D0084986

DATA AND COMPUTER COMMUNICATIONS

NETWORKING AND INTERNETWORKING

Gurdeep S. Hura
Mukesh Singhal

CRC Press
Boca Raton London New York Washington, D.C.

Library of Congress Cataloging-in-Publication Data

Hura, Gurdeep
Data and computer communications : networking and internetworking / Gurdeep S. Hura, Mukesh Singhal.
p. cm.
Includes bibliographical references and index.
ISBN 0-8493-0928-X (alk. paper)
1. Computer networks. 2. Internetworking (Telecommunications). I. Singhal, Mukesh. II. Title.

TK5105.5 .H865 2001
004.6—dc21 00-051866

Direct all inquiries to CRC Press LLC, 2000 N.W. Corporate Blvd., Boca Raton, Florida 33431.

Visit the CRC Press Web site at www.crcpress.com

International Standard Book Number 0-8493-0928-X
Library of Congress Card Number 00-051866
Printed in the United States of America 1 2 3 4 5 6 7 8 9 0
Printed on acid-free paper

Dedication

To my loving parents and sons, Devendra and Mandeep.
Gurdeep S. Hura

To my daughers, Malvika and Meenakshi.
Mukesh Singhal

About the Authors

Dr. Gurdeep S. Hura received his B.E. (B.S.) in Electronics and Tele-communication Engineering from Government Engineering College, Jabalpur University (India) in 1972, M.E. (M.S.) in Controls and Guidance Systems from University of Roorkee (India) in 1975, and Ph.D. from University of Roorkee (India) in 1985. He joined the faculty of Regional Engineering College, Kurukshetra (India) in 1974 as a Lecturer and became Assistant Professor in 1983. He has held positions at Concordia University, Montreal, Canada; Wright State University, Dayton, Ohio; and Nanyang Technological University, Singapore. Currently, he is Associate Professor at University of Idaho at Idaho Falls. His research interests include Petri net modeling and its applications, computer networks, software engineering, distributed systems, and real-time system design.

Dr. Hura is author/co-author of over 100 technical papers, which have been published in international journals and presented at conferences. He guest edited special issues of *International Journal of Microelectronics and Reliability* (*IJMR*) on "Petri Nets and Related Graph Models: Past, Present and Future" in 1991 and on "The Practice of Performance Modeling and Reliability Analysis" in 1996 and of *Computer Communication* on "Internet: The State of the Art" in 1998. He is an editor of *Computer Communications* and *International Journal of Electronics and Reliability* and Regional Editor of *Journal of Integrated Design and Process Control.*

He is a senior member of IEEE, was a General Chairman of IEEE EMS Singapore in 1995, is a reviewer of various IEEE Transaction journals (Software Engineering, Reliability, Computer), and is on the Program Committee of various IEEE Conferences (Distributed Computing System, System Integration, TENCON' Region 10). He has organized tutorials on computer networks, modeling and analysis, software engineering for various international conferences.

Dr. Hura has authored chapters on "Local Area Networks, Wide Area Networks, and Metropolitan Area Networks: An Overview" in *Computer Engineering Handbook*, McGraw-Hill, 1992, and on "Use of Petri Nets in System Reliability Evaluation" in *Current Issues in System Reliability Evaluation*, Elsevier, 1993.

Mukesh Singhal is a full professor of Computer and Information Science at The Ohio State University, Columbus. He received a Bachelor of Engineering degree in Electronics and Communication Engineering with high distinction from University of Roorkee (India) in 1980 and a Ph.D. degree in Computer Science from the University of Maryland, College Park, in May 1986. His current research interests include operating systems, distributed systems, computer networks, database systems, mobile computing, and computer security. He has published over 140 refereed articles in these areas. He has coauthored two books entitled *Advanced Concepts in Operating Systems*, McGraw-Hill, New York, 1994, and *Readings in Distributed Computing Systems*, IEEE Computer Society Press, 1993.

Dr. Singhal is currently serving on the editorial board of *IEEE Transaction on Knowledge and Data Engineering* and *Computer Networks*. He is a fellow of IEEE and a member of ACM. He served as the Program Chair of the 6th International Conference on Computer Communications and Networks, 1997, and of the 17th IEEE Symposium on Reliable Distributed Systems, 1998. He is currently the Program Director of the Operating Systems and Compilers program at the National Science Foundation.

Preface

This book presents a comprehensive treatment of the basic concepts of communication systems, signaling basics, data communications, transmission of digital signals, layered architecture of computer communication systems (local, metropolitan, and wide area networks), and well-known standards and protocols for networking and internetworking. Each of the layers of OSI-based local area networks is presented in more detail covering the services, functions, design issues, interfacing, protocols, etc. An in-depth discussion of integrated digital networks, integrated services digital networks, and very high-speed networks, including the currently evolved technologies such as asynchronous transmission mode (ATM) switching and their applications in multimedia technology is also presented with discussion on existing networking environments with appropriate applications.

This book is intended mainly for junior or undergraduate level students majoring in computer engineering or computer science and can also be used as a reference or handbook for graduate level computer engineering and computer science students, business information sciences and systems, management information systems, communication, and computer professionals. It can be used for both undergraduate and graduate level courses on, e.g., computer networks, data communications, protocols and interfacing, internetworking, and high-speed networks. It can also be used as a handbook for the computer professional particularly in the application areas of protocols and standards for both International Standards Organization (ISO) OSI-based and non-OSI-based networks and the internetworking of different types of networks (public, LANs, high-speed, Internet).

Multivendor networking has become a crucial need to most users of late; vendors have started working on the products that meet the customers' needs for easy, rapid, hardware-independent communication. There is a very large number of vendors of network products (proprietary) in the market, and it is very difficult (if not impossible) to understand the use of protocols and standards for networking and internetworking as each vendor defines and specifies the features, advantages, and applications of different types of network topology, control access, internetworking, security, data integrity, and many other performance-based issues for its own products. This book also serves as a practical guide on the technical details of many of the widely used protocols and standards for data communications, and networking and internetworking interfacing. It explains various technical details (insights of protocol primitives, parameters, data transfer, options, algorithms, etc.) of widely used protocols and standards.

This book is self-contained; it is hoped that the readers will become familiar with the basic and practical concepts of this rapidly growing and dominating technology in the data processing industry. This book provides a systematic approach to help the reader understand the various issues and problems of networking and internetworking during data communication over different network interconnectivity environments. This book is organized into six parts covering the following topics:

I. Computer network applications and standardization
II. Fundamentals of digital communication and signaling
    a. Basic concepts of communication systems
    b. Signal transmission basics
    c. Signaling techniques
    d. Digital communication
    e. Transmission media
    f. Telephone system
III. Local area networking and internetworking
    a. Local area network characterization
    b. Subnet (lower three layers of OSI-RM)
    c. Host-to-host (upper four layers of OSI-RM)
    d. Proprietary LANs (SNA, DNA, or DECnet)
    e. Internetworking devices and protocols
    f. LAN interconnectivity
IV. The OSI-RM architecture and protocols
    a. Layer functions and services of each layer of OSI-RM
    b. Layer protocol and interfacing of each layer of OSI-RM
    c. X.25 packet-switched public network protocol
    d. TCP/IP protocols
    e. LAN and WAN interconnectivity
    f. Network routing techniques
    g. Value-added networks and popular network vendors
    h. Network security and management
    i. Internet services and applications
    j. Details of popular services and their implementation
V. High-speed networking and internetworking
    a. Integrated digital networks
    b. Integrated services digital networks (ISDNs)
    c. ISDN protocols
    d. Fast packet switching
    e. Internetworking with different networks
    f. Evolution of broadband services
    g. Detailed discussion of services, protocols, formats of FDDI, SMDS, B-ISDN, frame relay, and asynchronous transfer mode high-speed networks
    h. High-speed network implementation for LAN environments
    i. ATM-based LAN and WAN interconnectivity
VI. Client–server LAN implementation
    a. Distributed computing architecture
    b. Client–server models
    c. Graphical user interface tools
    d. Client–server LAN implementations

For each part, a detailed description of issues, concepts, theory, etc. are highlighted and discussed in various chapters.

The first part starts with a brief introduction of network structure and its various applications. The services and functions of various layers of open system interconnection-reference model (OSI-RM) are discussed briefly, along with different layered protocols and the format of packets of the data information as it flows through the layered OSI-RM. The need to define standards for various layers and the advantages and disadvantages of OSI-RM are also explained. A detailed discussion of various international and national standards organizations, including standards and protocols defined by these organizations, is also presented. A discussion on current trends in communication networks briefly highlights various technologies.

The basic concepts and fundamentals of digital communication systems, signaling, and transmission basics are discussed in Part II. Chapter 2 presents the fundamentals of analog and digital signal transmission with emphasis on digital data communications, a comparison between analog and digital signals, and a representation of digital signals using Fourier series transformation. Different types of errors and noises (undesirable signals in the transmission of data signals) in data communications along with some remedies for their reduction are also presented.

The communication of signals (both analog and digital) across a larger distance requires modulation (a technique of raising the frequency band of the signal to higher level). Various modulation techniques for both analog and digital signals over synchronous and asynchronous transmission configurations are addressed in Chapter 3. The function of the modem (**mo**dulator/**dem**odulator), different types of modems and PAD (packet assembly and disassembly) interfaces for data communication, and various standards are also discussed. Different types of multiplexing and concentrators for data communication using classes of switching techniques over transmission media are introduced. A detailed discussion of each of the multiplexing techniques with its relative merits and demerits, their practical applications, and available vendor products are also presented (and will be very useful for understanding the working principles of these products). A detailed discussion of the implementation of the communication control interface using the universal asynchronous receiver and transmitter (UART) chip is presented.

Different modes of operations, e.g., simplex, duplex (half and full for the signal transmission from different types of terminals), and various error-control strategies are explained in detail in Chapter 4. Under error-control strategies, we present techniques and standards for both error-detection and error-correction controls along with well-known standards and a good reference for their application. A discussion of the different types of terminals should be very useful to the readers.

Various communication media for transmitting analog and digital signals are described in Chapter 5. The transmission media fall into two categories: wired and wireless. For each category of media, we have given an extensive technical presentation of all available media. Various channel accessing techniques are further discussed in this chapter. A detailed discussion of broadband LAN will provide insight of the network design hardware aspect of data communication.

To get an idea of how the telephone system came into existence, in Chapter 6 we present the evolution of the telephone system, different options available in telephone systems (e.g., leased or private telephone lines) different types of switched networks for analog data communications such as telephone networks, telex networks, switched networks, and a private branch exchange. A different type of carrier used exclusively for digital data communication is also discussed. In order to understand the workings of this type of carrier, we consider Bell System Carrier T1 which, while not a standard, is a very popular digital carrier and has been adopted by many countries, including parts of Europe. A detailed discussion of various digital transmission techniques will provide a strong foundation for understanding high-speed networks and also gigabit LANs.

Once we have an understanding of the basic concepts of layered architecture, protocol configuration, digital communications, transmission of digital signals, and transmission basics, we next introduce a detailed discussion of the conceptual design of layered architecture using OSI-based networks, its constituents and their roles in networking and internetworking in Part III. Chapter 7 of this part discusses various standard OSI-RM and non-OSI LANs, including local area networking, interfacing, and various aspects of internetworking. The characterization of LANs is discussed through various topologies (the physical connections of the network), commonly used communication media, accessing techniques (contention and token passing), and many other design and implementation issues.

A qualitative comparison between different types of OSI-based and non-OSI-based LANs is also provided to gain an understanding about these LANs, and is very useful in selecting a LAN with the appropriate standards and protocols for a particular application. It presents a detailed discussion of the recommendations of the IEEE 802 committee in defining different types of standard LANs. Further, a detailed implementation of each of these LANs will give readers insight into LANs with a view to compare them. The IEEE 802 committee's standard metropolitan area network (MAN) and a new concept in local area networking based on wireless LAN (IEEE 802.11) is discussed at the end of Chapter 8.

There are certain LANs which do not follow OSI standards, but are very popular. These non-OSI-based LANs are also discussed in detail in Chapter 9 to gain an understanding of the layering approach concept, which helps in implementing various aspects of networking. Further, data communications between the OSI-based and non-OSI-based LANs, and various internetworking devices with their applications in network interconnectivity in different types of networks, e.g., public networks, LANs, WANs, and different high-speed networks, are also discussed. Also presented are standard architectures, various standards, wireless spectrum allocation, and digital cordless standards.

After a detailed discussion of LANs in general, we next discuss each layer of OSI-RM LAN (Chapters 10–16) in detail in Part IV. Each chapter in this part is devoted to each layer of this model and provides a detailed description of the standards and protocols for data communication. For each of the chapters in this part, we look at the organization of the various sections by considering the following OSI specifications:

- Services and functions
- Design and implementation issues
- Standards and protocols
- Primitives and parameters of protocols

The first three chapters (10–12) present various design and implementation issues between subnet and host machines. The remaining chapters (13–16) present various design and implementation issues for data communications between host-to-host connections provided by a subnet and an understanding of LAN and WAN interconnectivity.

A detailed discussion of TCP/IP, internetworking interfaces and protocols, value-added networks, and internetworked LAN protocols for some of the private LANs is presented in Chapters 12, 13, and 17. A detailed discussion of security, public key cryptography, and secure socket layer protocol is presented in Chapter 15.

The detailed discussion of these concepts in Chapters 10–16, along with design implementation issues and limitations, makes this a useful and practical reference book for the professional in the networking and data processing industries. The last chapter in this part, Chapter 17, gives an overview of various applications of the Internet, the evolution of the Internet, various connections, Internet service providers, vendors, and other related implementation information.

In Part V, Chapter 18 discusses the evolution of integrated digital networks, integrated services digital networks (ISDN), functions, services, the architecture of ISDN, various standards, different data rates, protocols, their applications such as videoconferencing, teletext, telefax, and facsimile. A detailed discussion of the need for broadband services provides an understanding of the various technologies evolving for high-speed networking and LAN/WAN interconnectivity.

Chapter 19 presents a detailed discussion of high-speed networks with an emphasis on upcoming asynchronous transmission mode switches, and many other transmission techniques with a view to understanding their application in multimedia. The concepts, theory, design issues, limitations, standards, and protocols for high-speed networks are also discussed in this chapter. In particular, it includes the main components of B-ISDN and its capabilities, FDDI and FDDI-II and their applications, SMDS technology and its services and protocols, frame relay and its protocols, ATM and optical transmission (synchronous digital hierarchy), ATM networking (virtual paths, channels, resource allocation, etc.), user-network interface, B-ISDN protocol reference architecture, and B-ISDN equipment. A discussion of LAN/WAN interconnectivity based on ATM will provide a sound understanding of various technologies for various applications.

The last part discusses the importance of client-server models, various graphical user interface tools and a few applications. It also describes the ways in which client–server models have been used for LAN implementation.

We have also provided a compact definition of various terms (and their abbreviations) used in networking and internetworking in the glossary for ready reference.

The main object of this book is to present the practical concepts and fundamentals of networking and internetworking through a systematic and modular approach. This book covers the material found in the following courses currently offered (in various universities and organizations) in this subject area:

- Data communications (basic concepts of transmission, signaling, and digital data communications)
- Computer networks (applications and descriptions of different layers of OSI-RM)
- Protocol and standards (description of various well-known protocols and standards)
- Networking, internetworking, and interfacing (for both OSI-based and non-OSI-based networks)
- Digital communication systems (basic concepts of communication, signal transmission theory, and signaling)
- X.25 (packet-switched networks) and TCP/IP
- High-speed networking and its use in LAN/WAN interconnectivity
- Internet
- Client–server LAN implementation
- Client–server computing
- Broadband services

Gurdeep S. Hura
Mukesh Singhal

Contents

Part I Computer network applications and standardization

Part II Fundamentals of digital communication and signaling

part I

Computer network applications and standardization

chapter one

Computer networks and standardization

"One machine can do the work of fifty ordinary men. No machine can do the work of one extraordinary man."

Elbert Hubbard

1.1 *Network goals*

Computers have become an integral part of any organization, corporation, business, government agency, or university. These computers are at different locations within the organization. Each computer is performing a different set of tasks. For example, in a university with many buildings and campuses, generally there are computers at each location to keep different records such as student registration, courses offered, graduate students, employees, payroll, research records on funded projects, administrative strategies, guidelines, handbooks, and many others. Each of these computers works in a different environment without interacting with other computers. If these computers can be connected and managed properly, many of the overlapping operations can be avoided and a useful strategy can be developed for the university. Similarly, for organizations sharing resources, databases, programs, costs, communications, utilization for short- and long-term benefits may determine whether these computers should be connected or not.

A computer network connects several geographically dispersed computers so that they can exchange information by message passing and one computer can access resources available to other computers. If computers are connected by a **network**, then existing information on them can be shared.

A computer network provides the facility of information exchange among the computers connected to it. A network provides the ability of resource sharing to its users such that users of any computers (working under any environment) connected via a network can access the data, programs, resources (hard disks; high-quality, expensive laser printer; modems; etc.), peripheral devices, electronic mail (e-mail), software, etc., regardless of their physical locations. The physical distance between computers can be from a few feet to thousands of miles, but the users and computers exchange data and programs in the same way as they do locally. Regardless of the distance, a network creates a **global environment** for the users and the computers. Resource sharing is one of the main advantages of networks. Resources such as printers, files, peripheral devices, disks, computers, and even people can be connected together. This interconnectivity allows faster, better, more reliable mobile communications, such as cellular and wireless. Very large files, programs, software, graphic files, computer-aided design tools, and many other tools can be transferred over the networks. Sending and receiving e-mail via online services such as America Online (AOL), CompuServe, and many others over networks around the world adds a high degree of mobility.

One of the main disadvantages of networks is the cost of network equipment and various software operating systems and applications modules. Further, cost is also involved in training and certifying LAN engineers who will be responsible for designing, installing, maintaining, and upgrading the networks. There are three main LAN operating system designers — Microsoft, Novell, and Banyan.

Data communications have evolved from 300 B.C., when Heliograph was invented by Greeks. Some of the main inventions or events related to data communications are:

- The code for encryption (Francis Bacon in 1605)
- The difference engine (Charles Babbage in 1831)
- The first digital 0/1 communication device (Joseph Henry in 1831)
- The telephone (Philip Reise in 1854)
- Telegraph lines from New York to San Francisco (in 1861)
- Optical fiber communications (John Tyndall in 1870)
- The electromagnetic wave equation (James Clerk Maxwell in 1873)
- The first telephone switchboard (in 1885)
- American Telephone and Telegraph (AT&T) (formed in 1885)
- Radio communications across the Atlantic Ocean (Guglielmo Marconi in 1901)
- The U.S. Federal Communications Commission (FCC) (introduced by the Communications Acts of 1934)
- ENIAC, the first digital computer (in 1945)
- The transistor (invented by Bell Laboratories in 1948)
- The first integrated circuit (developed by Texas Instruments in 1959)
- The Systems Network Architecture (SNA) protocol (developed by IBM in 1972)
- The X.25 protocol for public networks (in 1976)
- Ethernet as a standard LAN (in 1978)
- AT&T broke up into AT&T and seven Regional Bell Operating Companies (RBOC) (in 1984)
- AT&T introduced a new telecommunications law allowing local carriers, long distance carriers, and cable television companies to compete with each other (in 1996) (and we are watching a similar situation with Microsoft)

More information can be found in Reference 7.

1.2 Network evolution

The Defense Advanced Research Project Agency (DARPA) research and development program of the U.S. Department of Defense (DOD) designed the Advanced Research Project Agency Network (ARPANET) in 1969. This network pioneered packet-switching technology and provided communication links between heterogeneous computers. The main aim of this program was to design an experimental network, which connected a few universities and organizations within the U.S., and allow them access to and use of shared computer resources. As ARPANET became successful, more functionality and applications were added to it. This technology was handed over to the Defense Communication Agency (DCA) to develop and improve it for evolving new advances. As a result of this, in 1983 the **Defense Data Network (DDN)** based on ARPANET technology was defined. DDN is a packet-switching network with dynamic routing capabilities and is able to adjust itself in the event of crash/failure without affecting services to the users. These networks have been defined and designed so as to provide survivability, security, and privacy.

There are basically three types of networks: local area networks (LANs), metropolitan area networks (MANs), and wide area networks (WANs). LANs are usually used for interconnecting computers within a range up to 10 miles, and they offer data rates from 10 Mbps to 1 Gbps (1 Mbps = 1×10^6 bits per second; 1 G = 1×10^9 bps). LANs allow the sharing of resources such as mainframes, file servers, printer servers, and other peripheral devices. MANs use optical fiber to connect computers in various offices and organizations within a metropolitan area. MANs have a typical range up to 70 or so miles and offer data rates of 1 Mbps to 150 Mbps. They are used mainly for financial transactions between banks in a metropolitan city. WANs connect computers located at very large distances (up to 800 miles or so) and offer data rates. Usually government agencies control these networks. According to some, the interconnection of WANs defines the Internet or a global area network (GAN), which covers distances of thousands of miles and offers data rates of 1 Mbps to 100 Gbps. The interconnection between these types of networks can be realized in a hierarchical way; i.e., LANs cover all the computers of a campus and may be connected to MANs in the metropolitan area. These MANs may be connected individually to WANs across the country, which may then be connected via GANs or routers (another form of GANs), to define the structure of the Internet at a very high level across the globe.

Early computer networks were dedicated to a limited number of operations and, as a result, only a limited number of connections were offered. These networks, by and large, were static and inflexible. Various features such as routing, switching, and intelligent functionalities were expensive to incorporate in the networks. Further, these networks were based on mainframe hosts and the online users were submitting their data requests to a centralized database monitored by a mainframe. This restricted the use for transmitting binary digital signals over a distance. Most of the PCs nowadays are connected by LANs. The number of LANs used has already approached over 2 million and the data rates which are offered by these LANs have been increasing.

In the 1980s, LANs were supporting data rates of 1 to 3 Mbps. In the 1990s, data rates of 10 to 16 Mbps became very common and standard. With new technologies in high-speed switching and transmission, data rates of 100 Mbps over twisted pair cabling **(copper distributed data interface (CDDI))** due to **fiber distributed data interface (FDDI)** LAN technology can be obtained over desktop transmission. This will now allow users to link with high-performance servers and workstations and access a variety of services of multimedia, video applications, virtual-reality applications, and many other applications requiring high bandwidth and high performance.

LANs are being used for information sharing and have proliferated quickly. According to a number of surveys, the number of LANs has already gone above 2 million around the world. With different types of traffic, the requirements of LANs have also changed. Now the requirements of LAN environments include a higher data rate for integrated traffic (data, video, voice, graphics, etc.), an acceptable quality of service, and a uniform network management system (using the same equipment). LAN environments have evolved from low speed (Ethernet, token ring, token bus, and some non-OSI LANs) to high speed, e.g., switched Ethernet, FDDI, 100-Mbps voice grade, gigabit Ethernet, and so on.

The switched Ethernet offers a very high data rate and dedicated bandwidth. The dedicated access to a resource causes starvation and does not support multimedia applications. FDDI offers superior performance compared to Ethernet and token ring, and it is currently being used as a backbone network in most organizations, universities, and other agencies. Although it is expensive and does not support isochronous (voice) traffic, a newer version of FDDI-II has been introduced which can handle the voice traffic and can also support Asynchronous Transfer Mode (ATM).

Inexpensive coax has been used in defining a new LAN as coax distributed data interface (CCCI), which may become an alternative to expensive FDDI. 100-Mbps voice-grade LAN is a new technology and seems to be better than FDDI. The ATM for LAN environments offers very high bandwidth (on demand) and much better performance with very high data rates for data, video, and voice. ATM can be used as a backbone, and routers associated with it offer interfaces to different LANs such as High Speed Serial Interface (HSSI), Digital Signal 1 (DS1) circuits, Data Exchange Interface (DXI), and so on. Details of LAN interconnectivity can be found in later chapters.

The frame relay was deployed in 1990, offers flexible bandwidth-on-demand services, and has nearly 200,000 ports supporting over 10,000 customers. Both Integrated Services Digital Network (ISDN) and Switched Multimegabit Data Service (SMDS) have seen very slow growth, but due to the Web, they have picked up in the market. ATM, on the other hand, offers a tremendous growth potential due to the fact that it supports different types of services, including frame relay. The use of asymmetric digital subscriber line (ADSL) modems with existing copper wires offers megabit/second speeds from the network to the users along a slower speed channel. This finds its application in accessing the Web.

1.2.1 *Public analog telephone networks*

The advent of **modulation** of binary digital signals with analog wave form made it possible to transmit digital signals across a greater distance through **public analog telephone network or system (PTN or PTS).** Sharing of resources over PTN was another feature. As a result of these advances, the network has evolved from dedicated resource usage to sharing of resources by a number of users.

Switching of message is one of the most important aspects for resource sharing across the network, and it lowers cost. The switching technique determines the **path** (or **route** or **circuit** or **logical channel)** over which the message is transmitted efficiently. The switching supports two types of services: **connection-oriented and connectionless**. Based on the type of switching (message, circuit, or packet), the following three categories of networks have been identified:

- Private line network (PLN)
- Circuit-switched network (CSN)
- Packet-switched network (PSN)

A brief introduction of each of these networks is given below. For details, please refer to the appropriate chapters.

Private Line Network (PLN): This network is also known as a dedicated network between two nodes and its structure is shown in Figure 1.1.

This network offers a fixed bandwidth which is known in advance. Network security and independence of protocol features are used when connected via a modem (MOdulation/DEModulation). It also offers transparency to data messages. A fixed monthly charge makes it more suitable for continuous, high-volume data transfer as resources are permanently allocated to the users. The fixed bandwidth and slow speed of modems are major drawbacks with this type of network, but high data rates may be achieved using high-speed modems (56,000 baud).

Figure 1.1 Private line network.

Circuit-switched network (CSN): This network is based on the concept of a normal telephone network. It allows sharing of resources uniformly among users and as such it reduces the cost. A user makes a request for the establishment of a logical connection over the link. Once the circuit is established, it becomes a dedicated circuit for the duration of its usage. After the data is transferred, the user makes a request for the termination of the connection. The user is billed for the duration of usage which includes the connection set-up time, data transfer time, and finally the termination time. These times are also collectively known as the dial-up connection time on switched telephone networks.

The circuit-switching connection is made up of a number of connections of switching nodes within the network. Noise and echo are inherent during the transmission and may cause high error rates if they are not properly handled. Special security is a major problem with this type of network, and this network alone can not be used as a backbone network for heavy usage.

A dial-up line usually offers the bandwidth for voice-grade of 300 Hz to 3400 Hz and as such the data rate is lower when compared to the private line. A private line may offer a data rate of at least 19.2 Kbps and higher. A dial-up modem (connected to a dial-up line for sending binary digital signals) is limited to 56Kbps. A digital private line offers 56 Kbps (Kbps = 1×10^3 bps) to 1.544 Mbps over copper wires, or high bandwidth over optical fiber.

Packet-switched network (PSN): The packet-switching technique was introduced by Paul Baran of Rand Corporation in the early 1960s while working on a U.S. Air Force project to study the requirements for a secure and fault-tolerant military communication network, intended for both voice and data communications. In this technique, a message (voice communication) is divided into slots which can be used to provide separation between audible signals into a packet of fixed size. Each packet is attached with a destination address, a sequence number, control information, etc., and packets may travel through different routes across the network (connections or links). These packets may arrive at the receiving side at different times via different routes, and they are reassembled so that the original message can be constructed. In the event of the failure of any node, the packets can be rerouted to prevent the loss of any packet. X.25, a widely used protocol standard that defines interface between host system and packet switched network, International Telecommunication Union-Tele-Communication (ITU-T), is an international standard defined by CCITT (located in Geneva, Switzerland) and and used in most of today's packet switching. It is universal and the telephone services in every country provide an X.25 network. It is slower and generally limited to 56-Kpbs transmissions.

1.2.2 Subnet interface

The packet-switched networks offer higher throughput than circuit-switched networks, because in packet-switched networks resources are allocated to the users on demand. The **ARPANET** is based on packet-switching, which defines the packet switches as **interface message processors (IMPs)**. The Defense Department's Advanced Research Projects Agency (ARPA) handed over the packet-switched network with other management and controlling functions to **Bolt, Barnak, and Newman (BBN)** of Cambridge, MA (U.S.) to deploy a fully operational network, which became known as ARPANET. This network is still being used for military applications and at present supports more than 100 nodes. The first network had only four nodes. England's National Physical Laboratory also defined its own experimental packet-switched network during the same time and for the first time, a packet of 128 bytes was defined and sent across the network. Some of the material presented is partially derived.[1,3,4,7,8,12]

1.2.3 Public switched data networks (PSDNs)

The first commercial public packet network was introduced in early 1970 as the McDonnell Douglas Tymnet network after the introduction of X.25 as a basic interface to a public network by CCITT. The initial version of Tymnet offered support to only low-speed asynchronous terminal interfaces which were connected to its nodes (located at main locations) via dial-up connections. This version was then updated to support X.25 and synchronous terminal interfaces, too. Tymnet uses a centralized routing technique which is based on the concept of a virtual circuit which defines a fixed route or circuit during the establishment of a connection. It is defined as a geographically distributed packet-switched digital network which provides services to the users.

1.2.4 Value-added networks (VANs)

A value-added network consists of communication processors known as computer nodes. The minicomputer running at a node is known as the engine. This engine is operated by software applications developed by ISIS International. Various services and popular network vendors offering value-added network services are discussed in Chapter 10.

The Tymnet provides interfaces to host computers, terminals with the networks, and offers the routing of packets and other control functions at the node. The packets can be transmitted over this circuit and all packets follow the same route. The circuit remains active until a request for its termination has been made by either the source or the destination node. This concept of a fixed route or circuit is also known as a **virtual circuit** and has become a part of the switching technique in X.25 networks.

Telenet, another public network based on ARPANET, was introduced in 1974 by GTE Telenet Communication Corporation (U.S.). It is a geographically distributed packet-switched network which uses digital lines at a data rate of 54 Kbps. One of its main objectives was to utilize resources effectively to allow users to access their remote computers at a reasonable cost. This network is a modified version of ARPANET and uses an integrated routing technique of distributed (used in ARPANET) and centralized (used in Tymnet and IBM's System Network Architecture (SNA)). IBM's SNA was introduced in 1974 and is a hierarchically structured LAN with centralized routing and other connection control facilities. Details on IBM's SNA can be found in Chapter 7.

1.3 Components of computer networks

A computer network comprises many components (hardware and software). The hardware components include computers, modems, interfaces, network cards, peripheral devices, networks, etc. The software components include operating systems (such as MS-DOS, Windows, and UNIX), protocols, and other software tools supporting the data communication across the network. The number of components (hardware and software) may depend on a variety of factors such as the type of network to be used, number of hosts, PCs, terminals, workstations, other peripheral devices to be connected to the network, data volume, functions of the systems, transmission speed, application programs, and services required. Next, we briefly describe the main hardware and software components which make up computer networks.

1.3.1 Hardware components

A typical configuration of a computer network is shown in Figure 1.2. Various hardware components include communication channels, modems, line interfaces, terminals, host, etc. In the following, we describe each of these hardware components in brief.

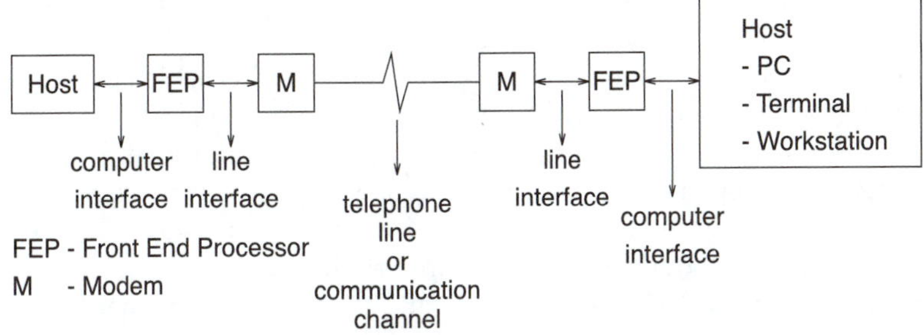

Figure 1.2 Components of computer network.

Communication channel: A communication channel, or link, provides a physical path for transmitting signals between computers connected to the network. Channels are usually provided by telephone common carriers and access the **public switched telephone networks (PSTNs)**, **switched networks**, or **leased private lines**. Communication channels have evolved with an emphasis on multiplexing multiple channels at higher data rates over the same channel.

Telephone lines were introduced in the late 1890s, and the concept of multiplexing (using a pair of telephone wires to carry more than one voice channel) was introduced in the 1920s. In the early 1940s, coaxial cable was used for carrying 600 multiplexed voice channels and 1800 channels with multiple link pair configuration. The total data rates offered by these 600 channels was about 100 Mbps. The number of channels which could be multiplexed over the same coaxial cable was continuously increasing (due to technological changes in switching and transmission techniques) and the data rates were also increasing for voice communication. Some of the popular number of channels over coaxial cable include 1860 voice channels (with data rates of about 600 Mbps), 10,800 voice channels (with data rates of 6 Gbps), and 13,200 voice channels (with data rates of 9 Gbps). (1 Mbps = 1×10^6 bps (bits per second), 1 Gbps = 1×10^9 bps.)

With the introduction of fiber optics in 1990, we can get data rates of over 10 Gbps with various advantages of fiber over cable, such as high bandwidth, low error rate, reliability, high capacity, and immunity to external noises and interferences. Another communication link, which is based on wireless transmission, is the microwave link which offers voice channels ranging from 2400 (in 1960) to over 42,000 (in the 1980s).

The channels support a variety of data communications options. In the literature, these options have been separated into three categories of communication: **narrow-band**, **voice-band**, and **broadband**. Narrow-band communication includes applications requiring data rates up to 300 Kbps. Voice-band communication usually defines the bandwidth of voice signals in the range of 300 Kbps–3.4 KHz. These voice signals can be transmitted above transmission speeds of 9600 bps. Broadband communication offers data communication rates higher than voice-band channels (up to several Mbps). Broadband communication channels are more reliable and efficient than the other two communication channels.

Modem (MOdulation/DEModulation): A modem allows the transmission of digital data over analog telephone lines. It converts the digital signal (pulses) coming out of a computer or terminal to analog signal acceptable for transmission over telephone lines. This conversion of digital signals into analog is performed by a device known as a **modulator**, which is **data circuit terminating equipment (DCE)**. It connects users' **data terminal equipment (DTE)**, e.g., personal computers (PCs) and terminals, in **wide area networks (WANs)**. The analog signal is modulated and transmitted over telephone lines. Another device known as a demodulator recovers the original digital signal from the received modulated signal by converting analog signals back into digital form. The interfaces between modem and DTE include EIA-232 and V.35.

There are different types of modems available on the market which provide support for different transmission modes, e.g., synchronous and asynchronous (thus defining synchronous and asynchronous modems, respectively). Synchronous modems are usually used with continuous data signals from devices such as magnetic tape and computers. Asynchronous modems are usually used with keyboards, terminal devices such as the teletype-writer, CRT displays, and terminals. Modems in general support all three transmission configurations: simplex, half-duplex, and full-duplex. More details can be found in Chapter 3.

There is another device known as a **coder/decoder (CODEC)** which does the inverse of a modem. It codes voice into digits by grading the amplitude at one of 256 different levels. The code is then defined as 8-bit (a byte or octet) and is transmitted. On the other end, the octet's bits are used as an index to the same scale of amplitude to reconstruct the signal.

The speed, or data rate, of modems is expressed in baud, or bits per second, and modems are available in a variety of forms with data rates of 1200 baud to 128,000 baud. Different types of modems and standard interfaces are presented in Chapter 3.

Acoustic coupler: An acoustic coupler accepts serial data from a data-processing device and modulates it in the voice-grade range such that the modulated signal is generated as an audible signal. This device may be used as an alternative to a modem with the only drawback being that it supports a low data rate. These are usually used with teleprinters and slow-speed terminals. Acoustic couplers accept signals from a telephone handset and provide a direct interface for acoustic signals at the mouthpiece. On the receiving end, the acoustic coupler accepts the audible signal from the telephone ear piece and demodulates it into serial form.

Line interface: This communication interface device provides a communication link between computers or terminals over a particular line. These interfaces are usually provided by the computer manufacturer for different combination of connections such as line-terminal or line-computer and remote computers/terminals. It conforms to ASCII (American Standard Code for Information Interchange) coding, RS-232 interface (physical layer DTE/DCE interface), and it also provides an interface for modems/acoustic couplers. Various line interfaces are discussed in Chapter 3.

Terminals: Terminals are used for processing and communication of data and provide an interface between users and computer systems. This interface provides a mapping between two types of representations: user-oriented and electrical signals–oriented frameworks. Various hardware devices can be interpreted as terminals, e.g., a printer, which provides mapping for processing data in the form of hard copy, and other input/output (I/O) devices. Terminals are a very useful component for processing, collecting, retrieving, and communicating data in the system and have become an integral part of application systems. The interface (provided by the terminal) for data communication takes input from input devices, such as a keyboard, card reader, magnetic tape, mouse, or any other pointing devices, and converts it into output devices, such as displays, paper, tape, screen, and other devices. Some I/O devices may have only input and no output; such devices are known as receive-only devices.

Input devices: Keyboards are a very popular input device and offer effective input capabilities to both terminals and computers. They include different types of keys, which in turn depend on the underlying applications. In general, keys can be grouped into three categories: **textual** (defining text, numbers, special symbols and characters, punctuation signs, etc.), **control** (dealing with various options during editing of text and its transmission), and **macro functional** (defined for a special sequence of operations which may include data processing, etc.). For converting the functions of keys into a serial bit stream for transmitting over a transmission link, or bus, various codes have been defined. The codes have a fixed number of bits for each key; e.g., ASCII defines 7-bit code, while EBCDIC (Extended Binary Coded Decimal Interchange Code) defines 8-bit code. ASCII is more popular than EBCDIC code.

Output devices: An output device accepts a bit stream from the terminal and generates equivalent representation symbols. The representation of the symbols can be defined in various forms such as regular full print-out and dot-matrix print-out. A regular full print-out (laser printers) offers better quality but is expensive. A dot-matrix printer, which is less expensive, does not offer good quality and characters are represented by combination of small dots. Other output devices include paper tape and display devices and will not be discussed here.

I/O devices that send/receive one character at a time are very slow. The speed of transmission of characters can be increased by using a buffer (a small segment of memory) which will store a block of characters, then transmit/receive it between input and output devices. These types of terminals (having buffers) are known as buffered terminals. In a buffered terminal, the capability to process data can be included to make it an intelligent terminal. An intelligent terminal is nothing more than a combination of micro-programmed computers (with stored programs in read-only memory [ROM]) and their attached input/output devices. A dumb terminal does not have any processing capability and is not controlled by software; it simply passes characters from one node to another. A detailed discussion of terminals, input and output devices with different categories, and their interfaces is presented in Chapter 4.

Internetworking devices: There are many devices that can be used to connect different types of LANs. A brief discussion of those devices is given below.

A repeater regenerates the signal strengths between various segments of similar LANs, thus increasing the total length of the LAN. It operates on the physical layer and does not provide any routing or management of packets/frames. If the packet/frame is lost or corrupted, we must look into another interconnecting device, known as a bridge. The bridge connects different types of LANs at the data link layer, where it looks at the destination address in the received frame and determines whether to keep it or pass it to another connected LAN. At the physical layer, the bridge acts like a repeater in the sense that the received signals are regenerated. It must use the same protocol on either end. This is useful only for small- or medium-size network systems. For a larger network system, another interconnecting device known as a router is used. The router is similar to a bridge for interconnecting different LANs except that it works with other types of networks, such as wide-area networks, FDDI, or any other high-speed networks. It operates at the network layer. It can be used to connect LANs to WANs, MANs to WANs, LANs to LANs, and so on. At the network layer, hardware devices do the routing for sending a frame to another connected network. Cisco Systems is the world's largest supplier of routers. A gateway is used to provide interconnectivity between LANs and other LANs, packet-switched networks, private networks, databases, videotex services, or any other services. The gateways provide the compatibility of protocols, mapping between different formats, and easy data communication across them. Our PC will send the packets to a gateway, which now becomes our address, which will send it to another network. In this way, a gateway basically becomes another node between networks that provides compatibility between the operations of these two different networks.

1.3.2 Software components

A typical connection between terminal and telephone line interface for data communication requires the following software components, shown in Figure 1.3. The connection between modem and hardware interface is controlled by various software modules such

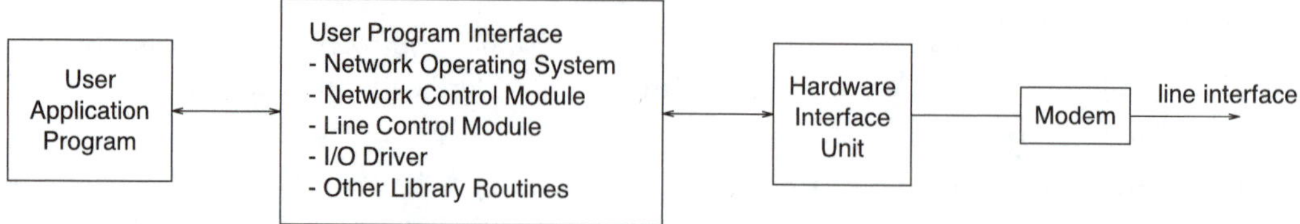

Figure 1.3 Software components of computer networks.

as a software I/O driver, a software module for line control, and a software module for network control. The I/O driver module controls various functions associated with data transfer, e.g., control and data transfer between the line control module and hardware interface and various status information about the data control.

The line control module allows the user application program to specify the type of line protocol procedure being used, modes of operation, code conversion options, error-control flow control, etc. and depends on the type of protocols, number of the devices used, and communication links attached to the interfaces. This module provides an interface network control module and software I/O driver. The network control module provides interpretation communication within the network and, as such, is responsible for defining logic paths, or links, between routing strategies over these links, acknowledging flow-control diagnostic capabilities, multiplexing, etc. All these modules are controlled and managed by the operating system. The user application program module is usually developed by users who use the operating system to perform various operations over the network.

1.4 Network structure

The general network structure includes hosts (computer, terminal, telephone, or any other communicating devices) and a communication subnet (also referred to as a network node, a subnet, or a transport system). Each host is connected to the communication subnets. The set of network nodes defines the boundary of the network and transfers the data from a **source host** to a **destination host** via transmission media (cable system, satellite, leased telephone lines, etc.) over communication channels across the network. The host provides various services to its users, while a subnet provides a communication environment for the transfer of data.

The communication subnet, in general, consists of switching systems and transmission links (also known as channels). The transmission links carry the bits from one computer to another through networks. The switching system (high-speed dedicated processing elements with large memory) on the other end is responsible for forwarding the data to its destination over the transmission links. The switching system is connected to the host via a transmission link which carries the data from its source to the system. Transmission links connect various switching elements of subnets and define the topology (physical connection) of the subnet. The subnet defines the three lower layers (physical, data link, and network) of Open System Interconnection-Reference Model (OSI-RM).

The switching system, after receiving data from its source, looks for a free transmission link between it and the switching element which is connected to the destination host. If it finds a free link, it will forward the data on to it; otherwise, it will store the data in its memory and try other route for the data. It will send the data to another switching element, which will again look for a free link until the data is delivered to its destination. The elements of a switching system have been referred to by different names, such as **interface message processor (IMP)**, **packet-switch mode**, **intermediate system**, and **data exchange**

system. These elements provide an interface between hosts and communication systems and establish a logic connection between hosts over transmission links. Various operations within communication systems are performed in dedicated and specialized computers. Some of the terms used here are derived from References 1–3.

Each host is connected to one or more IMPs. All the data originating from a source host first arrives at one of these connected IMPs, from where it is then delivered to the destination host through intermediate IMPs. The IMPs of a network may be connected by lines, cables, satellite links, leased telephone lines, etc. Based on the topology of the subnet, three types of communication services (offered by IMPs to the users) have been defined:

- Point-to-point communication
- Multicast communication
- Broadcast communication

1.4.1 Point-to-point communication

In a point-to-point communication, data from one host is transmitted over either **direct or indirect** links between IMPs (to which these hosts are connected). In a direct link, IMPs are directly connected via a physical communication medium during data transfer. In an indirect link, the data is transferred over intermediate IMPs until it reaches the destination host. Here, the data from an incoming line arrives at the IMP, which stores it and waits for a free link between IMPs. If it finds a free link, the data is sent over it; otherwise, it sends it over intermediate IMPs, and at each intermediate IMP, data is stored and forwarded to the next available IMP. This process is repeated until the data is received by the destination host. In the case of an indirect link, the objective is to always minimize the number of hops between hosts (a hop defines a simple path of length of one between two nodes, two hops defines a simple path of length of two, and so on).

A communication system supporting point-to-point communication is defined as **point-to-point**, **store-and-forward**, or **packet-switched** subnet.

1.4.2 Multicast communication

In a multicast communication, data can be sent to users of a selected group. The usual way of implementing this communication is to set the high-order bit in the address field (of data) to 1. The data will then be delivered to all the users whose high-order bit in the address field is set to 1. There exists only one channel or circuit which can be shared by all connected users' (of the selected group) hosts/IMPs. This type of communication suffers from the problem of **contention**. The problem of contention can be resolved either by using a centralized dedicated processor (which will decide as to which host can send the data next) or by using a distributed system (where each IMP resolves the problem independently). This obviously requires a complex protocol at each IMP.

1.4.3 Broadcast communication

A more general form of multicast communication is broadcast communication, where data is delivered to all hosts/IMPs connected to the network. A special code in the address field of the data is used to distinguish between point-to-point and multicasting communications. The data is sent on the network circuit and is received by every host/IMP connected to it. If the address of any connected host/IMP matches the address contained in the data, it can copy the data into its buffer. This type of shared communication typically defines the following network topologies: bus, satellite, radio, and televsion.

1.5 Data communication networks (DCNs)

Data communication networks can be classified as **local area networks (LANs)** and **wide area networks (WANs)** (also called **long haul networks (LHNs)**). Both classes of networks provide the same services such as sharing resources, programs, files, information (voice, data, and video), and peripheral devices. LANs are used for a typical limited range of up to 1 kilometer, such as a university campus or the premises of an organization. WANs, on other hand, are used for longer distances and provide services for transmitting different types of information (voice, data, and video) between universities and organizations. Users that are far apart can communicate with each other, send data in bulk, transfer files/reports, make changes in shared files, and allow other users to see it interactively. Some of the popular LANs are Ethernet, token ring, ARCnet, token bus, SNA, and FDDI. The wide area network (WAN) provides connection to various places around the world. Some examples include T1, X.25, System Network Architecture (SNA), Synchronous Optical Network (SONET), Integrated Services Digital Network (ISDN), metropolitan area network (MAN), and Fiber Distributed Data Interface (FDDI).

1.5.1 Applications of DCNs

One of the main applications of data communication networks is **electronic mail (e-mail)**, through which we can send information from our terminal, personal computer (PC), or workstation to anyone anywhere in the world (with the ability to receive it).

Other new technologies, such as digitization of voice, television pictures, video, and image patterns, can also be integrated into e-mail service. These services together provide a variety of applications, e.g., accessing an encyclopedia, language interpretation, learning at a distant location, pattern recognition, construction of patterns with different symbols and meaning at a different location, labeling the patterns/drawings, etc.

A closely related application of e-mail service is the **electronic bulletin board** system, which gained widespread popularity recently, particularly in universities and various corporations and organizations. In the university environment, for example, faculty members, graduate students, researchers, and others can access information about conference dates, funding and contracts, new publications, new products in different technologies, employment, and main news of different countries.

Digital technology has become an integral part of the new telephone networks because it's inexpensive and offers better quality of voice transmission over the networks. ISDN, which is based on the concept of a completely digital network, has proved that digital technology will not only provide fast and reliable service but will also help in introducing other technologies such as multimedia and virtual reality. The integration of different types of signals (audio, video, text, graphics, and many other application types) over high-speed networks is becoming an essential service of communication networks. The main philosophy behind digital technology for the integrated digital network is based on digital switching and transmission. The digital switching technique is based on digital **time division multiplexing (TDM)**, while synchronous TDM is used for carrier transmission between switching nodes. The full-duplex transmission of digitized voice signals is generally used between the subscriber and switching node. For control information regarding routing and other control-related parameters, **common channel signaling (CCS)** over public packet-switched telecommunication networks is being used.

1.5.2 Integrated digital service networks (ISDNs)

Integrated service digital network (ISDN) integrates voice and data signals. The advantages of ISDN are enormous, but due to continually evolving new technologies and demand for capacity, bandwidth, and higher data rates, somehow ISDN has not been widely accepted. In order to extend the concept of ISDN's integration of voice and data, broadband-ISDN (B-ISDN), a new definition of a unified network geared to revolutionize ISDN, has emerged to provide broadband services. It is hoped that this will meet the demands of various multimedia applications' requirements. Chapter 15 is devoted to discussing various issues of switching, ISDN services, the ISDN layered model, data rate interfaces, and ISDN protocols.

B-ISDN offers not only higher bandwidth, or capacity, but also higher bit data rates of hundreds of Mbps. One of the very interesting and powerful applications of B-ISDN is video conferencing and, in fact, it has become a standard service of B-ISDN. In this service, lectures and seminars can be transmitted over TV screens. B-ISDN may not be able to introduce other services such as point-to-point, file transfer, facsimile, or multi-point, but in pure video conferencing, there is no exchange of files, facsimiles, etc. Different conference places can be connected to receive point-to-point service. Video conferencing can also be implemented for multi-point configuration, where a small number of users can communicate with each other and discuss common documents, displaying the users' live pictures either one at a time or all together on the screen. All these options are usually handled by video service providers.

Chapter 17 discusses different types of high-speed networks and presents the following network in detail: **FDDI**; **SMDS**; **frame relay**; different hierarchies for **SONET**, **synchronous digital hierarchy (SDH)**, and **asynchronous digital hierarchy (ADH)**; different modes of transfer for **synchronous transfer mode (STM)** and **ATM**; **B-ISDN**; ATM-based networks; and various protocols and standards. In order to get an overview of networking and internetworking, the following paragraph gives a brief discussion of some of these technologies.

Frame relay was introduced in 1991 and is considered a modified version of X.25 packet switching. It offers data rates of 56/64 Kbps, 256 Kbps, 1Mbps, and 2 Mbps. It is primarily designed to interconnect LANs and offers a higher capacity with low delay, low overhead, and reliable data transfer over public switched networks. It does not support both audio and data. It offers flexibility in bandwidth and is cheaper than T1. It has found its applications in manufacturing, financial, state government, and similar areas. Various future plans by some carriers include working to ATM migration and interoperability, dial-up access to global frame relay networks and connectivity, internet-access services, adding more capacity on the existing T1 backbone networks, offering more bandwidth on demand, etc. The network administrators can reconfigure their domain by using low-speed dial-up frame relay access to the main office for the exchange of e-mails, documents, files, programs, etc. The frame relay supports data rates of 1.54 and 6 Mbps and is a virtual data network service.

Switched multimegabit data service (SMDS) is a virtual data network service which allows the addition of direct links without any additional cost. This improves performance, as information can be sent directly to the destination without going through a number of intermediate switching nodes. It is useful when a company has a large number of local nodes connected to it.

Asynchronous transfer mode (ATM), based on standards defined by ITU, was introduced in 1986 and is a packet-switched (cell relay) network. It transmits video, data, and audio at a data rate of 45 to 622 Mbps or more. It offers higher network speed and scalability. It supports the needs of both LANs and WANs of public and private networks. The ATM technology defines a cell of 53 bytes out of which five bytes correspond to a header. Although it is a packet-switched network, it supports the circuit-switching mode over packet-switched networks. One of the unique features of ATM is its ability to switch the transmission between audio, video, and data at various transmission speeds.

ATM Forum, a non-profit international consortium, was formed in 1991 and is responsible for ATM interoperability specifications, standardization of communications needs at the international level, and development of standard-compliant ATM products. IBM introduced ATM software architecture as broadband network services (BBNSs) in 1993 and subsequently defined a set of products for end-to-end ATM internetworking for backbones and desktops in 1994. The software architecture addresses issues like interoperability, transport and congestion-control protocols, bandwidth management, path-switching, and quality of service for video, data, and audio. As mentioned earlier, ATM is based on cells and a variable length packet. This architecture can be connected to Ethernet, token ring, frame relay, and other WANs. 2220 Nways' ATM switch containing BBNS software provides connectivity to LANs and WANs. A reader is advised to read Reference 8 for a more detailed discussion on broadband services and communications.

1.5.3 Internet services

The Internet can be defined as a collection of different types of computer networks and offers global connectivity for data communications between them. The communication between these computer networks is defined by a common language through a standard protocol, called **transfer control protocol/Internet protocol (TCP/IP)**. Although the introduction of the Internet is derived from ARPANET, it now includes a variety of networks such as NSFNET, BITNET, NEARNET, SPAN, CSNET, and many others.

Due to the emerging use and importance of ARPANET, it was decided (in 1983) that ARPANET should be divided into two types of networks (based on packet-switching), MILNET and ARPANET, and generic word was defined for the combination of these two networks: **Internet**. The Defense Communication Agency (DCA) decided to impose the use of TCP/IP on ARPANET and MILNET via packet-switching software and defined separate network numbers and gateways for each of these networks. The number of computer networks connected to the Internet expanded from 100 in 1983 to more than 2220 by 1990. The information regarding the number of computer networks connected to the Internet is maintained mainly by the **DDN Network Information Center (DDN NIC)**. Various services available on the Internet, accessing these services, detailed implementations of different services, and different browsers available for use for different applications are discussed in Chapter 15.

Many backbone networks connected to the Internet were also introduced during this time; e.g., CSNET was defined in 1984 (to provide interaction between computer and engineering professionals and researchers who were not connected to either ARPANET or MILNET), and NSFNET was defined in 1986 (to interconnect supercomputer centers for their use by university faculty members). NSFNET has become a backbone network for U.S. national research networks and uses TCP/IP. The main feature of NSFNET is that it connects a number of low-level networks, which are then connected to the universities and other commercial enterprises. In other words, NSFNET itself is like a mini-Internet.

The Internet is being used heavily for information sharing. Searching for particular information may be done with a number of tools (browsers or net surfers). These tools offer a listing of sources, or URLs (Uniform Resource Locators), from which the information can be accessed (reading, downloading, printing, etc.). Archie uses client-server-based application software such as file transfer protocol (FTP), which looks at databases to get file listings and stores them in database servers. A wide area information server (WAIS) provides the databases with keywords. Here we have to provide the keywords for searching a particular file/program/source of information. Gopher was introduced by the University of Minnesota and is friendlier than FTP. It offers a user-friendly menu to perform searches for files/resources and provides connections to those sites. Veronica acts as a search engine for WAIS.

The World Wide Web (WWW) has become an integral part of the Internet and there are a number of Web sites created by different vendors. Information about their products and catalogs, and other related information, are available. Online service is another step towards getting the information about the product, its pictures, specifications, and data sheets instantaneously. With the advent of e-commerce, it is possible to perform online shopping and pay using credit cards. There are a variety of encryption codes which provide protection from the misuse of credit cards. Commercial online services are available which provide various applications excluding Internet e-mail service. Some online service providers are America Online, Delphi (smaller online system patterned after CompuServe which offers FTP and e-mail service), Prodigy (a centralized commercial online service), CompuServe (oldest and very extensive online service), Genie (smaller commercial online service defined by General Electric), Fidonet (an international network composed of local computer bulletin board systems), MCI mail (e-mail network usually used by corporations), Applelink (an e-mail message network defined by Apple), and UUCP (similar to Fidonet; operates between computers running UNIX systems and uses UNIX-to-UNIX copy protocol UUCP).

UseNet is defined as a system that collects the messages from various newsgroups on a particular topic. The newsgroup is nothing more than a database of messages on different topics. There are many newsgroup servers available and newsgroups can be downloaded onto a user's PC using reader software. There are two types of newsgroups: moderated and unmoderated. In moderated newsgroups, the news on a particular topic is e-mailed to moderators who will store it on a server containing the messages of that topic. In unmoderated newsgroups, the message goes straight to a newsgroup server. The UseNet hierarchy newsgroup is based on the concept of the domain name system (DNS), where different domains, or groups, are defined starting from the right side using a major name, followed by sub-names, and so on. Newsgroups are organized into the same hierarchy by defining major topics, followed by subtopics, and so on.

A PC can be connected to a host via a modem using client–server software. The PC becomes a client that can use the graphical interface provided by the host. A PC can also run a terminal emulation program and allow itself to be a remote terminal to the host. A VT-100 is the most popular terminal to emulate and this configuration provides very limited Internet services (services provided by the host). A dumb terminal can also be used for Internet access by connecting it directly to a host. Its applications are usually found in libraries, government offices, departments, and other organizations. As said earlier, PCs can use dial-up access to access the Internet via Point-to-Point Protocol (PPP) or Serial Line Internet Protocol (SLIP). These protocols create a complete connection over telephone lines using high-speed modems. PPP ensures that packets arrive undamaged. In the event of a damaged or corrupted packet, PPP resends the packet, making it more reliable than SLIP. PCs and UNIX workstations can use either PPP or SLIP.

It is hoped that once these services/facilities become available at national and international levels, they can be accessed by both computer-oriented technical users and non-technical users around the world. These users not only can communicate with each other on technical matters but also can get to know the culture, problems, social life, and many other aspects of socioeconomic issues.

The implementation of different new applications on computer networks depends on a number of constraints, such as type of network to be used, distance, number of computers connected, reliability, costs, and many other performance-based issues. These applications of computer networks can be either time-constraint or non-time-constraint. In time-constraint applications, an acceptance/indication control signal is required from the remote (destination) host before a connection can be established between two hosts for the transfer of information between them. The applications in this category include airline reservation systems, bank autoteller systems, automation of equipment, customized software, control and guidance systems for engines, spacecraft, etc.

In non-time-constraint applications, less emphasis is given to the time when a reliable connection needs to be established between the hosts. Although time of transfer of data and control signals are not critical issues, high reliability and accurate transfer of data are important. Applications such as e-mail, resource sharing, office automation, etc., fall into the category of non-time-constraints.

According to Global Reach (a San Francisco–based firm), English is the native language of only half the online population. The researchers at Angus Reid Group in Vancouver, British Columbia, Canada estimate that the U.S. online population recently dropped below 50% of the world's total for the first time. The Internet Engineering Task Force (IETF), a U.S.-heavy standards committee, is looking at the alphabet with accent marks and non-Roman characters. It is estimated that by 2005, most of the Internet will be in languages other than English. At this time, the number of online users using English is about 72 million users and it is expected to rise to 265 million by 2005. Within the same period, non-English, Chinese, Spanish, and Japanese will go from 45, 2, 2, and 9 million to 740, 300, 100, and 70 million users, respectively.

1.5.4 Client–server LAN implementation

Due to a tremendous push for worldwide competition in recent years, financial and economic sectors are gearing towards globalization for balancing the traditional centralized corporate control. This trend has forced the business community to adopt new techniques for increasing productivity at lower operating costs. This has given birth to the concept of **re-engineering**, where the corporate-wide workflow processes are being redesigned instead of simply automating the processes within their organizations. Emerging technologies are being used to fulfill the performance and productivity targets of corporate re-engineering.

The local area network (LAN) has increasingly become a growing phenomenon in the enterprise network. LANs provide interconnection between workstations, PCs, and servers so that resources can be shared. They support various administrative applications such as word processing, spreadsheet, graphic, CAD/CAM, simulation, process control, and manufacturing. The standards for LAN came into existence in the mid-1980s. Integrated multi-vendor systems have captured the market, defined networks with more than 25,000 nodes, and are replacing minicomputers and mainframes.

Two popular standard LANs are Ethernet and token ring (IBM). In Ethernet LANs, all the stations are connected to a single cable and receive the messages, which can be sent in both directions and broadcast. An Ethernet LAN is a contention-based LAN and uses carrier sense multiple access with collision detection (CSMA/CD) protocol for its operation. The token ring is composed of stations (workstations, servers, PCs, etc.) connected by either unidirectional or bi-directional communication links. A free token (a unique bit pattern) is always circulating around the ring. A station wishing to send a message has to capture the token, append it, and send it back into the ring. The destination station receives the message and sends an acknowledgment to the sending station by inverting a bit in the token.

Ethernet provides high network availability, scalability, and low channel access delay for a light load. Due to poor efficiency and lower throughput, it is not suitable for real-time applications. Token ring LAN offers a higher throughput and efficiency under heavy load conditions. The fiber channels offer a very high data rate and have therefore been used as a backbone linking existing networks for greater connectivity. Distributed processing is another area where LANs are supporting processor communications and coordination.

Large corporations usually make the decisions for any future expansion or strategy based on data stored in central databases or files residing on mainframes or minicomputers. The off-loading of processing and manipulation of data from these expensive minicomputers and mainframes should be done on cheaper workstations which offer access to the host machine. This is exactly what is proposed in client–server computing.

A client is an application program which runs using local computing resources and can simultaneously make a request for a database or other network service from another remote application residing on a server. The software offering interaction between clients and servers is usually termed middleware. The client is typically a PC workstation connected via a network to more powerful PCs, workstations, mainframes, or minicomputers usually known as servers. These are capable of handling requests from more than one client simultaneously. A detailed discussion of client–server LAN implementation, different client–server models, network and client–server operating systems, various **graphical user interface (GUI)** tools, and toolkits customizing user interfaces, followed by a few case studies, is presented in Chapter 20.

1.6 Network architectures and layered protocols

Today, a large number of network products are available on the market from different vendors. It is time-consuming, confusing, and difficult (if not impossible) to connect heterogeneous systems (hosts) and to transfer data between them. In some cases, these systems are incompatible and unable to provide any communication between hosts. Another serious problem regards the architecture of networks. Vendor-customized network architecture may have different layers and different interfaces with common functions. Each manufacturer may have its own standards for the same product (hardware or software) and the replacement of any broken part may cause an incompatibility problem. This causes a lot of inconvenience and is an inefficient way of doing business. The standardization of products helps the customers through convenience and the manufacturers through profit, ease of upgrading, and new product development. There are different types of standards, such as proprietary and open. The former is controlled by one company and no other company can manufacture the product using that company's standard. Some examples include ARCnet and Apple MacIntosh computers. The open standard allows other companies to manufacture the product, sometimes charging a royalty fee from them, e.g., the personal computer (PC).

In order to overcome some of these difficulties, in 1977 the **International Standards Organization (ISO)** set up a committee to propose a common network architecture which can be used to connect various heterogeneous computers and devices.[9] It was also an initial step for defining international standardization of various standards and protocols to be used for different network layers.

ISO published a document in 1984 containing the specification about network architecture and protocols as ISO 7498 and defined it as the **open system interconnection-reference model (OSI-RM).** This name is given due to the fact that it will be used for those systems which are open, i.e., the systems that are open to communication with other vendor-customized or general system architectures. The customers have a choice of choosing a vendor, making investment, and communicating with others. One of the unique features of this is that it allows smaller companies to compete in licensing the products to other companies. Another international standards organization, the **Consultative Committee for International Telegraphy and Telephone (CCITT)**, which is now merged with the **International**

Communication Union-Telecommunications (ITU-T), also published similar and compatible standards as X.200. The United Nations (UN) created this organization for defining the technical aspects of the standards. The Canadian Standards Organization has also defined two standards organizations: the Canadian Radio and Television Commission and Industry Canada. The U.S. has defined a variety of industrial and government organizations such as the American National Standards Institute (ANSI), the Institute of Electrical and Electronics Engineers (IEEE), and the Electronic Industries Association (EIA). Xerox, Intel, and Digital Equipment Corporation (DEC) constituted the IEEE 802.3 committee to define the standard LAN as Ethernet. The U.S. government standard organization is the National Institute of Standards and Technology (NIST), formerly known as the National Bureau of Standards (NBS).

One of the main objectives of OSI-RM is to define/develop protocols to perform the functions of each layer and implementation detail of the services it is providing. The reference model concept is based on IBM's SNA (defined in 1972). The layering concept can be seen in most of the design products, buildings, or any organizational environment.

1.6.1 *Open system interconnection-reference model (OSI-RM)*

OSI-RM architecture of network is shown in Figure 1.4. The following paragraphs present a brief description of each of the layers. Detailed functions, services protocols, and interfacing and protocols for internetworking of these layers can be found in Chapter 7. A reader may read other publications.[1,3,4,10]

Physical layer: The physical layer is mainly concerned with electrical, mechanical, procedural, and functional aspects of transmission media for information transmission and receiving over the network. It specifies the details of connecting cables, processing of digital signals, interfaces to different media, etc. The most popular standard for this layer

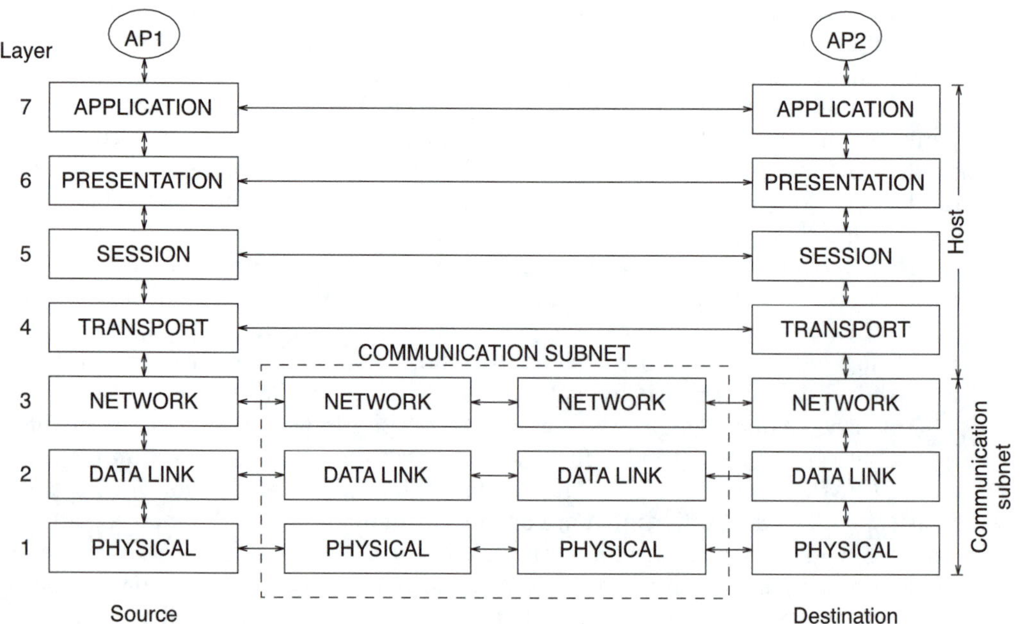

Figure 1.4 OSI reference model.

of digital network interface is X.21 (CCITT). Other standards for this layer include V.24 (CCITT) and RS-232 DCE/DTE interface (Electronics Industries Association (EIA)).

Data link layer: The data link layer is responsible for maintaining the integrity of data information between two sites. It offers a reliable channel for data transmitted over the transmission media. The protocols of this layer provide error recovery (error detection and error correction) to handle the errors caused by noisy media. Various standard protocols for the data link layer which provide reliable channel and error recovery are IBM's BSC, synchronous data link control (SDLC), ISO's high data link control (HDLC), and link access protocol-balanced (LAP-B) (CCITT's version of HDLC). This layer is further divided into two sub-layers known as media access control (MAC) and logical link control (LLC). For asynchronous transfer mode (ATM) switch (giving a very high data rate of a few hundred Mbps), this layer is divided into two sub-layers known as segmentation and reassembly (SAR) and convergence.

Network layer: The network layer provides communication between the user's PC and public or private networks. It defines addressing and routing (logical link) between source and destination sites. It also provides internetwork routing between two remote sites. This layer also makes sure that data packets are transmitted across the network correctly. X.25 is a standard for this layer; other proprietary networks providing an equivalent layer to this layer include IBM's SNA and DEC's DNA.

Transport layer: The transport layer offers network-independent service to higher layers and hides all details regarding the network being used for transmission. The upper layers have no idea about the type of network which is transmitting or receiving the data information. This layer breaks the message into smaller packets and provides the segmentation and reassembly. It also offers end-to-end error control and recovery.

Session layer: The session layer provides a session for data communication between two application processes. It also supports the synchronization between sites and defines checkpoints from which diagnostic tests can be performed in the event of a failure. It establishes the length of the session during which users log in and log out.

Presentation layer: The presentation layer represents the data information in appropriate form to be acceptable to the lower layers of the network. Two standard representation schemes for character representation are ASCII (plain text) and EBCDIC (IBM plain text), which are usually supported by most of the networks. The presentation layer provides different formatting styles to the data information and associated compatibility between various cooperating processes. It also converts the information (bitstream) into video, audio, and other formats. All the application entities of the application layer are also translated or mapped into suitable entities by this layer.

Application layer: The application layer provides an interface between application entities and the user's computer. This layer offers services to a variety of aspects of data communication between the user's computer and application entities including terminal handling, file handling, text interchange, job transfer, and manipulation. For each of these aspects, a number of standards have been defined, and a collection of these standards is known as the specific application service elements (SASE). Within each of these aspects of application, quite a number of standards have been defined and are continuously being defined to cater to different upcoming applications from various technologies. For example, a large number of terminal handling standards have been defined, such as basic class, form, graphics, and images. The application layer offers a variety of applications such as e-mail, data transfers, file transfer, digitized video, audio, data, remote login, and other Internet services.

1.6.2 Layered protocols

In OSI-RM, it is assumed that a lower layer always provides a set of services to its higher layers. Thus, the majority of the protocols that are being developed are based on the concept of a layered approach for OSI-RM and are therefore known as **layered protocols**. Specific functions and services are provided through protocols in each layer and are implemented as a set of program segments on computer hardware by both sender and receiver. This approach of defining/developing a layered protocol or functional layering of any communication process not only defines a specific application of communication process but also provides interoperability between vendor-customized heterogeneous systems. These layers are the foundation of many standards and products in the industry today. The layered model is different from layered protocols in the sense that the model defines a logic framework on which protocols (that the sender and receiver agree to use) can be used to perform a specific application. A frame (carrying the information) is defined as a bit pattern sent over the physical media (wires, cables, satellites, etc.). Within the frame, the boundaries for various fields are defined. Typically, a frame carries the data information of between 400 to 16,000 bits, depending on the type of network being used.

Functions of a layered protocol: Communication control procedures are defined as a set of rules and procedures for controlling and managing the interaction between users via their computers, terminals, and workstations across the network. The control procedures have been addressed by different names in the industry, such as protocols, line control procedures, discipline procedures, and data link control procedures. The protocols provide a communication logical link between two distant users across the network and the handshaking procedures provide the call establishment within the protocols.[1,3,4]

The following is a list of functions being offered by the layered protocols:

- Allows distant users to communicate with each other.
- Allows the sharing of media resources by both control and data packets and other facilities using interleaving or multiplexing. The control packets are used mainly to ensure the correctness of data transfer across the network.
- Supports serial communication, i.e., multiple communication transfer over a single channel.
- Supports point-to-point and multi-point connection configurations.
- Supports leased-line (dedicated) and public-switched connections.
- Supports orderly interleaving of control and data packets.
- Provides error-free communication between users in the presence of inherent nodes and errors introduced during the transmission.
- Provides data integrity.
- Supports the establishment of connection, data transfer, and termination of connection.

Layered protocol configurations: A layered protocol can be defined in a variety of ways and usually depends on the type or category of organization, application of data communications, etc. It has been standardized to define the client–server relationship for the layered protocols and, based on the type of services and operations required, different categories of configurations for layered protocols can be defined (discussed below).

1. **Point-to-point (non-switched):** A dedicated line or a private leased line defines this configuration. The protocol for this configuration defines one node as a controller performing various operations for data transfer across the line, while other nodes interact with each other on the basis of point-to-point connection or communication. The configuration is based on star topology and is also known as a **star-type network**.

2. **Point-to-point (switched):** In this configuration, the connection is established via dial-up operation. The protocol defines the procedures for connection establishment, data transfer, and connection de-establishment/termination phases. It uses device automatic calling units (ACUs). Any station can make a request at any time and the protocol must resolve the contention problem.
3. **Multi-point (non-switched):** This configuration is also known as a **multi-drop system**. Here, a communication channel is shared among users and all the users receive a message transmitted on the channel. Typically, one station controls all these operations and it also resolves the contention problem.
4. **Multi-point (switched):** This configuration is different from multi-point (non-switched) in the sense that users have to make an initial request for sending a message. Different phases for sending messages across the channel are handled by the appropriate protocols and set of procedures.
5. **Loop system:** This configuration uses a ring topology where the data is moving from one station to another over the ring (physical link). It requires nodes on the ring to include a set of protocols, which resolve various issues during the data communication over the ring. This configuration may define both the centralized and distributed categories of protocols.

1.6.3 Features of layered protocols

A layered protocol for OSI-RM should offer the following features:

- Decomposition of a large communication process into a smaller number of sub-processes (layers) using levels of abstractions at different levels should be defined.
- Each sub-process should define a specific set of functions and offer a specific set of services to its neighboring sub-processes.
- An appropriate interface (standard, protocol, etc.) between each sub-process should be defined through service access points (SAPs).
- Each sub-process should have the same function to provide interoperability for defining international standards.
- An attempt should be made to minimize the information flow across the interface.
- The number of layers should be enough to define the complete function with the minimum size of computer architecture.
- It should provide modularity and flexibility so that changes can be incorporated with little overhead.
- It should provide survivability and security for the computer communication networks.

1.6.4 Layered protocol representation

A representation scheme for a layered communication protocol is useful during its design and implementation, verification, and validation processes. It must show various time-dependent functions, the direction of data flow, various states of the communication system, etc. There exist various representation schemes for layered protocols, but the most common and widely used include **time-line graph**, **ANSI-graph**, **state transition graph**, **Petri nets**, **logical flow charts**, **queuing models**, etc. We briefly describe these schemes next.

Time-line diagram (TLD): The TLD defines various activities (message transfers) between two nodes over the time axis in sequential order. This diagram is useful and suitable for showing the sequence of all the characters of message transfers over a time axis, as shown in Figure 1.5.

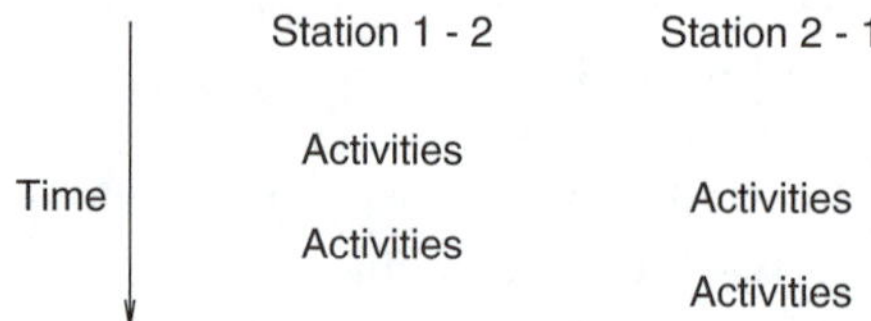

Figure 1.5 Time-line diagram.

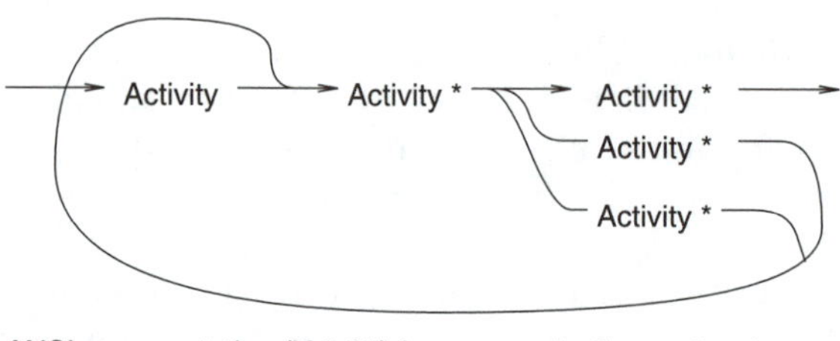

Figure 1.6 ANSI representation.

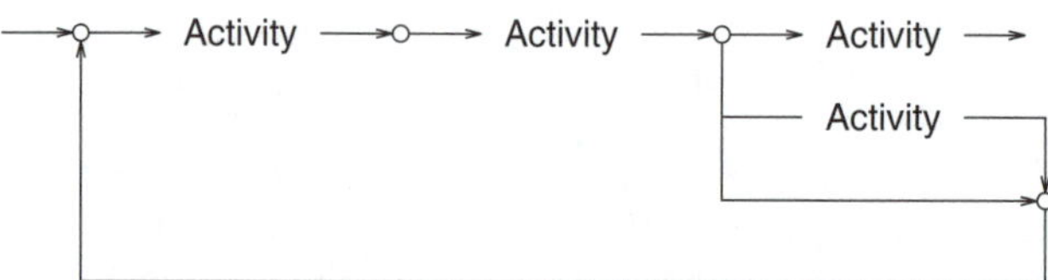

Figure 1.7 State transition graph for communication protocol representation.

American National Standards Institute X.3.28 (ANSI X.3.28): This representation scheme describes the activities of a protocol in a very compact and structured form and also depicts iterative transactions between nodes. This scheme is suitable for a half-duplex transmission system. A typical ANSI representation (ANSI recommendation: X.3.28) for a communication protocol between two nodes is shown in Figure 1.6. This scheme does not indicate any specific iterative flow between nodes. The symbol * represents line turnaround.

State-transition graph: A state-transition graph follows a different approach than the previous method; the states of the system are defined as nodes and various activities are defined as connecting links between the nodes, as shown in Figure 1.7. A change in state is defined by the occurrence of some activities between nodes, and after the completion of those activities, a new state (corresponding to the connected node) is defined to which those activities are connected.

Petri nets and logical flow charts: These methods basically describe the behavior of the system in the form of states.[5,6]

A layered communication protocol defines three phases for data communication:

- Connection establishment.
- Data transfer.
- Connection de-establishment/termination.

In the case of connectionless services, these three phases are not required and the packets to be transmitted contain the complete information regarding destination address and other control information.

The protocol depends heavily on **handshaking** for the connection establishment phase.

Layered protocols using the same connections ideally will allow different systems to communicate openly with each other without changing the logic of layers. Thus, the systems which were unable to communicate with each other due to their closeness should now be able to communicate with each other using layered protocols with slight changes in the software and hardware.

The reference model indicated earlier is a logic framework of different layers and provides specific functions and services to users. Layered protocols, which define a set of rules, are implemented on the framework of the reference model. The main purpose of the model is to provide interoperability and to define an international service on it. ISO defined the framework for the model in terms of layers describing services/functions that are needed in international computer communication networks. By having defined layers for the framework, designers/developers can implement layered protocols independently because the functions and services of each layer are very specifically defined and are accepted by the network community.

1.6.5 Protocols, layers, and interfacing in computer networks

There are a number of ways of defining communication systems between hosts. The usual approach is to define a communication system as a set of interacting processes (components) which provide functions and services to the users. This approach is not a useful one because it describes the function of communication environment in a very limited mode and it does not provide any means of interoperability for different types of networks.

The second approach is rather abstract in the sense that the functions/services of networks are defined as a collection of processes at various levels. At each level, we can define one layer for a particular process(es). Each layer in turn represents a specific function(s) and service(s) and can be different for different types of networks. However, each layer offers certain services to its higher layers. The implementation of these services is totally hidden from the higher layers. The services offered by the layers are provided through protocols and are implemented by both source and destination on their computers.

Protocols are defined in terms of rules (through semantics and syntax) describing the conversion of services into an appropriate set of program segments. Each layer has an entity process that is different on different computers, and it is this entity (peer) process through which various layers communicate with each other using the protocols. Examples of entity process are **file transfer**, **application programs**, and **electronic facilities**. These processes can be used by computers, terminals, personal computers (PCs), workstations, etc.

In OSI-RM, it is assumed that a lower layer provides a set of services to its higher layers. Thus, the majority of the protocols that are being developed are based on the concept of the layered approach for OSI-RM and are therefore known as the layered protocols. Specific functions and services are provided through protocols in each layer and are implemented as a set of program segments on computer hardware by both sender and receiver. This approach of defining/developing a layered protocol or functional layering of any communication process not only defines a specific application of communication process but also provides interoperability between vendor-customized heterogeneous systems. These layers are, in fact, the foundation of many standards and products in the industry today. The layered model is different from layered protocols in the sense that the model defines a logic framework on which protocols (that the sender and receiver agree to use) can be used to perform a specific application.[1]

A layer (*m*) defines two neighboring layers as layer *m*–1 (lower) and layer *m*+1 (upper). This layer (*m*) uses the services of layer *m*–1 and offers services to *m*+1. The *m*-entity

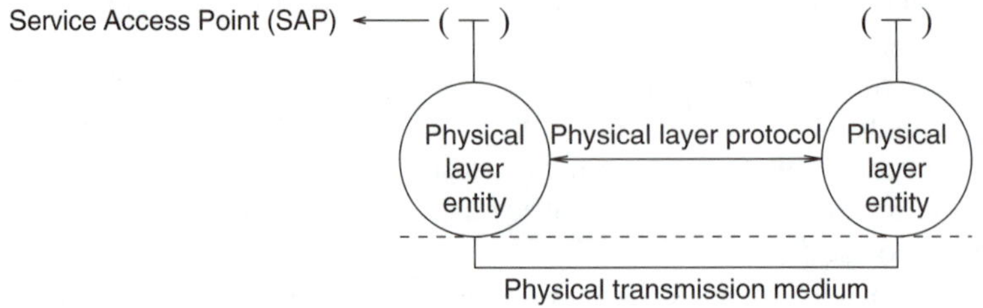

Figure 1.8 Communication link between physical layer entities.

performs functions within layer *m*. The same layers of two different networks define a communication between their respective entities via peer-to-peer protocol. Thus, we have a communication logical link between peer *m* layer entities of both networks as *m* layer peer-to-peer protocol. A protocol associated with each layer processes the data being exchanged between the peer layers. Each node (PCs, terminals, etc.) connected to a network must contain identical protocol layers for data communication.

Peer protocol layers exchange messages (which have a commonly understood format) that are known as a **protocol data unit (PDU)**. A PDU also corresponds to the unit of data in *m* level peer-to-peer protocol. The peer *m* entities communicate with each other using the services provided by the lower layer (*m*–1). The services of layer *m* are provided to layer *m*+1 at a point known as *m* service access point (SAP). This is the point through which services of layer *m* can be accessed by layer *m*+1. The communication link between physical layer entities is shown in Figure 1.8.

For a description of the interface between *m* and *m*+1 layers, the *m*-primitives are defined. The *m* service data unit (SDU) associated with *m* primitives is delivered to layer *m*+1 and, similarly, the layer *m*+1 also uses *m*+1 primitives to interact with layer *m*. The higher layer *m*+1 sends *m*+1 protocol data unit (PDU) to layer *m*. The *m*+1 PDU consists of *m*+1 protocol, control information, and *m*+1 user data. The *m*+1 control information is exchanged between *m*+1 entities.

Each layer passes message information (data and control), starting from the top layer (application) to its lower layer, until the message information reaches the lowest layer (physical) which provides the actual data communication. An interface is defined between layers, which supports the primitive operations and services the lower layer offers to its higher layer. The network architecture is then defined as a set of protocols and interfaces.

A **protocol** defines rules, procedures, and conventions used for data communication between peer processes. It provides data formatting, various signal levels for data and control signals, error control, synchronization, and appropriate sequencing for the data to be exchanged. Two networks will offer two-way communication for any specific application(s) if they have the same protocols. The layer *m* is known as the service provider for the layer *m*+1 and offers the services at SAP.

Private networks use a non-standardized reference model and, therefore, interface and protocols are unable to offer access to other networks. Although gateways may be used to connect these networks for data exchange, the implementation is very tedious and expensive. A variety of data transmissions within public data networks have been proposed. Some of these transmissions along with their data rates are:

- The start/stop mode (used in asynchronous transmission) offers data rates of 50–300 bits/sec.
- The synchronous operation mode offers data rates of 600, 2400, 4800, 9600, and 48,000 bps.

- The CCITT-recommended X.25 offers data rates of 1200, 2400, 4800, 9600, and 64,000 bps.
- The CCITT-recommended X.29 offers data rates of 50, 300, 1200, and 2400 bps.
- ISDN offers a data rate of 64 Kbps.

IBM's SNA offers LAN-to-LAN traffic and has been very popular to enterprises with large networks. It is managed by IBM's NetView (IBM's host-based network management system). The network management system can be shifted to a LAN from the mainframe using advanced peer-to-peer networking (APPN) that is based on the client–server framework. It allows users to connect LANs across an SNA backbone, providing simplified dynamic scalability without the intervention of a central computer and extra reprogramming. Peer-to-peer networking allows more than one station to communicate with each other and share resources, thus avoiding the use of a dedicated server. This type of architecture seems to be better than client–server, as it does not need any server for resource sharing because the stations can switch their modes (client or server) dynamically.

ISDN is based on a digitized telephone network, which is characterized as a 64-Kbps channel network. The 64-Kbps channel is derived from the 3.4-KHz voice channel which is required for voice (eight-bit sampling with a frequency of 8 KHz, offering a data rate of 8 KHz $\times$ 8 = 64 Kbps). ISDN is a circuit-switched network, but it can also offer access to packet-switched services.

1.6.6 Information coding

A coding scheme provides a unique representation to various symbols used in data communication. Symbols include letters, digits, and other special characters. These different types of symbols are also known as graphic characters. These characters are usually considered information characters for the coding scheme.

Early computers (first and second generation) used six-bit combinations for each of the character representation in binary form. This coding scheme was known as **binary-coded decimal (BCD)** and can represent 64 combinations. **ASCII** is a seven-bit code and can represent 128 combinations of different symbols. When it is modified to include one more bit for character parity, this coding scheme is known as EBCDIC and can represent 256 combinations of symbols. One of the main advantages of this code is that it provides error detection in addition to parity technique, and, also provides some control functions.

One of the main objectives of data communication systems is to allow users to transmit messages over a transmission media (usually noisy, susceptible to natural calamities, etc.) error-free. In other words, all the errors introduced during the transmission (due to transmission or media) must be detectable. The coding technique alone cannot detect all the errors all the time. A **redundancy** may be used for error detection. This adds extra bits in the message and will be used to detect all the errors at all times. However, this will affect transmission efficiency due to overhead. In order to improve the performance of the network, the redundancy must include a minimal number of bits for detecting the errors. There are a number of ways of achieving this.

In the simplest way, the receiver sends a **negative acknowledgment (NAK)** after detecting an error. For error detection and correction, the automatic repeat request (ARQ) method is used. In the ARQ method, the erred frame is re-transmitted until it is received error-free. Alternately, we can add enough redundancy into the messages so that the receiver is able to detect and correct the error itself rather than sending a NAK to the transmitter. The error detection and correction can be achieved by forward error-correcting (FEC) schemes. More details are given in Chapter 9.

1.6.7 Character error detection and correction

When we transmit a character over transmission media, a parity bit scheme (for redundancy) is used for detecting the error. The parity bit may be defined either as an even parity (total number of 1s in any character is even) or odd parity (total number of 1s in any character is odd).

Parity scheme: The parity scheme often requires the execution of some logic operations at both sending and receiving nodes. This type of redundancy can be achieved in both block coding and convolution coding techniques.

Block coding: In this scheme, a logical operation is performed to introduce redundant bits into the original message so that the message defines a fixed size of the format. For example, we translate a message of size *A* bits into a size *B* bits with *B-A* as redundant bits. There are a number of ways of implementing this translation process. The block coding technique is well-suited to ARQ techniques and can use either constant-ratio coding or parity-check coding. Two popular and widely used codes supporting the parity-check coding are geometric and cyclic.

Geometric coding: In this scheme, one bit is added to *A*-bit information to make it either odd or even code. This code can detect only either odd or even number of bit error. It cannot detect both errors together. The geometric code includes one parity to each character and one parity bit to each level of bits for all the characters in the block. The parity bit added to each character corresponds to **vertical redundancy check (VRC),** while the parity bit added to each bit level corresponds to **longitudinal redundancy check (LRC)**. This code is useful for ASCII code, but it can not detect all types of errors.

Cyclic (or polynomial) coding: This code requires a polynomial generator, which is sometimes very important because the number of bits in it ensures the detection of all errors of length less than the number of bits. This is more powerful and attractive than any other code for a given level of redundancy. This code is very easy to implement via shift registers and connections. The standard polynomial generator chips are available from Motorola Corporation for CRC. These chips are used in serial digital data communication for error detection and correction. The single chip can support more than one polynomial generator and it also offers additional functions.

Convolution coding: This coding is, in general, considered the most efficient for forward error-correcting (FEC) systems. This code is very useful for high-speed data communication.

1.7 Data communication in OSI-RM

The issue of data communication in OSI-RM deals with how the data actually flows through it, as shown in Figure 1.9. The OSI-RM looks at a computer network as a collection of sub-processes with a specific set of services and functions transformed into each sub-process (corresponding to layers). Within each sub-process, the services or functions of layers are obtained through entities (segment of software or a process, hardware I/O chips, etc.). Entities in the same layer on different systems are called **peer entities**.

The sender gives the data (to be transmitted for a particular application) to the application layer. For each of the applications, an entity carries on a dialogue with its peer entity at the same layer, i.e., the application layer on another computer. An application header *AH* containing this dialogue and other control information is attached to the beginning of the data and it then offers this as user's data to the presentation layer, as shown in Figure 1.9.

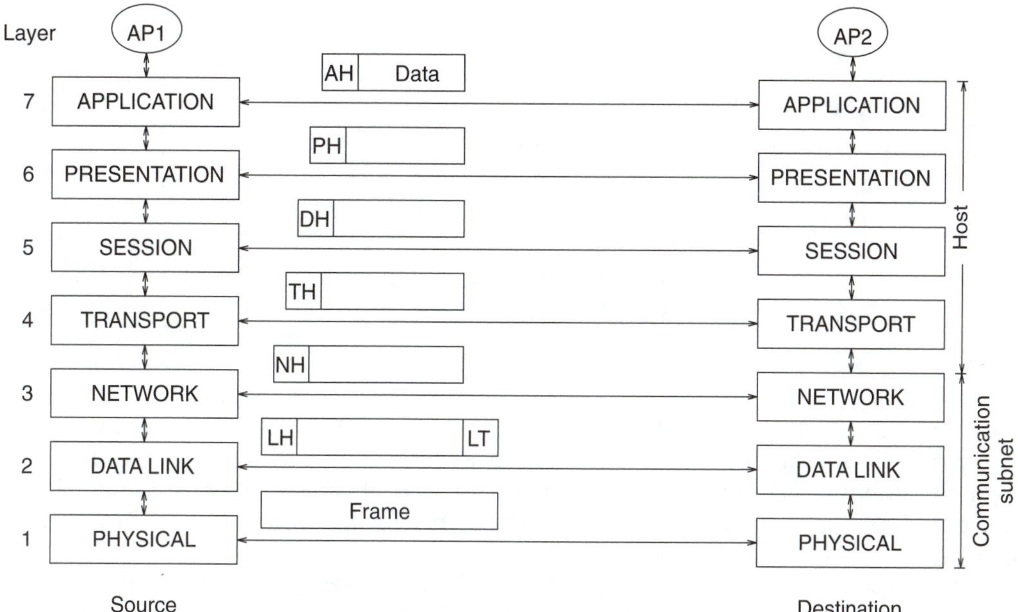

Figure 1.9 Data flow through OSI reference model.

The presentation layer selects an appropriate syntax for data transformation, formatting, encryption, etc., and adds its own header at the beginning as *AP* and presents this as the user's data to its session layer. The actual data at the application layer moves in this way through each of the lower layers until it reaches the physical layer. Up to the network layer, the user's data is contained in various packets. The data link layer makes a **frame** for each of the packets received from the network layer. These frames are transmitted over a communication medium to a host computer on the other side.

At the receiving side, the physical layer receives these frames and presents them to its data link layer. The data link layer takes its header from a frame and passes it on to its network layer. In this way, the original data reaches the application layer after going through the various layers and doing the opposite of what was done at the sending side. Similarly, if the receiver wants to send data or an acknowledgment, it goes through its OSI-RM in exactly the same way as it went through the OSI-RM at the sending side. It is important to note that it is only the data link layer which adds a header and trailer on each side of the packet (data or control) it receives from its network layer. The other layers add only a header at the beginning of the user's packet.

An advantage of defining peer entity is that these entities on OSI-RM first establish a connection before the transfer of the user's actual data. Depending on the type of protocol under consideration, we can have different primitives, but for a typical protocol, the following are the steps for establishing peer-to-peer connection using primitives on OSI-RM, as shown in Figures 1.10a and 1.10b. The upper layer (sender side) issues a **request** command to the protocol of lower layers of the network for establishing a connection with its peer entities and sends it to the upper layer entity of the other side (receiving) for its **indication**. If the connection is approved (optional), the receiver side sends the **response and confirmation** of the establishment of a connection between them. Once a connection is established, the upper layer entity of the sender will use the same set of primitives for sending the data to the receiving side which, in turn, uses the same set of primitives for confirming the acknowledgment of receipt of data. These four commands — request,

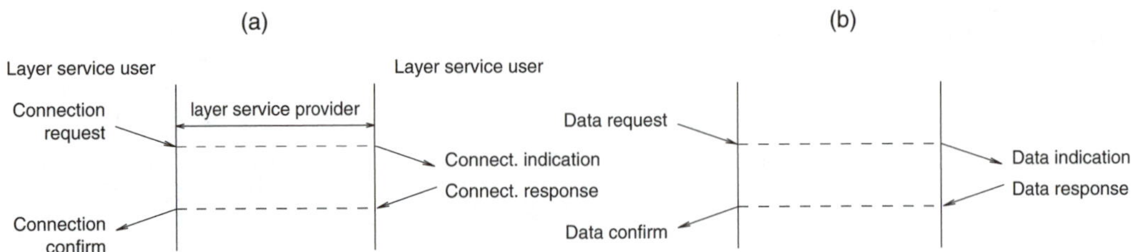

Figure 1.10 Peer-to-peer connection between source and destination of OSI reference model. (a) Connection establishment. (b) Data transfer.

indication, response, and confirmation — are known as attributes which are attached to every primitive issued by higher layers for its lower layers. Each layer defines its set of primitives and each primitive is expressed with four attributes. Each primitive is also associated with a number of parameters, which include the destination address, source address, quality of services, acknowledgment, etc.

As indicated earlier, layers of OSI-RM provide the interface between each other; i.e., the data link layer provides the interface between network and physical layers. Further, each layer has its own entity through which services on protocols can be performed/implemented. An entity of the application layer is known as **application entity**, that of the presentation layer as **presentation entity**, and so on. Higher layers are further defined as **service users**, while lower layers are defined as **service providers**. The services offered by higher layers are implemented by the entities of lower layers. That is, presentation layer entities contain all the services provided by the application layer.

The services offered by layers are available at service access points (SAPs) and can be accessed by higher layers at these points. Each SAP has a unique address (either hardware or software). Socket numbers (as in UNIX Berkeley) define these SAPs as sockets and the SAP addresses. An operating system can also create ports for these SAPs in the OSI-RM.

LAN environment: The typical LAN environment that exists today in universities, corporations, organizations, and government agencies uses FDDI as a backbone network which connects different types of LANs such as Ethernet and token ring. These LANs are connected to FDDI via a 10-Mbps link for Ethernet or a 4- or 16-Mbps link for a token ring network. Various servers are usually connected to FDDI via 10-Mbps links. The FDDI is also connected to routers via high-speed links of 50 Mbps or more. Routers are connected to the Internet or other routers via links ranging from 1.544 Mbps to 100 Mbps and provide access to the outside world.

ATM can be used for the existing LAN environment in a number of ways. In one environment, the FDDI can be replaced by an ATM switch and some of the servers may be connected via high-speed links. The FDDI can also be reconfigured as an access network for different LANs. The second environment may define an ATM hub (ATM-based) where ATM switches are connected via OC3 links and can be used as a backbone network in place of FDDI. Here the ATM switches will be providing interfaces (ATM adapters) and accesses to different types of LANs. These ATM switches will also provide interfaces to routers via DS3 or above links for access to the outside world. In video-based applications, the ATM can be connected directly to workstations or terminals via optical fiber. This allows a data rate of 100 Mbps with the option of transmitting video, voice, and data. A number of servers dealing with these types of traffic (video, voice, and data) can also be directly connected to ATM via ATM adapters.

1.7.1 OSI network management

The network management problem has been considered by a number of organizations, proprietary network vendors, etc. In the area of LAN management, the LANs interconnect a number of autonomous, multi-vendor, and different operating systems. The OSI committee has defined a number of management domains for future network management activities such as network management, LAN management, OSI management, and management dealing with loosely coupled interconnection and distributed computing aspects of LANs. LAN management deals mainly with lower layers and IEEE considers LAN management as system management.

One of the main objectives of LAN management is to provide continuous and efficient operation. Based on a number of recommendations, various issues are addressed in LAN management. These include configuration management, fault management, performance management, access control management, and accounting management. Configuration management deals mainly with controlling LANs from a remote node. This node should be able to broadcast to all the users on the LAN. Fault management deals with fault detection, diagnosis, and correction. Protocols must offer these services in the form of error reporting, confidence testing, repair (hardware and software), etc. Performance management allows the evaluation of the performance of the LAN system, collection of statistical data about the network traffic, etc. Access control management deals mainly with maintaining the integrity of the LAN system. Accounting management keeps a record of the number of packets and maintains statistics of packets for planning capacity, bandwidth, and other application-based updating.

The IEEE defined the two layers of the system management model as layer management and system management. Layer management is defined for each layer and each layer is responsible for gathering all the relevant information for layer management. Within each layer, a layer management entity is defined to collect the information of its layer. Layer management is coupled with the system management application process entity, and they interact with each other.

In wide area networks, typically end users communicate with each other over a logical connection defined over a series of switching nodes. The user data is divided into fixed, smaller-size packets and each packet is transmitted separately over different logical connections. There are a large number of wide area networks defined which offer connectivity within and across countries. In the U.S., ARPANET became the first experimental network connecting users for sharing of information. In the U.K., British Telecom has defined Packet Switch Stream wide area network. Wide area networks are managed by a number of organizations. In contrast to this, LANs are usually controlled by the organizations using them. Obviously, LANs offer higher data rates than WANs because LANs cover a smaller distance, typically a few kilometers. Further, LANs offer higher reliability and a wide range of interconnectivity.

Network management in WANs deals mainly with auditing and accounting. This is due to the fact that service providers or carriers usually provide these services to the users. The charges are calculated on the basis of the number of packets and also the number of switching nodes involved in the communication process.

The network management standard protocols have been defined for both LANs and WANs. In WANs, the switching nodes are interconnected and carry the packets from a source to a destination node. There are a large number of different types of WANs being defined and used in different countries. WANs are usually controlled and maintained by telephone companies. Some of the popular WANs include Catenet (U.S.), British Telecom's Packet Switch Stream (PSS), and Joint Academic NETwork (JANET). Network management in the case of WANs deals mainly with accounting and billing because users are getting these services from the same carriers. There are a number of software tools that can be used to manage networks, congestion control, charges for packet transmission, and many other related functions.

1.8 Standards organizations

There are a large number of network products and a variety of application-based network tools available from different vendors. It is very difficult (if not impossible) to connect even homogeneous systems (defined by different vendors) for the transfer of data from one system to another at a global level. In some cases, these systems are incompatible and there cannot be any communication between them. Another serious problem is in regard to the architecture of the networks. Vendor-customized network architecture may have different layers and interfaces for the common functions and services defined in OSI-RM.

In order to spread wide acceptance of using both communication and computer products (for interoperability) of different vendors for avoiding, or at least lowering, burdens on customers, the customers were required to develop protocol conversion software for their computers to communicate with different vendor computer communication products. This was achieved by defining several standards organizations so that the rift between the two communities (communication equipment and computer products) could be reduced and customers could use services offered by the networks by using these standards. The standardization process is one way to avoid incompatibility between vendor products and helps to provide interoperability between their products.

There is no well-defined definition of standards, but the widely used definition proposed by the National Policy on Standards for the United States of America in 1979 has been accepted widely and is reproduced below:

> A prescribed set of rules, conditions, or requirements concerning definition of terms; classification of components; specification of materials; performance, or operations; delineation of procedures; or measurement of quantity and quality in describing materials, products, systems, services, or practices.

Standards organizations are classified as government representatives, voluntary, non-treaty, and non-profit organizations. The standards organization is defined as a hierarchical structure-based or bottom-up model for reviewing and approving the standards. Various standards bodies have been identified to serve the needs of various regions, e.g., T1 (North America), ETSI (Europe), TTC (Japan), TTA (Korea), AOTC (Australia), CITEL (Latin America), etc. There are two levels of standards organizations:

1. International organizations;
2. National (domestic) organizations (U.S.).

1.8.1 International standards organizations

This level of organization is composed of three different and distinct international standards organizations, each performing a specific task(s).

- International Telecommunication Union (ITU)
- International Telegraph and Telephone Consultative Committee (CCITT)
- International Standards Organization (ISO)

International Telecommunication Union (ITU): This is an agency of the United Nations and provides coordination among various governments for the development of standards. It is a treaty organization formed within the U.N. in 1965. ITU has now become a special body in the U.N. Other organizations which are either part of ITU or interact with ITU are the **International Radio Consultative Committee (IRCC)** and the **International Telegraph and Telephone Consultative Committee (CCITT).** The former organization includes two subgroups; one group deals with international radio broadcasting, while the other group deals with telephone and data communication. The main objective of the CCITT is to provide technical recommendations on telephones, telegraphs, and data communications interfaces. The functions of the ITU include interconnectivity of network resources and services, international regulations, and growth of telecommunications in developing countries via the Technical Cooperation Department and the Telecommunication Development Bureau. Some of the useful ITU-T study groups are:

- Study Group (SG) I (MHS, fax, directory services, ICCN, PCS/UPT).
- SG II (network operations, numbering/routing country codes, fax quality).
- TG VII (data communication networks, MHS, data communications, X.25 for ICCN).
- SG IX (telegraph networks and telegraph terminal equipment).
- SG XVII (data transmission over the telephone network).
- SG XVIII (ISDN, broadband, UPT network terminology).
- SG XVIII (set up in 1985 to look into B-ISDN with an emphasis on architecture and services for high-speed networks)

More details can be found in Reference 8.

International Telegraph and Telephone Consultative Committee (CCITT): CCITT was a member of ITU, which itself is a treaty organization formed within the United Nations in 1965. The goal of CCITT (as mentioned in its chart) is "to study and issue recommendations on technical and tariff questions relating to telegraphy and telephony." Recently CCITT merged with ITU. CCITT submits recommendations to ITU on data communication networks, telephone switching standards, digital systems, and terminals. Since CCITT comes under ITU (now merged with ITU-T), various governments (voting members) are the members of CCITT. The Department of State (U.S.) is a representative of the government in this organization. Several other categories (non-voting members) at different levels have also been allowed, such as recognized private operating agencies (RPOA), scientific and industrial organizations (SIOs), and international organizations. Two ITUs known as ITU-T and ITU-R were given the task of coordinating global standards activities. ITU-R is mainly responsible for radio communication, while ITU-T deals with telecommunications.

The primary objective of CCITT is to define standards, techniques, and operations in telecommunication to provide end-to-end service on international telecommunication connections regardless of the country, region, or culture. This organization is dedicated to establishing effective and compatible telecommunications among its member nations. It has five classes of members, which are discussed below:

- **A Class** (National Telecommunication Administrations) includes government agencies and representatives of nationalized companies (duly approved by the government of the respective country). In general, these agencies and representatives are known as **Post, Telegraph, and Telephone administration (PTT).** The PTT is more popular in European and Asia Pacific countries. In the U.S., there is no PTT, so the government agency known as the Federal Communications Committee (FCC) under the Department of State represents the U.S. in this class. In France, the PTT is a representative of this class and many European countries are represented by their respective PTTs. Only A class members have voting power.

- **B Class** includes private organization administrations, which are duly recognized by their respective governments. For example, a private company, **American Telephone and Telegraph (AT&T)**, is a representative of the U.S. in this class.
- **C Class** includes scientific and industrial organizations.
- **D Class** includes other international organizations engaged in data communication standards for telegraph and telephones.
- **E Class** includes other miscellaneous organizations, which are not directly involved in the previous organizations but have interests in CCITT's work.

A standard X.25 defining three layers of OSI-RM (physical, data, and network) is a very popular protocol and is being used in all public networks around the world. The public networks are packet-switched networks. X.25 provides an interface between public networks and computers, terminals, workstations, etc.

Another standard defined by CCITT which is also widely accepted is V.24 (physical layer standard). This connector provides an interface between terminals and networks. Various pins, their placement, and meaning on the connector are defined by asynchronous terminals/computers. In the U.S., the V.24 is also known as an **EIA RS-232 DTE/DCE** interface.

CCITT (now merged with ITU) recommendations are often the result of collaborative efforts with other information standards organizations. Another agency helping ITU is the **Conference of European Postal and Telecommunication Administration (CEPT)**, chartered to facilitate cooperation among its 26 Postal Telegraph and Telephone Administrations (PTTs).

On behalf of the **European Economic Commission (EEC)**, the CEPT has undertaken the responsibility of standardizing devices in data communication, particularly terminals and other equipment connected to public networks. It influenced the integrated services digital network (ISDNs) standardization a few years ago.

International Standards Organization (ISO): ISO is a non-treaty, voluntary organization (based in Geneva) and has over 90 members who are members of the national standards organization of their respective countries. This organization was founded in 1946 and has defined more than 5050 standards in different areas. Although it is a non-treaty (non-government) organization, the majority of its members are either from government standards agencies or from duly recognized standards organization by public law. The main objective of this organization is to define/issue various standards and other products to provide exchange of services and other items at the international level. It also offers a framework for interaction among various technical, economic, and scientific activities to achieve its purpose. The members of ISO are typically the standards-making committees such as ANSI (U.S.), BSI (U.K.), and DIN (Germany). Both ISO and ITU-T work closely in the area of communications.

Various groups of **technical committees (TCs)** within ISO have been formed to define standards in specialized areas. ISO currently has nearly 2000 TCs, each dealing with a specific area/subject. Further, TCs are broken into sub-committees and subgroups that actually develop the standards for a very specific task. For example, **Technical Committee 97 (TC-97)** is concerned with the standardization of computers and information processing systems and ISDN. TC-97 has two sub-committees: SC-6 and SC-21. SC-6 deals with telecommunication and information exchange between computers, in particular the various layers of architecture and the interfaces between layers, while SC-21 deals with data access, data transfer, network management, and defining the OSI reference model which provides the basis of ITU-T's ISDN.

There are over 100,000 volunteers worldwide working in different sub-committees and subgroups. Volunteers include employees of the organizations whose products are under consideration for standards, government representatives proposing their ways of dealing and implementing technologies as standards in their respective countries, and academicians providing their knowledge and expertise for the standardization process.

Although both CCITT and ISO are separate international organizations, they cooperate in some areas such as telecommunication standards (ISO is a D Class member of CCITT), while in the areas of communication and information processing, they have defined a boundary between them. Per this agreement, ISO is mainly concerned with layers 4 through 7 of OSI-RM (defining the host part in the model), which deal with computer communication and distributed processing issues. The lower layers (1 through 3) deal with data communication and communication networks (defining the communication subnetwork part in OSI-RM) and will be the responsibility of CCITT. Further, IEEE is mainly concerned with the lower layers of OSI-RM for different LANs and has defined two sublayers of the data link layer of OSI-RM. The reference model defined by IEEE has eight layers as opposed to seven layers in OSI-RM.

With the integration of various technologies such as data communication, information and data processing, and distributed processing, there can be overlapping between the standards, but due to their mutual cooperation, they have so far not defined different standards for the same service/function.

As indicated earlier, ISO has members from various national standards organizations or organizations duly recognized by public law. American National Standard Institute (ANSI), which is a private, non-profit, non-government, domestic organization, represents the U.S. The members of ANSI are from manufacturing institutions, common carriers, and other interested organizations. Many of the standards recommended by ANSI have been defined/accepted as international standards. Great Britain is represented by the British Standards Institute (BSI), France is represented by AFNOR, and West Germany (now United Germany) is represented by DIN, while other countries are represented by their respective government agencies, in general, PTT. The ISO has defined standards mainly for LANs, which are based on OSI-RM, and non-OSI-RM LANs.

International Trade and User Groups (INTUG) was established in 1974 and includes various telecommunication organizations from several countries such as the U.S., the U.K., Australia, and Japan. It promotes the interests of its associations such as ITU, CEPT, PTTs, and ITU-T. It cooperates with the International Electrotechnical Commission (IEC) in a number of areas through the Joint Technical Programming Committee (JTPC). Both ISO and IEC have created a Joint Technical Committee on Information Technology (JTCI) for the standardization process within information technology. A detailed standard architecture, the relationship between various standards organizations, and other related information can be found in Reference 8.

1.8.2 U.S. standards organizations

For data and computer communication networks, there are two international standards organizations: CCITT and ISO (as discussed above). There are other standards organizations which are involved in defining standards for data communication networks but are not considered to be international standards organizations. Some of these organizations are in the U.S., while others are in Europe. The standardization process in the U.S. is managed by the U.S. State Department. It reviews the recommendations made by various standards organizations such as IEEE, T1 and T3-ANSI, and even some companies. After a careful review of the recommendations, they're forwarded to ITU-T. The following is a brief description of some of these organizations (typically national) along with their role in the standardization process:

American National Standards Institute (ANSI): ANSI is a non-government, non-profit, volunteer organization. It includes active members from manufacturing institutions, carriers, customers, and other interested organizations. It participates in an uncommonly large variety of standards activities. It plays a crucial role in the coordination and approval process of the voluntary American national standards. It has approved domestic standards in areas such as COBOL and FORTRAN programming languages, safety glass, screw thread dimensions, and bicycle helmets. It serves as the sole representative for the U.S. in the ISO. A standards committee, X3, was established in 1960 to study computer-related information system standards. A new accredited standards committee, ASC-T1, was formed in 1983 to set standards for the telecommunication industry. It helped CEPT to standardize ISDN. Other committees are ASC-X12, which deals mainly with business and **electronic data interchange (EDI)**, and X3.139, which defined the fiber distribution data interface (FDDI) specification for high-speed, 100-Mbps, token ring LANs. Due to its nature, it is also known as the National Clearing House (NCH), and various coordinating agencies provide their services for standards definition/implementation within the U.S. on a voluntary basis. It is a member of ISO and its main objective is to define and coordinate standards in data communication (office systems, encryption, etc.) for OSI-RM. ANSI formed an X.3 committee to define standards for data communication, programming, and magnetic storage media for OSI-RM. It is a voting member of ISO and has the same interests as that of ISO's TC-97.

Among its prominent contributing organizations is the **Electronic Industries Association (EIA)**, which started in 1924 as the Radio Manufacturers Association (RMA). At present, it has over 4000 industry and government members from semiconductor manufacturers and network integrators. A physical DTE/DCE interface RS-232 is a very popular standard defined by this organization and its main objective is to define standards for digital interface, signaling, and digital transmission.

Another of its contributors is IEEE, which defined the IEEE 802.x series of standard LANs in 1984. The IEEE 802.1 defines the MAC bridge (inter-LAN bridge) standard. Other contributors include MAP (Manufacturing Automation Protocol) and TOP (Technical Office Protocol) user groups (established in 1984), which provide standards for engineering, office, and manufacturing disciplines.

The National Institute of Standards and Technology (NIST): This organization was formerly known as the National Bureau of Standards (NBS). This organization is governed by the U.S. Department of Commerce and is very active in international standards committees. It coordinates all standards activities for the U.S. government and is an agency of the Institute for Computer Science, which has close ties with government and vendors. Its main objectives to define federal information processing standards (FIPS) used for the equipment bought by federal agencies, publish **U.S. Government Open System Interconnection Protocols (GOSIP),** and define protocols for the upper layers of OSI-RM standards. GOSIP is mainly concerned with the protocols to be used by U.S. government agencies. The U.S. Department of Defense (DOD) does not have to comply with FIPS or any other standards defined by NIST. NIST provides services to both CCITT and ISO. It sponsors the OSI implementers workshop, which is open to the government and commercial sectors. It also created and managed OSInet, an international research network, as a means of promoting the development of OSI. Some participants of this workshop include Boeing Computer Services, DEC, GM, HP, IBM, and NASA.

U.S. National Committee (USNC): This committee is the coordinating group for the U.S. in CCITT, while the U.S. State Department is the principal member of CCITT. There are four U.S. CCITT study groups which provide their input to the advisory committee groups at the international level. The **National Communication System (NCS)** is a consortium of various federal agencies which have telecommunication interests and capabilities. The main objective of the NCS is to coordinate with other organizations such as the Electronics Industries Association (EIA), ISO, and CCITT. It provides input to international standards organizations from federal agencies, and the majority of the input/recommendations are based on OSI-RM.

Electronics Industries Association (EIA): EIA is a national trade association composed of a vast number of electronic firms, it and is a member of ANSI. It is mainly concerned with the physical layer of OSI-RM. Its main job is to define hardware-oriented standards. It defined the best-known standard RS-232, a DTE/DCE interface. The first standard, EIA-232, was defined in 1962, subsequently modified to RS-232-C, and then renamed EIA-232-D in 1987. Some of the limitations of RS-232 were overcome in RS-449 (RS-422-A and RS 423-A) in 1975.

Defense Communication Agency (DCA): This organization deals mainly with communication-related military standards (MIL-STD) and issues the popular TCP/IP communication protocol.

Federal Telecommunication Standards Committee (FTSC): FTSC is a government agency which defines interoperability between government-owned networks and equipment. It adopts the existing standards of CCITT and EIA. FTSC and NBS try to work together. The lower layers of OSI-RM are being handled by FTSC, while the higher layers are handled by NBS.

The Institute of Electrical and Electronic Engineers (IEEE): This is the number one professional society, with various sections/chapters located throughout the world. It is a member of ANSI and is concerned mainly with local area networks (LANs) and other standards. The IEEE 802 committee has defined a number of local area networks and is primarily concerned with the lower two layers (physical and data) of OSI-RM standards. The various standards defined by IEEE 802 are:

IEEE 802.1	Higher Layer Interface Standards (HLIS)
IEEE 802.2	Logical Link Control Standards (LLC)
IEEE 802.3	Carrier Sense Multiple Access with Collision Detection LAN (CSMA/CD)
IEEE 802.4	General Motors (GM) Token Bus LAN
IEEE 802.5	IBM's Token Ring LAN
IEEE 802.6	Metropolitan Area Network (MAN) LAN
IEEE 802.7	Broadband Technical Advisory Group
IEEE 802.8	Fiber Optics Technical Advisory Group
IEEE 802.9	Isochronous LAN/Integrated Voice and Data Networks (ISDNs) Working Group
IEEE 802.10	LAN Security Working Group
IEEE 802.11	Wireless LAN Working Group
IEEE 802.12	Demand Priority Working Group
IEEE 802.14	Cable Modem Working Group
IEEE 802.16	Broadband Wireless Access Study Group

The IEEE 802.7 and IEEE 802.8 groups are working together on standards for broadband cable and fiber optics as transmission media which may be used in various LANs for higher data rates and bandwidths.

IEEE 802.9 was set up in 1986, and its main objective is to define the architecture and interface between desktop devices and LANs and ISDNs. The IEEE 802.10 standard for LAN security has defined the **Standard for Interoperable LAN and WAN Security (SILS)**, which will deal with the problems encountered during interoperability of LANs. According to IEEE 802.10 recommendations, it must include provisions for authentication, access control, data integrity, and confidentiality. The IEEE 802.3 committee has formed a high-speed study group to look into the high-speed ethernet LANs which offer data rates of over one gigabit (1000 Mbps).

1.8.3 Other national standards organizations

A list of other organizations around the world is given below:

The European Computer Manufacturers Association (ECMA): ECMA was founded in 1961 as a non-commercial organization to define standards for data processing and communication systems. It includes all European computer manufacturers, the European divisions of some American companies (IBM World Trade Europe Corporation, International Computers and Tabulators Limited, and many others). It is a non-voting member of CCITT and ISO and works closely with many technical committees of ISO and CCITT. Members of ECMA include AEG-Telefunken, Burroughs, Honeywell, IBM Europe, Siemens, Olivetti, Ferranti, ICL, Nixdorf, and NCR.

Conference European des Administration des Postes et des Telecommunications (CEPT): The European Community (EC), an integrated organization of the countries of Europe, has assigned the task of handling the issues of telecommunications, information industries, and new direction to the Directorate General (DG XIII). This organization deals with the standardization process for the free trade of goods and services within Europe. A white paper proposed setting up a single organization known as the European Standard Organization (ESO) to coordinate with other European standards organizations like CEN, CENELEC, and ETSI. The CEPT was formed in 1958 and has about 31 members throughout Europe and the European Free Trade Association (FETA). In January 1988 it set up an independent body known as the European Telecommunication Standards Institute (ETSI) to look into the standardization process.

European Telecommunication Standards Institute (ETSI): ETSI is funded by EC and EFTA and is responsible for the standardization of telecommunication, information technology (in cooperation with CEN/CENELEC), and broadcasting (in cooperation with the European Broadcasting Union). There are about 23 countries within ETSI, represented by EC and EFTA, and over 12 technical committees. Some of these include radio, equipment and systems (RES), group special mobile (GSM), paging systems, networks aspects, and transmission and multiplexing committees. ETSI coordinates with various standards organizations such as ITSC/GSC, ITU-T, ITU-R, and the Joint Technical Committee on Information Technology (JTCI). Both ITU-T and ITU-R are managed by ITU, while JTCI further coordinated with ISO and the International Electrotechnical Commission (IEC). Within Europe, ETSI coordinates with organizations such as CEN, the Joint Programming Group (JPG), and European bodies.

British Standards Institute (BSI): The BSI is a recognized U.K. organization for the formulation of national standards. It participates in ISO meetings through the membership of various technical sub-committees.

Deutsche Institut fur Normung (DIN): DIN is West German (now United German) national standards body analogous to BSI. The DIN standards are well recognized in areas of everyday use.

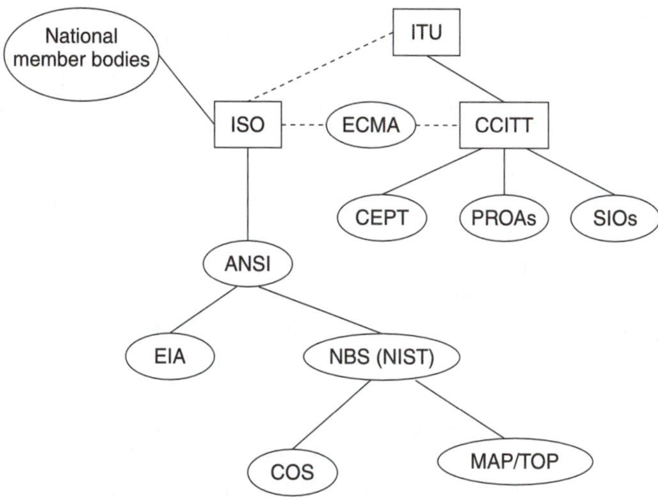

Figure 1.11 Standards organization hierarchy.

Association Francaise de Normalisation (AFONR): This is the French national standards body. It is very active within ISO.

International Electrotechnical Commission (IEC): The IEC is an independent international agency which specializes in the formulation of international standards in the area referred to by its name.

International Federation for Information Processing (IFIP): The IFIP is an international federation which includes technical groups with interests in providing information science and technology and assisting in research development education.

Canadian Standards Organizations: The industry standards organization, Industry Canada, and the government agency, Canadian Radio and Television Commission, are the two standards organizations in Canada. They play the same roles as the U.S. standards organizations.

These standards organizations (international and national/domestic) define standards for data communication networks. On the basis of these standards, vendors and customers have to agree on the products, which must comply with standards. Due to the spread of standardization throughout the world, many organizations have been formed during recent years. The main objective of these organizations is to provide acceptance between the vendors (supplying the products) and the customers (buying the products). The standards organizations hierarchy is shown in Figure 1.11.

1.9 Local area network (LAN) evolution

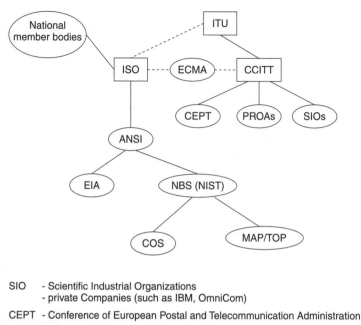

LANs are being used for information sharing and exchange of information and have grown very fast. According to a number of surveys, the number of LANs has already increased to more than 2 million around the world. Ethernet is a very popular LAN and has been adopted in many LAN protocols. It is available in a variety of data rates and supports different transmission media. Some of the material presented here is partially derived from References 7, 8, and 12 (see references for further reading).

The following is a list of Ethernet-based LANs which have been introduced:

- 10Base-2 (10-Mbps baseband with 200 meters coaxial cable — Thinnet)
- 10Base-5 (10-Mbps baseband with 500 meters coaxial cable — Thicknet)
- 10Base1-T (10-Mbps baseband with 100 meters twisted pair)
- 10Base-FL (10-Mbps baseband with 2km fiber)
- 100Base-T (100-Mbps baseband with 100 meters twisted pair).
- 100Base-FX (100-Mps baseband with 100 meters optical fiber)

Gigabit Ethernet is based on the concept of high-speed packet switching and uses the CSMA/CD protocol of the earlier, original Ethernet LANs (10-Mbps and 100-Mbps). It supports both 10Base-T and 100Base-T LANs. This network offers higher speeds and high performance in the LAN backbone. It is based on the same concept as Ethernet and uses the same Ethernet frame format, management framework, and collision detection and avoidance methods.

It uses the fiber channel signaling mechanism and can extend from 500 meters (multimode) to 2 kms (single mode). The existing twisted pair cabling can support only a few meters (20 to 100). Gigabit Ethernet provides LAN support at a data rate of 1 Gbps and is easy to manage, but it does not support WANs. ATM, on the other hand, provides LAN support at data rates of DS1 (1.5 Mbps) through OC-48 (2.5 Gbps), has built-in quality of service, and supports WANs.

Ethernet is the most popular cabling technology for local area networks. The most commonly installed Ethernet system was known as 10Base-T, which provided transmission speeds of up to 10 Mbps. With high bandwidth requirements for many multimedia applications, a new 100Base-T LAN, also known as Fast Ethernet, which could (theoretically) transmit at 100 Mbps, was introduced. Fast Ethernet uses the same cabling as 10Base-T, packet format and length, error control, and management information. Fast Ethernet was typically used for backbones, but due to ever-growing bandwidth requirements, it found its use in workstations. The Gigabit LAN that can (theoretically) provide a bandwidth of a billion bits per second is becoming very popular as a backbone. Applications in the modern enterprise networks are making heavy use of desktop, server, hub, and the switch for increased bandwidth applications. Megabytes of data which need to flow across intranets as communication within enterprises will move on from text-based e-mail messages to bandwidth-intensive real-time audio, video, and voice.

A very straightforward upgrade scenario is upgrading 100-Mbps links between Fast Ethernet switches or repeaters to 1000-Mbps links between 100/1000 switches. Such high-bandwidth, switch-to-switch links would enable the 100/1000 switches to support a greater number of both switched and shared Fast Ethernet segments.

The switched 10-Mbps and 100-Mbps Ethernets have already captured the market, and it is hoped that the introduction of Gigabit Ethernet may pose tough competition to ATM. Nearly 80% of all desktops are using Ethernet LAN. That high usage, different types of LANs (Fast Ethernet and Gigabit Ethernet), and their support for gigabit WAN transport position ATM as the dominating switching technology for high-performance LAN connectivity. For the applications (real-time, multimedia, video, and voice) that require quality of service in LANs, the ATM ensures end-to-end quality of service communications. At this time, real-time protocol and resource reservation protocol defined within IP are offering quality of services for Ethernet networks. The ATM provides a very useful and is the best LAN-to-WAN migration path. A server based on LAN can have a connection anywhere from 25 Mbps ATM to 622 Mbps ATM. The ATM has an already standardized OC48 (2.5 Gbps) supporting full duplex mode of operations and offers much higher data rates than Gigabit Ethernet (half duplex).

With different types of traffic, the requirements of LANs are also changed. Now the requirements of the LAN environment include higher data rates for integrated traffic (data, video, and voice), an acceptable quality of service, and a uniform network management system (using the same equipment). As mentioned earlier, the LAN environment has evolved from low-speed (Ethernet and token ring) to high-speed (switched Ethernet, FDDI, and 100-Mbps voice grade). The switched Ethernet offers a very high data rate and dedicated bandwidth. The dedicated access to a resource causes starvation and does not support multimedia applications. FDDI offers superior performance compared to Ethernet and token ring and is currently being used as a backbone network in many organizations, universities, and other agencies. Although it is expensive and does not support isochronous (voice) traffic, a newer version of FDDI-II has been introduced which can handle voice traffic and support ATM. Also, inexpensive coax has been used in defining a new LAN as Coax Distributed Data Interface (CDDI) which may become an alternative to expensive FDDI. 100-Mbps voice-grade LAN is a new technology and seems to be better than FDDI. No standards are available yet. The ATM for the LAN environment offers very high bandwidth (on demand) and much better performance with very high data rates for data, video, and voice.

ATM is a dedicated-connection switching technology that organizes digital data into packets and transmits them over a digital medium. Because ATM is designed to be easily implemented by hardware (rather than software), fast transmission speeds are possible. ATM, unlike Ethernet, is meant to accommodate real-time data transmission with varying priorities. It is fast, with top speeds of 622 Mbps today and 9.6 Gbps planned in the future.

The new technology, Virtual LAN, is based on the concept of LAN switching. The switching is mainly defined in Ethernet-based LANs (10 Base-T, 100 Base-T, and others) at layer 2. The switching at layer 2 provides control flow for the frames transmitted over the segment and, as a result, it may offer a higher bandwidth to the frames. LAN switching uses the look-up table for sending the frames to the appropriate segment of LAN. This is precisely what is done at the bridges, which also work at the data link layer and maintain a look-up table for routing the frames to the appropriate LANs. The VLAN may provide services to all the nodes connected to it. Instead of connecting bridges between segments of LANs, a switch is used for each segment. All these segments are then connected to another switch which is connected to LAN. In this way, we can get security, better performance, and better resource utilization. The security is achieved by the fact that the traffic in VLAN remains in the group of the connected nodes. Since the number of users per switch or segment is limited, better performance and higher bandwidth is expected. The network management seems to be easier because the physical connection between the nodes and switches can be performed through software.

1.10 Current trends in communication networks

LAN interconnectivity started in the late 1980s for connecting computers and PCs within a building or campus. The number of LANs currently being used is over 2 million and is expected to rise. LANs find their use in a variety of applications and offer different data rates. Some of the applications along with data rates are e-mail (9 K to 56 K), file transfer (56 K to 1.5 M), multimedia (3 M to 45 M), imaging (1.5 M to 45 M), and x-window system (56 K to 1.5 M). Computer-aided design traffic is usually local and the file transfers take place between mainframe and workstations that are located at the same place. The off-line file transfers typically are the remote traffic. Data rates of 56 K to 1.5 M may be sufficient for stand-alone applications but are not sufficient for these CAD/CAM tools.

An increasing number of enterprises are employing data warehousing for strategic planning, as it deals with a very high volume of data and low transmission latency. These warehouses may be composed of terabytes of data distributed over hundreds of platforms and accessed by thousands of users. This data must be updated on a regular basis to ensure that users access real-time data for critical business reports and analyses. Enterprise-critical applications will proliferate and demand ever-greater shares of bandwidth at the desktop. As the number of users grows rapidly, enterprises will need to migrate critical portions of their networks (if not the whole network itself) to higher-bandwidth technologies. Crossing over from Megabit to Gigabit Ethernet will be inevitable, starting at the backbone. At first we had companies, and then we had companies that made use of computers. Enterprise is a relatively modern term that includes any organization where computing systems replace traditionally used computers to increase productivity.

The current move toward high-speed networks, particularly in the area of asynchronous transfer mode (ATM) technology, seems to be revolutionizing information technology. In ATM technology, the information is broken into a small, fixed-sized cell of 53-byte packets and sent using connection-oriented services offered by ATM. On the receiving side, the cell packets are assembled and the original information is constructed. The routing is a crucial aspect in the design of any network. Routing may be an interesting and useful criterion for comparing LANs to ATM technology. LANs are usually considered non-scalable, which means that the LAN segment offers services for a limited distance to a limited number of users. Further, LAN protocols and access methods also operate for a limited distance, so LANs offer a limited capacity and physical size. See References 7, 8, and 12 for further reading.

It seems that the price difference between ATM switches and conventional LAN Ethernet switches is very small. Therefore, many of the major vendors are producing both the switches and other products. A typical configuration of Gigabit Ethernet includes a Gigabit switching hub that supports a number of servers via Gigabit Ethernet. It also supports other nodes with 100- and 1000-Mbps hubs that are linked to a Gigabit switching hub via a 1000-Mbps link and links to other workstations via 100-Mbps links.

The ATM network finds its application in LANs for a higher bandwidth and data rate than the entire ATM protocol and has moved from WAN groups such as ITU-T, Bellcore, and others toward being the possible LAN solution for every organization and university. Looking at the possible LAN solutions based on ATM, IEEE is moving toward defining 100-Mbps LANs (Ethernet and token ring). The existing 10-Mbps Ethernet will be developed as switched Ethernet, which will offer the entire bandwidth of 10 Mbps at each node rather than share the bandwidth in the traditional working of Ethernet. FDDI is already running over fiber at data rates of 100 Mbps. FDDI provides similar features as ATM in terms of support of data and time-based applications such as video and voice (isochronous channels). That means that in some applications, it may be cheaper than ATM. Other alternatives to ATM for possible LAN solutions may include 100-Mbps switched Ethernet, FDDI-II as a LAN, IEEE 802.6 as a MAN, packet transfer mode as a WAN, and SMDS and frame relays as public network services.

The 10-Mbps LANs may not be suitable if we have more than 80 users, but it will be more than enough for a single user. There are many that have introduced point-to-point link running at 10 Mbps by attaching each user to a hub segment of the LAN. The segment has only one user and one hub port and still uses CSMA/CD. The hub has linked input and output via a switched fiber. The hub having input and output linked can be used as an ATM switch offering a data rate of 10 Mbps. This type of switched Ethernet is essentially an Ethernet or 802.3 LAN but can be implemented in a variety of ways. The switched Ethernet can be realized in a duplex mode where we can expect a data rate of 20 Mbps supporting only data communication. The transmission of video and multimedia will still be avoided in this type of LAN. Both 100-Mbps and switched Ethernet are not suitable for video, multimedia, or graphic-based applications. In other words, these LANs do not support time-sensitive applications. FDDI-II, an extended version of FDDI, seems to be suitable for time-sensitive applications, as it supports variable-length data frames and uses circuit-switching techniques. The establishment of a network connection is needed for isochronous constant bit rate services. For more details, readers are advised to read Reference 12.

A typical LAN environment that exists today in universities, corporations, organizations, and government agencies uses FDDI as a backbone network which connects different types of LANs such as Ethernet and token ring. These LANs are connected to FDDI via a 10-Mbps link for Ethernet and a 4- or 16-Mbps for a token ring network. Various servers are usually connected to FDDI via 10-Mbps links. The FDDI is also connected to routers via high-speed links of 50 Mbps or above. The routers are connected to the Internet or other routers via different links from T1 (1.544 Mbps to 100 Mbps) and up, providing access to the outside world. ATM can be used for the existing LAN environment in a number of ways. In one environment, the FDDI can be replaced by an ATM switch and some of the servers may be connected via high-speed links. The FDDI can also be reconfigured as an access network for different LANs. The second environment may define an ATM hub (ATM-based) where ATM switches are connected via Optical Carrier level 3 (OC3) links and can be used as a backbone network in place of FDDI. Here the ATM switches will be providing interfaces (ATM adapters) and accesses to different types of LANs.

These ATM switches will also provide interfaces to routers via Digital Signal level 3 (DS3) or above links for access to the outside world. In video-based applications, the ATM can be directly connected to workstations or terminals via optical fiber, giving data rates of 100 Mbps with the options of transmitting video, voice, or data. A number of servers dealing with these types of traffic (video, voice, and data) can also be directly connected to ATM via ARM adapters.

The routers provide interconnection between different data link layers and offer a common network addressing that is independent of the details of the data link layer and also have the capability to convert from one data link format to another. The routers make decisions about forwarding incoming packets, based on the network address defined in each packet. The routers are also expected to perform off-line updating of route tables that are used to determine the route for the packet. The routers usually create and maintain their own routing tables dynamically which automatically adapt to the new network environment under traffic congestion, failed links or nodes, etc.

If we compare computer networks to telephone networks, we find that the data link layer is analogous to the analog or modem signal, while the telephone network does not use any routers. This is due to the fact that router-based networks are based on connectionless service, while telephone networks are based on connection-oriented service. Similarly, the ATM is based on connection-oriented service and is similar to telephone networks with the only difference being that the ATM provides a combination of high-speed and statistical multiplexing that allows it to transmit the bursty data multiplexed with PCM voice or video signals. As such, there is no equivalent to the router (software-based routing) needed in ATM. Here, the ATM switch offers a faster and inexpensive method of transferring the data through it (hardware-based routing).

The end-to-end delay in ATM is ten to hundreds times less than that provided by routers for an equivalent cost. Routing determination in ATM follows the same steps required as that in routers, e.g., exchange of network topology information, available routes, transmission costs, etc. These operations are important for an ATM switch to determine the minimum cost route for a particular session when the end nodes are requesting a connection. Further, ATM offers similar fault tolerance as is available with the routers. In a sense, we can say that ATM provides the scalable and hardware-based data transfer efficiency of the telephone networks with the high speed of LANs. It will offer a universal end-to-end data link service for local and remote connections. For more details on broadband services and various high-speed networks, you may read an excellent book on broadband communication by B. Kumar.[8]

Some useful Web sites

A number of USENET groups have been established that provide more discussion on the various issues of a particular topic. Some of the USENET groups are (others follow basically the same sort of addressing mechanism):

- Comp.dcom.lans
- Comp.dcom.security
- Comp.dcom.cell-relay
- Comp.dcom.frame-relay
- Comp.protocols.tcp-ip

There is a large number of Web sites that can be used to get more information on various aspects of networking, internetworking, standards, vendors, magazines, etc. A partial list of such Web sites is given below. There are many others that can be found using search engines such as Yahoo!, Alta Vista, Web Crawler, and Meta Crawler.

- http: / /www.internic.net
- Internet and IETF Data Communication Magazine-CMPNet *http://www.Tel.com*
- NCworld magazine-*http://www.ncworldmag.com*
- Network Marketing World and Network World Fusion *http://www.nmworld.com, http://www.nwfusion.com*
- Building Industry Consulting Services International (non-profit professional association) *http://www.creossbowen.co*
- Personal computing *http://www.tokyopc.org, http://www.viewz.com, http://wwwibm.com*
- Datamation *http://www.datamation.earthweb.com, http://intranetjournal.earthweb.com*
- Computerworld *http://www.computerworld.com*
- The Task Force on Cluster Computing *http://www.tu.chemnitz.de*
- A&T Network *http://www.atnetworks.com*
- Digital Network Associates *http://www.dna.com*
- Macintosh consultant at *http://wwmacn.com*
- Myriad Data Solution at *http://www.myriad-data.com*
- Hardware and software products *http://www.ronin.com*
- Mapping between various ISO and CCITT protocols along with RFC and complete description *http://www.cis.ohio-state.edu.*

Also see *http://www.cs.berkeley.edu* for network and computing infrastructure that contains information about various network products. There are Web sites which maintain the bibliography of computer professionals; one such Web site is *http://liinwww.ira.uka.de.*

Routers Enterprise routers and software products are available at: *http: support.3com.com, http://www.networking.ibm.com, http://www.zoomtel.com, http://www.playground.sun.com,* and many other sites.

For Gigabit Ethernet, a number of Web sites have been introduced, such as:

- Gigabit Ethernet Alliance *http://www.gigabit-ethernet.org*
- Gigabit Ethernet consortium *http://www.iol.unh.edu*, *http://www.3com.com*, *http://www.ciol.com*, *http://www.lan.net.au*
- WWW consortium *http://www.biomtrics.org*
- International Institute for Internet Industry Benchmarks *http://www.iiiib.org*
- Some of the vendors for Gigabit Ethernet *http://www.cisco.com*, *http://www.3com.com*, *http://www.onmitron-systems.com*, *http://www.allied-telesyn.com*
- Newsgroup for Ethernet *http://comp.dcom.lans.ethernet*.

References

1. Tanenbaum, A.S., *Computer Networks*, Prentice-Hall, 1989.
2. Day, J.D. and Zimmermann, H., The OSI reference model. *Proc. IEEE*, Vol. 71, Dec. 1983.
3. Stallings, W., *Data and Communication*, 2nd ed. Macmillan, 1988.
4. Hutchison, D., *Local Area Network Architectures*, Addison-Wesley, 1988.
5. Peterson, J., *Petri Net Theory and Modeling of System*, Prentice-Hall, 1981.
6. Hura, G.S., Petri nets, *IEEE Potential*, Oct. 1988.
7. Sveum. *Data Communications: An Overview*, Prentice-Hall, 2000.
8. Kumar, B., *Broadband Communications*, McGraw-Hill Series on Computer Communications, 1995.
9. Zimmerman, H., OSI reference model: The ISO model of architecture for open systems interconnection, *IEEE Trans. on Communications*, April 1980.
10. Jain, B. and Agarwala, A., *Open Systems Interconnection*, McGraw-Hill, 1993.
11. Halsall, F., *Data Communication, Computer Networks and Open Systems*, Addison-Wesley, 1996.
12. Goralski, W.J., *Introduction to ATM Networking*, McGraw-Hill International Edition, 1995.

References for further reading

1. 3Com, *Gigabit Ethernet on the Horizon: Meeting the Network Challenges*. 1997. gigabit-ethernet.org. *Gigabit Ethernet Alliance*. White paper, 1996.
2. Cohen, J., *ATM and Gigabit Ethernet Camps Collide*, Network World, Sept. 1996. Cells in Frame: *http://cif.cornell.edu*.
3. Jeffries, R., *Three Roads to Quality of Service: ATM, RSVP, and CIF*, Telecommunications, April 1996. IBM, *Desktop ATM Versus Fast Ethernet*, white paper, 1996.

part II

Fundamentals of digital communication and signaling

chapter two

Basic concepts of digital communication and signaling

"Mistakes are a fact of life. It is the response to the errors that count."

Nikki Giovanni

2.1 Introduction

Data communication deals with the transmission of analog and digital signals. This chapter introduces various aspects of communication systems starting from the basic concept of electricity, which basically carries electrons from one end to another over copper wire. Various concepts of alternating current waveforms such as frequency, bandwidth, representation of information or speech by waveforms, and transmission of signals over transmission media are explained. A communication channel is a very important component within communication systems, as it defines a path over the communication link for carrying the information from one node to another. The communication channel can be characterized by a number of parameters, such as modes of operations, capacity, speed of transmission, and different types of transmission. These are explained in this chapter. The basic components of communication systems and their relationships with each other are explained. Different forms of representing various signals are discussed.

2.2 Basic transmission signaling concepts

A communication system is made up of three subsystems, namely, a transmitter, a transmission medium, and a receiver. The transmitter at the source node transmits the voltage signal (corresponding to the user's information message) toward the receiving node over the medium (connecting them). Depending on the type of transmission, the signal may represent either baseband or broadband type of signals. From the point of view of transmission, the broadband signal requires modulation before transmission, while baseband, which is basically a digital signal, does not require modulation. The user information can be transmitted using either analog transmission or digital transmission. The user information can be expressed in either analog or digital form; as such, the transmission process is known as analog or digital transmission signaling of analog or digital data. In other words, the analog or digital signaling can be used for transmitting both types of signals (analog and digital). We discuss both types of signaling in the following sections.

2.2.1 Analog signaling

In analog transmission (using modulation), continuous variation of amplitude, frequency, and phase of modulating signals with respect to high-frequency carrier signal represent the user's information. The modulation is a process which increases the frequency of the modulating signals so that the signals can be transmitted over a larger distance with increased transmission capacity. The modulated signal is demodulated on the receiving side to recover the original user's information. In this signaling scheme, the content of user information is not important to modulation and can be represented in either analog or digital signals. The amplitude of the modulated signal gets attenuated as it travels over a long distance. This type of signaling is known as the analog signaling system.

In order to receive the analog modulated signal, it needs to be amplified at a regular distance. The amplifier raises the amplitude to the level determined by the amplification factor and then re-transmits it over the link. If any noise or unwanted signal is present in the modulated signal, it will also be amplified by the same amplification factor. Thus, the quality of the received signal depends on the **signal-to-noise** ratio, which sometimes becomes a main design consideration in analog transmission. Further, the number of amplifiers is generally large in the analog-signaling scheme.

Based on the variation of parameters of modulating a signal (amplitude, frequency, and phase), different types of modulation schemes have been defined as amplitude modulation (AM), frequency modulation (FM), and phase modulation (PM).

In AM, the amplitude of the sinusoidal carrier is varied in accordance with the instantaneous value of the modulating signal (user's information) by keeping the frequency of the modulated signal unchanged (same as that of the carrier). This type of modulation finds its use in applications like medium- and short-wave radio broadcasting systems, and it provides a reasonable bandwidth.

In FM, the frequency of the carrier signal is varied in accordance with the instantaneous value of the frequency of the modulating signal by keeping the amplitude of the modulated signal the same as that of the carrier signal. This type of modulation is immune to noise and finds its use in applications such as satellite link, microwave link, high radio frequencies, etc.

Phase modulation, AM, and FM are discussed in more detail in Chapter 3.

2.2.2 Digital signaling

If we consider the carrier as a series of periodic pulses, then modulation using this carrier is defined as **pulse modulation**. In this signaling scheme, the content of the user information is important, as it is expressed in logic 1s and 0s. The digital signals are not modulated (the digital form of the signal is defined only after modulation). This requires a repeater, which regenerates the logic 1s and 0s and re-transmits them over the link. Since the repeater simply regenerates the logics, this scheme does not include any noise, and the number of repeaters required to regenerate the digital signals for the same distance is much less than that of the number of amplifiers required in analog signaling. This type of signaling is used in a system known as a digital signaling system. If the amplitude of the carrier signal is varied in accordance with the amplitude of the user's signal, then we define pulse amplitude modulation. Similarly, other modulations can be defined.

These two forms of modulation (with periodic carriers or carriers with a series of periodic pulses) are defined in digital signaling and find their use in a variety of applications, e.g., voice communication, audio communications, video communications, optical communications, telemetry, etc.

Instead of considering the carrier as a series of periodical pulses, if we consider the sampling of the user's signal, then we get different types of modulations known as **shift keying modulations**. In these modulations, the carrier is considered to be a sinusoidal signal. These modulation schemes find their use in applications such as data and text transmission on analog channel/link, digital data transmission over wireless communication systems, optical communications, etc. These modulation schemes are discussed in more detail in Chapter 3.

A digital transmission transmits the signals in digital form. The input can be in any form at the transmitter; it will be converted into digital form after the modulation. The conversion of an analog signal into digital form is called **sampling**, whereas compression of an infinite number of analog signal pulses to a finite number is called **quantization**. The carrier signal is defined by a clock signal, which provides the bit rates of the digital transmission.

Digital signaling systems provide the following advantages over analog signaling systems:

- They allow the integration of the data and voice signaling scheme, thus defining the main principle and concept behind integrated services digital networks (ISDNs).
- They allow the regeneration of the digital signal in the presence of noise, offering better quality of service as compared to analog transmission systems
- They provide ease of encrypting the data.
- Due to inexpensive integrated hardware chips, different types of information (analog and digital) can be multiplexed, which improves the link utilization.

However, digital signaling transmission systems suffer from the following problems:

- They require a larger bandwidth. This is evident from the fact that in ISDN, the minimum data rate is 64 Kbps for each channel, compared to 4 KHz for voice communication in an analog system.
- Digital systems are dependent on synchronization between source and destination, which adds extra overhead in the transmission.
- Noise due to the sampling of the user's signals, phase jitter, high attenuation, etc., restricts the use of high-speed transmission media using digital transmission.

2.3 Electrical and voice signals: Basic definitions

In a communication system, the processing of information is carried out by electrical signals. The electrical signals (analog and digital) are produced by force corresponding to voltage differences between the ends of a conductor. The movement of electrons in a conductor (such as a copper wire) constitutes a current. Generally, all metals are good conductors. Silver is best, and copper is second best. Silver is very expensive, while copper is cheap and is, therefore, very commonly used.

The communication system between two stations usually has one wire for the current flow in one direction, while the earth is being used as a return path for the current. The value of signals on the line is measured in **volts (V)**, while current flow is measured in **amperes (A)**. An electrical current is defined as the rate of change of a charge flowing through a conductor.

The currents flowing through branches and voltage differences across the branches in the circuit can be used to define the values of currents and voltages (along with directions) in the analog circuits. They also represent some binary logic, codes, symbols, etc., in the case of digital circuits.

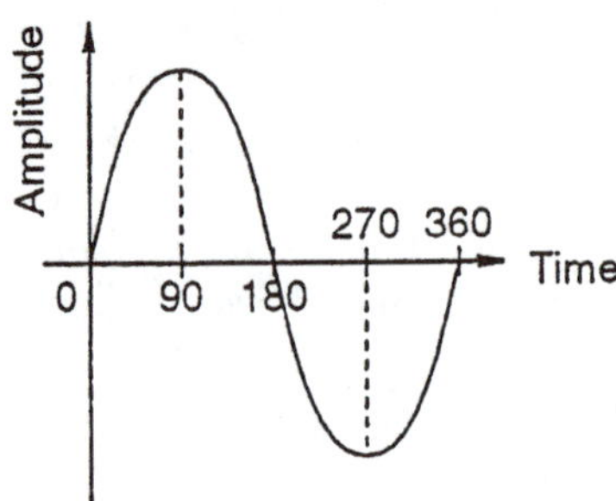

Figure 2.1 Analog signal.

2.3.1 *Signal waveform*

The electrical signals (currents or voltages) are typically defined in two ways:

Direct current (DC): The magnitude of current remains constant and the steady current flows in one direction. This current is produced by a steady or constant value of voltage obtained between two polarities (such as batteries). Direct current is used in a variety of applications such as cars, flashlights, electronic appliances, etc.

Alternating current (AC): Alternating current provides varying amplitude of signals between positive and negative polarities over a time axis (shown in Figure 2.1). Its amplitude assumes maximum value at two different polarity axes: one at 90 degrees (positive), the other at 270 degrees (negative). The waveform starts at 0 degrees (giving zero amplitude of signal), giving another zero amplitude at 180 degrees and 360 degrees, respectively. This waveform can represent both current and voltage signals.

2.3.2 *Basic definitions of sinusoidal signal waveform*

Cycle: One cycle of the waveform is defined as one complete oscillation (starting from 0 to 360 degrees) of the wave. The time for one cycle is known as one cycle time (T). The cycle is measured by wavelength.

Frequency: The number of cycles over a period of a second is defined as frequency (F). Thus, the frequency of a waveform is defined as the number of cycles per second and is commonly denoted as Hertz (Hz) or cycles per second (c/s). The signal (current or voltage) moves in one cycle time, and this movement is known as one wavelength. In other words, the time needed to travel one wavelength represents the time for one cycle, or T, as shown in Figure 2.2.

The waveform propagates through the conductor or atmosphere (in the case of wireless transmission) with a speed given by

$$v = f \times 1 \tag{2.1}$$

where v represents velocity in meters per second, f defines frequency in Hz, and 1 wavelength is the distance the waveform travels during one cycle time.

Electrical signals (particularly current) carry information or messages over communication circuits, and these signals can represent different forms of energies in the waveform. In other words, the waveform, as shown in Figure 2.2, can be used to represent any type of message/information and the speed of transmission of these waveforms over transmission media can be calculated using Equation (2.1).

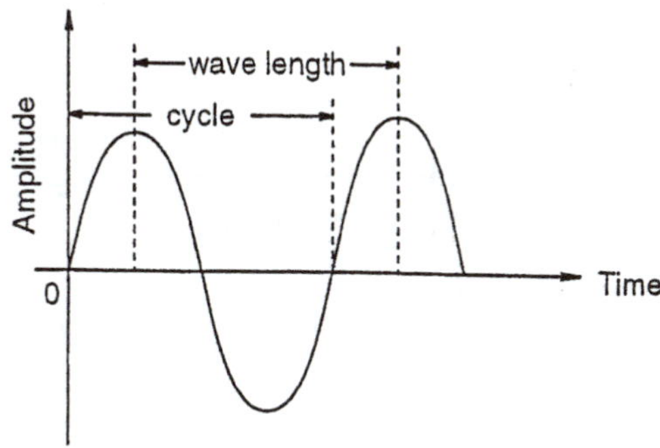

Figure 2.2 Analog waveform.

Bandwidth: Bandwidth is defined as the difference between two values of frequencies; i.e., if we choose two points on the frequency scale as $f1$ and $f2$, then the bandwidth is given by $f2 - f1$ in Hz.

Since a voice signal does not contain one unique frequency but consists of waveforms of different frequencies, it will be difficult to predict the exact frequency of the voice signal. Instead, we define a bandwidth for voice signals where bandwidth is defined as the difference of highest and lowest frequencies present in the signal.

The human ear can recognize sounds between 40 Hz and 18 KHz, but most people's voice/speech lies within the frequency range of 300 Hz to 3500 Hz. Due to this fact, the telephone system has a voice band (or voice grade) of 4 KHz. All the information or message in a person's speech can be accommodated in this frequency bandwidth. Different books use different bandwidth for voice band, but it has been widely accepted that the range of voice band in the telephone system is from 0 to 4 KHz. Further, it has been observed that the complete information found in a person's speech can be accommodated within a bandwidth of about 3.2 KHz. These voice-band channels are transmitted over the communication channel. In order to utilize the communication link effectively, these voice-grade bands are multiplexed and then transmitted over the media. These channels (typically 4 KHz each) are spaced over the frequency scale and are arranged within a predefined bandwidth. Since these channels are spaced over the frequency scale, the inter-channel interference becomes a serious problem. In order to avoid any electrical interference between consecutive channels, a guard band of 1 KHz is usually used between channels.

2.3.3 *Voice signal waveform*

In the following sections, we discuss the various forms of voice signals.

Analog signal waveform: Speech or voice signals coming out of the mouth can be represented by analog, continuous electrical signals (sinusoidal wave) as shown in Figure 2.1 and are similar to the waveform shown in Figure 2.2. The signal follows a continuous pattern and is known as an analog signal.

Digital signal waveform: Digital devices such as computers, terminals, printers, processors, and other data communication devices generate signals for various characters, symbols, etc., using different coding techniques. The digital (or discrete) signals can have only two binary forms or logic values 1 and 0.

If we have computers and machines which can generate discrete signals at output (as in old computational machines), then we don't have to represent the signals of computers

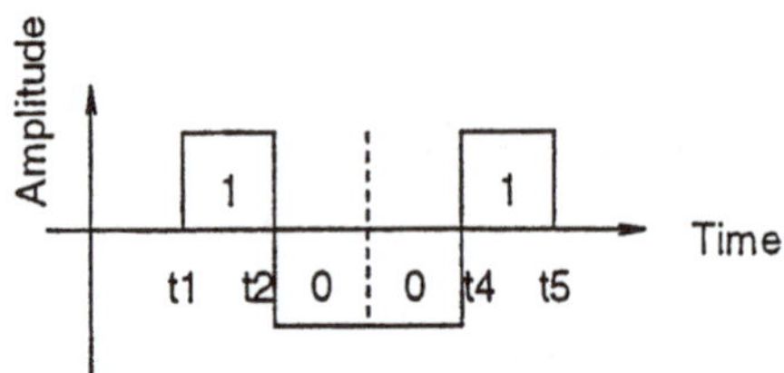

Figure 2.3 Digital signal.

and machines as discrete signals. But with the advent of **very large scale integration (VLSI)**, we can now represent any object, such as symbols, numbers, images, patterns, pictures, etc., as discrete signals. In some cases, we can also change the existing analog waveforms into suitable digital signals using different techniques. A waveform corresponding to a digital signal is shown in Figure 2.3.

In Figure 2.3, logic 1 is represented by the abrupt rise in amplitude at the discrete instant of time *t*1, remains the same until *t*2, goes down to 0, and changes its polarity to the other side at *t*2. It maintains the amplitude at a constant value to represent logic 0 until *t*3 and continues until *t*4 to represent another logic 0. We can assign different amplitude of signals (current or voltage) to logics 1 and 0. For example, high amplitude may represent logic 1 and low amplitude may represent logic 0, or vice versa.

Digital signals have the following advantages and disadvantages over analog signals:

Advantages

- Signal regeneration is cheaper and better than amplification.
- Digital signals provide for ease of encryption.
- Both data and voice can be integrated.
- They offer increased requirement for computer-to-computer data communication.

Disadvantages

- Digital signals require more bandwidth.
- They need synchronization.
- Quantization noise will be present with the signal.
- They have to avoid the problem of phase jitter.

Quasi-analog signal waveform: This is an analog signal that resembles a digital signal and allows digital signals to be sent on analog lines. These signals have been used in various communication devices using different types of modulation methods.

2.3.4 *Transmission basics*

The signals of data communication during transmission and reception may be in either analog or digital form. The digital signals travel from a communication device to the link via an **interface** or **line adapter** or **line driver**, while modems are used as an interface between a communication device and link for modulating signals (both analog and digital). These signals can be transmitted either serially or in parallel, depending on the type of transmission media being used.

Serial transmission: In this mode of transmission, message information is transmitted bit by bit over the link and the transmission speed of the transmitting site depends on the signaling speed. Both types of digital signals (data and control) are transmitted serially. The control signals are used for providing synchronization between two sites. The signaling rate is defined as the rate at which signaling per second is selected for the communication device and is usually expressed in baud. The time period of one signaling rate can be defined as the reciprocal of the baud. For example, if the signaling rate of a communication device is 300 baud, the time period of one signaling rate is 1/300 or .0033 sec. The communication circuit (containing electronic circuits like modulators/demodulators) must support the rapid change of the voltage levels (between logic 1s and 0s) during this time.

The digital signal bits are sent during the signaling period. The number of bits to be transmitted during the signaling period could be one or more, depending on the signaling rate. If the number of bits (0s or 1s) during the signaling period is one, then the signaling rate is the same as that of the transmission speed of bits/second (bps), but if the number of bits is more than one, then the baud may be more than the transmission speed of bps. The voltage values for describing logic 1s and 0s may be different depending on the signaling or coding methods used; e.g., we may have 0 and 5 volts (in the case of TTL IC), –15 and 15 volts, or –3 and 3 volts (in Ethernet LAN), etc.

Although this mode of transmission is slower, it is useful for transmitting signals over long distances and finds applications in communication systems such as facsimile, e-mail, satellite communication, microwave links, etc.

Parallel transmission: In this transmission mode, each bit is assigned a specific, separate channel number and all bits are transmitted simultaneously over different channels. For every bit, a channel is defined. Thus, for transmitting eight-bit data, eight channels are used. Both types of signals (data and control) are generated by different circuits and are also transmitted on different channels at the same time. The control signals provide synchronization between two sites and are sent on a separate channel to indicate different types of status (of transmitter and receiver) such as ready, busy, acknowledgment, etc. The number of data bits in the information to be sent by the data circuit defines the number of channel signals. The above-discussed basic definitions can be found in many books on networking.[1-4]

2.3.5 *Fourier-series representation of periodic waveform*

As discussed earlier, sinusoidal signals are continuous (analog) and periodic can represent different forms of analog quantities (pressure, speech, temperature, and so on), and are made up of a number of waveforms. The signal will have one fundamental frequency. Multiples of frequency are called **harmonics**. It has been observed that a typical voice waveform contains both types of harmonics (odd and even). These harmonics are what distinguish one person's voice or speech from another's. Periodic waveforms can have different shapes and waveform patterns for each of these shape repeats after each cycle. The various shapes of periodic signals are sinusoidal, triangular, sawtooth, square, etc. Any periodic signal can be expressed as a sum of infinite number of sine and cosine components given by a Fourier series, as shown below:

$$X(t) = 1/2C + \sum_{n=1}^{\infty} a_n \sin\left(2\prod nft\right) + \sum_{n=1}^{\infty} b_n \cos\left(2\prod nft\right) \tag{2.2}$$

where f is the fundamental frequency of the periodic waveform and a_n and b_n are the amplitudes of the nth harmonic components of sine and cosine functions, respectively. If period T is given and amplitudes are known, then the original function of time can be defined by using Eq. (2.2).

This type of representation of a periodic waveform in terms of an infinite number of harmonic terms was defined by French mathematician Jean Fourier in the early nineteenth century and is known as Fourier series. If these harmonic components are known, then the function can be redefined using Eq. (2.2). Periodic waveforms repeat the pattern after each cycle infinitely and hence can represent the signal (corresponding to any communication circuit) in the same way.

The harmonic amplitude can be computed using the following steps:

$$a_n = \int_0^T \sin(2\pi k f t)\sin(2\pi n f t) = 0 \text{ for } k \neq n \qquad (2.3)$$

$$T/2 \text{ for } k = n$$

Here, harmonic amplitude b_n disappears over the period of integration. Similarly, amplitude b_n can be computed using the same steps between range 0 and T and harmonic amplitude a_n will not be present in the integration.

The constant C can be obtained by integrating both sides of Eq. (2.3). The following are the values of a_n, b_n, and C of the Fourier-series representation of the periodic waveform:

$$a_n = 2/T \int_0^T X(t)\sin(2\Pi n f t)dt$$

$$b_n = 2/T \int_0^T X(t)\cos(2\Pi n f t)dt$$

$$C = 2/T \int_0^T X(t)dt$$

The periodic signals can also be represented in a frequency domain (a graph showing frequency and amplitude). The frequency representation of a pure sine wave (containing one frequency) shown in Figure 2.4(a) is given in Figure 2.4(b). The vertical line represents one frequency for all the amplitude of the signal.

Fourier series can be used to represent a variety of physical quantities such as temperature, pressure, vibrations, and sounds of any musical instrument, periodic waveform

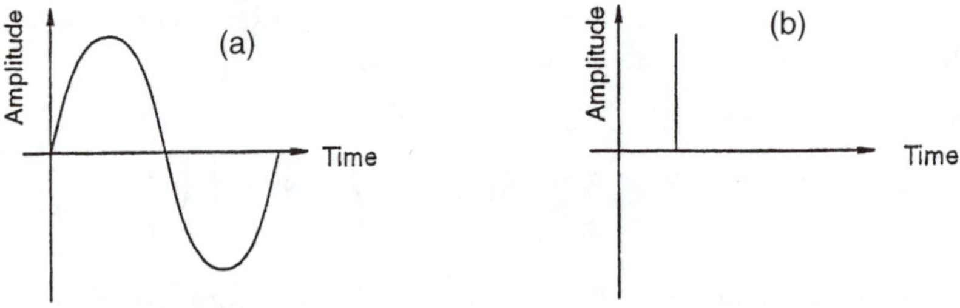

Figure 2.4 Pure sine wave (a) and its frequency representation (b).

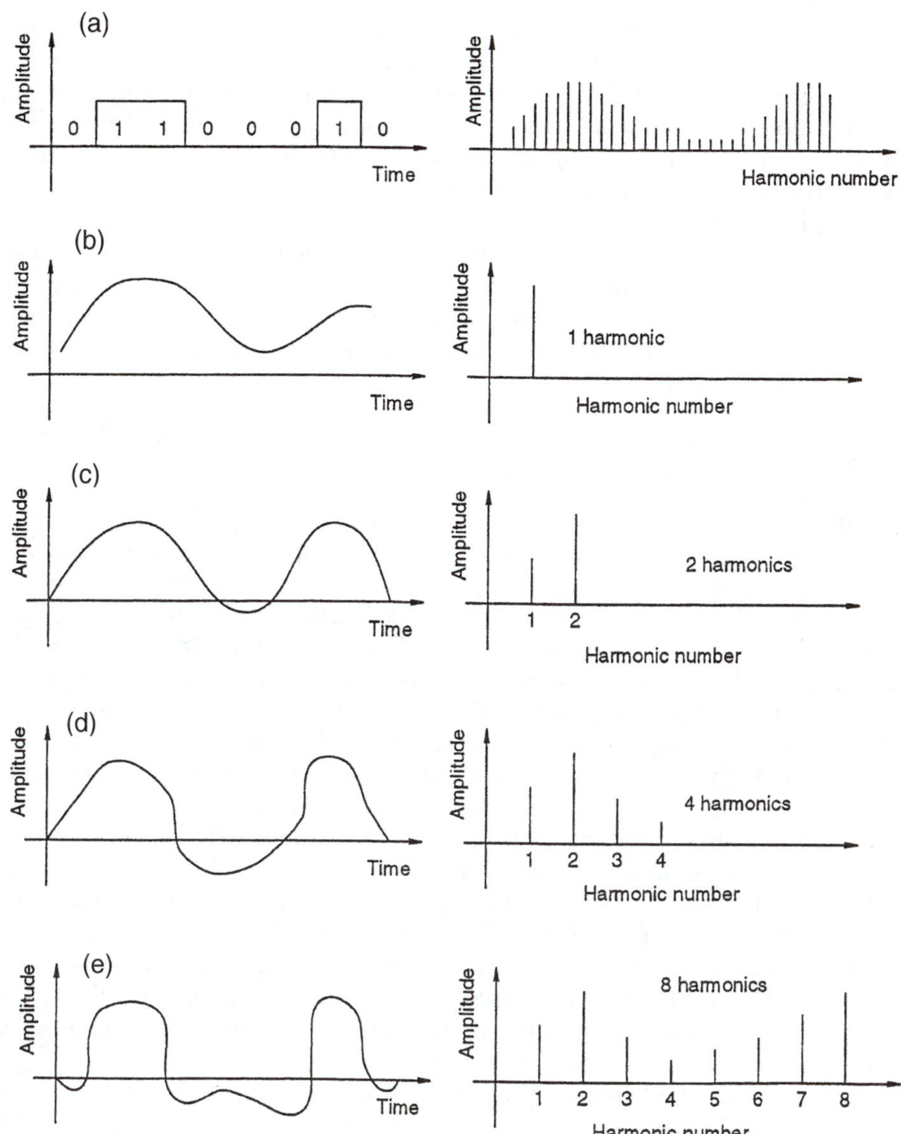

Figure 2.5 Fourier series.

(we discussed above) and any continuous analog signal in a communication system. Note that we can use Fourier series for representing digital signals. The use of Fourier series for data communication as discussed in Reference 1 is defined in Figure 2.5(a–e).

Consider the transmission of the eight-bit ASCII code for the letter b (01100010 with odd polarity). The Fourier analysis of the signal using Eq. (2.3) will give the coefficients a_n, b_n, and c. In the equation, the $X(t)$ defines the output voltage. Multiple harmonics can be blocked by choosing a low bandwidth filter where the bandwidth is lower, and, as a result, harmonics affect the accurate shaping of the transmitted letter b. But if we increase the bandwidth, the harmonics help in transmitting the accurate shape of the letter.

The harmonics, which are given by the sum of the root mean square of a_n and b_n, are shown on the right side of Figure 2.5(a–e). These correspond to the energy transmitted at the corresponding frequency. As the signals travel through media, they become attenuated. If all of the Fourier components equally disappear from the waveform, the output will represent the same shape of binary signal with less amplitude. The distortion is introduced due to the fact that different Fourier components are attenuated differently. In general, the amplitude of the signal components can be transmitted unchanged up to a cut-off frequency in such a way that the frequency components above this cut-off frequency are attenuated. This can be arranged in different ways by using different types of filters.

In the signal shown in Figure 2.5(a), the situation is that the frequency of the signal is so low that it transmits only the lowest frequencies. The Fourier analysis will offer the first few terms only and all the higher terms are absent. If we let a few harmonics go through the channel (fundamental frequency), the signals look like those shown in Figure 2.5(b). For higher-bandwidth channels, we get spectra and reconstructed functions as shown in Figure 2.5(c–e).

2.3.6 Data communication systems

A communication system (process) is an important part of data processing and data communication. It is characterized by five components: input devices, transmission, communication link, reception, and output devices.

Transmitter and receiver perform the transmission and reception, respectively. These are hardware units which include modulators/demodulators, multiplexers, line drivers, line adapters, etc. The transmitter performs the transmission of analog and digital signals over the communication channel. The analog signals are characterized by a continuously varying waveform and are usually in sinusoidal form. On the other hand, discrete pulse values or binary logic values 1 and 0 characterize a digital signal. Different values of voltage, current, or any other energy may be assigned to binary logics 1 and 0. The receiver receives these signals where these sides are connected by a physical link (transmission media).

If analog signals are used, then the transmission and reception require modulation and demodulation. These are obtained using a device known as a **modem (MOdulator/DEModulator)**. Modems have also been referred **as dataset or dataphone (AT&T modems)**. In order to transmit dial tones, a portable modem known as an **acoustic coupler (AC)** is used, coupled with a dial telephone set. In order to provide an interface to the signals (analog or digital) between the devices and communication channels, devices known as line drivers or line adapters are used.

The main function of a line driver is to provide a low-level interface between communication channels and devices. This is used in all the applications where we have to provide an interface between hardware devices (operating systems, protocols, programming languages, etc.). The line driver or adapter is different from the modem, as the modem is used for modulating and demodulating the signals (analog only) over the top of the driver, whereas the line driver is used for the transmission of analog or digital signals.

Digital signals do not required modems. In the case of digital transmission, the function of the line driver is to provide data communication control within computers. Other devices offering similar functions to that of line drivers are known as terminal service drivers, channel service drivers, etc.

Different types of communication links are available (telephone lines, coaxial cables, satellite link, microwave links, radio link, etc.). These links are usually based on two types of transmission: synchronous and asynchronous. These are discussed in brief here, and details are given in Chapter 5.

Synchronous communication: This type of transmission sends character symbols over the communication link. We have to provide synchronization between transmitting and receiving sites, whereby the receiver should know the start and end of the information. Because of this, control characters are always sent along with the character message information. The field corresponding to these control characters is defined as idle (IDLE) or synchronization (SYNC), which typically contains a unique control character or bit pattern and is attached before the message information. The characters are always in the form of coded blocks of information and are dependent on the type of information we are dealing with. For example, if the message is in text form, an ASCII scheme generates eight bits (actually, it defines only seven bits for each character, but the last bit is used as a parity) for each of characters transmitted. The message blocks are usually defined for groups of the characters and are of fixed size (eight bits in ASCII). This type of transmission is more efficient because it defines fewer bits in the control characters and error-detecting protocols can easily be used. Generally, the data block (packet) contains 16 to 32 bits plus the control bits. Modems and terminals for this type of transmission are more expensive and, in the event of errors, the entire block of data is re-transmitted.

Asynchronous communication (start-stop communication): In this transmission, a start bit is attached at the beginning of each text character (of eight bits) and represents the synchronization between transmitter and receiver clocks during the duration of the character. One or more stop bits then follow the character. When the channel is idle, the marking signal (logical 1s) is continuously transmitted over the link. In this way, this mode transmission allows the regular transmission of characters at varying or different instants of time. This type of operation is exactly the same operation being performed at the input devices (keyboard). Once a character is encoded with a start bit, eight bits (ASCII), and stop bits, the **least significant bit** (LSB: start bit) is transmitted first and the **most significant bit** (MSB: stop bit) is transmitted last. For calculating the speed of transmission in characters per second, we can divide the total bits transmitted per second by the average number of bits in the characters.

Asynchronous transmission systems are inexpensive and easy to maintain, but they offer low speed. Due to the low speed, channel capacity is wasted. In the event of errors, only lost characters need to be re-transmitted.

Various input devices include keyboards, transducers, analog-to-digital (A/D) converters, sensors, etc., while the output devices are teleprinters, printers, teletypewriters, etc.

2.3.7 *Communication channel properties*

One of the main objectives of data communication is to provide the transfer of messages between two communication devices connected by a communication link (transmission medium). This also defines a path between transmitting and receiving nodes over the link, and the signals (analog or digital) travel over it. The communication devices could be computers, terminals, printers, processors, etc. Messages are exchanged between the communication devices by electrical signals. The signals are generated by the transmitter at the sending side and are received by the receiver on the receiving side. The two communication devices are connected by transmission media (telephone lines, coaxial cables, satellite link, microwave link, etc.) which define a communication channel on different types of links. Obviously, we expect a communication channel to offer an error-free path for the transmission of signals. However, the channel introduces a number of distortions and noise into the signal, which affect the capacity of the channels (expressed as the number of bits per second) and the quality of the transmission.

Channel configurations: A communication must establish a route or circuit (logical path) between transmitting and receiving sites over which electrical signals are transmitted.

This type of connection may be defined over many transmission media, e.g., telephone lines, coaxial cable, etc. A combination of links may be defined as a communication channel over which message information can be transmitted and is characterized by various parameters such as channel operations, transmission speed, modes of operation, transmission configurations, etc. Various modes of operations in the communication channel are **simplex**, **half-duplex**, and **full-duplex**. More details on these configurations with respect to channels can be found in Chapter 4.

In a simplex configuration, information always flows in one direction, similar to a radio broadcast system, one-way traffic system, etc. This type of operation usually depends on the characteristics of the computer/terminals being used and is independent of the characteristics of the communication link being used for the transmission. Examples are send-only terminals, receive-only terminals, etc.

A half-duplex configuration allows the transmission of a signal in one direction at a time. It usually depends on the characteristics of the terminals, data communication link, and modulation method, but it allows the transmission in only one direction at any time. The end of a signal in one direction is recognized by the other side, which may switch its mode to the transmission state. The turnaround and overhead time sometimes become serious problems and affect the throughput of the communication system. Typically, the turnaround time is in the range of 20–200 milliseconds and depends on propagation, capacitance, inductance, and resistance of the lines. The majority of the terminals offer half-duplex operations.

A full-duplex configuration allows the transmission in both directions simultaneously. Since the turnaround time is eliminated (as the stations are not waiting for a response), the efficiency of the communication system is improved dramatically. Here we use two pairs of lines, and typically one pair of wires carries the data while another pair of wires carries the control signal. The operation depends on the characteristics of the communication devices being used. It is very useful in computer communication networks, especially with protocols of high-level data link layer control (HDLC), synchronous data link layer control (SDLC), link access control — balanced (LAP-B), link access control — D channel (LAP-D), advanced data communication control procedures (ADCCP), digital data communication message protocol (DDCMP), and many other vendor-customized software.

These configurations (simplex, half-duplex and full-duplex) have been referred to by different names by the standards organizations and countries where these are used. For example, CCITT has defined half-duplex and full-duplex operations as simplex and duplex operations, respectively. ANSI refers to these operations as one-way-only, two-way-alternate, and two-way-simultaneous, respectively. Both simplex and half-duplex operations have also been called a two-wire system, while full-duplex operation has been called a four-wire system. The details of these configurations, along with their applications in data communications, are discussed in Chapter 4.

Signaling speed of channels: Signaling speed to a degree represents the capacity of the communication channel. Signaling speed (the number of times it alternates the value in one second) and encoding method (such as ASCII) affect the transmission speed (time it takes to transmit the character). This speed may be defined for different devices; e.g., for a keyboard it may define the number of characters per second, for teleprinters it may be words per minute, for a communication channel it is baud, and so on. The transmission signaling speed or rate of any communication channel is characterized by the following:

- Baud rate
- Bit rate
- Throughput

Baud rate: Baud rate is the unit of signaling speed derived from the duration of the shortest code element. Baud rate = I/T where T represents the shortest pulse duration. A baud rate unit is known as baud.

Bit rate: Bit rate represents the number of bits transmitted in a given amount of time. Bit rate = total number of bits transmitted/total time. The unit of bit rate is given by bits per second (bps).

Throughput: Throughput is a measure of the efficiency of a communication system; it is the rate at which system can do work. This can be expressed in many ways, as given below:

- Number of information bits sent divided by the time it takes to transmit all bits, including overhead bits (bps)
- Number of information bits divided by total bits transmitted over a period of time (% throughput)
- Number of packets accepted divided by number of packets transmitted over a period of time.

Given a bit rate of r bits/sec, the time required to transmit eight bits (corresponding to the letter b in ASCII) will be $8/r$ seconds. The frequency of the first harmonic is $r/8$ Hz. We know that a telephone line under normal conditions will provide a voice band of 3 KHz. This means that a telephone line has an upper limit of 3 KHz, and, therefore, the number of highest harmonic frequencies which can pass through it are $3000/r/8$ or $24000/r$ Hz (24 KHz/r).

Channel capacity: A communication channel is characterized by its capacity to carry signals (voice or data). The signals represent different types of information, depending on their sources. Various sources of these signals include computer, sensors, terminals, etc. These signals are transmitted over the communication channel at different speeds. Based on the speed of the communication channels, different bands of signals have been defined as narrow-band (300 bps), voice channel or voice-band channel (600–188 bps), and wide-band channel (more than 19.2 Kbps). Other wide-band data rates of wide-band channels are 40.8, 50, 56, 230.4, 250 Kbps, etc.

The capacity of a channel (number of bits per second) is affected by bandwidth, which in turn is constrained by the characteristics of the transmitter, receiver, and medium.

Although there is no direct relationship between bandwidth and the capacity of a channel, various frequency ranges for different applications clearly demonstrate the relationship between them. Accordingly, the greater the bandwidth, the greater the capacity of the channel will be. A process known as multiplexing can obtain the transmission of a different combination of low-speed channels over a high-speed channel. In this process, the lower-band channels can be multiplexed to the maximum speed of the higher–data rate channel. There are two types of multiplexing: frequency division multiplexing (FDM) and time division multiplexing (TDM). In FDM, the channels are allocated frequency and are put together on the frequency scale, whereas in TDM the channels are placed together on the time scale.

The signals are classified as baseband and broadband. The baseband signals are typically digital signals and do not use modulation. They offer a limited bandwidth and use TDM for channel sharing. In contrast to this, broadband signals require modulation and are represented as analog waveforms. They offer higher bandwidth and use FDM for channel sharing.

The signals coming out from computers and terminals are usually square pulses, and a signaling scheme for these pulses can be defined in a variety of ways. In signaling, we define the voltage values for logic values 1s and 0s. In the simplest form of signaling, the logic 1 is represented by applied voltage, while logic 0 represents no voltage. In another scheme, different polarity of voltages is used for logic 1s and 0s (+ voltage for 1, – voltage for 0). In the return-to-zero (RTZ) scheme, the logic 1 is represented by voltage value, which returns to zero value during the time interval of transmission, thus providing a separation between consecutive pulses. In the differential encoding scheme, the difference of voltages is used to represent 1s and 0s. These schemes are used to provide signaling to logics 1 and 0 in such a way that it becomes easy to detect any error in the information during the transmission and also provide synchronization between the nodes. Further, these schemes transmit less DC component, which requires a lot of power and is therefore wasteful.

The pulses corresponding to information travel through a communication channel over communication links and experience resistance of link, inductive and capacitive effects of link, and some kind of loss or leakage. Due to these effects, the pulses are attenuated and, as such, lose their amplitude and other characteristics. In order to reduce these distortion effects on the signals, amplifiers (for analog signals) and regenerative repeaters are used at regular intervals from the communication link.

Channel bandwidth: As we have seen in the previous section, speech signals do not contain a unique frequency; instead, they consist of many harmonics. Similarly, signals corresponding to other communication circuits such as radio, television, and satellite also consist of a range of frequencies rather than just one frequency. Thus, these signals can be characterized as a bandwidth of signals which can be transmitted; e.g., a TV channel usually has a bandwidth of 6 MHz. Table 2.1 shows the frequency range of various applications.

Table 2.1 Frequency Range of Various Applications

Frequency	Applications
10^3 Hz	Telephone voice frequencies (with low to high speeds)
10^4 (very low frequency (VLF))	Telephone voice frequencies (with higher speeds)
10^6 (coaxial cable)	AM sound broadcasting (high-speed voice and data)
10^8 VHF (coaxial cable)	VHF sound and TV broadcasting (FM)
10^9 UHF	UHF TV broadcasting
$10^{12\text{-}13}$	Infrared transmission (local data transmission)
$10^{14\text{-}15}$	Optical fiber (very high-speed voice and data)
$10^{19\text{–}23}$	X-rays and gamma-rays

Wire transmission supports a frequency range of up to 100 kHz. Coaxial cable supports up to 100 MHz. The frequency range of up to 100 GHz is usually supported by the microwave medium, while waveguides use a frequency range greater than 100 10^{15} GHz and up to 100 THz (terahertz, 10^{12} Hz). The frequency range of over 10^{15} up to 10^{23} is defined for optical fiber communication. Laser offers a frequency of 10^{14} Hz.

The wavelength of any frequency range can be calculated by using Eq. (2.2), where c represents the velocity of radio and light waves in space and is equal to 3×10^8 m/s. These frequencies are allocated by the International Telecommunication Union (ITU) standards organization. The baseband bandwidth for AM is about 5 KHz, and for FM is 15 KHz. A 0.25-MHz frequency band for TV audio transmission is allocated. Each channel is assigned a different frequency band for these transmissions. It can be seen that this frequency is higher than that allocated by ITU because out of this frequency, some portion of the frequency band needs to be used for avoiding any interference between the channels.

The channel bandwidth for AM is 10 KHz, for FM is 200 KHz, for video is 4.5 MHz, and 6 MHz. The broadcasting bandwidths for AM transmissions are 535–1605 KHz and for FM transmissions are 88–108 MHz, 54–216 MHz, and 470–890 MHz. Looking at the bandwidth of each of the channels in these transmissions, the number of channels supported are 107 for AM, 100 for FM, 12 for VHF, and 70 for UHF.

2.4 *Filters*

The human ear cannot hear sound frequencies below 20 Hz or above 7.5 KHz. Thus, we can obtain low-band pass and high-band pass frequencies for the human ear as 20 Hz and 7.5 KHz. Similarly, telephone lines under normal circumstances have low-band passes of 200 Hz and high-band passes of 3.5 KHz, so the bandwidth of a telephone line is 3.2 KHz. Thus, for the reasons of economics, performance, etc., frequencies below 200 Hz and above 3.5 KHz are not transmitted over telephone lines under normal conditions.

When channels of voice signals are transmitted over a common medium, guard bands of 800 Hz separate these channels to 1 KHz. These guard bands prevent signal and electrical interference between the channels. Similarly, guard bands are also used in other communication circuits to prevent signal interference between channels. Sometimes, it is advisable (on the grounds of economics, performance, response time, and throughput) to select a band of frequency out of the channel and pass it while blocking other frequencies that are lower and higher than the lowest and highest frequencies of the channel, respectively. The selection/passing or rejection/blocking of a band of frequency can be obtained using a filter. There are four types of filters, shown in Figure 2.6(a–d), which can be used by communication systems for selecting or blocking a band of frequency.

A low-pass filter (LPF) passes lower-frequency signals, defined by its cut-off frequency, and blocks higher-frequency signals beyond the cut-off value. A high-pass filter (HPF) passes higher-frequency signals defined by its cut-off frequency and blocks lower-frequency signals beyond the cut-off value. A band-pass filter (BPF) passes a band of frequency (defined by lowest and highest frequencies) and blocks all the frequencies lower and higher than the lowest and highest, respectively. A band-elimination filter (BEF) passes all the frequencies lower and higher than the lowest and highest frequencies (defining a band) while blocking that band.

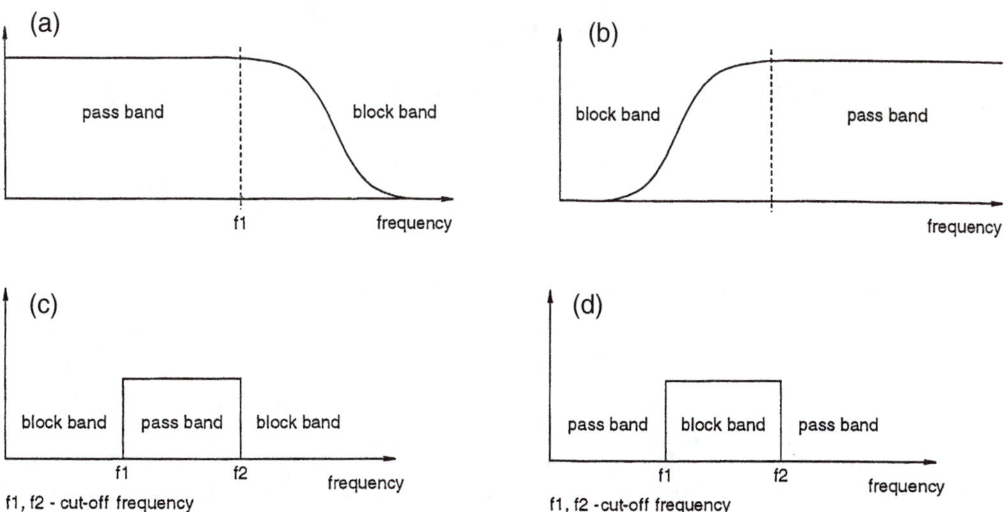

Figure 2.6 (a) Low-pass filter. (b) High-pass filter. (c) Band-pass filter. (d) Band-elimination filter.

2.5 Errors in data communication

Errors have become an integral part of any communication system. Human beings, signal distortions, noises, malfunctioning of electronic and electrical circuits, etc., may cause errors. The signal distortion may be defined as unwanted signal, which may be influenced by attenuation in the links, noises, and behavior of the various components of communication systems. On the other hand, a noise is unwanted signal generated by atmospheric disturbances and natural calamities and may be due to the communication circuits which are also transmitted with the data signal. Although there are different ways of grouping unwanted, undesired, and unpredictable signals, we will group them together into the following two groups:

- Signal distortion
- Noise

2.5.1 Signal distortion

Electrical signals transmitted over transmission media or noises, distances, attenuation, etc., generally influence communication links. Due to atmospheric noises, signal-to-noise ratio may become smaller at the receiving end, while signals may become weak as they travel over transmission lines over a long distance, or the communication media may offer significant attenuation (due to capacitive or inductive effects, resistance, etc.) to the signals. In all these cases, signals received at the receiving end may not provide us with the exact message information transmitted.

Signal-to-noise ratio: It is defined as the ratio of relative strength of the signal to noise and is expressed in decibels as shown below:

$$S/N = 10 \log_{10} S/N$$

If we have a number of 1,000, then S/N in decibel is given by

$$10 \log 1000 = 10 \log 10^3 = 10 \times 3 = 30$$

This ratio can also be used to define either gain or attenuation as given by

$$\text{Gain (db)} = 10 \log \mathit{Sout}/\mathit{Sinput}$$

$$\text{Attenuation (db)} = 10 \log \mathit{Sin}/\mathit{Sout}$$

where S represents signal strength as voltage, current, or power.

The wavelength is defined as

$$C = \lambda F$$

where C represents the speed of light as 3×10^8 m/s, λ represents the wavelength, and F denotes the frequency. The wavelength is defined as one period or one cycle of the signal. Frequency is defined as the ratio of c to wavelength, the inverse of period.

The wavelengths of some signals (using the above equation) are 300,000 m for voice grade (20 Hz to 20 KHz), 0.03 for satellite (10 GHz), and 3 for broadcast FM (100 MHz).

The following are the various types of distortions and unwanted signals (called noises) which contribute to signal distortion.

Attenuation: As the signal travels over the transmission media, the amplitude of the signal decreases (due to the resistance offered by the conductor, inductive and capacitive effects, and the length of the conductor), thus making the signals weaker.

High-frequency distortion: It is known that high-frequency signals are attenuated more than low-frequency signals when the transmitter is sending signals of different frequencies at different times. This type of distortion is also known as phase distortion.

Delay distortion: Lower-frequency signals travel faster than higher-frequency signals if a transmitter is sending more than one frequency signal at the same time. Thus, low-frequency signals have less delay than high-frequency signals. This type of distortion is known as delay distortion.

2.5.2 *Noise*

A noise is an unwanted signal and is transmitted along with the original signal over the transmission media. These signals are introduced/generated due to many atmospheric conditions, thunderstorms, rains, etc., and the equipment being used in the communication systems. There are a variety of noises that have been defined. Next, we discuss types of noises, the sources of the noises, and how these noises can be reduced in communication systems.

Electrical noise: This category of noise is basically a high-voltage generator, high-frequency unwanted interference on the link. The sources of this noise are transmission stations, lightning, heavy motors and generators, heating and cooling systems, etc. The high-frequency noise may cause the transmission of signals to wither over the links. In other words, the interference may be between the ground and the lines or between the lines themselves. This noise has a very significant effect on the quality of signal received if we use the frequency modulation scheme in the transmission.

Cross talk: In an open wire, one wire may pick up the signal transmission on an adjacent wire and thus cause interference in it. We experience this problem when we notice that our voice is not clear or we get many other voices together while using the telephone system. The interference created by open wires is called cross talk. Here the electromagnetic waves of the signals flowing through adjacent wires interfere with each other due to the inductive and capacitive effects created by other circuits near these lines, e.g., power lines, telephone lines, and other interference facilities. Cross talk may also be created by motors, generators, or even transmitters/receivers near the lines. This type of noise (distortion) is very common with open wire systems because telephone companies link the pairs of open wires as a carrier system using amplifiers accommodating 12 voice-grade channels using FDM.

When we use a parallel pair of wires (or cables), current flowing in each wire either in the same direction or in opposite directions will interfere with each other. This is due to the fact that the electromagnetic fields induced by the flow of current will interfere with each other, affecting the value of current and hence the quality of signals. In real life, sometimes we notice this type of noise (due to some other conversation) while using a phone. Either twisting the wires or putting a separator between parallel wires can minimize this noise, which reduces the interference effect between the electromagnetic fields.

Impulse noise (spikes): We know that the electrons constituting the current flowing in a metallic wire have a random movement and sometimes, due to this random movement, neighboring components of the electronic circuits experience thermal energy which causes the thermal motion of the electrons in the electronic circuit. This is caused due to the sudden increase in the amplitude of the signal to peak value within or outside the data communication circuit. The peak value of the amplitude of the noise (or internal impulse noise) is usually caused by poor connections and contacts, loose connections, improper grounding, etc., and can be noticed in the telephone system where we sometimes do not hear any tone. The external impulse noise is typically caused by atmospheric conditions such as thunderstorms, lightning, or even relay switchgear near the station. One can feel this noise in an AM receiver during bad weather, thunderstorms, or even rains. This noise is a form of random impairment.

We notice this type of noise **(hum)** when we turn off our radio or when it is not tuned to any station at a moderate volume. At low volume, we may not notice it, but if the volume is moderate, we will hear this noise. This noise has different names, such as **white noise**, **Gaussian noise**, **random noise**, etc. The most common name for this noise is **Gaussian (random)** as its amplitude follows the very popular Gaussian distribution function. The name white noise is derived from the fact that this noise occurs at all frequencies in more or less the same way. It is also known as thermal noise because the noise level varies linearly with the temperature of the system. This is due to the thermal agitation of electrons; this type of noise cannot be reduced and is independent of distance.

Transient noise: This is another form of random impairment noise and is usually caused by sudden voltage change (due to power equipment or heavy transformers), electrical storms, and even rotary motion during the dialing of rotary telephone systems, etc. Some people consider both noises (transient and impulse) the same and, as a matter of fact, this may be true, as some sources for these noises are common. The duration of transient noise may vary from a few milliseconds to a few seconds and depends on the magnitude of voltage developed by the sources. The loss of bits from bit rate transmission depends on the duration of transient noise. Shielding the wires may reduce this noise.

Phase jitter: This noise is usually caused by interfering with the phase shift of a transmitted signal and can be noticed at the receiver side. It has been shown that this noise is of the form of frequency or phase modulation and seems to have little effect on the transmission of voice-grade channel. It becomes one of the significant performance criteria for data communication as it limits the speed of the transmission. It usually causes errors in the data stream, which are transmitted over a frequency-division-multiplexed communication channel. In the single-sided-band amplitude modulation (SSB-AM), the carrier frequency is changed to a higher level (based on the amplitude of modulating signal) before multiplexing at the transmitting side, while the carrier frequency is changed to a lower value after de-multiplexing at the receiving side (to recover the original level). The carrier frequency must be the same at both ends (transmitter and receiver) of the communication link. Phase jitter will occur if the carrier frequency at the ends is not same. This difference of carrier frequency between two sides may be due to many factors such as ripple voltage noises in the respective oscillator circuits.

Phase jitter is usually measured as peak-to-peak derivation of the instantaneous value of the phase angle of the signal. In voice telephone communication, it may affect only the quality of voice received, but in the case of data communication, the loss of a large number of bits in a bit rate transmission becomes a serious problem. This noise may be reduced by using highly stable frequency circuits, multiplexers of the required specifications, etc. In most of the data and voice communications, the self-test tones may be transmitted to the receiver to measure the phase jitter between them and the received signals can be calibrated accordingly. This noise cannot be reduced and is independent of distance.

Harmonics distortion: This distortion is generated by different circuits (amplifiers, modulators, etc.) of communication systems and is defined as unwanted signals having integer multiples of amplitude frequencies of the original signal. The output of these circuits, in general, does not have a linear relationship with the input. The functions of the circuits provide nonlinear distortions and are collectively known as harmonic distortions. Using linearizing circuits to eliminate the effect of nonlinearity on the output transmitted signal may reduce this noise.

Other noises: Lightning storms and faulty equipment cause catastrophic noise that cannot be reduced, and it increases with distance. The higher-frequency signals lose their strength more than lower-frequency signals, which causes attenuation errors. Amplitude errors are caused by a sudden drop in power caused by faulty electrical connection or contacts, change in load, or switching errors. Errors which occur in burst are known as bursty errors. Intermodulation noise is similar to cross talk and is created by modems due to the interference between two signals. For more details on noise and other distortions, see Reference 4.

Echoes: This noise can be noticed during a conversation on a telephone system. The two-wire lines are connected to the network (typically four-wire lines) and the impedance of both two-wire and four-wire lines varies with the frequency in the communication circuits. Impedance is a generic word used to represent the effect of various components (inductance, capacitance, and resistance) with reference to frequency in the communication circuit. In addition to this, there may be a mismatch of impedance between two-wire and four-wire lines. The total effect of these causes is noticed by getting a portion of the signal reflected back to the source in the form of an undesirable signal, which is called an echo. Echoes on data lines cause intermittent errors.

A speaker may get an echo of his/her own voice if there is a mismatch at the transmission side. This may have a significant effect on the conversation. If a delay of the signal (going from the microphone to the junction and coming back) is small, the speaker's voice can be heard in his own receiver. If the delay is substantial, it has a significant effect on the receiver because the speaker has to repeat the conversation, speak louder, etc., causing dissatisfaction. This type of echo has little effect on the listener's side if the delay is small, but if it is large, the listener is unable to understand the conversation because it sounds like more than one person is talking simultaneously. Fortunately, this noise has little effect on data communication when we are using half duplex. For short-distance communications, echoes are generally tolerable, but for long-distance communication, impedance mismatching becomes a serious problem and echo suppressers, echo-cancellation techniques, etc., must be used.

The echo suppressor device reduces the problems of echoes caused by mismatching by comparing the transmitted and received paths for the signals. If the difference in the levels of signals over the paths is significant, it adjusts the signal levels over the paths to the levels by which the effect can be neutralized or reduced. It operates over four-wire lines, as shown in Figure 2.7(a). PADs define the adjustment of levels. When the node is transmitting signals over the transmitting path and the signal level on this path is much higher than the receiving path, the echo suppressor switches the PADs onto the the receiving path to reduce the effect of echo loss, as shown in Figure 2.7(b).

In the case where the node is in transmitting mode, it inserts the losses in the transmitting path so that we do not get any loss on the receiving path. The echo suppressor circuit must include minimal round-trip propagation delay in the telephone system; otherwise, the quality of the voice signal will be affected considerably. The echo suppressors used for the telephone system usually do not work on the tone frequencies (2000-2250 Hz), but once the connection is made, the signals can be transmitted over the link.

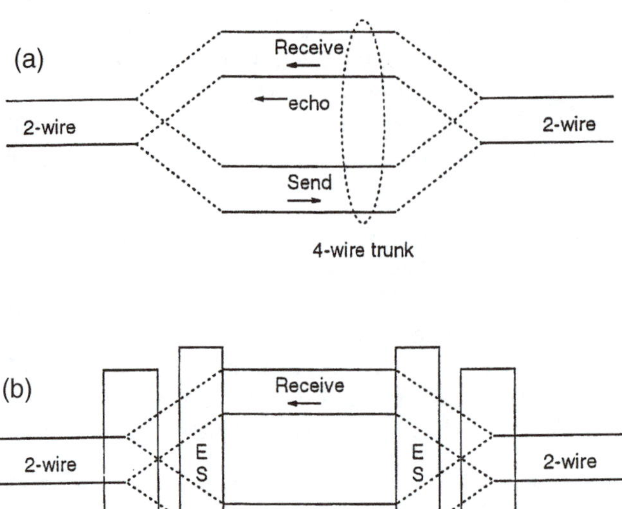

Figure 2.7 Echo suppressors.

Transmission distortion: These distortions are caused because transmitted signals take different times to reach the receivers. For example, in radio frequency (RF) transmission in the range of 3–30 MHz, the transmitted signals may be affected by magnetic storms, cosmic rays, weather conditions, etc. (ionosphere). The ultra-high frequency (UHF) transmission signals in the range of 300–3000 MHz may lose their strength due to fading (a phenomenon of troposphere).

Modulation (attenuation) noise: The behavior of the communication channel is affected by the modulation techniques we choose for the transmission of the signal. This noise is also referred to as attenuation noise and has been defined in two main classes: amplitude-frequency and phase-frequency attenuation distortions.

In the first class of this noise (amplitude-frequency), the level of transmitted signal depends on the frequency of the signal, transmission media, distance it travels, length of communication circuits, etc. The signal varies with the frequency over the bandwidth of the signal. This loss of level (decaying amplitude) of the transmitted signal is not the same for all frequencies, and this non-uniform variation of the transmitted signal across the bandwidth causes amplitude-frequency distortion. Installing inductive padding coils in the channel circuit at regular intervals can reduce this type of noise, which introduces lumped inductance over the channel. This scheme defines a uniform decay of the amplitude of the signal over the channel. Another method of reducing the noise is line conditioning, which is being used with leased voice-grade communication. In the case of switched networks, equalizers are attached with modems and they nullify the effects of the noise on the channel.

The second type of attenuation distortion (phase frequency) is caused by the fact that signals do not vary linearly with both frequency and phase shift and, as such, consist of a large number of frequency components. These frequency components have different phase shifts with respect to frequency, travel at different speeds, and arrive at the receiving side at different instants of times in different order. This type of noise is also known as **delay distortion** or **envelope delay, phase distortion**, or simply **delays**. Loaded cables, hybrid circuits, or filters (if used in the communication circuits) can cause the noise. Interestingly, these sources of noise also cause phase jitter. This noise is not serious for voice-grade communication but becomes a major problem with data communication. It can be reduced by using the same techniques as that of the previous noise (amplitude-frequency distortion), e.g., line conditioning for leased lines or attaching equalizers with modems.

Non-linear distortions: The above-described modulation (attenuation) distortions in general have deterministic behavior and affect the workings of communication circuits. In other words, these distortions can be analyzed and measured, and protective circuits may be designed to handle these noises. There are certain distortions which are nonlinear in nature and hence offer nonlinear behavior to the communication circuits. These distortions are usually caused by saturation effects in magnetic cores, nonlinear characteristics and behavior of the communication circuit itself, etc., and the effects become more significant if frequency-division multiplexing (FDM) is used in the systems. The difference in the carrier frequency of both transmission and receiving sides causes the harmonic relationship (frequency offset) and becomes a serious problem in data communications which use the narrow-frequency modulation scheme. In applications where the carrier signal at the receiving side is controlled or managed by the carrier signal generators of the transmitting side, the frequency offset may have little effect and can be controlled.

Another nonlinear distortion category is inter-modulation distortion, which is caused by a frequency-division multiplexing (FDM) system where a large number of voice-grade channels are multiplexed and transmitted over high-capacity channels (coaxial cable, satellite link, etc.). The signals on two different communication circuits may interfere with each other and produce a signal possessing a frequency band (via inter-modulation) which may have been assigned to signals going through another communication circuit. Although the interference between two circuits may produce a small frequency band value compared to the ones that are transmitted, sometimes it may become intolerable. Some of you may have heard unwanted conversation or dial tones in your receiver. This noise may become serious in data communication if the modem is transmitting a high-amplitude signal at a constant frequency. This noise can be reduced if signals are of a low amplitude level and of variable frequency.

Quantization noise: This type of noise occurs with a pulse-coded modulation (PCM) signal. In the PCM process, an analog signal is encoded in terms of quantization levels, which are obtained by sampling the signal (using Nyquist's Law). The number of bits which are included and transmitted usually determine the values of the samples. On the receiving side, during demodulation, this distortion may be a serious problem. Increasing the number of bits in the samples can reduce the noise, thus reducing the number of steps required for the quantization process to transmit the signal. This in turn reduces the number of steps required at the receiving side, too. Obviously, we can not increase the number of bits in the samples to just any value; it is constrained by the available bandwidth of the PCM channel.

Digital distortion: This type of distortion occurs in the DTE/DCE interface if the standard defined by the EIA is not properly and strictly followed. The noise is caused by the difference in the sample transition, different signal values at different transitions, different times to restore the decaying amplitude of the digital signal, etc. As a result of this, the transition from one logic to another takes different times and causes errors in data information. An appropriate logic error occurs at the receiving side if it takes more time than at the other side to reproduce it. These different timings of logic values may be caused due to a non-uniform sampling rate generated by a sampling clock. It is interesting to note that the same sources may also cause timing jitter noise.

Timing jitter: The synchronizing pulses generated by clocks sometimes are not stable and are affected by various factors such as instability in the electronic circuits (due to feedback), unwanted signals, etc. The signals generating high-frequency components obviously experience more distortion (attenuation) than the low-frequency components and, as a result, these components are going to be out of sync on the time scale. This type of distortion is called timing jitter.

Electronic circuits have many stages of operations. The output generated by one stage supplies clock pulses to the next stage electronic circuit, which in turn supplies pulses to the following, and so on. If signals are generated during each clock pulse, which will also be regenerated at the repeaters, this provides both signals: the actual data signal and the signal generated by the noise. A small number of signals generated during the clock pulses may have little effect on the synchronization, but in the case of excessive signals, both sides will be out of sync, causing the non-functioning of the clock. Letting the data link protocol recover so-called "lost packets" (during the excessive timing jitter) by bypassing the lost frame can reduce the noise. This way, the synchronization between transmitter and receiver will be maintained.

Baseband bias distortion: Signals from computers and terminals are square wave pulses. These pulses experience some losses due to the parameters of communication links. These parameters include resistance, inductive and capacitive effects, and shunt resistance. These parameters oppose the flow of signals and, as such, are distorted. The resistance offered by the link opposes the flow of signals, and the resistance offered by it is dependent on the frequency and distance. The baseband signals are generally used for low frequencies over a longer distance. The inductance opposes the rate of change of current flow through the link and, as a result, the changes in the logic (1 to 0 or vice versa) are distorted and the rise or fall of the pulses gets distorted. The opposition to the rate of change of flow of pulses corresponds to a resistance given by reactance of inductance. The capacitive effect opposes the rate of change of voltage levels of logic and, as a result, the original varying signal gets opposed by varying voltage in opposite direction. This distorts the pulses. The reactance of capacitance usually depends on parameters such as frequency, wire diameter, type of insulation, etc. The shunt resistance represents the losses due to current flowing into insulators or dielectric or hysteresis losses and is dependent on frequency, external environmental conditions, etc. These causes provide significant interference between the signal pulses, which are distorted, and this situation is known as bias distortion. This distortion does not include the effects of propagation delay of the link. Changing the sampling rate and its threshold value can change this type of distortion. A higher sampling threshold may reduce this distortion.

2.6 Repeaters and amplifiers

Electrical signals (analog or digital) lose their amplitude as they travel through the communication media due to resistance. Because of this, the behavior of the signals gets changed. This change is known as **attenuation**. Attenuation is caused by the communication media and depends on the length of the link. For example, the resistance offered by metal wire over its length, the impurities found in optical fiber, and the effect of atmospheric noise (rains, thunderstorms, clouds, smoke, lightning, etc.) on open wire/twisted wires all affect the quality of the signal. In the case where signals are

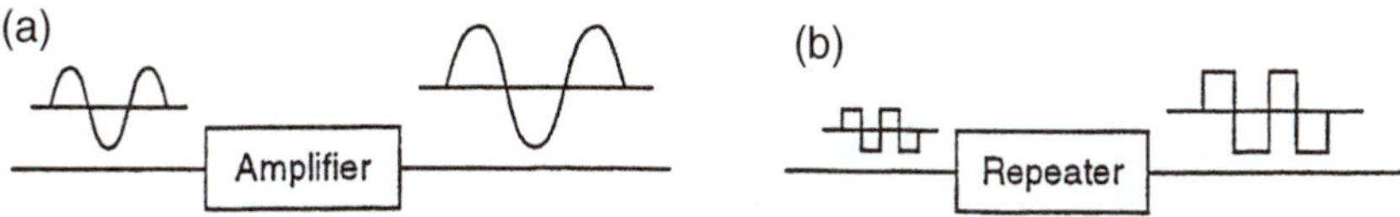

Figure 2.8 (a) Amplifier. (b) Repeater.

transmitted through the atmosphere, reflection and refraction attenuate (scatter) the signals. In order to achieve an unattenuated signal over a long distance, amplifiers are used for analog signals and repeaters are used for digital signals.

Amplifiers amplify the signal levels in one direction by an amplification factor, typically 70–100, and re-transmit the amplified signals over the media, as shown in Figure 2.8(a). The amplifiers have to be used at a regular distance (typically a few miles, e.g., 2–30). For signals flowing in opposite directions, separate amplifiers must be used. For digital signals, a repeater is used which regenerates the logic 1s and 0s and re-transmits this logic over the media, as shown in Figure 2.8(b). The repeaters have to be used at a regular distance, as well. Amplifiers are more expensive than repeaters and the number of repeaters for the same distance is much smaller than the number of amplifiers.

Some useful Web sites

For general information, see *http://www.usbuy.co*, *http://www.infoRocket.com*, and *http://www.ebay.com*. For more information on analog amplifiers, visit *http://www.analog.com*; for digital repeaters visit *http://www.e-insite.com*; and for network electronic components, modulators, amplifiers, etc., visit *http://www.lashen.com*. Information on data communication for network products can be found at *http://www.primusdatacom.com*.

References

1. Tanenbaum, A.S., *Computer Networks*, Prentice-Hall, 1989.
2. Techo, R., *Data Communications: An Introduction to Concepts and Design*, Plenum Press, 1984.
3. Loomis, M.E.S., *Data Communication*, Prentice-Hall, 1983.
4. Thurwachter, C.N., Jr., *Data and Telecommunications Systems and Applications*, Prentice-Hall, 2000.
5. Couc, L., *Digital and Analog Communication Systems*, Prentice-Hall, 1997.
6. Lathi, B.P., *Modern Digital and Analog Communication Systems*, Oxford University Press, 1998.

chapter three

Signal transmission basics

"People seldom become famous for what they say until after they are famous for what they've done."

Cullen Hightower

3.1 Modulation and demodulation: Basic concepts

A communication process is an essential component for both data processing and data communication systems. **Modem (MOdulation/DEModulation)** and transmission media together provide a platform for the transmission of data from computers/terminals, while the data from computers/terminals provide a communication interface between input devices and transmission media. We will use the terminology defined by ARPANET, and computers/terminals will be referred to as *hosts*. An example of input devices in our case will be a keyboard. The role of computer networks is to accept the data from the sender host and transmit it to the receiving host.

The data communication between data terminal equipment (DTE) can be defined in a number of ways and it depends on the distance. The DTE may represent PCs, terminals, workstations, etc. If the distance between them is small (i.e., on an organization's premises), the DTE may be connected by transmission media such as cable or twisted pair (shielded or unshielded optical fiber). If the distance between DTE is large, then we may use wireless media (e.g., microwave, satellite, or even lines of public telephone carrier companies). The telephone carrier companies offer switched circuits and dedicated (private) circuits. The establishment of circuits can be defined in a number of switched networks (public switched telephone networks, integrated services digital networks, and high-speed networks such as B-ISDN, ATM-based, etc.). The computer network is composed of hardware and software components which control and manage the transmission and reception of data, routing strategies, flow control, error control, statistical information, billing information, etc., between two hosts. The hardware components are generally defined by the transmitters, receivers, multiplexers, modems, etc., while software components are generally defined by protocols, customized software, operating systems, etc.

An electrical signal generated by a communication device (transducer, computer/terminal, etc.) may be in either analog or digital form. The signals so generated are termed **baseband signals** or **modulating signals**.

The analog signal is characterized by a waveform continuously varying its amplitude, while the digital signal is characterized by two discrete states (logics 1 and 0). Due to the presence of noises (caused by transmission, devices used, hardware failure, etc.) in the signals, the signals cannot be transmitted over the communication link or transmission line to a larger distance as the level of signal compared to noise signal level becomes very low (due to resistance offered by the links). Although amplifiers can be used to raise the level of the signals, they amplify both the signals with the same amplification factor. If we can define a process which can change the characteristics of signal of higher frequency by instantaneous value of amplitude, frequency, and phase of baseband (modulating) signal, then this signal can be transmitted to a larger/greater distance. This process is termed **modulation**. Similarly, on the receiving side, a process which can recover the original signal from the received signal is known as **demodulation** or **detection**. The processes of modulation and demodulation are performed by transmitter and receiver, respectively, and are both present in a modem.

The telephone network has a nominal bandwidth of 3 KHz. A bandwidth of 3 KHz has been chosen for the telephone network because it provides sufficient fidelity to recognize the voice of a person at the other end of the line. Obviously, if we have more bandwidth, the fidelity or quality of voice reproduction will be higher. Usually, a person speaking has his/her voice in the bandwidth of 3 KHz, while singers or announcers on a radio show may go up to a 10-KHz bandwidth. The **public switched telephone network (PSTN)** provides the services of conversation between users at a distance by making circuits or links between them over a medium or combination of media. Although the PSTN is primarily defined for voice communication, it can also be used for data communications using modems. The switched circuits are defined for all switched networks (PSTN, **switched data networks (SDN)** such as ISDNs (integrated services digital networks), B-ISDNs (broadband-ISDNs), etc.). The PSTN allows users to send their data packets over it via modem, while ISDN allows users to send their requests for call establishment and data transfer directly over it. The leased circuits (private or dedicated) are also defined in these networks and use modems when used with PSTNs, while leased circuits over switched data networks are all in digital.

If we transmit signals (voice or any other signal) of a bandwidth or from computers in the same way as they are coming out of the transducers, the transmission is known as **baseband transmission**. This transmission is possible only if an appropriate transmission medium is used. The generated signals from computers/terminals are generally square wave pulses. There is a variety of control baseband signal waveforms defined, and these are useful for error detection, maintenance of synchronizations, and reducing the **direct current (DC)** components in a signal in addition to normal representation of modulating signal. The baseband transmission is useful in telephone systems, television (video signals of 0–5 MHz produced by camera), and data communication signals coming out of terminal equipment. Similar discussions can be found in References 2 and 3.

Modulation: Modulation is an operation which translates a modulating signal (corresponding to information or message) into another signal using a constant carrier signal of high frequency. The modulation supports adaptation (to any unfavorable atmospheric conditions, noisy environment, etc.) for high quality of reception of the signals and allows multiple transmission of signals over a common transmission medium simultaneously.

Modulation can also be defined as an electrical or electronic process which in one way or another multiplies one signal modulation with another signal. Usually, people think that modulation is a linear process, but in fact it is a nonlinear process, since multiplication of time function is inherently a nonlinear process. Similarly, **demodulation** or **detection** is also a nonlinear process, although we might use linear circuits for linearizing due to obvious reasons (linear circuits offer stable response and are inexpensive and less complex than these processes).

One of the main advantages of the modulation process is that it overcomes noises and can transmit the signal over a larger distance and, hence, offers a good quality of transmission.

Further, several sub-channels can be transmitted over the communication link via another process termed **multiplexing**. Different types of multiplexing techniques based on time slot, frequency allocation, etc., allow the transmission of sub-channels of analog signal (modulated) and digital signal (coded) simultaneously over the transmission media. Various multiplexing techniques are discussed in Section 3.4.

In modulation, a baseband signal (modulating signal) is translated into a modulated signal using a carrier signal. Different types of modulation are defined based on how the instantaneous values of modulating signals are related to the characteristics of modulated signals. The operation of restoring a modulating signal is done by demodulation or detection. This restored signal is not exactly the same as that transmitted, as it may include distortions/interferences in the communication channels, and there may be a mismatch between modulator/demodulator circuits. In order to reconstruct the original signal from the modulated signal, different types of filter circuits are used, and these are discussed in this chapter.

There are two main classes of modulation techniques: **analog modulation (AM) and digital modulation (DM)**.

3.2 Analog modulation (AM)

This class of modulation requires the change of one of the parameters (amplitude, frequency, phase) of carrier analog signal in accordance with the instantaneous values of modulating analog signals. In this class, the contents or nature of information (analog or digital) is not modified. Instead, the modulating signal is mixed with a carrier signal of higher frequency and the resultant modulated signal typically defines the bandwidth in the same manner as that of the carrier signal. The carrier frequency is very high when transmitted over a telephone line, and this allows the frequency division multiplexing to utilize the link effectively.

Analog modulation can be further defined as either analog or digital, depending on (1) the carrier waveform (sinusoidal or pulse), (2) the type of transmission (analog or digital), and, (3) the type of modulation based on change of parameters (amplitude, frequency, phase of modulating signal). For digital data communication, analog modulation requires the digital data to first be converted to analog before the modulated signal is transmitted.

There are different categories of analog modulation, and each category has a different method of implementation. In the following sections, we discuss various techniques which are defined for these categories of modulation class. For more details on some of these modulation techniques, readers are referred to References 1 and 2.

3.2.1 Analog modulation for analog signals

An analog signal can be described by three main parameters or basic characteristics: **amplitude**, **frequency**, and **phase** in time domain, and, accordingly, we have three modulation techniques:

- Amplitude modulation
- Frequency modulation
- Phase modulation

3.2.1.1 Amplitude modulation (AM)

This is one of the earliest forms of modulation and one of the most commonly used modulation techniques. In this modulation, the amplitude of a carrier signal is changed

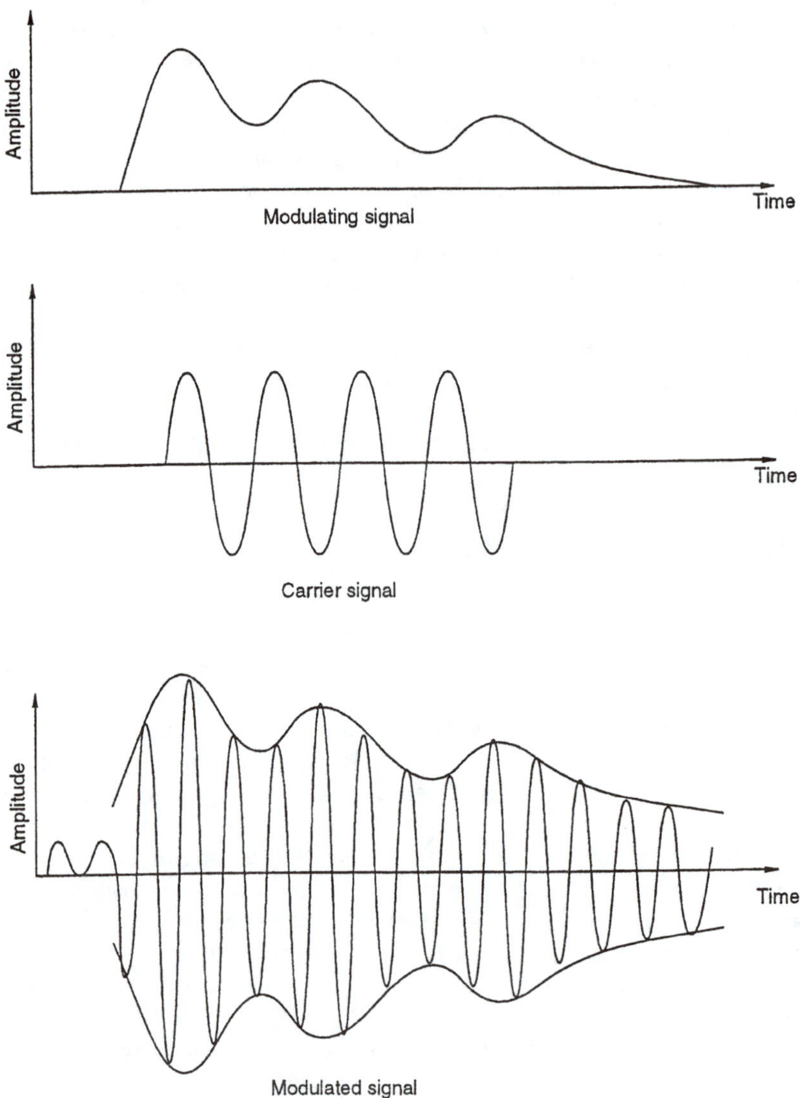

Figure 3.1 Amplitude modulation.

in accordance with the instantaneous value of lower-frequency modulating signals (Figure 3.1). The AM process generates a modulated signal which has twice the bandwidth of the modulating signal. It is obtained by multiplying a sinusoidal information signal with a constant term of the carrier signal, and this multiplication produces three sinusoidal components: **carrier**, **lower sideband**, and **upper sideband**. The bandwidth of each sideband is the same as that of the modulating signal. As indicated earlier, most of the information signals are made up of many Fourier components, and since each Fourier component is sinusoidal, each Fourier component present in the modulating information signal will produce a pair of sidebands in the manner just discussed.

The average power contained in the carrier is usually twice as much as in both sidebands for 100% modulation. The average power contained in the sidebands decreases with the decreased level of modulation, and at very low levels of modulation, sidebands may carry a fraction of the power transmitted while the carrier contains the majority of the power. It is important to note that the carrier does not carry any information — only sidebands carry the information. Thus, a standard AM signal is considered to be highly inefficient for radio frequency (RF) power to send the information through signals. In spite of this problem, AM is popular due to the ease with which AM signals can be demodulated or detected.

3.2.1.2 Versions of AM

As stated earlier, the carrier signal carries very little or no information, and the majority of the information is carried by sidebands. These sidebands are characterized as mirror images of each other. This means that information contained in one sideband is also contained in another sideband. In other words, we could send the information by transmitting one sideband only, but this method requires an expensive transmitter and receiver to select/receive a sideband from the signal, respectively. From the point of the resultant modulated waveform, AM offers a bandwidth which is near the frequency of the carrier (sum and difference of the frequency of carrier and modulating signals).

There are different ways of implementing AM modulation techniques, and the selection of an appropriate method depends on many factors such as cost, reliability, complexity of the circuits, and so on. The following implementation/version procedures of AM exist:

1. Transmit full **AM single envelope**.
2. Transmit only **single sideband (SSB)**.
3. Transmit both sidebands with carrier with reduced power **(double sideband suppressed carrier (DSBSC) modulation)**.

AM Envelope: AM modulation based on version 1 above is useful in radio broadcasting transmission because it provides moderate bandwidth and offers extremely simple and cheap demodulation of the signal.

Single Sideband-AM (SSB-AM): In single sideband suppressed carrier (SSBSC-AM) or, in short, SSB-AM, only one sideband is transmitted and the other sideband and carrier are suppressed. By sending only one sideband, half the bandwidth is needed and less power is required for the transmission. This is because the transmission of the carrier does not require power. SSB modulation finds application in telephony and also as an intermediate stage in constructing frequency division multiplexing of telephone channels before they can be sent using radio transmission (microwave link, satellite link), which uses another method for modulation. SSBSC modulation is unsuitable for direct application in transmission due to the requirements for isochronous (circuit-switching) reconstruction of the carrier at the receiving side; however, it is very useful as an intermediate stage in the implementation of SSB modulation.

Double Side Band-AM (DSB-AM): If the carrier signal (carrying no information) in DSB-AM can be stopped from the transmission of the modulated signal, the power wastage can be reduced but bandwidth remains the same. This is the working principle of double sideband suppressed carrier amplitude modulation (DSBSC-AM). Since the carrier is not transmitted, the phase reversal in the modulated signal (for transition from one level to another) becomes a problem. This requires a complex receiver which can detect this reversal in the received signal to demodulate the modulated signal.

Double sideband amplitude modulation (DSB-AM) is widely used for radio broadcasting (commercial). The frequency of the human voice (picked up by a microphone) varies from 30 Hz to 4 KHz and is given at the input of the radio transmitter. The radio transmitter provides a 750-KHz channel signal. It has two sidebands of the same bandwidth (the same as that of the modulating signal), and these carry the information of the modulating signal. The power distribution of the modulated signal varies linearly with the power distribution of the modulating signal, and as such there is wastage of power being carried by the carrier signal. The main advantage of a suppressed carrier is that this carrier signal can be used for other control information such as synchronization, timing, etc. Another category of modulation based on carrier suppression is known as **vestigial sideband** (discussed below) which uses one sideband and carrier signal with reduced power for transmission.

In all of these variations of AM, the bandwidth of the modulated signal transmitted is at least twice that of the modulating signal, and hence a larger bandwidth and greater power are required for its transmission. Further, the modulation index of AM modulation depends on the amplitude and frequency of the modulating signal, and the level of modulation changes the power in the modulated signal but the bandwidth remains around the frequency of the carrier.

For the transmission of digital data (defined as a low-frequency signal), we use two levels of modulation. In the first level, we use a low-frequency carrier for the modulation using the double sideband (DSB-AM) technique, and in the second level, modulation based on SSBSC-AM is used. This second level of modulation will transform the frequency of the signal to the desired level of carrier signal frequency. We can also use the frequency or even phase modulations in the first level of modulation.

3.2.1.3 Frequency modulation (FM)

In this modulation process, the carrier frequency changes in accordance with an instantaneous value of the baseband modulating (carrier amplitude remains unchanged), as shown in Figure 3.2.

Due to immunity to noises, FM is used for high-fidelity commercial broadcasting (FM radios). The immunity to noise can in fact be seen only at the receiving side, as the demodulation process maintains the amplitude of the signal at the constant value. The FM process contains a large number of sidebands and defines amplitude of modulating signals as derivation of carrier signal from allotted center frequency. (This is usually controlled by the FCC in the U.S.; in other countries, it is controlled by the respective government agency.) For example, a carrier frequency range of any commercial broadcasting station cannot be more than 150 KHz; i.e., frequency deviation on each side of the center frequency will be 75 KHz. If we have a modulating frequency of, say, 20 KHz, the channel bandwidth of that station for the broadcasting may be around 200 KHz.

This modulation process is very useful for radio-frequency (RF) transmission over wireless communication links (e.g., microwave link, satellite link, etc.) and can utilize the link bandwidth effectively by multiplexing the frequencies of a few gigahertz with a bandwidth of transmission in the range of 4 to 6 MHz.

FM modulation in high-radio-frequency transmission provides high-quality reception despite unfavorable conditions, noisy medium, adverse atmospheric conditions, etc. One of the most popular methods of FM demodulation is based on a **phase-locked loop (PLL)**, which extracts the instantaneous frequency from the modulated signal. The modulation index of FM depends on the frequency and phase of modulating signals.

3.2.1.4 Phase modulation (PM)

In this type of modulation, the phase of the carrier signal varies linearly according to the instantaneous value of the baseband modulating signal (Figure 3.3). The instantaneous frequency deviation is proportional to the derivative of the modulating signal.

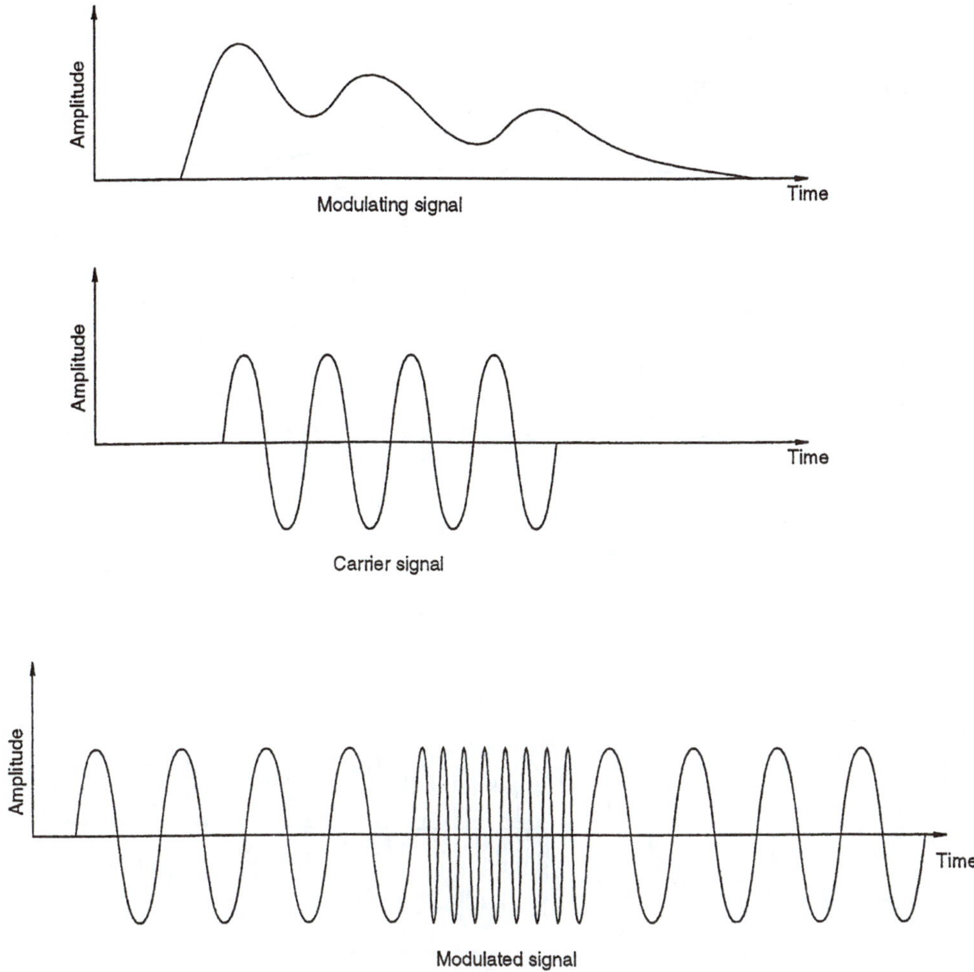

Figure 3.2 Frequency modulation.

FM modulation and phase modulation are suitable for applications in telecommunications in the following areas: microwave links, satellite links, mobile radio communication, audio radio broadcasting transmitters, and mobile radio telephony. In some books, these two types of modulation are collectively considered **angular modulation**. Features of angular modulation include

- Noise immunity at high modulation index.
- Larger available bandwidth than AM and available at all times.

FM and phase modulation are clearly similar, except for the difference in derivation of the modulating signal term. This indicates that an FM signal can be demodulated with an FM demodulator on the condition that we take a derivative of the reconstructed modulating signal which is performed afterward. Similarly, an FM signal can be demodulated on an FM demodulator by first demodulating it on an FM demodulator and then integrating this signal.

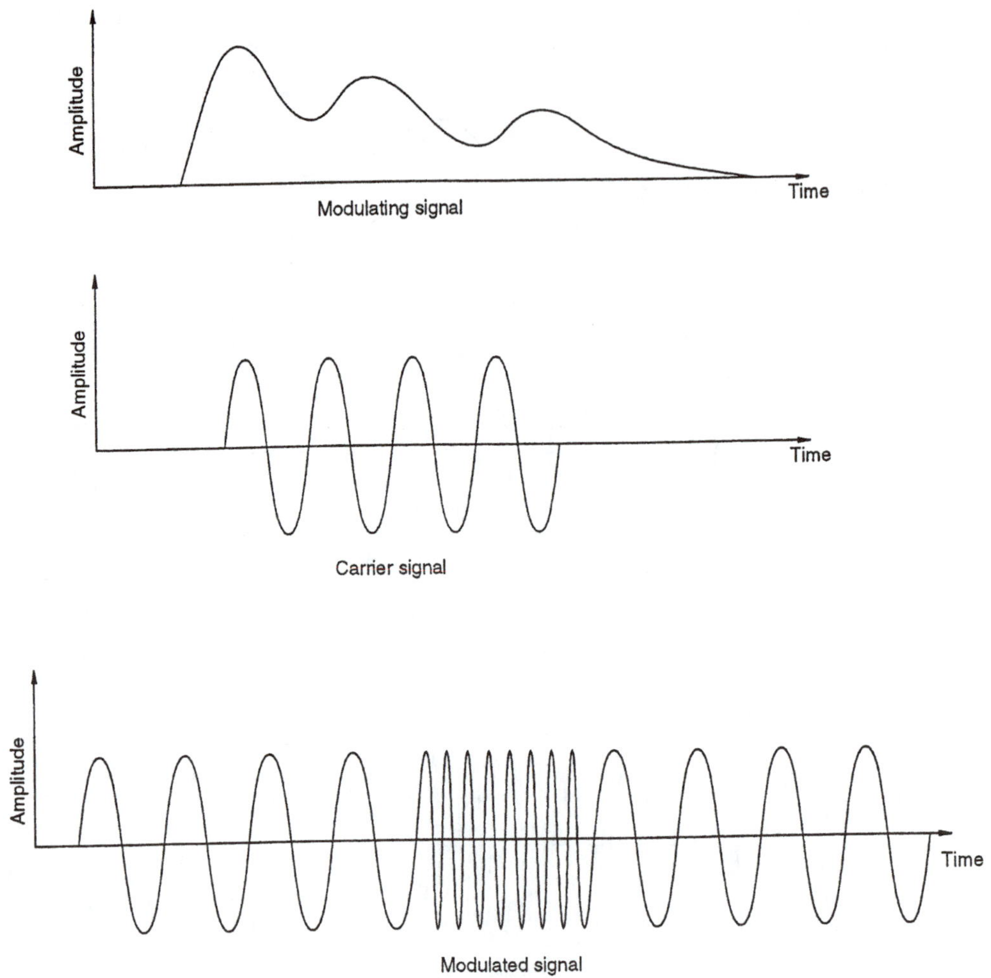

Figure 3.3 Phase modulation.

In applications where a stable frequency of carrier signal is required (e.g., telemetry, fixed bandwidth transmission, etc.), the phase modulation is preferred over the frequency modulation. This requires the transmitter to generate a stable frequency of the carrier, thus making the transmitter more complex and expensive. In a way, this helps during the demodulation process at the receiving side, as the frequency of the carrier signal is stable.

3.2.1.5 Vestigial sideband modulation (VSBM)

This type of modulation uses one sideband and a suppressed carrier carrying reduced power. We can consider this method of modulation a compromise between single-sideband and double-sideband carrier suppressed versions of AM (as discussed above). Television broadcasting transmission is different from radio broadcasting transmission, as it contains the following additional features:

- The bandwidth of a video signal (baseband) is approximately 5 MHz.
- Video signals contain very low frequencies (speech or voice).
- Transmission must comply with the scanning of images.

We cannot use AM for television (TV) transmission, as it defines two sidebands. The use of one sideband in SSB-AM may be advantageous, but the blocking of very low frequencies of video signals is very difficult. As discussed above, in the third version of AM modulation (DSBSC-AM), we don't transmit the carrier signal. This requires an isochronous demodulation, which is not an easy task. Thus, looking at these types of AM techniques, we find that none are suitable for television transmission. The VSBM, in fact, is a SSB-AM technique with a carrier, but we eliminate one sideband partially.

In television transmission, the video signal occupies in the baseband a bandwidth of 5 MHz (0 Hz {noninclusive} to 5 MHz). With VSBM, this can be transmitted in a modulated-frequency bandwidth of approximately 6 MHz. It also can be used in data communication at the speed of 48 Kbps in a bandwidth of 44 KHz. This bandwidth can include 11 analog telephone channels.

3.2.2 *Analog modulation for digital signals*

In this modulation process, the modulating (baseband) signal is in the analog form, but the carrier signal is in the form of a series of periodic pulses of high frequency. We can also say that this modulation transmits digital signals using analog signals. In this modulation, we vary the instantaneous value of parameters of samples of modulating signal. Various parameters of pulses are **pulse amplitude**, **pulse duration**, **pulse position**, and **pulse frequency**. This modulation type is sometimes also known as the **pulse modulation (PM)**.

Recall that in analog modulations, the characteristics of the carrier signal are changed by the instantaneous values of the parameters of the modulating signal. In pulse modulation, the modulating signal has to be sampled first; the values of the samples are then determined, and it is these values which are used with the pulse carrier signal.

For sampling rate, **Nyquist's law** is used: "The original signal can be obtained from the received modulated signal if the sampling rate is twice the value of highest signal frequency in the bandwidth." This sampling rate will be just enough to recover the original signal with acceptable quality. For example, if we consider a voice-grade signal of 4 KHz, 8000 samples each second are required to recover the original signal from the modulated signal.

3.2.2.1 *Pulse amplitude modulation (PAM)*

This type of modulation is obtained by sampling a modulating signal using a series of periodic pulses of frequency Fc and duration t and defining the instantaneous value of the amplitude of this signal for each of the samples, as shown in the modulated PAM signal in Figure 3.4(a). The value of the sample is used to change the amplitude of the pulse carrier. This type of modulation finds its use in time division multiplexing (TDM), discussed later in this chapter.

PAM is not useful for telecommunications, as it is sensitive to noise, interference, atmospheric conditions, cross talks (just as in AM), and propagation conditions on different routes through PSTN, but it has been used in some of AT&T's Dimension PBX systems. We can use PAM as an intermediate stage (for sampling) in other pulse modulation schemes (e.g., PDM, PPM, or PCM) that are discussed below.

3.2.2.2 *Pulse duration modulation (PDM)*

This type of modulation is obtained by varying the pulse duration t according to the modulating signal while keeping the amplitude constant, as shown in the modulated PDM signal in Figure 3.4(c).

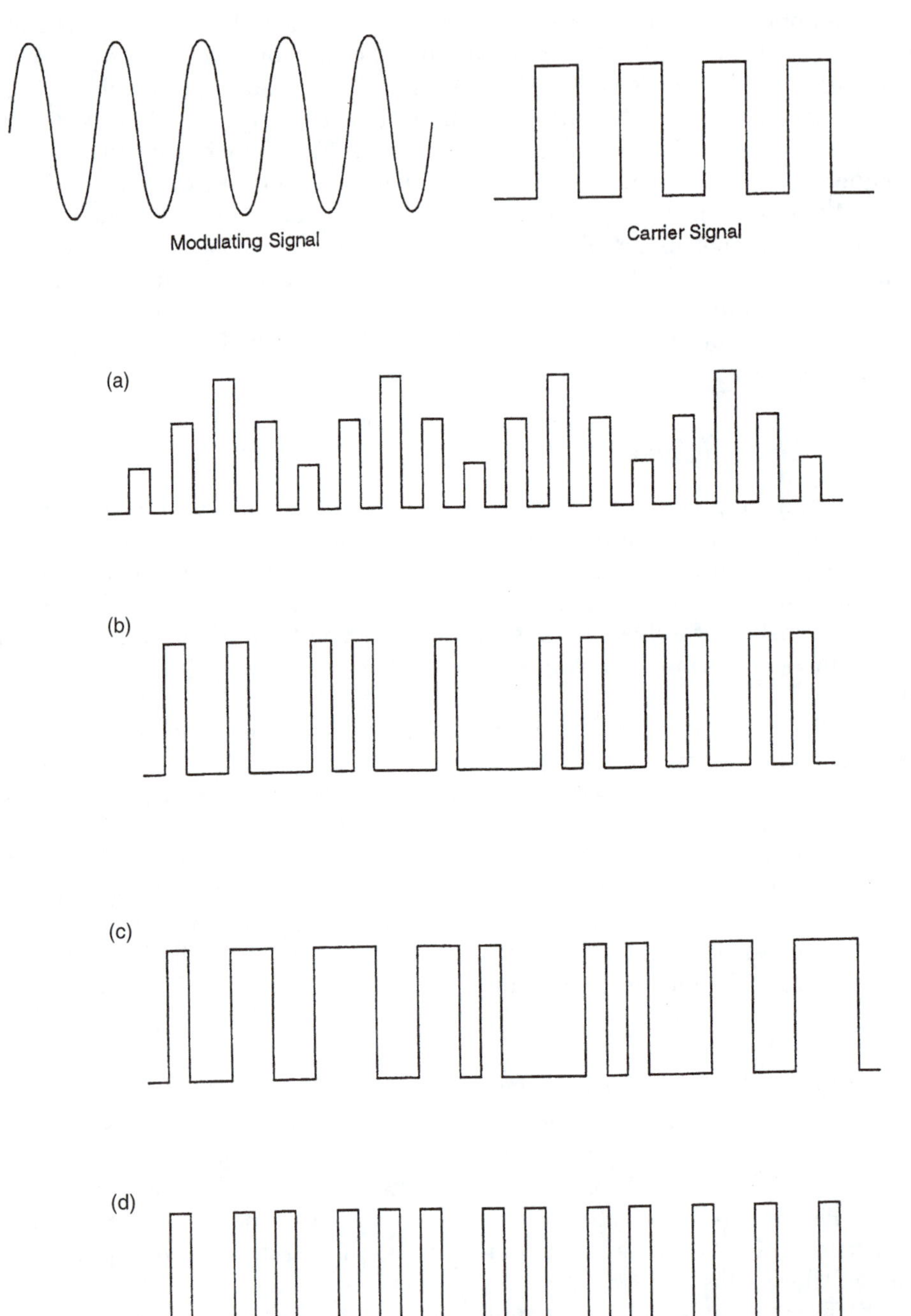

Figure 3.4 (a) Pulse amplitude modulated signal. (b) Pulse position modulated signal. (c) Pulse duration modulated signal. (d) Pulse frequency modulated signal.

3.2.2.3 Pulse position modulation (PPM)

In this type of modulation, the trailing edges of pulses represent the information and, hence, it is transmitted through the trailing edges only, as shown in the modulated PPM signal in Figure 3.4(c). The duration of the unmodulated pulses in this modulation simply represents a power that is not transmitted. Thus, PPM provides an optimum transmission, as only pulses of constant amplitude with very short duration are transmitted. This requires lower power and, hence, the modulation circuit is simple and inexpensive.

3.2.2.4 Pulse frequency modulation (PFM)

PFM is similar to phase modulation, and the instantaneous frequency for two consecutive modulated pulses is a linear function of modulating signals, as shown in the modulated PFM signal in Figure 3.4(d). This modulation can not be used for the construction of time division multiplexing (TDM).

The three modulation schemes of PDM, PPM, and PFM are also collectively known as **time modulation** and possess the following features:

- Immunity to noise.
- No amplitude linearity requirements.

3.2.3 Discrete (digital) analog modulation (DAM)

In this category of modulation, the modulating signals are defined by discrete values of signal, and these discrete or binary logic values carry the information. The carrier signal is varied between two or more values (typically between ON and OFF to represent digital signal) in accordance with the modulating signal, which is in the discrete form. The following three methods of this type of modulation have been considered and defined: **amplitude shift keying (ASK)**, **frequency shift keying (FSK)**, and **phase shift keying (PSK)**.

3.2.3.1 Amplitude shift keying (ASK)

The amplitude of an analog carrier signal varies in accordance with the discrete values of the bit stream (modulating signal), keeping frequency and phase constant, as shown in Figure 3.5(a). The level of amplitude can be used to represent binary logic 0s and 1s. We can think of a carrier signal as an ON or OFF switch. In the modulated signal, logic 0 is represented by the absence of a carrier, thus giving OFF/ON keying operation and hence the name **amplitude shift keying (ASK)**.

Like AM, ASK is also linear and sensitive to atmospheric noise, distortions, propagation conditions on different routes in PSTN, etc. It requires excessive bandwidth, and there is a wastage of power. Both ASK modulation and demodulation processes are relatively inexpensive. This type of modulation is mainly used in conjunction with phase modulation for leased lines and must use a very sophisticated modem. It can also be used to transmit digital data over fiber.

3.2.3.2 Frequency shift keying (FSK)

In FSK modulation, the frequency of the carrier is changed in accordance with discrete values of the signal, keeping the amplitude constant, as shown in Figure 3.5(b). Different frequencies (typically two frequency values for audio tone) are used to represent binary logic values 0 and 1. All the frequencies are within the voice-grade bandwidth. The binary values of 1s and 0s transmitted in opposite directions are represented by different frequency values, hence avoiding any overlapping of frequency values. Since separate frequencies represent 1s and 0s in both directions, FSK is well-suited for full duplex transmission and is also known as **digital FM (DFM)**. The FSK is usually used in low-speed asynchronous application of data communication. FSK supports 1200-bps or even higher bit rate modems for voice-grade channels with standard switched connections across PSTN. This is because bandwidth requirements here are very small and the modulation circuit is simple.

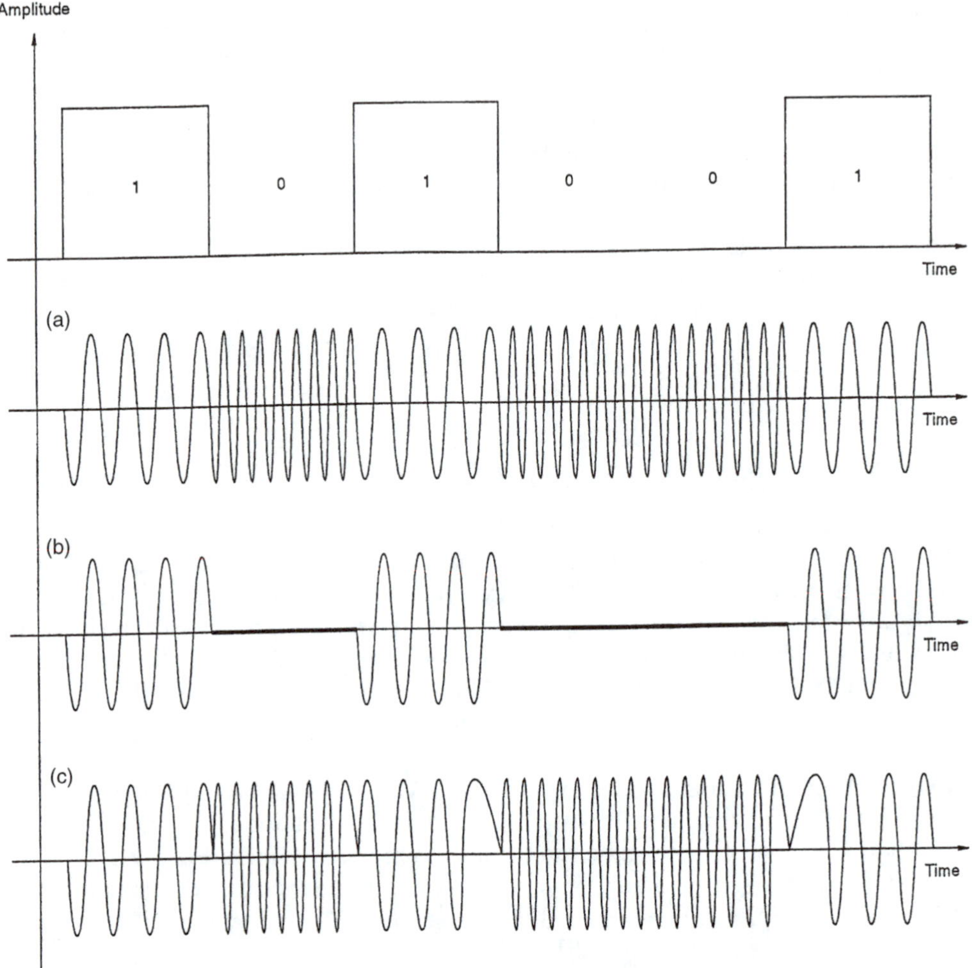

Figure 3.5 Discrete analog modulation. (a) Amplitude shift keying (ASK). (b) Frequency shift keying (FSK). (c) Phase shift keying (PSK).

FSK modulation is used in the following applications:

- Radio broadcasting transmission (2–35 MHz).
- Local area networks using broadband coaxial cables.
- Low-speed modems (up to 1200 bit/s).

FSK is less susceptible to noise than ASK. Due to a widespread acceptance of phase shift keying (PSK) modulation in all high-speed digital radio and network systems, the FSK is losing its grip on voice-grade communications.

3.2.3.3 *Phase shift keying (PSK)*

This modulation compares the phase of the current state with the phase of the previous state; thus, it utilizes bandwidth more efficiently than FSK, as this modulation assigns more information with each signal, as is shown in Figure 3.5(c). By changing the phase of the carrier signal, appropriate phase values can be used to represent binary logic 1s and 0s. Due to this feature, it is more efficient than FSK, as it utilizes bandwidth better than FSK.

3.2.3.4 *Differential PSK (DPSK)*

The phase angle of the carrier is changed by the value of binary signal. For example, if we use two bits 0 and 1, then 0 may represent 0 degrees while 1 may represent 180 degrees, giving rise to four-phase PSK modulation (usually written as **4-PSK**). With three bits, 8-PSK can be defined. This type of modulation is very susceptible to random changes in the phase of modulated signals. In order to reduce this problem of random changes in phase, a variation of the PSK modulation scheme is proposed as *differential PSK*. This scheme utilizes the relative position of the next bit to be transmitted with respect to the current for controlling the shifts in phase at each transition. In other words, a phase shift of 270 with respect to the current bit may represent a logic 1 while a phase shift of 90 with respect to the same current bit may represent a logic 0. On the receiving side, the relative magnitude of the logical signal as opposed to absolute magnitude needs to determined by the receiver. A disadvantage of PSK lies in the requirement of a complex system for the generation of modulated signals and also for demodulation of this signal.

The discrete analog modulations are useful in data transmission (telephone channel of carrier systems), optical transmission on fibers (ASK), radio-elective transmission of digital information by microwave links or satellite links, and other data communication systems. The device used for modulation and demodulation is known as a modem. Two main types of basic modems have been defined as **asynchronous** and **synchronous** to support appropriate transmission modes. The details about different types of modems, standard modem interfaces, data rates, and operations of modems are discussed in Sections 3.4–3.6.

3.3 *Digital modulation (DM)*

This class of modulation changes the analog form of a modulating signal into a digital modulated signal form. The modulated signal is characterized by bit rate and the code required to represent analog signals into digital signal.

For digital modulation class, the information is in discrete form but can be used on analog channels and is treated as analog modulation with a different name. If we choose the carrier as a series of pulses of frequency F_c, then the modulating signal will be sampled into samples whose amplitude is equal to the instantaneous values of the modulating signal over the duration of the pulses. The modulated signal also consists of a pulse train of frequency F_c. At procedural levels, both analog and digital modulation methods have some similarities and possess identical properties. The digital modulation used in switched circuits and leased circuits provides data and voice communication. This allows all the signals exchanged between switching exchanges of switched networks to be in digital form. The digital mode of signaling has also been adopted by consumers, who use ISDNs to send the data directly without using a modem.

The digital modulation process considers the carrier signal a periodic train of pulses, and the modulation process changes the characteristics of a discrete form of the signals. There are two methods of achieving digital modulation. In the first method, the characteristics of carrier pulses are changed in accordance with the instantaneous values of the modulating signals (digital form). The resulting waveforms may represent different types of modulated signals and are collectively known as **pulse modulation**. In the second method, we use binary symbol code to represent the signals and define a quantization process — this is termed **coded modulation**. The steps required in the quantization process include sampling of the baseband (modulating), signal quantization, and transmission of binary values of the samples.

3.3.1 Pulse modulation

In the pulse modulation scheme, the carrier is defined as a series of pulses and the characteristics of the carrier signal is changed in accordance with the instantaneous value of the baseband modulating signal. The characteristics of pulses are discrete in nature, and the modulated signal will have the following four types of modulation schemes: **pulse amplitude modulation (PAM)**, **pulse position modulation (PPM)**, **pulse duration modulation (PDM)**, and **pulse frequency modulation (PFM)**, as shown in Figure 3.6(a–d).

3.3.2 Coded modulation (CM)

In previous modulation methods, the carrier signal is changed for the transmission in accordance with the instantaneous values of the parameters of the modulating signal. In coded modulation, we take a different approach of generating a coded signal for an analog signal using binary coding representation schemes. We convert the analog signals contained in continuous signals into a sequence of discrete codes/signals generated by code or symbol generators.

The process of converting analog signals into discrete codes, in general, requires the following two steps:

1. Sampling analog signals at a certain instant of discrete time; it may not be a periodic signal.
2. Quantizing, i.e., mapping a large number of analog values to a finite number of codes/symbols (for representing these analog values).

Digital modulation/demodulation (in the coded modulation type) is primarily based on the following sequence of processes:

1. **Digital modulation:** A quantizing process accepts the analog continuous signal, quantizes it in the form of integers, and passes it on to the next process of coding (mapping) and modulating. This process chooses an appropriate code to represent values of integers, required bandwidth, etc., and transmits these as a modulated signal.
2. **Digital demodulation:** A decoding (mapping) and regenerating process will receive modulated signals as input and produce the same integer values (using some mutually agreed upon coding/decoding scheme). This set of integers is then mapped onto the original analog signal using a physical translation process.
3. **Quantizing process (QP):** This process usually assigns integers to the values of analog signals, which in turn are represented in the same form of code for the transmission. Quantizing is defined as a procedure for approximating the instantaneous value of an analog signal by the nearest value obtained from a finite set of discrete values, each assigned an integer. Each integer can represent a range of analog values called **quantizing intervals** which may be of different duration for different intervals of time.

Coded modulations are very important and useful, as they allow us to map analog signals into digital form in the communication circuits. These circuits are free from noises and distortions and offer high bit rate and low error rate. These modulations have become

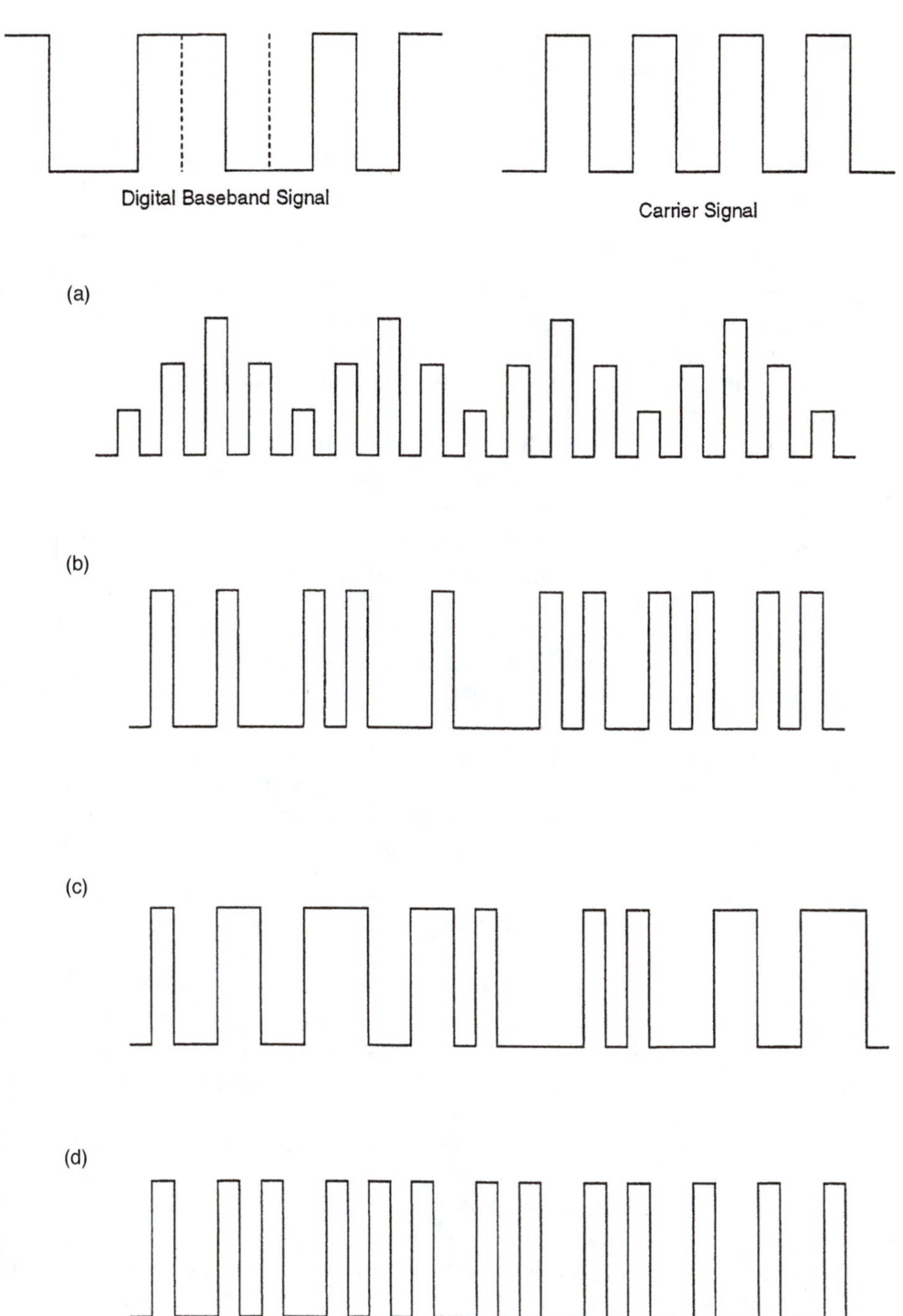

Figure 3.6 (a) Pulse amplitude modulation (PAM). (b) Pulse position modulation (PPM). (c) Pulse duration modulation (PDM). (d) Pulse frequency modulation (PFM).

the backbone of the upcoming digital data networks, integrated services digital networks (ISDNs), and many high-speed and intelligent networks.

Depending of the types of codes and mapping (between analog and digital signals), there are three forms of coded modulations:

1. Pulse code modulation (PCM)
2. Differential pulse code modulation (DPCM)
3. Delta modulation

The following sections will discuss in detail these forms of modulation and some of their applications.

3.3.2.1 Pulse code modulation (PCM)

The signal is sampled at regular time intervals at a rate which is greater than twice the original frequency of the modulating signal. These samples will include the entire information of the modulating signal. On the receiving side, low-pass filters may be used to construct the original signal from these received modulated samples.

As discussed above, digital modulation accepts a digital form of information as the input to the quantization process. This digital signal may have been derived from digitizing an analog signal, or it may represent information already in digital form (obtained from a computer or digital data device). It is important to clarify the difference between **pulse modulation** and **coded modulation**. In the former, we consider an analog discrete time system which produces analog signals; at the receiving side, we demodulate the modulated pulse train to get the original analog signals. In digital coded modulation, we start the modulation from a digital form of signal (representing information). This digital signal is then quantized into some binary coding representation scheme (e.g., ASCII). The coding techniques provide a fixed number of bits for each of the characters in the digital characters set.

The binary representation of four decimal numbers 0 to 3 requires 2 binary bits, 3 bits are required to represent decimal numbers 0 to 7, 4 bits are needed for decimal numbers 0 to 15, and in general, the number of bits required to represent x decimal numbers 0, $1 \ldots x - 1$ is given by $x = 2^y$, where y is a binary digital sequence $a_{y-1} \ldots a_0$.

Construction of PCM: The essential operations needed for constructing PCM are given below:

1. Determine the range measured in terms of volts of excursion of an analog information signal and obtain the samples of modulating signals at the rate of at least twice the highest frequency in the modulating signal (based on Nyquist's law).
2. Determine the quantization levels between the range by assuming a unit value of volts for the size between quantization levels, yielding the value of amplitude of each sample. This in turn depends on the range of excursions. For example, if the range of the analog information signal is from –5 to +5 volts, we may have ten quantization levels, each of size 1 V.
3. Locate each level on the scale at a point where the value of the level gives the nearest value of the analog information signal. This will determine the value of each of the quantization levels.
4. Assign the decimal number to these levels starting from the lowest value of the quantization level, i.e., starting from –5 or 4.5 V (whatever value we choose) and go up to a maximum of 5 V.
5. Now assign binary representation to each decimal code (assigned to its quantization level). The number of bits for binary representation depends on the number of codes considered.

If we define, say, eight bits for quantization, it will represent 256 levels and offer a data rate of 64 Kbps (at the rate of 8000 samples for eight bits). It can be observed that the rate

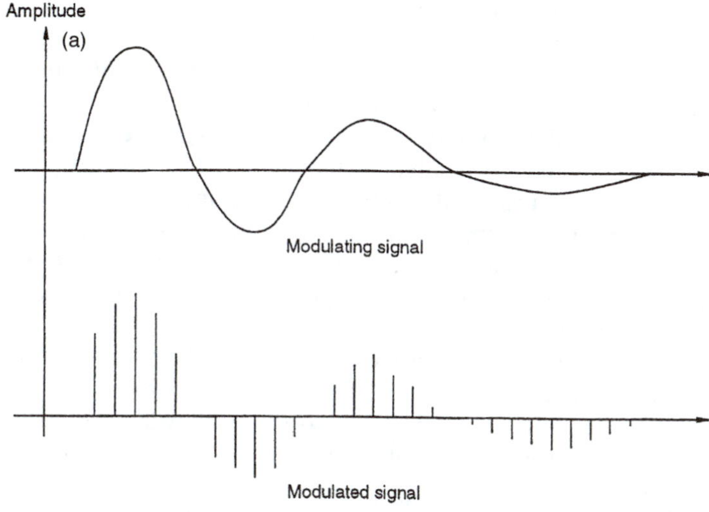

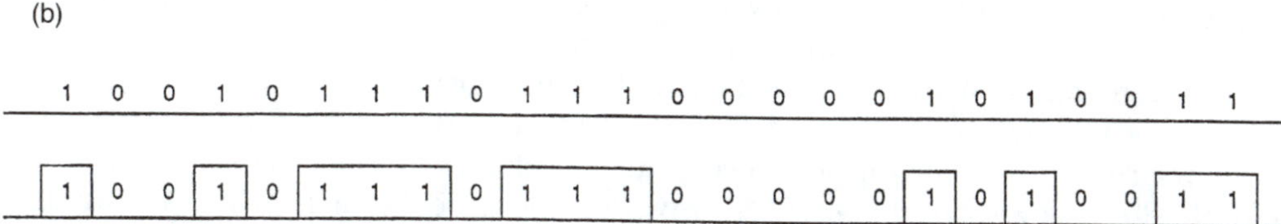

Figure 3.7 (a) Pulse code modulation. (b) Binary signal representation of quantization levels.

of change of amplitude of the signal will be different during different times of the cycle. Further, the signal changes faster at low amplitudes than at high amplitudes. If we quantize the signals linearly over the complete amplitude range, a quantization distortion is generated. This distortion can be reduced by defining eight levels before quantization, and the net result of this is to compress larger signals and expand smaller signals. This process is known as **companding.** This technique defines three segment code bits (defining eight segments) and four quantization bits (giving 16 quantization levels within each segment).

The binary representation of codes in PCM can be transmitted using either pulses or voltage levels, as shown in Figure 3.7(a).

Let's assume that we have eight quantization levels where 100, 101, 110, and 111 represent the positive side of the signal while 000, 001, 010, 011 represent the negative side of the signal. For the PCM signal shown in Figure 3.7(a), the binary signal transmitted is given in Figure 3.7(b).

Figure 3.7(b) shows how pulses can be used to represent and transmit codes (corresponding to quantization levels) over a transmission medium. We might use different types of interpretation of the pulses for representing the binary number. For example, binary bit (logic) 1 can be represented by a pulse while binary bit (logic) 0 can be represented by no pulse (absence of pulse). The pulses/no pulses are shown on different slots and are of the same duration. Another way of defining the pulses would be that in which a binary logic 0 is represented by a lower level of waveform while binary logic 1 is represented by an upper level of the waveform. This method has an advantage over the pulse method, as it does not have a DC component and hence DC power and its contribution to the reliability of the transmission is saved.

Quantization in PCM: The quantizing process maps the exact value of an analog information signal with an integer which represents the interval in which that value is defined/recognized/located. The translation of the integer into an appropriate logical binary signal which provides a binary information signal (corresponding to the nearest analog information signal) is known as **PCM word** and the translation process itself is called **coding**.

The coding does not affect the quality of modulation, but the selection of coding depends on many factors, such as technical advantages for implementing coding and decoding procedures, high data rate, compensation of DC component, frequency range, etc. Various codes based on binary code and its derivatives include **weighted codes** (serial coding/decoding), **Gray's code** (parallel coding/decoding), **folded code** (bipolar quantized), **iterative code** (coding/decoding), etc.

The PCM-based system consists of a sampler, an analog-to-digital converter (composed of a quantizer and encoder) at the transmitting side. A digital-to-analog converter (composed of a quantizer and decoder) and filter are used at the receiving side and, of course, a communication link between transmitting and receiving sides is provided by transmission media.

The quantized samples of the analog information signal obtained from the quantizer are applied to an encoder, which generates a unique binary pulse or voltage level signal (depending on the method used). The quantizer and encoder combined are known as an **analog-to-digital converter**, usually abbreviated **A/D converter**. The A/D converter accepts an input analog information signal and produces a sequence of code words where each word represents a unique train of binary bits in an arithmetic system. In other words, it produces a digitally encoded signal which is transmitted over the communication link.

Similarly, on the receiving side, a combination of quantizer and decoder is known as a **digital-to-analog converter,** or **D/A converter**, and produces a sequence of multilevel sample pulses from a digitally encoded signal received at the input of D/A from the communication link. This signal is then filtered to block any frequency components outside the baseband range. The output signal from the filter is the same as that of the input at the transmitting side, except that it now has some quantization noise and other errors due to noisy communication media.

A sampling frequency of 8 KHz has been allocated to telephone signals with PCM modulation by the international standards organization, the **Consultative Committee of International Telegraph and Telephone (CCITT)**. The coding schemes are different in different countries.

Some applications based on PCM modulation have the following sampling frequencies:

Music (transmission): 32 KHz
Music (recording): 44.1 KHz
Television (video signals): 13.3 MHz.

3.3.2.2 *Differential pulse-code modulation (DPCM)*

In PCM, we transmit the instantaneous value of a signal at the sampling time. If we transmit the difference between sample values (at two different sampling times) at each sampling time, then we get **differential pulse-code modulation**. The difference value is estimated by extrapolation and does not represent the instantaneous value as in PCM. If these difference values are transmitted and accumulated, a waveform identical to an analog information signal can be represented at the receiving side.

PCM transmitting difference values (as in DPCM) offers an advantage over PCM, as we make fewer samples; the difference values will also be much lower than the individual sampling/quantizing values. Further, the number of bits required to encode these levels is also reduced. Differential PCM (DPCM) can operate at approximately one-half of the bit rate of PCM, thus saving the spectrum space. This modulation finds its use in voice or video transmissions.

3.3.2.3 Delta modulation (DM)

This modulation is, in principle, similar to DPCM; the difference is in the encoding of the difference signal value and in its simple implementation. In DPCM, the difference signal is encoded into the number of bits present in the coding scheme considered, while in delta modulation, the difference signal is encoded into just one bit. This one bit, having two states (0 or 1), can be used to estimate the signal.

A Delta modulation system (based on linear delta modulation) consists of a comparator, sample and hold, up–down counter, and D/A converter. Both modulating and its quantized signals are applied to the comparator, which gives two outputs indicating whether the modulating signal is greater or smaller than the quantized value rather than different in magnitude. Depending upon the level of output, the counter also counts up or down accordingly, giving rise to the accumulation or addition of these difference values for their transmission. Differential PCM and Delta modulation have some unfavorable effects on high frequency signals. These effects have been removed in Delta-Sigma modulation which is based on DPCM and DM.

3.3.2.4 Adaptive delta modulation (ADM)

This is a modification over delta modulation where step size is not kept constant. It has more quantization error but provides a net advantage over delta modulation. ADM when operating at 32 Kbps offers similar performance to that of PCM at 64 Kbps. Another feature of ADM lies in its ability to operate at 16 Kbps (of course, at the cost of degraded performance). A variation of this modulation has been used in the space shuttle.

Quantization of samples has led to a number of digital modulation schemes, and although each one of these is independent of the others, PCM modulation has become the most popular and universal. It has been used in various communication networks on different transmission media such as coaxial cables, optical fibers, microwave links, and satellite links (radio communications), and other applications such as recording, signal processing, image processing, etc.

Delta modulation (DM) has a unique feature of representing a difference value of 1 bit, comparable to PCM at lower rates of 50 Kbps. At this low rate (for example, in telephone systems), for some applications in military transmission, lower quality of reception can be accepted. Further, DM may be useful in digital networks if we can digitally convert DM into PCM.

Differential PCM and ADM improve the performance significantly at the cost of complex and expensive systems. These modulation schemes provide a uniform range of values and, hence, can easily be used to transmit stationary images. For moving images, different variations of DPCM have been defined for use in digital television, videophony, etc.

The telecommunication services are usually provided by analog signals which, in fact, represent both voice and data communications. Typically, the digital signals from digital devices are converted into analog signals and then digitized by Codec before its transmission. During the transmission, various interconnecting devices such as repeaters are used where the signals are going through a number of transitions. Similarly, on the receiving side, the conversion from analog to digital form is defined to carry the signal to its destination. Instead of going through so many conversions, it is more appropriate to define an encoding or modulation technique where the modulated signal is defined in a higher frequency range, and this is exactly what is being done in the above-mentioned coded modulation techniques of PCM, APCM, delta modulation, adaptive delta modulation, etc.

3.3.2.5 *Higher data rate digital modulation*

In general, the **digital modulations** provide representation of analog information signals into digitally encoded forms for digital processing and are characterized by a variety of factors such as coding/decoding with security, recognition of original signal from noisy channels, and multiplexing (for switching/selecting routes of coded signals, storage of these signals, digital filtering, etc.).

To achieve higher bit rates, more bits have to be fit into each baud or signal pulse, and such modems usually use a combination of modulation methods as opposed to a single modulation method. Various combinations (of mixing modulation methods) are available for use in data communication, especially with modems, e.g., quadrature modulation, quadrature amplitude modulation, etc.

In **quadrature modulation (QTM)**, dibits (2-bits) are transmitted with single sideband amplitude modulation (SSB-AM), and this method is also known as two-phase–two-level amplitude modulation. Here two independent bit combinations are used on a single sideband amplitude modulation scheme using the same carrier, but the carrier signal in each combination has a phase difference of 90. These combinations are transmitted as four independent channels within the frequency bandwidth without any interference among each other.

In **quadrature amplitude modulation (QAM)**, each pulse is assigned with four bits and is transmitted at 2400-bps band, giving rise to a bit rate of 9600 bps over a voice-grade channel. These four bits define 126 combinations of amplitude, and each level of amplitude has three distinct values on it. This method is basically a combination of amplitude and phase modulations. The Bell system 209A modems use the QAM method.

3.4 *Modems (MOdulator/DEModulator) and modem standards*

A modem performs the conversion of binary signals into analog signals which can be transmitted over the telephone lines of PSTN. It performs a transformation between digital signal (used in network ports, terminals, hosts, etc.) and analog signals (used for transmission over telephone lines). It translates a square-edged pulse stream from computers or terminals into an analog-modulated signal to fit into the telephone channel frequency between 300 Hz and 4 KHz.

The modem accepts a serial bit string (from computer or terminal) and generates a modulated output waveform which, in turn, depends on the type of modulation used. We have already discussed different types of modulation for analog and digital signals in the previous sections. Amplitude modulation (AM), frequency modulation (FM), and phase modulation (PM) have been discussed for analog signals, while pulse modulation and coded modulation have been discussed for digital signals.

Modems differ in the way the digital tones are used to modulate some form of AC carrier which is transmitted across the transmission medium. One type of modulation is AM, where the amplitude of the carrier is changed between two levels 0 and 1. In data communication, this type of modulation is known as **amplitude shift keying (ASK)**. The shift keying is a Morse code term to describe using a key to shift between two signals, one for OFF and one for ON. A second method of altering AC signal is changing its phase from the reference value. A change to one phase from another indicates 1, and a change to a second phase represents 0. This method is known as phase shift keying.

The combination of phase shift keying and amplitude modulation is defined as **quadrature amplitude modulation (QAM)**, and it permits higher rates of data transmission.

The FSK modem contains a modulating circuit around a **voltage controlled oscillator (VCO)** whose output frequency changes as voltage applied to it changes. The first DC level represents 1 while the second level represents 0. The receiver portion of the modem accepts signals generated by a similar modem and then returns to DC levels. The circuit generally used to perform demodulation is known as **phase lock loop (PLL)**, which generates error voltage whenever an incoming signal differs in phase or frequency from a reference signal. The error voltage is used to establish a digital data level of 1 and 0, e.g., incoming tone patterns. In the case of FSK modems, the reference signal is usually taken at center or carrier frequency. The rate at which coded information is sent is defined as baud. It includes bit rate information, character size, and character transfer data rate. The least significant bit (LSB) is always sent first, and the most significant bit (MSB) is sent last.

The modem accepts the binary data stream, which is usually in coded form such as differential encoding, differential Manchester encoding, Gray coding, etc. (to facilitate the transmission with minimal errors). This data stream is then translated into a form which typically consists of a bit stream containing the 1s and 0s with equal probability. This type of representation of the original data stream is independent of the content and type of the data information intended to be transmitted. The bit stream is then modulated using modulation techniques (discussed earlier); e.g., for low speed (300, 600, and 1200 bit/s), frequency shift keying (FSK) is used, while phase modulation (PM) is used for speeds of 2400 bit/s and 4800 bit/s. The phased multi-amplitude modulation (PM-AM) is used for speeds of 7200 bit/s and 9600 bit/s. The amplitude level and frequency bands of the modulated signals are adjusted with the help of suitable filters and amplifiers according to CCITT recommendation V.22.

3.4.1 *Standard modem interface*

The standard hardware unit which provides an interface between a modem (DCE) and computer/terminal (DTE) is recommended to be a 25-pin cannon plug, as shown in Figure 3.8. The output of the modem is connected to different output devices. This standard is defined as **EIA-RS-232** and is the same as that defined by Electronics Industries Association standard RS-232, version C interface. The U.S. military data communication system has also adopted a standard MIL 188C similar to RS-232. A similar interface has also been defined by CCITT as V.24. This standard is very popular in European countries.

The telephone system transmits voice and data over a long distance, within a narrow bandwidth called a voice-grade channel/band (200 Hz to 4 KHz). This narrow band contains baseband signals and hence can be used only for a short distance with a low speed of transmission. Further, the telephone system poses problems of capacitive and inductive effects to digital signals which will change the features of digital signals, i.e., rise and fall times, and hence the waveform at the receiving side. If these problems can be reduced, then we can transmit the data at a speed of about 2 Mbps. In order to avoid these problems with digital signal transmission and utilize the narrow voice-grade band, we use a modem which converts baseband digital information (contained in voltages

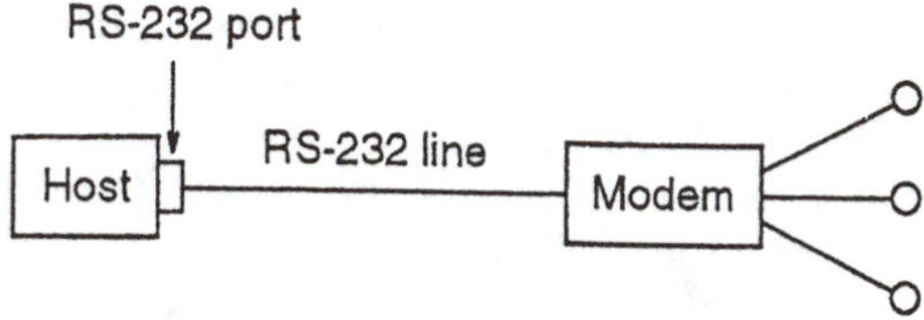

Figure 3.8 DTE/DCE interface. DTE = data terminating equipment (PC or terminal); DCE = data circuit–terminating equipment (modem).

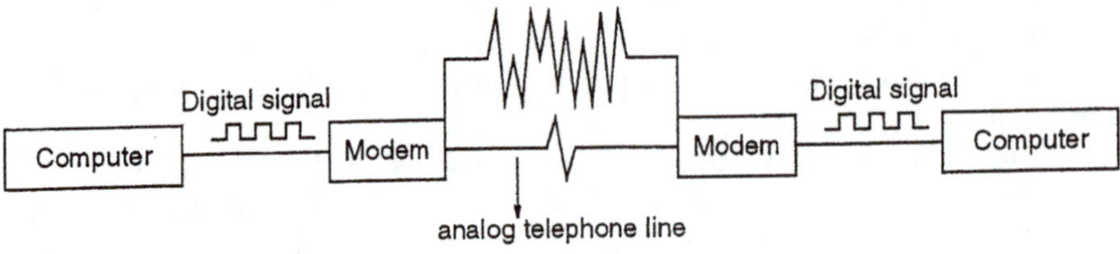

Figure 3.9 Modem connection.

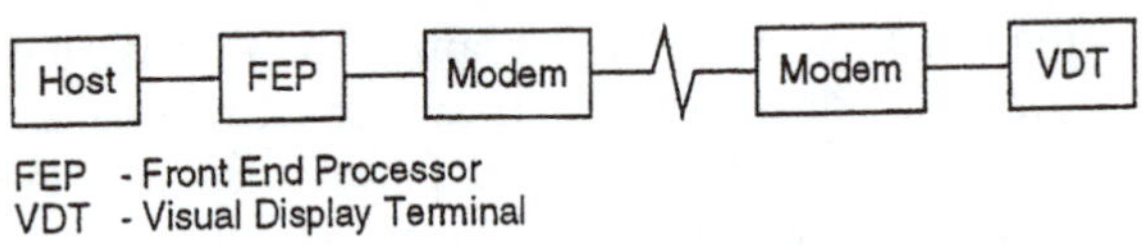

Figure 3.10 Modem interface.

representing binary logic 1s and 0s) into an analog signal within that band. On the receiving side, this signal is converted back into a baseband digital information signal, as shown in Figure 3.9.

A modem serves as an interface between devices which want to share data, and the transfer of data is usually performed by a transmitter and receiver. A computer or terminal provides digital information signals to a modem, which transmits the data from one side of the telephone line to another side, where another modem receives it and, after demodulation, passes it on to a computer or terminal. The modem translates the digital information signal into an appropriate analog signal which is compatible with the telephone line. Telephone lines transmit analog (continuously varying amplitude) signals. Thus, two computers can be linked together with modems on telephone lines for the transfer of data back and forth, as shown in Figure 3.10.

Figure 3.10 shows a simple and typical modem/terminal/network connection configuration. There are three interfaces in this type of configuration. The first is between the host and front-end processor (provided by the mainframe supplier). The second is between the front-end processor and modem (this interface is the same as that between a modem and a **visual display terminal (VDT)**). The third interface is between the modem and line and is the same at each end of the line.

3.4.2 *Modem/telephone line interface*

Modems provide an interface between computers/terminals and the telephone lines. They provide data communication over the telephone link via **two-wire lines** or **four-wire lines** via modem operation. In general, the majority of modems possess options of both two-wire and four-wire lines. The modems are connected to the telephone lines via a 1:1 transformer, as shown in Figure 3.11(a,b).

Two-wire-line modem operation consists of two circuits, **transmitter** and **receiver**, and can send and receive the data information at the same time. When the transmitter circuit is sending the bits, the receiver circuit is disabled and vice versa. The transmission and receiving of data is carried out on the same pair of wires, and the mode of operation is changed by a **switch** or **jumper**. Both the circuits cannot be enabled at the same time. This

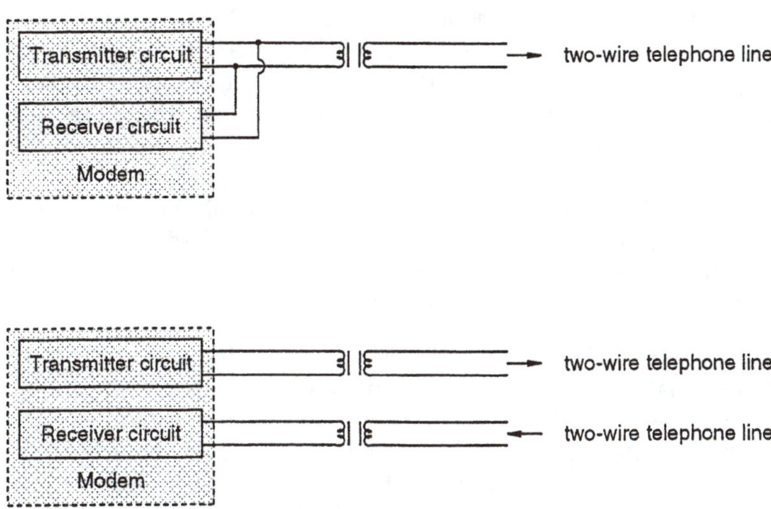

Figure 3.11 Modem/telephone line interface. (a) Two-wire-line interface. (b) Four-wire-line interface.

operation is very popular with asynchronous DTEs and is issued with packet-switched networks.

In four-wire-line modem operation, both transmitter and receiver can send and receive the data information simultaneously on different pairs of lines. This type of operation finds application with private or leased networks.

3.4.3 Modem characterization

Modems can be categorized on the basis of various parameters such as **two-wire** or **four-wire lines**, **synchronous** or **asynchronous**, **point-to-point** or **multi-point configurations**, etc. There are a variety of Bell modems available that provide different bit rates or transmission speeds. In addition to these normal modems, there are modems which can be used with multiplexers and are known as **multiplexer modems**. These modems operate on 1, 2, 4, and many other channels, providing different speeds of transmission.

Modems can be used over **dial-up lines** or **private (leased) lines**, and each category of modem can be further classified as using one pair or two pairs of wires. The dial-up line (a pair of wires) connecting a user's telephone set to a local/central switching office of the telephone company is known as a *local loop*. One line of this pair is grounded at the switching office while the other line carries the analog signal. The typical length of a local loop is about 8 km. We may use ordinary telephone lines or dedicated lines for dialing into these systems. For providing a continuous link for data communication between our computer and a remote computer, a modem over a standard dial-up line can be used, but it is very expensive.

The **leased line** or **dedicated line**, on other hand, offers a full-time permanent connection between two remote nodes and is billed on a monthly basis. The line is always connected and, hence, no dialing is required. It supports point-to-point and multi-point connection (using bridges). It provides one-loop (a pair of wires) or two-loop (two pairs of wires) modes of operation. The leased lines can be conditioned via special electronic circuits which improve the transmission speed with fewer errors. Most of the private lines are balanced and, as such, the lines are not grounded at the local switching office; instead, electrical grounding is provided at the local switching office.

The leased line modem gives a dedicated connection 24 hours a day at a flat rate and does not have any busy signals and toll charges. It supports both point-to-point and multipoint configurations. These modems are useful in applications such as banks, retail stores, organizations with central computing resource facilities, etc. These offer asynchronous data rates of up to 2400 bps, use four wires, and support half- and full-duplex modes of operation.

Modems are also characterized by their baud rates and bits per second, and it is worth mentioning here that both these parameters are not controlled by carrier frequency. In order to make this point clear, we redefine both baud rate and bits per second used in modems, as these are also defined with PSK techniques.

Baud rate of modem: The baud rate (BR) is the rate or frequency at which the carrier frequency is modulated per second. Every time the carrier frequency is modulated, one or two bits can be transmitted per modulation. The baud rate is a unit of signaling speed and can be defined as the reciprocal of the speed of the shortest character. **Bits per second (bps)** are defined as the number of bits that pass a particular point in a second of time and, as such, are considered a unit of information. Since a bit is usually represented by a pulse, the baud rate and bps are generally considered the same. With a two-phase (two-wire) modem, **bps = BR**.

For any number of phases in a modem, bps is given by the product of the baud rate and number of bits per baud. For example, we have a carrier frequency of modem of 1800 Hz and the baud rate (modulation frequency) of the modem is 1200. If we are sending one bit per modulation, it transmits 1200 bps. The baud rate will be 2400, if we are transmitting 2 bits/modulation. Hence, it is concluded that the baud rate is independent of the carrier frequency but depends on the number of bits/modulation. The number of bits per baud can be obtained by the following ratio: **bits/baud = $\log_2$(number of discrete signal levels that can be transmitted**). Thus, if we have 1200 baud and a four-phase modem, using this ratio we get bits/baud = $\log_2(4) = 2$. This ratio of bps to baud is equal to two and, hence, the bps is $2 \times 1200 = 2400$. In other words, this modem of 1200 bauds offers a data rate of 2400 bps. Thus, the bit rate in bits per second is given by a product of baud rate and bits/modulation.

The bit rate is independent of carrier frequency but depends on the number of bits/modulation. For a higher rate channel to be defined over PSTN, an appropriate modulation scheme is a very important factor. The baud rate as defined above also defines a **signaling rate** that corresponds to the number of times the characteristics (amplitude, frequency, and phase) of signals change their behaviors per second. It is important to know that although a higher signaling rate and higher baud rates may be defined by considering a large number of changes of signal per second, the bandwidth of the channel will always impose constraints on these baud rates.

3.4.4 PC communication protocols

The personal computer (PC) is now being used as a very useful tool for data communications across networks. The role of the PC as a data communication device has evolved from ASCII terminals, dumb terminals, RS-232C serial interface, and ASCII character set followed by asynchronous protocols. Before, PCs were dedicated mainly to word processing applications, but due to the availability of asynchronous serial ports and dial-up modems, PCs have now become a useful and effective device for data communication.

The Hay's AT command set has become a *de facto* industry standard which offers a variety of services for modems like dialing, variable speed setting, connection, disconnection facilities, etc. For accessing services of packet-switched networks, there are a number of software tools (protocols) available which support different types of communication schemes (synchronous or asynchronous).

Asynchronous communication protocol: This provides connection between asynchronous terminals and packet-switched networks and is available as terminal emulation, file transfer, etc. The terminal emulation enables the PC to emulate a dumb terminal using available emulation tools such as DEC VT 100, V T200, Data General D200, Televideo 920, etc. A few specialized emulation tools are also available from Hewlett Packard, Tektronix, etc. The emulation programs may become a useful window on the multi-tasking environment where users can easily perform many functions such as cut and paste, copy, delete, etc. The packet-switched network's X.3 parameters interact with the emulation program to configure on the screen.

The asynchronous protocol is widely accepted in both public and packet-switched networks and, as such, the global connectivity for desktop forced PC and Mac manufacturers to use this protocol as a device for data communication beyond the normal and traditional word processing applications. This has been well taken care of by introducing access to dial-up X.25 LAN gateways, and now users can easily access the services from both public and packet-switched networks. With the availability of serial ports operating asynchronously at data rates of up to 115 Kbps (PCs) and 57.6 Kbps (Macs), these devices can be used for wide-area communication applications with different types of modems which are offering data rates of 1.2 Kbps, 2.4 Kbps, 9.6 Kbps, etc.

For logging on to a host on the network, various steps are required in a particular sequence, such as dialing the number up and configuring X.3 parameters, user ID, passwords, etc. All these steps in the sequence can be automated via a file known as a **script file**. This file defines all the above-mentioned steps or procedures and allows the users to log on to the host by executing the file programs. PCs and Macs are usually equipped with predefined scripts in their communications packages. These scripts are generally popular for public databases (value-added networks, VANs) and e-mail service providers such as CompuServe, MCIMail, EasyLink, etc. Scripts may help in drawing windows on the screen, network security, protecting the user's identity (e.g., host address, phone number, etc.), text files to binary file conversions, etc.

Synchronous communication protocol: The synchronous communication protocol does not require any additional bits (as required in asynchronous communication protocols in terms of start, stop, and parity with every character). Instead, synchronizing bits are required before and after the block of data to maintain synchronization between sender and receiver. Due to this, it is expected that the synchronous communication protocol should offer faster response, higher throughput, and more efficient implementation than asynchronous protocols.

Most PCs (IBM-compatible) and Macs are based on asynchronous transmission and hence use asynchronous communication protocols; the synchronous communication protocols require a special type of hardware device/software unit to execute them. IBM has defined synchronous protocols such as 3270 SNA and 2780/3780 remote job entry (RJE), which run on the IBM-compatible boards. These protocols also require leased lines or direct-dial connections, which are more expensive than the ones which use asynchronous communication protocols.

A PC can be connected to a host via a modem using client–server software, where it becomes a client that can use the graphical interface provided by a host. A PC can also run a terminal emulation program and allow itself to be a remote terminal to the host. A VT-100 is the most popular terminal to emulate, and this configuration provides very limited Internet services (those provided by the host). A dumb terminal can also be used for Internet access by connecting it directly to a host. Its applications are usually found in libraries, government offices, departments, and other organizations. As said earlier, PCs can dial access to the Internet via PPP or SLIP protocols. These protocols, in fact, create a complete connection over the telephone lines using high-speed modems. The PPP makes sure that the packets arrive undamaged and is more reliable than SLIP. In the event of a damaged or corrupted packet, it resends the packet. PCs and UNIX workstations can use either PPP or SLIP.

3.4.5 *Modem line configurations*

The communication line between modems can be either a two-wire line or a four-wire line. In the case of two-wire lines, modems can derive two communication channels The modems generally find applications in **public switched telephone networks (PSTNs)**. Many common carriers define the limit of transmission speed for the dial-up networks. Before using dial-up connection communication, it is advisable to check with the common carrier (as it defines the availability of various transmission speeds) about these features.

The modem can operate in **half-duplex or full-duplex** operations in both modes of transmission (synchronous and asynchronous). Voice-grade channel modems using switched telephone networks usually work in half-duplex operation. This is due to the fact that end loops or subscriber loops, in general, use a pair of wires, and as such can only support half-duplex. However, new modems at 1200 bps and even higher–bit rate modems are available which can work with full-duplex operations.

Half-duplex modems have a serious problem of turnaround which sometimes becomes significant (especially when we are using high-speed transmission lines) compared to the overall bit rate of the system. The majority of modems operating at bit rates of up to 1800 bps are asynchronous and offer serial-by-bit operation for the transmission. The synchronous modems operate at higher bit rates (9600 bps to 19.2 Kbps). Asynchronous modems are cheaper than synchronous modems and are useful for low bit rates because they transmit additional start and stop bits with each character they transmit. These modems are described in the following section. A partial list of typical two-wire full-duplex modems (CCITT) is given in Table 3.1.

Two channels for full duplex at higher speed can be obtained through a four-wire system. The four-wire system is like two two-wires in parallel and is widely adopted in telephone networks. The modem performs a transformation between digital signals — which are acceptable to terminals, network ports, and host computers — and analog signals used for transmission lines. This translates a square-edged pulse stream from computer or terminal into an analog signal to fit into the telephone channel frequencies between 300 Hz and 4 KHz.

A four-wire modem providing two channels gives full-duplex capability in the line. This means that modems along with four-wire systems provide the transmission and receiving of data at both nodes simultaneously. In other words, both terminals and computers can

Table 3.1 Typical Two-Wire Full-Duplex Modems

Modem	Description
CCITT V.19	Asynchronous for parallel transmission — 12 characters/sec
CCITT V.20	Asynchronous for parallel transmission — 20 characters/sec
CCITT V.21	Asynchronous — 300 bps
CCITT V.22	Synchronous/asynchronous — 600/1200 bps
Bell 103	300 bps
Bell 212 A	1200 bps
Bell	2400–2400 bps

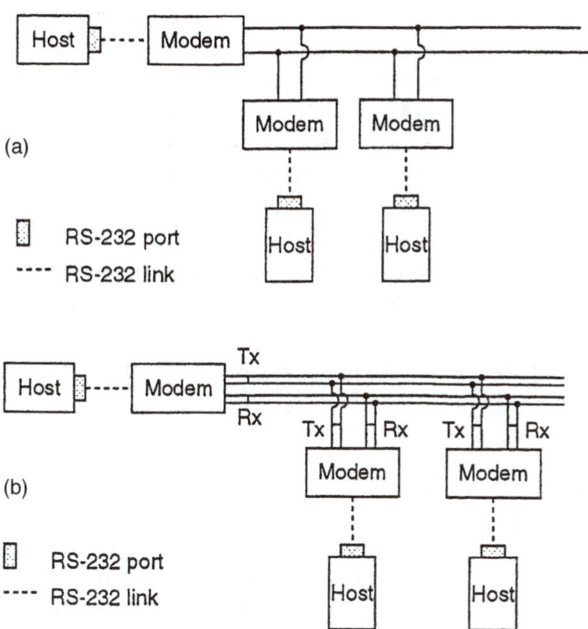

Figure 3.12 (a) Two-wire multi-point system. (b) Four-wire multi-point system.

transmit and receive the information between them through modem-line-modem simultaneously. If this does not happen, then they are connected through four-wire lines but offer half-duplex connection.

The utilization of communication lines can be improved by connecting two-wire or four-wire lines in a multi-point or multi-drop configuration. Terminals and computers can interact using a different framework/connection for each configuration, i.e., two-wire or four-wire systems using half- and full-duplex connections, as shown in Figure 3.12(a,b). There are many issues in this connection such as polling and selecting, which affect the performance of the system.

The point-to-point and multi-point connections can also be configured on leased lines using a **line splitter**. Usually, a line splitter is used at the exchange to provide the connection. By using a line splitter, one assigned line of exchange can be used by more than one location at the same time. The assigned line transmits information from a computer on the line to various terminals connected to different exchanges. Similarly, when these terminals want to transmit the information, it is combined on that line and delivered to computer.

On the other hand, a **digital splitter** allows more than one terminal to use a single modem port for transmitting and receiving information. One modem port can be shared by many terminals based on polling (by computers), which is controlled by a link-control protocol. After the polling, the selection is also made by a line-control protocol under the control of computers.

3.4.6 Operation modes of modems

A modem serves as an interface between devices which need to share data across the network and generally consists of transmitter and receiver. It transmits the data from a computer/terminal to one side of a modem which is connected to a computer/terminal while the other side is connected to a telephone line. It changes the characteristics of the signal so that it is compatible with the telephone channel. On the receiving side, another modem is connected between the telephone line and computer/terminal and it reconfigures the signal into original data formatted for a computer. One of the most important criteria in choosing a modem is whether the data terminal equipment (DTE) is synchronous or asynchronous. If there is a mismatch between the modem and DTE, then the transmission of data will generate erroneous data information. Thus, it is essential that we choose an asynchronous modem for a asynchronous DTE and synchronous modem for synchronous DTE.

In general, asynchronous modems offer low-speed transmission while synchronous modems offer high-speed transmission. There are modems which can offer the same speed for both types of DTEs.

Based on the modes of operation, four types of modems have been classified:

- Synchronous
- Asynchronous
- Synchronous/asynchronous
- CSU/DSU alternatives

Synchronous Modems: In synchronous transmission, the frames don't contain any start or stop bits/characters. There is no delay between the frames and, as such, this type of transmission offers a faster data rate than asynchronous transmission. It works on both dial-up (two-wire half-duplex on public-switched networks (PSTN) or leased lines) and leased lines with RS-232/V.24 interfaces and provided lines (four-wire full- and half-duplex), and it supports 4800-, 7200-, and 9600-based applications on point-to-point or multi-point connections. These modems on dial-up lines (4800-bps) or leased lines (9600-bps) are usually connected between **front-end processors (FEPs)** and a central telephone office. The front-end processors in turn are connected to synchronous devices such as IBM mainframes. Modems with automatic dialing and auto answer capabilities are also available for file transfer in synchronous devices (mainframes) and mini-environments.

Modems having both dial-up and leased connections together are also available and should be used initially as a dial-up modem and then as a leased-line modem at a later stage. Modems having the capabilities of loop-back testing of remote modems and local modems will improve turnaround time (typically a few milliseconds for synchronizing the host with its multi-drops). Interestingly, this value of turnaround time is much higher (typically 30–35 times) than in standard V.29.

These synchronous modems are used for synchronous data terminal equipment (DTE) devices. These modems break up the data into segments (or data blocks) and add error-correction codes with each transmitted segment. In order to provide synchronization between transmitting and receiving nodes, they also transmit a clock signal with each segment as a part of the segmentation scheme, as shown in Figure 3.13(a). The segments are long and sometimes may contain thousands of eight-bit data characters. These modems usually transfer the data at high rates more efficiently, but due to the large size of frames, a greater chance of error exists.

The segmenting and synchronization characters each begin with a synchronous data segment. These are followed by a special character **Start-of-TeXt (STX)**. Actual computer information occupies the next field in the synchronous-modem segmenting frame format, as shown in Figure 3.13(b).

Another special character **End-of-TeXt (ETX)** follows the data. The last field of the format contains a **block check character (BCC)** which detects error in the transmission. A BCC is similar to a parity bit and is a calculated character generated by both the sending and receiving modems based on the data transmitted in the long synchronous segments.

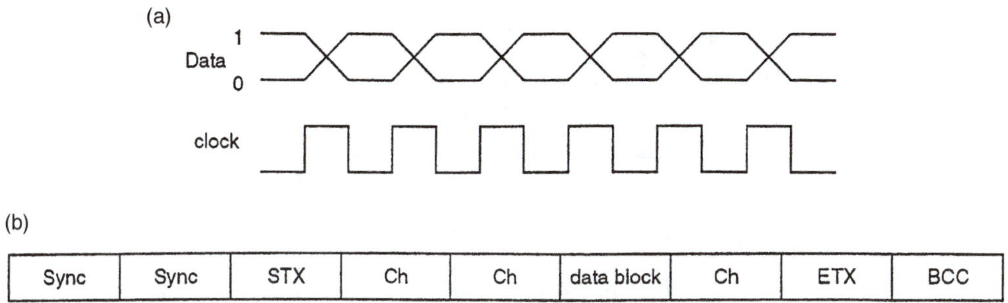

Figure 3.13 Synchronous modem. (a) Segmentation. (b) Segmentation frame format.

The receiving modem compares two BCCs. If they are the same, this indicates error-free transmission; otherwise, the block will be re-transmitted.

Synchronous modems offer a higher data rate and have been used in the following applications:

- Dial-up and leased line applications
- Manual dial/auto answer
- 4800-baud applications
- Point-to-point or multi-point configurations
- 3780/2780 bisync applications (file transfer in main and mini environment)

PC modem gateway cards connect PCs to synchronous mainframes or minis. They may include emulation or other options. These cards offer 2400 bps over two-wire PSTN, 4800 bps over two-wire PSTN or private lines, and 9600 bps over two-wire leased lines. Various emulation packages supported are SNA 3270 (protocol for interactive and file transfer communication), SNA 3770 software (protocol for 3770 RJE terminal emulation and micro-to-mainframe file transfer), SNA 5251 (interactive communication between remote PC and IBM systems, e.g., 34, 36, 38, AS/400™), bisync protocol (BSC, which provides micro-to-mainframe and micro-to-micro communication), dial-up software (menu-driven offering auto dial and auto answer to synchronous devices), etc.

Various capabilities (e.g., automatic dial/manual, auto answer) are available in emulation packages which run on PC-DOS and Microsoft PC-DOS and are compatible with SNA software protocols (e.g., interactive and file transfer, and communication between mainframe and remote PC), standard bisync protocols (micro-to-mainframe, micro-to-micro communication, etc.), and other dial-up software sync communication packages.

Asynchronous Modems: In an asynchronous mode of operation, the data to be transmitted is represented as frames of a fixed number of bits or characters. Typically, the length of a frame varies from 508 bits/character to some predefined value. In order to provide synchronization between sender and receiver, each frame is preceded and succeeded by two to three extra bits as **start** and **stop bits**. Due to the fact that there is a brief delay between each frame, the asynchronous modem speeds are relatively low (typically 200–2400 bps). FSK modulation is usually used for this modem. The transmission data rate depends on the digital signal and hence varies up to the maximum supported. The existing data rates are 100, 134, 150, and 300 bps.

In an asynchronous data block a **"start"** bit (mark pulse or positive pulse) at the beginning of character indicates the start of a new frame. After the character (digital information signal), a parity bit for error detection followed by a **"stop"** bit (usually one

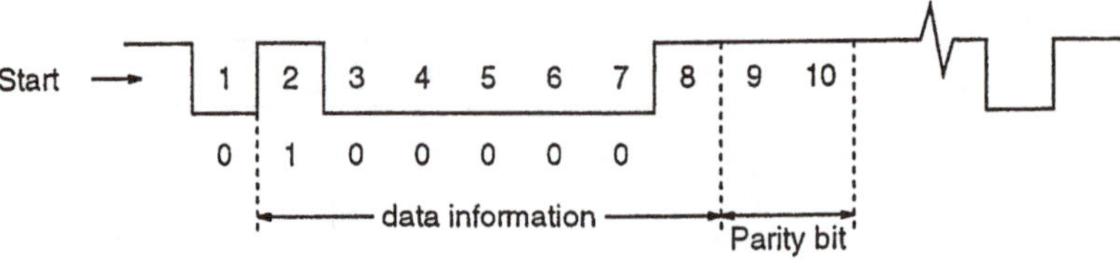

Figure 3.14 Asynchronous modem.

or two bits with space pulse or negative pulse) is included. In other words, for each character we are sending three or four extra bits with eight bits of computer data, as shown in Figure 3.14.

The parity bit may be either even or odd. In odd parity, transmitting modems will include 1 or 0 in the parity bit location (eighth data bit) to make the total number of 1s in the frame an odd number. When even parity is selected, 1 or 0 is included in the parity bit location (eighth data bit) to make the total number of 1s in the frame an even number. None/Off indicates that the eighth data bit is ignored.

A **stop** bit indicates the end of an asynchronous frame. An infinite **mark** signal follows (after the stop bit) before the next character frame is initiated by a start bit from the transmitting modem. **Mark** and **Space** represent logic 1 (high) and 0 (low), respectively.

Asynchronous modems are usually low-speed modems because an extra three or four bits are required with each character frame; further, they require overhead to deal with these extra bits, which is more than synchronous modems.

Each modem generates a carrier frequency which is used for modulating the incoming signals. This carrier frequency does not affect the transmission rate (in bits per second) of the communication media. The rate at which carrier frequency is modulated per second is defined as **baud rate**. In other words, baud rate of a modem indicates the number of bits which can be transmitted per modulation.

The transmission bit rate is the rate at which the modulated signals can be transmitted on media and is related to baud rate by the following expression:

$$\text{Bit rate}\left(\text{bps}\right) = \left(\text{baud rate}\right)\left(\text{bits/modulation}\right)$$

These modems are useful in connecting asynchronous devices to databases and bulletin board applications such as CompuServe, The Source, and Dow Jones, and in transfer of data between office and home or with anyone who has a compatible modem. These modems support two-wire full duplex over dial-up lines, full duplex over RS-232/V.22 interfaces. Various standards supported by modems are Bell 212, 103 CCITT's V.21, V.22 (data rate of 300 bps and 1200 bps); CCITT's V.22 bis (2400 bps). A portable modem (pocket-sized) that is battery operated to support 122, 2400 bps is also available which can be used on laptop computers.

These modems are usually compatible to Hayes modems to cover a variety of applications. These modems may be connected to an asynchronous host on RS-232 cable and dial-up line (1200 bps) to a central telephone office.

The protection and security of passwords and reconfiguration by calling the user back can be provided by a modem known as a **security modem**. Such a modem usually provides security at three levels. At level one, it does not offer any security and, as such, the modem works like a standard 2400-baud modem. At level two for password access, the modem stores and recognizes up to a few (typically 16) user passwords. At the third level, the password access along with call-back facility offers an efficient security provision. In this type of modem, it calls back the user at the phone number assigned to the user's password. The modem disconnects the user if the password and call-back number do not match. It is very useful in applications such as stock, payroll, inventory, and financial records. It can be called in by remote sites over a 2400-bps standard Bell 212A, which is characterized by asynchronous, full duplex over two-wire dial-up line, etc.

These modems (offering security) are connected to an asynchronous host to a dial-up line (2400 bps) to a common central telephone office. In this case, a standard dial-up line may be expensive, and the cost of this telephone circuit can be reduced by using a dedicated circuit all day at a flat rate. The modem for a leased line may be connected between an asynchronous host and the central telephone office. This leased-line modem offers point-to-point and multi-point, multi-drop connections to ensure data accuracy and fast transmission. This modem makes a segment or frame of 8- or 10-bit size out of the incoming data from a computer terminal. Both sending and receiving modems have to agree on a particular set up of frames before communications between them can begin.

Asynchronous modems can be used in a variety of applications:

- Accessing databases and bulletin boards such as CompuServe, The Source, Dow Jones, and many others
- Dial-up applications, auto or manual dial/answer
- Data protection/security
- Multi-point configurations
- Local or remote diagnostics

These modems provide operations on full duplex, two-wire full duplex over dial-up lines or leased lines, four wires, and half or full duplex over leased lines and offer data rates of 300, 1200, and 2400 bps. The traditional mainframes and minis such as Hewlett-Packard 1000 and AT&T 3B2 and 3B5 use asynchronous transmission. Also, PCs, ATs, and compatible DEC and Hewlett Packard Micro use asynchronous transmission.

Synchronous/Asynchronous Modems: Some synchronous/asynchronous (sync/async) modems (which support both synchronous and asynchronous data formats) are available in the market. These modems are compatible to Hayes modems and can use various popular communication packages such as CROSSTALK, XVI, and MTE.

Modems supporting asynchronous and synchronous data formats over Bell 103, CCITT's V.22, and V.22 bis standard on RS-232/V.24 interfaces are also available. These modems offer data rates up to 2400 bps for full duplex over two-wire dial-up or leased line and support point-to-point or multi-point connections. In general, in the initial stages of a company or business, one might opt for a dial-up line, but later on when the business grows, one might think of opting for a leased line, as here a flat rate is charged as opposed to the charge per connection in the case of the dial-up connection. Modems supporting asynchronous and synchronous operations are usually available with an asynchronous-to-synchronous converter and can be used with dial-up or leased lines, two wires or four wires for half duplex or full duplex.

These modems have the same applications as those of asynchronous and synchronous modems individually and have the following features:

- Automatic baud-rate conversion
- Auto-dial/auto-answer
- Data accuracy
- Compatible with 4800 or 9600 baud modems
- Data rates of 300, 1200, 2400, 4800, and 9600 bps

(a)

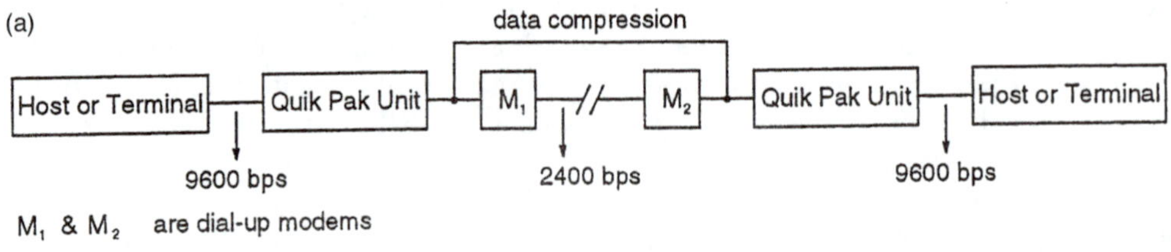

(b)

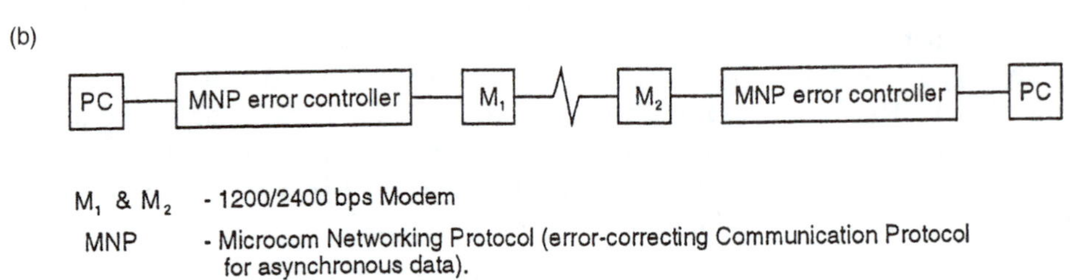

Figure 3.15 (a) Typical configuration for high-speed 1200- or 2400-baud modem. (b) Typical connection using MNP error controller.

In addition to these modems, there are modem support products in the market which are used with modems for improving the performance of communication over transmission media:

- **Quick-Pak** uses a data compression algorithm to increase throughput by converting the baud rate of the host to a slower modem. It allows the PC to send the data at a normal rate of 9600 bps and adjusts this rate to the transmission rate of a 1200/2400 dial-up modem automatically. In other words, it maps the 9600-bps data rate of a PC to match the slower dial-up modem rates of 1200/2400 bps. This modem product uses a standard ACT data compression technique. A typical connection using this unit is shown in Figure 3.15(a). It supports asynchronous transmission (ASCII) and full duplex and is compatible with different classes of **MNP (microcom network protocol).**
- **Microcom Network Protocol (MNP) Error Controller** is an error-correcting communications protocol for async data (interactive and file transfer applications). It can be used over dial-up lines via 1200/2400-bps modems. A typical connection using this device is shown in Figure 3.15(b). It conforms to OSI model's data link layer (for providing reliable data transfer). Traditional modems cannot provide error-free data transfer because the transmission includes noise and distortions introduced by voice-grade telephone circuits. This protocol has become the *de facto* standard of the industry. It has defined five versions (classes) of protocols, and each version or class of MNP interacts with each other and offers efficient operations over the media.
- **Class I** uses an asynchronous byte-oriented block method of data exchange. It is seldom used.
- **Class II** adds full duplex and offers throughput of 2000 bps from 2400-bps modems.
- **Class III** uses a synchronous bit-oriented full-duplex data packet. It eliminates start and stop bits and offers throughput of 2600 bps from 2400-bps modems.
- **Class IV** adds adaptive size packet assembly and data phase packet format optimization and offers 2900 bps from 2400-bps modems.
- **Class V** adds data compression, which uses a real-time adoption algorithm to compress the data and offers 4800 bps from 2400-bps modems.

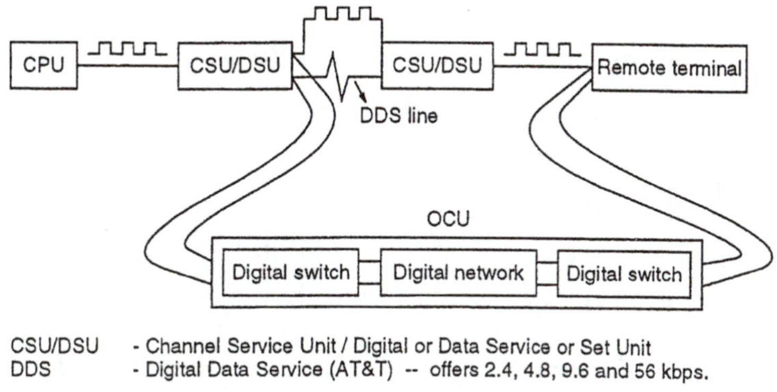

Figure 3.16 Configuration for CSU/DSU.

Channel Service Unit (CSU)/Digital or Data Service or Set Unit (DSU) Alternative: Many people think that this is a modem, but it is not. This provides a high-speed alternative for a digital transmission on digital networks. The Federal Communications Commission (FCC) advised AT & T and Bell Operating Companies (BOC) not to install equipment containing customer premises equipment (CPE) and advised customers to use CPE on their own. The CPE basically provides digital communication between computers or terminals (DTEs) on digital computers (Figure 3.16).

In the early 1980s, a device containing two components known as a **combined channel service unit (CSU)** and **data service unit (DSU)** was introduced by Western Electric 500 A. This pair of CSU and DSU provides an interface between DTEs and digital channels. The DSU converts a signal from computers or terminals (DTEs) to a bipolar digital signal. It also provides clocking and signal regeneration on the channel. The CSU, on the other hand, provides line equalization, signal reshaping, and loop-back testing (between the CSU and the network carrier's office channel unit OCU).

Most of the CSU/DSU alternatives support synchronous and asynchronous data formats and can handle a data rate up to 56,000 bps in point-to-point and multi-point configurations. Leased digital lines are required for these alternatives, but now with the reduction in cost of hardware components and increased use of very large-scale integration (VLSI) technology, the cost of digital lines is comparable to that of analog leased lines.

3.4.7 Bell modems

Bell modems are very popular and widely accepted in North America and other countries. In fact, physical layer specifications in North America are defined on the basis of Bell modem specifications. For automatic dial-and-answer modems, Bell 103/212A specifications are widely used by different vendors, while Bell 103, 113, 201 C, 208 A/B, and 212 A specifications have been adopted by many manufacturers. Each of these modems have different data rates and baud rates and use different types of modulation techniques.

Modems produced by Hayes Microcomputer Products, Inc. have become the *de facto* standard for low- and medium-speed auto-dialing modems. These modems may be used in both full- and half-duplex communications. Popular Hayes series modems are Smart modems (300, 1200, 2400, 9600 bps).

A brief discussion of the Bell 103 F modem provides further insights on the working principles and concepts of modems.

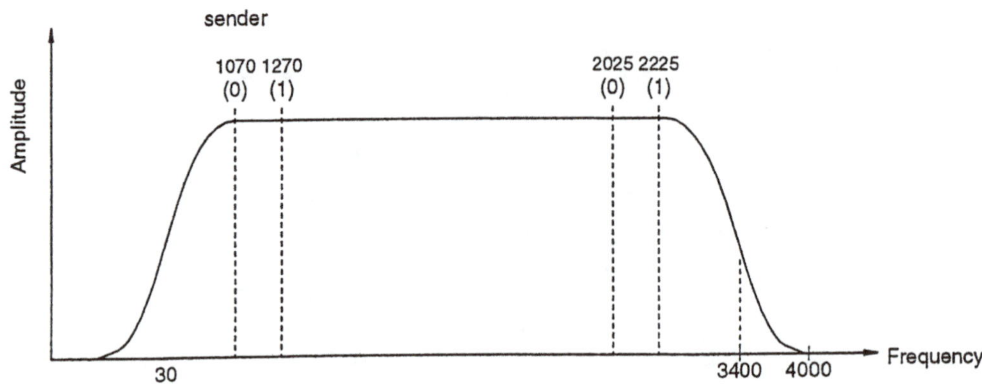

Figure 3.17 Bell 103 F modem.

Bell 103 F modem: This is one of the simplest modems of Bell series modems. It uses a pair of wires for full duplex communication at 300 bps. It uses voice-grade baud (300 Hz). It generates two sub-channels out of this band. One sub-channel represents a band of 1400 Hz (300–1700 Hz) for sending the information, while another sub-channel represents a band of 1300 Hz (1700–3000 Hz) for sending the response. If two devices are connected via these modems on a telephone line, then one device will be allocated with one sub-channel while the other will get another sub-channel. The frequency allocated to 0s and 1s between these devices will be different for each way of communication; i.e., the frequencies allocated to 0s and 1s from one device to another may be 1070 and 1270 Hz, respectively, while for the opposite direction of communication, 2025 and 2225 Hz may be allocated to them. Similarly, the receiver on each side will accept the frequency band (transmitted from the other side). All these four frequency components will be present on the medium at the same time but will not interfere with each other (Figure 3.17).

3.4.8 *Turnaround time of a modem*

The turnaround time is a very important performance parameter for a modem and depends on a number of timing factors. We will consider a typical modem which is used on half duplex and for which the further channel also uses an echo suppressor. The following is a description of those timing factors which must be considered for determining the turnaround time:

1. The ready-to-send (RTS) signal from the data terminal is asserted and a clear-to-send signal is generated which typically takes 10–240 ms after assertion of RTS.
2. One-way propagation delay (depends on length and transmission speed of the network). For example, a terrestrial link may offer this delay in the range of 2–15 ms, while a satellite may offer around 300 ms.
3. One-way propagation signal transmission delay through transmitter and receiver equipment; may have values in the range of 3–16 ms.
4. The receiving data terminal equipment takes a finite time for responding to the RTS signal.
5. The time taken to send acknowledgment for the receiving data terminal at modem bit rate (typically 4 to 12 characters). Each character typically consists of 6 to 11 bits depending on the coding and modes of transmission(synchronous or asynchronous).
6. The transmitting data terminal takes a finite time to process the acknowledgment it has received from the receiving data terminal, evaluate it, and send its reply as a request-to-send signal (a few milliseconds).

While calculating the turnaround time, the timing factors corresponding to 1, 2, and 3 above must be considered twice, as these are defined for one-way. Thus, the turnaround time is given by the sum of all these timing factors, as discussed above.

3.4.9 *Types of modems*

Modems and related devices can be classified into the following categories. Some of the material is partially derived from References 1, 2, and 7.

Low-speed modems: The maximum data rate of these modems is 600 bps and they can be used over voice-grade, dial-up, or private lines for unlimited distances. These modems, in general, use frequency shift keying (FSK). The sender modem usually uses low frequency values for 1 and 0, while the receiver modem uses higher frequency values for its 1 and 0 to send acknowledgment to the sender, as shown in Figure 3.17. The low-speed modems are useful for keyboard terminals. These modems are compatible to CCITT V 21.

Medium-speed modems: These modems provide data rates between 1200 and 3600 bps and can be used for the same applications as those of low-speed modems. These modems also use frequency shift keying (FSK) and support a bit rate in the range of 1200–3600 bps; however, 1200 bps is very common.

High-speed modems: These modems are used for the same applications as low-speed and medium-speed modems. These are very popular with the RS-232 physical layer interface. The transmission bit rate is between 4800 bps and 19.2 Kbps.

Wide-band modems: These modems use more than one transmission line to send the data. The bit rate of each line is added to provide the total bit rate of over 19.2 Kbps on each side of the telephone line.

Limited-distance modems: These modems include **short-hand modems (SHM)** and **line drivers**. These modems can be used for short-range communications and are available in both synchronous and asynchronous versions, covering a wide range of host applications. The asynchronous version transmits up to 2 miles at 9600 bps, while the synchronous version transmits up to 0.8 miles at a maximum speed of 19.2 Kbps. These modems can easily replace RS-232 twisted-pair cable (due to distance limitation and high cost of bulk quantities) in the same building over a campus or industrial complex. These modems handle short-range communication quickly, easily, and inexpensively with the help of in-house unshielded twisted pairs or local private lines. These modems are useful in point-to-point configuration by offering data rates up to 19.2 Kbps for synchronous and 9600 bps for asynchronous. Further, the use of optical isolation greatly reduces the electrical interference in the transmission. SHM modems support point-to-point on four-wire applications. Limited distance modems (standard V.35) are used for the transmission of the bulk of data within a local distance of a few miles (typically less than 3 miles); limited-distance modems such as V.35 are very useful for high-speed applications. They support point-to-point or multi-point configurations and offer data rates up to 64 Kbps.

High-performance modems: High-performance modems for transmission of data on telephone lines are also available and are expensive compared to ordinary modems. For any application, we have to make a tradeoff between the cost and traffic volume. For example, for heavy traffic on point-to-point links, we will be better off using high-performance modems, as the cost of these models will be much less than the link itself. A V.32 modem supports full-duplex operations on switched networks and offers a bit rate of 9600 bps. Modems with a bit rate of 14,400 bps or higher are available in the market, and since the bit rate is approaching the maximum value of bit rate as defined by Shannon's theorem, it seems unlikely that new modems supporting higher bit rates than given in Shannon's theorem will be available in the market.

Long-haul modems: The modems for long-haul transmission systems offer a frequency band of 48 KHz and use frequency-division multiplexing. CCITT's recommendations V.35 and V.36 define modems for synchronous transmission with a bit rate of 48 Kbps. In addition to this bit rate, V.36 also supports bit rates of 56, 64, and 72 Kbps. The V.37 supports 76, 112, 128, and 144 Kbps for synchronous transmission. V.29 series modems are based on CCITT's V.29 specifications.

Modem Eliminators: Sometimes the distance between host terminals is so small that even small-distance modems or line drivers may not be used; instead of a pair of modems for providing communication between host terminals, the local connection for synchronous devices running on limited-distance modems (V.35) can be extended to thousands of feet apart without any line booster. The modem eliminator offers various mapped pin connections (clocking for interfaces, etc.) between a synchronous host machine, terminals, or peripheral devices (e.g., printers), thus eliminating use of a pair of modems. It supports a typical data rate of 56 Kbps over a distance of a few thousand feet. It supports a data rate of 64 Kbps on full or half duplex for RS422/RS449 or V.35 interfaces.

A typical connection using a modem eliminator is shown in Figure 3.18. Here, the eliminator offers synchronous operations. A synchronous modem eliminator may be used between central computing services with, say, an IBM mainframe which supports a number of synchronous devices placed at a distance of up to 15 m. The modem is connected with a mainframe over an RS-232 cable connection. It offers transparent operations. Similarly, an **asynchronous modem eliminator (AME)** can be used between any asynchronous devices (PCs, terminals, or printers) in any way. This eliminator supports a distance of up to 15 m on RS-232/V.24 and offers transparent operations and data rate.

If we are transmitting data on a point-to-point configuration over a very short distance, then we do not need a modem or line drivers. We can use an inexpensive modem eliminator in place of the two modems that otherwise would be required for the short distance. Also, if we are using synchronous devices running on V.35 or RS-422 interface, we can increase the distance up to 2000 ft (V.35) or 4000 ft (RS-422) without any line booster just by using modem eliminator. It offers connections up to 2000 ft at the typical 56-Kbps transmission rate (used in most domestic V.35 applications) and also the standard overseas rate of 48 or 64 Kbps. Similarly, it offers connections up to 4000 ft (specified by RS-422) and supports up to a 64-Kbps transmission rate. These modems provide all required clocking for their respective interfaces.

If we have to connect two local asynchronous devices such as mainframes, terminals, printers, etc., **null modem (cross-over)** cable consisting of both RS-232 cable and a modem eliminator can be used for a range of 50 ft to 500 ft.

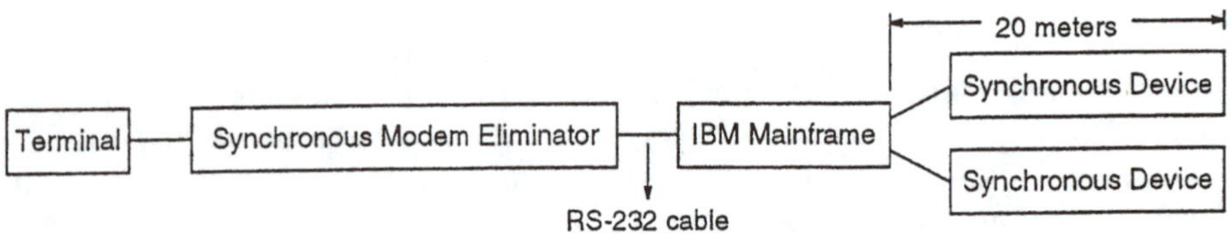

Figure 3.18 Modem eliminator.

Similarly, a synchronous modem eliminator can be used to connect two units of synchronous data terminal equipment (DTE) such as IBM minis, mainframes, and terminals up to 50 ft. In addition, it enhances the asynchronous environment in the sense that for changing from a synchronous to asynchronous environment, the same eliminator can be used.

Advanced intelligent modems: Modems with the capability of keeping the status of the various types of signals (analog or digital) are known as **intelligent modems** and may be used for network management systems. A microprocessor-based system performing these functions is built in with these modems. These modems typically operate at a normal data rate up to 19.2 Kbps for data transfer and are called a **primary channel**. In order to maintain the network management systems, a secondary channel with a rate up to a few bits per second is used between modems which communicate with a management system to provide the necessary data for the management of the network. Although a secondary channel is defined out of voice-grade bandwidth, it occupies very little bandwidth, while the primary channel occupies the majority of bandwidth; hence, the quality or fidelity of voice is not affected. The network management system basically monitors analog signals and related parameters and also the characteristics of modems.

Self-testing modems (loop-back): One of the options available in industrial modems is a **loop-back testing** facility which is also known as a **buy-back option.** This option allows the user to detect any fault or trouble in the modem before the transmission of data on the telephone line. This option in modems generates a test bit stream (after the modem is turned on) which travels from the host to the modem and returns to the host, making a self-looping configuration. The connecting lines provide a loop between host and modem as shown in Figure 3.19(a). The returned test bit stream is compared with the transmitted test bit stream for any fault in the modem. Data from the host will be transmitted to telephone lines via modem only if this test gives no fault in the modem. If there is a fault, then transmitted data from RS-232 of the host will be echoed back to the host by the modem and will not be transmitted on the line. This type of test is known as a loop-back test and checks the RS-232 interface between host and modem.

Another type of loop-back test is known as remote loop-back and is used to test the line. Here also a test bit stream is sent from one side to the remote side, as depicted in Figure 3.19(b). In the case of any fault, data will be echoed back to the transmitting side. Although modems at both of these sites are sending and receiving bit streams to and from each other, data transfer can take place only if there is no fault.

Null Modem: As discussed above, asynchronous devices (mainframes, terminals, PCs, printers) may be connected by a modem eliminator (eliminating the use of a pair of

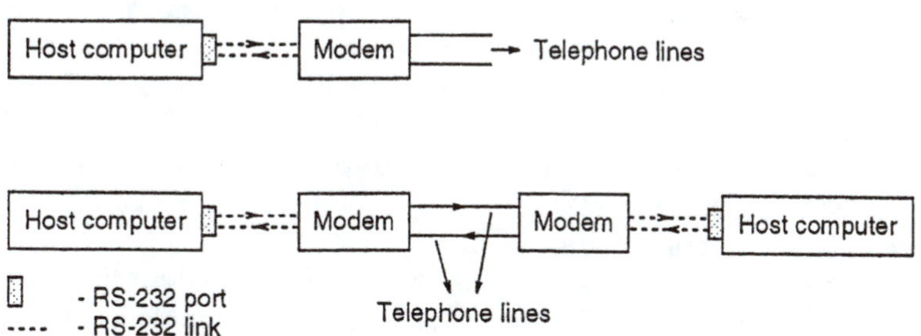

Figure 3.19 (a) Local loop-back testing of modem. (b) Remote loop-back testing of modem.

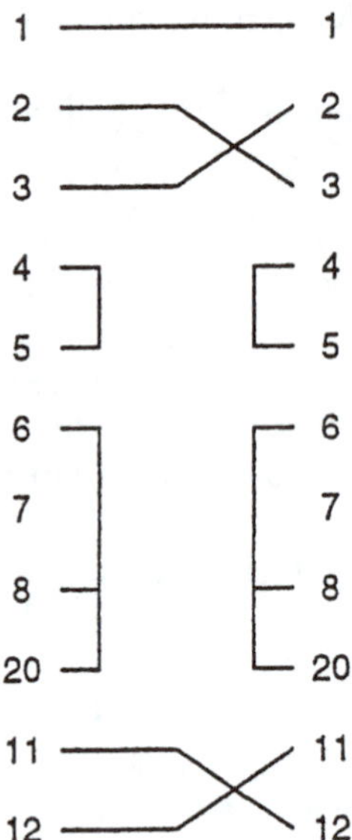

Figure 3.20 Null modem.

modems). The devices are connected by cables. The **null modem (crossover)** cable includes both modem and cable and can support a length of 15 m for local applications. The cable is usually immune to noises and is not available for a higher distance than the standard 15 m.

Modems which support both synchronous and asynchronous operations can be used to connect synchronous data terminal equipment (DTE) such as IBM mainframes, terminals, etc., up to a distance of 15 m offering a maximum data rate of 38 Kbps over full- or half-duplex operation for RS-232/V.24 interfaces. They can also be switched to asynchronous mode to connect asynchronous devices with extended-length RS-232 cable, and a maximum distance of 150–160 m can be obtained. The standard pin assignment of null modems is shown in Figure 3.20.

Fiber-optic modems: These modems are used with fiber-optic links only and can transmit the data of bit rate in the range of 56 Kbps to 2 Mbps. The standard physical interface RS-232-C and V.35 can be used with these modems.

Line drivers: Line drivers are used to increase or boost the distance for transmitting data over transmission media. For short distances, low–data rate line drivers can be used. For longer distances, the line drivers are usually very expensive. Instead of using expensive line drivers, we can use line drivers with compression techniques, and there are many line drivers with the facility of data compression available in the market. These line drivers provide an interface between RS-232 and short-haul modems and convert RS-232 signal levels to low-voltage and low-impedance levels. Typically, SHMs get their power from RS-232 interface, which gets it from the DTE. These modems do not convert digital signals into analog signals (as is done in other modems).

On the other hand, synchronous modems are usually defined for faster transmission, boosting data rates to 2 Kbps for a distance of 0.8 miles. These modems are also available with a data rate of 600 bps to 2 Kbps, covering a distance of 5 miles to .8 miles. The fiber driver can be plugged into an RS-232 port of a PC or terminal, and fiber cable of length 2.5 miles that offers a data rate of 76.8 Kbps can be connected between the PC and

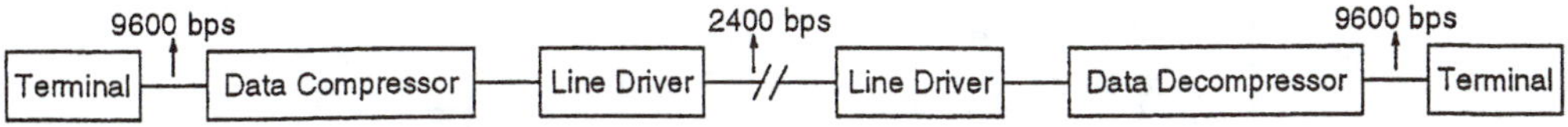

Figure 3.21 Line drivers.

transmission media. A fiber PC card is also available which can be used for file transfer applications and is compatible with fiber drivers.

The average price of asynchronous or synchronous modems is in the range of a few hundred U.S. dollars. The distance and data rate for shielded cable will be lower than for unshielded cable by 30–37% of the values shown above.

The following section discusses some of the widely used line drivers for asynchronous and synchronous transmission.

i) Asynchronous line drivers: As said earlier, line drivers are used to boost the distance over which data can be transmitted. Although LAN data can be used to transmit the data a few hundred feet or a few miles, it has a limitation on the distance. Asynchronous line drivers can be used to transmit the data over inexpensive twisted-pair (four-wire on half or full duplex) wires between various locations within organizations or across buildings within some complexes (within a distance of a few miles). These modems usually support RS-232, V.24, and V.28 interfaces.

The twisted-pair wire can be used to connect all asynchronous devices within the complex. Using data compression techniques, the line drivers can be used to transmit a large volume of data which will occupy a small area at slower speed. The line drivers offer various data rates ranging from 100 bps to 64 Kbps, covering a distance range of 40 miles to 0.8 miles on different sizes of wires. For example, a lower AWG number (e.g., 19) will cover more distance than a higher AWG (e.g., 27).

The line drivers may be connected as shown in Figure 3.21. Here, the terminal is connected via a 9600-bps link to a device which provides data compression and sends compressed data to a line driver. This device also provides speed conversions from 9600 bps to 2400 bps or even 1200 bps. On the receiving side, the compressed data is converted back into the original format and speed of transmission.

The line drivers allow us to transmit the data over a short distance at a data rate of up to 64 Kbps and in some cases even up to 128 Kbps. They support point-to-point configuration and also can connect more than one station to the same line via multi-point line drivers.

The asynchronous multi-point drivers support point-to-point or multi-point configurations and usually offer data rates of 64 Kbps and are compatible with RS-232. They usually support 64 users or more at data rates of up to 64 Kbps. There is no need to buy expensive modems or even use telephone lines. The data transfer can take place over in-house twisted-pair cable within the organization offices in a complex. These drivers support various DCE/DTE interfaces such as RS-232 and RS-485 ports.

ii) Fiber drivers: A higher data rate of 76.8 Kbps for a distance of 2.5 miles can be obtained using a fiber driver and plugging it into an RS-232 port of PC or terminal. Obviously, we have to use fiber-optic cable in our campus or building. These drivers can be plugged into RS-232 ports of PCs and the data transmission can take place on the optic cable at a very high unprecedented speed (data rate). The fiber drivers are immune to noises, electrical interferences, lightning motors, high-frequency signals, etc. In contrast to fiber drivers, other line cable drivers such as copper-based drivers generally use a surge protector to reduce the effect of noise. The asynchronous fiber drivers support data rates of 7608 Kbps

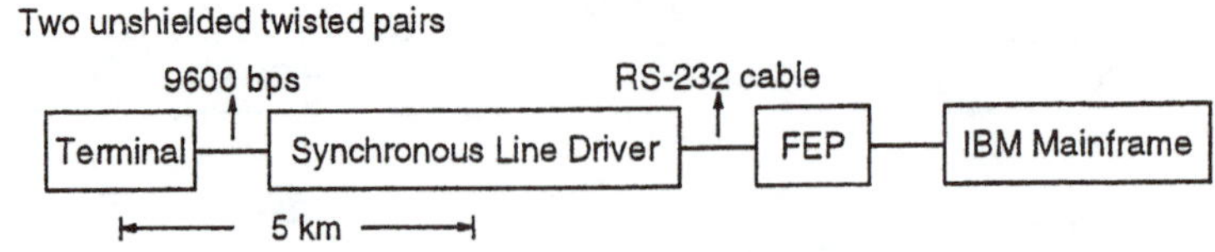

Figure 3.22 Synchronous line driver.

on RS-232 DEC/DTE interface fiber optics. The fiber-optic PC card is suitable for file transfer applications while fiber drivers are suitable for terminals. These drivers usually support data rates ranging from 100 bps to 76.8 Kbps for a distance range of 32 miles to 2.5/3 miles.

iii) Line boosters: This device increases the transmission distance for asynchronous data over an RS-232 interface. As we know, RS-232 cable length goes up to 15.5 km, while by connecting one line booster between two RS-232 cables, the signals now can travel to a distance of 30 m. One of the applications of line boosters would be to place a printer at a distance from a PC/workstation.

iv) Synchronous line drivers: These drivers offer higher distance than asynchronous line drivers. They use the twisted-pair line or telco line which can be leased from the telephone company and supports RS-232/V.24 interfaces. These drivers offer a data rate of 1200 bps to 128 Kbps for a distance range of six or seven miles to 0.5 miles (different for different gauges) on two-wire half duplex, four-wire half/full duplex, or four-wire full duplex. The general configuration of these line drivers includes a mainframe and front-end processor (FEP) for point-to-point or multi-point models, as shown in Figure 3.22.

3.4.10 Packet assembly and disassembly (PAD) interface

The main objective of PAD is to define a packet from the characters it receives from character-oriented asynchronous terminals. These packets follow the format of the packet defined by the X.25 public switched packet network (PSPN). PAD, after receiving these packets, disassembles them back into character form and sends the characters to their respective terminals one by one. In other words, PAD offers all the services of X.25 (call establishment, data transfer, flow control, etc.) for asynchronous terminals. CCITT defines X.25 as the specification for a network interface and uses the protocols of three lower layers (physical, data, and network). Many people think that X.25 is a protocol; rather, it is a specification for the network interface. According to CCITT's recommendation, X.25 is an interface between data terminal equipment (DTE) and data circuit-terminating equipment (DCE) for terminals which are functioning in the packet mode in public data networks (PDN).

A typical X.25 uses a generic function for DTE (PCs, terminals, etc.) which may be a network-attached, end user device or any other communication device working in the packet mode. For example, a minicomputer or any other communication device running at X.25 specification is a DTE, the front-end processor loaded with X.25 code is a DTE, etc. An alternative to these DTEs is packet assembly/disassembly (PAD), which allows the users of X.25 or non-X.25 to communicate with public or private packet-switched networks.

The PADs are usually used for display terminals but have also been used with computers, printers, teletype machines, terminal servers, etc. The available PAD devices support asynchronous, IBM bisynchronous, IBM SNA, DECnet, HP, Burrough, and Poll/select types of protocols. The PADs do not provide any protocol conversion between different end-system communication devices. The PADs are always required for X.25 interfaces (with similar protocol on the network) for synchronous communication mode. Another type of DTE is a dumb terminal, which uses asynchronous protocols for its operations. This terminal transmits a character at a time (as opposed to a block of characters in other character- or byte-based terminals). This terminal uses start–stop bits with every ASCII 7-bit character for providing synchronization between sender and receiver.

The CCITT has defined a separate non-X.25 network interface specification for these terminals (also known as start–stop DTEs). A pair of DTE and DCE is needed to use X.25, and for this reason, it has been defined as standard interface specification by CCITT. Other possible combinations such as DTE–DTE and DCE–DCE interfaces have not been defined or standardized.

The information on the X.25 packet network is in the form of a packet/frame. A terminal (which is typically an asynchronous non-buffered/buffered device) cannot be connected to X.25 directly through DCE (modems or multiplexers) and hence cannot access the packets from the network. A PAD interface device can be connected between terminals and an X.25 network. The PAD gets characters from terminals and makes a packet out of these characters in block mode. Thus, the input to PAD is a string of characters and the output is a block of characters defining a packet. The packet is transmitted to the X.25 network. On the receiving side, PAD receives the packet from the network and disassembles it into characters, and these are sent to terminals. Terminals at different speeds are supported by PAD.

Features of PADs: PADs can support multiple packet networks which will provide permanent or switched network connection. The data link of PAD can be individually set for different networks.

- PADs can be used for point-to-point communications between terminal and host with increased throughput and fallback capabilities.
- PADs can provide higher data rates (19,200 bps) with higher throughput and improved fault tolerance (in case any data link fails).
- PADs can support various asynchronous standard speeds of 50, 75, 110, 134.5, 150, 300, 600, 1200, 1800, 2000, 2400, 3600, 4800, 7200, and 9600 bps and the speed of a 4800-bps computer port.
- PADs are used in a variety of applications. Some of these applications will be discussed briefly here. They can be used for local switching where both local and remote terminals can be connected to various local resources and other terminals. These PADs are programmable and as such can adapt different data rates automatically. The adaptive speeds may range from 110 bps to 1200 bps.

PAD CCITT standard protocol: As indicated earlier, users can access X.25 via a dial-up connection over the public telephone network or via X.25 packet-mode terminals. Users having leased lines or asynchronous mode terminals (start–stop) cannot access X.25 networks. The CCITT defined three X recommendations — X.3, X.28, and X.29 — which offer a PAD option on PSDNs to users having leased lines or asynchronous terminals. These three protocols are sometimes called the **triple X protocol**.

i) X.3: It defines the functions and various parameters for controlling the operations of PADs. There are typically 22 parameters maintained for each of the PADs connected to a terminal. PADs may be connected to more then one terminal.

ii) X.28: It defines the protocol between terminal and PAD device. Two phases of operation for X.28 have been defined. The first is the data transfer phase in which the exchange of data takes place between the terminal and host, while the second is the control data phase in which the exchange of data takes place between the terminal and PAD in the form of characters. Various attributes such as desired baud rate between terminal and PAD, opening of a connection with remote DTE or host, etc. may be obtained with the help of parameters defined in X.3. Thus, this standard offers access to PAD, sets the terminal parameters to required values, establishes a virtual connection between terminal and PAD, offers data transfer between them, and de-establishes the connection.

The X.28 protocol defines the procedures and primitive language between the dial-in terminal (start–stop) or local DTE and X.3 PAD (sitting in PSDN). It defines two types of primitive languages: **PAD commands** and **PAD responses**. The primitive language requests initiated by DTE for PAD such as call setup are known as PAD commands, while requests such as clearing are known as PAD responses. X.28 protocol defines the procedure for controlling the functions of X.3 PAD port parameters over the network for its transmission to remote DTE (X.25 host).

X.28 commands allow the users to communicate with PAD directly and also to submit their requests. The users have to dial up the local PSDN to make use of X.28 commands. The X.28 offers nine PAD commands, and these commands allow the user to clear the call, read or change the X.3 parameter for any configuration, and also check the call.

iii) X.29: It defines the interface and a protocol between PAD and host. The PAD–host protocol defines a list of X.29 messages which are used between PAD and host for data exchange. Typically, PADs support 4, 8, 24, and up to 48 RS-232 ports on their input. PADs meet certain requirements of CCITT to support different types of networks.

PADs can be used in a variety of applications on public networks (domestic) such as Accunet, Telenet, Tymnet, Uninet, and CompuServe, and also international networks such as Datapac, Datex-P, DCS, DNI, and PSS Transpac.

A variety of DTEs are available from different vendors, and these devices define or support various options provided by all three layers of X.25 protocol. An attempt is being made to define a universal type of unified DTE which must provide not only various options offered by X.25 but must also support different types of networks. It is important to note that most of the existing networks such as Telenet (U.S.), DDX (Japan), Datapac (Canada), Transpac (France), and many others do not support each other; i.e., these are totally incompatible with each other. Except for one or a few services, these networks have altogether different formats and implementation schemes. One of the compatible services is Virtual Circuit, which supports packets 128 bytes long. This size of packet is supported by all the existing networks. A detailed discussion of how the PADs can be used to interconnect with X.25 and other implementation details can be found in Chapter 12.

3.5 Multiplexers

Telephone systems have become an integral part of various communication systems. We can transmit digital or data signals (via modems) on these lines in addition to voice communication. Usually, communication lines have much higher capacity than the number of bits/characters which can be sent by keyboard terminals, which can typically send and receive only a few hundred bits per second. These bits are exchanged only during the session with computers and, hence, a line is utilized. There are applications where the link is not continuously utilized during the session (e.g., waiting for acknowledgments) and, hence, the link is idle during that time. This is a very poor utilization of the link. Obviously, this becomes worse if we use high-speed lines for the transmission of data. Further, the cost and maintenance of low-speed and high-speed communication lines between two switching sites are comparable. In order to solve this problem, a method known as multiplexing has been devised where these low-speed terminals (through which they are connected to the network) can provide higher data speeds so that the line can be utilized fully at all times. In this method, various low-speed data coming out of voice, terminals, user applications, etc., can be combined into a high-speed bit stream which can utilize the high-speed line fully.

3.5.1 Advantages of multiplexers

A **multiplexer** is defined as a process which combines the small-capacity or low-speed sub-channels into a high-capacity or high-speed channel and transmits the combined channel over the communication link/transmission medium. On the receiving side, each of the sub-channels is separated from the high-speed channel and sent to respective receiving devices operating at different speeds. This technique of combining low-speed sub-channels into one high-speed channel and transmitting over a communication link/transmission medium improves the link/medium utilization and also the throughput of the system. The use of multiplexers offers the following advantages:

1. The response time of various hosts connected via the multiplexers is not affected.
2. They are relatively cheap and reliable.
3. The high volume of the data stream can be easily handled by them without any interference from other hosts.
4. Various send and receive primitives used for the transfer of data are not affected by the multiplexers but instead are dependent on the protocols under consideration.

Multiplexers accept multiple low-speed data (voice, data signal, telephone, terminals, etc.) and combine them into one high-speed channel for transmission onto a telephone line. On the receiving side, a de-multiplexer separates the incoming high-speed channel into appropriate low-speed individual signals. Thus, we are not only utilizing the high-speed lines but also using the same link for multiple channels containing a variety of applications (voice, data, video, etc.) Both the input and output of multiplexers/de-multiplexers can be in either digital form or analog form, depending on the distance and configurations. Multiplexing is an expensive technique for sending more than one voice-grade communication over a single channel simultaneously. The channel can be realized by any of the transmission media discussed in Chapter 5, such as coaxial cable, twisted-pair (unshielded or shielded), microwave, satellite, optical fiber, radio, or any other medium. Some of the material is derived partially from References 1, 2, and 7.

Figure 3.23 shows a simple example of how a multiplexer can be used to send three RS-232 low-speed signals on one RS-232 high-speed line for a short distance.

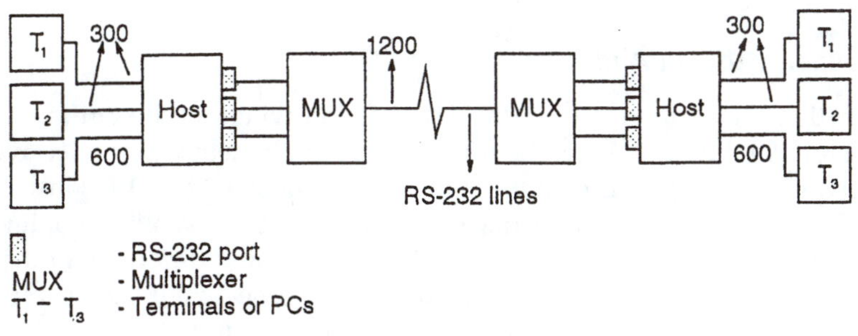

Figure 3.23 Simple multiplexer RS-232 line connection.

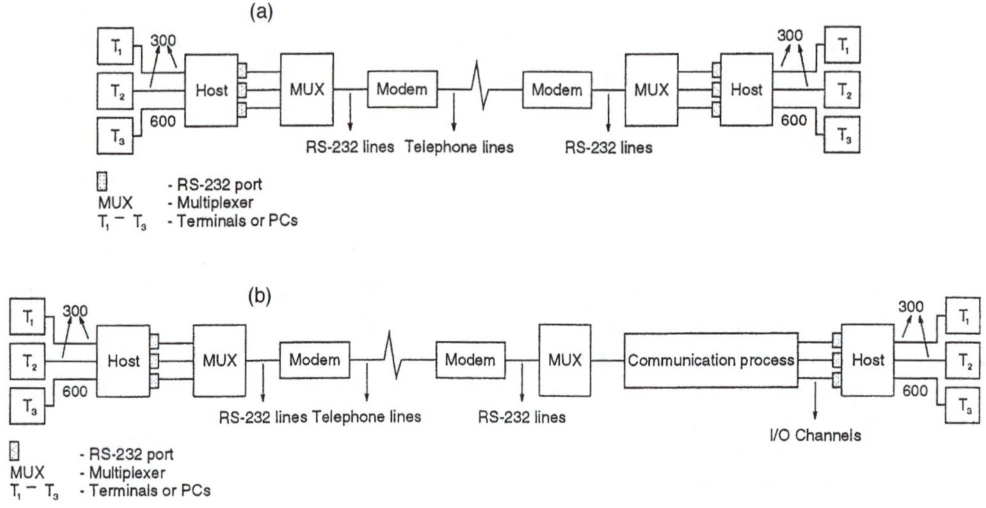

Figure 3.24 (a) Multiple lines via modem. (b) Multiple telephone line connection.

Multiplexers and **concentrators** (same as multiplexers with added storage and processing capabilities) are used to connect customer devices, time-sharing terminals, remote job-entry systems, C-based systems, etc., and are a part of a network interface. These can operate at a maximum speed equal to that of the high-speed lines connecting nodes, but they usually share high-speed lines with a number of other low-speed lines (circuits) and thus operate at considerably lower rates.

If we use a telephone line, then these low-speed terminals are multiplexed (as discussed above), but now the output of the multiplexer is connected to the line via modem, as shown in Figure 3.24(a).

In Figures 3.23 and 3.24(a), we showed that multiplexers are directly connected to mainframe host machines through communication lines. In general, these host machines are not connected to these lines but instead are connected via front-end processor communication controllers to handle low-speed communications, as shown in Figure 3.24(b). High speed I/O lines are used for connecting the communication controllers to the host machines. Communication processors have features such as buffering, serial-to-parallel transmission of data between line and host, parallel-to-serial transmission of data between host and line, etc., and provide low-level communication, thus allowing the load-sharing of host machines.

The multiplexing process can be divided into two main classes: frequency-division multiplexing (FDM) and time-division multiplexing (TDM).

3.5.2 *Frequency-division multiplexing (FDM)*

The FDM method is analogous to radio transmission where each station is assigned a frequency band. All the stations can be transmitted simultaneously. The receiver selects a particular frequency band to receive the signal from that station. FDM consists of mixers and filters. A mixer mixes two frequency components and is usually a nonlinear device. Due to nonlinearity, the output is proportional to the sum and difference of the frequency components. Each of the slots in FDM includes a predefined frequency component. The band pass filter is a very common device, which allows the upper side band to the mixer. It is based on an analog-signaling technique and finds its application in cable television,

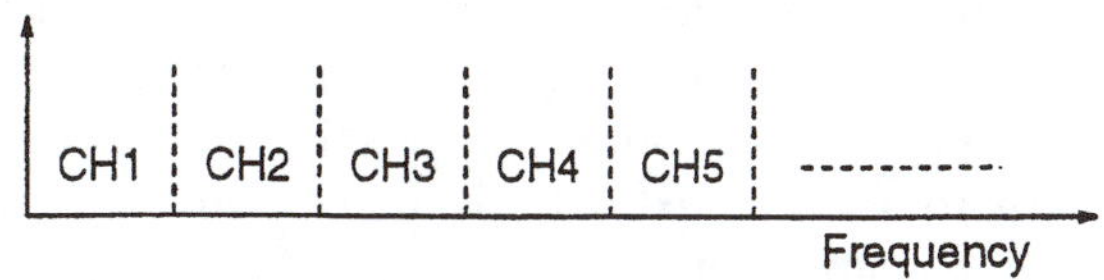

Figure 3.25 Frequency-division multiplexing (FDM).

which multiplexes a number of TV channels of 6 MHz onto one coaxial cable. A frequency-spectrum analyzer may be used to troubleshoot the FDM.

Various operations defined in this multiplexer are shown in Figure 3.25. Here the entire bandwidth (transmission frequency range or channels) is divided into smaller bandwidths (sub-channels) where each user has exclusive use of the sub-channel. Each sub-channel can be used for sending different signals such as data, voice, and signals from low-speed devices, and represents a lower frequency band. These sub-channels need not all terminate at the same time. Further, the terminals may be at a distance apart from each other.

The modulation process translates these sub-channels to their assigned band from their modulating signal-frequency values. Similarly on the receiving side, these sub-channels have to be converted back into the respective modulating signal-frequency values. FDM has been used in a variety of applications like AM radio broadcasting, telephone systems, CAble TeleVision (CATV), multi-drop networks, etc. In AM broadcasting systems, we have about a 1.2-MHz bandwidth for transmission which accommodates various commercial stations (in the range of 500 to 1400 KHz). Each commercial station (sub-channel) has been allocated a frequency band for its operations. Unused frequencies in bandwidth (guard-band) are used for providing non-interference between these sub-channels. Each channel can be used as a separate communication system of lower bandwidth. These sub-channels can further be divided into logical sub-channels: music and advertising. Both these logical sub-channels use the same frequency band but at different times in an alternate fashion.

In the case of multi-drop networks, the frequencies can be dropped off at a host (terminal or processor) and the remaining frequencies can be transmitted to more distantly located hosts. This type of mechanism of dividing sub-channels at two different times is based on **time-division multiplexing (TDM)** (discussed later in this chapter).

3.5.2.1 Guard-band carrier

Frequency bandwidth of sub-channels is very low and, due to the guard band for preventing electrical signals and noise interferences of a sub-channel with its neighbors, the actual bandwidth allocated to the user after FDM is further reduced. This is still enough to accommodate voice signals and signals from other low-speed devices.

Each sub-channel usually transmits data at a rate of 50–500 bps, while the typical transmission rate of a channel can go up to 4800 bps. A frequency modulation scheme is used to divide a channel into sub-channels. There are different types of channels (corresponding to different signals like voice, data signals, low-speed devices, etc.). In order to take advantage of the larger bandwidth of various communication media, e.g., twisted-pair, telephone trunk, coaxial cable, microwave, satellite, fiber optics, etc., we can group similar types of signals into one channel. A separate carrier is generated for each channel and is assigned a different frequency. The frequencies of these carriers are separated by a **guard band**.

The frequency band for a guard band is a significant portion of the total bandwidth of sub-channels. For example, voice-grade telephone sub-channels typically have only

3 KHz of usable bandwidth out of 4 KHz of allocated frequency band. This is because 1 KHz of frequency band is divided into two 500-Hz bands for each side of a sub-channel, and this is used as a guard band for reducing electrical interference between sub-channels and hence wasted. Filters are used to extract 3 KHz from an allocated 4 KHz frequency band. In spite of these guard bands, there is always going to be some interference/overlapping between two consecutive sub-channels. This overlapping can be seen in terms of spikes (at the end edge of one sub-channel onto the beginning edge of a neighboring sub-channel), which are non-thermal noise.

3.5.2.2 FDM standard

A popular and widespread standard for FDM is a frequency band of 60 to 108 KHz. It contains 12 400-Hz voice frequency channels (out of which only 3 KHz is usable with two guard bands of 500 Hz each on each side of the channel). These 12 voice frequency channels are multiplexed using FDM within a frequency band of 60–108 KHz. Each frequency band is called a *group*. A frequency band of 12–68 KHz (another group) is sometimes also used. Five groups (giving 60 voice channels) can be multiplexed to give a supergroup, and five supergroups, when multiplexed, give a *mastergroup*. CCITT has defined a mastergroup composed of five supergroups, while Bell System's mastergroup is composed of 10 supergroups.

Another standard defining a group of 230,000 voice channels is also defined and is being used. FDM can also be used to transmit voice and data simultaneously at the existing telephone network exchange within a building or organization. Filters will be required to separate an analog voice-grade band (300 Hz to 4 KHz) and also for data signals (500 KHz to 1.5 MHz). Crosstalk at higher bandwidth becomes a problem which can be reduced by another filter in this frequency range. The frequency range of 60 KHz to 108 KHz is being adopted in the U.S. for 12-channel voice frequency channels at the lowest level of its FDM hierarchy. It uses different modulation and multiplexing techniques at different levels in the FDM hierarchy.

FDM is useful in carrier systems where a number of voice-grade channels are multiplexed in various ways for transmitting over the communication link or transmission medium. This technique is popular for low-speed asynchronous terminals performing modem functions and also as an intermediate node in multi-dropped lines. The use of a guard band restricts the use of this technique to 2400 bps on leased lines for voice-grade transmission.

FDMs are generally used on private-line, dial-up applications. Another useful application of FDM is to create several low-speed, full-duplex sub-channels for teletype over a pair of lines for communicating signals over a wide geographic range. FDM modems combine both processes of multiplexing and modulation into one process. These modems have been used extensively in telephone networks, switched-data networks, dial-up networks, etc. One such application of an FDM modem is shown in Figure 3.26.

One of the major drawbacks with FDM is its inability for extension; i.e., once we have decided on the number of sub-channels on the link, it becomes very difficult to add any sub-channels, as this requires reassignment of all frequencies of other sub-channels. This is because the frequency band for FDM is defined and then the sub-channels are allocated with the frequency band which can accommodate the required data (e.g., a voice-grade channel has a fixed frequency band of 4 KHz).

3.5.3 Time-division multiplexing (TDM)

TDM allows users to share a high-speed line for sending the data from their low-speed keyboard computers. For each of the ports, a time slot and a particular device is allocated,

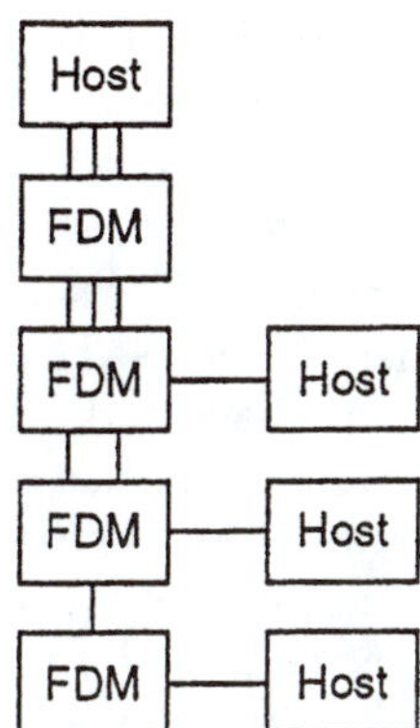

Figure 3.26 FDM modem.

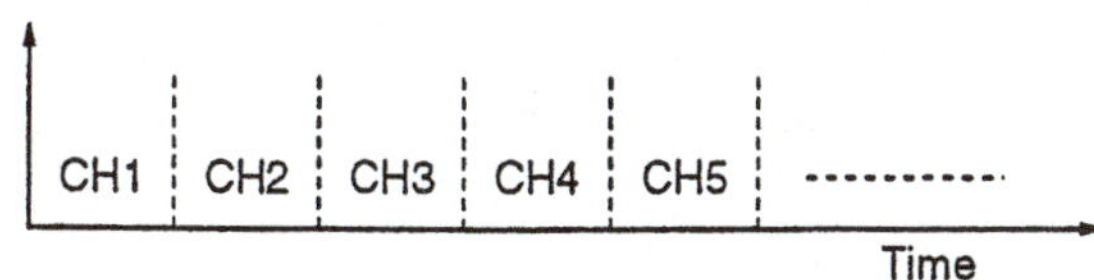

Figure 3.27 Time division multiplexing (TDM).

and each port when selected can provide bits, bytes, characters, or packets to the communication line from that device. At that time, the user can utilize the entire bandwidth and capacity of the channel. TDM scans all these ports in a circular fashion, picks up the signals, separates them out and, after interleaving into a frame, transmits them on the communication link, as depicted in Figure 3.27.

It is important to note that only digital signals are accepted, transmitted, and received in TDM. If a particular device does not provide any signal to its connected port, then during its allocated time slot, the TDM, while scanning, waits for that time interval at the time slot before it moves on to the next port. Thus, there is a wastage of channel capacity. The combined transmission data rate of these ports cannot be more than the link capacity.

TDM is better than FDM, as it allows the use of the entire bandwidth and the channel capacity. We can multiplex devices of different varying speeds (at ports) and hence more time can be assigned to the slots of higher speed than the slots assigned to slow speeds. TDM is generally more efficient than FDM because it has more sub-channels, and these can be derived. It has long been used with digital signals and may allow the sharing of a line by 32 low-speed terminals. It is simple here to add channels after the link has been established provided time slots are available.

TDM accepts data from both asynchronous terminals and synchronous devices. FDMs typically accept the data from asynchronous terminals. TDM is easier to implement and can easily multiplex the devices which are transmitting the data continuously, but some overhead is incurred with it. The assigned time slots must be defined to separate various data formats.

The TDM supports point-to-point and multi-point configurations. In the case of point-to-point configurations or networks, the TDM accepts data from a number of asynchronous hosts and delivers to a number of asynchronous terminals.

In the case of multi-point configuration (networks), the TDM accepts the data from an asynchronous host (going through a modem) and then transmits over a common telephone line. Different modems can receive the data from telephone lines via modems and deliver the data to various asynchronous devices.

3.5.3.1 Drop and insert TDM channel

The concept of **drop and insert** TDM channels has been applied where hosts are connected over TDM channels through intermediate hosts. The intermediate hosts typically gather data from either of the hosts and transfer back and forth between them. The configuration of this interconnection of hosts over TDM channels requires the use of TDM at each host or intermediate hosts. The data from any host is received by intermediate hosts as multiplexed data. The intermediate hosts simply combine the data from different ports of hosts located at two different locations. This combined multiplexed data is received by the host and has to be de-multiplexed there. This may increase the burden on one host.

An alternative configuration may add one more TDM at each intermediate host for each of the ports it is interested in. This data is de-multiplexed at the intermediate hosts and re-multiplexed and then transmitted to the host. The host thus does not need TDM for this data, as drop and insert operations have been performed at the intermediate hosts — hence the name **drop and insert TDM** channels. These channels are dropped at the hosts and then inserted at intermediate hosts.

3.5.3.2 Interleaving in TDM channel

As stated above, TDM can multiplex bits, bytes, characters, and packets. Based on this, there are two modes of operation in TDM. In the first mode, **bit multiplexing**, a bit from a channel is assigned to each port, which in turn represents a preassigned time slot. This mode is also termed **bit interleaving**. In the second mode, **byte multiplexing**, a longer time slot (corresponding to a preassigned port) is assigned to a channel. This mode is also known as **byte interleaving**.

Depending on the applications, these bit and byte (character) interleaving techniques of TDM are used. The modulation techniques which use one bit per sample (e.g., delta modulation) should use bit interleaving. Although we can use byte interleaving too, it will cause unnecessary delay (more number of bits for samples are added) and wastage of power for transmitting the samples (these do not carry any information). Synchronous devices commonly use bit interleaving. Byte interleaving will be useful in applications where we can represent the signals in eight-bit scheme (like ASCII-code scheme), e.g., digitized voice systems, terminals, computers, etc.

Communication between two TDMs is achieved by sending and receiving TDM frames. Each TDM frame usually contains fields like link header, link trailer control information, and a fixed slot (preassigned to a particular port and hence to a device). Link headers, trailer, and control (if it is available) can be grouped into one slot as a synchronization slot. This slot provides synchronization and also control to remote TDM so that the data can be delivered to the appropriate port of a different data rate, as these ports may not have the same data rate as that of transmitting ports. TDMs used as end user multiplexers find applications in carrier transmission systems. TDM modems combine the processes of multiplexing and modulation and have been used in telephone networks, switched networks, etc. One such connection for this modem is shown in Figure 3.28.

3.5.3.3 Cascading of TDM channels

In order to provide communications between remote ports on the host and various branch terminals located at a distance, we typically should require the number of lines for each terminal to be equal to the number of ports on the host. This number of lines can be reduced to one and still provide communication between them. TDMs can be cascaded.

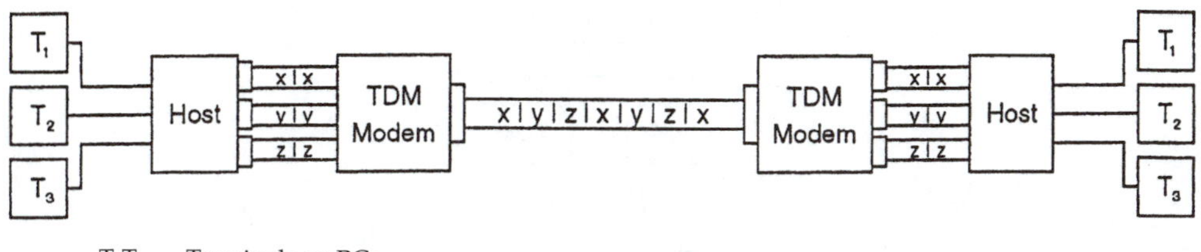

Figure 3.28 TDM modem.

This configuration will provide communication between the host and remote terminals on one pair of lines. It offers various functions like conversion of speed code and formats of the data information, polling, error control, routing, and many others, depending on a particular type of concentrator under consideration. Similar discussions may be found in References 1 and 7.

The cascading of TDMs can reduce the number of telephone lines. Consider a situation where a particular university has a large number of small and large campuses. The main campus (M) has a host which is assigned with different channels. Let's assume that there are two large campuses (L1 and L2) and there are three small campuses (s1, s2, and s3). We further assume that at each campus (small and large), there are various terminals connected to the main computer of the campus. If all the campuses have to be connected to the main campus host, then the use of TDM will be difficult, as TDMs supporting a large number of channels of different speeds may not be available. Further, in real-life situations, not all the campuses need to be directly connected to the main campus host. Instead, we can cluster them in two or three sites such that each site has a TDM which gets the data from its allotted campuses, multiplexes them, and transmits to another, larger campus site. This larger campus site receives the data messages via modems on the high-speed lines. The larger campus has its own channels of different speeds and passes on the data messages of smaller campuses to the host on its TDM via modem, as shown in Figure 3.29.

3.5.3.4 *TDM standards*

The TDM defines a time slot, and each time slot can carry digital signals from different sources, with the multiplexed TDM channel offering much higher data rates than individual channels. The digital signal is defined by sampling analog signals at the rate of 8000 samples per second and it offers an eight-bit sample once every 125 µs. The total data rate at the output of a multiplexer depends on the number of channels used. The number of channels defined for this type of multiplexer is different for different countries; e.g., North America and Japan define 24 voice channels, while European countries which comply with CCITT define 30 channels.

The total data rate of TDM with 24 channels will be given by 1.544 Mbps ($8000 \times 24 \times 8 + 8000$), while 30 channels will offer a data rate of 2.048 Mbps ($8000 \times 30 \times 8$). In order to define the boundary of the frame and call setup (signaling), a few control bits are also used. For frame boundary, North America uses one bit at the start of each frame (consisting of 24 channels) which alternates between 1 and 0 for consecutive frames. For signaling, the last bit of time slot 6 and 12 leaving seven user bits in these slots thus requiring an additional 8000 bits. For CCITT recommendation, the first time slot (zero number) is defined for frame synchronization. The 16 time slots carry the control information for signaling, giving 16×8 Kbps as control bits. These TDM circuits are also known as E1 links. Table 3.2 lists data rates defined for these two systems.

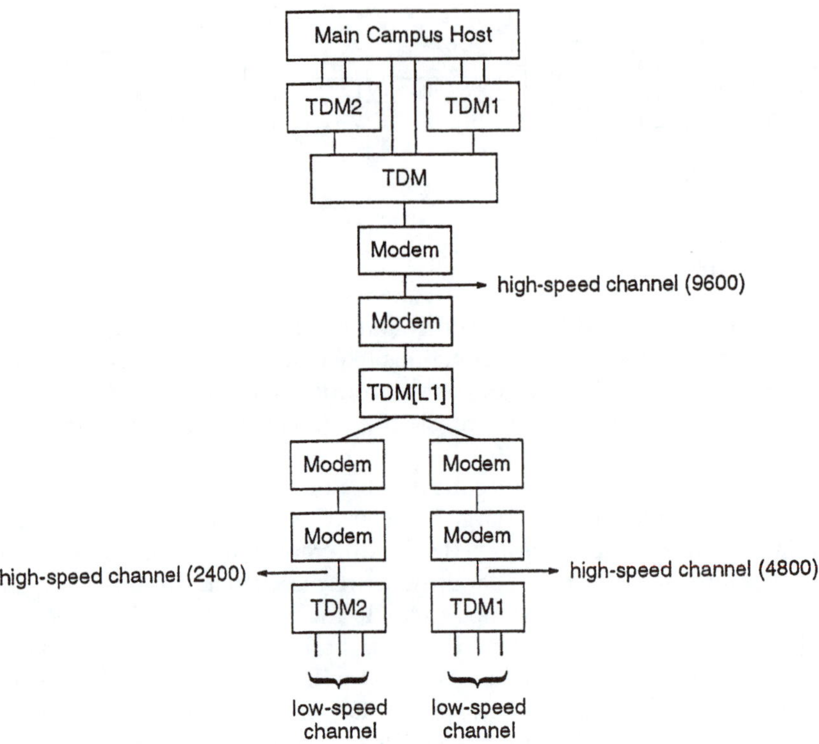

Figure 3.29 Cascaded TDM.

Table 3.2 Data Rates for North American and European Systems

Circuit name	Mbps	Voice/data channels
	North America	
DS1	1.544	24
DS1C	3.152	48
DS2	6.312	96
DS3	44.736	672
DS4	274.176	4032
	CCITT(Europe)	
E1	2.048	30
E2	8.448	120
E3	34.368	480
E4	139.264	1920
E5	565.148	7680

There are quite a number of local asynchronous multiplexers which offer different data rates (to channel) and composite rates (for all channels). The local multiplexers are connected by different transmission media and are supported by a number of ports of the host. Each of the ports is connected to a channel. These multiplexers carry the multiplexed data over the medium and deliver to various asynchronous and synchronous devices such as terminals, printers, PCs, etc. Each channel offers a data rate specified by the multiplexer's specifications, and the composite data rate is defined for all the channels together.

Table 3.3 Data Rates and Composite Data Rates for Various Media

Medium	Data rate of channel	Composite data rate
Twisted-pair cable	Up to 9600 bps	About 76.8 Kbps
Four-wire cable	Up to 9600 bps	About 830 Kbps
Two twisted pairs	Up to 9600 bps	About 1.544 Mbps
Dual fiber-optic cable	Up to 19.2 Kbps	About 14 Mbps (up to 5 km)

The asynchronous multiplexers provide connection between asynchronous devices over media with different data and composite data rates, as shown in Table 3.3. The data rates in the table are typical and do not represent a particular vendor. The composite data rates are based on the number of channels the multiplexer supports.

Asynchronous/synchronous wideband/voice multiplexers are also available which can carry integrated signals of data and voice, and the modems are equipped with voice cards. The asynchronous/synchronous multiplexers accept the data from asynchronous/synchronous hosts/terminals and also from PBXs. The integrated signals are transmitted and delivered to PBXs or asynchronous/synchronous hosts/terminals.

TDM is cheaper and requires more bandwidth than FDM. FDM is an analog method, while TDM is digitally based. Bandwidth is cheaper over wire transmission media and very expensive in wireless communication. The U.S. Federal Communications Commission (FCC) and the United Nations handle the frequency allocation to different channels. TDM uses digital signaling, and hence a digital analyzer may be used to troubleshoot TDM.

3.5.4 *Statistical time-division multiplexing (STDM)*

In conventional TDM, each port is assigned a time slot, and if there is no information on a particular port, the TDM waits for that preassigned time at that port before going to the next port. In other words, this technique scans these ports on the basis of static allocation of time slots. This causes a wastage of bandwidth and also waiting time, as shown in Figure 3.30. If we can allocate the time slots to the ports in a dynamic mode, i.e., if any particular port is empty, the scanning process will slip and go to the next port and so on, so that we can prevent the wastage of bandwidth and waiting time can also be reduced. This is the working principle of **statistical time-division multiplexing (STDM)**.

The control field of an STDM frame is used to identify the owners of the ports on the transmitter side. It is used on the receiving side to de-multiplex the data and then reassemble similar data signals of each channel and pass them on to appropriate ports. STDM uses the ports on the basis of variable-length-of-time slots (depending upon the information on the ports) as opposed to fixed-time slots in the case of TDM. Thus, STDM offers flexibility, better utilization of the line, and reduced waiting time for various ports. It is interesting to note here that the transmission data rate in STDM may exceed the transmission data rate of the line, provided all ports are not active at the same time, for a long time. This situation, in general, happens in the application of digitized voice, remote job entry (RJE), etc. This

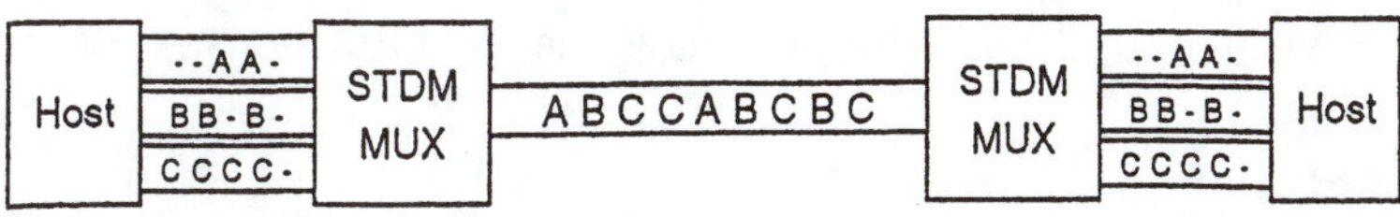

STDM — Statistical Time Division Multiplexing
MUX — Multiplexer

Figure 3.30 STDM multiplexer.

method is ideal for the situation where multiple terminals (specifically asynchronous) are accessing a remote host over lines. This is based on the fact that human operators cannot use the terminals at their full capacity (bps) at all times. The length of time slots depends on the length of data on that port. Synchronous devices can also be used on STDMs.

As discussed in the previous section, two TDMs communicate through a TDM frame, and the frames contain control and synchronization bits and also define the time slots. Similarly, STDMs use an STDM frame to convey the following control messages:

- Channel which is sending data through frames.
- Device which is transmitting the data and also the length of bits, bytes, etc., in it.
- Sequencing numbers, error-control, flow-control for making sure that error-free data (composed of frames) may be delivered to the destination computers/terminals. The flow control maintains the constant flow of frames in spite of the different speeds of transmission at transmitting and receiving sites.

STDMs provide a very powerful and flexible method of multiplexing the incoming data to utilize the full channel capacity using dynamic allocation of time slots. Based on this technique of multiplexing, vendors have built systems which are being used in applications like PBXs (private branch exchanges), message/packet-switching front-end processors, satellite communications, etc. The available STDMs typically support multiple channels (8, 16) over the same telephone line with a control configuration facility at each channel. These multiplexers offer data rates of 2400 and 9600 bps for switched and leased operations over both classes of transmission (synchronous and asynchronous). Some of the leading vendors include Universal Data Systems, ITRON, Micom Communications Corporation, and ARK Electronic Products.

STDMs are preferred over FDMs and also over TDMs for the following reasons:

1. STDMs provide extensive error-control techniques in the communications (very much the same as those provided by the data link layer of OSI model). These include error detection, error correction, flow control, and buffer management.
2. Some STDMs contain a built-in modulation interface which allow them to be connected directly into analog networks (without modems).
3. The adaptive method results in improved line utilization when traffic is bursty. For uniformly time-distributed traffic, ordinary TDM may be cheaper and may offer the same utilization as that of STDM. STDM will require more complex time-sharing devices along with additional overhead for keeping a record of source and destination addresses and the traffic units.

Sometimes, people call STDM a **"concentrator,"** as both terms represent the same function. A concentrator provides an interface between n input and m output lines such that the total data capacity of all n input lines should be equal to the total data capacity of m output lines. If input data capacity exceeds the output capacity, concentrators prevent some terminals from transmitting the data signals. The only difference between a concentrator and STDM is in the number of output lines. STDM has only one output line, while a concentrator has m output lines, but both have identical functions.

3.5.5 Concentrators

If we add the storage and processing capabilities to an ordinary multiplexer, it becomes a concentrator. It also includes software control which provides storage, queuing, and other statistical information to the hosts which allows the low-speed lines to share high-speed lines. In other words, we can define a hierarchy of different sets of lines of varying speeds which can be shared by various hosts. The data from low-speed lines can be multiplexed over available high-speed lines, which again can be multiplexed over still-higher-speed lines and so on.

The concentrators work on the principle of the store-and-forward concept, where large amounts of data can be stored and transmitted at later time. In this way, terminals which are not sending any data packets can be spared. On the other hand, the packets can be reassembled by concentrators and delivered to the destination stations when they are free. This principle finds a very useful application in low-speed synchronous terminals which do not have any storage buffers.

The concentrators can also be clustered into various groups such that any data will not automatically go to the host computer; instead, it will be decided by these clusters which one should get this packet. For smaller networks, this may not be a feasible and cost-effective approach, but in large networks with many terminals, it not only will reduce the response time but will also reduce the overall cost of the networks.

If we add processing capability to the concentrators along with storage buffering, the program residing in the concentrators can be executed and provide interface between the incoming data on low-speed lines onto data information of high-speed lines. It also provides an interface between terminals and computers and, as such, is responsible for converting the formats, code, and even speed of the incoming data to outgoing data on that particular high-speed line between the terminals and computers.

A variety of concentrators are available in the market and not all offer the functions described above. Offerings depend on a number of factors, such as characteristics of terminals, type of protocols, model of computer, the processing capabilities of concentrators, storage buffering, and many other parameters. In many cases the concentrators may even provide error control, switching, routing, etc. and, in general, they provide processing communications. They do not process the contents of the data but instead provide conversion between incoming and outgoing data of the concentrators so that incompatible terminals and computers can exchange the data between them. A multi-drop line is another way of improving the link utilization where all the terminals and communication devices are connected. Only one terminal/device can be active at any time in the network. A network employing this type of link/line is known as a multi-drop network.

There exists a variety of network concentrator systems for use in wide area networks (WAN). These systems offer point-to-point, multi-point, drop-and-insert, and multi-link hub network configurations. These devices typically include the following features:

- Diagnostics for high-speed links and channels
- Network status monitoring
- Performance and flexibility
- Error-free transmission using ARQ protocols, and CRC algorithm with inherent automatic retransmission of the packets in the event of error detection
- Flow control (eliminates any loss of data between device and asynchronous terminals)
- Support of multiple channels by a single network over wideband links

A device similar to a concentrator is a **front-end processor (FEP)**, which provides interface between computers and computer networks. The FEPs may have storage buffering or auxiliary storage which store the data packets which will be delivered to the hosts when they are available. One of the main functions of FEPs is to share the communication processing of the host. Unlike concentrators, they also provide conversion of codes, formats, and link characteristics between hosts and networks. The exchange of data generally takes place at data rates of 56 Kbps, while offering data rates of 9.6 Kbps (asynchronous transmission) and 19.2 Kbps (synchronous transmission).

3.5.6 Difference between multiplexers and concentrators

In TDM, a predefined time slot is allocated to each of the ports, which are connected to terminals or computers. Theoretically speaking, the speed of transmission on the input side of a multiplexer is equal to the capacity of the channel.

The main problem with TDM is the wastage of a slot if its port does not have data. In practice, we send what we call a "dummy character" for that slot. If we skip that slot, then we have to provide additional control bits to indicate to the receiving node the address of terminals, length of characters, etc. Generally, the synchronization between transmitting and receiving nodes is provided by the sequence of time slots. For very light digital information signals, TDM wastes bandwidth and channel capacity.

In order to send only data signals and avoid the transmission of dummy characters, the sum of the input speed of transmission of lines may be greater than the channel capacity. This utilizes the channel capacity fully but requires sending of two characters for each of the input data addresses of the terminals and the actual data. Concentrators using this method are sometimes called asynchronous time-division multiplexers (ATDMs) as opposed to STDMs.

Concentrators are better than TDMs or even STDMs if the amount of data on the input slots is moderate. But if all ports become active and start sending the data at maximum speed, then some of the data will be lost, and the channel capacity will not handle this situation. That is why concentrators always have enough storage (buffers) to handle these unexpected situations. The amount of storage may become a critical issue, since if it is too little then more data will be lost, while if it is too much then it is expensive. Thus, choosing enough storage for concentrators is a very important issue and should be considered by collecting statistical data over a period of time. Even when chosen carefully, a concentrator still may not give optimal performance.

3.5.7 Inverse multiplexing

In inverse multiplexing, low-speed lines are combined to give high speed and bandwidth, as depicted in Figure 3.31. This figure shows how inverse multiplexing can be used to increase the transmission data rate on full-duplex transmission. Consider the case in which there are two nodes transmitting data back and forth on a leased voice-grade line which provides a 9600-bps full-duplex transmission data rate. This means that each node is using a different pair of half-duplex lines for transmitting and receiving. If we want to send data at a higher rate than 9600 bps, say at 19,200 bps, and if the telephone line supporting that rate is not available, then we have to choose the next available line. If we are using Telco 8000 series lines, then after a 9600-bps data rate, the next available data rate in the 8000 series line is 56,000. If we use this line, then the total capacity of channels used will be 19,200 × 2 = 38,400 bps for each transmission, thus offering an unused bandwidth of 56,000 – 38,400 = 17,600 bps. It is economical to use two pairs of 9600 bps each for transmission

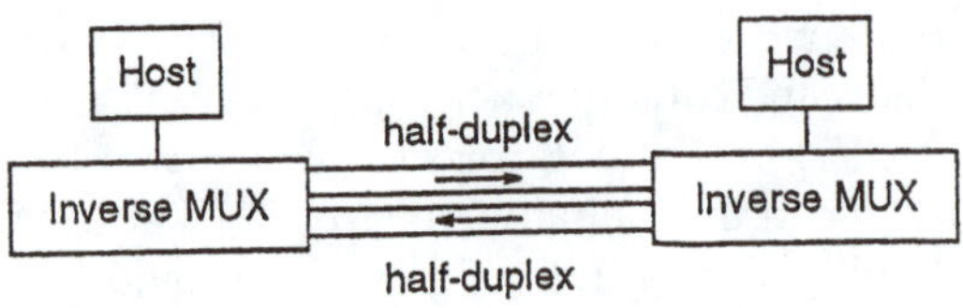

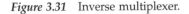

Figure 3.31 Inverse multiplexer.

and reception connected through an inverse multiplexer. The total bits per second of these two lines combined gives 19,600 bps in each direction.

If we compare the multiplexers with point-to-point configured networks, we may not feel any difference except that in the case of multiplexers, we notice a time delay. Sometimes, this delay may become significant, especially when we are considering the response time of the network. But, in general, it has been seen that the delay caused by multiplexers is not significant and may not cause a bottleneck in the overall performance of the networks. The multiplexers do not modify the data while it is being transmitted; they simply help in increasing the link utilization and also in decreasing the cost of the links in the network.

3.6 Switching techniques

A computer network may be defined either as a set of terminals connected to a computer or as an interconnection between computers. In general, computer networks include both computers, terminals, transmission links, hosts, etc. This interconnection between computers allows the computers to share the data and interact with each other. Sometimes the computer network thus defined is also known as a **shared network**. In shared networks (e.g., telephone networks, packet-switched networks, etc.), switching for routing the data packets performs the following functions:

1. The routing of information between source and destination across the network.
2. The link-utilization techniques which can transmit the traffic equally (depending on the capacity of the links) on all links.

These switching functions can be performed on interconnected computers by the following switching techniques:

- Circuit switching
- Message switching
- Packet switching

These three switching techniques allow users to achieve maximum utilization of resources by using suitable resource allocation strategies. The networks using these techniques are appropriately known as circuit-switched networks, message-switched networks, and packet-switched networks, respectively.

3.6.1 Circuit switching

In circuit switching, a complete circuit (route or path) between source and destination nodes is established before the data can be transmitted. The establishment of the entire circuit is obtained through many intermediate switching offices which provide logic connection between input and output lines of those intermediate offices. The circuit between source and destination can be established on any communication link/transmission medium (telephone lines, coaxial cable, satellite link, microwave link, etc.). The establishment of the circuit is based on the bits contained in the data frame. A line contention delay is a major problem in circuit-switched networks.

The following three steps are required to establish connections:

1. **Setup of link:** This detects a subscriber's request for service, identifies the terminal, searches and grabs a circuit, informs the required terminal, receives its response, and, after the data is transmitted, de-establishes the circuit (connection).

2. **Hold-up of link:** The established link is held during the transmission of data between source and destination and sends out the billing information (depending on the distance and duration of the connection) to the subscribers.
3. **Release of the link:** After the communication is completed, the link channels and shared devices are released.

As circuit-switching systems require a dedicated circuit before the data can be sent, this technique is inefficient and uneconomical. In the first phase, we have to send a signal all the way from the source to the destination to establish the circuit and, once the circuit is established, the data is sent from source to destination. For short distances, the propagation times of signals may be small, but for long distances, sometimes the total setup time (including propagation) may become significant. Thus, this technique is not useful for the applications which require minimum setup time (e.g., credit card identification, police recording and retrieving of social security number) but has found its application in very large Tymnet systems.

Circuit-switched networks transmit signals over different types of channels for long-distance communication links. It gives a feeling to the users that switching of the lines from more than one line is taking place for the establishment of the communication channel. The established channel is available to the users during the time when they are active. It has been decided that the total number of available channels should always be much less than the number of users.

It is very unlikely that all the users will use the telephone network at the same time. The network guarantees the existence of established paths and does not present any delay to the user once it has been set up. The users can send the message information at the rate defined by the capacity of the communication link. The only problem with this type of switching is that sometimes the setup time may become relatively longer. The circuit-switched network is useful for telephone networks and, looking at the behavior of call requests (traffic), the patience of the users, and many other features of the network, sometimes this relatively longer setup time may be negligible. The network is not useful for teleprocessing traffic applications, as these users always remain connected to the network and, as a result, waiting time before the transmission of any message information may become a serious problem.

3.6.2 *Message switching*

In message switching, an individual message is separately switched at each node along its route or path from source to destination. A circuit (or path) is not established exclusively for a message; instead, messages are sent using the store-and-forward approach. Each message is divided into blocks of data by users based on the capacity of the networks, and these blocks are transmitted in a sequence. The receiver, after receiving these blocks, constructs the original message from the blocks (by putting them in the same sequence as when it was transmitted) and sends an acknowledgment to the source.

The first block of the message contains control information regarding routing, quality of service parameters, etc. If the destination node does not accept the message (due to incomplete message, errors, or some other failure), the source node will transmit the same message again on the same link. The intermediate switching nodes (IMPs) will store the block of messages in the event of circuit failure, heavy traffic, broken links, loss of acknowledgment, etc. These intermediate nodes have secondary devices (disks) and direct access storage devices for storing these blocks of messages.

The switching nodes store the blocks of messages and look for the free link to another switching intermediate node. If it finds any free node, it sends one block at a time to that node until all the stored blocks are sent. If it does not find any free node, it will store the blocks and keep on trying to find free links until it finds one and transmits the block to it. The intermediate nodes are predefined between source and destination on the basis of certain criteria (e.g., shortest route, fastest route, heavy traffic, shareable files and programs, real-time data, etc.). If it does not find any intermediate node within the constraints, it will re-route the blocks on a different alternative route. Each block of messages is received as a complete frame and errors are checked in each block separately. If an error is found in any block, then that block will be re-transmitted. The transmission in this mode of switching is typically unidirectional (simplex operation). Autodin (U.S. Department of Defense) is a digital network that uses this switching technique for sending messages across networks.

The message-switching technique is less expensive than circuit switching and is widely used as a technique for interconnection. However, it does not support dynamic load balancing and, as such, heavy traffic on a particular node will be handled by itself. There is no limit on the size of the blocks of messages, which presents two obvious problems: (1) the switching node must have enough storage capability for storing the blocks, and (2) if a free link between source and destination is busy for a long time, then waiting and response time for the blocks to be transmitted may be significant. Thus, the technique seems to be useful for batched messages of longer duration but not suitable for interactive messages of shorter duration.

Once a route is established in a message-switched network, it does not present any delay to the transmission of the message except the transmission delay, propagation time, etc. Thus, the transfer time for the message depends on these parameters and it also may depend on the amount of traffic in the network. In other words, if the network is experiencing very heavy traffic, the nodes of the network will also have a large queue of traffic and, as such, waiting time may become another parameter for the transfer time of the message information. Thus, it can be said that the transfer time in the network (under such conditions) will always be greater than that of circuit-switched networks. This network allows the efficient utilization of a communication link. Each of the nodes connected in the network share the load on equal probability and hence the channels are utilized to their maximum capacity level.

3.6.3 Packet switching

In principle, packet switching is based on the concept of message switching with the following differences:

- Packets are parts of messages and include control bits (for detecting transmission errors).
- Networks break the messages into blocks (or packets), while in message switching, this is performed by the users.
- Due to very small storage time of packets in the waiting queue at any node, users experience bidirectional transmission of the packets in real time.

In message switching, a message is divided into blocks and the first block contains the control information (routing across the switching nodes of the network). Since there is no limit on the size of the block, blocks can be of any variable size, and this requires the switching node to have enough buffer to store these blocks. Further, it reduces the response time and throughput of the networks, as the access of these blocks from secondary devices requires significant time.

In packet switching, the message is divided into blocks (or packets) of fixed size and, further, each packet has its own control information regarding the routing, etc., across the network. This means the packets may follow different routes across the networks between source and destination and that they also may arrive at the destination in a different sequence. The receiver, after receiving the packets out of sequence, has to arrange the packets in the same order as they were transmitted from the source.

Due to fixed or limited size of packets and also due to different routing of the packets, the switching nodes store the packets in the main memory as opposed to a secondary storage device (disk) in message switching. The time to access or fetch a packet from main memory takes much less time than from disks and, further, the packets can be forwarded from any switching node to any other switching node before the complete arrival of the packet there. These two features together reduce the delay and, as such, the throughput (the number of packets transmitted per unit of time) is also improved.

In packet-switched networks, if any node receives a garbled packet, it will request the sending node to transmit the same packet again. When the destination node receives the last packet of the message, it sends an acknowledgment (defined as a typical bit pattern packet message) to the source node. Similarly, if the destination node does not receive all the packets within the specified time (after request for the connection has been confirmed), it sends a request for the data (instead of acknowledgment) to the source node about the missing packets. After the receiving node receives the missing packet, it does not send the acknowledgment.

The packet-switching technique allows the switching nodes to transmit the packets without waiting for a complete message and also allows them to adjust the traffic they have, thus minimizing the resource requirements of the nodes. If any particular node is already heavily loaded, it will reject the packets until its load becomes moderate.

Packet switching can, in fact, be considered as a mode of operation rather than switching, as it allows optional use of bit rate by transmitting packets between source(s) and requested destination(s) in a dynamic multiplexed mode, hence utilizing the link capacity effectively. This technique may pose a problem if it is used in digital telephone networks, as packets will be made out of PCM-coded samples. Further, these packets reach the destination with unpredictable delay.

Packet-switched networks basically work on the same principle as that of message-switched networks, except that here the message information is broken into packets which contain a header field along with the message information. The header field includes a packet descriptor, source and destination addresses, and other control information. The packets can be routed on different routes independently. Due to the fixed size of the packets (compared to any size of the message block in the case of message-switched networks), there is an improvement on the performance of the network, and also the utilization of the communication link is improved. This network can be used in two different modes of operation: **datagram** and **virtual circuit**. In the former, the users can transmit packets of fixed size back and forth independently, while in the later mode, the network manager performs the operations of segmenting, reassembling, and maintaining the path/route until the packets are delivered to the destination.

Advantages and disadvantages of packet switching: The packet-switched network offers the following advantages over circuit switching (also message switching):

1. The link utilization is improved, since the link can be shared by a number of different packets. Each node has its own queue storing those packets which are going to use the node. After the communication link is established, the packets (of different or the same data messages) may be transmitted during the established connection.
2. In the case of packet-switched networks, stations with different data rates can communicate with each other, and the necessary conversion between different data

rates is done by the network, while in the case of circuit-switched networks, both stations must have the same data rate.

3. In circuit-switched networks, the packets may be lost, as the network will not accept them in the event of a busy network, while the packets will be accepted in the case of a packet-switched network but there may be some delay in their delivery.
4. The delay in the transmission of packets at any node may become a problem in the case of packet-switched networks, while in the case of circuit-switched networks, a dedicated circuit has been established and hence there is no delay in the transmission.
5. Priorities can be assigned to the switching nodes in the packet-switched networks such that nodes with higher priorities will transmit the packets waiting in their respective queues before the lower-priority nodes. In the case of circuit-switched networks, there is no concept of priority.
6. Circuit-switched networks generate the ordered delivery of the packets, while packet-switched networks do not give any guarantee for the ordered delivery of the packets.

3.6.4 Comparison between switching techniques

Circuit switching establishes a dedicated circuit/route before a message can be transmitted. Once the switching node gets a circuit, the entire bandwidth of the communication circuit is used by the user. If there is any unused bandwidth on the established circuit, it will be wasted. In packet switching, the circuit and bandwidth are allocated only when they are required, and after the use, they are released.

In circuit switching, messages are transmitted and received between source and destination in the same sequence (no recording of blocks or packets). No other messages can be transmitted on this circuit, and a sudden rise in the number of messages on any switching node can be handled easily without losing them. In packet switching, packets are received from different circuits in different sequences and these packets have to be arranged in the same sequence to rebuild the original message. The switching nodes have limited storage capability, and hence in the event of a sudden rise in the number of packets, the packets may be lost.

Circuit switching is basically defined for plain old telephone service (POTS), while packet switching is being used for data communication over interconnected terminals, PCs, and other nodes. The circuit-switching technique has been modified to derivatives such as multicare circuit switching, fast circuit switching, and finally ATM for a fixed packet size. On the other hand, packet switching has been refined into derivatives for use in frame relay, SDMS, and ATM.

In circuit switching, a dedicated circuit is established in advance and hence various parameters such as data format conversion, code exchange, speed incompatibility, etc., also have to be decided in advance. In packet switching, switching nodes provide these parameters depending on the traffic, available protocols, routing tables, etc., in a dynamic mode. Further, in packet switching, error control strategies may be used to detect errors in the packets. In circuit switching, no error-control strategies are used.

In circuit switching, the customers are charged on the basis of time (for a dedicated circuit) and distance (between source and destination), and this is independent of the number of messages (to some extent, it is similar to a long-distance telephone call!). Here, the time does include the setup time. In packet switching, customers are charged on the basis of the number of bytes (in the packets) and connect time (it does not include setup time, as there is no setup time in this technique) and it is independent of distance (like e-mail, facsimile, etc.).

The following is a brief summary of various parameters in different switching techniques:

- Delay:
  Message-switching (second to minutes)
  Packet-switching (microseconds to second)
  Circuit-switching (few microseconds)
- Memory requirement en route:
  Message-switching (high)
  Packet-switching (low)
  Circuit-switching (none)
- Applications:
  Message-switching (modern telex)
  Packet-switching (telematics)

3.7 *Communication control interfaces*

For each type of transmission (asynchronous and synchronous), hardware products in the form of integrated chips (IC) are available which perform appropriate functions. For example, the IC for asynchronous transmission generates clocking (bit), character synchronization, and cyclic redundancy check polynomial (Motorola has defined an IC which can offer different types of standard polynomial generators (CCITT)). For synchronous transmission, ICs are also available which perform character synchronization, generation of parity codes for error detection, and transmission of character-oriented messages. Other ICs are available which offer functions such as stuffing, checking of functions for character-oriented and bit-oriented messages, etc. Encoding and decoding at different data rate transmissions have also been implemented in ICs, and these chips are commercially available. The majority of these chips are available from different vendors like Motorola,[6] Intel, etc. Most of these chips are programmable in the sense that they offer different types of operations like transmission configuration, polynomial generators, parity or other error detection, etc. For each of these options, the users have to select via selector switches. For this reason, these chips are known as universal transmission control interface circuits. The following is a partial list of such chips, which are used for communication interfaces during data communication over PSTNs, PSDNs, and high-speed networks.

- **Universal asynchronous receiver and transmitter (UART):** UARTs offer clock (bit) synchronization, character synchronization, generation of parity, and error checking for each character. They support asynchronous transmission and convert parallel transmission from a computer/terminal into serial transmission over telephone lines. This IC chip is discussed in detail in Section 3.7.1.
- **Universal synchronous receiver and transmitter (USRT):** A USRT supports character synchronization and generates bit rate clock (bit) synchronization, parity and error checking of each character, idle character synchronization, etc.
- **Bit-oriented protocol (BOP):** BOP is mainly used for stuffing. It offers insertion and deletion of bits (flag), zero bit insertion and deletion, generation of CRC and idle bit pattern, etc.

In addition to these chips, there are a few other IC chips available such as the universal synchronous/asynchronous receiver and transmitter (USART), universal communication interface control circuits, etc. These chips can be programmed as one of the above chips.

For example, USART can be programmed to operate as either a UART or USRT and offers a variety of other features.

3.7.1 Universal asynchronous receiver and transmitter (UART)

The data transmission between two DTEs usually takes place in a serial mode, while the CPU within the DTE processes the data in a parallel mode in a power of one byte (eight bits). The peripheral interface adapter (PIA) is a device used to provide a parallel transfer of data between an I/O device and microprocessor. The transfer of data takes place over an eight-bit bus through eight parallel peripheral lines at a time controlled by the same clock. This device is basically useful for a short distance, e.g., keyboard and computer. This device will be expensive for a longer distance due to cabling, maintenance, etc. For transferring the data over a longer distance, the approach described below is always preferred. In this approach, the parallel data information is converted into a serial stream of data which can be transmitted over a single line. We have seen that the majority of data and voice communication takes place over telephone networks; this serial stream of data (digital form) can be transmitted over telephone lines by first converting parallel data (from computer) into a serial form by using a **universal asynchronous receiver and transmitter (UART)** device. This serial data stream (containing different DC levels for logics 1 and 0) is converted to analog signal using audio signals. These analog signals are transmitted over telephone lines. The device which converts data into analog signals is known as a modulator/demodulator (modem).

When digital signals are converted into AC signals, these signals can be transmitted over the telephone lines over a longer distance. The frequency range for these signals is usually between 300 Hz to 3.4 KHz. This bandwidth is considered a practical bandwidth for telephone networks. The higher value of frequency represents logic 1, while the lower value of frequency represents logic 0. The FSK modem contains a modulating circuit based on a voltage-controlled oscillator (VCO) whose output frequency depends on the voltage at its input. The first DC level represents logic 1, while the second level represents logic 0. On the receiving side, the signals are received by modem, which converts them back into DC levels. The demodulation based on this is obtained by a circuit known as a phase lock loop (PLL). The PLL generates an error voltage at its output whenever the frequency or phase of an incoming signal is different from the reference signal value. The error voltage is used to establish a digital data level of 1 and 0. In the case of an FSK modem, the reference signal is the center or carrier frequency, which is usually defined by a value between the frequency values of 1 and 0.

The UART converts the parallel data of a terminal or PC into serial data which will be used by a modem to translate into analog modulated signals and be transmitted over the telephone lines. The parallel data (from parallel bus) from terminals or PCs are first loaded at the input of UART. In UART, the data are received by shift register which checks for start, stop, and parity bits associated with each character coming out from terminals (asynchronous transmission). After receiving these bits, the shift register inserts them into their correct locations. These characters are then shifted out along a serial data line or bus. After the first character has been shifted, the register takes the next character and repeats the process until the last character of data is transmitted, as shown in a generic and typical configuration given in Figure 3.32.

On the receiving side, the serial data are accepted only after the receiver detects a start bit. These characters are received by the shift register, which checks the parity bits one by one. If the parity bits are correct, it will separate the character bits from the data stream and store them in a buffer. The computer then fetches the parallel data output from the buffer register as the next character is shifted into the shift register. The CPU accepts the

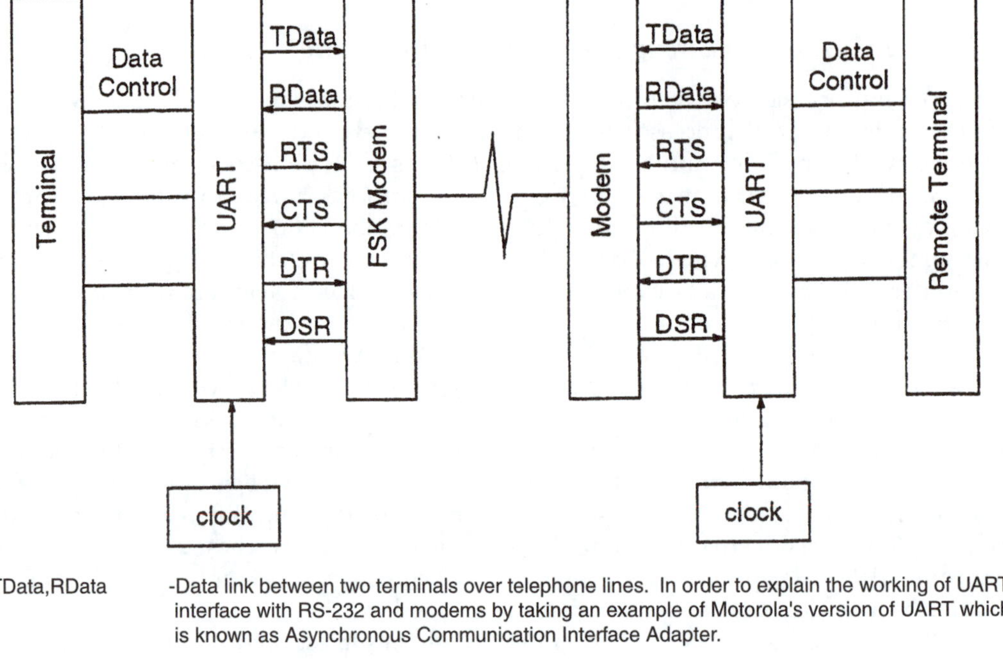

TData,RData -Data link between two terminals over telephone lines. In order to explain the working of UART interface with RS-232 and modems by taking an example of Motorola's version of UART which is known as Asynchronous Communication Interface Adapter.

D_0 - D_7 -Standard 6800/68000 family data lines that carries the data between up and ACIA

CS_0, $\overline{CS}_1$, CS_2 -ACIA's chip selector used for address decoding

RS (Register Select) & R/W (Read/Write) -These two control signals continue to select which of four registers is accessed when ACIA is addressed. These are normally connected to lower address and/or $\overline{\text{VDS}}$ or $\overline{\text{LDS}}$

Figure 3.32 Typical UART configuration.

serial data from the DTE and after processing sends it back to the DTE. There must be a device between the DTE and CPU which will convert incoming serial data into parallel (for processing by the CPU) and then after processing convert it back into serial mode for its transmission to the DTE. The interface between the DTE and CPU is a special integrated circuit (IC) chip called the universal asynchronous receiver and transmitter (UART). The device is programmable and can be configured accordingly by the designers or users. The programming of a UART requires certain controlling data signals which can be specified by the users, and the device will perform various functions such as conversion of incoming serial data into parallel, detecting parity and framing errors, and many other features.

Consider Figure 3.33, where two DTEs are connected to each other's RS-232 interface. Each computer (DTE) has a serial I/O interface card which is connected to the bus. The UART (placed within the serial input/output [SIO] board) is connected to an RS-232 interface port. The UART receives a byte from a CPU in parallel mode via the system bus. The CPU performs read and write operations on UART in exactly the same way as it does on memory. For example, when the CPU wants to write (i.e., transmit) a byte to UART, it will perform a write operation on UART and one byte (eight bits on a PC) at a time will be transmitted to UART in a parallel mode. The UART will convert the mode of operation by sending the bits one by one (depending on the data rate) back to the CPU (in the event of response) or to the serial data link between two DTEs.

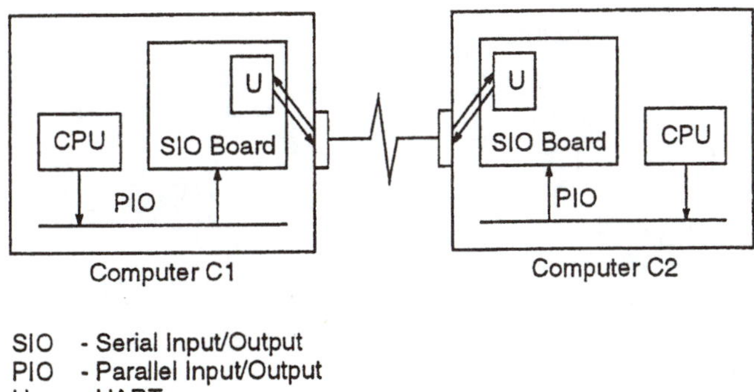

Figure 3.33 SIO board.

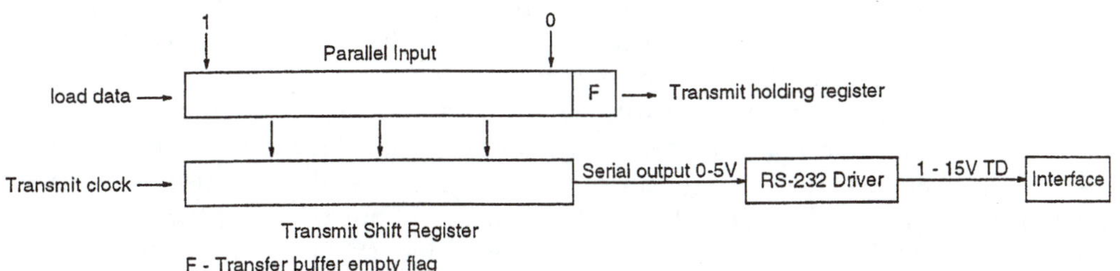

Figure 3.34 Transmitter section of UART.

3.7.2 *Transmission mechanism in UART*

The UART defines two registers as *transmit holding* and *transmit shift*. The CPU loads a character (to be transmitted) in parallel mode onto a holding register. The characters can be distinguished by inserting start bit "0" at the beginning of the character followed by a stop bit "1" at the end of the character within the holding register, as shown in Figure 3.34. These bits are transferred to the shift register whenever it is available, after which the bits are transmitted serially (at the selected data rate) to the transmit data (TD) of the RS-232 interface. The main reason for providing two registers is to allow the transmission at full speed. The UART chip is a TTL logic chip and, as such, describes the signal waveform in the voltage range of 0–5 V while RS-232 describes the data and control signals in the voltage range of +15 to –15 V). In order to match these different voltages, the SIO board offers this conversion between the outputs of UART and RS-232 drivers (also known as EIA drivers). The EIA driver produces the RS-232 voltage, which is then fed to the RS port.

The UART uses a STATUS bit (also known as transmitter buffer empty, TBE), which indicates the availability of the holding register. When this bit or flag is ON, it accepts the characters from the CPU and, after receiving the characters, it is set to OFF. After the transmission of the character, it is again set to ON. The UART can either be polled by the CPU or it can be programmed to generate an interrupt after a TBE is set. We have to be careful while programming UART that we do not try to load the holding register if it is not set; otherwise, it will overwrite the contents of the holding register and the previous contents will vanish. In general, most of the UARTs do not show this error, known as overrun.

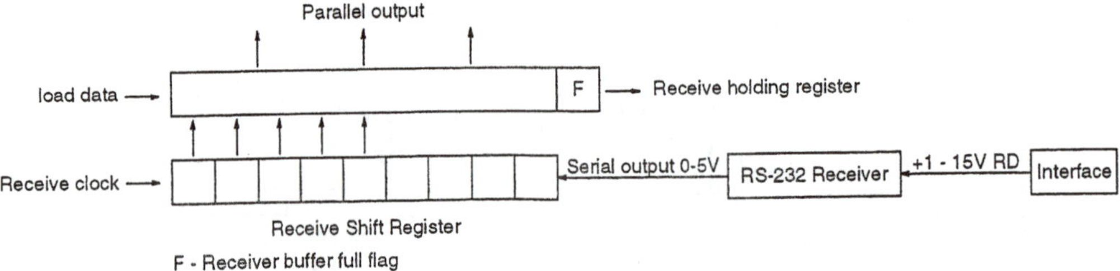

Figure 3.35 Receiver section of UART.

3.7.3 *Receiving mechanism in UART*

In this mode, the receive data (RD) pin of the RS-232 port is connected to an RS-232 receiver (EIA driver), which now converts the +15 to –15 V range into the 0–5 V suitable to TTL logic (UART). This input is sent to a receiver shift register, which is driven by a receiver clock, as shown in Figure 3.35. The register will accept the data signal after detecting a start bit and start shifting the bits until it detects a stop bit. As soon as it detects a stop bit, it sets the **receiver buffer full (RBF)** flag to the ON state. Again, this can be either polled (through software) or programmed via interrupt signal generation. After the CPU has read the characters, the state of RBF is reset to the OFF state. Here also, if we have a situation where the reading operation is performed without checking the status of the RBF flag, the data signals either may be read inaccurately or may be unpredictable.

3.7.4 *Error correction in UART*

As mentioned above, the UART also detects the errors in parity, framing, and data overrun. In the case of a **parity bit (PB)** error, this bit is set in the event of any mismatching of parity bit. In the case of **framing (FE)** error, if a stop bit is 0 the UART detects a framing error, as it should always be set to 1. This error is mainly caused by a mismatch between the speeds of the transmitter and receiver. In the case of a **data overrun (OV)** error, the previous data contents of the holding register will be lost if the CPU did not remove it from the register and another character arrives at the register. The status of the register, such as holding and shifting, is defined in terms of flags or status bits, usually stored in another register known as a line status register.

3.7.5 *Timing signaling in UART*

The UART circuit includes a data sampling click whose speed is in general very large (typically 16–18 times the data rate of an RD line). The line is sampled at a faster rate in order to detect a 1-to-0 transition in the event of receiving a start bit, as shown in Figure 3.36. As soon as the circuit detects a 1-to-0 transition, it assumes that it is a start

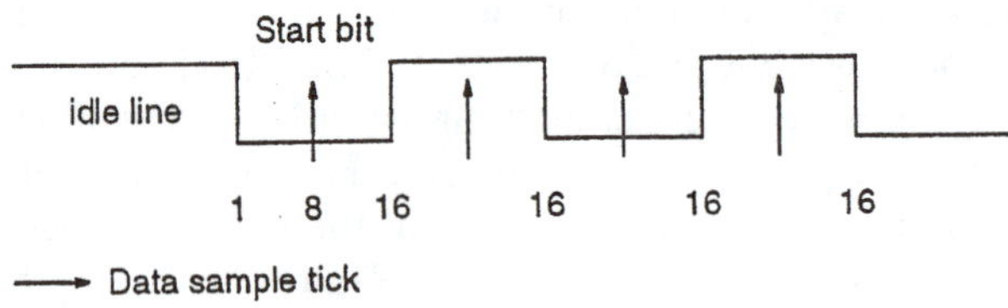

Figure 3.36 Start bit detection and data bit sampling.

bit and starts counting the next eight clocks. After each eight clock, it samples the bit and if the bit is 0, it assumes that it is a valid start bit; otherwise, if it gets 1, it assumes that the first transition was due to some noise on the line and it will not take any action. This event of detecting the noise and taking action is known as *spike detection*. The UART enables a counter which divides the clock by 16 and generates a sampling clock tick after every 16th tick of the original clock. In order to reduce the error in binary representation, usually the tick is adjusted at the middle of the bit and the data bits are counted onward from there. As soon as the UART detects a stop bit, it goes back to the state of waiting for a start bit.

The programming of the UART has to be carried out carefully, as otherwise — due to mismatching between transmitter and receiver clocks and sampling at a rate different from the speed in bits per second — the data signals will be sampled at different sampling clocks and hence the data signals will be corrupted. For example, consider the case where the receiver is mistakenly set to twice the speed (bps) of the transmitter side's speed. As usual, the 1-to-0 transition is detected for the start bit and the UART waits for another eight clocks and then samples the start bit to check its validity. If it is a valid bit, it will generate a bit-sampling clock which will now be sampling at twice the speed (at the receiving side), so that the first sampling takes place within the start bit and thereafter the sampling of each data signal is taking place twice.

If the UART is being used for, say, seven-bit ASCII, then UART samples the seven bits as 0110011 with 0 as parity bit and 0 as stop bit. If the parity is even, then a framing error flag is set, while if the parity is odd, it will set both parity error and framing error flags.

In order to explain the working principle of UART, we consider in Table 3.4 a Motorola UART chip 6850 known as an asynchronous communication interface adapter (ADIA). The pin connections with RS-232 and modems are shown in Figure 3.37. For more details, please refer to Reference 6.

The most common UART used in IBM PCs is the 8250 IC. The interface between the I/O bus of this chip and the RS-232 interface is shown in Figure 3.38.

Table 3.4 Working Principle of UART

Control signals/pins	Functions
D0–D7 (15–22)	Standard 6800/68,000 family data bus to carry the data between microprocessor and ASCI.
CS0, CS1, CS2 (8–10)	ACIA's chip selector used for address decoding.
RS (register select) (11) and R/W (read/write) (13)	These control signals combined provide the selection for a register (out of four registers) to be selected for accessing when ACIA is addressed. They are normally connected to a lower address bus and/or UDS or LDS.
Enable (E) (14)	Used by ACIA for internal timing. Connected to E output clock.
IRQ (interrupt request) (7)	Sends an interrupt request to 68,000 interrupt priority logic when ACIA interrupts are enabled.
TxData and RxData (6, 2)	Actual RS-232C designated serial data lines that handle data transfer between ACIA and modems.
TxClock and RxClock (4, 3)	sets up the bit rate of transmitted data.
Data link control leads	Control leads based on the RS-232C standard.
RST (5)	Generated by computer and sent to the local modem.
CTS (24)	Modem, after a short delay, sends the reply back to its computer.
DCD (23)	The last in a sequence of handshaking to establish data communication over telephone lines.

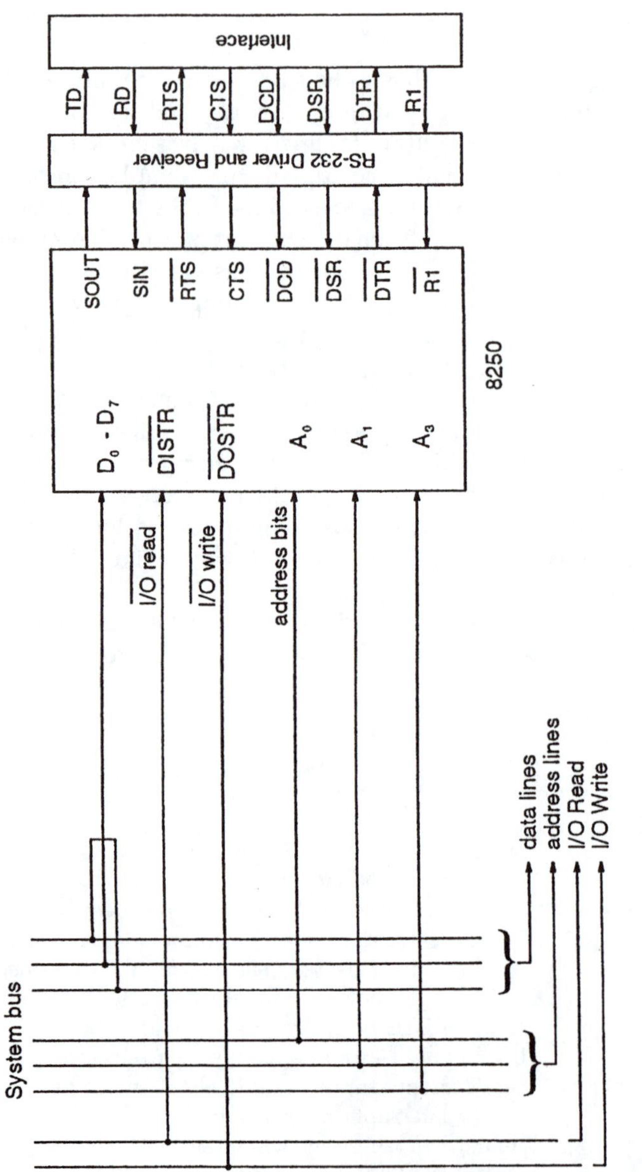

Figure 3.37 8250's I/O bus and RS-232 interface.

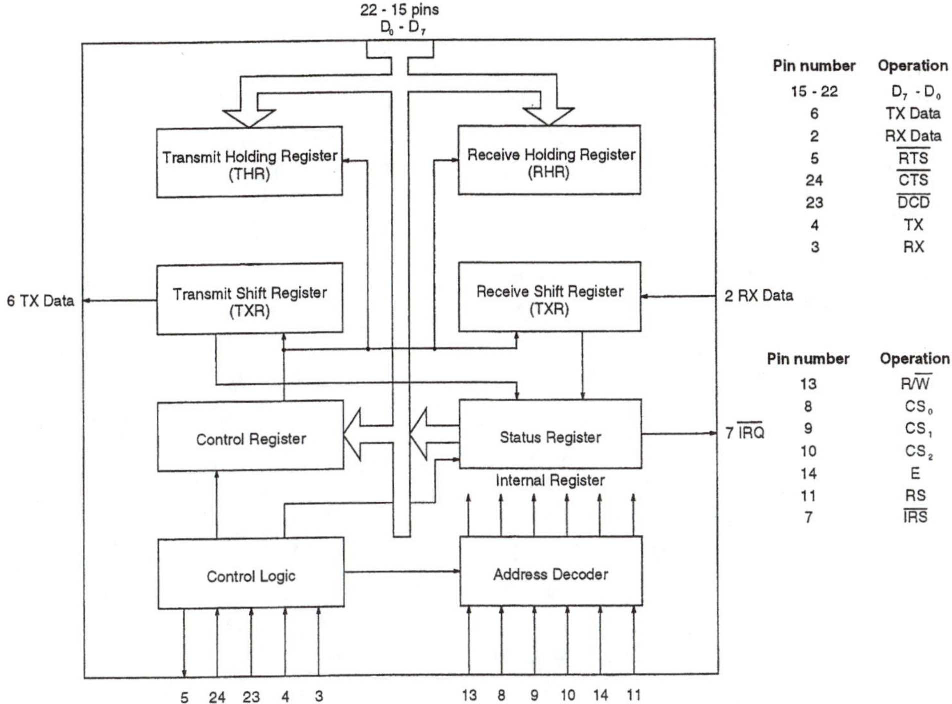

Figure 3.38 6850 ACIA block diagram.

Some useful Web sites

Information on network electronic components, modulators, amplifiers, etc., can be found at *http://www.lashen.com*; on modulators and related vendors at *http://www.eecs.unm.edu*; on cable modem suppliers at *http://www.datacomnews.com*; on modems, wireless LAN products, HomeLAN, broadband access cables, ADSL, analog modems, etc., at *http://www.zoomtel.com*; on modems at *http://www.modems.com* and *http://www.56k.com*; and on concentrators at *http://www.his.web.cern.ch*.

References

1. Techo, R., *Data Communications: An Introduction to Concepts and Design*, Plenum Press, 1984.
2. Loomis, M.E.S., *Data Communication*, Prentice Hall, 1983.
3. Technical reference, *Black Box Catalogue*, Pittsburgh, USA.
4. Technical reference, *ARK Electronic Products*, Inc., USA.
5. Technical reference, *Universal Data Systems*, Inc., USA.
6. Miller, M.A., *The 68000 Microprocessor Family: Architecture, Programming and Applications*, Merrill, an imprint of Macmillan Publishing Company, 1992.
7. Black, U., *Data Networks Concepts, Theory and Practice*, Prentice Hall, 1989.

chapter four

Modes of communication channel

"Recovering from failure is often easier than building from success."

Michael D. Eisner

4.1 Introduction

A communication channel defined over transmission media or links carries the data signal (analog or digital). The communication channel is a very important component within a communication system, as it defines a path over a communication link or a combination of links for carrying the information from one node to another. In the previous chapters, we briefly characterized a channel by a number of parameters such as different forms of representing various signals, modes of operations of a channel, channel capacity, speed of transmission, different types of transmission, etc.

In this chapter we discuss some of these issues in more detail. We also discuss the representation of the symbols and data by a scheme known as *coding*. Different types of symbols and characters used in graphic display and different types of terminals are discussed. Data communication between terminals connected by a physical link involves a number of issues, such as communication control procedures, protocols, error control, error detection and corrections, error recovery, handshaking, line control, data-link control, etc.[1-7]

4.2 Communication channel

One of the main objectives of data communication is to provide the transfer of message information between two communication devices connected by a communication link (transmission medium). A path is defined between transmitting and receiving nodes over the link and the signals (analog or digital) travel over it. The communication devices could be computers, terminals, printers, processors, etc. The message information is carried between the devices by electrical signals. The signals are generated by the transmitter at the sending side and are received by a receiver on the receiving side. The two communication devices are connected by transmission media (telephone lines, coaxial cables, satellite link, microwave link, etc.) which define a communication channel on different types of links. Obviously, we expect the communication channel to offer an error-free path for the transmission of signals, but the fact is that it offers a number of distortions to the signal which affect the capacity of the channels (expressed as the number of bits per second) and also the throughput of the systems.

The communication channel offers a variety of services on different systems. These services are available on private networks (usually owned by private organizations or corporations and governed in the U.S. by the Federal Communications Commission, FCC), private leased networks (usually known as common carriers to provide leased communication services and governed by the FCC), and public switched networks (known as communication service providers for communication services). Various aspects of communication channels are discussed below.

4.2.1 *Channel modes of operation*

The communication channel is defined over a transmission medium or combination of transmission media and offers a path between the nodes. The signals (analog or digital) travel over this path. The communication channel may be used as a service in data communication and can be established over different media within the communication systems. In some books, the transmission media are also known as *communication links*.

There are three modes of operation which provide communication between two nodes on different channel configurations, namely, simplex, half-duplex, and full-duplex. In other words, these three communication channels are defined for the transmission of signals between nodes.

Simplex: In this channel mode of transmission, the signals flow in one direction from node X to Y, and there cannot be any transmission or reception from Y to X (Figure 4.1(a)). A separate transmission channel has to be used to transmit signals from Y to X. This mode of transmission is useful in commercial transmission applications such as radio broadcasting, television, video displays of airline arrivals and departures at airports, sending of information from a terminal to a main office, etc., and generally is not considered a good transmission mode in data communication applications. Node X could be a computer or terminal, while node Y could be an output device such as a printer, monitor, etc. The advantages of this channel mode include simplicity (including software), inexpensiveness, and easy installation. The simplex mode has restricted applications, as it allows transmission or reception in only one direction. A typical simplex channel mode (over a telephone line) can be defined over a simplex line, as shown in Figure 4.1(b). In this mode, the entire bandwidth is available for transmission of data.

Half-duplex: In this channel mode of transmission, signals flow from X to Y in one direction at a time (Figure 4.2(a)). After Y has received the signals, it can send signals back to X at another time by switching its state from receiving to transmitting (when X is not sending to Y). In other words, signals flow in both directions between X and Y in such a way that signals flow in only one direction at any time. Therefore, there will be only one transmission on a transmission medium at any time. Both the nodes have both transmitting

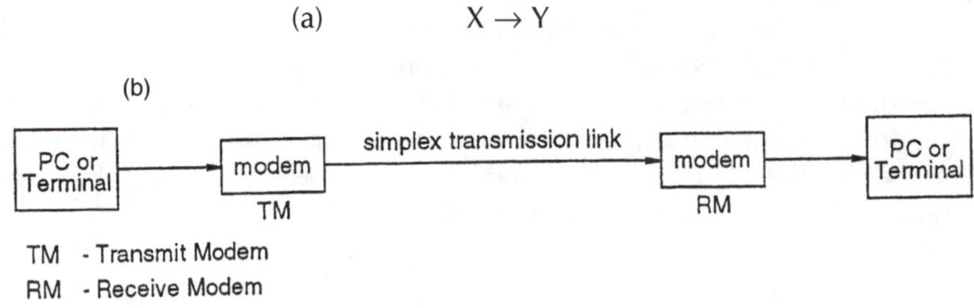

Figure 4.1 (a) Signal flow in one direction from X to Y. (b) Simplex channel mode over simplex line.

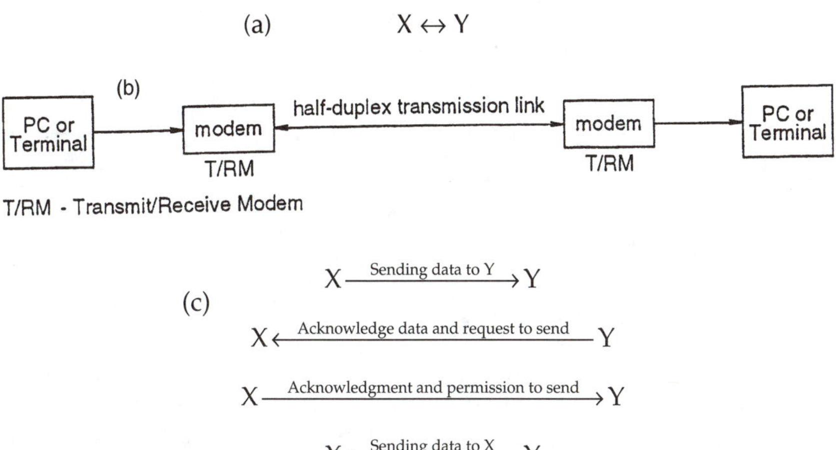

Figure 4.2 (a) Signal flow in one direction at a time, from X to Y or Y to X. (b) Half-duplex channel mode. (c) Sequence of data exchange between X and Y.

and receiving circuits. Half-duplex mode is sometimes also known as the **two-way-alternate (TWA)** mode of transmission (U.S. telecommunication industry standard). CCITT calls this mode of operation "simplex," and it uses two wires.

The half-duplex channel mode can be used in walkie-talkie-type system applications such as police-radio patrolling, air-traffic controllers, emergency 911 number systems, radio communication systems, ham radio systems, etc. In general, many computer terminal systems usually operate in this mode where the main computer (mainframe) sends messages to these terminals and then waits for its reply from these terminals (connected in primary–secondary, polling/selection configurations, etc.). Here, both nodes X and Y could be computers or terminals and must be using protocols which allow them to switch their status for exchange of data back and forth with a constraint that only one can transmit at any time. Both nodes are using the same pair of wires for transmitting the data. A typical half-duplex channel mode configuration (over telephone lines) using a modem is shown in Figure 4.2(b).

The change of status for transmitting the data by Y follows the following sequence of operations, as shown in Figure 4.2(c). The transmission medium between X and Y in this case is a telephone line. The modems (connected between the device and telephone lines at both nodes X and Y) used in this configuration have both transmission and receiving capabilities and are connected by one pair of wires known as a duplex connection. The modem accepts a digital signal from a digital device and converts it into an analog signal so that it can be transmitted over telephone lines. The modem of X sends the data to the modem of Y and, after the modem of Y sends an acknowledgment to X, it also sends a request with it to X for sending the data. After the modem of Y receives acknowledgment from the modem of X, it sends the data to X. After the data has been transferred to X, any node that must send data can place itself in the transmitting mode. In this mode, the entire bandwidth is available for transmission in either direction.

Full-Duplex: This channel mode of transmission allows the signals to flow in both directions between X and Y simultaneously (Figure 4.3(a)). This provides the flow of signals in both directions on the media at any time. This mode is sometimes known as a **two-way simultaneous (TWS)** mode of transmission (U.S. telecommunication industry standard) or four-wire transmission. CCITT calls this mode of operation "duplex" and uses four wires. The RS-232 is an ideal interface (for DTE/DCE) which works in full-duplex

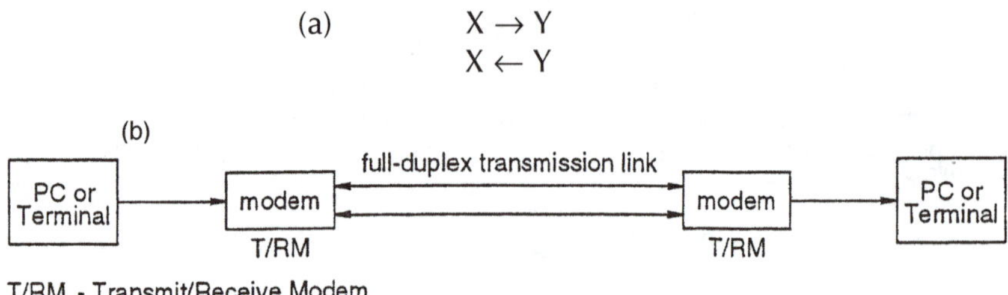

Figure 4.3 (a) Signal flow in both directions simultaneously. (b) Full-duplex channel mode.

mode. It constitutes two identical transmission circuits between X and Y. Each node has a transmitter and receiver which are basically transducers connected by two pairs of wires (four-wire). Communication between the pair of transducers takes place in both directions simultaneously over different pairs of wires. The various pairs of transducers are microphone–receiver, keyboard–printer, camera–screen, etc.

The communication channels may operate in one of these modes and can be defined over communication links (two-wire or four-wire). It is important to note here that selecting a particular mode of operation over a communication link requires that it be supported by the devices being used and their interfaces. For example, if we want to have a full-duplex facility in our communication systems (consisting of modem connections), both terminals and modems must support full-duplex capabilities (in terms of hardware and software support).

In a typical mode configuration (based on telephone lines), modems possessing both transmission and receiving capabilities are connected by two pairs of wires defining full-duplex connection, as shown in Figure 4.3(b). In this mode, the modem can send information over one pair of wires and send a request command as an interrupt which is regarded by another computer as a high-priority request, and the information may be sent to you over another pair of wires. Thus, in this mode, both nodes can receive (listen) and send (talk) at the same time (in contrast to either sending or receiving at any time in half-duplex mode). The software for this channel mode of transmission is more complex, and this mode is more expensive than other modes.

4.2.2 *Channel configurations*

As discussed above, data communication over a transmission medium, particularly a telephone line, can be defined in three modes (simplex, half-duplex, and full-duplex). The telephone lines can be either **dial-up switched** lines (used in residences, organizations, etc.) or **leased lines** (with higher data rates and low error rates). The data communication is defined for the transfer of bits or characters between X and Y. The bits or characters can be transmitted in two different configurations: parallel and serial.

Parallel configuration: In order to send data signals of n bits between X (computer/terminal) and Y (computer/terminal or external device) for transmission, we need to have n separate lines, as shown in Figure 4.4(a). The n bits will be transmitted over these lines simultaneously. The transmission of characters can take place in both directions with the constraint that only one transmission in any direction can take place at any time. Each bit is transmitted over a separate line. This type of configuration has a limited distance (maybe a few meters) due to the number of lines and hence can be used in applications such as transmission of signals between measuring instruments and computer (or terminal) which are close to each other or also within various modules of computer architecture such as

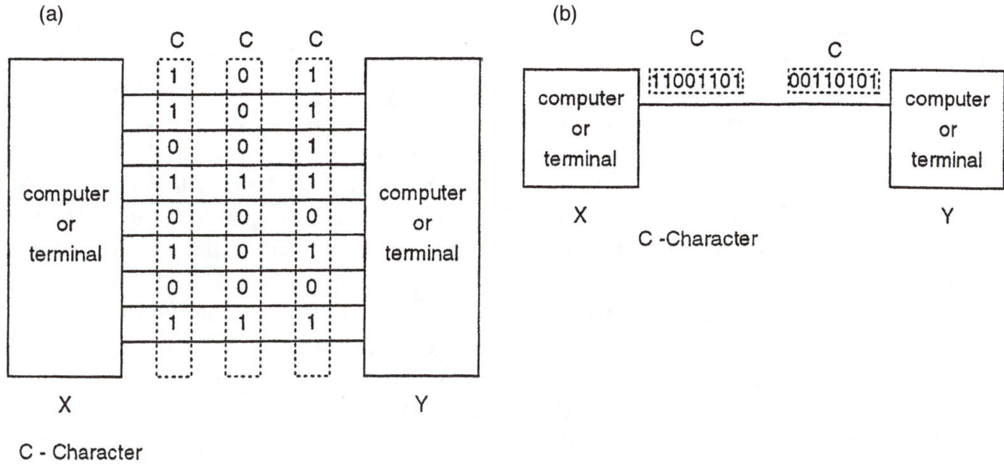

Figure 4.4 (a) Parallel configuration. (b) Serial configuration.

a central processing unit (CPU) and bus, etc. If we want to send the data for a longer distance, the amplitude of signals becomes weak as it travels down the line, and also the signals get distorted due to electromagnetic interference between the wires. Although the amplifiers or repeaters may be used to amplify or regenerate the signals, it becomes expensive. In the case of analog transmission, amplifiers are used to amplify the signals, while in the case of digital signal transmission, repeaters are used to regenerate the digital signals, which are then transmitted back onto the transmission line. These devices have to be used at a regular distance over the transmission media.

Although this configuration is generally used for a short distance, e.g., between computer and printer, it offers a higher speed of transmission. It offers a higher data rate and requires a separate control circuit for providing synchronization between the two nodes. This control circuit uses different primitives between transmitter and receiver nodes to provide the status of the receiving nodes to the receiver:

READY — An indication by transmitting node to receiving node that it is ready to transmit the data
BUSY — Line is busy
ACKNOWLEDGMENT — Indication of receipt of the data sent to the transmitter by the receiver.
STATUS — Status of receiving node.

The transmission of characters between X and Y is in fact obtained by transmitting all the bits of characters over different lines simultaneously. The characters can be transmitted over the lines one after another, either by using extra synchronizing bits between consecutive characters to identify the bits of different characters on each line (to be provided by the transmission mechanism discussed below) or by being transmitted at different instants of time.

Serial configuration: In this channel configuration, there is only one line between two nodes (X and Y) and the transmission signal bits are transmitted bit by bit (as a function of time) or sequentially over the transmission media, as depicted in Figure 4.4(b). The signaling rate is controlled by baud (rate of signaling per second on the circuit). The reciprocal of baud gives the time period of signaling for the transmission. The time period of signaling is a very important parameter, as during this period only, data signals are transmitted. We can send one, two, or more bits during this period, giving rise to different methods of coding the binary signals, which in turn define the data rate in bits per second (bps).

The character composed of bits can also be sent such that the bits are transmitted sequentially. We can also send more than one character one after another over the same line; in this case the identification of the characters is determined by a transmission mechanism discussed below. This configuration offers a low speed of transmission but can be used at greater distance. This configuration, in fact, has become the most common mode for the transmission of data. Here also the data can be sent in both directions over the same link, but only one transmission will take place at any time.

Each signaling method offers different performance which, in turn, may reflect the complexity of the circuit for timing the transmission of bits over a link. The bits generated by a CPU (the most expensive resource in the digital computer) are usually transmitted in parallel mode. But if we want to send a digital signal from a digital computer to a channel (telephone line, which is much slower than a CPU), a communication I/O port is defined and different vendor communication I/O boards are available. The communication I/O board accepts the bits in parallel on a bus (connected to a CPU) and converts them into a serial bit stream. This serial bit stream is transmitted over the channel to its maximum speed (selected by the communication I/O board). The communication board on the receiving side which accepts the serial bit stream from the channel converts it back into parallel mode and sends it on the bus (connected to the CPU). Although this mode of transmission is slower, it is very useful for transmitting signals over long distances and is used in communication systems such as facsimile, e-mail, satellite communications, microwave links, etc.).

The preceding sections discussed two types of channel configurations. In the following section, we discuss different classes of channel configurations which have been defined for the RS-232 interface (DTE/DCE). This interface is a standard defined for the physical layer of OSI-RM.

4.2.3 Channel configurations for RS-232 interface

Information can be shared exclusively by a system's users by accessing the channel which provides a framework under which different channel configurations can be defined. The following sections describe channel configurations based on RS-232 interfaces and telephone lines.

Point-to-point configuration: This configuration establishes a communication circuit between two nodes X and Y which are connected through their RS-232 interface (by a cable). This configuration allows the sharing of access to the channel between these pairs of nodes only, e.g., PC and printer, terminal and mainframe, etc., as shown in Figure 4.5(a). The communication between two nodes takes place one node at a time; i.e., when node X is transmitting, node Y will be receiving and vice versa. In other words, the channel can be accessed by any node at any time. This configuration between two nodes can also be defined on telephone lines, as shown in Figure 4.5(b–d).

As stated above, a point-to-point configuration can be established between two nodes either through their RS-232 port connections or through modem connections. The lines between them can be either telephone leased line or dedicated line. There are two lines used between the nodes. For telephone leased lines, two-wire modems or four-wire

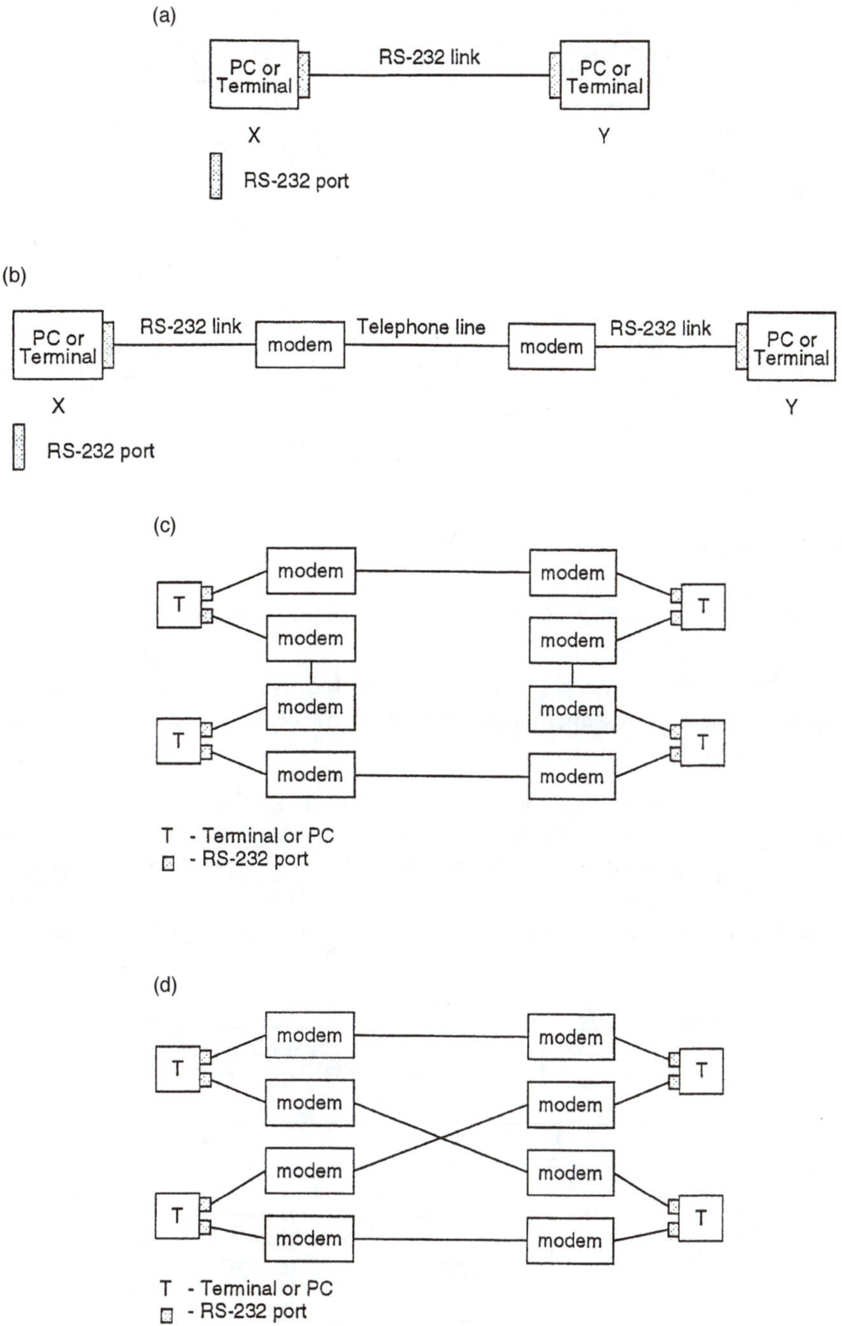

Figure 4.5 (a) Sharing access to the channel between nodes X and Y only. (b) Point-to-point lines between X and Y. (c) Point-to-point lines between many nodes. (d) Point-to-point configurations.

modems are used to provide a channel between them, as shown in Figure 4.6(a,b). Signals travel from X to Y on one line and return on the second line from Y to X to form a closed loop of the circuit.

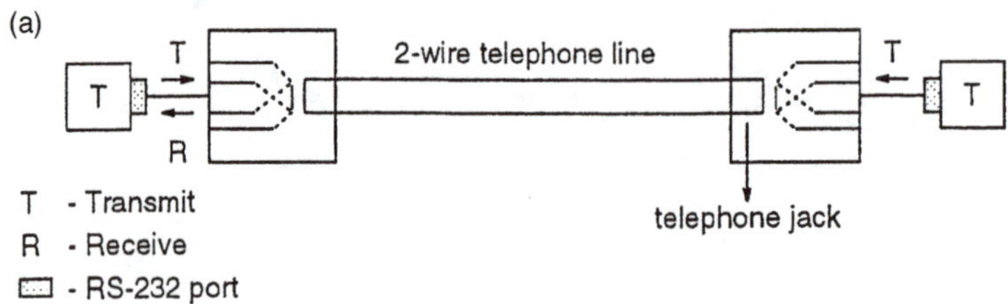

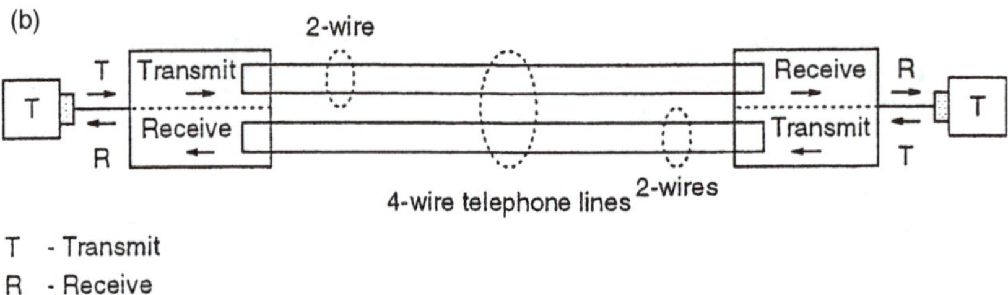

Figure 4.6 Communication channels. (a) Two-wire modem. (b) Four-wire modem.

In Figure 4.6(a), only one message can be transmitted/received on the line at any time, while in Figure 4.6(b), two messages can be transmitted/received simultaneously between X and Y.

Multi-point configuration: In this configuration, a link is established between more than two nodes at any time; i.e., more than two nodes can access the channel simultaneously, as shown in Figure 4.7(a). We can also allow more than two stations to access the telephone lines or local area networks, as shown in Figure 4.7(b), which depicts the connection between various terminals or personal computers to mainframe or host machines. When

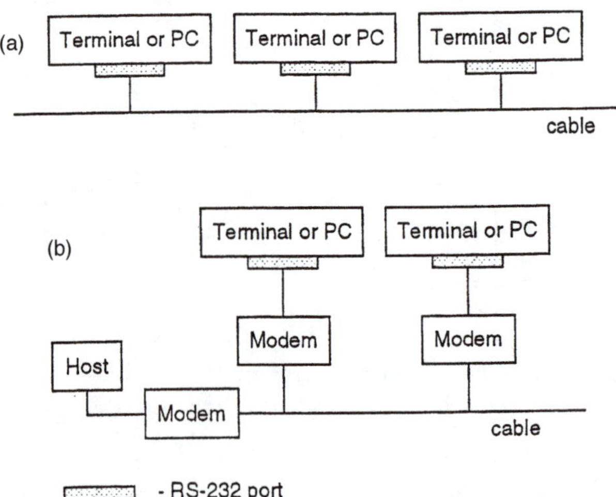

Figure 4.7 Multi-point configurations. (a) More than two nodes can access the channel simultaneously. (b) Connections between terminals and host.

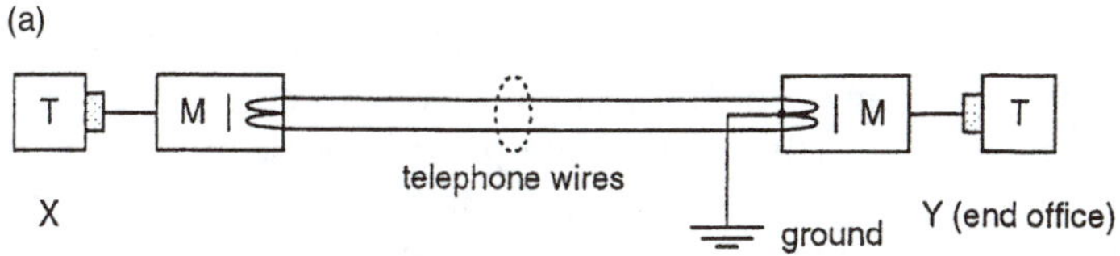

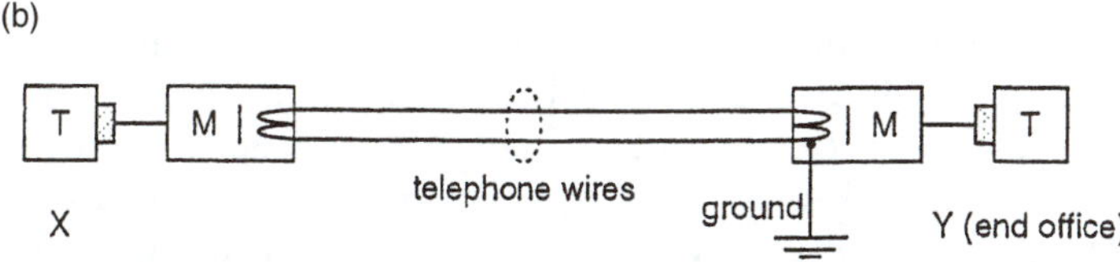

Figure 4.8 Channel circuit connection. (a) Balanced circuit (leased lines) termination. (b) Unbalanced circuit (dial-up lines) termination.

any particular node is transmitting a message, the remaining nodes connected to it will be accessing it simultaneously. Only one message can be sent on it at any time. This configuration is used for conference calls, video conferencing, e-mail services, etc.

Unbalanced and balanced channel configurations: In these configurations, one of the lines in a two-wire modem is grounded. This grounded line provides the return path for the signals. The telephone lines are connected to the modem either in balanced or unbalanced circuit mode, as shown in Figure 4.8(a,b). In general, balanced circuit connections are used for leased lines, while unbalanced circuit connections are used for dial-up lines. A power supply of 48 V is used with these lines.

The above was a discussion on the configurations of channels over which data can be sent. One of the main problems with these configurations is the identification of characters being sent, and this problem becomes more serious in the case of serial configurations, as the bits are being sent continuously. On the receiving side, the node has to identify the bits for each of the transmitted characters. Further, the receiving node must be informed about the transmission of data from the transmitting node. In many books, these two problems have been referred as the "synchronization problem," in which complete mutual coordination (synchronization) must be provided between X and Y for data communication without any problem. In the next section, we discuss various issues associated with synchronization and also how this problem is solved in communication channels.

4.2.4 Synchronization in transmission

Synchronization can be provided in a number of ways. First, we have to make sure that both X and Y know the encoding technique being used by the other for serial configuration. This is usually defined by the transmission rate expressed in baud or sometimes in bits per second (bps). Note that baud and bits per second are not the same. Both X and Y must have modems of the same baud and also the same bits per second. In the case of different bits per second, the conversion must be done by a suitable circuit between the X and Y nodes. These nodes have to use some synchronizing bits between the characters, not only to identify the boundary of each character but also to provide synchronization between them. When the receiver receives the synchronizing bits, it knows when to start considering the first and last bit of the character. If the clocks at both X and Y are running at the same speed, both nodes are synchronized for the transmission of characters between them.

Another way of providing synchronization between X and Y is through the selection of a transmission mechanism which inherently provides synchronization for serial transmission between the nodes. The transmission mechanism for data communication is usually defined as either asynchronous or synchronous, and each of these can be implemented in their respective modes. For example, asynchronous transmission is based on sequential bit-by-bit or character-by-character transmission (which is again based on bit by bit within each character), while synchronous transmission is based on parallel-by-parallel transmission of characters (one character after another).

Asynchronous communication channel: This transmission mechanism is based on serial-by-serial transmission of bits or characters. Character-by-character transmission is usually performed on the keyboards or teleprinters where each character is encoded by start and stop bits. If we want to transmit digital signals from a computer/terminal to another computer/terminal or mainframe, the digital signals are transmitted in an asynchronous form and there is no fixed time interval between the text characters. On the receiving side, we have to provide some sort of synchronization mechanism which will indicate the start of data characters, beginning and ending of messages, number of characters in the message, etc. A regular time relationship between characters is lacking, and as such each character needs to be synchronized separately.

In order to provide synchronization between transmitting and receiving nodes, we send a start bit (preceding each character) on the communication circuit which will begin a clocking system to provide clocks during the interval of a single character and also initialize timing functions before the character arrives. This transmission method is also known as **start–stop transmission**, as it consists of a start bit as a prefix bit to the information and a stop bit (one or two) as a suffix to the information.

The start bit indicates the start of information, while the stop bit indicates the end of information. The nodes also know how many bits are defined in the characters (depending on the coding technique used, e.g., in ASCII, we use eight bits per character, out of which only seven bits are used, while the eighth bit is used as a parity for error detection) and accordingly store that number of bits (by counting via counter) of characters without control bits into their buffers. In case of no transmission over the communication link, a continuous 1s or 0s signal is always flowing through it. Each character is enclosed by start and stop characters, and these characters can be sent either at regular or irregular intervals, depending on their transmission from the keyboard.

For example, if we use ASCII characters, then the start pulse will provide clocking during eight-bit duration. In the absence of transmission (when no data signal is transmitted), we usually send logic 1s constantly over the line. This situation indicates that the line is ideal and is in *marking* state. At the end of each character, we send one or two stop bits, depending on the type of parity error-control technique chosen. The last bit of ASCII is usually represented by a parity bit and logic 0s define the spacing state, as shown in Figure 4.9. To summarize,

Logical 1 (marking)
Logical 0 (spacing)

Consider an ASCII character whose binary representation is given by 01110011 (it may not be any valid character). It is preceded by a start bit (**least significant bit, LSB**) represented by 0s and will be sent first to the receiving side. As soon as the receiving side

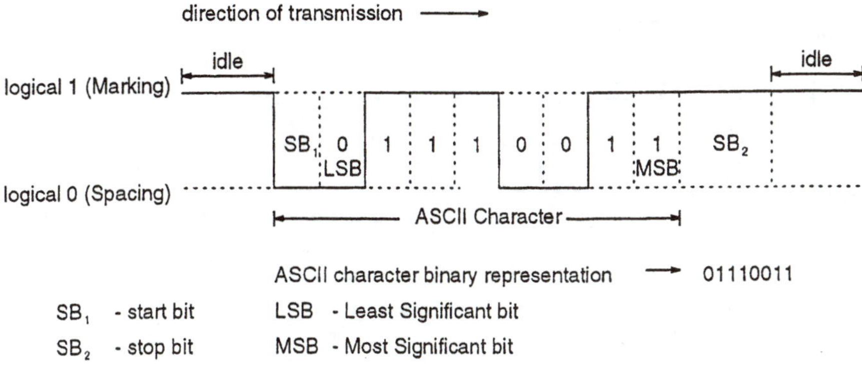

Figure 4.9 Asynchronous transmission.

sees the 0s bit, it begins preparation to receive the first bit of the text. After the last bit of the character (usually a parity bit), a stop bit is received. The receiving side extracts the actual text by neglecting start and stop bits. Thus, for each ASCII character, we are sending two or three extra bits which provide synchronization between the transmitting and receiving nodes. If we know the data rate of the line, and if we decide to use one start and one stop bit with ASCII characters (usually have eight bits with the **most significant bit (MSB)** representing the parity bit), the speed of transmission of characters per second over the communication channel can be calculated. The parity bit is used for error detection. For example, if we have a 4800-bps line, then 480 characters per second can be transmitted on an asynchronous channel (number of characters per second = bits per second/number of bits in each character). The asynchronous transmission can be used for a number of codes such as Baudout, ASCII, etc.

Asynchronous transmission uses two types of parity: even and odd. As indicated above, the last bit (MSB) of an ASCII character usually represents the parity bit. Let's take the example of an ASCII character used in this type of transmission, as shown in Figure 4.10(a). For even parity, we have to substitute 0 at the MSB position (since in even parity, the number of 1s including parity bit should be even), as shown in Figure 4.10(b), while for odd parity, we have to substitute 1 at the MSB position (since the number of 1s must be odd), as shown in Figure 4.10(c). The parity bit can be used for error control (error detection and in some cases error correction), using either even or odd parity. This type of error control and detection is also known as **parity checking** or **virtual redundancy check (VRC)**.

A parity bit can detect only one bit error, so if there are more than one bit error, these cannot be detected by parity check or VRC. Instead, we have to use additional parity bits for detecting more than one bit error, and this is handled by techniques such as **block check character (BCC)** or **longitudinal redundancy check (LRC)**, which are discussed in the following sections. These techniques of detecting errors are usually used in asynchronous transmission where burst errors in adjacent characters have more effect than non-adjacent characters. The LRC technique, in general, is suitable for a data rate of less than 4800 bps, while for higher rates other techniques such as the spiral redundancy check (SRC), cyclic redundancy check (CRC), etc., are used.

Asynchronous transmission interfaces are very popular and are widely used as DTE/DCE interfaces on computers/terminals, as they are very cheap and as the majority of the computers/terminals use asynchronous interfaces. These systems can be used only for lower speeds of communications.

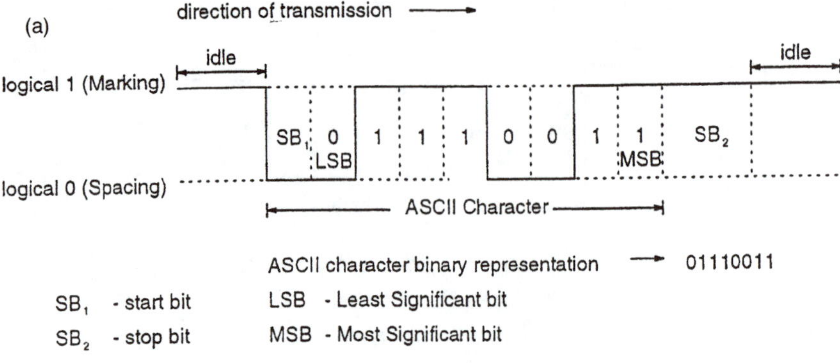

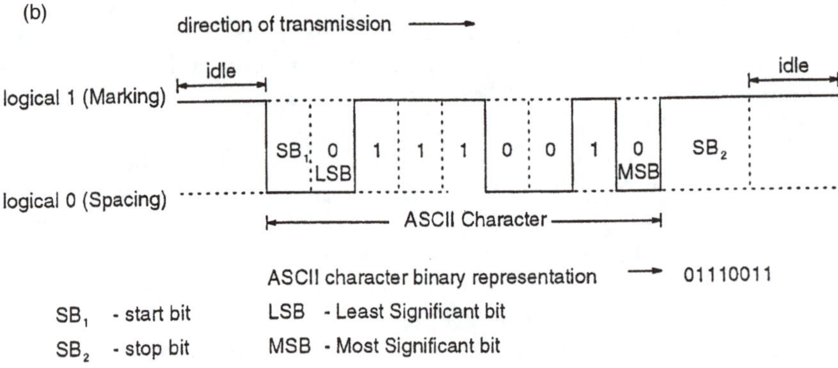

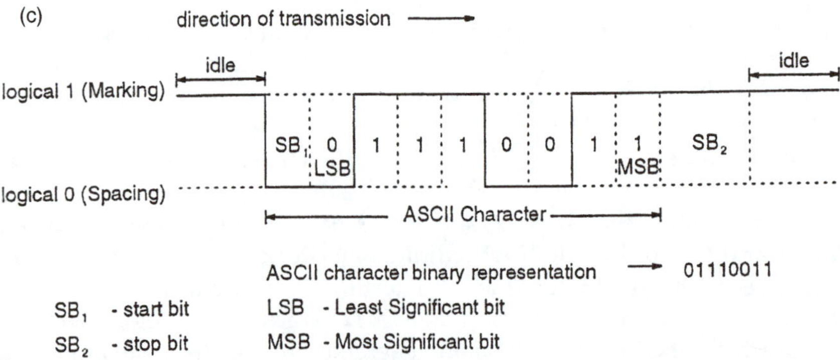

Figure 4.10 (a) ASCII character used in asynchronous transmission. (b) Even parity. (c) Odd parity.

Synchronous communication channel: In this transmission method, we transmit text characters contiguously without any delay between consecutive characters. There exists a regular relationship between characters, and the transmitter and receiver are always synchronized. Synchronous transmission does not require any start or stop bits to provide synchronization as required in asynchronous transmission. If we consider the ASCII format, then the eight-bit characters (seven-bit representation and the eighth bit as a parity) will be transmitted one after the other on the transmission circuit (media). With this transmission, a problem of synchronization exists which must be taken care of, as the receiver has no knowledge about the start and end of the character bits. This synchronization problem is solved by using a synchronizing bit pattern at the start of transmitting a block of data rather than individual characters (as in asynchronous transmission). A block of data may contain more than one character. In synchronous transmission, it is required that characters are always transmitted over the communication channels at a regular rate even if there are no information characters being sent. The **synchronization (Sync)** or **idle (Idle)** characters are defined and are being sent continuously over the communication link at all times.

There are different protocols available to handle synchronization problems in synchronous transmission. The size of the synchronizing bit pattern is different for different protocols. For example, if we consider two eight-bit synchronizing bits, the transmitted signals look like

Sync 8bit | Sync 8bit | character1 |character 2| .... Sync 8bit| Sync 8bit

The synchronizing bits are of a special bit pattern and hence indicate the receiving node about the start and end of the block of data composed of characters (of fixed size). The receiving node extracts the block of data from received signals by deleting sync bits. In synchronous transmission, characters, bytes, and bits can be sent over the transmission medium and for each of these types of data, the synchronization control characters or bits are defined. Based on the type of data transmission, we have different types of protocols to provide synchronization between nodes for data communication, such as character-oriented, byte-oriented, and bit-oriented protocols.

In **character-oriented protocols** (based on standard alphanumeric codes), two sync characters are used as control characters for the synchronization. After recognizing the first two characters as control characters, the receiver considers the remaining bits as the bits of characters defined as data characters.

In **bit-oriented protocols**, there are no control or data characters. Instead, a special bit pattern known as a *flag* or *preamble* (01111110) is used as an initial synchronizing bit (similar to Idle in asynchronous transmission). The following is a general format of a synchronous frame with appropriate fields used in synchronous transmission:

| flag | header | data | trailer |

The length of the header and trailer (defining the boundaries of block of data) is different for different protocols. When there is no transmission, flag bits are continuously sent on the line. The header field initializes the clocking system and timing functions of the receiving node, while the trailer indicates the end of a block of data. These two fields are used to define the size of a block which may have more than one character. These are different from start–stop bits (as in asynchronous transmission) in the sense that start–stop bits are recognized for each character while a header/trailer is recognized for an entire block of characters.

Synchronous transmission provides communication between computers or devices which have buffering capabilities and is not suitable for terminals without buffering capabilities. Synchronous devices are more expensive than asynchronous devices. Further, these devices usually prefer high data rates (2400 bps or more), while asynchronous devices are preferred for data rates of 1200–1800 bps due to the speed limitations of their respective modems.

In order to understand the differences between these two types of transmission (synchronous and asynchronous) in terms of implementation and efficiency, we consider the following example (let's assume we want to transmit 1000 ASCII characters):

In the case of asynchronous transmission, for each of the characters, we require at least two extra bits (start and stop), giving rise to $1000 \times 2 = 2000$ bits. The total number of bits transmitted is thus given by $1000 \times 8 + 2000 = 10{,}000$ bits. The overhead (representing the ratio of extra bits to the actual bits transmitted) is given by $10{,}000/2000 = 50\%$.

For transmitting the same number of characters in synchronous transmission, the total number of synchronizing fields may be four (two synchronizing eight-bits on each side of the block of 1000 characters), giving rise to 32 bits. Thus, the total number of bits transmitted is $1000 \times 8 + 32 = 8032$ bits. The overhead (representing the ratio of extra bits to the actual bits transmitted) is given by $32/8032 = 0.398\%$ (less than 0.4%). Thus, the selection of an appropriate transmission is based on the tradeoff between cost (overhead) and efficiency.

For isochronous (data and voice) transmission, the start–stop bits may or may not be used for providing bit synchronization.

4.2.5 Channel access techniques

So far we have discussed various modes of operation (simplex, duplex, etc.) and channel configurations (point-to-point, multi-point). The configuration and modes, in fact, provide a framework or connection for sending the messages between connected nodes. The nodes (computers, terminals, hosts, printers, etc.) have to use techniques to access the channel for transmitting/receiving the message from the lines. The channels can be accessed by the following two widely used techniques:

- Contention (random access)
- Polling/selection

Contention: In this technique, the channel is shared by various nodes and only one node can transmit a message at any time. The transmitted message can be accessed simultaneously by other connected nodes to the lines. This method is generally useful for a multi-point configuration on either half-duplex in full-duplex mode of operation. Remember, only one message by any node can be sent on the line at any time. This obviously raises a number of questions regarding the controlling of transmitting and receiving nodes, their turns, and so on. One very popular and effective method of solving this problem is through a primary/secondary relationship, as shown in Figure 4.11(a). One node acts as primary while other nodes act as secondary. Secondary nodes can transmit messages only when they are signaled by the primary to do so. In multi-point, the primary node can send messages to secondary nodes, and secondary nodes can send messages back to the primary. But the secondary nodes cannot send messages directly to each other. These messages are transmitted via the primary node, as shown in Figure 4.11(b). This type of primary–secondary relationship can be expanded in the form of a tree structure. Multi-point configurations on half-duplex and full-duplex operations are shown in Figure 4.11(c,d).

Another way of solving the contention problem in channel accessing is through a *distributed relationship*. Instead of primary or secondary nodes, all the nodes are the same and can access the channel whenever they want to. Each node uses a protocol (of contention) to access the channel. If a node wants to send a message, it will listen to the channel, and if it is available, it will send the message on it. But if it is not available, it will wait for a random time and then again listen to the line. If two nodes (after listening to the line) have sent their messages on the line, a collision will take place and both nodes will

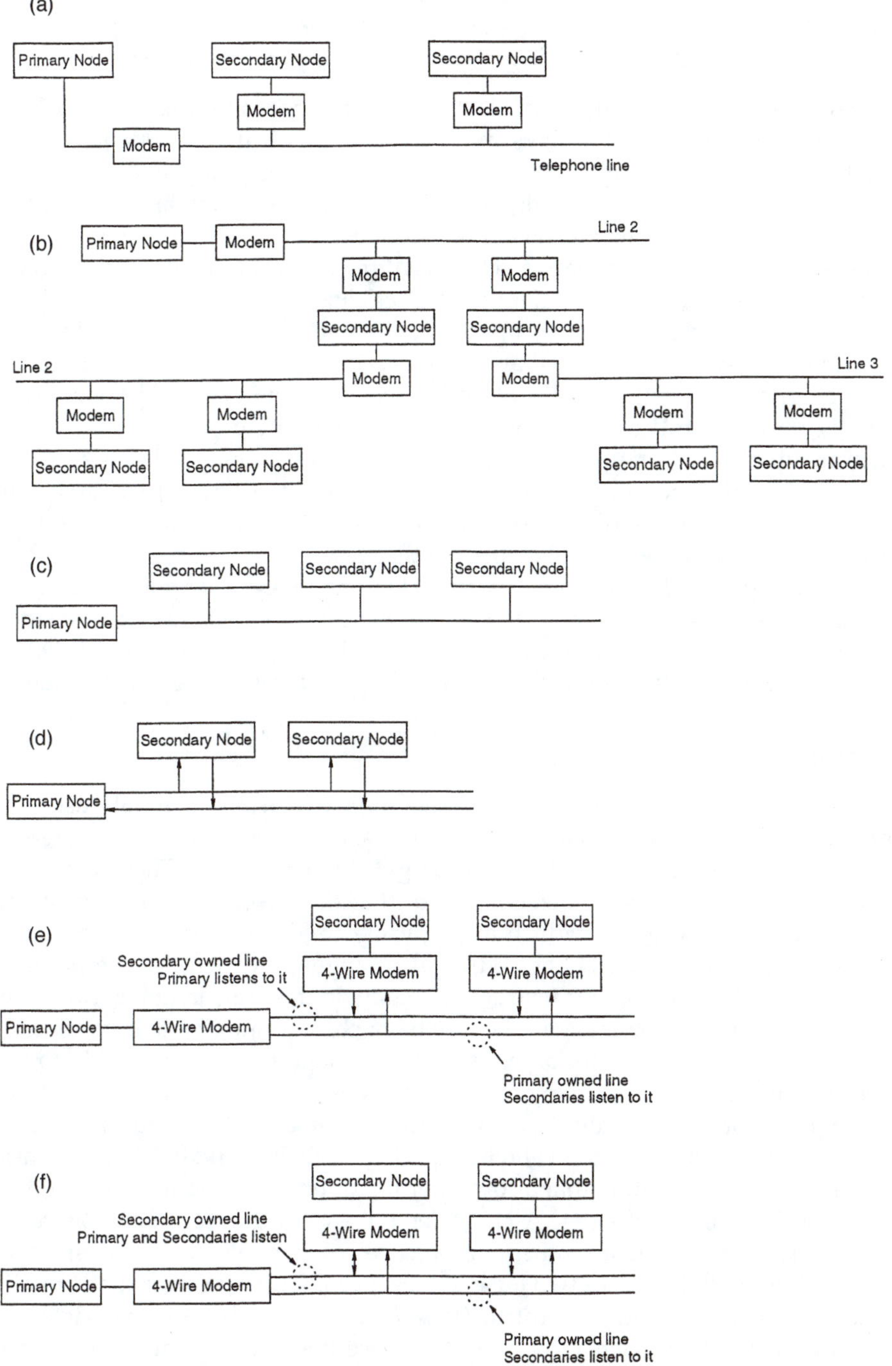

Figure 4.11 (a) Primary and secondary nodes. (b) Secondary node messages transmitted via the primary node. (c) Two-wire multi-point half duplex. (d) Four-wire multi-point full duplex. (e) Roll-call polling. (f) Hub polling.

have to withdraw their messages and re-transmit after a random time. That is why this method is sometimes called the *random access contention method*. This method is very useful in Ethernet local area networks but is never used in wide area networks.

Polling/Selection: This technique can also use either a primary–secondary relationship or a distributed environment for accessing the channel. A primary node sends a **"poll"** message to secondary nodes in a circular fashion. If any secondary node has a message to transmit, it will send an **acknowledgment (Ack)** to the primary, which then allows that secondary node to transmit its message. If a secondary node does not have any message to be transmitted, it will send **negative acknowledgment (Nak)**. If a primary receives Ack from more than one secondary node, it will "select" one secondary at a time. Also, when a primary wants to send a message to a particular secondary, it will send a **Select** message to it and, after receiving Ack from that secondary, it will transmit the message to it. In this case, the primary maintains a list of polled secondary and selected nodes. This type of polling is known as **roll-call polling (RCP)**, as shown in Figure 4.11(e).

There are number of variations and applications of this technique. For example, in one variation, the primary node may send both the Select and the actual message together (if the message is small) to a particular secondary node(s) without waiting for Ack(s). This type of selection is known as **fast selection**. In other cases, it will send Select and then wait for the acknowledgment. This technique is useful for long messages. Another type of polling is **Hub** polling, as shown in Figure 4.11(f), where the primary polls the farthest secondary first. If that secondary has any message to send, it will send it to the primary. Otherwise, the farthest secondary will poll the closest secondary, and this process is continued.

4.3 Information coding

A coding scheme provides a unique representation of various symbols used in data communication. The symbols include letters, digits, and other special characters. These different types of symbols are also known as **graphic characters**. These characters are usually considered as information characters for the coding scheme. Another category of characters is **control characters**, which are typically used for controlling terminals and computers. We can also consider the coding as a language which the computer understands and processes based on our instructions, again in the same language, and provides us the results in the same language. The language chosen must be able to represent all the characters, symbols, and special signs. This language must also be able to support error control, flow control, and many other features required for data communications.

The language or code usually transforms these symbols into the form of bits, bytes, and characters. A bit is a binary digit and has two logic values, 1 and 0. A byte is considered the smallest unit of information that a computer manipulates as a unit and includes eight bits. It is also known as an octet. A character represents different symbols (alphabet, numbers 0 through 9, and many other special types of symbols) and is again expressed in bytes. The binary digit defines two logic values, 1 and 0, and it corresponds to different states, such as on/off, low/high, and true/false in different applications of data communications and networking. Generally, computers use two types of numbering schemes: octal and hexadecimal. The octal representation of numbers uses a base of 8, while hexadecimal uses a base of 16 (0 to 9 and then A to F).

Early computers (first and second generation) used six-bit combinations for each of the character representations in the binary form, and the coding scheme was termed **binary-coded decimal (BCD).** It can represent 64 combinations. Modern computers use new coding techniques (mainframes, minicomputers, PCs, etc.) and these are discussed next.

4.3.1 American Standard Code for Information Interchange (ASCII)

Most microcomputers use the American Standard Code for Information Interchange (ASCII). ASCII is a seven-bit code for representing a character and can represent 128 combinations of different characters and symbols. The last bit (eighth bit position) includes a parity bit (used for error detection). The binary representation of any ASCII character has seven bits as defined in the ASCII code table. The first three bits corresponding to the column heading give the higher-order bits, with the left-most corresponding to the most significant bit (MSB). The remaining four bits, as obtained from the row heading, represent the lower-order bits, with the right-most representing the least significant bit (LSB). The parity bit is defined at the eighth bit position and after adding it to the seven bits, the code defines this added parity bit as the MBS. A variation of ASCII code is known as **Data Interchange Code (DIC).** This code includes non-printing control characters (different from the ones defined in ASCII) and offers only odd parity. This code has been successful for providing communication between computers and is also known as computer-to-computer communication code.

4.3.2 Extended Binary Coded Decimal Interchange Code (EBCDIC)

This code is called the *Extended* Binary Coded Decimal Interchange Code (EBCDIC) (eight-bit) because it is modified to include one more bit as character parity and can represent 256 combinations of symbols. This coding is defined for IBM's large mainframes. When we get messages from IBM networks or IBM mainframes, the data is expressed in EBCDIC. One of the main advantages of this code is that it provides error detection in addition to the parity technique and also provides some control functions.

It is worth mentioning here that systems using these different codes (ASCII and EBCDIC) will not be able to interact directly, as the codes have different meanings for the last bit of the code. However, with the help of an ASCII/EBCDIC translator or converter (available on the market), the conversion between these codes is done inherently by the system and is considered to be part of communication protocols.

One of the main objectives of a data communication system is to allow users to transmit messages over a transmission medium (usually noisy, susceptible to natural calamities, etc.) error-free. In other words, all the errors introduced during the transmission (due to transmission or medium) must be detectable. The coding technique alone cannot detect all the errors all the time. A **redundancy** technique may be used for error detection. This technique adds extra bits in the message and will be used to detect all the errors at all times. However, this will affect transmission efficiency due to overhead. In order to improve the performance of the network, the redundancy must include the minimum number of bits for detecting the errors, and there are a number of ways of achieving this.

In the simplest method, the receiver sends a negative acknowledgment (Nak) after detecting an error. For error detection and correction, the automatic repeat request (ARQ) method is used (details of these techniques can be found in Chapter 13). This method of error detection and correction has defined three protocols: stop-and-wait, go back N, and selective repeat ARQ protocols. In the ARQ method, the frame containing the error is re-transmitted until it is received error-free. Alternately, we can add enough redundancy to the messages so that the receiver itself will detect and correct the errors rather than send a Nak to the transmitter. In such a case, the error detection and correction can be achieved by forward-error-correcting (FEC) methods. Some versions of FEC schemes are discussed briefly in the next section.

4.4 Character error detection and correction coding

When we transmit an asynchronous character over transmission media, a parity bit scheme (for redundancy) is used for detecting the error. The parity bit may be defined as either an even parity (i.e., the total number of 1s in any character is even) or odd parity (i.e., the total number of 1s in any character is odd). A **parity scheme** often requires the execution of logic operations at both the sending and receiving nodes. This type of redundancy can be achieved in either of two coding techniques: **block coding** and **convolution coding**. These techniques will be described briefly in the following subsections and in greater detail in Section 4.5.

4.4.1 Block coding

In this technique, a logical operation is performed to introduce redundant bits into the original message such that the message defines a fixed size of the format. For example, we translate a message of size A bits into a size of B bits with B – A as redundant bits. There are a number of ways of implementing this translation process. The block-coding technique is well-suited to ARQ techniques and can be either constant-ratio coding or parity check coding. Different versions of block-coding techniques are available for these two techniques, but the majority of the block-coding techniques are based on parity-check coding, in which we define codes of an A-bit block which can be obtained by performing logical operations on B bits of information to the B – A parity bit (redundancy). Two popular and widely used codes supporting parity-check coding are **geometric** and **cyclic** (or polynomial).

4.4.1.1 Geometric coding

In this technique, one bit is added to the A-bit information to make it either an odd or even code. This code can detect only an error of either an odd or even number of bits. It cannot detect both types of error together. The geometric code includes one parity bit to each character and one parity bit to each level of bits for all the characters in the block. The parity bit added to the character corresponds to **vertical redundancy check (VRC)**, while the parity bit added to each bit level corresponds to **longitudinal redundancy check (LRC)**. A geometric code is useful for ASCII code, but it cannot detect all types of errors.

4.4.1.2 Cyclic (polynomial) coding

This code requires a polynomial generator which sometimes is very important, as the number of bits in it ensures the detection of all errors of length less than the number of bits. This is more powerful and attractive than any other code for a given level of redundancy. This code is easy to implement via shift registers and connections. Very simple algebraic techniques are used for decoding which can be also be implemented easily by the hardware configuration. The standard polynomial generator chips are available from Motorola Corporation for cyclic redundancy check. These chips are used in serial digital data communication for error detection and correction. The single chip can support more than one polynomial generator, and it also offers additional functions.

4.4.2 Convolution coding

Convolution coding is different from block coding and in general is considered to be most efficient for forward-error-correcting (FEC) systems. This code is useful for high-speed data communication.

4.5 Error-control techniques and standard protocols

An error present in transmitted signals affects various parameters of the communication system, e.g., quality of transmission, data rate, response time, throughput, efficiency, etc. This form of error is caused by different types of noise, including thermal noise, impulse noise, cross talk, echoes, intermodulations, attenuation, phase jitter, timing jitter, radio signals, fading delay, etc. Some of these noises are predictable while others are unpredictable. Of the possible types of noise, thermal noise, impulse noise, cross talk, and echoes are the major contributors to error. A detailed discussion on these types of noise, their effects on data communications, and various methods of reducing them can be found in Chapter 2.

For error control, techniques have been developed to reduce errors introduced in the transmission or to reduce the effects of noise on the signals. These error-control techniques usually include codes for checking the errors and for correcting them. Additionally, nearly all the error-correcting codes consist of some code for checking the error. As mentioned in the previous section, there are two widely used codes for checking errors: block code and convolution code.

A **block code** consists of information bits and redundant bits, and various code words (based on Hamming code) are defined to detect and correct the errors. These code words can be implemented by combinational circuit logic, are based on the value of information bits, and hence can be considered as memory-less. The general form of these codes is given by (n, k) where n represents the total number of bits in the block and k represents k bits of information.

Convolution codes also generate code words and depend not only on the current value of information bits but also on the previous value of information bits. These codes have memory and hence can be implemented by sequential circuit logic.

Various error-control techniques based on block or convolution codes have been implemented and are being supported by many commonly used protocols. A cyclic redundancy check (CRC) is one technique which is based on block coding.

4.5.1 Error detection and correction

As discussed previously, there are four main types of noise which contribute to transmission error: thermal, impulse, cross talk, and echoes. This error sometimes becomes significant, especially in the case of digital signal transmission. For example, if we are sending digital signals at a data rate of 2400 bps and if the transmission error occurs as spikes of duration of 10 ms, then these spikes, which sound like clicks to the human ear, will produce a large sound of 24 bits on a line of 2400 bps. An impulse noise is a major source of noise for data communication and occurs as impulse clicks of very short duration.

Techniques (codes) for handling error detection and error correction are different. The *error-detecting* technique introduces codes with a transmitted block of data, which the receiving node uses for the detection of error. If error occurs, it requests the transmitting node to re-transmit the same block of data. In the *error-correcting* technique, the receiving node tries to reconstruct/extract the original signal from received signals. If it cannot regenerate/reconstruct, then it requests the transmitting node to re-transmit the same block again. When digital signals are transmitted toward the receiving node and if the receiving node detects the error, the popular method for error recovery, automatic-repeat request (ARQ), is used. In this technique, after the error is detected at the receiving side, the transmitting node is requested to re-transmit the same digital information signals. There are different versions of ARQ that have been applied in the protocols for error recovery in local area networks and wide area networks, and these will be discussed in more detail in Chapter 9.

4.5.2 Error detection techniques

There are various techniques for error detection and these are implementable in computer hardware and software. The following discussion on error-detection and error-correction techniques can be found in References 1–7 and many other publications. A reader is advised to read these references. Some of these techniques have been mentioned above, but we discuss their implementation details in the following sections.

Parity-check techniques: We have discussed parity check in asynchronous transmission where the last bit (the most significant bit) usually represents the parity in the seven-bit ASCII format. This method can be used to detect an odd number of bit errors. A parity bit defines even or odd parity, depending on whether the logical sum of 1s (including the parity bit) is even or odd in a seven-bit ASCII format.

For even parity, the sum of bits in modulo 2 (without carries) is zero and the parity bit is zero. If the sum of the seven-bit ASCII is one and the parity bit is set to one, then modulo 2 of this sum will be zero. This means that this technique will not detect an even number of bits. This technique is also known as the **vertical redundancy check (VRC)**.

For example, we want to send a ASCII character with data bits 1101010 using odd parity for error detection. Since the number of 1s here is 4, we add one more 1 at the eighth position to get 11010101. When this bit combination of the character arrives at the receiving side, the receiver knows that the information is defined by the first seven bits while the last bit is parity. Now if the data received is, say, 11010100, the receiver knows that this is not correct, as the number of 1s is even while mutual agreement between them was for odd parity. It will then send a negative acknowledgment (Nak). This method of error detection may detect a one-bit error only. It will not be able to detect the error if the bits within the combination change their positions. Further, the even errors will not be detected if odd parity is chosen. This technique cannot detect burst errors, which are usually generated during data communication.

Longitudinal redundancy check (LRC): This technique is a modified form of the parity check or VRC. Here, seven bits of ASCII characters are arranged vertically. The eighth bit is defined as the **vertical redundancy check (VRC)** bit. The bit values contributing to LRC characters are eight bits long (one byte) and are arranged longitudinally. The bit position in each ASCII byte may be used to calculate LRC. For example, the first position of each ASCII byte is used for the first LRC calculation, and the second bit position is used for the second LRC calculation. LRC and VRC are used to detect errors in rows and columns, respectively, and hence the combined LRC and VRC will detect all possible errors of any bit length. This method enables the parity of each character in the frame to be checked and hence is used to detect errors in a block of data. The error-detection capabilities may be enhanced by combining LRC with other parity patterns. An even number of bit errors cannot be detected by LRC and, as such, some even bit patterns and errors in the same positions on an even number of characters will not be detected.

Cyclic redundancy check (CRC): This technique is based on block codes; hence, it is memory-less and can be implemented by combinational circuit logic. It uses a check character at the end of a block of data. The vertical parity and longitudinal parity (as discussed above) are calculated by the sender side using a mutually agreed-upon polynomial code. This parity character is sent at the end of a block of data in the check character field. The receiver side calculates the proper parity character from the received block of data. The calculated parity is compared with that sent in the character-check field. If these two parity characters are the same, there is no error and the receiver side sends an acknowledgment. But if these two parity characters are not same, it sends a negative acknowledgment indicating the occurrence of error in the transmitted data. It uses bit stuffing and is usually realized by a register. Due to polynomial code used in these checks, CRC is sometimes also known as **polynomial check**. The data-link layer protocols use these parity-check characters and have different formats for parity characters; e.g., the IBM Binary Synchronous Protocol (Bisync) uses block-check character (BCC) while a frame-check field (FCF) is defined in Synchronous Data Link Layer Control (SDLC), and both these parity characters use CRC format.

A polynomial code is formed as an algebraic expression consisting of terms with coefficients of 0 and 1 only. If we have a *k*-bit frame, it will be expressed as a coefficient expression for a polynomial with the same bits, i.e., *k* terms having decreasing power-of-coefficient terms from X^{k-1} to X^0. This expression has a degree of $(k-1)$. The decreasing power-of-coefficient terms represent the highest bit with a degree of $(k-1)$ on its left; the next-highest bit with a degree of $(k-2)$, etc., to define a polynomial generator PG(X). A special bit pattern generated on the transmitting node known as a frame-check sequence (FCS) or block-check count (BCC) appends the original contents of the frame. The frame actually transmitted from the transmitting node contains the original frame and an FCS at the end of the frame such that the total number of bits in the frame is divisible by a mutually-agreed upon polynomial generator either with a known remainder or without a remainder. On the receiving node, checksum is computed (using the same polynomial generator) for the received frame. The division operation giving a nonzero or unknown remainder (not predefined or mutually agreed upon between transmitting and receiving nodes) will correspond to the presence of error.

Various types of errors such as single-bit errors, two-bit errors, an odd number of errors, etc., can be detected by CRC, and detection of each of the errors requires some conditions to be satisfied.

CRC is used for detecting an error in a long frame, and it usually computes (using CRC algorithms) and appends the CRC at the end of the frame at the transmitting node. When the receiving node receives the frame, it again computes the CRC (using the same algorithm) to determine if any error has been introduced during transmission. The algorithm for CRC requires the mutually agreed-upon polynomial generator and is based on division operations. The following polynomial codes for error detection represent international standards protocols: CRC-12, CRC-16, and CRC-CCITT. CRC-CCITT detects errors of lengths up to 16 bits, and most errors of length greater than 12 bits are detected. CRC-16 detects most of the errors of more than 16 bits. Of these two polynomial codes, CRC-16 is a very common protocol. CRC-12 is used for six-bit-long characters, while CRC-16 is used for eight-bit-long characters.

Another error-correcting code very popular in LANs is a 32-bit frame-check sequence (FCS) which is similar to CRC. Although it requires a 32-bit-long character (as opposed to six or eight bits in CRCs), it has better error-detection capabilities than those of CRCs.

Computation of checksum: The following steps are required to compute the checksum.

1. Construct a polynomial generator PG(X) such that both high-order and low-order bits must be logic 1. For example, a bit pattern of 110001 is represented by the polynomial as

   $$PG(X) = X^6 + X^4 + 1$$

   As indicated earlier, the coefficient of each bit in the expression can have a value of either 0 or 1. The high-order polynomial coefficient is on the left-hand side and it decreases until the coefficient has a degree of 0 toward the right-hand side.

2. The information frame is appended by a number of 0s equal to the degree of PG(X). This new appended frame is now divided by PG(X) using modulo 2. PG(X) is one bit longer than FCS. Here, it is assumed that there are no carries and borrows for additions and subtractions, respectively.
3. For subtractions and additions, EXCLUSIVE-OR logic rules are used (no carries and borrows) as shown below:

```
Subtraction:   1    1    0    0
              -1   -0   -1   -0
              ---  ---  ---  ---
               0    1    1    0

Addition:      1    0    0    0
              +1   +0   +1   +0
              ---  ---  ---  ---
               0    1    1    0
```

   Division operation follows the same concept as that in binary division, with the only difference in the subtraction. Here the subtraction is based on modulo 2 (stated above). If the dividend contains the same number of bits as that of the divisor (PG(X)), then the divisor can be divided into the dividend.
4. Continue the division operations until no further division can be carried out. The remainder represents the FCS bit; it is placed at the end of the information frame and transmitted. All zeros which appended the information frame are now replaced by the remainder (FCS).
5. On the receiving node, the same division (with the same divisor) is performed on the received frame which contains the information frame and the FCS bit.
6. If the remainder is zero or some predetermined value (mutually agreed upon), then the transmission of the frame is considered to be error-free.

An example:
To clarify the way CRC works, consider the following example:

Information frame (to be sent): 1100011
Polynomial generator PG(X): 110001 ($X^5 + X^4 + 1$). Information frame appended with five (degree) 0s as 110001100000
Division operation is performed to calculate the checksum, as given below:

```
              10000001
       110001 110001100000
                1100001
             -------------------
                 000000100000
                       110001
                       ----------
                       010001
```

The transmitted frame has two blocks: an information frame and an FCS at the end of the information frame

On the receiving frame, the receiving node performs the division operation with the same PG(X) on it, as shown below:

```
          1000001
   110001 110001110001
          110001
          -----------------
           00000110001
                110001
                ----------
                 00000 → zero remainder
```

Since the remainder is zero, the transmission is error-free. For division operations on both nodes, the quotient is ignored. The division operation can be realized using EXCLUSIVE-OR (EX-OR) logic functional gates and register. The output of EXCLUSIVE-OR gate is 0 when both inputs are the same and is 1 if the inputs are different. The polynomial generator defines the setup of registers.

4.5.3 Forward error-correcting techniques

Forward error correction is another way of handling error recovery after the receiver detects an error. In this technique, the protocol on the receiving node tries to reconstruct the original digital information signal (after an error is detected). If it can reconstruct the signal, then all is well; otherwise, the receiving node requests the transmitting node to re-transmit the same digital signals again. This technique is quite complex and is generally used in expensive links and also for links which provide significant propagation delay between sending and receiving of the signals. It is worth pointing out here that even with inexpensive links, the choice of whether to correct the error at the receiving node or to re-transmit the signals may depend on many factors, such as efficiency, overhead, cost, delay, interference, etc.

We must determine the relationships between check bits (required to detect the errors) and the information bits. Obviously, data security and protection can be achieved if there are as many check bits as bits in the information. This method may be useful in applications such as banks and other financial institutions, where a high error rate in data such as customer account numbers, balances, withdrawals and deposits, transfer of accounts, etc., is not acceptable. The error-detection and error-correction software in these applications is very complex.

It is obvious that error-detection software will be cheaper than error-correcting software. Both types of software may add their own errors, and the possibility of error increases with the increase in complexity (due to addition of layers). Also, the delay introduced by error-correcting software may become a serious problem for time-critical applications.

In applications where there is a long delay on each side of the transmission (e.g., satellite communications, defense communications applications, microwave-link communications, etc.), it will be easier and cheaper to have FEC code implemented at the transmitting node. This node enables the receiving node to correct the error at the receiving node itself rather than ask the transmitting node to re-transmit, eliminating a significant delay. Further, interference may also be an important issue which might force use of FEC rather than error-correcting codes (ECC). This FEC can be either hardware or software, but due to lower costs of the hardware, there are various FEC controller circuits available, although the hardware implementation requires a very complex algorithm for error correction.

Error detection and correction: Error-correcting codes are useful in increasing the transmission rate over a channel by maintaining a fixed rate. These codes reduce the bit-error rate while maintaining a fixed transmission rate. In general, these codes help us in designing a communication system which specifies transmission rate and bit-error rate separately and independently, but both are subject to bandwidth and channel capacity constraints. Thus, the objectives of these codes should be to transmit at channel capacity with the smallest possible error rate and to provide efficiency of code which allows the system to approach Shannon limits.

In order to understand how the errors are handled, we first consider that the error is present in the frame and then discuss the ways to detect and correct this error. These steps are also explained in References 1, 2, 4, and 6.

Consider a frame consisting of k binary bits containing the data. Assume that the information source generates M equally likely message pieces. Each piece is represented by k bits with $2^k = M$. If we add r redundant bits (parity check bits) to k bits in the frame, then each piece can be expressed in a code word of length n bits with

$$n = k + r$$

The total number of possible n-bit code words is 2^n, and the total amount of possible information is 2^k. Thus, the number of n-bit words which do not represent data is given by $(2^n - 2^k)$. The code word is defined as a block code of length $(n - k)$, as we have added r redundant bits to the information bits. The code words are usually designed as (n, k) codes. For example, a code word (5, 3) has three information bits giving rise to $2^3 = 8$ possible information. There are five bits in each code word such that each code word has two redundant bits. If we assume that the time required to send a code word with n bits and k bits of information should be the same and if we have the duration of coding and encoding the words as Tc and Tn, respectively, then the following condition must be satisfied:

$$Tc \times n = Tn \times k$$

We define the rate of code (to be transmitted) as:

$$Rc = k/n$$

Hence, $Tc/Tn = k/n = 1/Rc = fn/fc$, where fn and fc are the frequencies of codes and are the reciprocals of the respective duration.

R. N. Hamming of Bell Laboratories in 1950 introduced a code for forward error correction which detects and corrects multiple errors in received signals at the receiving node. The error may be due to any type of noise; for the present discussion it is irrelevant. The difference in the number of bit positions in any two consecutive codes is defined by the *Hamming distance* in error-correcting code.

Consider two code words, Cm and Cn, in a particular block. The number of bit positions difference between C1 and C2 defines the Hamming code, and we represent the difference in bit position as $d12$. Thus, dmn in the following two code words is 4 (Hamming distance):

$$Cm = 10110011$$

$$Cn = 01111010$$

Similarly, we can obtain *dmn* between other possible pairs of code words. Now, we define the minimum Hamming distance *dmn*(min) of all the *dmn*s obtained. This minimum value of Hamming distance for given code words defines an upper limit on the effectiveness of a code.

The following are two important properties of *dmn*(min):

1. For *E* errors in a received code word, it can be detected that the code word received is not a valid code word (and also an invalid word) provided

$$E < dmn(\min) < -1$$

   In other words, to detect *E* errors, a Hamming distance of *E* + 1 code is required.
2. If there are *T* errors in a received word, then the received word is not a valid word and *T* errors can be corrected provided

$$2T + 1 < dmn(\min) < 2T + 2$$

   In other words, to correct *T* errors, a Hamming distance of 2*T* + 1 code is required.

For example, say we have *dmn*(min) = 5. This means that we can detect an invalid word which may have up to four errors and also that it will correct the errors if not more than two errors have been introduced. The above conditions for detecting and correcting errors represent worst-case situations; i.e., in some situations, we may detect or correct more errors than obtained by these conditions.

In general, these codes can detect and correct even-parity bits. For a given Hamming code, say (5, 3), we have eight distinct bits of information and two redundant bits, and we can construct a code word of five-bit length by associating even-parity checks as follows:

C1I1C2I2I3

where I1, I2, and I3 are the information bits and C1 and C2 are even-parity check bits. C1 is for I1, while C2 is for I2 and I3.

For a 1-bit error, comparison of C1 and C2 will indicate which bit is in error. If both C1 and C2 are in error, then the information bit common to them is incorrect; in the example, it is I2.

Other detecting and correcting codes based on the new concept of forward error detection (FED) include Trellis coded modulation (TCM) used in the CCITT V.32 modem, which provides a data rate of 9600 bps on full-duplex (dial-up, two-wire) lines. Modems with a higher data rate of 19.2-Kbps links based on TCM are also available on the market. Due to its accuracy and amount of errors it can reduce, TCM is becoming popular; many data link protocols may be able to use these codes for error control in the networks.

4.5.4 Communication link control procedures

In order to provide error-free transmission of data over the physical link in any communication system, errors introduced during transmission by either the link, the communication channel, hardware failure, or any other reason must be handled properly and immediately. Otherwise, the errors will disrupt the communication and the data will not be delivered to the destination node accurately. The data sequencing and synchronization have to be restored for communication between nodes again. This problem of providing synchronization and error handling is usually performed by a set of rules and procedures known as **communication link control procedures**. These procedures provide interaction between transmitting and receiving (remote) nodes and also between computers and terminals in the network. The set of procedures controlling the interaction between them is implemented as a protocol and is responsible for providing correct sequencing, synchronization and data integrity during data communication. The data and control characters or information are sent over the same physical link; therefore, it is expected that the protocol will distinguish between control and data information. The control procedures have been referred to in the literature as **handshaking protocols, line control protocols, link control procedures, communication link control procedures, link discipline procedures, data link control procedures or protocols**, etc. All the functions described above are provided by a second layer of OSI-RM (data link layer).

In order to understand the working of different types of protocols defined for the data link layer, we will discuss various functions of protocols and also the design considerations for their implementation. It is worth mentioning here that, although we have discussed some of these issues in the preceding sections, the following discussion is aimed to provide us with a complete design procedure for defining our data link control procedures, so we may again mention some of these issues to maintain the continuity.

Functions of a protocol:

- To distinguish between data and control characters or information.
- To support different types of channel configurations over RS-232 and telephone lines.
- To control the data transfer.
- To provide coding to the information.
- To support different types of communication links (leased lines or switched-line-based systems).
- To provide error control (error detection, error correction, and error recovery).
- To provide synchronization between sending and receiving computers.
- To provide information and communication facilities transparency to the users.
- To provide handshaking and bootstrapping.
- To offer efficient utilization of a communication link.

In the following section, we discuss these functions in detail to clarify the requirements of link control procedures.

Data and control characters: The distinction between control and data characters is usually provided by the information codes. The information is presented to the communication link in a defined format which typically is defined in terms of fields. For example, the standard format of data information (to be transmitted) defines three main fields as (1) a header or control field (containing the control characters or information), (2) a user's information field (containing the message or text to be transmitted), and (3) a trailer or error-checking field (containing error-control procedures). There are some special control characters which are used by some data link protocols to delimit these fields, including start of header (SOH), start of text (SOT), end of text (EOT), etc. The EOT character indicates that the information following is control information for error control. The protocols distinguish between control and data characters using appropriate information codes.

Information format: The header field includes the following control information: addresses (destination and source), frame sequencing, control flags, and facilities for acknowledgment. The frame sequencing makes sure that the frames are neither lost nor duplicated. The control flag indicates the type of information (control or data) and also the status of length of data information (defined in terms of blocks: first block, last block, etc.). The control information provides a sequence of operations between the sender and receiver defining who will be sending next, who will be receiving, and acknowledgments (for every block or a group of blocks, lost blocks, duplicate blocks, etc.). These options of control information can be implemented by considering appropriate procedures by a process known as *handshaking*. The trailer field contains the control information regarding error control (error detection, error correction, and error recovery). The acknowledgment control information deals with the status of data information being received by each side in opposite directions.

Channel configurations: The link control procedures must support point-to-point connections between nodes over RS-232 and telephone links. They should also support multi-point configurations where a number of nodes are connected to the network, and they must be able to identify the address of a particular node or group of nodes to whom the data information is to be delivered. The protocols must provide interaction between these nodes on a multi-point configuration in an orderly manner, including acknowledgments from the nodes for proper data communications between them. These protocols are also expected to support private lines (based on a master–slave configuration), multi-drop systems (sharing of the channel by various nodes and accessing based on poll/selection), dialed calls (based on the method of connection establishment and termination), dialed conference (based on multi-point configuration), loop systems (based on multi-point configuration in a loop), etc.

Information coding: As we noted earlier, terminals and computers communicate with each other via a common computer language (information coding). The information code defines the syntax for different characters, symbols, and control characters such that these are represented in the same language for proper processing of the data by computers. The information coding provides a unique representation of each of the characters in a fixed number of bits. The characters (data or control) used for data communication are typically divided into **graphic characters** (representing symbols) and **control characters** (for controlling the terminals and computers).

To review, among the codes used in data communications are the seven-bit code with an additional bit for parity known as the American Standard Code for Information Interchange (ASCII) and IBM's Extended Binary Coded Decimal Interchange Code (EBCDIC), which is primarily defined for IBM mainframes and contains eight bits for each character, hence enabling more combinations (255) than ASCII (128).

The link control protocols must support connections on different types of systems which are provided by the carriers, e.g., leased lines, switched lines, private lines, and dedicated lines. They must support different types of services over connection-oriented and connectionless connections. Further, different types of transport services (connection-oriented, unacknowledged connectionless, and acknowledged connectionless) must also be supported by these procedures.

Error control (error detection, error correction, and error recovery): It is expected that a communication link should offer error-free communication between nodes. The communication system as a whole is error-prone and the link control procedures must reduce the effects of these errors on data communications. The link control procedures define some **check characters** and these characters are sent along the data information. These check characters or bits are known as **block check characters (BCC)**. The BCC is included in the trailer part of the information format. This is mutually agreed upon between the sender and receiver and is generated by a checking algorithm at the transmitting side and sent to the receiving side with data information. The checking algorithm is used on data information.

On the receiving side, this error checking is performed in a number of ways, depending on the information code and its functions. One such checking method, mentioned earlier, is vertical redundancy checking (VRC), which checks the parity on a character-by-character basis as the data is received. Other methods are longitudinal redundancy checking (LRC) and cyclic redundancy checking (CRC), which check block by block as the data is received. In all these checking techniques, a positive acknowledgment (Ack) is sent if there is no error in the received data and a negative acknowledgment (Nak) is sent if an error occurs in the received data. The Nak also expects re-transmission of the same data information until it is received error-free, so that until the receiving side has received error-free data, that particular data will be occupying the buffer at the transmitting side. Obviously, the aim should be to minimize the number of re-transmissions of the data.

Vertical redundancy checking (VRC) is a technique for detecting error based on parity (even or odd). This technique is performed on a character-by-character basis and, as such, defines a particular position of parity in the binary representation of characters. If we are using an ASCII character, then the first seven bits represent the binary representation of the character, while the eighth-position bit can be used for checking, thus making the binary representation of a character of one byte long. This checking is also known as *parity check* and can be defined as either even or odd parity checking. In general, the LRC is used with VRC to provide increased error-detection capabilities within the link control protocols.

The **longitudinal redundancy checking (LRC)** technique checks the entire data information or block of data. At the transmitting side, we define an EX-OR operation on the entire data information bit by bit with a checking algorithm. The result of the EX-OR operation is defined as a BCC and is transmitted in the trailer field of the information format. On the receiving side, a similar operation (EX-OR) is performed on the received data information and a BCC is computed. This BCC is compared with the one received with the information data format in its trailer field. If these are identical, it indicates that there is no error; otherwise, error exists. The receiving node will send Nak to the sender side.

Like LRC, the **cyclic redundancy check (CRC)** technique of error detection is also carried out on a block-by-block basis, but it is more complex than LRC. It defines a polynomial generator which is used for polynomial division (using modulo 2) on the data information. The CRC polynomial is preset while the 1s and 0s represent the coefficients of the dividend polynomial. The subtraction modulo 2 (without a carry) is used for division operation, and the division operation is continued until we get a remainder known as CRC. This remainder is sent in the trailer field of the information data format. On the receiving side, a similar division operation is performed on the received data information and the process is continued until a remainder (CRC) is obtained. This remainder is compared with the one received in the trailer field. If the two remainders are the same, it indicates an error-free condition; otherwise, error has occurred. In the case of error, it will send Nak and wait for re-transmission of the same data information.

Popular versions of the CRC error-checking algorithms are CRC-16 (which uses a polynomial generator of $X^{16} + X^{15} + X^2 + 1$) and CRC-CCITT (which uses a polynomial generator of $X^{16} + X^{12} + X^5 + 1$). Both versions generate 16-bit BCCs.

The error-checking methods discussed above also handle sequencing errors, and different protocols handle this problem in different ways.

Synchronization: Synchronization between a transmitter and receiver becomes crucial when we are sending a continuous data stream. Such synchronization can be provided in a number of ways. The most efficient and popular method is to use a synchronizing pulse (unique bit pattern) or control characters in front of the data. The receiver always looks for this special bit pattern or control characters and, as soon as it finds one, it gets synchronized with the transmitter by initializing its variables, counters, and other data structures of the link control procedures. The control characters used in ASCII may also become a part of the data information. In order to use these control characters as part of the data, a *stuffing* (bit or character) technique has been suggested. The protocols differ in the number of synchronizing bits or control characters and also in the information coding used.

Link utilization: Link utilization is affected by the structure of link control procedures and usually depends on factors such as direct utilization, acknowledgment handling, support for a number of nodes, overhead for providing these options via control characters, etc. The direct utilization defines a direct physical link between nodes and can be either simplex, half-duplex, or full-duplex. The control characters are used in header and trailer fields of the information data format. The number of control characters should be neither too large nor too small. The ratio of the information bits to the total number of bits defines the link utilization and must be calculated for each way of transmission. This ratio should be as small as possible, as a greater number of control characters will simply lower the link utilization. The number of control characters for header and error checking required by the protocols should be minimal. The link utilization and its efficiency are also affected by the way information and communication facilities are providing transparency.

Acknowledgment handling: The acknowledgment is also considered a part of control characters, and a number of ways have been suggested to reduce the overhead for acknowledgments. If we want acknowledgment for all information, the overhead will be very high. This can be reduced if we decide to have acknowledgments after a certain number of blocks of data. This can be further reduced if we use it only for reporting the errors and so on.

Communication transparency: The protocol must support different types of modes such as character or bit modes. When the characters are transmitted, we need to indicate the start of the message (character message) and where also it should end. Similarly, in bit transmission, the start and end of the bit message must be indicated by the link control protocols. The protocols usually use a method for providing a boundary of messages in different modes (characters or bits) known as **transparency**. In some books, it is also known as **transmission in transparent mode**. The protocols used for character-based transmission typically use an escape control character (e.g., data link escape, DLE) to indicate the transparent transmission. A pair of special control characters (DLE-STX) is used to indicate the start of the message (transparent text) followed by another pair of control characters (DLE-ETX) to indicate the end of the transparent text message and also the end of transparent transmission. If the data information also contains the same characters as a data character, then this problem is solved by a technique known as **stuffing**. Whenever the control characters DLE-ETX appear, the protocol stuffs by inserting an extra DLE character into the message. The transmitter scans the message to be transmitted in transparent mode, and it inserts an extra DLE whenever it finds a DLE in the message. This is known as **character stuffing**. The receiver scans two characters at a time and whenever it finds two DLEs, it deletes one DLE. But if it finds the first character as DLE and the second one as ETX, it assumes this a normal termination of the transparent mode of transmission.

Similarly, for bit transmission, the protocols use a bit stuffing which is implemented in the following way. These protocols use a unique bit pattern 01111110 as a flag. This control byte indicates the transparent mode of bit transmission. On the transmitting side, a 0 is inserted after every five consecutive 1s. The receiver, after having received the unique bit pattern of 01111110, will automatically delete 0 after every five consecutive 1s. This is known as **bit stuffing**.

Information transparency: The same protocol must provide transparency for various other options during transmission for different modes of transmission (asynchronous or synchronous) such as serial or parallel, serial asynchronous, serial synchronous, etc. The majority of link control procedures are designed for one type of configuration but can be used for other configurations as well, depending on how the transparency is implemented.

Bootstrapping: Computer systems usually have their software loaded at the end of lines of link control procedures and are then restarted via communication line. The starting of a system which was inoperative is defined as *bootstrapping*. Generally, the bootstrapping program is considered a part of link control procedures or can be embedded in the text files.

4.6 Terminals

Terminals are used for the processing and communication of data and provide interface between users and the system. This interface, in fact, provides a mapping between two types of representations: users and electrical signal–oriented platforms. Various hardware interfaces can be called terminals, e.g., a printer which provides mapping for processed data in the form of hard copy and other I/O devices. Terminals are a very useful component for processing, collecting, retrieving, and communicating data in the system and have become an integral requirement of any application system. The main function of terminals is to convert characters of the keyboard into electrical signal pulses expressed in the form of code which is processed by computers. For example, a keyboard-based terminal based on ASCII will generate seven-bit electrical pulses corresponding to the code of that key which was hit/pressed. On the receiving side, the character corresponding to the pressed key on the transmitting side will be printed.

The interface (provided by terminals) for data communication takes input from input devices such as a keyboard, card reader, magnetic tape, or any other input device and converts it into output devices like display devices, paper tape, etc. Some of the I/O devices may have only input and no output; such devices are known as *receive-only devices*.

The choice of a terminal for a particular computer application is a very important issue in the design of communication systems. For example, computer applications such as query response, data collection, interactive systems, and database systems consider the terminal a remote device for I/O operations. Terminals are used to handle practically any communication function in either a controlled or uncontrolled mode. In uncontrolled mode, data from the terminals is controlled by the terminal's operator, while in controlled mode, data from the terminals is controlled by the device (i.e., computer) on the other side.

Most unbuffered asynchronous terminals operate in uncontrolled mode in the sense that as the characters are entered via the keyboard of the terminals, they are transmitted over the line. If a computer is connected to the terminals, then it offers point-to-point line configuration to each of the terminals connected to it (operating in uncontrolled mode). Since the computer has no control over the data, it can accept the data from each of the terminals on the buffers of each of the lines.

Most controlled terminals are usually buffered, which allows them to run on high-speed shareable lines between a number of terminals. The terminals in one building/floor can be connected to one line through a cluster controller or daisy-chain (concatenated) connection to the computer. The terminals (buffered or unbuffered) have become an integral part of the requirements of any application system. Various applications of terminals include teleprinter, remote job entry, facsimile, banking data collection, credit/debit, etc.

4.6.1 Input–output devices

The keyboard input device is very popular and offers effective input capabilities to both terminals and computers. It includes different types of keys which in turn depend on the application. But in general, the keys can be grouped into three groups: **textual** (defining text, numbers, special symbols and characters, punctuation signs, etc.), **control** (dealing with various options during the editing of the text and its transmission) and **macro** functional (defined for special sequences of operations which may include data processing, etc.). For converting the functions of the keys into a serial bit stream for transmitting over transmission link or bus, various codes have been defined. The codes have a fixed number of bits for each key; e.g., ASCII and EBCDIC define eight-bit code, but ASCII has become a more popular code than EBCDIC.

Output devices accept a bit stream from the terminal and generate the symbol. The representation of the symbols can be defined in various forms: regular full printout or dot matrix printout. The regular full matrix printout offers better quality of characters but is expensive.

The above-mentioned input and output devices can send/receive one character at a time, making the devices very slow. The speed of transmission of characters can be made faster by using a buffer (a small segment of memory) which will store a block of characters and transmit/receive it between input and output devices. This type of device (having a buffer) is known as a **buffered terminal**. In buffered terminals, the processing capability of data can be included to make intelligent terminals. An intelligent terminal, in fact, is nothing more than a combination of a microprogrammed computer (with a stored program in ROM) and attached input/output devices. In general, these intelligent terminals may be characterized by various features such as buffer storage, error-detection capability, automatic repeat request (ARQ) facilities, support of stuffing (bit or character), automatic answering, data compaction, clustering of various terminals, back-up storage, etc.

As we don't have a standardized definition of intelligent terminals, different vendors are using different criteria for describing some of the features of their products. For example, from one vendor's point of view, intelligent terminals are more expensive than dumb terminals, while another vendor may say that intelligent terminals are buffered terminals associated with an input device (keyboard), teleprinter, an output device, and other relevant components. Further, these terminals are monitored by software while dumb terminals are not controlled by software; instead, they simply forward the message from one node to another node. Based on this, we also see intelligent or smart facsimile (fax) machines available from different vendors in the market.

In any event, intelligent terminals possess a variety of features such as buffering, functional key, processing capabilities, software-supported keyboards, paging, automatic transmission of blocks of characters, etc. These terminals have found applications in word processing, time sharing, electronic mail services, data entry, query, etc. The storage capabilities of intelligent terminals can be enhanced by introducing auxiliary storage (e.g., floppy diskettes or tape devices).

4.6.2 Classes of terminals

There are varieties of terminals, and it is very difficult to define classes of terminals. A discussion on different types of terminals can also be found in References 5–7. For the sake of knowing their functionality and type of transmission mode, we divide them into two classes: functionality-based and transmission mode–based.

4.6.2.1 Functionality-based terminals

There are three versions of functionality-based terminals: character-oriented, high-speed, and programmable.

Character-oriented terminals: These are inexpensive, offer no data-handling capability, have little or no buffering, and operate at very low speed. These are asynchronous and generally operate in uncontrolled mode. They are not useful for applications such as in-house lines, local area dial-up lines, etc.

High-speed terminals: These terminals, in general, have more logical functions than character-oriented terminals. These have storage (such as tape drive) and operate either synchronously or asynchronously at high speed. These provide security (useful in applications such as banking, allied fields, etc.) and can be used for off-line data entry (with printers, tape drives, etc.).

Programmable terminals: These terminals have computing capability which can be programmed. Most of these terminals communicate just like simple unbuffered character-oriented terminals. They have data-handling capability and are commonly used in remote data entry, detecting data entry errors, handling of entry errors, etc.

4.6.2.2 Transmission mode–based terminals

There are two categories of transmission mode–based terminals: synchronous and asynchronous. These terminals are low-speed devices for sending the characters to a mainframe and are connected to it through RS-232 ports. The mainframe, in general, supports both types of transmission modes for sending and receiving, therefore providing asynchronous and synchronous channels between terminals and mainframe.

Asynchronous terminals: As discussed previously, in asynchronous transmission, each character is preceded by a start bit and followed by stop bits. For this reason, these terminals are also known as start/stop terminals. At any time, a binary signal logic 1 (marking) is continuously transmitted from an asynchronous terminal's RS-232 port to RS-232 interface on the mainframe. When a character key is pressed, that particular character is transmitted to the mainframe via an RS-232 interface port. Similarly, the characters can be transmitted from the mainframe port to terminals. The data rate of asynchronous terminals and mainframe RS-232 should be the same. Since asynchronous terminals do not have any buffer, one character is sent between the terminals and the mainframe at any time.

Synchronous terminals: These terminals have a buffer capability and can store a block of data (constituting more than one character). When there is no digital signal, a flag bit pattern of 01111110 is continuously transmitted from terminals to mainframe. When a character is transmitted from the keyboard (by hitting a particular key), the synchronizing bits are also transmitted and the end of the block of data is recognized by the return key. Since the buffer stores a block of data, these terminals are also known as **block mode terminals (BMTs)**, e.g., IBM 3270, etc.

Some useful Web sites

For information on error-correcting codes, visit *http://www.amazon.com*. Courses on error-detecting codes are being offered in various universities such as *http://www.systems.caltech.edu*, *http://www.syr.edu*, and *http://www.theory.lcs.mit.edu*. For error-detection techniques, visit *http://www.ironbark.bendigo.latrobe.edu*, *http://www.engr.subr.edu*, *http://www.shakti.cs.gsu.edu* (also deals with wireless LANs and ATMs), and *http://www.fokus.gmd.de* (summary of error-recovery techniques in IP-based systems).

References

1. Tanenbaum, A.S., *Computer Networks*, Prentice Hall, 2nd ed., 1989.
2. Stallings, W., *Handbook of Computer Communications Standards*, vol. 2, 2nd ed., Howard W. Sams, 1990.
3. Markeley, R.W., *Data Communications and Interoperability*, Prentice-Hall, 1989.

4. Black, U., *Data Networks: Concepts, Theory and Practice.* Prentice-Hall, 1989.
5. Spragins, J.D., *Telecommunications: Protocols and Design.* Addison-Wesley, 1991.
6. Techo, R., *Data Communications: An Introduction to Concepts and Design.* Plenum Press, 1984.
7. Loomis, M.E.S., *Data Communication.* Prentice-Hall, 1983.

chapter five

Transmission media

> "Each year it takes less time to cross the country and more time to get to work."
>
> Mary Waldrip

5.1 Introduction

In order to transmit analog data (voice, speech, or any other analog signal) and digital data (from computers, terminals, or a digitized signal) between a sending node transmitter and a receiving node receiver, a transmission medium defines a physical link between them. The frequency bandwidth offered by a medium represents the frequency range of the signal transmitted and is expressed in cycles per second (Hertz). The bandwidth and quality of a signal transmitted on the medium depend on a number of its components such as resistance, capacitance, inductance, distance, and other parameters such as atmospheric noise, interference, etc. The loss of magnitude of signal due to these factors is known as *attenuation* (expressed in decibels) while the quality of signal is measured through *signal-to-noise ratio*. Since the objective of transmission media is to transmit an information signal from one node to another, it should provide a communication channel that is error-free and has greater speed, larger bandwidth, reliability, and many other features.

For transmitting analog and digital information from one node to another node, a communication system consists of three blocks: (1) transmitting equipment, (2) communication medium (for providing a logical connection over the physical link and path between them), and (3) receiving equipment.

Transmitting equipment accepts input information from different types of transducers and converts it into an appropriate electrical signal, which can travel over the transmission medium. The transmission medium may be a physical wired connection or logical wireless connection. Receiving equipment receives the electrical signals from the medium and reconverts them back into the original information. These three parts of a communication system have to be compatible with each other from a physical connection point of view and also from the point of view of behavior and properties of these parts.

The communication channel over transmission media plays an important role in a telecommunication system and has become an integral part of any communication system (to be used on land, in water, or in space). In this chapter, we review the essential characteristics and properties of different types of transmission media from the points of view of telecommunication systems, which are using them for data transfer over communication channels.

5.2 Communication channels

A communication channel or link provides physical paths for transmitting the signals between computers connected to a network. The channels are usually provided by telephone common carriers, and these channels access the **public switched telephone network (PSTN)**, switched networks, or even leased private lines.

Over the years, communication channels have evolved with great emphasis on multiplexing of multiple channels at higher data rates over the same link of channel. Telephone lines were introduced in the late 1890s, and the concept of multiplexing was used over a pair of telephone wires to carry more than one-voice channels in the 1920s. In the early 1940s, the coaxial cable was used for carrying 600 multiplexed voice channels and 1800 channels with multiple-link pair configurations. The total data rate offered by these 600 channels is about 100 Mbps. The number of channels that can be multiplexed over the same coaxial cable has continuously increased with the technology changes in switching and transmission techniques, and hence the data rates are also becoming higher for voice communication. Some of the most popular numbers of channels over coaxial cable are 1860 voice channels (with a data rate of about 600 Mbps), 10,800 voice channels (with a data rate of 6 Gbps), and 13,200 voice channels (with a data rate of 9 Gbps).

With the introduction of fiber optics in 1990, we can get data rates of over 10 Gbps, with such advantages of fiber over cable as high bandwidth, low error rate, greater reliability, immunity to external noises, etc. At the same time, another communication link was created based on wireless transmission through microwave link. This offers a number of voice channels ranging from 2400 in 1960 to over 42,000 in the 1980s and still rising.

5.2.1 Channel bands

The channels support a variety of data communications options that can be grouped into three categories: **narrow-band**, **voice-band**, and **broadband**. Narrow-band communication includes applications requiring data rates of up to 300 Kbps. Voice-band communication usually defines the bandwidth of voice signals in the range of 300 Hz to 3.4 KHz. These voice signals can be transmitted above a transmission speed of 9600 bps. Broadband communication offers data communication higher than a voice-band channel (up to several Mbps). Broadband communication channels are more reliable and efficient than the other two communication channels.

5.2.2 Channel services

Communication channels are used to access both switched and leased services of networks. **Switched services** are being offered on public switched networks and sometimes are also known as **dial-up services**. Switched services are controlled and managed by AT&T, Data-Phone system, and Western Union broadband channel. These companies send billing information based on the use of switched services (duration of usage). A public switched network may have a long connect time, disconnect time, and busy time, as it is based on a connectionless configuration. **Leased services**, on the other hand, are available on dedicated lines or private lines. The leased or private lines are always physically connected and available to the users. The billing information is based on a 24-hour-a-day and 7-day-a-week basis. The response times are relatively shorter than switched network services.

Common carriers offer a variety of services that are being used in data communication networks. Some carriers include AT&T, Western Union, Datran, and MCI WorldCom. MCI WorldCom uses a microwave link for providing physical connection, while Datran uses the switched network for the transfer of data. Each of the these physical connections provided by carriers possesses some minimum features and maximum data transfer speed. For example, if we consider leased lines, the tariff is usually divided into the following categories (depending on the type of application): teletype, sub-voice, voice, and wide-band. The teletype (or telegraph-grade) communication channels usually provide a maximum transmission speed of 75 bauds (i.e., 100 words per minute). In order to send digital signals, high-speed lines, digital-to-analog, and analog-to-digital conversion is required at transmitting and receiving sides, respectively. For low speed, the direct current (DC) carrier avoids the use of digital-to-analog signals.

5.2.3 Channel standard interfaces

A transmission speed of 180 baud (15 characters per second) over leased lines is available on sub-voice-grade channels. A higher transmission speed is available on voice-grade lines. The voice-grade channels can offer as much as 10,000 baud without distortions, but usually they are used in systems with a transmission speed of 2400 baud. The wide-band channels offer the highest transmission speed: 60,000 to 120,000 bauds.

The modem device used with telephone lines allows the transmission of digital data communication over analog telephone lines. It converts the digital signal (pulses) coming out of a computer or terminal to an analog signal acceptable for its transmission over telephone lines. A modulator performs this conversion of digital signals into analog. A demodulator recovers the original digital signal from the received modulated signal at the receiving side by converting the analog signals into digital form. Modems available on the market provide support to synchronous and asynchronous transmission modes and are thus known as **synchronous** and **asynchronous modems**, respectively.

Synchronous modems are usually used with continuous data signals coming from devices such as magnetic tape, computers, etc., while asynchronous modems are usually used with keyboards, terminal devices such as teletype-writers, CRT displays, terminals, etc. In general, modems support all three transmission configurations: simplex, half-duplex, and full-duplex. Details about these are provided in Chapter 3.

The speed or data rate of modems is expressed in baud or bits per second; modems are available in a variety of forms with data rates of 1200, 2400, 4800, 9600, and 56,000 baud.

The following sections detail different types of transmission media that have been defined for computer networks. More details on these media and their characteristics can be found in References 1–5.

5.3 Open-wire lines

An open-wire line consists of two uninsulated pairs of conductors or wires hung in air and supported by poles. Since wires are exposed to open air, they are affected by atmospheric conditions, interference from high-voltage power lines, corrosion, etc., which affect the quality of signal transmitted over them. The attenuation coefficient of this medium typically ranges from 0.04 to 0.1 dB per 1.6 km at a frequency of 1 KHz. Thus, for long-distance transmission, amplifiers are needed at a regular distance of 40–50 km over the distance. Mutual interference of electromagnetic waves between two wires causes **cross talk**, and it becomes significant and affects the quality of signals.

A number of technologies have been used to reduce cross talk, such as transpositions (crossing the wires of each line periodically), twisting the wires around each other, etc. Due to the second method of reducing cross talk, these wires are known as **twisted-pair wires/cables.** The twisting of wires reduces the interference if two conductors carry equal currents in opposite directions. The magnetic fields created by these two conductors will cancel each other out. Twisting the wires can also reduce the coupling effect caused by stray capacitance between adjacent wires.

5.4 *Twisted-pair wires/cables*

In twisted-pair wires or cables, two wires are twisted together to make a first pair of twisted wires. Now we twist together two pairs of twisted wires (as obtained above) to form a **quad**. Cross talk is further eliminated by changing the direction of twist in each pair. Usually, a number of pairs of these wires are put together into a cable. The cable may contain more than a hundred pairs of wires for long-distance communications. The wires usually have a diameter of 0.016 to 0.036 in.

Twisted-pair wires are the most common media in a telephone network. These wires support both analog and digital signals and can transmit the signal at a speed of 10 Mbps over a short distance. The existing telephone lines in a building/organization can be used for data communication inexpensively. They have been used in a number of applications, such as a subscriber loop connecting the telephone system (home or office) to the exchange or end office, an in-house private branch exchange (PBX) connected to telephones within the building, a connection to a digital switch or digital PBX within the building for a data rate of 64 Kbps, PC LANs within the building, etc. With PC LANS, a data rate of between 10 Mbps and 100 Mbps can be achieved. For a long-distance PC LAN, a data rate of 4 Mbps or more is possible. It can be used for transmitting both analog and digital signals. In analog transmission, an amplifier is usually used after every three to four miles, while repeaters are needed for every one to two miles in the case of digital transmission. The twisted pair has a limited data rate, distance, and bandwidth. Interference, attenuation, noise, frequency interference, etc. can affect it.

The twisting of wires with different twisting lengths reduces the effect of cross talk and low-frequency interference. It also offers a different bandwidth for different types of signal transmission. For example, a bandwidth of about 270 KHz for point-to-point analog signaling and a bandwidth of a few megabits per second for long-distance point-to-point digital signaling is possible. For a short distance, a higher bandwidth can be achieved in digital signaling.

Twisted pair (TP) copper wires are mainly used between the central office (CO) of a telephone exchange and a subscriber's handset, either at the office or home, and have been installed in most buildings and offices. They are also being used in some LANs. The TP is typically a pair of solid wires (American Wire Gauge numbers 22 and 26). The gauge number indicates the diameter of the wire. Each wire in TP is insulated separately. The pair of wires or a group of pairs can be either a **shielded twisted pair (STP)** or **unshielded twisted pair (UTP)**.

UTPs are cheaper, more flexible, and easier to install. They provide enough support for telephone systems and are not covered by metal insulation. They offer acceptable performance for a long-distance signal transmission, but as they are uninsulated, they are affected by cross talk, atmospheric conditions, electromagnetic interference, and adjacent twisted pairs, as well as by any noise generated nearby. The majority of the telephone twisted pairs are unshielded and can transmit signals at a speed of 10 Mbps.

STPs are usually used for about 1 MHz, but with different types of distortion-reduction techniques, they may be used for several megahertz. Coaxial cable, on other hand, can be used for about 600 MHz, and this range can be extended to gigahertz with special distortion-reduction devices. Both media are good for low-frequency applications and can be used for any frequency application with reasonable attenuation tolerances. This attenuation tolerance concept at higher frequencies is being used in new media such as advanced digital subscriber line (ADSL).

Some of the material below is derived partially from References 1, 3, 4, and 8.

Shielding by metallic braid or using sheathing can reduce interference. The Electronics Industries Association (EIA) published Standard EIA-568 for UTPs and STPs for data communications. This standard offers a data rate of 1 to 16 Mbps over LANs within buildings and offices. The EIA-568-A standard defined 150-Ω shielded twisted-pair and 100-Ω unshielded twisted-pair cables in 1995. EIA-568-A has defined the following five categories of UTPs. A brief discussion of each of these categories is presented below:

Category 1 UTP: This is the oldest type of copper-wire telephone cable. These cables use paper and rubber and offer a very low speed of data communication (about 2400 bps).

Category 2 UTP: These wires are made of copper and have been used for telephone systems since the early 1980s. IBM used these wires in its IBM token ring LAN that offers a data rate of 4 Mbps.

Category 3 UTP: This category is defined mainly for voice communication (voice-grade twisted pair) and is already in existence in many old buildings. This standard wire was introduced basically for popular IEEE 802.3 and 802.5 LANs. The 802.3 LAN is 10 B-T and offers a data rate of 10 Mbps, while the 802.5 offers a data rate of 16 Mbps. The 802.5 can also be used for 100 B-T LANs but has not been recommended (Category 5 has been recommended for these high-speed LANs). It is a 100-Ω twisted-pair cable and is being used extensively with LANs. This medium offers a limited data rate and distance and a bandwidth of 16 MHz. It has fewer twists per unit of length and offers a data rate of 16 Mbps.

Category 4 UTP: This cable offers data rates of up to 20 Mbps. Four-pair cable can, of course, be used for high-speed LANs (100 B-T). Again, it has been recommended that Category 5 be used for high-speed LANs. This medium has more twists than Category 3 and offers a bandwidth of 20 MHz.

Category 5 UTP: This category is defined mainly for data transmission at a rate of 100 Mbps over a short distance. It is more expensive than categories 3 and 4, as it has more twists per unit of length, but it offers better performance than those UTPs. Category 5 is becoming very popular in new buildings and offices. This cable has been defined for use in high-speed LANs offering data rates of 100 Mbps. This cable is of unshielded type and can be used for data rates over 500 Mbps.

Shielded twisted-pair (STP): Shielded open-wire lines are covered by metal insulation and hence are immune to cross talk and atmospheric conditions. By putting many pairs of insulated wire into one protective sheath (or insulating jacket), we can construct a cable. Insulated wires are twisted so as to reduce the inductive coupling between the pairs. These wires use copper as a conductor, although aluminum (lighter and cheaper than copper) can also be used. Four insulated wires are twisted together to form a **quad**. A sheath provides external electrical protection to these quads. Wired cable using wood pulp insulation can house 600–3600 pairs of wires per cable. If it uses plastic insulation (also known as **polyethylene insulated cable (PIC)**), it can house 6–1800 pairs per cable.

Plenum cable: The installation of TP is typically in ventilation airways or ducts or plenum, as plenum goes to all rooms in buildings. The National Electric Code (NEC) has defined a specific insulation for a higher resistance for plenum cable. It reduces smoke inhalation injuries, and the insulation in plenum cable is Teflon in the U.S.

As mentioned earlier, TPs can also be used with LANs and are available as 10BASE T or 100BASE T LANs. The number before "BASE" represents the data rate, while "T" means the twisted pair. BASE means one frequency. The IBM STP uses a different standard known as IBM Type 1 STP. This is mainly defined for a token ring LAN. This type consists of two pairs of STP per cable, and the entire cable is insulated.

The typical wire-pair cable system used in the U.S. is based on a K-carrier and N-carrier which can carry 12 voice channels over a distance of 30 to 290 Km. The wired cable usually operates at a low-frequency range of up to 1 MHz.

There are two types of twisted open-wire lines: **balanced-pair lines** and **unbalanced-pair lines**. In balanced-pair lines, we use two wires twisted around each other, while in unbalanced, one wire carries the current while the other wire is grounded to provide a return path for the current. Due to ground disturbances such as stray capacitance, soil conditions, etc., unbalanced wire lines are affected by ground distortions and hence usually are not used.

With balanced-pair lines, amplifiers/repeaters can be used for analog/digital signals. With analog signals, an amplifier is used to boost the amplitude of the analog signal at a regular interval of every 6–8 km, while for digital signals, repeaters are used at a regular distance of 3–4 km. A twisted pair can carry 24 voice channels (each of 4 KHz), giving a total bandwidth for analog communication of 96 KHz. Twisted pairs can be used for star, ring, bus, and tree topologies (discussed in Chapter 7) of LANs within a building or at a maximum distance of 10–15 km. They are very cheap, but installation increases the cost.

Open-wire lines have been used in high-voltage power transmission lines, telephone conversations, telemetry, and even data communication. Wires are usually defined by their diameter in terms of the **American Wire Gauge (AWG)** system. A smaller gauge number indicates that the wire is of larger diameter. Further, the larger diameter represents less resistance of the wire, which in turn will increase the bit transfer rate on the medium. In general, smaller-gauge wires are used for long-distance transmission, while the majority of the twisted pairs of wires in telephones are of larger-gauge size (typically 24 or 26). A local loop usually employs a gauge size 22 or 24.

The twisted-pair cable can be used as a communication medium in the following applications:

1. Long-distance transmission at low frequencies.
2. Subscriber cables with 2400 or 3400 pairs for low frequencies (telephone conversation) or musical broadcasting (160 to 355 KHz).
3. Urban interchange cable with hundreds of pairs.
4. Analog signal (bandwidth of 4 KHz per channel for 24 voice channels, giving a bandwidth of 268 KHz) with amplifiers at a regular distance of 6–8 km.
5. Digital signal using a modem and supporting a data speed of up to 9600 bps for 24 channels.
6. Bell T1 (digital carrier) with 24 PCM voice channels, giving a data rate of 1.544 Mbps.
7. Local area networks.

Twisted pairs using copper telephone lines have a very high error rate at higher speeds and poor security. They are used to connect home or office telephones with the end office or central office and carry the analog signal. The telecommunication industries convert these analog signals into digital signals (using an analog-to-digital converter) before they can be sent over PSTN. On the receiving side, the digital signals are converted back into analog form (using a digital-to-analog converter) before they can be sent to a subscriber's phone. In some books, both A/D and D/A converters are collectively known as (COder/DECoders) (codecs). Codecs are usually eight-bit. Japan and North America use

Mu-law-based codecs, while Europeans use A-law-based codecs. The Mu-law codec gives higher gain at a low-input voltage and less gain at a higher input voltage. The A-law codec does just the opposite.

Codec: A coder/decoder (codec) provides coding of voice into digits, as required in digital lines. It looks at an analog sample from analog circuits and maps the amplitudes in one of 256 different levels (as it uses eight bits). On the receiving side, these bits are used as an index to the same scale of amplitudes to reconstruct the signal. Usually the scale used to define the amplitude does not provide a linear relationship, so we need to define a scale that is characterized by gradations that represent a better replication of a speech signal. The gradation points are closer together for lower amplitudes and are further apart for higher amplitude. The technique using this concept is known as *companding*. The sender uses COMPression technique for amplitude signals while the receiver expANDs the signal. There are two standards defined for companding technique. In North America, the companding scale is based on Mu-law, while a scale based on A-law is used in Europe.

5.5 Coaxial cable

With an increase in frequency, the flow of current in a metal conductor develops a tendency to go toward the shallow zone on the surface, thus causing an increase in the effective resistance of the wire. This is known as the **skin effect**. In order to reduce the skin effect, coaxial cable has been designed as a skin with a center conductor that is suitable for high frequency. Coaxial cable (coax) consists of two concentric copper conductors. The inner conductor is surrounded by insulating sheets to provide support to it, and both of these are then covered by another copper or aluminum conductor which supports them through a cylindrical architecture. A protective cover (sheath) is then wrapped around the entire unit to prevent interference from signals from other media, as shown in Figure 5.1. The inner conductor is of very small gauge size and hence can be used for long-distance transmission, local area networks, toll trunk, and carrier systems (10,800 telephone channels per pair at maximum frequency of 60 MHz with a data rate of a few Mbps). Coaxial cables are not useful in applications which have frequencies lower than 60 KHz. They usually transmit the signal in one direction, which is why they are available in pairs.

Coaxial cables are used as unbalanced circuits where an inner copper conductor carries the current while an outside shield (insulation) is grounded. Due to this, unbalanced coaxial cables have more alternation than balanced wire pairs, but they are cheaper and simpler to install and maintain. They offer larger bandwidths for long distance and, as such, they are very popular and widely used. In particular, the advent of **integrated digital networks (IDNs)** and **integrated services digital networks (ISDNs)** has given a boost to

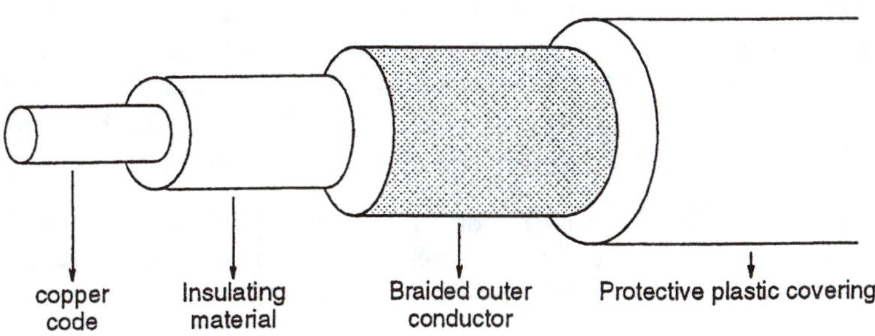

Figure 5.1 Coaxial cable.

coaxial cable as the leading medium for providing digital services to subscribers over a long distance. It can be used to transmit both analog and digital signals. Due to shielding, it is less susceptible to interference and cross talk than twisted-pair cable. For a long-haul analog or digital transmission, amplifiers and repeaters are used every mile or so to get higher frequency and data rates.

Coaxial cable is immune to electrical/noise interference (due to shielding), and the shielding is effective at higher frequencies (>1 MHz). Voice, data, and video signals can be transmitted simultaneously on it. The design of construction of coaxial cable reduces channel noise and eliminates cross talk. The self-shielding property prevents any interference from external signals and, in fact, is effective only at higher frequencies (higher than 1 MHz). Due to this property of shielding, the capacitance, inductance, and conductance are usually considered independent of frequency. Here, the resistance changes due to the skin effect and is proportional to the square root of the frequency. Coaxial cables are used in television systems, long-haul telephone networks, short-distance high-speed connectivity within computer systems, LANs, etc. Cable TV, which started as community antenna television (CATV), offers entertainment and communication services to remote locations serving homes and offices over long distances (a few hundred miles). Typically, a TV channel requires a bandwidth of 6 MHz, and cable carries hundreds of TV channels. Between 10,000 and 15,000 voice-grade channels (each of 4 KHz) can be transmitted over cable simultaneously. Other media in great demand are optical fiber, satellite, and microwaves, which carry many more voice-grade channels and TV channels than coaxial cables.

5.5.1 Classes of cable transmission

For each type of electrical transmission (baseband and broadband), we have a separate coaxial cable. **Baseband transmission** supports analog signals and uses coaxial cable of 75-Ω resistance. **Broadband transmission**, on the other hand, is a digital electrical transmission and uses broadband coaxial of 50 Ω. This supports data transmission at a speed of more than 100 Mbps up to 400 Mbps.

Depending on the type of LAN being used, we may expect that LANs will support connection of a few tens to tens of thousands of devices (PCs, workstations, terminals, peripheral devices, printers, etc.). LANs may be characterized on the basis of a number of parameters, such as topology, access-control technique, transmission media, etc. The bus or tree topology–based LANs are one of the earliest defined and implemented LANs. These LANs include Ethernet (based on baseband bus), MITREenet (based on bus/tree broadband) and many other popular low-cost, twisted-pair LANs available for PCs. The difference between LANs, in fact, is defined in terms of a number of parameters such as topology, access media control, switching, signaling, etc., and as such we have seen different types of standard LANs (IEEE 802 recommendations).

Baseband and broadband are the switching techniques defined over transmission media and are distinguished by the type of signaling (digital or analog) used for the communication. The baseband switching technique is supported by twisted-pair and coaxial cable, while broadband coaxial cable uses analog signaling using an RF modem. Standard LANs may be categorized as baseband or broadband depending on the switching and signaling techniques, as discussed in the following section.

Baseband signaling for LANs: A baseband LAN uses digital signaling. It does not use any analog carrier. It requires a bus topology and the digital signals are transmitted using Manchester or differential Manchester encoding techniques and get the entire bandwidth of the medium. For this reason, it does not use frequency division multiplexing (FDM). It uses twisted-pair cable, coaxial cable, fiber optics, and microwave transmission media.

It offers one channel and supports a variety of topologies such as ring, bus, and tree. It transmits both analog and digital signals in their original forms without using any modulations. It is bidirectional, i.e., uses only one cable, and signals travel in both directions. The only problem with baseband LANs is that they offer a limited capacity over a limited distance.

In this type of transmission, the signal is directly impressed onto the underlying media in binary form (via Manchester encoding). In this encoding scheme, logic 1 is described by low-to-high transition, while logic 0 is described by high-to-low transition; therefore, it does not require any extra clock pulse for synchronization. Manchester encoding is very popular with Ethernet LANs.

An equally popular and widely used encoding technique is differential Manchester encoding. In this encoding scheme, transition is always defined at the center of each interval of the pulse in such a way that logic 1 is represented by no transition at a pulse boundary while logic 0 is represented by a transition at the pulse boundary. This scheme is widely used in the IEEE 802.5 IBM token ring LAN. It is simpler and very inexpensive, as the layout of the cable is very simple.

Baseband-based LANs may use any cable (coaxial, twisted pair, etc.), and no modulation is required because they support only very short distances (typically a few kilometers). They use 50-Ω cable and support multiple lines. For multiple access to the channel, baseband LANs use time division multiplexing (TDM). They do not divide the bandwidth for data communication but instead allocate the entire bandwidth to the channel, thus offering higher data rates. Only one datum can be transmitted at a time. A device known as a *transceiver* usually provides the access method to a baseband system (usually Ethernet). This device receives the data from the node and defines the packets for transmitting over the bus. The packet contains a source address, destination address, user's data, and some control information. Baseband LANs are inexpensive and very simple. The shielded twisted-pair cable offers more capacity than the twisted-pair baseband, but it is more expensive. An unshielded twisted pair may allow users to use the existing wiring system (telephone wiring, etc.).

Some vendors of baseband LANs include Datapoint Corporation (Texas), Starnet Data Systems (Colorado), and Xerox Corporation (Connecticut). Typically, the baseband LANs from various vendors include the following specifications:

Topology: bus, tree, logical ring, star, twin ring
Access method: CSMA/CD, CSMA/CA, token passing
DTE interface: RS-232-C, RS-449, IEEE-488, RS-423
Number of connections: varies between 32 and 16,000

Broadband signaling for LANs: These LANs use analog technology and a high-frequency modem to define carrier signals for the channel. They are based on the concept of broadband systems for transmitting data, voice, and video signals through cable television (CATV, or community antenna TV). They support and carry a wide variety of traffic and offer very high capacity and data rates. The carrier signal generated by high-frequency modems is used for modulating the user's digital data. A digital-signaling switching technique is used for data communication and allows the entire bandwidth to be used by the data signal. Typically, it offers a channel having a bandwidth greater than a voice-grade channel (4 KHz). In broadband transmission, the multiple channels are modulated to different frequencies using FDM and need a headend. The analog signals are usually in the radio frequency range of 5–400 MHz and typically use 75-Ω cable.

In broadband LANs, a single communication channel is shared by a number of nodes (machines) over the networks. A packet transmitted from any node contains the destination address and also allows the multicasting (addresses of a group of nodes over the network) or broadcasting (addresses of all the nodes over the network) of the same packet. The broadband LANs contain the path device and interface unit. The interface unit contains a radio frequency (RF) modem, which accepts the data from the device and converts it into the RF of the inbound band. The network interface packetizes the digital data into analog radio frequency signals.

Another form of broadband is known as a **single-channel broadband** or **carrierband**. Here the entire frequency spectrum of cable is allocated to a single transmission circuit for analog signals. It supports bidirectional transmission and uses bus topology. It does not require any amplifiers or any headend. It offers very low attenuation at low frequency and uses some form of FSK modulation.

CATV systems provide a frequency bandwidth of 300 MHz to 400 MHz and support 39 to 58 television channels each of 6 MHz. The cable used by CATV is of 75-Ω impedance and provides modulated signals for larger distances. The digital signals need to be converted to analog signals (via radio-frequency modem) to be transmitted over CATV. Here, the entire bandwidth is divided into channels each of 6 MHz, which are further divided into sub-channels of smaller bandwidths. These sub-channels are multiplexed using either TDM or FDM.

Most of the broadband LANs use the CSMA/CD access technique and are based on bus topology (used in standard Ethernet LANs). 3Com has defined a broadband LAN based on bus topology using a token-passing access method (which is basically defined for token ring LANs) called LAN/1. Datapoint initially proposed the token-passing method for token ring LANs. The concept of broadband transmission can be achieved on two types of cable: **single (split channel)** and **dual cable**.

The **single-** or **split-channel cable** carries the network traffic in both directions on one cable and supports different bands of frequencies, e.g., **forward band, return band,** and **crossover band**. In the forward band mode, it carries the signals from the headend to the network in the frequency bandwidth of 160–300 MHz. The return band mode is opposite to the forward and carries the signals from nodes of networks to the headend in the frequency bandwidth of 5–110 MHz. The headend device receives RF-modulated signals and distributes them to various connected receiving nodes. In a restricted sense, this can be considered a repeater for providing frequency conversion. Finally, the crossover band mode of a single cable acts as a guard band to avoid any electrical interference in the frequency bandwidth of 110–160 MHz.

There are three main schemes for split channel available: **subsplit**, **midsplit**, and **highsplit**. Each of the schemes defines a different frequency range for inbound and outbound channels.

The dual cable provides different segments of cable for the transmission and reception of signals in opposite directions. The forward cable segment is used basically as a receive cable or line in the frequency range of 5–300 MHz, while the return cable segment is used for transmitting signals in the same frequency range of 5–300 MHz. This type of cable does not need headend but may use amplifiers. One of the main advantages of dual cable broadband mode is that there is a savings of frequency bandwidth, as the same frequency range can be used on different cable segments. Since we have to use two separate cables, it is very expensive, because two ports are required on computer systems (terminals, PCs, or workstations) to use dual cable.

A partial list of broadband vendors includes Concord Data Systems (Massachusetts), Gould, Inc. (California), and Proteon Associates, Inc. (Massachusetts). Typically, the broadband LANs offered by these vendors include the following specifications:

Topology: bus, logical ring, tree, physical ring
Access Method: CSMA/CD, CSMA, token passing, FDM
DTE interface: RS-232-C, RS-449, V.35, IEEE-488
Supporting protocols for DTE: BSC, HDLC, TTY
Number of connections: varies between 32 and 62, and 535.

5.5.2 Control signaling

Control signaling provides a means for managing and controlling various activities in the networks. It is being exchanged between various switching elements and also between switches and networks. Every aspect of networks' behavior is controlled and managed by these **signaling functions**. With circuit switching, the most popular control signaling has been **inchannel**, in which the same physical channel is used to carry both control and data signals. This certainly offers advantages in terms of reduced transmission setup for signaling. Inchannel signaling can be implemented in two different ways: **Inband** and **out-of-band**.

In inband inchannel signaling, the control signaling uses the same physical channel and frequency within voice-grade range. In this way, both control and data signals can have the same properties and share the same channel and transmission setup. Both signals can travel with the same behavior of the network anywhere within the network at any location. Further, if any conversions (analog-to-digital or digital-to-analog) are performed in the networks, they will be used by both signals in the same way over the same path.

In out-of-band inchannel signaling, a separate frequency narrow band is defined within voice-grade range (4 KHz) for control signals that can be transmitted without even sending data signals. In this way, the monitoring of the channel is continuously performed. Due to lower frequency for control signals, the transmission speed of signaling is lower and, furthermore, the logic circuit to separate these signals is also complex.

Both of these methods of inchannel signaling suffer from the fact that the circuit switched networks always introduce delay and the frequency bands for control signals are also very low. A new signaling technique known as **common channel signaling (CCS)** has been introduced which alleviates these drawbacks. The main philosophy behind CCS is to define two separate circuits for control and data signals (the details of CCS can be found in Chapters 6 and 18).

Broadband systems use two channels for transmitting data. One channel is known as the **out-of-band** channel (54–400 MHz) and is used for data leaving the headend toward the network, while the other channel, **inbound** (5–30 MHz), is used to carry the data from the network toward the headend. The headend can be considered a repeater, which provides frequency conversions of signals. It is an analog translation device, which accepts the data defined in inbound band channels and broadcasts it to outbound band channels. All the nodes send their data on inbound band channels and listen for their addresses on outbound band channels. There is a small band known as a **guard band** that separates these two bands. These two bands define a number of channels in their respective bands. Each channel can carry different types of signals, e.g., voice, video, data, etc., at different rates. No multiplexing of voice mail signal with video signal will be carried out. The channels work on bus and tree topologies and are suitable for multi-drop operations.

A headend is typically a signal frequency converter, which provides conversion between signal frequency channels. The data is always transmitted on an inbound channel and received on an outbound channel. Both channels may be defined in either one cable or two separate cables. Headends are used in bus and tree topologies of LANs and cover only a limited distance (a few kilometers). They use FDM for defining different channels corresponding to video, data, and audio signals over one circuit and always operate in one direction. Headends are used only in broadband transmission due to their unidirectional amplifiers. Typical applications of broadband include telemetry, integrated video and data, manufacturing, simultaneous operations in instrumentation, etc.

Plenum cable: This cable does not use PVC sheathing for insulation but instead uses Teflon. Plenum cable is used to carry the signals through ducts, as Teflon does not become toxic when burned. This cable is very expensive. On the other hand, non-plenum cables use PVC sheathing that becomes toxic when burnt.

Connectors: There are different types of connectors for cable available in the market. A brief description of some of these is given below. For details, please refer to Reference 8.

- **F-connector:** Used in TVs and VCRs; supports low UHF range of 500 MHz.
- **RCA plug and jack:** Available in audio equipment; connects phonograph or CD audio channels from phone/CD unit to the amplifier.
- **BNC connector:** The most popular connector for electronic equipment used in electronics laboratories, e.g., oscilloscope, signal generators, spectrum analyzers, and many others.
- **SMA connector:** Smaller than the BNC connector. Expensive. RCA, BNC, and SNA connectors have common applications.
- **N-connector:** Used in laboratory equipment dealing with UHF and SHF frequency bands. Larger than other connectors and very expensive.

5.6 *Optical fiber communication*

In the early days of communications, light was used to carry messages through the atmosphere from one point to another. Atmospheric conditions alternated the light and changed the direction of its transmission (as light passed through various layers of atmosphere), and the signals were degraded, affecting the quality of transmission. But if we propagate the light through a channel (free of atmospheric noises) which possesses known and stable characteristics and provides controlled reflections along the channel to the light, it can be used to carry the entire message (without any loss) from one node to another. This is true because the channel will provide a loss-free environment to the light and the total reflection offered by the channel can provide optimal transmission of light through it. The channel, tube, or dielectric guide is known as **fiber**, which has a refractive index greater than atmosphere. Light propagating through a glass-fiber channel (optical fibers) can be used to carry different types of signals such as voice, data, and video. An optical-fiber cable is of cylindrical shape and consists of three layers: core, cladding, and jacket. The core is the innermost layer consisting of a few thin strands or fibers (plastic or glass). The second layer, cladding, provides coating (glass or plastic) for each of the fibers. The jacket (the outermost layer) provides a covering for one or a bundle of cladded fibers. The main function of the jacket is to provide protection against moisture, environmental interference, crushing, etc. More details can be found in References 5 and 7.

As we have seen, metallic wire–based media (open wire, coaxial cables) have one problem in common: capacitive or inductive interference with electromagnetic radiation. Further, these media have a limitation on distance and also require the use of amplifiers and repeaters at a regular distance. Wireless technology defined a variety of communication media such as radio waves, microwaves, satellites, light waves, etc. The signals transmitted from these media in the atmosphere experience alternation (because they travel through various layers of atmosphere, they are reflected and refracted, so that energy contained in the signals gets spread in a wider area).

If we can define a medium (other than the atmosphere) which does not have atmospheric noises and provides a controlled optimal transmission of signals, then signals can be transmitted with less alternation over a long distance. For example, high-frequency signals such as radio signals and microwave signals can be transmitted in wave guides, light signals in optical fibers, etc. Impurities in wave guides and optical fibers pose a serious problem.

Fiber optics has been used as a medium of transmission in all broadband communication protocols. Two standards for specifications and interconnection of different network systems through fiber optics have been defined by international standards organizations: synchronous optical network (SONET, used mainly in North America and Canada) and synchronous digital hierarchy (SDH, used mainly in Europe). Both standards provide direct synchronous multiplexing to different types of traffic over their higher data. There is no need for converting the traffic signals into any format before they can be multiplexed; therefore, SONET/SDH networks can be interconnected directly to any network. Both SONET and SDH can be used in different types of networks such as long-haul, LANs, and loop carriers. CATV networks also use these standards for video transmission. One of the unique features of SONET/SDH is their ability to provide advanced network management and maintenance, since these standards allocate about 5% of their bandwidth for these functions.

The SONET standard is set at a 51.84-Mbps base signal for a new multiplexing hierarchy known as synchronous transport electrical signal 1 (STS1) and its equivalent, optical signal level 1 (OC1). This is equivalent to 1 DS 3 and 28 DS 1. Other optical levels are OC 3 (155.52 Mbps), OC 12 (622.08 Mbps), OC 24 (1.244 Gbps), OC 48 (2.488Gbps), OC 192 (9.6 Gbps), and so on. These are discussed in more detail in the following sections.

Features of optical fiber: Optical fiber used as a communication medium in telecommunication systems offers the following features (or advantages):

1. It supports a very large bandwidth (50–80 MHz in multi-mode fibers (MMF), several GHz in single-mode fiber (SMF)), allowing a high bit rate of transmission over thousands of miles. It offers a very high capacity compared to hundreds of megabits per second over a mile or so for coaxial cable and a few megabits per second for less than a mile of twisted pair. For example, 30,000 telephone voice channels can be transmitted simultaneously at a bandwidth of 500 MHz. It offers a bandwidth of 10^{14} to 10^{16} Hz.
2. It offers immunity to electromagnetic interference such as cross talk and electrical sparks and, thus, can be used in a very highly noninterference-type of electromagnetic environment.
3. Due to internal reflection inside the fiber, it has a minimal alternation coefficient, which requires fewer repeaters than that of coaxial cables for the same distance. The spacing between repeaters depends on factors such as data rate and type of fiber being used. Typically, spacing will be in the range of 20–40 miles as opposed to 2–4 miles in the case of coaxial cable. The Bell Systems T1 carrier uses coaxial cable and supports 32 channels at 1.544 Mbps, while the General Telecommunications and Electronics carrier supports 24 channels at 1.544 Mbps and uses coaxial cable. AT&T has developed a fiber-based communication system offering a data rate of over 3.5 Gbps over a distance of nearly 200 miles.
4. It has a very low error rate of 10^{-10} to 10^{-12}, as compared to 10^{-5} to 10^{-7} in coaxial cable.
5. It has very small dimensions (offering small size and volume), minimal insight, and higher mechanical flexibility.

6. It has very low sensitivity to range of temperature (low and high) variation and can easily be installed and operated.
7. The rapid and continuing decline in the cost of hardware chips (due to VLSI technology) will make this medium useful and popular.
8. It possesses a long life and long-term reliability.
9. It is impossible to detect or tap fiber-optic signals and, hence, it is highly secured communication.
10. Well-timed integration of optics and opto-electronics seems to have a great impact on economic conditions with respect to telecommunication services, speed of transmission, etc.

Various applications of optical fiber include long-haul communication trunks, metropolitan communication trunks, rural-exchange trunks, LANs, and subscriber loops.

The *long-haul telephone route* covers a large distance (a few hundred miles) in a telephone network and offers a very high capacity (20,000–65,000 voice-grade channels). In most countries, optical fiber or satellite from microwave and coaxial cables is slowly replacing the existing transmission medium. As compared to the long-haul telephone route, the *metropolitan telephone route* offers a short distance of a few miles and accommodates over 10,000 voice-grade channels. Typically, the routes are installed underground and are connected to long-haul telephone routes. The *rural-exchange telephone trunks* offer a route of a few miles covering a small village or town and accommodate over 5000 voice-grade channels. Typically, these trunks use microwave links. With *LANs*, optical fiber offers a higher data rate of 100 Mbps and allows a large number of stations within a building or office to share resources at that rate. The *subscriber loop* connects a subscriber phone with an exchange (central office) and offers a higher data rate for video images and audio compared with the twisted-pair cable currently being used.

Problems with optical fiber: In spite of these features/advantages, fiber optics suffer from the following problems:

1. The development of opto-electronics transducers, their optical interface with fiber, and reliability.
2. Electrical insulation between emitter, fiber, and receiver, and also the problem of supplying direct current power to intermediate repeater stations.
3. Impurities in fiber, mismatching, poor alignment of connectors, operating constraints, reliability issues, etc.
4. Difficulty in coupling of light signals from one fiber to another fiber.
5. Difficulty of installation and repair.
6. Expense of replacing the existing equipment for introducing and using fiber optics.
7. Lack of standardization.

In 1977, General Telephone and Electronics (GTE) introduced the first commercial optical fiber system with a 5.6-mile link providing highly secured communication. A fiber-optic-based LAN has been developed and is standardized as fiber-distributed data interface (FDDI), providing speed up to 100 Mbps. It has also been used in metropolitan area networks (MAN), the IEEE 802.6 standard LAN for a metropolitan area.

Synchronous optical network (SONET) hierarchy: With digital technology, most of the networks are shifting toward digital transmission over existing cables. The digital links so defined support the addition of overhead bytes to the data to be transmitted. This payload is a general term given to a frame (similar to the one defined in LANs). The frames used in LANs carry a lot of protocol-specific information such as synchronization,

Table 5.1 SONET Hierarchy

SDH	SONET	Data rate (Mbps)
	STS-1/OC-1	51.84
STM-1	STS-3/OC-3	155.52
STM-3	STS-9/OC-9	466.56
STM-4	STS-12/OC-12	622.08
STM-6	STS-18/OC-18	933.12
STM-8	STS-24/OC-24	1244.16 (1.244 Gbps)
STM-12	STS-36/OC-36	1866.24 (1.866 Gbps)
STM-16	STS-48/OC-48	2488.32 (2.488 Gbps)

error control, flow control, etc. In the case of ATM networks, the data is enclosed in these frames (if available). The frames are generated at a rate of 8000 frames per second, and digital links offer a very high data rate. The commonly used digital link in the U.S. is a synchronous optical network (SONET). The basic rate is defined as STS-1, which offers 51.84 Mbps, while STS-3 offers 155.52 Mbps.

The signals represented in electrical form are known as **Synchronous transport signals (STS)**. An equivalent optical representation is known as an **optical carrier (OC)**. SONET was formerly known as synchronous digital hierarchy (SDH). Both SDH and synchronous transport module (STM) are popular in Europe. The STM is defined within SDH. SONET, STS, and OC are all popular in the U.S. The SONET hierarchy is given in Table 5.1.

In this hierarchy, each row represents a level of multiplexing. For example, if we want to move from STM-4 to STM-5, the number of multiplexed data signals in STM-5 will be five times that of the STM-4 signals.

In order to provide internal reflections, a cladding with a lower refractive index covers the core of the fiber. Both core and cladding are usually made up of extremely pure quartz, transparent glass (silica), or plastic. The fibers are then inserted into a protective tube. The light signals traveling through optical fiber are usually in the form of binary pulses (ON/OFF).

Fiber communication system: The fiber-optic-based communication system is implemented using the following three components, as shown in Figure 5.2:

1. Modulation (light source)
2. Light transmission (fiber)
3. Demodulation (detector)

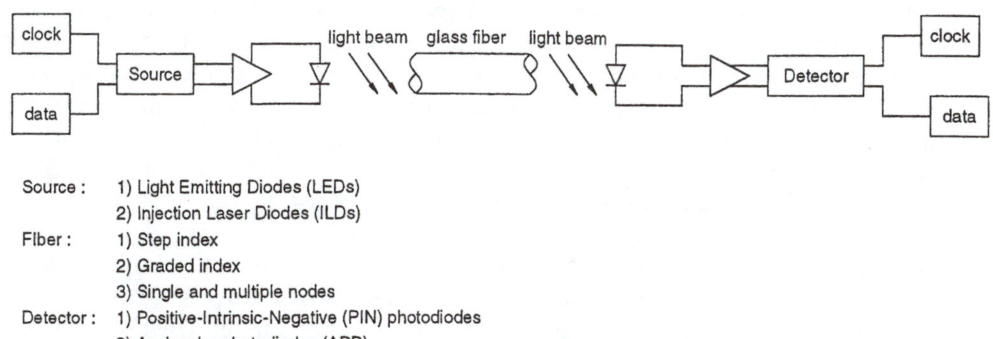

Figure 5.2 Optical fiber communication system.

Modulation consists of electro-optical transducers and opto-electronic transducers at both transmitting and receiving nodes. These transducers convert the electrical signals into light signals and light signals into electrical signals, respectively. These transducers are interfaced with fiber via an optical adapter. Two popular sources of fiber optical communications are **light emitting diodes (LEDs)** and **injection laser diodes (ILDs)**. Both are semiconductor devices that emit a beam of light with the application of volts. LED is cheaper, has a longer temperature range of operation, and has a longer operational life. ILD, on other hand, is more efficient (uses laser) than LED and offers greater data rates.

Various types of modulation (PCM, PPM, PFM, and FM) can be used to transmit the electrical signals, which will be input to an electro-optical transducer. Any amplification (analog signals) or regeneration (digital signals) has to be done before the transducer at the electrical signals level. LEDs and ILDs are used for producing a light beam in infrared wavelength (10^{12} Hz) range, while **positive intrinsic negative (PIN) photo diodes** and **avalanche photo diodes (APD)** are used for detecting the light generating appropriate electrical signals.

LEDs generate an incoherent light beam into fiber for optical communication of frequency up to 50 Mbps. LEDs are independent of temperature changes and are very cheap. LEDs consume a very small amount of power, typically on the order of a few microwatts. These devices suffer from intermodal dispersion (caused by distortions within the mode of operations). This is due to the fact that various segments of incoherent light travel with different speeds on different paths and cause pulse smearing.

ILDs, on other hand, generate a coherent light beam, and dispersion is eliminated. They offer high data rates of gigabits per second (Gbps). In contrast to LEDs, these devices are dependent on temperature changes and are more expensive. Due to their electrical characteristics, they dissipate higher power than LEDs.

Types of fibers: There are four types of optical fibers: step index, graded index, single mode, and multi-mode fibers.

The multi-mode step-index type of fiber contains a core of large size so that it can support multiple modes of light rays. It offers a clear distinction between core and clad and provides an easy interface to the source (LEDs, ILDs). One of the main problems with this fiber is that it suffers from inter-modal and intra-modal dispersions. The inter-modal dispersion is due to the fact that the light beam rays are traveling over different paths and that it offers dispersion between modes (i.e., this fiber may support different modes of operations but suffers from the distortions between those modes). The pulses generated are also affected by the smearing effect. The intra-modal dispersion is basically caused by the fact that the light beam rays are traveling with different speeds and that this fiber offers distortion within the mode of its operations. It also causes pulse smearing.

In summary, we can say that this type of fiber is hampered by different speeds and different paths taken by light beam rays and also by the distortions caused by different modes and within modes. This fiber is not suited for a high quality of transmission.

The **single-mode step index** type of fiber consists of a core that is just enough to support a single mode of operation. It eliminates inter-modal and intra-modal dispersions but is very expensive to manufacture.

The **multi-mode graded index** type of fiber does not distinguish between core and cladding and reduces the inter-modal dispersions.

Both single-mode and multi-mode fibers can support various wavelengths and use either a laser or LED light source. In a typical optical fiber, the light travels in three different wavelengths: 600 MHz, 1000 MHz, and 1500 MHz (in infrared range). The loss of light is lower at higher wavelengths, offering a higher data rate over a long distance. LED or laser sources with wavelengths of 1300 nm offer higher data rates. An even higher date rate can be achieved with 1500-nm laser sources.

The **positive-intrinsic negative (PIN)** diode is very inexpensive and has a very long life. It possesses low sensitivity and offers a relatively slower response time. In contrast to this, the **avalanche photo diode (APD)** possesses more sensitivity and a reasonable signal-to-noise ratio. It has a short lifetime and is very expensive. It offers a very fast response time.

For long-haul communications, a communication system based on optical communication uses an ILD as source, a single-mode fiber, and an APD detector. In general, the fiber is said to possess the following losses during the transmission: absorption, scattering, radiation, and coupling. These losses are self-explanatory.

Optical modulation: Optical fiber supports the following modulation schemes, known as optical modulation. These techniques are based on the variation of optical power according to electrical signals applied to the electro-optical transducers.

1. **Continuous:** In optics, we usually call this **intensity modulation (IM)**, similar to analog modulation (AM), with the only difference in the nature of the optical carrier signal (non-sinusoidal).
2. **Discrete:** This technique is known as **optical on–off keying (OOK)**, similar to shift keying with a sinusoidal carrier. The modulated signal is composed of pulses (ON/OFF) of constant amplitude.

It is recommended to use frequency pulse modulation (FPM) for analog transmission of a television channel by optical fiber. For more channels to be transmitted, PPM, which allows time-division multiplexing, can be used. The most effective use of optical fiber links is in digital transmission, as it offers a very large bandwidth with minimal error.

The features of optical fibers are very useful for a long-distance network and, in particular, the digital systems support implementation in optical form. CCITT has defined the following standards: 34 Mbps (480 channels), 140 Mbps (1920 channels), 565 Mbps (7680 channels), or multiples of these rates. These are four-wire systems with each transmission performed by a separate fiber. Fiber optics is used in local broadband networks, audio program broadcasting, duplex data transmission of one or more television channels, PCM form of analog voice channels, etc.

5.7 *Wireless communication*

As mentioned earlier, electromagnetic signals and light signals carry information in the atmosphere. In order to transmit and receive the signals in the case of unguided media, we need an antenna. The antenna transmits the electromagnetic signals into the atmosphere (usually air) and the receiver receives the signals through the antenna. The atmosphere has been divided into three regions: troposphere, stratosphere, and ionosphere.

- **Troposphere:** This region is at altitudes lower than 15 km and is characterized by clouds, winds etc.
- **Stratosphere:** This region is at altitudes from 15 to 40 km and is devoid of water vapor.
- **Ionosphere:** This region is at altitudes from 40 to 500 km, which is characterized by ionized layers, refraction, and reflection phenomena, and is useful for sky waves.

Electromagnetic signals are transmitted using different frequency ranges; allocation to a particular mode of operation and its use is made by the **International Radio Communications Consultative Committee (CCIR)** and **International Frequency Registration**

Table 5.2 Frequency Bandwidths and Their Applications in Telecommunications

Category	Frequency
Direct current (DC)	0 Hz
Alternating current (AC)	60 Hz
Low frequency (LF)	4 KHz
AM radio	500–1600 KHz
Medium frequency (broadcasting)	1.5 MHz
High frequency (broadcasting)	10 MHz
Very high frequency	100 MHz
Detroit police radio	2 MHz
Short wave radio	1.6–30 MHz
Highway mobile services	35–47 MHz
Ham radio	50–54 MHz
VHF-TV (channels 2, 3, 4)	54–72 MHz
VHF-TV (channels 5, 6)	76 MHz
FM radio (broadcasting)	88–108 MHz
First Bell system service	150 MHz
Mobile cellular radio (wireless communication)	105 MHz to 1 GHz
VHF-TV (channels 7–13)	174–216 MHz
Ham radio	430–450 MHz
UHF-TV	500–800 MHz
Cellular communication	825–890 MHz
Microwave terrestrial	1–500 GHz
Satellite (telecommunication)	1–50 GHz
Microwave radio	2–12 GHz (4, 6, 11GHz)
Visible light	1–4 THz (10^{12} Hz)
Optical fiber (data communication)	100–1000 THz
X-ray, gamma-ray, nuclear radiation	1 MTHz
Infrared	100 GHz to 10 THz
Ultraviolet	1–100 MTHz

Board (IFRB). Table 5.2 shows various frequency bandwidths and their applications in telecommunications. Similar tables can be found in References 1 and 10.

The following are the frequency bands allocated by the CCIR and IFRB to radio broadcasting (audio and video).

Long wave (150–285 KHz): In this transmission, the modulated signals travel close to the ground between sender and receiver.

Medium wave (525–1605 KHz): In this band the propagation takes place by ground waves traveling close to the ground between transmitter and receiver. Ground waves are affected by parameters such as curvature of the earth, earth conditions (water, temperature, noise from large equipment), etc.

Short wave (4–26 MHz): Ground waves possess less alternation for a short distance, while for a long distance, high-frequency signals can be transmitted through the ionosphere. This offers a worldwide broadcasting system. It is based on high-frequency (HF) transmission, which is typically of frequency range 3–30 MHz and is usually divided into two types of transmission: skywave and groundwave. Groundwave transmission is typically used for a distance of about 700 km, as the transmission travels over the surface of the ground (earth), while skywave transmission travels through the ionosphere and covers a

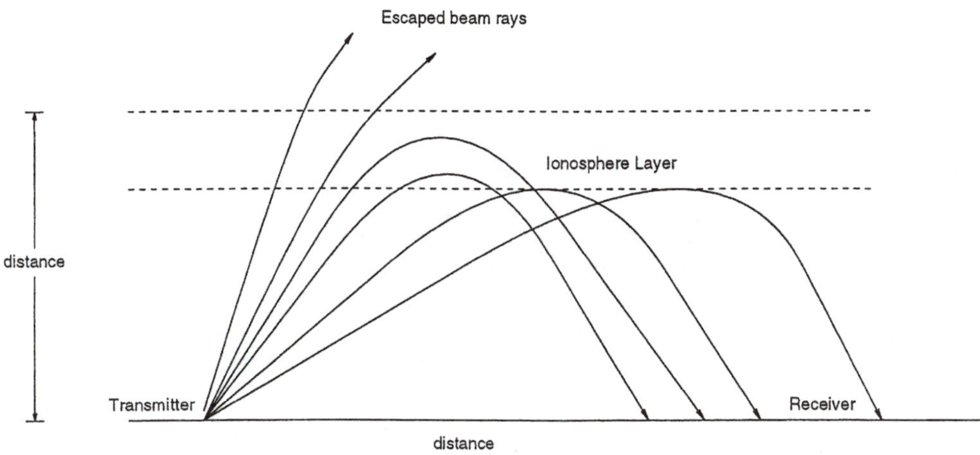

Figure 5.3 High-frequency refraction.

larger distance of up to 700 km. The main difference between these two transmissions is that in groundwave, a larger number of amplifiers are used.

Skywave allows the transmission of both data and voice through the ionosphere by refracting and reflecting the signal off the ionosphere, as depicted in Figure 5.3. The ionosphere layer, the region of atmosphere over the earth, contains ionized gases. The HF transmission is used in ground and sea communication, mobile transmission, packet radio transmission, etc. In general, we experience a periodic blackout from the ionosphere due to the fact that the lower layers are affected by solar flares and the signals lose strength due to multi-path propagation. The higher frequencies are usually disrupted due to these effects. These frequency bands for audio radio broadcasting are used nationally and internationally and, in general, use an AM technique for their transmission.

For high-frequency communication (900–5000 MHz), another form of transmission is known as *troposphere*, where a narrow beam is sent to high-density spots in the troposphere (above the ionosphere). The beam is scattered and propagates back to the earth. In this process of scattering, there will be lost scatter, forward scattering, and back scattering. The forward scattered beam carries the signals, as shown in Figure 5.4. This type of transmission covers a distance of 300–600 km and is characterized by a narrow transmit beamwidth. It requires high transmitter power (due to high losses, typically of 60–90 dB) and is very expensive. It offers low fading (in contrast to high frequency) due to the fact that the changes in atmosphere at that layer are minimum. Due to multi-path propagation, a loss of amplitude of signals (typically 20 dB) occurs quite often, but it is much less than that of HF transmission. This type of loss of amplitude or attenuation is known as **Rayleigh fading**.

Ultra-short wave: Five different frequency bands have been allocated for video broadcasting in the range of 41 MHz to 96 GHz. One band is reserved for radio broadcasting (using the FM technique) while the remaining four are for television broadcasting (using vestigial sideband modulation, VSB).

Radio waves also find applications in many areas other than broadcasting (discussed above), and for each of these applications, the CCIR and IFRB committees have allocated a frequency band. For example, **mobile radio communication** is used for short distances and uses the frequency bands of 80 MHz, 160 MHz, 460 MHz, and 900 MHz. Other applications, along with their frequency bands, are given in the following sections.

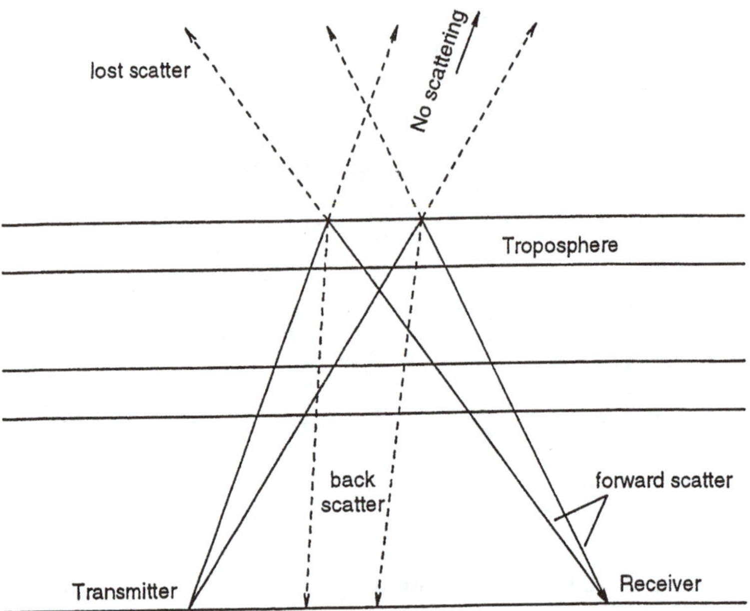

Figure 5.4 Troposphere scatter propagation.

5.7.1 Microwave communication

Microwave communication works on the principle of **"line of sight,"** which dictates that the antennas at both transmitter and receiver be at the same height (this means that the height should be great enough to compensate for the curvature of the earth and also provide line-of-sight visibility). It generates highly directional beams and provides point-to-point communication. It is also used for satellite communication. It carries thousands of voice channels over a long distance and operates at a frequency range of 1.7 GHz to 15 GHz (in some books, this frequency range has been stated as 250 MHz to 22 GHz). But this transmission operates in the range of a few gigahertz (1 GHz = 10^9 Hz). It can carry 2400 to 2700 voice channels or one TV channel.

AT&T has defined three types of microwave systems: TD, TL, and TH carrier systems. In this transmission, the channels are first frequency-division-multiplexed using SSBSC-AM (discussed in Chapter 3). This offers a frequency mapping for the channels. Then this FD-multiplexed signal is transmitted using FDM in the gigahertz band over microwave transmission. Microwave links usually multiplex the channels for their transmissions. There are a few single communication applications where only one communication is sent at a time on the link, e.g., radiotelephone, radiotelegraphy, etc. The parabolic dish has been the most popular and common antenna, and its size is typically in the range of a few feet in diameter. It is fixed rigidly and, using the line-of-sight concept, it transmits a narrow beam toward the receiving antenna. Both the antennas should be installed at a height so that they can point at each other. There should not be any obstacles (trees, buildings) between them. Microwave relay towers can be used to provide point-to-point communications over a long distance.

Microwave communication is used in long-haul telephone networks, television transmission, LANs, and point-to-point communication across buildings. It requires fewer amplifiers and repeaters than do coaxial cables. It can also be used as a closed-circuit TV. A direct connection between local long-haul telephone networks and a company can be established for analog or digital communication, thus bypassing the local exchange.

Microwave transmission suffers from attenuation (absorption) caused by solid objects such as earth, fog, rain, snow, etc., reflection from flat conductive surfaces (water, metal, etc.), and diffraction from solid objects (rain, snow, etc.).

Microwave links can be used to transmit multiplexed voice channels, video signals (television), and data. These are very useful in television transmission, as they provide high frequency and multiplexing for different television channels. The size of the antenna for television transmission depends on the wavelength of the transmission, which is defined as a reciprocal of frequency in hertz. The unit of wavelength is meters. It has been shown that the size of antenna will be smaller for smaller wavelengths (larger bandwidths). Microwave links are a very effective means for transmitting signals across countries and, in particular, are being used extensively for speech/voice channel transmission. Two types of transmission systems are possible with a microwave link: digital microwave link and analog microwave link. Basically, in both the links, we first multiplex the signal and then convert signals into high-frequency range transmission.

Analog microwave link: In the first step, we use frequency-division multiplexing analog telephone channels and then convert into high-frequency range (microwave) by using phase modulation.

Digital microwave link: In the first step, we use time-division multiplexing of digital telephone channels or data, and then in the second step these are converted into microwave tray range using discrete analog modulation of the carrier with PSK, amplitude shift keying, etc. At high frequency, a higher data rate can be obtained; e.g., 2–3 GHz may give a data rate of about 13 Mbps, while a still-higher frequency of 20 GHz may give a data rate of 280 Mbps.

A typical communication system based on this type of transmission includes antennas of 70-m length which are spaced about 50 km apart. At each antenna, a radiolink repeater (regenerative repeater or amplifier) is used. For analog signals, the amplifiers simply amplify the signals and transmit back onto microwave transmission. In the case of digital signal communication, the signals are demodulated into baseband using regenerative repeaters. The channels now either may be dropped (if these belong to the destination) or added (the ones which have to be transmitted further). A transmitter modulates these channels over to another antenna site. A *hop* is usually defined between a pair of antenna sites as a direct link between them. For long-haul transmission, we may define a large number of microwave links. Such a configuration supports the majority of voice and message traffic and is used in telephony, TV, and data communications. It offers long-haul terrestrial communication at lower costs.

The antenna defines a line-of-sight radioless link between the nodes and, as such, includes the functions of radiation (provided by the radiator). There are two types of radiators used in this transmission: **active** and **inactive**. An active radiator offers transmission on both sides by using forward feed (cassegrain) and reverse feed. An inactive radiator, on other hand, offers redirective reflector and periscope antenna.

The repeater is used to regenerate the signal (in the case of digital signal) and retransmit the regenerated signal back onto the link. The repeater is required to extend the loss shot, contain spreading and absorption losses, and also counter the curvature of the earth. There are two types of repeaters: **passive** and **regenerative**. A passive type of repeater simply repeats the incoming signals, while a regenerative type of repeater repeats and regenerates the incoming signals and also modulates the signals over the link. The spreading loss is also known as *free space loss* and is given by the following equation: Spreading Loss == 10 Log $(4\ PD/l)^2$, where D represents distance in meters, l represents wavelength in meters ($l = 3 \times 10^8/F$, F represents a frequency). It has been observed that this loss remains constant for transmission.

The absorption loss is caused by water vapor in the atmosphere, which in fact absorbs the power. It increases with frequency and also with the intensity of rain. A typical value of this loss is given (at 15 GHz) as 0.15 dB/km (for light rain) and 2 dB/km (for heavy rain). Repeaters and amplifiers may be placed at a typical distance of a few miles (5–75). Various frequency bands are regulated by the U.S. Federal Communications Commission (FCC), ranging from 2 to 40 GHz. Out of these bands, the most common bands for long-haul telephone networks are 4–6 GHz. The 11-GHz band is being considered for future long-haul networks. The cable TV system uses a band of 12 GHz. One useful application of this link is to provide TV signals to local CATV installations. The transmission from local CATV to various users is achieved via coaxial cable. The 22-GHz band is typically used for point-to-point communication. At a higher microwave frequency band, the size of the antenna may be small, but the attenuation will be higher.

Antennas: Different types of antennas for different applications in wireless communications are described below. For more details on these antennas, a reader is advised to read Reference 8.

1. **Half-wave dipole:** This is the simplest form of antenna that can be designed using coaxial or thin wire and is useful in the design of other antennas.
2. **Folded antenna:** This antenna is a modified version of the half-wave dipole and finds the same use as that of the half-wave.
3. **Quarter-wave antenna:** This antenna is used in automobile AM, FM, and CB antennas.
4. **Array antenna:** This antenna is used in the design of such popular antennas as parasitic or Yagi and log periodic antennas.
5. **Parasitic array and Yagi antenna:** This antenna is useful in FM and TV antennas and VHF and lower UHF (30 MHz to 1 GHz) applications. They are very easy and inexpensive to build.
6. **Log periodic array:** This is a very popular antenna and finds its applications in VHF TV and FM antennas and VHF and UHF (30 MHz to 1GHz) applications.
7. **Loop antenna:** This antenna is useful for UHF band, television reception, AM radios, and direction finding.
8. **Helical antenna:** This antenna is used in car telephones.
9. **Parabolic antenna:** This antenna is commonly used as a satellite-receiving antenna in the range of 4–6 GHz.

5.7.2 Satellite communication

Satellite communication consists of a microwave link beyond the atmosphere, usually assigned a pair of frequencies at 4 and 6 GHz, 11 and 14 GHz, or 20 and 30 GHz. The satellite is in fact a microwave relay station, which connects a few ground-based microwave transmitters/receivers (known as earth stations or ground stations) and uses the concept of line of sight. One band of frequency is used to transmit the signals to the satellite, which amplifies or repeats the signal and transmits it on another frequency band. A satellite working on bands of frequency is known as a *transponder*. It supports point-to-point and relay communications. In point-to-point communications, the satellite provides communications between two different ground-based antennas pointing toward the satellite. In relay communications, the satellite receives the signal from one ground-based transmitting antenna, amplifies or repeats it, and transmits it to a number of ground-based receiving antennas.

In order to maintain a line of sight between the antennas and the satellite, the satellite must be stationary with respect to its position over the earth. This means that the satellite must rotate at the same speed as that of the earth — at a distance of 22,300 miles. In order to reduce any interference between satellites in orbit, the current standard defines an angular displacement of 4° spacing in a band of 4 to 6 GHz and a spacing of 3° for a band of 12 to 14 GHz. Satellite communication is used in long-haul telephone communication, TV distribution, private business networks, etc. The Public Broadcasting Service (PBS) uses a satellite for distributing its television programs. Other commercial networks such as CNN, CBS, and others use satellite and cable television systems. Due to ever-growing demands for television programs, satellites now deliver television programs of different countries to homes via the direct broadcast satellite (DBS) system. The advantage of DBS is that the size of the antenna is considerably reduced (a maximum of 2 ft in diameter).

Thus, we have two distinct types of point-to-point communication within a country or across countries and also in local area networks: cable links or wired (balanced pair or coaxial), and wireless or radio (microwave wave, satellite wave, short-wave telecommunication) links. Radio waves are a directorial broadcasting system (radio and television) and can be extended to an international platform using short wave. Cable link communication can also broadcast information on cable networks and offers advantages over radio waves in terms of its immunity to electromagnetic interference, extended programs capability, no antenna at individual receiving stations, etc.

Satellite link: These links provide a large number of voice-grade channels placed in microwave frequency bands that then can be accommodated on a satellite station (at a distance of 22,300 miles above the earth). Due to their ability to cover a large geographic area, they are used for broadcast transmission of analog signals, data, and also video signals, as shown in Figure 5.5.

Usually, satellites at low altitudes are known as *moving satellites*, which rotate rapidly with respect to a satellite earth station point on the earth's surface. At higher altitudes (22,100–22,500 mi), satellites rotate at the same speed that the earth rotates in 24 hours, and hence these satellites are fixed with respect to the satellite earth station on the earth's surface. These satellites are known as *geostationary satellites*. The material presented below is derived partially from References 1, 3, 8, and 9.

The former U.S.S.R. and the U.S. launched satellites Sputnik and Explorer in 1957 and 1958, respectively. These two satellites were not defined for any type of communications. The first telecommunication satellite, SCORE, was launched by the U.S. Army in 1958. The first active satellite Telstar I (for transmitting transatlantic television signals at altitudes between 600 and 3600 miles) was launched in 1962, while the first geostationary commercial satellite, Early Bird or Intelsat I, was launched in 1965 and contained 240 telephone channels. Other geostationary satellites include Intelsat II, III, IV, IV A, V, Symphony, OTS,

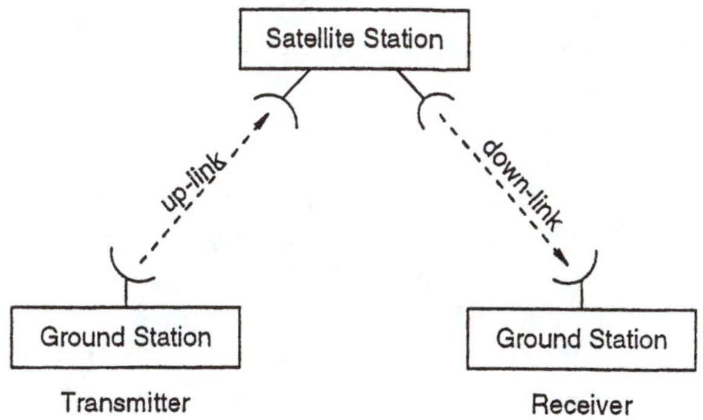

Figure 5.5 Satellite communication system.

WESTAR, COMSAT (Vs domestic satellite), etc., and are characterized by increasing performance (240–15,000 channels).

A satellite link provides the following features to data communication:

- High speed
- Broadcast capability
- High reliability
- Low cost

We can send data with a speed of more than 100–200 Mbps, which is relatively higher than other existing communication media but less than that of fiber optics. Satellites are characterized by higher capacity and also by broadcast capability (multiple databases can be sent with one transmission). Satellites links are inexpensive with respect to distance. The total propagation delay between an earth station and a satellite is usually in the range of a few hundred milliseconds, since microwave signals can travel at the speed of light (186,000 mi/s). This delay may introduce error into data transmitted on half-duplex communication protocols. The satellites can be used effectively if we use full-duplex communication protocols for point-to-point links, since the loop delay does not have an effect on the number of links.

Many of the satellites are at a distance of about 22,300 miles and rotate around the earth at a speed of about 7000 mi/h. These satellites are in *synchronous orbit*, covering large geographic areas and transmitting signals at 6 and 4 GHz. They offer a variety of business applications; a number of groups per channel are defined and leased to users. A user having a dish (antenna) can use it for a private network at a lower data rate. Large organizations can be allocated different bandwidths ranging from 56 Kbps to 256 Kbps per channel. These signals will be received by a hub at the receiving side, processed by a front-end processor, and allow the users to share information from databases, documents, files, etc. A low-cost **very small aperture terminal (VSAT)** system provides such an application; users having VSAT antennas can share the information at a very reasonable lease of a few hundred dollars per month per VSAT.

Satellite frequency bands: There are different classes of frequency bands for satellite links, such as C band (6 and 4 GHz), Ku band (11/12/14 GHz), and K band (30/20 GHz). Each country is assigned a frequency band and the type of geostationary satellite it can use by the United Nations International Telecommunications Union (ITU). For example, in the U.S., the FCC uses C band for commercial communication and has divided this band into two distinct microwave frequencies: 3.7–4.2 GHz for downward microwaves and 5.925–6.425 GHz for upward waves. Higher bands can also be used but are more susceptible to atmospheric conditions such as rain, thunderstorms, and clouds, which may provide significant deviation in the reflection of the transmitted signal — and they also require a complex and expensive setup for the transmission and reception of signals.

The C band, which provides commercial communication, is a civilian satellite link and transmits a large number of channels (voice, data, or video communication). In order to prevent interference between channels, the communication uses a 250-MHz guard band for each direction of wave propagation (upward and downward), giving rise to a total bandwidth of 500 MHz for the guard band alone. This band usually requires a distance of about 500 miles between two satellite stations in orbit and a large-sized antenna. Other frequency bands Ku and K require less distance between satellites, and also the size of the antenna is very small. But as discussed above, these bands experience alterations from atmospheric conditions, especially raindrops, which not only reflect the propagation of waves but also absorb energy from it. Further, the equipment required to transmit and receive signals through a satellite is very complex and expensive.

It has been shown that C band is superior to Ku and K bands as it has much less alteration to atmospheric conditions than higher bands. Alterations can be reduced by using boosting power instruments both at the satellite and the satellite earth station. In spite of these problems, satellite communication has revolutionized the communication industry, transmitting voice, data, and video signals between any part of the world. Also, the smaller size of the antenna has made it very popular for television communications. Antennas or dishes of 2–5 ft in diameter can be installed on the premises (organizations, motel industries, universities, and homes) to receive television channels.

Currently, there are about 1500 satellites rotating in orbit and providing various services to subscribers. Different types of signals can be sent to a satellite earth station via different types of media, e.g., cable, fiber optics, radio communications, etc., and they can be linked to a satellite link for transmission. In some applications, the existing communication facilities may be bypassed and the satellite carrier signals may be used directly to link them with a satellite communication system.

As indicated above, the FCC has allocated different frequency bandwidths for upward and downward transmission of waves. Usually, more power is required to transmit signals from an earth station to a satellite in orbit; that is why a higher-frequency bandwidth is used for upward transmission. **Transponders** receive a signal in a frequency bandwidth and amplify and convert the signal in another frequency bandwidth. A satellite system contains large number of transponders. The bandwidth of 500 MHz (guard band) accommodates a large number of channels of 36- or 40-MHz bandwidth in such a way that each channel is allocated to one transponder.

The signal transmission below 1 GHz will introduce different types of solar, galactic, atmospheric, and human-made noises. Due to the long distance, satellites suffer a significant propagation delay during transmission and reception of signals between one ground station and another ground station. They do not support error control and flow control and also do not provide secured data communications.

5.7.3 Cellular radio communication

Another wireless communication medium is **cellular radio**. The purpose of cellular radio is to provide communication between two subscribers separated by thousands of miles. No satellite or microwave link is established between them, but a frequency band of 800–900 MHz (assigned by the FCC) is used for the transmission of radio waves. Cellular radios enhance the use of mobile radio-telephone systems. The main purpose of cellular radios is to provide voice communications in a mobile system.

The cellular system uses either analog or digital signaling. The first mobile radio system to work with telephone systems was introduced in 1946 and known as a mobile telephone system. This system, a centralized system supporting half-duplex communication, consisted of a powerful transmitter covering a radius of about 20 miles. The first cellular system, the mobile telephone switching office (MTSO), is based on the concept of frequency reuse. Here, many users are using the same frequency with minimum interference between them. Each cell can support a number of channels; cells using the same frequencies are usually placed close to each other at a sufficient distance to reduce the interference between them. A cell site is connected to an MTSO, which is connected to an end office and other MTSOs. An MTSO controls a number of cells. The connections between MTSOs and end offices can be provided by a variety of media, such as twisted-pair cable, coaxial cable, fiber, microwave, satellite, etc. The MTSOs are then connected to PSTN (Public Switched Telephone Network) via end offices. The MTSO computers keep a record of all the users. Most of the cellular systems are full-duplex and use the same concept of satellite communications via uplinking and downlinking mechanisms. For each of these operations, a different frequency standard value is used. The uplinking frequency component sends the signal from mobile to base while downlinking sends the signal from base to mobile.

Another communication medium for short-distance transmission is the transmission of infrared frequencies. It provides a high data rate (a few Mbps) and is inexpensive and immune to microwave interference.

5.8 Typical broadband LAN configuration

Broadband local area networks (LANs) use a broadband cable which has a sleeve of extruded aluminum of diameter wider than a half-inch and which is more expensive than baseband. Modulating radio frequency (RF) carrier signals with the modulating signals to be transmitted send the broadband information. Many devices can share the information signals by sharing the cable's bandwidth by using frequency division multiplexing (FDM). Each FDM channel can be divided into sub-channels, which can correspond to different access techniques of LANs, e.g., random contention, token passing, etc.

5.8.1 Components of broadband LANs

Broadband LANs consist mainly of two components: **headend** and **distribution networks**. The headend receives modulated RF signals and redistributes them to the receiving nodes. This can be considered a repeater, which provides frequency transformation. The distribution network typically consists of coaxial cables, splitter, joins, amplifiers, and many other components and is responsible for carrying the information to the destination nodes. Broadband cable may use a single-cable or dual-cable system.

Single cable layout: A typical broadband single cable layout system is shown in Figure 5.6. Here, a single cable is used to carry transmitted and received signals. The total bandwidth of the cable is split between transmission and reception of the signals. Half of

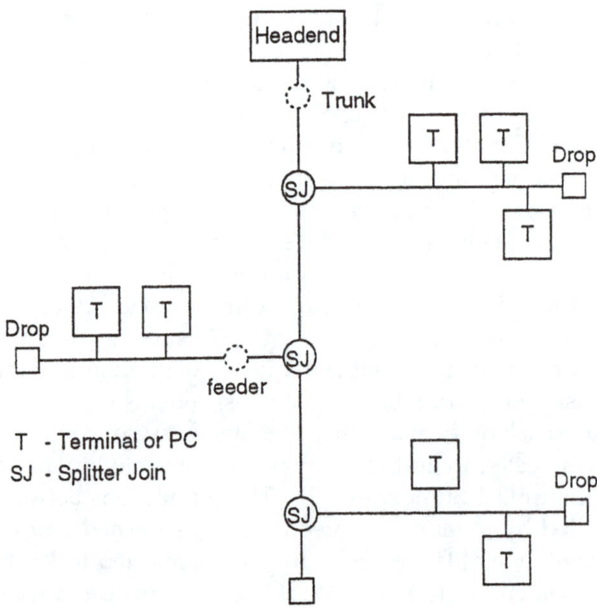

Figure 5.6 Single-cable broadband layout.

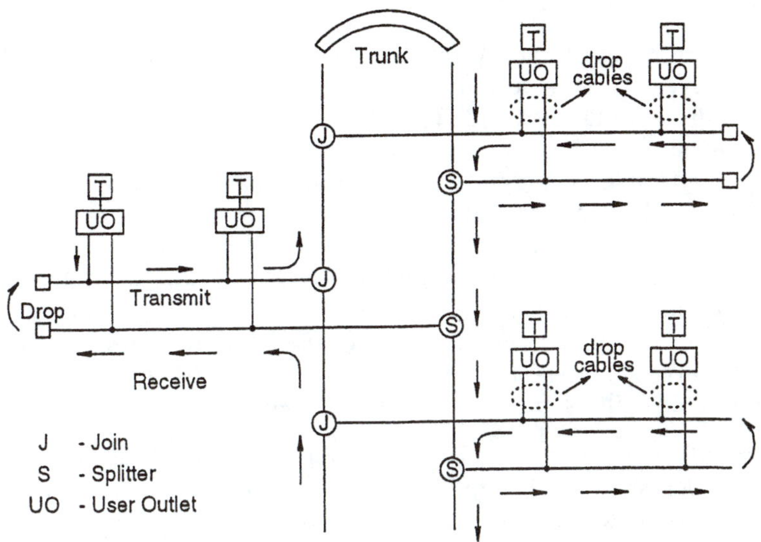

Figure 5.7 Dual-cable broadband layout.

the bandwidth is used for forward (transmit) signals, while the other half is used for backward (receive) transmission. The bandwidth of the cable is divided into three sections: return, forward, and guard.

The return band carries signals returning to the headend. Signals forwarded by the headend to receiving devices use the **forward** band. The **guard** band does not carry any signal but is used to separate the forward and return signals from each other.

Dual cable layout: A typical dual broadband layout for LANs is shown in Figure 5.7. It uses two separate cables: one for transmission and another for reception. The cable distribution is provided by a main cable known as a **trunk cable**. At the midpoint of the cable (active/passive), a point known as a transmit–receive crossover (network loop) divides the network into two sections: **transmit** and **receive**. The signals transmitted from any node propagate in only one direction along the transmit line defining the distribution tree toward the network loop. Signals arriving at the network loops cross over to the receive line distribution tree and propagate back to individual nodes. The transmit cable (line) contains the *join points*, while the receive cable (line) defines *split points*, and these lines are physically separated by dual cable distribution and electronically isolated by unidirectional propagation characteristics of LAN cable components.

The dual cable broadband layout uses return and forward bands but does not use any guard band, as separate cables are used for carrying the signals for transmission and reception. The scheme defined to separate the bands for transmission and reception is known as **band splitting**, and there are three splitting schemes defined for dual cable broadband cable, as shown in Table 5.3. The **sub-split** scheme is used in CATV systems. The **mid-split** scheme seems to be very popular at this time, while the **high-split** scheme is a new innovation and offers more bandwidth.

Table 5.3 Band-Splitting Schemes in Dual Cable Broadband Cable

Scheme	Transmit (return) frequency	Receive (forward) frequency
Sub-split	5–30 MHz	54–400 MHz
Mid-split	5–116 MHz	168–400 MHz
High-split	5–174 MHz	323–400 MHz

5.8.2 *Network hardware design*

Network hardware design typically uses CATV-type broadband cable components as discussed above. Networks can be defined as active or passive. Some of the material below is derived partially from References 1, 2, 3, and 10.

In **passive design**, the main trunk and distribution cables interconnect CATV-type signal splitters and cable taps. Splitters provide multiple cable branches from a single cable run. Taps provide unidirectional paths for the signals to pass from the transmit connector on a user outlet to a distribution transmit line or from a receive distribution line to a receive connector on a user outlet. Normally, a passive design may meet the requirements of small installations, but larger installations usually require an **active design**. This design scheme uses the following components: network user outlets, coaxial cable, trunk coaxial cable, drop cable, and passive directional components such as splitters, combiners, multi-taps, filters, power supplies, and network headend.

Network user outlets: These are similar to the CATV jack-type wall-mounted user outlets shown in Figure 5.8. They can be connected to a particular branch of a network by means of a **drop coaxial cable**. One drop cable is attached to the branch transmit line by a combiner tap. At the same location, the second drop cable is connected to the branch receive line by means of a distribution tap. The user outlet is then attached to the free ends of the drop cables. The two connectors have some sort of identification so that the receive and transmit sides can be easily identified. If the connectors are not attached to a device, two terminator (50/75 Ω) resistors are attached to the unused connectors.

Figure 5.8 Network user outlet.

Coaxial cable: Coaxial cable provides the center conductor of the cable and the outer shield with a common axis. The main purpose of coaxial cable is to serve as a medium for RF signals. These cables are generally color-coded. For example, the transmit lines may contain blue color bands after every 10–30-ft interval and the receive lines may have red bands. Components of this cable are as follow:

- **Center conductor:** The center of coax consists of solid copper or copper-clad aluminum wire. Outer conductor is composed of aluminum.
- **Dielectric insulation:** Keeps the position of the center conductor and is composed of foamed polyethylene.
- **Flooding compound:** A gel which provides protection in case the jacket is ruptured.
- **Jacket:** A coating used to provide a weather-tight seal.

Trunk coaxial cable: Typically, trunk lines are available in six types ranging from 0.412-in. to 1-in. diameter. These cables have attenuation ranging from 1.6 to 0.5 dB per 100 ft at 300 MHz. Generally, trunk lines should be five inches or more in diameter, as shown in Figure 5.9. The trunk cables are used outside the building and may have to go through weather stress. To avoid this interference, the cables are usually jacketed. Cables which are used in conduits or buried underground also have a corrosion-resistant gel.

Distributed/feeder/branch cable: These cables are normally 0.5 inches in diameter and are used for indoor distribution. They are available as jacketed or non-jacketed.

Drop cable: These are outlet cables connected to the distribution cables and typically 10–15 ft in length. These cables range from RG 11 and RG 6 to RG 59. Each type has foil and shielding to prevent pickup of noise, radiation, and RF signals.

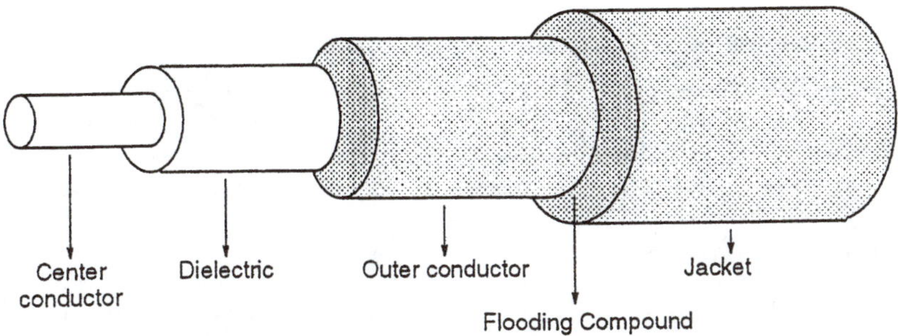

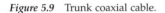

Figure 5.9 Trunk coaxial cable.

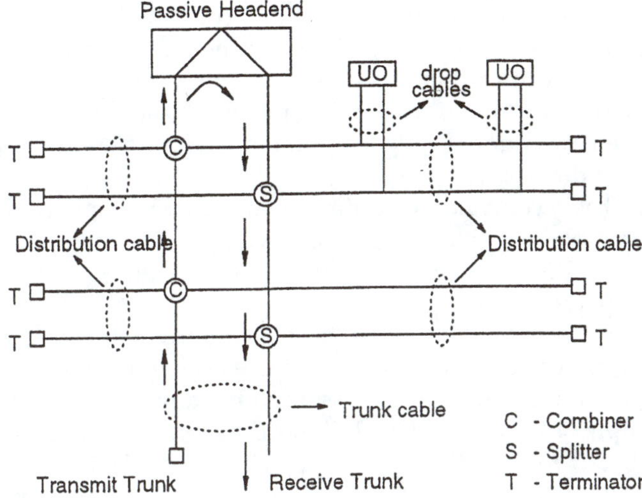

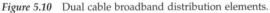

Figure 5.10 Dual cable broadband distribution elements.

Passive directional components: In the transmit lines, the signals propagate toward the network loop or headend, while in the receive lines, they propagate away from the headend, as shown in Figure 5.10. To establish this unidirectional connection between transmit and receive lines, a special component is installed to pass (couple) signals in only one direction. These unidirectional coupling elements are installed at every location in the network cabling where the cable is divided to create one or more additional distribution branches or drops. The passive components are the ones that require no power to operate and include connectors, splitters and combiners, directional couplers, multi-taps, filters, and terminators. Each of these components introduces signal loss, known as passive loss or insertion loss. The passive components are described in brief below.

1. **Connectors:** 75-Ω connectors are made up of all types of coaxial cable. Connectors must be installed carefully, as 70% of network failure is due to bad connector installation.
2. **Splitters and combiners:** Both halves of the trunk are divided into identical branches. As a result of this, a transmission tree is formed on transmit lines (half of the network) and on receive lines (the other half). All transmit line components are positioned physically adjacent and parallel to all receive-line components. Thus,

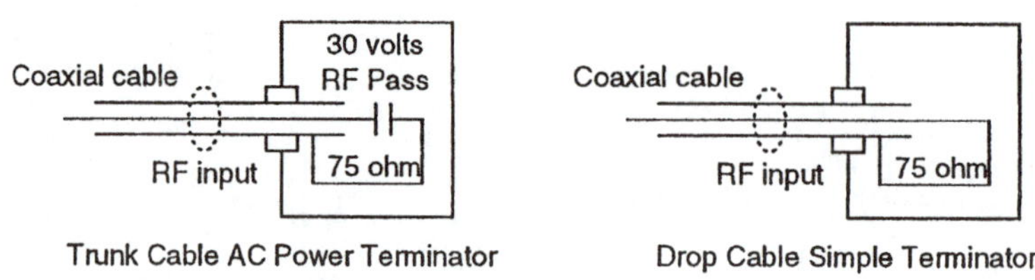

Figure 5.11 Broadband terminators.

the receive distribution tree is a mirror image of the transmit distribution line. These components are used for either splitting the signals from the receive line or joining/combining them on the transmit line. Each combiner or splitter can couple more than one drop cable. The opposite end of the cable is terminated at the connector on a user outlet.

3. **Directional coupler:** This device divides and combines RF signals while maintaining 75-Ω impedance. It also makes sure that the signals being transmitted by any device will go only toward the headend and minimizes the reflection of RF signals back to their source. It has three parts: trunk input (placed toward headend), trunk output (placed away from headend), and tap. The couplers should not be installed in a backward direction.
4. **Multi-taps:** These combine a directional coupler with a signal splitter and connect several drop cables to feeder cable. This device provides more than one-user outlet drop cables. Multi-taps may contain two, four, or eight simultaneous tap connections, called **ports**.
5. **Filters:** Filters are normally used at the headend to combine or separate frequency bands. They are used in amplifiers for single-cable mid-split-type systems. There are five types of filters: **low band pass**, **high band pass**, **band pass**, **band elimination**, and **diplex (diplexers)**. The diplexer directs/splits signals in high- and low-frequency bands to the bands' processing equipment. When used with amplifiers of mid-split systems, signals from the headend are passed by diplexers to the forward/receive amplifier and isolated from the return/transmit amplifiers. These allow signals of 5–4000 MHz.
6. **Terminators:** These devices are placed at the end of trunk lines, feeder lines, and unused outlets to limit reflections. They convert RF energy to heat energy and are of 75 Ω. Terminators are available in different varieties for indoor and outdoor applications. Details of these components and their connections are shown in Figure 5.11.
7. **RF modems:** Most companies define a number of data channels inside a single 6-MHz bandwidth. Within this bandwidth, data communication is provided by RF modems. This 6 MHz provides a common reference for allocating frequencies for RF modems. The number of sub-channels within this band is different for different companies. To accommodate up to 9600 bps, 60 full-duplex communications, 120 channels with 300-KHz channel spacing is used. Three types of RF modems have been defined: **fixed frequency**, **point-to-point** or **multi-point**, and **frequency agile**. RF modems can also be divided on the basis of the following parameters: required bandwidth, tuning ability of frequency agile modems, and access method used (CSMA, token passing, etc.).

A typical mid-split RF modem will have the following features:

- Manual switching up to 60 channels over a 6-MHz band
- Transmit frequency between 72.1 MHz and 89 MHz.

- Receive frequency between 264.35 MHz and 281.25 MHz
- Channel spacing of 300 KHz
- Modulation technique FSK (frequency shift keying)
- Medium access CSMA/CD or token passing

Point-to-point RF modems operate only on their assigned channels. Their assigned frequency is fixed. Only one transmitter is allowed over the frequency. The device at the headend has to translate the transmit frequency from one modem to a fixed receive frequency of the receiving modem, thus offering four frequencies instead of two. These modems have continuous access to a network. Some modems may use CSMA/CD over a 300-KHz band, while others may continuously generate their transmit frequency, just like modems used at home.

Multiplexed RF modems do not need continuous access to the network and can be multiplexed together via RF multiplexer to share a single sub-channel. All the devices can monitor the channel, but only one can transmit at any time. Some companies have allocated a 300-KHz bandwidth for each sub-channel and have allocated 20 of these 300-KHz channels in a 6-MHz bandwidth. Each of these channels can support up to 200 asynchronous ports, each port up to 9600 bps. This way, 4000 asynchronous ports can be allocated over this 6-MHz bandwidth. Some modems permit a user to select any one of the 60 channels.

Modems can transmit and receive 64,000 bps. The interface to DTE is RS-449 ports. Certain companies allocate 16 full-duplex channels out of a 6-MHz. bandwidth.

Some RF modems can interface two DTEs to the LAN. They use a card known as an **asynchronous interface unit (AIU)**. This card contains two RS-232-C interfaces, one z80 microprocessor, and RAM. Units are available to support 16 AIUs, allowing a total of 32 interfaces. These cards can be replaced or removed.

Active network hardware design: In active design, CATV-type amplifiers are used to boost signals on the cable. Companies may use different power levels or sources. For example, 30-V AC may power all amplifiers. This power is distributed throughout the network. One or more amplifier power supplies are connected to the network through AC power inserters. Power inserters block network signals from the output of each power supply. Cable taps isolate user outlets from the 30-V AC supply voltage. A typical functional block diagram of a network headend is shown in Figure 5.12.

The main difference between active and passive design lies in the network headend. A passive design holds a frequency standard generator and the network loop, while an active design headend has one or two frequency standard generators, two amplifiers, attenuator pads, amplifier power supply, and network loop. The following cable components are common to both designs: cable, amplifiers, directional couplers (splitters and combiners), equalizers, taps, terminators, and connectors. The active elements used in active design of broadband LANs include amplifiers, power converters and power inserters, and network headend.

The **amplifiers** are inserted at calculated intervals to compensate for signal attenuation caused by all passive in-line components. Signals produced at the amplifier output side are restored to the desired trunk or distribution operating levels.

The gain of amplifiers is expressed in decibels, which are defined as a ratio of output to input signal level. The gain level of amplifiers can be adjusted by gain control, which can be either manual or automatic. **Automatic gain control (AGC)** amplifiers are more expensive. **Manual gain control (MCG)** amplifiers are usually used for short distances, especially on feeder cables. Automatic gain control produces a constant gain in decibels, regardless of variations in environmental conditions, within a range of ±3 dB. The output

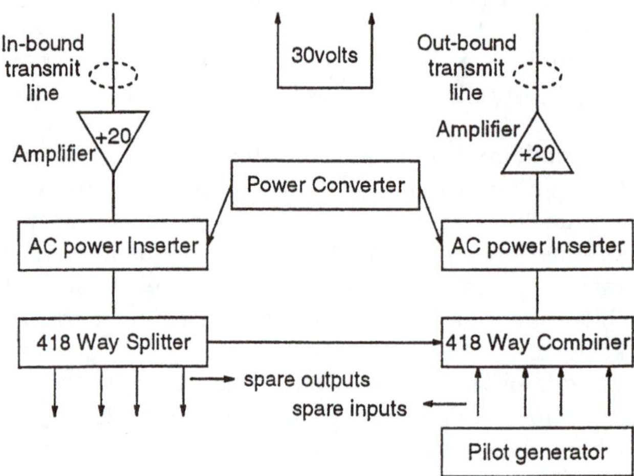

Figure 5.12 Functional block diagram of network headend.

level of the signal lies in the RF range. The amplifiers can also produce noise of their own over RF signals. Distortion of signals is also present. For a large LAN, where many amplifiers have been cascaded, both manual and automatic gain control amplifiers are installed for improved overall performance. Typically, after the third or fourth MGC cascaded amplifier, an automatic gain control amplifier is installed. This scheme considerably reduces spike-like noise problems, which can result in a large amount of signal or data loss for several seconds.

The different types of amplifiers are trunk, bridge (bridger), line, and distribution.

- **Trunk amplifiers** typically provide a 22-dB gain for input levels of about 10 dB mV and output levels of 30 dB MV for 35-channel systems. About 20 of these trunk amplifiers can be cascaded. Each amplifier in the cascade usually reduces the output by 3 dB or less.
- **Bridge amplifiers** provide signals for individual distribution of feeder cables. These receive signals from directional couplers or taps connected to the trunk cable. From these bridges, up to four feeder cable outputs are possible. The maximum output of a bridge is +47 dB mV. With a good combination of bridging and trunk amplifiers, the signal level on trunks can be maintained at CATV standards, while bridging amplifiers will feed the distribution cables.
- **Line amplifiers** provide a higher level of signals if the bridging amplifier is not enough. These are more expensive than trunk and bridging amplifiers, which are low-quality amplifiers. A maximum of three to four amplifiers should be used (preferably).
- Internal **distribution amplifiers** use a 110-V power supply and are not cascaded, as they offer very high gains. These have high gain signal distribution capacity. They are used, for example, where feeder cable runs through a whole building going through different floors.

Power supply converters normally produce 30-V AC cable signal amplifiers. Typically, one converter is located at the headend of the network. Additional power converters can be connected at other locations in the network. The current-handling capability of passive in-line cable components determines the location of each power converter. Usually each amplifier draws about 600 mA from power converters and supports 10 amplifiers. The AC power is put on coax by a device known as a **power combiner**. Power combiners put

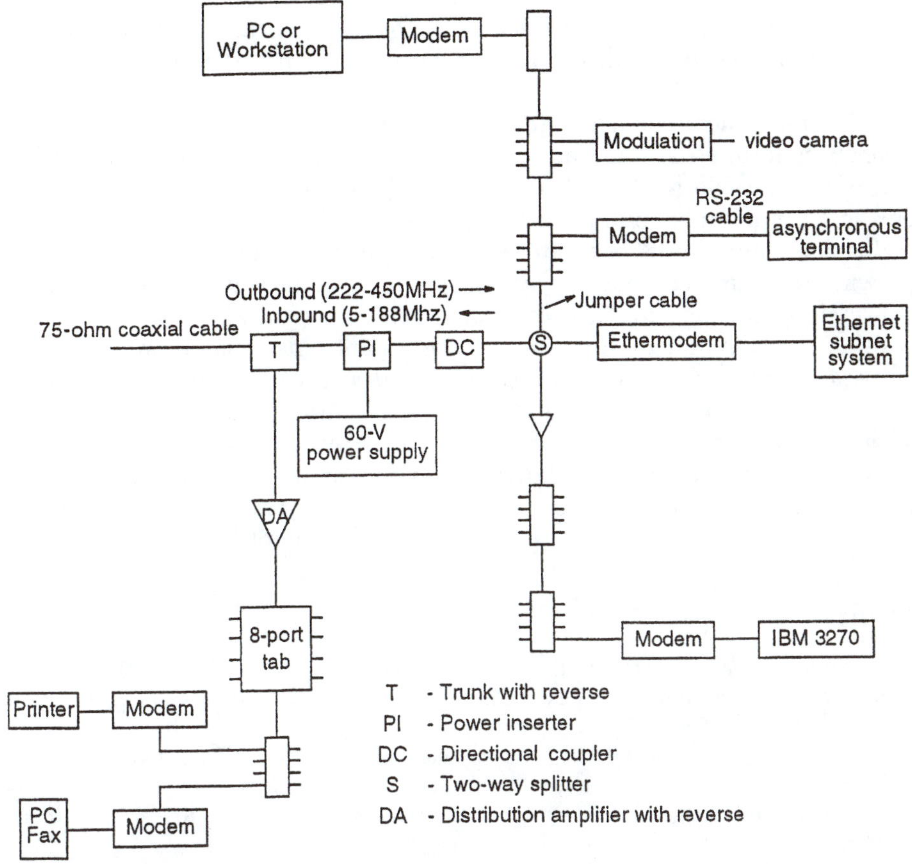

Figure 5.13 A typical broadband cable layout for LANs.

AC power in either one or both directions, with little effect from RF signals. 30-V AC is used for older LANs, and 60-V AC is used for new versions of broadband LANs. Multiple power supply converters are also available and can be used.

Network headend: A typical broadband cable layout for LANs as derived from Reference 6 is shown in Figure 5.13.

Some useful Web sites

For coaxial cables and cellular radio systems, an encyclopedia is found at *http://www.encyclopedia.com*, and a telecommunication dictionary is at *http://www.ctechsolutions.com*.

One of the largest global cable product suppliers is at *http://www.cable-link.com*. Two theses on high-speed communications on twisted-pair wires are found at *http://www.opubl.luth.se*. A course on twisted-pair wires (both shielded and unshielded) is available at *http://www.w3.sba.oakland.edu*.

Microwave links for multi-megabit data communications are at *http://www.tapr.org*.

Various hardware and software components of microwave links are available at *http://www.radioville.com* and *http://www.comsearch.com*. Cellular radiotelephone systems are discussed at *http://www.amazon.com* and *http://www.encyclopedia.com*. A cellular radiotelephone service fact sheet and other documents are maintained by the FCC at *http://www.fcc.gov*.

Optical fiber communication products and components are available at *http://www.fiberoptek.com*, *http://www.fibrecomms.co.uk*, *http://www.try1css.com*, and *http://www.analysys.co.uk*. Optical standards can be found at *http://www.siemon.com*, *http://www.bicsi.com*, and *http://www.ansi.gov.*

SONET digital hierarchy standards are available at *http://www.ansi.gov.* The SONET interoperability forum (network and services integration) is at *http://www.atis.org*.

A very useful cable modem glossary can be found at *http://vidotron.ab.ca*, and cable modems in general can be found at *http://www.cabel-modems.org*.

A glossary about standards organizations is maintained at *http://www.itp.colorado.edu*.

For wireless communications, services and products are available at *http://www.ems-t.com*, *http://www.ameranth.com*, *http://www.cellnetcell.com*, *http://snapple.cs.washington.edu*, and many other sites. A list of industries, subscribers, and related information is at *http://www.wow-com.com*. For reference sites, visit *http://www.ornl.gov.* Wireless communications standards and mobile satellite and related standards are found at *http://www.msua.org*. Wireless LANs represent a new technology and have already been used in various applications. The following sites contain information about the products of LAN, installation, and their uses: *http://www.rhowireless.com*, *http://www.wlana.com*, *http://www.hpl.hp.com*, *http://www.cis.ohio-state.edu*, and many others.

References

1. Freeman, R., *Telecommunication Transmission Handbook,* Addison-Wiley, 1991.
2. Bellcore (Bell Communication Research), *Telecommunications Transmission Engineering,* 3rd ed. (three volumes), 1990.
3. Gibson, J., *Principles of Digital and Analog Communications.* Macmillan, 1989.
4. Stallings, W., *Handbook of Computer Communications Standards,* Macmillan, vol. 2, 1990.
5. Zanger, H. and Zanger, C., *Fiber Optics: Communication and Other Applications,* Macmillan, 1991.
6. Technical reference, Anixter Bros. Inc., 1988.
7. Park et al., 2.488 Gbps/318 km repeaterless transmission using barbium-doped fiber amplifier in direct-detection system. *IEEE Photonics Technical Letters,* Feb. 1992.
8. Thurwachter, C.N., Jr., *Data and Telecommunications Systems and Applications,* Prentice-Hall, 2000.
9. Reeve, W., *Subscriber Loop Signaling and Transmission Handbook,* IEEE Press, 1995.
10. Black, U., *Data Networks Concepts, Theory and Practice,* Prentice-Hall, 1989.

chapter six

Telephone systems

"Silence is a text easy to misread."

A. A. Attanasio

6.1 Introduction

A telephone system is an analog telecommunication system that carries voice signals. The telephone set uses a transducer to convert human speech into electrical signals for transmission by a transmitter; the signals are reconverted into voice on the receiver side by another transducer. The traditional private automatic exchange (PBX) allows voice communication among different telephone sets within organizations and also provides interfaces with public switched telephone networks (PSTNs) and many other telecommunication networks. Digital transmission has opened a new era in accessing telecommunication services from the digital private branch exchange (DPABX), to which users can connect their computers and terminals along with their telephone sets to access services from existing PSTNs and also from new, emerging networks, such as integrated services digital networks (ISDNs). The services available on these networks include voice, data, text, images, video, motion pictures, etc.

6.2 Evolution of telephone system

In the first-generation telephone system (early 1880s), an operator needed to make a direct connection (local) between the calling and called parties by inserting jacks and cords into proper sockets on the switchboard. If the direct connection was not local, then the operator sent a request for connection to another operator, who then connected the two parties by inserting jacks and cords on the switchboard. This indirect connection was established through the operator and hence could be heard or be misused.

The manual switchboard containing sockets and switches was replaced by a *strowger switch*, which provides step-by-step connection in the exchange, thus eliminating the manual operator. This switch establishes connection automatically between calling and called parties and is used in automatic exchanges. Telephone systems using these switches became the second-generation system — plugs, jacks, and sockets were replaced by relays, electromagnetic strowger switches, etc. The strowger switch (being slow) was then replaced by a crossbar switch for faster establishment of connections in a telephone system.

A dialing process by which customers could dial the numbers was also an important part of the automatic exchange. Nowadays we have a rotary dial or push buttons for dialing a required number. The rotary dial provides two signals: one that corresponds to

Figure 6.1 Connection of network-terminating units (NTUs).

the number dialed and one that dialing (when the rotary dial is coming back to original position) provides a delay signal to the exchange. In push-button dialing, a similar concept is used in a different way. For each of the buttons, two tones of different frequencies are generated. The frequency for these two tones ranges from 697 Hz to 1477 Hz, and this range is within a voice-grade bandwidth of 4 KHz. Telephone systems carry analog signals. If we want to send digital data signals on telephone systems, these signals have to be converted into analog and then can be sent on the telephone network.

Data communication takes place between two computers over telephone systems. Two computers can send data back and forth by providing a communication channel between them over transmission media through telephone systems. If the distance between them is small, we might connect them by cables, but if the distance is large, we have two choices.

The **first choice** is to design and install new facilities which allow the users to send the data from their computers to other computers across countries. This seems to be reasonable but requires much money and time, and it may so happen that by the time it is operational, new technology will have come into existence.

If we have repeaters at suitable distances, the distorted signals can be regenerated and sent to the telephone network. The speed of transmission usually depends on the distance between repeaters. Even if the new facilities for digital signals use pure digital transmission, we have to use some kind of interface between the digital signals coming out of terminals or computers and digital signals going into the digital network. This interface is usually called the network terminating unit (NTU). This unit is used between computer (host) and digital data network and also between terminal and digital data networks, as shown in Figure 6.1.

The **second choice** is to use the existing public network for the transfer of data between computers across countries. Although PSTNs were developed for sending voice between two nodes, by including some transducers and other equipment, data can be sent on this network between two computers anywhere in the world.

Although the so-called new digital network connecting computers via cable may give us better transmission speed and less error rate compared to the existing telephone system, it is still worthwhile to find ways for transmitting data efficiently on PSTNs rather than waiting for new technologies.

6.3 Hierarchical model of telephone systems

The telephone system has become an integral part of every country, and almost all countries are connected by telephone systems. Different communication media at various hierarchical levels in the systems are used to establish connections between any two customers (telephone subscribers) across countries.[1,2] A hierarchical model of a telephone system is shown in Figure 6.2, where each level is defined as a class:

Class I Regional Center
Class II Sectional Center
Class III Primary Center
Class IV Toll Center

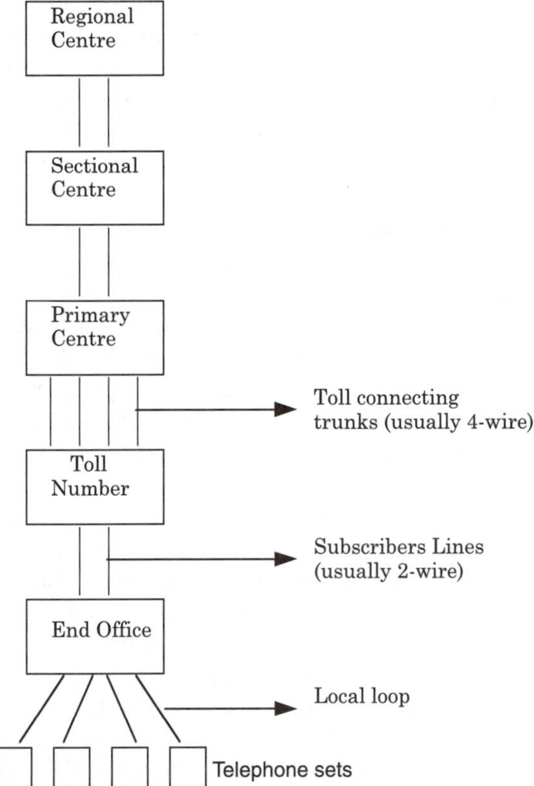

Figure 6.2 Hierarchical model of a telephone system.

Class V End-Office/Local Canter Office/Local Exchange

Each customer (telephone subscriber) has a telephone set which connects him/her to a Class V end office through a two-wire or four-wire (copper conductor) system and is known as **local loop**, **subscriber loop**, or **end office**. The distance between the customer's telephone set (either at home or office) and the end office is typically a few miles. Each country has thousands of end offices, and these are connected by Class IV **toll offices** (tandem center) to provide connections between the customers of different end offices. All the subscribers connected to the same end office are connected by **direct connection** through switching switches. However, a subscriber of one end office is connected to a subscriber of another end office through a toll office, and the end offices are connected to the toll office through lines known as **toll-connecting trunks**. Similarly, subscribers of the same toll office are directly connected by toll-connecting trunks. The subscribers of different toll offices are connected indirectly by appropriate offices of Class III, Class II, and Class I via **intertoll trunks**, **intersectional trunks**, and **inter-regional trunks**, respectively. Each of the trunks at Class III, Class II, and Class I have higher bandwidths.[1,2]

A subscriber communicates with the end office using analog voltage signals on twisted-pair lines. End-to-end connectivity between any two subscribers goes through these levels for the duration of a call. The circuits are established during the establishment of the call, while at the termination of the call, the circuits are released and used with other connections.

As indicated above, local loops usually use **two-wire** (copper conductor) systems for providing an interface with the switching equipment at the end office. In two-wire systems, current flows from a subscriber's set to the end office on one wire and returns on another wire. When the distance between subscriber and end office is in the range of a few miles, we usually prefer two-wire systems. The two-wire system is good for low-frequency signals but are not used for high-frequency signals due to various problems. An interface between a two-wire system and a four-wire system is provided through a circuit known as **transformer** or **hybrid**, which maps electrical and other characteristics of the two-wire system to a four-wire system.

The four-wire system uses a separate transmission path for each direction. The two-wire circuit is connected to a four-wire circuit, and a interface connection circuit needs to be defined at the junction to provide the calls between them. This interface-connecting circuit is known as a *hybrid balancing circuit*. This circuit provides a complete matching between the signals coming out in two-wire and four-wire systems. The signal coming from a local subscriber is divided into two signals at the junction to be transmitted over two pairs of wires (incoming and outgoing). The incoming pair of wires carries an attenuated signal, while the outgoing pair carries the signals to be transmitted. This balance circuit is used on both sides.

Further, the impedance of two-wire or four-wire systems varies with frequency and is not the same at the junction of two-wire and four-wire systems or between wires and the network. This mismatching of impedance at the junction causes *reflection*. In order to reduce this reflection (echo) due to unequal impedance at the junction, echo suppressers are used. (Details on echo suppressers are given in Chapter 2).

As we go from the end office to higher-level offices, the delay becomes greater and greater, as offices at each level are connected to offices of higher levels by switching centers. Usually Class I offices have the shortest path configuration and hence require fewer switches, thus reducing the delay time. The number of trunk lines and their configuration depend on the number of connections (to be made) and vary from country to country. Telephone companies follow the hierarchical model for establishing connection between subscribers, and going through all intermediate switching centers (Class IV through Class I) causes delay in setting up the call, sometimes significant delay. The switching offices are connected by different communication media such as insulated copper wires, coaxial cables, microwaves, satellites, and fiber optics. Due to the very large bandwidth offered in fiber optics, many cables can be replaced by one channel which will carry the same amount of data as those cables would.

6.3.1 Advantages and disadvantages of the hierarchical model

The hierarchical model offers many advantages such as reliability, quick connection, and many new facilities at a much cheaper rate. However, the routing strategies at various classes are predefined for a static and fixed topology. This means that sometimes call setup time and other delays may be significant and customers often get a busy signal or have to wait for a long time.

In order to reduce these problems, many telephone companies have started using a new routing strategy known as **non-hierarchical routing (NHR)**, based on a non-hierarchical model to be discussed in the following section. NHR allows these companies to include a set of predefined routing paths at each office which will handle heavy traffic efficiently, such that customers get fewer busy signals and are connected faster than in fixed topology, as shown in Figure 6.2. It is interesting to note that this does not define a new model, but new routing strategies are implemented over the hierarchical model of fixed topology. It has been shown that this new hierarchical routing strategy provides not only faster connection but also reduces the cost of telephone systems.

6.3.2 *Forms of non-hierarchical routing (NHR)*

Variations of NHR strategies are being used by telephone companies of different countries. For example, AT&T's telephone system is based on a decentralized approach where switching offices of each class contain a set of predefined routes (paths). This information is used in a proper sequence, which in turn depends on the particular day and time.

Bell Canada uses a centralized dynamic technique in its telephone system. The routing strategies depend on the amount of traffic and accordingly adapt to it, yielding a "load-balancing concept" to utilize the trunks efficiently. Some countries (mostly European) offer the choice of a particular path to the end office and provide connection based on it.

6.3.3 *Telephone system standards*

The carrier signals (used for modulation) to different countries are assigned by a U.N. treaty organization known as the International Telecommunications Union (ITU). Usually, countries or their representative governmental agencies/departments are the members of ITU. The Department of State represents the U.S. in the ITU.

The **ITU-T** is an organization that deals with telephone, telegraph, and data communication standards and makes recommendations to the ITU. It meets every four years. Another branch of the ITU, **ITU-R**, is mainly concerned with radio communications. In the 1989 Plenipotentiary, ITU-T and ITU-R were advised to coordinate global standard activities. The carriers assigned by ITU to various countries are usually controlled/managed by the **Post, Telephone, and Telegraph (PTT)** department of the respective country's government. **International Record Carriers (IRC)** are international organizations which include various telephone companies, such as International Telephone and Telegraph (ITT), World Communications, Radio Corporation of America (RCA), Global Communication, Tropical Radio and Telegraph Corporation Telecommunications (TRT).

Common carriers offer basic communication and transportation services for voice and data communications. Telephone companies are examples of common carriers. Telephone companies offer carrier services such as microwaves, satellite links, fiber optics, coaxial cables, or any other transportation media for providing end-to-end communication. In the U.S., telephone companies such as MCI, Sprint, and AT&T are regulated by the FCC.

6.4 *Telephone network systems*

Each country has its own telephone network, managed either by the government or by a private company. In order to understand the infrastructure of telephone networks, we consider the U.S. Telephone companies typically offer services for voice communications such as call forwarding (switching calls to another phone), camp on (setting a phone on hold for an incoming call), call waiting (alerting the receiver to the waiting call), auto switch to extension, auto redial, etc. Telephone network services are provided to customers by seven **Regional Bell Holding Companies (RBHCs)** which control various long-distance carriers such as AT&T, MCI, Sprint, etc. The main duties of RBHCs are to install and operate about 180 **local access and transport areas (LATAs)**, which are usually located in metropolitan areas. Inter-LATA services are provided by RBHCs with equal access to all long-distance carriers.

Seven local telephone services companies known as Regional Bell Operating Companies (RBOCs) were formed following the government-mandated breakup of Bell Systems. A number of small carriers that provide services in a specific region work with these companies. The seven original companies of RBOC were Ameritech Corporation, Bell Atlantic Corporation, BellSouth Corporation, NYNEX Corporation, Pacific Telesis Group,

Southwestern Bell Corporation, and U.S. West, Inc. Some of these have merged, leaving only Bell Atlantic, BellSouth, Southwestern Bell, and U.S. West. At this time, there are over 180 local access and transport areas (LATAs). Each company of RBOC manages between 20 and 30 LATAs, depending on the number of subscribers and size. One carrier, via local-change carrier, offers the local wired telephone services within each LATA. For long-distance telephone services, interexchange carrier providers include AT&T, Sprint, MCI, and a few others.

Each long-distance carrier company has its own communication backbone network and is known as a **common carrier (CC)**. The types of services, prices, etc., are defined in a document called a **tariff**. The FCC approves this document for interstate tariffs. These long-distance carrier companies must provide commercial telecommunication services to the RBHCs. Further, these companies must provide a remote concentrator or private branch exchange (PBX) to a university, organization, or company for use on their premises. This will group a large number of telephone calls and send them to the central office of the PBX or remote concentrator (small switching office), which is connected to the trunks directly.

Telecommunication services can be divided into four main classes: analog, digital, satellite, and value-added. The *analog services* are basically offered in the form of analog channels and are used for voice and data communication. Some popular analog services include **direct long-distance (DDD) calling** using PSTN, **wide area telephone services** (WATS, a dial-up service), **telex** (a low-speed switched network), **data phone** (high-speed switched data service of 50 Kbps), and many private leased channels offering a voice band of 300–3400 KHz. The DDD establishes calls over a network and is also known as **long-distance calling (LDC)**. The telephone system could be a dial-up network or public switched network.

Digital services, on other hand, represent services at different data rates, such as 2.4 Kbps, 4.8 Kbps, 9.6 Kbps, 56 Kbps, etc., and a very low error rate of 109 to 1014. AT&T offers **data phone digital service (DDS)** to cities around the U.S. DDS is defined for transmission of digital data, usually to analog options and facilities. *Satellite services* are usually in the form of analog or digital channels.

Value-added services are different from the previous services in the sense that they offer end-to-end services for the transfer of data instead of only communication channels. Value-added carrier services are available on a larger network, which utilizes expensive resources (e.g., transmission lines) and a message-switching concept. This network offers a variety of services via the user's subscribers based on the volume of traffic requested (for tariff charges). The value-added carriers can request a large transmission capacity (depending on the traffic load) and, as such, are very flexible in adapting to new changes and types of services, as these are available from established common carriers. The value-added networks use a packet-switching technique and are better than existing data communication networks and many private data networks.

Telephone connectors: The FCC has defined a number of connectors. The registered jack (RJ) series is very popular and offers a variety of connectors. Some of these are discussed here in brief; for more details, see Reference 4.

RJ-11: A modular plug that connects the telephone set to the wall socket, used for voice and data communications and also for LAN connections.

RJ-14: Has the same function as that of RJ-11 but is more expensive and is not recommended for applications of RJ-11.

RJ-22: A compact version of RJ-11; can connect a telephone set to the telephone base.

RJ-45: Used in DTE/DCE interfaces and LANs (10 and 100 B-T).

RJ-48: Used in Ds-1 or T1 circuits.

6.4.1 Communication channels

A communication channel or link provides physical paths for transmitting the signals between computers connected to a network. The channels are usually provided by telephone common carriers, and these channels allow the users to access the **public switched telephone networks (PSTNs)**, **switched networks**, or even **leased private lines**. Communication channels have evolved with more emphasis on multiplexing of multiple channels at higher data rates over the same link or channel.

6.4.2 Telephone lines or links

Telephone lines were introduced in the late 1890s, and the concept of multiplexing was used over a pair of telephone wires to carry more than one voice channel in the 1920s. In the early 1940s, coaxial cable was used for carrying 600 multiplexed voice channels and 1800 channels with a multiple-wire-pair configuration. The total data rate offered by these 600 channels is about 100 Mbps. The number of channels which could be multiplexed over the same coaxial cable has continuously increased with technology changes in switching and transmission techniques, and hence data rates for voice communication are also increasing. Popular numbers of channels over coaxial cable include 1860 voice channels (with data rates of about 600 Mbps), 10,800 voice channels (with data rates of 6 Gbps), and 13,200 voice channels (with data rates of 9 Gbps).

With the introduction of fiber optics in 1990, we can get data rates of over 10 Gbps with such advantages of fiber over cable as high bandwidth, low error rate, greater reliability, etc. In addition, another communication link based on wireless transmission has been introduced, the microwave link, which offers a number of voice channels ranging from 2400 in 1960 to over 42,000 in the 1980s.

6.4.3 Data communications options

Communication channels support three categories of data communications options: narrow band, voice band, and broadband. **Narrow-band communication** includes applications requiring data rates up to 300 Kbps. **Voice-band communication** usually defines a bandwidth of voice signals in the range of 300 Hz to 3.4 KHz. These voice signals can be transmitted at a speed of 9600 bps. **Broadband** communication offers higher data communication rates than voice-band channels (up to several Mbps). Broadband communication channels are more reliable and efficient than the other two communication channels.

Common carriers (e.g., AT&T, Western Union, Datran, and MCI WorldCom) offer a variety of services used in data communication networks. MCI WorldCom uses microwave links for providing a physical connection, while Datran uses a switched network for transfer of data. Each of the physical connections provided by carriers possesses some minimum features and offers maximum data transfer speed. For example, if we consider leased lines, the tariff is usually divided into the following categories (depending on the type of application): teletype, sub-voice, voice, and wide-band.

Teletype (or telegraph-grade) communication channels usually provide a maximum transmission speed of 75 baud (i.e., 100 words per minute). In order to send digital signals, high-speed lines with digital-to-analog and analog-to-digital conversion are required at the transmitting and receiving sides, respectively. For low speeds, a direct current (DC) carrier avoids the use of digital-to-analog signals.

A transmission speed of 15 characters per second at a data rate of 180 baud is available on **sub-voice**-grade channels. A higher transmission speed is available on **voice**-grade lines. The voice-grade channels can offer a data rate as high as 10,000 baud without distortions, but they are usually used in systems with transmission speeds of 2400 baud. The **wide-band** channels offer the highest transmission speeds of 60,000 to 120,000 baud.

6.4.4 *Types of telephone network services*

Network services can be obtained in two classes: **leased** and **switched**.

Leased or private lines are always physically connected and available to the users. Users are billed on a 24-hour-a-day and 7-day-a-week basis. The response times are relatively shorter than for switched networks. **Switched** services are offered on public switched networks and are also known as dial-up services. They are controlled and managed by AT&T Data-Phone system and Western Union broadband channels. Users are billed based on the use of switched services (duration of usage). Public switched networks may have a longer connect time, disconnect time, and busy time, as they are based on a connectionless configuration.

6.4.4.1 *Leased/private lines*

As mentioned above, dedicated/leased/private lines provide a permanent full connection between two nodes in the telephone network. Since the line is always connected, no dialing and hence no call-setup time is required. Further, with leased lines, customers will never get busy signals, even if all the circuits are busy. From a cost point of view, sometimes it is cheaper to transmit data on a leased line than on a switched line (especially when a customer has huge amounts of data to be sent every day). Billing is based on distance only and does not depend on the amount of usage time.

As discussed in Chapter 4, there are two types of termination circuits: unbalanced and balanced. In unbalanced termination circuits, one of the two wires at the switching office is grounded and hence represents a reference voltage with value 0 V (there is some voltage, in the range of 0.1 to 0.2 V, but for practical purposes we always assume it to be 0 V). The other line uses 48 V and carries the data between a user and a switching office.

In contrast to this, a balanced termination circuit has two wires between the user and end office. Only the electrical center of the circuit at the end office is grounded. Both lines have the same voltage, 48 V (of course, with opposite polarity) with respect to ground. The user's telephone sets are connected to the end office through an unbalanced termination circuit, and this connection is sometimes called a **dial-up circuit (DUC)**. The private/leased lines are connected to the end office through balanced termination circuits.

There are two methods of processing incoming multiplexed messages over the link: multi-point and multi-drop. In the **multi-point** method, on the receiving side there is one de-multiplexing receiver which, after de-multiplexing the messages, passes them onto appropriate terminals operating at different speeds. In the **multi-drop** method, the entire message goes over all the terminals, and the incoming multiplexed message keeps dropping part of its data at the nodes until it reaches the final destination. In other words, all the destination nodes will get their respective message over the same link or circuit. The main difference between these two techniques is that in multi-point, the transmitter considers only one destination, while in the multi-drop, the transmitter considers each of the destination addresses as a separate address.

Telephone lines originally were designed for transmitting analog signals (particularly human voice/speech), and telephone companies were asked to provide the services to the customers. These companies provide different services for both dial-up and private lines in the following ranges of bandwidth.

Slow speed — Narrow bandwidth with a speed of 30–80 bps
Medium speed — Voice band with a speed of 1200–4800 bps
High speed — Wide band with higher speeds

A telephone system, by its nature, deals with analog signals and hence the above-mentioned services are represented and processed via analog signals, providing what we may call **analog data service (ADS)**.

If we can map digital signals (digital bits of 0s and 1s) in analog signals, then digital data service (DDS) can also be provided on these telephone lines. This requires a modem device, which converts digital signals into analog signals, which are modulated and transmitted by the transmitter over the link. On the receiver side, another modem converts modulated signals into appropriate digital signals. On four-wire private lines, the services providing for digital signal communication are point-to-point and multi-point, with data rates of 2, 4, 4.8, 9.6, and 56 Kbps. In order to send digital signals on these lines, the subscriber has to use a **channel service unit/digital or data service unit (CSU/DSU)** (discussed earlier) for interfacing computers to the lines without using modems.

6.4.4.2 Switched networks

In the terminology of telephone systems, we use the word *switching* as a synonym for voice grade, and switching basically deals with voice communication on voice-grade channels. Telephone networks allow subscribers to establish a point-to-point connection automatically.

There are applications where subscribers (from telephone set terminals) want to use telephone lines for a short time each day. Although telephone companies can provide private lines to subscribers, it is sometimes cheaper to use various switched networks for sending both voice and data signals. Switched networks provide a point-to-point connection between a subscriber's telephone set and terminals and are charged to subscribers based on the duration of the connection, amount of data transmitted, distance, etc.

There are three main categories of switched networks:

1. Circuit-switched networks
2. Message-switched networks
3. Packet-switched networks

The telephone networks allow subscribers to establish point-to-point connection automatically by dialing the requested number from the telephone system. The public switched telephone network (or dial-up network) is used for data communications. The telephone network is usually a two-wire system, which allows half-duplex operation at a data rate of 9600 bps. Modems can be used to achieve full-duplex operation at a speed of 2400 bps or 9600 bps. Terminals and computers are connected to the network via modems and other switching circuits. Usually the switching circuit is built into the modem and also is available as a **switched network adapter (SNA)**.

Circuit-switched networks: The circuit-switching technique was used in the first generation of telephone systems where an operator inserts a jumper cable into input and output sockets to establish the connection between subscribers. A complete circuit or route (or path), between source and destination nodes is established before the data or voice information can be transmitted. The establishment of an entire circuit is obtained through many intermediate switching offices which provide logic connection between input and output lines of those intermediate offices. The circuit between source and destination can be established on any communication link/transmission medium (telephone line, coaxial cable, satellite link, microwave link, etc.). The switching function for dialing equipment nowadays is performed automatically. The telephone system is an example of this type of switching.

The dialing process in a computer network is usually manual, and it is used mostly with remote terminals for interactive communications. A user dials a sequence of bits to get access to a particular system. If he/she gets it, a dedicated circuit is established and the data information can be transmitted on it. If he/she does not get access to the computer network, he/she terminates the connection and tries to get access by redialing the sequence of bits later on, perhaps establishing a different circuit for the same computer. Some networks provide automatic circuit switching where the establishment of a circuit is based on the bits contained in the data frame. A line contention delay is a major problem in circuit-switched networks.

This network is generally bidirectional, with the same channel for transmitting the data in both directions (two-wire line or four-wire line system). For each of these systems, the following three steps are required to establish connections:

- **Set up the link:** The system detects a subscriber's request for service, identifies the terminal, searches and grabs a circuit, informs the required terminal, receives its response, and, after the data is transmitted, de-establishes the circuit (connection).
- **Hold up the link:** The established link is held during the transmission of data information between source and destination and sends out billing information (depending on the distance and duration of the connection) to the subscribers.
- **Release of the link:** After the communication is completed, the link channels and shared devices are released.

As a circuit-switching system requires a dedicated circuit before the data can be sent, this technique is inefficient and uneconomical. In the first phase, we have to send the signal all the way from the source to the destination to establish the circuit and, once the circuit is established, the data is sent from source to destination. For short distances, the propagation times of signals may be small, but for long distances, sometimes the total setup time (including propagation) may become significant. Thus, this technique is not useful for applications that require minimum setup time (e.g., credit card identification, police recording and retrieving of social security numbers, etc.) but has found its applications in very large Tymnet systems.

The circuit-switched networks transmit signals over different types of channels for long-distance communication links. It gives a feeling to users that switching of the lines from more than one link is taking place for the establishment of the user's communication channel. The established channel is available to users during the time when the user is active. It has been decided that the total number of available channels should always be much smaller than the number of users.

It is very unlikely that all the users will use the telephone network at the same time. The network guarantees the existence of established paths and does not offer any delay to the user once it has been set up. The users can send the message information at the rate defined by the capacity of the communication link. The only problem with this type of switching is that sometimes the setup time may be relatively longer. The circuit-switched network is useful for telephone networks and, looking at the behavior of call requests (traffic), the patience of the users, and many other features of the network, sometimes this relatively longer setup time may be negligible. The network is not useful for teleprocessing traffic applications, as the users here always remain connected to the network and, as a result, waiting time may become a serious problem before the transmission of any message information. Further, the allocation of the entire communication resources and capacity to the users during the connection may become a problem in some applications. At the same time, it may be advantageous to have these facilities in some applications.

Two popular networks are based on circuit-switching: public switched telephone networks (PSTNs) and telex networks.

In **PSTNs**, a subscriber's telephone set is connected to the end office through a two-wire system or four-wire (typically unshielded twisted pair) system and should contain some standard form of carrier signal to provide loop carrier signaling between the set and the end office for recognizing various control signals. This connection is known as a local or subscriber loop. The end office is also known as the local exchange, and the local loop carries analog signals or audio tones from a modem. Various tones include **hook-off and hook-on conditions, dial tone, busy tone**, and, of course, **dialing** of a number. Early analog voice telephone systems (public switched network) used a 4-KHz bandwidth for sending all the control information by generating different frequency signaling for different controls. The hook-off condition represents the lifting of the handset (20-mA current flows between the end office and the user's telephone), while hook-on represents the resting condition of the handset (no current flows from the end office through the user's telephone).

When a user dials a number, the end office searches for a direct connection (trunk) between itself and the destination end office. If it finds one, it establishes the routing for the call. If it does find one, it will search via a higher level of **toll center**. If it finds a direct routing between itself and a toll center, the call follows that route; otherwise, it will search for a direct link via its higher level, and so on. AT&T uses a hierarchical system starting with end loop to toll center to primary center to section center to regional center. There are over 19,000 end offices in the U.S. **Local access and transport areas (LATAs)** usually handle the code numbers, and a local telephone company is known as a telco or local exchange carrier. Popular long-distance carriers (e.g., AT&T, Sprint, MCI WorldCom, U.S. West) are usually known as **interexchange carriers (IECs)** or **interXchange carriers (IXCs)**.

Different types of telephone sets such as rotary-dial and touch-button basically generate current pulses for each of the control signals for establishing a connection between two subscribers. In the old rotary-dial system, different control signals are transmitted to the end office at a defined rate of transmission (pulses per second). In touch-button systems, two different frequency signals are generated for each of the buttons. After the required number has been dialed, these signals are processed either by the end office for direct connection or by a higher-level office for inter-office trunk signaling for indirect connections. This type of system is known as **in-band signaling (IBS)** and has been in existence for a long time. This scheme has many problems and drawbacks, such as misuse of bandwidth for sending both control and voice signals, misuse of frequency signaling for different controls, restrictions to direct connection only (without operator), etc.

In order to eliminate these problems and other drawbacks of an in-band signaling system, a new scheme known as **common channel inter-office signaling (CCIS)**, based on CCITT Signaling System No. 6 (SS6), was approved by telephone companies. It was designed and installed in 1976 by AT&T as a packet-switching network to provide signaling (different from the public switched network). This network runs at 2.4 Kbps (as opposed to 4.8 Kbps in CCITT SS-6) and is known as an **out-of-band signaling (OBS)** scheme. It separates signaling from voice and uses a separate digital circuit. Since one digital signal can be shared by more than one voice channel, the name "common channel" is used.

When a user dials a number, CCIS checks the status of the destination end office. If it is busy, it will signal to the sender end office, which in turn informs the user. The busy tone, in fact, is coming to the user from the local end office. But if the destination end office is not busy, it will establish a voice circuit over the needed trunks (connecting different centers at different levels) and send a ringback signal to the sender end office, which in turn sends a 20-Hz ringing signal to the user's telephone. The message format includes 28 bits (eight bits for error checking, while the actual information is defined within 20 bits). CCIS offers various options to the subscriber for making and paying for the telephone calls, such as coin-operated, calling-card-operated, operator-assisted, collect, etc. Some calls are processed by digitized voice; there is no human operator involved during the establishment of connection or routing of calls, as it is a completely automated system.

The end office in CCIS chooses a trunk on the public switched network and sends a packet (containing the route, channel selected, etc.) to the CCIS. After this, CCIS has complete control over the routing and choosing of intermediate nodes within CCIS until the connection is established. Thus, CCIS gets complete control and management and subscribers/operators have no access to this information (routing path, channels selected, etc.); therefore, there is less likelihood of any intruder misusing the system.

CCIS offers the following advantages:

- Call setup and takedown are faster due to faster transmission of signals.
- Subscriber's post-dialing delay is reduced.
- The holding time for equipment and trunks is reduced.
- It has more capacity than in-band systems.
- The control signals flow during conversation on full-duplex signaling.
- CCIS is flexible in the sense that control messages and software can be included to add new services.

In CCIS, packet-switching network concepts (packets for call setup and takedown) are used. A brief description of CCIS is as follows: a signal unit (SU) contains a piece of information and consists of 28 bits (eight bits are used for error detection). A packet composed of 12 SUs is transmitted, out of which 11 SUs contain signaling information; the last one is an acknowledgment SU (ASU). All SUs within a packet have the same destination. A block diagram of a CCIS signaling link is shown in Figure 6.3.

Data circuit–terminating equipment (DCEs) or modems are used for converting digital signals into analog and back into digital. Signaling terminals (STs) are used for storing messages, error control (detection and re-transmission), monitoring data carriers for link outage, and selection of an alternative link in case of failure in a link. In CCIS, a link is a four-wire dedicated system. A terminal-access circuit (TAC) allows the processor to access

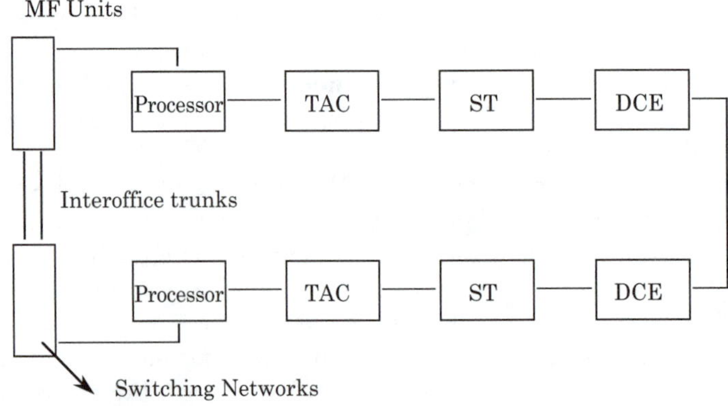

TAC: Terminal-Access Circuit
ST: Signaling Terminal
DCE: Data Circuit–Terminating Equipment
MF: Multi-Frequency

Figure 6.3 CCIS-signaling link.

signaling links and select a particular link (based on a procedure executed by the processor). The processor acts as a link controller, receives CCIS messages, selects a link, initializes appropriate procedures, selects alternative links (if needed), and provides the instruction to TAC.

CCIS became very popular in the late 1970s, and there were a number of commercial packet-switching networks on the market based on CCIS. The reason for their wide use and popularity was the ability to use them for applications such as remote databases, remote file access, time-sharing systems, etc. In these applications, a physical connection is not required for a long duration, and hence cost of the call basically depends upon the amount of data and not on the connect time.

The carrier's equipment for the telephone system is composed of three different types of networks: **analog public switched networks** for voice signals, CCIS for controlling the **public switched network**, and **packet-switching networks** for digital signals. These three networks together allow the subscribers to send voice and data (from digital computers, terminals, telephones, facsimile, or any other digital device) from one node to another within a country or across countries. CCIS utilizes analog carrier signals for providing various controls between a subscriber's set and the end office.

A new signaling scheme recommended by CCITT is based on digital signals and is known as **Signaling System No. 7 (SS7)**. SS7 has been widely accepted and implemented by various telephone companies. This scheme was the first signaling scheme used in integrated services digital networks (ISDNs) and has already been accepted as a standard specification of signaling systems. It contains very complex coding and caused many problems during installation into the electronic switching system (ESS) computer in New York City. The software has been revised, tested, and installed and is now providing support to the Federal Aviation Administration's Air Control System.

SS7 overcomes some of the limitations of earlier schemes (SS6 and CCIS), such as support of data speeds of 2.4 Kbps or 4.8 Kbps, limited size of service units (SUs), etc. Many telephone companies have already decided to switch to SS7. AT&T and Telecom Canada have implemented message transfer part (MTP), which is ISDN user part (ISDNUP) and signaling connection control part (SCCP) based on the SS7 signaling scheme. The system is also known as CCITT Signaling System No. 7, or **CCSS7**. It offers better monitoring, maintenance, and network administration.

CCSS7 uses network topology similar to that of CCIS and contains digital 56-Kbps trunks. Signaling transfer points (STPs) cover all seven regions on the network topology, and STP control signals manage and operate CCIS and CCSS7 protocols.

SS7 provides the following functions and services to telephone systems:

1. Message transfer part (MTP) (Q.701–Q.709).
2. PABX (or computerized branch exchange: CBX) application (Q.710).
3. Signaling connection control part (SCCP) (Q.711–Q.714).
4. Integrated services digital network user part (ISDN UP) (Q.761–Q.766).
5. MTP monitoring and measuring (Q.791).
6. Operations and maintenance application part (OMAP) (Q.795).

A set of procedures is defined in SS7 for various operations, such as setup of a call, transfer of data, termination, clearing of a call, etc. These procedures also provide a mapping between telephone control information and signals of SS6 telephone exchange and SS7 signaling transfer points (STPs). There are a number of specialized dedicated computers performing the switching, and the software can be easily be replaced and updated. Lucent Technologies has designed electronic switching systems known as 4ESS and 5ESS computers. The 4ESS computer usually handles over 1 million calls per hour, while the 5ESS handles fewer than a quarter of a million calls per hour. AT&T uses about 140 4ESS computers in the U.S.; Northern Telecomm (formerly known as Normtel; a Canadian company) manufactures these computers. The SS7 circuits monitor the signaling, while trunks interconnecting the computers carry data, audio, and video.

Along with PSTNs, another type of network based on circuit switching is the **telex network**. This network is similar to a telephone network, with the only difference being in the devices, which can be connected. In telex, teleprinters are connected using point-to-point communication and can send typed messages on the telex network (switched-type network). New services that have been introduced in these networks are store-and-forward capability, automatic conference calls, setup facility, etc.

Telex networks provide a speed of 40–60 bps (baud of code), while American TWX networks provide 100 bps (ASCII code). Many countries are connected to telex networks, and there are a few million telex installations around the world. Telex networks, although used by many countries, are very slow and unreliable, and they produce very poor-quality output. Telex communication has been replaced by teletex communication, which offers higher speed and better quality of output (using higher-quality printers) on word processors. This communication can provide a transmission speed of 2400 bps, 4800 bps, 9600 bps, and so on.

Due to incompatibility between various word processors (by different vendors), CCITT, in 1982, defined an international standard for telex communication. This standard is based on ISO open-system interconnection seven-layered architecture model and includes protocols, definition of communication rules, character sets, error-control functions, etc., collectively known as **telex protocol (TP)**. This protocol provides interface between different types of word processors and also allows conversion of formats between telex and teletex networks.

Different countries use different types of teletex networks, such as packet-switched networks, public switched telephone networks, and circuit-switched teletex networks. The telex networks can be characterized as the networks which provide high-speed transmission of text and high quality of output and allow subscribers to send text from old text or a word processor to any part of the world. Data stored in databases can be retrieved using videotex, which can be either a public or private network. Terminals may be connected to a videotex network via modem and can also update data stored in a database. Telegram networks are another type of network providing services in telephone networks.

Message-switched networks (MSNs): In the message-switching technique, an individual message is separately switched at each node along its route or path from the source to the destination. In other words, there is no direct physical connection between the users. A circuit is not established exclusively for the message, but instead messages are sent using the **store-and-forward** approach. Sometimes these networks are also known as store-and-forward switched networks. Users, based on the capacity of the networks, divide each message into blocks of data, and these blocks of messages are transmitted in a sequence. The receiver, after receiving these blocks, constructs the original message from the blocks (by putting them in the same sequence as when they were transmitted) and sends an acknowledgment to the source. MSNs offer a multiple-receiver capability in the sense that a user may send the same message to a number of locations simultaneously. Further, the message-switching technique allows assignment of priority to messages which will supersede the low-priority messages, while this is not supported in circuit switching.

The first block of the messages contains control information regarding routing, quality of service parameters, etc. If the destination node does not accept the messages (due to incomplete message, errors, or some other failure), the source node will transmit the same message again on the same link. The intermediate switching nodes will store the block of messages in the event of circuit failure, heavy traffic, broken links, loss of acknowledgment, etc. These intermediate nodes have secondary devices (disks) and direct access storage devices for storing these blocks of messages.

The switching nodes store the blocks of messages and look for the free link to another switching intermediate node. If it finds a free node, it sends one block at a time to that node until all the blocks stored are sent. If it does not find any free node, it will store the blocks and keep on trying to find free links until it finds one and transmits the block over it. The intermediate nodes are already predefined between source and destination on the basis of certain criteria (e.g., shortest route, fastest route, heavy traffic, shareable files and programs, real-time data, etc.). If it does not find any intermediate node within the constraints, it will reroute the blocks on a different alternative route. This offers advantages in keeping a fairly constant error rate (although lower on the average) for all traffic, while in the case of circuit switching, the error rate may go up during heavy traffic. Each block of a message is received as a complete frame, and errors are checked in each block separately. If an error is found in any block, then that block will be re-transmitted. Transmission in this mode of switching is typically unidirectional (simplex operation).

The MSN technique is less expensive than circuit switching and widely used as a very useful switching technique for interconnection. However, it does not support dynamic load balancing and, as such, heavy traffic on a particular node will be handled by itself. There is no limit to the size of the blocks of messages, which poses two obvious problems: (1) the switching node must have enough storage capability for storing the blocks, and (2) if a free link between source and destination is busy for a long time, then waiting and hence response time for the blocks to be transmitted may be significant. Thus, the technique seems to be useful for batched messages of longer duration but not suitable for interactive messages of shorter duration.

Once a route is established in a message-switched network, it does not present any delay to the transmission of message information except the transmission delay, propagation time, etc. Thus, the transfer time for the message information depends on these parameters and also may depend on the amount of traffic in the network. In other words, if the network is experiencing very heavy traffic, the nodes of the network will also have a large queue of traffic and, as such, waiting time may become another parameter for the transfer time of the message information. Thus, it can be said that the transfer time in the network (under such conditions) will always be greater than that of circuit-switched networks. Circuit-switched networks allow efficient utilization of communication links. Each of the nodes connected in the network share the load with equal probability and, hence, the channels are utilized to their maximum capacity levels.

Packet-switched networks (PSNs): In principle, packet-switching is based on the concept of message switching with the following differences:

- Packets are parts of messages and include control bits (for detecting transmission errors). Each packet has complete control and identification information associated with the user's data and is handled separately.
- Networks break the messages into packets (or blocks), while in message switching, the users break the messages.
- Due to very small storage time of packets in the waiting queue at any node, users experience bidirectional transmission of the packets in real time.

In packet-switching, the message is divided into packets (or blocks) of fixed size; further, each packet has its own control information regarding routing, etc., across the network. This means the packets may follow different routes across the networks between source and destination and they may also arrive at the destination at different sequences. The receiver, after receiving the packets out of sequence, has to arrange the packets in the same order as they were transmitted from the source.

Due to the fixed or limited size of packets and also due to different routing of the packets, the switching nodes store the packets in the main memory, as opposed to a secondary storage device (disk) in message switching. Accessing or fetching a packet from main memory takes much less time than with disks, and further, the packets can be forwarded from a switching node to any other switching nodes before the complete arrival of the packet there. These two features together reduce the delay time in this technique and, as such, the throughput (the number of packets transmitted per unit of time) is also improved.

In packet-switched networks, if any node receives a garbled packet, it will request the sending node to transmit the same packet again. When the destination node receives the last packet of the message, it sends an acknowledgment (defined as a typical bit pattern packet message) to the source node. Similarly, if the destination node does not receive all the packets within the specified time (after request for the connection has been confirmed), it sends a request for the data (instead of an acknowledgment) to the source node about the missing packets. After the receiving node receives the missing packet, it does not send the acknowledgment.

The packet-switching technique allows the switching nodes to transmit the packets without waiting for a complete message and also allows them to adjust the traffic they have, thus minimizing the resource requirements of the nodes. If any particular node is already heavily loaded, it will reject the packets until its load becomes moderate.

Packet-switching can, in fact, be considered a mode of operation rather than switching, as it allows optional use of bit rate by transmitting packets between source(s) and requested destination(s) in a dynamic multiplexed mode, and hence it utilizes the link capacity effectively. This technique may pose some problems if it is used in digital telephone networks, as packets will be made out of PCM-coded samples. Further, these packets reach the destination with unpredictable delay.

The packet-switching technique was used in ARPANET and also in X.25 networks and finds its application in WAN protocols. This type of switching offers a limited capacity and speed of transmission. Recently, a modified version of packet-switching based on transfer modes (transmission and switching) and multiplexing, known as fast packet-switching (FPS), has been used in applications such as integrated services digital networks (ISDNs) and its extended form, broadband integrated service digital networks (B-ISDN); asynchronous transfer mode (ATM)-based high-speed networks; etc. These networks, along with their architectures, services offered, and various standards, will be discussed in Chapters 18 and 19.

6.5 *Private automatic branch exchange (PABX)*

PABX provides connections between telephone subscribers of an organization or university and other subscribers (within city, state, country, or across countries). The main function of PABX is to transmit and receive telephone calls. With the advent of digital signal technology, digital **PBX** (early 1970s) was introduced for office data communications, automation, and other telecommunication services such as local area networks, electronic mail, bulletin board messages, file transfer, etc.

The transmission link between a PABX subscriber's phone system and the end/central office supports only voice-grade channels occupying a frequency range of 300 Hz to 3.4 KHz and offering a very low data rate of 56 Kbps. Internet traffic needs much higher rates, as it includes graphics and video, which may take a longer time to download on

PCs. The links between the Internet and end offices provide data rates as high as 2.5 Gbps. Although modems offering 56 Kbps and even higher rates have been available on the market since 1996, the bottleneck continues to be the twisted-pair cables connected between the user's phone and the end office. A number of new technologies are being considered to provide a high data rate for Internet traffic. Some of these include ISDN, cable modems, and asymmetric data subscriber line (ADSL). Telco has suggested the removal of loading coils from twisted-pair cable as a means of increasing the data rate at the expense of reduced bandwidth. Cable modems provide a very large bandwidth (due to the cable medium).

ADSL modems provide much higher data rates than ISDN. Here also, Telco suggests a higher data rate by removing the coils from unshielded twisted-pair cable (being used between a subscriber's set and the end office). ADSL modems offer a higher data rate for downward and a lower data rate for upward communication. The telephone link is being used for both data communication and the telephone system at the same time. The modem is connected between the twisted pair and the end office. On the receiving side, it uses a plain old telephone service (POTS) splitter which provides two different paths on the same link between an ADSL modem and the user's PC and between the splitter and telephone system.

Hughes Electronics, Inc. offers high-speed Internet access over its Hughes' Galaxy IV geosynchronous satellite system. The downward (from satellite to user's PC) data rate is 400 Kbps. The user has to go through the Hughes Satellite Control Center via PSTN (as is done in a normal course for Internet access). The request for any file or program is sent to the satellite, which then relays it to the satellite receiver. The files are then downloaded from the receiver onto the user's PCs. The downward communication uses a frequency band of Galaxy IV 11.7 to 12.2 GHz. The data communication in this frequency band is transmitted using the TDM technique.

Digital PBX is capable of handling computer data and digital signals from word processors, office machines, etc. Some well-established PBX suppliers and computer mainframe suppliers have announced joint ventures for computer communication system development. Digital PBX offers the following features and applicability:

1. Voice signals can be converted into a 64-Kbps data stream for switching, thus avoiding the use of modems. If public networks can be converted into integrated ISDN transport carries, then digital transmission will be used between subscribers, and a digital transmission path of 64 Kbps can be established between any two ports/sockets.
2. It provides flexible switching and control functions. For lower-speed applications such as file transfer, local area networks can be used which will not use the entire 64 Kbps; however, this rate is enough for the majority of office applications.
3. The wiring should not be a problem for digital PBX, as the existing wiring facilities can be used to carry digital signals at transmission speeds of 300 to 600 Kbps and also can be extended beyond it. Digital PBX provides communication between various devices and equipment and, as such, offers appropriate switching between them. Both data and voice can be transmitted simultaneously.
4. With the upgrading of common carriers on telephone networks for transmitting digital signals (like ISDNs), it will be possible to connect digital transmission links running at 1.5 or 2 Mbps and even greater from ISDN to PBX and provide the equivalent of 30 voice channels.

In the U.S., a Bell System T1 carrier supports 24 channels at a speed of 1.544 Kbps, while in Europe, CCITT's standard 2.048 Mbps supports 30 channels running at 64 Kbps.

The majority of the networks are public networks and are required to provide services to different users. In the U.S., these networks fall into the following classes: the local exchange carriers (LECs) that provide local access services (e.g., U.S. West) and inter-exchange carriers (IECs) that provide long distance services (e.g., AT&T, MCI, Sprint, etc.). Regional Bell Operating Companies (RBOCs) operate the LECs. IECs also provide inter-state or inter-local access transport areas (LATAs). In most countries, government-controlled post, telegraph, and telephone (PTT) providers provide these services.

Asynchronous transfer mode (ATM) is used in different classes of public networks. For example, it can be used to replace the end office and the intermediate office above it by connecting them with DS1/DS3 links. As we know, the subscriber's phone is connected to the end office via a twisted pair of wires which must be replaced by coax/fiber. The public network system and CATV can use ATM for providing video-on-demand services. This may require the use of compression techniques such as asymmetrical digital subscriber loop (ADSL) that uses the MPEG2 coding method.

There are so many LANs that internetworking has become a serious problem. Some of these LANs include SNAs (mainly for corporations) for sharing the mainframe data, PBXs (for voice services), point-to-point voice channel networks for PBXs, videoconferencing, and many other types of networks. Each of these networks provides a specific type of service, and it is quite obvious that for any of the services, higher data rates, higher bandwidths, and minimum delay are expected. For some applications (cable TV, voice-grade communication), constant bit rates are needed, while in some applications (digital data, videoconferencing, multimedia, graphics, etc.), variable data rates are needed. The use of ATM is expected to offer higher bandwidths and data rates to LANs (FDDI, high-speed Ethernet, etc.) and LAN connectivity and bandwidth required for graphics, images, and multimedia applications.

The existing packet-switched network X.25 provides interfaces to routers, which are connected to routers, bridges and LANs. The interface offers a dial-up connection via a 56-Kbps link from LANs. The function of the X.25 protocol is to provide the functions of the lower three layers of OS-RM. Since the transmission media are not reliable, the X.25 must support error control and error detection. Further, a packet going through three layers requires processing time, and hence the network is very slow. There are a number of ways an ATM can be used in X.25 networks, which are typically used for data communication. In one X.25 switched network environment, each of the X.25 switches is replaced by frame relay (connection-oriented services). This will increase the speed from 56 Kbps to T1 (1.544 Mbps) speed. For connectionless services, switched multimegabit data service (SMDS) switches offering higher speeds than T1 are used. They can offer T3 (45 Mbps) speed and provide public broadband services. The underlying frame relay may become an access point to an SMDS network. In another X.25 environment, ATM may be used as a backbone network.

6.5.1 PABXs vs. LANs

In the hierarchy of LANs, the higher-level LANs typically become a sort of backbone LAN for the lower-level LANs. There may be a situation where PABXs may be used rather than LANs, as these offer services similar to those of LANs. The PABX basically offers switching nodes for transporting the telephone services of voice communications over a transmission medium. It is basically considered a first generation of integrated digital-switching and telephone-exchange systems. The PABX is derived from the earlier version, private branch exchange (PBX), which was defined for providing telephone interconnections within the premises of private organizations. PBXs offer access to public telephone systems. PABXs support analog signals and also provide support for data connection over a modem.

PABXs can accept digital data and also can access LANs or can be connected to LANs. In other words, PABXs are being used for both voice and data communications. The only problems with PABXs may be their limited bandwidth. Due to widespread use of digital technology in data and voice communications in ISDN technology, as well as the use of pulse coded modulation (PCM), which is an inexpensive device for sampling communication signals, selection between PABX and LAN has become a sensitive issue.

For limited applications, it may be more appropriate to use PABX over LAN. Although both LANs and PABXs offer similar services, they each have their own merits and demerits, and in no way will these two technologies replace each other. Instead, they will complement each other in the area of data and voice communications; however, the main services with PABX will still be voice communications. Digital switching, in fact, is derived from time-division switching, which was defined for analog signals, i.e., voice communications. The switching functions remain independent of the signals being used. Space-division switching defines a signal circuit for each of the signals and is divided in space; i.e., each signal connection travels over a dedicated circuit through a sequence of switches.

Space-division switching is usually implemented by a crossbar configuration with N full-duplex I/O lines with N inputs and N outputs, where N represents the number of switches. These switches provide connections between single input and output lines. Due to the size of crossbar connection points, these switches are now replaced by multi-stage switches which provide connections between groups of lines (in contrast to input and output lines separately). The second generation of digital exchange is **computerized branch exchange (CBX)**, which supports analog voice traffic and is used in codecs for digital switching. The new or third generation of digital exchanges is based on integrated voice and data communications.

The switching concept of asynchronous transfer mode (ATM) is finding its use in PBXs, where it will become a server (providing various telephone services) in ATM LANs. Some PBXs have already started using ATM interfaces (Siemens and ROLM products). PBX vendors have already introduced ATM-ready PBX products that will be replacing ATM LAN switches. Three APIs have been defined to access the features of calls and other control voice calls such as telephone API, telephony services API, and WinSock 2.0. Voice traffic can be generated from different sources such as human voice, fax data, modem signals, and recorded audio signals. Two methods or services of sending voice over ATMs are known as constant bit rate (CBR) and real-time variable bit rate (rtVBR) and are discussed in more detail in Chapters 18 and 19.

6.6 Digital carrier systems (DCS)

When a telephone subscriber wants to send his/her voice on the telephone system, the voice (continuous analog signal) of 4-KHz bandwidth is modulated (using an analog modulation technique such as amplitude modulation (AM), frequency modulation (FM), and phase modulation (PM)) over another continuous signal of high frequency, and this modulated signal is sent over the lines (or any communication medium) to another subscriber. The telephone system accepts the analog signal at an end office and transmits an analog-modulated signal on the medium. If we want to send a digital signal (from computer or terminal) on the telephone system, we have to convert it into a modulated analog signal using a modem on the transmitting side and another modem which reconverts the analog-modulated signal into a digital signal on the receiving side.

6.6.1 Digital networks

Digital private networks, in general, consist of two components: **transmission** (leased circuits) and **switching centers** (circuit, message, or packet-switching) adapted to a particular need of data switching. These networks can be considered an alternative to the use of switched circuits across the public telephone or telex networks. These networks are limited to a group of subscribers and are generally incompatible with each other. The digital public networks (analogous to public telephone networks) have to provide effective, economical shared transmission, switching, and information storage services at both the national and international levels. Further, these networks have to meet the needs of subscribers belonging to businesses, universities, and the general public.

There is a trend toward digital technology in network data transmission systems and exchange systems; many organizations have already started going toward digitalization for transmission, multiplexing, switching, etc., and some organizations have chosen a different approach of interfacing with digital public networks via their digital PBXs. The analog PBXs are using digital technology due to competitive tariffs and new services and, in fact, many private networks have already started offering services through their digital exchanges. The introduction of integrated digital technology is a step in this direction. Here PBXs are linked to digital private circuits. This shift of analog exchange into digital offers various advantages, such as elimination of expensive multiplexers and expensive analog-to-digital interfaces, and it also allows both data and voice to use the same digital circuits for their communications.

The common channel signaling has been a useful concept in digital transmission, as now both data and control signals can be transmitted over the same channel (as opposed to transmission of control signals over a speech circuit path). The private networks gearing toward digital transmission find their applications in offering a wide range of features. These networks use digital switching and transmission and digital devices such as digital multiplexers, etc., and they offer interesting features which are coupled with stored program control. The digital PABXs offer services to both voice and data over the same network with advantages of reduced cost and increased flexibility, reliability, and availability.

Packet-switching is widely used in these networks. Packet assembly and disassembly (PAD) can be performed either at a switching office or at front-end processors, as shown in Figure 6.4. Although circuit switching can also be used in these networks, it does not provide flexibility and efficient use of links in the network.

In a typical configuration, the data from our asynchronous terminals cannot be sent directly to a public switched network. A device known as a packet assembly and disassembly (PAD) proposed by CCITT can be used between the terminal and public network. PAD is defined by the CCITT X.3 standard that provides the assembly and disassembly of the packets into data and control fields. The X.28 standard defines an interface between a terminal and PAD. The X.29 standard defines an interface between PAD and a public

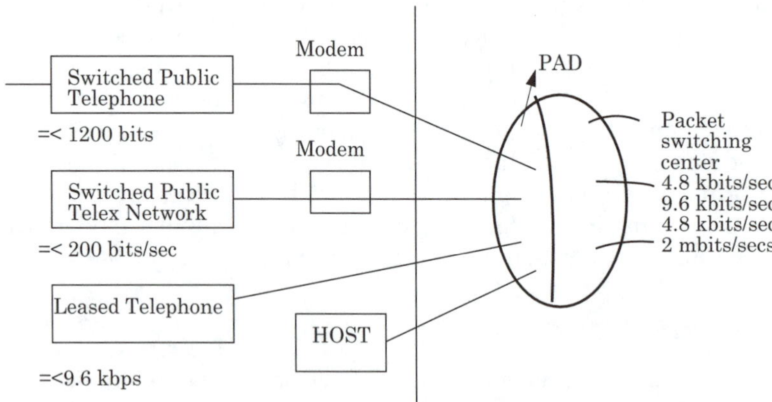

Figure 6.4 Packet-switching networks.

switched network. The packet from a terminal goes to the public network through PAD, and the packets from the public network go to another public network or networks. The X.75 standard defines the interface between various public networks like Tymnet, Uninet, and Telenet (all U.S.), Datapac (Canada), PSS (U.K.), and so on. For all these networks, the X.25 standard may provide a unified interface. The public network acts as a DTE to a second net's DCE. The CCITT proposed three possible configurations. In the first configuration, the X.25 DTE will be connected to a public network that in turn communicates with another X.25 DTE via the public network. In the second configuration, the DTE will interact with an X.25 LAN, which will send the packet to another X.25 DTE via public network. In the third configuration, the end X.25 DTE communicates with public networks that communicate with each other through private networks. The public networks can be accessed either via public access ports or private ports. The public port is usually dial-up or wide-area telephone service, while the private ports can also be dial-up or leased lines.

Switching in digital networks is different from that in analog networks and offers the following advantages:

- Binary representation allows the easy interconnection of any input and output of an exchange by using simple logical operations between input and connection-request signal (command).
- Time division multiplexing (TDM) allows the establishment of different communication on the same medium at a different instant of time, which reduces the cost and complexity of the networks.
- Digital transmission offers a high-speed data rate with a low error rate.

Although the advantages of digital transmission are substantial, the day when a new facility will be built entirely for it seems remote at this point. Since there already are huge facilities for analog transmission, a compromised approach for using these together has been widely accepted, i.e., integration of digital switching and analog transmission for a new network known as an integrated digital network (IDN).

IDNs are based on the concept of integrating switching and transmission, and this integration can be obtained in two different ways:

- Convert digital signals into analog for analog switching and integrate with an analog network. The resulting network will use digital transmission on an analog network.
- Convert analog signals into digital for digital switching and integrate with an analog network for analog transmission on it.

Both of these methods of integration offer the following advantages:

- Improvement in the quality of transmission of analog information (by avoiding quantization errors).
- Total digital transmission from one node to another with a total transparency of code and data rate.
- Analog-to-digital converters (PCM codecs) are used only at the boundary of the network.

6.6.2 *Integrated networks*

The switching techniques discussed above also find their applications in different types of data and integrated networks. In the integrated environment where both voice and data can be sent over the same channel, the data networks (based on packet-switching) will handle (process, transfer, etc.) only the data while PSTN (based on circuit switching) will handle voice communications. The networks supporting both data and voice are known as integrated digital networks, offering narrow-band and broadband services.

Integrated services digital networks (ISDNs): Integrated networks offering both data and voice are known as integrated services digital networks (ISDNs). These networks use both types of switching techniques for offering integrated services. For voice communications, circuit switching involves both channels B and D for their protocols, such that the B channel is defined for exchange of a message or user's information, and the D channel is concerned with the exchange of control information between the network and users. The control information deals with the establishment of calls, termination of calls, accessing of networks, etc. The switching nodes receive control information from the common channel signaling network (CCSN) regarding the control information for connection of circuits and its use for transmitting the user information. This CCSN uses the **common channel signaling system No. 7 (CCSS7)** scheme for ISDN as defined by CCITT. Packet-switching is used on channel B, which is using circuit switching for accessing the services of the ISDN, and also on channel D.

Packet-switched networks, as we noted earlier, work on basically the same principle as that of message-switched networks, except that here the message information is broken into packets which contain a header field along with the message information. The header field includes a packet descriptor, source and destination addresses, and other control information. The packets can be routed on different routes independently. Due to the fixed size of the packets (compared with the variable size of the message block in the case of message-switched networks), there is an improvement in the performance of the network and also the utilization of the communication link. This network can be used in two different modes of operation: datagram and virtual circuit. In the former mode, the users can transmit the packets of fixed size back and forth independently, while in the latter mode, the network manager performs the operations of segmenting, reassembly, and maintaining the path/route until the packets are delivered to the destination.

High-speed networks (HSNs): Due to availability of a fiber-optic transmission medium for high-speed transmission, a new version of packet-switching (fast packet-switching) is used in new services from the packet-switched networks. For example, the broadband ISDN is defined as relay-based, using fast packet-switched networks. The **frame relay network** provides packet-switching services over channel B such that the D channel is used for control signaling and channel B is used for user information as discussed above. Layer two of ISDN supports multiplexing of virtual circuits on the B channel and then the frames are relayed. The functions of layer three, such as sequencing and error control, which require much processing at the switching nodes, must move toward the end systems so that the switching node may be allowed to perform more functions of switching and transmission. This, in turn, will offer greater speed of transmission and also higher throughput. This is exactly what is being done in the frame and cell relay–based networks. It is considered a streamlined form of X.25 packet-switched data networks offering the network services at higher data rates.

The functions performed by the frame relay include framing and switching with multiplexing (using a data link connection identifier). The frame relay–based network is responsible for routing these frames to the destination node via intermediate switching nodes and also offers error control (error detection only). When a frame containing error is received, the frame relay network will simply discard the frame; it does not allow re-transmission of frames containing errors. The option of re-transmission or flow control has to be offered by the end systems and is not performed by the network. There is a possibility of including some control information for congestion-control functions by defining suitable bits in the frame header. The frame typically consists of a flag field, address field (usually three or four bytes), user's data field, and another flag (details of these fields can be found in Chapter 17, High-Speed Networks). This frame is defined at the MAC sublayer. As usual, the user's data field includes end-to-end LLC control information.

The **cell relay**, on other hand, defines an ATM cell of fixed size (53 bytes) that is relayed by layer one. The switching nodes in both of these fast packet-switched networks (frame and cell relay) do not provide error control and flow control and are concerned only with routing and switching the user information. The end systems must provide error and flow control. In other words, transferring these functions to end systems, thus offering a very high speed of switching and transmission, lowers the overhead from the switching nodes. The ATM cell offers switching and multiplexing within a fixed size for a variety of signals such as data, voice, video, etc., and these services are available over public switched data networks. The routing information is defined in a cell header and offers functions similar to that of frame relay. Control information such as address, routing (not the exact one), and other relevant control information are defined within five bytes of the cell header. ATM cell-based networks multiplex multiple logical connections over the same physical channel (like X.25 networks, frame relay networks, etc.). For more details, see Chapter 19.

In addition to these communication networks, there are other types of communication systems/carriers, such as value-added networks (VANs), which provide basic communication components such as modems, cables, multiplex various channels and send them to the networks. Message-switching and packet-switching networks are examples of VANs.

6.6.3 Digital communication circuit

If we have a digital end office in a telephone system, then a signal transmitted from a local loop (subscriber's telephone set) is an analog continuous signal. A codec (coder-decoder) which produces a seven- or eight-bit number will convert this signal into a digital signal at the end office. Codecs convert an analog signal into a digital signal and are therefore known as analog-to-digital (A/D) converters. Modems, on the other hand, convert digital signals into modulated analog signals and modulated analog signals back into digital signals. The Nyquist theorem is used in codecs for converting analog signals (4-KHz bandwidth) into digital signals. According to this theorem, it is sufficient to reproduce all the information from a 4-KHz bandwidth analog spectrum if it is sampled at the rate of 8000 samples per second (duration of each sample = 125 μs). This technique of sampling analog signals had become very popular and is being used in pulse-coded modulation (PCM), which was discussed earlier.

There is no agreed upon technique for converting analog signals into digital at an end office based on the above method. This has resulted in the development of a number of "converters" which are used for changing one into another, due to incompatible schemes.

6.6.4 Bell System carrier T1

One widely used digital communication circuit is Bell System's T1 carrier. The circuit includes a pair of wires (one for transmitting and another for receiving digital signals) and offers a bandwidth of 1.544 Mbps on full-duplex connection. Although it can transmit voice data and video signals simultaneously, its primary use is in voice transmission. The T1 carrier supports 24 voice channels; each channel consists of eight bits (seven bits for data and one bit for control). These 24 voice channels are multiplexed, and analog signals for each channel are sampled on a round-robin basis using one codec. The output of the codec represents digital signals for each of the input multiplexed voice channels. Each of the channels composed of eight bits is grouped into one frame known as a T1 frame having

$24 \times 8 = 192$ bits, to which one extra bit is added for defining and providing the synchronization of the frame. Thus, each T1 frame has 193 bits. Each frame has a duration of 125 μs, or we can also say that 8000 T1 frames are generated each second, giving a data rate of $193 \times 8000 = 1.544$ Mbps.

As discussed earlier, it is sufficient to reproduce voice from analog signals at the rate of 8000 samples per second; each voice channel is sampled at the rate of 8000 samples per second, yielding $7 \times 8000 = 56{,}000$ bps for data and $1 \times 8000 = 8000$ bps for control information.

On the receiving side, T1 frames are de-multiplexed into 24 channels (digital), and the analog signal for each of the channels is reproduced. Both sides can transmit and receive 24 channels on a T1 carrier simultaneously at a data rate of 1.544 bps on a pair of wires, making a full-duplex connection.

Looking at the success and widespread acceptance of Bell System's T1 carrier, CCITT finally defined two incompatible variations of the standards. It defined eight-bit data for the channel and, therefore, quantization samples of 256 combinations instead of 128 in the T1 carrier. The following two incompatible variations of standards are proposed by CCITT:

1. **Common Channel Signaling:** In the T1 carrier, an extra bit for framing and synchronization of a unique bit pattern 01010… is attached at the end of the frame, while CCITT recommends that it should be added at the beginning of the frame and should have a new bit pattern of 101010… in the odd frames and contain signaling information for all the channels in the even frames.
2. **Channel-Associated Signaling:** This carrier system uses PCM at 2.048 Mbps. It has 32 samples (each of eight-bit data) in one frame of duration 125 μs. Out of the 32, 30 channels are used for carrying information while two channels carry signaling. A group consisting of four frames will have eight channels, giving 64 bits of signaling. Out of these 64 bits of signaling, 32 bits are used for channel-associated signaling, while the remaining 32 bits are used for frame synchronization or an option specified by the country.

Each channel defines its own signaling sub-channel, known as a private sub-channel. One of the bits (from eight-bit data) from each sixth frame is used for frame signaling. This means that out of six frames, five frames will have eight-bit data while another one will have seven-bit data. This carrier at 2.048 Mbps is very popular and is being used widely outside North America and Japan. Although there is still some disagreement as to how these channels should be multiplexed, it is a useful carrier for voice transmission.

6.6.5 Voice digitization

The sampling of analog signals using the Nyquist theorem is obtained using a previously discussed modulation technique known as pulse amplitude modulation (PAM). The efficiency of transmission lines can be improved by multiplexing digitized voice traffic with other digital data. Digital data offers features such as error control, data encryption, and error detection (forward) and also uses less transmission medium, as digitized speech uses a compaction technique over voice-grade lines. This technique produces samples of analog signals of different values that are then encoded using seven bits (since seven bits of data are used in each channel). In other words, each of the samples corresponding to a signal value is represented by an eight-bit number (from 0 to 127 unique combinations). A unique combination is assigned to these signal values such that these values are closest to the actual value of the signals of the samples. This encoding of unique combinations to signal values is known as *quantization* and the difference between the value of a signal and the assigned value is known as *quantization error*. An attempt should be made to make this error as small as possible. The complete process of sampling and encoding is the working principle of pulse-coded modulation (PCM).

The voice can be digitized a number of ways, but two techniques are very popular. One is the use of a *vocoder*, which uses filers and defines signal-frequency ranges and then analyzes the pitch (expressed in terms of phonemes: a unit of spoken language) of the voice and a few parameters of the vocal tract which produce the sound. The second method of digitization is known as *waveform analysis*, which depends on the digitization of analog waves. It uses a statistical approximation technique (e.g., Fourier series). The most common method of waveform analysis is defined as pulse code modulation (PCM), which uses codec. In both techniques of digitization, the main philosophy is to reproduce low-frequency sounds more accurately than high-frequency sounds.

A T1 frame contains 193 bits where the 193rd bit is used for framing and synchronization. It consists of a typical bit pattern of 010101…. A receiving node on each side initially checks this bit pattern. If it finds the same bit pattern as transmitted, then it is in sync with the transmitting node; otherwise, it keeps checking for this bit pattern until it finds it and then get resynchronized. The CCITT also defined equivalent standards known as the DS series: DS-1 (1.544 Mbps), DS-2 (6.312 Mbps), DS-3 (44.736 Mbps), and DS-4 (274.176 Mbps).

6.6.6 Digital subscriber line (DSL)

This new technology is proving to be very hot in the communications industry. *x*DSL is a family of technologies that is based on the DSL concept. The variable *x* indicates upstream or downstream data rates. The upstream data rate is provided to the user for sending the data to the service provider, while the downstream data rate is provided to the service provider for sending the data to the user. For more details, readers are advised to read Reference 3 and Maxwell's paper (see References section at the end of this chapter for further reading). Different DSLs with data rates and applications as defined by the ATM Forum are given below:

- Digital subscriber line (DSL) — 160 Kbps, ISDN voice services
- High data rate digital subscriber line (HDSL/DS1) — 1.544 Mbps, T1 carrier
- High data rate digital subscriber line (HDSL/E1) — 2.048 Mbps, E1 carrier
- Single-line digital subscriber line (SDSL/DS1) — 1.544 Mbps,T1 carrier
- Single-line digital subscriber line (SDSL/E1) — 2.048 Mbps, E1 carrier
- Asymmetric digital subscriber line (ADSL) — 16–640 Kbps (upstream), 1.5–9 Mbps (downstream), video on demand, LAN access
- Very high data rate digital subscriber line (VDSL) — 1.5–2.3 Mbps (upstream), 13–52 Mbps (downstream), high-quality video, high-performance LAN

The ATM offers network services over XDSL, as these basically provide an access technique for public networks. This means the existing unshielded twisted pair can be used to provide ATM services for video (TV channels) and data (teleconferencing and Internet access), and local-exchange carriers manage voice (dial-up services) and all the services. In other words, instead of going to different service providers, we can get all these services (ATM and non-ATM) over ADSL by the same carrier or service provider. The use of an analog modem has been very popular to access Internet services at a lower speed, but recent advances in remote LAN access has moved toward the use of 128-Kbps ISDN access. It is quite obvious that the use of *x*DSL will offer much higher data rates than that of 128-Kbps ISDN access modems. The use of ATM will certainly offer virtual connections at a specified bandwidth with confirmed quality of service.

Cable modems offer data rates of up to 30 Mbps (full-duplex), which is the same as those of *x*DSL modems. Vendors of cable modems include COM21, Motorola, DEC, Bay Networks, Media General, and others. Both types of modems provide high downstream data rates, i.e., from service providers to the users. It is important to know that these rates are not available between users or for peer-to-peer across the network. A fee of about $50 per month per connection can be expected for these two modems. The same fee may apply for an ISDN connection but at a much lower speed of 64 or 128 Kbps.

The ATM-service networks offer a variety of resident services available to the users through a home area network (HAN) that is defined within the ATM network. The ATM Forum has set up a working group known as Residential Broadband that introduces the ATM into HANs.

The ATM networks use ATM switches at both the central office and customer premises equipment (CPE). The majority of ATM networks use SONET as a backbone in North America and SDH for global services. The ATM-based networks can be accessed in a variety of ways: via ATM user-to-network interface (connected to CO switch), ATM DXI (connected to CSU/DSU switch), ATM DS1a and nDS1 inverse multiplexing, frame relay interface (frame relay interworking functional switch), SMDS interface (connectionless server), or even Internet (IP interworking functional switch). ATM offers the following classes of services: point-to-point or point-to-multi-point connections, switched and permanent virtual connection, variable bandwidth and reservation, virtual path or virtual channel connections, SMDS access over ATM, voice over ATM, Internet access to IP via ATM, ATM cell-relay service, etc.

A number of vendors are developing a modem analyzer. The HP 79000 ADSL test station is being introduced for testing ADSL standard ANSI T1.433 and POTS standard ITU T0.133. This analyzer tests both local modem (end office ADSL transceiver ATU-C) and remote-end ADSL transceiver modem ATU-R. For more details, see References for Further Reading publications.

Application: A typical configuration depicting the reference model of DSL includes all the devices (PC, ISDN, TV) accessing the premises distribution network via different service modules. The service modules may use a variety of transport modes, e.g., a complete ATM, STM, STM/ATM, packet, packet/ATM, and so on. The premises distribution network is interfaced with the telephone lines to the public network via an *x*DSL modem. The modem contains three classes of channels: high upstream data rate of 1.5–52 Mbps, one medium with a data rate of 16 Kbps to 2.3 Mbps, and one telephone channel. The analog splitter is used between the *x*DSL modem and public switched telephone network. It acts as a filter, which separates the analog signal from these channels. The remaining two channels are connected to an access node via an *x*DSL modem. The access node provides interfaces to different types of networks, such as ISDN, B-ISDN, packet networks, etc.

The data rates offered by the *x*DSL modems are the same as those of American and European digital STM hierarchy. The access nodes are typically digital subscriber loop access multiplexers/modems, which offer ADSL subscribers line termination and also perform the functions of multiplexing and subscriber lines in ATM networks. Vendors who have developed these access nodes include Alcatel, Ariel, Dagaztech, Motorola Semiconductor, and others.

6.6.7 CCITT DS-Series carrier systems

The DS-1 transmission format, which is used in North America and Japan and has been accepted as a CCITT standard, multiplexes 24 channels with eight bits per channel. A frame consists of 24 channels. It also defines one bit as a framing bit, bringing the total number of bits to $24 \times 8 + 1 = 193$ bits. The DS-1 format is used for both voice communications and digital data services. When used for voice communication, a voice signal can be sent in each channel and is usually called a **PCM word**, as the voice signal is digitized using the PCM technique. The PCM defines 8000 samples of 4-KHz voice-grade signals, and these samples are defined at the rate of 8000 samples per second for each channel carrying voice signals. The size of one frame, composed of 24 channels (each of eight bits sampled at a rate of 8000 samples per second) and one bit for framing, offers a data rate of 8000 per frame (193 bits) per second, thus offering 1.544 Mbps. One eight-bit PCM is used between five out of every six frames, and for every six frames, every channel used in the frame contains a seven-bit PCM word and the eighth bit is a signaling bit (used for routing and network control information).

When DS-1 is used to provide digital services, the same data rate is available, with the only difference being in the number of channels available for data services. Here, only 23 channels are used for data while the 24th channel is basically used for synchronization. Also, within the channel, only seven bits are used for data while the last bit (eighth bit) is used to indicate the status of the channel (data or control). Also, in this application, the sampling rate remains as 8000 samples per second, but due to seven bits, the data rate per channel is reduced to 56 Kbps ($8000 \times 7 = 56$ Kbps) rather than 64 Kbps as in the case of all eight bits being used as data ($8000 \times 8 = 64$ Kbps).

The DS-1 format can also offer lower data rates by using one bit from the channel for other control information. For example, if we use one bit out of seven bits for, say, a multiplexing operation within the same channel, then we now have six bits, giving us a data rate of 48 Kbps (8000×6). The channels can also define sub-channels within them, with the only constraints being that the total sum of the data rates of the sub-channels should not exceed their data rate. Let us consider the situations where we are getting incoming signals from 9.6-Kbps modems. A channel of 48 Kbps of DS-1 can accommodate five sub-channels each of 9.6 Kbps. Similarly, for other low-speed modems, a number of sub-channels can be accommodated within this channel.

The DS-1 can carry both data and voice signals within the frame, where different channels may have different types of signals. The higher data rate formats are DS-2 (6.312 Mbps), DS-3 (44.736 Mbps), and DS-4 (274.176 Mbps). DS-1 may accommodate four DS-1 formats, with the remaining unused bandwidth available for control purposes. The digital service offered by DS-1 is known as data phone digital service (DDS). This service does not need any modem and is available at a user's devices over twisted-pair lines (customer premises) operating at data rates of 2.4 Kbps to 56 Kbps.

The Bell System and CCITT have defined incompatible standards, as given in Figure 6.5.

As discussed earlier, the Bell System T1 carrier multiplexes 24 voice channels, and each channel consists of eight bits (seven bits of data and one bit for control), giving a data rate of 64 Kbps (8000 samples times the number of bits in each channel). We have to remember that this data rate also includes control information transmission at 8 Kbps. Now, if we are sending pure voice signals on the channel, then the loss of one bit from each channel or even from one or two channels from a T1 frame will not affect the quality of voice signals. But if we are using data signals also in the T1 frame, then the loss of one bit or frame(s) may destroy the data packet. That is why, for data communication, we prefer 56 Kbps over 64 Kbps, while 64 Kbps may be used for voice signals.

There are different types of slots available for plugging in cards for each channel on the input of multiplexers. By inserting these cards into slots, a particular channel may be used for transmission. These cards are available for both voice and data communication. The plug-in cards designed for data communication can further multiplex a number of computers or terminals for sending the data on a T1 carrier. For example, a number of computers can be sub-multiplexed (on plug-in cards) on a local loop, which can be inserted

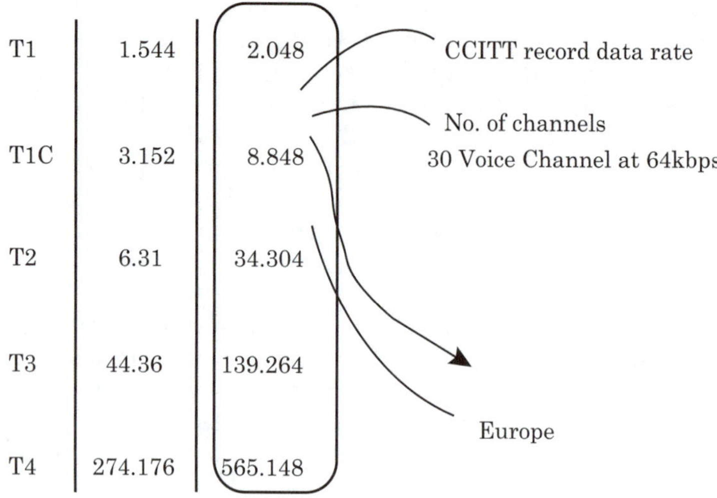

T1
1.544
2.048
CCITT record data rate
No. of channels
T1C
3.152
8.848
30 Voice Channel at 64kbps
T2
6.31
34.304
T3
44.36
139.264
Europe
T4
274.176
565.148

Figure 6.5 Bell System and CCITT recommendations.

into an appropriate slot on the main multiplexer board at the public carrier office. On the receiving side, we have one plug-in card assigned for receiving the data on it, which will be de-multiplexed on the local loop to the appropriate computers. The advantage of this scheme is that we can use the digital end office directly without using modems. Further, this scheme is useful in applications where a 56-Kbps bandwidth can be shared by terminals and computers.

A T1 carrier is usually used for short distances, typically a few hundred miles. For larger distances, the T1 carrier uses communication media such as microwave, radio, fiber optics, etc. A T1 circuit usually accepts bipolar digital signals from a multiplexer. In bipolar encoding, 0s and 1s are represented by voltage between ±2.75 and ±3.3 volts. In order to avoid any jitter error, consecutive 1s are represented by the same voltage with opposite polarity. Repeaters are used to regenerate like digital signals in a T1 carrier and usually are placed at a distance of a few thousand miles (equal to the distance between manholes, as the T1 carrier goes through manholes).

6.6.8 *Telex communication*

In order to transmit information (such as text, pictures, etc.) in digital form, services are offered through different forms of communication links. Each of the links is suited to a particular application and is obtained through appropriate devices, such as telegraphy (transmission of alphanumeric text using electrical signals), telex (transmission and routing of alphanumeric text using a teleprinter), teletex (compatible to a word processor), telecopying (facsimile transmission of pictures or text), and telecontrol (concerned with telemonitoring, remote control, telemetry, etc.). These communication links provide services to the data communication from the point of view of integrated telecommunication computer-based communication. These are also known as *telecomputing* or *telematics* (videotex and teletex).

Videotex is an interactive text and graphics service and offers text and graphics over a videotex terminal or a TV with a special decoder, over cable TV, or even over telephone lines attached to a PC via modems. In contrast to videotex, the **teletex** does not allow any interaction the text and graphics services; it is available over a cable TV channel, a part of an FM broadcast signal, or even an unused portion of commercial TV signals plus decoder.

The videotex vendor usually has contracts with information providers and makes that information available to the users.[2,5] Most of the information providers for videotex offer information retrieval or transaction services, e.g., travel services, bank services, weather reports, stock prices, database queries, entertainment services. Various services offered on videotex include request for any balance, transfer of funds, payment, purchase, database queries, purchase of tickets, travel documents, interactive games.

The above services are also available as teletex — news summaries, sports news, financial newsletters, travel listings, stock market reports, etc. — with the only difference being that in videotex, the users can select the use of services interactively by user interfaces (provided by information providers), while teletex service is one-way communication; we can see the information but cannot perform any operations on it.

Various codes have been defined for transmitting text, pictures, numerals, etc. Some of these codes are alphanumeric codes, Morse code (used in telegraphy), five-bit telex code (for telex communication), seven-bit error-detecting code (CCITT seven-bit code number 3), seven-bit ISO code, etc. Seven-bit code has been defined and used differently by various organizations (e.g., International Standards Organization (ISO) Code, CCITT Code No. 5, and American Standards Code for Information Interchange (ASCII)).

Based on seven-bit codes, telex can be used for data transmission using the following circuits:

1. **Switched circuit:** This is similar to a switched telephone network and provides communication during the established connection on the network.
2. **Dedicated circuit:** This is just like a private line between two nodes and provides connection at all times. It offers a speed of 30–400 bps on telegraphic circuits or 2400–9600 bps on telephone circuits.
3. **Packet-switched circuit:** This offers a connection between two nodes using either circuit switching or packet-switching on the digital network.

Telex communication can also transmit voice channel telegraphy on a telephone network. A carrier frequency band has to be assigned for telex, and various telex channels (accommodated in a voice channel of 4 KHz) can be multiplexed and transmitted. Various band rates for different multiplexed telex channels can be obtained. For example, with 24 telex channels multiplexed, we can get 50 band, while 12 channels multiplexed will give rise to 100 band, and so on.

6.6.9 Cellular communication

The cellular telephone is based on the *capture* concept of the frequency modulation technique. In this modulation technique, if two signals use the same frequency, then the resulting mixing of these signals produces interference between those signals. But, if we change the amplitude of one signal to be much greater than the other, it will not produce any interference between them. The receiver will receive the stronger signal and will not pick up the weak signal. This capturing concept enables use of the same frequency in nearby cells. This is precisely the main working principle of cellular systems. The transmitted signals may be reflected or blocked by mountains, tall buildings, hills, etc. The receiver receives part of the signals over a direct path, while part of it is being received after getting reflected from those obstacles. Due to this, the signals usually lose their strength but are still within the amplitude limitations of FM receivers.

As we noted in Chapter 5, the cellular system uses either analog or digital signaling. The first mobile radio system to work with telephone systems was known as a mobile telephone system and introduced in 1946. This system consisted of a powerful transmitter covering a radius of about 20 miles and a centralized system supporting half-duplex communication. The first cellular system, the mobile telephone switching office (MTSO), is based on the concept of frequency re-use. Here, many users are using the same frequency with minimum interference between them. Each cell can support a number of channels; cells using the same frequencies are usually placed close to each other at a sufficient distance to reduce the interference between them. A cell site is connected to an MTSO, which is connected to an end office and other MTSOs. An MTSO controls a number of cells. The connections between MTSOs and end offices can be provided by a variety of media, such as twisted-pair cable, coaxial cable, fiber, microwave, satellite, etc. The MTSOs are then connected to PSTN via end offices. The MTSO computers keep a record of all the users. Most of the cellular systems are full-duplex and use the same concept of satellite communications via uplinking and downlinking mechanisms. For each of these operations, a different frequency standard value is used. The uplinking frequency component sends the signal from mobile to base, while downlinking sends the signal from base to mobile. For further information, see Reference 3 and References for further study at the end of this chapter.

There are four standards defined for cellular communication:

1. **Advanced Mobile Phone Service (AMPS)** is the first analog system used in North America. It uses frequency division multiplexing (FDM).
2. The digital system **Interim Standard (IS) 54** uses digitized voice data using vocoders and supports both types of multiplexing: FDM and TDM. It uses a differential quadrature phase shift modulation technique. The current analog-based AMPS is slowly being replaced by this standard. It offers greater capacity, uses the same frequency of 30 KHz, and offers a data rate of about 13 Kbps.
3. **Global System for Mobile Communications** is used mostly in parts of the world other than North America. It uses digital signaling and uses both TDM and FDM. It provides interface with ISDN. The frequency values for two-direction communications are 890–915 MHz (uplinking) and 935–960 MHz (downlinking), with a frame bandwidth of 200 MHz.
4. **Interim Standard (IS) 95** uses a digital spread spectrum known as code division multiple access (CDMA) for providing a higher capacity.

The cellular frequencies in the U.S. are as follows:

- Transmit signal (uplinking) — 824–849 MHz; 25 MHz is divided between two providers
- Receive signal (downlinking) — 869–896 MHz with 25 MHz divided between two providers
- Mobile receive frequency — mobile transmit frequency + 45 MHz

Since each channel is of 30-KHz bandwidth, the total number of channels is 25 MHz/30KHz = 832 channels. Each provider gets 416 channels; of these, 21 channels are used for signaling, and there are one signaling channel per cell and 395 voice channels per provider.[3]

Some useful Web sites

More information about cellular radiotelephone systems can be found at *http://www.amazon.com* and *http://www.encyclopedia.com*. The cellular radiotelephone service fact sheet and other documents are maintained by the FCC at *http://www.fcc.gov.*

PSTN and Internet internetworking information can be found at *http://www.ietf.org*. For asymmetric digital subscriber line, visit the ADSL forum at *http://www.adslforum.com*. A telephone report on *x*DSL is found at *http://www.telechoice.com*. Vendors of DSL products can be found at *http://www.ariel.com*, *http://www.dagaztech.com*, *http://www.efficient.com*, and *http://www.motorola.com*.

Bell System T1 carrier and related documents are available at *http://www.Nunetsolutions.com*, *http://www.santafeterminal.com*, *http://www.oreilly.com*, *http://www.webclasses.net*, *http://digitalcreators.com*, *http://webopedia.internet.com*, and many others.

ADSL Forum is at *http://www.adsl.com*, and some of the vendors manufacturing or providing digital subscriber line services can be found at *http://www.alcatel.com*, *http://www.3com.com*, *http://www.zyxel.com*, *http://www.virata.com*, and many others.

Telephone system information can be found at *http://www.softcab.com*, *http://www.cascomdvi.com*, *http://www.fcc.gov*, and other sites. Some manufacturers include Adtran (csu/DSU, CPE, telco lines) at *http://www.adtran.com* and American Technology (ATM, frame relay, and others) at *http://www.tli.com*. Some telephone service providers include AT&T, Cisco (routers, switches, gateways, etc.), 3Com (network solutions), Intel (hardware chips), and Lucent (one of the largest network equipment manufacturers in the world).

References

1. Tanenbaum, A.S., *Computer Networks,* Prentice-Hall, 1988.
2. Black, U., *Data Networks, Concepts, Theory and Practice,* Prentice-Hall, 1989.
3. Sveum, M.E., *Data Communications: An Overview,* Prentice-Hall, 2000.
4. Thurwachter, C.N., Jr., *Data and Telecommunications Systems and Applications,* Prentice-Hall, 2000.
5. Marney-Petix, V.C., *Networking and Data Communication,* Reston Publishing, 1986.

References for further reading

1. Wright, D., Voice over ATM: an evaluation of network architecture alternatives. *IEEE Network,* Sept./Oct. 1996.
2. "ATM report," *Broadband Networking News,* May 1997, April 15, 1996.
3. Maxwell, K., Asymmetric digital subscriber line: interim technology for the next forty years. *IEEE Communication Magazine,* Oct. 1996.

part III

Local area networking and internetworking

chapter seven

Introduction to local area networks (LANs)

"The world is extremely interesting to a joyful soul."

Alexandra Stoddard

7.1 Local area networks evolution

The first report describing the use of LANs to increase an employee's productivity, efficiency, and sharing of expensive resources (including use of obsolete computers/resources) was given in Metacalfe's Ph.D. Dissertation at MIT in 1973. Later on, he — along with David Boggs and others working at Xerox Corporation — developed the first LAN, known as **Ethernet** (a trademark of Xerox Corporation). This Ethernet was adopted by various companies, and Intel built a single-chip controller for it. Thus, Ethernet became a *de facto* standard for LANs and is characterized as a non-deterministic LAN, known commercially as **DIX** (derived from the specification document jointly published by Digital, Intel, and Xerox in 1980). DIX is based on the ALOHA Radio Network developed at the University of Hawaii. It operates over coaxial cable, fiber-optic cable, and twisted-pair cable and offers a throughput of 10 Mbps. It is based on a bus topology and uses the CSMA/CD access method to access the network. The Ethernet LAN is considered the basis for the International Standards Organization (ISO) to define various standard LANs, resulting in the introduction of OSI-RM followed by IEEE 802 standard LANs.

In 1982, General Motors (GM) of the U.S. decided to increase the productivity of its assembly plants through factory automation and competitiveness in the market and, as such, was looking for a LAN that could connect all of its factories and dealers so that GM could supply/ship the customer's orders from anywhere in the world. Further, this network needed to be able to provide a fixed rate of movement of various parts for assembly by robots connected to the network. The first CSMA LAN (Ethernet) developed was a *non-deterministic* (based on contention) one. GM created its own LAN, known as a **token bus**, which works on the basis of round-robin allocation to various machines, robots, programmable devices, and resources connected to it, thus giving a *deterministic* response to them. This LAN provided acceptance to both vendors and customers, and later on it became one of the standards of IEEE. From the vendor's point of view, this provided specification to the network products for factory use which were acceptable to the organization. From the customer's point of view, productivity of manufacturing and meeting the requirements of its clients were significantly improved. The **manufacturing automation protocol (MAP)** was defined as a standard and can be used in the automobile industries/factory environment. Further, it offered a large market for the products that comply with those standards. There are many companies and organizations which have participated in a MAP environment and have benefited from the mass production of their products, particularly those used by the automobile industry.

While GM was busy selecting/defining a token bus LAN for its factory automation, International Business Machines (IBM) developed its own LAN known as a **token ring**. This LAN later became a standard; its main objective was to provide high reliability and availability and maximum utilization of the resources and devices that were connected to the LAN.

A similar attempt (based on factory automation by GM) to define its own LAN was made for office automation by Boeing Company (U.S.).[2–4,34] The main objective of defining its own LAN was to support office automation and engineering aspects in a real-time domain to Boeing 747 fleets. Boeing created its own LAN and also a set of protocols for office automation known as **technical office protocol (TOP)**. It is defined as an office application counterpart of the industrial-based MAP. It is also based on OSI-RM and uses the access control of IEEE 802.3 Ethernet. TOP is now under the purview of the Society of Manufacturing Engineers (SME). GM and Boeing worked together to provide interoperability between their networks.

Ethernet is also considered a suitable LAN choice for many engineering graphics and other high-demand applications environments. Ethernet supports more than 100 trunks or series of twisted pairs of concentrators. A variety of Ethernet products is available, such as network interface cards for host computers, bridges, and other interconnecting devices.

As LANs were being designed by different organizations with different standards, the international standards organization IEEE was asked to form a standardization oversight committee and, as a result, the **IEEE 802** committee came into existence. The main objective of this committee was to define standards for all types of LANs and also provide interoperability among them, since each LAN was defined for a specific environment and application and therefore the architectures were different. The standards defined for each of the LANs were based on OSI-RM.

In early 1986, a non-profit joint venture of more than 60 manufacturers/suppliers of data processing and communication products was formed, known as the **Corporation for Open Systems (COS)**. Its main objective is to provide interoperability for multi-vendor products and services offered by OSI-RM LANs and ISDNs and related international standards which can convince the customers and vendors to concentrate on an open network architecture model. It also deals with the definition and development of a set of testing criteria, facilities, and certification procedures. Thus, vendors are obliged to certify that their products comply with international standards and hence will be acceptable to customers.

During the late 1980s, the **U.S. Government OSI User Committee (USGOSIUC)** was formed by the U.S. government. Its main objective is to provide OSI-related standards on the world market. The committee offers OSI requirement specifications and also coordinates with government agencies and industries that are working for OSI-based standards.

All the above-mentioned organizations are working for common objectives: to provide OSI-based standards, widespread development products and protocols for OSI-RM, and interoperability between various OSI-based incompatible products, LANs, and non-OSI-based LANs.

All the LANs proposed as individual standards are based on OSI-RM in one way or another. For each of the LANs, protocols are different for different layers and are collectively known as a **set of protocols**. The only difference between the types of LANs lies in their lower layers, but still these networks can communicate with each other in the same way as two similar networks via gateways, bridges, and routers (to be discussed later).

In Chapter 8, we discuss each standard LAN (IEEE 802.3, IEEE 802.4, IEEE 802.5, IEEE 802.6, FDDI). Non-standard proprietary LAN (IBM's SNA, DEC's DNA or DECnet) are discussed in Chapter 9. For each of the LANs, a complete description is presented starting from the topologies, access techniques, design issues of access techniques (useful during implementation), lower-layer formats (IEEE 802 recommendation), frame formats of MACs, design issues in MAC-accessing techniques, interoperability, etc. A detailed discussion on various interconnecting devices (bridges, routers, gateways) is also presented. A section on wireless LANs is also presented.

7.2 LAN definitions and standards

The computer network may be defined as a set of autonomous computers interconnected for exchange of data. The connection may be defined by different transmission media, e.g., wires, coaxial cables, microwaves, satellite link, laser, etc. It encompasses both transmission media and communication software. A computer network defines the network architecture with associated protocols for different topologies and different access techniques and offers various applications. Various network activities are defined in terms of a hierarchy of layers, with each layer performing specific functions. Further, the system architectures conforming to the standards should be able to communicate with each other. The network is used to provide connectivity, high reliability, a powerful communication medium, and cost savings for sharing of expensive resources. The connectivity usually corresponds to providing access to information and processing power that are situated a distance apart. The distance typically classifies the type of networks as either **local area networks (LANs)** or **wide area networks (WANs)**.

LANs are typically used for a short distance (a few kilometers), while WANs cover a distance of thousands of kilometers. The expensive resources that are shared among the users include information-processing devices (computers, printers, file servers, etc.) and interconnection bandwidth. The sharing of resources is important, as the interconnection of bandwidth (frequency spectrum) is shared among the users. For each new application, appropriate servers are being shared by users over the networks. An example of sharing can be found in such applications as the multimedia server shared by a number of users to access multimedia applications over the networks.

Computer professionals try to establish a relationship between **distributed system** and **computer network**. In fact, the difference between these two types of systems lies in the way the software is implemented. For example, in a computer network, the network manager (a software module) alone provides explicit access to the remote machine and offers services such as submission of remote jobs, file transfer, etc. In the case of distributed systems, the operating system decides the processor to be used for a particular task and then accordingly allocates the memory to it. All these operations provide a total transparency to the users, as users always feel as if these tasks are running on their local machines. We can consider the distributed system a special case of computer network.

Multimedia applications include video, images, text, audio, etc., with real-time and non-real-time delivery requirements. Within LANs, a variety of configurations can be defined based on accessing techniques; e.g., a **point-to-point** network is defined as a set of point-to-point links interconnected by switching nodes. A workstation accesses a point-to-point network either directly to the switching node or through intermediate access points that concentrate traffic from several workstations and multiplex them for sending to switching nodes. Examples of this class of network include the Internet, ATM-based networks, and many PSDNs. The multiplexed data can be accessed in different ways.

De-multiplexing encompasses different types of configurations such as multi-point and multi-drop. If we have one receiver that de-multiplexes the entire data stream and routes the data to respective terminals, PCs, or any other computer node, this type of accessing corresponds to **multi-point** configurations. If the receiver, after accepting the data stream, goes to each of the destination nodes separately and drops a part of the data at one node and then moves on to the next one, drops a part of the data there, and keeps repeating until it reaches the last node in the network, this accessing scheme is known as a **multi-drop** configuration. The main difference between these two accessing schemes is basically the way the sender looks at the receiving side. In the case of a multi-point configuration, the accessing scheme sees only one destination and the sender also sends the message as one multiplexed message as opposed to messages for all the end points separately.

Another accessing configuration is known as a **multi-point line connection**, which is different from the point-to-point accessing configuration, as here several terminals and computer nodes are connected to the same line. In this configuration, an efficient method of establishing topology requires determination of the shortest path between nodes on the line. This path is known as a **minimal spanning tree** of the network. Two terminals on a multi-drop line cannot use the line at the same time. Therefore, a procedure to determine which terminal may use the line must be determined (contention problem). This **contention problem** may be reduced by providing two lines for a computer so that balanced loading may be established over the shortest path connection. For more information on accessing configurations, see References 2, 3–7, 10, 14, 33, 34, 40, and 41.

7.2.1 LAN services

LANs provide a hardware and software interface for communication devices. The interface device and its interfacing details with the LAN are known as a **network node.** An intelligent interface may include a **packet assembly and disassembly (PAD)**. The main objective of PAD is to provide an interface between public networks (X.25) with character-based asynchronous terminals. The devices which are connected to LANs include PCs, terminals, workstations, gateways, servers (disk, printer, file, etc.), peripheral devices, etc. Each device has its own interface. A general configuration of a LANs may connect the servers (file, printer, fax, multi-function, etc.) to different users and devices. Each of the connected devices access or share the transmission facility over the LAN. LANs provide sharing of information, accessing of remote information, communications with remote users, various applications, network resources, etc. For sharing of information, we can collaborate with each other electronically, and each of us can perform a specific task on our PC and share it with others over the network. For more information about its functions, standards, and applications, see Chapter 1.

Various resources which can be shared by a number of users over a network include programs, data, databases, expensive devices, etc., regardless of their locations. In the event where one central processing unit (CPU) goes down, the relocation of files and programs on other CPUs and replication of these on more than one machine are some of the features of computer networks. A powerful PC or workstation is usually used to keep the data on a shared file server which can be shared by many machines over the network, thus offering cost-effective sharing of resources. The updating of system performance can be obtained at an incremental rate by adding the resources (processors, servers, etc.) over the existing configuration. Typical applications of networks include military, banking, air-traffic control, and other applications which are solely dependent on the hardware (the failure of which will collapse the whole system). The accessing of remote information includes stock prices and airline schedules via public information service facilities, or we can also dial up access to our organization's databases of various applications. New applications such as remote programs, databases, and value-added communication facilities can be added economically, e.g., accessing various data and other integrated services via a network over the existing telephone links — data can be transferred from any node to any other node anywhere around the world. Computer networks also offer paperless communications via e-mail services. Sharing of database program files, application programs, etc., on storage devices can be achieved via networks. They also allow users to send files to the network printers and continue with other tasks.

If we look at the evolution of data, we notice that in all these technologies, one of the main objectives has been to define a high-speed network for both short and long distances. The evolution of data communication actually started with telegraphy and then telex. Other communication systems which were developed and used are as follows: low-speed data transmission (teletype), circuit-switched data networks, medium-speed data networks, high-speed data networks, digital data transmission, packet-switched networks, public switched networks, electronic mail, directory systems, electronic data interchange (EDI), integrated services digital networks (ISDNs), broadband-integrated services digital networks (B-ISDNs), Internet services, etc.

LAN interconnectivity started in the late 1980s for connecting computers or PCs within a building or campus. The number of LANs currently being used is over 2 million and is expected to rise. LANs are used in a variety of applications and offer different data rates. Some applications along with approximate data rates are as follows: e-mail (9 K to 56 K), file transfer (56 K to 1.5 M), multimedia (3 M to 45 M), Imaging (1.5 M to 45 M), X-Window system (56 K to 1.5 M), etc. Computer-aided design traffic is usually local, and the file transfers take place between mainframe and workstations that are located at the same place. The off-line file transfers typically are the remote traffic. The data rates of 56 K to 1.5 M may be sufficient for stand-alone applications but are not sufficient for these CAD/CAM tools.

LAN-to-LAN communication always puts some constraint on the data rate and bandwidth. T1 channels offers 64 Kbps but may not be sufficient even for DOS applications. Full-screen information of a DOS application is typically 25 × 80 characters, and each character is represented by 16 bits. The number of pixels and number of bits to represent the number of colors and data in a graphical user interface such as Windows is so large that it will take a long time to load it. Images, graphics information, multimedia, and many other applications require a lot of bandwidth. For example, an X-ray report may occupy about 120 Mbps. Compression algorithms can be used to compress the images, but some information is lost during decompression. With ATM connectivity with LANs, it is possible to achieve a data rate of over 155 Mbps, which will reduce the transmission time significantly. The ATM-based network uses a cell of fixed length and transfers the information whenever it is generated. The asynchronous concept, in fact, is a transmission with requested bandwidth-on-demand philosophy. The ATM switch uses mainly the packet-switching system and is being used as a switch in ATM networks.

For the general needs of any organization, the 10-Mbps LANs may not be suitable if we have more than, say, 80 or so users, but they will be more than enough for a single user. There are many that have introduced a point-to-point link running at 10 Mbps by attaching each user with a hub to a segment of the LAN. The segment has only one user and one hub port and still uses CSMA/CD. The hub has linked input and output via a switched fiber. The hub having input and output linked, in fact, can be used as an ATM switch offering a data rate of 10 Mbps. This type of switched Ethernet is essentially an Ethernet or 802.3 LAN but can be implemented in a variety of ways.

The switched Ethernet can be realized in a duplex mode, where we can expect a data rate of 20 Mbps supporting only data communication. The transmission of video and multimedia should still be avoided in this type of LAN. Thus, neither 100-Mbps nor switched Ethernet are suitable for video, multimedia, image, or graphic-based applications. In other words, these LANs do not support time-sensitive applications. FDDI-II, an extended version of FDDI, seems to be suitable for time-sensitive applications, as it supports variable-length data frames and uses a circuit-switching technique. The establishment of a network connection is needed for isochronous constant bit rate services.

The ATM network is used in LANs for a higher bandwidth and data rate, so that the entire ATM protocol has moved from WAN groups such as ITU-T and Bellcore toward possible LAN solutions in every organization. Looking at possible LAN solutions based on ATM, IEEE is moving toward defining 100-Mbps LANs (Ethernet, token ring). The existing 10-Mbps Ethernet will be developed as switched Ethernet, which will offer the entire bandwidth of 10 Mbps at each node rather than share the bandwidth as in the traditional working of the Ethernet. FDDI is already running over fiber at a data rate of 100 Mbps. FDDI provides features similar to those of ATM in terms of support of data and time-based applications such as video and voice (isochronous channels); in some applications, it may be cheaper than ATM. Other alternatives to ATM for possible LAN solutions may include 100-Mbps switched Ethernet, FDDI-II as LAN, IEEE 802.6 as a MAN, Packet-transfer mode as a WAN with SMDS, and frame relays as public network services.

7.2.2 Open system interconnection-reference model (OSI-RM) and LAN standards

The international standard defines a reference model composed of different layers (open system interconnection). It does not define any services or protocols for OSI but, instead, provides a framework for coordinating the development of various standards for interconnecting different systems. According to OSI, a system is defined as a collection of computers with their associated software and attached peripheral devices configured so that the information can be transferred across its connected devices. Obviously, a system which conforms to the OSI reference model is known as an open system, as shown in Figure 7.1.

The aim of OSI-RM[1] is to define a platform for different vendors that conforms to the reference model and associated standards so that these different vendors can communicate with each other over the network. The OSI-RM architecture is defined in terms of seven layers, where each layer performs a specific function(s). Each layer further offers a set of services to its higher layers. The information processing required for any application (over the open system) requires an element known as an application process (AP). The implementation of each of the layers is shielded or hidden from its lower layer via an abstraction mechanism and, as such, the implementation of entire networks provides total transparency to all the users.

For each layer of OSI-RM we have a specific set of functions and services. Each layer has its own element known as an **entity**, which basically interacts with the entity of its lower or higher layers. The peer entity is defined for the same layer of two network systems following OSI standards. For each of the layers, we have a set of rules and procedures known as **layer protocol**. The services offered by any layer (through its entity) to its higher layers (to its entity) are usually accessed at **service access points (SAPs)**. When a user's request for any service is invoked at the highest layer (application), the data communication through the model takes the user's request all the way to the lowest layer. Each layer adds its own control information and address with the message until it reaches the data link layer. This layer defines a frame by adding a header and trailer around the packet

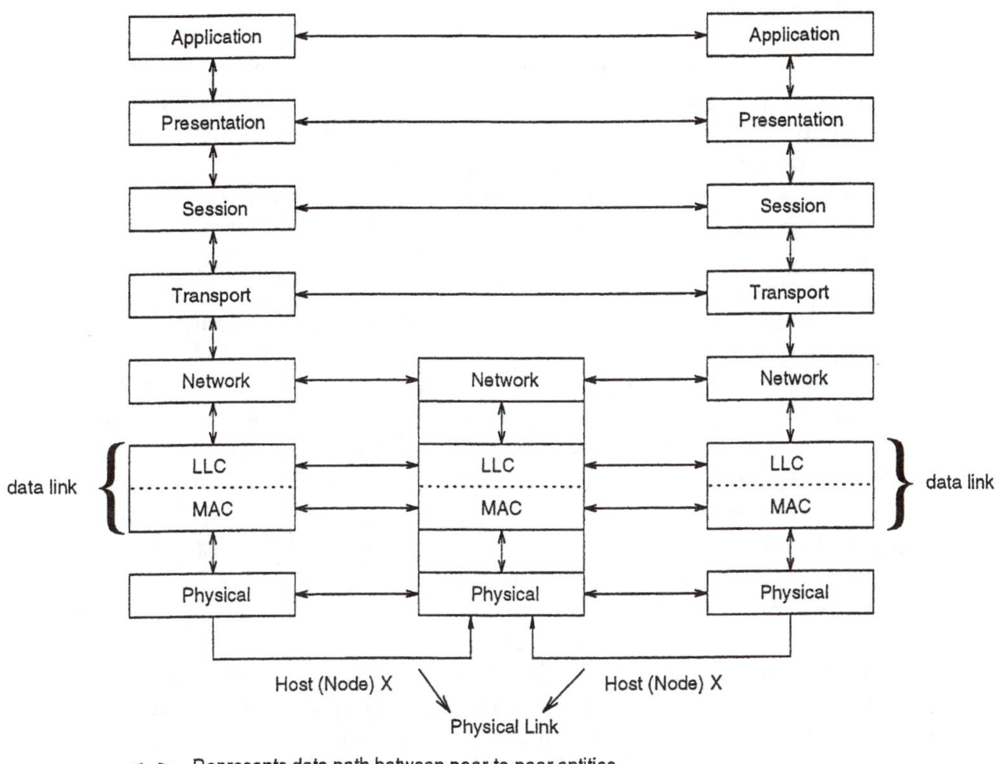

Figure 7.1 OSI-RM layering concept.

(which it receives from the network layer). This process of taking the user's request onto the medium is termed **frame construction**. On the receiving side, the frame goes all the way to the highest layer (application) from the lowest layer (physical) in the same way as it traveled on the transmitting side but in the reverse direction. Each layer removes the header and control information from the frame when it sends the frame to its higher layer. This process is termed **frame reduction**. In Section 1.6 (Chapter 1), we described the functions and services of each of the layers of OSI-RM in brief to get an idea of the total functionality of OSI-RM and appreciate the advantages of a layered approach for defining a model of a computer network. The functions, services, and standards of each of the layers can be found in References 1–7, 10, 14, 18, 32–36, 40, and 41.

7.2.2.1 Physical layer

The physical layer provides an interface between a data link and transmission medium circuit. This layer is responsible for providing transparent transmission of data across the physical link system and includes various functions required to offer this link or connection for data communication. It connects the network to the transmission medium and also generates the electromagnetic signals. It offers two types of transmission configurations: point-to-point and multi-point. It supports different types of line configurations: half-duplex and full-duplex in serial or parallel modes. The data link entities receive the services from the physical layer for establishing a logical connection to the destination, ordered delivery of data stream, etc. After the connection through switches has been established, the switching nodes are hidden from the upper layers by the physical layer.

The characteristics of this layer are as follows:

1. The physical behavior is defined by a **data terminal equipment (DTE)/data circuit terminating equipment (DCE)** interface where DTE can be a computer, terminal, or workstation while DCE can be a modem, multiplexer, controller, etc. The standard DTE/DCE interface offers interchange circuits.
2. Mechanical behavioral function is concerned with physical properties of the interface-to-transmission medium (size, dimension of plugs, allocation of pins, configuration, etc.). 25-pin (RS-232), 37-pin, and 9-pin DTE/DCE interface connectors (ISO/DP 4902) are available.
3. Electrical behavioral function is concerned with the representation of bits (voltage levels) and data-transmission media. Standard V.28 (CCITT) describes electrical behavior of unbalanced double-current interchange circuits (DTE/DCE).
4. Functional behavior is concerned with functions offered by physical interface between the system and transmission media, e.g., the interpretation of signal voltage on the media, value of the voltage, etc. The X.24 standard (CCITT) includes definitions of interchange circuits between DTE and DCE on public data networks (PDNs).
5. The procedural function deals with rules defined for various functions, sequence of the events, etc. It is typically handled by a protocol as to which bit streams are exchanged across the transmission media. The X.21 standard (CCITT) defines a general-purpose interface between DTE and DCE for synchronous operations on PDNs.
6. Activation/deactivation of a physical link is usually concerned with physical connections, control of the physical link and preparation of transmission media to pass bit stream in both directions, termination at the end of transmission, etc. It defines a **physical service data unit (PSDU)** as its function and also as a service to the data-link layer entities. It provides transparent transmission of bit streams between data-link entities.
7. Maintenance and management (MM) is concerned with the transmission of bits and management activities related to the physical link.
8. Fault condition notification signifies that the data link entities are notified of fault conditions in physical links.
9. Quality of service is usually determined by characterization parameters of quality of transmission paths.
10. The sequencing and identification function defines the sequencing for PSDUs, as well as the labeling with identification of various data circuits established between two endpoints.
11. The layer supports half-duplex and full-duplex transmission of bit streams.

The design of the physical layer is concerned mainly with mechanical (pin connection of the interface), electrical (value of voltage to represent logic 1s and 0s, duration of the bits), and procedural (establishment and termination of connections between sender and receiver), aspects, as well as the characteristics of the physical medium under the physical layer. These characteristics provide access to the physical medium in addition to transmitting the stream over the medium/circuit.

Standards:

EIA-232-D — 25 pins
EIA RS-449 (RS-422-A and RS-423-A) — 37 pins
EIA-530
EIA-366
CCITT X.21-1 (circuit-switched network) — 9 pins

CCITT X.21 bis (for PSDNs)
CCITT X.25-1 (packet-switched network)
CCITT V.24 (list of definitions for interchange of circuits between DTE and DCE — for PSTNs)
5.28 (electrical characteristics for unbalanced)
5.21 (general-purpose interface between DTE and DCE) — for Ethernet, token bus (UNIX)
ISO 8802.3, 8802.4, and 8802.5 (for LANs only)
CCITT I.430 and I.431(for ISDNs)

7.2.2.2 Data link layer

This layer provides an interface between physical and network layers. It offers procedural and functional support for establishing, maintaining, and terminating a data connection among network entities. The data link layer services data units transferred over this connection. It offers timing and controls the transmission and access to the network medium. It hides the details of transmission media from the higher layers. It provides error detection and correction in order to offer reliable transmission for higher layers, independent of the data being transmitted.

The characteristics of this layer are as follows:

1. Manages the data link connection during the transfer of data between sender and receiver and also manages the data interchange by connecting network entities via **data link service data unit (DSDU)** mapping. The data link connections are defined on the underlying physical connection. It does not provide any blocking. It simply provides mapping between **network service data unit (NSDU)** and **data link protocol data unit (DPDU)** based on data link protocols used.
2. Provides synchronization and delimiting between sender and receiver. Suitable mechanisms are used by protocols to deal with situations where the transmitters and receivers have different speeds with limited buffer space, thus providing continuous traffic flow across the physical layer.
3. For reliable and error-free transfer of data across the physical layer, it breaks the incoming data up into blocks of data or frames of fixed size (typically a few hundred bytes or octets); includes various synchronization, error-control, error-detection, and flow-control bits with the data; and then transmits the frames to the receiving side using frame-sequencing techniques, e.g., **frame sequence control (FSC)**, for error detection.
4. Processes the acknowledgment frame received to provide link-flow control.
5. Handles the recovery from abnormal conditions in the events of lost frames, bit-error recovery, duplicate frames, damaged frames, etc.
6. Defines the start and end of frames, the length of data defined in the frame, various control information, etc. Once the data link layer makes a frame, it passes it on to the physical layer. The physical layer has no idea about the data structure, contents of the frame, etc. It simply accepts the bit stream and transmits it on the physical medium. Special bit patterns are used to define boundaries of the frames, while special techniques are used to send acknowledgments for the received frames in both directions between sender and receiver.
7. Performs data link management; other functions managed by this layer include control of data circuit interconnection, exchange of parameters and identifications, quality of services, error notifications, etc.
8. May define various parameters for quality of service (QOS) (optional), including mean time between detected error, residual error rate (due to loss of packet, duplication, etc.), transit delay, and throughput.

Standards:

High-level data link control (HDLC: bit-oriented)
ANSI's X3.28 (character-oriented)
IBM's binary synchronous communication (character-oriented)
CCITT X.21-2 (circuit-switched network)
CCITT X.25-2 (packet-switched network)
CCITT X.212/222 (for PSDNs)
CCITT T71 (for PSTNs)
ISO 8802.2
DEC's digital data communication message control (DDCMP: character-oriented)
Protocols for HDLC:
- ISO 3309 HDLC procedures (frame structure)
- ISO/DIS 4335 HDLC procedures (elements of procedures)
- ISO/DIS 6159 HDLC unbalanced classes of procedures
- ISO/DIS 6256 HDLC balanced classes of procedures — Ethernet, logical (UNIX)

CCITT I.440 and I.441 (for ISDNs)

7.2.2.3 *Network layer*

This layer provides an interface between the data link and transport layers. It offers routing and addressing of the message data and also directs it to the destination network and node. It provides procedural and functional support for maintaining network connections for exchange of network service data units between transport entities. It offers a transparent transfer of all the data it receives from its transport layer and transport to any of the transport layer entities defined in the layered model (OSI-RM). The transport layer has no idea about the interconnections being defined. It offers a point-to-point connection between two network addresses and between a pair of network addresses. More than one network connection may be defined (generally implemented by the transport layer). It provides its supported services to the transport layer during network-connection establishment.

The following are characteristics of this layer:

1. It establishes, maintains, and terminates network connections between sender and receiver addresses.
2. It provides routing, relaying, and switching for the packets to be transmitted over the transmission medium.
3. It provides the packet interface through segmenting and blocking of the data information.
4. It offers network connections and network connection multiplexing for data transfer, including expedited data transfer.
5. It provides sequencing and flow control at packet level.
6. It handles error recovery for the lost/damaged frames (or packets) in the network and offers error notification.
7. It keeps an account of the number of bits, packets, size of the packets, etc., that are sent by each sender so that billing information can be sent to the senders.
8. It handles addressing schemes of different networks such that the packets can be sent back and forth across heterogeneous connected networks.
9. It handles different types of communication configurations such as point-to-point, multi-cast, and broadcast and supports resetting facilities.
10. It performs network layer management: it manages service selections, reset options, and sequencing.

The design issues of the network layer include routing strategies (both static and dynamic), addressing schemes, controlling of the communication subnets, adoption of different vendor protocols, controlling of traffic congestion, etc. With this layer, we can include algorithms for determining the optimal path between sender and receiver and also the next host computer to which packets should be sent in the case of overlooking on a certain computer.

Standards:

CCITT's X.25 includes lower three layers (physical, data link, and network), usually known as levels 1, 2, and 3
X.21 has been defined for level 1, high-level data link control (HDLC) for level 2, and packet level standard for level 3
CCITT X.213 (circuit-switched network) (for PSDNs)
CCITT X.25-3 (packet-switched network)
CCITT T30 (for PSTNs)
ISO 8880/8473/9542/10589 (for LANs only)
ISO 8203/8881(for LANs)
ISO connectionless mode network Internet Protocol (IP) (UNIX)
CCITT I.450 and I.451 (for ISDNs)

7.2.2.4 Transport layer

This layer provides an interface between the network and session layers. It offers continuity in the transmission and reliability of data communication between two nodes. It provides a transport connection between two session entities for the transfer of data between them, independent of their locations. The transport entities support its services on an end-to-end basis, which means it must map its address to the network address. It is responsible for optimal usage of resources, maintaining a guaranteed quality of service (QOS). The protocols of this layer transparently offer end-to-end communication and are not aware of the underlying protocols, topology of the network, etc. This layer offers two duplex-based transport connections between transport addresses and does not support the multi-point transport connections.

The following are the characteristics of this layer:

1. It provides an error-free, reliable, and cost-effective communication link between sender and receiver for each application process in the application layer.
2. It provides mapping between different networks and provides transparent addresses to the network layer address. It also manages concatenated networks.
3. It offers three phases for data transfer: establishment, data transfer, and release phase. In each phase, various operations are defined; e.g., the data transfer may include functions such as sequencing, error control, multiplexing, flow control, concatenation, segmenting, expedited data transfer, transport service data unit, etc.
4. It provides end-to-end error control, reporting of error, and overseeing of lower-level controls.
5. It breaks a long message or file into smaller blocks (due to limitation on packet size of different networks) at the sender side and reassembles at the receiver side.
6. It supports multiple network connections, multiplexing, and flow control of information. These features affect the cost of a network connection as well as throughput and, of course, response time of the network.
7. It provides sender-to-receiver direct conversation using end-to-end layer connection.
8. It offers identification service (transport address and connection, endpoint identifiers, class of service requested, etc.).

The design issues of this layer include breaking up of the user's data packet into small blocks of data; delivering them in proper order to the receiver efficiently; selecting suitable algorithms to obtain cost-effective network connections, higher throughput, and minimum response time; an error-free, point-to-point link for delivering the information; sender-to-receiver end-to-end layer communication for establishing connection between two networks; multiple connections for multiprogrammed computers; multiplexing capabilities; etc.

Class of services:

- Unacknowledged connectionless
- Acknowledged connectionless
- Connection-oriented

Connection-oriented (virtual) service is based on the telephone system. Here the service user establishes a connection and transfers the data over this connection, which is followed by the termination of the connection. In this service, the in-order delivery of a data message is guaranteed. Examples of this service include Telnet and file transfer protocol (FTP).

Connectionless (datagram) service is based on the Postal Service. Here, the message, including the destination address, is routed over different systems and routes; therefore, it does guarantee in-order delivery of the message. Examples of this service include network file system (NFS) and junk mail.

All the layers above the transport layer usually provide the same functionality. The IEEE 802 introduced connectionless service. Three types of network connections are defined as Type A (usually for LANs), Type B (usually for X.25), and Type C.

Standards:

ISO 8072 transport service specification (for LANs)
ISO 8073 transport layer protocol (for LANs)
CCITT X.214/X.224 (for PSDNs)
CCITT T70 (for PSTNs)
Classes of protocols:
Class 0 Simple class
Class 1 Basic error recovery
Class 2 Multiplexing
Class 3 Error recovery and multiplexing
Class 4 Error detection and recovery
CCITT transport layer protocol
Transmission control protocol (TCP) (UNIX)

7.2.2.5 *Session layer*

This layer provides an interface between the presentation layer and transport layer. It offers a means for presentation entities to cooperate for organizing and providing synchronization between dialogue and data exchanges. It defines and manages the sequence of interaction between communicating devices for data communication between them. It provides a session connection and manages the data transfer over it. It also interacts with presentation entities to support different modes of transmission (full-duplex, half-duplex, or simplex).

Various characteristics of this layer are as follows:

1. It provides/establishes a dialogue control session between sender and receiver. This dialogue session includes transfer of the user's data packets in one or both directions, control data packet exchange for priority traffic, interaction and synchronization, etc.

2. It provides administrative services (not well-defined by ISO) such as binding connections, unbinding connections, token management, etc. In a binding connection, a session is set up between sender and receiver allowing ordinary data transport between them (in addition to the services offered by the transport layer). It allows the users to log in on a remote computer and also transfer a file between two computers. Token management basically avoids the access/execution of the same operations at the same time by two computers, as this will corrupt communication between them. Any computer holding a token can execute that particular operation.
3. It keeps a record of error statistics.
4. It provides synchronization for either introducing checkpoints or remembering already sent or lost frames. This will avoid sending duplicate frames on the network in the event of a network crash.
5. It offers mapping between session and transport connections. Further, various operations during session connection, recovery, and release are provided by this layer.
6. It offers transport entity interaction; it supports different types of modes of transmission for session connection by interacting with transport entities with guaranteed quality of service.

Design issues for this layer include establishment of session between two computers, continual transfer of data in the case of disconnected transport layer connection, synchronization for avoiding the duplicate sending of frames, avoidance of execution of the same operation by two computers at the same time, etc.

Standards:

ISO 8326 session service specification (for LANs)
ISO 8327 session layer protocol (for LANs)
CCITT X.215/X.225 (for PSDNs)
CCITT T62 (for PSTNs)
Remote procedure control (RPC) library (UNIX)

7.2.2.6 Presentation layer

This layer provides an interface between the top-most layer (application) and the session layer. It provides mainly mapping to different formats and syntax used for data transformation and selection of syntax used for various application processes defined in the application layer for network transmission. It possesses the following characteristics:

1. It provides a suitable syntax for data transmission; i.e., it defines how the transformation/translation of the data into a suitable format for transmission can be obtained. Also, it provides some way to select the syntax or transformation technique for data transformation.
2. It provides a common representation of information that application entities can communicate with and also use during their dialogue session.
3. It provides a means of data formatting (text compression, etc.); data encryption for security, privacy, and authentication; and structuring of data for improving the reliability and efficiency of the data transfer rates.
4. It allows the establishment and termination of sessions between sender and receiver.
5. It provides efficient data transfer across the layers of the networks.
6. It offers presentation connection for application entities and supports addressing and multiplexing.
7. Other functions related to presentation connections are maintained by presentation layer management.

The design issues for this layer include selection and acceptance of suitable transmission syntax, data transmission, formatting and other types of transformation (if needed), data encryption, code conversions, etc.

Standards:

ISO 8822 presentation service and syntax notation (for LANs)
ISO 8823 presentation layer protocol (for LANs)
ISO 8824, ISO 8825 (for LANs)
CCITT X.216, CCITT X.226 (for PSDNs)
CCITT T50, T51, and T61 (for PSTNs)
XDR (UNIX)

7.2.2.7 Application layer

This layer forms an interface between user applications and the presentation layer. It is the highest layer of OSI-RM. It provides various network services required by user applications, e.g., file service. It serves as a window between application processes which are using OSI-RM and defines its aspect for various application entities. It defines application entities corresponding to application processes. The application processes can be either system functions (control of various operations of the system) or user functions (data processing). It possesses the following characteristics:

1. It provides a window for a variety of tasks between the user's applications and the OSI model.
2. It provides mapping between various incompatible terminals and also real terminals. In order to provide this type of mapping, virtual terminal software containing network virtual terminal, editors, and software are supported by this layer.
3. It provides various services comprehensible to the users, e.g., identification, availability of resources, authentication, agreement on syntax, agreement on acceptable quality of service, etc.
4. The layer management function provides the determination of cost-effective allocation strategies, agreement on error recovery, data integrity, data syntax protocols, etc.
5. It provides communication between open systems using the lower layers of OSI-RM.
6. It provides various applications to the users, e.g., file transfer, electronic mail, access to remote resources (job entry, printing, accessing a remote file or program, etc.), access of directory, unique user-oriented services, and many other applications.
7. The application layer protocols define communication between application processes dealing with various issues during the data communication and, as such, we have different categories of protocols, e.g., system management, enterprise-specific, industry-specific, application, etc.

Different classes of applications such as terminal-handling, file-handling, job transfer and manipulation, etc., have been standardized and are usually termed **specific application service elements (SASEs)**. The terminal-handling application, concerned mainly with basic, graphics, images, and other types of terminals, is defined by a **common application service element (CASE)** which offers common services to all application protocols. An application process (AP) within OSI-RM is defined as an **application entity (AE)** and is available at the **service access point (SAP)**. The AP may contain more than one AE which is basically composed of application service elements and associated user elements. The user element acts an interface into application processes. The application service elements are protocols defining the service applications from CASE and SASE.

The design issues for this layer range from application — allowing different incompatible terminals, PCs, workstations, etc. working under different environments to use the networks — to layer management and communication between open systems using lower layers of OSI-RM.

Groups:

Group I	System management protocols
Group II	Application management protocols
Group III	System protocol
Group IV	Industry-specific protocols
Group V	Enterprise-specific protocols

Standards:

CCITT X.400 message handling system (MHS) (for PSDNs)
CCITT FTAM (file transfer access and management) (for PSDNs)
CCITT X.420 interpersonal messaging (for PSDNs)
CCITT X.500/520 directory services (for PSDNs)
CCITT TTX telex services (for PSTNs)
CCITT T(0-4) facsimile (for PSTNs)
CCITT T100, T101 videotex (for PSTNs)
Common management information protocol (ISO 9595/6)
ISO VTP (virtual terminal protocol) (for LANs)
ISO 9579 remote database access (for LANs)
ISO 10026 distributed transaction processing (for LANs)
ISO 8571(1-4) job transfer access and management (JTAM) (for LANs)
ISO 8649, ISO 8650 common application service elements (CASEs) (for LANs)
ISO 9594 directory system (for LANs)
Network file system (NFS), file transfer protocol (FTP), remote login (Rlogin), remote procedure call (RPC), simple mail transfer protocol (SMTP), simple network management protocol (SNMP), and others are provided by TCP/IP communication interface (UNIX)

7.2.3 Protocols, layers, and interfacing in LANs

In OSI-RM, it is assumed that a lower layer always provides a set of services to its higher layers. Thus, the majority of the protocols that are being developed are based on the concept of a layered approach for OSI-RM and hence are known as **layered protocols**. Specific functions and services are provided through protocols in each layer and are implemented as a set of program segments on computer hardware by both sender and receiver. This approach of defining/developing a layered protocol or functional layering of any communication process not only defines a specific application of a communication process but also provides interoperability between vendor-customized heterogeneous systems. These layers are, in fact, the foundation of many standards and products in the industry today. The layered model is different from layered protocols in the sense that the model defines a logic framework on which protocols (that the sender and receiver agree to use) can be used to perform a specific application on it. Some of the definitions and concepts in this subsection are partially derived from References 1–7 and 33–36.

A layer m defines two neighboring layers as $m - 1$ (lower) and $m + 1$ (upper). This layer m uses the services of layer $m - 1$ and offers services to $m + 1$. The m-entity performs functions within the m-layer. The same layers of two different networks define a communication between their respective entities via a peer-to-peer protocol. Thus, we have a communication logical link between peer m-layer entities of both networks as m peer-to-peer protocol. A protocol associated with each layer processes the data being exchanged between the peer layers. Each node (PCs, terminals, etc.) connected to a LAN must contain identical protocol layers for data communication. Peer protocol layers exchange messages which have commonly understood formats, and such messages are known as **protocol data units (PDUs)**. PDU also corresponds to the unit of data in m peer-to-peer protocol. The peer m-entities communicate with each other using the services provided by the lower layer (m – 1). The services of layer m are provided to m + 1 at a point known as m-service access point (SAP). This is the point through which services of m-layer can be accessed by layer m + 1. For a description of the interface between m- and (m + 1)-layers, the m-primitives are defined. The m-service data unit (SDU) associated with m-primitives is delivered to (m + 1)-layer and, similarly, the layer (m + 1) also uses (m + 1)-primitives to interact with the m-layer. The higher layer (m + 1) sends an (m + 1)-protocol data unit (PDU) to the m-layer. The (m + 1)-PDU consists of an (m + 1)-protocol, control information, and (m + 1)-user data. The (m + 1)-control information is exchanged between (m + 1)-entities.

The communication control procedures are a set of rules and procedures for controlling and managing the interaction between users via their computers, terminals, and workstations across the network. The control procedures are known by different names in the industry, such as protocols, line-control procedures, discipline procedures, data link control procedures, etc. The protocols provide communication logical links between two distant users across the network, and the handshaking procedures provide the call establishment within the protocols.

Each layer passes message information (data and control), starting from the top layer (application) to its lower layer, until the message information reaches the lowest layer (physical) which provides actual data communication. An interface is defined between layers which supports the primitive operations and services that the lower layer offers to its higher layer. The network architecture is then defined as a set of protocols and interfaces.

There are a number of ways of defining a communication system between hosts. The usual approach is to define a communication system as a set of interacting processes (components) which provide functions and services to the users. This approach is not a useful one, as it describes the function of the communication environment in a very limited mode and, further, it does not provide any means of interoperability for different types of networks. The second approach is rather abstract in the sense that the functions/services of networks are defined as a collection of processes at various levels. At each level, we can define one layer for a particular process(es). Each layer, in turn, represents a specific function(s) and service(s) and can be different for different types of networks. However, each of the layers offers certain services to its higher layers.

The implementation of these services is totally hidden from the higher layers. The services offered by the layers are provided through protocols and are implemented by both source and destination on their computers. Protocols are defined in terms of rules (through semantics and syntax) describing the conversion of services into an appropriate set of program segments. Each layer has an **entity process** and is different on different computers, and it is this entity (peer) process through which various layers communicate with each other using the protocols. Examples of entity processes are file transfer, application programs, electronic facilities, etc. These processes can be used by computers, terminals, PCs, workstations, etc.

A protocol defines rules, procedures, and conventions used for data communication between peer processes. It provides data formatting, various signal levels for data and control signals, error control, synchronization, and appropriate sequencing for the data to be exchanged. Two networks will offer two-way communication for any specific application(s) if they have the same protocols. The layer *m* is known as a service provider for the layer ($m + 1$) and offers the services at SAP.

The ISO has defined the following terms used for data communication through OSI-RM. Each layer defines active elements as entities, and entities in the same layer of different machines connected via networks are known as **peer entities**. The entities in a layer are known as *service providers* and the implementation of services used by the higher layer is known as a *user*. The **service access point (SAP)** is defined by a unique hardware or port or socket address (software) through which the upper layer accesses the services offered by its lower layer. Also, the lower layer fetches the message from its higher layers through these SAPs. The SAP defines its own address at the layer boundary and is prefixed to identify the appropriate layers; for example, TSAP is a transport layer SAP. The **service data unit (SDU)** is defined as message information across the network to peer entities of all the layers. The SDU and control information (addressing, error control, etc.) together are known as the **interface data unit (IDU)**, which is basically a message information sent by an upper-layer entity to its lower-layer entity via SAPs. For sending an SDU, a layer entity partitions the message information into smaller-sized messages. Each message is associated with control information defined by the appropriate layer as header and sent as a **protocol data unit (PDU)**. This PDU is known as a **packet**.

7.2.4 LAN hierarchy

In the hierarchy of LANs, we may identify different types of LANs based on the data rates being offered by them. For example, high-speed LANs are expected to offer data rates of 100 to 500 Mbps, while other LANs may offer data rates of 4–100 Mbps. As indicated earlier, the **Ethernet LAN** is considered the LAN of choice for many engineering graphics and high-demand applications. The Ethernet LANs offer various advantages, and some of these are listed below:

- Simplicity and reliability
- Low cost of installation
- Compatibility
- Ease of maintenance
- Support of interconnections between PCs to hosts, PCs to PCs, hosts to hosts, etc.
- Approved LAN standard by ISO, IEEE, EMCA
- Supports backbone and broadband LANs

Ethernet is available in three versions: **standard Ethernet, ThinNet, and twisted-pair Ethernet.** The standard Ethernet and ThinNet are coaxial cable-based LANs, while twisted-pair Ethernet uses hierarchical star topology using a concentrator bus configuration. The standard Ethernet is more expensive and difficult to install than ThinNet, but it offers a larger distance coverage for a large number of users. Ethernet is based on a 10 BASE T specification, which allows the use of two pairs of existing pair drop cable wiring for LANs and is based on star or hub topology. Other cable media are also supported in the standards documents: 10 BASE 2 uses thin wire coaxial cable with a maximum segment length of 200 m, 10 BASE 5 uses thick wire with a maximum segment length of 500 m, 10 BASE F uses optical fiber drop cables and is based on star topology (similar to 10 BASE T).

Based on the data rates, applicability, and interoperability, the LAN hierarchy is usually categorized in two groups as **PC-LAN** and **telephone-LAN**. These two groups are generally considered as complementing each other for providing a better communication environment through which users can access a variety of services.

PC-LANs: The PC LANs usually offer data rates of 1–5 Mbps. A PC LAN offers sharing of resources such as storage, output devices, files, programs, etc., and as such lowers the burden on a single computer system. Networking of PCs is more effective and offers an efficient communication system for sharing of resources among the networked PC or workstation users. An interface between PC and LAN is known as a *co-processor,* which runs on the user's PC in the background but provides all the necessary functions of the network. This interface is available as a network card. The standard LANs, such as the **carrier sense media access/carrier detection (CSMA/CD)**, or even token ring network protocols, support both baseband and broadband transmission of data rates between 1 and 10 Mbps. A minimum of 2-Mbps data rate is very common among these networks without extending the network beyond the flexibility limit. PCs generally run under operating systems such as MS-DOS, PC-DOS, OS/2, etc.

In addition to the Ethernet, other LANs have been developed such as Starnet, ARC-NET, etc.[18] The **Starnet LAN** was developed by AT&T as a low-cost-per-user network based on flexible star topology, e.g., star, daisy chains, or both, and it offers a modular structure. It is easy to install and uses a CSMA/CD access technique, and it operates over twisted pair (unshielded). It offers a throughput of 1 Mbps. This LAN can be internet-worked with other Ethernet LANs via bridges (available in the market). By using Novell Advanced NetWare protocol, it provides interconnection to other networks through file servers.

Datapoint Corporation's **ARCNET** LAN is based on token-passing access. It offers a throughput of 2.5 Mbps. Due to its support for token-passing access, 50–75 users can be connected to one segment of the LAN, and it offers excellent reliability. It operates over coaxial cable, offers low cost per user, uses distributed star topology, and is easy to install, expand, and modify.

Most PCs are asynchronous and usually require an asynchronous-to-synchronous adapter between PCs and modems which offers 9.6 Kbps for full duplex using dial-up asynchronous transmission. PCs can be connected to a mainframe (using mainframe communication assistance), workstation (using a workstation emulation program), and other interfaces and emulators. There are many application programs available which offer file transfer on PCs, including XModem, YModem, Blast, Kermit, MNP, etc.

The **XModem** program allows two PCs to communicate with each other by sending a Nak (negative acknowledgment) to the transmitter. The data blocks (128 bytes) are exchanged one at a time. The receiver performs checksum and sends back Ack or Nak. There is a limit of nine tries and if the transmitter fails to send, it times out by sending cancel (Can). The transmitter waits for 10 seconds to listen to Ack and if it does not hear within this time, it re-transmits the data again. The **YModem** is similar to the XModem but it offers additional features such as abort of file transfer after two Cans, support of CCITT CRC-16 error checking, transmission of multiple files at one time, 1-Kb block size, etc. For more information on **Kermit**, see Reference 35.

In this hierarchy of LANs, the higher levels of LANs typically become backbone LANs for the lower levels of LANs. There may be a situation where **private automatic branch exchanges (PABXs)** are used rather than LANs, as these offer similar services. For a discussion of issues surrounding the use of PABXs vs. LANs, see Section 6.5.1 of Chapter 6.

Resource sharing among PCs over a LAN can be obtained using a variety of interfaces providing PC networking. These interfaces or protocols provide communication logical links between users of PCs over a LAN. These interfaces or protocols include **Network Basic Input/Output System (NETBIOS), Novell's Advanced NetWare, 3Com networks, IBM PS/2 LAN interface, Macintosh AppleTalk networking interfaces**, etc. A brief discussion of each of these interfaces or protocols is given below.

NETBIOS protocol offers peer-to-peer communications between machines without being connected to a host. It defines system commands in the form of **network control block (NCB)**. It allows users to make a request for a resource of another machine by sending a resource request and name to each machine on the network; if it is found on the network, it establishes the connection for the users to use it. The broadcasting facilities require additional overhead but are supported by the protocol.

Novell Advanced NetWare protocol provides communication with PC hardware, e.g., disk drivers running under DOS and other operating systems. The vendors have to provide their own disk drivers, as they cannot use DoD disk drivers. It offers a reasonable throughput and response time to access services over the LANs.

3Com networks are based on CSMA/CD protocols and are easy to install and use. They offer the following networking application products for specific functions:

Product	Function
EtherShare	Disk management
EtherPrint	Print sharing
EtherMail	Electronic mail
EtherMenu	Customized menu lists
EtherTerm	Terminal emulation
EtherBackup	Disk-to-tape backup
Ether3270	3270 emulator

The **PS/2 LAN interface** is a collection of products defined for different types of networks (baseband, broadband, token ring, etc.). IBM's baseband PC network adapter/A allows the PC to interface and communicate with baseband LANs. IBM's broadband PC network adapter II/A allows PCs to interface and communicate with broadband LANs. IBM's token ring network adapter/A offers an interface for a PC into an IBM token ring and supports 4 Mbps and 16 Mbps (with a 9-pin connector).

The **AppleTalk** interface, defined for Macintosh, connects file servers and printers to PC-based workstations and supports up to 32 computers and peripherals. It uses shielded twisted-pair cable and a physical layer standard which is a variation of EIA RS-449. It runs SDLC and CSMA/CA link-level protocols.

The two resource-sharing technologies, LANs and PABXs, can be interfaced via gateways or adapters, allowing their users to communicate with each other as if they are connected to the same LAN or PABX. The users of PABX may use the resources of LAN or communicate with its users and also may provide a direct link between two different LANs.

A high-performance and high-speed LAN (e.g., token ring, FDDI) may be used as a data communication link connecting different PABXs in a ring topology (using coaxial cable with broadband transmission). This type of configuration creates a ring topology using a high-bandwidth LAN for connecting PABXs (as high as 64) as the nodes of the LAN. It is very attractive and has been installed in a number of organizations. It uses two token ring cables providing 50 Mbps (using time-division multiplexing access, TDMA) for circuit-switched voice and data communications and 10 Mbps for data communications and some control information packets. Thus, this type of configuration can be considered as fully integrated and distributed, as both data and voice communications are taking place over the same channel simultaneously.

A complete integrated and centralized configuration based on PABX may provide interface between various types of devices such as workstations, PCs, public switched telephone networks (PSTNs), modems, hosts, and even public data networks. The PABXs offer interfaces to PSTNs such as X.25, making them more useful in providing a direct connection between LANs and X.25 or any other public data networks.

Telephone LANs: Telephone LANs are defined for private branch exchange (PBX) or PABX and use a star topology. Both data and voice signals can be transmitted over PBX, which is basically a switching device. The PBX offers connectivity similar to that of baseband or broadband LANs but with a slower data rate. If we try to compare PBX or PABX with standard LANs, we may notice that both provide connectivity to both synchronous and asynchronous signals. In terms of wiring and installation charges, the digital PBX may be cheaper than standard LANs. However, LANs will be preferred over digital PBX for the speed of transmission, as standard LANs can offer data rates of 1–100 Mbps as opposed to 56 Kbps or possibly 64 Kbps in the case of PBX. Further, in some cases modems may be required. The PABX may not need modems. It offers data rates of 1200 and 2400 bps for asynchronous data. The broadband and baseband LANs offer data rates of 4800 and 9600 bps, and up to 19.2 Kbps.

7.3 LAN characterization

The LAN can be characterized in a variety of ways by considering different parameters such as topology, access control, application domain, architecture, transmission media, standards, characteristics, communication data rate, LAN functionality, etc. There may be overlapping of some of the parameters for characterizations. Some of the parameters can be grouped into one generic group known as **LAN technology**. This may contain parameters such as access control, topology, data rate, transmission media, and standards, which are provided by the lower layers of OSI-RM. The OSI-RM defines seven layers while the IEEE 802 committee (responsible for defining standards for LANs) defines an eight-layer LAN model. This eight-layer model is known as the LAN model. The primary purpose of the IEEE 802 committee was to concentrate on the lower two layers of OSI-RM and, as such, the data link layer was partitioned into two sublayers: logical link control (LLC) and media access control (MAC). In the following sections we describe the lower layers of OSI-RM in more detail and compare the frame formats and media-access techniques of different standard and non-standard LANs.

LANs are used in both offices (typically allows the sharing of expensive resources, files, databases, etc.) and industry (typically performs distributed real-time control applications). The requirements for LANs in office applications usually include a high-speed bandwidth which will be enough to provide to the users more than 9600 bps on V.24 lines. No real-time constraints are defined, and, as such, the protocols are very simple and inexpensive. The underlying LAN provides flexibility whereby the nodes can be deleted or added, and also these reconfigure to the changes made in the network requirements. The LANs in industry-based applications offer a number of time constraints, high speed of transmission, and deterministic responses with an acceptable quality of service. In these applications, important issues include reliability, availability, fault analysis, and other options for error control.

7.3.1 LAN (network) topology

The topology of a network defines the physical location of the devices and their connections with the transmission media, which in turn defines the configurations of the network. It is considered to be a physical makeup of the network describing how the nodes or users are interconnected. It also defines a connection between the adapters in the communication subsystems which may have different concepts, cabling, accessing techniques, etc. Different topologies for the network can be defined and, in fact, new technologies are always coming out with different topologies, encouraging states of change, new transmission media, new switching techniques, growth, and flexibility in the network architecture. The networks can be classified as centralized or distributed.

A typical example of a **centralized network** may have a mainframe host that performs all the data processing, while the **distributed network** may have a mainframe as well as a remote host (mainframe or other computer such as PC, workstation, etc.) that also processes the jobs or tasks for the end users. Interestingly, the distributed network can also have centralized or distributed data processing.

Various topologies for networks have been introduced, and the topologies defined for LANs will be discussed in the following paragraphs. Some of these topologies are also used in IEEE 802.6 metropolitan area networks (MANs). The LANs cover a limited distance of a few kilometers while MANs cover a distance around the city and offer higher data rates. As with LANS, the MANs can be owned or controlled by a single organization, with the only difference being that MANs typically have other agencies and organizations involved. The transmission medium used in MANs is either optic fiber or coaxial cable and may present relatively higher delay and bit error rates than for LANs. The topology has been useful for defining or identifying a type of LAN. These topologies are discussed in References 1–7, 10, 14, 18, 32–36, 40, and 41 and can also be found in many other books.

In general, the LAN topologies can be grouped into three categories: star, bus, and ring. The **star topology** is used basically by PBX and PABX. This category includes topologies supporting the following configurations: point-to-point, multi-point, strongly connected, hierarchical. **Bus topology** defines configurations for standard LANs, e.g., IEEE 802.3 (CSMA/CD) and IEEE 802.4 (token bus) LANs. It also includes a tree-structure topology. **Ring topology** defines configurations for the standard LAN, e.g., IEEE 802.5 (token ring), slotted ring, Cambridge ring (British Standard Institute, BSI), and fiber distributed data interface (FDDI).

As mentioned above, the LAN topology defines the physical connection for various nodes (host or mainframe, PCs, databases, etc.). Various specifications to compare different LAN topologies include data rate, maximum segment length, total length of the network, maximum nodes per segment, minimum node spacing, physical media, coax impedance, IEEE specification, topology, signaling technique, and others. The three popular LANs are Ethernet, token ring, and token bus. Ethernet LANs are usually denoted by 10 Bx, where 10 indicates the data rate and x denotes the spacing between the repeaters in meters. 10 B5 was the first Ethernet system that used 10-mm-diameter coax connected via vampire tap. 10 B2 uses 7-mm-diameter coax. The coax is connected to a network interface card by a BN tee connector. 10 BT, on the other hand, uses American Wire Gauge (AWG) #24 unshielded twisted pair, which is connected to the hub NIC and the hub via RJ-45 telephone connector. The maximum length for 10 BT is 100 m, but it can be extended to 500 m using repeaters. It uses the bus topology where all the nodes are connected to a hub.

7.3.1.1 Star topology

Star topology is defined by a central or common switching node (working as a primary or master) providing a direct connectivity to other nodes (known as secondary or slave). It defines one primary or central node which provides communication between different computers or secondary nodes. Each of these computer or secondary nodes are connected directly to the primary node on a separate transmission link and are usually termed *secondary nodes*, as shown in Figure 7.2. Each secondary node sends a data packet to the primary node, which passes it on to the destination secondary node. The data packet (frame) contains the destination node address. It defines a **centralized network.**

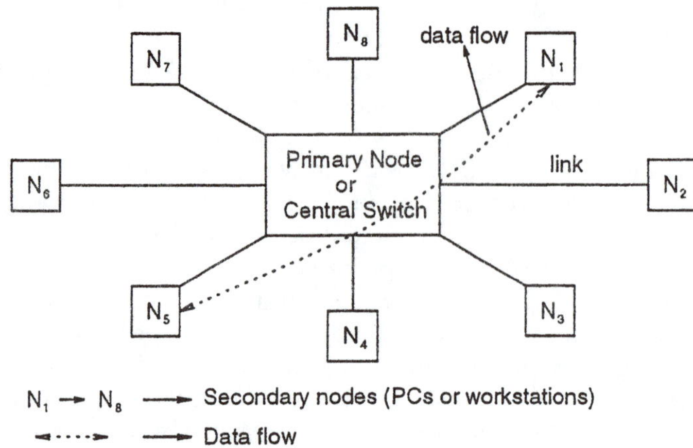

Figure 7.2 Star network.

The primary node controls the entire network and sometimes is also known as the **controller of peripheral (COP)**. Any secondary node that wishes to communicate with another secondary node in the network has to go through the central node. The communication between any nodes takes place via circuit switching being used by the central node (hub) over a pair of unshielded wires or cables. Each node is connected to the hub via two pairs of wires; one pair is used for transmission while the second pair is used for receiving. Any node wishing to communicate with another node must request the central node, which establishes the connection or circuit between it and the destination node. The data transfer will take place only after this connection or circuit has been established. The central node or hub basically acts as a repeater, accepting a data signal from one pair of lines and, after repeating on it, passing it on to the requested node.

Although the star topology is defined by physically connecting each node by a pair of wires with the hub, it also looks like a bus topology (logically). During the transfer of data over the circuit, the nodes are connected by a direct and dedicated circuit providing a **point-to-point link** between them. It uses an unshielded twisted pair of wires. Multiple levels of the central node or hub can be defined in such a way that the main hub provides a direct connection to either nodes or intermediate sub-hubs. These sub-hubs in turn may be connected to either nodes or other intermediate hubs.

Star topology provides easy expansion capabilities inexpensively (flexibility) and supports only a small number of connections. In other words, it offers an expandable structure and requires a smaller length of cables for any expansion of the network. Only two hops (number of direct connections defined by links over a pair of nodes) are required to provide a logical connection between those secondary nodes for data communication. Interestingly, this number of two hops is the maximum one can get in this topology for data communication between any pair of secondary nodes. Further, it is very easy to implement this type of topology. A failure of any secondary node does not affect the working of the network as a whole.

One of the major problems with this topology is that the entire network goes down when the primary node fails. Also, this topology is suitable for small networks but becomes very expensive for large networks (due to cabling, connectors, etc.). Further, the centralized protocols may become vulnerable, as the topology does not offer any local processing; it only provides the connection between two secondary nodes. **PBX** telephone systems use this topology, where a PBX switch works as a primary node while the subscriber's telephone set works as a secondary node and the connection between requesting and

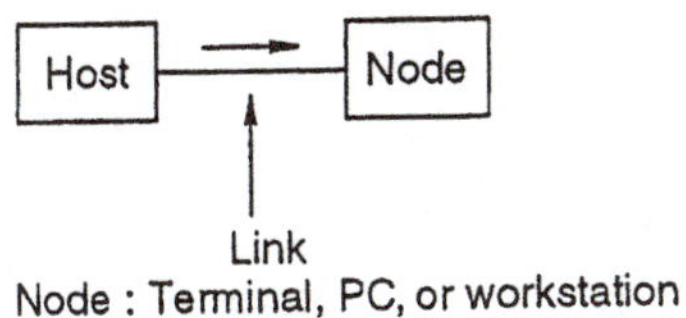

Figure 7.3 Point-to-point topology.

requested nodes is established by an operator. When this topology is used with PABX, various services we may expect from it include call forwarding, call waiting, auto redial, etc.

The central node is very complex, as it has to maintain concurrent circuits established between different pairs of secondary nodes. This certainly lowers the burden on the nodes of the network, as the central node provides the necessary switching and transmission. Each of the secondary nodes use appropriate logic circuits for requesting connections, receiving connections, and communication logic for point-to-point configurations. The star topology is useful for transmitting digital data switches and PBX switches. Thus, the star topology supports both digital or voice/digital services in a variety of applications.

Point-to-point: This is the simplest topology which consists of a host, a communication link (transmission medium), and a terminal or PC or even workstation, as shown in Figure 7.3. Any type of terminal (batch or interactive) can be used. This topology is very simple and can be easily implemented; further, it supports different types of hosts (mainframe, minicomputer, microcomputers, etc.). One or more terminals can be attached to a host and the number of links depends on how many terminals can be supported by the host. Each terminal communicates with the host on the basis of point-to-point transmission. This topology defines a **centralized network**. It is interesting to note that point-to-point communication can also be defined by other topologies such as star, ring, tree, symmetrical, etc. We can use any one of these topologies for point-to-point networks or subnets. LANs usually use symmetric or regular topologies (as discussed above), while WANs use asymmetric or irregular topologies (due to geographic coverage).

Multi-point: This topology is an extension of point-to-point topology and uses multiple remote hosts which are connected via independent transmission links to the different types of hosts (terminals, PCs, or workstations). Further, these nodes interact with a remote computer over a single link via multiplexing, as depicted in Figure 7.4. In both point-to-point and multi-point topologies, the remote nodes are characterized by the amount of data processing, number of tasks, etc., being performed. Some LANs support this topology and do not control the operation of the topology, as the interfaces at each node or point possess some form of intelligence to decide whether to accept the packet or forward it to other nodes in the network architecture. This type of topology supports transmission of a data packet to multiple nodes simultaneously.

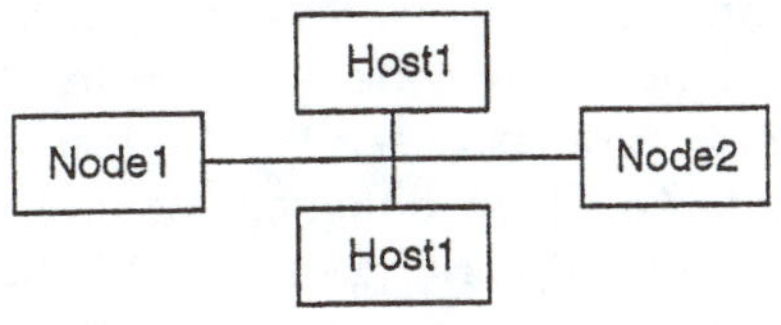

Figure 7.4 Multi-point topology.

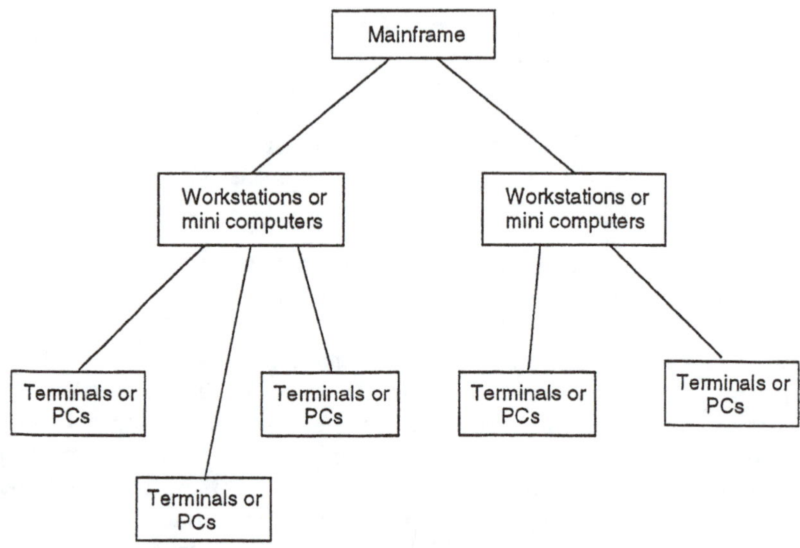

Figure 7.5 Hierarchical topology.

Hierarchical: In this topology, the user's computers or nodes in the same geographic area are grouped together as a cluster and then connected to the network. It defines a fully distributed environment where the computer nodes pass the information to computer nodes at lower levels, which in turn pass the information on to lower-level computer nodes, and so on. The root node has complete control of the network. Usually, the root node consists of a general-purpose computer with a few **front-end processors (FEPs)** attached to interface it with other nodes of the network. Typically, nodes at lower levels include inexpensive devices of fewer processing capabilities and less intelligence as their logical distance from the root node increases. The computer nodes at each level provide data processing facilities. In a typical distributed architecture, at the top-level computer node (root), a computer with high processing power such as a mainframe is used, while the lower-level computers are placed in order of their decreasing processing power (minicomputers, microcomputers, PCs, and terminals), as shown in Figure 7.5.

Use of a hierarchical topology is very common in **WANs**. The actual topology of the network may not follow the hierarchy as defined; instead, it may be altered to suit the application. In other words, this topology is application-dependent, and failure of any link or node will affect the overall function of the network. Depending upon applications, the hierarchical topology offers different levels of processing within the application; e.g., in the database management system, the data transaction may take place at lower levels by terminals or PCs, while at higher levels, other operations such as processing the data, sorting, updating the database, etc., are required before the transaction can be completed.

Strongly connected (Fully connected): In this topology, all the nodes are connected to each other by a separate link. Since each node is connected to the others directly, the number of hops (a simple path or route between two nodes) is always one between any pair of nodes, as shown in Figure 7.6. This topology sometimes is also known as **crossbar** and has been very popular in various networking strategies for fast response requirements. It is very fast, as there is only one hop or a simple path between each pair of nodes, and as such it offers very small (negligible) queue delays. It does not suffer from any routing problems, is very reliable, and survives under any traffic conditions (heavy or light). If

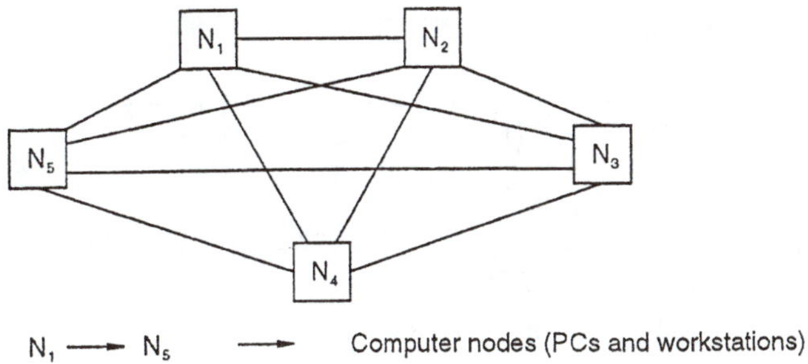

Figure 7.6 Fully connected topology.

any particular link or node fails, only the particular node or nodes connected to that failed link are affected, and it does not affect the entire network in any way.

Due to direct interconnection between each pair of nodes, this is a very expensive topology, as links (interfaces) are very expensive. This may not be a problem with a small network, but it is certainly not suitable for large networks (due to the large number of links), as the number of links is always n $(n - 1)$ where n represents the number of nodes in the network.

7.3.1.2 Ring (loop) topology

In this topology, all the nodes are connected to a circular ring link or closed loop via repeaters and a point-to-point configuration is established between different pairs of nodes, as shown in Figure 7.7(a,b). A *repeater* is a hardware device that regenerates the incoming data and transmits it as new data which does not contain any noise or attenuation, as such disturbances have been nullified by the regeneration actions defined by repeaters during the transmission (timing jitter). The data signals are represented using appropriate encoding (e.g., **Manchester** and **differential Manchester encoding**) which can introduce *clocking*. The signals represented in these encoding techniques define some form of clocking inherent in it at the expense of lower bandwidth. The clocking is detected by the repeater at each node and it indicates at what time the repeater must sample the incoming data signals to recover the bit data, and it also uses the clocking with the data signals to be transmitted from the repeater of that node.

The operations performed in repeaters are different from those in amplifiers where the signals are actually amplified in one direction at a time. In the case of repeaters, each logic 1 and 0 is regenerated such that noises or attenuation, if present, are compensated or nullified in the new values of logic (1 and 0) so generated by it. Repeaters can be **unidirectional or bidirectional**, while amplifiers work only in unidirectional.

Each node connected to the ring has adjacent nodes at its right and left and, as such, each repeater works for two links at any time. This topology enables unidirectional transmission of data to all logically connected nodes; i.e., data is transmitted in only one direction around the ring. The repeaters accept the data from one node bit by bit and pass it on to the next connected node. The repeaters do not define any buffer and, as such, the data can be received and transmitted at a faster speed, with the added advantage of regeneration of digital signals at each node. The computers at the nodes are usually microcomputers and are useful for geographic local networks. Since the nodes form a ring, no routing is required.

The bidirectional loop structure can also be defined where one loop or ring carries the data information in one direction while the other loop or ring carries the data in the opposite direction. The number of hops required depends on the physical location of the

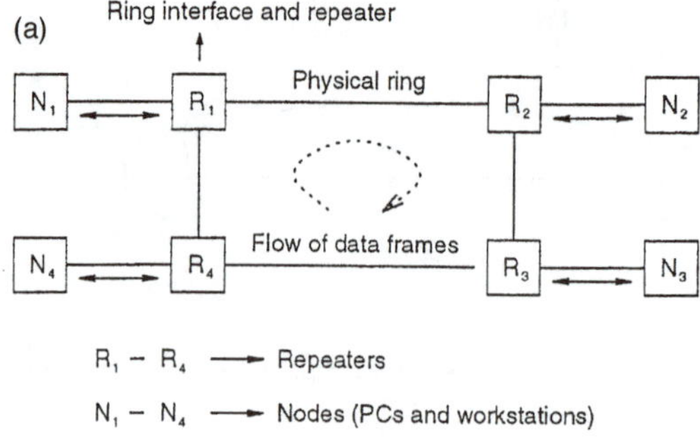

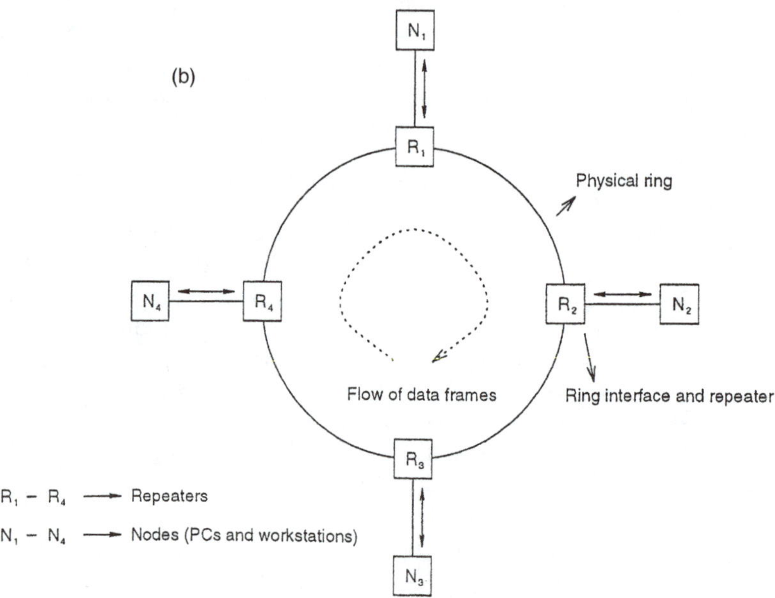

Figure 7.7 Ring topology.

nodes and, in the worst case, will be $(n - 1)$ where n represents the number of nodes connected to the ring.

A node sends the data packet (frame) to another node on the ring — via a repeater connected between them — which includes the data and destination address. When the data packet goes around the ring, and after a destination node recognizes its address defined in the frame, it copies the data on its buffer. The data packet may have any number of bits (from one bit to several bits). The data packets are transmitted as blocks or frames known as **protocol data units (PDUs)**. Each frame contains the user's data, control information, and destination address. One of the designated nodes in the ring topology works as a monitor node which is responsible for monitoring and controlling access to the ring and various operations over the token within the ring topology (e.g., who creates the token, who is taking away the token from the ring, who should get the token next, etc.). This in turn depends on the access logic technique being used (in general, it is based on token passing). The protocols used in the network usually are implemented in a **distributed control environment.**

This topology defines a **distributed network (DN)** and supports high-speed data rates. It usually supports a small number of connections (or nodes) and can be expanded easily with minimal cabling, connections, etc., and supports the **token passing method** for accessing the data over the network. This method determines which node gets an access right to the ring for data communication. In this method, a token of unique fixed bit pattern is defined and this token always circulates around the ring topology. Any node wishing to send the frame must first capture the token. After it has captured the token, it attaches the data packet (frame) with the token and sends it back onto the ring. The destination node receives this token along with the frame, copies the frame onto its buffer, and changes the bit position in the token frame to indicate its acknowledgment to the sender node. At each node the incoming bits go through repeaters which regenerate the digital signal. After the sender receives the acknowledgment, it sends back the token onto the ring. Due to this action, this topology is also known as **token passing topology.** It is not considered as survivable. The network interface devices are simply repeaters and, as such, the nodes require less processing of functions needed to access the network (in contrast to star topology, where complex processing of functions is needed to access the network at each node). The only requirements in ring topology are that the nodes have an access control logic and also are responsible for framing.

Here, the data packet (including control and data) is transmitted around the ring, and each station requires a ring interface to be connected to the ring which is supplied by the network vendors. The transmission media used in ring topology include twisted-pair cable, coaxial cable, or fiber-optic cable. The communication protocol for ring topology allows the users to transmit or receive a frame of fixed size, and it also monitors various operations such as tokens, errors, loss of tokens, duplication of tokens, etc. The frame contains a fixed size and the data rate defines the number of bits going around the ring per second. For example, if a ring LAN defines a data rate of 1 Mbps, one bit over a 1-km ring will travel a distance of 200 m and, as such, only five bits can be present on the ring at any time.

Based on this scheme of accessing in ring topology (as discussed above), we have such standard LANs as IEEE 802.4 (token bus), IEEE 802.5 (token ring), BSI's Cambridge ring, and FDDI, which use this topology in different forms.

One of the main drawbacks with this topology is that the entire network goes down (collapses) or fails when either a node or a link fails (crashing or failure of a node, broken link, or failure of both node and link). Further, the waiting time and transmission time for sending the data packet may become significant, as these depend on the number of nodes connected and the access control technique being used. In this topology, the data packets always travel in one direction and, as a result, even if the destination node is a neighbor to the source node, the packets will have to travel around the ring network.

The problem of a broken link or node can be alleviated by considering multiple rings. In general, a combination of two rings has been very common with token ring LANs, FDDI LANs, and other high-speed networks based on token passing topology. In this scheme, a relay is used with every pair of rings. The combination of two rings is connected in such a way that the failure of a node or link does not affect the normal working of the LAN, as a relay in that particular failed node is actuated and bypasses that node or link by connecting each node via a **wire concentrator**, as shown in Figure 7.8. Each node is physically connected to a wire center by a pair of twisted wires in such a way that one wire is used to carry the data to the node while another wire brings data out of the node.

In each pair of twisted-pair wires attached with every node, one wire is attached with a relay. In normal working this relay is intact, as the entire current corresponding to the signals flowing in or out of the node is also flowing through it. When any link or node is broken, the relay does not get sufficient current to remain intact, and it gets released from

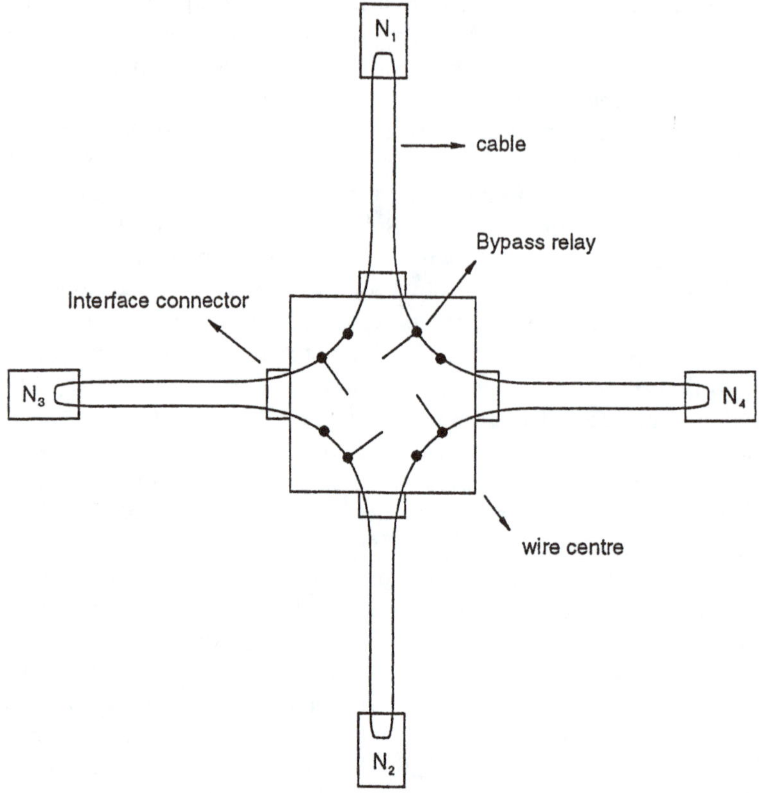

Figure 7.8 Wire concentrator.

the circuits, causing a bypassing of the failed node or link to which it was connected. The relay makes a connection with a wire of another node and allows the current to flow to the next node. It continuously works by bypassing that failed node or link until it has been repaired or corrected.

In this topology, the data packets go through the nodes of the ring until the destination receives it. After receiving, it indicates within the token the action it has taken (recognizing its address defined in the data frame, copying it on its buffer, changing the status of the frame status (FS) field defined within the token frame format, etc.). The source node removes the data packets from the ring and submits a new token back onto the ring. The minimum number of hops required is one, while the maximum number of hops required will be $n - 1$, where n defines the number of nodes connected in the network. A **line interface coupler (LIC)** connecting various PCs in this topology is available from various vendors.

The ring topology offers a point-to-point configuration and supports higher data rates over a fiber-optic medium. The maintenance for point-to-point configurations is much simpler than for multi-point configurations.

Plait: There is another topology known as plait (based on ring topology) where the main emphasis is given to the reliability. This topology is derived partially from the work of Meijer and Peters.[6] Here, if a central node fails, there is an alternate node which can be used for this node or any other node in the network. Similarly, if any link fails, there is another link which will be used for this failed link. This is exactly what is done in reconfigurability, where in the event of any failed node or link (in the event where the

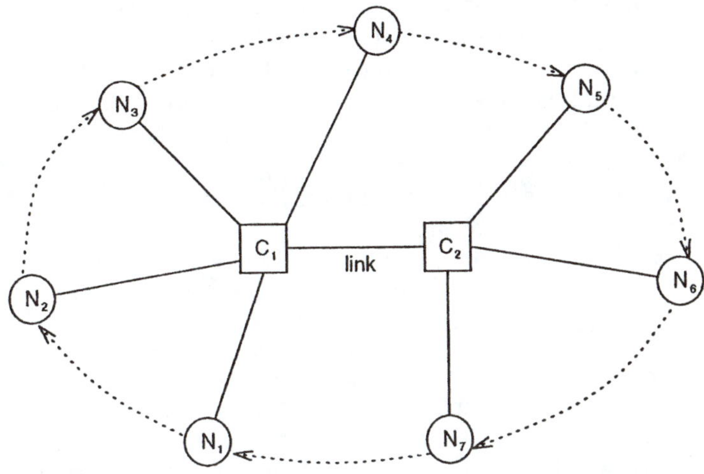

Figure 7.9 Star-shaped ring topology.

requirements are changed), the network topology offers acceptable behavior (if not exactly the same) by bypassing that failed node or link. The reconfigurability is further enhanced by connecting the nodes not only to its neighbors but also to its successors. Although the number of links and nodes may become large, any failed node or link will not affect the routing strategies of data messages over other links or nodes. This topology encompasses fewer connections (cables) and, as such, the network does not go down if the cable goes down. Further, if one of the concentrators goes down, it has little effect on the network. A **concentrator** is a device which carries multiplexed data packets, similar to a multiplexer, with the only difference being that it offers storage capability while a multiplexer does not. Both of these devices offer the same operations and applications. The plait topology offers cost-effective, variable capacity and supports a number of media. It has been used in IBM token ring and FDDI LANs. One of the main drawbacks with this topology is that the latency time tends to increase with the load.

Another way of enhancing reliability of the network is via a **star-shaped ring topology**, a new configuration where the LAN nodes are not directly connected to each other; instead, they are connected via concentrators, as depicted in Figure 7.9. The nodes connected to a concentrator use star topology, and the concentrators are connected via a ring topology. The concentrator is an intelligent multiplexer which possesses a storage buffer and a minicomputer or microcomputer processor. The main difference between a multiplexer and concentrator lies in the way the input and output relationship is defined. A multiplexer is basically a hardware device; the input and output are directly related through data rates. On the other hand, in the concentrator, the low-speed input channels can share a small number of high-speed channels at its output via software control. The allocation and queuing of available channels are provided by the storage capabilities of concentrators. The multiplexer does not know the structure of data, while the concentrator offers intelligent and sophisticated access logic. One of the concentrator's main applications would be to provide a direct coupling or interface between low-speed devices (asynchronous, terminals, etc.) and high-speed devices (synchronous over high-speed wide-band circuit).

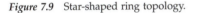

Logical ring topology (based on ring topology as discussed above) uses a token passing control scheme and passes it from one address to another predefined address, giving each successive address permission to send the data packets to the next in the line. All data must travel through a central switch (as in star topology). This topology is easily expandable at both levels: concentrators and star connection for concentrators. One of the main features of this topology is that the logical ring does not allow the central switch to control the transaction. This is because it combines the inexpensive feature of star topology with that of greater flexibility of token passing ring topology. It is flexible and inexpensive for expansion. It does not offer any disruption in the event of failed cable or failed nodes and supports different transmission media. One of the main drawbacks with this topology is that if the concentrators are down, the network goes down.

Multiple-connected topology: A multiple-connected topology (based on ring topology) provides at least two paths between a pair of nodes in the network. Although it does not offer any better advantages over ring topology, this type of redundancy is certainly useful in enhancing the reliability of communication networks. It is used in geographically dispersed networks.

7.3.1.3 Bus topology

Bus topology connects the nodes to a common link (transmission medium) via connecting taps or hardware-connecting devices (defining the interfaces) and, as such, allows easy addition or deletion of the nodes onto the network. In contrast to ring topology, it provides a direct link (or bus) between the nodes, as shown in Figure 7.10. It does not require any switches or repeaters. Since only one link or bus is used as a transmission medium among connected nodes, each node first has to listen to the bus; if the bus is free, it will send the data packet (frame) over it to the destination. If the bus is busy, it will wait and try again to send the same data frame at a later time. All the nodes are treated equally and any node can request access to the network at any time. Bus topology typically uses coaxial cable and fiber-optic transmission media. At the ends of the bus cable, a device known as a terminator, controller, headend, absorber, or connector is connected. The main function of this device is to remove the data frame from the network after it has been transmitted by any node and received by the destination node. Further, it defines an upper limit on the length of the topology, as the propagation time around the ring is useful data for predicting throughput and other performance measures. At any time, only one node is allowed to transmit the message over the network and can be considered a **master node**; other nodes are not allowed to send any message during this time. An arbitration mechanism is needed to handle and avoid situations of contention or conflict.

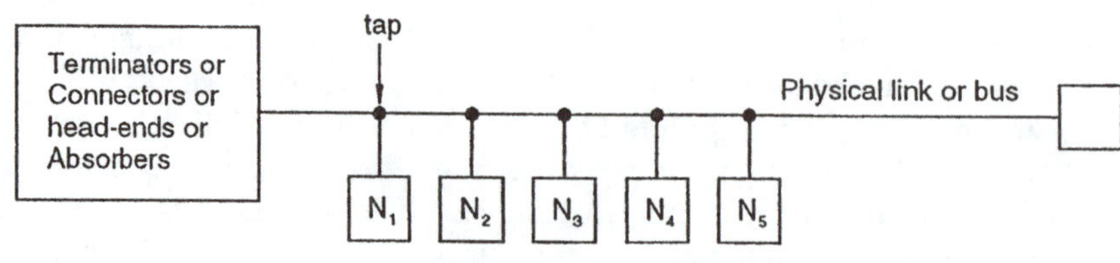

Figure 7.10 Bus topology.

The media used in this topology are connected in multi-point configuration. Either one or two wires are used for transmitting the data across the network. If one wire is used, then the same wire is used for transmitting the data in the forward direction and receiving a reply/acknowledgment in the reverse direction. In the case of two wires, one wire is used for forward data transmission and the second for reverse transmission. The most widely used transmission medium with this topology is coaxial cable. The access control logic and framing must be supported by the nodes (as is done in ring topology) and, as such, the underlying network may have very small processing of the network functions. Usually, one node acts as a **network controller** which manages the networks. The controller is based on the concept of **polling and selection (PAS)**. In this concept, the controller polls the available nodes (which are requesting to send the data packet) and, based on some criteria, selects one node out of the polled nodes to access the bus or link (also known as a channel).

This topology (using either a single wire or two wires) is very useful and popular with LANs, broadband LANs, and transmission (CATV) systems. It is useful for a large number of devices and offers higher capacity. It is easy to implement and expand and it offers local control of the protocols to the users. The addition and deletion of nodes can easily be accomplished by just plugging into the bus dynamically (without affecting the normal working of the network; i.e., the network is still in operation). Since we are using one bus, the failure of this link will cause failure of the entire network. The tapping of the network by intruders is a very common problem with this topology, and sometimes the waiting time and stability considerations may cause a bottleneck. The baseband bus is simple and does need any bridge or ring wiring concentrators. It supports both baseband and broadband transmission. In baseband transmission, it uses a digital signal and is simple and inexpensive. In broadband transmission, analog signals in the RF range used, and it supports multiple channels.

This topology is considered a **distributed network** and is very popular for resource sharing, file transfer, etc., in LAN interconnection. When used with cable television systems, a headend processor converts low-frequency signals (received from the I/O devices) into high-frequency signals and then transmits these signals over the bus (channel). In contrast to the stations in a ring topology, which send the messages around the ring and in fact are connected in a circular or closed-loop configuration, the stations in bus topology are connected in an open loop or linear configuration of the cable (bus). Each station is connected to the network via a network interface which incidentally controls access control, too. **Line interface couplers (LICs)** connecting PCs to a bus in this topology are available from different vendors.

A variation of bus topology is known as **multi-bus topology**, which basically consists of two or more buses. These buses are connected to each other by an inter-bus link known as a **bridge.** Typically, a bus topology alone can support as many as 100 devices and, with this topology, the length or range of the network can be increased. The only requirement for this type of topology is that every bus must represent the same network (LAN).

By comparing bus and ring topologies with respect to suitable parameters and model assumptions (for analytical techniques), it has been established by a number of experimental results that the ring topology is better than bus for real-time applications as long as the ring topology–based LANs use the token access control technique to access the media. One of the reasons for this is the absence of conflict in the accessing technique of ring topology. The broadband subnets normally use bus, satellite, ground radio, and ring topologies and support multicasting by dividing the nodes into groups such that the users can use either one or more than one group.

Tree: This topology is derived from bus topology and is characterized as having a root (headend) node (in contrast to the bus where all the nodes are the same and make a request

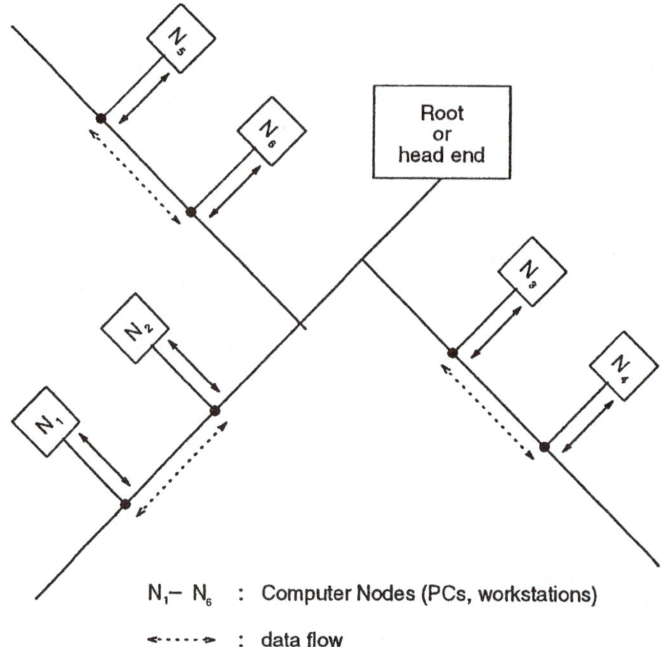

Figure 7.11 Tree topology.

to access the network at any time). We can also say that this topology is a generalization of bus topology without any loop or star-shaped topology. It is characterized as a tree structure which defines certain intermediate nodes for controlling the information flow. There is one root node which is connected to other nodes over respective bus topologies based on a hierarchy of branches and sub-branches coming out of the root node, as shown in Figure 7.11.

Tree topology uses a pair of coaxial cables for connecting all the nodes, where one cable carries all the transmitted data packets (frames) while the second cable provides a link to the data coming out from the receiving nodes. Data is transmitted over the entire medium and can be received by any node in the topology; hence, this is a multi-point configuration. Since all the nodes are sharing the same medium, only one data packet can be transmitted at any time. The nodes in this topology also must have access control logic and framing capabilities, and the network itself has less processing of network functions.

Tree topology allows flexibility where sub-modules corresponding to the branches of the tree can be added or deleted to the topology, and it is very easy to implement in a highly modular way. The routing strategies are simple and the maximum number of hops is equal to the length of the tree plus one. The throughput may be affected (lowered) for large networks, as sub-module failure may lead to a big bottleneck due to the number of interconnections.

Tree topology is mainly useful for broadband transmission (e.g., CATV) and is now being used in broadband LANs, as well. It offers various branches and sub-branches and hence can be useful for providing connectivity to broadband transmission within large buildings of a organization. It offers the same advantages and disadvantages as that of bus topology and is being used as a **centralized network**, as the mapping and conversions are defined only at the root node.

Another topology based on this concept is known as a **mesh topology**, which connects different types of nodes (hosts, computers, etc.) operating under different environments supporting different sizes of packets so that these different devices can exchange information. These nodes are not peers; i.e., the nodes represent different types of computers (PCs, workstations, mainframes, etc.). There is no set pattern for this topology. This topology was used in the Department of Defense (DoD) Internet. It usually is useful for WANs.

In addition to these main topologies of LANs, there are other types of topologies and each of these topologies can be either derived from these existing ones or defined as extensions or generalizations of these topologies. Some other topologies introduced include rings with and without a mediator, direct transport without routing, indirect transport, regular and irregular structures, rootless tress, etc.

7.3.2 Topology design issues

Design issues in LAN topology: The performance and efficiency of a LAN topology can be predicted by considering the following parameters.[7]

- **Modularity:** This parameter deals with the behavior of the network when nodes are added or deleted. The addition and deletion of nodes should not affect the behavior of the network, and these operations must be carried out independently. Each addition and deletion may add to the cost, which is known as modularity of cost, and it is expected that it should be minimal.
- **Logical complexity:** This parameter defines the internal architecture of the information and also the processes (logic) required to handle the transfer of this information. Obviously, the smaller the number of processes required to transfer the data, the less complex the logic.
- **Flexibility:** This parameter deals with the ability to add or delete nodes in the network; it is expected that the network will have this ability. This parameter is useful for defining modularity, as it is used to predict behavior, cost of the network, etc.
- **Throughput:** This parameter indicates the number of packets which can be delivered over a period of time. It is a very important parameter, as it helps in determining utilization of the network. It depends on a number of factors, e.g., access control technique, number of stations connected, average waiting time, propagation time delay, connection, etc.
- **Capacity:** This parameter indicates the maximum number of bits which can be transmitted over the network over a unit of time (typically the smallest unit, e.g., a second). Thus, a network offering a capacity of, say, 10 Mbps transmits 10 Mbps over the link. It also indicates the transmission speed or data rate of the link (network).
- **Stability:** This parameter deals with the behavior of the network in the case of a failure (cable or node). The network is expected to provide similar or uniform behavior under these conditions (failed node, broken cable, hardware error in any communicating devices, etc.) and is continuously monitored for network operations and connecting cables.
- **Interdependence level:** This parameter defines the number of cables which can go down without the loss of data communication link among the users.
- **Reconfigurability:** This feature or behavior of the network will offer an alternate network environment in the event of failed network components (failed node, failed cable). The network is expected to provide acceptable behavior (if not exactly the same) for any added or new requirements occurring due to these changes.

The topology of the network has a great effect on the stability, reliability, and efficiency of the network. The efficiency of the network further depends on a number of parameters such as minimum response time, smaller waiting time, higher link utilization, higher degree of flexibility, and other related parameters. Although some of these parameters may not have a direct effect on efficiency, it is certain that they have at least an indirect effect.

The number of cables, their lengths, and connections are usually defined by the topology; these parameters also affect the routing strategies within the defined topology of the network, because the network must provide at least one link between any two nodes in the event of failure of all the remaining links. The transmission media at this point do not have any effect on the performance parameters of the network. Various network topologies are being used in different types of networks (LANs, WANs, and high-speed networks). Each of these networks support broadcasting or multicasting configurations. Usually these configurations can be implemented in static or dynamic mode.

In **static mode**, the time is divided into discrete slots which are assigned to various nodes (machines). A node can transmit during its allocated time slot. This wastes the channel capacity of the network. The **dynamic mode** can be defined as either centralized or decentralized (distributed). In the centralized mode, one entity determines which transmits next, while in the decentralized mode, each node decides whether to transmit and they interact with each other regularly.

Link performance parameters: With all these topologies, there are various performance parameters associated like connectivity, reliability, availability, etc. These parameters are discussed below.

The **connectivity** can be defined for both link and node. The link connectivity is defined by the minimum number of links which must be removed from the network topology to disconnect two nodes or users under consideration. Similarly, the node connectivity is defined as the minimum number of nodes or users which may be removed from the network such that the two nodes or users under consideration are disconnected from the network. If two nodes are connected by a link, then the node connectivity of the network may be expressed as $n - 1$ where n represents the number of nodes in the network. Usually, node connectivity is used for performance evaluation of the network, as it helps in identifying the number of links required to provide minimum acceptable response time in the network.

Reliability is a performance measure of the network assessing whether it will provide a minimum level of acceptable service over a period of time. This is affected by node or link connectivity and is directly proportional to the connectivity. Higher reliability can be achieved by considering higher node or link connectivity. Another method of improving reliability is through some kind of **redundancy** (used in error detection and correction in data link layer protocols). The reliability is directly proportional to the redundancy; we have to make a compromise between reliability and redundancy that usually depends on a number of parameters such as communication traffic and cost, overhead, data rate, transmission codes, etc.

Availability is yet another performance measure to assess whether a network system is usable over a certain period of time. The availability of the network depends on the capacity of the links used in the network (the capacity expresses the maximum number of bits a link can carry over a predefined time). For example, if we say that the capacity of any link is B bits/sec, it means that the traffic over this link can go up to a data rate of B bits/sec. Once this is achieved, the link is now unavailable for other traffic. Thus, if we have greater capacity of the link, we have greater availability of the network.

The maximum capacity of the link can be determined by finding the *minimum cut*, which basically is a group of links connecting two nodes with least capacity weights. The bandwidth or capacity of any network depends on the physical media (e.g., wires, coaxial cables, fiber, etc.). Different media offer different transfer and transmission characteristics and behavior and, as such, define their data transfer (or transmission speed) or capacity in terms of bits per second or frequency range. For example, coaxial cable offers a transmission speed of 300–400 MHz. Twisted pairs of wires (telephone) offer 1–2 MHz, while optical fiber offers 2–6 GHz.

As discussed above, different LANs offer different data rates, and sometimes we take for granted that when we use these LANs, we will certainly get these data rates. Actually, we never get the upper limit of data rates of LANs as suggested by vendors in their specifications. For example, consider a LAN offering a data rate of 10 Mbps. The data link layer consumes nearly 5 Mbps, while higher layers also add overheads offering throughput of around 2 Mbps. Similarly, for internetworking protocols, we may expect around 4 Mbps, the transport layer may pass around 3 Mbps, and the application may offer throughput of around 3 Mbps. The following are typical bandwidths of some of the categories of LANs:

Backbone LANs	100 Mbps
Broadband ISDN	400 Mbps
Fiber private LANs	1 Gbps

A backbone LAN is usually used to interconnect different types of independent LANs used in various departments or divisions within a university or organization. These different types of LANs are connected to the backbone LAN via bridges, offering a high data rate (only for transit of data) between LANs and WANs connected by it.

There are two types of coaxial cable, baseband and broadband, used as transmission media in LANs. Baseband cable is a carrier wire surrounded by a woven mesh of copper, while broadband cable (used for CATV) has a sleeve of extruded aluminum. Baseband cable offers an impedance of 50 Ω and is typically used for digital signaling (digital transmission). Broadband coaxial cable offers an impedance of 75 Ω and is typically used for analog signaling with FDM in community antenna television (CATV). Broadband cable is also used for analog signaling and digital signaling (high-speed networks) without FDM. With analog signaling, it supports a bandwidth of up to 400 MHz and carries data, video, audio, etc. (defining a bandwidth of 6 MHz per TV channel) over the baseband cable.

Channels carry analog or digital data signals at a data rate of 20 Mbps. The bus or tree topologies use twisted-pair cable offering 2 Mbps and allowing 10 taps or interfaces. Baseband coaxial offers 50 Mbps with 100 taps, while broadband coaxial cable offers up to 500 Mbps with nearly a thousand taps.

Baseband cable is typically 3/8 in. in diameter and costs about 20 cents per foot, while broadband is wider and costs 45 cents per foot. The broadband information is transmitted by modulating radio frequency (RF) carrier signals, and use of FDM allows sharing of devices and the cable's bandwidth.

Broadband cable is used in two types of LANs: single-cable and dual-cable. In single-cable broadband LAN, the bandwidth is split into two directional signals, while in dual-cable broadband LAN, two separate cables are used for signals in opposite directions. The single-cable broadband LAN uses the schemes of frequency splitting shown in Table 7.1.

Table 7.1 Single-Cable Broadband LAN Frequency-Splitting Sequences

Scheme	Transmit (return) frequency	Receive (forward) frequency
Sub-split (used in CATV systems)	5–30 MHz	54–400 MHz
Mid-split (popular scheme)	5–116 MHz	168–400 MHz
High-split (recent innovation)	5–174 MHz	232–400 MHz

Transmission media parameters: The physical connection between various nodes of any topology is obtained by different physical media which provide a signal path between them. The signals are transmitted over it bit by bit from the transmitter (source node) to the receiver (destination node). The commonly used media for LAN topologies are twisted pair, coaxial cable, and optical fiber. These media provide point-to-point configurations (ring topology) and multi-point configurations (bus, tree). Each of these media have different characteristics which include different parameters. These parameters may include features of transmission (analog/digital, modulation techniques, bandwidth, capacity, etc.), configurations (point-to-point or multi-point), the range or length of the link supported, number of nodes, noise immunity, etc.

Twisted-pair cable can be used to transmit both analog and digital signals. It can accommodate 24 channels each of 4 KHz, giving rise to a total bandwidth of 268 KHz. For transmitting digital signals over it, a modem of 9600 baud using phase-shift keying (PSK) modulation may give rise to 230 Kbps over 24 channels.

Digital or baseband signaling can be performed over twisted-pair cable and, in fact, Bell's T1 digital carrier uses twisted pairs which support 24 voice-grade channels (PCM) offering a data rate of 1.544 Mbps. It is used in both point-to-point and multi-point configurations.

The ring-based LANs offer shared access for point-to-point and multi-point configurations and, as such, these are also useful for broadcasting applications (just like bus/tree-based LANs). These LANs offer a longer distance of transmission (due to repeaters at nodes around the ring) for both baseband and broadband transmission, higher data rates over optical fiber transmission media (due to their deterministic behavior), easy maintenance of links (point-to-point), simpler error recovery and fault isolation, data integrity, and many other features. One of the nicest features of ring-based LANs is their ability to detect duplicate addresses within the LAN. In the case of bus/tree-based LANs, a very complex access logic protocol needs to be defined to detect duplicate addresses (two nodes having the same address).

In a ring-based LAN, when the frame is circulating around the ring, it will be received by a repeater of the destination node (after its address is matched with the address of the destination node defined in the frame). When this frame is circulating around the ring, the duplicate node will not be able to accept it, as its status has already been changed and it is only the source node which can remove it from the ring. In spite of all these features, ring topology suffers from various problems such as failure of the ring or repeater, the limit on the number of repeaters, and token monitoring (dealing with generation of token, removing it from the ring, and controlling various operations of the token, etc.).

The ring topology supports all three types of transmission media. In the case of twisted-pair cable, unshielded twisted pair offers 4 Mbps and can support up to 70 repeaters, while shielded-pair wire offers 16 Mbps and support for 250 repeaters. Baseband coaxial cable offers 16 Mbps with support for 250 repeaters. (The ring topology supports only digital signaling and, as such, does not support broadband coaxial cable.) Optical fiber offers a data rate of 100 Mbps and supports 250 repeaters.

7.3.3 Switching techniques

Switching plays an important role in offering speed in the network. It can be classified as circuit switching and store-and-forward switching. Details of these types of switching are available in Chapter 4. In order to understand the concepts within the framework of LAN technology, we describe these switching techniques briefly here.

In **circuit switching**, the entire circuit or link is established before the transfer of data over it. This circuit is defined in terms of various intermediate switching nodes between sender and destination nodes. Circuit switching can be used for either analog signals

(**PSTNs**) or digital signals (**digital data networks, DDNs**). Further, circuit switching for digital communication uses a multiplexing technique for transmitting channels over the same transmission media. There are mainly two types of multiplexing techniques being used in digital communication based on circuit switching: **space division multiplexing (SDM) and time division multiplexing (TDM)**.

Store-and-forward switching defines the circuit between sender and receiver on demand; i.e., the packet containing all the user data, control information, and destination address is transmitted to the next switching node en route to the destination. This node looks for a free link to the next switching node, which repeats the process of looking for a free link until the packet is delivered to the destination node. It can be used in either **message** or **packet** data.

One of the major differences between these types of switching (circuit and store-and-forward) is the size of the data message. In the case of message switching, there is no restriction on the size of message blocks, while in the case of packet switching, the entire message information is partitioned into packets of fixed size (depends on the type of underlying network). Each packet includes the source, destination addresses, and some control information and is transmitted separately.

Packet-switching (PS) can be extended to support both connection-oriented (virtual circuit) and connectionless (datagram) services. In the **connection-oriented services**, we have three phases for data transfer: connection establishment, data transfer, and connection de-establishment/termination. In the case of **connectionless service**, each packet is transmitted over different circuits or links without establishing any connection in advance. Packet-switching is used in **PSDNs** and high-speed networks.

The first experimental network, known as ARPANET, was based on packet-switching and network management. Another experimental packet-switched network was developed at the National Physics Laboratory, England. It was here that the term "packet" was first used to describe a 128-byte block. The X.25 standard became a basic interface to public packet networks. The first commercial public packet network was developed in 1970 and is known as Tymnet. It uses a centralized routing policy under the control of a network supervisory computer. It was designed to support low-speed asynchronous terminals via dial-up connection to the nodes located in major metropolitan areas. Later it also provided support for synchronous terminal interfaces. In 1974, Telenet started as an enhanced version of ARPANET and adopted another approach of defining its routing strategies by combining distributed routing (as defined in ARPANET) and centralized routing (as defined in IBM's SNA[19]). For more information on switching techniques, see References 1–4, 6, 7, 10, 32, and 34.

7.3.4 Access control

As we pointed out earlier, OSI-RM defines seven layers for a local area network. But if we divide the data link layer of OSI-RM into two sublayers, **media access control (MAC)** and **logical link control (LLC)**, then this reference model defines eight layers and is known as the **OSI reference model for LANs** (defined by the IEEE 802 Committee). The physical layer provides an interface between the transmission media and data link layer, while an access cable composed of cable, physical media attachment, and media sockets provides interface between the physical layer and transmission media. Every transmission medium has its own media attachment but provides the same services to the physical layer.

As mentioned above, the data link protocols maintain the virtual connections for sending and receiving data over the network. The virtual connection is basically a logical connection defined by a data link protocol over the existing transmission media and may support different data rates. One very important service offered by the data link protocol is to provide and control access to transmission media, as this resource is being shared by all the connected stations with a restriction that only one station can use it at any time.

The MAC sublayer is mainly concerned with media access strategies, and these are different for different LANs and support different types of transmission media at different data rates. The LLC provides an interface between the network layer, and MAC offers common **LLC service access points (LSAPs)** to all the MAC sublayer protocols. All the MAC sublayer protocols are supported by a common LLC sublayer standard protocol, IEEE 802.2. More detailed discussion on this layer is given in Chapter 13.

There are mainly two types of access control techniques, namely carrier sense and token passing (as discussed above). We have two main categories of MAC media access protocols, known as **carrier sense multiple access (CSMA)** and **token passing**. The CSMA protocol is used by Ethernet over bus topology, while the token passing protocol works over either ring or bus topology. The token passing access control technique is used in token ring LANs (token bus, token ring, Cambridge, FDDI). The Cambridge ring LAN (standard adapted by BSI) uses the concept of *empty slot* for its MAC sublayer protocol.

The first carrier sense access technique was known as ALOHA, developed for centralized packet radio transmission at the University of Hawaii. In this access method, a node waits for the data packet and sends it to the network immediately after it receives it from the user. After sending, it listens for the acknowledgment (Ack) packet from the destination node. The duration of listening must be equal to the maximum value of round trip propagation time delay of the network topology. The transmission is said to be complete when it receives the Ack. If it does not receive the Ack or time-out, it will re-transmit the same packet after a random amount of time (typically greater than round-trip propagation time).

A variation of the ALOHA access technique is known as slotted ALOHA, where the node, after getting the data packet, waits for the time slot (allocated to it). After the node transmits the packet during its slot, it waits for Ack from the destination node. If it does not receive the Ack, it will re-transmit the same data packet. Since the node is transmitting the data packets during allocated time slots, collision is reduced considerably (in contrast to ALOHA), but it needs an overhead for maintaining synchronization between two nodes periodically. More details on ALOHA can be found in Reference 2.

7.3.4.1 Carrier sense (random or contention) access

In this access technique, a node listens to the transmission medium (channel) before sending any data packet (message). If the channel is free, it can transmit the message in the form of fixed size of frame (containing control and data information). But, if the channel is not free, it will wait for a fixed amount of time (dependent on the type of protocols used) and retry to see if the channel is free. A situation may occur where, at the time of listening to the channel, it finds it free, but after it transmits the message, there is another message by another node also in the channel. This causes a collision and the messages have to be withdrawn from the channel and transmitted again after some time. This does not guarantee access or offer any permission to any node before it starts transmitting; i.e., any node can start transmitting at any time. This scheme on bus topology uses a carrier sense multiple access (CSMA) technique for accessing the media.

The carrier sense access control technique is used mainly in CSMA and token bus LANs for bus or tree topologies (very useful for an office environment). Under average network traffic conditions, in most cases, the data frames reach the destination node. The size of the data frame depends on the physical distance the frame has to travel from one end to another (propagation time). The propagation time is defined as the time taken by a frame to go from one end of the LAN to the other. The round-trip propagation delay is twice this.

If the transmission time of the frame is too short compared to the propagation time, collisions may not be detected. In the event of collision (where two stations, after finding the channel free, transmitted their messages), the first station which detects the collision sends a jam signal (a predefined bit pattern sequence of 32 bits) into the channel. This unique bit pattern signal over the channel is continuously flowing and indicates to all the connected stations that a collision has occurred in the channel. Further, after the occurrence of the collision, both the stations have to withdraw from the channel and these stations retry after a random time to listen to the channel. It is possible, but very exceptional, that a particular station will try many times to transmit a message unsuccessfully, which may become a serious problem in that situation.

In general, this problem of unnecessary longer delay is handled by discarding the message, using an efficient transport layer protocol ISO 8073 class 4 service, or just informing the higher-layer protocols via various data unit primitives. This access control technique defines the MAC (lower sublayer of the data link layer). The destination node must send its acknowledgment; otherwise, the source may send the frame again. More details on this access technique will be given in Section 8.2.3 of Chapter 8.

CSMA/CD and CSMA/CA: CSMA access control protocol offers a set of utilities for different characteristics. This sharing of channels by a number of stations is derived from the original access technique used in pure ALOHA networks. In ALOHA networks, any station which wants to transmit the message uses the common channel and it gets an acknowledgment from the receiving station. This works well for light traffic (offering fewer collisions), but for heavy traffic, the maximum utilization of ALOHA will be reduced. The utilization in pure ALOHA is improved by introducing slotted time slots, and the messages are transmitted only at the beginning of the time slots. This type of network is known as *slotted ALOHA*. The CSMA access technique yields many collisions and, as such, network utilization is affected significantly (particularly under heavy traffic conditions).

An improved version of this access technique is known as **listening before talking (LBT)** or **carrier sense multiple access (CSMA)**. In this technique, a node listens or senses the channel before sending any message over the channel or network. This technique offers advantages over the one used in slotted ALOHA. The number of collisions is reduced, which improves network utilization and throughput since now the channel can carry more messages. Further, the synchronization mechanism is not required, thus reducing the overhead (in contrast to slotted ALOHA). In the event of collision, both stations have to withdraw from the network and retry after a random amount of time (depending on the algorithms being used for listening to the channel). During the collision, both messages are corrupted and therefore have to be withdrawn from the network.

There are versions of CSMA protocols where the collisions can be detected by special hardware devices and appropriated utilities of protocols. Further, the collisions can be avoided by using special types of hardware and software techniques. In these versions of protocols, the access control techniques are appropriately termed CSMA/(collision detection, CD) and CSMA/(collision avoidance, CA). Most manufacturers have developed variations of CSMA as CSMA/CD and CSMA/CA protocols.

In **CSMA/CD**, the network interface listens to the channel and, after it finds it free, puts the frame on the channel (as discussed above). In **CSMA/CA**, the network interface avoids any attempts to eliminate collision detection but instead senses the channel two times and makes necessary arrangements before it sends the data packets over the channel. Thus, it may not need any jam signal to be transmitted in the event of collision, as collisions are less likely to occur. Although the number of collisions is not reduced, the efficiency of this CSMA/CA scheme is not lower than that of CSMA/CD.

The physical layer of Ethernet LANs handles the actual transmission of data over coaxial cable, which is connected to the transceiver by 50-Ω cable (trunk). Each transceiver is positioned whenever the network access is required. A length of interface cable (drop cable) joins each transceiver to a controller/terminal. The interface cable consists of four twisted pairs and carries the following signals on each cable: transmit, receive, collision, and power.

Ethernet and IEEE 802.3 are well-known CSMA/CD bus systems. Ethernet offers a data rate of 10 Mbps and is being manufactured by a large number of manufacturers and vendors. The earlier versions of Ethernet used the baseband transmission technique on coaxial cable (transmission medium) with its bus topology. Now Ethernet uses the broadband transmission technique, fiber optics, etc. and offers higher data rates. IEEE 802.x standards were announced in 1980. IEEE 802.3 has a seven-byte preamble and one byte of start frame delimiter. Ethernet has a preamble of eight bytes. The Ethernet-type field defines higher-level (level 3 and above) protocols in the data field; some use TCP/IP, while others use XNS, Xerox Network System.

Versions of carrier sense protocols: There are different types of strategies for sensing and retrying the channel before and after the occurrence of the collision: **non-persistence (NP)** and **p-persistence protocols (p-P)**. There exists a variety of vendor protocols for each of these classes. A number of CSMA algorithms have been defined for handling the situation of a busy medium and letting the nodes send their data messages. Each of these classes of protocols will be discussed in detail below. Some of these techniques have been derived partially from References 1–6, 14, 33, 34, and 36.

In non-persistence (NP) carrier sense protocols, the protocol senses the channel and, if it is free, transmits the message. In the event of occurrence of collision, it waits for a random amount of time before it re-listens or re-senses the channel. If it finds it free, it will transmit; otherwise, it keeps on trying by waiting a random amount of time until it is able to transmit the data message. In other words, this access algorithm waits for some time (after it finds the medium busy) before listening to the medium again. This type of protocol is not efficient because it wastes a lot of bandwidth, as during the time that the particular station (which has experienced an collision) is waiting, no stations, including that one, may be trying to sense the channel. The steps required in this version of protocol are shown in Figure 7.12. This offers an easy implementation but wastes the bandwidth of channel and may offer low throughput.

p-Persistence carrier sense protocols are based mainly on continuous monitoring of the channel in the hope of effectively using the bandwidth of the channel by reducing the number of collisions. There are two versions of p-persistence protocols available: 1-persistence (1-P) and p-persistence (p-P).

In the **1-persistence (1-P)** category, the protocol senses the channel continuously until it finds it free and then transmits the data message. In the event of occurrence of collision, it waits a random amount of time and again senses the channel continuously. In this way, it has to wait a random amount of time only in the event of occurrence of collision, while in other cases it is sensing the channel continuously. This protocol is certainly better than non-persistence, as it offers higher throughput and is more efficient. In other words, this access algorithm listens to the medium continuously (after finding that medium is busy) until it finds media free and then transmits the data packets immediately. All the required steps for this protocol are shown in Figure 7.13.

The **p-persistence (p-P)** class of the access control technique can be considered the one which tries to reduce the number of collisions so that the bandwidth of the channel can be used effectively. A station listens to the channel and, if it is busy, it keeps on monitoring it until it becomes free. As soon as it finds a free channel during the first time

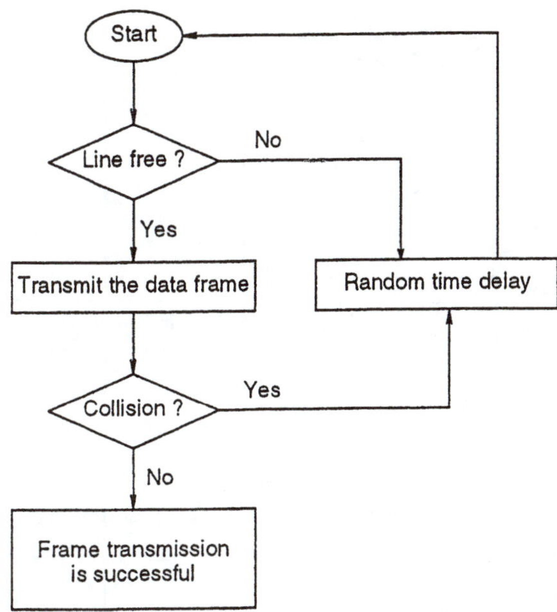

Figure 7.12 Non-persistence CSMA.

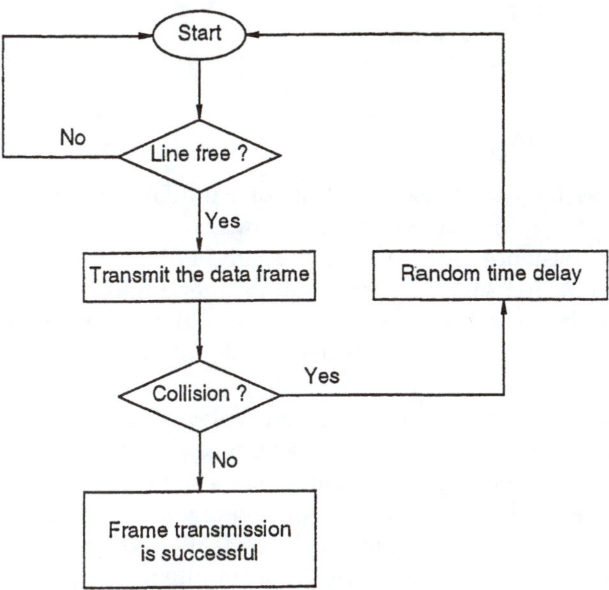

Figure 7.13 1-Persistence CSMA.

slot, it transmits the data message with a probability of p or it can wait for another time slot with a probability of $(1 - p)$. The time slot represents maximum propagation delay of the network. As usual, in the event of occurrence of collision, the protocol waits a random amount of time, but it keeps on monitoring the channel. In other words, in this access algorithm, the node transmits the data packet with a probability of p (after it finds the medium free).

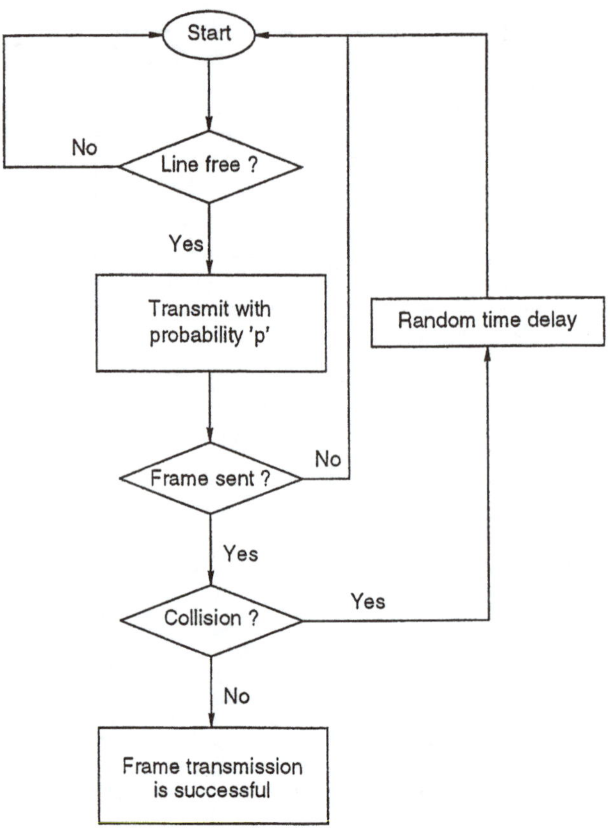

Figure 7.14 p-Persistence CSMA.

If the medium is a bus, it will listen continuously until it is free and then try to transmit with the probability of p. All the required steps of this version of protocol are shown in Figure 7.14. The 1-persistence protocol may be a subset of this protocol, as the probability associated with the data message being transmitted after it finds a free channel is always 1, and as such the number of collisions may be large. In the case of p-persistence, not only is the number of collisions reduced but the bandwidth of the channel is also utilized more efficiently than in 1-persistence.

The main issue in p-persistence is the value of p. If the value of p is large, more collisions are likely to occur. In the worst case, where p is equal to 1, the number of collisions may be very large (1-persistence). If we select a low value of p, the average waiting time for transmitting any data message by any station is greater and may affect the throughput, but the number of collisions is certainly reduced. We can determine the number of stations which are waiting to transmit simultaneously, denoting it by N. Then, in order to expect acceptable service from the network, the product of N and p must be much smaller than 1 (i.e., $N \times p \ll 1$). If this condition is not satisfied, then it is possible that the large number of stations waiting to transmit at peak times may collapse the network and no service will be available to the stations.

The above classes of protocols were mainly defined for an access control technique based on CSMA or LBT. This access control technique has been modified to a new category of access control technique known as **CSMA/collision detection (CD)** or **listen while talking (LWT)**. This access control technique is very popular in Ethernet LANs and offers a suitable access logic interface for connecting the node to a network. Each node is

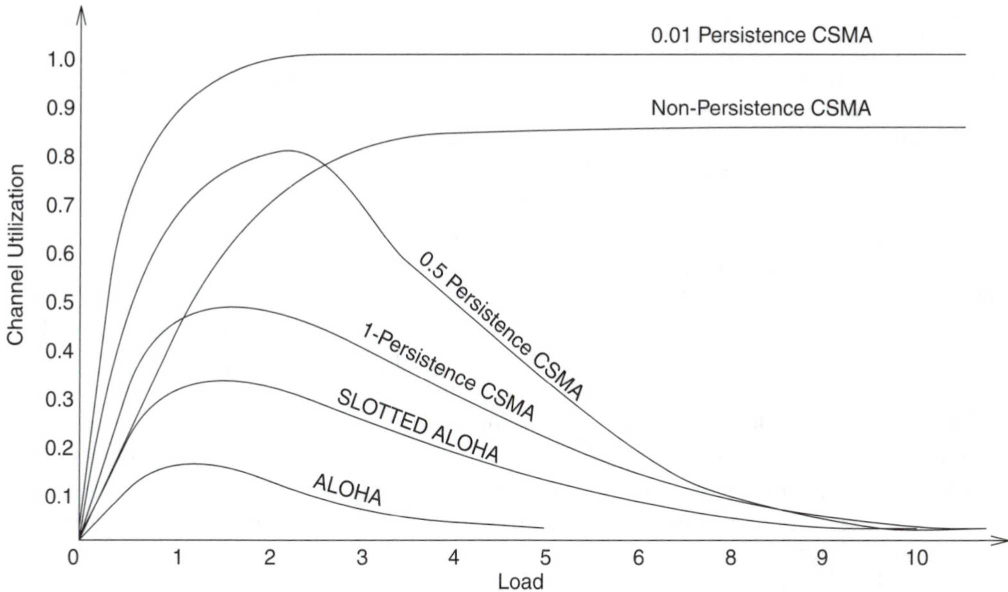

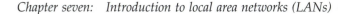

Figure 7.15 Channel utilization vs. load for various access techniques.

connected to the channel via a **LAN computer interface card (LANCIC)** which, in fact, is a combination of a node and its associated access logic and is also known as a **controller**. Each node has a unique address on the network and is defined as a hardware address which is implemented as a hardwired address plug or set of switches. This is the version of protocol which is being used in Ethernet and IEEE 802.3 LANs.

A graph showing how the above-mentioned access techniques help in the utilization of LANs for different network traffic is shown in Figure 7.15.

CSMA protocol design issues: In the previous section, we briefly introduced the CSMA/CD access techniques. Now we will discuss various design issues that are encountered during their implementation. There are a number of design issues underlying the working concepts of CSMA accessing techniques that one must understand, as these will be useful during the implementation. The CSMA/CD access control technique–based protocol senses the channel before sending the data over the network. If the channel is not free, it will transmit the message; otherwise, it will wait and try to sense the channel again. If it finds a free channel, it waits for some time to let the network reset all the counters, shift registers, and other hardware devices and then begins to transmit the data. During this time, before it transmits the data, it will again sense the channel to see if any collision has occurred.

The collision corrupts both the messages, and one of the stations involved in the collision must send a jam signal on the network which can be listened to by other stations which are planning to transmit messages. The station transmitter checks the data on the channel with the data it has transmitted and after it finds that a collision has occurred, it stops the transmission and instead transmits a jam signal on the channel. This jam signal is of a specific bit pattern (typically a 32-bit unique bit sequence) which is transmitted for a random amount of time before re-transmission can resume from the station transmitter.

In the situation where two stations want to transmit messages at the same time, it is possible for one station not to listen at the time of transmission and, after it has transmitted the messages over the channel, a collision occurs. The nodes involved in collision invoke a **binary exponential backoff (BEB)** or **truncated binary exponential backoff (TBEB)** algorithm before they try to re-transmit. In TBEB, the number of time slots (equivalent to round-trip propagation delay) is reduced. If a packet has m unsuccessful transmissions, the next transmission time is delayed by n slots. This integer n is selected from the uniform distribution function in the range of $0 < n < 2^k$ where $k = m(1,0)$. In these cases, we have to consider the propagation time, in particular the round-trip propagation time.

For example, consider this situation. If station A is transmitting the data message to station B, the message will take time (depending on transmission media, data rates, etc.) to travel down to station B. If station B also wants to send a message over the channel, there will be a collision. But we can design the protocol in such a way that when the last bit of the message sent by A is going back to station A, at that time B can start sending the message, as by the time it reaches A, the last bit has already come back and no collision will be detected. The propagation time so considered is defined **slot time (ST)**, which is usually used for handling the collisions in the network.

For deciding the length of the time slot, we consider the round-trip propagation time (twice the propagation time, i.e., the time a signal takes to go from A to B and come back to A). This time slot must be greater than the round-trip propagation delay; otherwise, in the worst case (when two extreme end stations are sending messages back and forth), collisions may occur.

In addition to these access control strategies, there are other access control techniques which are based on different concepts. The **demand access control–based protocols** may be either *central* control (used in polling and selection scheme) or *distributed* control (used in token passing and slotted-ring LANs). The adaptive access control–based protocols are very complex in nature, due to their working principles, as they have to adapt to the changes of the networks' characteristics, traffic loads, updating, etc. Although for heavy traffic this scheme may take the form of polling or even **time division multiplexing (TDM)**, in light traffic it seems that it will still be using some kind of random access control strategies.

Fast Ethernet: There are three fast Ethernet LANs: 100 Base T, 100 Base VG (voice grade) or AnyLAN, and fiber distributed data interface (FDDI). The 100 Base T can use unshielded twisted pair (categories 3, 4, and 5) and optical fiber. It is based on CSMA/CD, as it uses the same MAC sublayer and also is compatible with 10 Base 10. 100 Base VG or AnyLAN was introduced by Hewlett-Packard and it uses twisted pair (categories 3, 4, and 5) and optical fiber. It supports the transmission of audio, data, and video. FDDI is based on the token concept and allows many frames to travel simultaneously over the media. It uses dual ring and hence offers fault tolerance. The American National Standards Institute (ANSI) introduced FDDI; it uses copper twisted pair and coax. It supports synchronous traffic (audio, video), asynchronous traffic (data), and isochronous traffic (data, video).

7.3.4.2 Token passing access

LANs using token passing access control techniques were initially popular in Europe, but now they have gained some popularity in the U.S. There are four LANs which are based on token passing protocols: token bus (IEEE 802.4),[24] token ring (IEEE 802.5),[25] FDDI,[17] and Cambridge ring LAN (BSI). IBM introduced a token ring network in early 1980 which operates at a data rate of 4 Mbps. In 1988, IBM introduced a 16-Mbps token ring network. The maximum distance is about 300 m and it uses coax for connection. It supports over 270 devices. With unshielded twisted pair, it supports fewer devices and a smaller distance. All the nodes are connected to a multi-station access unit (MAU) which does not have any amplifier or relay.

In the **token passing LAN**, the network operates in a deterministic manner where the stations are defined in a logical ring topology (using token bus) or they are physically connected to the ring topology. The token bus LANs working over a bus topology (coaxial cable) defines a logical ring for the stations which wish to transmit the message among themselves irrespective of their physical locations. Each node is given an identification number which is attached with the message data packet and transmitted to the node with that address in the logical ring topology. Each node knows the address of its logically defined **predecessor** and **successor** stations. A node will receive the data packet from its predecessor and, after performing required functions on the packet, pass it on to its successor station.

A *free token*, defined as a certain number of bits (in a unique bit pattern), always circulates around the logical ring. A station which wants to transmit any data message data packet waits for this free token. After it receives this free token, it attaches the data packet with it and transmits it to the successor station. After the data packet has been received by a particular station, it makes the token a free token and passes it on to the next station. This scheme of accessing the channel or network is based on a round-robin technique where each station has an equal right to access it. Further, the priority for any station can be determined at the beginning and that station always gets the token first and then transmits the data packets. The assignment of priority is independent of the stations' physical locations.

A similar concept is being used in token ring LANs where a token of fixed and small length (3 bytes) circulates around the physical ring to which stations are connected. Any station wishing to transmit the message data packet has to capture the token. After getting the token, it appends (attaches the data message with token) it and sends it back onto the ring. The destination station, after receiving the message, copies it, changes the status bit of the token, and sends it back onto the ring. The sending station looks at the bits in **frame status (FS)** of the token ring MAC frame for the acknowledgment. If the status bit is not set, the token, along with the message data packet, keeps on circulating around the ring until it is either received or removed by the sender station (in the event of a failed or inactive destination station or if it goes over the maximum limit of time for letting the token circulate around the ring).

A token does not carry any data. The starting delimiter (SD) provides synchronization and indicates the destination node about the token or frame. Initially a node with the highest address will act as an active monitor node. It generates a token and sends it to the next node by not setting T bit. That node may not have the data to send but will check the token for any error. If it finds no error, it will regenerate the token and send it to the next node. If the next node has the data to send, it will capture the token, append it, and rest A and C bits. When a frame including a token goes over the first node (which generated the token), it will again check it for any error. If no error is found, it will pass it to the next node until it reaches the destination. The destination node also checks for any error, copies the data, and, after setting C, returns it to the sender. After receiving the frame and checking that the destination has received the data, it will generate a new token and send it to the next node. In the event of error in the frame, the node will set the E bit and pass the frame on to the destination. The destination node, after seeing the E bit set, will not copy the data and send it back to the source node.

As mentioned above, in this type of LAN, the nodes are connected to the ring media via repeaters and allow transmission in one direction only, offering a closed loop. The repeaters accept the data bit by bit from a source node and, after regenerating this data, send it to another repeater until it reaches the final destination node. The repeaters also provide interface to various nodes and offer the following main functions: **data transmission, data receiving**, and **data removing**. These three functions are required for data communication around the ring topology. The data message to be transmitted is usually given as blocks or frames, and these frames carry the user's information, control information, and destination address. When the frame is moving around the ring, it goes over the repeaters, and as soon as the address of any node matches the address of the destination included in the frame, the frame contents are copied in the buffer associated with that node.

A ring-based LAN (token bus, token ring, FDDI, Cambridge ring) typically includes two main components for providing access to the LANs: **ring interface** and **connection.** The ring interface allows the nodes to access the LAN, while the connection between nodes is usually provided by twisted or coaxial cable and defines point-to-point configurations between a pair of nodes.

Design issues in token ring: The physical length of bits circulating around the ring plays an important role in the design of efficient LAN protocols. In the following paragraphs, we will discuss some of the conceptual design issues which must be understood before the implementation of protocols for this type of LAN.

Consider a ring-based LAN which operates at a data rate of A Mbps. The time to transmit one bit is given by the reciprocal of this data rate (1/A Mbps/sec). The ring LAN transmits the data frame bit by bit. We assume that the transmission speed of bits over the cable (ring) is B m/sec. Each transmitted bit which has taken a time of 1/A will be over the ring for a distance of B × 1/A m. If we assume that the length of the ring is 1 km and the transmission speed is 200 m/sec, the number of bits present in the ring at any time is given by (length of ring)/(distance traveled by one bit). Let's consider the data rate of 1 Mbps; the distance traveled by one bit is 200 m. Thus, the number of bits present at any time in the ring of length 1 km is given by 1000/200 = 5. The number of bits at any time over the ring will be five.

The repeater attachment for ring-based LANs is different from the taps or adapter required in bus/tree-based LANs. The repeater attachment in ring topology has to offer the above-mentioned functions. In the case of bus/tree, a **terminator** or **headend** is used which absorbs the data being transmitted over it even though the destination node has not received it. In other words, the bus/tree-based LANs are available for the next transmission immediately after they have transmitted the data packets. They do not wait to sense or know whether the data packets have been received by the destination node, as the data will be absorbed or removed by the terminators or headend. In the case of ring-based LANs, the frame will be circulating around the ring until it has been received by the destination node, after which it will still circulate around the ring. The source node has to remove this frame from the ring and issue a new token to the ring so that the network can be used by other nodes for data transmission.

The token provides a provision of assigning a priority to the stations (via an access control field defined in the token frame) which are attached with a **ring station interface (RSI)** or **network access unit (NAU)**. These RSIs or NAUs perform various functions such as copying the message data packets onto a buffer, making a request for a token, requesting a priority, changing the status of a token to indicate the acknowledgment, etc., and are based on a round-robin technique. These interfaces usually provide two modes of transmission: listen and send (Transmit), as shown in Figure 7.16(a,b).

In the **listen mode** of transmission, a station is ready to either modify the token or copy the message data packet and change the bit status of the token (defined in the frame status field of the token MAC frame) to indicate the acknowledgment. The ring interface introduces a one-bit delay in the information it receives at the input, after which it simply copies to the output and the node can listen to it. In other words, it introduces a one-bit delay into the transmission.

The station in this mode can only receive the message (if it belongs to it), copy it, and send the token back into the ring. Here each bit received by the repeater is re-transmitted with a small delay (typically one bit). During this delay, the repeater is required to perform all of its functions, including recognition of frame, copying all the bits received (bit by

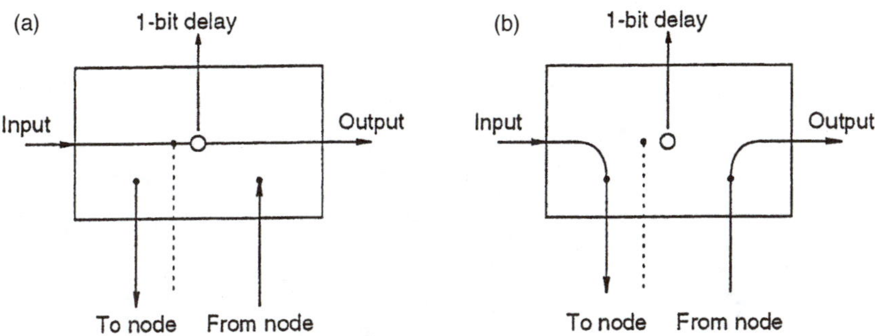

Figure 7.16 (a) Listen mode. (b) Send–transmit mode.

bit), re-transmission of the bits, indication of acknowledgment by changing the status of a bit, change of mode, etc. All these functions are performed by a repeater during the delay without affecting the bit-by-bit re-transmission of the frame.

In **send (transmit) mode**, a free token is captured by the node interface. Once the token is captured, the node can either copy the message or send a message into the ring. The repeater of that node has obtained permission to transmit via its access logic circuit interface. After this, the repeater goes into transit mode from listen to send during the one-bit delay and it must have its message data packet ready for the transmission waiting in the buffer. In this mode, the repeater receives bits from the node and re-transmits the frames bit by bit on its outgoing link or circuit. During the transmission, the bit of frame may appear on its incoming signals, as these frames are circulating around the ring. After the message data has been transmitted, the sender station interface looks at the last bit of its message (attached with the token) and, as soon as this bit is received by it, it changes its send mode into listen mode. The token is sent back into the ring.

When the message data packet (frame) is circulating around the network, the sender station keeps a copy of the transmitted message on its buffer and compares it with the transmitted one to check for any error. When the repeater is still transmitting the frames from that node, the bits of transmitted frame may have come back to the node, which typically will happen if the bit length of the token ring is smaller than the length of the frame. The repeater, after receiving the bits of its transmitted frame, will send them back to the ring after checking the status (acknowledgment from the destination node). If it receives the same bit it is transmitting, it will continue with the transmission, but if it finds it is getting bits it is not transmitting, it will accept them and store them in its buffer without interrupting the transmission. After it is done with the transmission, it will transmit the bits stored in the buffer by changing the status and mode appropriately.

These two modes provide all the necessary functions of the ring-based LANs. The repeaters are usually equipped with relay switches which are activated in the event of any failure or malfunctioning of a repeater or broken link. In any of these events, the affected node, its attached repeater, and both links are bypassed after the relay switch is activated. The ring-based LAN does not collapse, but that particular node is bypassed from the ring topology. We may consider this situation a third mode — bypassing mode — in addition to the listen and send modes as discussed above. During this bypassing mode, the bits bypass the bad node and its repeater and are transmitted to another repeater. Since they are bypassing this node, there will not be any delay introduced by this node, as the bits were not received at all by it. A typical connection of repeaters using a bypass relay switch is shown in Figure 7.17.

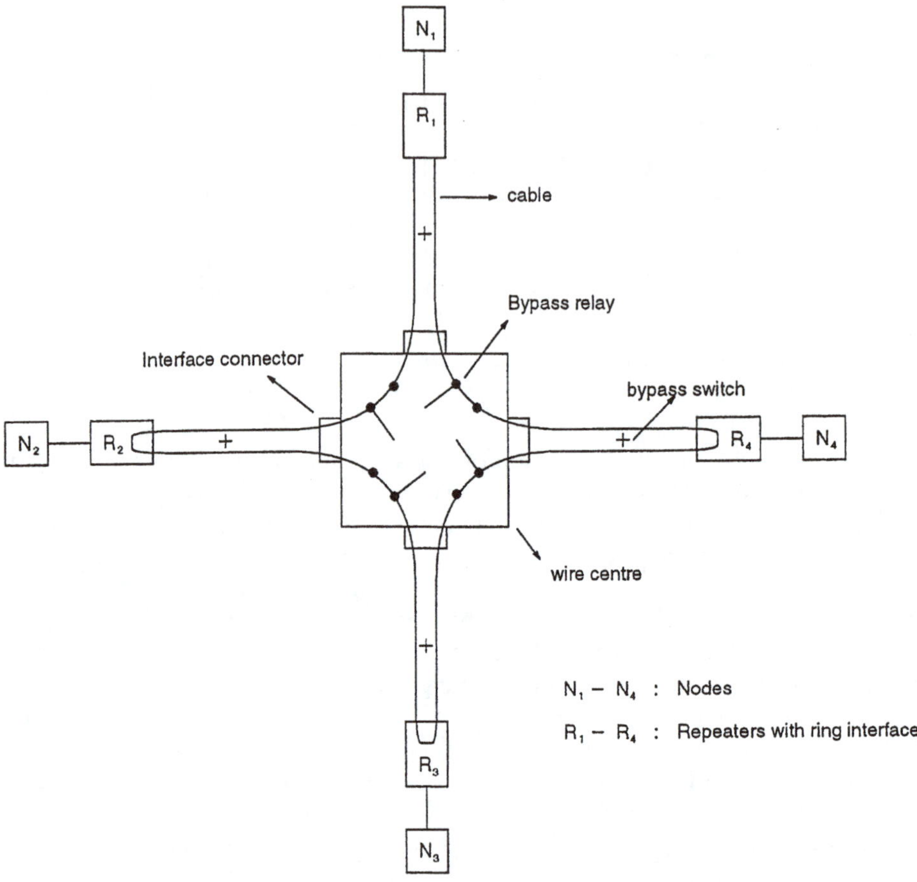

Figure 7.17 Bypass relay switch.

The ring-based LAN in normal conditions usually works in a highly decentralized way, but there are certain problems (related to token) which the nodes cannot solve by coordinating with each other. In these cases, the problem is usually solved by considering a **central** or **monitor** node. This node handles all the problems or faults related to tokens. These faults include loss of token, removal of token from the ring, duplication of the token, monitor going down, etc. A detailed discussion on these faults and how they can be avoided or controlled in ring-based LANs can be found in Section 8.2.5. Some of the material discussed above has been derived from References 1–4.

Contention ring: In a traditional ring system, there is only one token circulating around the ring all the time, and this may become a problem in terms of significant waiting time in the case of light network traffic applications. A different approach is adopted in ring-based LANs known as a **contention ring**, where no token is circulating. If any node wishes to transmit a data frame, it waits to see if any data frame with a token has passed over it. If there is none, it will send the data and append a free token frame at the end of the data frame. The source node will either remove the data frame from the ring after it has been received by the destination node or let the frame go around the ring for a few more times (depending on the protocols being used, it will remove the token).

During the time the data frame and appended (new) token are circulating around the ring, the token may become busy or remain free. If it becomes busy, it will carry the data from that node (requesting) and the LAN will behave like a token ring network. But, if it is free, it will come back as a free token. The source node removes it from the ring along with the data frame. Now, if any node wishes to transmit the data frame, it has to redefine the token contention configuration by creating a token and appending it at the end of its data frame. This scheme may be comparable to that of a token ring LAN except that it is intended for light network traffic applications and the selection between these two types of ring (token ring and contention ring) depends on parameters such as collision frequency, waiting time, network traffic, etc.

Slotted ring: Another version of a ring-based system is **slotted ring**, where time slots of fixed size or length are defined. These slots are represented as carriers for the data frames around the ring. Each slot has its own status indicator and, in the case of a free slot, the data frames are copied into the carriers of fixed sizes. If the length of the data is greater than the length of the carrier, it will be partitioned into packets of the size of the carriers (slots) and transmitted. For each slot, a bit can be set/reset to indicate the acknowledgment, and the source node must make it free for further transfer of data frames.

7.3.5 Media-based LANs

There are a variety of LANs available on the market, and not all of these LANs are standard. Each LAN has a different application and usage.

Depending on the type of LAN being used, we may expect that a LAN will support a connection of a few tens to tens of thousands of PCs, workstations, terminals, peripheral devices, printers, etc. LANs may be characterized on the basis of a number of parameters such as topology, access control technique, transmission media, etc. The bus or tree topology–based LANs are one of the earliest defined and include Ethernet (based on baseband bus), MITREnet (based on bus/tree broadband), and many other popular low-cost twisted-pair LANs available for PCs. But, in general, bus/tree-based LANs can be grouped into two classes: baseband and broadband (based on type of switching used). The baseband and broadband types of cables have already been discussed in Section 7.3. Now we will discuss two very popular LANs based on these types of cables, along with their applications. We will also describe some vendor LANs with their specifications for each category of cables.

Baseband and broadband are switching techniques defined over transmission media and are distinguished by the type of signaling (digital or analog) used for the communication. The baseband switching technique uses digital signaling and is supported by twisted-pair and coaxial cable, while broadband coaxial cable uses analog signaling using an RF modem. The remainder of this section highlights the main differences between these two switching techniques.

The difference between LANs, in fact, is defined in terms of a number of parameters such as topology, access media control, switching, signaling, etc., and as such we have seen different types of standard IEEE LANs. These standard LANs may be categorized as baseband or broadband LANs, depending on the switching and signaling techniques used.

Baseband LANs: A baseband LAN uses digital signaling. It does not use any analog carrier or frequency division multiplexing (FDM). It requires a bus topology, and the digital signals are transmitted using Manchester or differential Manchester encoding and get the entire bandwidth of the medium, thus explaining why it does not use FDM. It uses twisted-pair cable, coaxial cable, fiber optics, and microwave transmission media. It offers one channel and supports a variety of topologies such as ring, bus, and tree. It transmits both analog and digital signals in their original forms without using any modulations. This

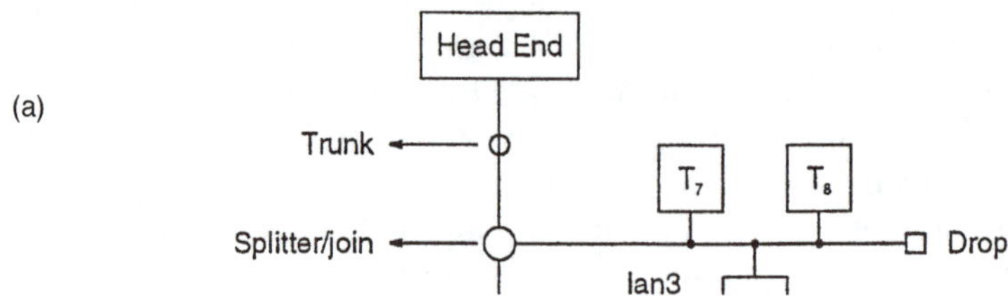

Figure 7.18 Baseband (bidirectional) or single-channel broadband.

is a bidirectional communication system (signals travel in both directions) and uses only one cable.

In baseband LAN transmission, the signal is directly impressed onto the underlying media in binary form via **Manchester encoding**. In this encoding scheme, logic 1 is described by low-to-high transition while logic 0 is described by high-to-low transition and, as such, it does not require any extra clock pulse for synchronization. This type of encoding is very popular with Ethernet LANs, IEEE 802.3 for baseband coaxial cable, and twisted-pair CSMA/CS bus LANs, and also for shielded twisted-pair bus LAN (MIL-STD 1553B). Another encoding technique which is equally popular and widely used is **differential Manchester encoding**. In this encoding scheme, transition is always defined at the center of each interval of the pulse in such a way that logic 1 will be represented by no transition at a pulse boundary while logic 0 will be represented by a transition at the pulse boundary. This scheme is widely used in IEEE 802.5 LANs using shielded twisted pair. It is simple and very inexpensive, as the layout of the cable is very simple. The only problem with baseband LANs is that these LANs offer a limited capacity over a limited distance. Other encoding schemes defined for digital signals include **non-return-to-zero-level (NRZ-L)** (high level for logic 0 and low level for logic 1), **non-return-to-zero-inverted (NRZI)** (logic 0 is represented by no transition at the beginning of an interval while logic 1 is represented by transition at the beginning of an interval), **bipolar with 8 zeros substitution (B8ZS), high density bipolar 3 zeros (HDB3), bipolar-AMI (alternate mark inversion)**, etc. The NRZ code usually is used for generating or identifying the digital signals of terminals and devices and is used in digital magnetic recording. The encoding scheme bipolar-AMI is a form of multilevel binary representation and is useful for error detection.

The baseband-based LANs may use any cable (coaxial, twisted-pair, etc.), and no modulation is required due to its support for only a very short distance (typically a few kilometers). It uses 50-Ω cable and supports multiple lines. For multiple access to the channel, it uses time division multiplexing (TDM). It does not divide the bandwidth for data communication but instead allocates the entire bandwidth to the channel, thus offering higher data rates. Only one piece of data can be transmitted at a time. The access method to a baseband system (usually Ethernet) is usually provided by a device known as a **transceiver**. This device receives the data from the node and defines the packets for transmitting over the bus. The packet contains a source address, destination address, user's data, and some control information. These LANs can only handle the data and are cheap and very simple. The shielded twisted pair offers more capacity than twisted-pair baseband, but it is more expensive. An unshielded twisted pair may allow the users to use the existing wiring system (telephone wiring, etc.). A general bus topology for baseband or single-channel broadband LANs is shown in Figure 7.18.

Some of the vendors of baseband LANs include Datapoint Corporation (Texas), Starnet Data Systems (Colorado), and Xerox Corporation (Connecticut). Typically, the baseband LANs from different vendors include the following specifications:

Topology: bus, tree, logical ring, star, twin ring
Access method: CSMA/CD, CSMA/CA, token passing
DTE interface: RS-232-C, RS-449, IEEE-488, RS-423
Number of connections: varies between 32 and 16,000

A brief description of some of the vendor-baseband LANs is given below.[40]

An **Omninet LAN** is a baseband LAN developed by Corvus Systems, Inc. In this LAN, after the occurrence of collision, each sender node waits a random amount of time before sensing or listening to the channel. The amount of waiting time before it tries again is known as the **backoff**, and it increases exponentially as the traffic increases. As a result of this, the frequency of occurrence of collision also increases. It uses the CSMA/CA variation of the CSMA protocol at a data rate of 1 Mbps over a segment of 33 m serving 64 nodes. Repeaters can lengthen the size of the network up to 1.2 km. An **intelligent network interface (INI)** known as a **transporter** (describing the lower four layers of OSI-RM) accepts the data from the device, packetizes it, and transmits it through sockets. In order to reduce the burden on memory management systems, it uses direct memory access (DMA) instead of using the computer's main memory.

An **Ethernet LAN** implements the collision detection version of CSMA (CSMA/CD) in hardware and uses an exponential backoff algorithm after the occurrence of collision. It offers a data rate of 10 Mbps over a segment of 500 m that can be extended to 2.3 km via repeaters, and it supports 1024 nodes. It uses coaxial cable. It has a very complex cable installation and requires expensive maintenance. Further, the average propagation time delay for any node in broadband LANs is twice that of baseband LANs and, due to this, the efficiency and performance of these LANs are reduced.

Broadband LANs: This type of LAN uses analog technology and a high-frequency modem to define carrier signals for the channel. Broadband systems are defined for transmitting data, voice, and video signals through cable television (CATV). A broadband LAN supports and carries a wide variety of traffic and offers very high capacity and data rates. The carrier signal generated by high-frequency modems is used for modulating a user's digital data. A digital signaling switching technique is used for data communication and allows the entire bandwidth to be used by the data signal. Typically, it offers a channel having a bandwidth of greater than voice-grade channel bandwidth (4 KHz). In broadband transmission, the multiple channels are modulated to different frequencies using FDM and, hence, the system needs a **headend**. The analog signals usually are in a radio frequency range of 5–400 MHz and typically use 75-Ω cable.

The control signaling provides a means for managing and controlling various activities in the networks. It is exchanged between various switching elements and also between switches and networks. Every aspect of the network behavior is controlled and managed by these signals, also known as **signaling functions**. With circuit switching, the popular control signaling has been **in-channel**. In this technique, the same physical channel is used to carry both control and data signals. It certainly offers advantages in terms of reduced transmission setup for signaling. This in-channel signaling can be implemented in two different ways: as **in-band** or **out-of-band** in-channel signaling.

In **in-band in-channel signaling**, the control signaling uses the same physical channel and frequency within voice-grade range. In this way, both control and data signals can have the same properties and share the same channel and transmission setup. Both signals can travel with the same behavior of the network anywhere within the network at any location. Further, any conversions (analog-to-digital or digital-to-analog) used in the networks will be used by both signals in the same way over the same path.

In **out-of-bound in-channel signaling**, a separate frequency narrow band is defined within voice-grade (4 KHz) for control signals and, as such, can be transmitted without even sending data signals. This way, the channel is continuously monitored. Due to lower frequency for control signals, the transmission speed of signaling is lower and the logic circuit to separate these signals is also complex.

Both of these methods of in-channel signaling have the drawbacks that the circuit-switched networks always introduce delay and the frequency bands for control signals are also very low. A new signaling technique known as **common channel signaling (CCS)** has been introduced which alleviates these drawbacks. The main philosophy behind this is to define two separate circuits for control and data signals. (The details of CCS can be found in Chapter 19).

The broadband systems use two channels for transmitting the data. One channel, known as an **outbound** channel (54–400 MHz), is used for data leaving the headend toward the network, while the other channel, an **inbound** channel (5–30 MHz), is used to carry the data from the network toward the headend. The headend can be considered a repeater which provides frequency conversions of signals. It is an analog translation device which accepts the data defined in inbound band channels and broadcasts it to outbound band channels. All the nodes send their data on inbound band channels and listen for their addresses on outbound band channels. There is a small band known as a **guard band** which separates these two bands. These two bands define a number of channels in their respective bands. Each channel can carry different types of data, e.g., voice, video, data, etc., at different rates. No multiplexing of voice mail signals with video signals will be carried out. The channels work on bus and tree topologies and are suitable for multi-drop operations.

The **headend** is basically a signal frequency converter which provides conversion between signals in a frequency channel. The data is always transmitted on an inbound channel and received on an outbound channel. Both channels may be defined in one cable, or two separate cables can be used. The headend is used in bus and tree topologies of LANs and covers only a limited distance (a few kilometers). The headend uses FDM for defining different channels corresponding to video, data, and audio signals over one circuit. It always operates in one direction. Typical applications of broadband include telemetry, integrated video and data, manufacturing, simultaneous operations in instrumentation, etc. The headend is used only in broadband transmission because the amplifiers' work is unidirectional.

In the broadband LANs, a single communication channel is shared by a number of nodes (machines) over the networks. A packet transmitted from any node contains a destination address and also allows multicasting (addresses of group of nodes over the network) or broadcasting (the addresses of all the nodes over the network) of the same packet. Broadband LANs contain a path device and interface units. The interface unit contains a radio frequency (RF) modem which accepts the data from a device and converts it into the RF of the inbound band. The network interface packetizes the digital data into analog RF signals.

Another form of broadband is known as a **single-channel broadband** or **carrier band**. Here the entire frequency spectrum of cable is allocated to a single transmission circuit for analog signals. It supports bidirectional transmission and uses bus topology. It does not require any amplifiers or any headend. It offers very low attenuation at low frequency and uses some form of FSK modulation.

CATV systems provide a frequency bandwidth of 300–400 MHz and support 39 to 58 television channels each of 6 MHz. The cable used by CATV is of 75-Ω impedance and provides modulated signals for larger distances. The digital signals need to be converted to analog signals (via a radio-frequency modem) to be transmitted over CATV. Here the entire bandwidth is divided into channels of 6 MHz, which are further divided into sub-channels of smaller bandwidth. These sub-channels are multiplexed using either time division multiplexing (TDM) or frequency division multiplexing (FDM).

Most of the broadband LANs use the CSMA/CD access technique and are based on bus topology (used in standard Ethernet LANs). 3Com has developed a broadband LAN based on bus topology using a token passing access method (which is basically defined for token ring LANs) known as LAN/1. The token passing method was initially proposed by Datapoint for token ring LANs. Broadband transmission can be achieved on two types of cable: single (split-channel) and dual cable. The **single or split-channel cable** carries network traffic in both directions on one cable and supports different bands of frequencies, e.g., forward band, return band, and crossover band. In the **forward band mode**, it carries the signals from the headend to the network in the frequency bandwidth of 160–300 MHz. The **return band mode** is opposite to the forward and carries the signals from nodes of networks to the headend in the frequency bandwidth of 5–110 MHz. Finally, the **crossover band mode** of a single cable acts as a guard band to avoid any electrical interference in the frequency bandwidth of 110–160 MHz. Some of the material presented here has been derived from References 1–6, 10, 18, 33, and 34.

A typical broadband LAN single-cable layout consisting of coaxial cable, splitter, join amplifiers, feeders, and other components is shown in Figure 7.19(a). In this LAN, the single cable carries the transmitted and received signals over the same channel. The bandwidth of the cable is split into two frequency signals: forward (receiving) and return (transmission). In order to avoid any interference between transmitted and received signals, the entire bandwidth is divided into three sub-bands: return, forward, and guard. The return band allows the device to transmit its signals toward the headend. The return is used to indicate that the signal is returning to the headend. The devices receive the signals on the forward band and forward the signals to the receiving device; hence, the name forward is derived. The guard band carries no signal and is basically used to separate the return bands from each other. The dual-cable broadband LANs do not need the guard band, as separate cables are used for transmitting and receiving the signals.

There are three main schemes for split channel available, known as **subsplit**, **midsplit**, and **highsplit**. Each of the schemes defines a different frequency range for inbound and outbound channels.

The **dual cable** provides different segments of cable for the transmission and reception of signals in opposite directions. The cable distribution begins with a main trunk cable. At the midpoint of the cable, a passive/active transmit–receive crossover point (network loop) divides the network into a **transmit half** and a **receive half**. The forward cable segment is basically used as a receive cable or line in the frequency range of 5–300 MHz, while the return cable segment is used for transmitting signals in the same frequency range of 5–300 MHz. This type of cable does not need a headend but may use amplifiers. One of the main advantages of dual cable (broadband mode) is that there is a savings of frequency bandwidth, as the same frequency range can be used on different cable segments.

Since we have to use two separate cables, the expensive is very high, as we now need two ports on our computer systems (terminals, PCs, or workstations) and more amplifiers to use dual cable. Signals transmitted from any node propagate in only one direction along the transmit line distribution tree toward the network loop. Signals arriving at the network loop cross over to the receive line distribution tree and propagate back to the individual nodes. The transmit and receive signals are physically separated by dual-cable distribution and are electronically isolated by the unidirectional propagation characteristics of LAN cable components. A general LAN architecture of dual-cable broadband is shown in Figure 7.19(b).

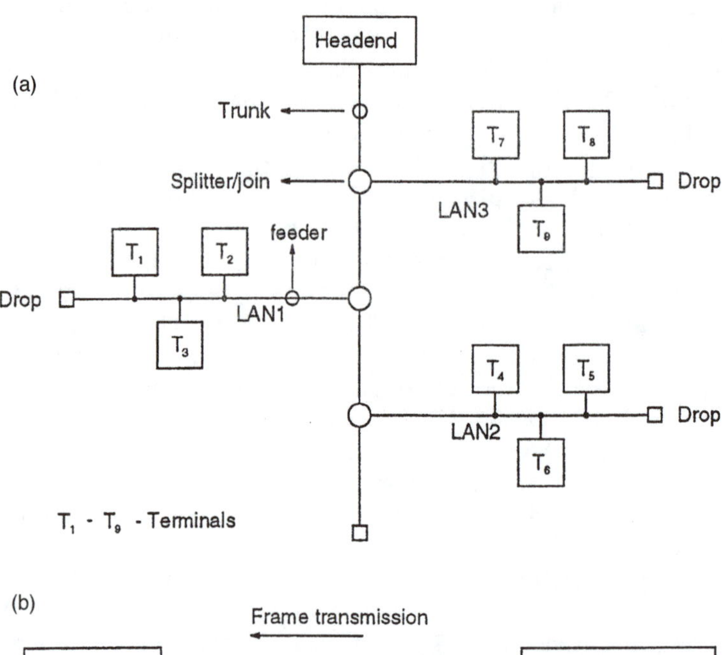

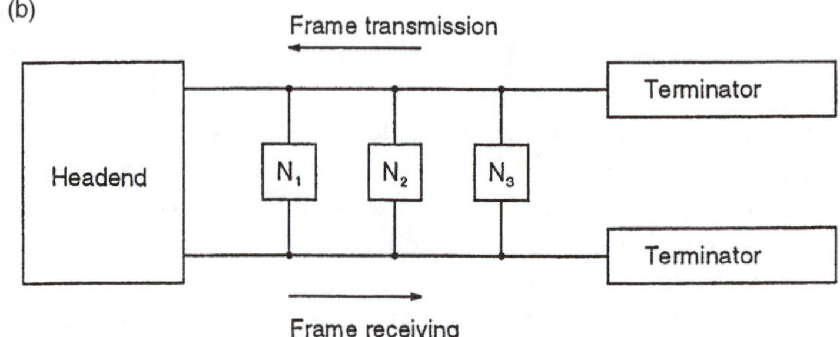

Figure 7.19 (a) Single-cable broadband LAN. (b) Dual-cable broadband LAN.

It is worth mentioning that a TV channel occupies a 6-MHz bandwidth and, as such, 35 channel networks will require a 300-MHz coaxial bandwidth, while 52 channels will require a 400-MHz coaxial bandwidth.

Typically, the broadband LANs offered by various vendors include the following specifications:

Topology: bus, logical ring, tree, physical ring
Access method: CSMA/CD, CSMA, token passing, FDM
DTE interface: RS-232-C, RS-449, V.35, IEEE-488
Supporting protocols for DTE: BSC, HDLC, TTY
Number of connections: varies between 32 and 62, 535

We briefly discuss two broadband LANs below.[40]

The **Net/One** network was developed by Ungermann-Bass and provides an inbound band of 59.75–89.75 MHz and outbound band of 252–282 MHz. These two bands carry up to five channels and offer a data rate of 5 Mbps. The **network interface unit (NIU)** includes an RF modem and can connect 24 devices. It conforms with the RS-232, RS-449, and V.35 serial. Each channel translator can handle up to 300 NIUs per channel, so the maximum number of devices which can be connected to the network is 36,000 (300 × 24 × 5).

The **Local/Net** LAN was developed by Sytek and defines an inbound band of 40–106 MHz and an outbound band of 196–262 MHz. It offers a 128-Kbps capacity channel which is suitable for low-speed, low-traffic devices, such as PCs. It supports both synchronous and asynchronous devices and covers a distance of 35 miles one way, supporting 55,000 nodes.

Some useful Web sites

LAN standards can be found at *http://standards.ieee.org*, high-speed LANs are at *http: www.zyxel.com*, and faster LAN/WANs and other products and links are available at *http://ww.wkmn.com*. Wireless LANs, WANs, and MANs standards are available at *http://www.wireless-nets.com*, *http://web.syr.edu*, *http://www.host.ots.utexas.edu*, *http://www.iol.unh.edu*, *http://www.rhowireless.com*, *http://www.wlana.com*, *http://www.hpl.hp.com*, *http://www.cis.ohio-state.edu*, and many others. FDDI LAN standards are available at *http://www.server2.padova.ccr.it*.

Some of the LAN vendor sites include *http://www.lucent.com*, *http://www.fcw.com*, and *http://www.3com.com*. Links to various manufacturers of wireless LANs can be found at *http://www.wlana.com*, and *http://www.tansu.com.au* (for security issues).

Links related to Ethernet and high-speed Ethernet LANs are found at *http://www.ots.utexas.edu*. For gigabit Ethernet, a number of Web sites have been introduced. Gigabit Ethernet Alliance is at *http://www.gigabit-ethernet.org*; a Gigabit Ethernet consortium established at an interoperability lab has sites at *http://www.iol.unh.edu*, *http://www.3com.com*, *http://www.ciol.com*, and *http://www.lan.net.au*; WWW consortia like *http://www.biometrics.org*, and the International Institute for Internet Industry Benchmarks are at *http://www.iiiib.org*. Some of the vendor sites for gigabit Ethernet include *http://www.cisco.com*, *http://www.3com.com*, *http://www.onmitron-systems.com*, and *http://www.allied-telesyn.com*. A newsgroup for Ethernet is at *comp.dcom.lans.ethernet*.

Enterprise routers and software products are available at *http: support.3com.com*, *http://www.networking.ibm.com*, *http://www.zoomtel.com*, *http://www.playground.sun.com*, and many other.

Bridges, routers, and protocols associated with these are available at *http://www.globalknowledge.com* and *http://www.datatech.com*. Books covering these in more detail are at *http://www.booksoncomputing.com*, *http://www.bookspool.com*, and other sites.

Gateways, internetworking devices, and service providers are available at *http://www.oneworld.com*, *http://www.ngn99.com*, *http://www.cisco.com*, and *http://www.techweb.com*. Also, the IETF charter for PSTN and Internet internetworking has information at *http://www.ietf.org*.

References

1. ISO 7498, *Open System Interconnection — Basic Reference Model*, Information Processing Systems, 1986.
2. Tanenbaum, A.S., *Computer Networks*, Prentice-Hall, 1989.
3. Stallings, W., *Handbook of Computer Communications Standards*, Howard W. Sams and Co., vol. 1, 2nd ed., 1990.
4. Black, U., *Data Networks: Concepts, Theory and Practice*, Prentice-Hall, 1989.
5. Schlar, S.K., *Inside X.25: A Manager's Guide*, McGraw-Hill, 1990.
6. Meijer, A. and Peeters, P., *Computer Network Architecture*, Computer Science Press, 1982.
7. Kauffles, E.J., *Practical LANs Analyzed*, Ellis Horwood, 1989.
8. Digital Equipment Corp., *DECnet Digital Network Architecture, General Description*, May 1982.
9. Digital Equipment Corp., *DNA Digital Data Communication Message Protocol (DDCMP), Functional Specification*. 1982.

10. Markeley, R.W., *Data Communication and Interoperability,* Prentice-Hall, 1990.
11. Mollennauer, J.F.,"Standards for metropolitan area network. *IEEE Communication Magazine,* April 1988.
12. Hemrich, C.F. et al., Switched multi-megabit service and early availability via MAN technology. *IEEE Communication Magazine,* April 1988.
13. ISO 8802.6 standard, Metropolitan Area Network (MAN).
14. Minoli, D., *Telecommunications Technology Handbook.* Artech House, 1991.
15. Valovic, Metropolitan area networks: A status report, *Telecommunications,* July 1989.
16. IEEE standard 802.6, Distributed queue dual bus (DQDB) subnetwork of MAN. 1991.
17. Ross, F.E., An overview of FDDI: The fiber distributed data interface. *IEEE Journal on Selected Areas in Communications,* 7(7), Sept. 1989.
18. Hutchison, D., *Local Area Network Architectures.* Addison-Wesley, 1988.
19. IBM, Systems network architecture: technical overview. IBM Corp., Research Triangle Park, NC, No. GC30-3073, 1985.
20. Institute of Electrical and Electronics Engineers. Computer Society, *IEEE Draft 802.1 (part A), Overview and Architecture.* IEEE, Oct. 1990.
21. Institute of Electrical and Electronics Engineers, *IEEE Standard 802.1: MAC Bridges.* IEEE, Sept. 1990.
22. Institute of Electrical and Electronic Engineers, *Local Area Networks — Logical Link Control (LLC).* IEEE 802.2 (ISO 8808.2).
23. Institute of Electrical and Electronics Engineers, *Carrier Sense Multiple Access with Collision Detection (CSMA/CD) Access Method and Physical Layer Specification.* ANSI/IEEE 802.3 (ISO 8802.3), 1990.
24. Institute of Electrical and Electronics Engineers, *Token Passing Bus Access Method and Physical Layer Specifications,* ANSI/IEEE 802.4 (ISO 8808.4), 1990.
25. Institute of Electrical and Electronics Engineers, *Token Ring Access Method and Physical Layer Specifications,* ANSI/IEEE 802.5 (ISO 8808.5), 1989.
26. Institute of Electrical and Electronics Engineers, *IEEE 802.5 Appendix D; Multi-Ring Network (source routing),* 1991.
27. CCITT recommendation X.3, *Packet Assembly/Disassembly (PAD) in a Public Data Network,* CCITT, 1984.
28. CCITT recommendation X.25, *Interface between Data Terminal Equipment (DTE) and Data Circuit-Terminating Equipment (DCE) for Terminals Operating in Packet Mode and Connected to Public Data Network by Dedicated Circuit,* CCITT, 1984.
29. CCITT recommendation X.75, *Terminal and Transit Call Control Procedures and Data Transfer System on International Circuit between Packet-Switched Networks,* CCITT, 1984.
30. CCITT recommendation X.121, *International Numbering Plan for Public Data Network,* CCITT, 1984.
31. Huffman, D.A., "A method for construction of minimum redundancy codes," *Proceedings of IRE 40,* Sept. 1952.
32. Dijkstra, E.W., "A note on two problems in connection with graphs," *Numerical Mathematics,* 1, 1959.
33. Spragins, J., Hammond, J., and Pawlikowski, K., *Telecommunications Protocols and Design,* Addison-Wesley, 1991.
34. Stallings, W., *Local Networks: An Introduction,* Macmillan, 1987.
35. Schwartz, M., *Telecommunication Networks: Modeling and Analysis,* Addison-Wesley, 1987.
36. Halsall, F., *Data Communication, Computer Networks and Open Systems,* Addison-Wesley, 1992.
37. IEEE Document P802.11-93/028, Standards Working Group IEEE P802.11, Wireless Access Method and Physical Layer Specifications, March 1993.
38. Olivett reference manual, "The cordless LANs for enterprise," 1990.
39. AT&T reference manual, "Wireless Personal Communication," 1990.
40. Marney-Petix, V.C., *Networking and Data Communication,* Reston Publishing Co., 1986.
41. Simonds, F., *LAN Communication Handbook,* McGraw-Hill, 1994.

References for further study

1. Naugle, M., *Local Area Networking,* McGraw-Hill, 1996.
2. Bertsekas, D. and Gallager, R., *Data Networks,* Prentice Hall, 1992.
3. Kadambi, J., Crayford, J., and Kalkunte, M., *Gigabit Ethernet,* Prentice-Hall, 1998.
4. Seifert, R., *Gigabit Ethernet,* Addison-Wesley, 1998.
5. Spurgeon, C., "Ethernet configuration guideline," *Peer-to-Peer Communication,* 1996.

chapter eight

IEEE LANS

8.1 IEEE 802 recommendations (OSI-RM)

Most of the LANs available on the market comply with the international standards organizations, and the Institute of Electrical and Electronic Engineers (IEEE) has been responsible for defining various LANs and standards for the lower two layers of OSI-RM. A list of the standards defined by the IEEE 802 committee is given in Figure 8.1 and is discussed below. Each of the LANs has been derived from appropriate IEEE 802 subcommittees (CSMA/CD, token bus, token ring, MAN). The standardization of LANs has made LAN products available on the market at a reasonable price.

The IEEE 802 committee was established in 1980 with an objective to define standards for local area networks (LANs), as there were different types of LANs available on the market and it was a difficult task for users and vendors to choose a particular LAN for their specific applications. After a rigorous study and detailed survey, the IEEE 802 committee defined four standards in 1985. These standards were accepted by the American National Standards Institute (ANSI) in 1985 as **American National Standards (ANS)**. The committee also submitted these standards to the International Standards Organization (ISO) in 1987, which finally approved these as international standards in 1989 as ISO 8802 and defined an open standards interconnection (OSI) reference model (a layered architecture) for LANs.

The IEEE 802 committee also approved two types of LANs: **carrier sense** (bus/tree topology) and **token passing** (token ring topology). Within bus topology, two access control techniques known as CSMA/CD and token bus were approved for both switching transmissions (baseband and broadband). In order to get some idea about the standardization process and various tasks defined for each of the subcommittees, we discuss the roles and functions of each of the subcommittees under IEEE 802 in the following section. Each of the subcommittees was asked to study and propose a specific function within OSI-RM, and their subsequent proposals were approved by the IEEE 802 committee as a universal architecture for all types of LANs.

8.1.1 IEEE 802.1 high-level interface (HLI)

This standard offers high-level interface and is common to all LANs.[1] It defines internetworking, network architecture, and network and system management, and it gives an overview of other standards and explains their relationship with the OSI-RM. It also discusses higher-layer protocols, various aspects of internetworking and management, and associated protocols defined for different LANs.

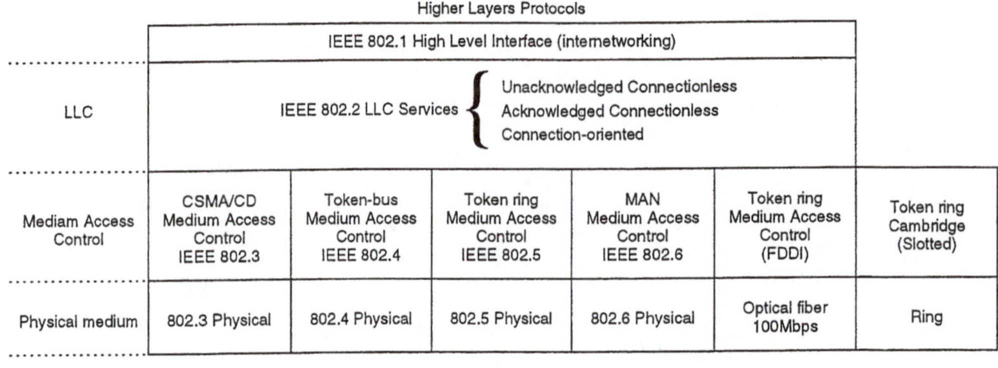

- 802.3 Physical Medium :
  - Baseband Coaxial: 1, 5, 10 and 20 Mbps
  - Unshielded Twisted Pair: 1, 10 Mbps
  - Broadband Coaxial: 10 Mbps
  - Bus/Tree Topology
- 802.4 Physical Medium :
  - Baseband Coaxial: 1, 5, 10 Mbps
  - Carrierband (broadband) Coaxial: 1, 5, 10 Mbps
  - Optical Fiber: 5, 10, 20 Mbps
  - Bus Topology
- 902.5 Physical Medium :
  - Shielded Twisted Pair: 1, 4 and 16 Mpbs
  - Ring Topology
- 802.6 Physical Medium :
  - Optical Fiber: 5 Mbps
- FDDI Physical Medium :
  - Optical Fiber: 100 Mbps
  - Ring Topology
- Cambridge Ring Physical Medium :
  - Shielded Twisted Pair: 10 Mbps
  - Ring Topology

Figure 8.1 IEEE LAN standards.

8.1.2 IEEE 802.2 logical link control (LLC)

As we mentioned earlier, the OSI-RM is composed of seven layers, while the IEEE 802 committee defined two sublayers of the data link layer (second layer of OSI-RM) within the model as logical link control (LLC) and media access control (MAC).[2] Each of these sublayers has specific functions, but these two sublayers together provide the entire functions of the data link layer. One of the main reasons for dividing the data link layer into two sublayers was to provide support for different transmission media and offer different data rates. It also allows control accesses to share channels among autonomous computers and provides a decentralized access scheme. Further, it offers a more compatible interface with WANs, since the LLC is a part of the HDLC superset, as it has a similar control field. By dividing the data link layer of OSI-RM into two sublayers, MAC and LLC, the LAN becomes a model of eight layers. The MAC sublayer encompasses standard IEEE LANs 802.3, 802.4, and 802.5.[3-5] The LLC sublayer, on other hand, includes the IEEE 802.2 standard.

The MAC sublayer accepts the data from LLC and calculates a cyclic redundancy check (CRC) and then transmits the encapsulated data (frames). The transmit media access management sends the serial bit stream to the physical layer, which listens to the medium. It halts the transmission if it senses a collision and tries to re-transmit it; at the same time, it sends a jam signal over the medium. It inserts PAD fields if the LLC field length is less than the minimum length. When it receives a frame from another node, it first computes CRC and accepts the frame after its address (DA) matches the one defined in the frame and sends it to LLC. The receive media access management receives the serial bit stream from the physical layer and discards if it is an invalid frame (if the length of the frame is less than the minimum length).

The LLC standard corresponds to the upper sublayer of the data link layer of OSI-RM. It uses an asynchronous balanced mode and supports connectionless services by using an unnumbered information frame and connection-oriented services by using two unnumbered frames. It also allows multiplexing via its SAPs. The data link layer, in fact, provides a communication architecture and, as such, this may be considered the top sublayer of communication architecture of LANs. It defines the control procedures used in both types of service, connection-oriented and connectionless, which can be used for any MAC. This is the lower sublayer of the data link layer of OSI-RM and FDDI. It provides a platform for exchanging the data between LLC users over different MAC-based links (like bus/tree, ring topologies).

The LLC standard is common to all LANs and offers three types of service: **unacknowledged connectionless**, **acknowledged connectionless**, and **connection-oriented**. It supports three classes of transmission configurations: point-to-point, point-to-group, and broadcast. In connectionless services, the datagrams are transmitted over the network without setting up any logical connection. In contrast to this, the connection-oriented services require three phases: connection establishment, data transfer, and connection deestablishment or connection termination. The class of transmission configuration is supported and established by the data link layer.

As mentioned above, the LLC is common to all LANs, while each LAN has its own MAC which depends on the access control procedure being used. Each of the LANs is standardized on the basis of its MAC protocol.[6-13]

Different transmission media offer different characteristics and data rates for different distances. For example, twisted wire offers a data rate of 50–60 Kbps over a distance of 1 km. Baseband coaxial cable offers a data rate of 50 Mbps over a distance of 1–3 km, while its broadband version offers a data rate of 350 Mbps over a distance of 10 km. Fiber-optic cable offers 1 Gbps over a distance of 10 km. Here the distances mentioned for the transmission media represent one segment without any repeater or amplifier and any intermediate nodes.

8.1.3 IEEE 802.3 CSMA/CD LAN

This subcommittee developed a CSMA/CD bus standard that is very similar to Ethernet.[3] This standard defines a MAC interface for Ethernet LANs and also its own operation for accessing the LAN. It defines the physical layer for baseband coaxial operating at data speeds of 1, 5, 10, and 20 Mbps and also specifies the electrical and physical details about these media required for data communication. In fact, it includes both data link and physical layers. As discussed earlier, the data link layer consists of two sublayers: LLC and MAC. This standard, in fact, defines an interface for LLC with its higher layers. Physical signaling is used as one of the top layers of the physical layer which provides a common interface between MAC and different media-dependent interfaces.

This standard protocol is used with Ethernet and supports both baseband and broadband at the physical layer. Baseband transmission offers data rates of 1, 5, and 20 Mbps, while broadband offers data rates of up to 10 Mbps and CATV transmission speeds. The physical layer is connected to different types of transmission media and, as such, it must include a number of interfaces and attachments. The physical signaling (interface point for physical layer) component is interfaced with a **physical media interface attachment (PMIA)** via a physical interface device known as an **attachment unit interface (AUI)**. This PMAI (different for different media) is connected to a medium via an appropriate **medium-dependent interface attachment (MDIA)**, as shown in Figure 8.2.

The PMIA and MDIA together may define a media attachment interface unit (MAIU) which contains the necessary hardware and software support for these interfaces between

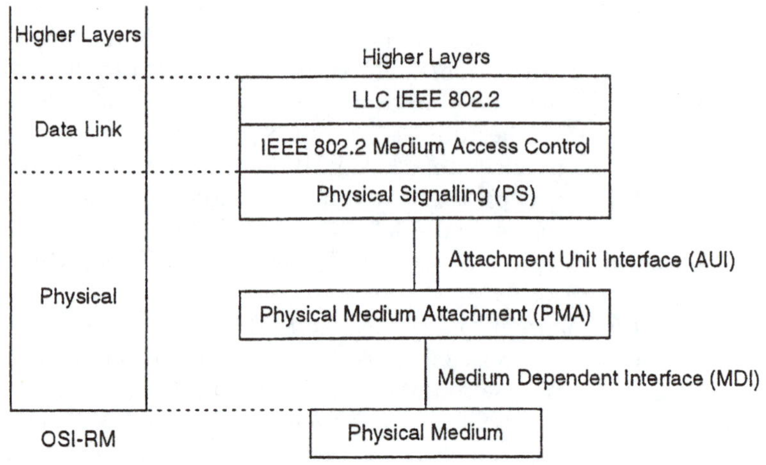

Figure 8.2 IEEE 802.3 architecture.

the media and physical layer. The physical layer also provides common interfaces (for all the media) with the MAC sublayer which simply provides a means for exchange of bits and some timing controls between them.[8,12] The specifications of the media attachment and other information for CSMA/CD are given below:

Cable media:	50-Ω coaxial
Bit rate:	10 Mbps
Slot time:	512-bit times
Interframe gap:	9.6 μs
Jam size:	32-bit
Maximum frame size:	1518 bytes (octets)
Minimum frame size:	512 bytes (octets)

Various topologies supported by this LAN include linear, tree, segmented, etc. The CSMA/CD bus LANs are usually used in technical and office environments and are also known as Ethernet (as described earlier). This LAN typically offers a 10-Mbps data rate and operates over coaxial cable. Other cables are also supported by it; some of the popular cables used in this LAN include 10 BASE 2 (thin wire), 10 BASE 5 (thick cable), 10 BASE T (twisted-pair drop cables and mainly used for hub over star topology), and 10 BASE F (optical-fiber drop cables and mainly used for hub over star topology). Specifications for 10 BASE 2 and 10 BASE 5 cables are given in Table 8.1.

Gigabit Ethernet: This network offers higher speeds and high performance in an LAN backbone. It is based on the same concept as that of Ethernet, including its frame format, management framework, and collision detection and avoidance methods. It uses a fiber

Table 8.1 10 BASE 2 and 10 BASE 5 Cable Specifications

Parameter	10 BASE 2 thin Ethernet	10 BASE 5 thick Ethernet
Maximum cable segment length	185 m	500 m
Maximum number of nodes	30	500
Maximum node connecting spacing	0.5 m	2.5 m
Attachment	BNC T or standard transceiver cable and BNC-equipped transceiver	Transceiver and transceiver cable; length of cable = 50 m

channel signaling mechanism and can extend from 500 m (multimode) and 2 km (single mode). The existing twisted-pair cabling can support only a few meters (20 to 100). It provides LAN support at a data rate of 1 Gbps, which is easy to manage, but it does not support WANs. ATM, on the other hand, also provides LAN support at data rates of DS1 (1.5 Mbps) through OC-48 (2.5 Gbps), built-in quality of service, and support of WANs. It seems that the price difference between ATM switches and conventional LAN Ethernet switches is very small, and hence many of the major vendors are producing both switches and other products.

The switched 10-Mbps and 100-Mbps Ethernets have already captured the market, and it is hoped that the introduction of gigabit Ethernet may pose tough competition for ATM. Nearly 80% of all desktops are using Ethernet LANs, and that high usage includes different types of LANs (fast Ethernet, gigabit Ethernet). Their support for gigabit WAN transport makes ATM the dominant switching technology for high-performance LAN connectivity. For applications (real-time, multimedia, video, and voice) that require quality of service in LANs, the ATM ensures end-to-end quality of service communications. At this time, the real-time protocol and resource reservation protocol defined within IP are offering quality of services for Ethernet networks. The ATM is the best LAN to use for a WAN migration path. Thus, a server based on LAN can have a connection anywhere from 25-Mbps ATM to 622-Mbps ATM. The ATM has already standardized OC48 (2.5 Gbps) supporting a full-duplex mode of operations, offering much higher data rates than gigabit Ethernet (half-duplex).

8.1.4 IEEE 802.4 token bus LAN

This standard LAN uses the token passing access technique to access the medium and data and is based on bus topology.[4] It also uses different types of media-dependent interfaces which offer different data rates, and each one of these interfaces uses the common analog signaling provided by the physical layer. As usual, the IEEE 802.4 MAC services are defined for its LLC and higher layers. These services provide a platform for transmitting and receiving the protocol data units from higher layers and provide support for error control and flow control. The physical layer provides one common interface between it and the MAC layer for all the types of media interfaces and supports different media-dependent interfaces. The media-dependent interfaces typically include hardware and software. The physical attachments for CSMA/CD, token bus, and token ring LANs can be found in References 8–10. In token bus–based LANs, the physical media are connected to a physical layer via a **drop cable** and **cable coupling unit attachment (CCUA)**, which provides physical connection with the cable, as shown in Figure 8.3.

Another connection based on drop cable is known as a **multi-drop cable** connection, which is different from the drop cable connection (providing point-to-point). In a multi-drop cable connection several nodes are connected to same link. The efficient way of

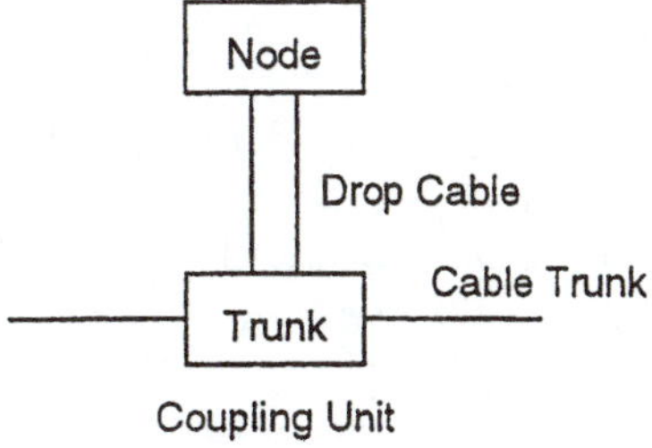

Figure 8.3 Coaxial cable connection.

establishing topology is to simply compute the shortest path between nodes on the link. This path is known as the **minimum spanning tree** of the network. Two nodes on a multidrop connection cannot use the link at the same time. In order to utilize the link efficiently, a suitable algorithm needs to be used which will determine the next node to use the link. In general, this problem of contention is reduced by providing two lines to the computer; further, the load-balancing algorithm may provide an efficient way of utilizing the link.

This LAN also offers baseband and broadband transmission at physical layers. The baseband transmission data rates of 1 and 5 Mbps and broadband transmission data rates of 1, 5, 10, and 20 Mbps are supported by the physical layer. A logical ring is defined by the MAC protocol, and the transmission of packets over this logical ring is controlled by a token bus protocol. This LAN is flexible, as the stations can be added and deleted at our discretion and the priority can also be assigned, thus providing a different type of topology of LANs irrespective of the locations of the stations. The algorithms for managing and controlling such LANs are quite involved and complex.

The priority in token bus LANs is assigned to the frames, and the frames with the highest priority use the maximum capacity of the network. The lower-priority frames get the network only when it has enough capacity. There are four classes of priority levels: in decreasing order, 6, 4, 2, and 0. When a node receives the token, it transmits the data from frames of the highest-priority queues (i.e., class 6) until the transmission becomes equal to the **token hold time (THT)** or there is no data to transmit. If the **token rotation time (TRT)** is less than the token rotation time of class i, then the data is transmitted in the order of priority. The token cycle time is defined as follows: If $n \times$ token hold time is greater than $n \times$ maximum token rotation time, then only class i data will be transmitted. But, if $n \times$ token hold time is less than $n \times$ maximum token rotation time, the class 6 data is assured of the requested capacity while other class data will get the remaining capacity.

The token maintenance protocols deal with a number of functions such as inserting a node, deleting a node, inactive node, etc., and each of these functions are implemented by separate commands defined within the protocols.

These LANs also provide interaction between MAC and LLC via primitives and parameters (request, indication). The current standard (IEEE 802.4) offers one common physical layer interface with MAC (independent of media) and four MAC media-dependent interfaces providing support to only four categories of broadcast media:

1. Baseband coaxial cable (75-Ω) at 1 Mbps over bus topology
2. Baseband coaxial (75-Ω) at 5 and 10 Mbps over bus topology
3. Broadband coaxial (75-Ω) at 1, 5, and 10 Mbps over directional bus/tree topologies
4. Optical fiber at 5, 10, and 20 Mbps over optical fiber

8.1.5 IEEE 802.5 token ring LAN

The IEEE 802.5 standard is based on token passing (defines token ring access control (MAC)) and uses a ring topology.[5] It defines a physical layer for use of a token on a sequential medium (twisted pair at 1, 4, and 16 Mbps) and differential Manchester signaling. This LAN defines both the physical and data link layer together as one unit. The IEEE 802.5 specifies the services offered by LLC to its higher layers and the services offered by the physical layer to its MAC sublayer. The interface at the LLC layer provides a platform for transmission of receiving of protocol data units from higher layers and supports error control and flow control. The physical layer offers a common interface between the physical layer and different MACs for exchanging of bit streams between them, signaling method (encoding, etc.), and data rates. On other hand, for each medium, a separate attachment or interface (describing the electrical, mechanical, and functional behavior of the medium) is connected to the common interface attachment.

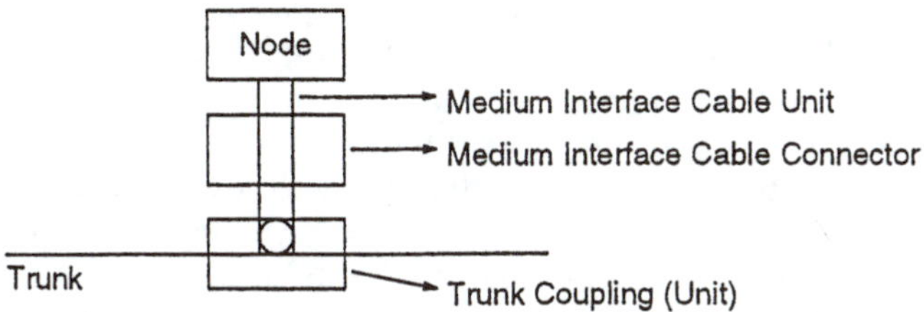

Figure 8.4 Token ring cable connection.

When we compare token ring LANs with CSMA (Ethernet) LANs, we find the following advantages of token ring over CSMA LANs:

- Token ring LANs offer deterministic behavior for the users. If we know the number of nodes, the maximum size of the packet, and the maximum propagation time around the ring, it is possible to determine the maximum time interval between two consecutive frames, and hence we can determine the time slot for the next frame to be sent. In CSMA, due to collision, it is not possible to determine the time between consecutive transmissions.
- Token ring LANs offer a point-to-point configuration for data communication, and as such it is easy to use optical fiber, as it offers taps or sockets for attaching the nodes. The CSMA does not offer a point-to-point configuration and, as such, a high-speed link such as optical fiber cannot be used with it.

The IEEE 802.5 LAN provides one physical ring (in contrast to many logical rings in the case of the token bus LAN). It supports baseband transmission over twisted or coaxial cables. Twisted pair cable offers data rates of 1, 4, and 16 Mbps, while coaxial cable offers 4, 20, and 40 Mbps.

The physical layer is connected to the cable at the **cable coupling unit interface (CCUI)** via a hardware **medium interface connector attachment (MICA)**. This interface provides different segments of cable on each side, known as physical-to-MICA and MICA-to-CCUI, as shown in Figure 8.4. It is important to know that the cables and various coupling units and attachments are not a part of the standards but provide a physical connection between a node and cable coupling unit. The connection typically consists of two 150-Ω shielded twisted-pair cables between the physical layer and CCUI. The signal transmission takes place over different pairs of cable in opposite directions. Various functions of token ring LANs are performed by repeaters attached to each node using **listen** and **transmit** modes. In the event of failure of repeaters or broken links, the LAN switches to bypassing mode, where this function is provided by cable coupling unit interfaces (CCUIs). Eight classes of priority access levels are defined in this LAN. It uses twisted-pair and duplex optical-fiber cables. A special node known as an active monitor node defined by the protocols makes sure that all the functions of the token ring network are performed properly, such as network initialization, token generation, removal of frame circulation continuously around the ring, avoiding duplication of tokens, etc. IEEE 802.5 has also defined a multi-ring topology for LANs.[14]

8.1.6 IEEE 802.6 metropolitan area network (MAN)

The standard LANs discussed above (CSMA/CD, token bus, and token ring) provide data communication within a limited range, but the length of LANs can be enhanced by using bridges between identical LANs. The LANs are usually defined for data communication, resources sharing, etc., within a bandwidth of 10 Mbps (Ethernet), 4 or 16 Mbps (token ring), or 100 Mbps (FDDI). The Ethernet and token ring LANs are typically used for data communication and do not offer enough bandwidth and speed to video or voice communications. Although one may try to compress the data and send them over these LANs, this will restrict the use of the LAN for other data and will cover only a very small distance. For providing both audio and video services over a long distance in metropolitan cities, IEEE defined another standard LAN ISO 88802.6,[15] known as a metropolitan area network (MAN). A chapter on MANs can be found in Reference 16. This LAN provides media access protocols for sharing of resources to users over a large distance within a metropolitan city using coaxial cable or fiber transmission media. The media access protocols of existing LANs (Ethernet, token ring, etc.) by themselves cannot cover a greater distance, and further, the speed of transmission of data over these links is limited to only 10 Mbps or 16 Mbps. The MAN provides a transmission speed of data over 100 Mbps and supports data and voice over the same link. For definitions, architecture, and standards of MAN, readers are referred to References 11, 15, and 17–20.

LANs are usually controlled by a single user and, as such, can be considered dedicated to one user. These LANs have a limited geographic distance within the premises of an organization or university. In order to make LANs cost-effective, the LANs of different organizations may be connected to each other via MANs, making those LANs into public networks which will offer services to those interconnected organizations over a larger geographic distance within metropolitan cities. MANs consist of dedicated circuits which are distributed throughout the metropolitan area or city at various organizational locations.

For data communication within premises or organizations, LANs are suitable, but for voice communications, we may prefer private branch exchange (PBX). Further, the PBXs of each of the organizations may be interconnected over shared media. Although dedicated links DS1 or DS3 may also be used, these links are very expensive. Thus, the MAN provides both integrated voice and video data and, as such, may be used for interconnecting different types of mainframes, WANs, and PBXs for both public and private networks. The interconnection of mainframes provides a channel-to-channel link for data communication, while interconnection of PBXs provides private network-based applications.

The MAN can transmit such data as medical images and data, graphics, CASE tools like CAD/CAM, and digital video signal (compressed) used in teleconferencing, etc. For voice communication, the delay introduced by the transmission or network should be minimum, and the IEEE 802.6 standard sets a maximum limit of this delay for voice communication as 2 ms in MAN. The protocols defined for MAN must deal with this delay to provide proper synchronization between sending and receiving sites for voice and video signals with some kind of security.

Internetworking can also be achieved between WANs, mainframes, etc. Based on distance and applications, the concept of MAN (high-speed network providing interconnections to LANs, WANs, etc. and supporting data, video, and audio information) has been extended to different types of networks developed lately, including **fiber distributed data interface (FDDI)**, **distributed queue dual-bus (DQDB)**, **broadband-integrated services digital networks (B-ISDN)**, and **switched multi-megabit data services (SMDS)**. The FDDI has been used as a backbone network in the majority of the campus environments and hence can be considered a private network providing high-speed interconnections to LANs, WANs, etc.

The DQDB is defined by IEEE 802.6 Working Group[15] and is based on the switching used in the concept of queued packet and synchronous circuit exchange (QPSX) and used as a MAN/LAN standard. It supports isochronous, connection-oriented, and connectionless services simultaneously. The DQDB MAN uses two buses in such a way that two signals flow in opposite directions. It is independent of physical media and supports the transmission of plesiochronous digital hierarchy (PDH) systems (34, 45, and 145 Mbps) and has synchronous digital hierarchy (SDH) based transmission and data rates (of gigabits per second).

The communication switching in DQDB MAN integrates video, voice, and data and transmits these over circuit-switched and packet-switched links; it is used mainly for communication over a larger geographic distance/area. It also offers sharing of telecommunication resources and facilities to its users and, as such, is also known as a high-speed public communication network within that area. The information within this network is transmitted within slots. A slot consists of a header of five octets and an information field of 48 octets (ATM cell of 53 octets: 48 octets of information and five octets of header is derived from the DQDB slot). The slots are identified by inserting VCI values in the slot header and are controlled by a slot generator. There are two types of slots defined in DQDB: pre-arbitrated and queue-arbitrated. The non-isochronous information is transported with queue-arbitrated slots.

8.1.7 Other IEEE 802 recommendations

The IEEE 802.7 and IEEE 802.8 groups basically work together for broadband cable systems and use fiber optics as transmission media which may be used in various LANs for higher data rates and bandwidths.

IEEE 802.9 was specified in 1986, and its main objective is to define an architecture and interface between desktop devices and LANs or ISDNs.

The IEEE 802.10 standard for LAN security is known as the **Standard for Interoperable LAN and WAN Security (SILS)** which deals with the problems encountered during interoperability of LANs. According to IEEE 802.10 recommendations, it must include provisions for authentication, access control, data integrity, and confidentiality.

With **authentication**, users will not be allowed to access the data frames which are addressed to a different station; i.e., users will be able to access only those frames which are addressed to them. This is usually provided only in conjunction with integrity services. **Access control** defines the control for limited use of the resources (files, servers, bridges, gateways, remote services, etc.). This is also provided in conjunction with integrity and authentication services.

The **data integrity** (for connectionless) deals with the situation where the data frame cannot be modified by anybody before it reaches the destination node and usually depends on external key management services. **Confidentiality** is achieved by using encryption to mask the data information frames to the users by an appropriate key pattern. It offers multiple confidentiality algorithms and depends on the external key management service.

8.1.8 Other proprietary LANs (non-IEEE)

The above was a discussion on OSI-RM-based LANs and IEEE 802 recommendations. There are proprietary LANs which are based on a layered approach but do not follow the OSI-RM. One of the reasons is that the OSI-RM does not provide a well-defined model and also does not use some of the modern concepts of structured programming. The following are some of the criticisms of OSI-RM which have been identified by computer network professionals:

- The session layer does not contribute any useful functions in most applications.
- The application layer is not well defined and, as such, is nearly an empty layer.
- The data and network layers are very complex and perform most of the network functions, and, as such, these layers need to be split.

- OSI-RM is a very complex model and offers confusing operations, which makes the implementation inefficient.
- OSI-RM includes duplicate functions and services (e.g., addressing, error control, flow control, etc.) being offered or supported by more than one layer within the model.
- The original OSI-RM does not provide connection-oriented and connectionless services, but these are supported by the newer model.
- The features provided by some of the layers are not quite obvious; e.g., virtual terminal handling was originally associated with the presentation layer but has been assigned to the application layer; data security and encryption and network management was not considered at all in the original model but are now being supported by it in one way or another.
- The original OSI-RM was geared more toward communication professionals without mentioning any relationship between computing and communication; also, efficient implementation concepts such as interrupt-driven data structures were not included.

The OSI-RM, when used under UNIX operating, offers only four layers (with communication protocols of TCP/IP): application, transport layer (TCP), network (IP), and data link layers. The data link layers are connected by a physical link of cable.

Cambridge token ring: In the mid-1970s, the Computer Laboratory at Cambridge University, England, developed a LAN based on slotted ring by Wilkies and Wheeler. It connects a number of time-sharing computers and a number of shared servers (name server, printing server, file server, etc.).[17] Different types of terminals can also access the services provided by a number of terminal terminators connected to the ring. The connection between users' terminals and computers and sharing of various services are usually handled by a bank of processors connected to the ring.

This LAN, based on ring topology, defines an empty slot (in contrast to a fixed size of token used in token ring and token bus LANs) and offers a data speed of 10 Mbps. Each node is assigned with a slot for sending only a single packet; i.e., the slot can be used by the node only once for sending one packet, after which it has to pass the control to the next node. The second time slot must be passed unused. The third time slot will be used again. This means that a node can use at most one out of three slot rings at any time. In general, we can say that a node will use a ring slot once in $n + 2$ in an n-slot ring, provided no other node is making a request to use it. In the case of more than one node requesting the ring slot, each will get it at least once in $m + n$ slot, where n represents the number of slots while m represents the number of nodes making requests at the same time.

The Cambridge token ring LAN has been standardized by the British Standard Institute (BSI) only and is used widely in companies and organizations within the U.K. A proposal for standardization by an international organization has also been under consideration for a while.

The MAC protocol used in this LAN is based on an **empty slot** concept. In this technique, a small packet is continuously circulating around the ring topology. At the beginning of the packet, a bit is reserved for indicating the status of the packet (full or empty), similar to the bit used in token ring LANs for indicating acknowledgment. Any station wishing to transmit the packet will capture the token, change the status of this bit to full, and append the data into the token; i.e., the data packet is attached to the token. No station can use the token more than two times; after capturing it once, it must send the token back onto the ring by making it an empty token. Then, if the station wants to send the data packet again, it has to capture the token again. This MAC protocol supports a number of transmission media such as twisted-pair wires and fiber-optic cable (offering 100 Mbps).

Various hardware chips for Cambridge token ring LANs are available on the market. There are a number of Cambridge implementations available.

Fiber distributed data interface (FDDI) token ring: The standard for FDDI was initially proposed by the Accredited Standards Committee (ASC) X3T9.5 and accepted by ANSI and defines a MAC sublayer and physical layers of LANs for optical fiber LANs. This is defined for high-speed ring LANs. It also uses token ring algorithms similar to those of IEEE 802 token ring LANs, with minor differences. In FDDI, the node generates and places the token immediately after the frame has been received, and, as such, we may have more than one frame in the ring at any time. This is in contrast to token ring, which generates and puts the token onto the ring after the leading edge of its transmitted frame. As a result of this, the FDDI offers higher efficiency for larger ring LANs. Further, the FDDI offers flexibility to support a variety of high data rates for both continuous stream and bursty traffic. It offers excellent security, and it is virtually impossible to wiretap without detection. An excellent overview of FDDI is given in Reference 21.

As stated above, all the standard LANs use a common standard LLC sublayer defined by IEEE 802.2, and optical fiber uses the same LLC for data communication. FDDI is expected to offer a data rate of 100 Mbps and support of 500–1000 nodes over a distance of 100–200 km. The FDDI access control technique is similar to the IEEE 802.5 token ring protocol with a minor difference in the way the free token is generated by the source node after the transmission is complete.

FDDI uses ring topology and offers a high speed of transmission. The control regarding various functions of an FDDI LAN working in the ring at each node (up to the MAC layer from the PMD) is provided by a separate layer known as layer management (LMT), which basically resides within the MAC and both sublayers of the physical layer. Various other controls for overall functions of the networks are defined in station management, which includes LMT as well and provides control for LLC and higher layers of an FDDI LAN. The ring interface includes both the physical layer and MAC sublayer. A typical ring topology for an FDDI LAN can be seen in Figure 8.5.

The FDDI LAN offers a few common functions with IEEE 802.6. It does not support priority. The FDDI uses a very small size of token which is circulating around the ring. Any node wishing to transmit the message waits until it detects a token. It captures the token by terminating the token transmission over the ring. When the token is completely captured, the frames are transmitted over the ring. The node puts the free token back onto the ring immediately after sending the last data frame within the maximum transmission duration period. In contrast to this, in the IEEE 802.5 token ring, the free token can be put back onto the ring only after the source node receives its transmitted data frame back at the node. In other words, in FDDI several data frames can be sent by stations having tokens, while token ring allows only one frame at any time.

Design issues in FDDI LANs: The FDDI allows multiple frames onto the ring at any time, and this feature is implemented by using two timers: token rotation timer (TRT) and token holding timer (THT). The TRT defines the time between arrival of the free token as a measure of the network's traffic. The THT offers limits of duration within which frames can be sent while the node holds the free token. It is given by the difference of the TRT and a fixed portion of the total circulation time. This time is usually derived from the internode initialization process, and a part of the overall bandwidth is reserved for synchronous traffic.

During the time the token is captured and the token transmission is terminated, the ring does not have any token, and, as such, no node will be able to transmit until a node

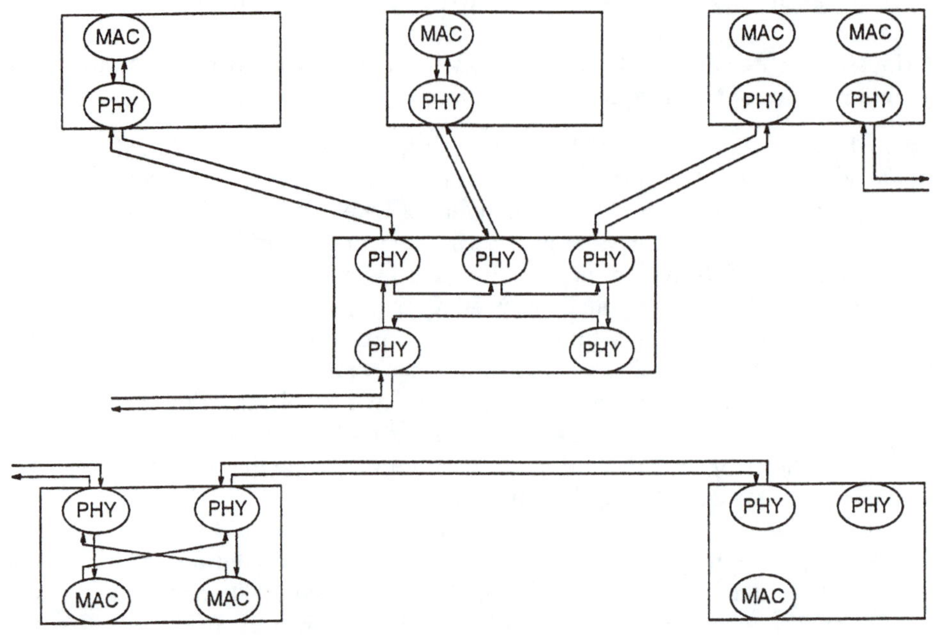

MAC — Medium Access Control
PHY — Physical Protocol Layer

Figure 8.5 FDDI ring topology.

captures the token. After the transmitted frame is received by the destination node, it is removed from the ring by the transmitting node. It then issues or inserts a token back into the ring. The FDDI does not define any busy token. The concept used in FDDI fundamentally differs from that of the token ring (IEEE 802.5) in the sense that here the bit length of the ring is always greater than the length of frame transmitted from the station, and, as such, the token can be issued and inserted by the transmitting node even before it receives the last bit of the transmitted frame for its removal.

Typical FDDI specifications include the following:

- Optical-fiber channel offers a data rate of 100 Mbps, 1300-nm signal
- One optical-fiber ring can accommodate up to 1000 nodes
- Nodes can be spaced up to 2 km apart
- Ring circumference can be defined up to 120 miles
- Offers features similar to IEEE 802.5 token ring
- Permits multiple users to occupy the ring at a time

In general, a node will take off its own transmission from the ring and establish a free token. Typically, the normalized throughput of FDDI LANs is 60 (for a distance of 200 km) to 70 (for a short distance of 1 km) Mbps out of 100 Mbps. FDDI uses IEEE's LLC protocols and is usually used as a backbone for heterogeneous LANs (i.e., a flexible local network supporting premises-wide data communication service at a high speed with high capacity). This backbone LAN offers resource sharing among devices which may be placed at a distance in the premises. Although a single LAN interconnecting all the data-processing devices within premises can be defined as a backbone LAN, this type of LAN may pose problems of reliability, cost, capacity, etc. Instead, we should use low-cost LANs within departments or divisions and then connect these low-cost LANs with high-cost, high-speed backbone LANs within the premises.

The FDDI LANs offer high data rates and capacity and lower cost; they have been adopted as backbone LANs in many organizations and universities. They connect different LANs which can operate in parallel and offer greater processing efficiency. FDDI typically filters network traffic and forwards only those messages which are destined for other LANs. It covers a greater distance and offers higher bandwidth. Due to the requirements of high bandwidth and long-distance transmission, optical fiber or microwave transmission media for backbone networks are preferred. The connection to backbone networks may require bridges, routers, or gateways, depending on the applications and architecture of the LANs being used, and it can be built with leased wideband digital carriers such as T1 (1.5 Mbps) and T3 (45 Mbps). The FDDI-II (a second version of FDDI) divides the FDDI capacity into 16 dynamically programmable 6.144-Mbps channels which support T-carriers (popular in the U.S.).

8.2 Design issues in standard LANs

In order to understand the detailed design of standard LANs, we present a brief overview of the service primitives and various LLC services which are common to LANs. More details on these can be found in Chapter 11. As discussed earlier, standard LANs (CSMA/CD, token bus, token ring, MAN, and FDDI) have a common LLC but different MACs. In fact, due to the different working principles of media access techniques, we have LANs in the names of those access techniques, as shown in Figure 8.6. In this section, we discuss the details of various media access design issues, physical layer and MAC sublayers, and frame formats of different standard and non-standard LANs.

8.2.1 Layer service primitives

The OSI-RM's second layer (data link) has been divided by the IEEE 802 committee into two sublayers, logical link control (LLC) and media access control (MAC). The layers or sublayers of the LANs interact with each other via service primitives through **service access points (SAPs)** between them. Each layer's service is defined as a set of primitives.

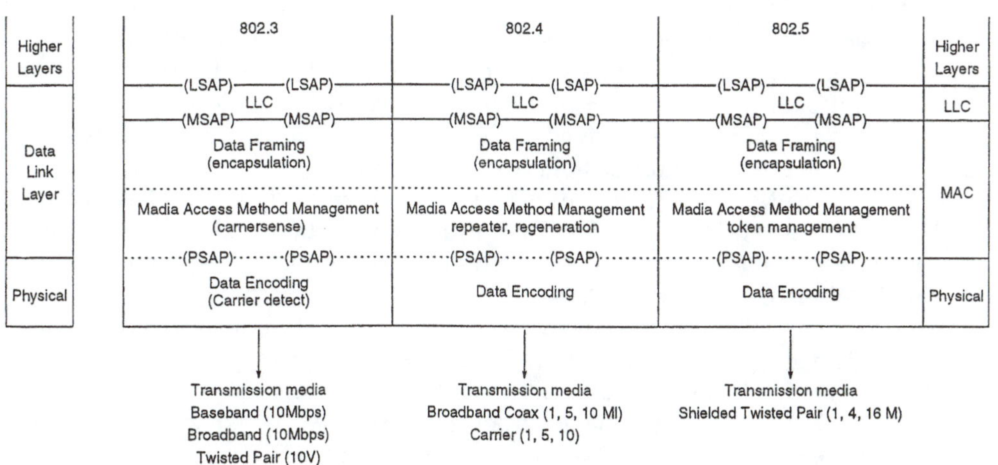

Figure 8.6 Access technique-based standard LAN.

These primitives are concise statements containing the service name and associated information to be passed through SAPs. We discuss the primitives and parameters provided by physical signaling for the MAC sublayer and also the primitives and parameters offered by MAC to the LLC sublayer in more detail in Chapter 11. These primitives can also be found in References 6–9, 11, 12, 19, 22, and 23.

The LLC defines its own set of primitives and parameters for the network layer. LLC supports three types of services (unacknowledged connectionless, acknowledged connectionless, and connection-oriented services), and it defines a different set of primitives and parameters for each of these services. There are four standard service primitives types which are used by different layers for both control and data transfer: **request**, **indication**, **response**, and **confirmation** (already discussed in previous chapters). We will describe in brief the functions of each of these below.

The **request primitive** is initiated by the users to invoke a primitive of a service layer. It uses an application of an underlying provider layer or remote peer protocol layer. The service layer passes an event to the user about the receiving of the request. The **indication primitive** to the user defines an acknowledgment by passing some event regarding the invoking of some procedure. It alerts the user through provider about incoming message. Finally, the service layer passes the **response** of the initiated request by passing the appropriate event to the user via **confirmation** (indicating end-to-end configuration). In other words, the remote user has successfully processed requested operations.

In the **eight-layer LAN model**, the confirmation primitive type simply indicates the confirmation of delivery of the underlying service, and there is no guarantee that the request has been processed. Similarly, the response attribute primitive type in OSI-RM indicates end-to-end acknowledgment of a successful operation. By looking at these conceptual differences between OSI-RM and eight-layer LAN models (IEEE 802), we can say that the OSI-RM is typically a connection-oriented network while the LAN model is a connectionless network. However, both of the models support both of these services (connection-oriented and connectionless).

To understand the exchange of service primitives between layers via SAP, we consider the LLC, MAC, and physical layers. The LLC sublayer, after receiving a packet from its network layer at its LSAP, sends a service primitive type MA-DATA.request to its MAC sublayer via its MSAP for transfer to its peer LLCs. This primitive has enough control information and the LSAP address. The MAC sublayer defines a frame for each packet it receives from its LLC by attaching a destination address, length of fields, etc., and sends a primitive type MA-DATA.confirm to its LLC via the same or different MSAPs indicating the success or failure of the operations over the request. For receiving any frame from a remote node, it uses another service primitive type MA-DATA.indication to indicate to the LLC sublayer a frame it has received from another MAC address and also the type of addressing, i.e., individual, multicasting, or broadcasting.

Similarly, the MAC sublayer uses service primitive types Ph-DATA.request, Ph-DATA.indication, Ph-DATA.confirm, Ph-CARRIER.indication, and Ph-SIGNAL.indication to interact with the physical layer via physical layer service access points (PhSAPs). For more details, refer to Chapter 10.

These primitives with the above-mentioned attributes can be implemented in a variety of ways. The service primitives can be used in different ways to support various services offered by the model. In some cases, confirmation from the local MAC sublayer is sufficient, while in other cases, confirmation from a distant MAC is required. Confirmation from a distant service provider can also provide confirmed services.

The peer protocol layers in different layers within the model communicate with each other by exchanging **protocol data units (PDUs)**. These PDUs, which are defined between each pair of layers, are implemented differently across the interface between them. For example, at the physical layer the bits are encoded using appropriate encoding techniques which will be understood by both source and destination nodes. At the data link layer and also at higher layer levels, the packets or frames are handled (instead of bits) by respective protocols. Each of the layer protocols, when passing the information (service data unit) to lower layers, adds its header (containing its address and control information) and sends it as a PDU to its lower layer, which accepts this as its SDU. When the frame arrives at the receiving side, it goes through the same layers, starting from the lower side to the upper layers, and at each of its peer layer protocols, the protocol of the layer removes the headers from the PDUs until it reaches the destination node. It is important to know that the data link layer protocol is the only protocol which adds both header and trailer to the PDU it receives from the network layer and defines this as a **frame**.

In order to distinguish between the data passed from lower layers to higher or vice versa with its peers, the data passed from the higher layer to its lower layer is known as a service data unit (SDU). PDUs are exchanged across peer layers while SDUs are packaged with the header and passed down as SDU.

8.2.2 LLC interfaces

The LLC sublayer offers a common interface to various MAC sublayer protocols and, in fact, is independent of the media access control strategies. The LLC PDU has the formats shown in Figure 8.7.

The LLC PDU includes destination service access point (DSAP), source service access point (SSAP), a control field, and the users' data. Each address field consisting of eight bits uses seven bits for addressing and one bit for control. In the case of DSAP, the first bit is a control bit and defines the type of address; e.g., if it is 0, it corresponds to one MAC address (of that node). If its value is 1, it represents a group of broadcasting addresses of MAC nodes. Similarly, the first bit of SSAP indicates whether the PDU is a **command** or **response** frame. The control field identifies a particular and specific control function. It is either one or two octets, depending on the type of PDU. The primitives between higher layers and the LLC are always given a prefix of **L**, while the primitives between MAC and LLC sublayers are given a prefix of **M**.

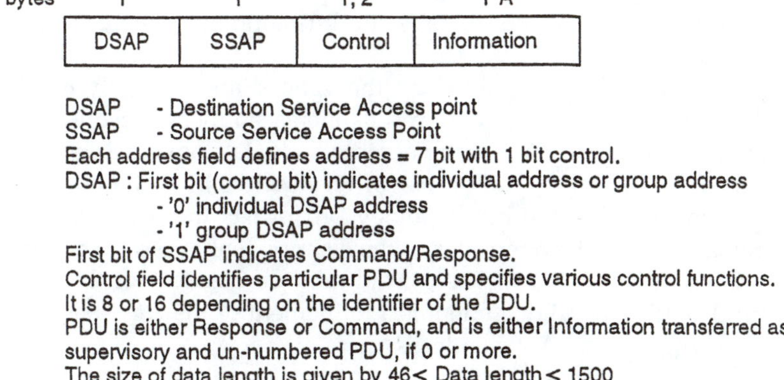

Figure 8.7 LLC protocol data unit formats.

The set of primitives and parameters for connection-oriented services is divided into three categories: **connection establishment**, **data transfer**, and **connection deestablishment**. For each phase, we have a set of primitives and parameters which are used for the exchange of the data between the LLC entities with appropriate peer LLC entities. It is worth mentioning that out of four primitive attributes (request, response, indication, and confirmation), the **response** attribute is not used in the latest recommendations.

These primitives will discard all the frames, and LLC does not support the recovery of these discarded or lost data units. The higher-layer protocols have to provide for recovering these lost or discarded frames. These primitives use parameters such as source address, destination address, and the amount of data the LLC entity can handle, and these can be set to any value. In the event of a zero value, no data will be accepted by the LLC entity. Likewise, the network layer defines its own primitives for controlling the data it is receiving from the LLC sublayer.

8.2.3 IEEE 802.3 (ISO 8802/3) CSMA/CD vs. Ethernet

The original Ethernet LAN (still used in many applications) developed by Xerox PARC is based on the concept of carrier sense multiple access with collision detection (CSMA/CD) and uses coaxial cable for its bus topology. The specification document published jointly by Digital, Intel, and Xerox (DIX) introduced Ethernet to the manufacturer by encouraging the adoption of a *de facto* industry standard. It offers a data rate of 3 Mbps. The bus uses a impedance terminator or reflector at its ends to remove the data from the bus after it has been received by the destination node. Each segment of coaxial cable is terminated by a terminator at its ends and corresponds to one Ethernet LAN segment connecting the devices and PCs via taps.

The LAN topology defines the physical connection for various nodes (host or mainframe, PCs, databases, etc.) Various specifications to compare different LAN topologies include data rate, maximum segment length, total length of the network, maximum nodes per segment, minimum node spacing, physical media, coax impedance, IEEE specification, topology, signaling technique, and others. Ethernet LANs are usually denoted by 10 B*x*, where 10 indicates the data rate and *x* denotes the spacing between the repeaters in meters. 10 B5 was the first Ethernet system to use 10-mm diameter coax, and the coax is connected via vampire tap. 10 B2 uses 7-mm diameter coax. The coax is connected to a network interface card by BN tee connector. On the other hand, 10 BT uses American Wire Gauge (AWG) #24 unshielded twisted pair, which is connected to the hub NIC and to the hub via RJ-45 telephone connector. The maximum length for 10 BT is 100 m, and it can be extended to 500 m using repeaters. It uses the bus topology where all the nodes are connected to a hub.

The length of a LAN is determined by the vendor and typically can be of a few kilometers. The length of a LAN segment can be increased by connecting it with another Ethernet segment via repeaters. A repeater accepts the data from one segment and passes it on to the next Ethernet segment. The use of terminators at both ends of an Ethernet segment simply ensures that the bus topology broadcasts the message data to all the Ethernet segments connected via repeaters, which simply enhances the range of an Ethernet LAN, as shown in Figure 8.8(a). The coaxial cable connection for this LAN is shown in Figure 8.8(b). The IEEE 802.3 document is a refinement of the DIX specification, which brings it into line with other LAN standards.[3]

IEEE 802.3 CSMA/CD LANs: The Ethernet LANs were introduced as a *de facto* industry standard offering a data rate of 10 Mbps. With the advent of optical-cable transmission, a data rate of 100 Mbps is possible. In fact, a few attempts of defining experimental LANs

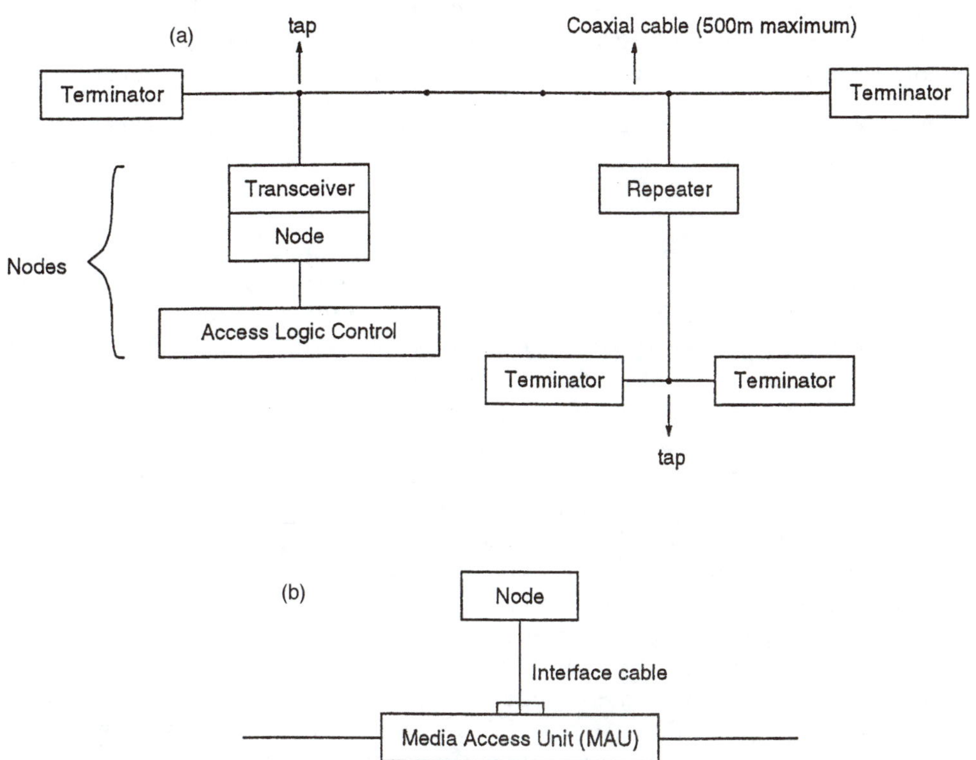

Figure 8.8 (a) Ethernet topology. (b) Coaxial-cable connection.

based on CSMA concepts using fiber have resulted in LANs such as **Fibernet**, **light bus system (LIBSY/PERNE)**, and **Ethernet**. In each of these LANs, the existing switching and connecting components in the transmission have already been replaced by the optical components. The Fibernet LAN uses optical cable and offers a data rate of 150 Mbps. LIBSY LAN uses a unidirectional ring topology and is based on the token access technique. It offers a data rate of 128 Kbps to 1 Mbps.

Ethernet is a very popular LAN and has been adopted in many LAN protocols. It is available in a variety of data rates and supports different transmission media. The following is a list of Ethernet-based LANs which have been introduced:

10 Base 2 (10-Mbps baseband with 200-m coaxial cable, Thinnet)
10 Base 5 (10-Mbps baseband with 500-m coaxial cable, Thicknet)
10 Base T (10-Mbps baseband with 100-m twisted pair)
10 Base FL (10-Mbps baseband with 2-km fiber)
100 Base T (100-Mbps baseband with 100-m twisted pair)

In the existing Ethernet LANs, drawbacks have been identified such as poor installation and poor performance with CSMA under heavy traffic conditions. The use of optical transmission can offer a higher data rate up to a distance of 4 km without using any repeaters. The ends of Ethernet LANs are connected by optical transceiver interfaces which support up to eight nodes.

Another highly advanced LAN using CSMA concepts is WAGNET. This LAN offers a minimum of 10 MHz and a maximum of 400 MHz. It uses double coaxial cable (send and receive) in tree-like structures. It uses both FDM and TDM for its medium. The FDM divides the segments up to a 390-MHz bandwidth to support various WAGNET services, while the segments are allocated to these services based on TDM (CSMA/CD polling method). We have already discussed a few other LANs, such as Starlan and ARCNET.

Performance parameters: We will now discuss the widely accepted performance parameters (cost per user, performance, ease of installation, interconnectivity) which can be used to compare LANs.

1. **Cost per user:** This parameter is based on hardware expenses, primary cables, network interface cards, and other charges related to installation and administrative activities. Looking at the working of these LANs, it has been found that IBM token ring has the highest cost per user, followed by twisted pair, standard Ethernet, ThinNet, Starlan, and ARCNET (in decreasing order).
2. **Performance:** This is a very generic parameter representing the bit rate and access methods used. Token passing is certainly suited for a large number of users at peak traffic, thus offering high percentage utilization of a network's capacity. The CSMA/CD, on the other hand, offers efficient utilization of network capacity at low traffic. Ethernet (all types) offers data rates of 10 Mbps (CSMA/CD); IBM token ring, 4 Mbps (token passing access); ARCNET, 2.5 Mbps (token passing access); and Starlan, 1 Mbps (CSMA/CD).
3. **Ease of installation:** This parameter is concerned with the topology, which determines the ease of installation. The star topology being used in twisted pair, Ethernet, Starlan, and distributed star topology of ARCNET allows easy deletion and addition of nodes into LANs. The standard Ethernet and ThinNet use bus topology, while IBM token ring uses a physical star logical ring topology (flexible and fault-tolerant).
4. **Interconnectivity:** This parameter is concerned with the ease with which particular LAN vendor products are available for connecting similar or different types of LANs. The Ethernet offers the highest connectivity, followed by IBM token ring and Starlan. ARCNET offers the lowest connectivity.

Working of CDMA/CS LANs: In this type of LAN, each node is connected via a transceiver and a tap (defined on the cable). The transceiver is a hardware-connecting device which has an electronic circuit which provides the necessary logic to access the cable. It accepts the data from the cable and broadcasts to all the connected nodes and also receives the data from the cable and passes it on to the node. In both cases (transmission and receiving), the data is handled serially. The tap and transceiver together provide a physical and electrical interface and connection between the nodes and the cable. The CSMA/CD protocols provide access to these hardware interface attachments. Each node has its own hardwired physical address used by the protocols.

The CSMA/CD protocol offers resource sharing among the connected users and, at any time, only one node can use the LAN for sending the frame over it. Any node wishing to transmit a frame has to first listen or sense the channel (cable) and transmits the frame only when it is free. If it is not free, it will wait and keep retrying (after a random amount of time) to see if the channel is free. During this time, it is possible that another node has also found that the channel is free. If both nodes try to transmit their frames at the same time, there will be a collision in the channel.

Due to this collision, the original data packets are corrupted and the protocol withdraws both the frames from the channel and tries to re-transmit the same frame(s) after a random amount of time. A transmitter which has noticed the collision (after it notices that the data on the channel is not the same as it has transmitted) will send a jam signal (a unique bit pattern of 32 bits) over the channel. This jam signal being transmitted by one of the transmitters is an indication to other nodes (about the collision) which are trying to transmit data frames by listening to the channel. The access control logic circuits recognize this jam signal and, as such, do not try to transmit the data frames. Instead, these nodes wait for a random amount of time. The random amount of waiting time for retrying is usually generated by suitable algorithms which also resolve the problem of contention.

CSMA/CD MAC sublayer: As discussed above, in CSMA/CD LAN, a station listens to or senses the channel before sending any message data and, after it finds a free channel, it transmits the message data over the channel. Each station is equipped with a device known as a **network access unit (NAU)** or a **media access unit (MAU)** (in the terminology of the IEEE 802.3 sub-committee), as shown above in Figure 8.8(b). It senses or listens to the channel to see if there is any traffic on the network. It is possible that two NAUs may try to send the message data over the network at the same time after finding it free at the time of listening or sensing the network. In this case, a collision will occur which will corrupt both the messages.

CSMA media access design issues: When a collision has occurred, both nodes withdraw their data frames and, after some time, have to sense or listen to the media again for transmitting the same frame(s). During this time, the channel is not being used and, hence, its bandwidth is wasted and remains unused until these two collided frames are re-sent or another node tries to send a frame. If the bit length of the frame is longer than the propagation time, then it is obvious that the wastage of bandwidth may become significant, as it affects the throughput of the LAN. The *propagation time* is the time an electrical signal takes to reach the upper length of the cable. The following are two typical situations which may arise when two nodes (A and B) are connected at opposite ends of the channel and try to transmit data frames at the same time.[6–11,13,22,24]

1. Both A and B are transmitting at the same time. It is likely that collision may occur halfway through their transmission, i.e., after half of the propagation time. This will corrupt the frames, and a jam signal of 32 bits (a unique bit pattern) will be transmitted over the media to inform other nodes about the occurrence of collision. The total time for any node to know about the collision is given by the round-trip propagation time (which is double the propagation time in one direction from one end to another of the medium).
2. A has transmitted the data frame and, after it has arrived very near B, node B detects a collision after it has listened to the channel as the last bit of the frame is about to be received. This collision is heard by node B immediately, while node A has to wait until the propagation time for the jam signal to go back to A. We can say that node A will notice the collision only after twice the propagation time (round-trip propagation time). Thus, in the worst case, the time to detect collision will be twice the propagation time of the medium.

In order to handle collision, we must define a **time slot** (characterizes the Ethernet) with minimum and maximum values so that we may get reasonable throughput of LAN. Let us consider different situations for understanding this time slot.

For the second situation described above, node A knows about the collision after twice the propagation time (round-trip time). We also have to add some finite amount of time during the collision and also after it is detected, when one of the transmitters involved in the collision generates a jam signal which travels over the channel. Thus, the time a node knows about the collision is given by round-trip time (twice propagation time) plus time wasted during collision plus MAC layer jam time. This MAC jam time depends on the coding and other functional and electrical behaviors of the physical layer. Thus, the time slot becomes a very important parameter for handing collision in Ethernet. The bit length of a data frame (MAC) should be at least this length.

The round-trip propagation time also helps in determining the **time slot window**. This can be explained by considering the second situation above. Node A has transmitted a frame and is not sure if node B has detected its data packet and, hence, is waiting until the round-trip time has elapsed. After this time, node A gets possession of the medium and starts sending the frame. In other words, after node A has transmitted the frame and while waiting for another to send, it must keep listening to the channel. Thus, the round-trip time also affects the upper limit of the size of data packets (partially transmitted) and hence the time slot.

In the case of **baseband transmission**, the time required to detect collision (in the worst case) is equal to the round-trip time. In the case of **broadband transmission**, the use of a headend will double the time for detecting the collision, as now the signal has to travel to the headend node, too. In other words, the bit length of the frame must be at least equal to the time required to detect collision in the case of broadband transmission, which is four times the propagation time. Further, this type of access control technique is usually used on short-distance LANs because, if we use it for longer lengths, the time slot window also gets bigger, which offers more collisions. Typically, a longer propagation delay must be used with shorter frames and, as such, the nodes will then be getting more chances to use the channel than if we use longer frames.

As mentioned earlier, each node uses an access control algorithm which listens and transmits the data frame over the channel. This algorithm is very complex, as it has to deal with all the problems mentioned above. In addition to these problems, it has to deal with the following situations.

When a collision occurs, both the affected nodes withdraw their data frames and retry sending the same frame after a random amount of time. Now assume that the same node is involved in the collision again and the same node keeps on getting kicked out due to collisions, thus causing a starvation to that node. This problem may be resolved if we can assign the random amount of time the nodes must wait for retry from a fairly uniform probability–distributed time in such a way that the number of collisions is reduced.

The number of collisions here presents a cumulative effect, since collisions generate additional traffic and increase traffic on the medium. If the traffic on the medium increases, there are more chances of collisions due to additional re-transmission, which in turn adds extra traffic, and so on. This problem is very serious and affects the throughput of the LAN considerably, as more collisions will not transmit the data frames but will keep the channel busy resolving this collision problem. This problem can be alleviated if we assign a waiting time for the node after collision, known as a **truncated binary exponential back-off (TBEB)**. This time allows the nodes affected by collision to wait for a longer time than those which are less involved with collisions. The back-off time is calculated using back-off algorithms.

With TBEB, a node which is involved in a collision for the first time waits for an amount of time equal to an integral number of slot time before it tries to sense the channel again for re-transmission. The integral number of the time slot is usually expressed in the range of 0–2, giving the three values 0, 1, and 2. After this, if the same node is again involved in a collision, the delay will be expressed as an integral value in the range 0–4 before re-transmission, and in the range 0–8 after the next collision, and so on. The back-off algorithm usually limits the number of times the delay should be computed; typically, this limit is 10 attempts. The media access protocol defines the delay as the integral times of slot time before 10 transmission times, after which it aborts the transmission by defining this event as an error to the LLC sublayer.

The nodes which are involved in many repeated collisions must wait for a longer time before they can retry than those not involved in many collisions. This will certainly hamper those nodes which have waited for a longer time with more collisions.

Due to the features of this type of MAC sublayer protocol, the network using this protocol sometimes is also known as **probabilistic** or **non-deterministic**, as there is no guarantee that a particular station which has been involved in a number of collisions will ever be able to transmit its message data over the network. If there is a large number of collisions taking place in the network (in particular, during a heavy traffic load), it is possible that the transmission of data messages will be reduced and hence the link or channel or network utilization will decrease. This will further reduce the throughput of the network. This type of network is more suitable for light traffic and unsuitable for heavy traffic. Further, this network may not be useful for real time–based applications.

In a typical CSMA/CD LAN configuration, each node is attached with a transceiver (with plug connector) and a controller, and these two are connected by a transceiver cable. The ends of the cable are connected by a connector or terminator and the cable possesses a load capacity of 50 Ω at 1 W. Cable segments of 500 m may connect 100 transceivers and are usually 2.5 m apart.

CSMA/CD physical layer: The physical layer of CSMA/CD has a 802.3 bus topology of passive shared configuration. The coaxial cable uses a baseband transmission scheme which does not require any modulation of a high-frequency carrier. The physical layer interface, known as a **media access unit (MAU)** in the terminology of IEEE 802.3, uses Manchester encoding-decoding signaling. This encoding-decoding scheme is better than ordinary binary representation, and it establishes a voltage transition at the midpoint of every logic bit (1 and 0). The Manchester coding generates a steady stream of 0s, 1s, or even a combination of 0s and 1s and looks like a square wave giving rise to a constant stream of signals. The advantage of using Manchester signaling is that it generates a carrier (similar to radio signal transmission) and, in fact, this is the signal which is detected by nodes when they listen to or sense the channel.

CSCM/CD protocol primitives: As explained earlier, the physical layer provides a common interface between LLC and all the media. Each transmission medium has its own interface which defines the medium-dependent services offered by it to the MAC sublayer. It also provides signaling services (common to all the media) and supports various functions of repeaters. The physical layer specification provides the following services to the MAC sub-layer of an IEEE 802.3 LAN, offered by physical signaling and defined in terms of primitives and parameters. These primitives include **Ph-DATA.request**, **Ph-DATA.indication**, **Ph-DATA.confirm**, **Ph-CARRIER.indication**, and **Ph-SIGNAL.indication**.

CSMA/CD MAC services[3,25]**:** The services offered by MAC to LLC allow the LLC entities of both nodes (LLC service users) to exchange the data with respective peer entities where these two nodes are a distance apart. The exchange of data between them is provided via primitives and parameters of MAC.

MAC defines a frame for the data packets it receives from LLC and sends the bit stream to the physical layer for its onward transmission over appropriate media. When it receives any frame (as bit stream from a distant physical layer), it sends it to the LLC, which may send it to the destination address mentioned in the frame or broadcast it to all the connected nodes. In the case where the frame does not belong to it, the frames are discarded and sent to the network management facility for appropriate actions on these frames. The MAC waits for transmitting frames when it finds the medium is busy. It adds a frame check sequence (FCS) value to each frame being sent out from the node, and for a received frame, it checks this FCS for any errors.

As mentioned earlier, in the event of collision, both frames are withdrawn from the channel. The MAC sublayer protocol, after noticing a collision, halts the transmission and tries to sense the channel for re-transmission after some random time. In order to check the validity of the frame, the MAC checks the size of the frame and, if the size of the frame is less than defined, it discards the frame as the frame is invalid. In order to avoid this situation, if the MAC finds a frame of length less than the minimum value, it adds a packet assembly and disassembly (PAD) field to make it a valid frame. When a frame is transmitted, various control fields are inserted into it and, similarly, when the MAC receives a valid frame, it removes those control fields (discussed below) before it sends it to LLC.

For providing synchronization between the nodes over a channel, the MAC uses a sync signal, which is also known as a preamble. It provides appropriate encoding (using self-clocking coding) at the transmitting side for the data frame and decodes at the receiving side using some decoding technique (Manchester, etc.) for recovering the original signals. It also generates an access signal which allows the nodes to access by sensing or listening to the channel and starts sending the user's frames if it finds the free channel. It further provides support for detecting any collision on the channel so that the user's data can be withdrawn from the channel. The frames defined by the MAC sublayer containing the user's data information, control information, and many other related control fields are shown later.

The primitives offered by MAC include **MA-DATA.request**, **MA-DATA.indication**, **MA-DATA.confirm**, and **MA-DATA.status.indication**.

The MAC and LLC sublayers interact with each other via PDUs through service access points (SAPs). The LLC uses a primitive MA-DATA.request to transfer the data units to peer LLC(s) through a data link service access point (DLSAP). The MAC sublayer constructs a frame from LLC data units, as it contains all the relevant information such as destination address, length of data, and other options requested. The MAC sublayer, after receiving the LLC data units, constructs the frame and indicates the success or failure of its request by using the primitive MA-DATA.confirm via another DLSAP. It uses a primitive MA-DATA.indication to indicate to its LLC sublayer (through available DLSAP) about a frame it has received which has as its destination this MAC address. It also indicates if the frame has a broadcasting address or multicast address in the destination address field.

The parameters associated with this primitive include source address, destination address, media access control service (MACS), transmission status, service class, and MAC service data unit. The service primitives and parameters defined by MAC are different for different LANs, as these functions are not defined by a common LLC sublayer. The MAC sublayer offers the following functions: assemble the data packets (LLC) into a frame with address and error fields for transmission, disassemble the frame into LLC data packets, Recognize the address and error fields on the receiving side, and provide management for the data communication over the link. Some of the material presented above has been derived partially from References 6–9, 11, 12, 19, 23, and 24.

CSMA/CD MAC control frame format: The MAC sublayer defines a frame which carries the user's data, destination address, control information regarding the boundaries of the frames, error detection, requested service, etc. IEEE 802.x standards were announced in 1980 while Ethernet was developed earlier, and as such the frames of these LANs are slightly different. The preamble for 802.3 has a seven-byte preamble and one byte of start frame delimiter. Ethernet has a preamble of eight bytes. The Ethernet-type field defines higher-level (level 3 and above) protocol in the data field where some use TCP/IP while others use XNS (Xerox Network System). The format of IEEE 802.3 CSMA/CD is shown in Figure 8.9.

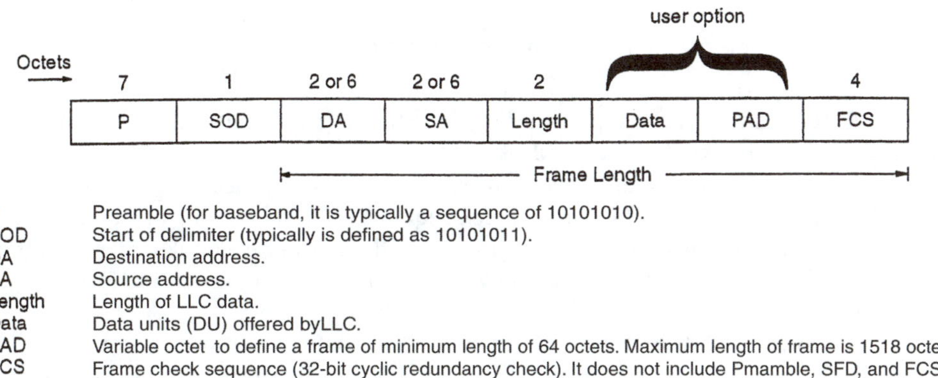

P	Preamble (for baseband, it is typically a sequence of 10101010).
SOD	Start of delimiter (typically is defined as 10101011).
DA	Destination address.
SA	Source address.
Length	Length of LLC data.
Data	Data units (DU) offered byLLC.
PAD	Variable octet to define a frame of minimum length of 64 octets. Maximum length of frame is 1518 octets.
FCS	Frame check sequence (32-bit cyclic redundancy check). It does not include Pmamble, SFD, and FCS.

Figure 8.9 CSMA/CD MAC control frame format.

The seven-octet (byte) preamble establishes the same unique bit sequence for each octet and is used for providing bit synchronization between nodes for physical layer signal-coding circuits, and medium initialization and stabilization. The least significant bit (LSB) in each octet is transmitted first. The preamble typically follows the bit pattern of **10101010** for every octet.

The next field contains one octet of **start frame limiter (SFL)**, which is used for defining frame synchronization and indicates the boundary of the frame. In particular, it represents the start of the first frame. Here again, the LSB bit is transmitted first. It has a unique bit pattern of **10101011**.

IEEE 802.3 allows two classes of addresses, two-octet or six-octet, for both source and destination addresses. The **destination address (DA)** field may include a single node, a group of nodes, or all the connected nodes (broadcasting). For broadcasting, the MAC frame includes all nodes in the DA field.

Each node connected to the LAN checks this field to establish whether it belongs to it. The first bit of this field (DA) identifies the type of address being defined. If the first bit is 0, it indicates a single address of MAC at that node. If it is 1, it indicates a group of destination nodes (a group of MAC addresses). A generalization of this is a broadcasting address where all the bits of the DA field (six bytes) are 1. In general, if it represents one node, it defines the physical layer address (MAC), but if it is a group of addresses for broadcasting, it may represent controller addresses, which usually include 8, 16, 32, or 64 nodes. The Ethernet uses six octets for a destination address.

The two-octet-length field identifies the lengths of the LLC (number of octets) data field and PAD field. The LLC data field is of variable length and contains the user's data. If the LLC data field contains data of length less than the maximum length, the PAD field is filled with extra octets so that the length of frame may be at least equal to the minimum length (64 octets) requirements in IEEE 802.3.

The user's field includes four subfields: **DSAP, SSAP, control information**, and **user's data**. The **destination service access point (DSAP)** field of the LLC frame uses the following bits:

1/G D D D D D D D

where 1/G defines the following address configuration: 1/G = 1 (group address), 1/G = 0 (individual address), and D represents the actual address and has a special bit pattern for broadcasting addressing of 1111111 (broadcasting).

The source service access point (SSAP) field of the LLC frame uses the following bits:

C/R S S S S S S S

where C/R defines the frame type as follows: C/R = 1 (response), C/R = 0 (command), and S represents the actual source address.

It is important to note that the length of a frame is measured from the destination address field up to the FCS. The frame check sequence (FCS) is a four-octet field used to detect errors and contains CRC-like checking algorithms. In some books, these two fields are combined into one field known as the *user's option field* containing data and PAD and of variable length.

The maximum length of the MAC frame is 1518 octets. The maximum length of a packet is 1526 octets, out of which the data is 1500 octets. Similarly, the minimum length of the packet is 72 octets, of which the data occupies only 46 octets. In order to have a valid frame of minimum length of 64 octets, the length of the user's option field must be at least 46 octets. If we use six octets for each of the address fields, two octets for the length, and four octets for FCS, the remaining 46 octets (64 – 18 = 46 octets) must be assigned data and PAD fields together.

The CRC is calculated for the entire frame except the preamble, SFD, and FCS itself. The 32-bit cyclic redundancy checksum (CRC) is produced by the following polynomial generator (denoted as $G(x)$):

$$G(x) = \left(x^{32} + x^{26} + x^{23} + x^{22} + x^{16} + x^{12} + x^{11} + x^{10} + x^{8} + x^{7} + x^{5} + x^{4} + x^{2} + x + 1\right)$$

The first bit of DA is considered to be the information polynomial with the highest exponent. The information data frame polynomial is divided by $G(x)$ and the remainder is expressed as $R(x)$. The highest term in $R(x)$ is the first bit transferred in the CRC field. CRC is implemented by considering a linear register which contains only 1s at the beginning. The contents of the register are inverted (after the transfer of the last bit) and are then transferred as CRC. On the receiving side, the information data polynomial and remainder are added and divided by $G(x)$. If there is a remainder, it indicates an error; otherwise, there is no error.

In order to see the difference between the original Ethernet and its current standard (IEEE 802.3), we show the MAC control frame format of the Ethernet in Figure 8.10.

The CSMA/CD LAN offers the following features:

- It is very simple, inexpensive, and reliable.
- The protocol offers a fair access to the nodes and all the nodes are allocated over the entire bandwidth. For light traffic, it offers small delay and higher throughput.
- Collision is the main cause of poor throughput and delay time under heavy traffic. Also, collision detection sometimes becomes difficult; this imposes constraints on the size of the frame used in CSMA/CD LANs. The protocol offers poor performance under heavy traffic. It is suitable for long transmission (e.g., large data files).

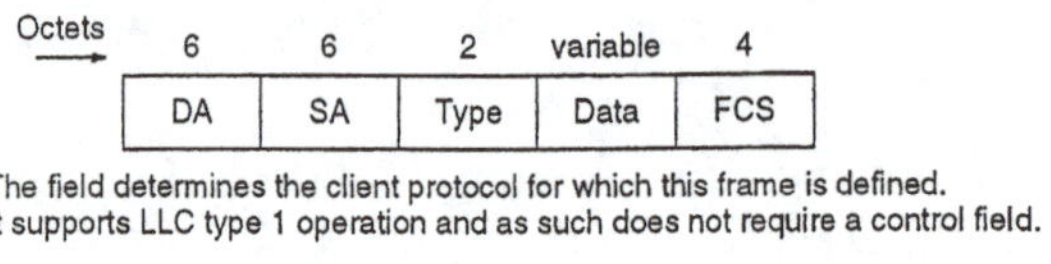

Octets →	6	6	2	variable	4
	DA	SA	Type	Data	FCS

The field determines the client protocol for which this frame is defined.
It supports LLC type 1 operation and as such does not require a control field.

Figure 8.10 Ethernet MAC control frame format.

For evaluating the performance of CSMA/CD LANs, some of the parameters to consider are as follows: the minimum inter-packet delay is defined as the minimum rise and fall time of transmission and is usually 9.6 μs. The round-trip delay, which is defined as double the propagation time (time for a bit to travel from one end to another) is 51.2 μs.

The IEEE 802.3 committee has formed a High-Speed Study Group to look into the very high-speed Ethernet LANs offering data rates in the gigabits (1000 Mbps). This gigabit Ethernet is based on the concept of high-speed packet-switching and uses a CSMA/CD protocol of original Ethernet LANs (10 Mbps and 100 Mbps). It supports both 10 Base T and 100 Base T LANs. A typical configuration of gigabit Ethernet includes a gigabit switching hub that supports a number of servers via gigabit Ethernet. It also supports other nodes with 100- and 1000-Mbps hubs that are linked to the gigabit switching hub via a 1000-Mbps link, and links to other workstations via 100-Mbps links.

Ethernet is the most popular cabling technology for local area networks. In the early days of Ethernet, the most commonly installed Ethernet system was known as 10 Base T, which provided transmission speeds of up to 10 Mbps. With high bandwidth requirements for many multimedia applications, a new LAN — 100 Base T, also known as fast Ethernet — which could (theoretically) transmit at 100 Mbps was introduced. Fast Ethernet uses the same cabling as 10 Base T, as well as packet format and length, error control, and management information. The fast Ethernet was typically used for backbones, but the ever-growing bandwidth requirements have made it useful in workstations. The gigbit LAN that can (theoretically) provide a bandwidth of a billion bits per second is becoming very popular as a backbone. Applications in the modern enterprise networks include a heavy-use desktop, server, hub, and the switch for increased bandwidth applications. Megabytes of data need to flow across intranets, as communication within enterprises will move on from text-based e-mail messages to bandwidth-intensive real-time audio, video, and voice.

An increasing number of enterprises are employing data warehousing for strategic planning, as it deals with a very high volume of data and low transmission latency. These warehouses may be composed of terabytes of data distributed over hundreds of platforms and accessed by thousands of users. This data must be updated on a regular basis to ensure that the users access real-time data for critical business reports and analyses.

Enterprise critical applications will proliferate and demand ever-greater shares of bandwidth at the desktop. As the number of users grows rapidly, enterprises will need to migrate critical portions of their networks (if not the whole network itself) to higher-bandwidth technologies. Crossing over from megabit to gigabit Ethernet will be inevitable, starting at the backbone.

A very straightforward upgrade scenario is upgrading 100-Mbps links between fast Ethernet switches or repeaters to 1000-Mbps links between 100/1000 switches. Such high-bandwidth, switch-to-switch links would enable the 100/1000 switches to support a greater number of both switched and shared fast Ethernet segments.

8.2.4 IEEE 802.4 (ISO 8802/4) token passing bus LANs[4]

The token bus standard LAN using a token passing access technique[4] offers uniform, guaranteed access to the media. It supports different types of physical layer standards offering different data rates of 1, 5, and 10 Mbps (with a bandwidth of 1.5, 6, and 12 MHz) and over coaxial cable (broadband of 75 Ω) for bus/tree topologies. The frequency allocation to different channels is defined by North American channel specification definitions. For example, for 10 Mbps, two channels are defined, while for 1 Mbps, a 1.5-MHz sub-channel is defined within one channel recommended by the standard channel. The single-channel broadband (offering the entire bandwidth for one transmission) is also supported by IEEE 802.4 using a phase-continuous FSK scheme (1 Mbps) or phase-coherent FSK scheme (5 and 10 Mbps).

Optical fiber is basically an analog medium (due to its inherent working principle) and the data is transmitted in the optical range. For using optical fiber, we need to transform the digital data into analog signal form (using a digital-to-analog converter) before it can be transmitted. As we mentioned earlier, both single-channel and broadband switching in optical fiber use different forms of frequency shift keying (FSK), while the IEEE 802.3 and IEEE 802.4 broadband systems use different forms of phase shift keying (PSK). These forms of FSK or PSK will not work in the case of high-speed switching; therefore, another technique based on amplitude shift keying (ASK) is being used for optical transmission. These modulation techniques are discussed in Chapter 3.

The IEEE 802.4 optical fiber specification works on the bus topology (as long as a logical bus topology is defined). Here also, if two nodes are trying to transmit at the same time, a collision will occur. Due to the high cost of bus topology and other connecting devices (low-loss taps, access logic unit, and connecting cables), the active or passive star topology is highly recommended as a suitable topology for optical transmission. The optical-fiber transmission medium using Manchester coding offers data rates of 5, 10, and 20 Mbps. Some of the material in this section is partially derived from References 6, 11, and 19.

Token bus LAN: In IEEE 802.4, the nodes define a logical ring with each node assigned a number as its address. The token determines the right for the nodes to access the bus. The token frame contains the destination address. In order to transmit the data, a node must first get the token, and the node interface makes sure that the node gets the token completely before it can use it. The node, after getting the token, can either pass (as token frame) it on to another node or send the data frame to its defined destination node. Each node gets a predefined fixed time to keep the token, during which it can either transmit the frames or pass the token to the next node.

There is a finite time for any node to receive the token which typically is equal to the propagation time. In the worst case, it will include a number of other timing parameters such as the propagation time of the signal to go from one end to another over the bus topology, the time required by each node interface to provide necessary functions (time to pass the token to another node and time to transmit the frames), and the time for all the nodes to transmit the data frames. The node must pass the token to another node if it does not have any data frame to transmit. In the case when it has the data frame to send, it will send all the frames. It will send the control token frame to the next node either after it has sent all the frames or after the predefined allocated time has expired. It is important to note that the logical ring defined in this LAN is independent of the locations of nodes and may be defined at our choice, which in turn depends on the required nodes for the application. A typical coaxial-cable connection for IEEE token bus LAN architecture is shown in Figure 8.11.

Token bus MAC sublayer: In the token passing MAC protocols, the networks operate in a deterministic manner where the stations are defined in a logical ring topology (token bus) or they are physically connected to the ring topology. The token bus LANs working over a bus topology (e.g., coaxial cable) define a logical ring for the stations which wish to transmit the message among themselves irrespective of their physical locations. Each station is defined by an identification number, which is attached with the message data packet and transmitted to the station with that address in the logical ring topology. Each station knows the address of its logically defined predecessor and successor stations. It will receive the data packet from its predecessor and, after performing required functions

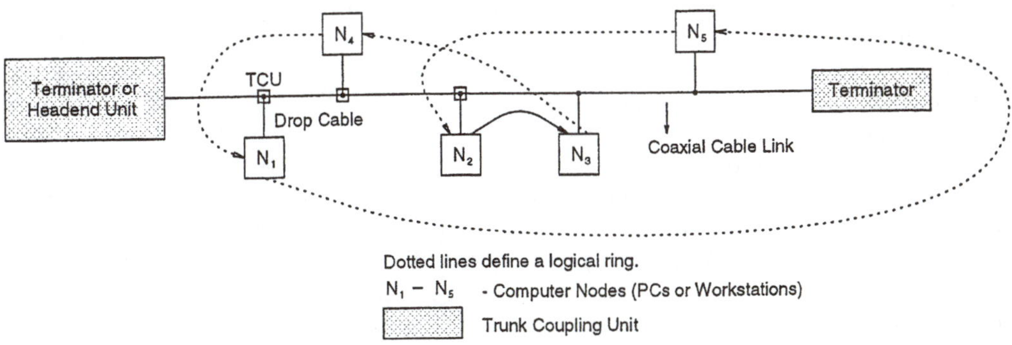

Figure 8.11 Token bus IEEE 802.4 LAN architecture.

on the packet, will pass it on to its successor station. A *free token* is defined as a certain number of bits in a unique bit pattern, and it also circulates around the logical ring.

A station which wants to transmit any data message packet waits for this free token. After it receives this free token, it attaches the data packet to it and transmits it to the successor station. After the data packet has been received by a particular station (destination), it makes the token a free token and passes it on to the next station. This scheme of accessing the channel or network is based on a round-robin technique where each station has an equal right to access it. Further, the priority for any station can be defined at the beginning so that it always gets the token first and transmits the data packets. The assignment of priority is independent of the stations' physical locations.

The maximum length of a message data packet in the token passing LANs (physical or logical) can be determined by looking at the number of stations connected. Once the maximum length of a packet is known, the waiting time, propagation time, and other time delays can be calculated by assuming that every station is sending the maximum size of the packet each time (worst-case situation). The average waiting time can be adjusted to any predefined value (based on statistical data) to receive an acceptable quality of service from the token passing LAN. Due to this, the token passing LANs (token bus and token ring) are also known as **deterministic LANs** and are useful for real-time applications, in contrast to CSMA/CD, which is a non-deterministic LAN.

The deterministic LANs are useful for heavy traffic where we may expect nearly 100% network utilization. These LANs are not suitable for light data traffic due to the fact that the average waiting time in this case may become significant. The only problem with token passing LANs is that a token may be affected by any noise or other transmission error, an error caused by duplicate tokens, or any other error. In these events, the token access time may be severely affected. Although token-recovery procedures have been defined and standardized, these LANs still do not provide optimal resource utilization for real-time applications.

Token bus media access design issues: In IEEE 802.4, the token (a fixed unique bit pattern sequence) circulates around the logical ring. This logical ring is defined by the users and is independent of the physical location of the nodes. Each node is assigned with a number as its address and the token always goes a higher-number to a lower-number address node. The concept of a logical ring is different from a token ring (IEEE 802.5) in the sense that in an IEEE 802.4 token bus, only those nodes which have initialized the transmission over the ring will be connected to the logical ring. In the case of the token ring, all the nodes connected to it are always connected and do not require any transmission initialization process.

Further, in token bus, after the node has sent the token frame to its successor, it waits for a valid frame (following the token). It knows that the token has reached the destination node and starts transmitting the data frames. If it does not hear the valid frame following the token, it knows that the destination is not active. It may establish this node by defining it as its successor. In token bus, there may be nodes which are connected to the logical ring and may listen to the data frames being circulated over it but will not be able to transmit the frames. After a node has finished the transmission, it sends the token to the next node (with a lower-number address and defined as its successor) by passing a token control frame.

In order to maintain the logical ring, each node keeps a record of its predecessor and successor. When it receives a token from its predecessor, it knows its own address and also the address of its successor where the token is to be sent. This scheme of defining a logical ring is implemented dynamically to ensure that there is one logic ring at any time. The implementation for accessing the logical ring is more complex than CSMA/CD, as the access logic control at each node must possess the following features:

1. In the event of restarting (for broken ring) or initialization of logical ring (next day), a node from the ring must initialize the ring and include other nodes.
2. A node may be included in or deleted from the logical ring, and it should not affect the normal working of the token bus, as in each case, the node must assign its address and inform its predecessor and successor nodes. A node may be added to the logical ring by using the following steps:
   - A node, after getting the token, issues a new frame known as solicit-successor and inserts the node address as a number between its own number and the number of the successor node. It waits for a time slot (twice the propagation time of the bus).
   - If it gets any response, it will be a set-successor frame from a requesting node, and the node holding the token will change its successor address. It then sends it to the requesting node, which sets its address and sends it to the next node. But if there is no response from any node, it will transmit the token to its designated successor node. Similarly, when a node is to be deleted, the node after receiving the token sends a set-successor frame to the predecessor defining its successor in the token.
   - In the event of multiple responses, another procedure is invoked to resolve the contention. It must provide token recovery in the event of a lost or duplicate token from the logical ring. It must also support priority to the nodes in terms of more time or capacity, depending on the requirements.

As defined in token ring access control technique, the destination node changes the status of a bit to indicate acknowledgment to the source node; a similar concept is being used in token bus. Here, after the node has possession of the token, it may send predefined bits as **request-with-response (RWR)** in the data frame. This particular bit is recognized by the destination node, which sends a response to it by changing its status back to the source node without waiting for the token to arrive at the node. In this scheme, unacknowledged and acknowledged connectionless services can also be supported. Further, IEEE 802.4 allows these services to be included with the following service class. Each class of service is assigned a priority on the basis of transmission and other related modes, as shown below:

Class 6 — Synchronous
Class 4 — Asynchronous urgent
Class 2 — Asynchronous normal
Class 0 — Asynchronous time-available

The token holding node provides control and maintenance for priority services which are usually implemented via timers.

For more details on token bus media design access issues, see References 8, 9, 12, 19, 22, and 23.

Token bus physical layer: The physical layer provides services for MAC in terms of primitives and parameters. The primitives defined by physical-layer services include Ph-DATA.request, Ph-DATA.indication, Ph-MODE.invoke, and Ph-NOTIFY.invoke. Various parameters associated with these primitives include symbols and modes.

The MAC sublayer services provided for LLC are in the form of primitives which include MA-DATA.request, MA-DATA.indication, and MA-DATA.status.indication, and associated parameters including destination address, source address, MAC service data unit (MSDU), and quality of service (QOS).

Structure of token bus MAC sublayer: The IEEE 802.4 committee has defined the following modules of the MAC sublayer, as shown in Figure 8.12.

The **interface machine (IFM)** provides mapping between LLC and MAC for the quality of service and also performs address checking on the received LLC frames. The **access control machine (ACM)** interacts with the IFM and, based on the mapping, determines at what time the frame should be sent to the bus and offers a decentralized algorithm for sharing of a common bus between various ACMs of different stations. In other words, these ACMs of different nodes share the common bus. Other functions of this module include initialization of logical ring, error detection, maintenance of logical ring, controlling of addition or deletion of nodes from the ring, fault recovery, and flow control. The IFM creates LLC frames out of the packets it receives from the network layer and passes them onto ACM.

The other two modules **transmit** and **receive machine (TM** and **RM)** are used to transmit the frames to the physical layer and receive frames from ACM, respectively. The TM module accepts the frames from the ACM and defines a **MAC protocol data unit (MPDU)** by adding various control fields such as start of delimiter (SOD), frequency check

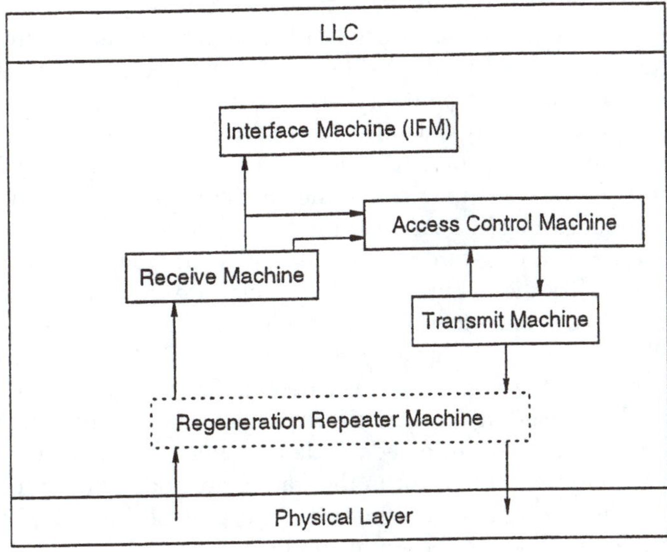

Figure 8.12 Modules of MAC (IEEE 802.4). (Modified from Black, U., *Data Networks: Concepts, Theory and Practice,* Prentice Hall, 1989.)

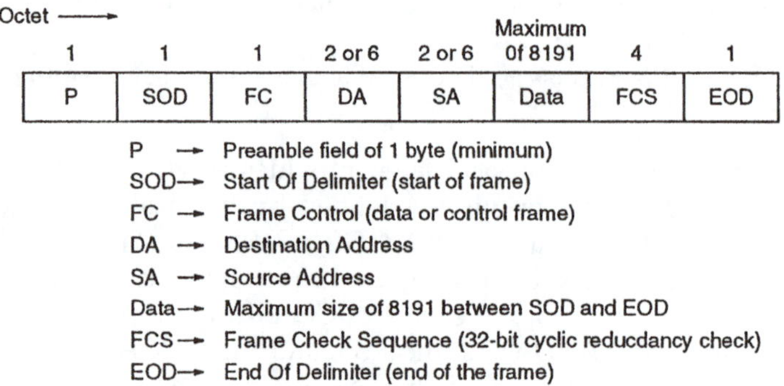

Figure 8.13 IEEE 802.4 token bus MAC control frame format.

sequence (FCS), preamble, and end of delimiter (EOD). Each of these control fields will be discussed in the following section on MAC control frame format. The RM module accepts the data from the physical layer and, after checking the error-free transmission (by checking FCS) and identifying the boundary of the valid LLC frame, it passes it on to the IFM. The IFM indicates the arrival of this frame and passes it on to the LLC sublayer. The sublayer sends it to the network layer, from where it goes through higher layers via a common interface (IEEE 802.1) to an end-user application or another layer.

Token bus MAC control frame format: The MAC control frame format of IEEE 802.4 token bus is shown in Figure 8.13. The preamble field defines one- or more-octet unique bit patterns used to provide bit synchronization between two nodes. The start of delimiter (SOD) field of one octet defines the starting boundary of the frame and hence provides frame-level synchronization between the nodes. The signaling bit pattern used in this field is always different from the data. A typical combination of SOD may have a non-data bit at an appropriate position in the field, and the placement of this non-data bit depends on the encoding technique being used for the transmission media. The frame control field of one octet defines the type of frame (control or LLC data, token, etc.) being sent. Different operations on the token are defined by different combinations of control frame, which basically handle these operations on the token.

The token does not carry any data. The start of delimiter (SOD) provides synchronization and indicates the destination node about the token or frame. Initially, a node with the highest address will act as an active monitor node. It generates a token and sends it to the next node by not setting T. That node may not have the data to send but will check the token for any error. If it finds no error, it will regenerate the token and send it to next node. If the next node has the data to send, it will capture the token, append it, and rest the A and C bits. When a frame including token goes over the first node (which generated the token), it will again check it for any error. If no error is found, it will pass it to the next node until it reaches the destination. The destination node also checks for any error, copies the data, and, after setting C, returns it to the sender. After receiving the frame and checking that the destination has received the data, it will generate a new token and send it to the next node. In the event of error in the frame, the node will set the E bit and pass the frame on to the destination. The destination node, after seeing the E bit is set, will not copy the data and send it back to the source node.

As we discussed in CSMA/CD LANs, the MAC sublayer supports two categories of addresses, two or six octets, and similar options are available with the token bus MAC frame.

The destination address may be defined for a particular node, a group of nodes (multicast group), or globally (broadcasting to all connected nodes). The use of either a two-octet or six-octet address depends on the application and implementation and must be the same for all the same nodes of a LAN. Each node is assigned a unique hardware physical address.

The data field of variable size contains the user's data (LLC data) related to control operations and must be greater than or equal to zero. This field is always limited between the start of delimiter (SOD) and the end of delimiter (EOD). The frame check sequence (FCS) field uses a 32-bit cyclic redundancy check (CRC), and it is a 32-bit CRC-like error-detection sequence. The CRC is computed for the entire frame except the preamble, SOD, EOD, and FCS itself. The EOD field of one octet indicates the end of the frame. A typical bit pattern for this field looks like NN1NN1IE, where N represents a non-data symbol and I represents an intermediate bit and indicates the status of the frames transmitted. If its value is set to 1, it indicates the transmission of the last frame from that node. The E-bit represents an error bit, which is usually set by the repeater in the event of any detected error by FCS.

In summary, we can list the following features of a token bus LAN:

- It is a deterministic-type LAN where an upper limit for waiting time can be determined or defined. This is possible since each node in the logical ring can keep the token for a predefined time, after which it has to pass it on to its successor node.
- It offers excellent throughput for a data rate and, as data rates increase, the throughput tends to rise. The throughput and performance of token bus LANs are not affected significantly by increasing the length of the bus topology (by increasing the number of nodes).
- It does not cause any collision between the data frames, as only one data frame can be transmitted over it at any time.
- Due to its deterministic behavior, this LAN is used in real-time applications and process-control applications.
- It provides a fair access to the nodes (as done by CSMA/CD for light traffic), not only for light traffic but also for heavy traffic.
- It supports priority and provides a guaranteed fixed value of bandwidth via a token, as required in applications needing voice, data, and audio signal transmission.

The main disadvantages of token bus include the complexity, which adds overhead. Further, the noise generated during the transmission may damage the token or may lose it. It is not suitable for applications requiring light traffic, as the waiting time for the nodes to get a token and transmit the data frame may cause a bottleneck.

Derivative protocols of IEEE 802.4 standard: As we pointed out earlier, IEEE 802.4 is a deterministic type of LAN and, as such, can be useful for real-time applications, process control, etc. When this LAN was defined, its main use was for applications dealing with industry and factory-based production lines. This is evident from the fact that it has successfully been used in the following three applications related to these areas: **process data highway (PROWAY)**, **manufacturing automation protocol (MAP)**, and **technical office products (TOP)**. For each of these applications, a standard protocol has been defined. The following section describes each of these in brief.[7–9,11,17,19,23]

For monitoring processes and various applications, the International Electromechanical Commission (IEC) defined a user-driven protocol for LANs known as **process data highway (PROWAY)**. The main aim of this protocol was to provide a framework of interconnecting systems of different vendors' process-control applications. This framework of interconnection must offer high reliability and predefined guaranteed access times over a LAN for data communication. The Technical Committee of the Instrument Society of America (TCISA) defined a LAN for its industrial control applications in 1976, and in 1983, the standards document Extensions for Industrial Control (EIC) defined compatibility between IEEE 802.4 and PROWAY LANs. The industrial-control applications should provide support to both continuous and discrete processes and must provide an optimum feature and behavior of the industrial control systems. The PROWAY LAN will be required to provide high availability, high data integrity, and proper operations of the systems in the presence of electromagnetic interference and other sources of noise (due to voltage differences in ground voltages, etc.). It must provide a high information transfer rate among the processes within a geographic distance.

The standard supports 100 devices/equipment over a distance of two km, and the devices are connected via a dual bus medium. The average length of a packet is 128 bits. The specifications of this standard from different vendors usually provide a minimum value of access time to access the LAN (typically 20 ms) for a maximum number of nodes (100). In this LAN, it is expected that nearly half the nodes are trying to send at the same time, and as such the length of packets is defined as a very small size (128 bits or 14 octets) compared to most of the standard LANs.

The committee's recommendation was for a single-channel token passing bus MAC, as it offers immunity to low-frequency noise typically present in the factory environment. This channel uses analog signaling and is less expensive than broadband cable (but with smaller capacity). Further, due to deterministic behavior and priority options available with the IEEE 802.4, it became a suitable candidate for these applications as it can support a real-time environment of factory or process control in a dynamic mode.

The PROWAY provides five functions at the user highway boundary, which is analogous to the boundary between LLC and MAC defined in standard IEEE LANs: **send data with acknowledgment (SDA), management of PROWAY (MOP), global send data (GSD), request data with reply (RDR)**, and **remote station recovery (RSR)**. Each of these functions is self-explanatory. PROWAY offers connectionless services and it does not support connection-oriented services.

The **manufacturing automation protocol (MAP)**, the second industry-related application for IEEE 802.4, was developed by General Motors (automobile industry, U.S.) in 1962 for providing compatibility between various communication devices used in the manufacturing processing environment. The communication devices include terminals, resources, programmable devices, industrial robots, and other devices required within the manufacturing plant or complex. This attempt to develop such a local area network was transferred to the Society of Manufacturing Engineers (SME). These devices connected to local area networks are used for various manufacturing processes, assembly line operations, and many other factory floor operations.

The main idea of developing such a LAN was to provide a backbone network through a factory which connects management and manufacturing control–based computers and other manufacturing floor cells–based networks, computers, and devices. The low-speed and low-cost proprietary networks may be connected to this backbone LAN via repeaters. To get an idea, an automotive assembly plant may have thousands of programmable controllers, thousands of robots, and a large number of intelligent devices, and all of these devices may be from different vendors. The intent of MAP was to provide a common communication interface for these communicating devices such that these incompatible devices from different vendors could communicate with each other in the assembly plant.

A number of standards and options within standards already in use within the factory and based on the open system interconnection (OSI) model were developed by GM. The main objectives of GM were to develop its own standard for application-to-application

Table 8.2 MAP Layers Compared with OSI-RM Layers

OSI-RM (IEEE 802)	MAP
Application layer (7)	ISO CASE kernel, 4 ASEs: File transfer and management (FTAM), directory service, network management (NM), manufacturing messaging format standard (MMFS)
Presentation layer (6)	ISO presentation kernel, ISO 8822, 8823, 8824, 8825
Session layer (5)	ISO session kernel, ISO 8326, 8327, 8326/DAD2, 8327/DAD2
Transport layer (4)	ISO 8072, and ISO 8073 transport class 4
Network layer (3)	ISO network connectionless Internet 8473
Data link layer	IEEE 802.2, different classes:
LLC	Unacknowledged and acknowledged connectionless services
MAC	IEEE 802.4 token bus with 48-bit address, no options required
Physical layer	IEEE 802.4 broadband (10 Mbps) and single-channel or carrier band (5 Mbps)

communication and define a format standard for application-functions messages. It was also decided to recommend that these protocols be used with MAP. MAP has various layers conforming to the layers of OSI-RM, as shown in Table 8.2.

The network layer offers connectionless service, while the transport layer provides class 4 service. In this way, we can build on the connectionless lower layers to give a full connection-oriented service. The MMFS protocol is a MAP-based format which provides encoding of messages and limited support for file transfer to manufacturing nodes (controllers, robots, etc.). The EIA has also defined a standard for the MMFS, and it is renamed **manufacturing message service (MMS)**.

The MAP supports connectionless service only. This is evident from the fact that it uses an HDLC-type unnumbered frame for the data link, while Internet protocol (IP) at network layers supports connectionless only. Since these lower-layer (2 and 3) protocols support connectionless services, the higher layers will also handle the connectionless services. The adoption of the token by LAN (IEEE 802.4) for MAP also offers finite and predictable access times to the nodes, and communication is said to be simple and reliable. It offers application-to-application messaging and gateways to other non-MAP networks. The MAP-compatible products with another protocol architecture, technical and office protocol (TOP), and other components have made a significant advancement in the development of products by different vendors. In fact, there is a MAP group which has been established, and various companies are members of this group. A new trend of defining an end system for communication with other networks has promoted development of various products for interconnectivity with different LANs, e.g., PROWAY/mini-MAP networks, MAP bridges, gateways to connect MAPs with non-MAP networks, etc.

The third industry-related application for IEEE 802.4 is the **technical office products protocol (TOP)**. The Boeing Company wanted to use a local area network which can support the office automation processes for its computer services. Also, it had to offer services similar to MAP and be compatible with it. This LAN uses an OSI-RM layered approach, and the TOP standard has protocols in common with OSI-RM for data communication in its applications for office automation and computer services (Table 8.3).

If we look carefully at MAP and TOP, we will notice that TOP does not use many of the protocols for its application, e.g., manufacturing messaging format standard (MMFS), common application service elements (CASE), or network management and directory services, while these are supported by MAP. Further, the CSMA/CD MAC is defined for TOP but is not defined for MAP.

Table 8.3 Top Layer Compared with OSI-RM Layers

OSI-RM	TOP
Layer 7	ISO file transfer access and management (FTAM)
Layer 6	Not required
Layer 5	ISO 8327 session layer protocol
Layer 4	ISO 8473 transport layer protocol
Layer 3	ISO 8473 connectionless Internet
Layer 2	IEEE 802.2 LLC (ISO 8802/2)
Layer 1	IEEE 802.3 CSMA/CD (ISO 8802/3)

8.2.5 IEEE 802.5 (ISO 8802/5) token ring LANs

The IEEE 802.5 subcommittee proposed a LAN (initially developed by IBM) for connecting various devices in a ring topology where the devices can transmit the data frame back and forth and, also, a priority can be assigned to the devices.[5,14] This LAN uses a token access control technique to access the medium.

Token ring LAN: IBM introduced token ring networks in early 1980, which operate at a data rate of 4 Mbps. In 1988, IBM introduced a 16-Mbps token ring network. The maximum distance is about 300 m and it uses coax for connection. It supports a maximum number of over 270 devices. With unshielded twisted pair, it supports fewer devices and a smaller distance. All the nodes are connected to a **multistation access unit (MAU)** which does not have any amplifier or relay.

In a token ring LAN, every node is physically attached to two adjacent nodes in such a way that it receives a data frame from one node and sends it to another node. As discussed earlier, the IEEE standard LANs usually encompass both physical and MAC sublayers into it; similarly, the IEEE 802.5 encompasses both physical layer specifications and MAC sublayer specifications. This has been implemented by a number of vendors, including IBM's IEEE 802.5-compatible token ring LANs (for use in its PCs). It uses a unidirectional ring topology, and each node regenerates and repeats bits to keep the LAN active at all times.

Token ring physical layer: The services provided by the physical layer are expressed in the form of Ph-DATA.request and Ph-DATA.indication. The parameters associated with these primitives are usually symbols (binary form for data and non-data symbols), and a single entity MAC delivers the data to all the MAC entities defined within the same local area network. The data from one MAC entity initiated as Ph-DATA.request is received by various MAC entities via another primitive, Ph-DATA.indication.

As discussed earlier, the physical layer offers a common interface to various transmission media and is known as a **medium-independent specification**. This specification supports data rates of 1 Mbps and 4 Mbps and the standard uses differential Manchester encoding for a signaling technique. In this encoding technique, for logical 0, the polarity of the leading signal is opposite that of the trailing element of the previous symbol, and for logical 1, the polarity of the leading signal is the same as that of the trailing element of the previous signal. The synchronization between bits is provided by the mid-bit position which is inherent within encoding.

Further, it has been traditional that the non-data symbols are usually used in the start of delimiter (SOD) and end of delimiter (EOD), which provide synchronization between frames. In order to distinguish between the data and non-data symbols, the non-data symbols (usually J and K symbols are used) do not represent any mid-point transition. Thus, we represent the J symbol by one which has the same polarity as the trailing part of the preceding symbols, while the non-data K symbol has a polarity just opposite that of the J symbol.

The physical layer also offers medium-dependent specifications for different media which basically is a connection between the node and trunk coupling unit (TCU). This connection is typically provided by a pair of 150-Ω shielded twisted wires which carry the transmission in opposite directions for each side. The MAC services allow the LLC entities to exchange the data (LLC protocol data units) between their peer LLC entities where these LLCs are at different locations on different networks.

Token ring media access design issues: The IEEE 802.5 LAN specifies a three-byte token for transmitting the data frames around the ring topology. Any node wishing to transmit the frames must first capture the token, and the associated interface with the node decides whether it has to send the data frame or pass the token to the next node. Each node introduces a one-bit delay for each time a token passes over it, usually known as **node latency**. During this delay, each node must decide whether to pass the token or send the frame with it. The total time around the ring topology is termed **ring latency** or **ring walk time.** The ring latency time can also be determined by adding the latency times of all the connected nodes and the propagation time around the ring. The walking time will change with an increase in the number of nodes in the ring.

The three-byte token is composed of three fields: **start of delimiter (SOD)**, **access control (AC)**, and **end of delimiter (EOD)**. The SOD and EOD fields represent the boundaries of the frame. This token frame is recognized as a control frame because it has a field for access control (AC) of one byte which performs various functions for the nodes to capture the token for transmitting the frames. It also allows the nodes to indicate their priority. When a node is ready with the frame (control or data) with the appropriate address and other frame-control information for its transmission, it waits for the token to arrive. The details of each of these fields are discussed below.

The MAC protocol changes the status of the access control field into the data frame control; i.e., earlier the AC was representing these three bytes as a token, and now the AC is changed to the data frame control (indicating the transmission of a frame from the node rather than a token). During the transmission of frames, the node may receive its own frame over it and checks with the data frame it has sent. If it finds that the data frame it has received is not the one(s) it has transmitted, it will terminate the communication. In the case that it finds that the received frame is the same as it transmitted, it will keep sending the frame, and at the end of all the frames, it will compute a 32-bit FCS and append it at the end of the frame.

All the nodes listen to this frame, and if any node finds its address matches the address defined in the frame, it copies it into its buffer and submits the frame back onto the ring by changing the status of a bit in the token. When the sending node sees this bit, it gets an acknowledgment from the destination that it has received the frame (by looking at the frame status field of one octet in the MAC frame), and now the sending node must remove this frame from the ring. After this, it must issue the new token (attached by it at the end of frame transmission) and send it to the ring.

The sending node when transmitting the frames may receive frames from another node. It discards all these incoming frames but will not discard its own transmitted frames. At the end of its frame transmission, it attaches a new token and then waits for frames from another node or even a token (if it wishes to transmit frames). Thus, the node, after capturing the token, rechecks the address and, based on the above-mentioned situations, repeats the signals for its onward transmission.

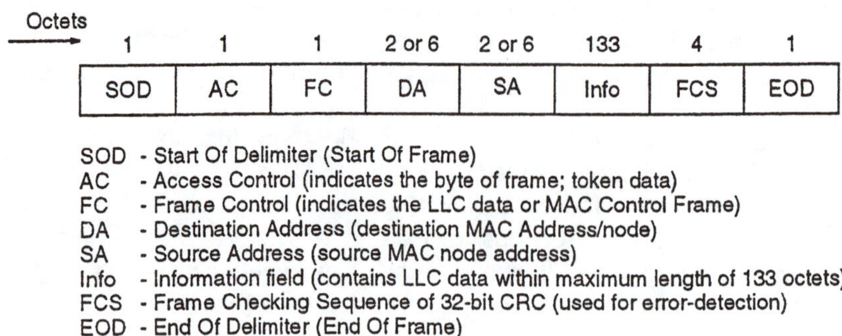

Figure 8.14 IEEE 802.5 token ring MAC control frame format.

Sometimes due to either time-out, failure, or error-generated tokens, a node in the ring must regenerate a new token and circulate it around the ring. The node which generates the new token becomes the monitor for the token and transmits a token frame with SOD, EOD, and AC fields, as if it is transmitting this frame after capturing the token. After the token has been defined and is circulating over the ring, it stops generating further new tokens.

As stated previously, the token does not carry any data. The start of delimiter (SOD) provides synchronization and indicates the destination node about the token or frame. Initially, a node with the highest address will act as an active monitor node. It generates a token and sends it to the next node by not setting T. That node may not have the data to send but will check the token for any error. If it finds no error, it will regenerate the token and send it to next node. If the next node has the data to send, it will capture the token, append it, and rest A and C bits. When a frame including the token goes over the first node (which generated the token), it will again check it for any error. If no error is found, it will pass it to the next node until it reaches the destination. The destination node also checks for any error, copies the data, and, after setting C, returns it to the sender. After receiving the frame and checking that the destination has received the data, it will generate a new token and send it to the next node. In the event of error in the frame, the node will set the E-bit and pass the frame on to the destination. The destination node, after seeing the E-bit set, will not copy the data and send it back to the source node.

Token ring MAC control frame format: The token ring MAC format is shown in Figure 8.14. The **start of delimiter (SOD)** field of one octet indicates the start of the frame. The **access control (AC)** field of one octet indicates the type of frame (token, or other data, etc.) and defines various operations of the token. The **frame control (FC)** of one octet distinguishes between the type of frame (MAC control or LLC data). It is also used in establishing the priorities between LLC peer entities. The **destination address (DA)** and **source address (SA)** fields define the destination and source nodes in the ring and can be either two-octet or six-octet addresses. The information or user's data field indicates the user's data, and the maximum value of octets which can be defined in this field is 133 octets. The **frame check sequence (FCS)** field is of four octets and detects error using a 32-bit CRC-like error-detection technique or sequence. The **end of delimiter (EOD)** field of one octet indicates the end of frame and also provides an error-detected bit to the sending node.

The **frame status** field of one octet is usually used as an acknowledgment byte by the receiving node to inform the sending node about its matching address with the one defined in its frame and copying of the transmitted frame. We mentioned earlier that the destination node, after copying the frame, changes the status of bits to indicate the acknowledgment and, in fact, this is the field which provides the acknowledgment facility to the

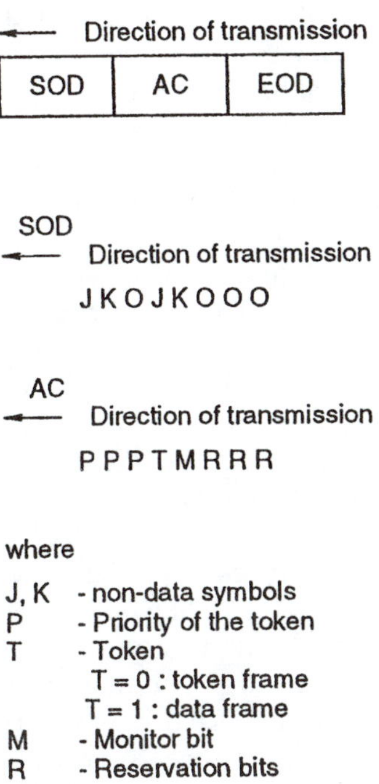

Figure 8.15 Token control frame format.

sending node. This type of technique of indicating acknowledgment is also known as a **piggybacking** type of acknowledgment, where we do not use a separate frame for the acknowledgment but, instead, change the status of bits in this field defined in the same frame (circulating over the ring back to the sending node) to indicate the acknowledgment to the sending node.

The token frame consists of three bytes: start of delimiter (SOD), access control (AC), and end of delimiter (EOD) (as discussed above), as shown in Figure 8.15. The purpose of the two delimiters (SOD and EOD) is to provide the boundaries for the frame and synchronization for frame transmission. It also indicates the beginning and end of frame transmission over the ring LAN. The standard bit format of the SOD field is **J K 0 J K 0 0 0**, where J and K represent non-data symbols and are identified by the absence of normal mid-bit transition in their representation by differential Manchester encoding.

The access control (AC) field has four sub-fields as shown below:

bit →	**0**	**1**	**2**	**3**	**4**	**5**	**6**	**7**
	P	**P**	**P**	**T**	**M**	**R**	**R**	**R**

where the **priority (P)** field of three bits indicates priority of the token and also indicates which nodes will be accessing the token ring LAN. The node which has the data frame to transmit and the highest-priority frame on the ring gets to access it first. The values of these priorities are also stored in a register known as a **priority register (PR)**. The value of priority defined in the token circulating around the ring corresponds to the current token ring priority service.

The **token (T)** field of one bit distinguishes between token and data frames; i.e., if T = 0, it represents the token frame and if T = 1, it represents the data frame. The **monitor (M)** field of one bit makes sure that the token with the highest priority always circulates around the ring. The **reservation (R)** field of three bits is used by the nodes with highest priority to make a request for using the next token. In other words, the node having a data frame with the highest priority makes a request for the token to be issued to the node at high priority. The priority value defined in the token frame is also stored in a **reservation register (RR)**.

As indicated in the previous sections, the ring-based LANs usually work in a very highly decentralized way but, since problems dealing with the tokens cannot be easily solved by cooperative nodes, in general, a central or monitor node is required which can solve different problems (errors or faults) in the token. These token faults typically include loss of token, duplication of token, etc. The monitor node must handle these faults for the correct operation of a token ring LAN. In the following paragraphs we discuss, in brief, the schemes used to overcome these faults occurring with the token.

In the case of a token being lost or busy, the monitor node checks that a token passes over it at certain time intervals. The maximum time a token should take to pass depends on the propagation time (for the token to go around the ring) and the maximum length of time for a data frame. If the token does not go over it after the expiration of this time, it switches into send mode and removes any information frame circulating around the ring. It then transmits a sequence of bits followed by a token back into the ring.

If a busy token is going around the ring more than once, the monitor node, after realizing that the busy token has passed over it more than once, removes it from the ring and submits a new token into the ring. The free and busy tokens are distinguished by one bit, and as such, if the free token has been sent out for a busy token (by mistake), it will also be handled in the same way that the busy token was handled.

After a node has transmitted a data frame onto the ring, it waits to receive its own address on the frame. If it finds that the address defined in the frame is not the same as its own, it knows that other nodes are on send mode. It will immediately terminate the transmission and will not generate any token. Now the monitor node provides its services, as explained in the previous cases.

If the ring is functioning well without any problem with the monitor, all is well, but if the monitor node itself has gone down (crashed), then one of the potential monitor nodes after a period of time (greater than or equal to the total propagation time multiplied by the number of nodes plus the time for a token to go around the ring) looks for a free token. If it finds one, it uses it; otherwise, it will generate a new token and become a new monitor of the ring-based LAN.

Table 8.4 provides is a quick reference for MAC formats for different types of IEEE standard LANs. In order to compare these formats, we define them within a standard general LAN frame format. This general frame format includes three fields: header, user's data, and trailer. In Table 8.5 we identify the header and trailer fields for each of these LANs.

Table 8.4 MAC Formats for IEEE Standard LANs

				IEEE 802.3 CSMA/CD					
Octets	7	1	2 or 6	2 or 6	2	Variable	4		
	Preamble	SOD	DA	SA	Length	MAC DU	FCS		
				IEEE 802.4 Token Bus					
Octets	2	2	1	2 or 6	2 or 6	<8174	4	1	
	Preamble	SOD	FC	DA	SA	MAC DU	FCS	ED	
				IEEE 802.5 Token Ring					
Octets	1	2	1	2 or 6	2 or 6	Variable	4	1	1
	SOD	AC	FC	DA	SA	MAC DU	FCS	EOD	FS

Table 8.5 Header and Trailer Fields for IEEE Standard LANS

Field	IEEE 802.3	IEEE 802.4	IEEE 802.5
Header	Preamble and SOD	Preamble, SOD, and FC	SOD, AC, and FC
Trailer	FCS	FCS and ED	FCS, EOD, and FS

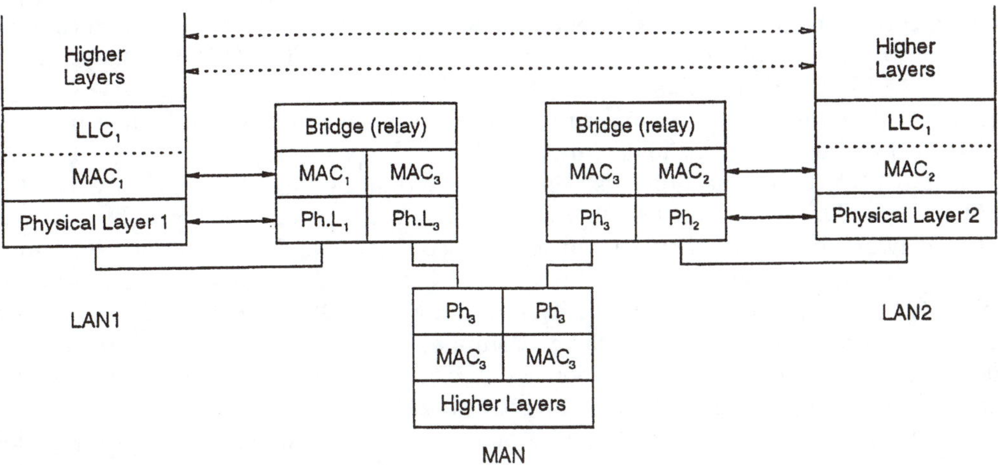

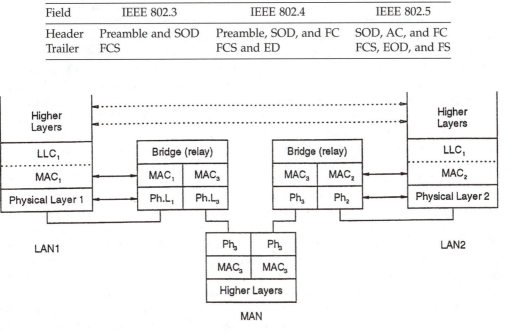

Figure 8.16 Interconnection of LAN1 and LAN2 via MAN.

8.2.6 IEEE 802.6 metropolitan area network (MAN) LANs

The MAN LAN protocols support interconnection to various LANs, WANs, and other high-speed networks (providing channel-to-channel transmission) and support data, video, and audio.[15,18-20] They offer these services over coaxial cable or fiber transmission media. Similar LANs can be interconnected by MANs using bridges in such a way that they accept the data frame from one LAN and translate it to another LAN, as shown in Figure 8.16.

In Figure 8.16, LAN1 is connected to LAN2 via relay bridges. The MAN including the relay bridge is a part of the carrier's equipment and provides transport for the data communication at high rates (100 Mbps). This type of bridging is defined at the MAC sublayer for identical LANs. Since the MAC sublayer defines bridging operations, it is transparent to the layers and users. The data from one LAN will be accepted by another LAN only when it is addressed to it. Similarly, the bridge defined at the MAC sublayer will forward the data to the MAN only if it is addressed to that particular MAN. In other words, the packet from any LAN (corresponding to the access network) will be transmitted to the user of another LAN only if is addressed to the MAN which provides interconnection with it. The MAN along with the relay bridge forms a carrier network known as a *transport network*. The transport network provides interconnection between any LANs (based on bus, token, or any other topology) or private LANs.

Distributed queue dual bus (DQDB): During the standardization process for shared media and access techniques for MAN, the IEEE 802.6 committee developed a distributed queue dual bus (DQDB) that had similar features to those of the ATM advocated by ITU-T.[15] The DQDB, as the name suggests, uses a two-bus architecture (similar to FDDI) and offers data rates from 34 M to 155 M. It supports both types of circuit- and packet-switching techniques and hence services.

A DQDB consists of two unidirectional buses carrying data in opposite directions. Since these two buses are independent, the capacity of the network is doubled with respect to single bus architecture. The entire bus architecture is defined as fixed-length slots. Each node connected to the network is equipped with an access unit (AU) with which the read and write operations can be performed on the allocated slots. IEEE 802.6 has defined the protocols for the physical layers and MAC sublayer. The physical layer of DQDB performs the same functions as that of the physical layer of OSI-RM and contains a physical layer convergence protocol (PLCP) that provides services to the DQDB layer. It supports a wide range of physical media and speed of data communication. The DQDB is the same as MAC in OSI-RM and supports different types of services, such as connectionless (datagram), connection-oriented (virtual circuit) using packet-switching, and connection-oriented isochronous (audio) using circuit switching

In 1987, Telecommunication Australia defined a dual-bus architecture over which the IEEE 802.6 MAN standard, the DQDB, was used. This concept was developed into a MAN-based network by a company known as Queued Packet and Synchronous Exchange (QPSX). The DQDB network provides communication services within a metropolitan area. The MAN provides interconnection between various DQDB sub-networks via bridges, gateways, routers, etc. The entire capacity of sub-networks can be used by various services such as connection-oriented, connectionless, and isochronous. The DQDB offers a data rate of 1.5 Gbps on fiber and 45 Mbps on coaxial cable.

MAN uses two contra-directional buses and offers unique access control protocols for both integrated isochronous (circuit-switched) and non-isochronous (packet-switching) traffic in the network. The isochronous service is based on the arbitrated access method; the fixed number of slots of fixed length can be defined by the TDM frame. The non-isochronous service is based on the queue-arbitrated access method and hence is supported by STDM (statistical TDM). The idea of defining a **single integrated broadband communication network (SIBCN)** was to offer various services to users at a reasonable price, and this is being accomplished with the introduction of B-ISDN and associated interfaces for LANs and PBXs. The MAN (private or public) offers an easy integration with SIBCN, as both MAN and B-ISDN use the same transmission medium and switching technique.

The MAN offers integrated information services (data, video, audio) and communication services can be implemented over a variety of platforms like FDDI, DQDB, B-ISDN, ATM, etc. Each of these platforms may provide support for synchronous, asynchronous, or isochronous services. The ATM-based platform for MAN implementation offers various advantages such as dynamic allocation of bandwidth, variable data rates, etc. The ATM provides support to asynchronous transmission (e.g., burst traffic, high-speed data, etc.) and synchronous transmission (e.g., regular traffic at fixed and regular interval, such as PCM audio and video).

The DQDB architecture defining the structure of MAN includes two contra-directional buses. The MAN provides a typical access control where the packets are assigned with priority numbers based on whether the packets are isochronous or non-isochronous communication. It also defines an upper limit on the capacity which can be allocated to the packets with higher priority.

MAN topology: The network topology of MAN is defined by two contra-flowing unidirectional buses I and II. The stations are connected to the buses via **access units (AUs)** by OR-write connection over contra-flowing unidirectional buses in MAN, as shown in Figure 8.17. Each of these operations (read and write) is performed in opposite directions. MAN supports two types of topologies: open and looped bus. The partitioning of the bus bandwidth into fixed-length slots is performed by a slot synchronization unit (SSU)

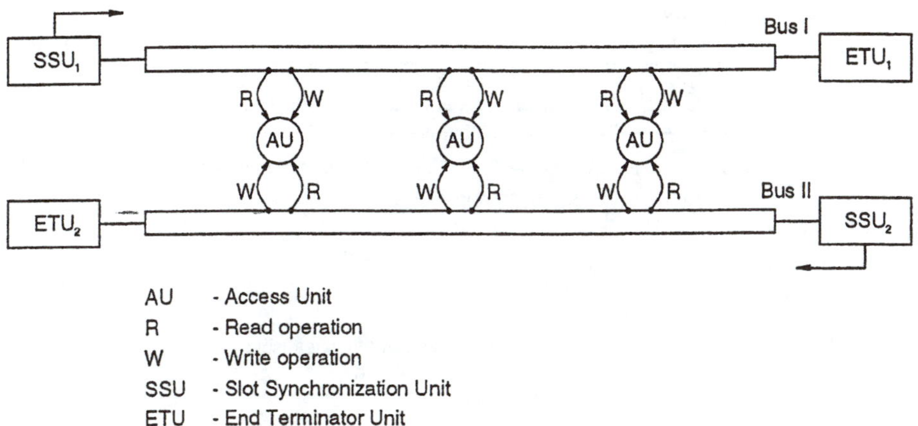

Figure 8.17 Contra-flowing unidirectional buses in MAN.

attached at the end of the bus while the other side of the bus is attached by an end terminator unit (ETU).

The DQDB may be connected to any LANs via gateways, and it provides support to a number of nodes connected to these LANs. A separate node to generate a frame may be considered in MAN. The MAN switch provides integrated circuit switching (for both voice and video communications) and packet-switching (for data communications). The bandwidth allocation to synchronous traffic is done dynamically, and any unallocated bandwidth is then shared by packet-switched traffic. It uses a frame of 125 μs which defines the number of slots between isochronous and non-isochronous communications.

The earlier version of MAN used a slot length of 32 octets, but a slot length of 69 octets is now being used with B-ISDN and ATM networks. The first octet of the slot provides the access control (AC) field, which is typically used for slot identification and protocol control. Out of 68 octets in the slot, the first four octets are used as a cell header, while the remaining 64 octets are available for isochronous channels (non-arbitrated frames). The cell header further assigns two octets for message identification (logical link) and two for service and segment header check. The segment type of message length is defined by two octets. Thus, out of 69 octets, 62 octets are available for user information. A voice channel of 64 Kbps is defined by one octet. The allocation of isochronous channels is performed by a frame generator of MAN which is a centralized node providing access to one or several channels to the nodes upon their requests.

The non-isochronous slots (queue-arbitrated) basically transfer the asynchronous message segment which is defined by the previous message. The message segmentation uses three types of messages: beginning of message, end of message, and more message.

AUs carry the user's information which may be based on isochronous (voice) or non-isochronous (high-speed data) communications. The slots defined by SSU are allocated with a view to maximizing the bandwidth and minimizing the overhead (particularly in the case of synchronous headers). The structure of a single slot is shown in Figure 8.18.

The first octet in the slot represents access control (AC) which is basically used by AUs. One of the main objectives of AU is to write the user's information into a free slot and read the user's information from the slot whenever it is busy or full. Figure 8.19 shows the description of bits of an AC field, where S represents synchronizing bits used for synchronization, set by SSU. The AU sets a busy bit (B) to indicate that it is carrying the information. The type of communication, isochronous (voice) or non-isochronous (high-speed data), is determined by a type (T) bit set by SSU. The request bits (R) are used by

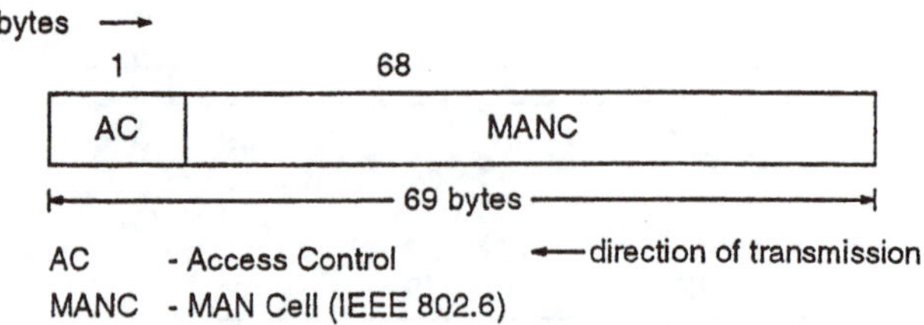

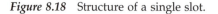

Figure 8.18 Structure of a single slot.

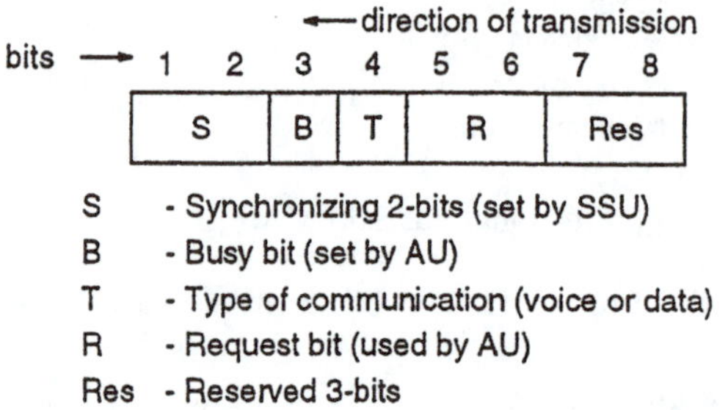

Figure 8.19 Description of AC field byte.

AUs to request voice or data communications. The remaining two bits are reserved bits (Res). The remaining 68 octets are defined for the cell of MAN slot (IEEE 802.6 standard). Thus, the total length of the slot becomes 69 octets. Some of the material above is partially derived from References 15 and 16.

8.2.7 Cambridge ring LANs

As mentioned above, the bandwidth in a Cambridge ring LAN is divided into ring slots which circulate around the ring continuously. The ring slots either carry the data or are available for carrying the data. Each node is allowed to use one slot for sending one small packet (known as a mini-packet) and then wait for a third ring slot to send any packet. There is some bit gap between head-to-tail. A typical cable connection for this LAN looks similar to a token ring LAN, as shown in the Cambridge ring LAN topology illustrated in Figure 8.20. For a ring LAN of 10 Mbps, the data field is approximately 40% of the mini-packet capacity, giving about 38 bits per mini-packet. Out of this, one third of the bandwidth for point-to-point configuration will offer nearly 1.3 Mbps for one slot. If there are n ring slots, the bandwidth in any slot will be $4/(n + 2)$ Mbps, and if there are m nodes trying to access the network at the same time, it will be $4/(n + m)$ Mbps.[26]

The nodes are connected to the ring via repeaters and, out of these, one node acts as a monitor node. Other nodes connected via repeaters are connected to other devices such as a station unit and access box.

Various control signals travel between repeaters and a node access logic unit attached by a network attachment interface unit (NAIU) to each node, as shown in Figure 8.21. As we discussed in token ring LANs, the repeater is connected between two nodes, accepts the data from one node, and, after regenerating it, sends it to another node. Here also, the repeater performs the regeneration functions over the data it receives from the node. There are five lines (two data and three control) in the interface between the node and repeater, and these are described below.

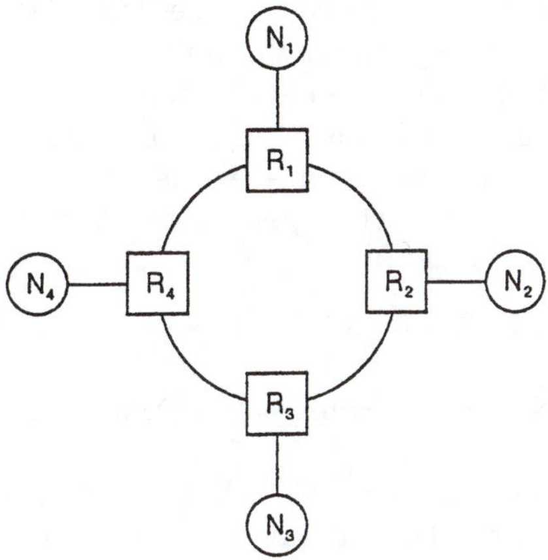

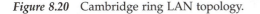

Figure 8.20 Cambridge ring LAN topology.

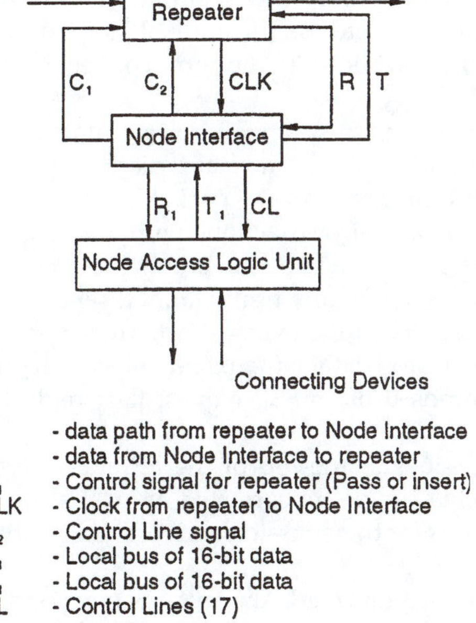

Figure 8.21 Ring structure and node structure.

The data from the repeater flows continuously over a **receive (R)** line. The data passed from the node to a repeater for its onward transmission travels over the **transmit (T)** line. These two lines are known as data lines. The control signal over line C1 indicates the condition for data to pass from the ring to another node or inserts the data of a node into the ring. The C1-line defines the clocking signal from a repeater to the node. The third control line C2 comprises three types of registers: **transmit (T)**, **receive (T)**, and **source select (SS)**. Various other control operations performed on this line include framing, parity checking, transmission and receiving of mini-packets, recognizing the acknowledgments, etc.

The T-line detects an empty mini-packet, defines the data, and transmits it on the ring. The response bits are defined as 11 after the transmission. The R-line, on other hand, monitors the data mini-packets addressed to it continuously. The R-register will store this mini-packet, provided it is cleared. If it is not clear, it will send a busy response by setting the response bits to 00. The returning mini-packets are detected by a suitable detection circuit used in this control circuit. The interface between node unit and access box is composed of two local 16-bit data buses.

Mini packet format: The format of a mini-packet (38 bits) is shown in Figure 8.22.

S F U Dest Source Data Data Res Par

where S field of one bit is used to provide synchronization. The F-bit indicates the status of the mini-packet (full/empty). The monitor node uses one bit to remove any erroneous mini-packet. The destination and source address fields of one byte define 255 addresses for destination and source, respectively. The two fields of data of two bytes carry the user's data. The two-bit response field indicates the response of the packet, while the parity field of one bit is used for error detection. We can have a number of mini-packets circulating around the ring at any time.

Ring protocols: The protocols for this LAN conform with the OSI-RM and include three higher-level protocols: **packet protocol (PP)**, **single shot protocol (SSP)**, and **byte stream protocol (BSP)**. The first two protocols (PP and SSP) support connectionless services and, as such, define datagrams, while BSP supports connection-oriented service and hence defines a virtual circuit.

The PP usually is used for transmitting large amounts of data within the mini-packet in the form of a block or packet. The PP format typically consists of a header, router, user's data, and checksum, and the length of the packet is two octets.

The SSP allows a number of transactions within a single packet over the ring. The packet may carry the data from one node to another and can also carry the single-packet response in the form of results for any transaction. It sends the data and also receives the results and any error message in the event of occurrence of an error. In other words, the return single packet may carry either the results or an error indication. The SSP packet format of 16 bits is composed of an SSP request flag, reply port, function number, and data for request.

The BSP protocol resides at the top of the PP and offers synchronization between nodes. The errors detected by PP are corrected by it. It also provides flow control. Each packet contains commands for transmission and receiving, control information, and user's data.

The specification for a 38-bit ring LAN contains two parts: **Cambridge Ring 82 interface specifications** (dealing mainly with the ring hardware and device interfaces) and **Cambridge Ring 82 protocol specifications**. These two parts are also known as **CR82**. An attempt was made to make this specification compatible with LLC, MAC, and the physical layer of OSI-RM, and after a number of drafts were exchanged between BSI and the IEEE 802 Committee, finally in 1984 two standards for Cambridge LANs were established: CR82, which deals with hardware, and a second specification that deals with IEEE LAN MAC sublayer protocols. The name of the Cambridge ring was changed to slotted ring LAN and conforms to ISO standards hierarchy 8802/7 defining services for both MAC/physical and LLC/MAC interfaces. The following primitives were accepted for these interfaces: P-PACKET.request, P-PACKET.confirm, P-PACKET.indication, M-DATA.request, M-DATA.confirm, and M-DATA.indication. The parameters are defined in a similar way as for standard LANs.

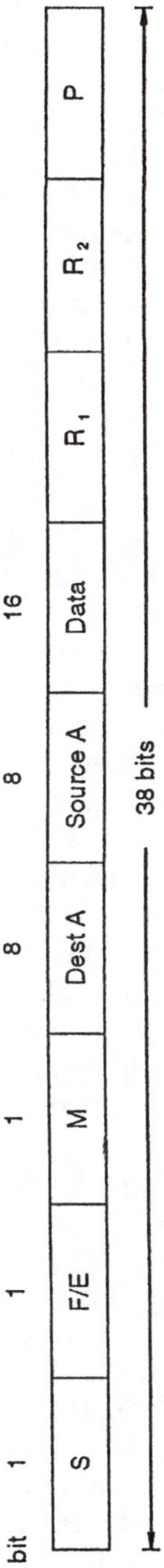

bit	1	1	1	8	8	16			
	S	F/E	M	Dest A	Source A	Data	R_1	R_2	P

38 bits

S - Synchronization (for minipacket)
F/E - Full/Empty (Status of minipacket)
M - Monitor bit used by monitor node
Dest A - Destination Address of 1 byte
Source A - Source Address of 1 byte
Data - Data field of 2 bytes
R_1, R_2 - 2-bits for reponse
P - Parity bit for error-detection

Figure 8.22 Mini-packet format.

8.2.8 Fiber distributed data interface (FDDI) LANs

FDDI uses a timed-token protocol that provides fairness to all the nodes to access the token ring.[21] The FDDI is based on fiber optics and covers a distance of nearly 150 miles. It uses two dual-counter rotating rings and a token passing MAC frame (IEEE 802.5 token ring standard). It offers a data rate of 100 Mbps and supports both synchronous and asynchronous data services by allocating the bandwidth required dynamically. The following is a list of ANSI/ISO standards that have been defined.

- Hybrid ring control (HRC) — ANSI X3.186, ISO 9314-5
- Physical layer medium dependent (PMD) — ANSI X3.166, ISO 9314-3
- Single mode fiber physical layer medium dependent (SMF-PMD) — ANSI X3.184, ISO 9314-4
- SONET physical layer mapping (SPM) — ANSI T1.105
- Token ring layer protocol (PHY) — ANSI X3.148, ISO 9314-1 PHY-2
- Token ring medium access control (MAC) — ANSI X3.139, ISO 9314.2 MAC-2

FDDI supports two types of nodes: dual-attachment and single-attachment. A dedicated node providing connection with fault tolerance known as a concentrator is also used. Realizing the need to send video, images, audio, and data over the network, the FDDI was extended to another version known as FDDI-II that meets the requirements at a higher data rate.

The dual counter–rotating ring is composed of primary and secondary rings. The primary ring usually carries the data and the secondary ring is used as a standby ring. In the event that a fault on the primary ring occurs, the secondary ring takes over and restores the functionality of the ring by isolating the fault. Now the network is running over a single ring alone. After the fault is repaired, the network operates with dual rings. If more than one fault occurs in the network, then only those nodes that are still connected by a ring can communicate with others Various states of FDDI ring operations can be defined as ring initialization, connection establishment, ring maintenance, and steady state. The steady state represents the existence of an operational FDDI where the nodes connected to the ring can exchange data using a timed-token protocol. FDDI supports both synchronous and asynchronous services. As mentioned earlier, the synchronous frames are transmitted whenever the agreed upon bandwidth is available, e.g., video and audio messages. On the other hand, asynchronous frames are transmitted even if the bandwidth is not available, e.g., datagram-based messages.

The FDDI specifies MAC services to its LLC and higher layers. The LLC accepts these services and offers further services to higher layers. The interface between MAC and LLC provides support for the exchange of protocol data units (PDUs) between them through primitives and parameters. As in other LANs, the physical layer offers a common interface with the MAC which is independent of the transmission media. Another interface offered by the physical layer is dependent on the media being used, as it depends on the electrical, mechanical, and functional behavior of the media. In FDDI, the physical layer is composed of two sublayers: **PHY (physical protocol)** and **physical medium dependent (PMD)**. The PMD sublayer defines the behavioral characteristics (electrical, mechanical, and functional) and also other relevant components of attachment interfaces, receivers, drivers, cable connectors, etc. These interfaces over the FDDI-layered model are shown in Figure 8.23. This model uses a pair of rings offering opposite directions for transmission of information, as shown in Figure 8.24(a). Various types of nodes used in FDDI reference models are shown in Figure 8.24(b).

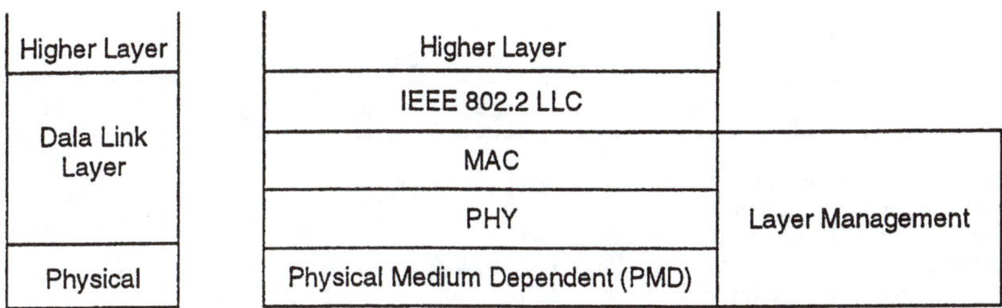

Figure 8.23 FDDI architecture model.

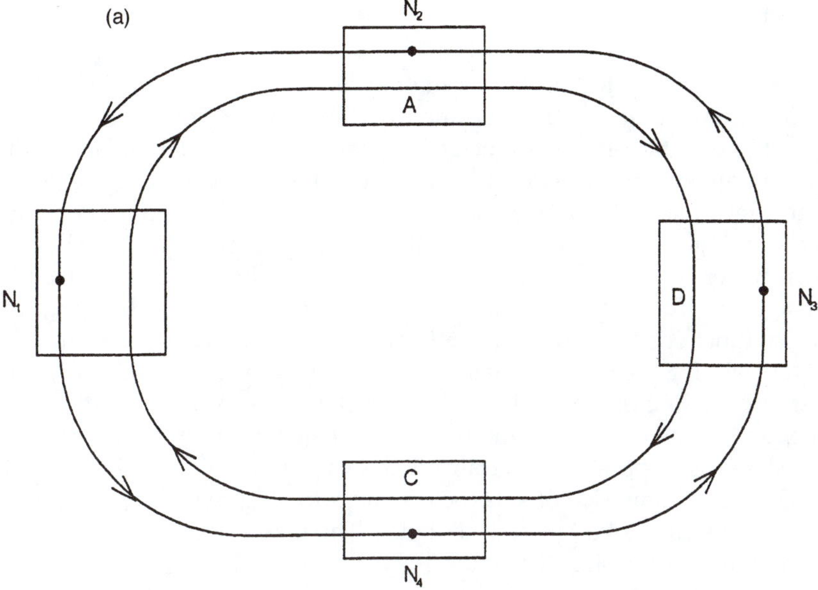

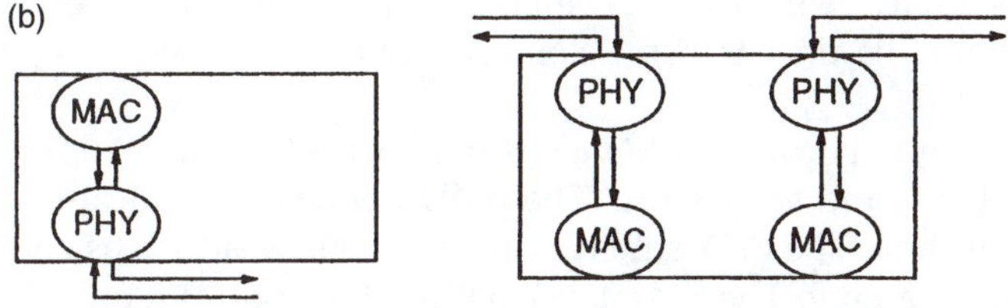

Figure 8.24 Types of nodes in FDDI LANs. (a) Rind node. (b) Reference model nodes.

There are a few differences between IEEE 802.6 and FDDI in the way of capturing the token and acknowledging the frame(s). As we explained earlier, the FDDI uses a very small token and, after transmitting frame(s), it issues and inserts the token immediately even before the arrival of the last bit of its transmitted frame(s). Further, the changing of the status of a bit required for acknowledgment is not needed here (as is done in IEEE 802.5 LAN).

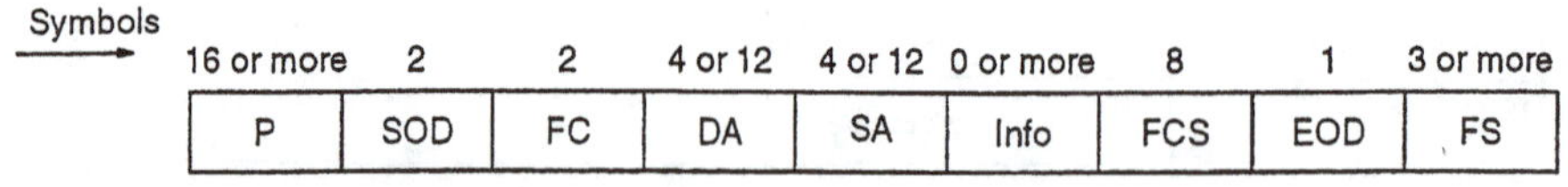

P - Preamble (for providing synchronization between frame and clock)
SOD - Start of delimiter (start of frame)
FC - Frame control (different control function)
DA - Destination address of destination node of 4 symbols (2 octets) or 12 symbols (6 octets)
SA - Source address of sending node of 4 symbols (2 octets) or 12 symbols (6 octets)
Info - Information field contains the LLC data units (to be transmitted) related to control operation
FCS - Frame checking sequence defines 32-bit cyclic redundancy check (CRC) on frame containing FC, DA, SA, and Info field only
EOD - End of delimiter (end of frame)
F S - Frame status field of 3 or more symbols for error detection, address recognition, frame copying control, and other control indication can be defined.

Figure 8.25 FDDI MAC control frame format.

These differences are due to the FDDI's high speed of transmission. The FDDI does not support priority, since a node may issue and insert the token before it has actually received the last bits of its own transmitted frame(s). The main aim of FDDI is, in fact, the high-speed transmission of different types of data (bursty, voice, video, etc.) requiring higher bandwidth and also the transmission of multiple frames over the ring topology simultaneously.[7–9,12,15,19,22]

The monitoring of various functions within ring topology is provided in a distributed environment where each node must make sure that the ring is not initialized for invalid conditions created by either the same node or any other node in the ring. In order to provide these functions, error-detection and error-correction techniques must include the processes for claiming the token, initialization of the ring, and error recovery (in the event of a broken link, failure of any node, etc.). The FDDI offers a data rate of 100 Mbps over optical fiber under ring topology. It defines its own FDDI frame and supports a distributed clocking system (as opposed to centralized clocking and IEEE 802.5 frame). The FDDI allows a maximum frame size of 4500 octets and supports two categories of source and destination addresses of two bytes or six bytes. The token ring, on the other hand, does not have any limit for the size of a frame.

FDDI MAC layer: The standard IEEE 802.6 also includes the first two layers of OSI-RM, and the media access control (MAC) sublayer of the data link layer offers more or less the same services as in the token ring LAN (IEEE 802.5 token ring), with minor differences. It provides connectivity at 100 Mbps. The FDDI LAN provides support for both synchronous and asynchronous data transmission. The error rate is very low (typically in the range of 10^{-9}).

The MAC sublayer is concerned mainly with access to the medium, error checking, framing, flow control, and addressing. The PMD defines the optical-fiber link and other components of this link. The PHY sublayer deals with issues such as encoding, decoding, clocking, and data framing. The SMT layer defines the control required for proper operation of stations on the ring and services such as station management, configuration management, fault analysis, error recovery, routing management, etc.

The FDDI MAC control frame format is usually defined in terms of symbols of four bits each, and these symbols are exchanged between MAC entities (Figure 8.25). It is important to note that the MAC entity always handles the data frames bit by bit for its physical layer. The fields in the frames are as follows:

- **Preamble (P):** This field is used for providing synchronization between the FDDI frames and a node's clock. The source node determines 16 symbols which are basically non-data and typically a unique set. The actual sequence of symbols depends on the encoding schemes being used with the media.

- **Start of delimiter (SOD):** This field of two symbols (one octet) indicates the start of a frame. This field contains the non-data symbols, which are different from the data symbols. It consists of J and K non-data symbols where each symbol is assigned four bits.
- **Frame control (FC):** This field of two symbols (one octet) is used for defining the status of various frames defined by MAC. It consists of a typical unique symbol pattern given as CLFFZZZZ, where the C-bit represents whether the frame is synchronous or asynchronous, the L-bit indicates the choice between two-octet or six-octet addresses, and the FF-bits distinguish between the LLC data frame and MAC control frame. If the frame is a MAC control frame, then the remaining bits (ZZZZ) represent the type of MAC control frame.
- **Destination address (DA):** This field is of either four symbols or 12 symbols and provides the physical address of the destination node. This address may be unique for one node, a multi-cast group of selected addresses (nodes), or a broadcast address (all the nodes). The FDDI ring may have both 16-bit and 48-bit addresses.
- **Source address (SA):** This field of four symbols (two octets) or 12 symbols (12 octets) defines the address of the source node.
- **Information (user's data):** This field contains a variable length of symbols (starting from 0 symbols) and represents the LLC data related to the control operation defined in the MAC control frame.
- **Frame check sequence (FCS):** This field of eight symbols (four octets) defines a 32-bit cyclic redundancy check (CRC), and its computation is performed on the frame containing only FC, DA, SA, and user's data fields.
- **End of delimiter (EOD):** This field of one symbol (four bits) indicates the end of the frame. This field represents a four-bit terminate (T) symbol for a data frame (indicating its end) and an eight-bit token frame containing two T-symbols (indicating the end of a token frame).
- **Frame status (FS):** This field is of three or more symbols and is used to indicate the status of the frame after it has been transmitted. It contains an error (E) bit, and address (A) recognized and frame copied (C) indicators. Each of the indicators is defined in terms of symbols reset (R) and set (S). The symbol R will have **Off** and **False** conditions, while S will have the conditions **On** and **True**.

FDDI token ring protocols: The token ring protocols of FDDI work more or less in the same way as IEEE 802.5 token ring protocols, with some minor differences. In FDDI, a token (fixed number of bits) control packet circulates around the token ring continuously. The **FDDI token frame** contains 22 symbols.

The first field of this frame is a **preamble (P)** of 16 symbols. The next three fields, **start of delimiter (SOD)**, **frame control (FC)**, and **end of delimiter (EOD)**, are of two symbols each. The token frame provides the right to transmit the data packets to the destination node.

Any node which wants to send the data needs to capture the token from the ring. It removes the token from the ring and transmits the data frame (which includes a token at the end of the data frame). The data frame is copied by the destination node and during this copying time, the token (released by the sender) may be captured by another node which will start sending its data frame around the ring. At any time, we may have more than one frame on the ring, one after the other (with no collisions, as these are spaced apart over the time axis). Once the data frame has been copied by the respective destination node, the data frames still going around the ring will be removed from the ring by the source node only. In IEEE 802.5 token ring LANs, only one frame can be on the ring at any time, as opposed to multiple data frames in FDDI LANs.

Out of two rings, one ring is based on multi-mode fiber which provides fiber connection up to two km, while the second ring is based on single-mode fiber supporting a fiber connection of up to 100 km. In other words, the multi-mode fiber provides a network tolerance for up to two km, while the single-mode fiber provides such tolerance for up to 50 km, giving rise to a total fiber cable distance of 100 km (perimeter path per fiber cable). It supports 500 stations and requires 1000 physical connections, yielding a total fiber length of 200 km (a pair of fibers is being used).

The ring topology is suitable for optical transmission and, further, FDDI is intended to provide point-to-point communication. Other topologies such as bus, star, and symmetric may also be used, but these topologies have a large number of source points and also do not provide point-to-point configurations (although they may support them indirectly). Since FDDI is using two rings for ring topology, the failure of any connecting device or link may automatically reconfigure the network without affecting the normal working and performance of the LAN.

FDDI is used as a backbone network within a campus or the premises of an organization. It connects different types of LANs being used in various departments and divisions. Those LANs usually offer data rates of 4 and 16 Mbps (token ring LAN) and 10 Mbps (Ethernet), and when connected with FDDI, a data rate of 100 Mbps is achieved. Typical average throughput with a reasonable number of LANs can be as high as 80 Mbps. Although it can carry isochronous (audio) messages, it still uses the public telephone network for carrying audio over PBXs.

Differential encoding in FDDI: One of the main objectives of ANSI Committee X.3T9.5 (established in 1982) was to define a standard for a high-speed data network which supports a packet-switched network at a very high speed over a variety of fibers. This resulted in FDDI, which meets the requirement of providing interconnections to LANs, mainframes, mini-computers, sharable resources, peripheral devices, etc., at a very high speed. It uses a token ring (similar to IEEE 802.5 token ring LAN) over twisted-pair or fiber-optic cable, with the only difference that here a pair of fiber-optic cables is used as a unidirectional looped bus. Each looped bus (or ring) offers a data rate of 100 Mbps. The FDDI ring so defined provides interconnections to LANs independent of protocols being used by them and is used as a high-speed backbone LAN. At the same time, the FDDI can be used as a high-speed LAN (similar to existing LANs) for workstations. Thus, the FDDI LAN offers high speed, high bandwidth, noise immunity (due to optical transmission of signal through fiber), security, low-error data rate (due to NRZI encoding 4B/5B standard), high reliability, etc. The 4B/5B code is used for timing and clocking on the network. Some of these encoding techniques[6–9,12,13,22,24] have already been introduced and explained earlier; we now discuss them in more detail.

In the **non-return-to-zero (NRZ)** encoding technique, the logic 1s and 0s are represented by low and high values of levels, as shown in Figure 8.26, and during the encoding in digital transmission, two steps are followed. Other encodings based on this are return-to-zero (RTZ) and non-return-to-zero-inverted (NRTZI). In RTZ, logic 0 is represented as a zero signal for full one-bit duration, while logic 1 is represented by one signal for the first of the half-bit duration and then zero for the second half of the duration. In NRZI, logic 0 is encoded by no change in transmitted value, while logic 1 is encoded by inverting the previous transmitter value. In **Manchester encoding**, the transition is defined at the middle of each bit period. The middle-bit transition serves as a clock and also as data. The logic 1 is represented by a high-to-low transition, while logic 0 is represented by low-to-high transition. In **differential Manchester encoding**, the middle-bit transition is used

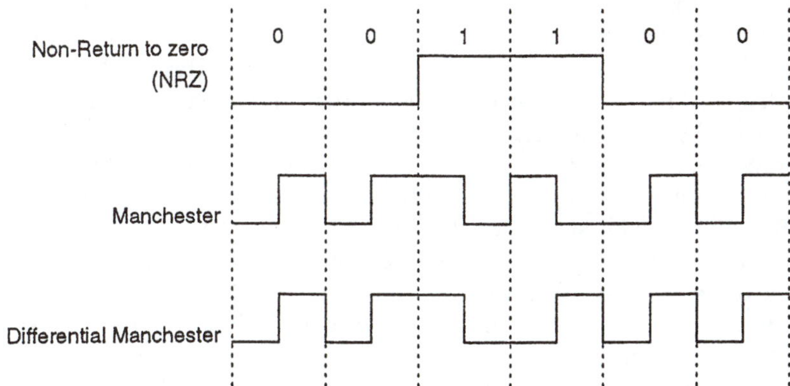

Figure 8.26 Encoding techniques.

to provide a clocking signal only and it does not represent data (in contrast to Manchester). The logic 1 is represented by the absence of transition at the beginning of the bit period, while logic 0 is represented by the presence of transition at the beginning of the bit period, as shown in Figure 8.26.

The differential Manchester encoding technique offers 50% efficiency, as each logic 1 and 0 data is represented by a transition in signal and, as such, it occupies twice the bandwidth for each data bit. The optic fiber transmission medium is less susceptible to noise, and hence two transitions to represent and recognize the data bits are really not required. To avoid this problem of using 200 MHz, the FDDI defined a new encoding technique as **4B/5B NZRI group encoding**. By using this encoding technique, a throughput of 100 MHz with a 125-μs rate can be achieved, instead of 200 MHz (needed in differential Manchester encoding). One of the main objectives of 4B/5B NRZI is to transmit the signal from a network into symbols. Here, four bits of data are converted or translated into a five-band value, and then this signal translated into symbols is transmitted over the network.

The NRZI encoding also follows two steps. In the first step, the encoding is defined by considering four bits at a time; i.e., each set of four bits is encoded as five bits. Thus, each data component is considered a binary value and encoded using NRZI (here, the logic 1 is represented with a transition at the beginning of the bit period, and no other transition occurs). This type of encoding via NRZI offers certain advantages: the representation of data as four bits gives 16 combinations, while a five-bit pattern gives 32 combinations. This offers users the opportunity to select a particular set of combinations to be used, and other unused combinations are invalid. The selection of the code combinations follows certain guidelines; for instance, a transition is present at least twice for each five-bit combination code. In other words, in terms of NRZI, no more than three 0s are allowed in any combination, since the logic 0 is represented by the absence of transition. The 4B/5B NRZI encoding does not support clock recovery while, due to the large number of pulses used in Manchester encoding, clock recovery is not a probability. Out of 32 combinations, 16 combinations are used for data: three for delimiters (two start and one end), two for control, three for hardware signaling, and eight unused. This saves the bandwidth required on the network, as now we can get 100 Mbps on FDDI in a frequency band of merely 125 MHz. Some popular network-based optical-fiber media are **S/Net, Fastnet, Expressenet, Datakit,** etc.

FDDI LANs: The FDDI LAN defines two types of nodes as Class I and Class II. Two rings used in FDDI are known as primary and secondary rings, providing data communications in opposite directions. The Class I node is connected to both primary and secondary rings, while Class II is connected to only one ring. The Class I node can connect several Class II nodes and, hence, can be used as a **concentrator** or **controller** in FDDI networks. The FDDI has been used mainly as a backbone network for interconnecting LANs, WANs, etc. and also as a feeder plant, intelligence data communication network. Its application in the area of high-speed LANs for workstations has not been promoted by vendors or users due to its cost, complex hardware, and software support for the transmission medium. Also, the FDDI is not well suited for the applications of broadband LANs which carry full analog video (6 MHz). If we use a digitization technique for an analog video signal of 6 MHz, the digitized channel may require a data bandwidth somewhere between 45 and 90 Mbps. The compression technique may result in a bandwidth on the lower side of the above range (i.e., 45 Mbps). Although multiple channels of integrated video, audio, and data can be defined in FDDI, the above example shows that the FDDI offering 100 Mbps may be fully reserved for one digitized 6-MHz TV channel only.

The bandwidth requirement in high-definition television (HDTV) is usually over 1.5 GHz per channel. The compression technique may reduce the bandwidth requirements per channel to around 200 Mbps. HDTV transmission is becoming very popular in B-ISDN, and a standard of 600 MHz has been approved which can accommodate three channels (compressed), each of 150 Mbps. As we mentioned earlier, the FDDI can also be used for resource sharing, and, further, it supports both synchronous and asynchronous and, in general, handles uncompressed data. Clearly, the FDDI is not suitable for video application (uncompressed) and other applications requiring higher bandwidth.

There are three fast Ethernet LANs: 100 Base T, 100 Base VG (voice grade) or AnyLAN, and FDDI. The 100 Base T can use unshielded twisted pair (categories 3, 4, and 5) and optical fiber. It is based on the CSMA/CD, as it uses the same MAC sublayer and also is compatible with 10 Base 10. 100 Base VG, or AnyLAN, was introduced by Hewlett-Packard and uses twisted pair (categories 3, 4, and 5), and optical fiber. It supports transmission of audio, data, and video. FDDI is based on the token concept and allows many frames to travel simultaneously over the media. It uses dual ring and hence offers fault tolerance. The American National Standards Institute (ANSI) introduced it; it uses copper twisted pair and coax. It supports synchronous traffic (audio, video), asynchronous traffic (data), and control information.

8.2.9 OSI LAN management protocol primitives and parameters

In the area of LAN management, the LANs interconnect a number of autonomous, multivendor, and different operating systems. The OSI committee has established a number of management domains for future network management activities such as network management, LAN management, OSI management, and management dealing with loosely coupled interconnection and distributed computing aspects of LANs. LAN management deals mainly with the lower layers, and the IEEE considers LAN management as system management. One of the main objectives of LAN management is to provide continuous and efficient operation of LANs.

Based on a number of recommendations, various issues which are addressed in LAN management include configuration management, fault management, performance management, access control management, and accounting management. *Configuration management* deals mainly with the controlling of a LAN from a remote node; this node should also be able to broadcast to all the users over the LAN. *Fault management* deals with fault detection, diagnosis, and correction, and the protocols must offer these services in the form of error reporting, confidence testing, repair (hardware and software), etc. *Performance management* enables evaluation of the performance of a LAN system, collection of statistical data about the network traffic, etc. *Access control management* deals mainly with maintaining integrity of the LAN system. *Accounting* keeps a record of the number of packets and maintains statistics of packets for planning capacity, bandwidth, and other application-based updating.

The IEEE specified two layers of a system management model: layer management and system management. Layer management is defined for each layer, and each layer is responsible for gathering all the relevant information for layer management. Within each layer, a layer management entity is defined to collect the information of its layer. Layer management is coupled with the system management application process entity. These entities of layer and system management interact with each other.

The layer management entity of a particular layer provides its services to the system management entity, while the layer entity can get information about other layers via system entities. The data communication between these two layers follows the same steps as for normal data, and each of these management layers defines its own set of primitives. Interaction between the system entity and the layer entity is regulated by a layer management interface (LMI), while the system layer entities interact with each other via a system management data service interface (SMDSI). As we have discussed in OSI-RM, each layer of OSI-RM defines its own set of protocol primitives and parameters; similarly, these two interfaces also define their own set of primitives and parameters.

The SMDSI interface defines the primitives through which it interacts with the layer management entities as **SM-DATA.request** and **SM-DATA.indication**. SMPDUs are used with primitives to carry the protocol data units of the system management layer which are transmitted to remote system management layer application processes. Similarly, SMPDUs are received from the remote management layer application processes. The management layer application processes offer interaction between various protocol entities.

The LMI interface, on the other hand, is used by the management layer application process for making a request to the layer; the application process informs the layer about the event. The layer informs the application process after the occurrence of the event and, finally, the layer makes a request to the application process. The following is a list of associated primitives which have been defined for LMI:

LM-SET-VALUE.request
LM-SET-VALUE.indication
LM-COMPARE-AND-SET-VALUE.request
LM-COMPARE-AND-SET-VALUE.indication
LM-COMPARE-AND-SET-VALUE.confirm
LM-COMPARE-AND-SET-VALUE.response
LM-GET-VALUE.request
LM-GET-VALUE.indication
LM-ACTION.request
LM-ACTION.indication
LM-ACTION.response
LM-ACTION.confirm
LM-EVENT.indication

System management layer protocol data unit (SMPDU): SMPDUs are used to exchange information between peer layers, and each protocol data unit is associated with a set of parameters. The following eight SMPDUs have been defined: PrivatePDU, LoadPDU, RequestPDU, ResponsePDU, EventPDU, EventAckPDU, TraceRequestPDU, and TraceResponsePDU. A part of the above discussion is derived partially from Reference 24.

8.2.10 LAN specifications and performance measures

Specifications of popular standard LANs: In order to compare the performance and specifications of different LANs, we consider the specifications of some of the popular LANs which are useful during the installation of LANs.

Standard Ethernet:

Maximum length of coaxial cable segment:	500 m
Maximum length of coax between nodes:	1500 m (includes three tapped segments with two repeater connections)
Maximum length of transceiver cable:	50 m
Maximum number of nodes:	1024
Maximum delay for point-to-point segment:	2.75 μs

Standard Ethernet (single segment coax cable):

Maximum cable segment:	500 m
Grounding at a single point to building's ground system	
Each end is terminated with 50-Ω resistor	
Maximum number of transceivers per segment:	100
Minimum distance between transceivers:	2.5 m
Maximum transceiver cable length:	50 m

Large-scale Ethernet system (copper):

Cable length:	500 m
Each segment is terminated at each end by 50-Ω resistor	
Maximum number of transceivers:	100
Maximum number of stations per network:	1024
Maximum distance between nodes:	2.5 km
Maximum number of copper segments:	5
Maximum number of repeaters:	4
Maximum transceiver length:	50 m
Maximum distance between two transceiver nodes:	2.5 m

ThinNet:

Maximum length of coax cable segments:	185 m
Maximum length of coax between two nodes:	925 m
Minimum length of coax between two nodes:	0.5 m
Maximum number of nodes:	1024

Standard Ethernet (10 Base 5):

Maximum length of coax cable segment:	500 m
Maximum length of cable:	2.5 km
Maximum length of transceiver cable:	50 m
Minimum distance between transceivers:	2.5 m
Maximum number of transceiver nodes per segment:	100
Both ends per segment are terminated by 50-Ω resistor	

Thin-wire Ethernet (10 Base 2):

Maximum length of ThinNet coax cable segment:	185 m
Maximum length of cable:	4 km
Maximum length of transceiver cable:	50 m

Minimum distance between transceivers:	0.5 m
Maximum number of nodes:	30
Both ends of coax cable segment are terminated by 50-Ω resistor	
T-connectors are used to connect devices	

Twisted-pair Ethernet:

Maximum transceiver coax cable length:	50 m
Distance between concentrator-to-concentrator or concentrator-to-transceiver:	110 m
Network cable is unshielded two twisted pairs	
Maximum cable segment length:	110 m
No direct connection between transceivers	
Repeater or bridge between twisted-pair transceiver and standard Ethernet	

Token ring:

Maximum length of ring:	366 m
Maximum distance of one 8-node MAU on ring:	5 m
Maximum number of MAUs:	33
Maximum number of nodes per ring:	255
Maximum distance between MAU to node:	100 m
Maximum length of ring (with repeater):	760 m
Maximum length of ring (with fiber):	4 km

ARCNET

Maximum number of nodes:	255
Distance of node from passive hub:	600 m
Distance of node from active hub:	31 m
Install PC hub in server or workstation (never power off)	
Addition and deletion of nodes without powering off	
Active hubs may be daisy-chained for increased connectivity	
Passive hub cannot be daisy-chained	

The multiport transceivers can be used to expand the capacity of the transceivers; i.e., more nodes can be added to the transceivers. This reduces the number of segments of LANs. In general, the multiport transceivers can support eight nodes. Thus, these nodes are not connected to cable but instead are connected a multiport transceiver which is connected to cable. The multiport transceivers can also be cascaded, thus offering greater reduction in the number of transceivers. This reduces not only the number of transceivers but also the cost.

Similarly, the multiport repeater can be used to connect different types of Ethernet LANs. These repeaters may be used to connect six to eight LANs (e.g., six ThinNet, two standard Ethernet, or some other combination). These repeaters also give a choice of topologies; e.g., ThinNet segments can be connected in a star topology or one or two standard Ethernet cable systems. These carry preamble regeneration, fragment extension, and signal regeneration. Some multiport repeaters provide an automatic partitioning and reconnection of the repeaters in the event of a faulty segment.

Twisted-pair transceivers are used to link PCs in the network to a concentrator. One side of the twisted-pair transceiver is connected via its 15-pin transceiver cable to a PC Ethernet board, while the other end is connected to twisted-pair telephone cables. The transceivers can be used to connect a variety of Ethernet-compatible devices, such as IBM, Sun, Apple, and DEC devices.

Fiber-optic transceivers are used to connect a variety of DTEs or PCs to optic-fiber baseband 10-Mbps CSMA/CD LANs. These transceivers include two ports (transmit and receive) that can connect the transceivers to duplex fiber-optic cable. The fiber-optic cable covers a greater distance and can be used to connect a number of LAN segments via fiber-optic and fiber-optic multiport repeaters.

Performance measures of LANs: We have discussed a number of criteria of performance parameters such as performance measurement, performance based on topology, and performance based on links, and all of these performance parameters help us in comparing LANs to suit our needs. These parameters are computed or evaluated using various measures. There are many ways of comparing LANs, and one of the criteria for comparing them is by performance measure of the network. Different books have defined different measures of performance, but in general, the performance measures used in LANs or in any network are characterized by parameters such as delay, waiting time, and throughput. There is a possibility of some overlapping between these measures, but these measures are enough to measure the performance of LANs. Some of these measures also depend on the transmission media, MAC protocols, and other characterization of LANs (discussed earlier).

The **throughput** of a LAN is defined as the number of data packets it sends over a period of time. This is always measured as a function of total traffic offered by that LAN. The **delay** is defined as the time a station has to wait to access the network after the frame is ready to be sent. This measure is also expressed as a function of total traffic a LAN offers. The delay is usually expressed as a range having a lower and upper limit. The lower limit of delay, in fact, represents transmission of a greater number of frames, which in turn offers a higher throughput. The lower limit of delay as an average delay for any LAN is acceptable in a few applications, but for time constraint–based applications, the upper limit of the delay is more important for the design of the systems. The **waiting time** (also known as response) is the time a station is waiting in the queue to transmit the frame. Although this time may have some relationship to the delay time as defined earlier, the difference between these two times lies in the fact that the delay time is caused by transmission media, while the waiting time is due mainly to the type of MAC protocol being used.

In general, these performance measures depend on a number of other related parameters and are considered for determining the performance of the network. These parameters include capacity (transmission link), propagation delay (transmission media), frame length (MAC, LLC sublayer protocols of LAN), number of stations (depends on the LANs), maximum amount of data traffic (handled by transmission media), access protocols (MAC sublayer protocol), error rate of transmission media, data rates (depends on the MAC protocols and transmission media), etc.

The **bit error rate (BER)** is a performance measure defined as the ratio of the number of bits received to the total number of bits transmitted over a period of time. Bit errors may be a single error or multiple (burst) error. A single error is usually caused by noise and non-synchronized transmitter and receiver and represents one error. A multiple error, on the other hand, is caused by impulse noise and factors due to improper management of the bits during transmission.

A **packet error rate (PER)** is defined for packets, as opposed to BER, which is for bits. The PER is the ratio of the number of packets received to the total number of packets transmitted over a period of time.

Two additional performance measures have been established for packets and are known as **packet loss rate (PLR)** and **packet insertion rate (PIR)**. The **PLR** is the ratio of the number of packets lost to the total number of packets transmitted over a period of time. The loss of a packet is basically due to traffic congestion, hardware failure, or misrouting. The **PIR** is the ratio of the number of packets inserted during the transmission to the number of packets transmitted over a period of time. The packets which are inserted are those which are accepted as valid packets by the nodes to which these packets do not belong.

Bit errors are usually detected by the transmission system, as it deals with bits only, while packet errors are detected by switching and multiplexing techniques, as these systems deal with the packets.

The majority of these parameters are usually handled by the network configuration, except the error rate of the media, which is usually very low for any type of LAN (single owner and complete control and management of the network locally), but it may not be low for WANs which may work over leased lines or packet-switched telephone networks or any other switched network (as there is no local and single control).

The **capacity** of the network or media is defined as the maximum rate in bits per second which can be transmitted over it. Since stations are transmitting the frames, these frames are being transmitted at this data rate over the network or medium.

The **propagation delay** is defined as a time a frame takes to travel down the length of the medium or network. In general, most of the transmission speeds of the available transmission media used in LANs are more or less the same; it may be possible to determine this delay by knowing the length of the network. It has been observed that this speed is typically two-thirds of the speed of light (2×10^8 m/s).

The **frame length** used in LANs is the number of bits in the frame and is defined or constructed at the data link layer. The amount of data information in the frame is generally expressed as bytes or octets (one byte or octet is equal to eight bits) and typically depends on the type of MAC protocol being used and the support it is getting from physical layer protocols. It is interesting to note that the type of MAC defines the type of LAN, as all the LANs have a common LLC protocol. The LANs support different types of communication media. In some cases the size of a frame defined by individual LANs is also different, as destination addresses of two values (two or six bytes) are used.

The maximum **number of stations** which can be connected to LANs is usually determined by the protocols used at the physical layer and MAC sublayer. Note that it is possible to have a large number of stations connected to LANs, as it may improve the throughput. The only drawback in this technique is the reduction of the bandwidth for each station, as the total bandwidth (expressed as bits per second) is now shared by these stations. If we assume that all the stations want to use the LAN at the same time with equal probability of accessing it, each station will get one-*N*th of the total bandwidth or capacity of the network.

The **network traffic** offered depends on the access control technique used and is usually calculated on the basis of average traffic offered by all the stations simultaneously. As expected, we will assume that all the stations are transmitting at the same time and are sharing a one-*N*th bandwidth for the application. It is possible that the total traffic at any time may exceed the upper limit of the capacity (an ideal condition where all stations are sharing a maximum of one-*N*th bandwidth of the network). It is easy to calculate the average traffic of deterministic LANs (such as token bus, token ring, and Cambridge ring), but CSMA/CD LANs will not allow calculation of the average traffic, as it is non-deterministic due to its random access control strategies (MAC protocol).

The **access control technique** is another important parameter for the performance evaluation of LANs. There are different techniques for performing studies to determine the performance of networks, such as analytical models, simulations, etc.

Of the above-mentioned parameters, it is quite clear that the data rate of any LAN is fixed, while other parameters such as frame length, number of stations, and length of the

LAN are variable. In order to compare the performance of the LANs, various combinations of these parameters can be used. For example, one may try to assess the traffic offered by the networks by considering both sizes of frames (different source and destination addresses) and fixed number of stations. For each combination of frame length and fixed number of stations, the average traffic and length of the networks may be determined.

Performance measures of communication link: In the previous section, we discussed a number of parameters which are useful for predicting the performance behavior or measures of the networks. These measures include delay, throughput, and waiting time. We further discussed a list of other parameters which affect these measures in one way or another. The efficient design of a network considers these measures and parameters. In addition to these parameters, there is another measure (ANSI X3.44 standard) known as the **transfer rate of information bits (TRIB)**, which is generally used for comparing the efficiency of various protocols being used. The choice of a particular protocol depends on a number of parameters such as coding, size of frame, and options offered by the protocols for error control, control information, and flow control. In this section we introduce TRIB and, with the help of a suitable example, show how this can be calculated for a particular protocol.[11]

The coding technique plays an important role in encoding the characters of a data message being transmitted, as it represents the number of bits to be transmitted. Further, different coding techniques use different methods for error control and other control information. Thus, the efficiency of a code can be defined as a ratio of the number of bits per message or character to the actual number of bits transmitted. The actual number of bits will include the bits required to represent a character and also the bits required for error control and other control information. Obviously, it is best to have a minimum number of bits for control information and error control.

The TRIB, defined by ANSI X3.44, is used for comparing the efficiency of protocols and depends on a number of time delays defined during data communication. It is given by a ratio of the total number of bits transmitted to the total time to transfer these bits from sender to receiver. The number of bits transmitted includes data bits and control bits required for providing synchronization, error control, and flow control. The total time to transmit these bits includes a number of time factors, described below.

The bits experience some finite delay at the transmitting and receiving interfaces, propagation time (depends on the electrical behavior of the medium and its transmission speed), processing time required at the interfaces, time to receive an acknowledgment, time to fetch the data bits from buffer or disk, and also finite time for their storage. Out of these different time factors, the TRIB uses only four time factors: propagation time (t_{prop}), bit message–processing time (t_{proces}), circuit-delay time at the receiver (t_{rec}), and time (t_t) to transmit the bits. Thus, the total time to transmit the data message is given by the sum of these times.

The total time depends on the transmission speed of the channel. As we discussed earlier, there are three modes of transmission — simplex, half-duplex, and full-duplex. Further, we can include an option of acknowledgment, which will take the same amount of time to travel from the receiving side to the transmitting side.

Let's assume that the transmitting rate or speed of a communication channel is M bits per second. The time to transmit the number of bits in a message (say N bits, which includes data bits and control information) is given by N/M seconds. Similarly, if we want to include acknowledgment bits (say ACK), then the time for transmitting and receiving the acknowledgment is given by $(N + ACK)/M$ seconds.

This calculation of time can be made for each of the modes of transmission separately. For each of the protocols used in different standard LANs, some of the time factors like interface time, interframe time, etc., are defined in their specifications and can also be calculated by looking at the size of the frame being considered. In the above case, we are assuming that no error occurs, and hence TRIB defines the ratio for an error-free situation. But if we assume the occurrence of error, then the time for transmitting the data is given by the ratio of time to send the data message (without occurrence of the error) to the probability that there is no error introduced during data communication. This probability of no error can be calculated by using the following formula:

$$\text{Probability of no error} = (1 - \text{Bit Error Rate})^N$$

where N represents the number of bits in a block and Bit Error Rate defines the bit error rate of the channel.

Consider the example of CSMA/CD LANs. For evaluating the performance of CSMA/CD LANs, some of the parameters considered are given below:

- The minimum inter-packet delay is defined as the minimum rise-and-fall time of transmission and is usually given 9.6 μs.
- The round-trip delay, which is defined as double the propagation time (for a bit to travel from one end to another end), is given as 51.2 μs.
- The total time for transmitting a frame includes the following time factors: access time (t_{acc}), time for one frame (t_{fra}), and time gap between frames ($t_{interfra}$).
- To include the random access features of this LAN, the access time must be calculated on the basis of probability of successful transmission by one station at any time. The probability that one station transmits successfully is given by

  $$\text{probability} = N\,p(1-p)^{N-1}$$

  where p is the probability that a station wants to transmit during the next time slot and N is the number of stations. The average time required for recovery from the collision depends on the time of sensing collision and jam time. This time can be calculated by considering the round-trip propagation time in the worst case, and it may be given as half the round-trip propagation time (denoted as time slot). This time corresponds to slot time, which does not include the jam time.

The above equation must be optimized for calculating the probability that one station has transmitted successfully. From here, we can predict the number of re-transmissions. Then we can calculate the average time required for re-transmission by multiplying this probability by the time slot and then add the average time before collision is detected (half the round-trip propagation time); this will give us the access time. The slot time or round-trip propagation time in the case of CSMA/CD LAN is 51.2 μs. The t_{fra} can be calculated, as the data rate of this LAN is 10 Mbps and the size of one frame can be calculated. The control information includes the preamble and start of delimiter (SOD) fields. The actual data message to be transmitted will include all the fields (defined in frame format) except the preamble and SOD. The $t_{interfra}$ is given in the specification.

For a token passing LAN, the access time is determined from the token time, which is defined as the time taken by a station to pass the token to the next station. The access time can be calculated for the worst case by considering t_{access} = token time × N, where N is the number of stations. Further, in the worst case, when all stations are transmitting, the token will service them in sequence, and hence access time will be simply the sum of the token time-passing time for N stations and the average time for each of the other stations to transmit:

$$t_{\mathrm{acc}} = \text{Token time} \times N + (1 - N) \times t_{\mathrm{averfram}}$$

where t_{averfram} is the average time for a frame.

For a token ring LAN, t_{acc} is given by

$$t_{\mathrm{acc}} = \text{Token time} \times N + t_{\mathrm{interface}} \times N$$

where $t_{\mathrm{interface}}$ defines a one-bit delay at each interface and this access time is based on the worst-case situation where one active station has to wait for a token to go around the ring. Further, when all the stations are active (transmitting), then the access time in worst case is given by

$$t_{\mathrm{acc}} = \text{Token time} \times N + t_{\mathrm{interface}} \times N + (1 - N)t_{\mathrm{averfram}}$$

where t_{averfram} represents the average time for each of the stations and $t_{\mathrm{interface}}$ represents a circuit delay of one bit at each interface.

Internetworking: In any business or organization, two types of communication systems are very common: voice communication and data communication. Voice communication is handled by PBX, which also handles facsimile, while data communication is handled by LANs. As we have seen in the previous sections, a number of LANs have been introduced into business or organization premises, and different types of standard and non-standard protocols are being used in these communication systems. The existing standard LANs typically offer data rates of up to 16 Mbps over a typical distance of about 10 km. Typically, coaxial cable is used as a transmission medium in these LANs, and most of these LANs use ring or bus topology through star configuration. The available LANs include CSMA/CD, token bus, token ring, etc. A high-speed LAN (like MAN, FDDI, DQDB) offers a data rate of 100 Mbps. These high-speed LANs are used for interconnecting existing LANs and also for high-speed data communication required by workstations and file servers. These high-speed networks are generally used over a larger geographic distance, such as a metropolitan area of 100-km diameter, and support as many as 1000 users for data exchange. In these LANs, optic fiber is used as a transmission medium, and typically the high-speed LANs are classified into FDDI and DQDB.

Due to increasing communication requirements, interconnections must be made between different types of LANs, between MANs and B-ISDNs, and between LANs and private MANs. A hardware device known as an **internetworking unit (IWU)** is used to provide interconnections between these networks. Two networks (any) can be connected by this IWU if the distance between them is small. If the distance between the networks is large, then the networks are interconnected by intermediate sub-networks. The IWUs provide interconnection between two different types of networks and, as such, must deal with the problems of addressing, naming, routing, congestion control, flow control (due to different speeds), segmentation and reassembly (due to different sizes), etc.

The IWU is known as a repeater if two similar LANs are interconnected at layer 1 (physical). It is known as a bridge if different LANs are interconnected at layer 2 (data link) and a router if the networks are interconnected at layer 3 (network). If the networks are interconnected at a higher layer (normally at transport or application layers), the IWU is known as a gateway.

Brief descriptions of different interconnecting devices follow:

- **Repeater:** Regenerates the signal strengths between various segments of similar LANs, thus increasing the total length of the LAN. It operates at the physical layer. It does not provide any routing or management of packets/frames. If the packet/frame is lost or corrupted, we must look into another interconnecting device known as a bridge.
- **Bridge:** Connects different types of LANs at the data link layer. At the data link layer, it looks at the destination address in the received frame and determines whether to keep it or pass it to another connected LAN. At the physical layer, the bridge acts like a repeater in the sense that the received signals are regenerated. It must use the same protocol on either side. This is useful only for small or medium-sized network systems, but for larger network systems, another interconnecting device known as a router is used.
- **Router:** Similar to a bridge for interconnecting different LANs, except that it allows other types of networks such WANs, FDDI, or any other high-speed network. It operates at the network layer. It can be used to connect LANs to WANs, MANs to WANs, LANs to LANs, and so on. At the network layer, the routing for sending a frame to another connected network is done by hardware devices. Cisco Systems is the world's largest supplier of routers. The Cisco Series 7000 provides connectivity to high-speed networks using ATM and T/E1 systems and, in fact, is being used in Internet applications for connecting different types of networks (see Chapter 17 regarding the Internet). The routers in Internet processes ensure the timely delivery of frames and reduce the loss of frames. They handle Internet protocol (IP). The network nodes in high-speed networks are usually implemented via switches, while the nodes are implemented via repeaters in Internet processes. Routers provide connectionless service and use hierarchical routing using IP. The nodes of high-speed network switches, on the other hand, are usually telco and connection-oriented; use X.25, frame relay, and ATM; and are based on a fixed address.
- **Gateway:** Connected between the Internet and the destination network (target network). It receives Internet e-mail and forwards it to the destination. As said earlier, the message is broken into IP packets. The gateway receives it and uses TCP to reconstruct the IP packet into the complete message. It then provides mapping to the message into the protocol being used at the destination.

There exists a variety of tools for managing the TCP/IP suite of protocols. Every gateway connected to the Internet defines and maintains a database known as a management information base (MIB), which contains information regarding the operations of gateways, e.g., the operating system used in gateways, TCP, UDP, ARP mapping, IP, statistics such as the number of datagrams received and forwarded, number of routing errors, datagram fragmentation, reassembly, IP routing table, etc.

LANs offer data rates of up to 16 Mbps, while ISDN offers only 64 Kbps. If we have to interconnect LANs for higher data rate services, then these are interconnected to public MANs via dedicated links (attached with IWU). MANs provide connectionless services, but in the future we may expect connection-oriented and isochronous services.

LANs can be interconnected with public MANs, which are then connected to B-ISDN nodes. The B-ISDN nodes are connected to each other. This type of interconnection offers LAN users access to wider areas with flexibility, low delay, and high throughput.

The LANs can be interconnected via inter-switch link switches, e.g., the IEEE 802.10 standard and ATM LAN emulation standard. The LAN emulation standard protocol allows the ATM host and server to be internetworked with different types of LANs. As mentioned earlier, the data link layer of OSI-RM consists of two sublayers: logical link control (LLC) and media access control (MAC). The LLC provides interface between the network layer and service access point (SAP). The function of these SAPs is to support multiplexing within a single host over a single MAC sublayer address/port. Different types of MAC standards have been developed: IEEE 802.3 (10 Mbps, coax and fiber, frame size of 1500 bytes), IEEE 802.4 (1, 5, 10 Mbps, coax, frame size of 8191 bytes), IEEE 802.5 (4 and 16 Mbps, STP, frame size of 5000 bytes), IEEE 802.6 (34, 44, 155 Mbps, coax and fiber, frame size of 8192 bytes), IEEE 802.9a (isoEthernet, 16 Mbps, UTP and STP), IEEE 802.12 (100 vg-AnyLAN, 100 Mbps, fiber), and so on.

An interesting application of ATM supports many ATM switches connected at a central location that are connected to various frame relay switches at different locations of the same organizations. These locations connected via frame relay generally require an efficient access speed.

Virtual LAN is a new concept which uses ATM networks. In this environment, the users are connected to different LANs which are connected to other networks (LANs, WANS, etc.). The users can access the virtual LAN resources located and connected to remote LANs as if they are local resources. Virtual LAN supports multiple MAC layer protocols.

The TCP/IP protocol is generally used for LAN interoperability. This pair of communication protocols supports a variety of operating systems such as AT&T UNIX, VAX/VMS from DEC, various IBM mainframes, and PC systems. A client application defined within TCP/IP allows users to access any application software on the host, transfer files between any hosts irrespective of their locations, use electronic mail facilities, and perform many other tasks. The TCP/IP allows users to run their terminal sessions with Telnet, transfer files with FTP, use the electronic mail system with SMTP, and perform many other functions. Many TCP/IP-based software applications for workstations for different network operating systems and servers are available on the market.

The broadband Ethernet LANs are simpler to implement and install and are very reliable. These are available with bandwidths of 12 or 18 MHz. These LANs operate at a full 10-Mbps CSMA/CD capacity with 100% collision detection. These LANs are transparent to non-standard LANs (like DECnets), TCP/IP, and many other configurations and other higher-layer protocols. The baseband Ethernet usually operates at 120 Mbps with 100% collision detection and enforcement. Broadband Ethernet bridges are also available for bandwidths of 12 and 18 MHz to provide interconnection between baseband and broadband Ethernet for increased coverage and capacity. These bridges offer various features such as high performance for 10 Mbps throughput, redundancy and loop detection for reliable network operations, flexibility, etc.

Well-known LAN protocol standards: Figure 8.27 shows a list of protocol standards defined for the layered model by different standards organizations.

CCITT is mainly concerned with defining access protocols to the public communication data networks. It has defined data terminal equipment (DTE) as packet-switched networks that provide data circuit-terminating equipment (DCE). It has defined X.25 addressing schemes for the network layer. Layer 2 (data link) has been redefined as link access protocol-balanced (LAP-B), which is analogous to ISO's high-level data link protocol (HDLC) and ANSI's advanced data communication procedures (ADCCP). A corresponding standard X.21 has been defined for circuit-switched networks.

CCITT has also defined services for telematics, which typically cover user-oriented information transmission services like teletex (communication among office word processing systems), videotex (interactive information retrieval), and facsimile (images and text document transmission).

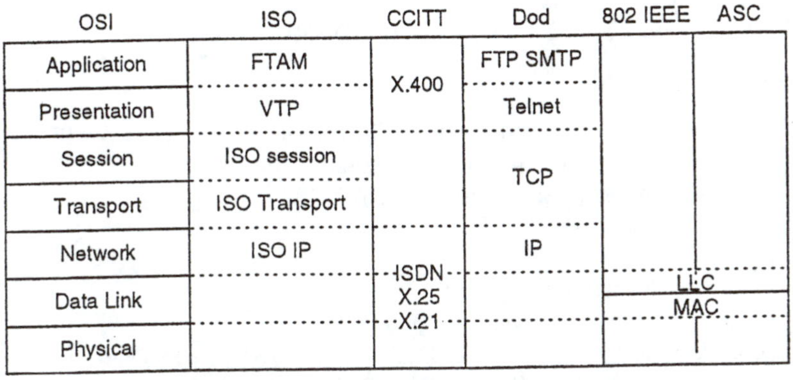

OSI	ISO	CCITT	Dod	802 IEEE	ASC
Application	FTAM	X.400	FTP SMTP		
Presentation	VTP		Telnet		
Session	ISO session		TCP		
Transport	ISO Transport				
Network	ISO IP		IP		
Data Link		ISDN X.25 X.21		LLC MAC	
Physical					

CCITT - mainly concerned with defining access protocol to public communication network.
- has developed Data Terminal Equipment (DTE) to packet-switched network that provides Data Circuit-terminating Equipment (DCE)
- X.25 address layer 3
- Layer 2 is refined as Link Access Protocol-Balance (LAP-B) and is anologous to ISO's High level Data Link Protocols (HDLC) and ANSI's Advanced Data Communication Procedures (ADCCP).

Figure 8.27 Some well-known protocol standards.

The Defense Communication Agency (DCA) promulgates communication-based and related standards, e.g., MIL-STD and standards defined by DoD. It issues these standards for military and DoD. It has developed a well-known communication interface standard for internetworking known as transmission control protocol (TCP) (equivalent to transport layer) and Internet Protocol (IP) (equivalent to network layer).

The IEEE 802 committee defined two layers for a data link of OSI-RM, thus making an eight-layer model for LANs. The IEEE 802 committee is concerned only with the lower two layers of OSI-RM.

ANSI's subcommittee, ANS X3T9.5, has developed standards for a type of local area network known as high-speed LAN (HSLAN).

The standards organizations are involved in the development of a variety of standards for the OSI model which can be used with different types of networks such as LANs, PSTNs, PSDNs, ISDNs, and high-speed networks (B-ISDN) based on ATM switching. The high-speed networks are discussed in Chapter 19, while ISDN is discussed in Chapter 18.

The ISO and CCITT have defined a series of standards for each of the layers of OSI-RM, given below.

ISO standards:

- Application Layer: Remote database access ISO 9579
  Distributed transaction processing ISO 10026
  Job transfer access and management ISO 8571(1-4)
  Virtual terminal ISO 9040/1
  Common application service elements ISO 8649/8650
  Directory system ISO 9594
- Presentation layer: ISO 8822(3-5)
- Session layer: ISO 8326/7
- Transport layer: ISO 8073/3
- Network layer: ISO 8880/8473/9542/10589, ISO 8208/8881

- Data link layer: ISO 8802.2
- Physical layer: ISO 8802.3, ISO 8802.4, ISO 8802.5 (for LANs)

CCITT standards:

- Application layer: Common management information protocol (ISO 9506/6)
  Interpersonal message X.420
  Message-handling system X.400
  Telex service TTX
  Videotex T100/1
  Facsimile T (0,4,5)
  Directory system X.500/520
- Presentation layer: X.216/226 (for PSDNs), X.50/51/61 (for PSTNs)
- Session layer: X.215/225 (for PSDNs), T62 (for PSTNs)
- Transport layer: X.214/224 (for PSDNs), T70 (for PSTNs)
- Network layer: X.213 and X.25(for PSDNs), T30 (for PSTNs), I450/451 (for ISDNs)
- Data link layer: X.212/222 (for PSDNs), T71 (for PSTNs), I440/441 (for ISDNs)
- Physical layer: X.21 and X.21bis (for PSDNs), V.24 (for PSTNs), I430/431 (for ISDNs)

References

1. Institute of Electrical and Electronics Engineers, *IEEE Standard 802.1: MAC Bridges*, IEEE, Sept. 1990.
2. Institute of Electrical and Electronic Engineers, *Local Area Networks — Logical Link Control (LLC)*, IEEE 802.2 (ISO 8808.2).
3. Institute of Electrical and Electronics Engineers, *Carrier Sense Multiple Access with Collision Detection (CSMA/CD) Access Method and Physical Layer Specification*, ANSI/IEEE 802.3 (ISO 8802.3), 1990.
4. Institute of Electrical and Electronics Engineers, *Token Passing Bus Access Method and Physical Layer Specifications*, ANSI/IEEE 802.4 (ISO 8808.4), 1990.
5. Institute of Electrical and Electronics Engineers, *Token Ring Access Method and Physical Layer Specifications*, ANSI/IEEE 802.5 (ISO 8808.5), 1989.
6. ISO 7498, *Open System Interconnection — Basic Reference Model*, Information Processing Systems, 1986.
7. Tanenbaum, A.S., *Computer Networks*, Prentice-Hall, 1989.
8. Stallings, W., *Handbook of Computer Communications Standards*, vol. 1, 2nd ed., Howard W. Sams and Co., 1990.
9. Black, U., *Data Networks: Concepts, Theory and Practice*, Prentice Hall, 1989.
10. Schlar, S.K., *Inside X.25: A Manager's Guide*, McGraw-Hill, 1990.
11. Markeley, R.W., *Data Communication and Interoperability*, Prentice-Hall, 1990.
12. Spragins, J., Hammond, J., and Pawlikowski, K., *Telecommunications Protocols and Design*, Addison-Wesley, 1991.
13. Stallings, W., *Local Networks: An Introduction*, Macmillan, 1987.
14. Institute of Electrical and Electronics Engineers, *IEEE 802.5 Appendix D; Multi-Ring Network (source routing)*, 1991.
15. IEEE standard 802.6, Distributed Queue Dual Bus (DQDB) Subnetwork of MAN, 1991.
16. ISO 8802.6 standard, Metropolitan Area Network (MAN).
17. Hemrich, C.F., et al., Switched multi-megabit service and early availability via MAN technology, *IEEE Communication Magazine*, April 1988.
18. Mollennauer, J.F., "Standards for Metropolitan Area Network," *IEEE Communication Magazine*, April 1988.
19. Minoli, D., *Telecommunications Technology Handbook*, Artech House, 1991.

20. Valovic, "Metropolitan area networks: A status report," *Telecommunications,* July 1989.
21. Ross, F.E., "An overview of FDDI: The fiber distributed data interface," *IEEE Journal on selected areas in communications,* 7(7), Sept. 1989.
22. Halsall, F., *Data Communication, Computer Networks and Open Systems,* Addison-Wesley, 1992.
23. Simonds, F., *LAN Communication Handbook,* McGraw-Hill, 1994.
24. Schwartz, M., *Telecommunication Networks: Modeling and Analysis,* Addison-Wesley, 1987.
25. Institute of Electrical and Electronics Engineers, Computer Society, *IEEE Draft 802.1 (part A), Overview and Architecture,* IEEE, Oct. 1990.
26. Hutchinson, D., *Local Arae Network Architectures,* Addison-Wesley, 1988.

References for further study

1. Naugle, M., *Local Area Networking,* McGraw-Hill, 1996.
2. Bertsekas, D. and Gallager, R., *Data Networks,* Prentice Hall, 1992.

chapter nine

Nonstandard LANS and internetworking

9.1 Non-OSI-RM LANs

In the preceding chapter, we discussed various standard IEEE LANs, and these LANs are widely used in various organizations, industries, universities, etc. A user organization may also develop its own LAN, which basically will include hardware and software components from different vendors. Obviously, such LANs are customized to meet the requirements of the organization's data processing and data communication technologies. These LANs are known as *private LANs*. There are various advantages and disadvantages of private LANs. Some of the advantages include complete control over the network, access, and flexibility. The main disadvantage of private LANs is that the user's organization has complete control over the LAN and, as such, is responsible for design, implementation, and maintenance. A number of computer vendors have proposed private LANs and designed their own network architectures, with functional descriptions of various aspects of hardware, software, and protocols used for data communications. These private LANs do not follow OSI-RM but instead define their own layers, with functions similar to the layers of OSI-RM. The following is a partial list of some private LANs:

- Burroughs Network Architecture (BNA) by Burroughs
- DEC Network Architecture (DNA) by Digital Equipment Corporation
- Distributed Network Architecture (DNA) by National Cash Register (NCR), now merged with AT&T
- System Network Architecture (SNA) by IBM
- Distributed System Network (DSN) by Hewlett-Packard
- Expand Nonstop Network (ENN) by Tandem

Of these private LANs, we will discuss IBM's SNA and DEC's DNA in the following sections, as these are more popular than other LANs. For details on proprietary LANs (IBM's SNA, DEC's DECnet) see References 1–6; references used in preparing this section include 1–7.

9.1.1 IBM System Network Architecture (SNA)

IBM has developed its own non-OSI LAN for interconnecting PCs, known as System Network Architecture (SNA).[3,6] Although SNA is a non-OSI architecture, its upper layers perform the functions of the application layer of OSI-RM. In this architecture, users can define their private networks on a host or even on sub-networks. The SNA architecture is a complete model of layers, and each layer has specific specifications and implementation guidelines. The main philosophy behind it is to provide a uniform framework for communication management. This framework offers users a variety of functions, some of which are not available in other networks, including DECnet. SNA is centralized, works on a host machine, does not lose data, is connection-oriented, is difficult to maintain, and is very complex. In contrast to this, TCP/IP is highly distributed, works on a network, loses data, is connectionless, is easy to maintain, and is less complex.

SNA uses a network control program (NCP) which manages all the peripherals and terminals. The host runs a virtual telecommunication access method (VTAM) that keeps a record of all the connected machines, devices, communication logical paths, etc. In contrast, TCP/IP is highly distributed, does not run on one host machine, and does not keep any routing information.

The operating system for SNA is usually multiple virtual system, which manages all the resources of the host. The communication over the network is managed by another operating system known as VTAM, while a subset of VTAM known as system services control point (SSCP) manages the operation, session, configuration, error control, and other related network management services. Other software modules are network addressable unit (dealing with input/output port and residing in the host) and logical unit (deals with resource allocation, activation and deactivation of network).

The SNA architecture has evolved over the years, following its inception in September 1974. In the first version of SNA, a host was used as a centralized root of the tree-structured network. The terminals connected to the host were known as **leaves**. Here each leaf has point-to-point communication with a centralized node. This version of the LAN had to control the entire network with one large host computer. Subsequent versions of SNA were supported by multiple hosts. One of the main objectives of SNA was to provide device independence so that the physical configuration considerations could be separated from application programs and the logic elements of the network. A typical and general configuration of SNA follows a hierarchical configuration, where it consists of a root node (host) which is connected to front-end processors via communication channels. These channels are then connected to terminals, clustered terminals, PCs, workstations, and a remote controller which may be connected to other controllers, etc. The clustered terminals are usually controlled by programmable cluster-control units and have been used in banking, retailing, supermarkets, etc.

The later versions allowed communication between leaves of different trees via their root nodes. The root nodes of the trees were connected together, providing connection between the leaves (terminals). The later versions of SNA included more general forms and categories of communication links, e.g., public networks, satellite links, local area networks, etc.

SNA architecture is designed to provide a structural framework which can be compared with OSI-RM. The lower layers of SNA and OSI-RM seem to have common functions, but the higher layers are different. The protocols used in the SNA model provide a basis for communication between nodes of a homogeneous communication architecture network. The SNA architecture is defined in terms of logical levels which are not analogous to the layers of OSI-RM. The levels in SNA are defined as **logical units (LUs)**. The levels are of different types, and we can have LUs for different layers.

Logical units (LUs): The LU may be defined as a gateway which provides communication between end users via the SNA network. The end users can access the functions provided by the centralized manager of the network. In other words, the LU provides an interface between it and the centralized manager and also between it and other LUs of IBM networks.

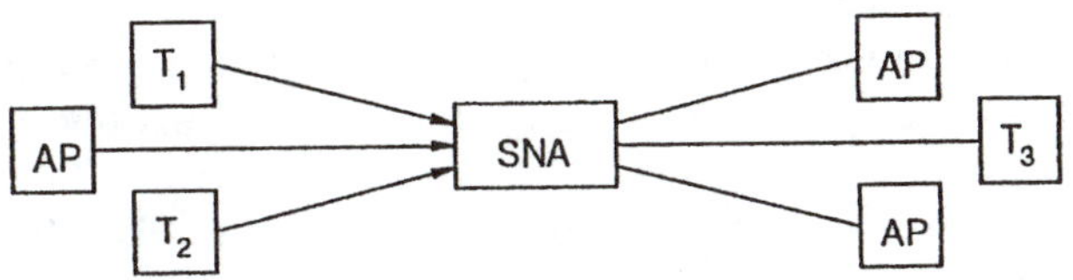

Figure 9.1 Terminals, application programs, and SNA network.

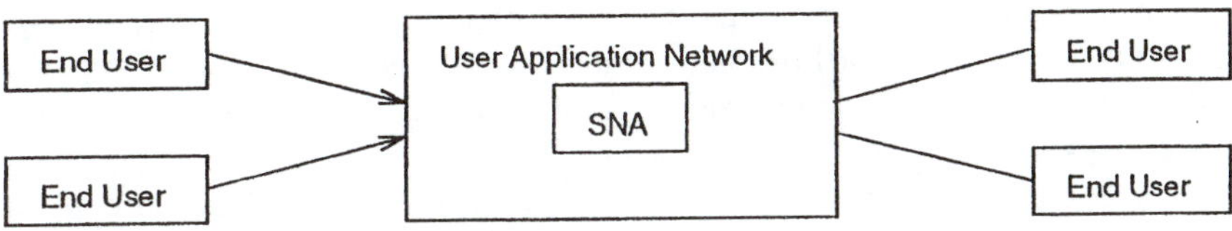

Figure 9.2 End users, user application network, and SNA network.

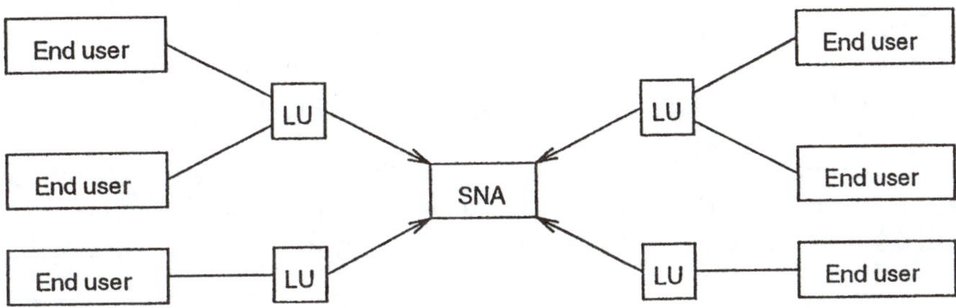

Figure 9.3 Logical units (LU) represented as end users.

The LUs support a variety of nodes within the SNA architecture. Each node, in fact, represents a machine. The machine may represent terminals, hosts, controller nodes, etc., depending on the characteristics of the **physical unit (PU)**.

A typical SNA network may have terminal users and application programs being shared by the users, as shown in Figure 9.1. In another situation, the end users may be connected to the SNA network which becomes a part of a user application network, as shown in Figure 9.2. The end users are connected to the SNA via LUs, nodes with a central manager system service control point (SSCP) which includes LUs along with APs and PUs, nodes without SSCP at PU, and peripheral nodes at LUs, all connected to individual terminals to cluster management units, as shown in Figure 9.3. The SNA usually operates in domains, and a *domain* is usually defined as an environment which includes a central manager SSCP, PUs, LUs, and various resources connected to SNA. The terminal node may represent a data station with less intelligent capabilities. The IBM 3270 was considered a terminal node in an SNA network.

The hierarchy of an SNA network contains subsections. Each subsection consists of a subsection node to which peripheral and terminal nodes are attached. The subsection node receives the messages from either its own connected nodes or some other subsection node and forwards then to the appropriate destination node of the network. The peripheral or terminal nodes can either send information to or receive information from the subsection node. The SNA consists of four types of nodes: **host**, **communication**, **cluster**, and **terminal**. Each of these nodes has a specific function in the overall SNA network. The functions provided by these nodes along with standards are described in brief here.

The **host node** is a physical unit type 5 and is typically a system consisting of a CPU with an operating system, an access technique, and routines, and supporting a few relevant application programs. The operating systems supported by this node are OS/MVS and DOS/VS. The virtual telecommunication access method (VTAM) is a widely used access method at the application layer. In the case of a host node of an SNA network, usually this node uses the advanced communication facility version of VTAM, which is basically an extension of VTAM. The host node, in general, uses the system service control point (SSCP).

The **communication node** is used to lower the processing burden on the CPU and is also known as a front-end processor (FEP). This node is of PU type 4 and is connected directly to the host node. The management of data interchange over the channel connection or SDLC connection, error-handling techniques, block checksum generation, checking of transmission counters, and communication between channel manager and network address managers for transmission of information are some of the important functions of this node. The SSCP is not required by this node; instead, it uses an **advanced communication facility/network control program (AFC/NCP)** which provides transfer of information between two managers.

The **cluster node** usually provides a balanced distribution of functions and processing between it and host nodes. It is the node through which the host node can be remotely accessed. The application programs usually residing on the host may be run on the cluster nodes, as well, to provide a better utilization of the resources of the SNA network. In summary, we can say that this node can also be used as another host node for running various application programs.

The **terminal node** is a PU type 1 and is usually considered an intelligent terminal. It can be considered a type of cluster node which may allow the execution of programs of the host node on it, but it has fewer capabilities than the actual cluster node.

9.1.1.1 SNA layered architecture model vs. OSI-RM

The SNA network is a seven-layered model, similar to OSI-RM. Although each of the layers has specific functions (as in OSI-RM), these are different from their counterpart layers of OSI-RM. In other words, although SNA and OSI-RM are totally different, they have some common functions at lower layers. Further, the main difference between these models lies in the type of network, communication, architectural products, and protocols.

The OSI-RM is defined to provide a unified system for communication between different types of networks (heterogeneous) and, as such, provides more emphasis on creating an environment of understanding between these networks. In contrast, the SNA provides a communication system between homogeneous networks and offers the complete system which takes into account the management, data integrity, control, adaptability, speed, maintenance, etc. These issues are not well defined and implemented in OSI-RM, due to the fact that in OSI-RM, most of the architecture components for higher layers are not available and are usually implemented as application programs. The application programs require the operating system, access techniques, etc., to support these issues. In essence, although both the models are totally different (with the exception of some common functions at lower layers), they have been defined to provide communication between networks.

It is very difficult to compare the SNA network with OSI-RM using a one-to-one correspondence. However, there are a few layers which are similar to and provide the same functions as those of OSI-RM. The names of the layers are not the same, but overall, the functionality of the SNA network has some similarity with OSI-RM, with a difference of main emphasis. In the following paragraphs, we compare the layers of SNA and OSI-RM.

Figure 9.4 shows the various layers of the SNA architecture model. It is important to note that the names of the layers in SNA are different in different books. For this reason, we use generic names for these layers in our present discussion.

ISO	SNA
Application Layer	End User Layer
Presentation Layer	NAU Services Layer
Session Layer	Data Flow Control Layer
Transport Layer	Transmission Control Layer
Network Layer	Path Control Layer
Data Link Layer	Data Link Control Layer
Physical Layer	Physical Layer

Figure 9.4 IBM SNA layered architecture.

End user layer: This layer is analogous to the application layer of OSI-RM and supports functions such as application programs, user processes, and even users' hardware devices to be interfaced with the top layer. Further, it supports all the services offered by its lower layer, **network access service (NAS)**. Typically, an application program requires two types of tasks: execution of the application program and distributed application processing. Each of these tasks is defined as a sequence of processes or sub-tasks which are performed on different computer systems. In general, an appropriate load-balancing technique is implemented which can perform these sub-tasks on different systems concurrently and cooperatively. The application program uses distributed sub-tasks-processing environments for files, programs, etc. In distributed application processing, the user's requests are performed by different applications running over different systems. These two tasks can share the resources to process various applications more effectively if these are connected to SNA.

Network access service (NAS) layer: Although this layer is analogous to the presentation layer of OSI-RM, it offers the functions of the presentation and session layers. The functions offered by the presentation layer of OSI-RM include text formatting, syntax transformation, text compression, code conversion and coding, etc., and various services offered by the session layer of OSI-RM include establishing a session and logical connections, management of the connection, etc. The network access service (NAS) layer of SNA provides all of these functions. It also supports the services offered by the lower layer, known as the data flow layer.

In OSI-RM, the application layer is not well defined and standardized and does not have specific functions, but it provides a variety of services to the users. In an SNA network, the functions are defined in the NAS layer. In some books, this layer has been compared to the presentation layer.

The logical unit (LU) can be considered a port through which end users communicate with each other. Although in most cases the interfaces, including LU, are not defined specifically, these need to be implemented as LUs, as the LU offers the functions of the session layer of OSI-RM. The network operation, including transportation, is performed by interconnecting LUs. The SNA network defines different types of units/controls which are collectively known as network addressable units (NAUs). These units are **physical units (PUs)**, **logical units (LUs)**, and **system service control points (SSCPs)**.

The LUs first establish what we call a **semi-session** with their own units' SSCPs. The LU may be thought of as having functional, data flow, and transmission layers, and these are well supported by NAU service managers (services offered at the top layer). The functional layer includes the services that support end users, and each of the services is session-specific. The next layer, data flow function, provides conversion of protocol primitives and has the function of an interpreter of the contents of headers between LUs. These two layers basically exchange information regarding request and response primitives before a session can be maintained by the lower layer (transmission layer). The LU provides routes to the packets received from the path control network. It is also responsible for making headers of the primitives being used and interprets the headers it receives from the path control network. The path control network provides two types of controls: **path control (PC)** and **data link control (DLC)**. Path control is mainly concerned with the construction of a path over the available links, while data link control is mainly concerned with the actual transmission of packets.

Data flow layer: This layer is analogous to the session layer of OSI-RM but supports its functions as services for the NAS layer. The function of the data flow layer is to maintain and regulate the half session (seen by an LU). It checks the working of various primitives, generates a sequence number, maintains a table for the request response relationship of individual users and groups of users, etc. The services provided by the lower layer known as the transmission control layer (offering transport-connection management) are supported by this layer.

The services offered by OSI-RM's session layer include session-connection controls, quality of service parameters, synchronization, exception and error reporting, etc. In an SNA network, an SNA session is established via a start data traffic (SDT) request unit, which is also sent to its lower layer (transmission control). For data exchange, the normal request and response units are used. The synchronization is provided in the **request control mode (RCM)** and **response control mode (RCM)**.

Transmission control layer: This layer is analogous to the transport layer of OSI-RM. The main function of this layer is to provide a transport connection between two sites. Other functions of this layer include data flow control (after a session is established), interface to the higher layers (independent of the type of network under consideration, in particular the transport connection), multiplexing/de-multiplexing for higher layers, etc. In some cases, it also supports encoding and decoding of signals. This layer includes the higher-level task of identifying the endpoints of a session between users and must perform functions of interaction at the endpoints. In general, these endpoints are the host and terminals. It is implemented by different modules of the **virtual telecommunication access method (VTAM)** which reside in each host computer. The layer offers implementations for front-end processors as **network control program/virtual storage (NCP/VS)**. This converts the data code, stores it in buffers, transfers data between hosts and terminals, handles links, performs error recovery, adds and deletes the header information, etc.

The services of the transport layer in OSI-RM include mapping of addresses, connection control, data transfer services, etc. The SNA network address of NAS is equivalent to the transport address and is always suffixed with NAU for session establishment. The data flow control layer (lower layer) does not get the transport connection establishment services directly; instead, the NAU service layer sends a request to the SSCP. The SSCP requests the transmission control layer to send an appropriate request unit **(BIND)** to the corresponding correspondent. The transport service data unit, in fact, is a request/response unit coming out from the data flow control layer. This request unit is used during connection establishment. Similarly, **UNBIND** will be used during the connection termination.

Various features during the data transfer (after the establishment of connection) in the transmission control layer include sequencing, contention, segmenting, and error recovery. It multiplexes multiple sessions (SNA) to the same virtual route initialized by different users in the path control layer. It gives a preference to the expedited facility for data transfer in the sense that these expedited request units are not queued with other request units and are controlled by a connection point manager (CPM).

Path control layer: This layer is analogous to the network layer of OSI-RM and offers some of its functions. It offers the functions of routing and congestion control. Other functions of this layer include conversion of transmission layer units into path information units; allowing users to share remote resources, programs, and files; and the transformation of information units between NAS and path control layers over the network as well as the routing information of each of the information bits. This control exists for every node of the network.

The path control layer's **routing functions** define a logical channel and are responsible for delivering the packet to the destination. The addressing scheme is defined for NAU, which is a unique address assigned by the SNA network. The path control uses this address for the selection of path and routing of the packet. In addition to this address, it has a logical name address which is internal to the network. Each NAU address has two fields and a node address. The node address of each NAU is unique, while the subsection address of NAU is used mainly for routing of the messages between subsection nodes. The routing of packets across the network is performed in three steps as control functions: **virtual route**, **explicit route**, and **transmission route**.

System services control point (SSCP) functions: As stated above, the path control layer provides a reliable session between end users, while the actual transport connection is established by the transmission controller. The main function of the data flow control layer is to provide user-oriented services to the users over this transport connection. In a normal operation, a user process requests its SSCP to establish a session connection, and, as such, a process (NAU) defines a session with SSCP in its domain and uses this session for sending any type of request unit. If this request unit pertains to a local site, it goes to the appropriate address. But if it pertains to a remote site with a symbolic name, this name is converted into a suitable network address by the local SSCP.

The SSCP provides two local services through which it controls the resources in the control domain: **physical unit (PU)** services and **logical unit (LU)** services. The PU services reside in every node of the network, as they define the function of the set of resources attached to that node. The request from the user to the SSCP in a session is passed as an action unit to its local physical unit services on the same channel. Similarly, the LU services are present in every LU and are controlled by the SSCP in the same session. The end user can also make a request for services to the SSCP through its LU services.

The SSCP provides network services to various NAUs via its control function and can request these services from NAUs, as well. The SSCP uses a set of **network service request units (NSRUs)** to communicate with NAUs over the session established between them. This session needs to be established between SSCP and all the NAUs with which SSCP wants to communicate. Each NSRU performs a specific function, and the sessions for sending NSRUs are established by activating the network, which keeps the session active for the entire life of the network.

The services provided by the SSCP and request packets from the NAU are classified as **configuration**, **session**, **maintenance**, **measurement**, and **operator**. The **configuration** services deal mainly with the physical characteristics and configuration of the network. The control functions are required to configure, modify, and update the network. The **session** services are usually responsible for using the function of activation/deactivation of session between LUs, SSCPs, and request units. The **maintenance** services deal with the maintenance management of the network. The control function for these services deals with testing, tracing error information, recording these errors, etc. The **measurement** services define a set of measurements which are to be performed on the network. These measurements are mainly for the performance evaluation of the network. In general, these measurement services are implementation-dependent services. The **operator** services, as the name implies, are the services offered by the SSCP to the users or operators regarding the use of the network, calling a particular function, and other control information of the network.

Data link layer: This layer is analogous to the data link layer of OSI-RM and offers more or less the same functions as in OSI-RM, including error control, flow control, etc. Higher layers are not aware of each of the functions provided by this layer, except that these layers may be notified about the errors detected. Error control, including error detection, correction, and recovery, is performed solely by this layer. It provides communication links over which the data communication takes place. The data link protocol synchronous data link control (SDLC) provides the common line configurations for data communication across the network.

Physical layer: This is the lowest layer of the SNA network and is analogous to the physical layer of OSI-RM. The function of this layer is basically the same as that in OSI-RM: to accept the raw stream of data from the data link layer and transmit it over the transmission link/medium.

9.1.1.2 SNA protocols

The network addressable unit (NAU) is application software which allows users to access the network. The NAU, in fact, defines the address through which users can interact with each other and is considered a socket or port similar to that in a UNIX operating system. The LUs allow the application programs to communicate with each other, and the user processes can access the devices connected to the network. On the other hand, the PUs are associated with each node and the network activates/deactivates the nodes; i.e., these nodes are connected to the network online or off-line. The addresses of devices connected to the network are defined by these PUs, which do not have any idea about the processes which are going to use them.

LANs have reduced the importance of a centralized host network, and, as such, IBM introduced a new distributed networking system known as Advanced Peer-to-Peer Networking (APPN) that allows the use of the same peripherals, terminals, and other devices. It is compatible to SNA.

The centralized manager node or system services control unit (SSCU) is usually implemented by the hosts (PU type 5). These three units (LUs, PUs, SSCUs) in general are subsets of NAU.

The SSCP carries knowledge about other nodes of types 1, 2, and 4 and also controls the operation on these nodes, connected to the host. In the terminology of IBM, each domain of the SNA network consists of SSCP, PUs, LUs, and other resources/devices, and the controlling and management of these devices are provided by SSCP. The SNA network may operate in more than one domain.

Extensions to SNA LANs: The APPC LU6.2 is being used by IBM as a higher-level protocol which provides support for various categories of architecture, e.g., distributed processing, office automation, interchange architecture, etc. IBM has developed the following products which are supported by APPC LU6.2 at higher layers, and these products are collectively known as **transaction services architecture (TSA)**.

- SNA distribution services (SNADS)
- Document interchange architecture (DIA)
- Document control architecture (DCA)
- Distributed data management (DDM)
- Low entry networking (LEN)

SNA distribution service (SNADS): This is a kind of complete communication system in which the packet information can be stored in the buffer and forwarded to the user at a later time when he/she is available. The packet information may include the actual message, a request for the resources, or even files, documents, or programs. This service supports the concept of store and forward to all types of messages (request to access resources remotely, electronic mail, texts, etc.). The service sends the packet information asynchronously. In some cases, it also supports the routing service, whereby it looks at the routing table available at each node and, based on the routing information, may select the most appropriate route for the packet information. The most appropriate route may depend on a number of factors: priority, security required, capacity of intermediate data service units (DSUs), data rate of the channel, etc.[2,3,7,8]

The DSU is a device which converts the digital signals of the DTE into bipolar digital signals. The functions of repeater (i.e., regeneration of digital signal) and clocking (for synchronization) are also provided by this device. This device, along with another device known as a **channel service unit (CSU)**, provides an interface with the digital channel. The combination of **CSU/DSU** provides an interface between digital DTEs and the digital network via a T1 channel. This device is discussed in Section 3.3.

The DSU is connected to the user DTE via RS-232 or V.35, while the CSU is connected to the digital network via a digital switch known as an office channel unit over a four-wire digital channel and local loop. The OCU is connected to the digital network over a T1 channel. The CSU performs the functions related for the signals traveling over the channels, e.g., equalization or conditioning of pulses, reshaping of pulses, etc. It also provides the loop-back testing between the CSU and OCU. Other features related to digital signals may be attached to the CSU when used with more complex digital systems. The newer digital systems offer a variety of features not available in older ones. The use of a CSU/DSU with a T1 system (capable of transmitting both voice and data) provides analog-to-digital conversion and multiplexing. This device is discussed in more detail in Chapter 3.

The intermediate DSUs are connected by channels of fixed capacity of the DSU nodes, and the packet information delivery to the DSU by the SNA is controlled by SNADS. The nodes connected to the same machine do not use SNADS for the transfer of packet information, but SNADS will be required if the packet information is delivered to different DSUs. In other words, for the same DSU, there is no need of SNADS; instead, another protocol known as document interchange architecture (DIA) is used. This protocol is discussed below.

Document interchange architecture (DIA): This protocol provides document exchange service between the office system node (OSN) and the source/recipient node (SRN). As stated above, if the users are on the same machine using one DSU, the DIA is used to provide communication between them. But if the users are on different machines and use different intermediate DSUs, the SNADS provides communication between them. The services provided by DIA over documents include processing, formatting, priorities, storing, searching, retrieval, file transfer, etc. All these services are provided by different layers defined in DIA (based on the layered approach used in OSI-RM).

The DIA is defined by the following layers: user (the top layer), DIA with source/recipient node (SRN), office system nodes (OSN), and SNADS and DSU at the lowest layer. The lowest layers of two nodes are logically connected by peer-to-peer entities. The user of the sending node makes a request for services on the documents and the OSN layer determines the source or destination node. If the sending node OSN is a source destination, it gives a unique name to each document request it receives from DIA SRN, stores it in the buffer, determines the route (if available) and stores it, provides status information to the DIA SRN, etc. On the receiving node, these documents requests are received and stored in the buffer.

The operating system for SNA is usually a multiple virtual system which manages all the resources of the host. The communication over the network is managed by another operating system known as the virtual telecommunications access method (VTAM), while a subset of VTAM known as a system services control point (SSCP) manages the operation, session, configuration, error control, and other related network management services. Other software modules are the network addressable unit (deals with input/output port and resides in host) and the logical unit (deals with resource allocation, activation and deactivation of network).

As stated above, a typical physical configuration of an SNA network includes different types of nodes which are connected by different types of links. This configuration offers device independence in the sense that the physical configuration requirements for different application programs are different and the logical communication is also different. A host node is connected to a front-end processor by a communication channel which in turn is connected to other nodes (communication controllers, cluster controllers, or even remote controllers, terminals, or PCs) via communication lines. The communication controllers may also be connected to cluster controllers via communication lines. In this configuration, there is a single host computer that provides centralized control over the management activities, while other nodes are mainly concerned with the routing and transmission of data. The terminal nodes provide input and output to the data via the terminal devices.

This configuration is defined for a single host. The multiple-host configuration of an SNA network can be defined by networking the host nodes via their respective **front-end processor (FEP)** nodes. The FEP of host number one is connected to the FEP of host number two, which is connected to another host, and so on. The transmission between FEPs can be carried over any transmission medium and is usually supported by **digital dataphone services (DDS)**. This digital facility is becoming popular in the transmission of digital signals over a digital framework. The advantage of this framework is that it does not require a line interface.

The transmission of a signal between sending and receiving terminals over the DDS facility is entirely in digital form, and both digital and analog signals (modulated) can be transmitted over it. A very popular technique of analog-to-digital modulation is pulse-coded modulation (PCM). As mentioned above, different links are used between different types of nodes, e.g., data channels between host and FEPs, while common carrier lines, on-site loops, etc. are used to provide links between other nodes. Each node is considered a logical structured entity with a specific set of well-defined tasks and a common set of functions such as transmission and application. The transmission function is a mainly concerned with routing strategies, formatting of messages, message scheduling, etc., and cannot interpret or examine the contents of the message.

The SNA network provides three types of communication between its various nodes. A data link can be established between any two nodes, a complete path can be defined between source and destination nodes, and a session can be established between end users across the network. Each of these communication links is defined in a hierarchy, whereby if a session communication link is defined between the host and, say, a terminal, it will also include the other two communication links, since otherwise, there is no session between source and destination. The session link is established over various intermediate nodes between which a data path is defined. In other words, all the communication links are defined in the session link between the end users of an SNA network.

The information regarding these links or connections is handled by the transmission management layer. This layer is implemented on different nodes in a different way. For example, at the host node, this is implemented as a VTAM protocol, while at the front-end processor node, it is implemented as **network control program/virtual storage (NCP/VS)**. The main function of NCP/VS is to provide conversion of data codes, lines, error recovery, interpretation of control information headers, etc., for messages between the host and terminals.

Document contents architecture (DCA): This extension of SNA deals mainly with the internal structure of information to be transmitted. It offers an interface between the user and the network and controls the transmission of information over the network. This scheme has been implemented over different platforms; various software packages such as PC Tex3, PC-Office (for PCs), and PROFS (for mainframe) are available on the market from different vendors.

9.1.2 DEC's Digital Network Architecture, or DECnet

Digital Equipment Corporation (DEC, now merged with COMPAQ Computers), has also developed its own LAN conforming to OSI-RM, and it is hoped that the DEC products will be fully compatible to OSI-RM. The LAN developed by DEC is known as Digital Network Architecture (DNA) or DECnet.[3-5] DNA is another layered model specification that can be implemented differently under different environments and, in fact, products are available for each of the different environments.

DECnet specifies a set of hardware and software products allowing Digital's operating systems to work in the computer network environment. The network consists of nodes (terminals, computers, PCs, workstations, etc.) which are connected by transmission media. Depending on the type of connection (characterized by media, topology, distance of networking, etc.), DECnet can be connected as a LAN (based on Ethernet), satellite network (based on satellite communication), or X.25 network (based on a packet-switching network). In each network, there are usually two types of nodes: user node and network manager node.

A **user node** uses the network for various applications such as file transfer, resource sharing, etc., while a **network manager node** basically ensures the correct operation of the network by using various diagnostic and trouble-shooting utilities. The nodes of the network correspond to DEC machines such as VAX, MICRO VAX I and II, PDO-11, DEC SYSTEMs 20 and 10, Professional 350 PCs, IBM-PC/XT/AT, etc., and each of these machines has a different operating system; for example, VAX has VAX/VMS, ULTRIX while MICRO VAX I and II use Macro VMS, VAXELN, ULTRIXm. The IBM-PC/XT/AT uses standard DOS while DEC SYSTEMs 20 and 10 use TOPS 20 and 10, respectively.

DECnet can be implemented or configured in a variety of ways, depending on its applications. For example, if the distance between computers is small, a LAN configuration using Ethernet can be defined for DECnet where a DECnet node can communicate with other nodes connected to the network. Each node has a unique address and transmits and receives the message to and from the network, respectively. DECnet offers various services, including that of Ethernet (LAN), that are summarized below:

- Resource sharing (expensive peripherals, laser printers, etc.)
- Data transfer between nodes
- Services of other networks
- Computations by powerful nodes (i.e., nodes having workstations or high computing machines)
- Gateways for different OSI-based and non-OSI-based networks
- Gateways for public network X.25

In addition to these services, many new services for different applications can be defined and implemented. Some of the material in this subsection is derived partially from References 2–5, 9, and 10.

ISO	DNA
Application	User
	Network Management
Presentation	Network Application
Session	Session Control
Transport	End-to-end Communication
Network	Routing
Data Link	Data Link
Physical Link	Physical Link

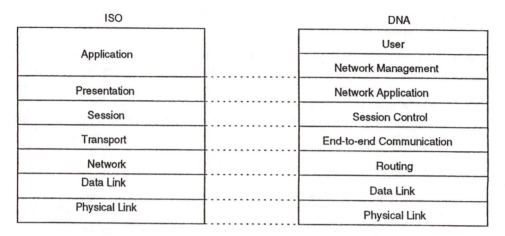

Figure 9.5 DEC DNA layered architecture.

9.1.2.1 DNA layered architecture model vs. OSI-RM

As stated above, the DECnet consists of hardware and software components, and for hardware components, we need an architecture. The DEC defines a layered model architecture for DNA or DECnet which conforms with OSI-RM, as shown in Figure 9.5. In DNA, the specifications for the management of various hardware and software components and interaction between them and other interfaces are defined. Each of the layers in DNA has a specific set of services and functions to be offered to its higher layers, and the functions are usually implemented by protocols of those layers. It is worth mentioning here that different authors have defined different numbers of layers with different names, but one thing is clear: these proprietary LANs (IBM's SNA, DEC's DECnet, etc.) are based on a layered approach as used in OSI-RM. These LANs provide interconnectivity with other LANs (OSI and non-OSI). The software component of DNA, in fact, consists of all the protocols of the layers of DNA architecture. Various layers of DNA (seven-layer model) are discussed below.

User and network management layer: This layer consists of two sublayers: **user and network management sublayers**. The top sublayer (user layer) defines various services offered by DNA as routines. Some of the network program controls of DECnet are defined in this layer. It has direct access to lower layers of DNA such as network management, network applications, and session control layers. Similarly, the second sublayer, network management, also has access to all the these layers. The user layer of DNA defines two functions in terms of its two sublayers: **access service sublayer** and **network service sublayer**. The first sublayer, access service, consists of various program services such as access service and network service. It allows the user to access the network. The second sublayer, network service, provides support to various tasks of users, applications, and global management. Widely used services offered by this layer are file transfer, database management, and network management (local and global).

The second sublayer (network management) gets all the relevant information about the performance and quality of service parameters from the lower layers. The functions of this sublayer include planning, controlling, managing, and maintaining the network; down-line loading of remote nodes; and network connection loop testing (local and remote). Various network operations are controlled and managed by this layer, e.g., down-line loading of resources, files, testing of channels, and status of channels and nodes of networks. In some books, this layer is called the network application layer.

When a DEC system is used as an Ethernet LAN, the Ethernet coaxial cable provides an interface for the VAX/VMS host and terminal drivers, local area transport protocol (LATP), end-to-end communication layer, transport layer, and the Ethernet driver underneath these layers. For a VAX/VMS host, the protocols for the application layer, terminal access driver, two drivers (one for a LAN port connected to an Ethernet coaxial cable while the second is connected to an appropriate driver for the terminal) are similar. This configuration indicates how we can implement a local area transport protocol on the Ethernet LAN.

The channel established can be used for carrying out the heading commands in both ways for receiving and transmitting. In the event of malfunction, a facility termed **upline dumping (UD)** immediately transmits an extract of the image system to a larger system attached with it. After the execution of this option or utility, the target node is then repaired and restored. The down-line loading of computers is very useful during diagnostic testing, backing up of the files, and other comprehensive system testing.

Network application layer: This layer consists of different types of application modules which are being supported by the application layer of OSI-RM. Although the application modules in the layer may not have any similarity with the functions of the application and presentation layers of OSI-RM, they do provide services for the higher sublayers of the top layer. The functions offered by the network application layer are data access, remote file access, file transfer, remote terminal access capabilities, communication services to the users, accessing of X.25 network access to SNA gateways, and data access protocols (DAP). The remote terminal–accessing capability also includes virtual terminal protocol. Some of these functions are explained below briefly.

1. **Remote terminal accessing (RTA):** Offers accessing capabilities via the network virtual terminal protocol to a support terminal attached to a host node of DECnet to interact with a remote host node of DECnet.
2. **Data access protocol (DAP):** Supports file access and allows the transfer of files independent of their I/O structures, thus offering a universal I/O language within DNA. This further allows the transfer of files across the heterogeneous system.
3. **Loop-back mirror protocol (LMP):** Concerned with the use of network management access modules for transmitting and receiving data messages across the network.
4. **Packet net access interface (PSI):** Allows users to send messages from DECnet to non-DECnet systems across the public network (X.25). The X.25 PAD provides interface between DECnet nodes and remote DECnet nodes.
5. **SNA gateway:** Provides an interface between DECnet nodes and IBM's SNA nodes within SNA LANs.

Session layer: This layer is concerned mainly with the mapping of addresses, layer management, data transfer, and interprocess communication (based on the underlying system). We can say that it also includes all the functions of an OSI-RM session layer.

End-to-end data communication layer: The modules residing in each node perform a specific function such as creation and controlling of logical connections, data transfer, remote file accesses, etc. The nodes have different operating systems and may have been implemented differently in different programming languages. In order to avoid placing a burden on the users to develop and design their protocols independent of programming languages, DECnet system has defined various processes/tasks for establishing end-to-end communication tasks between the terminal and remote server nodes or other nodes:

- User request for a logical connection
- Receiving of user's request by the remote node
- Acceptance/rejection of user's request for connection

- Establishment of logical connection
- Transfer of data from terminal
- Detection of end of data
- Sending of interrupt data
- Receiving of interrupt data
- User's request to release the logical connection
- Termination of logical connection
- Acknowledgment

Other services/functions like data integrity, error control, flow control, etc., are being provided by other subroutine modules. Further, the subroutine modules provide support for distinguishing between a user's node, addresses of source and destination nodes, logical connection, and other parameters of the control field.

For remote file accessing, the subroutine modules for file access reside in the remote node and allow the users to access files of the remote node, perform various operations on it, and also transfer the files and records between them. Various other options available in remote file accessing are transfer of files from terminal to remote node, deletion from the files at remote node, listing of directory of the remote node, printing of files by remote printers, etc. Each of these options is implemented by appropriate DECnet calls which are supported by a **network file transfer (NFT)** utility.

A typical DECnet configuration for the transfer of files via DECnet file access subroutines may include the following sequence of actions through DNA of local and remote server nodes:

1. User at user layer of local node may initialize the accessing program (NFT) utility via VAX/VMS command.
2. The DAP speaking routine of the network application layer will use RMS in the local file system (peer-to-peer routine between local node and server node).
3. The lower layers simply prepare the data packets until they reach the lowest layer of the DNA. This layer sends the data packet to the DECnet system via a return routine, which then transmits the data message to a remote server node. The packet goes from the physical link layer up to the session control layer of DNA in the same way that it came from the session control layer to the physical link layer at the local node side. The network application layer will have DAP-speaking routines and is in listening mode by a file access listener module which executes the RMS utility.

Routing layer: DECnet also offers flexibility in changing the topologies of LANs and WANs using a DECnet routing protocol. This protocol avoids the requirement of having direct connections between pairs of nodes. It also allows the network manager to reconfigure the topology of the network. Although the pairs of nodes need not be connected directly, the DECnet routing ensures the communication between every source and destination node. If DECnet routing goes down, the adaptive routing algorithm is initialized, which tries to find an alternate route between nodes. DECnet has many phases (Phases I–IV) and has been successfully used in networking (both LANs and WANs). The phases offer capabilities for supporting different servers of LANs, baseband and broadband transmissions, and other DECnet LANs and WANs. Each phase has a unique capability and is due to the addition of a new feature in the networking. Various common options available with DECnet routing phases are listed below. Depending on the phase version of DECnet routing, additional facilities may be available.

- The routing strategies for local and remote destinations are different. In the case of a local destination address, the routing strategy sends the packet to the end-to-end communication layer. If the packet is addressed to a remote address, the routing strategy transmits the packet to the next available node, which in turn searches for an available node (if the packet does not belong to that node) and so on, until the packet reaches the destination.
- DECnet routing offers flexibility of reconfiguring the topology, link circuits, or channels, updating of routing strategies of other nodes, etc. In the event of a failed node, the routing strategy finds an alternate node for routing of the packets. Different types of data link channels exist (point-to-point or multi-point on DDCMP, X.25 virtual circuits, Ethernet). The changes in the routing strategy are broadcast to all other nodes.
- If a node is down and a packet is addressed to it, the routing returns the packet, but the contents of packet are never lost, even if the packet itself is lost.
- The memory requirements for packets and the buffer requirements at the nodes should avoid any overloading on the nodes. Other information required for nodes includes the number of packets transmitted via the node, a time limit on the processing of the packets at the nodes, management information for looking at overloading problems at the nodes, etc.

A typical DECnet LAN configuration includes the following devices connected to Ethernet coaxial cable:

- Terminal server
- DECnet-SNA gateway (which is connected to IBM SNA network)
- DECnet router-X.25 gateway (connected to public packet-switched network X.25)
- DECnet VAX
- DECnet-RSX

Data link layer: This layer is analogous to the data link layer of OSI-RM and provides similar functions. It establishes a large number of error-free routes between various nodes (local and remote). The location of these nodes depends on the type of link being used. For example, if we are using two LANs connected by an Internet working unit (IWU), only local and remote nodes are defined. If we are using a gateway for connecting two nodes over X.25, then we have local, gateway, and remote nodes. This layer supports different types of links such as X.25, Ethernet, and DEC's digital data communication message protocol (DDCMP).

Physical layer: This layer is analogous to the physical layer of OSI-RM and offers similar functions and services. Typical functions of this layer include channel signaling, hardware communication link interfaces, device drivers, notification of any hardware or other communication errors to higher layers, etc.

9.1.2.2 DECnet protocols

The **network virtual terminal protocol (NVTP)** developed by DEC provides a common higher-layer protocol for providing communication supporting different operating systems across the network. This protocol offers a variety of peer-to-peer services across the network and is divided into two modules (sublayers): **command terminal protocol module (CTPM)** and **terminal communication protocol module (TCPM)**.

If a DECnet user uses a standard LAN such as Ethernet, the terminals need to be connected to a terminal server. In this way, any terminal can be logged on to any computer connected to LAN. The NVT protocol may be used at the application layer, but it does not require all of the functions of it. Instead, another protocol known as a local area transport protocol (LATP) has been defined. This protocol resides on a terminal server. The LATP port driver is attached to each host machine. Any terminal may be connected to any host machine within an Ethernet LAN via a sublayer CTPM. A terminal may interact with any host outside the Ethernet by converting LATP to CTPM.

Data access protocol (DAP): This protocol resides in the network application layer and provides an interface between the file access listener (FAL) utility and the subroutine doing the accessing. The FAL utility (if resident on the node) is responsible for accessing the requests and then processing them accordingly (as discussed below). Users cannot access DAP directly but can be involved by external subroutines to use it. DAP controls network configuration, data flow, transportation, formal conversions, and monitoring of the programs. As stated above, this protocol is concerned mainly with the managing and controlling of remote file access, file transfer, etc. The main function of DAP is to perform different types of operations on files, e.g., reading and writing of files onto I/O devices, deletion and renaming of files, directory of remote files, execution of files, support of different data structures of files (sequential, random, and indexed file organizations) and their access techniques (sequential, random, and indexed), etc.

The user makes a request for a remote file via I/O commands, which are transformed into DAP messages. These messages are transmitted over a logical connection established between the terminal and remote server. The interpretation of DAP messages into actual I/O operations on a file is carried out by the server. The server sends back to the user an acknowledgment indicating the status of the file. Initially, when the message is sent from user to server, it is required that this message contain information about the files (file system organization, size of the file for buffer, record size, etc.). The DAP is composed of very complex software handling the situations of varying sizes of files and data records, data structures of files, databases, etc.

The DEC system includes the following protocols for remote file access and file transfer. These protocols are based on the concept of DAP.[2,3,7]

1. **File access listener (FAL):** This utility targets programs of a file access operation, accepts a user's request, and processes it on the node where the FAL is residing. This protocol resides at the remote server node and responds to user I/O commands for establishing a listening session between the nodes.
2. **Network file transfer (NFT):** This protocol resides at the top layer (user layer of DNA) and maps the DAP-speaking process into the appropriate DAP function. It offers file transfer and various operations on the files.
3. **Record management services (RMS):** The majority of the DEC operating systems use this scheme for their file management systems. The DAP messages are transmitted over a logical connection by the file system (RMS). The DAP message includes a user request for remote file, remote node address, and possibly access control information. This message is transmitted to a remote server node where FAL, after receiving the message, performs the operations.
4. **Network file access routines (NFAR):** This protocol is a set of routines which is a part of the user process (when making an I/O request). These subroutines, in fact, allow the user to access remote files and are executed under the control of FAL. These subroutines reside at the application layer of DNA at the server node.
5. **VAX/VMS command language interpreter:** Various commands for file access and file transfer are interfaced with RMS for allowing users to access remote files and perform operations on them; therefore, NTF is included in VAX/VMS.
6. **Network management modules (NMM):** These modules are used to provide services like down-line loading of remote files, transfer of up-line dumps for storage, etc.

The DAP messages are exchanged between users for each of the above facilities (defined by the protocols). In order to understand the DAP messaging scheme, we describe a few of the DAP messages; such as configuration (capability and configurations like operating system, exchanged after the establishment of a link), attributes (structuring of data in the files, including information on file organization, data type, format, device characteristics, etc.), access (file name and type of access), continue transfer (recovery from errors), data (file data transfer over logical link), summary attributes extension (character of allocation), and name (for renaming a file or obtaining file directory data).

9.2 LAN-to-LAN interconnection

LAN-to-LAN communication always puts some constraint on the data rate and bandwidth. A T1 channel offers 64 Kbps but may not be sufficient even for DOS applications. Full-screen information of a DOS application is typically 25 × 80 characters, and each character is represented by 16 bits. The number of pixels and bits to represent the number of colors and data in the graphical user interface of Windows is so large that it takes a long time to load it. Images, graphic information, multimedia, and many other applications require significant bandwidth. For example, an X-ray report may occupy about 120 Mbps. Compression algorithms can be used to compress the images, but some information is lost during decompression. With ATM connectivity with LAN, it is possible to achieve a data rate of over 155 Mbps, which will reduce the transmission time significantly. The ATM-based network uses a cell of fixed length and transfers the information whenever it is generated. The asynchronous concept, in fact, is a transmission with a requested bandwidth-on-demand philosophy. The ATM switch uses mainly the packet-switching system and is being used as a switch in ATM networks.

There are so many types of LANs that internetworking has become a serious problem. Some of these LANs include SNA and DECnet (mainly for corporations) for sharing the mainframe data, PBXs for voice services, point-to-point voice channel networks for PBXs, videoconferencing, and many other types. Each of these networks provides a type of service, and it is quite obvious that for any of the services, higher data rates, higher bandwidth, and minimum delay are expected. For some applications (cable TV, voice-grade communication), constant bit rates are needed, while in some applications (digital data, videoconferencing, multimedia, graphics, etc.), variable data rates are needed. Use of ATM is expected to offer solutions to higher bandwidth and data rates of LANs (FDDI, high-speed Ethernet, etc.) as well as LAN connectivity and bandwidth required for graphics, images, and multimedia applications.

As stated above, the services are described as modules, and modules having the same or equivalent functions usually reside in different nodes and communicate with each other with the help of protocols. The protocols play an important role in defining an environment for communication and information interchange. The user layer usually does not have any protocol and, as such, does not contain a certain routine or programs to interpret all the data messages received. Since the idea of DNA was to make a more widely acceptable network, many of the layers of DNA provide different categories of function and include different types of protocols/standards. For example, the data link layer of DNA provides interfaces for DDCMP, Ethernet, and public network X.25. It is important to know that the DECnet system does not enforce the use of DEC's protocols or standards, but rather other protocols (either standard or designed by users) can also be used, provided these are consistent with the DEC system's layered architecture. The architecture of DECnet is very flexible and allows the user to upgrade and improve the system with advances in transmission technology or higher-level compatibility. In summary, the existing configuration of nodes, including equipment, need not be changed in the event of upgrading the system.

A typical DAP message exchange for accessing a file requires the transmission of configuration information (file system, buffer size, DECnet phase number, etc.) as a message to the remote node server. The server node now defines the characteristics of the file (type, block and record sizes, etc.) to the server node, along with an access request. Both of the messages (file characteristics and access request) are returned with attribute messages after successful operation where the file is opened.

9.2.1 Internetworking DECnet and SNA LANs

The interconnection between DEC's DECnet and IBM's SNA has the following characteristics:

- SNA network allows a direct access between DECnet and the IBM host.
- SNA considers the DECnet host as a logical system residing in the host.
- Various services and features are offered by transmission control.
- The emulated SNA physical unit type 2 (PU2) (to be discussed later) offers communication between DECnet and SNA. The emulated PU2 resides on the DECnet system.
- The emulated PU2 of the DECnet system may be used to access the resources of the SNA network.

Internetworking between DECnet and SNA can be obtained by the **DECnet/SNA gateway** and **VMS/SNA software**.

DECnet/SNA gateway: This consists of DECnet nodes and provides communication with the SNA network consisting of IBM system nodes. The DEC's gateway is used as an SNA type 2 physical unit (PU2) and can be implemented by two approaches.[3,7,8]

In the first approach, the DECnet SNA gateway provides point-to-point connection between it and the DECnet host system and usually is called a DX.24 gateway. The DX.24 gateway allows the DECnet nodes to be connected in sequence to it and the SNA COMC. It basically works as a controller which is connected to various SNA nodes (like IBM host, SNA PC, SNA disk system, etc.). The SNA PC mode may also be connected to the SNA host. The DX.24 gateway offers more than one communication channel with different data rates for DEC nodes. The nodes support a variety of connections like modems (low-speed CCITT or high-speed Bell type), RS-232 interfaces, V.24/V.28 interfaces, coaxial connection, modem eliminator, or any equivalent connection.

In the second approach, the Ethernet LAN is directly connected to the DECnet host. This configuration is defined as a DECSA gateway. The DECSA gateway is connected to the LAN (consisting of DECnet nodes) and SNA COMC (connecting to SNA nodes, IBM host, etc.). Interestingly, in both approaches the gateways remain transparent to the users.

VMS/SNA software: The previous method of internetworking between DECnet and SNA is defined via gateways (hardware) and supported software. In the VMS/SNA software method of internetworking, the gateway operation is offered via DEC's Micro VAX system itself. The VMS/SNA software offers the communication link between DEC and IBM, bypassing DECnet. The Micro VAX acts as a physical unit type 2 (PU2) to the SNA host. It offers both functions of communication between them and support of applications. The DECnet/SNA gateway provides the only connection between them, as the gateway is merely a communication processor device.

9.2.2 Internetworking with X.25 and LANs

There are a number of ways of providing internetworking between LANs and WANs via gateways. Various configurations for internetworking of X.25 LANs have been recommended in ISO 8881 (Annex A). The following four configurations are the possible connections for internetworking.[1-3,7,9,11-13]

In the first configuration, two DEC DNAs are connected via a X.25 gateway. In other words, two different LANs are interconnected via one gateway. Both LAN1 and LAN2 have peer-to-peer entities for protocols of all layers of DEC DNA (user layer, transport layer protocol, X.25 packet layer protocol (PLP), LLC/MAC).[1] The gateway includes two tables within it. In table 1, the packet goes through the LLC/MAC to X.25 PLP, while table 2 includes X.25 PLP, LAPB, and X.25 PL. The gateway interface provides mapping between these two tables. In other words, the routing of the packets maps the packets from LLC/MAC mode to LAPB and X.25 PL. This packet is received by LAN2, which offers X.25 PL and LAPB at the lower layers. The gateway offers services of mapping, interfacing between LANs, etc., and is termed an internetworking unit (IWU).

In the first configuration, Figure 9.6(a), one IWU provides a direct link and routing between LANs. In the second configuration, Figure 9.6(b), the LANs are connected to individual IWUs, which are then directly connected to each other. This configuration uses two gateways between LANs. The IWU provides mapping between two X.25 PLPs across the network. There is a fundamental difference between the working of the IWU and any other wide area network connecting LANs, and that lies in the functioning of the interface between them. In the case of the WAN as an interface between LANs, the mapping across the network assumes both LANs to be of the same type (DCE), while in the case of the IWU, the terminal side has an emulated DEC while the network side has an emulated DTE. The main functions of the IWU are to receive the call setup, clear and reset packets from both sides of the gateways, and map between the format of packets across it.

During the connection setup, the incoming call packets come out either from LANs or X.25 networks. These packets, in fact, are sent by an X.25 PLP which has provided mapping to the request from the user who created a call request packet and sent it to the X.25 PLP (packet layer protocol). The IWU determines the size of the packet and options requested by looking at the header of the packet and matching it with the size of the packet on the other side of the gateway. The IWU may have an option of segmenting/reassembly of packets. In this case, the packets of fixed size (supported) are defined by mapping MAC addresses of the LAN at its interface (LAN) and X.25 addresses of packets at its interface (X.25). The X.25 interface defines the X.25 address, address extension fields, etc. If the IWU does not support any option for segmentation/disassembly, then it will provide mapping between the formats of packets on each side of the IWU.

After identifying the destination address, the IWU maps this address across it. It looks for a free channel on the other side of the IWU, selects a logical channel, and groups it with the channel on the left side. Once this is done, the packets for LANs or X.25 are transmitted over the channel group (on each side) to the destination during the data transfer phase. The call request packet goes to the other side, where it is mapped onto an incoming call packet. If the call request is accepted by the receiving node, the IWU receives a call-connected packet which is mapped onto a call-accepted packet for the sending node at the interface. It uses the same channel which was used for sending the call packet. Other events such as reset, clear, etc., are performed at the IWU for the calls which were not accepted by the receiving node. In summary, we can say that one or two gateways usually provide an interface between sending and receiving LANs or X.25 for the packet in both directions.

Figure 9.6(a,b) demonstrated how internetworking between two LANs can be obtained via one IWU or two IWUs. In these two configurations, a node of LAN (left-hand side) can communicate with another node of LAN (right-hand side); i.e., these two nodes are not on the same LAN but, instead, are on different LANs and still communicate with each other via the IWU(s). The IWU offers internetworking between them by providing

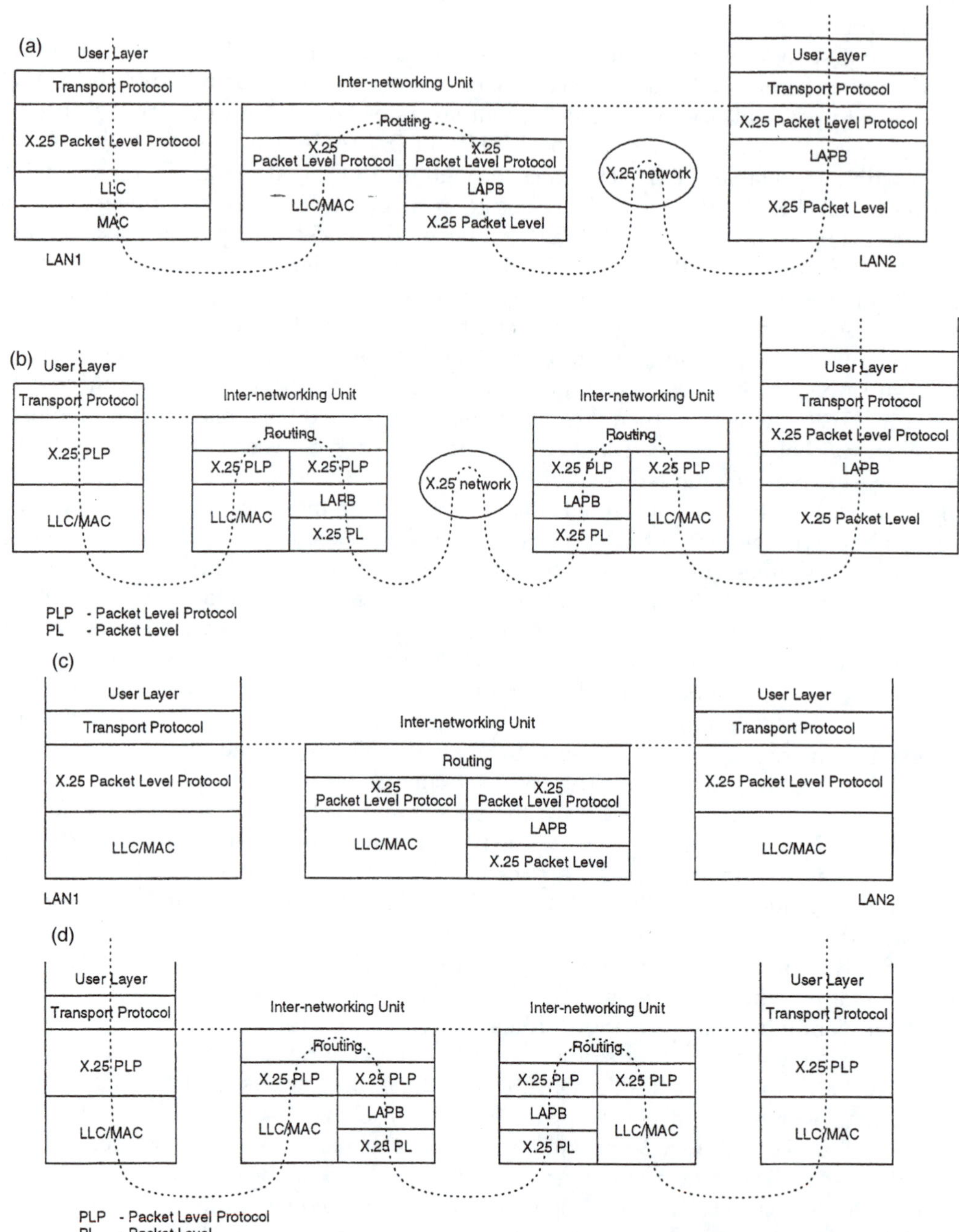

Figure 9.6 Internetworking configurations. (a) One IWU provides a direct link and routing between LANs. (b) Two interconnected IWUs connect LANs. (c) One IWU provides an interface between LAN and WAN. (d) Two IWUs provide an interface between LAN and WAN.

the function of a gateway node. As we can see in Figure 9.6(a,b), the protocols at the data link layer of LANs are different but still communicate with each other via the IWU. The standard ISO (Annex) recommendations discuss how the X.25 PLP on both sides are interfaced and communicate with each other via the IWU. This recommendation describes the connection between X.25 PLPs of both LANs which is, in fact, being managed and controlled by the IWU.

Similarly, the interface between the LAN and WAN requires either one or two IWUs (Figure 9.6(c,d)). A typical configuration for internetworking includes a LAN on each side with peer-to-peer protocols, with the only difference being in the protocols of the data link layer. The left-hand LAN sends a call request packet to IWU via its left-side protocol hierarchy. The routing of this packet goes through the same protocol.

ATM-based LANs: LANs are widely used for information sharing and exchange. The number of LANs around the world has already gone above 2 million. With different types of traffic, the requirements of LANs are also changed. Now the requirements of a LAN environment include higher data rates for integrated traffic (data, video, voice), an acceptable quality of service, and a uniform network management system (using the same equipment). As mentioned earlier, the LAN environment has evolved from low speed (Ethernet, token ring) to high speed (switched Ethernet, FDDI, 100-Mbps voice grade). The switched Ethernet offers a very high data rate and dedicated bandwidth. However, the dedicated access to a resource causes starvation, and it does not support multimedia applications. FDDI offers superior performance compared to Ethernet and token ring and is currently being used as a backbone network in many organizations, universities, and other agencies. Although it is expensive and does not support isochronous (voice) traffic, a newer version of FDDI — FDDI II — has been introduced which can handle voice traffic and can also support ATM. Also, inexpensive coax has been used in developing a new LAN known as coax distributed data interface (CDDI), which may become an alternative to the more expensive FDDI. 100-Mbps voice-grade LAN is a new technology and seems to be better than FDDI. No standards are available yet.

The ATM for LAN environments offers very high bandwidth (on demand) and much better performance, with very high data rates for data, video, and voice. ATM can be used as a backbone, and routers associated with it offer interfaces to different LANs such as HSSI, DSI, DXI, and so on. Different types of ATM hubs can be established, such as router-based, ATM-based, or a combination of these. In the router-based hub, all the routers are connected and ATM provides interfaces between different LANs and routers. The ATM contains interfaces for different types of LANs and also interfaces with routers and, hence, becomes the access point for LANs.The routers so connected form a router-based hub. In an ATM-based hub, all the ATM switches are connected, and the routers provide interfaces between different LANs and ATM. Here the router provides interfaces to various LANs and hence becomes the access point for LANs. A combination of these two types of hubs can be obtained by combining these two classes of hub architecture and is usually used in big corporations.

A typical LAN environment that exists today in universities, corporations, organizations, and government agencies uses the FDDI as a backbone network which connects different types of LANs such as Ethernet and token ring. These LANs are connected to the FDDI via a 10-Mbps link for Ethernet, or a 4- or 16-Mbps link for a token ring network. Various servers are usually connected to FDDI via 10-Mbps links. The FDDI is also connected to routers via a high-speed link — 50 Mbps or above. The routers are connected to the Internet or other routers via different links ranging from T1 (1.544 Mbps to 100 Mbps), providing access to the outside world. ATM can be used for the existing LAN environment in a number of ways. In one environment, the FDDI can be replaced by an ATM switch and some of the servers may be connected via high-speed links. The FDDI can also be reconfigured as an access network for different LANs. The second environment may define an ATM hub (ATM-based) where ATM switches are connected via OC3 links and can be used as a backbone network in place of the FDDI. Here the ATM switches will provide interfaces (ATM adapters) and access to different types of LANs. These ATM switches will also provide interfaces to routers via DS3 or above links for access to the outside world. In video-based applications, the ATM can be directly connected to workstations or terminals via optical fiber, giving a data rate of 100 Mbps with the option of transmitting video, voice, and data. A number of servers dealing with these types of traffic (video, voice, and data) can also be directly connected to ATM via ARM adapters.

ATM-based WANs: WAN interconnectivity is slightly more complex than that of LANs. The WAN, as said earlier, is used for greater distances than for LANs. The LANs use their own link to provide services, while WANs use the public carrier or a private line. The LANs and WANs can communicate with each other through an interconnecting device known as a channel service unit/data service unit (CSU/DSU). This device provides a mapping between LAN protocol and network digital protocol corresponding to T1 or any other link in the same way that the bridges provide two types of LANs at the MAC sublayer. The function of a router is to provide interconnectivity between LANs located within a building or at different locations (via WAN). The LANs are connected to WANs via CSU/DSU devices which provide mapping between the protocols of LANs and that of a public carrier (such as T1 or any other link) or private lines. The protocol used on public carrier or private lines use the TDM technique to accommodate the data from different sources. Unfortunately, the existing carrier supports data traffic only. Other types of traffic such as video or voice can also be handled by a TDM device via codec and PBX. For WAN interconnectivity, the ATM hub can be defined by connecting ATM switches via OC3 links, and then the ATM switches of the hub can provide interconnectivity with LANs through the ATM LAN connected via OC3 links. The ATM switches from the hub can be connected to WANs via OC3/DS3 links for access to the outside world.

A typical WAN environment includes various TDM devices connected together via T1 carrier lines defining a TDM hub. The TDMs are connected to various routers located at different locations miles apart. The routers are used as access points for the LANs (as discussed above). There are a number of ways of providing WAN interconnectivity. In one WAN interconnectivity method, each TDM may be replaced by an ATM switch, and these switches are connected via either T1 links or T3 (DS3) links (45 Mbps). The ATM switches transmit cells between them for the traffic coming out of the routers. The ATM hub supports different types of traffic (video, data, and voice). Here the ATM hub, in fact, defines a MAN/WAN backbone network.

In another type of WAN interconnectivity, the ATM switch may be connected via OC3 links and also can be directly connected to various routers and video services through video servers. It is interesting to note that with OC3 links, the optical signals now carry information which may allow us to use the public services available on an ATM-based backbone public network. These networks are connected to ATM switches via OC3 links.

ATM is used in different classes of public networks. For example, it can be used to replace the end office and intermediate office above this by connecting them with DS1/DS3 links. As we know, the subscriber's phone is connected to the end office via a twisted pair of wires which must be replaced by coax/fiber. The public network system and CATV can use ATM for providing video-on-demand services. This may require the use of compression techniques such as asymmetrical digital subscriber loop (ADSL) that uses the MPEG2 coding method.

The existing packet-switched network X.25 provides interfaces to routers, which are connected to routers, bridges, and LANs. The interface offers a dial-up connection via a 56-Kbps link from LANs. The function of the X.25 protocol is to provide the functions of the three lower layers of OS-RM. Since the transmission media are not reliable, the X.25 must support error control and error detection. Further, the packet goes through three layers, which requires processing time, and hence the network is very slow. There are a number of ways an ATM can be used in X.25 networks, which are typically used for data communication. In one X.25 switched network environment, each of the X.25 switches is replaced by frame relay (connection-oriented services). This will increase the speed from 56 Kbps to T1 (1.544 Mbps) speed. For connectionless services, switched multi-megabit data service (SMDS) switches offering higher speeds than T1 are used. SMDS can offer T3 (45 Mbps) speed and provide public broadband services. The underlying frame relay may become an access point to the SMDS network. In another X.25 environment, ATM may be used as a backbone network, offering virtual circuits with much higher bandwidth between T1 and OC12 (622 Mbps). We can use frame relay and SMDS as access networks to ATM. We can transmit voice data and video over an ATM backbone using circuit switching. We can also use the complete B-ISDN network using B-ISDN protocols and interfaces (ATM/SONET).

9.2.3 ISO internetworking protocols

ISO has defined a series of protocols for different layers of LANs and WANs which provide interconnection between different combinations of LANs and WANs.

Application: For LANs and WANs, file transfer and access management (FTAM), transaction protocol (TP), virtual terminal protocol (VTP), X.400 message-handling system (MHS), X.500 directory services, job transfer and maintenance protocol (JTMP), network management, etc.

Presentation: For LANs and WANs, ISO 8822, ISO 8823 with ISO 8824 and ISO 8825 (ASN.1/BER), kernel, abstract syntax notation (ASN), etc.

Session: For LANs and WANs, ISO 8326 and ISO 8327, basic combined subset (BCS), basic synchronized subset (BSS), basic activity subset (BAS).

Transport: For LANs and WANs, ISO 8072 and ISO 8073 types 0, 1, 2, 3, and 4.

Network: For LANs, connectionless network protocol, ISO 8473, Internet protocol (IP). For WANS, connection-oriented, mode, X.25 packet layer protocol (PLP).

Data link: For LANs, IEEE 802.2 logical link control (LLC) common to all LANs, IEEE 802.3 media access control (MAC), IEEE 802.4 MAC, IEEE 802.5 MAC. For WANs, high-level data link control (HDLC), link access control-balanced (LACB).

Physical: For LANs, baseband (IEEE 802.3 MAC), broadband (IEEE 802.4), twisted pair (IEEE 802.5). For WANs, X.21 bis (RS-232D), X.21, V.35.

9.3 Internetworking devices for LANs

Internetworking allows users working on different machines under different operating systems to interact with each other and use the services of remote networks as if these were local networks. Internetworking can be provided for both LAN and WANs. For internetworking between similar LANs, repeaters and bridges are used, while a gateway is used to internetwork dissimilar LANs. Internetworking defined by higher layers is obtained by protocol converters.

As digital signals travel along the cable, the amplitude of these signals decreases gradually. If the communication stations are widely separated from each other, then the signals must be regenerated gradually along the length of cable, and the device which performs this regeneration is known as a *repeater*. A repeater copies the bits from one segment of the LAN and passes it on to another connected segment. Obviously, both segments belong to the same category of LAN and, as such, the repeaters here are used to enhance the length (cable length) of the LAN. For example, the transceiver chip used in CSMA/CD LANs cover a distance of 500 m, but the use of repeaters can extend the length of the LAN up to 2.5 km. The repeaters provide internetworking at the physical layer. In some implementations, network stations themselves provide the operation of a repeater at their network interfaces (token passing networks).

Server: A server is either a hardware device or special software running on a general-purpose computer.

Socket: The socket defines an address through which the data transfer between PC and network takes place. There may be a few sockets within an interface unit. The received packet is partitioned into two fields: information and control. The message goes to the correct process in the memory of the user's PC. The socket must be unique. The CCITT has defined the socket address of 14 bits. The first four digits represent the segment or a particular network. The remaining 10 bits are used to define the network station and unique socket within it.

Repeater: A repeater provides interconnection between two identical LANs. The bridge is used when LANs have different physical transmissions and different protocols at the physical layer. The protocols higher than the physical layer must be the same. A bridge may be connected between more than one LAN, and all the nodes must have the same address format. Address format conversion is not provided by a bridge. The repeater also provides temporary storage for the messages that it has forwarded to another network and fetches the messages from the storage in case a re-transmission is requested. Multiple bridges may be used for connecting multiple LANs, and there must be one route to every node connected, as the bridges do not provide any routing. The repeater regenerates the signal strengths between various segments of similar LANs, thus increasing the total length of the LAN. It accepts a data signal at its input and regenerates the signal via some kind of error correction. The regenerated signal is then re-transmitted. This device works at layer 1 of OSI-RM and can be bidirectional for bidirectional cables. In the case of token ring networks, it is a unidirectional device. A repeater is a very inexpensive device. It does not provide any routing or any management of packets/frames. If the packet/frame is lost or corrupted, we must look into another interconnecting device, such as a bridge.

9.3.1 Media access control (MAC) bridges

LANs can be interconnected by bridges (at the data link layer), routers (at the network layer), and gateways (at higher layers). LANs which have the same protocols for the physical layer and the IEEE MAC sublayer are connected by a **bridge**. The bridge connects different types of LANs at the data link layer by looking at the destination address in the received frame and determining whether to keep it or pass it to another connected LAN. At the physical layer, the bridge acts like a repeater in the sense that the received signals are regenerated. It must use the same protocol on either side. This is useful only for small- or medium-sized network systems, but for a larger network system, another interconnecting device known as a router is used. A bridge operates at layers 1 and 2 of OSI-RM and is commonly used to either segment or extend LANs running the same LLC protocols.[11,14-16] In other words, the bridge can be considered a **store-and-forward internetworking unit** between similar or different LANs. It does not modify the frame formats (including destination address field). A bridge connecting two LANs must have those LAN boards, and it can be connected to more than two LANs. The exchange of MAC headers between two LANs takes place over the physical layer (based on differential Manchester encoding). It is closest to the hardware device and offers excellent price/performance internetworking.

The bridges accept the frames from one LAN, listen to them, and pass them on to appropriate LANs. If the frame belongs to a LAN connected to it, it will accept it and pass it on to the destination node on that LAN; otherwise, it will pass it on to another connected LAN. Both LANs are identical, and as a result these MACs offer minimal processing. Thus, the function of a bridge is to transfer the frame data from one LAN to another LAN; it is equivalent to a repeater, which also provides a link between identical standard LANs. The repeater accepts the frame data from one LAN and transfers it to another LAN after regenerating the data signal (equivalent to amplification in analog signal), thus extending the length of the LANs. It does not provide any mapping or routing. On the other hand, the bridge offers the following main functions to the frames during internetworking:

- Address mapping
- Routing information
- Relaying of the frames
- Buffer space for providing flow control

The features offered by a bridge include reliability (by inserting bridges at critical nodes), security (by programming bridges not to forward sensitive traffic, e.g., examination papers, marks, etc.), connecting LANs of different floors of buildings, partitioning of a load on a single LAN, connecting segments of LANs to avoid cabling, etc.

Let's consider a case where two LANs, LAN1 and LAN2, are connected by a bridge. All the data frames transmitted from LAN1 (or LAN2) are received by their respective LLC sublayers which define LLC headers. The packet from LAN1 is passed to its MAC sublayer, which adds an IEEE 802 header and trailer (depending on the type of standard LAN) and defines frames. These frames are transmitted over the cable and are received by the MAC of the bridge, which strips off the IEEE 802 header and trailer and sends the packet with the LLC header to its LLC. The LLC then passes this packet to the MAC sublayer of LAN2 at the bridge, which adds its header and trailer (of LAN2). This frame, which is now defined for LAN2 format, goes through cable up to LAN2. In this case, we are assuming that the bridge is between LAN1 and LAN2 and, as such, it contains MACs of these two LANs only. Thus, this bridge allows the transfer of frames between LAN1 and LAN2. In other words, the bridge will receive only those data frames from LAN1 which are addressed to its connected LAN2. After it receives these frames, the MAC sublayer protocol will re-transmit them over to LAN2. Similarly, the same bridge will accept the data frames from LAN2 which are addressed to LAN1 and its MAC sublayer protocol will re-transmit them over LAN1. If a bridge is used to connect N different types of LANs, it will have N different types of MAC sublayers and also N different types of physical layers for each one. It is important to note that the LLC sublayer remains the same for internetworking via bridges.

The bridges usually do not change the contents of frames and also do not change their formats, as both LANs are identical. In fact, the higher layers above the LLC sublayer use the peer protocols and, hence, are identical. Each node on any LAN is defined by a unique address (MAC), and the use of a bridge gives us a feeling that we are transmitting the frames to the destination node on the same LAN. This means that the bridge also extends the length of the LANs. The relaying function is based solely on the MAC address in the frame and is transparent to higher-layer protocols. This allows different protocols to interact with each other. The bridge also allows the network more flexibility in its management, as operational configurations can always be changed dynamically by changing the status of bridge ports only.

In addition to the above, a bridge offers other functions corresponding to the status of the bridge and also the control mechanism associated with it. The status of the bridge typically indicates the state of the bridge such as busy, root node identification, port identification, etc., while the control mechanism is mainly responsible for reducing the congestion problems with the buffer of the bridge. The standard defined for bridges by the IEEE 802 committee is known as **MAC level relay**. As indicated above, the bridge accepts the frames, stores them, and then forwards them appropriately. This operation at the bridge causes delay, which may be significant compared to the operation of a repeater. The MAC sublayer does not provide support for flow control and, hence, the bridge may impose traffic congestion during heavy traffic. Since MAC formats of different LANs are different, a frame-check sequence at the bridge for forwarding the frame from one format to another will introduce errors.

MAC architecture: As mentioned above, the bridges are connected at the MAC sublayer level; the bridge is composed of the physical layer and the MAC sublayer. Since the LANs to be connected by a bridge are the same, the architecture of the bridge (MAC sublayer and physical layer) will have the same MAC sublayer protocol, while the physical layer may use different protocols depending on the media used. A typical interconnection between LANs via bridge can be seen in Figure 9.7.

Since the bridge is connected at the MAC sublayer, the end user node is defined at the MAC level. In other words, the LLC submits the packets at **MAC service access points (MSAPs)**, as shown as A, B, C, and D in Figure 9.7. Both LAN1 and LAN2 use the same LLC and MAC protocols. After the MAC receives the packets from its LLC, it defines a frame and defines the destination node in the DA field of its MAC frame. At this point, this node does not know the address of the LAN to which that particular destination node is connected. Due to this, when the bridge receives the frame, it stores it in its buffer. After looking at its address table, it re-transmits the frame to another bridge via its connected LAN. If there is only one bridge between two LANs, the session between these LANs is defined by its peer entities of protocols of two end user nodes, and we really do not require an LLC sublayer in our bridge.

The data communication through bridges follows the same pattern we discussed regarding the data communication through layers of OSI-RM. The packets are received by the LLC sublayer from its network layer, and it adds its header, which includes the source and destination addresses (LSAPs), control information, and user's data, and passes it on to the MAC sublayer via MSAPs. The MAC sublayer defines a frame for each of the packets it receives from its LLC sublayer by attaching a header and trailer around every packet separately. These MAC frames are then transmitted over the transmission medium.

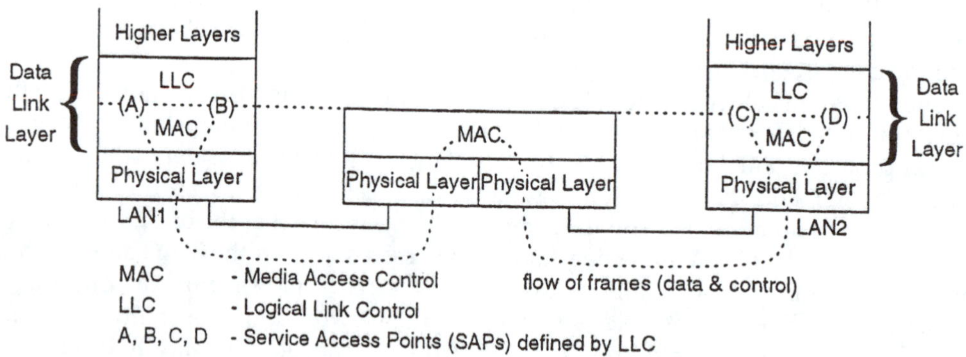

Figure 9.7 Interconnection between LANs via a bridge.

The MAC destination address defined in the MAC header is read by the bridge, which stores it in the buffer. After looking at the address table, the bridge either sends it to its connected destination node or passes it to another bridge connected via its connected LAN. Both LANs connected use the same standard protocols of layers above the MAC sublayer. The bridge will contain the MAC sublayers of these two connected LANs. The protocols at higher layers may be from ISO or other vendors.

For ISO standard protocols, the following is a list of protocols for these layers: application (undefined), presentation (ISO 8823), session (ISO 8327), transport (ISO 8073 Type 4), network (ISO 8473), LLC (ISO 8802.2), and MAC (ISO 8802 and depends on the type of LANs).

The bridge represents only up to the MAC sublayer; as such, the LANs will have the same LLC while the MAC and physical layers of both LANs and bridge use the same protocols. The bridge can also provide interconnection with packet-switched networks. Here again, for top layers up to the LLC layers, identical protocols are used. The bridge defines lower layers (physical and MAC) of two connected LANs. This LAN bridge is then encapsulated within X.21 (physical layer of packet-switched network), which also provides MAC conversion. The physical layers of the packet-switched networks are connected and provide peer-to-peer entity connections, as both packet-switched networks use the same protocols at all three layers. The network layer uses X.25 PLP, the data link layer uses ISO 7776 LAP-B, and the physical layer uses X.21 standard.

Transparent bridge: If more than one bridge is connected between sender and receiver for defining the routes between them, the routing strategies for the packets are solely defined by the bridges, and the LANs have no control over these routing strategies. Due to this, the implementation of the bridge and various routing strategies are transparent to LANs; hence, the name **transparent bridge** is derived. If there are two LANs connected by a bridge, then the implementation of this type of interconnection is very simple and does not require any kind of routing strategies. The LAN segment is connected to a bridge at its physical port, and typically a bridge defines two ports. Multiport bridges define connected ports, which are basically defined for segments.

The normal working of a bridge is based on **promiscuous mode**, which defines the implementation steps required for realizing the bridge. When a bridge receives a frame, it stores it in the buffer (controlled by MAC sublayer) and makes a request to the **bridge protocol data unit (BPDU)** for processing. The bridge includes a subroutine for handling a port and another for controlling the bridge protocol entity. The protocol used in the bridge transfers the memory locations of the frame to the protocol entity and also handles the multiple frames coming to the port and leaving the port.

If the LANs are connected to more than one bridge, then the implementation of the bridges becomes complex, as it has to determine not only the destination node but also the least expensive route between source and destination nodes. This is due to the fact that now the data frames are received by more than one bridge, and each bridge has its own address table and uses its own routing strategies. Each bridge has to recognize the MAC address separately, and if any one bridge finds it in its connected LAN, it passes it to that node; otherwise, all the connected bridges will forward the data frames to their respective connected LANs through which they are received by other bridges. Those bridges will repeat this process until the data frame is delivered to the destination node.

The routing strategy adopted by the bridges in this situation typically uses **tree structure** for its implementation. At each level of the tree, the frames are received by the bridges, which pass them on to lower levels of the bridges until they arrive at the lowest level of the tree, which corresponds to the destination node. At each level, the bridges have a choice of either storing the data frames or forwarding them to bridges of lower levels. It is obvious that as the number of levels of the tree increases (i.e., the destination node is further from the source node), the complexity of the protocols to be used in the bridges becomes greater.

Initially, when the bridges are plugged in, the hash table maintained for the routing table is empty. When it receives any frame, it looks at the source address and, depending on its destination, it will either forward the frame or discard it. Further, it notes the time the frame arrives and makes an entry in the table. The next time the frame arrives, it simply updates the entry in the table with the current time of arrival. In this way, it defines the routing table in a dynamic mode about the current status of the frame. This type of routing does not offer an optimal path and supports connectionless services. The errors are handled by bridges; hence, the bridges are very complex, locating the errors transparently in a backward learning.

The frames which arrive at any bridge node identify both source LAN and destination LAN, and the bridges perform the following set of rules for handling them. If both source and destination LANs are the same, the frame is discarded. But if both source and destination LANs are different, the frame is forwarded. For each of these situations, hardware units (chips) are available on the market. Bridges support a variety of routing strategies like static routing, source routing, adaptive routing, etc., and do not support flow control.

Gateways: This internetworking device is used to provide interconnectivity between LANs, packet-switched networks, private networks, databases, videotex services, or any other services. Gateways provide the compatibility of protocols, mapping between different formats, and easy data communication across them. The user's PC will send the packets to the gateway, which now becomes the address, which will send it to another network. In this way, a gateway basically becomes another node between networks that provides compatibility between the operations of these two different networks.

Spanning tree algorithm: A spanning tree algorithm initializes all the bridges to exchange the frames among themselves, and these frames are known as bridge protocol data units (BPDUs). This algorithm is based on tree structure, and each bridge is assigned a unique identifier. A bridge-sending initialization control is chosen as a root bridge dynamically and is assigned the highest priority. This bridge keeps sending a BPDU indicating its status as a root bridge to other bridges. The other bridges respond to this BPDU by determining which port assigns the least cost to the root bridge, and this is termed the *root port*. There may be more than one root port, and the root bridge will decide on one root port for data communication. All the BPDUs from the root bridge will be sent through this root port. This port may not be physically near the root, but it is chosen based on the lowest-cost path between itself and the root bridge.

The link data rate determines the lowest cost for the root based on the port root which offers higher data rates. Based on the lowest-cost path, the root port will designate one root port through which all the frames are forwarded to each segment of the LAN. The bridge selected as the root port will always receive the frames from the root bridge and can perform two functions: **forwarding** and **blocking**. The root port bridge cannot transmit any frame. Initially, all the bridges are in forwarding mode. Once the root and root port bridges have been identified, they are set in forwarding mode while other bridges are set to blocking mode. The transition between forwarding and blocking basically goes through two states: **listening** and **learning**. Each mode and transitional state will receive only a class of BPDU, depending on its functionality.

When a port bridge is not active, it will receive management BPDUs and will process them. In blocked mode, the port bridge will receive configuration BPDUs and will process them. When the port bridge is in forwarding mode, it will receive all BPDUs and also the user's message frame, and each one of these is processed by the port bridge. In the listening state, it will receive BPDUs, while in the learning state BPDUs are processed while a learning process is invoked to learn about the user's message frame without forwarding them.

Initially, all the bridge ports are active when switched on. The BPDU entity is invoked either when a BPDU is being received or when the time-out corresponding to that bridge protocol is expired. In both cases, the bridge protocol entity will be invoked to cause a transition between various modes and states, as mentioned above, and the transition between modes and states depends on the type of BPDU being received by the bridge protocol. The port goes to blocking mode when a management BPDU for enabling the port is received by the network manager. The network manager can also make any port bridge inactive or disabled.

The bridge protocol changes the blocking mode to listening state after the root and root port bridge have been decided, and the bridge goes into listening state and a timer is set. The transition from listening to learning takes place in case the timer expires, if during this time the root and root ports are the same, and the timer is reset. This procedure is continued until the time-out is expired. The ports which are still root and root port will be changing their states to *forwarding* mode during this time. If during this time, the ports are neither root nor root port, then these ports will be returned back to *blocking* mode.

Source routing bridges: Packet processing is relatively faster than routers and is usually best suited for extending local network systems. The source routing strategy for bridges is very useful in token ring LANs, while the CSMA/CD and token passing LANs use the above-mentioned routing technique. In this technique, the end nodes perform the routing functions. In other words, the node determines the route in advance for any frame to be delivered to a destination before the frame is actually transmitted. This information is defined in the header field of the frame. The bridge looks at this field to determine whether this frame needs to be forwarded. The routing information includes a sequence of all the segment bridges and their identifiers.

On the receiving side, the bridge looks at the routing field (a part of the header field) to match its own identifier with the one defined in the header. If it finds its identifier, it then checks the next segment identifier defined in the field connected to its output ports and forwards it. If it does not find the above information, it simply will not forward the frame. If the frame is forwarded, its address is recognized in the A-bit of the frame status (FS) field and the frame copied is shown in the C-bit of the FS field. This is an indication to the sender node that the frame has been received (forwarded) by the destination bridge. It is simple to install. The source routing defines a path for routing the packet and is expressed in terms of the LAN address. Each LAN is assigned a unique 12-bit number, while each bridge is assigned a unique four-bit number.

9.3.2 Internetworking IEEE LANs

The bridge may have the same number of different LANs but will have a different number of the same LANs. Thus, the path is defined in terms of sequence of bridge, LAN, bridge, and so on. It offers optimal routing for connection-oriented services, and the errors are usually handled by the hosts (making it more complex). The location of errors is implemented by discovering the frame and is not transparent.

Since there are three standard LANs defined by the IEEE 802 committee, the internetworking between these LANs via bridge offers nine possible combinations. Due to different formats of MAC frames, different access techniques, and support for different media, there are problems which must be identified and solved during the design of bridges. The following is a partial list of some of the general problems encountered during internetworking for standard LANs via bridge.

- Different formats for copying require CPU time
- New checksum calculation
- Detection of errors introduced by bridge
- Different data rates (needs buffers)
- Loss of bandwidth during collision in IEEE 802.3
- Value of timers in the higher layers (difference in speed may time out and packets will be lost)
- Different frame lengths (802.3 — 1518 octets, 802.4 — 8191, 802.5 — no upper limit, except that a node cannot transmit longer than token holding time, and this puts a limit on the length of 5000 octets with default value of 10 ms)
- No provision for segmenting the frames
- Support for priority (802.3 does not support priority)
- Semantics of the bits are changed by the bridge

9.3.3 *Bridge protocol data unit (BPDU)*

The bridges exchange information via protocol data units (PDUs) at service access points (SAPs) as described in OSI-RM, and these PDUs are known as **bridge PDUs (BPDUs)**. The BPDUs contain addresses, control information, and users' data. Since the routing strategies in a bridge follow a tree structure, a bridge, after receiving the data frame, defines itself as a root node in its BPDU which is transmitted over the media. This BPDU is received by all LANs connected to it. A LAN which claims to have the address of either the bridge which has the node's MAC address or the node itself will accept it and pass it on to its connected bridge. This information about the claim is broadcast to all the nodes to indicate its connection with the root node (as one hop). Meanwhile, the first bridge will regularly transmit the BPDU, indicating its status as a root node. This process of broadcasting the BPDUs is continuously carried out along with their respective addresses with respect to the node, which has sent it to the LAN with its address with respect to the original root node in terms of the number of hops. The process terminates as soon as the destination node receives the BPDU.

The IEEE 802.1 has defined the control frame format of BPDU as follows[7,14,15] (also see Figure 9.8).

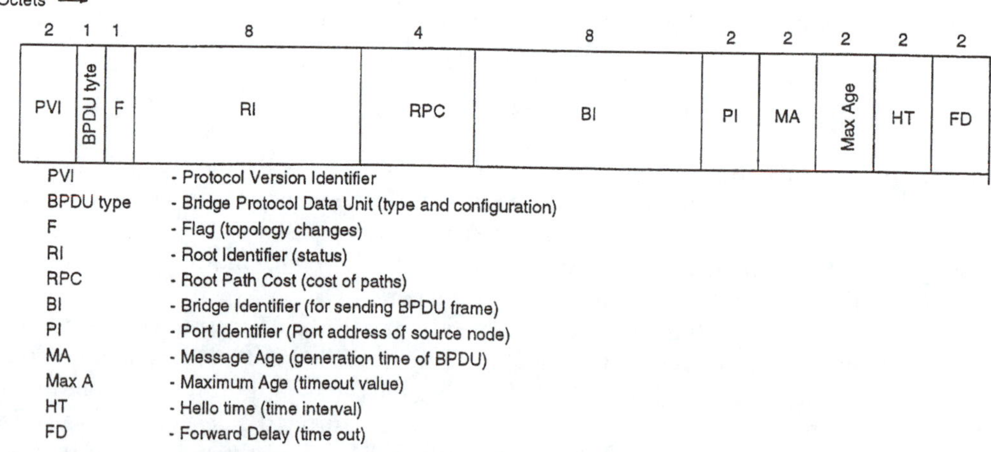

Figure 9.8 BPDU control frame format (IEEE 802.1).

- **PVI:** The protocol version identifier (PVI) field of two octets defines the spanning tree algorithm and protocol defined by IEEE 802.1. Initially this field is set to all 0s.
- **BPDU:** The bridge protocol data unit (BPDU) field of one octet indicates the type and configuration of the BPDU; initially all the bits are set to 0s.
- **Flag:** The flag field of one octet indicates the topology change.
- **RI:** The root identifier (RI) field of eight octets indicates the status of a root node to all connected bridge nodes.
- **RPC:** The root path cost (RPC) field of four octets indicates the cost of path for transmitting a BPDU from one bridge to another bridge and is identified by the root identifier (RI).
- **BI:** The bridge identifier (BI) field of eight octets defines a parameter which helps the bridge to decide whether to send the BPDU (if two or more bridges are connected) based on the cost paths to the root.
- **PI:** The port identifier (PI) field of two octets indicates the port address of the transmitting bridge.
- **MA:** The message age (MA) field of two octets indicates the time at which the BPDU configuration is generated by the root node. Any BPDU received whose age exceeds the maximum age defined by it must be discarded by the bridge.
- **Max A:** The maximum age (Max A) field of two octets indicates the time-out value to be used by all the bridges. This indicates that each bridge defines a consistent value against which to test the age of stored configuration information.
- **HT:** The hello time (HT) field of two octets defines the time interval between the generation of configuration BPDUs by the root node. This parameter is not used in a spanning tree algorithm.
- **FD:** The forward delay (FD) field of two octets defines a time-out to be used by all bridges and is set by the root node.

The bridges are required to maintain status information such as direction, receiving port, source nodes, next nodes, next ports, mapping, etc.

The bridges used in token ring LANs must introduce minimal delay and, as such, should transmit and receive the frames at the same time. As we stated earlier, the token ring LANs use a pair of rings, and the hardware used in one ring for transmitting must be coupled properly with the hardware of another ring for receiving the frames. The multiplexing is also used by the bridges for transmission and receiving in the same way that it is used in the transmission.

A number of mini-packets can be multiplexed and transmitted over a number of destinations. Each of these mini-packets is assigned the ring number for both source and destination addresses, thus allowing the bridges to accept packets from a number of rings. The bridges can also accept mini-packets for a set of destination nodes by maintaining a table look-up. The major drawback with the mini-packets is the question of how to preserve the response bits, as these are actually used to provide flow control within the transmission of data frames.

The transmission of a number of data frames over a number of token ring LANs to a number of destination nodes requires some kind of global address. This global addressing scheme includes both source and destination node addresses, including ring number and the address of the node. These are included in a routing table so that the frames can be forwarded to the destination or the next bridge on the route. Here also, frames may come in a different order, and re-sequencing has to be offered by the host software. This software deals with address sizes and performs part of building the address route for internetworking LANs and many other internetworking environments. The transport session between internetworked LANs is achieved by an interface architecture developed by Xerox PARC, known as **PARC Universal Packet (PUP)**. This interface offers transport for connectionless service (datagrams).

The internetworked protocol for this interface (PUP) resides in the host machine, which can support different types of networks. The gateway nodes usually accept the packets from one host and deliver them to another host and also provide routing (in contrast to bridges). In the ring bridge terminology, the datagrams are in fact PUPs. Due to the fact that the PUP offers connectionless transport service (which is usually not reliable), the higher layers must provide protocols for reliable transport. The PUP offers a packet transport service at lower levels (network-dependent), while at the middle level, it offers Internet datagrams independent of media. Various applications such as file transport and byte stream application protocols are provided at higher levels.

9.3.4 Performance design issues in bridges

As indicated above, the length of a LAN can be extended via repeaters, and there is a limit on the number of nodes which can be connected to each segment. The MAC also limits the number of repeaters which can be used for a particular LAN. The throughput and response time can be calculated by considering a number of parameters such as the number of segments, number of nodes in each segment, length of segments and total length of the LAN, type of traffic, type of LAN, etc.

There are a number of issues which must be considered while discussing performance design of bridges, and these performance parameters include forwarding delay (rate at which the frames are processed at the bridge port and, based on the address, either blocked or forwarded to another port; sometimes this is also related to throughput), buffer capacity (defined by bridges for storing the frames for queuing for their forward transmission or blocking), delay caused by cascading of bridges (where the bridge ports are connected to form a daisy chain via two- or multi-port bridges and the frames have to go through a number of intermediate bridge ports before they reach the destination bridge port of the LAN segment), and multi-port bridge (typically, this category of bridges can support between 4 and 14 ports and is very useful when connecting a number of LAN segments of different departments within an organization). The frames can be transmitted to and from a few ports, while a few ports can be allocated for providing access services of memory for transmitting and receiving the frames.

Multi-port bridges with processing times of 10,000 to 35,000 frames per second are available on the market. A remote bridge is used in connection with leased lines which link various nodes and other devices located geographically apart. The remote bridge will cause delay in terms of queuing of frames and processing time, and other delays with frames include transmission delay, propagation delay, etc. The remote bridges can be used only with non-switched networks (like dedicated or private), while interconnection between switched networks can be obtained by other interconnection devices such as routers.

9.3.5 Routers

A bridge provides interconnection between two similar LANs. A router provides interconnection between two different networks. The router, an internetworking unit (device), is compatible with the three lower layers. Unlike bridges, it supports at least three physical links (in general, it supports other links, too). This device is similar to a bridge for interconnecting different LANs, except that it is used in other types of networks such as

WANs, FDDIs, and other high-speed networks. It operates at the network layer. It can be used to connect LANs to WANs, MANs to WANs, LANs to LANs, and so on. A message frame transmitted over a LAN goes to all the nodes. At the network layer, hardware devices do the routing for sending a frame to other connected networks. By looking at the address defined in the frame, each node determines if the frame belongs to it. If so, the router accepts this frame.

The router defines the route for a frame to be transmitted to a destination. It is possible that the router will define more than one route for any frame to a destination, and the frame has to go through a number of routers. Each frame must contain two addresses: the destination address and the address of the next node along the route. The second address changes as it moves from one router to another. The routing strategies deal basically with the determination of the next node to which the frame must be sent. Routers are most commonly used for interconnecting networks from a single vendor or interconnecting networks which are based on the same network architecture. Here, the physical and data link layer protocols may be different, but higher-layer protocols must be the same.

A router is used at layer 3 of OSI-RM and, as such, connects different types of networks that have the same network layer protocol. It accepts the packet from one network, finds the best route (based on a routing algorithm), and uses the functions of the data link layer defined within the MAC address of the bridge (another internetworking device). Routers are used in LAN/WAN interconnectivity and provide different categories of ports. Usually, they support a few ports of each category. The port for a LAN typically supports 100 Base T, but it may have ports of other LANs, as well. These ports are available in the form of a card. In addition to providing ports for LANs and WANs, routers also provide interface to various serial devices such as CSU/DSU. The router maintains a routing table giving the information regarding the routes a packet will take going from its LAN to WAN and finally to the destination LAN. The decision regarding the routing is based on the routing algorithm being used by the router. There are a number of protocols that provide routing of the packets through internetworking: **interior gateway routing protocol**, **exterior gateway protocol**, **border gateway protocol**, **intermediate system to intermediate system**, etc.

Cisco Systems is the world's largest supplier of routers. The Cisco Series 7000 provides connectivity to high-speed networks using ATM and T/E1 systems and is used in the Internet for connecting different types of networks (see Chapter 17). The routers in the Internet ensure the timely delivery of frames, reduce the loss of frames, and handle Internet protocol (IP). The network nodes in high-speed networks are usually implemented via switches, while the nodes are implemented via repeaters in the Internet. Routers provide connectionless service and use hierarchical routing using IP. The nodes of high-speed network switches, on the other hand, are usually telco and connection-oriented; use X.25, frame relay, and ATM; and are based on fixed addresses.

9.3.6 Gateways

A gateway is used to interconnect different networks and, as such, must offer high-level protocol conversion. It must offer message format conversion, as messages from different networks have different formats, sizes, and coding. It must provide address translation, as different networks use different addressing schemes. Finally, because these networks are using a different set of protocols for their layers, the gateway must provide conversions for different functions (implemented differently in different networks), such as flow control, error control, and error recovery. Since gateways provide interconnection between different networks, they are flexible, expensive, and complex. The conversions of protocols have to be performed on the basis of layers.

A gateway provides interface between two different networks having different protocols and provides mapping between their protocols. It operates at layer 7 of OSI-RM. Typically, a PC may be connected through a gateway to a mainframe or other computer having different protocols so that the packets can be sent to it and the mainframe resources can be used.

For an incoming packet, a gateway determines the output link. It offers connection-oriented configuration-based protocols (like X.25), and the decision to route the packets is made only after the connection is established. It defines an internal path during the duration of a call. In the case of connectionless service, the address of every incoming packet must be examined. Routing determination makes routing strategies and its processing cause overhead over connection-oriented configurations. Since it operates at the network layer, it can easily transform or map the address of one LAN to another one, making it slower. This device for internetworking is usually used in WANs, where response time is slow and it is not required to handle more than 10,000 packets per second. Internetworking between dissimilar LANs can be defined for both connection-oriented and connectionless services.

A gateway connected between the Internet and the network of the destination (target network) receives the Internet e-mail and forwards it to the destination. As said earlier, the message is broken into IP packets. The gateway receives it and uses TCP to reconstruct the IP packet into the complete message. It then provides mapping to the message into the protocol being used at the destination.

The gateway for connection-oriented networks defines a virtual circuit at the network layer and is usually managed by different organizations (as opposed to bridges, which are managed by the same organization). In the architecture of a MAC bridge, as shown in Figure 9.7, we notice that the bridge contains physical and MAC layers of each LAN on each side. In the case of a gateway, the Internet sublayer (sitting on top of the network layer) is common to different sublayers (subnet access and subnet enhancement) of the network. The gateway is partitioned into two parts, attached with each host, and connected by a communication link. Each part of the gateway consists of two sublayers — LAN to Internet and Internet to LAN; thus, the gateway at node 1 will have LAN1 to Internet, while node 2 will have LAN2 to Internet and Internet to LAN2. Each partition of the gateway is known as a **half-gateway (HG)** and is controlled by another organization. The half-gateway uses CCITT's X.75 protocol for data communication over the network.

The X.25 protocol builds up an internetworking connection by concatenating a series of intra-networks and half-gateways to half-gateway virtual circuits. Each connection consists of five adjacent virtual circuits known as VC 1–5. The VC1 is defined between the source node and half-gateway (also known as **signaling terminal**) in the local network. The VC2 connection is defined between the half-gateway (of source) and half-gateway of intermediate networks. The VC3 and VC5 are intranet just like VC1, while VC4 is another form of intranet just like VC2.

Internetworking supporting connectionless services defines datagrams. Here, the gateways typically consist of Internet and transport packet format protocols and the formats of frames of networks which are connected by gateways. The formats of frames include data link layer headers and trailers around the Internet packets. Each gateway contains the formats of those networks which are internetworked by it. The Internet packet format and transport packet formats remain the same for all types of networks.

Due to a number of parameters involved in internetworking across different types of networks, such as operating systems, buffer sizes, protocols, size of packets, error control, flow control, etc., the internetworking devices are getting very complex. For some protocols, the maximum size of packets is known, and this may pose a problem when packets of a smaller size need to be transmitted. For example, HDLC offers variable size, IEEE 802.4 offers 65,528 bits, X.25 offers 32,768 bits, and ARPANET offers 1008 bits.

9.4 Wireless LANs

In most of the standards of LANs defined by IEEE or by non-IEEE private or proprietary products, the transmission medium (coaxial cable, twisted pair, or even optical fiber) plays an important role, as it not only provides the physical link or circuit across LANs, but it also defines the capacity and bandwidth (data rates) of data frames transmitted across the networks. For the LANs used on the premises of organizations, universities, etc., the cabling or wiring may sometimes become messy and expensive and need to be redone in the event of relocation of resources and other communication devices. Further, the cabling sometimes poses a serious problem in cases where updating is performed frequently on the premises for relocation of offices, resources, etc. Quite often, the cable installation and length of cables connecting these devices may become cumbersome and inconvenient.

The problem of installing and connecting cables has been addressed in another category of LANs which are based on the concept of data communication over a *wireless transmission medium*. The medium through which data communication takes place is air, and, as such, we need to define an air interface. This does not require any cabling to connect devices to LANs. This concept of wireless transmission has been used in voice communication and radio-frequency transmission, e.g., **radio frequency (RF)**, microwave link, satellite link.

In wireless LANs,[17-19] we may expect the following advantages:

- The cabling installation and cost is reduced.
- Since no physical link is used, there is no electrical interference and, hence, no noise.
- Support is offered for portability and flexibility.
- The reorganization or relocation of office devices does not require any additional cost in the configuration or cabling, or moving of devices.

Wireless LANs are used where cabling installation may become a problem, e.g., old buildings, temporary installation, restricted or prohibited areas, unhealthy environment, and even in historical buildings.

9.4.1 Wireless LAN layout configuration

A typical wireless LAN layout configuration is shown in Figure 9.9. It is important to note that this is a generic configuration and does not correspond to a specific vendor-based wireless LAN product.

In Figure 9.9, each terminal interacts with the hub node by using a radio-frequency band of 1.88–1.90 GHz over a distance of less than 200 m. A typical hub may support a few radio devices or units. The wireless LAN configuration as defined above offers a new type of star topology which supports radio connection to terminals and is based on an

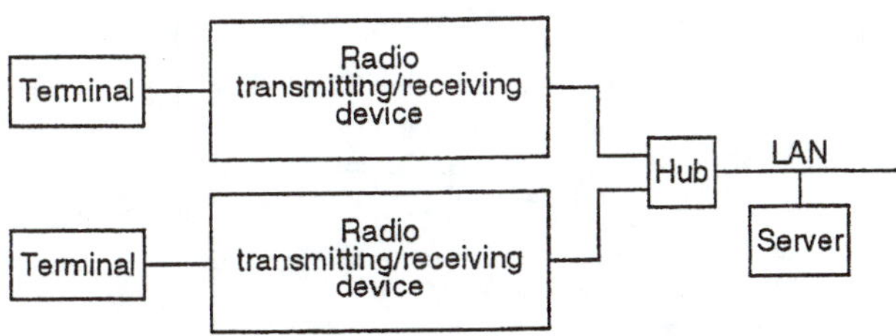

Figure 9.9 A typical wireless LAN layout.

intelligent hub and several cell management units. The hub offers network management services, and the wireless workstations communicate with each other transparently.

The topology follows a client–server implementation which, due to its features, offers high performance and highly efficient configuration. The hub is a wired LAN system and the client unit is composed of a **network interface card (NIC)** and software. The servers are connected to wired LANs which define the hub. The servers must manage wireless connections to all PCs via radio units and bridge the standard with an Ethernet segment of LAN. A typical hub contains a radio unit, Ethernet controller and board, and other software units.

The radio units are installed near the hub, about 200 m away. The client PCs must be near the region for providing a wireless communication link with the hub (typically a few hundred meters). The radio unit typically is a small board with two antennas (omnidirectional) coming out through holes. A cable connects a controller, while the PC bus provides electrical supply to this unit. The radio unit usually operates within a frequency band of 1880–1900 MHz, which is divided into ten channels with a spacing of 1.728 MHz between them.

The software required for the client wireless PC includes three modules: LAN manager, network operating system, and installation. The LAN manager provides an NDIS-compliant interface for the LAN manager Protocol Stack (NETBEUI), and the NetWare version protocol provides an OSI-compliant interface for an IPX protocol. The hub can be a dedicated PC which offers functions such as communication with wireless clients and a communication link between Ethernet and DECT (via a bridge).

9.4.2 IEEE 802.11 standard architecture

There are a number of proposals for defining an architecture of wireless LANs pending with the IEEE 802.11 subcommittee.[17] The proposed architecture model that is widely accepted among researchers and professionals working in this area is shown in Figure 9.10. Since this architecture model has not yet been standardized, the functions of various layers are not standardized, but the conceptual functional architecture can be understood.

The physical layer is composed of three sublayers: media-independent, media-dependent, and convergence sublayers. The **media-independent sublayer** represents a uniform interface to MAC and also to MAC management sublayers, as this sublayer supports a variety of physical layers defined by different media-dependent sublayers. This sublayer

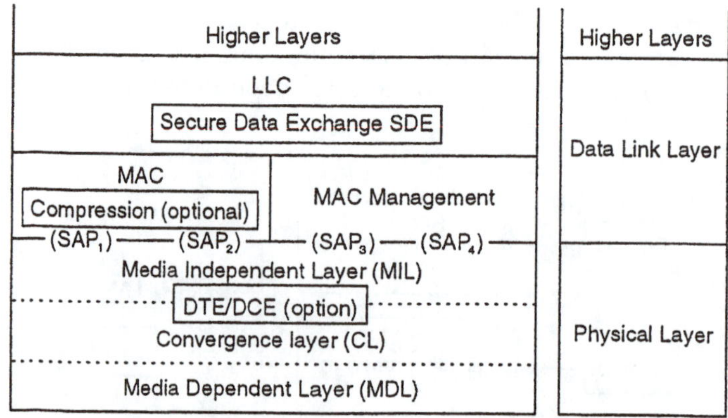

Figure 9.10 IEEE 802.11 proposed architecture.

has to coordinate with MAC and MAC management sublayers to provide synchronization for data communication (frame by frame). It also offers a DTE/DCE interface (optional) which provides interaction and coordination between MAC and the physical layer. This does not require any additional implementation and, in fact, uses a minimum number of pins and signals to minimize the implementation costs. The **media-dependent sublayer** represents all the characteristics and behavior of a particular medium and offers an interface between it and the media being used. The **convergence sublayer** accepts the information from different media-dependent sublayers and offers a mapping between media-dependent and media-independent sublayers.

The MAC layer is divided into two sublayers: MAC and MAC management. The **MAC sublayer** offers the same functions as that of conventional MAC sublayers (already discussed above). This sublayer is mainly concerned with data communication, reception, controlling of data and control frames across the medium, etc. It also offers compression technique as an optional service. The **MAC management sublayer**, on the other hand, is concerned with providing an effective interface between radio media and quality detection and control transmission at the packet level. This offers ISO-style management in the sense that it offers access to the MAC and physical layers.

Both MAC and MAC management sublayers offer separate and independent **MAC service access points (SAPs)** to the media-independent sublayer. The interface between MAC and the media-independent sublayer carries data and timing in the same way as the traditional LAN interface. It also carries the frames from the physical layer at the bit level; i.e., it carries bit-by-bit information from the physical layer. This information is useful for the MAC management sublayer, as it creates a map of the transmission characteristics between each pair of wireless LAN nodes.

For providing security, the LLC sublayer includes a **secure data exchange (SDE)** module which uses the following primitives to interact with the MAC sublayer. The primitives associated with these services are defined in the SDE Services Specification Document:

- **UNITDATA.request (source address, destination address, MAC SDE)**
- **UNITDATA.indication (source address, destination address, MAC SDE)**

The data communication through this proposed wireless model (IEEE 802.11) is shown in Figure 9.11. The SDE protocol may affect the framing and other frame management problems, but these implementations are provided transparently. The SDE frame format is shown in Figure 9.12.

The standardization process of the IEEE 802.11 subcommittee is still in the development phase, and this committee is looking at a number of relevant problems, e.g., security procedures (which layer should performs this?) and authentication procedures. The IEEE 802.11 subcommittee has made the following recommendations for time-based services

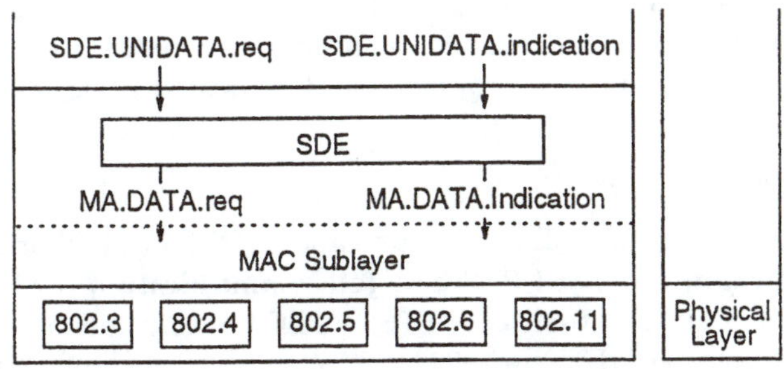

Figure 9.11 Communication through wireless IEEE 802.11 model.

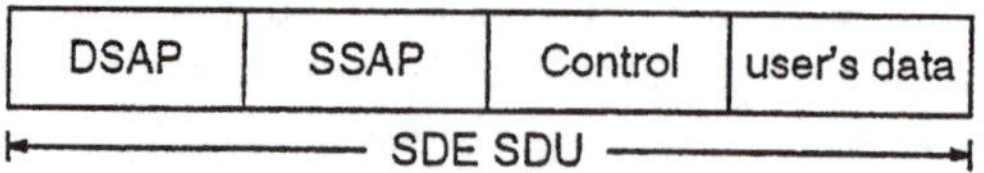

Figure 9.12 SDE frame format service data unit.

such as voice, video, industrial automation, and multimedia. A number of new concepts in switching and transmission have been proposed recently for providing integrated voice and data communications.

It seems that the ATM cell has been widely considered and accepted as a suitable switch for future high-speed integrated networks such as B-ISDN, IEEE 802.6, MAN, and many others. With this notion, the ATM-based backbone network will provide the backbone network environment for offering a variety of existing and new services at high data rates. The wireless LAN will provide a local distribution service to users. The ATM cell has been chosen for all future high-speed networks because switching and formats for cells are independent of whether the content is a packet or connection type of service. Further, the physical medium and access methods are decoupled from most of the details of providing connection-type services and public network compatibility. The IEEE 802.6 has created and adopted a standard based on a multiplexed isochronous medium over which ATM cells can be transmitted. In order to accommodate the ATM cells, slots are defined to match the cell length and access methods, and the state of slots (busy or available) is also specified. The medium is a dual bus created from a two-ended cable between drops. The IEEE 802.6 is a backbone interface adopted by Bellcore's SMDS.

IEEE 802.11 has recommended regular periodic slots which are organized into frames and integer multiples of 125 μs for wireless media. The wireless LANs are being considered an important alternative for general communication. Increased versatility and flexibility of transmission media are considerable incentives to involve asynchronous switching for offering a range of services to users at a transfer rate of 16 Mbps and higher. This results in distribution of wired and wireless communication for all types of communication systems, such as metropolitan, office, factory, and residential areas for PBXs, LANs, and other service providers.

Many new digital modulation schemes have been introduced and defined for high-speed integrated networks. A **non-return-to-zero smooth transition modulation (NRZ-ST)**[16] has recently been proposed. In this modulation scheme, the transmitted signal is generated by overlapping pulses of the same or opposite polarities. The pulse shape must have the property of adding to a near constant value when repeated and added at the desired pulse rate. The pulses are differentially pre-coded so that the information can be carried in the same polarity as that of the current pulse relative to its previous pulse.

The IEEE 802.9 subcommittee has produced a draft standard which proposes a telephone twisted pair of wires as a communication link between hub and workstations. The satiation lines are dedicated to the served station and operate at 4.096 Mbps or 20.480 Mbps. The multiplexed isochronous frames in the medium usually have a common header with frame configuration control fields and dedicated basic rate interface (BRI).

9.4.3 *Wireless LAN standards*

The Digital European Cordless Telecommunication (DECT) standard was developed under the **European Conference of Post and Telecommunication (CEPT)** and **European Telecommunication Standard Institute (ETSI).** DECT was published as the European Telecommunication Standard (ETS) in 1992 by ETSI and is a result of cooperation between 20 telecommunication companies and over 18 countries.

DECT is an open, non-proprietary standard defined for point-to-point communication between the base station and portable telecommunication station. It was initially developed for a wireless PBX application and has become a general standard for cordless telecommunication in Europe. It is a standard for air interface rather than a communication network standard. Its applications are in wireless PBX, domestic cordless telephone, and cordless LAN (CLAN), and it offers up to 11 Mbps for CLAN. The following are features of DECT:

- Performs speech and data communication
- Has a variable bandwidth channel to cover bit rate segment of speech and data
- Accommodates the total traffic capacity requirement
- Offers speech quality (same as in wired telephone)
- Provides a data rate of 1 Mbps to each terminal
- Provides overall throughput in the range of Ethernet LAN data rate (10 Mbps)
- Offers a highly protected radio-frequency band for data communication
- Offers compatibility with existing standard LANs
- Offers high portability and flexibility
- Offers reasonable security

9.4.4 *Frequency bands for wireless communication*

There are three bands of frequency which have been proposed as a communication link for wireless LAN: spread spectrum range, microwave, and infrared. For the sake of comparison, Table 9.1 presents various categories of frequency bands, along with their frequency ranges within the electromagnetic spectrum.

Spread spectrum technology (SST): SST is defined basically for military applications and is immune to interferences.[18,19] This technique offers various advantages but does not guarantee a secure environment. It is now being used in LANs. Here the SST propagates over a wide band of spectrum. There are two main techniques of defining and transmitting SST data over the LAN: direct sequence (DS) and frequency hopping (FH). DS is based on amplitude modulation (AM), while FH is based on frequency modulation (FM). The DS modulates the original signals with a signal that has a wide spectrum and allows the number of users to share the same frequency by using time division multiplexing access (TDMA), which is used in microwave wireless LANs. The sharing of the same frequency band by a number of users in DS is also known as **code division multiple access (CDMA)**, which is becoming a *de facto* standard for wireless LANs at the radio end of the spectrum. It allocates to each user a sequence number which is unique and recognizable by the receiving antenna. CDMA is becoming more popular than TDMA in many countries, particularly the U.S.

Microwave: This band of frequency for implementation of LANs offers data rates of Ethernet (10 Mbps) and above. The frequency band around 18 to 19 GHz is used (somewhere between radio wave and infrared). These LANs operate at the highest frequency end of the band.

Infrared: This frequency band is on the lower side of visible light. It is immune to electromagnetic interferences. Laser transmits infrared in a very light beam that can be modulated up to 16 km. This particular scheme of transmitting the signal is used in outdoor applications. It can also use light-emitting diodes (LEDs), which offer a short range of distance and are ideal for indoor applications. The light source is typically from fluorescence and not the open air. Direct exposure to sunlight may flood the transmitter with messages and, hence, these transmitters are never kept under direct sunlight.

Table 9.1 Frequencies of Various Bands

Category	Frequency
Direct current (DC)	0 Hz
Alternating current (AC)	60 Hz
Low frequency (LF)	4 KHz
AM radio	500–1600 KHz
Medium frequency	1.5 MHz
High frequency	10 MHz
Very high frequency	100 MHz
Detroit police radio	2 MHz
Short wave radio	1.6–30 MHz
Highway mobile services	35–47 MHz
Ham radio	50–54 MHz
VHF-TV (channels 2, 3, 4)	54–72 MHz
VHF-TV (channels 5, 6)	76 MHz
FM radio	88–108 MHz
First Bell System Service	150 MHz
Mobile cellular radio	105 MHz to 1 GHz
VHF-TV (channels 7–13)	174–216 MHz
Ham radio	430–450 MHz
UHF-TV	500–800 MHz
Cellular communication	825–890 MHz
Microwave terrestrial	1–500 GHz
Satellite	1–50 GHz
Microwave radio	2–12 GHz (4, 6, 11 GHz)
Visible light	1–4 THz (10^{12}Hz)
Optical fiber	100 TGHz to 1000 THz
X-ray, gamma-ray, nuclear radiation	1 MTHz
Infrared	100 GHz to 10 THz
Ultraviolet	1–100 MTHz

Source: Derived from References 7 and 17–19.

The microwave and infrared techniques have a common drawback of line of sight, and as a result of this they are usually not used. SST offers different frequency bands below 900 MHz for different applications such as FM radio and cellular telephones; however, these frequency bands may not provide enough bandwidth required for the data. Just to get an idea, these bands of frequency may provide less than a 2-Mbps data rate for Ethernet (10 Mbps). To avoid the interferences from other frequency bands, the 1.8–1.9 GHz portion of the frequency band is considered a suitable candidate for offering enough bandwidth for the data with the flexibility of reusing it. The advantages offered by these bands of frequency for wireless LANs include excellent propagation characteristics within 100–200 m and general availability.

9.4.5 *Wireless spectrum allocation*[18,19]

Cordless phone (CT-1)	46 MHz and 49 MHz
Cellular phones	824–849 MHz
	869–894 MHz
SM band	902–928 MHz
Tele-point (CT-2)	940–941 MHz

PCN	1850–1990 MHz
CT-2	864–868 MHz
Cellular phones	890–915 MHz
	935–960 MHz
DECT	1880–1900 MHz
Cellular handy phone (Japan)	1000–3000 MHz

9.4.6 *Digital cordless standards*

The CEPT entered the digital cordless telephony market via two different directions. In the first approach, the time division multiple access/time division duplex (TDMA/TDD) was introduced for cordless communication in an office environment. The second approach is based on frequency division multiple access (FDMA), which partitions the entire bandwidth into channels and is used in cordless telephony.

The first product was a cordless handset which has a base station and uses radio-channel and analog transmission. These telephone sets are unreliable, have a short range, and provide poor quality. The second generation of digital cordless telephony is based on a digital standard (CT-2 — CAI, common air interface) offering a high-quality, secure replacement for CT-1 with increased range and capacity. The CEPT accepted the CTMA/TDD approach and recommended an allocation of 1880 to 1900 MHz rather than 900 MHz for cordless telephony. This was accepted by companies like Alcatel, Fujitsu, Mitel, Northern Telecommunication, Phillips, etc.

The Cellular service providers in the U.S. include McCaw Cellular, PacTel, Southwestern Bell, Nynex, GTE Mobilnet, Bell South Mobile, U.S. West, American West, Bell Atlantic Mobile, American Cellular, U.S. Cellular Corporation, and many others. The approach divides the bandwidth in the frequency domain, multi-carrier (MC), and then divides these carriers into time slots in the time domain. Each voice time slot has a capacity of 32 Kbps, and we can have a higher capacity up to 384 Kbps in both directions. At present, 120 basic channels are available.

The DECT frame format is shown in Figure 9.13.

Radio interface architecture: DECT system users get available channels by **cordless portable port (CPP)**. The base station, known as the radio fixed part (RFP), offers channels for paging and synchronization. All the RFPs page out an incoming call, which is picked up by the CPP that responds to a paging message on a channel of the CPP's choice. CPPs have the capability within DECT specification to poll carriers and always look for stronger RFP. The radio interface between RFP and CPP is composed of four layers (physical, MAC, LLC, and network).

The **physical layer** creates the means for a CPP to communicate with the RFP. Ten RFP carriers are generated via modulation/demodulation with a defined bit stream rate of 1.544 Mbps. It transports the user's channel frequency and, within this channel, signaling and speech/data information is included.

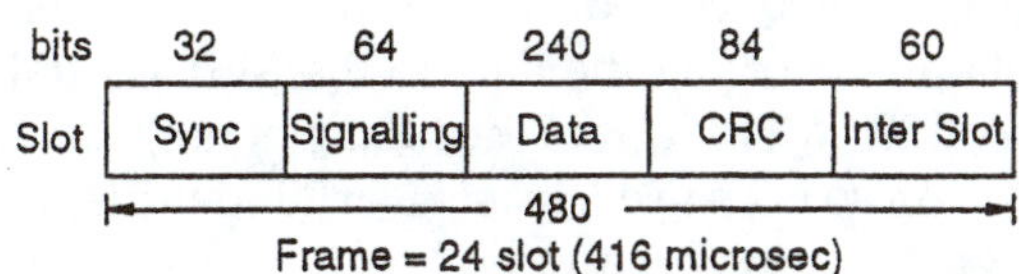

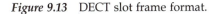

Figure 9.13 DECT slot frame format.

A combination of time-slot pair and RF channel is known as a *channel*. The maximum number of channels is 120. The physical channel contains three logical channels: **32-bit synchronization (sync) channel**, **64-bit control channel**, and **information channel**. The control and speech/data information is carried within the control channel. The physical layer interfaces with MAC through these two logical channels. The sync channel is used by the physical layer itself to synchronize transmission. The physical channel is assigned to one user. The user's physical channels are put in an RF envelope by TDMA/TDD multiplexing. DECT offers 12 duplex channels (24 time slots) per frame. Similarly, control and information channels are mixed into the physical channel. The sync channel is always present and transmits the same 32 bits in each burst. It helps the fixed and portable stations to monitor several bursts within one frame at a time.

Radio interface acts as a bit pipe transporting bits of control speech/data through the C channel and me channels. The physical layer performs all the multiplexing and control functions. This is the reason to choose the MC/TDMA/TDD.

The **data link layer** ensures reliable data communication over the bit pipes provided by the physical layer. MAC and DLC and higher provide error protocols to evaluate the quality of service and the means to instruct the layer to use another radio channel whose header defines the type of layer 2 messages. One header is for paging and identification.

The **network layer**, the hangover function in DECT, is fully defined in layers 1 and 2. The DECT provides access to other networks, ISDN (basic rate), X.25, public switched telecommunication networks (PSTN), GSM (global system for mobile communication), and all LANs.

A LAN is used to transport data from the workstation to server. DECT is a suitable way to set up a CLAN (Cordless LAN) when complemented and integrated with the layer above DLC with protocols like Novell's NetWare and Microsoft's LAN manager. It offers 11 Mbps for CLANs.

9.4.7 Cellular system architecture interfaces

The data from a public switched telephone network (PSTN) is interfaced with the mobile switching center (MSC) via an appropriate PSTN-to-MSC interface. The call processing typically goes through the following steps: switch connection with PSTN and then establish switched connections between mobile subscribers. The use of radio voice channels is monitored by an appropriate layer within the architecture. The cell site receives the data from the MSC onto two different channels, as voice and data links are transmitted in the area. Various functions provided by the cell site include RF radiation and reception, data communication with the MTS central office and mobiles, routine maintenance testing, relevant equipment control, reconfigurable functions, voice processing, call setup, call supervision, and call termination.

The information from mobiles is picked up by an RF distribution module via various antennas installed at various locations. This module performs the functions of a cell site, e.g., voice radios, set up radio, locate radio, and maintenance and testing of various components used in the network. It also provides paths for each of these signals; e.g., all the voice signals are sent over the voice trunk to the MTS central office via voice trunk interfaces, while the remaining functions are received by the cell site controller and are sent over data link interfaces to the data link carrying the data to the MST central office. Most countries use the frequency band of 900 MHz in their cellular phones, but the frequency bands of 450 MHz and 400 MHz have also been used. In all, over 20 countries provide cellular services.

The radio-frequency band provides a radio link between users, and these signals are picked up by the terminal via antenna. The switching hierarchy within a mobile telephone system follows the same sequence of connection as is defined in the existing telephone system; e.g., a call message is originated from a local office and goes to a toll office. The operating company operates it over its connection with the central office of the MTS, which broadcasts it over the radio link via antennas. The mobile carrying the antenna picks up the message via its antenna.

The CEPT adopted WARC 79 recommendations for using the frequency bands of 890–915 MHz and 935–960 MHz for mobile telephony systems on land.[17] The GSM was created to define the standard for mobile telephony, it was generally agreed upon by all the European countries and also included digital technology standards dealing with TDMA, speech coding, channel coding, and digital modulation techniques. The following is a list of GMS teleservices defined:

Speech transmission
- Telephony
- Emergency calls

Short message services
- Mobile terminating for point-to-point configurations
- Mobile originating point

Message handling and storage services
- Video text access
- Teletext transmission
- Facsimile transmission
- Alternate speech

The GSM system typically offers the following functions for the above-mentioned categories of services: number identification, call offering, and call completion. A typical reference model of GSM[17] is shown in Figure 9.14. The reference points are shown between different modules. The reference point between MSC and the base station system (BSS) supports two types of configurations: integrated and distributed. In the integrated configuration, all of the base transceiver stations (BTS) get the same message from their base station controller (BSC) to a common antenna, while in the case of distributed configuration, the message goes to an individual BTS and to its respective antenna. The mobile service switching center performs all the switching functions for mobile communication and procedures required for hangover and location registration.

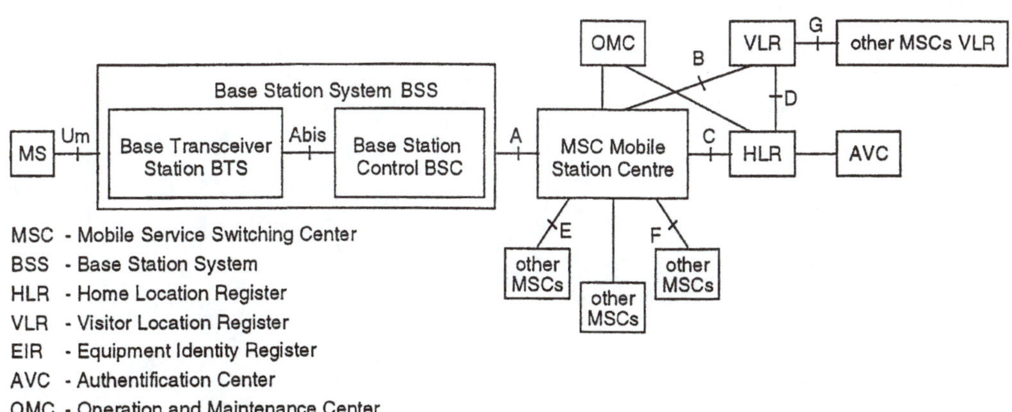

Figure 9.14 GSM reference model.

References

1. Schlar, S.K., *Inside X.25: A Manager's Guide,* McGraw-Hill, 1990.
2. Meijer, A. and Peeters, P., *Computer Network Architecture,* Computer Science Press, 1982.
3. Kauffles, E.J., *Practical LANs Analyzed,* Ellis Horwood, 1989.
4. Digital Equipment Corp., DECnet *Digital Network Architecture, General Description,* May, 1982.
5. Digital Equipment Corporation, *DNA Digital Data Communication Message Protocol (DDCMP), Functional Specification,* 1982.
6. IBM, "Systems Network Architecture: Technical Overview," IBM Corp. Research Triangle Park, NC, No. GC30-3073, 1985.
7. Black, U., *Data Networks: Concepts, Theory and Practice,* Prentice Hall, 1989.
8. Hutchison, D., *Local Area Network Architectures,* Addison-Wesley, 1988.
9. Markeley, R.W., *Data Communication and Interoperability,* Prentice-Hall, 1990.
10. Minoli, D., *Telecommunications Technology Handbook,* Artech House, 1991.
11. Tanenbaum, A.S., *Computer Networks,* Prentice-Hall, 1989.
12. Marney-Petix, V.C., *Networking and Data Communication,* Reston Publishing Co., 1986.
13. Simonds, F., *LAN Communication Handbook,* McGraw-Hill, 1994.
14. Stallings, W., *Handbook of Computer Communications Standards,* vol. 1, 2nd ed., Howard W. Sams and Co., 1990.
15. Spragins, J., Hammond, J., and Pawlikowski, K., *Telecommunications Protocols and Design,* Addison-Wesley, 1991.
16. Halsall, F., *Data Communication, Computer Networks and Open Systems,* Addison-Wesley, 1992.
17. IEEE Document P802.11-93/028, Standards Working Group IEEE P802.11, Wireless Access Method and Physical Layer Specifications, March 1993.
18. Olivett reference manual, "The cordless LANs for enterprise," 1990.
19. AT&T reference manual, "Wireless personal communication," 1990.
20. Huffman, D.A., "A method for construction of minimum redundancy codes," *Proceedings of IRE 40,* 1098–1101, Sept. 1952.

References for further study

1. Bantz D. and Bauchot, F., "Wireless LAN design alternatives," *IEEE Network,* March/April, 1994.
2. Pahlavan, K., Probert, T., and Chase, M., "Trends in local wireless network," *IEEE Communication Magazine,* March 1995.

part IV

The OSI-RM architecture and protocols

chapter ten

Physical layer

"What is a vision? It is a compelling image of an achievable future."

Laura Berman Fortgang

"Layer one, the lowest layer, in the OSI seven layer model, concerning electrical and mechanical connections to the network. The physical layer is used by the data link layer. Example physical layer protocols are CSMA/CD, token ring and bus."

The Free Online Dictionary of Computing,
Denis Howe, Editor (http://foldoc.doc.ic.ac.uk/)

10.1 Introduction

The physical layer, the lowest layer of the seven-layer OSI-RM model, provides an interface between the communication media and the data link layer. It deals with the bit-by-bit transmission of data in the form of electrical voltage and current over the transmission media connecting two adjacent nodes. This layer includes protocols and various interfaces such that computers which are physically connected to each other through the communication media can send the data back and forth between them. This physical link between the computers is fully maintained and controlled by the physical layer. This layer also provides mechanical, electrical, functional, and procedural functions and services to establish, control, and maintain the physical link during data communication.

Various vendors have developed different interfaces for computers or terminals — **data terminal equipment (DTEs)** — with modems — **data circuit-terminating equipment (DCEs)** — and the interfaces allow different vendor computers to interact with each other. These interfaces are provided by physical layer protocols and often are termed DTE/DCE interfaces. These interfaces are physically connected to communication media.

The physical layer accepts data from the link layer as **frames** and passes it on the communication media bit by bit (sequentially). The physical connection between two hosts for point-to-point or multi-point configuration (for broadcasting) provides the physical connection between various devices. Similarly, it receives data from communication media and passes it on to the data link layer. Other services, such as connection-oriented or connectionless services, and error control, are provided by the data link layer and supported by higher layers, while parameters like data rate, error rate, propagation delay, transmission speeds, etc., are determined by the physical connection circuits established over communication media. Thus, the characteristic properties of communication media should be properly interfaced/matched with the physical layer, as this layer accepts the bit stream, whether it is a transmitter or receiver of the data. In the case of the IEEE model of LANs, the physical layer provides an interface between **media access control (MAC)**, one of two sublayers of the data link layer of OSI-RM (another sublayer is **logical link control, LLC**), and the transmission media. The services offered by the physical layer are received by the data link layer as a **physical service access point (PhSAP)**. The physical layer fetches the frames from the data link layer through the PhSAP.

10.2 Layered protocol and interfacing

In OSI-RM,[8] it is assumed that the lower layer always provides a set of services to its higher layers. Thus, the majority of the protocols that are being developed are based on the concept of a layered approach for OSI-RM and, hence, are known as **layered protocols**. Specific functions and services are provided through protocols in each layer and are implemented as a set of program segments on computer hardware by both sender and receiver. This approach of defining/developing a layered protocol or functional layering of any communication process not only provides a specific application of a communication process but also provides interoperability between vendor-customized, heterogeneous systems. These layers are, in fact, the foundation of many standards and products in the industry today. The layered model is different from layered protocols in the sense that the model defines a logic framework on which protocols (that the sender and receiver agree to use) can be used to perform a specific application. Some of the material presented in this section is partially derived from References 1–3 and 16.

Layered protocols: The communication control procedures are a set of rules and procedures for controlling and managing the interaction between users via their computers, terminals, and workstations across the network. The control procedures are known by different names in the industry such as protocols, line control procedures, discipline procedures, data link control procedures, etc. The protocols provide a communication logical link between two distant users across the network, and the handshaking procedures provide the call establishment within the protocols.

Each layer passes message information (data and control) starting from the top layer (application) to its lower layer until the message information reaches the lowest layer (physical), which provides actual data communication. An interface is defined between layers which supports the primitive operations and services that the lower layer offers to its higher layer. The network architecture is then derived as a set of protocols and interfaces.

The implementation of these services is totally hidden from the higher layers. The services offered by the layers are provided through protocols and are implemented by both source and destination on their computers. Protocols are defined in terms of rules (through semantics and syntax) describing the conversion of services into an appropriate set of program segments. Each layer has an **entity process** that is different on different computers, and it is this entity (peer) process through which various layers communicate with each other using the protocols. Examples of entity processes are file transfer, application programs, electronic facilities, etc. These processes can be used by computers, terminals, personal computers (PCs), workstations, etc.

A protocol defines rules, procedures, and conventions used for data communication between peer processes. It provides data formatting, various signal levels for data and control signals, error control, synchronization, and appropriate sequencing for the data to be exchanged. Two networks will offer two-way communication for any specific application(s) if they have the same protocols. The layer m (where m represents any layer within OSI-RM) is known as a service provider for layer $m + 1$ and offers the services at SAP.

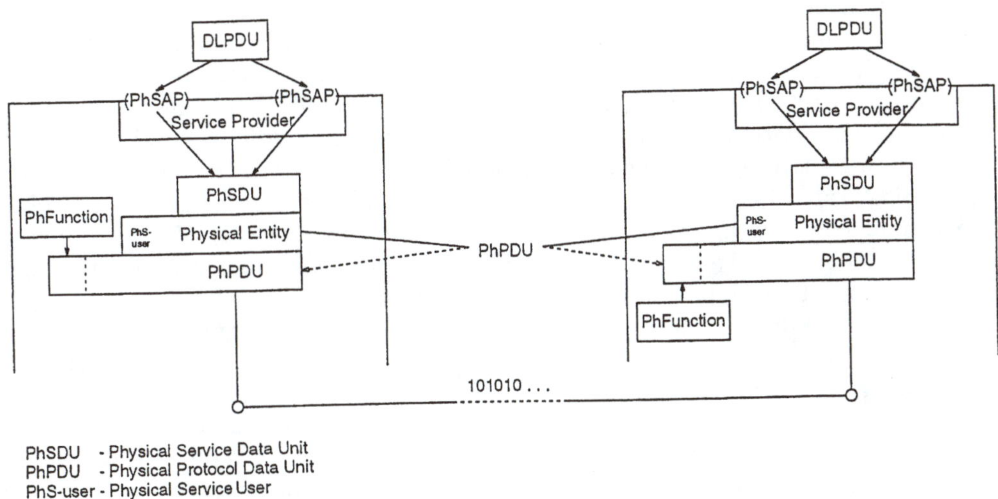

Figure 10.1 Relationship between units of the physical layer.

Interfacing: ISO[8] has defined the following terms used for data communication through OSI-RM. Each layer has active elements known as entities, and entities in the same layer of different machines connected via network are known as **peer entities**. The entities in a layer are known as service providers. The **physical service access point (PhSAP)** is defined by a unique hardware or port or socket address (software) through which the data link layer accesses the services offered by it. Messages from the data link layer are also fetched through these PhSAPs. The PhSAP has its own address at the layer boundary and, as such, is prefixed defining the appropriate layers. The **physical service data unit (PhSDU)** is defined as message information across the network to peer entities of all the layers. The PhSDU and control information (mechanical and electrical signaling) together compose the **physical protocol data unit (PhPDU)**, which is basically a frame which is received from the data link layer and transmitted bit by bit. The relationship between these units of the physical layer is shown in Figure 10.1. In this book, we use the terms defined by ISO. After the data link layer has defined a **frame** (by associating header and trailer to the NPDU), it simply forwards it to the physical layer entity for its transmission.

10.3 Physical layer functions and services

The physical entities are connected via a physical medium which carries the bits between two nodes. The physical layer establishes a physical connection between data link entities for bit transmission. The physical layer accepts the bit stream from the data link layer via its PhSAP. The data circuit is established between the two physical entities, which provides a communication path for bit transmission. These data circuits may be used to provide physical connections via relaying functions defined by the physical layer. The PhSAP is also known as a physical connection between endpoints and the physical entity which defines data circuits and physical connections (using data circuits via relaying functions).

ISO functions: As discussed earlier, the physical layer provides four main functions: mechanical, electrical, procedural, and functional. Each of these functions encompasses a specific set(s) of attributes being implemented in the protocols.

The **mechanical function** provides attributes such as behavioral parameters of interface to the communication medium, number of wires (for data control and signaling in the interface), number of pins on the connectors, mode of configurations, etc. The **electrical function** is concerned with attributes such as the magnitude of voltages (or currents), level signals for binary numbers, timing relationship between various signals (data, control, synchronization, clocking, etc.), and other electrical properties, e.g., duration of the signal, noises, capacitive and inductive effects, rise and fall times of the pulses, data rates, propagation delay and speeds, etc. The **procedural function** offers a set of procedures (or actions) which transfer the bit stream through appropriate connectors to the interface and to the communication media. The **functional function** deals with specific functions which will be performed (by that protocol) on the interface (between computer and communication media), and these are expressed in terms of parameters, e.g., control signal, data signal, timing signal, etc.

In addition to these four main functions, the physical layer maintains the physical connection (over a physical link) and also allows disconnection of the physical connection after data has been transmitted/received and physical layer management. It supports both asynchronous and synchronous transmission of physical service data units (PhSDUs) where PhSDUs are basically defined as bit streams. It also deals with physical layer management, which is basically a subset of OSI layer management.

ISO services: Various services offered by the physical layer to its data link layer include physical connection (across the data link entities for bit transmission), physical service data unit (PhSDU) which can be transmitted in a serial mode (containing one bit in PhSDU) or parallel mode (containing *n* bits in PhSDU), physical connection endpoints (defined by identifiers), data circuit (defined by identifiers), sequencing, quality of service parameters (error rate, service availability, transmission rate, transit delay, etc.), and fault condition notifications to data link layer entities.

10.4 Physical layer interface standards

The physical layer standard should provide a compatible interface to different types of DTEs and DCEs. Further, these interfaces should support different types of encoding, modulation, demodulation, and decoding techniques. For each of these techniques, interchange circuits and standards have been defined which should be added to the standard interface. According to OSI standards, the physical layer is concerned with synchronous or asynchronous transmission for the data transmission rate. Accordingly, there are different views about the contents of physical layer protocols.

CCITT has prepared a document known as *V series physical level standards* which defines this protocol as an interface between DTE and DCE and also between DCE and DCE. In other words, this protocol not only defines an interface between DTE and DCE but also defines signaling between two DCEs. Another standard defined by the Electronics Industries Association (EIA) is an EIA-232 interface which provides an interface mainly between DTE and DCE but has also been used to provide signaling between two DCEs by various vendors (following CCITT V series physical-level protocols).

The physical layer offers two types of transmission service: **private** (or leased or dedicated) and **switched-circuit** service.

In private service, a physical connection between hosts (computer or terminal) is always established and users of these connected hosts can transmit data without caring about the connection. There is no problem of de-establishing or disconnecting the physical connection. For hosts in switched-circuit service, we have to establish and de-establish a physical connection between hosts, and the data can be transmitted only after the establishment of connection. Thus, the physical layer in this case must include the procedures for connecting and disconnecting the connection; also, the same communication media can be used for providing connection to other computers or terminals.

While comparing these two services, we find that private service is more expensive and should be used for sending a bulk of data for a long duration, while switched-circuit service is useful for short-duration transmission and can be shared by a large number of users.

As pointed out earlier, DTE is a computer, terminal, or controller, while DCE is a modem or converter. In some books, multiplexers and data service units are used to represent DCE. The main function of a **modem (MOdulator/DEModulator)** is to convert digital signals into analog signals and vice versa. At the transmitting side, a modem (connected to computer/terminal) accepts the digital signal and modulates (using modulation techniques) it into an analog signal for transmission over a telephone line, while at the receiving side, the received modulated analog signal is converted (demodulated) back into a digital signal.

There are different types of modems available on the market, e.g., limited distance, medium- and high-speed, synchronous and asynchronous, **channel service unit/data service unit (CSU/DSU)** alternating, wide-band, intelligent, self-testing, and remote-testing. Details on these modems can be found in Chapter 3.

Although electrical function of the physical layer deals with various attributes such as data control, signal synchronization, clock signals, timing signals, etc., the synchronization and clock signals are more important than the others, as these provide the proper communication between two nodes. These functions are designed with a view to handling these concerns during data communication; e.g., what time should the sender transmit the data to the receiver? what is the size? etc. The timing and synchronization signals are provided by the mode of transmission being used. For example, in the case of asynchronous transmission, start, stop, and parity bits are used to provide the timing and synchronization, and parity bits can also be used to detect error on the receiving side.

In synchronous transmission, various techniques can be used to provide the synchronization and timing signals over the physical circuit between nodes. For example, before the data is sent, a clock signal on a separate line can be sent to initialize and synchronize the clock at the receiving side; once this is done, the data can be sent on the other line. This method is suitable and inexpensive for short distances, e.g., synchronous interface for terminal/computer and modem/multiplexer. As another example, clock and synchronization signals can be transmitted as data encoded at the transmitting node which can be decoded at the receiving node to initialize its clock. This data is transmitted at regular intervals on the line to maintain synchronization. This technique uses a digital phase locked loop to provide the clock and synchronization signals as part of the data and maintains the synchronization at bit level.

10.4.1 Interchange circuits (DTE/DCE interface)

The physical layer provides an interface between the terminal, computer, or controller and circuit interface equipment such as a modem (or multiplexer or transducer or connector) for the transmission of data, as shown in Figure 10.2. According to CCITT, the

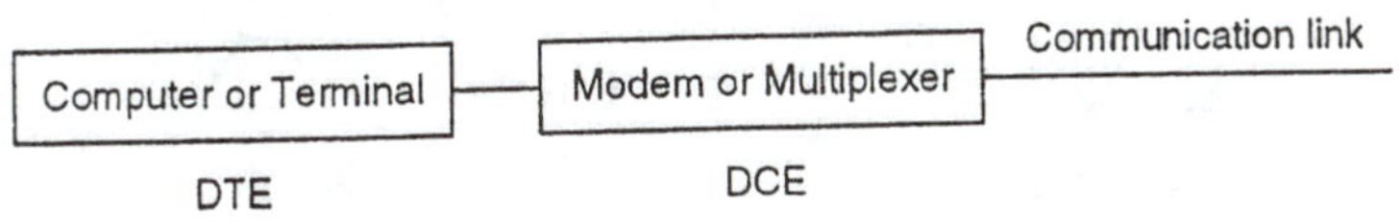

Figure 10.2 Interchange circuit.

terminals or computers are known as **data terminal equipment (DTE)** while the circuit interface equipment/devices are known as **data circuit terminating equipment (DCE)**. The modem is connected to communication media. Once the connection is established between the terminal/computer and modem, modem and transmission media, transmission media and modem, and modem and terminal/computer, various types of signals (data, control, clock, timing, and synchronization) can be transmitted between the modems.

Signals corresponding to digital data coming out of DTEs can be transmitted for a short distance with a very low speed. These signals are known as non-return-to-zero (NRZ), where zero voltage or negative voltage represents logic 0 while high voltage or positive voltage represents logic 1. DTEs are never connected directly to the communication media; also, these devices cannot be connected to all types of media. The most popular and common way of connecting DTEs to communication media is through a DCE (modem). A DTE sends data control information to a DCE over a pair of lines which provides appropriate conversion format between DTE and DCE, and due to this the DTE is often known as an **interchange circuit**. The DCE then converts this information into a bit stream and transmits it over transmission media using a serial bit transmission scheme. In this scheme, all the bits are sent serially over the same link or circuit. Another transmission scheme is parallel bit transmission, which requires a number of circuits equal to the number of bits to be transmitted between the DCEs. In other words, each bit is transmitted on a separate link or circuit. The transmission between DTEs can be effective and efficient if both use the same interface which provides identical functions and services.

There are various international standards for the physical layer interface (commonly known as DTE/DCE interface) on the network market. The most common and important interfaces are EIA RS-232, EIA RS-530 (with RS-422-A and 423-A), and X.21. Each of these standards will be discussed in more detail in the following sections.

10.4.2 EIA RS-232 DTE/DCE interface

The Electronic Industry Association (EIA)[9] has defined the RS-232 standard interface of the DTE with DCE which is connected to the analog public telephone system, as shown in Figure 10.3. Since these lines or related lines with telephone requirements and specifications are nowadays heavily used for data communication, EIA RS-232-D is considered one of the most effective and widely accepted DTE/DCE interfaces in the world.

The first version of the RS-232 standard was approved by EIA in 1962, while the third version, known as RS-232-C, was approved in 1969. The fourth and latest version, RS-232-D, was approved in 1987.[9] This version of RS-232-D is compatible with RS-232-C and also with other standards such as CCITT's V.24 (functional and procedural functions) and V.28 (electrical function) and ISO's 2110 (describing the mechanical function).

The RS-232 DTE/DCE interface supports both transmission (synchronous and asynchronous) schemes for data communication and can also be used for other types of interfacing. It contains a 25-pin, D-shaped connector which can be easily plugged into the

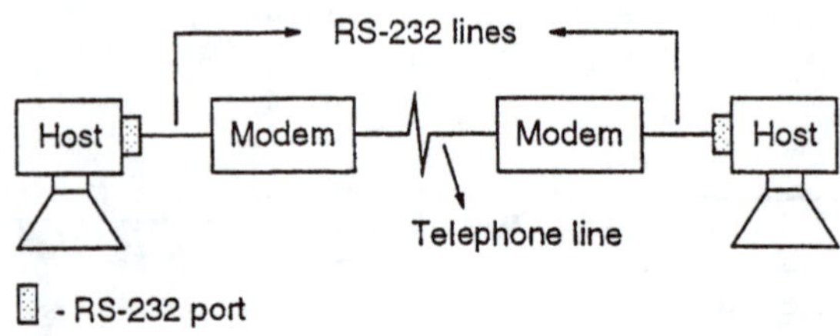

Figure 10.3 RS-232 DTE/DCE interface.

modem socket and which is also known as a DB-25 connector. Some standards, pin diagrams, and other electrical circuits for the physical layer are partially derived from References 1–6.

The standards organizations have defined and published documents on mechanical–electrical connections, communications, specifications, etc. The CCITT V series defines the connections, specifications, etc. It is ideal for the data transmission range of 0–20 Kbps and 15.2 m and employs unbalanced signaling. Serial data exists through a RS-232-C port via the transmit data (TD) pin and arrives at the other side's RS-232-C port through the received data (RD) pin. These TD and RD pins correspond to the receive and transmit lines which carry the data. The setting and resetting connection functions for switched connections with a public switched telephone network (PSTN) are defined by control and timing lines connected to appropriate pins. Each pin performs a specific function on a specific signal and defines a circuit for establishing connection between DTE and DCE. The connection between DTE and DCE is obtained over a pair of wires and is an interconnecting circuit. Here, fewer pins (fewer than half) are used to define interconnecting circuits for establishing connection. Details of these standards, along with their pin numbers, are discussed below.

10.4.3 EIA RS-232-D DTE/DCE interface

The RS-232-D interface[9] provides all functions (described earlier) for establishing connection between DTE and DCE, namely, initialization of the interconnection circuit for establishing the link between DTE and DCE, transmission of data and control across the circuits, timing and control signals to provide synchronization between them, maintenance and controlling of data transfer between them, etc. It establishes an unbalanced connection between DTE and DCE, and, as such, voltages and currents flow through circuits in one direction and return through the ground (return path). Their values are defined with respect to the ground (which is common to both transmitting and receiving circuits at two nodes). Further, the RS-232-D always establishes a circuit for data transmission in only one direction at a time; hence, it supports half-duplex operation. All the signals (data, control, timing, etc.) flow through the same circuit (primary). If we can provide a second circuit (secondary at lower speed) between DTE and DCE, then full-duplex operation can also be supported by RS-232-D.

Due to the common ground (return path) between transmitting and receiving nodes, noise is introduced into the signal, which limits the distance and the rate of signaling. For a larger distance between DTE and DCE, the ground voltage may be different, and this causes a ground loop current which affects the transmitting as well as receiving sites. Further, the noise limits the distance and the data rate of the interface. The typical value of the distance supported is 10–15 m at a signaling rate of 20,000 baud. The signaling rate controls the data rate: a higher signaling rate will give a higher data rate at a shorter distance, while a lower signaling rate provides a lower data rate at a longer distance.

The binary values consist of conventional voltage values defined in CCITT's V.28; according to this document, voltage more negative than –3 V with respect to common ground represents logic 1 (data signal), while voltage more positive than +3 V represents logic 0 (data signal). A voltage more negative than –3 V represents logic 0 or OFF (control signal), and a voltage more positive than +3 V represents logic 1 or ON (control signal). This indicates that the voltage values for the same logic in data and control signals are logically opposite.

In general, it has been observed that transmitting and receiving sites have different behaviors and characteristics of their equipment and, as such, may have different logic for the return path, i.e., ground. This will result in different values of voltage appearing at the DTE/DCE interface. If the difference of the voltage at the interface is small, then

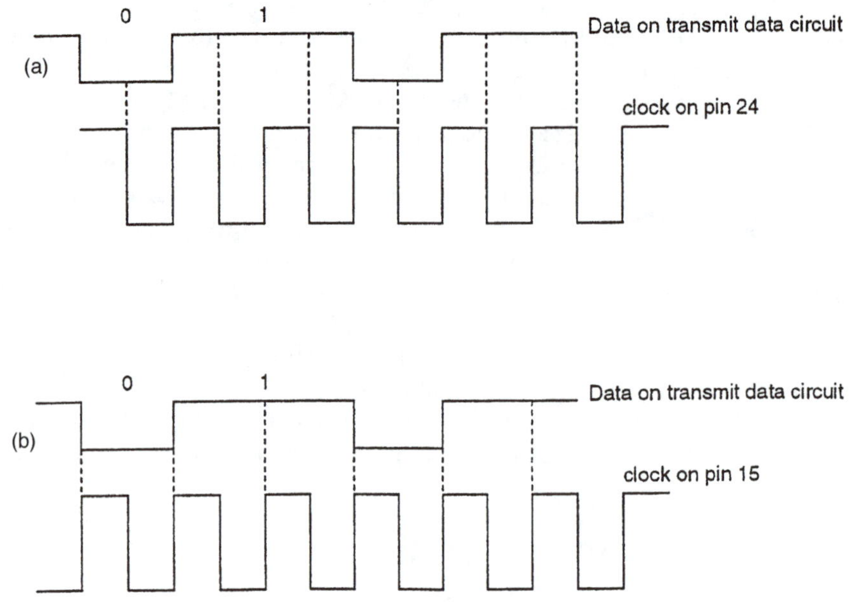

Figure 10.4 (a) Clock signal generated by DTE. (b) Clock signal generated by DCE.

the voltage with respect to this difference on the line will be very small and hence will not cause any error. But if the difference at the DTE/DCE is large, then the voltage with respect to this difference will be large enough to represent a different logic from the actual logic for the signal on the line and it will cause error. In addition to cross talk, impulse noise, and other types of noises (discussed earlier), errors may also be caused by capacitive and inductive effects of the line. This may change the response time of different logic, and the duration of the binary values may also be different.

Types of signals for DTE/DCE interface: There are four types of signals (as defined in the CCITT V series): data, control, timing, and ground. These signals together define an interchange circuit between DTE and DCE. The number of these circuits is different for different versions of RS-232. As an example, we consider RS-232-D here. It has 15 control circuits which provide transmission (asynchronous/synchronous) over both primary and secondary circuits and loop-back circuits. These control circuits are listed in Table 10.1. There are four types of circuits:

- **Data circuits:** Data signals from a terminal (DTE) are transmitted to a modem (DCE) under some conditions over these circuits.
- **Control circuits:** Control signals are transmitted from DTE to DCE on these circuits. The main functions of these circuits are to establish connection, check the readiness of DTE and DCE, send and receive the data, control the flow of data, etc.
- **Timing circuits:** Clock signals provide synchronous transmission across the interface and can be generated at either DTE or DCE. When DTE generates a clock signal (via DTE transmitter signal element timing pin 24), ON and OFF transmissions are defined at the middle of the data bit pulse on the transmit circuit, as shown in Figure 10.4(a). On the other hand, when a modem generates a clock signal, it sends a demand clock to DTE (via DCE transmitter signal element timing pin 15). After DTE demands a clock from DCE, it transmits data in such a way that the transition

between bits is synchronous with the ON/OFF transition of the demand clock, as shown in Figure 10.4(b).

- **Ground circuit:** This circuit provides common ground reference to all the circuits of an interface. No signaling is involved in this circuit, and it is simply for providing protection and safety to the interface.

There are four data circuits (two for primary and two for secondary circuits) in RS-232-D. Three timing signals provide suitable clock signals for transmitting and receiving equipment in the case of synchronous transmission. Finally, two ground signals establish the common ground to all circuits and also shield for various cables at DTE for eliminating electromagnetic interferences. Some of the circuits defined in RS-232-D were also defined or included in RS-232-C but have different names for their operations. The procedural function of the RS-232-D interface, in fact, defines the sequence of activating the circuits for the transfer of data between DTE and DCE and has a different sequence of operation for different types of transmission (synchronous and asynchronous).

Pin connector for DTE/DCE interface: The standard advises use of a 25-pin connector for cable to interface DTE and DCE, as depicted in Figure 10.5. This connector is also used for CCITT's V.24 and ISO 211 standards, as well as RS-232-D. The EIA RS-232-D is also available in a 9-pin connector (EIA-449). The DB 25 standard (discussed above), does not mention the type of connector, but this has become the standard for 9- and 15-pin connectors used in certain computers. The standard, however, mentions the assignment of circuits to specific pins in these connectors (Table 10.1).

The direction of data and control signals across the interface (i.e., between the DTE and DCE) and the standard connection between them are shown below.

DTE		DCE
TD (2)	0 ——————→ 0	TD (2)
RD (3)	0 ←—————— 0	RD (3)
RTS (4)	0 ——————→ 0	RTS (4)
DTR (20)	0 ——————→ 0	DTR (20)
CTS (5)	0 ←—————— 0	CTS (5)
DCD (8)	0 ←—————— 0	DCD (8)
DSR (6)	0 ←—————— 0	DSR (6)

Half-duplex DTE/DCE interface: In order to understand the workings of a DCE/DTE interface, we consider a half-duplex configuration which uses an asynchronous short-distance transmission modem. Various control circuits between A and B are shown in Figure 10.6.

The circuits and the sequence of establishment of the circuits between the two sites A and B are shown below. Initially, we assume that no site is communicating and, hence, there is no carrier on the line. All the control circuits are in an OFF state. As soon as the

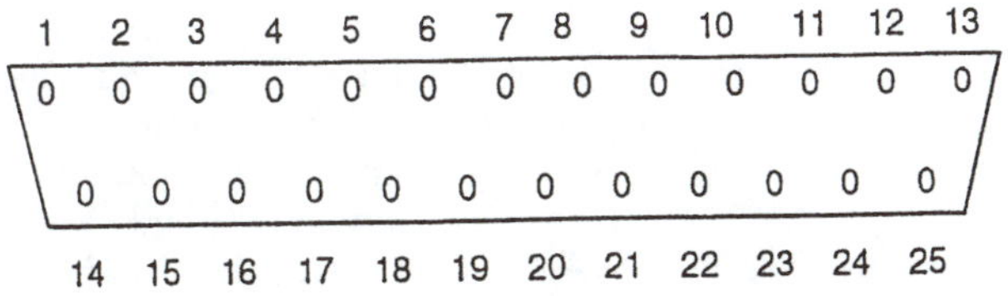

1 2 3 4 5 6 7 8 9 10 11 12 13
0 0 0 0 0 0 0 0 0 0 0 0 0
0 0 0 0 0 0 0 0 0 0 0 0
14 15 16 17 18 19 20 21 22 23 24 25

Figure 10.5 EIA RS-232-D pin number assignment.

Table 10.1 Assignment of Circuits to Pins on EIA RS-232-D Connector

Pin number (CCITT equivalent)	Name of the signal	Comments
2 (103)	Transmitted data	Output signal generated by local DTE for DCE
3 (104)	Received data	DCE (after receiving signal from DTE/DCE) at other node generates a signal at this pin
14 (118)	Secondary transmitted data	DTE generates output signal on secondary circuit for DCE
16 (119)	Secondary received data	DTE (after receiving signal from DTE/DCE of other node) generates a signal at this pin
4 (105)	Request to send	DTE sends signal to local DCE, which turns on the carrier of the modem
5 (106)	Clear to send	DCE informs DTE that the carrier is ON and is ready to receive its data
6 (107)	DCE ready (data set ready in RS-232-C)	DCE indicates to its DTE that the circuit connection is ready and it is ready to operate (circuit connection means establishment of telephone connection)
20 (108.2)	DTE ready (data terminal ready in RS-232-C)	TE indicates to its DCE that it is ready to operate and appropriate equipment is ready
22 (12)	Ring indicator	DCE on the other side indicates to its DTE that it is receiving the ring signal on communication/telephone circuits
8 (109)	Received line signal detector	DCE indicates to its DTE that it is receiving/detecting carrier signal from distant DCE
21 (140/110)	Signal quality detector	A signal on this pin at DTE indicates that error has been detected in received data
23 (111/112)	Data signal rate selector	A signal on DCE allows selection of higher data rate
23	Data signal rate selector	A signal on DTE allows selection of higher data rate
19 (120)	Secondary request to send	DTE indicates to its DCE for transmission on secondary circuit
13 (121)	Secondary clear to send	DCE indicates to the DTE that it is ready to receive the data on secondary circuit
12 (122/112)	Secondary received line signal	DCE indicates to its DCE that it is ready to receive the data on the detector secondary circuit
2 (140/110)	Remote loop back (not in RS-232-C, only CG interpretation)	A signal on this indicates to distant DCE to loop back signals to originating DCE
18 (141)	Local loop back (unassigned in RS-232-C)	A signal on this indicates to local DCE to loop back signal of its local DTE
25 (142)	Test mode	A signal on this indicates to DTE that local DCE is in a test mode
24 (113)	Transmitter signal element timing	A signal on it at DTE provides a clock signal (DTE four-bit synchronization) to DCE, which turns ON or OFF during the middle of 0s and 1s
15 (114)	Transmitter signal element timing	A signal on it at DCE provides DCE signal back to DTE
17 (115)	Receiver signal element timing	A signal on it at DCE provides information regarding received timing signal (from DTE) back to DTE
7 (102)	Signal ground	A signal on it provides common ground for all the circuits between DTE and DCE
1	Shield (same as protective ground RS-232-C)	A signal on it provides shield to several cables connected at DTE

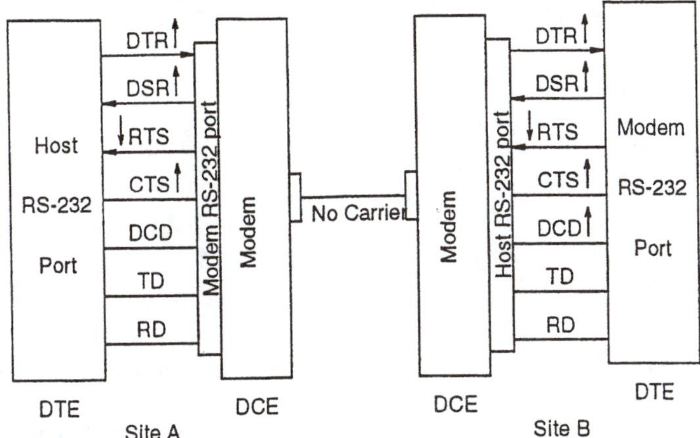

Figure 10.6 No site transmitting.

computer is on, the DTR is asserted (ON), and when the modem is ON, the DSR is asserted (ON), while other circuits are still OFF. In the idle line, a mark (1111) is continuously following for an asynchronous mode, while a synchronous byte 01111110 is continuously following through the line for a synchronous mode of operation.

	DTE	DCE		DCE	DTE
RTS	OFF	0 → 0		0 ← 0 OFF	**RTS**
DTR	ON	0 → 0		0 ← 0 ON	**DTR**
CTS	OFF	0 ← 0	**No carrier**	0 → 0 OFF	**CTS**
DCD	OFF	0 ← 0		0 → 0 OFF	**DCD**
DSR	ON	0 ← 0		0 → 0 ON	**DSR**
	Site A			**Site B**	

Let us assume that A wants to send data to B. The DTE of A asserts RTS high (ON), is recognized as high by its modem, and generates a carrier. Now the line has the carrier (ON), i.e., all 1s.

	DTE	DCE		DCE	DTE
RTS	ON	0 → 0		0 ← 0 OFF	**RTS**
DTR	ON	0 → 0		0 ← 0 ON	**DTR**
CTS	OFF	0 ← 0	**Binary 1s**	0 → 0 OFF	**CTS**
DCD	OFF	0 ← 0		0 → 0 OFF	**DCD**
DSR	ON	0 ← 0		0 → 0 ON	**DSR**
	Site A			**Site B**	

Site B receives the carrier and, after a delay (250 μs to 30 ms), the DCD of B is asserted (ON) and at the same time the DCE of A asserts its CTS (ON); i.e., the DTE of A can send the data.

	DTE	DCE		DCE	DTE
RTS	ON	0 → 0		0 ← 0 OFF	**RTS**
DTR	ON	0 → 0		0 ← 0 ON	**DTR**
CTS	ON	0 ← 0	**Binary 1s**	0 → 0 OFF	**CTS**
DCD	OFF	0 ← 0		0 → 0 ON	**DCD**
DSR	ON	0 ← 0		0 → 0 ON	**DSR**
	Site A			**Site B**	

Now A wants to stop. The DTE of A will assert RTS low (OFF), which will stop generating a carrier; i.e., the CTS becomes OFF. This will make the DCD OFF at B.

Full-duplex DTE/DCE interface: In the case of full-duplex transmission, we can use either a pair of wires or two pairs of wires (private/leased line). In both types of transmission, we are establishing a point-to-point link between A and B, and all the control circuits are asserted high (ON) for data communication between them in both directions over different pairs of wires, as shown below.

	DTE	DCE		DCE	DTE
RTS	ON	0 → 0		0 ← 0 ON	**RTS**
DTR	ON	0 → 0		0 ← 0 ON	**DTR**
CTS	ON	0 ← 0	**Binary 1s**	0 → 0 ON	**CTS**
DCD	ON	0 ← 0		0 → 0 ON	**DCD**
DSR	ON	0 ← 0		0 → 0 ON	**DSR**
	Site A			**Site B**	

In the case of a telephone system (where A and B are connected by a telephone network), two additional circuits, ring terminal ready and ring indicator, are used. A modem asserts high (ON) on its ring indicator after the ring signal is generated, which in turn is initialized by dialing the number. The DTE is asserted (ON) in the same way as in the case of half-duplex or full-duplex point-to-point communication. At the other site, the DTE will detect the carrier by asserting DCD (ON), and if there is no carrier, then the DTE of this site will assert DTR (OFF). Other circuits are asserted using the above-mentioned sequence for establishing the link, transferring the data, and de-establishing the link.

Null modem: The above was a discussion on how a DTE/DCE interface can be used for the different types of transmission modes: point-to-point communication, multi-point communication, telephone network, and so on. Here, both transmit and receive data between terminals and computers are transmitted over the same lines. This is because the same modem is being used at both nodes for both transmitting and receiving. The DTE/DCE interface is primarily an interface between character-oriented devices (like printer, terminals, etc.) and computer. The modem must be able to distinguish between these devices and computers. The function of the modem remains the same; it is not practical to provide emulation between the devices and computers, and this can be avoided if we change the pin connections between the nodes.

Further, sometimes these devices (DTEs) are close to each other and don't require connection by modem. In other words, these devices can directly provide signals to each other. This type of configuration is known as *null modem,* where DTEs are still connected through an RS-232 connector and modem, but the modem does not accept signals from the DTE and hence does not give any signal to the DCE on the other side. For asynchronous transmission, pins 2 (transmitted data), 3 (received data), and 7 (signal ground) of both RS-232 compatible DTEs can provide a direct link for the transfer of data between them. In this case, the modem is not connected in the link. Additional circuits have to be established to maintain the normal connection via the DCE, as shown below. Pins 2, 3, and 7 of site A DTE are connected to pins 3, 2, and 7 of site B DTE to provide asynchronous transfer of data, while additional circuit connections are shown below.

	DTE		DTE	
Shield (1)	0	———	0	Shield
TD (2,BA)	0		0	TD (2,BA)
RD (3,BB)	0		0	RD (3,BB)
RTS (4)	0		0	RTS (4)
CTS (5)	0		0	CTS (5)
DSR (6)	0		0	DSR (6)
C-GND (7)	0	———	0	C-GND (7)
DCD (8)	0		0	DCD (8)
DTR (20)	0		0	DTR (20)
RI (22)	0		0	RI (22)

The control pins corresponding to RTS and CTS of each side are connected together and are connected to DCD. Similarly, the DSR and RI are connected together to the DTR of the other side. The null modem, as shown above, provides interface between terminals and computers or other devices. A null modem for synchronous transmission is usually very complex and requires timing circuits in addition to the circuits in the case of an asynchronous null modem.

10.4.4 EIA RS-232-C DTE/DCE interface

The RS-232-C interface, now renamed RS-232-D, defines the electrical and mechanical characteristics of the interface between the DTE and the DCE using serial binary communication. In this version of the interface, the prefix D (after the interface was renamed RS-232-D), in fact, indicates its conformance with CCITT V.24, V.28, and ISO 2110. The new version of the interface includes new features like local loopback, remote loopback, test mode interchange circuits, etc., and has already been discussed above (Table 10.1). It was also decided to rename the earlier data communications terminal (DCT) as data circuit–terminating equipment (DCE), which is used in RS-232-D interfaces.

The RS-232-C interface has the following characteristics:

- Most common DTE/DCE interface
- One of the most successful standards
- Unbalanced electrical transmission
- Bipolar 3–25 V
- 50 ft maximum distance
- 20-Kbps maximum data rate
- 21 interchange circuits
  - Data
  - Control
    - Timing
    - 25-pin connector

Here, logic 1 is represented by voltage of less than –3 V, while logic 0 is represented by more than +3 V (a range of 6 V). The length of cable is recommended to be less than 15 m and offers a data rate of 20 Kbps. Different data and control circuits of RS-232-C are shown below.

	DTE		DCE
Send data (SD)	0	⟶	0
Signal ground	0	—	0
Receive data (RD)	0	⟵	0
Request to send (RST)	0	⟶	0
Clear to send (CTS)	0	⟵	0
Data set ready (DSR)	0	⟵	0
Data terminal ready (DTR)	0	⟶	0
Data carrier detect (DCD)	0	⟵	0
Ring indicator (RI)	0	⟵	0
External clock	0	⟶	0

In order to understand the difference between RS-232-C and RS-232-D, we present the pin diagram of RS-232-C and various associated signals (Table 10.2).

It is important to note that both RS-232-C and RS-232-D have the same pin numbers except that RS-232-D has additional pins as 9, 10, and 11 (without any equivalent CCITT or EIA numbers), 18 (141/LL for local loopback), and 25 (142/TM for test mode). Other details on RS-232 interfaces and their applications can be found in References 4, 5, and 16.

10.4.5 RS-449 (RS-422-A and RS-423-A) interface

As mentioned above, the RS-232 interface can be used for only a limited distance (less than 15 m) with a maximum speed of 20 Kbps due to noise (ground looping currents, shielding, capacitances, inductances, etc.). In addition to these factors, the quality and reliability of the circuit also affect the actual data rate. A new standard known as an RS-449 (RS-422-A and RS-423-A) interface overcomes this problem by using different distance electrical circuits RS-422-A (balance circuit) and RS-423-A (unbalanced circuit) between DTE and DCE, making it useful for a longer distance with a higher data rate (up to 2.5 Mbps).

The RS-449 interface[12] has the following characteristics:

- A new family of physical interface (DTE/DCE) standards
- Defined by RS-449, RS-422-A, and RS-423-A
- RS-449[12]
  - Defines mechanical, functional, and procedural functions
  - Has 30 interchange circuits
  - Has 37-pin connector (uses ISO 4902 mechanical connector) (Figure 10.7(a))
  - Offers a bit rate of up to 2.5 Mbps
  - Has plan for transition from 232-C and 232-D
  - Compatible with CCITT V.24, CCITT V.54, CCITT X.21 bis
- RS-422-A[10]
  - Compatible with X.27
  - Balanced electrical transmission
  - Can be used for a distance of up to 1.5 km
  - 100-Kbps data rate at 1.5-km distance
  - 10-Mbps data rate at 15 m

Table 10.2 Pin Assignments for RS-232-C and Associated Signals

	Pin/Circuit	Data Signal From		Control Signal From		Timing Signal From		
		DCE	DTE	DCE	DTE	DCE	DTE	CCITT/EIA
1	Protective ground							101/AA
2	Transmit data		x					103/BA
3	Receive data	x						104/BB
4	Request to send				x			105/CA
5	Clear to send			x				106/CB
6	Data set ready			x				107/CC
7	Signal ground common return							102/AB
8	Receive line signal detector			x				109/CF
9	Reserved for data set setting							
10	Reserved for data set setting							
11	Unassigned							
12	Secondary received line signal detector			x				122/SCF
13	Secondary clear to send		x					121/SCB
14	Secondary transmitted data	x						118/SBA
15	Transmission signal element timing (DCE)						x	114/DB
16	Secondary received data	x						119/SBB
17	Receiver signal element timing (DCE)						x	115/DD
18	Unassigned							
19	Secondary request to send				x			120/SCA
20	Data terminal ready			x				108.2/CD
21	Signal quality detector				x			110/CG
22	Ring indicator				x			125/CE
23	Data signal rate selector			x	x			111/112 (CH/CI)
24	Transmit signal element timing (DTE)			x				113/DA
25	Secondary clear to send				x			

Note: DCE: data communications equipment, DTE: data terminal equipment. Table partially derived from References 1 and 2.

- Less noise-sensitive (as half of the actual signal flows in each wire on the transmitting side; on the receiving side, the signals are inverted and combined to get the original signal, which cancels the noise present on the wire)
- RS-423-A[11]
  - Unbalanced electrical transmission
  - Can be used for a distance of up to 1.2 km
  - 50-Kbps data rate at 1100 m
  - 300-Kbps data rate at 10 m

Figure 10.7(a) also represents the pin number assignment of V.24 and ISO 4902. The pin number assignment of EIA-449 is shown in Figure 10.7(b).

Differential signaling on both the conductors of balanced transmission makes them less noise-sensitive, as signals on the conductors are equal in magnitude but opposite in polarity for representing logics 0 and 1. This scheme allows a constant voltage difference between these conductors. On the receiving side of RS-423-A, the difference voltage is the only signal recognized by it; other voltages are ignored.

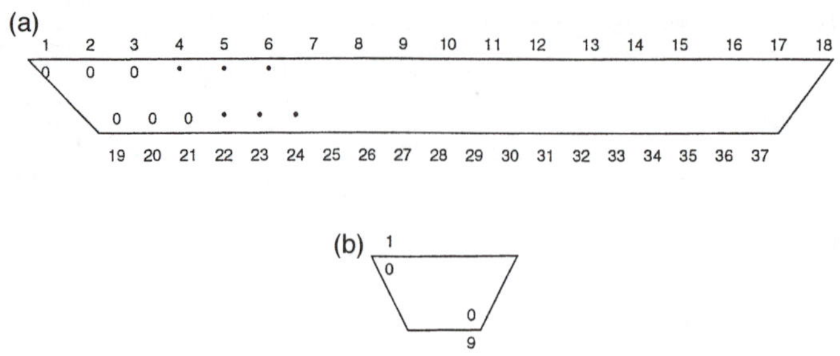

Figure 10.7 (a) Pin number assignment of RS-449. (b) Pin number assignment of EIA-449.

RS-422 balanced interface uses a dual state for the drivers. RS-485 resembles RS-422, with the only difference being that it uses tri-state driver capability. It is useful in a multipoint configuration, where one central computer may be used to control various devices. RS-485 can be used for controlling up to 65 devices.

The main purpose of defining a new standard RS-449 (in 1977) was to provide a long-distance capability and also a higher data rate. The idea was to overcome the limitations of RS-232-C and RS-232-D by defining a new standard RS-449 on a 37-pin connector and replace the existing RS-232 series completely by RS-449. But due to the fact that it is a 37-pin connector, it requires more space and fewer input and output connections. A new standard, RS-530, was introduced in 1987 by EIA. RS-530 uses the same 25-pin connector used by RS-232-C and RS-232-D and operates at a data rate of 20 Kbps to 2 Mbps. Thus, RS-530 has now completely replaced RS-449 and offers the same functional, mechanical, and procedural functions (on a 25-pin connector) as RS-449 (on a 37-pin connector).

For electrical functions, RS-449 and RS-530 use two other EIA standards, RS-422-A and RS-423-A. They transmit serial binary data (for synchronous and asynchronous modes) at a higher data rate. Interestingly, RS-530 and RS-232 are not compatible. Due to the constraints of the two conductors, this transmission is useful in coaxial or for a very short distance like the 15-m zone (RS-232-C can be used for a maximum distance of 15 m).

RS-422-A uses a balanced transmission for each of the signals in the interface RS-530. In balanced transmission, we use a pair of conductors. Currents (corresponding to signals to be transmitted) flow in one conductor from one side to the other and come back on another conductor. The advantage of balanced transmission lies in its higher data rate of 100 Kbps at 1100 m and 10 Mbps at 13 m.

Balanced transmission produces less noise compared to unbalanced transmission and even if noise is present, it will have the same value on both conductors and a combined effect will cancel the noise and, hence, not affect the value of voltage and current across the circuit. For digital transmission, this mode of transmission is based on the differential signaling technique and the binary signal value is determined by the voltage difference between the two conductors. The value of voltage difference for binary signals is very low compared to that in RS-232-C. Here logic 0 is represented by a voltage greater than 0.2 V while logic 1 is represented by –0.2 V, a difference of 0.4 V (as opposed to 6 V in RS-232) .

RS-423-A offers an unbalanced circuit (similar to RS-232). It uses the common ground of the circuit for transmitting the signal and a signal wire as a return path for the signal. This eliminates the ground looping current, since the return is not through the ground but through a wire. RS-530 provides an alternate interface for RS-232-D but uses an RS-232-D mechanical connector (25-pin). It offers a data rate in the range of 20 Kbps to 2.5 Mbps and can be used on private lines dedicated with a two-wire or four-wire circuit on point-to-point or multi-point operations. Both synchronous and asynchronous transmission are supported by this interface. This interface is basically defined for wide-band circuits and offers point-to-point configuration for synchronous transmission with data rates of 48–168 Kbps. The CCITT has defined three standards (equivalent to RS-449) as V.35 (48 Kbps), V.36 (48–72 Kbps), and V.37 (92–168 Kbps).

The logic 0 is represented by a positive voltage, while logic 1 is represented by negative voltage and a difference of 4 volts is used. It offers 300 Kbps at a distance of 15 m and 3 Kbps at 1.1 km, compared to 20 Kbps at distance of about 15 m in RS-232.

RS-449 provides 10 additional circuits for controlling and testing the signals. Usually it is available as a 37-pin connector and 9-pin auxiliary connector. Both connectors have their own ground and other common signals. The two classes of interconnection can be obtained on RS-449 (using RS-422-A and RS-423-A circuits) (see Table 10.3). One class (RS-423-A) provides unbalanced transmission on RS-449 with a minimum number of wires and low cost. The second class (RS-422-A) offers better performance.

RS-422-A uses the following 10 circuits on RS-449 and offers a data rate of over 20 Kbps.

- Send data
- Receive data
- Terminal timing
- Send timing
- Receive timing
- Request to send
- Clear to send
- Receiver ready
- Terminal ready
- Data mode

10.4.6 RS-530 interface

One of the main purposes of the RS-530 interface is to define mechanical interface characteristics between DTE and DCE. It operates in conjunction with RS-422-A and RS-423-A and is intended to replace RS-449. It is a 25-pin connector. In fact, it is an interface to RS-232-D and, as such, it uses the RS-232-D mechanical connector and offers data rates higher than RS-232, i.e., rates between 20 Kbps and 2 Mbps. This interface can be used with both synchronous and asynchronous systems. It supports two-wire and four-wire circuits and different line configurations (point-to-point, multi-point). It can be used with switched, non-switched, dedicated, or leased lines.

The interchange circuits, including the pin numbers of RS-530, are shown in Table 10.4.

Another interface, EIA-366, has been defined for an automatic call system and DTE through modem. This interface manages the operation, connection, and disconnection of the circuits.

10.4.7 CCITT V series standard interface

The physical layer standard should provide compatible interface to different types of DTEs and DCEs. Further, these interfaces should support different types of encoding, modulation, demodulation, and decoding techniques. For each of these techniques, interchange circuits and standards have been defined which should be added to the standard interface to achieve these. The standards for these are known as CCITT V series or Bell standards. The CCITT provides two sets of standards under the V series to be used with telephone networks and wide area networks. The standard for telephone networks is known as V.24, which conforms to RS-232-C. The V.24 standard can be used on either two-wire (half-duplex) or four-wire

Table 10.3 Pin Assignments for RS-449 and Associated Signals

Pin number (equivalent in RS-232-D)	Signal
1 (1)	Shield
2 (23)	Signaling rate indicator (A)
3 (11)	No connection (spare)
4 and 22 (2)	Send data (A) and send data (B)
5 and 23 (15)	Send timing (A) DCE source and send timing (B) DCE source
6 and 24 (3)	Receive data (A) and receive data (B)
7 and 25 (4)	Request to send (A) and request to send (B)
8 and 26 (17)	Receive timing (A) and receive timing (B)
9 and 27 (5)	Clear to send (A) and clear to send (B)
10 and 14 (10)	Local loop back (A) and remote loop (B)
11 and 9 (6)	Data mode (A) and data mode (B)
12 and 30 (20)	Terminal ready (A) and terminal ready (B)
13 and 31 (8)	Receiver ready (A) and receiver ready (B)
14 and 10 (10)	Remote loop back (A) and local loop (A)
15 (22)	Incoming call (A)
16 (23)	Signaling rate selector (A)
17 and 35 (24)	Terminal timing (A) and terminal timing (B)
18, 28, 34 (18)	Test mode (A), term in service (A), and new signal
19 (7)	Signal ground
20 (9)	Receive common
21 and 3 (11)	No connection (spare)
22 and 4 (2)	Send data (B) and send data (A)
23 and 5 (15)	Send timing (B) DCE source and send timing (A) DCE source
24 and 6 (3)	Receive data (B) and receive data (A)
25 and 7 (4)	Request to send (B) and request to send (A)
26 and 8 (17)	Receive timing (B) and receive timing (A)
27 and 9 (5)	Clear to send (B) and clear to send (A)
28, 18, 34 (18)	Terminal in service (A), test mode (A), and new signal
29 and 11 (6)	Data mode (B) and data mode (A)
30 and 12 (20)	Terminal ready (B) and terminal ready (A)
31 and 13 (8)	Receiver ready (B) and receiver ready (A)
32 (12)	Select standby
33, 12, 30 (20)	Signal quality (A), terminal ready (A), terminal ready (B)
34, 18, 28 (18)	New signal, test mode (A), term in service (A)
35 and 17 (24)	Terminal timing (B) and terminal timing (A)
36 (25)	Standby/indicator
37	Send common

(full-duplex) systems. The standards which work over two-wire systems include V.23, V.26, V.27, etc., while the standards for four-wire systems include V.21, V.21 bis, V.22, V.22 bis, V.29, V.32, etc.

The standard for wide area networks is known as V.35, which conforms to RS-449 and offers point-to-point synchronous configuration between data rates of 48 and 168 Kbps. The standards under this class include V.36 and V.37. The V.35 defines an interface between a DTE and a high-speed synchronous modem (48–168 Kbps). Various interchange circuits between DTEs and DCEs are defined in the V.24 standard but with data rates lower than 20 Kbps. Table 10.1 shows the equivalent interchange circuit number (as defined in V.24) for various pins of RS-232-D (see Figure 10.5). The following is a brief description of various standards defined in the CCITT V series.[13,14]

Table 10.4 Pin Assignments for RS-530 and Associated Signals

Pin number	Signal interchange circuit
1	Shield
2, 14	Transmitted data
3, 16	Received data
4, 19	Request send
5, 13	Clear to send
6, 22	DCE ready
20, 23	DTE ready
7	Signal ground
8, 10	Received line
15, 12	Transmit signal element timing
17, 9	Receiver signal element timing
18	Local loopback
21	Remote loopback
24, 11	Transmit signal element timing
25	Test mode

V.1 definition of binary and other signals
V.2 definition of signal level, power level, etc.
V.5 definitions of bps over half-duplex and full-duplex configurations
V.6 definitions of bps/dedicated/leased link
V.10 definitions of electrical function of an unbalanced interchange circuit in physical layer interface
V.11 definitions of electrical function of balanced interchange circuit in physical layer interface.
V.21 interface for PSTN (two-wire, half-duplex, switched-circuit, 300 bps)
V.22 bis interface for PSTN (two-wire, half-duplex, switched-circuit, 2400 bps)
V.23 interface for PSTN (two/four-wire, half/full-duplex, leased-circuit, 600–1200 bps)
V.24 definitions of all interchange circuits
V.25 definitions of four interchange circuits for automatic dial-and-answer interchange between DTE and DCE
V.25 bis definitions of one interchange circuit for automatic dial-and-answer interchange between DTE and DCE
V.28 definitions of electrical function for an unbalanced interchange circuit
V.29 interface for PSTN (two/four-wire, half/full-duplex, leased-circuit, 4800/9600 bps)
V.32 interface for PSTN (4800, 9600 bps, full-duplex)
V.35 interface for PSDN (synchronous, point-to-point, 48 Kbps)
V.36 interface for PSDN (synchronous, point-to-point, 48–72 Kbps)
V.37 interface for PSDN (synchronous, point-to-point, 96–168 Kbps)

The CCITT defined the pin number assignments shown in Table 10.5 for the V.35 standard, which is available as 37- and 15-pin connectors. The standard V.35 supports both unbalanced (RS-232-C and V.24) and balanced (RS-422-A) operation.

10.4.8 CCITT X series standard interface

The main philosophy in EIA physical interfaces has been to include a large number of circuits to handle all possible situations. Sometimes this approach becomes expensive,

Table 10.5 Pin Assignments for CCITT's V.35 Standard and Associated Signals

Pin number	Description
1	Shield
2	Transmitted data
3	Received data
4	Request to send
5	Ready for sending
6	Data set ready
8	Receive line signal detect
15	Transmit signal element timing
17	Receive signal element clock

particularly nowadays, when other hardware costs are decreasing rapidly. An alternative approach could be to use fewer circuits (i.e., pins) in the interface and include a greater number of logics for communication at the DTE and DCE level. This is the approach used in defining the CCITT X.21[13] physical layer interface, which is an interface between the DTE and DCE of public data networks (X.25 packet-switched, circuit-switched networks). X.21 is more popular in European and Asian countries than in North America.

The X.21 interface was defined as an international standard in 1976 for sharing the public data network (using synchronous transmission for data transfer). It is a 15-pin connector and has the interchange circuits shown in Table 10.6. It consists of two circuits for transmitting (T) and receiving (R) the data across the interface which defines DCE/DTE interchange circuits for a digital circuit-switched communication network. The flow of control and data signals in DTE/DCE interchange circuits using X.21 is shown below.

T	0	⟶	0	T
C	0	⟶	0	C
R	0	⟵	0	R
I	0	⟵	0	I
S	0	⟵	0	S
B	0	⟵	0	B
G	0	—	0	G

Table 10.6 Interchange Circuits of CCITT's X.21 Interface

Pin	Pin function
Signal Ground (G)	Provides common signal ground between DTE and DCE
Transmit (T)	Signal on this pin provides DTE to DCE data signals and network call control signals and depends upon control (C) and indication (I)
Receive (R)	Signal on this pin provides DCE to DTE data signal and network call control signals for distant DTE/DCE
Signal element timing (S)	Signal on this pin of DCE provides bit signal timing to its DTE-
Control (C)	Signal on this pin of DTE indicates the circuit condition to DCE, i.e., whether data is being transmitted
Indication (I)	Signal on this pin of DCE indicates the call establishment and start of data transmission (ON) to DTE
Byte timing (B)	An optional interchange circuit for DCE to indicate to DTE for eight-bit (byte or octet) timing (or synchronization)

The data could be either the actual data bit stream signals or control signals, and both these types of signals (T or R) are transmitted on these two circuits. This is the reason why we use fewer pins and why not all the pins are used for the transmission of data across the interface. Other circuits, control (C) from DCE to DTE, and indication (I) from DCE to DTE do not carry any data and simply represent the control and status information through ON/OFF states.

As indicated earlier, this interface can be used to share the public switch network, and a signal element timing (S) circuit provides bit signaling (for synchronization to DTE). As discussed above, these circuits are enough to provide transfer of data between two sides and can be used in a variety of applications.

10.4.8.1 CCITT X.21 standard interface

X.21 supports both modes of transmission (balanced and unbalanced) and offers the same speed and distance capability as that of EIA interfaces. In most of the applications, a balanced mode of transmission in all circuits is needed due to obvious reasons. It is more flexible than RS-232-D and RS-449 and less expensive.

CCITT X recommendations are also known as data communication network interfaces. The standard X.24 defines the description of interchange circuits. The X.26 (used in RS-422-A) and X.27 (used in RS-423-A) define the electrical function of X.21. Both X.26 and X.27 are used with integrated circuit equipment and are equivalent to V.10 and V.11, respectively.

The development of integrated services digital networks (ISDNs) has affected the use of X.21 on digital networks. Limited networks support X.21, but still it is a useful and widely accepted physical layer interface for terminals accessing the X.25 public packet-switched network.

Another standard, **X.21 bis**[14] (similar to RS-232), has also been defined by CCITT for accessing a X.25 network with DTE. This standard is considered a kind of interim and will not be useful for digital circuit-switched networks. This provides an interface to duplex V series modems. The X.25 provides an interface between DCE and DTE for asynchronous transmission on public networks. Although the X.21 bis interface is more efficient than RS-232 or RS-449 (RS-422-A and RS-423-A), due to ISDN, it looks as though this may soon disappear from the market or be replaced by another interface with still fewer circuits. It can also be used to provide interface between DTE of a synchronous modem and DCE of public data networks.

Various procedures defined for X.21, such as call establishing, disconnecting the connection, network congestion, etc., over a circuit-switched network involve a series of operations. These procedures are not provided by physical layer protocols but have to be provided by the status of DTE and DCE and are explained below.

- **Ready:** When DTE wants to send the data, it will first show a ready state by making T = 1 and a control OFF (logic 1) on C. At this time, on the receiving side, receive R gets 1 and its I is OFF (logic 1).
- **Call-request:** DTE makes T = 0 and C = ON (logic 0) while DCE remains unchanged.
- **Proceed-to-select:** DTE transmits ASCII symbols to DCE indicating its readiness and then transmits ASCII address.
  DTE T = O C = ON
  T = ASCII address
  DCE R = ++++++ I = OFF
- **Military standards:** These standards are specifically for military applications. The following is a partial list of some of these standards:

MIL STD 188-C is equivalent to RS-232-C and needs an interface for communication.
MIL STD 188-114 balanced standard is similar to RS-422-A and provides all the features of RS-422-A. Similarly, MIL STD 188-114 unbalanced is similar to RS-423-A and offers all the features of RS-423-A.

A detailed discussion of different types of protocols and standards for the physical layer can be found in Reference 6.

10.5 Terminals

Terminals are used for the processing and communication of data and provide interface between users and the system. This interface, in fact, provides a mapping between two types of representations: user's and electrical signal–oriented. Various hardware interfaces can be interpreted as terminals, e.g., a printer which provides mapping for processed data in the form of hard copy, and other I/O devices. Terminals are very useful components for processing, collecting, retrieving, and communicating data in the system and have become an integral part of the requirements of any application system. The interface (provided by terminals) for data communication takes input from input devices such as a keyboard, card reader, or magnetic tape of any other input device and converts it into output through display devices, paper tape, etc. Some of the I/O devices may have only input and no output; such devices are known as *receive-only* devices.

The keyboard input device is very popular and offers effective input capabilities to both terminals and computers. It includes different types of keys which, in turn, depend on the application. But in general, the keys can be grouped into three groups: textual (defining text, numbers, special symbols and characters, punctuation signs, etc.), control (dealing with various options during the editing of the text and its transmission), and macro-functional (defined for a special sequence of operations which may include data processing, etc.). Various codes have been defined for converting the functions of the keys into a serial bit stream for transmitting over transmission link or bus. The codes have a fixed number of bits for each key; e.g., both ASCII and EBCDIC define an eight-bit code, but ASCII has become the more popular code.

The output devices accept a bit stream from the terminal and generate the symbol. The symbols can be represented in various forms: regular full print-out or dot matrix print-out. The regular full-matrix print-out offers better quality of characters but is expensive. Laser printers (Hewlett-Packard) are popular nowadays; a nice laser printer may cost between U.S. $200 to $600. The dot matrix printers, on the other hand, do not offer good quality — characters are represented by a combination of small dots — but these printers are very inexpensive. Panasonic Kx series printers are popular and cost about U.S. $250.00. Other output devices include paper tape and display devices, and we will not discuss these here.

The input and output devices in the terminal can send/receive one character at a time, making the devices very slow. The speed of transmission of characters can be made faster by using a buffer (a small segment of memory) which will store a block of characters and transmit/receive it between input and output devices. A device having a buffer is known as a **buffered terminal**. In buffered terminals, the processing capability of data can be included to make it an intelligent terminal. An intelligent terminal, in fact, is nothing but a combination of micro-programmed computer (with stored program in ROM) and attached input/output devices.

Terminals (buffered or unbuffered) have become an integral part of the requirements of any application system. There are a variety of categories of terminals available on the market, and the selection of a particular terminal depends on the application, such as airline reservations, online banking system, data collection, inventory system, banking and brokerage transaction, processing, general purpose, etc. For each of these applications, the criteria for the selection of a terminal typically include the following parameters: speed, acceptable minimum error rate, input/output environment, coding scheme formats, storage, etc. Depending on the applications, terminals may be classified as teleprinter, video-display, remote job entry, facsimile, banking data collection, credit/debit, etc.

10.5.1 Categories of terminals

The terminals used with a computer network generally fall into three categories: typewriter-oriented, cathode ray tube (CRT) display, and intelligent terminals. The first two terminals do not possess any processing capabilities and also do not have their own memories.

1. **Type-oriented terminals** typically include a keyboard, printer, mouse, and appropriate control and interface cards. These terminals operate at a faster speed, support a variety of I/O formats, and are equipped with buffers and line control units. These terminals are used in applications such as time sharing, file manipulation, data retrieval, format and message conversions, etc.
2. **CRT display terminals** are used to record and display data over a big, TV-like display screen. Typically, these terminals can display 1000 characters simultaneously on the screen at a transmission rate of 100–1600 characters per second. These terminals are used mainly in applications such as an electronic bulletin board to provide display of the data stored in the computer system. Other applications include online reservations, arrival/departure information at airports, banking systems, inventory control system, brokerage system, etc. These terminals offer higher speed, an ability to display selected data quickly, and many other features.
3. **Intelligent terminals** have their own processing and memory and can be configured to perform specific functions. They avoid the use of host processes to provide basic services to the function. The services to the function typically include listing, formatting, editing, communication control and management, batch processing, etc. These terminals are also known as *swat* or *programmable terminals* and support a variety of devices such as printers, disk drives, etc. These terminals have found applications in word processing, time sharing, electronic mail services, data entry, query, etc. The storage capabilities of intelligent terminals can be enhanced by introducing an auxiliary storage (like floppy diskettes or tape devices).

As we don't have a standardized definition of intelligent terminals, different vendors are using different parameters for describing some of the features of their products. For example, from one vendor's point of view, intelligent terminals are more expensive than dumb terminals, while another vendor may say that intelligent terminals are buffered terminals associated with an input device (keyboard), teleprinter, output device, and other relevant components and that these terminals are monitored by software, while dumb terminals are not controlled by software but instead simply form one node to another. Based on this, we also see intelligent or smart facsimile or fax machines in the market.

Intelligent terminals possess a variety of features such as buffering, functional key, processing capabilities, software-supported keyboards, paging, and automatic transmission of blocks of characters.

Connector types for interfacing: Table 10.7 gives connector types for various interfaces.

Table 10.7 Connector Types for Various Interfaces

Connector type	Interface
DB25 (4, 12, or 24 pin)	RS-232 (V.24), RS-530, IBM parallel
DB37	RS-449, 442, 423
DB50	Dataproducts, Datapoint, UNIVAC
DB15	Texas Instruments, NCR Ethernet
DB15 (high-density)	IBM PS/2
DB9	Video interface
5-pin Din	IBM PCs
Mini 6-pin Din	PS/2 keyboard
Mini 8-pin Din	Apple Macintosh
M/34	V.35
BNC, BNC, and TNC	Coaxial (BNC or TNC)
Telco	Telephone (voice and data)
RJ-11	Voice telephone
RJ-45	Data telephone
Modular jack	DEC 423 DEC Connect

10.5.2 *Terminal interfacing standards*

There are different ways a terminal can be interfaced with the communication network, and all depend on the bit rate of the terminal being used. Based on these, there are two main classes of terminal interfaces: pollable and non-pollable. For a *pollable interface*, terminals must have buffering capabilities on a selectable ID. These terminals cannot send messages until they are polled by the primary and, hence, are useful for a multi-point configuration. For a *non-pollable* interface, terminals are non-buffered, use asynchronous transmission (start/stop type), and can be used only for point-to-point configuration between hosts.

Pollable and non-pollable device interfaces: The pollable interface can be used for polling the distant pollable terminals via an analog bridge (Figure 10.8(a)) or pollable terminals (which are within RS-232 distance range) via digital bridge (Figure 10.8(b)).

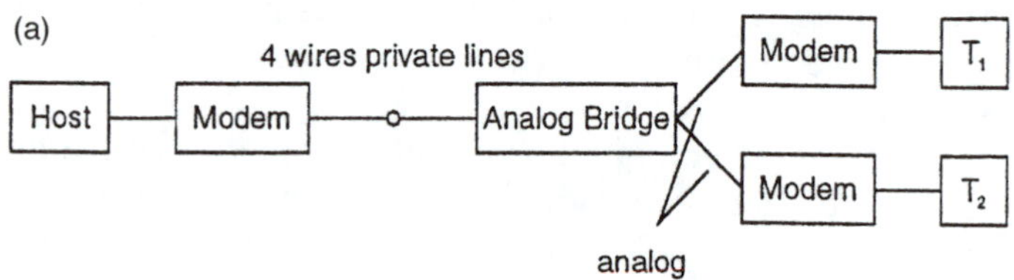

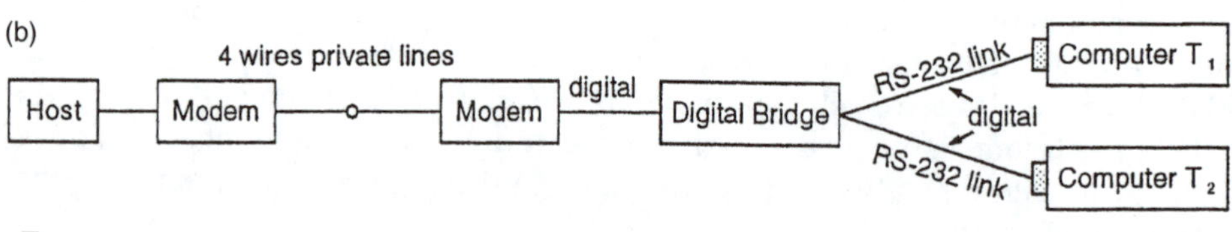

Figure 10.8 (a) Analog bridge. (b) Digital bridge.

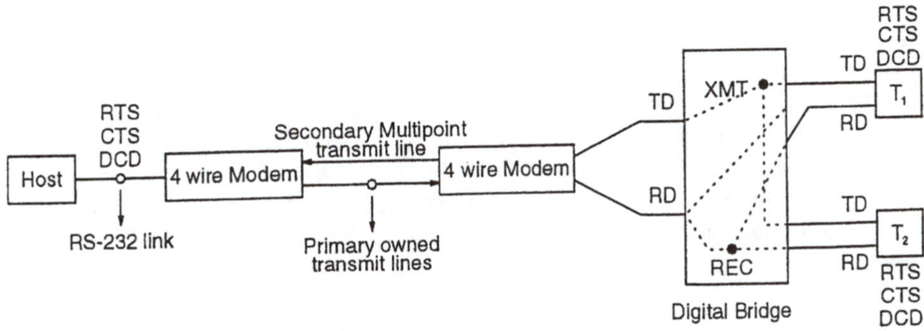

Figure 10.9 Full-duplex multi-point interface.

The analog bridge allows polling of devices which are not directly connected to the host but are connected by analog bridge at its secondary ports. This bridge can support eight to ten secondary ports. A request signal from the host (primary) is broadcast by analog bridge to all the secondary devices, and their response is OKed and sent back to the host. After receiving a response from the analog bridge, the host then decides which device it has to use and then makes a request to the analog bridge for establishing the connection between it and a specific port. On the other hand, if the devices are within the range of RS-232, a digital bridge again broadcasts the request to secondary devices. It then selects a particular port and establishes a connection between the host and the device.

Multi-point interfaces: A pollable interface based on the digital bridge can also be used to provide a multi-point link for both half-duplex and full-duplex communication.

For a **full-duplex multi-point interface**, the primary and secondary nodes are connected via a four-wire full-duplex set. A typical configuration is shown in Figure 10.9. The primary (host) has its own half-duplex pair of wires for transmitting the data to a secondary device. This pair of lines can be used only by the primary node. There is another half-duplex line over which a secondary node can send data to the primary. Both pairs must have the same data rate to define a full-duplex set of one data rate. The second half-duplex link is usually shared by more than one secondary node an RS-232 distance and is connected to the multi-point link via a digital bridge through RS-232 cables. When a primary node sends any message (including a pollable message), the digital bridge simply broadcasts the message to all the pollable terminals through the two slave ports.

Since the primary has its own half-duplex sending line, the RTS and CTS control circuits are always ON. This means that there will be a carrier all the time over the line owned by the primary node and the secondary nodes are always receiving this carrier signal, which makes their DCD control circuit always ON.

At the time when the RTS control circuit of the primary node is ON and the DCD circuit of both secondary nodes is ON, there is no transmission between them. The primary wants to poll one of the secondary (say T1), and this message is broadcast by the digital bridge to both the secondary nodes. The secondary node T1, after recognizing the message, will assert an RTS control signal The bridge connects the TD circuit of the secondary T1 to the modem. The modem generates a carrier which will assert DCD on the primary side. Now the transmit line from the secondary to the primary is allocated to T1, and it can send the message. The primary and secondary can now communicate back and forth. After the session, the secondary T1 de-asserts its RTS, and the bridge will disconnect its link with the transmit line and drop the carrier, which results in the drop of DCD on the primary.

In the case of a **half-duplex multi-point interface**, primary and secondary nodes are connected by a half-duplex channel, and only one node can send the message over it at any time. The purpose of the digital bridge is to connect one of the secondaries to the channel at any time when it is ready to send the data.

Initially, when no node is transmitting, the RTS, CTS, and DCD control circuits of these nodes are OFF. It is important to note that both DTR and DSR circuits are always ON. Let's say that the primary wants to poll the secondary T1; it will assert its RTS circuit to ON, which will generate a carrier. The bridge listens to this carrier and propagates the same carrier to both secondary nodes. After polling, the primary will let the secondary talk by dropping its RTS circuit to OFF, which will make its CTS circuit OFF. The modem connected to the primary will drop the carrier, which in turn will drop the DCD circuit to OFF for both secondaries.

Although the polling message was received by both secondaries, it was meant for the T1, and the secondary node T1 will assert RTS ON while the T2 will keep its RTS OFF. The bridge scans the RS-232 lines continuously, finds the secondary node T1's RTS is ON, and connects the TD line of the T1 to the TD line of the modem and, hence, connection is established.

Figure 10.8(b) shows the connection between a primary and secondaries connected via digital bridge. Since it is a half-duplex transmission, only one at a time (either primary or secondary) can transmit the data over a pair of lines. Further, all sites are connected via their RS-232 interfaces, and each of the sites has its RTS, CTS, and DCD control circuits initially OFF. Let us assume, for the sake of illustration, that DTR and DSR for all sites are ON. If the primary wants to poll any of the secondaries, it will make its RTS ON and the digital bridge will broadcast it to both the secondaries. Although the signal is received by both secondaries, only that secondary for which the request signal is sent will make its RTS ON. The digital bridge scans the secondaries continuously and, after it finds RTS of a particular secondary ON, connects this secondary with the primary. Once the primary has polled and sent the data, it will drop its appropriate circuits to OFF to enable others to use it.

The host is connected to a digital bridge via two pairs of lines. One pair of lines is reserved for the primary (host) to communicate with the bridge while the second pair of lines is used by the secondaries to send their requests.

As mentioned above, an automatic answering modem provides dialing, answering, and other operations automatically; it can also be used in full-duplex communication. The automatic answering modem will always keep its CTS ON (if it is powered ON). A ring signal will be heard on an R1 circuit of a distant automatic answering modem, which generates an interrupt at DTE. The ring signals are counted by the software of DTE and, depending on the type of software, DTE makes its DTR ON. This indicates to the modem to go OFF hook or answer the call state. The modem then generates a carrier signal which will be detected at the originating site. This site will make the DCD circuit ON, and now both DTE sites can transmit and receive the data as full-duplex.

The flow of data and control signals across the modem terminal interface is shown in Table 10.8.

10.6 Physical layer protocol primitives and parameters

The main function of the physical layer is to provide an interface between the communication media and data link layer of the OSI model (or interface between transmission media and media access control (MAC) in the IEEE LAN model). We have discussed in the preceding sections that the physical layer standards provide an interface between the physical layer and communication media. The interface between the physical layer and data link layer has been defined by services offered by the physical layer, including the establishment of physical connection, exchange of data over the connection, support for

Table 10.8 Flow of Data and Control Signals across the Modem Terminal Interface

Pin	Name	From DTE	From DCE	Type	Function
1	FG			G	Protective ground frame
2	TD	X		D	Transmit data
3	RD		X	D	Receive data
4	RTS	X		C	Request to send
5	CTS		X	C	Clear to send
6	DSR		X	C	Data set ready
7	SG			G	Signal ground
8	DCD		X	C	Data carrier detect
15	TSET	X		T	Transmit signal element timing (clock)
17	RSET		X	T	Receive signal element timing (clock)
20	DTR	X		C	Data terminal ready
21	SQ		X	C	Signal quality detect
22	RI		X	C	Ring indicator
23	DRS	X		C	Data rate select
24	TEST	X		C	External transmitter clock

different types of transmission (synchronous and asynchronous), mode of operation (half-duplex, full-duplex, simplex), configurations (point-to point, multi-point), etc. The standard protocol of the physical layer provides these services to the data link layer and, depending on the services required, an appropriate physical connection is configured which will map various links and modes of operation into an actual physical circuit. The configuration of physical connection between the physical layer and data link layer is independent of switching services and connection-oriented or connectionless link, and it provides a single physical connection for each of these categories.

The services offered by the physical layer are defined by ISO physical layer definitions (ISO 10022)[7] and are being provided by the **PH.ACTIVATE.request** and **PH.ACTIVATE.indication** primitives. The PH.ACTIVATE primitive is used to activate the data transfer operation along with its direction (useful for half-duplex configuration). PH.ACTIVATE.request activates the transmission from site A to site B and initializes PH.ACTIVATE. indication primitive of B. These two primitives establish the connection between the sites. The **PH.DATA.request (PH-user-data)** and **PH.DATA.indication (PH-user-data)** primitives are used to transfer data. For half-duplex, data can move in one direction at a time. Thus, when A wants to send data to B, these primitives transfer the data (bits or stream of bits) from A to B only. After the transfer of data, the **PH.DEACTIVATE.request** and **PH.DEACTIVATE.indication** primitives de-establish the connection between A and B. Now, if B wants to send the data back to A, the B-site has to establish the connection using the above primitives. In some cases, DEACTIVATE primitives may be used to change the direction of transmission of signals between the sites. This will ensure the establishment of a connection between them for as long as the sites wish to keep it.

Some useful Web sites

Some relevant sites on the physical layer are found at *www.whatis.com/phylayer.htm*, *www.atmforum.com/atmforum/library/53bytes/backissues/others/53bytes-0795-5.html*, and *www.ee.siue.edu/~rwalden/networking/physical.html*.

Tutorials on RS-232, RS-422, and V.35 interfaces can be found at *www.sangoma.com/signal.htm* and *www.arcelect.com/rs232.htm*. For information on RS-232, go to *http:/www.exclbr.com/101/rs232.htm*.

References

1. Stallings, W., *Handbook of Computer Communications Standards,* vol. 1, 2nd ed., Howard W. Sans and Co., 1990.
2. Black, U., *Data Networks: Concepts, Theory and Practice,* Prentice-Hall, 1989.
3. Markeley, R.W., *Data Communications and Interoperability,* Prentice-Hall, 1990.
4. Campbell, J., *The RS-232 Solutions,* Sybex, Inc., Berkeley, CA, 1984.
5. Seyer, M.D., *RS-232 Made Easy,* Prentice-Hall, 1984.
6. McNamara, J.E., *Technical Aspects of Data Communication,* 2nd ed., Digital Press, 1982.
7. International Standards Organization, Physical Layer Definitions, ISO 10022.
8. International Standards Organization, Open System Interconnection-Basic Reference Model, Information Processing System, ISO 7498, 1984.
9. EIA 232-D, Interface between data terminal equipment and data circuit-terminating equipment employing serial binary data interchange, Electronics Industries Association (EIA), 1987.
10. EIA RS-422A, Electrical characteristics of balanced voltage digital interface circuits, EIA, 1978.
11. EIA RS-423A, Electrical characteristics of unbalanced digital interface circuit, EIA, 1978.
12. EIA RS-449, General purpose 37-position and 9-position interface for data terminal equipment and data circuit-terminating equipment terminating employing serial binary data Interchange, EIA, 1977.
13. CCITT Recommendation X.21, Interface between data terminal equipment (DTE) and data circuit-terminating equipment (DTE) for synchronous operations on public data networks, CCITT, 1984.
14. CCITT Recommendation X.21 bis, Use on public data networks of data terminal equipment (DTE) which is designed for interfacing to synchronous B-series modem, CCITT, 1984.
15. ITU, The X.25 protocol and seven other key CCIIT recommendations, X.1, X.2, X.3, X.21 Bis, X.28, and X.29, International Telecommunications Union.
16. Black, U., *Physical Layer Interfaces and Protocols,* IEEE Computer Society Press, 1996.

chapter eleven

Data link layer

"Doubt is often the beginning of wisdom."

M. Scott Peck

"Layer two, the second lowest layer in the OSI seven layer model. The data link layer splits data into frames for sending on the physical layer and receives acknowledgment frames. It performs error checking and re-transmits frames not received correctly. It provides an error-free virtual channel to the the network layer. The data link layer is split into an upper sublayer, Logical Link Control (LLC), and a lower sublayer, Media Access Control (MAC)."

The Free Online Dictionary of Computing
Denis Howe, Editor (http://foldoc.doc.ic.ac.uk/)

11.1 Introduction

The data link layer offers an interface between the network layer and the lowest layer (physical) and provides a reliable connection for the transfer of data between two adjacent nodes (computers, IMPs, etc.) which are connected by a communication medium. Since communication media may experience noise in one way or another, the data link layer must provide a means to offer error-free data transmission to the network layer. This requires the data link layer to use techniques to detect and correct the errors introduced into communication media by atmospheric (or ground) conditions that will occur in the circuits of the physical layer. Further, it should support different types of communication media and accept the services offered by the physical layer protocols.

According to the OSI document, the main objective of the data link layer is to provide functional and procedural functions to establish and maintain a data link connection between the data link layer and network layer in response to a primitive request from the network layer to the data link layer. The data link layer, in general, defines the data into a block known as a *frame* of fixed size (which may vary from network to network). In OSI documentation, the frames are known as *data link protocol data units* (DPDUs). This layer accepts the data from network entities and constructs a frame which is then transferred to physical layer entities that finally transmit the data over the transmission media sequentially. The set of procedures defined in data link protocols detect errors (transmission, format, operational, etc.) occurring in the configured physical connection and try to recover from these errors. In this chapter, we discuss different data link protocols, services, and protocol primitives, along with their parameters and various error detection and correction protocols which are required for reliable data communication over the media.

11.2 *Layered protocols and interfacing*

In OSI-RM,[1] it is assumed that the lower layer always provides a set of services to its higher layers. Thus, the majority of the protocols that are being developed are based on the concept of a layered approach for OSI-RM and, hence, are known as *layered protocols*. Specific functions and services are provided through protocols in each layer and are implemented as a set of program segments on computer hardware by both sender and receiver. This approach of defining/developing a layered protocol or functional layering of any communication process provides not only a specific application of the communication process but also interoperability between vendor-customized heterogeneous systems. These layers are, in fact, the foundation of many standards and products in the industry today. The layered model is different from layered protocols in the sense that the model defines a logic framework on which protocols (that the sender and receiver agree to use) can be used to perform a specific application on it.

11.2.1 *Layered protocols*

A data layer has two neighboring layers: physical (lower) and network (upper). This layer uses the services of the physical layer and offers services to the network layer. The data-link entity performs functions within it. The same layers of two different networks communicate between their respective entities via a peer-to-peer protocol. Thus, we will have a communication logical link between peer data link layer entities of both networks, known as a data link layer peer-to-peer protocol. A protocol associated with it processes the data being exchanged between the peer layers. Each node (PCs, terminals, etc.) connected to the network must contain identical protocol layers for data communication. Peer protocol layers exchange messages which have a commonly understood format, and such messages are known as protocol data units (PDUs). Thus, the data link layer corresponds to the unit of data in a data link peer-to-peer protocol. The peer data entities communicate with each other using the services provided by the network layer.

The services of the data link layer are provided to the network layer at a point known as a data service access point (DSAP). This is the point through which services of the data layer can be accessed by the network layer. For a description of the interface between the data and network layers, the data link services and protocol primitives are defined. The data service data unit (DSDU), associated with data primitives, is delivered to the network layer and, similarly, the network layer uses its primitives to interact with the data link layer. The higher-layer network sends its protocol data unit (PDU) to the data link layer. The network layer PDU consists of a network protocol, control information, and its user data. The network layer control information is exchanged between network layer entities.

The communication control procedures are a set of rules for controlling and managing the interaction between users via their computers, terminals, and workstations across the network. The control procedures have been known by different names in the industry, such as protocols, line control procedures, discipline procedures, and data link control procedures. The protocols provide a communication logical link between two distant users across the network, and the handshaking procedures provide the call establishment within the protocols.

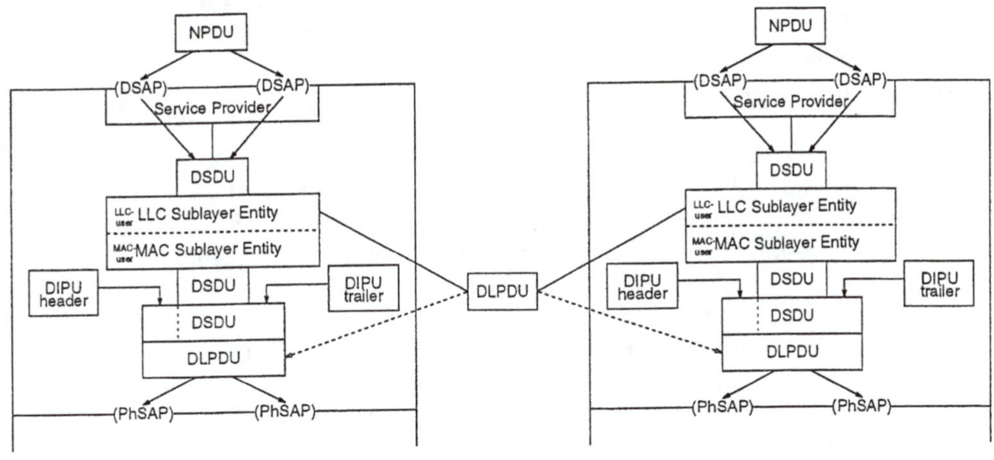

Figure 11.1 Relationship between DLPDU, DLSDU, and the data link layer.

11.2.2 *Interfacing*

A protocol defines rules, procedures, and conventions used for data communication between peer processes. It provides data formatting, various signal levels for data and control signals, error control, synchronization, and appropriate sequencing for the data to be exchanged. Two networks will offer a two-way communication for any specific application(s) if they have the same protocols. The layer m is known as the service provider for the layer $m + 1$ and offers the services at the SAP.

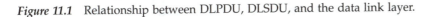

A service access point (SAP) is a unique hardware or port or socket address (software) through which the upper layer accesses the services offered by its lower layer. Also, the lower layer fetches the message from its higher layers through these SAPs. The SAP defines its own address at the layer boundary which is prefixed to define the appropriate layers. Thus, the data link layer DSAP is a data link layer SAP. The data link layer service data unit (DSDU) provides message information across the network to peer entities of all the layers. The DSDU and control information (addressing, error control, etc.) are known as an interface data unit (IDU), which is basically message information sent by an upper-layer entity to its lower-layer entity via DSAPs. For sending a DSDU, a layer entity partitions the message information into smaller-sized messages. Up to the network layer, each message is associated with control information defined by the appropriate layer as a header and sent as a protocol data unit (PDU) to its lower layer. This PDU is known as a packet. Thus, the network sends its PDU as an NPDU to the data link layer. We will be using herein these terms defined by ISO. The relationship between DLPDU, DSDU, and the data link layer entity is shown in Figure 11.1.

11.3 *ISO data link layer services and functions*

The data link layer is responsible for defining data link connections between network entities and for transferring data link service data units over it. These connections are defined via functional and procedural components of data link protocols. The data link connection is made over either one physical connection or multiple physical connections. It also supports the controlling of interconnection of data connections (within physical layer) by its higher layer (network).

The data link layer offers the following features:

- Establishes a data link connection between network and data link layer entities. Here the data link connection is established after the physical layer connection has been initiated. It supports a multi-point connection where it identifies the data link connections for physical connections defining multiple endpoints.
- Initializes various parameters, variables, and procedures in the protocol.
- The link control data link defines the sender and receiver on a half-duplex configuration. At the end of transmission in one direction, it changes the direction of transmission to the other direction through a specific field in the frame.
- Synchronizes between sender and receiver via special SYN characters or a specific field in the frame.
- Controls rate of receiving frames by software at the receiving side. A process known as handshaking is defined between sender and receiver; it simply acknowledges the receipt of data and readiness of receiver. Thus, it will prevent the sender from pumping more frames onto the media than the receiver can receive. The network entity can control the rate at which it wants to receive the data link service data units from the data link connection. The data link layer usually defines this rate as it corresponds to the receiving of data link service data units by the data link connection endpoint.
- Detects, corrects, and recovers from errors over the established connection between two sites. There are various techniques (and protocols) available which can be used to detect and also correct errors. It also notifies the network layer entity of errors. The error-detection capability plays an important role in defining the size of the data link service data unit and it also depends on the error rate supported by physical connection.
- Re-transmits the data in the event of error or transmits the same data after a predefined time (time out) in the event an acknowledgment that was sent is lost.
- Maintains a link between sender and receiver, identifies nodes, type of control and data, etc.

The parameters for quality of service (QOS) can also be selected for the duration of a data link connection. Quality-of-service parameters depend on many factors such as response time (in the event of errors), residual error rate (due to loss, duplication, or damage to data link service data units), throughput, transit delay, mean time between detected errors, error rate, undetected error rate, etc.

The above was a discussion of various features offered by the data link layer. There is a difference between functions and services. The *functions* are the procedural and functional aspects of the layer within the layer, and the *services* using these functions provide attributes for defining a number of facilities to its higher layer. Each layer offers its set of services to higher-layer entities. The higher layer is responsible for representing the use of all the services it is getting from its lower layer. When a lower-layer entity cannot provide all the requested services to its higher-layer entity, it communicates with its other entities to complete the services. These services are different from the services it is getting from its lower layer. The entities of the lowest layer (physical) always communicate with each other via the transmission media connecting them. The following are the functions and services of the data link layer.

- **ISO functions:** Establishment of data link connection, mapping of data link service data units, delimiting and synchronization, sequence control, error control, flow control, exchange of parameters of QOS, data link layer management.
- **ISO services:** Data link layer connection, data link service data units, sequencing, flow control, QOS parameters, error notification.

Note that there is an overlap of services and functions offered with the services of other layers.

11.4 Data link control functions and protocols

In this section we discuss different types of protocols defined for the data link layer. There are two models available for LANs: OSI-RM (composed of seven layers) and IEEE 802 (composed of eight layers). This latter model defines two sublayers for the data link layer and is very common with all standard LANs. We will briefly discuss the two sublayers of the data link layer defined in the IEEE 802 LAN model.

11.4.1 Functions of sublayers of the data link layer

As we pointed out earlier, OSI-RM defines seven layers for local area networks. But if we divide the data link layer of OSI-RM into two sublayers, **media access control (MAC)** and **logical link control (LLC)**, then this reference model has eight layers and is known as the OSI reference model (OSI-RM) for LANs (defined by the IEEE 802 Committee). One of the main reasons for dividing the data link layer into two sublayers is to provide support for different transmission media and offer different data rates. It also allows control accesses to shared channels among autonomous computers and provides decentralized access schemes. Further, it offers a more compatible interface with WANs, since the LLC is a part of the HDLC superset, as it also defines a control field similar to that of HDLC.[10,11]

Media access control (MAC) sublayer: The MAC sublayer is mainly concerned with media access strategies and is different for different LANs. It supports different types of transmission media at different data rates. The MAC sublayer accepts data from the LLC, calculates a cyclic redundancy check (CRC), and then transmits the encapsulated frames. The transmitting media access management procedure sends the serial bit stream to the physical layer, which listens to the media. It halts the transmission if it senses a collision and tries to re-transmit it, and at the same time it sends a jam signal over the media. It inserts PAD fields if the LLC field length is less than the minimum length. When it receives a frame on another node, it first computes the CRC and accepts the frame after its address (DA) matches with the one defined in the frame and sends it to LLC. The receive media access management receives the serial bit stream from the physical layer and discards it if it is an invalid frame (if the length of the frame is less than the minimum length).

Logical link control (LLC) sublayer: The LLC provides an interface between the network layer and MAC and offers a common service access point (SAP) to all the MAC sublayer protocols. All the MAC sublayer protocols are supported by a common LLC sublayer standard protocol IEEE 802.2. This standard (LLC) corresponds to the upper sublayer of the data link layer of OSI-RM. It uses asynchronous balanced mode and supports both connectionless services by using an unnumbered information frame and connection-oriented services by using two unnumbered frames. It also allows the multiplexing of frames via its DSAPs. The data link layer, in fact, provides a communication architecture and, as such, this may be considered the top sublayer of the communication architecture of LANs.

It defines the control procedures used in both connection-oriented and connectionless services which can be used for any media access control (MAC) (the lower sublayer of the data link layer of OSI-RM and FDDI). It provides a platform for exchanging the data between LLC users over different MAC-based links (like bus/tree, ring topologies).

The LLC sublayer interacts with its MAC sublayer via its primitives and parameters (request, indication) for transmitting its protocol data unit (PDU). The physical layer boundary of each of the standard LANs defines its own PhSAPs between the physical layer and MAC sublayer. Similarly, the MSAPs are defined between MAC and LLC sublayers, while the LSAPs are defined between the LLC and network layers, as shown in Figure 11.2.

This standard is common to all LANs and offers three types of services: unacknowledged connectionless, acknowledged connectionless, and connection-oriented. It supports three classes of transmission configurations: point-to-point, point-to-group, or broadcast. In connectionless services, the datagrams are transmitted over the network without setting up any logical connection. In contrast to this, the connection-oriented services require three phases: connection establishment, data transfer, and connection de-establishment or connection termination. The class of transmission configuration is supported and established by the data link layer. Some of the material in this section is partially derived from References 2–6.

As mentioned above, the LLC is common to all LANs, while each LAN has its own MAC which depends on the access control procedure being used. Each of the LANs is standardized on the basis of its MAC protocol. Different transmission media offer different characteristics and data rates for different distances. For example, twisted wire offers a data rate of 50–60 Kbps over a distance of 1 km. Coaxial cable (baseband) offers a data rate of 50 Mbps over a distance of 1–3 km, while its broadband version offers a data rate of 350 Mbps over a distance of 10 km. Fiber-optic cable offers 1 Gbps over a distance of 10 km. Here the distances mentioned for the transmission media represent one segment without any repeater or amplifier or any intermediate nodes.

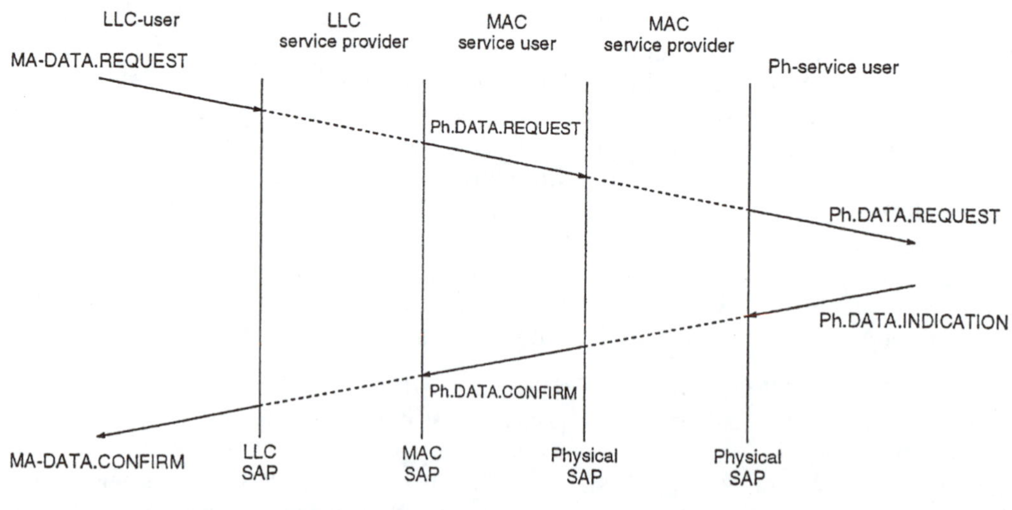

Figure 11.2 Service access points in the MAC and LLC layers.

11.4.2 Data link service and protocol data units (DSDUs and DPDUs)

The entities in a layer are known as service providers (DS-provider) and the implementation of services used by its higher layer is known as service user (DS-user). The data link service access point (DSAP) is defined by a unique hardware or port or socket address (software) through which the network layer accesses the services offered by it. Also, the data link layer fetches the message from its network layer through these DSAPs. A data link service data unit (DSDU) is defined as message information across the network to peer entities of all the layers. The DSDU associated with its primitives is delivered to the network layer and, similarly, the network layer uses its own primitives to interact with the data link layer. The DSDU and control information (addressing, error control, etc.) is known as interface data unit (IDU), which is basically message information sent by the network layer entity to its entity DSAPs.

Each node (PCs, terminals, etc.) connected to a LAN must contain identical protocol layers for data communication. Peer data link protocol layers exchange messages which have commonly understood formats known as **data link protocol data units (DPDUs)**. These also correspond to the units of data in the data link layer peer-to-peer protocol. The peer data link layer entities communicate with each other using the services provided by the lower layer (physical). The network layer sends its **network protocol data unit (NPDU)** to the data link layer. The NPDU consists of its protocol, control information, and its user data. The network control information is exchanged between network layer entities. Each message is associated with control information defined by an appropriate layer as a header and sent as an NPDU. This NPDU is termed a *packet*. The data link layer is the only layer in OSI-RM which adds a data link header at the beginning of the packet and a data link trailer at the end of the packet. The packet so encapsulated by the data link header and trailer is termed a *frame*. It is this frame which is passed to the physical layer for its onward transmission over transmission media. We will use herein the terms defined by ISO and used also in other publications.[2-6]

The communication control procedures are a set of rules and procedures for controlling and managing the interaction between users via their computers, terminals, and workstations across the network. The control procedures are known by different names in the industry, such as protocols, line control procedures, discipline procedures, and data link control procedures. The protocols provide a communication logical link between two distant users across the network, and the handshaking procedures provide call establishment and other control procedures within the protocols.

The physical layer provides an interface between transmission media, and the data link layer accession technique composed of access cable, a physical media attachment, and media is used to access the physical layer protocol data units. Each layer passes message information (data and control) starting from the top layer (application) to its lower layer until the message information reaches the lowest layer (physical) which provides actual data communication. An interface is defined between layers which supports the primitive operations and services that the lower layer offers to its higher layer. Every transmission medium has its own media attachment, but it provides the same services to the physical layer.

Recall that the data link protocols maintain virtual connections for sending and receiving the data over it. The virtual connection is basically a logical connection defined by the data link protocol over the existing transmission media and may support different data rates. One of most important services offered by the data link protocol is to provide and control access to transmission media, as this resource is being shared by all the connected stations, with a restriction that only one station can use it at any time.

11.4.3 Data link control protocols

The physical media carry the data between computers across the network. Various issues related to data communication such as data sequencing, synchronization, and data integrity between transmitter and receiver have to be dealt with appropriately in the presence of errors. The data link protocols which deal with these issues not only provide the proper sequencing, synchronization, and data integrity between computers across the network, but they also handle these issues between computers and terminals within the network. They also provide orderly and efficient communication between them. The physical link between computers carries both data and control signals and, as such, the data link protocol must be able to distinguish between the data and control fields within the user message code.

ISO functions of protocols: In order to understand the workings of protocols for the data link layer, we first identify various functions to be performed by the data link protocols defined by ISO.[10,11] The following paragraphs will discuss each of these functions in brief to get a feel of their importance and also to describe how each of these functions must be implemented. Some of the discussions in the following paragraphs are derived from References 2–7.

Control and management of data transfer: There are many ways of controlling data transfer, but the following three methods of controlling the data transfer have been widely accepted:

1. Data description
2. Control description
3. Handshaking procedures

In the data description method, various fields, their positions, etc., are defined for a message packet or block (containing information to be transmitted). The packet or block must include control information for error control, flow control, sequencing, addressing, acknowledgment, etc. The blocks usually have either two or three fields within them, as shown in Figure 11.3(a,b).

In Figure 11.3(a), the control information field includes all the above-mentioned information in one place. It can also be placed at two different locations around text data as header and trailer (Figure 11.3(b)). The header field includes information regarding the boundaries of data, control messages, number of frames to be sent, acknowledgment, etc., while the trailer field includes only information regarding error control. The header field uses a set of control characters (from the ASCII table) which provide the boundary, length of frame, etc. These control characters may also be used in the data field (using a stuffing technique). The available data link protocols use the following control characters: SOH, STX, and ETX, where SOH is the start of the header, STX is the start of text, and ETX is the end of text.

The header also includes other control information such as addressing, block sequencing, control flags, and acknowledgment. The addressing of both source and destination

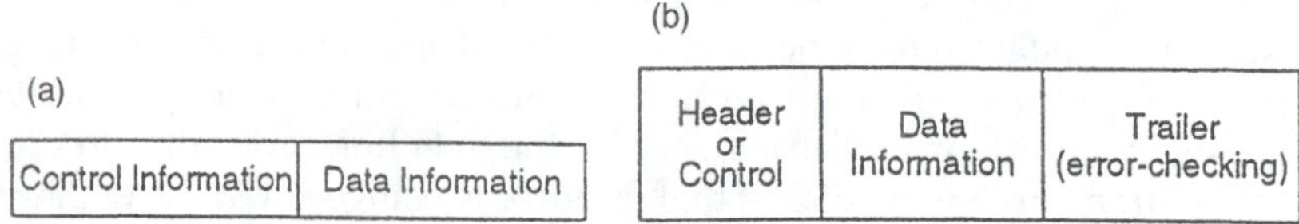

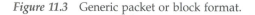

Figure 11.3 Generic packet or block format.

is also included in the packet. The sequencing for packets or blocks is concerned with the arrival of all the blocks in the same sequence as they were transmitted and also ensures that no blocks are lost or duplicated. The two most widely used codes for representing different combinations of characters, symbols, etc. are the American Standard Code for Information Interchange (ASCII) and the Extended Binary Coded Decimal Interchange Code (EBCDIC, defined by IBM). The CCITT defined its own code known as International Alphabet Number 5, IA5. The ISO has defined its own code as ISO 646.[18] CCITT's IA5 and ISO's 646 are both based on ASCII. ASCII uses seven bits for coding with one bit for parity and represents only 128 combinations or words, while EBCDIC uses eight bits for coding, representing 256 combinations.

The control flags define different types of the messages: data or non-data message, non-data message only, the location and numbers of various blocks in the sequence, the number of blocks in the message, etc. The acknowledgment control is concerned mainly with the handling of message blocks, keeping a record of source and destination addresses, transmission of positive acknowledgment (Ack), abortion of data transfer, informing of failure (hardware or software), etc. Each of these situations is a separate procedure, and together they are known as handshaking procedures. Some of the procedures may be used in error recovery, as well.

Error control: As we discussed earlier, data communication between two DTEs is based on serial transmission and it is very likely that error will be introduced into the data frames. This is a serious problem when the distance between these DTEs is large, as the data frame will include errors during its transmission and will be corrupted. The errors introduced will cause the voltage signals represented by binary logic to give different values. There are a number of schemes for error detection and error recovery which ensure the receipt of error-free data frames on the receiving side.

The simplest technique is **parity**, which is applicable to asynchronous transmission, as here each character is transmitted separately and extra bits can easily be attached with a bit combination of each character. On the receiving side, the similar parity technique is used to detect the parity bit. If the received parity bit and transmitted parity bit are the same, it indicates the absence of error. The presence of error is denoted by different parity values. In order to compute a parity bit for error detection, two techniques have been developed: even-parity and odd-parity. In the case of even parity, the parity bit is added at the end using modulo 2 such that the total number of 1s in the bit representation of characters (ASCII or EBCDIC) including parity is even. Similarly, odd parity requires the addition of a parity bit such that the number of bits should have an odd number of 1s. The implementation of these techniques can be realized by using simple hardware logic gates such as Exclusive-OR logic operations.

In **error checking and recovery**, various line control procedures include construction of an error-control algorithm, transmission of test signals, and re-transmission of lost or duplicated frames. Each of the procedures is defined as a special string of bits. The string of bits is usually termed **block check characters (BCC)** and is associated as a trailer field to the data information field. An appropriate algorithm (mutually agreed upon between the users) is executed to generate the string of bits and applied over the information message at the source node.

On the receiving side, the data information is checked for any error using different methods which usually depend on the coding technique and function involved. A number of error-detection techniques are being used in data link layers, and these techniques are defined as error-control procedures within the protocol. An error-checking technique which checks the parity bit of each of the characters in the data is known as **vertical redundancy checking (VRC)**.

Another error-checking technique which checks the block of data received at the receiving side is known as **longitudinal redundancy checking (LRC)** or **cyclic redundancy checking (CRC)**. This technique is well suited for burst errors and offers a reliable technique for error detection. It is based on polynomial codes which are basically used with the transmission of frames or blocks. The polynomial codes are used to define a single set of check bits (CRC) which is attached with every frame at its tail side. The value of this CRC depends on the contents of the frame. On receiving, a similar function is performed and compared with the transmitted check CRC bits. When both CRCs are the same, there is no error; otherwise, errors are present. The CRC is also known as a frequency check sequence (FCS), and 16 or 32 bits for calculating FCS are widely used.

In each of these error-checking techniques, after checking the data information, a positive acknowledgment (Ack) is sent back to the transmitting site if the data information is error-free; otherwise, a negative acknowledgment (Nak) is transmitted. The acknowledgment may either be defined as a special control character message (Ack or Nak) or may be sent in the "response field" defined in the header of the message information to be sent in opposite directions (based on piggybacking). After receiving Nak from the receiver, the message block containing error will be re-transmitted until it is received error-free at the receiving side, after which the buffer (at the transmitted side) containing the erred frame will be released.

VRC supports both parity types (even and odd) and the error checking is performed on each character separately. This requires a separate parity bit for each character that is treated as a check bit. If a character is represented by eight bits (ASCII), then the representation of the character is usually defined by seven bits while the last bit is considered as a parity bit. If the total number of 1s or 0s in the eight-bit character is even or odd, we have even or odd parity in our error-control procedure.

In the case of longitudinal redundancy checking (LRC), the entire message or block of data is checked for the error. An Exclusive-OR logic operation is performed on the entire message or block of data, and the resulting character, known as a block check character (BCC), is transmitted as the last character in the block or message (similar to a parity bit). On the receiving side, the same algorithm is executed independently and a BCC is computed. It compares its own BCC with the one it received. If these two BCCs are the same, it indicates the absence of error; if not, an error has occurred. In the event of an erred message or block, a Nak will be transmitted to the transmitting site.

It is quite obvious that by executing both LRC and VRC, the error detection options can be made more effective within the system. The CRC algorithm is more complex than LRC, as it involves polynomial division of the data stream by a CRC polynomial. The CRC polynomial and the coefficient of the dividend polynomial are defined by 1s and 0s of the message information. The division process of message information by a polynomial uses subtraction modulo 2 (ignoring carries) and the remainder is considered CRC. On the receiving side, the remainder is computed using its own algorithm. If it is the same as that of the received one, it indicates error-free receipt of the frame. But if these two remainders are not the same, there is an error in the frame.

Two of the most widely accepted and popular protocols for error checking and recovery are CRC-16 and CRC-CCITT (used in WANs). The former protocol uses a polynomial $X^{16} + X^{15} + X^2 + 1$, while the latter uses a polynomial of $X^{16} + X^{12} + X^5 + 1$. Both the protocols generate a 16-bit BCC. These protocols also support the checking of sequence error in a different way, and that depends on the type of protocol being used. The CRC-32 (used in LANs) is also used, and its polynomial is given by

$$X^{32} + X^{26} + X^{23} + X^{16} + X^{11} + X^{10} + X^8 + X^7 + X^5 + X^4 + X^2 + X + 1$$

The polynomial for CRC-16 is given by 1 1000 0000 0000 0101.

The Motorola universal polynomial generator (UPG) is commercially available as MC 8503.[21] This is used in serial digital rate handling systems for error detection and correction. The serial data stream is divided by a selected polynomial and the division remainder is transmitted at the end of the data stream as a cyclic redundancy check (CRC). On the receiving side, the same calculation is performed to see if the remainder is zero (for error-free) or not (erred frame). This UPG offers four common polynomial generators for error-detection techniques, has a typical data rate of 5 MHz, and can be used in floppy discs, cassettes, and data communication.

Available polynomial generators are:

CRCC-16 (forward) → $X^{16} + X^{15} + X^2 + 1$
CRCC-16 (backward) → $X^{16} + X^{14} + X + 1$
CRCC CCITT (forward) → $X^{16} + X^{12} + X^5 + 1$
CRCC CCITT (backward) → $X^{16} + X^{11} + X^4 + 1$
LRCC-16 → $X^{16} + 1$
LRCC-8 → $X^8 + 1$

Flow control is another function offered by data link protocols. When two DTEs are communicating with each other over public switched data networks or any other data network, due to the limited size of the buffer offered by these intermediate networks, the networks will accept the data message only within the limited space. It is also likely that the receiving side may not have enough space. In both of these cases, the data message is likely to be lost. Further, the loss of data frames is caused by different data rates of the transmitter, receiver, and underlying networks and, as a result of this, there might be traffic congestion, resulting in loss of data frames during the transmission.

Information coding: The formatting of message information in a proper, acceptable form requires information coding. The coding provides a uniform interpretation for code structure of bits, characters, and message syntax to define the information message in terms of characters and various control characters (required for data communication between users for exchanging information messages). There are different techniques of coding of character in the message based on the number of bits used for the characters. The characters may represent graphics, symbols, data information, control information, etc., and are typically used to control functions of computer terminals and the network. The data transfer between two DTEs normally takes place in a serial mode, whereby each character is transmitted either as a byte or a block of eight bits. The data transfer may be accomplished using either synchronous or asynchronous transmission, and, as such, the number of bits transmitted per character is also different. Both types of transmission (asynchronous and synchronous) carry the data serially.

It is interesting to note that, within DTE itself, the data and control transfer takes place in parallel mode (as discussed above). The data flowing within DTE in parallel mode has to be converted to serial mode before it can be transmitted over the channel or circuit. Similarly, on the receiving side, the data is received serially and has to be converted back into parallel mode before it can go through a DTE device. In both types of transmission (synchronous and asynchronous), synchronization must be provided at the bit, character, and frame levels, and different schemes of defining synchronizing control signals have been defined for each of these transmissions (discussed in the following section and also in Chapter 2).

The coding scheme defined by the U.S. Standards Institute and accepted as the U.S. Federal Standard technique for transmitting the binary data transparently is the American Standard Code for Information Interchange (ASCII). As we noted earlier, the ASCII code is an eight-bit code, out of which one bit is reserved for parity. In other words, the characters are effectively represented by seven bits. Special characters are reserved as non-data for dealing with control activities like synchronization, message headers, controlling, etc.

A different code based on parity is the **data interacting code (DIC)**, which is a variation of ASCII. It supports odd parity and uses different control characters (in particular, for printing).

As we noted earlier, a coding scheme similar to ASCII is the **Extended Binary Coded Decimal Interchange Code (EBCDIC)**, defined by IBM. It uses an eight-bit code for the characters, does not include any parity (in contrast to ASCII), and can represent 256 characters. This code was defined mainly for IBM mainframes and IBM-based machines.

The encoding scheme, including timing control, establishes different voltage signals for binary logics 1 and 0. Each bit is represented by suitable timing signals, also known as *bipolar encoding*. In the simplest encoding scheme, the logics can be represented by positive or negative voltage, and in both cases the voltage representation of these logics starts from zero level and returns to zero level; this scheme is also known as **return-to-zero (RTZ)**. In this scheme, we typically consider three states: positive, negative, and zero. An encoding scheme in which only two states are defined is known as **non-return-to-zero (NRTZ)**. The Manchester and differential Manchester encoding schemes fall into this category. We discuss these techniques in brief here; for further details, please read Chapter 3.

In Manchester encoding, the logic 1 is represented by low-to-high transition, while 0 is represented by high-to-low transition, where transition is defined at the middle of a bit period. The middle-bit transition serves as a clock and also as data. In differential Manchester encoding, the middle-bit transition is used to provide a clocking signal only and does not represent data (in contrast to Manchester). The logic 1 is represented by the absence of transition at the beginning of the bit period, while logic 0 is represented by the presence of transition at the beginning of the bit period. In other words, the transition is still defined at the middle of the bit, but a transition at the beginning of the bit will occur only if the next bit to be encoded is 0. These encoded signals do not contain any DC components and hence can be coupled with the data link via a transformer.

Another encoding technique is based on the assumption that a constant and stable clock is available at the receiving node and is always synchronous with incoming bit signals. In this technique, the synchronization between the clocks of the transmitter and receiver is initiated at a regular interval to make sure that these two clocks are synchronous. The binary logic 1 is represented by no change in transmission, while 0 is represented by a change in the transmission. This type of scheme is known as non-return-to-zero-inverted (NRZI). This scheme is more popular in WANs, while Manchester and differential Manchester schemes are more popular with LANs. One of the main reasons for this is the distance coverage of the networks. Pseudoternary encoding scheme represents logic 1 by no line signal and logic 0 by positive or negative signal, alternating for successive zeros.

In order to detect errors in the encoding techniques, multiple levels are generally suggested, and, as such, we have a number of schemes defined for WANs, such as **alternate mark inversion (AMI)**, **bipolar with 8 zeros substitution (B8ZS)**, and **high density bipolar 3 (HDB3)**. The AMI scheme defines a three-level code for binary logic and the signal transition is initiated by positive binary (mark) logic 1. B8ZS is based on the concept of AMI, with the only difference being that here the encoding of 00B0VB0V is used to indicate a string of eight zeros. The B represents a normal transmission of opposite polarity, while V represents a violation transmission of the same polarity. HDB3 is more popular outside North America and is based on B8ZS. Here, the string of four consecutive zeros is represented by three zeros.

Link utilization: The link or physical circuit or channel established between users is managed and controlled by link control procedures and is usually affected by various parameters such as link usage, directional link usage, acknowledgment, and the number of nodes supported by the link.

The established link (physical) between two nodes may be used for data transmission in different ways, e.g., simplex, half-duplex, and full-duplex. In simplex operation, one node sends while the other receives. In half-duplex operation, the message information is transmitted in one direction at a time, alternately. Full-duplex operation allows simultaneous transmission in both directions between nodes. For controlling and managing the simultaneous transmission of messages between two nodes over a full-duplex physical link, a full-duplex link control procedure is needed.

As stated above, the message information is transmitted along with different types of control information (control bits, error-checking bits, sequencing bits, etc.). All of these bits of control information are known as *overhead information*. The link utilization may be defined as the ratio of useful data message information packets to control information, as control information simply carries the data message information packets from one node to another. It is quite obvious that if we have more control information bits, the link utilization will be lowered. The transparency of information and communication may also affect the link utilization and efficiency.

Another parameter which affects link utilization is the way the acknowledgment facility is implemented. Each acknowledgment (Ack) is considered a regular message information packet and, as such, it also includes extra overhead (control information). The main objective for acknowledgment implementation in the line control procedure must be to reduce the overhead attached with the Acks. There may be a number of ways for reducing the overhead caused by Acks, such as avoidance of transmission of a separate Ack for each message information packet, few characters in control information, one acknowledgment packet for multiple blocks of message information, and piggybacking.

The line control protocols support both point-to-point and multi-point or multi-drop configurations, and the number of stations can be increased if the link utilization seems to be lower than the minimum or acceptable level. This lower-link utilization is defined for message blocks including control information.

Synchronization: In order to have proper data communication between nodes, some form of proper coordination or cooperation is required to provide synchronization between transmitter and receiver for message information blocks and control blocks. This is obtained by introducing a unique bit stream of bits before the message information blocks. This unique bit stream is termed *synchronization sequence bits* or *sync*. The receiver looks for this sequence and, after recognizing it, initializes various flags, counters, ports, etc., and ensures that it is properly synchronized with the transmitter.

The synchronizing sequence, in fact, is not produced with the message block but is usually supported by the transmission coding scheme and the number of synchronizing characters used. Due to these reasons, we have different types of data link control procedures and protocols. In coding schemes where all the combinations of symbols are used, an extra synchronizing sequence will be required.

Transparency for information communication: Users should not feel a difference when working on different types of channels (serial asynchronous, serial synchronous, or parallel). This can be achieved by using the same protocol for these channel configurations.

The line control procedure may include the bootstrap program as a part of it, or it may be embedded in the text field of the message information to provide transparency for information communication. The same set of link procedure controls and protocols can be used for these configurations.

11.4.4 Types of data link control protocols

Depending on the type of transmission, there are two types of data link controls:

- Asynchronous data link controls/protocols
- Synchronous data link controls/protocols

11.4.4.1 Asynchronous data link control protocols

In asynchronous transmission of signals, start and stop bits are sent with each character at the beginning and end of the code, respectively. The start bit (space polarity, i.e., logic 0) which precedes the data is detected by a process known as start bit detection at the receiving site. Various procedures are initialized such as synchronization, counting the number of frames received, receiving of data, the type of demodulation, sampling, etc. In an idle situation where no node is transmitting any character, a marking polarity of logic 1 is continuously flowing over the link (idle condition). When the receiver receives the data, it stores it in temporary storage (buffer) and later transfers it to the appropriate DTE memory for notification to the user. Similarly, a stop bit (usually 1 or 1½ bit duration) of marking polarity (logic 1) is detected by the stop bit detection process and indicates to the receiver the end limit of the data. Following this stop bit (logic 1), the link between the two goes to idle state (binary 1). Whenever the idle line changes its state from 1 to 0, this is an indication to the receiver that data is coming. In this method of transmission, there is no timing signal, as it is a part of the data (start/stop signal); hence, no synchronous bytes (signals) flow between sender and receiver. In general, all low-speed terminals and the majority of personal computers and teletypes or teleprinters are based on this mode of transmission.

11.4.4.2 Synchronous data link control protocols

In synchronous transmission, no extra bits (start/stop) are used with the data; instead, a defined bit pattern for preamble and postamble with the entire data is used (as opposed to start/stop bits with each character of data). These bit patterns are usually one byte (eight bits) each and are known as *synchronizing bits* (Syn). Preamble and postamble bits define the boundary of data to the receiver and also provide synchronization between sender and receiver. This synchronization is provided at two different layers: clock synchronization at the physical layer and identification of type of control, data, user's address, etc., at the data link layer. There are different types of synchronous data link protocols such as character-oriented (or byte-oriented), count-oriented (block-oriented), and bit-oriented. These protocols are discussed below. Some of the following information may also be found in References 3 and 5–8.

Character (byte)–oriented control protocols: This is one of the earliest link controls; it was developed in the early 1960s and is still in use. In fact, a very popular link control family known as binary control synchronous controls (BISYNC) is based on the character-oriented framing technique.

The character-oriented technique is code-dependent, wherein frames use a specific code (e.g., ASCII, EBCDIC) for making a distinction between data and control characters. This forces the use of the same code at both sites. In order to use different codes at different sites, a code-conversion process has to be used at one of the sites. In this technique, the boundary of a frame is defined by special ASCII sequence characters (*DLE, STX* and *DLE, ETX*) at the beginning and end of the frame (just like start/stop), respectively. The functions of these characters are given below.

DLE STX A STX B C DLE DLE STX

(a) Original data.

DLE DLE STX STX A STX STX B C DLE DLE DLE STX

(b) Stuffed data transmitted.

DLE STX A STX B C DLE STX

(c) Unstuffed data passed to network layer at the receiver.

Figure 11.4 Character stuffing.

DLE → data link escape (used for code transparency)
STX → start of text (defines text mode on the line)
ETX → end of text

Consider a message composed of A B C characters to be transmitted. This message will be sent as

DLE STX A B C DLE ETX

If, at any time, the receiver is not in synchronization with the sender (due to any transmission error or any hardware malfunction), the protocol looks at the sequence DLE STX and DLE ETX to recognize the boundary of the frame. This method of framing has a serious problem. What will happen if one of the control characters is part of the data? The easy solution is to add one more similar character to the data and transmit the control character within the data two times. On the receiving end, only one of these two will be considered as part of the data. This technique is known as *character stuffing*. The example in Figure 11.4 shows how the data is stuffed, transmitted, and then de-stuffed.

This version of the character-oriented synchronous data link protocol is popular and is being used in many systems. Although various vendor products based on this are still on the market, this technique is useful for a fixed-size code like ASCII or EBCDIC.

Count (block)–oriented control protocols: In this protocol, a frame has a header field which defines the number of characters in the frame. The header, in one way, defines the boundary of the frame, as it defines the starting point for the frame and also defines the number of characters in the frame. This field also precedes the frame. This technique is very simple but causes the receiver to be out of synchronization with the sender in the event of any transmission error, hardware error, or any other error in the header, as it will change the contents of the header. Another serious problem with this method occurs when the receiver recognizes that an error has occurred (checksum series error: check algorithm gave the wrong value from the one transmitted using the same algorithm at the sender), yet the receiver is unable to detect the starting point of the frame and, hence, cannot inform the sender about the exact starting point of the particular frame(s) for its re-transmission. In other words, in this situation, synchronization between sender and receiver has to be re-established before any data can be sent.

Bit-oriented control protocols: As discussed above, the count-oriented protocols can be used only for a fixed size of code (i.e., has a fixed number of bits or characters). Bit-oriented protocols can be used only for a fixed size of code, but with a variable length of data within the code. Bit-oriented protocols do not depend on the fixed-size codes but allow a variable length of code; i.e., data frames can have an arbitrary number of bits and the characters can also have an arbitrary number of bits. A special bit pattern of 01111110 is attached at the beginning and end of each frame. This bit pattern defines the boundary of the frames. In this method, we may have the same problem as character-oriented protocols; i.e., this bit pattern may also be part of the data. In order to eliminate this, we use the *bit stuffing* technique (similar to character stuffing) on the transmitting side using the following rules.

Data link layer protocols insert a 0 bit after every five consecutive 1s in the data (excluding the bit patterns at the beginning and end of the frame). Consider the following data to be transmitted:

0111110110

This data will be contained in the frame as shown below:

01111110 0111110110 01111110

Using bit stuffing, the transmitted frame will look like this:

01111110 01111100110 01111110
↑
bit stuffing

On the receiving side, the data link layer protocol checks the bit after every five consecutive 1s. If the bit is a 0, it deletes it; i.e., it unstuffs the bit and passes it on to the network layer.

This method has an advantage over character-counting protocols in terms of both code and data transparency. It is independent of the code being used for transmitting data; it only looks at this special bit pattern at the beginning and end of the frames.

The binary-oriented synchronous link protocols have been accepted as a powerful standard throughout the world, and different vendor products based on this protocol are available on the market and are being used in industry. IBM's product series 3270 and 3780 are based on bisync or binary synchronous control (BSC). We will discuss some of these protocols in detail at the end of this chapter.

11.4.5 Design considerations of data link control protocols

In using any of these synchronous data link protocols, the response from the receiving site is always defined/communicated in terms of two primitives: Acknowledgment (Ack) or Negative acknowledgment (Nak). If the transmitting site receives Ack, this indicates that the transmitted frame is received error-free. Nak indicates that the received frame has an error, which means that the transmitting site has to re-transmit the same frame to the receiving site. Although these protocols provide end-to-end multiple links for data communication, they don't provide end-to-end acknowledgments (or negative acknowledgments). This type of service is provided at different layers, which in turn depend on the systems. In OSI model, this service is provided by the network layer.

In order to provide the above-mentioned services by data link protocols, different timers are used in the protocols during their implementation. Basically, these timers define the state of the event occurring due to the initialization of a particular primitive within the layer. The value of each timer depends on a number of factors affecting that particular event and services offered by the layer. For example, when a frame is transmitted, a timer is initialized and puts that state into transmitting mode after it receives the acknowledgment from the receiving site. The value of the timer will show the total time taken to transmit the data with the acknowledgment. Another timer may be used for time out at the sender. This timer is set to a predefined value as soon as a frame is transmitted. If the sender does not get a response from the receiver within this predefined value, another (duplicate) frame is automatically transmitted, and this keeps going on until a response is received at the sender. We have to use another timer to determine how many times the frame should be re-transmitted (after the predefined value of the previous timer is expired). The predefined value of a timer depends on various factors and is calculated very carefully. The factors which affect the value are as follows: total round-trip propagation time (sender to receiver and back to sender), access time, processing time, transmission time, local and remote memory access time, etc.

11.5 Data link control configurations

Data link protocols can be used on different types of communication channels or links, topologies, configurations, etc., and are distinguished by the way these are implemented. Further, the implementation of protocols, depends on many issues such as the number of terminals/computers to be connected, cost of each connection, throughput, response time, priorities, management of communication link, etc. The following subsections will discuss some of the important issues which affect the implementation of protocols.

11.5.1 Topology

If two sites are connected directly by one link, then this link provides point-to-point communication. If more than two sites are connected by one link, then this link provides multi-point communication. The standard computer/terminal connection can be obtained by using point-to-point or multi-point communication. Point-to-point communication between computer and each terminal requires an I/O socket or port and a separate link (see Figure 11.5(a)). The number of terminals which can be connected to a computer depends on the number of I/O ports present in that computer. If we use multi-point communication between a computer and terminals, then we require only one link and one I/O port. All the terminals are connected to the computer via this link, and they use the same link to send the data (see Figure 11.5(b)).

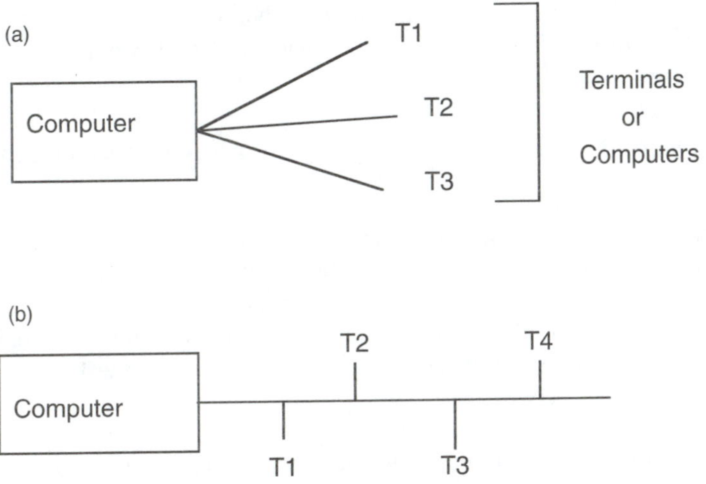

Figure 11.5 Types of connection configurations. (a) Point-to-point. (b) Multi-point or broadcast.

The communication control protocol between a computer (host) and terminals includes the following three steps:

1. **Establishment of link:** A logical link is established between two sites; i.e., their addresses are determined and both sites are ready for transmission. The sender site is ready to transmit the data while the receiver site is ready to receive the data.
2. **Transfer of data:** The transfer of data takes place between the sender and the receiver and, depending on the configuration and type of protocols, acknowledgment or negative acknowledgment will be sent by the receiver to the sender.
3. **De-establishment of link:** The logical link between sender and receiver is de-established or terminated and no further transfer of data takes place between them.

These three steps are found in one way or another in all the protocols being used in different configurations, transmission modes, or communication media.

11.5.2 *Transmission line configuration*

Point-to-point link connection: Different line configurations (simplex, half-duplex, and full-duplex) can be used for the types of connection configurations shown in Figure 11.5. In simplex configuration, the signal flows in only one direction. This is useful in applications such as I/O communication, where an input device may send data to a host and the data from the host can be sent to an output device. Simplex will not be useful for applications where an Ack or Nak is required (Figure 11.6(a)).

A half-duplex configuration requires one pair of lines, while full-duplex requires two pairs of lines. These configurations can be used in both analog and digital signaling. The type of configuration for each type of connection depends on the situation and requirement; e.g., in analog signaling, the frequency determines the type of configuration (Figure 11.6(b)). It allows the transmission of signals in both directions at different times. When one site is transmitting, the other site has to receive; after the transmission is over, the statuses of these sites can be interchanged. Both sites cannot transmit or receive simultaneously. A full-duplex configuration, on the other hand, allows transmission of signals in both directions simultaneously (Figure 11.6(c)).

Multi-point link connection: Multi-point link connections for various configurations are shown in Figure 11.7. In this category of connection, we can have either multiple sender single receiver (MSSR) or single sender multiple receiver (SSMR). In the case of half-duplex configuration, the flow of information will take place in one direction at any time between sender and receiver.

11.5.3 *Implementation and management of data link control configurations*

11.5.3.1 *Types of nodes*

ISO 7809[13] discusses the types of nodes, link configurations, and data transfer mode of operations of HDLC which, in most cases, satisfy a variety of requirements of different users. The data exchange between nodes (PCs, terminals, etc.) takes place over the transmission media, and the data link protocol controls various operations during data communication. There are different types of nodes which can be connected in different types of modes for various protocols.

- **Primary Node:** This node controls the operations of a data link and sends the command frames to all of its secondary nodes. It also receives frames (response)

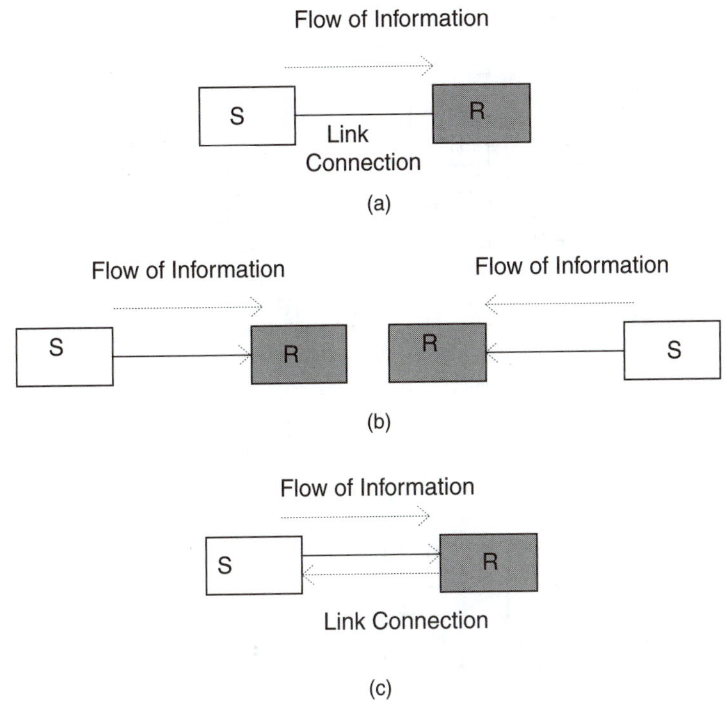

Figure 11.6 Flow of information in a ,point-to-point link. (a) Simplex configuration. (b) Half-duplex configuration. (c) Full-duplex configuration.

from the secondary node(s) and maintains a list of session(s) or logical link(s) with each node(s) attached to the link.

- **Secondary Node:** This node sends a response to the primary node and is not concerned with the controlling of link(s). One secondary node cannot communicate directly with another secondary node; they have to go through the primary node.
- **Combined Node:** This node can send and receive both commands and responses from another node of the same type. This type of node performs the functions of both primary and secondary nodes.

11.5.3.2 *Protocol configurations*

The implementation and management of communication links (point-to-point or multi-point) for various configurations (half-duplex, full-duplex) can be obtained using two important protocols: primary/secondary and peer-to-peer control.

Primary/secondary control: In this technique, one site (usually the host computer) acts as a primary while the other sites (terminals, controllers, devices, etc.) act as secondaries, as shown in Figure 11.8. The primary site controls all the secondaries and can provide point-to-point communication or multi-point communication to them. The primary/secondary protocols can be implemented as pollable systems or non-pollable systems. In pollable systems, the primary polls and then selects a secondary; the transfer of data then takes place between the selected secondary and the primary. The processes *polling* and *selection* usually use Ack or Nak for establishing a link and can be implemented in a variety of ways. The polling process, in fact, broadcasts the message to all secondaries and, once the primary gets Acks or Naks from any one of the secondaries, it selects an appropriate secondary for the transfer of data. If the primary gets a large number of Naks, then the

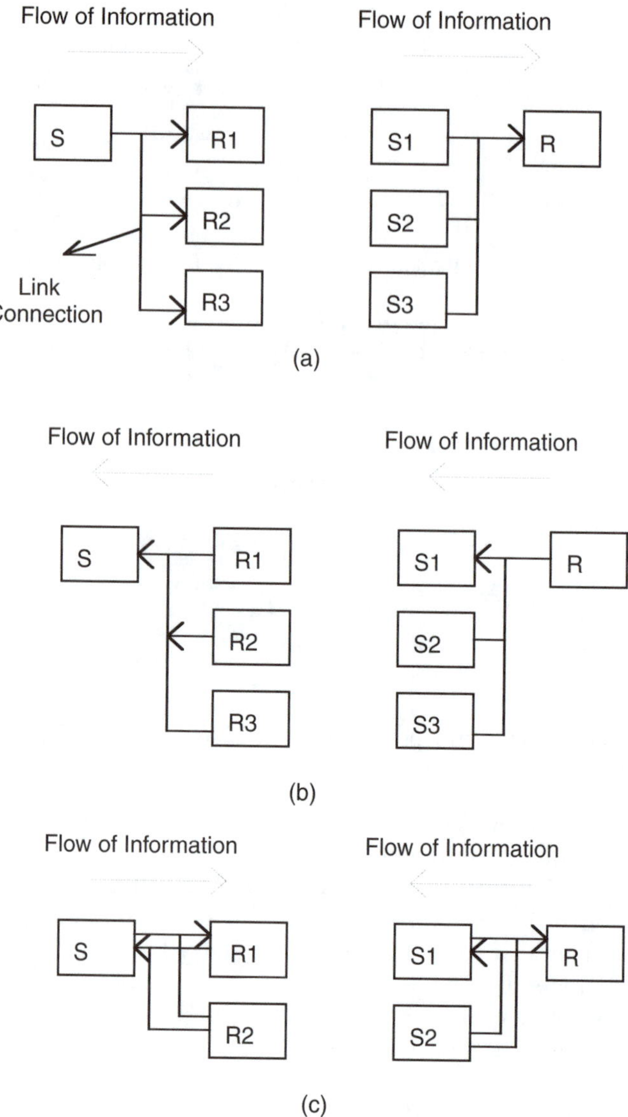

Figure 11.7 Multi-point link connections for different line configurations.

response time and throughput of the protocol is lowered, and also the overhead increases because the Naks use the resources.

In non-pollable systems, the polling and selection processes are carried out using a round-robin principle or scheme based on **time-division multiple access (TMDA)**. One site acts as a primary while the other sites act as secondaries and send requests to establish the link to the primary site. This request is contained in a special field of the frame sent out to the primary. The primary site assigns time slots to each of the secondaries (which wish to send the data), and during those time slots, the links are established between the primary and that secondary slot. The non-pollable system can be implemented in different manners, which in turn depend on the type of network, amount of traffic, configuration, and many other parameters.

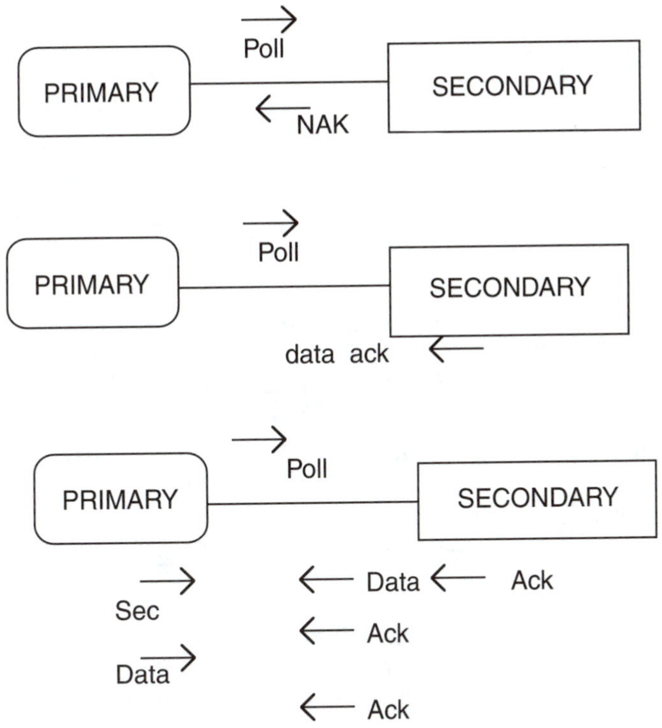

Figure 11.8 Primary/secondary control protocol.

Peer-to-peer control: In this scheme we don't have a primary/secondary relationship between the computer (host) and terminal (devices). All sites have equal access to the channel or link. Once the link is established between any two sites, a point-to-point communication exists between them. Similarly, the same channel can be used for multi-point communication (broadcast to all terminals connected to the channels). Peer-to-peer protocols can be used in different local area networks and depend on the characteristics of the networks; e.g., some networks provide priorities while some do not. Depending on this, these protocols can also be classified as priority-based or non-priority-based protocols. As we know, Ethernet LANs (IEEE 802.3) do not support priorities; hence, these protocols are known as non-priority peer-to-peer protocols (or contention protocols) and support both peer-to-peer and multi-point configurations. Token bus (IEEE 802.4) and token ring (IEEE 802.5) LANs support priorities and, hence, these protocols are known as **priority peer-to-peer protocols**.

While using these protocols, we have to provide the addresses of the sender and receiver. For point-to-point communications, in some cases, we have to provide the address of the receiver, while in other cases we don't have to provide the address, as establishment of the link can take place only between those sites.

In the case of a primary/secondary system, we have to provide the address of the receiver (in both pollable and non-pollable protocols). The distinction between these lies in the process of selecting the receiver site. Both the addresses (sender and receiver) have to be provided in the case of peer-to-peer communication protocols (for both priority-based and non-priority-based protocols).

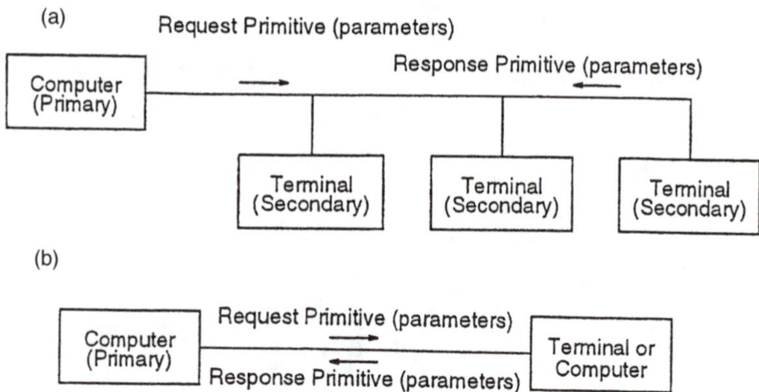

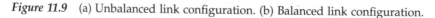

Figure 11.9 (a) Unbalanced link configuration. (b) Balanced link configuration.

11.5.3.3 Link configurations

The following three link configurations are used to configure primary, secondary, and combined nodes in HDLC.

Unbalanced configuration: This configuration supports one primary and one or more seondary nodes which may operate in either point-to-point or multi-point operation on both half-duplex and full-duplex connections. All the secondary nodes connected are controlled for link connection by one primary node (Figure 11.9(a)).

Balanced configuration: This configuration supports the connection between two combined nodes connected in point-to-point mode using half-duplex or full-duplex connection only. Each node has equal responsibility for sending frames and controlling the links between them. In some cases, these combined nodes may use their own commands (Figure 11.9(b)).

Symmetrical configuration: This configuration defines two nodes (primary and secondary) as independent point-to-point unbalanced configuration at each site. The primary node of one site controls the link operations for the secondary of a second site, while the primary of the second site controls the link operations of the secondary of the first site. The commands and responses from these nodes from each site are communicated on a single channel. This configuartion is rarely used, due to the cost and complex structure of the configuration.

The above-mentioned configurations have no resemblance to the attributes of the physical layer, since the physical layer provides unbalanced and balanced electrical behavior of transmission media. Although different layers may use the same attributes, in reality, the protocol for each of the layers has no knowledge about the attributes, as the link controls will have no way of detecting these.

11.5.3.4 Modes of operation

For the transfer of data, the communication between two sites uses one of the following modes of operation. The discussion of these modes is derived partially from References 3, 5, and 6.

Normal response mode (NRM): This mode is useful for multi-point communication. The primary site polls the connected sites and requests a secondary site to send the data. The secondary site will send a response (which may contain the data) to the primary site. After the last frame has been transmitted, the secondary site waits for another polling from the primary site. This mode of operation supports an unbalanced configuration. The connected sites may be be terminals, personal computers (PCs), controllers, peripheral device controllers, etc. This mode may also be useful for point-to-point communication where only one device is connected to the computer.

Asynchronous balance mode (ABM): This mode uses the polling and selection processes for the transfer of data and is useful for multi-point communication and not efficient for point-to-point communication. ABM mode allows combined nodes to transfer the data between them in a point-to-point communication mode of operation. The delay and overhead are reduced considerably, and combined nodes can start sending the data without waiting for the request from other nodes. This mode supports the balanced configuration and provides efficient use of point-to-point communication of the full-duplex transmission mode.

Asynchronous response mode (ARM): This mode allows secondary sites to send the frames without getting any request from the primary site; due to this, overhead is reduced. The frames may contain data and control information (status of secondary site). A secondary site can start transmission by detecting the idle state of the channel in half-duplex, or at any time in a full-duplex channel. The primary site, on the other hand, will still maintain link control such as setup, channel connection and disconnection, error recovery, and other related control information. This is a typical mode and is useful in hub polling or special situations where secondary sites may start transmission without being polled by the primary site.

This mode uses an unbalanced configuration. An asynchronous operation in ARM is not at all related to asynchronous transmission supported by the physical layer. The ARM operation represents the setting up of transmission between primary and secondary sites, while asynchronous transmission in the physical layer defines how the binary data can be transmitted on the transmission media with appropriate control functions (e.g., synchronization, flow control, etc.) included in the transmission. Each of the transmissions (from primary to secondary or vice-versa) is defined by the frame(s), and the frame(s) contains the data as well as the control information. The frames used in the data link layer define an entity of data (also known as a protocol data unit, PDU) between two sites.

11.6 Data link protocols

In order to understand the working of the protocols, we consider two situations: noise-free channel and noisy channel. In each of these situations, we have non-idealistic assumptions, and we will highlight the significance of each of the assumptions in the actual protocols in the following sections, with a view to understanding the possible situations which are useful during the implementation of data link procedures and protocols. This discussion is derived partially from the ideas used in References 2–6.

11.6.1 Noise-free link/channel protocol

We assume that a noise-free (error-free) channel is already established between the sender and the receiver. Data can be transmitted in only one direction, i.e., from sender to receiver. The receiver has unlimited space and the sender sends the data frames continuously on the channel. Since the channel is noise-free, we assume that the frames will be neither corrupted nor lost. We also assume that the frames will arrive at the receiver in the same order that they were transmitted by the sender and that they contain no errors. Without caring about performance, we can also assume that the time for processing at each site is

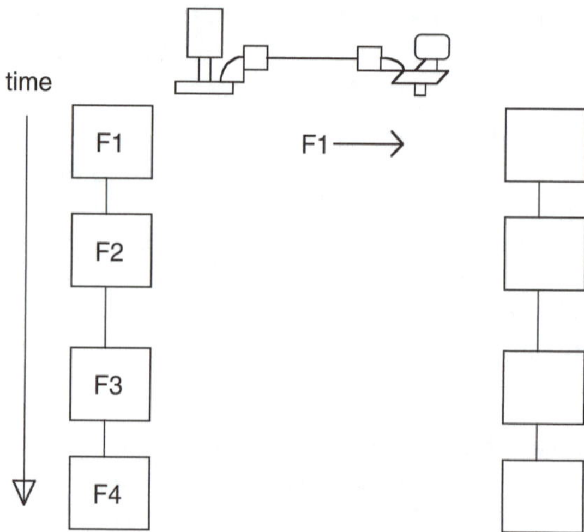

Figure 11.10 Frame transmission over a noise-free (error-free) channel.

not considered and that the delay between the transmitted and received frames is negligible. However, in a realistic situation, there will be a finite delay due to propagation, capacitance, inductance, etc., and the links between them.

On the sender's site, the data link layer gets a data packet from its network layer and constructs frames for it. These frames are then transmitted on noise-free channels to the other side, as depicted in Figure 11.10. Data link layer and network link layer protocols communicate with each other through a set of primitives and parameters (of the network layer), and the logical connection between them is established. During this process, the following three procedures are employed:

1. The data packets from the network layer are fetched by the data link layer at their data service access points (DSAPs), which are also known as **sockets** or **ports**.
2. Once the data link layer has gotten the data packets from the network layer, it makes a frame(s) for these packets which includes the header, trailer, checksum (error control), and destination and source addresses defined by the data link layer.
3. These frames are then handed over to the physical layer for transmission over the channel.

These three procedures are collectively known as the *sender process*, which is always in an infinite loop and executes these procedures in the same sequence. A suitable data structure for each of the procedures may be defined.

On the receiver side, we employ the same set of procedures as in the sender process, but with different functions. When a frame (also called a data packet) arrives at the receiving site, it is received by the physical layer and then is passed on to the data link layer. The data link layer removes the header and other control fields from the frame and passes it to the network layer. Thus, the receiver process executes these procedures in sequence and, after the execution of the last procedure (i.e., passing on the data packet to its network layer), it again waits for the arrival of the frame; i.e., the receiver process goes in an infinite loop. Both sender and receiver processes are in infinite loops, changing their original states to new states and then coming back to the original states after executing the process, and they will continue in the loop until the channel remains established.

We will now include more practical and realistic situations in our initial assumptions and configurations and then discuss different types of protocols to be considered to handle those typical situations.

11.6.2 Stop-and-wait link protocol

In this protocol, we assume that the channel is error-free; however, the sender and the receiver do not have infinite buffer space. In the previous protocol, there is no need for acknowledgments; the frames which are transmitted will definitely be received by the receiver, as it has infinite buffer space. However, in the stop-and-wait protocol, the receiver has a limited buffer space and may drop an incoming frame if the buffer is full. We use acknowledgments to control the flow of frames between the sender and the receiver.

When the receiver receives a frame, it sends an acknowledgment back to the sender informing it of the receipt of the frame (error-free). After the acknowledgment has been received by the sender, it sends another frame and waits for its acknowledgment. Thus, in this protocol only one frame can be sent on the channel at a time, as shown in Figure 11.11. The acknowledgment used here is usually a dummy frame or a few bits or a character in the frame.

Sender and receiver processes are similar to the ones employed in the previous protocol, with the only difference being that once the sender process has transmitted a frame, it is in the wait state, which will be changed to the state corresponding to the fetching of the new packet from its network layer only by an acknowledgment frame from the receiver site. The sender process is not concerned with the data in the acknowledgment frame. We assume here that acknowledgment frames and data frames are neither damaged nor lost during transmission.

If we use the stop-and-wait protocol in a noisy channel (a more realistic situation), it poses the problems of an acknowledgment frame. Consider a situation where a frame is sent out and the receiver calculates the checksum value, finds everything is all right, and accordingly sends an acknowledgment back to the sender. Let's consider the situation where the acknowledgment is lost (due to hardware failure or a broken link or some other reason). If the receiver receives a damaged frame, it will simply reject it and not send any response. The sender, after getting no response from the receiver, will ultimately time out, send the same frame again, and wait for the acknowledgment. This process of sending the same frame to the receiver is continued until the receiver gets an error-free frame.

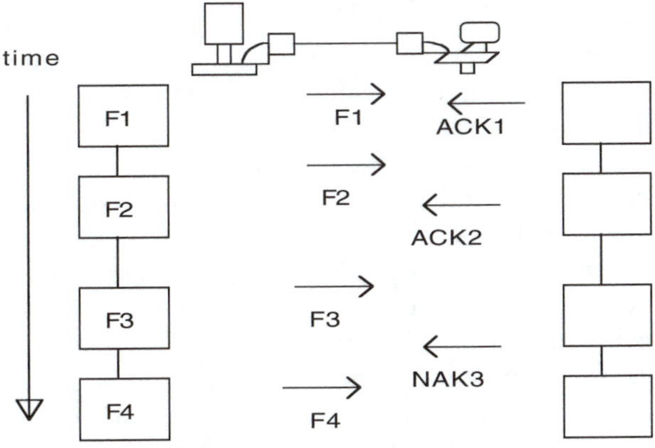

Figure 11.11 Flow control in the stop-and-wait protocol.

Obviously, this poses many problems, of which two are serious and are addressed below with a possible way of handling them.

1. The sender sends a frame to the receiver, does not get acknowledgment, and does not know whether the receiver received a damaged frame or received the frame correctly but the acknowledgment is lost. In this case, the problem is how many times should the sender send the same frame to the receiver site?
2. The receiver sends an acknowledgment to the sender and the sender, not getting any response (acknowledgment is lost) from the receiver site, times out and sends the same frame again. The receiver site is surprised because it knows that it has already sent the acknowledgment and, further, it has to decide which version of the frame will be selected for passing on the corresponding frame to its network layer.

These two problems can be overcome if we use a sequence number with each frame; also, the acknowledgment should include the sequence number of the received frame. Thus, when the frames are transmitted from the sender site to the receiver site, the receiver can recognize whether this is a new frame or a duplicate frame. The following question arises: what should be the length of the sequence number? That is, how many bits should be used to define the sequence number? In many similar situations, a somewhat similar problem may arise. For example, assume that the receiver, after receiving a particular frame, sends an acknowledgment which becomes lost, then receives another frame and sends another acknowledgment which is received by the sender. Now the sender looks at this number and may interpret the receipt of the earlier frame, as well (which is not true). This problem can be solved if we consider a one-bit sequence number 0 or 1 for the frames. A frame with sequence number 0 or 1 is sent to the receiver site, which expects the frame with sequence number 0 or 1. After the receiver gets the frame with the correct sequence number, it changes the sequence number to 1 or 0 using modulo 2. Now the receiver is expecting the frame with the sequence number 1 or 0. If at any time the receiver does not get the frame with the correct sequence, it will simply reject the frame.

This protocol requires an acknowledgment from the receiver before the sender can transmit the next frame. As indicated earlier, we may use two types of acknowledgment, positive acknowledgment (Ack) and negative acknowledgment (Nak). If the sender receives Ack, it sends the next frame, but if it receives Nak, it re-transmits the same frame (bearing the same frame sequence number). This allows the data to travel in only one direction. In the event of a lost acknowledgment or frame, the sender, after timing out, re-transmits the same frame. The time-out interval or duration associated with the sender's transmitter depends on many factors, e.g., propagation time (round-trip), access time, processing time, transmission time, etc., and should be long enough to avoid the transmission of duplicate frames (worst-case value).

The initialization of the protocols at sender and receiver does not take place exactly at the same time before each of them goes into an infinite loop. In other words, there is always some finite delay in the initialization process, as the receiver protocol depends on the occurrence of an event at the sender site. When the sender transmits a frame, it goes to wait state, and the event corresponding to either Ack or Nak will change the state of the sender's protocol, which will be executed accordingly (depending on Ack or Nak). On the receiving site, it can receive either a damaged frame or duplicate frame.

In any event, the data link layer will pass only the correct frame to its network layer. A lost or duplicate frame will not be passed on to the network layer. The above discussion is partially derived from References 2–5.

11.6.3 Piggybacking link protocol

In the previous protocols, we have assumed that the actual data frames are transmitted in one direction, while acknowledgment (dummy frame) is sent on the same circuit in the opposite direction. In real situations, we would like to transmit the data in both directions. There are a number of ways of implementing this scheme. One way is to use a full-duplex configuration, defining one simplex channel for each direction. Actual data is sent on one channel, while the acknowledgment data is received on another channel. This technique seems to be simple, but it does not fully utilize the channel capacity, particularly the second channel, which carries the acknowledgment.

Another way of transmitting the data in both directions is to use the same channels for sending data and acknowledgment on the same frame. We can define a special field known as a kind of header in the frame. This field can differentiate between data and acknowledgment; if both sites are having the same capacity, then protocols based on these concepts can provide a better utilization of channel capacity.

The above technique can be improved using the piggybacking concept. In this technique, we don't send a separate frame for the acknowledgment; instead, we attach the acknowledgment to the data frames to be transmitted from the receiving site. This does not require the transmission of a separate acknowledgment frame. Similarly, an acknowledgment from the sending site can be attached with the data frames to be transmitted toward the receiving site. The acknowledgment can be sent only if the data link layer gets packets from its network layer (Figure 11.12).

What happens if there are no frames to be transmitted at a site? In such a case, the protocol times out and sends an acknowledgment frame. Thus, this piggybacked link protocol offers advantages in terms of sending the data frame together in one frame. It also reduces the cost of transmission. The major drawback with piggybacked link protocol is that if the receiver (or sender) has no data frames to be transmitted, then the acknowledgment frames have to either wait until they fetch the packets from their respective network layer or transmit the acknowledgment frame separately. This obviously makes the protocol inefficient.

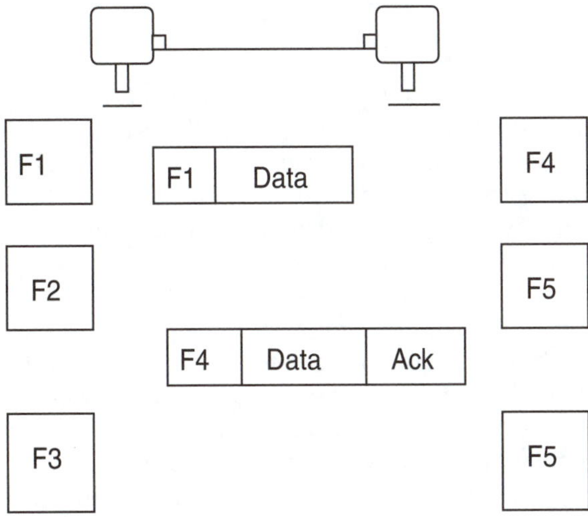

Figure 11.12 Piggybacked acknowledgment.

11.6.4 Window link protocol

Although the previous protocol allows the transmission of data and acknowledgment frames in both directions, the common point in them remains the same; i.e., only one frame can be in the channel at a time. Further, a message is encoded into a number of frames and the network usually has a fixed-size frame (e.g., ARPANET provides a frame with a maximum of 8063 bytes). If the bit length of the channel is greater than the maximum size of the frame, then the transmission and resource utilization will be highly inefficient. Thus, the techniques may not be useful in a satellite communication system, as it will limit the throughput and the transmission rate (due to a higher propagation delay in the satellite link).

In order to improve the efficiency of the protocol, we can send more than one frame over the channel at any time. The concept of a *window* allows more than one frame to be on the channel at a time. In this method, we assume that both the sending and receiving sites have sufficient buffer space to accommodate, say, m frames. The sending site transmits m frames at a time toward the receiving site without waiting for an acknowledgment. The receiving site, after receiving the frames, sends the acknowledgment by sending the next expected sequence number of the frame to the sender. This type of acknowledgment to sender indicates not only the receipt of a frame with the previous sequence-numbered frame but also that the receiving site is ready for the next sequence number in m frames. For example, if we consider $m = 10$, then the sender transmits frames with sequence numbers 0, 1, 2, …, 10 to the receiving site, which does not send the acknowledgment for each frame but, instead, sends the acknowledgment with the expected sequence number 11. This confirms the receipt of all previous frames 0 to 10, and the receiving site is ready to accept another m frames.

Both sending and receiving sites maintain a list indicating the allowed sequence number and expected sequence number, respectively. Intuitively speaking, each list details the range of sequence numbers that is defined as a window of the frames; this technique, based on the window concept, is known as *sliding-window flow control* and provides a sliding-window protocol for flow control between two sites. The list maintained at each site determines the size or range of windows, and these are known as sender's and receiver's windows, respectively.

11.6.5 Sliding-window link protocol

The earlier protocols assumed only one frame could be sent or received at a time on/from the channel. In terms of windows, we can say that the window size of both the sender and the receiver is one. As indicated above, the efficiency of transmission can be improved if we send more than one frame at a time without waiting for the acknowledgment. In this case, the size of the window will be more than one and can have a value which depends on the maximum size of frames that a network can support. A window size greater than one defines a new concept known as *pipelining*, where more frames can be transmitted at any time without waiting for acknowledgment. In this case, the number of frames can be equal to the maximum number of frames a network can handle; this helps in increasing the transmission rate on the network. The acknowledgment from the receiving site will be sent only after these frames, corresponding to the sender's window size, have been received. Similarly, the receiver's window size may also have any value higher than one and depends on the protocol being used, as shown in Figure 11.13.

Both sender and receiver maintain a list of the frame numbers. The sender's list contains information such as the outgoing frame number Ack or Nak and the movement of the pointer to the next expected frame number after it receives Ack (otherwise, the

A sends frames 0 and 1 to B

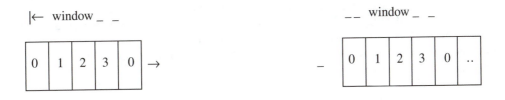

B recieves both frames 0 and 1 and updates its window

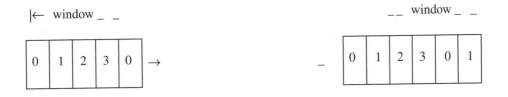

After receiving acknowledgement for frame no.1, A also updates its window.

Figure 11.13 Pipelined sliding-window protocols with sliding-window size of four.

pointer indicates the same frame number). Window size depends on the length of the frame number and should be considered carefully to avoid any overlapping. Further, the acknowledgment for the frames depends on the protocols being used: some protocols support Ack or Nak (in the event an error has occurred) after every frame received, while some protocols allow the acknowledgment after the last frame number (in mutually agreed upon sequence numbers), which by default will indicate to the sender that all the frames up to that number in the sequence have been received. In the latter case, if the sender receives a Nak, then all the frames starting with that faulty frame up to the last number of the frame in that sequence will be re-transmitted.

The size of the sender's and receiver's windows does not affect the delivery of the packets by the data link layer to its network layer; these data packets will be delivered to the network layer in proper order. It is assumed that the window size does not represent the number of frames that can be transmitted; it simply indicates the transmitting/receiving of the frames in a particular sequence. Further, initially the size for both windows is the same, but as frames are transmitted and received, the boundary of the windows also changes.

11.7 Error-control protocols

The above was a discussion of different types of data link layer control procedures for noise-free channels. We started with a non-realistic situation and slowly introduced realistic situations into it, and for each of these situations, various options for implementing these control procedures were discussed. In the following section, we extend further our discussion by considering a noisy channel and discuss various error-detection and error-correction strategies to be adapted within data link protocols. The sliding-window concept introduced for deriving various error-detection strategies has been derived partially from References 2, 3, 5, 6, and 8.

11.7.1 Sliding-window error control (automatic repeat request (ARQ)) protocols

This section discusses protocols based on the sliding-window error control concept. These protocols use the following parameters: frame transmission positive acknowledgment (Ack), negative acknowledgment (Nak), and re-transmission of frames in the event of either Nak or time out. Ack here represents the situation where the frame has been received at the receiving site error-free, while Nak represents the receipt of the frame with error and, hence, the frame is to be re-transmitted. A timer at the transmitting site intializes the transmitting process to re-transmission of the frame after the elapse of the preset value of time. The protocol providing these parameters is known as an **automatic repeat request (ARQ)** protocol. The following sections describe three versions of ARQ protocols.

1-bit sliding window or stop-and-wait ARQ protocol: This protocol is the same as the stop-and-wait protocol in the sense that the sending site transmits one frame and then waits for either Ack or Nak. Depending on the parameters discussed above, an appropriate action will be taken. The maximum size of the window of both the sender and receiver is 1. In normal operation, a sender transmits a frame and waits for a response (Ack or Nak) from the receiver (Figure 11.14). If it receives Ack, it transmits the next frame and again goes to the waiting state for the response. If it receives Nak, it will re-transmit the same frame until it receives Ack.

In a typical situation, if the receiving site, after receiving a frame, transmits Ack and it gets lost, the sending site, not having received either Ack or Nak, times out (after a preset value in the timer) and transmits the same frame until it receives Ack/Nak. This will cause problems to the receiver, as it has received duplicate frames. This problem is solved by using 0s and 1s for numbering the frame. When a frame with a number 0 is received at the receiving site, the receiver sends Ack with 1 indicating to the sender that a frame number 0 has been received correctly and now it is expecting the next frame with number 1. If it receives the frame with a number other than expected, it discards that frame. When both sites are attempting to transmit the frames back and forth simultaneously, only one site can transmit the frame at any time and, in the worst case, whenever

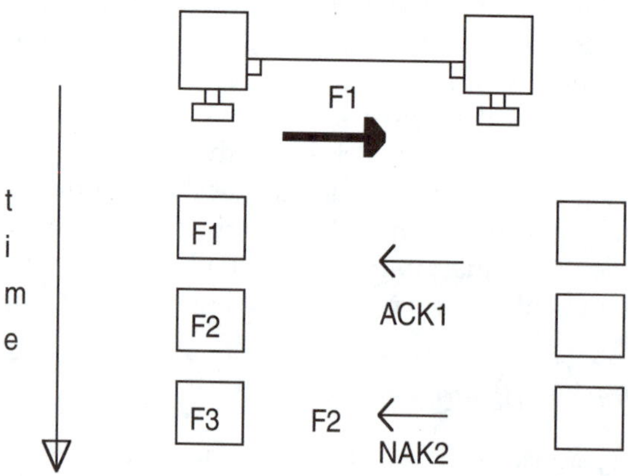

Figure 11.14 Stop-and-wait ARQ.

both sites are initialized to send the frames at the same time, then this protocol will provide duplicate frames on each site. As a result, the throughput of this protocol may not be more than 50% during heavy traffic of the data.

This scheme is simple to implement but very inefficient, particularly in situations where the delay time (propagation delay of the media, access time, transmission time and other delays associated with sites and communication media) is large. An example of this is found in satellite communication where during the time the frame is in transit, the sender process is in a waiting state and, further, while the receiver's Ack or Nak is in transit, both the receiver and its processes are idle. Thus, the utilization of these processes is reduced considerably.

Go-Back-N or Go-Back-N ARQ protocol: This protocol removes the restriction on the maximum size of the window (as was mentioned in the previous protocol) and allows a variable size of the window for both sender and receiver. As mentioned in the preceding section, the size of the window depends on many factors and should be just enough to give sufficient time for the frame to arrive at the receiving site and also for the Ack or Nak to come back to the sending site. Once the window size is determined/decided, then the sender can start transmitting the frames continuously up to the maximum window size without waiting for an Ack or Nak. If the window size is calculated properly, then theoretically at the time when the last frame has been transmitted, an Ack or Nak for the first frame should arrive; thus, the sender can send another set of frames continuously. The continuous transmission of frames is sometimes also known as *pipelining*.

In normal operation of the protocol, the frames are transmitted continuously over the channel. In other words, there is always more than one frame on the channel at any time. If the receiving site finds an error in any frame, it will send Nak for that frame, and the sending site has to re-transmit all frames preceding and including the frame containing error until the frame is received error-free. If the receiving site has any frame out of order, it will simply send Nak. This situation occurs where the first frame transmitted by the sender is lost in transit and the receiver waiting for this frame receives the next one, as depicted in Figure 11.15.

Consider a typical situation: a frame is received, and the receiving site sends Ack or Nak. If Ack is lost, what should the sending site do? There are a number of things which can happen. In one case, if there is no Ack, the sending site times out and re-transmits the frame with the same number. In the second case, the Ack number will indicate the receipt of all previous frame numbers prior to the Ack number frame. If this Ack number arrives before the time-out, then duplication of frames can be avoided and the receiving site expects the next frame from the transmitting site with a number associated with Ack. If Nak is lost, then again the transmitting site will time out and re-transmit the same frame until it receives either Nak or Ack.

A frame has a sequence number field, and the number of bits in this field represents the range of sequence numbers; e.g., if we have n bits, then the range of sequence numbers is given by 2^n. If we use a window size of 2^n, then there is a possibility of overlap between the last number of the sequence of the first set of 2^n frames and the first number of the sequence of the second set of 2^n frames, in the event of lost or piggybacked Ack. In order to avoid this situation, we should always consider the window size of $2^n - 1$.

In this protocol, more than one frame is in transit at any time. Hence, during the implementation of this protocol, for each of the frames, we have to use a timer which will be interrupted by the machine clock. A complete link list structure for each of the nodes (corresponding to frames) is maintained which is scanned periodically by the machine clock. During each clock tick, the timers are gradually decreased until their time-out values. Each time-out causes the re-transmission of that particular node.

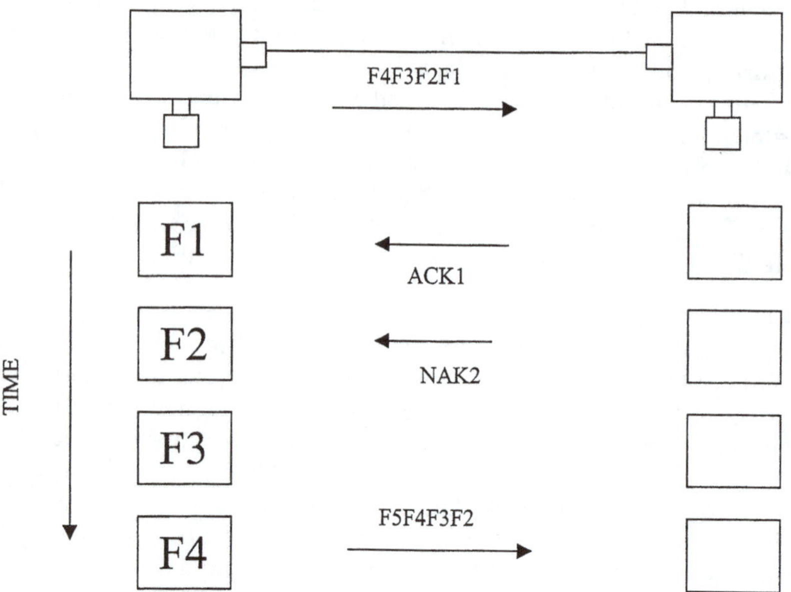

Figure 11.15 Go-Back-N ARQ.

Selective-repeat or selective-reject ARQ protocol: This is another form of pipelining which removes the main problem with go-back-n protocol. In go-back-n ARQ, if the sender receives a Nak, then all the frames preceding this error frame will be re-transmitted, and hence there is a wastage of the communication channel. Similarly, in *selective-reject ARQ protocol*, more than one frame is in transit and the receiving site buffer stores these frames. If any frame has an error, a Nak is transmitted to the sending site, and only that error frame will be re-transmitted until it is received error-free; other frames will be fetched from the buffer one at a time and passed on to the network layer only when their respective previous frames have been received error-free, as shown in Figure 11.16.

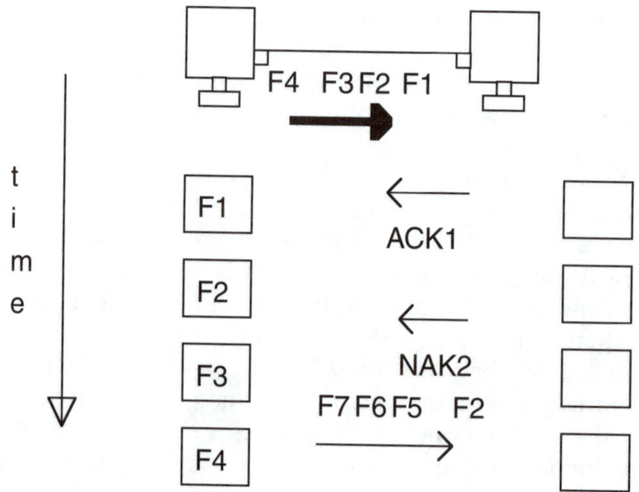

Figure 11.16 Selective-repeat ARQ.

This protocol requires fewer re-transmissions and hence is more efficient than the go-back-n protocol. It has several disadvantages, such as the facts that the receiving site must have enough buffer for storing all the frames until their previous frames have been received error-free, and the transmitting site must have an appropriate logic circuit which should keep a record of the Acks and Naks and take necessary action accordingly. Also, the window size is much smaller than that of the go-back-n protocol. As stated above, the maximum size of the window in go-back-n protocol is $2^n - 1$, while here it should not be more than half the range of the sequence number considered, i.e., $2^n/2$. This restriction on window size will avoid overlapping and loss of ACKs. Initially, both windows are set to the same value and, each time a frame is sent, the sender window decreases its window size by one while the receiver window advances by one. In this way, both sender and receiver are synchronized.

11.7.2 *Performance of sliding-window error-control protocols*

In this section, we discuss the performance of the sliding-window error-detection protocols.[2-6] The propagation delay affects the performance of protocols, as, during this time, both sites (sender and receiver) are idle. The propagation time (pt) is defined as the ratio of the distance of the communication link (d) and the speed of the electrical signal (s).

$$\text{pt} = d/s$$

The electrical signal usually travels at a speed of 0.2M km/s, while light travels at a speed of 0.3M km/s. In other words, the propagation delay between the sender and receiver sites is proportional to the distance, assuming that the electrical signal has a constant speed. The transmission rate of the channel is defined in bits per second (bps). For example, if we have a transmission rate of 1 Mbps and the length of the link between sender and receiver is 1 km, then the time taken to propagate the signal over the link is 5 μs (using speed of signal 0.2M km/s). During this time, there will be five bits on the link (using the transmission speed of 1 Mbps), which will give the length of transmission media five bits. Thus, we can transmit five bits with a propagation delay of 5 μs. If we increase the transmission speed to, say, 10 Mbps, we will have 60 bits on the link of the same length during the same propagation delay of 5 μs, giving rise to the length of the link as 60 bits. If we make it 100 Mbps, then the number of bits on the link (of the same length) will be 500 during the same propagation delay, giving rise to the length of the link as 500 bits. The satellite link typically has a propagation delay of 540 μs (round-trip delay). If we use a 1-Mbps satellite link, then we can transmit 270,000 bits during the propagation delay of 270 μs in one direction. Thus, the length of the satellite link is given by 270,000 bits. Hence, the length of any communication medium in bits is given by

$$\text{Length of the medium (bits)} = \text{Transmission rate } (r) \text{ (in bps)} \times \text{pt}$$

It is important to note that the transmission rate in bits per second has also been considered *channel capacity* (in bps) or *efficiency* parameters in many books. The product of r and pt will be the same for two different transmission rates over different lengths of a link but with the same propagation delay of the link. Further, the propagation delay is different for different media. Both the length of media and channel capacity represent the number of bits on the link per second. If the size (or length) of a frame is l bits and the transmission or channel capacity is c bps, then the time required to send one frame is l/c seconds. This is the time during which the link is busy. We have already defined the propagation delay of the link earlier. When a frame is transmitted, the last bit of the frame is received after pt, where pt represents propagation delay in one direction and the round-trip propagation is given by 2 pt. Once the last bit of a frame is received, an acknowledgment frame takes the same propagation time pt to reach the sender site. The utilization of a link can be defined as the ratio of the time required to transmit the frame and the actual time before the next frame can be transmitted. The time required to transmit a frame also depends on the window size and the type of protocol being used. For the sake of illustration, consider the time that of the window size; then link utilization (lu) can be written as

$$\text{lu} = \text{Window size}/(2m+1)$$

where m = propagation time delay (pt/time to transmit one frame). The value of m is a very important parameter in determining the efficiency and the utilization factor. Although other parameters exist, such as processing time, access time, waiting time, etc., they are relatively less important than m and, hence, are usually ignored as compared to m.

As discussed earlier, there is some relationship between the maximum size of a window and the maximum number in the sequence. We say *some* relationship, as we don't have an equational relationship, but looking at how each of these sliding-window protocols has been defined, it has been suggested that the maximum size of the windows should always be less than the number in the sequence. If the number of bits for representing sequence number is k, then in the case of the stop-and-wait protocol, the maximum size of the window should be 1 while the numbers in sequence are 2 (0 and 1 bits). In the case of go-back-N, the maximum size of the window should be less than $2^k - 1$, while in the case of the selective-reject protocol, it should not be more than half the maximum number in the sequence, i.e., $(2^k - 1)/2$. The relationship between the maximum size of the window and the maximum number in the sequence so defined for these protocols basically avoids overlapping between sets of the frames. For stop-and-wait sliding-window protocol, lu is given by

$$\text{lu} = 1/(2m+1)$$

For go-back-n protocol, lu can give us 100% utilization if the maximum window size is equal to the maximum number in the sequence. This 100% utilization is an ideal situation and is never achieved practically. Similarly, lu can be calculated in the case of selective-repeat sliding-window protocols by assuming that the maximum size of the window should not be more than half of the maximum numbers in the sequence.

The value of lu also depends on the size of the frame, and using the probability theory, we can determine the lus for each of the sliding-window protocols. Let us assume that the probability of a frame being received and damaged is p and the probability of receiving the damaged frame is given by $1 - p$. The probability of getting an undamaged frame depends on the size of the frame and the error rate.

Consider a transmission line having an error rate of 10^{-5}, which indicates that there will be one bit error out of 10^5 bits transmitted. If we are transmitting a frame of size 10^5 bits, then the probability that the receiver will receive an undamaged frame is zero and the probability that a damaged frame will be received is 1. If we reduce the size of the frame to, say, 0.5×10^5 bits, the probability that the undamaged frame is received will be half of p, as now we have two frames for the same data. If we further reduce the size of the frames, then the probability of receiving undamaged frames will be higher. Thus, we conclude from this that for small error rates, frames of larger size, and for large error rates, frames of smaller size will give a high probability of receiving undamaged frames. The line utilization based on probability theory (the probability that a damaged frame will be received) for each of the protocols can thus be defined as follows:

a. Stop-and-wait: $\text{lu} = (1-p)/(2m+1)$

b. Go-back-n: $\text{lu} = (1-p)/(2mp+1)$ for $N > 2m+1$

$= N(1-p)/(2n+1) \times (1-p+Np)$ for $N < 2m+1$

c. Selective-reject: $\text{lu} = (1-p)$ for $N > 2m+1$

$= N(1-p)/(2M+1)$

where N represents the number of frames transmitted.

11.7.3 Comparison of sliding-window error-control protocols

From the preceding sections, we get a complete idea of the working, implementation details, and relative merits and demerits of these three types of sliding window protocols. Based on these observations, we can conclude the following:

- Stop-and-wait protocol is inefficient for communication media and systems which have a long propagation delay.
- Go-back-n protocols are good for lower error rates but are inefficient for higher error rates. Also, these are efficient for small propagation delay and low data rates but will not provide higher throughput and efficiency for long propagation delay and high data rates.
- Selective-reject protocols offer very good throughput and efficiency, as the parameters (discussed above) don't depend on the propagation delay. The only drawback with these protocols lies in the requirement of enough buffer space at the receiving site and intelligent/complex logic on the transmitting site to maintain a list of the frames for which Acks have been received or the frames for which Naks have been received.

In spite of very good efficiency and throughput, go-back-n protocols are more popular than selective-reject, as the majority of the systems usually deal with low data rates and low propagation delays.

11.8 Examples of data link protocols

The data link layer provides an interface between the physical layer (characteristics of physical link, mode of operations, and configurations) and network layer (routing strategies, error control, connection control, etc.). Various characteristics and configurations of physical links (point-to-point or multi-point configurations), communication operations, and protocols (e.g., half-duplex, full-duplex, primary, secondary, and peer-to-peer communication protocols) will be supported by data link protocols. These protocols should be efficient and useful for high utilization of communication links with appropriate error control and flow control strategies. Also, the protocols should be code-independent; i.e., both byte-oriented and character-oriented approaches (for monitoring frame synchronization) should be available to the users. Each approach determines the length of frame in a different way. For example, a byte-oriented (or bit-oriented) approach provides a header in front of the frame, and the header includes the length of the data field. Thus, the receiving site can identify the length of the data field by calculating the location address from the contents of the header and hence does not depend on the control characters. Bit-oriented protocols are the same as that of byte-oriented, except that bit-oriented supports only frames of fixed lengths and, as such, contains a flag to terminate frames of variable lengths. Character-oriented protocols, in general, use available conventional symbols for the characters using ASCII or EBCDIC codes. There are various schemes for sending characters using these symbols, and these have already been discussed in earlier chapters.

The following subsections discuss the general requirements and formats of these protocols in brief and an international standard bit-oriented protocol in detail. Some of the material is derived from References 2–6.

11.8.1 *Character-oriented (binary synchronous communication (BSC)) protocols*

Character-oriented protocols use control characters with every character for providing synchronization between sender and receiver. The control characters include start of delimiter, end of delimiter, and error control, and when used with different protocols provide support for point-to-point or multi-point configurations. The simplest type of protocol in this category is Kermit, which has been used between DTEs over point-to-point configuration. This protocol is used for the transfer of files over a data link. This link can be either a public switched telephone network or a dedicated link (e.g., twisted pair). This protocol uses simple commands for sending files in one direction, such as KERMIT, CONNECT, SEND, EXIT, RECEIVE, END OF FILE, and END OF TRANSMISSION. This list of commands may be different for different versions of Kermit.

International Business Machines (IBM) in 1966 defined a **binary synchronous communication (BSC/BISYNC)**[22,23] protocol for providing communication between computer and terminal and also between computer and computer. This is one of the most widely accepted protocols in the industry and was defined for data communication between IBM machines and batch and video display terminals. It is a half-duplex protocol and supports point-to-point and multi-point communication configurations. It can also be used in both switched and non-switched channels. BSC is code-dependent, as it uses different types of special characters to differentiate between data, control, and synchronization characters. These defined characters are used to provide frame synchronization, character synchronization, and control of transmission of frames over the physical links. IBM has developed three versions of BSC protocols which are dependent on line configuration (point-to-point private, point-to-point lease or dial-up, and multi-point). BSC and other BCPs are popular when supported by large installed bases and are typically used for mainframes.

The general format of a binary synchronous communication (BSC) frame is shown in Figure 11.17. The number of fields for different BSC protocols are different and are usually

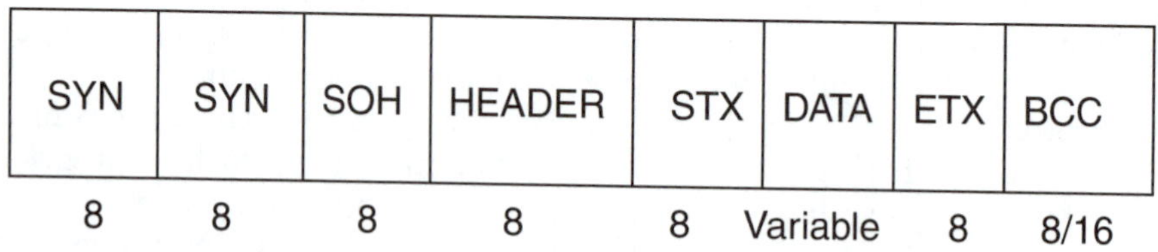

Figure 11.17 Format of a binary synchronous communication (BSC) frame.

considered optional. Some of the control characters are present in all protocols. In order to provide synchronization (for frame and character) between sender and receiver and also to keep the communication channel (or link) between them active (or ON), we use two synchronous bits (SYN) in front of the data frame. This control character is used to delimit the fields. Start of header (SOH) and start of text (STX) are the control characters and define the start of the header and text fields. These control characters are found in ASCII as well as in EBCDIC codes.

The header is optional. If it is used, start of header (SOH) will precede it and the header will be followed by start of text (STX).The users have to define the contents of the header, but in some cases such as testing of frames, these are provided by the system. The end of text is defined by another control character, end-of-text (ETX) or end-of-transmission block (ETB). In other words, these control characters SOH, STX, and ETX define the boundary of the frame, and we can have a variable length of data inside it.

The data field of variable length may also contain transparent data. For transparent data, delimiters are defined by a pair of DLE STX and DLE ETX (or DLE ETB). These control characters are non-data characters. DLE stands for data link escape. A detailed BSC frame format of a subframe is shown in Figure 11.18. The abbreviations for the fields are the following:

SYN — Synchronous idle (makes the channel active or retains the idle condition via synchronous transmission control)
SOH — Start of header (indicates the start of header)
STX — Start of text (initializes the text mode and terminates the header mode)
ETX — End of text (indicates the end of text message)
ETB — End of transmission block (indicates the end of a block or frame)
EOT — End of transmission (re-initializes the channel mode, terminates the connection)
ENQ — Enquiry (requests remote node for a response, useful for polling/section configuration)
ITB — end of intermediate block
ACK — Acknowledgment (indicates the receipt of message from sender)
NAK — Negative acknowledgment (indicates negative response from the receiver)

In this protocol, the VRC/LRC or CRC algorithms are used for error detection and correction in the transmission. If the transmission uses the ASCII coding scheme, then the VRC algorithm is used to check each character separately and the LRC is used to check the error in the entire block of the message.

The CRC typically used is the line control protocol if EBCDIC or the six-bit Transcode coding scheme is used. The six-bit Transcode uses CRC-12 with polynomial $x^{12} + x^{13} + x^{14}$ and generates a 12-bit BCC. The CRC-16 is used for EBCDIC coding and generates a 16-bit BCC.

The computed BCC at the receiver side is compared with the transmitted BCC. A positive acknowledgment is sent back in the case where both BCCs are the same, while a negative acknowledgment is sent in the case where both are different (resulting in an error). The BSC will re-transmit the error frame and usually checks for sequence errors by alternating positive acknowledgments to successive message blocks, with successive acknowledgments for even-numbered and odd-numbered blocks of messages.

ASCII, EBCDIC, and six-bit Transcode coding techniques are all supported by BSC. As a result, various control characters such as SOH, STX, ELX, ITB, ETB, EOT, NAK, DLE, and ENQ, and two-character sequences (e.g., ACK0, ACK affirmative Ack, RVI, and TTD) are being used as non-data characters.

SYN - Synchronous idle (makes the channel active or retains idle condition via synchronous transmission control).

SOH - Start of Heading (indicates the start of the header).

STX - Start of Text (initializes the text mode and terminates header mode).

ETX - End of Text (indicates the end of text message).

ETB - End of Transmission Block (indicates the end of a block or frame).

EOT - End of Transmission (re-initializes the channel mode, terminates the connection).

ENQ - Enquiry (requesting remote node for a response, useful for polling/section configuration).

ITB - End of Intermediate Block.

ACK - Acknowledgment (indicates the reciept of message from sender).

NAK -Negative Acknowledgment (indiactes negative response from the receiver).

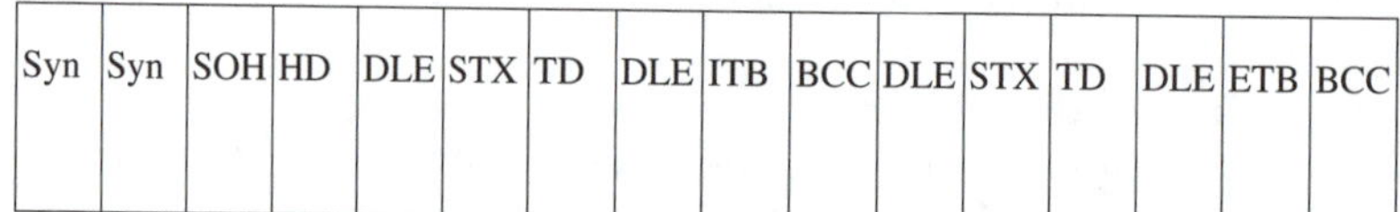

Syn	Syn	SOH	HD	DLE	STX	TD	DLE	ITB	BCC	DLE	STX	TD	DLE	ETB	BCC

HD: Heading, TD: Transparent Data

Figure 11.18 BSC frame format of a subframe.

The use of DLE STX at the beginning of the text field provides a total transparent mode to the user in the sense that during this mode, all the control characters must be preceded by DLE characters. In the case where DLE is used as data within transparent data, a DLE is inserted to allow the link control procedure to transmit DLE as data. This technique is known as *character shifting*; at the receiving side, the DLE from the transparent data is discarded.

The BSC system is a half-duplex transmission over both point-to-point and multi-point configurations. A minimum of two synchronous characters (SYN) are used at the beginning of the message block or control block and are usually defined as a unique bit pattern in each of the coding schemes. BSCs are defined for synchronous lines. Being a character-based protocol, BSC can implement both synchronous and asynchronous channels.

The BSC protocols support both ASCII (when CRC is used for error control) and EBCDIC (when a record character checking is used for error detection). In addition to these two codes, either code with an appropriate error detection algorithm is supported. The number of bits for each character is fixed for the code, e.g., eight bits in the case of ASCII. Thus, the length of the data is always an integral number of the eight bits (or

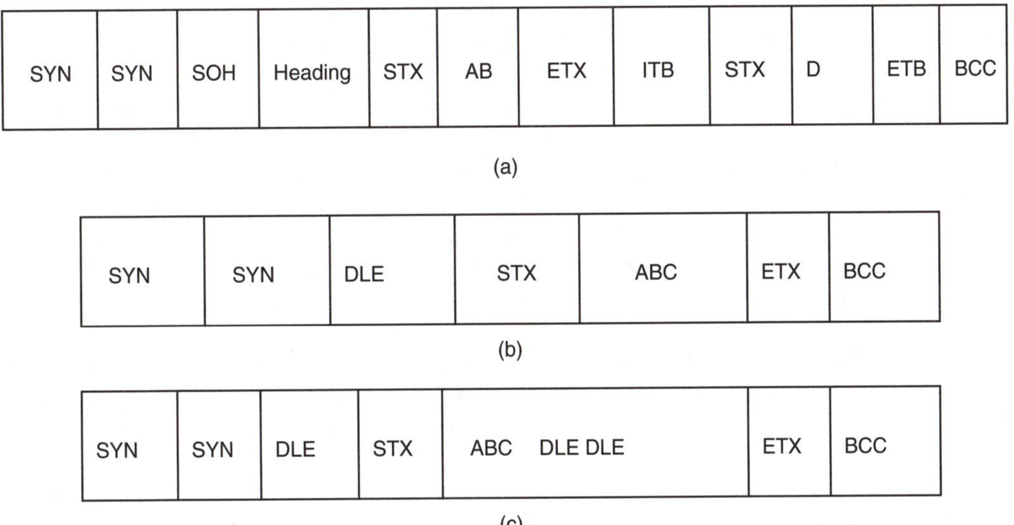

Figure 11.19 Character stuffing.

octets). Binary data can be sent as long as the data length is a multiple of the octet or eight bits.

A serious problem may arise when a user defines ETX or ETB in his data field and the receiver cannot differentiate between the same control characters in two different domains (data and control). As soon as it sees ETX in the data field, it stops reading the frame and waits for another frame. This results in an error which may be detected by an error-control algorithm used in the protocol. This problem can be overcome by the **character stuffing** technique (discussed earlier), where for each of the control characters used in a data field, we add another control character, data link escape (DLE) before ETX or ETB (Figure 11.19(c)). This character-stuffing technique provides data transparency. Similarly, DLE is added before STX if it is used in the data field. The receiver site identifies a 16-bit pattern for one control character in the data field and discards the first eight-bit pattern (corresponding to DLE).

If DLE is used in the data field, then another DLE is added to control character DLE. BSC uses a combination of parity and LRC in the case of ASCII code (eight bits) or cyclic redundancy check (CRC) in the case of EBCDIC (16 bits) for detecting and correcting the error in the frame during transmission. These are defined in the last field of the frames as a block check character or count (BCC). The content of the BCC defines the number of characters from the SOH through ETX or ETB. If the receiver gets a different calculated BCC value than was transmitted, it sends a Nak to the sender for ARQ error control. If both values of BCC are same, it means that there is no error in the frame, and the receiver sends an Ack to the sender.

Figure 11.19(a–c) shows how data transparency can be obtained by using character stuffing, as discussed above.

The above discussion of the BSC frame clearly demonstrates that this frame sends both control and data characters and, hence, encompasses two modes of operation: control and data (or character or text). The control character establishes the link between the nodes for the transfer of text. In general, the control mode is performed by a centralized node known as a *master node*, while the text mode is performed by other nodes (*slave*) under the control of the master node. The master node sends the control character (for polling or selection of frames) to various nodes and, after receiving the responses (polling data) from slave nodes, selects a particular node. This establishes the link between them; the slave node sends the data to the master node in text mode through frames. This is based on the poll/select scheme, which ensures that only one data message can be transmitted over the link at any time. The master node interacts with each of the secondary nodes by using unique identifiers attached to them and uses two types of messages — data and control.

The master (primary) node uses poll and select control primitives to get the message response from the slave (secondary). The secondary nodes use no-message and ready-to-receive control primitives to respond to the primary's control primitives. The poll/select scheme causes excessive overhead, and sometimes it becomes a bottleneck for data link protocols. This problem of excessive overhead is reduced by using a cluster controller and front-end processor (FEP). The cluster controller is used to poll and select a group of secondary nodes connected to it and thus reduces the response time, while the FEP is used to handle only the polling and selection procedures, leaving the burden of processing entirely on the CPU. The FEP is basically a dedicated processor which is programmed to handle only these control procedures.

The control characters contained in the frames define the text mode. At the end of transmission, the slave node sends the control end-of-transmission (EOT) character to the master, thus putting the channel back into the control mode. When the BSC channel is in text mode, no other node can send the message on the channel. The control character used in the frame for establishing the BSC channel in the control mode is an ENQ, along with the address of the node. This character is sent by the sender. When a receiver site sends an Ack to this request, then the BSC channel is established between the sender (master) and receiver (slave); thus, it provides contention transmission on a point-to-point configuration. After the messages have been transmitted (text mode), the EOT control character puts the channel into the control mode. Thus, the ENQ character provides various functions: polling, selection, and point-to-point configuration. The distinction between polling and selection function of the master node is defined through the lower or upper case of codes.

BSC protocol works similarly to stop-and-wait ARQ protocols on half-duplex only. Only one bit is sufficient for the acknowledgment. The various line control codes used in protocols are given below:

- ACK0: Positive acknowledgment to even-sequenced blocks of data or acknowledgment to selection or point-to-point connection request.
- ACK1: Positive acknowledgment to odd-sequenced blocks of data.
- WACK: Wait before transmit (positive acknowledgment); useful for reversing the direction of transmission on half-duplex.
- RV1: Reverse interrupt — indicates that the station will interrupt the process for sending the data.
- DISC: A request for de-establishing the link; useful for switched lines.
- TTD: Temporary text delay — this code indicates that the sender has established a connection and is in text mode but will transmit the message after a delay.

The sequence of these control characters is important and provides the sequence of operations which will take place; e.g., if we transmit "ENQ SYN SYN," this sequence of control characters indicates that a node (based on point-to-point communication configuration) requests the establishment of communication links. If two nodes are sending the sequence at the same time, then one of them will be asked to retry. A sequence "ACK0 SYN SYN" will indicate that the node is ready to receive the message. Other possible sequences are given below. The sequence of BSC line-control protocol is shown in Figure 11.20.

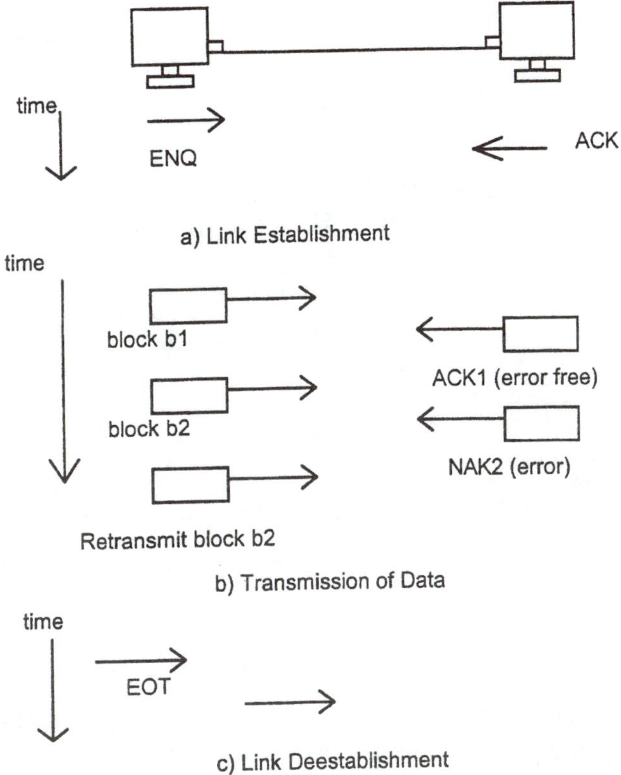

Figure 11.20 Sequence of line codes in BSC line-control protocol.

- ACK0/1 SYN SYN — node Ack transmission and/or is ready to receive message.
- NAK SYN SYN — node Nak transmission and/or is not ready to receive message.
- WACK SYN SYN — node requests other node to try again later on.
- RV1 SYN SYN — Acks the message and requests node to suspend transmission.
- ENQ STX SYN SYN — node is in temporary text delay (TDD) mode.
- EOT SYN SYN — negative response to poll.

The BSC protocol has been considered a very powerful and useful character-oriented protocol, and even today it is probably the best-known protocol and is being used throughout the world. Looking at the press and demand of this protocol, ISO prepared products and standards similar to BSC. These products and standards are not the same as those of BSC but provide some of its functions and services. These are ISO 2111, ISO 1745, and ISO 2628. The ISO 2111 deals with the management of the data link between two nodes (establishment, control of link, termination, etc.) through control characters (same as STX, DLE, and ETB in BSC frame). Various link operations (synchronous or asynchronous transmission) and configurations (half-duplex and full-duplex configuration, point-to-point, or multi-point) are managed by ISO 1745, while the error control (error recovery, interrupts, etc.) and timers are described in ISO 2628.

The full-duplex character-oriented protocols operate over point-to-point duplex links which are established between nodes. The transmission of frames takes place in both directions simultaneously over different links. This type of protocol has been used in ARPANET for controlling the exchange of data frames (packets) across the links. The subnets known within ARPANET as **interface message processors (IMPs)** define different

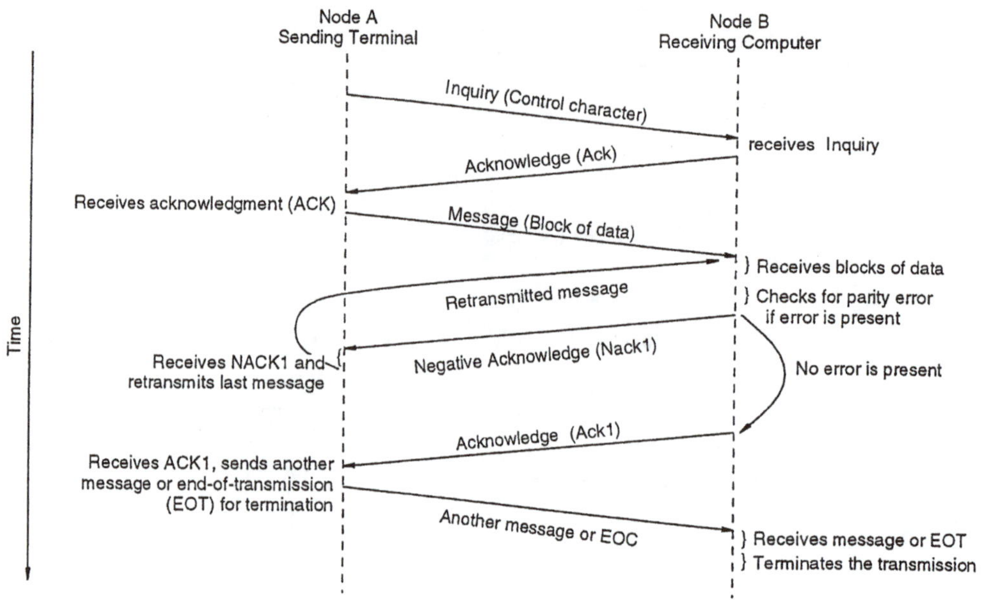

Figure 11.21 A typical data exchange across the network between two nodes using BSC protocol.

links among them based on full-duplex configurations, as described in Chapter 1. This configuration establishes different circuits between different intermediate nodes (IMPs) until a complete routing between sender and receiver is achieved. There may be more than one routing between the IMPs, and packets will be delivered to the destination via different routes. The protocols for the topology of IMPs use a full-duplex configuration which treats each of the frames as a separate frame and processes each one of these separately. Each data frame contains all the control information (destination address, error control, etc.). The protocol uses 24-bit CRC for error detection.

A typical data exchange across the network between two nodes using BSC protocols is shown in Figure 11.21.

11.8.2 *Byte-oriented (digital data communication message protocol (DDCMP)) protocol*

Digital Equipment Corporation (DEC) developed a byte-oriented protocol known as digital data communications message protocol (DDCMP) in 1974. It defines the number of characters in the header field which specifies the length of the data field in the frame and, as such, does not require any technique to solve the problem of data or text transparency. This is one of the major problems in the character-oriented protocols, where the code transparency has to be provided by the DLE control characters due to a number of problems with the DLE (as discussed above). The efficiency and overhead of character-oriented protocols are the major concerns. The DDCMP can be used on half-duplex or full-duplex configuration using both synchronous and asynchronous transmission. The length of the data field is given in bytes (or octets) and is specified in the header. The end of the frame is not defined by control codes but instead is determined by the protocol. This provides complete data transparency for all of the information in the data field (multiple octets). DDCMP supplies serial and parallel channels under both synchronous and asynchronous (start–stop) modes of transmission.

The DDCMP frame is simpler than that of BSC. Each DDCMP frame is assigned a unique number, and CRC-16 protocol is used for error detection. The ARQ technique (discussed previously) is used for error recovery (i.e., error correction). The maximum size of the data field is 16,383 octets and, depending on the situation, usually smaller-size frames are transmitted. The frame number is defined within a byte field; the maximum number of frames which can be transmitted is 255. The DDCMP provides three sub-frames: data frames (by start of header, SOH), control frames (control field defining the control message, e.g., error recovery, link management primitives, etc.) and maintenance frames (by DLE for testing and debugging the link in off-line mode). Figure 11.22(a) shows the general format of the DDCMP frame.

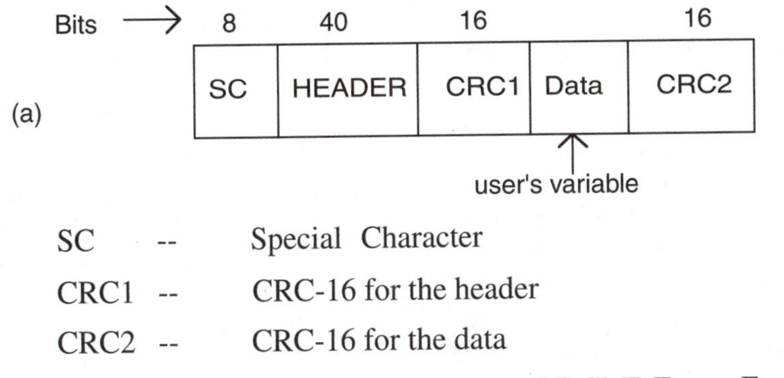

(b)

bits → ← direction of transmission

Syn	Syn	SOH	Count	Flag	Response	Sequence	Address	CRC1	Information variable	CRC2
			4	2	8	8	8	16		16

SOH - Start Of Header
Information field - any number of 8-bit characters

(c)

SOH	Count	Flags	Response	Number	Addr.	HBEC	Data	DBEC	CRC
8	14	2	8	8	8	16	user's variable	16	16

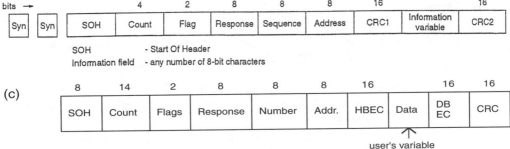

SOH -- Start of Header (indicates the data information)
Count -- Length of the Data information in the frame
Flags -- Used for line control
Response The last sequence number received correctly (generally useful for piggybacking and pipelining)
Number Sequence Number of data frame sent
HBEC -- Header Block Error Check (header field has its own CRC-16)
DBEC -- Data Block Error Check (data block has its own CRC-16)
CRC -- CRC-16 for the frame

Figure 11.22 (a)DDCMP frame format. (b), (c) Details of DDCMP frame format.

More detailed DDCMP message format versions for data frame headers containing fields (for defining various functions) are shown in Figure 11.22(b,c). It is worth mentioning that these two versions of DDCMP frame formats can be found in many books and are widely accepted and used.

The control characters SOH, ENQ, and DLE are used to distinguish between data control and boot slap message blocks.

The header includes the information characters of eight bits in the information field, control flags response field for positive acknowledgment, message sequence (modulo 256), and an address. The header has its own CRC. The information field is of variable length (up to 16,383 bytes). The last field is a BCC field, which contains CRC-16 (computed remainder). The positive acknowledgment is not sent for all message blocks, but instead the sequence number for positive and negative acknowledgment can be sent together either in a response field or in the respective fields for acknowledgment. The presence of error in the message block is not sent to the transmitter by the receiver, but the transmitter detects the error by looking at the response field.

As stated above, the various ASCII control characters such as start of header (SOH), enquiry (ENQ), and data link escape (DLE) are used to distinguish between different types of message blocks. The header is of fixed length, and the count in the header defines the length of the information field (up to a maximum of 16,383 bytes). The link utilization may be improved in a variety of ways (as discussed in BSC) in both half- and full-duplex modes of operations including no separate acknowledgment when traffic is heavy, multiple acknowledgments per Ack (one Ack for every 255 messages), etc.

In the frame, the special character (SC) defines the type of sub-frame being used. Each type of sub-frame has the same length for its header. The header includes the protocol CRC-16 (CRC1) for controlling and maintaining the link, while the data field defines its own CRC-16 protocol (CRC2) for controlling the user data; the user data field is a multiple of an octet.

As discussed above, the DDCMD frame contains three sub-frames: **data**, **control**, and **maintenance**. Each of the sub-frames is identified by a special control character, as described below. The data sub-frame header containing the fields (for defining the various functions) is shown in Figure 11.22(b,c).

- The starting of the sub-frame header is defined by SOH.
- The length of the data field in the sub-frame is defined by the Count field (in the range of 1–16,383 octets).
- The synchronization and selection are defined by two types of flags: SYNC flag and Select flag. The SYNC flag defines the transmission of SYN characters, while the Select flag defines the line communication configurations such as half-duplex, multi-point, and other functions.
- An acknowledgment of frame-received data (error-free) in piggybacking and pipelining is defined by the number in the response field, while the number in the number field represents the sequence number of the transmitted data.
- The address of the particular node on a multi-point line configuration is defined in the Address field.
- A check on a sub-frame header is defined in the CRC1 field, which ensures that the count field is received error-free and is not damaged, while CRC2 checks the sequence of data in the data field. The data field contains the user data.

The control sub-frame header, which performs functions such as establishment of links, management of link functions, etc., is shown in Figure 11.23.

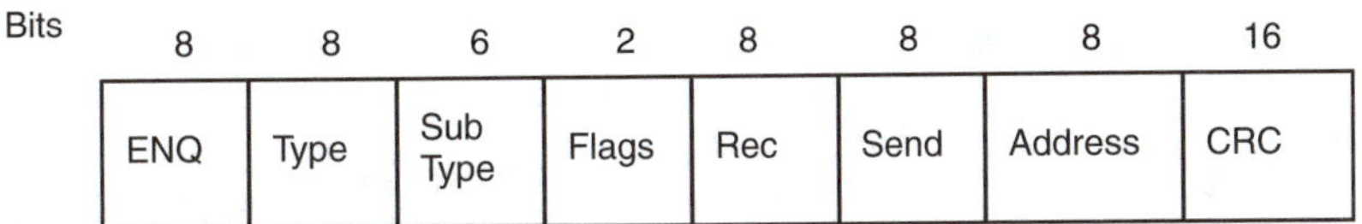

ENQ = Control character defining control frame
Type = Type of control frame
Subtype = Indicates reasons for NAK and other control information
Rec = Receiver (indicates some specific control information); from receiver to sender after receiver receives the frame
Send = Sender (field indicates some specific information from sender to receiver)

Figure 11.23 DDCMP control frame.

The start of the control sub-frame header is defined by the ENQ control character. Control frames don't define numbers (as in the case of data sub-frame headers). Various tasks defined in the link task (type) are discussed below.

- An acknowledgment message (Ack) (type = 1) indicates the receipt of error-free data (i.e., CRC check has been computed). This Ack is different from that in ASCII or EBCDIC codes, while similar to the response field in the data sub-frame header.
- Negative acknowledgment message (Nak) indicates the error in received data sub-frame, and the reason or cause of the error can be defined in the details of link tasks (sub-type) field
- The status of frame at the receiving site is defined by the REP in type field. The message in this type basically defines three situations: data message has been sent, Ack for data message has not been received, or it has timed out.
- An initial setup for establishing the link is defined by STRT in the type field and it represents the initial contact and synchronization for establishing the link.
- A response regarding STRT from transmitting node is defined by STACK which is sent not only to the transmitting node but also to other nodes on DDCMP.
- To check the correctness of both sending and receiving, sites may use these receiver and sender fields to inform each other about the number of the last data frame received error-free and the number of the last data frame transmitted, respectively.

Off-line testing and other maintenance functions are represented in the third maintenance sub-frame header, which starts from a DLE control character.

Various link functions in DDCMP are performed by three processes: **framing**, **link management**, and **data exchange**. The *framing process* basically identifies the start and end of a message. It supports both types of transmission (synchronous and asynchronous) and identifies start/stop bits for such synchronization on an asynchronous link while it recognizes a bisync character for synchronization in an asynchronous link. Further, it recognizes different types of sub-frames by identifying SOH, ENQ, and DLE control characters.

The *link management process* is concerned with link functions such as data flow, addressing, control flow, transmitting and receiving of data messages, etc. The *data exchange process* is responsible for sequencing error recovery (error checking and recovery). This process is invoked as soon as the framing process is initialized.

As explained above, DDCMP supports sliding windows, pipelining, and piggybacking with acknowledgments; the control sub-frame initially is used for setting up the link. STRT control characters provide initial contact and synchronization for the link and transmit

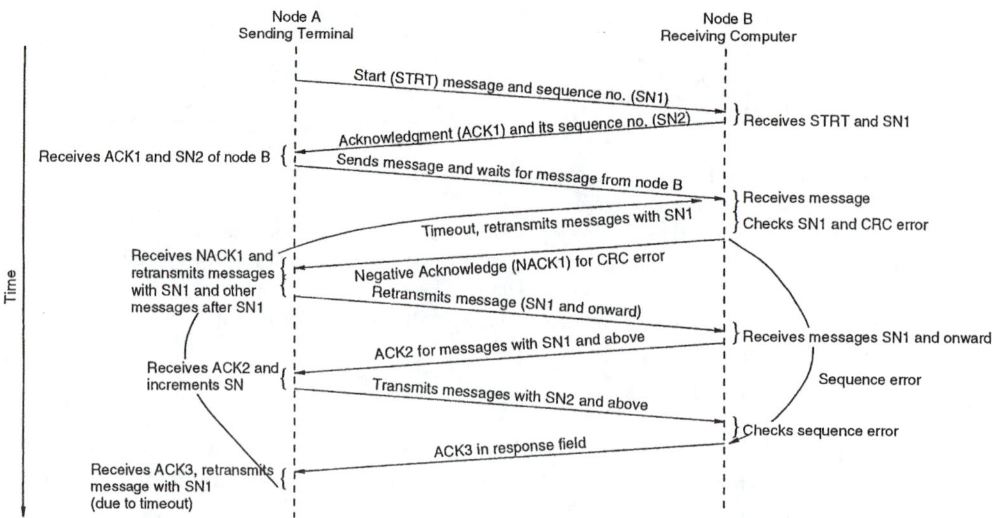

Figure 11.24 A typical data exchange between two nodes across a network using DDCMP.

on a DDCMP multi-point-link connection. A node on DDCMP receiving it responds by transmitting STACK and resets the frame number. An Ack is then sent to the receiving node (which sent STACK) indicating that it is ready to transmit the data. It can be used with synchronous and asynchronous transmission systems.

A typical data exchange between two nodes across the network using DDCMP protocols is shown in Figure 11.24.

BISYNC and other BCP are popular when supported by large installed bases and are typically used for mainframes. The byte-controlled protocols and IBM's synchronous data link control (SDLC) protocol are based on this. The SDLC uses an unbalanced normal response mode and is a superset of HDLC. In SDLC, secondary stations cannot send to each other. All communications take place through the primary station. The address file contains only the secondary addresses. The primary can send a series of frames to different secondaries by varying the address fields. The secondary station can respond only during SDLC polling. It does error checking on all frames (control and data). This means that both SDLC and bit-oriented protocols offer higher reliability and a lower error rate than BISYNC.

11.8.3 Bit-oriented ISO high-level data link control (HDLC) protocols

Previous protocols (character- and byte-oriented) have the following drawbacks:

1. **Code-dependence:** This prohibits the use of any control character (other than those defined) or bit pattern for data transmission. Further, users have to use DLE to obtain code transparency in the protocols.
2. **Efficiency:** The efficiency of these protocols is low due to the use of inefficient error-recovery and flow-control strategies and the overhead for supporting these.
3. **Flexibility:** These protocols possess very low flexibility for supporting efficient strategies for error detection and error recovery.

These drawbacks are removed in bit-oriented synchronous data link protocols. Instead of using control characters for providing synchronization and defining the boundary through character delimiters, bit-oriented protocols have frame-delimiting, unique bit patterns for defining the boundary of a frame (start and end) and the length of the frame in a header field, and they include some bits for encoding. Due to these advantages, bit-oriented protocols have been widely accepted throughout the world and have been approved as an international standard known as **high-level data link control (HDLC)** by the International Standards Organization (ISO 3309 and ISO 4335).[10,11] HDLC supports both point-to-point and multi-point (multi-drop) data link configurations.

The American National Standards Institute (ANSI) has defined its own standard based on bit-oriented protocol, **advanced data communication control procedures (ADCCP)**, which has been accepted by the U.S. National Bureau of Standards (NBS) and also by the Federal Telecommunications Standards Committee (FTSC). NBS uses it for procurement purposes in U.S. government agencies, while FTSC has considered it the standard for the national-defense-related National Communications System (NCS).

CCITT has also defined an international standard bit-oriented protocol known as link access procedure balanced (LAP-B) protocol, which is used as a process in the X.25 CCITT packet-switched network standard.[9] Another form of LAP standard is defined for ISDN as LAP-D, which is discussed in CCITT I.440 and I.441.[17,18]

Another protocol based on the bit-oriented approach has been developed by IBM and is known as **synchronous data link control (SDLC)**, which is very popular and widely accepted but is not a standard. Details of this protocol are discussed in Reference 24 and in Section 11.8.4 of this chapter.

Ideally speaking, there is not much difference between these bit-oriented protocols, and all the protocols have been derived from HDLC with minor changes. Different vendors may have a different implementation and name for HDLC, so we have to be very careful when purchasing HDLC.

Users may have different requirements, and these requirements have to be satisfied by these protocols. ISO 7809[13] discusses the types of nodes, link configuration, and data transfer mode of operations of HDLC which, in most cases, satisfy a variety of requirements of different users. HDLC can be used on half-duplex and full-duplex and supports point-to-point and multi-point communication with different channels (switched and non-switched). The following subsections describe the characteristics, frame format, control frame, and operations of HDLC.

ISO HDLC frame format: The frame format for HDLC by ISO 3309 and ISO 4335[10,11] is discussed below and shown in Figure 11.25(a). It has seven fields (including the user's data):

1. Flag field (F, eight bits)
2. Address field (A, eight bits but extendable)
3. Control field (C, eight or 16 bits)
4. Data field (D, variable)
5. Information field (I, variable length, some frames do not use it)
6. Frame check sequence field (FCS, 16 or 32 bits)
7. Flag field (F, eight bits)

Each frame is preceded and succeeded by flag fields. The flag field is defined by a particular bit pattern of 01111110 and is continously transmitted over the link between the sites for keeping the link active. The fields F, A, and C before the data are defined as the header of the frame, while fields I (optional), FCS, and F after the data are defined as the trailer of the frame.

The flag pattern 01111110 determines the start and end of the frame. All the connected sites are looking for this flag pattern and as soon as any site receives it, it is synchronized

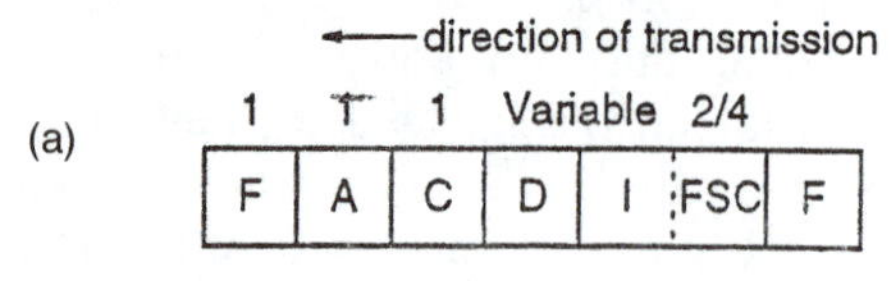

F - Flag
A - Address
C - Control (1 or 2 bytes)
D - Data (variable; user's option)
I - Information (variable length, some frames do not use it)
FSC - Frame Sequence Control (2 or 4 bytes)

(b)

	0	1	2	3	4	5	6	7
I	0	N(S)			P/F	N(R)		
S	1	0		S	P/F	N(R)		
U	1	1		M	P/F	M		

I - Information frame
S - Supervisory frame
U - Unnumbered
N(S) - Sequence number of frame to be transmitted
N(R) - Sequence number of frame to be received
S - Supervisory bit
M - Unnumbered bit
P/F Poll/Final bit

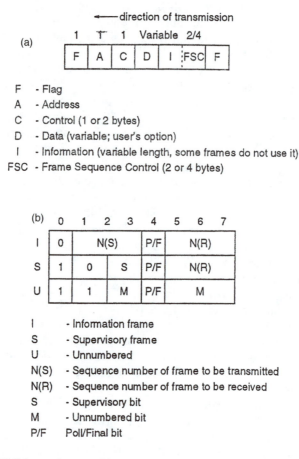

Figure 11.25 (a) HDLC frame format. (b) HDLC control frame.

with the frame and determines the start of the frame. Similarly, it also looks for the same bit pattern during data communication and as soon as it finds one, it knows the end of the frame. In order to avoid the synchronization for determining the start and end of the frames (we may have the same flag pattern in other fields of the frame), many techniques such as bit stuffing or non-return-to-zero-inverted (NRZI) technique can be used which will keep the receiver site's clock synchronized. In the bit-stuffing technique, a 0 is inserted after every five consecutive 1s of all the fields in the frame except the flag field itself. When the receiving site receives the flag field, it identifies the start of the frame.

For all the fields (except the last flag), it looks at the sixth bit. If the sixth bit is 0, it unstuffs (deletes) it, but if the bit is 1, it represents two situations: to abort or accept the frame. If the seventh bit is 0, it accepts this as a valid flag field; otherwise, if both the sixth and seventh bits are 1, then it indicates that the sending site has identified an abort situation. This type of bit stuffing (using any bit pattern) provides code and data transparency. The protocol based on bit stuffing is concerned only with a unique flag pattern which is mutually agreed upon between two users. The bit stuffing may cause error in the frames when used in a different way in the flag pattern.

Sometimes bit stuffing is used in conjunction with encoding schemes for providing clock synchronization between receiver and transmitter. There are different binary encoding techniques, but the non-return-to-zero-inverted (NRZI) technique is widely used where a line transition represents 1s, while no change represents 0s. Different derivatives of this type of scheme include **alternate mark inversion (AMI)**, **bipolar with 8 zeros substitution (B8ZS)**, **high density bipolar 3 (HDB3)**, etc.

The address field (A) defines the addresses of primary and secondary nodes which are going to transmit and receive the frames. Although this field will not define any other addresses it is used to define both nodes for the sake of using the same format for all the addresses. The length of the address field is eight bits, and the last bit (least significant bit) may be used for extending the length of the address field (having been mutually agreed upon), thus using only seven bits for addressing. All 1s in the address field represents a broadcasting mode, and as such all the connected secondary nodes will receive the frame simultaneously.

The address field (A) defines different addresses for command and response frames of transmitting (primary) and receiving (secondary) nodes; e.g., in an unbalanced configuration, the secondary node address is defined by both command and response frames, while for a balanced configuration, the response frame defines the primary node address and the command frame defines the secondary node address.

ISO HDLC control frame: The ISO HDLC control frame is shown in Figure 11.25(b). The **control field (C)** defines the frame fields for command, responses, and sequence number of frames for maintaining data-flow control between primary and secondary nodes. These fields may vary with the HDLC frame being used. Three types of frames defined by HDLC with different control frame fields are discussed below:

- **Information frame (I-frame)** contains an information field which carries the end user data to be trasnmitted between two nodes and is defined as user data. This frame may also include ARQ error and flow-control strategies, piggybacking on I frame, etc., for acknowledgment, and also functions for polling of the nodes.
- **Supervisory frame (S frame)** provides the flow-control functions like acknowledgment of the frames, re-transmission of frames, etc. It supports both ARQ and piggybacking mechanisms.
- **Unnumbered frame (U frame)** provides link-control functions, e.g., link initialization, link establishment, link de-establishment, link management, etc.

The frame type (I, S, and U) is usually defined by the first bit or first two bits of the control field. Five bits are used to define 32 commands and 32 responses in appropriate format, depending on the type of HDLC frame. S- and I-frames usually use a three-bit sequence in their control field and can be extended to a seven-bit sequence using appropriate set-mode command.

The format of HDLC is determined by the usage of these three frames and also the coding techniques used for them. These frames are being exchanged between a primary and secondary node or between two primary nodes which, in turn, defines the operation of the HDLC frame.

The I-frame defines the sequence number N (S) which is attached with the transmitted frame and also the piggybacking acknowledgment. The acknowledgment further defines the sequence number of the next frame, which the receiving node will be expecting from the transmitting node. The size of the window is defined by the bits in the control field, and the maximum-size window of the control field is 127 bits (as the first bit or first two bits define the type of frame).

The I-frame also includes a poll/final (P/F) bit. The primary node issues a command by setting the bit to poll (P) while the final (F) bit is used by the secondary node for the response. The P/F-bit is set to 1 by the respective node for command and response on the I-frame for appropriate action in normal response mode (NRM). The same P/F-bit is used for controlling the exchange of other frames (S and U) between nodes in asynchronous response mode (ARM) and asynchronous balanced mode (ABM).

The **supervisory (S-frame)** defines error and flow control between nodes. It supports go-back-N ARQ and selective-reject ARQ protocols. The P/F-bit in the S-frame is used by primary and secondary nodes for polling and sending responses between them, respectively. The I-frame uses I as command and response N(S) and N(R). The S-frame uses the following commands and responses[2–6]:

- **Receive ready (RR):** Both primary and secondary nodes can use it for receiving/acknowledging the frames, and it also is used by the primary to poll by setting the P-bit to A.
- **Receive not ready (RNR):** A node indicates a busy condition, sends a positive acknowledgment to the transmitting node on the N(R) field, and can clear a busy condition by sending RR. This frame can be used both as a command and response.
- **Reject (REJ):** A negative acknowledgment indicates the transmission of frames with the starting frame where the sequence number is defined in N(R). This also indicates the receipt of all frames with numbers (N(R) – 1). Go-back-N technique can be used by the REJ frame. This frame can be used both as a command and response.
- **Selective-reject (SREJ):** A negative acknowledgment indicates the transmission of only that frame whose number is contained in the N(R) field. This type of acknowledgment defines the receipt of frames with (N(R) – 1), as discussed above in the REJ frame. After the SREJ frame is transmitted, all the frames following the error frames will be accepted and stored for re-transmission. SREJ is reset by transmitting the I-frame when N(S) is equal to V(R). Only one SREJ frame can be sent at any time, and for each frame transmitted after the error frame is detected, the SREJ frame will be transmitted. This SREJ frame can be used for both command and response.

The commands of the unnumbered frame format are discussed below:

1. **Unnumbered information (UI):** This frame format provides exchange of both types of control data (in an unnumbered frame). It does not have N(S) and N(R) and hence can be considered a connectionless configuration protocol.
2. **Exchange identification (XID):** This defines the status and type of message (request/report) being exchanged between two nodes.
3. **Test:** This provides the testing strategy between two nodes by exchanging the contents of similar fields before any data can be transmitted from one node to another node.

All three command frame formats can be used as both commands and responses. There are a few frame formats which can be used only as commands, while other formats can be used for responses only. The command frame formats which provide commands include:

Set normal response/extended mode (SNRM/SNRME)
Set normal asynchronous response extended mode (SARM/SARME)
Set asynchronous balanced/extended mode (SABM/SABME)
Set initialization mode (SIM)
Unnumbered poll (UP)
Reset (RESET)

The frame formats for response are as follows: unnumbered acknowledgment (UA), disconnect mode (DM), request disconnect (RD), request initialization mode (RIM), and frame reject (FRMR).

The three frames — I-frame, S-frame, and U-frame — are exchanged between primary and secondary nodes, or even between two primary nodes. The above-mentioned frame formats for commands and responses are used to define the operation of HDLC. The I-frame contains the user data, sequence number of the transmitted frame, piggyback acknowledgment, and poll/final (P/F) bit. Flow control and error control are managed by the S-frame which uses go-back-N ARQ and selective-reject ARQ protocols.

Various functions such as commands and responses for mode setting, information transfer, recovery, and other functions are defined in unnumbered frames. Two classes of HDLC timers, T1 and T2, have been defined. These timers represent the link operations and how they are implemented. In T1, a primary node, after sending a poll bit, waits for a predefined time (waits for F-time) which is controlled by T1. T1 is being used in the majority of vendor products. T2 is used to control the wait for N(R) time to check the receipt of acknowledgment in ARM mode.

Derivative protocols of HDLC: HDLC supports both balanced and unbalanced link modes. In the unbalanced link mode, two formats, normal response mode UN and asynchronous response mode VA, are being supported, while in balanced mode, asynchronous balanced mode BA is supported. There are other data link protocols which have been derived from HDLC and are different from HDLC in terms of capabilities, support of UN, VA and BA formats, etc. The following is a brief description of available standard data link protocols which are being derived from HDLC.

1. Logical link control (LLC)
2. Link access procedure (LAP)
3. Multi-link procedure (MLP)

The data link layer in OSI-RM of a local area network is composed of two sublayers: **logical link control (LLC)** and media access control (MAC). IEEE has defined LLC as a part of the IEEE 802 family of standards for local area networks. This standard allows the local area network to interact with different types of local area networks and also wide area networks. LLC is defined as a subset of HDLC and is used in balanced mode with number BA-2, 4. LLC is different from HDLC on multi-point link control, as LLC supports peer multi-point link connection where there are no primary or secondary nodes. Each node connected to the network has to include both primary and secondary addresses with the data unit. The ISO 8802.2[14] discusses various services and functions of this sublayer.

Various services supported by LLC are unacknowledged connectionless, UI-frame, acknowledged connectionless using AC-frame, and acknowledged connection-oriented using I-frame. Each of these services uses a different set of commands and responses (as discussed above) for transmitting the data between nodes connected to the network.

The **link access procedure (LAP)** standard has been derived from HDLC and uses the HDLC set asynchronous response mode (SARM) command on an unbalanced configuration. It has been designated as VA 2, 8 and is being used to support X.25 networks. The primary and secondary nodes use SARM and unnumbered acknowledgment (UA) commands, respectively, for establishing a link between them. The extended version of this protocol is known as LAP-B.

The **link access procedure–balanced (LAP-B)** standard is a subset of HDLC which controls the data exchange between DTE and public switched data networks or even private networks and establishes a point-to-point link between them. It controls the data frame exchange across a DTE/DCE interface and offers an asynchronous balanced mode of operations across the interface. All of the data frames coming out at the interface are treated as command frames by LAP-B, as it considers a DTE/DCE interface one of the nodes within the switched data network or private network. This standard has been used in both private and public networks including X.25. In X.25, it defines the data link protocol. The DTE of X.25 is used to establish a LAP-B link for communication. This standard is also derived from HDLC and has been designated as BA2, 8, 10. Here these numbers represent the options which are available. It is very popular and widely used worldwide.

A version of LAP for the modem has also been defined for the error-correction protocol modem. The modem V.32 accepts asynchronous data from the DTE. It then converts this asynchronous (start–stop) data into synchronous mode using the bit-oriented protocol of HDLC. It also supports error correction. This version of LAP defined for a modem is composed of two modules known as the user DTE interface (UDTEI) and error-correcting (EC) modules. The UDTEI is primarily responsible for transmitting the characters or bytes across the V.24 interface and maintains flow control. The EC module, in fact, includes the LAP for error correction. Both of these modules communicate with each other via a predefined set of service primitives. The EC modules of the DCE across PSTNs mutually agree on the parameters before the data transfer between DTEs. The UDTEI issues a request for a set of parameters in the protocol data unit primitive as L-param.request which is forwarded by the EC to the remote EC over the PSTN. The remote UDTEI responds using L-param.response back to UDTEI.

These primitives use all four attributes, discussed previously: request, indication, response, and confirm. After mutual agreement over the parameters, appropriate primitives are used for establishing connections, transfer of data, and, finally, termination of connection (similar to one discussed in Section 8.2 of Chapter 8). The UDTEI receives the characters from the DTE over a V.24 interface, assembles it, and, after defining a block or frame, passes it on to the EC using the L-data.request primitive. The EC encapsulates the data in an information frame (I-frame) defined in the error-correcting procedure of HDLC and transmits it over PSTN.

11.8.4 IBM's synchronous data link control (SDLC)

The SDLC protocol was developed by IBM as a data link protocol for its LAN (SNA). In SDLC, secondary stations cannot send to each other. All communications take place through the primary station. The address field contains only the secondary addresses. The primary can send a series of frames to different secondaries by varying the address fields. The secondary station can respond only during SDLC polling. It does error checking on all frames (control and data). This means that both SDLC and bit-oriented protocols offer higher reliability and lower error rate than BISYNC.

LAP-B is derived from IBM's SDLC and, as such, offers similar services and functions. The data link layer LAP-B protocol defines the procedures between DTE and DCE (at local site) that allow them the link access. When LAP-B is used with the X.25 network, the LAP-B frame typically has a maximum size of 135 octets where the information field (defined within the frame) includes the packet it receives from the network layer. LAP-B is an OSI-based, state-driven protocol, and it supports the features of bit-oriented, synchronous, and different link configurations. Other derivatives which have working principles similar to that of SDLC are ANSI X.66-1979, **advanced data communication control procedures (HDLC)**, **logical link control (LLC)**, and **LAP-D**.

The execution of a particular state defined by LAP-B usually follows these steps:

1. It receives the selective frames while in the receiving state.
2. During the processing of frames, it discards the frames not belonging to that particular state. Other frames belonging logically to that state will be transmitted/stored accordingly, depending upon the command of the link protocol. For each such state, a sequence of actions is defined for its respective activities.

LAP-B supports a maximum of seven frames for a frame sequencing in a pipelined configuration. The error control is handled by a frame-sequencing check (FSC), which is a 16-bit frame similar to CRC-16. Here the error correction is implemented by a combination of acknowledgment and re-transmission of the frame. In the event of error, the frame is re-transmitted.

As indicated above, the LAP-B link can be established by either local DTE or local DCE (if using the X.25 network). The link can be established by any station using appropriate commands. Initially, before the link is established, the station must clear the traffic and set an appropriate mode of operation. A flag is then used to establish connection between stations, where the flag provides the synchronizing bit pattern and is attached at the beginning of the frame. Further, it can be used as the indicator of the end of the frame. In other words, we can use a single flag for indicating the end of one frame and also the start of another frame. The LAP-B general frame format is shown in Figure 11.26(a).

The address field defines two types of frames: command and response. In command frames, we have two types of sub-frames: one contains the sender's information or data, while the other contains the acknowledgment. These sub-frames, in fact, together define the address field. In other words, each address field has a sender's command sub-frame followed by a response sub-frame. The sub-frames contain the address of either DTE or DCE, depending on the type of sub-frame.

The control field includes the sequence number of the frame in addition to the command and response sub-frames. The control field allows three versions of the formats in the control frame:

1. Numbered (N) information transfer
2. Numbered supervisory (S) function
3. Unnumbered (U) control functions

The information field contains the sequence number of the frame, the polling bit, and the acknowledgment number of the frame, as shown in Figure 11.26(b). In the **information frame (I) control field**, N(S) indicates the sequence number of the information frame which has been transmitted. N(R) indicates the sequence number of the information frame which is next in the sequenced information (I) frame pool and is expected by the receiver. The fifth bit position within N(R) indicates the least significant bit. The P/F-bit is set to 1 by the sender, indicating that its command sub-frame is being transmitted.

The supervisory (S) frame defines the acknowledgment of I-frames and performs this operation by requesting the re-transmission of the I-frames. In the meantime, it will halt the transmission of I-frames. The command sub-frame has the following control fields, as depicted in Figure 11.26(c).

- The **P/F-bit** has the same function as that in the information (I) frame. The two-bit supervisory function offers the following operations:
  - **Receive ready (RR):** This operation indicates that the sender is ready for both receiving and acknowledging the I-frame, whose number is defined in N(R) and is represented by bits 00.
  - **Receive-not-ready (RNR):** This operation indicates that the sender is busy and hence cannot receive any I-frames; it is represented by bits 10.
  - **Reject (REJ):** This operation indicates that either DCE or DTE wants the re-transmission of I-frames whose sequence number is defined in N(R) and is represented by bits 01.
- The last bit combination 11 of the two-bit supervisory frame is not used.

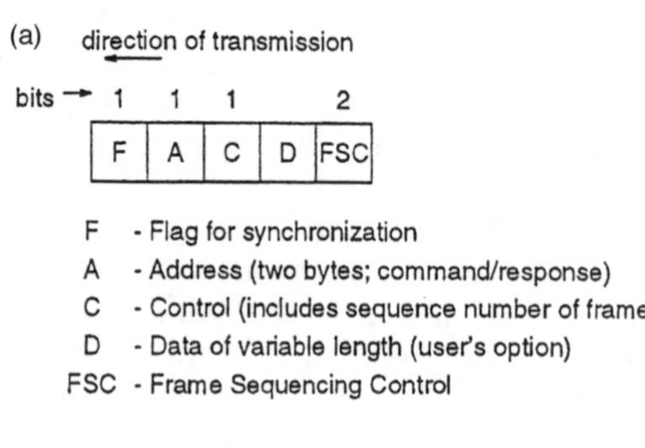

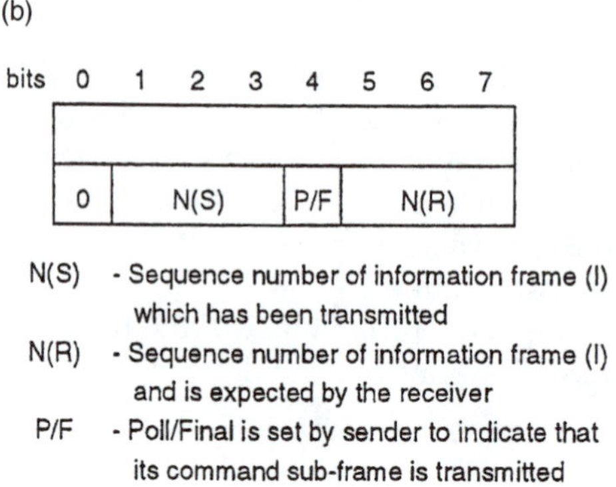

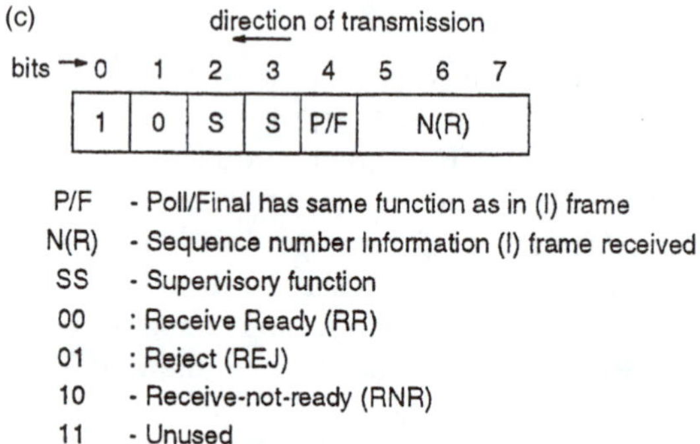

Figure 11.26 (a) LAP-B frame format. (b) Information (I) frame format. (c) Supervisory command sub-frame format.

The unnumbered sub-frame controls the operations dealing with the initialization of transmission, indicating variable lengths of sub-frames defining the transmission configurations, etc. Two unnumbered command sub-frames which have been defined for U-frames are (1) set asyncronous balanced mode (SABM) and (2) disconnect (DISC). Earlier versions of the LAP protocol did not use REJ or RNR frames as command frames, as they were working under asynchronous response mode only.

The HDLC protocols allow data frame exchange across a single channel, and for this reason, this type of protocol is sometimes also termed *single link procedure* (SLP). It also supports the establishment of multiple links or circuits, and the extension of LAP-B which controls these multiple links is termed *multiple link procedure* (MLP). The MLP can be

F	SAPI	TEI	CONTROL	I	FCS	Flag

F - A typical bit pattern of (01111110) for synchronization
SAPI - Service Access Point Identifier. It also includes a bit representing command/response category as well as an extension bit for addressing.

Figure 11.27 Format of LAP-D frame.

considered a list of physical links between LAP-B which resides in the data link layer. The protocol defines a field at the header side to indicate the services of multiple links between nodes. This field is known as multiple link control (MLC) and is transparent to SLP. The SLP frame treats the MLC and frame contents as an information field, and it adds its own address (A) and control (C) fields in front of the frame. The information field of SLP includes SLP and LAP-B. The flag (F) field is given, as usual, at the beginning of the frame as a header, while the frame check sequence (FCS) followed by another flag (F) is used as a trailer of the SLP frame. The MLC field uses two octets and a 12-bit sequence number, giving a maximum-size window of 4096.

Link access protocol-D channel (LAP-D): The data link layer protocols of OSI-RM provide framing synchronization, error control, flow control, and connection of data link layers. The majority of the protocols are subsets of HDLC. We discussed LAP-B earlier, which is also a subset of HDLC. **LAP-D** (ISO I.440 and I.441[17,18]) is a subset protocol for the data link layer of integrated services digital networks (ISDNs). The D-channel of an ISDN rate or access interface is used mainly for signaling and offers connectivity to all the connected devices to communicate with each other through it. Due to this, the protocol for this service is LAP-D, which requires a full-duplex bit transport channel and supports any bit rate of transmission and both point-to-point and broadcast configurations. The LAP-D frame includes user information and protocol information and, in fact, transfers the user's data it receives from the network layer. It offers two types of services to ISDN users: **unacknowledged** and **acknowledged services**.

The *unacknowledged service* basically transfers the frame containing the user's data and some control information. As the name implies, it does not support acknowledgment. It also does not support error control and flow control. The network layer (terminal equipment) uses the primitives of NT2 and ISDN (corresponding to the layer) as D-UNIT-DATA with appropriate attributes (request, indication, response, and confirm) to provide connectionless data communication between peer protocols (layer 3). This service is basically used for transfering management-based messages across the network.

The *acknowledged service*, on the other hand, is very popular and commonly used. It offers services similar to those offered by HDLC and LAP-B. In this service, the logical connection is established between two LAP-Ds before the data transfer can take place (as happens in connection-oriented services). Here, the standard primitives of the data link layer are used to implement all three phases of connection-oriented services (connection establishment, data transfer, and connection termination). It offers a transfer of call messages between DTE and the local exchange and also offers error control. The format of the LAP-D frame is shown in Figure 11.27, and various definitions are given below.

- **F** represents a typical bit pattern of 01111110 (for synchronization).
- **SAPI** is the service access point identifier. It includes a bit representing the command/response category and an extension bit for addressing. The user sends commands by setting this bit to 0 and sends responses by setting it to 1. The network responding to the user's request uses the opposite polarity of the bits for commands and responses.

- The **terminal endpoint identifier (TEI)** defines the number of terminals in the extension bit for addressing. Both SAPI and TEI have an EI bit which, in fact, offers more bits for addressing.
- The **I-field** represents the information field.
- The **FCS-field** controls frame check sequencing (FCS) and has the same functions as discussed in HDLC and LAP-B protocols.

The LAP-D protocol also offers various modes of operation (similar to HDLC), such as numbered, supervisory, information transfer frame, etc. The control field distinguishes between the formats of these modes (information, supervisory, unnumbered). In LAPD, flow control, error control, etc., are usually not supported, and a faster data transfer may be expected. These options are provided in HDLC by the unacknowledged information frame (UIF). Various services and protocol specifications of LAP-D are defined in CCITT I.440 and I.441, respectively. These standards are also known as **Q.920** and **Q.921**. More details about LAP-D protocol are discussed in Chapter 18.

As mentioned above, the SDLC protocol has procedures identical to HDLC in various procedure classes and has been considered a subset of HDLC or a version of the HDLC superset. It uses various procedure classes of HDLC, for example, UNC1, 2, 3, 4, 5, 12, and part of 1 (XD). The SDLC uses the unnumbered normal response mode (UNRM) and defines I-frames (within the frame) as a multiple of eight bits, while HDLC allows an I-frame (within the frame) of variable length.

In SDLC, link management is provided by the primary node and defines several command sub-frames which are not specified in HDLC products or standards. Due to these additional commands, the SDLC offers a useful facility for multi-point links which can be used in ring topology for ring-polling operations. This facility is not compatible with HDLC.

The SDLC is based on the ring topology system. Here, we have one primary station (also known as a central station) and a number of secondary stations. The primary station transmits frames containing the addresses of secondary stations connected in the ring configuration. The secondary station receives the frame, decodes the address, and transmits the frame to the following secondary station. If a frame is addressed to a particular secondary station, it decodes the address and also performs operations on it while it is sending to the following secondary station. The primary station can send the frame to one or more secondary stations, but secondary stations cannot send the frames among themselves directly or indirectly. Instead, they can send the frames to the primary station only.

At the end of the last frame, the primary station sends a minimum of eight consecutive 0s followed by a sequence of consecutive 1s. This scheme is a control transfer sequence known as go-ahead (GA), in which secondary stations are usually operating in daisy-chain fashion on the inbound channel/link. The GA scheme uses a frame consisting of one 0 followed by seven consecutive 1s (01111111). After this frame, the primary station changes its mode to receiving (from transmitting) while continuing to send 1s.

The secondary station has to receive a supervisory (S) frame with P-bit set to 1 or an unnumbered poll frame with P-bit set to 0 to receive a frame from the primary station. After receiving the frames, the secondary station waits to transmit the frame. As soon as it receives the control transfer sequence, it changes the last bit to 0, which, in fact, changes the status to a frame delimiter. This delimiter is then used with the rest of the frames yet to be sent. After the last frame has been transmitted, the secondary station changes to the repeater mode so that it can transmit 1s which it has received from the primary station. The last frame (flag) sent by the secondary is defined as 01111110. The last bit 0 of this frame, along with 1-bits re-transmitted by the secondary station, becomes a control transfer signal to be sent to the following station.

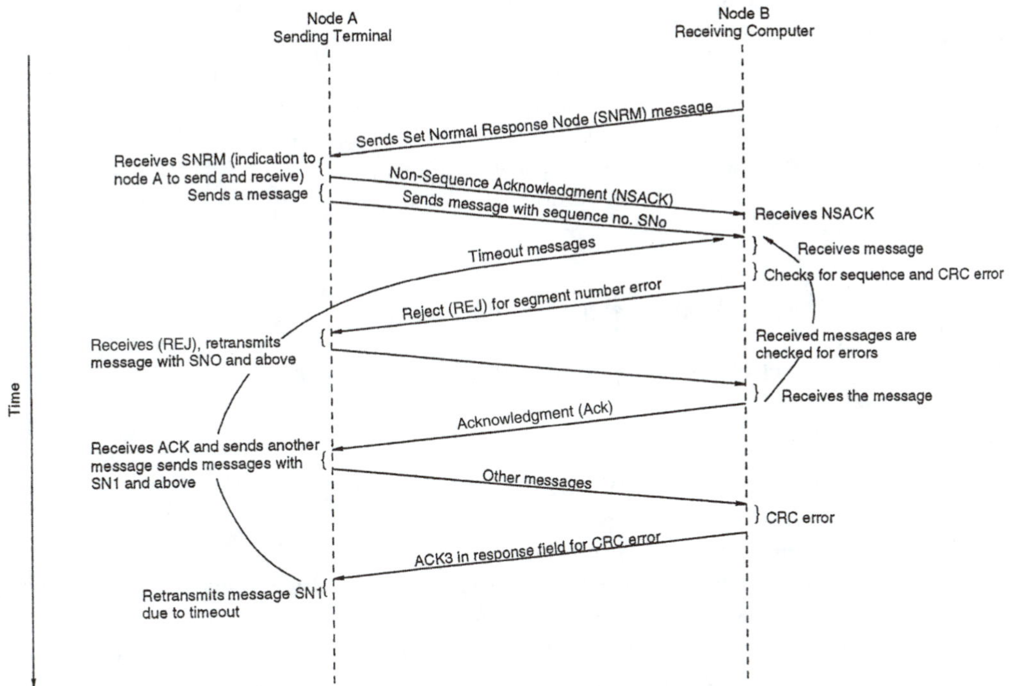

Figure 11.28 A typical data exchange between two nodes across the network using SLDC protocol.

A typical data exchange between two nodes across the network using SDLC protocol is shown in Figure 11.28.

In addition to the above-mentioned popular data link layer protocols, many organizations such as the American National Standards Institute (ANSI) and the International Standards Organization (ISO) have developed other forms of data link layer standard protocols which also include procedures and codes. ANSI's standard protocol is known as advanced data communication control procedure (ADCCP), while ISO's is known as high level link protocol (HDLC). Both of these protocols and SDLC have some common features. In summary, there are five popular standard data link layer protocols, listed below, and each of these has common features and characteristics.

- DEC's digital data communication message protocol (DDCMP)
- IBM's binary synchronous communication (BSC) and synchronous data link control (SDLC)
- ANSI's advanced data communication control procedure (ADCCP)
- ISO's high level data link control (HDLC)
- CCITT .212 and CCITT .222 for SPDNs
- CCITT T71 for PSTNs

11.9 Data link service protocol primitives and parameters

ISO has developed various data link service definitions (for both connection-oriented and connectionless services) in DIS 8886.[16] These services and various functions are usually defined by a set of primitives. The peer protocol layers in different layers within the model communicate with each other by exchanging **data protocol data units (DPDUs)**. These PDUs, which are defined between each pair of layers, are implemented differently across the interface between them. For example, at the physical layer the bits are encoded using appropriate encoding techniques which will be understood by both source and destination nodes. At the data link layer and also at higher-layer levels, the packets or frames are handled (instead of bits) by respective protocols. Each of the layer protocols, when passing the information (data service data unit) to lower layers, adds its header (containing its address and control information) as DPDUs. When the frame arrives at the receiving side, it goes through the same layers, starting from the lower to the upper layers, and at each of its peer layer protocols, the layer removes the headers from the DPDUs until it reaches the destination node.

11.9.1 *Primitives and parameters of data link layer protocols*

The following is a set of service primitives used by the data link layer protocols.

11.9.1.1 *Connection-oriented services*

A set of primitives along with parameters for these services is given below:

- Connection-establishment primitives
  - Dl.connect.request (destination address, sender address, expedited data selection, quality-of-service parameters)
  - Dl.connect.indication (destination address, sender address, expedited data, quality-of-service parameters)
  - Dl.connect.response (destination address, expedited data, quality-of-service parameters)
  - Dl.connect.confirm (sender address, expedited data, quality-of-service parameters)
- Data-transfer primitives
  - Dl.data.request (user data)
  - Dl.data.indication (user data)
  - Dl.expedited-data.request (user data)
  - Dl.expedited-data.indication (user data)
- Termination of connection
  - Dl.disconnect.request (originator, reason)
  - Dl.disconnect.indication (originator, reason)
- Reset primitives
  - Dl.reset.request (originator, reason)
  - Dl.reset.indication (originator)
  - Dl.reset.response (responder)
  - Dl.reset.confirm (responder)
- Error-reporting primitive
  - Dl.error.report.indication (reason)

11.9.1.2 *Connectionless services*

The following two service primitives, along with their parameters, have been defined for this class of service.

Dl.unidata.request (sender address, destination address, quality-of-service parameters, user data)

Dl.unidata.indication (sender address, destination address, quality-of-service parameters, user data)

11.9.2 *Sublayer Service Primitives and Parameters*

The OSI-RM's second layer (data link) consists of two sublayers termed **logical link control (LLC)** and **media access control (MAC)** by the IEEE 802 committee. The layers or sublayers of the LANs interact with each other via service primitives through **service access points (SAPs)** between them. Each layer's service is defined as its set primitives. These primitives are concise statements containing service name and associated information to be passed through the SAP.

The LLC also defines its own set of primitives and parameters for the network layer. Since LLC supports three types of services (unacknowledged connectionless, acknowledged connectionless, and connection-oriented), it has a different set of primitive and parameters for each of these services. There are four standard service primitives attributes/types which are used by different layers for both control and data transfer. These service primitive attributes/types are **request**, **indication**, **response**, and **confirm** (already discussed in previous chapters). It is worth mentioning here that the OSI-RM defines four primitives as **request**, **indication**, **response**, and **confirmation**. The reference model of LAN defined by IEEE 802 (based on OSI-RM) has eight layers, as opposed to seven in OSI-RM. This is because the data link layer is partitioned into two sublayers, LLC and MAC. This reference model of LAN uses only three primitive attributes: **request**, **indication**, and **confirmation**.

The services offered by MAC for LLC allows the LLC entities of both nodes (LLC users) to exchange data with their respective peer entities where these two nodes are a distance apart. The exchange of data is provided via primitives and parameters of MAC.

The MAC sublayer defines a frame for the data packets it receives from LLC and sends the bit stream to the physical layer for its onward transmission over appropriate media. When it receives any frame (as bit stream from the physical layer), it sends it to the LLC, which may send it to the destination address mentioned in the frame or may broadcast it to all of the connected nodes. In the case when the frame does not belong to it, the frames are discarded and sent to the network management facility for appropriate action on these frames. The MAC waits for transmitting frames when it finds the medium is busy. It adds an FCS value to each frame being sent out from the node, and for a received frame, it checks this FCS for any errors.

To understand the exchange of the service primitives between layers via SAP, we consider the LLC, MAC, and physical layers. The LLC sublayer, after receiving a packet from its network layer at its LSAP, sends a service primitive type MA-Data.request to its MAC sublayer via its **MAC SAP (MSAP)** for transfer to its peer LLCs. This primitive has enough control information and it also includes the LSAP address. The MAC sublayer defines a frame with destination address, length of fields, etc., and sends a primitive type MA-Data.confirm to its LLC via the same or different MSAP indicating the success or failure of the operations over the request. It uses another service primitive type MA-data.indication to inform the LLC sublayer about a frame it has received from another MAC address and also about the type of addressing, i.e., individual, multi-casting, or broadcasting.

Similarly, the MAC sublayer uses service primitive types Ph-Data.request, Ph-Data.indication, Ph-Data.confirm, Ph-Carrier.indication, and Ph-Signal.indication to interact with the physical layer via PhSAPs.

These primitives can be implemented in a variety of ways. The service primitives can be used in different ways to support various services offered by it. In some cases, confirmation from the local MAC sublayer is sufficient, while in other cases, confirmation from the distant MAC is required. Confirmation from a distant service provider can also provide confirmed services.

11.9.2.1 Primitives and parameters of LLC protocols

The LLC sublayer offers a common interface with various MAC sublayer protocols and, in fact, is independent of the media access control strategies. The LLC **protocol data unit (PDU)** includes the following fields: destination service access point (DSAP), source service access point (SSAP), a control field, and user's data.

Each address field consisting of eight bits uses seven bits for addressing and one bit for control. In the case of DSAP, the first bit is a control bit and defines the type of address. For example, if it is 0, it indicates that it corresponds to one MAC address (of that node); if its value is 1, it indicates that it represents a group of broadcasting addresses of MAC nodes. Similarly, the first bit of SSAP indicates whether the PDU is a **command** or **response** frame. The control field identifies particular and specific control functions. It is either one or two octets, depending on the type of PDU.

The primitives between higher layers and the LLC are always defined by a prefix of L, while the primitives between MAC and LLC sublayers are defined by a prefix of M. The following is a list of the primitives defined by LLC which are used in different types of services provided by it.

- **Unacknowledged connectionless:**
  L-Data.request and L-Data.indication — For each these primitives, the parameters used include source address, destination address, data units (defined as link service data units), priority, and service class.
- **Acknowledged connectionless:**
  L-Data.ack.request and L-data.ack.indication — This service was introduced in 1987 in the IEEE 802.3 recommendations. For each of these primitives, the parameters used include source address, data units (link service data units), priority, and service class. Here, no connection is established before sending of any data, but acknowledgment is required (L-data.ack.request) for the receipt of the data by the source node from the destination node (L-data.ack.indication).
- **Connection-oriented:** The set of primitives and parameters for connection-oriented services is divided into three categories: connection establishment, data transfer, and connection de-establishment. For each phase, we have a set of primitives and parameters which are used for the exchange of the data between the LLC entities and appropriate peer LLC entities. It is worth mentioning here that, out of four primitive attributes (request, response, indication, and confirm), the **response** attribute is not used in the latest recommendations.
  - **Connection establishment phase:**
    L-Connect.request, L-connect.indication, L-connect.response, and L-data.confirm — The parameters associated with these primitives include source address, destination address, priority, service class, and status (only in indication and confirm).
  - **Data transfer phase:**
    L-Data.request, L-Data.indication, L-Data.response, L-Data.confirm — Various parameters associated with these primitives include source address, destination address, data units (link service data units), and status (in confirm only).
  - **Connection De-establishment (termination):**
    L-Disconnect.request, L-disconnect.indication, and L-disconnect.confirm — The parameters used are source address, destination address, reason (in indication only), and status (in confirm only).

- **Link reset:** The LLC connection can be reset by the following primitives:
  L-Reset.request, L-reset.indication, L-reset.response, and L-Reset.confirm — The parameters used here are source address, destination address, reason (in indication only), and status (in confirm only).
- **Link flow control:** The packets received from the network layer are controlled by using the following LLC primitives:
  L-connection-flowcontrol.request, L-connection-flowcontrol.indication — These primitives will discard all the frames and do not support the recovery of these discarded or lost data units. The higher-layer protocols have to provide a provision for recovering these lost or discarded frames. They use parameters such as source address, destination address, and the amount of data the LLC entity can handle, and this can be set to any value. In the event of its zero value, no data will be accepted by the LLC entity. Likewise, the network layer defines its own primitives for controlling the data it is receiving from the LLC sublayer.

The data link layer protocol is the only protocol which adds both header and trailer to the PDU it receives from the network layer and defines the frame. In order to distinguish between the data passed from lower layers to higher or vice versa with its peers, the data passed from the higher layer to its lower layer is known as the service data unit (SDU); PDUs are exchanged across peer layers, while SDUs are packaged with the header and passed down as SDU.

11.9.2.2 Primitives and parameters of MAC protocols

The frames defined by the MAC sublayer containing the user's data information, control information, and many other related control fields are discussed below:

The primitives offered by MAC include MA-Data.request, MA-Data.indication, MA-Data.confirm, and MA-Data.status.indication.

As indicated earlier, the MAC and LLC sublayers interact with each other via MAC service access points (MSAPs). The LLC uses a primitive MA-Data.request to transfer the data units to peer LLC(s). The MAC sublayer constructs a frame from LLC data units, as it contains all the relevant information such as destination address, length of data, and other options requested. The MAC sublayer, after receiving the LLC data units, constructs the frame and indicates the success or failure of its request by using the primitive MA-Data.confirm via another SAP. It uses a primitive MA-Data.indication to inform its LLC sublayer (through available SAP) about a frame it has received which has the destination of this MAC address. It also indicates if the frame has a broadcasting address or multi-casting address defined in the destination address field.

The parameters associated with this primitive include source address, destination address, media access control service (MACS), transmission status, service class, and MAC service data unit. The service primitives and parameters defined by MAC are different for different LANs, as these functions are not defined by a common LLC sublayer. The MAC sublayer offers the following functions: assemble the data packets (LLC) into a frame with address and error fields for transmission, disassemble the frame into LLC data packets, recognize the address and error fields on the receiving side, and provide management for data communication over the link. The above discussion is partially derived from References 2–6.

Some useful Web sites

More information on HDLC can be found at *www.jyu.fi/~eerwall/tausta.html* and *www.erg.abdn.ac.uk/users/gorry/eg3561/dl-pages/hdlc.html*. Cisco solutions to WAN data-link switching can be found at *www.cisco.com/warp/public/cc/cisco/mkt/iworks/wan/dlsw/index.shtml*. More information on the data link layer can be found at *www.rad.com/networks/1994/osi/datalink.htm*. Information on data link protocols can be found at *www.cs.panam.edu/~meng/Course/CS6345/Notes/chpt-4/node19.html* and *www-dos.uni-inc.msk.ru/tech1/1994/osi/l2examp.htm*.

References

1. International Standards Organization, *Open Systems Interconnection — Basic Reference Model*, Information Processing Systems, 1986.
2. Tanenbaum, A.S., *Computer Networks*, 2nd ed., Prentice-Hall, 1989.
3. Stallings, W., *Handbook of Computer Communications Standards*, vol. 1, 2nd ed., Howard W. Sams and Co., 1990.
4. Markeley, R.W., *Data Communications and Interoperability*, Prentice-Hall, 1989.
5. Black, U., *Data Networks: Concepts, Theory and Practice*, Prentice-Hall, 1989.
6. Spragins, J.D., *Telecommunications: Protocols and Design*, Addison-Wesley, 1991.
7. Jennings, F., *Practical Data Communications*, Blackwell, 1986.
8. Schwartz, M., *Telecommunications Network Protocols and Design*, Addison-Wesley, 1987.
9. Schlar, S.K., *Inside X.25: A Manager's Guide*, McGraw-Hill, 1990.
10. International Standards Organization, *High Level Data Link Control Procedures — Frame Structure*, ISO 3309, 1986.
11. International Standards Organization, *High Level Link Control Procedures — Consolidation of Elements of Procedures*, ISO 4335, 1986.
12. International Standards Organization, *Multilink Procedures*, ISO 7478, 1986.
13. International Standards Organization, *High Level Data Link Control Procedures — Consolidation of Classes of Procedures*, ISO 7809, 1986.
14. International Standards Organization, *Local Area Network-Part II: Logical Link Control*, ISO 8802.2 (ANSI/IEEE 802.2).
15. International Standards Organization, *High Level Data Link Control Procedures — General Purpose XID Frame Information Field Content and Format*, ISO 8885, 1986.
16. International Standards Organization, *Data Link Services Definition for Open Systems Interconnection*, ISO 8886, 1986.
17. CCITT I.440, *ISDN User-Network Interfaces Data Link Layer-General Aspects.*
18. CCITT I.441, *ISDN User-Network Interface Data Link Layer Specification.*
19. CCITT X.212 and X.222.
20. CCITT T71.
21. Motorola, "The X.25 protocol controller (XPS)," Motorola Technical Summary no. MC 68605, Phoenix, AZ, 1986.
22. IBM, Binary Synchronous Communications, GA27-3004-2, IBM Corp., 1970.
23. IBM, System Network Architecture, GC30-3072, IBM Corp., 1981.
24. Martin, J. and Chapman, K.K., *SNA IBM's Networking Solution*, Prentice-Hall, 1987.

References for further reading

1. A. Rybczynski, "X.25 interface and end-to-end virtual circuit service characteristics," *IEEE Trans. on Communication*, April 1980.
2. IEEE, "X.25 and related protocols," *IEEE Computer Society*, 1991.
3. Black, U.D., *Data Link Protocols*, Prentice-Hall, 1993.

chapter twelve

Network layer

"There's nothing so dangerous for manipulators as people who think for themselves."

Meg Greenfield

"The third lowest layer in the OSI seven layer model. The network layer determines routing of packets of data from sender to receiver via the data link layer and is used by the transport layer. The most common network layer protocol is IP."

The Free Online Dictionary of Computing
Denis Howe, Editor (http://foldoc.doc.ic.ac.uk/)

12.1 Introduction

As discussed in previous chapters, the physical layer provides physical connection and services to the data link layer, while the data link layer is concerned with the construction of frames, error control, flow control, and the transmission of frames over the physical circuit connection between computers or even between different types of networks. The frames (containing user's data) may have to go through a number of nodes (computers, networks) which may be connected in different topologies. The network layer entity uses the services of the data link layer and provides services (routing of packets, relaying, congestion control, etc.) to its higher layers (transport) and also to different types of intermediate networks or sub-networks. The primary functions of the network layer are routing, selection of appropriate faults, controlling of overloading or congestion of traffic on communication lines, efficient use of various communication lines, etc. This layer manages network layer connections and provides an end-to-end reliable packet delivery mechanism. This layer provides user-to-user connection for exchange of data between them.

This chapter discusses various services offered by the network layer to the transport layer entity and data layer entity within OSI-RM, including routing and flow control of user's data (via packets) through nodes of the same network and also through sub-networks over different modes of operation (connection-oriented and connectionless), establishing connection between different types of networks (internetworking) for sending the user's packet over them.

The network layer in the OSI model is also known as the **subnet**, which may run on different networks (e.g., X.25, ARPANET, NSFNET, and others) where the network layer runs on **interface message processors (IMPs)**. The transport layer of the OSI model runs on the hosts. The control and management of the network layer are provided by different organizations in different countries, such as PTT, AT&T, etc. PTT is very popular in European countries and undeveloped countries. Thus, the network layer not only provides services to the transport layer and different sub-networks but also defines an interface and connection between different hosts and different types of networks.

12.2 Layered protocols and interfacing

As mentioned in previous chapters, the layers are, in fact, the foundation of many standards and products in the industry today. The layered model is different from layered protocols in the sense that the model defines a logic framework on which protocols (that the sender and receiver agree to use) can be used to perform a specific application.

12.2.1 Layered protocols

Network layer protocols provide a logical communication link between two distant users across the network. Each node (PCs, terminals, etc.) connected to a network must contain identical protocol layers for data communication. Peer protocol layers exchange messages which have commonly understood formats, and such messages are known as **protocol data units (PDUs)**. PDUs also correspond to the unit of data in a network peer-to-peer protocol. The peer *m* entities communicate with each other using the services provided by the data link layer. The network layer offers its services to the transport layer at a point known as a **network service access point (NSAP)**. For a description of the interface between it and the transport layer, the network primitives are defined. The network service data unit (NSDU) associated with *m* primitives is delivered to the transport layer and, similarly, it uses its primitives to interact with the network layer. The transport layer sends its transport protocol data unit (TPDU) to the network layer. The TPDU consists of the transport protocol, control information, and its user data. The transport control information is exchanged between its entities.

12.2.2 Interfacing

A protocol defines rules, procedures, and conventions used for data communication between peer processes. It provides data formatting, various signal levels for data and control signals, error control, synchronization, and appropriate sequencing for the data to be exchanged. Two networks will offer a two-way communication for any specific application(s) if they have the same protocols. The network layer is known as a service provider for the transport layer and offers the services at the NSAP.

ISO has defined the following terms used for data communication through OSI-RM. Each layer includes active elements known as entities, and entities in the same layer of different machines connected via network are known as **peer entities.** The entities in a layer are the service providers, and the implementation of services used by its higher layer is known as the user. The **network service access point (NSAP)** is defined by a unique hardware or port or socket address (software) through which the transport layer accesses the services offered by it. The NSAP defines its own address at the layer boundary. The **network service data unit (NSDU)** is defined as message information across the network to peer entities of all the layers. The NSDU and control information (addressing, error control, etc.) is known as the **interface data unit (IDU)**, which is basically message information sent by the transport layer entity to its lower layer entity via NSAPs. For sending an NSDU, a layer entity partitions the message information into smaller-sized messages. Each

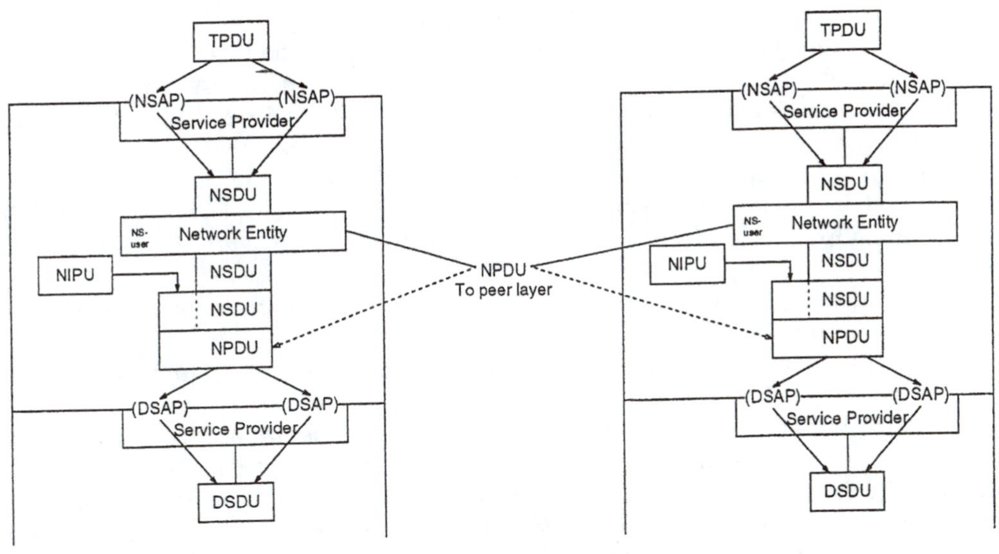

Figure 12.1 The relationship between NSDU, NPDU, and the network entity.

message is associated with control information defined by an appropriate layer as a header and sent as a **protocol data unit (PDU)**. This PDU is also known as a **packet**. We will be using herein the terms defined by ISO. The relationship between NSDU, NPDU, and the network entity is shown in Figure 12.1. Some of these definitions can be found in many publications and are presented here as derived from References 2–4, 7, and 28–30.

12.3 ISO network layer services and functions

According to OSI document ISO 8348,[10] the network layer must possess the following features:

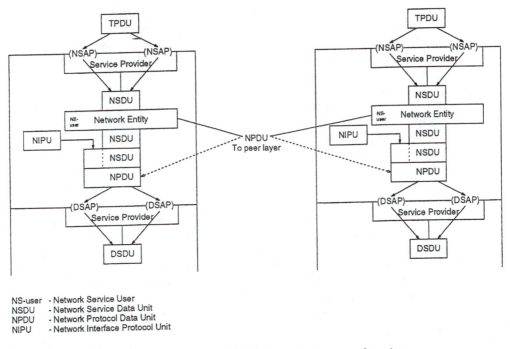

1. Offers services to various types of sub-networks. The sub-network may be defined as a set of one or more intermediate nodes which are used for establishing network connections between end systems and provide routing and relaying of frames through it.
2. Offers uniform addressing and services to LANs and WANs. The network layer defines network addresses which are being used by transport entities to access network services. Each transport entity is a unique network connection endpoint identifier defined by the network layer which may be independent of addressing required by underlying layers. The network entities define network connections and may also include intermediate nodes which usually provide relaying. The network connection over intermediate subnets is known as a **subnet connection** and is usually handled by the underlying protocols.
3. Establishes, controls, and maintains a logical connection between the transport layer entities for exchanging the data across the networks. The network connection offers point-to-point connection and more than one network connection may be defined between the same pair of network addresses. The network service data units

(NSDUs) are transferred transparently between transport entities over the network connection and generally do not have any upper limit on their size.

4. The transport layer can adapt to different types of sub-networks (it is independent of the characteristics of sub-networks, e.g., routing strategies considered, topologies and types of the sub-network).
5. Offers an acceptable quality of services (QOS) during the network connection, which in turn is based on parameters such as residual error rate, service availability, throughput, reliability, etc. It also reports any undetectable errors to the transport layer.
6. Provides sequencing for NSDUs over the network connection when requested by transport entities.
7. Provides both connection-oriented and connectionless services to the users.

In connection-oriented services, a logical connection is set up between two transport entities, including agreement for the type of service desired, parameters, cost, priorities, transfer of data (in packets) in both directions with appropriate flow control, and, finally, the termination of the connection. For each of these operations, the network entities use a primitive which is associated with different parameters. These primitives are also associated with various attributes such as request, indication, response, and confirm. Various network service primitives along with their parameters are discussed in the following section.

On the other hand, in connectionless services, network layer primitives SEND and RECEIVE are being used to send and receive the packets across the networks. The users are responsible for flow control and error control on their hosts. This is very common among the network users that these services (flow control) should be provided by the user's host. Thus, in connectionless mode, duplication of flow control and error control can be avoided. Each packet contains all of the information (i.e., address of destination, routing, etc.) and is processed separately. This type of mode typically is used for e-mails, bulletin boards, etc., while the connection-oriented mode is used for reliable transmission of packets across the network.

The difference between connection-oriented and connectionless services has been discussed in many books, and analogies between these services and other systems have been derived. For example, the service provided by a **public switched telephone system (PSTS)** is analogous to connection-oriented services, as it requires the same steps for transferring the data between users as in connection-oriented services. The dialing sets up the end-to-end connection (if it not busy), the transfer of data (spoken/verbal) messages takes place between the users, and, finally, the connection is terminated by either one of the users by hanging up the handset. Although the telephone system itself may be very complex from the inside, it provides connection-oriented service and offers end-to-end (user-to-user) connection along with other features such as flow control and error control. The connectionless service is similar to the postal service where the user has to provide the entire address on the envelope. Various control information associated with this type of service may include the type of transmission (ordinary post, registered post, insured, next-day delivery, printed-matter category, etc.).

Both services (connection-oriented and connectionless) are either provided or supported by the OSI model from the physical layer to the application layer. The routing in the network is implemented differently for different services (connection-oriented or connectionless). In other words, connection-oriented and connectionless networks will have different routing strategies. The connection-oriented network is analogous to a telephone system, circuit-switched physical network, or virtual circuit network, while the connectionless network is analogous to the postal service, packet-switched network, or datagram network.

The above was a description of features of the network layer. As indicated earlier, the network layer offers services to the transport layer and at the same time receives the services offered by the data link layer. In addition to the services offered by each of the layers, each layer also defines its set of functions. There will be overlapping between the functions and services of any layer. In order to distinguish between the functions and services of the network layer, we present the functions and services separately, as follows:

- **ISO functions:** Routing and relaying, network connections, multiplexing of network connections, segmentation, error control (detection and recovery), sequencing, flow control, expedited data transfer, service selection, reset, network layer management, and part of OSI layer management.
- **ISO services:** Network addresses, network connection, network connection identifiers, quality of service, error reporting, sequencing, flow control, reset, expedited transfer of NSDUs, and release and termination of network connections.

As explained earlier, each layer accesses the data through its **service access point (SAP)** above it. The top layer (application) of the OSI model does not have any SAP, as it provides the user's services directly to the lower layers. Similarly, connection-oriented and connectionless services are provided by the data link layer all the way up to the application layer. The function of the physical layer is to accept raw bits from the data link layer and pass it on to the physical media. It is important to note that these services are defined and implemented only in network and transport layers, and each top layer provides support for this service.

The network layer protocols combine control information and data into a unit known as a **network protocol data unit (NPDU)** in OSI-model terminology. The network services exchanged between transport entities are known as **network service data units (NSDUs)**. The NDSU also includes the network address, which indicates the address with which a connection is to be established for the transfer of data. Other services provided by the network layer to the transport layer are cost and **quality of service (QOS)**. The QOS is usually defined by various parameters. In addition to services offered by the network layer to the transport layer, the network layer also possesses routing and relaying information. The routing and relaying information is usually implemented in a tabular form and stored at each node. Thus, the network layer protocol is very complex, as it has to have various routing and relaying information across the sub-networks and also provide services to the transport layer and sub-networks of different characteristics.

12.4 *Network routing service concepts*

When the transmitted frame arrives at the computer of the receiving side, the physical layer of the receiving side accepts the frame and passes it on to the data link layer. The data link layer protocol checks the frame for errors. If there is no error in the frame, the data link layer removes its header and trailer from the frame and passes the packet on to the network layer. In the event of error, the data link layer sends negative acknowledgment (Nak) to the sender's side and discards the frame. In the case of an error-free frame, the network layer protocol looks at the network header of the packet. If the address pertains to the same network and computer, it will accept it; if it pertains to another computer on the same network, it will route the packet accordingly. If the packet does not belong to the network, it will send the packet to the appropriate sub-network(s) with information about the possible physical circuit connection, and this process is repeated until the packet is delivered to the designated destination. The ISO PDTR 9575[18] discusses the OSI routing framework.

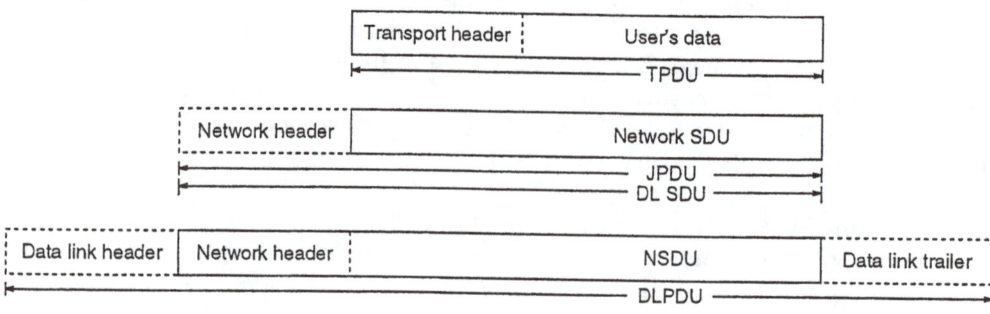

Figure 12.2 Data communication in OSI-RM.

12.4.1 *Data communication in OSI-RM*

The network layer, after receiving the packet from the transport layer, adds its own header to the user's data and passes it on to the data link layer. The network header includes the source address, destination address, and control information. The data field of the packet, in fact, represents the network service data unit (NSDU), as this offers services to the transport layer. The packet delivered to the data link layer contains the source destination addresses of NSAP and also an indication of establishing a connection with another computer, as shown in Figure 12.2.

12.4.2 *Connection-oriented networks*

In connection-oriented networks (e.g., circuit-switched network), in the user's data packet, a virtual channel number is stored in the destination address field. A packet from the transport layer is sent as a **virtual circuit unit (VCU)**. Full destination and source addresses are usually not required. This means that the size of the network layer header is very short, which helps in transmitting large amounts of data/files (more data in packets) efficiently. These networks have one major drawback: if any sub-network/computer node in the network is down or crashed, then all of the virtual circuit packets have to be re-transmitted, which requires fresh requests for the connection, re-establishment of connection, and re-transmission of all these packets that were lost during that time. The transport layer protocols perform these operations in the event that the computer goes down.

12.4.3 *Connectionless networks*

In connectionless networks (packet-switched networks), the packets from the transport layer are sent as a **datagram unit (DU)**. Each datagram unit has the complete destination and source address. The network layer makes the decision as to which datagram unit is to be sent next after the datagram units arrive there. The physical layer may have a pre-established physical connection and datagrams may have been there, but the decision will be made by the network layer. Thus, the datagram units in these networks may arrive either in sequence or out of sequence or may not even arrive. These networks are useful for short messages, and the datagram units can be sent more efficiently than virtual circuit units. The only problem with this service may be the time required to establish connection between two computers. If any node is down, it will not affect the operation, as the paths for each of the datagram units can easily be changed. This, of course, requires an additional protocol in the transport layer which must provide a reliable service to upper layers at the cost of extra overhead.

As discussed above, the network layer makes a logical connection between transport entities of the sub-networks. If the sub-networks are the same, then the logical connection between transport entities provides the network connection, provided both the sub-networks use the same protocols and offer the same services with the same QOS (quality of service). This then provides end-to-end connection between transport entities. But if the sub-networks are different and provide different services with different QOS, then the network layer protocol will be responsible for providing an acceptable QOS between those connected sub-networks. These sub-networks are usually connected in series or tandem to provide internetwork connection.

12.5 Sub-network (subnet) interface

As stated previously, the standard physical layer interface for the OSI model is DTE/DCE. This interface also provides end-to-end communication between two open systems. The open system, in fact, consists of two components: **user's host** and **sub-network (subnet)**. The lower three layers (physical, data link, and network layers) define/form the sub-network, while upper layers from the transport layer up to the application layer define/form the host. The communication between open systems is provided by sub-networks, and the end systems are attached to communication devices DTE/DCE. Depending on the type of networks, the communication devices may be used in a variety of ways to define the different configurations for end-to-end communication between open systems. Some of the widely used network configurations for open systems are discussed below.

Packet-switched data network (PSDN): In packet-switched networks, DCE is defined by the sub-network of the OSI model which is considered a switching node in the network. The end system is defined by the host, including its own sub-network that is attached with its switching node. The transport layer provides end-to-end service, while the network layer provides routing for data through the sub-network switching nodes. In other words, the transport layer provides DTE and DTE-oriented service, while the network layer provides DCE and DCE-oriented service. This is due to the fact that the network layer is the first layer of the sub-network which provides interaction with DTE. The network layer interacts with the transport layer to get information regarding priority, etc., and after it receives the packet, it provides information regarding the destination address to the sub-network which, in turn, based on this information, determines the routing for the data to the destination DTE address.

DTE-DCE oriented service requires the protocols between the lower three layers of DTE and the three layers of the sub-network's DCE. The set of protocols for local DTE-DCE service will be entirely different from those in the destination DCE-DTE service. In other words, the routing at the local DTE-DCE interface will be different from the actual route for data packets across the sub-network switching nodes, thus providing total transparency for DTE-DCE interactive and internal operation to the user at the local node.

A set of protocols for the DTE-DCE interface has been recommended; it is known as the X.25 standard (discussed in the next section). The top layer (network) of this standard is defined as a **packet-level protocol**. The services offered by the network to the transport layer of the OSI model are defined in ISO 8348,[10] connectionless service is defined in ISO 8348/DAD1,[11] and network addressing is included in ISO 8348/DAD2.[12]

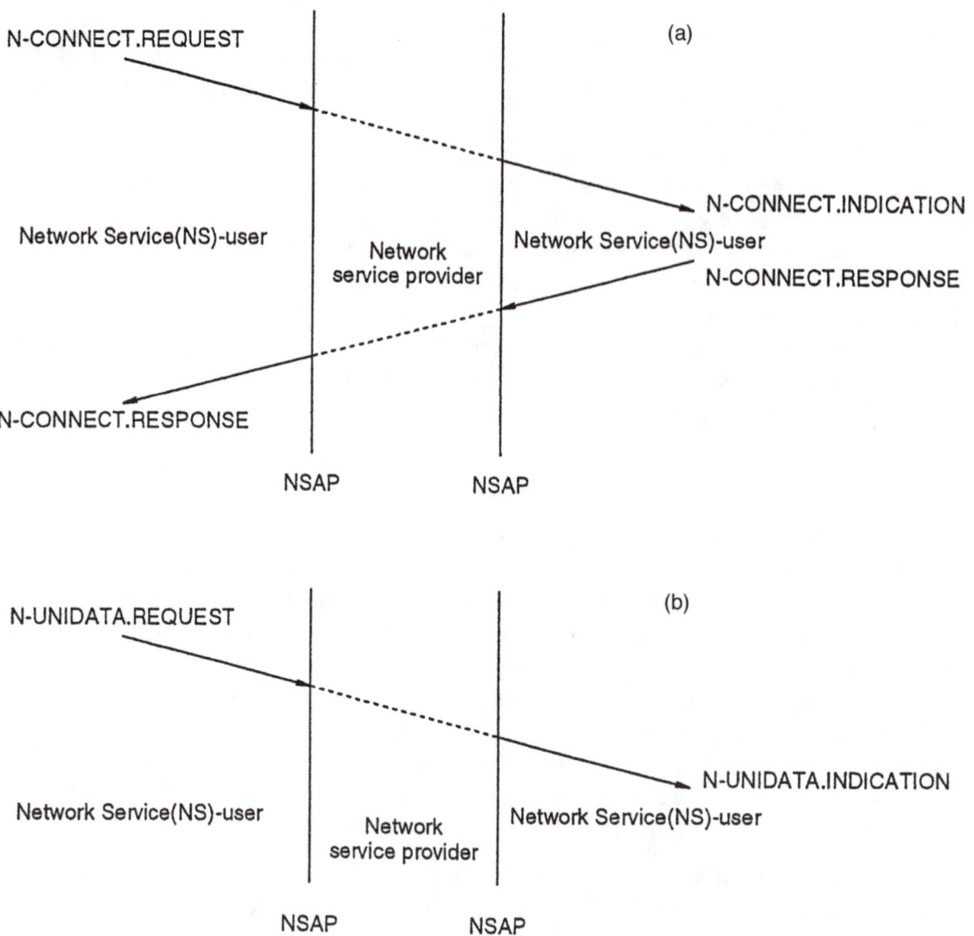

Figure 12.3 (a) Connection-oriented network connection operation. (b) Connectionless data transfer operation.

These services are implemented as protocol entities which are used to provide a logic connection between layers for exchange of data packets. The transport layer sends a primitive, **N-CONNECT.request**, to the network layer, which in turn sends another primitive, **N-CONNECT.indication**, to the transport layer entity of the destination network. The network service user (transport entity) uses a request primitive for the network service parameter (network entity). This primitive is received by the remote service provider as **N.CONNECT.indication** primitive. The remote transport layer entity sends back another primitive, **N-CONNECT.response**, to the sender network. This layer then sends another primitive, **N-CONNECT.confirm**, to its transport layer, thus establishing a logical connection between two nodes (sender and receiver) for connection-oriented service. Figure 12.3(a) shows the sequence of these primitives being exchanged between two nodes for establishing a logical connection between them for data transfer. Figure 12.3(b) shows the sequence of the primitives exchanged between nodes for exchange of data.

Point-to-point data network (PPDN): This type of network configuration uses a standard physical layer interface to provide end-to-end communication between open systems.

Here, the DTE (PCs, terminals, etc.) interacts with the DCE (e.g., a modem) at the physical layer only. The communication between two DCEs (modems) can be obtained by a physical layer protocol, depending on the type of transmission media considered. Layers 2, 3, and 4 provide end-to-end, DTE-DTE-oriented service, while the physical layer provides DTE-DCE-oriented service. The physical layer simply accepts the raw bits from DTE and transmits them on the communication media. In this configuration (point-to-point), the network layer does not provide any information regarding a destination address for routing, priority, etc. A packet-level protocol X.25 DTE-DTE has been recommended and defined for this configuration by CCITT (discussed in the next section).

Circuit-switched data network (CSDN): This network configuration is similar to a point-to-point configuration, with the only difference being in the establishment of a connection between DTEs. In a point-to-point configuration, connection is always established between DTEs, while in the case of a circuit-switched network, a link or circuit has to be established between source and destination, which requires the establishment of a circuit along the intermediate sub-networks. Here again, the physical layer accepts the raw bits from the data link layer and transmits them on the established circuit. The frames can be sent to the destination only when a circuit all the way from source to destination has been established. The network layer does not provide any information regarding routing, priority, etc. The layers 2 (data link), 3 (network layer), and 4 (transport layer) provide DTE-DTE-oriented service, while layer 1 provides DTE-DCE-oriented service.

Local area networks (LANs): In this network configuration, DTEs of open systems are directly connected, thus avoiding DCEs. Layer 2 (data link) protocol provides end-to-end service between connected communication devices (DTEs). The physical layer performs the same function as in the two previous cases (i.e., it accepts raw bits from the data link layer and transmits them directly to the destination DTE). DTEs use a shared sub-network medium. Here also, the network layer protocol does not provide information regarding routing, priority, etc. The standards for layers 1 (physical layer) and 2 (data link) of OSI-RM can be found Section 10.4 and 11.4, respectively. These standards are also presented in Section 8.5 of Chapter 8.

12.6 Packet-switched data network (PSDN) (CCITT X.25)

In the previous section, we discussed four classes of configurations for establishing connection between open systems. Out of these, the packet-switched data network (PSDN) has been more popular and important than others, as it offers connection-oriented service, which is one of the characteristics of all ISO-based model protocols.[20] Further, the packet-switched network interface standard X.25 has been adopted for interfacing different vendor products and also has been used in different systems. Due to its widespread acceptance and use, the network management and maintenance costs may be improved. X.25 provides a connection-oriented service in packet-switched networks that conforms to the network services and functions defined in the ISO 8348 document.

The virtual circuits defined in X.25 also support a packet multiplexing which multiplexes the packets from different sources and destinations over a link (in contrast to TDM). The host machine on the receiver side de-multiplexes the packets by looking at the multiplexing ID (e.g., virtual circuit number in X.25). It is worth mentioning here that the X.25 offers a virtual connection which remains active until the session (virtual circuit) is terminated, and it can be used for routing packets. For datagram packets, the packets are routed over different links defined by different sets of switching nodes, as these packets contain complete source and destination addresses.

CCITT recommended in 1974 a standard X.25 for providing interfacing to different types of sub-networks with a packet-switched network. This standard was revised in 1976, 1980, 1984, and 1988. Due to many revisions and modifications, it has become one of the most important standards for providing an interface to the open system OSI model with a packet-switched network (wide area packet network). It is also known as a DTE-DCE interface, which encompasses the lower three layers (1, 2, and 3) of the OSI model. The physical layer interface between DTE and DCE is defined in another standard, X.21 (see Section 10.4.8).

The packet-switched networks define various access points and establish virtual circuits using these access points. They also define a direct link for **value-added networks (VANs)**. These networks offer certain added services with additional qualities; e.g., after we have sent our message in the form of packets to access points, the network will reroute the packets in the event of any error without contacting the user's node. The X.25 offers a standard interface to VANs. Detailed information regarding VAN services, various VAN connections, and different-sized vendor networks offering these services is given in Sections 1.2.4 and 12.10 of Chapters 1 and 12.

The access points may be connected by a number of links, such as leased lines (not suitable for long distance), switched long-distance phone lines (expensive, as transfer of data over it is based on usage per call), and, finally, packet-switched data networks. They offer a central site for the networks and define virtual circuits for data transfer. The access points can be located at a corporate office, or users may access these points via conventional telephone lines. The access points are connected to different switching nodes which define a virtual circuit or path for the packets to travel through the network until they reach the switching node on the destination node. This node then routes the packet to destination-based access points.

12.6.1 Evolution of X.25 networks

The first X.25 recommendation was issued in 1976 and had few facilities (like call barring, link access procedure, etc.). The X.75 recommendation was defined in 1978. The 1980 version of X.25 introduced other facilities such as datagram, fast select, LAP-B, etc. In 1984, the recommendation for the datagram facility was dropped and new facilities such as call redirection, multi-link procedure, OSI address extension, etc., were defined and added. Also, OSI quality of service (QOS) and X.32 recommendations were defined.

The 1988 version of X.25 included call deflection facility and long address formatting. At this point, many vendors have defined PSDNs and other related products incorporating all of the facilities of the 1988 version of X.25, and they are available to the users.

12.6.2 X.25 standards

It is important to note that X.25 does not define any packet-switching specifications for the public networks; instead, it is a network interface specification for the network. The X.25 standard offers standard interfaces to VANs, as well. The X.25 does not provide any information regarding routing, etc., but offers a communication environment between DTE and DCE for the exchange of data between them, as shown in Figure 12.4. Here, the DTE is a normal user node (terminal or PC), while the DCE is considered a network node. The interface between DTE and DCE was initially termed an interface between **data terminal equipment (DTE)** and **data circuit terminating equipment (DCE)** for terminals operating in the public packet-switched data networks (PSDNs).

As stated above, the X.25 packet level specification standard corresponds to the standard of the three lower layers of OSI-RM (physical, data, and network layers).[20] The

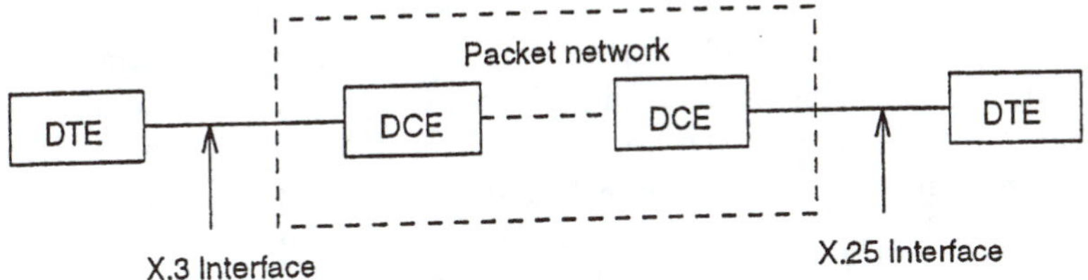

Figure 12.4 X.25 communication interface.

Table 12.1 Physical Layer Interfaces EIA-232 and V.24 Used with the X.25 Network

Control/data signal	EIA-232	V.24
Send data	BA	103
Receive data	BB	104
Request to send (RTS)	CA	105
Clear to send (CTS)	CB	106
Data set ready (DSR)	CC	107
Data terminal ready (DTR)	CD	108
Carrier detect (CD)	CF	109

physical layer standard provides interface between the DTE and DCE, and it offers an electrical path for the transportation of packets. The physical layer uses X.21 or X.21 bis, while the data link layer uses LAP or LAP-B protocols. It also supports the use of EIA-232 or V.24/V.28 physical layer interfaces (discussed in Section 10.4) and, in fact, is one of the useful features of the X.25 protocol as the X.21 interface is being used in many countries at the physical layer of the networks. The X.25 recommendation terms this physical layer interface (X.21) as X.21 bis in its recommendation; X.21 bis can also operate with other physical layer standard interfaces (e.g., RS-449, V.35, etc.).

The communication between X.25 and the network node can take place only when the V.24 circuits (105, 106, 107, 108, and 109) are in active states. These circuits represent the state of the physical layer. If these circuits are inactive, then X.25 cannot communicate with the data and network layers. The data exchange takes place in circuits 103 and 104. Table 12.1 shows the circuits of physical layer interfaces EIA-232 and V.24 when used with the X.25 network. For each of the circuits, the control and data signals are shown to be equivalent. The table does not show the ground reference and other timing signals (for details about these, please read Chapter 10 on physical layer).

12.6.3 X.25 layered architecture model

The X.25 packet-level standard is widely accepted and is being used by different vendor products around the world. It provides DTE-DCE interface across the network where DCE offers access interface to the packet-switched network. It is defined basically for the third layer of OSI-RM, but it also encompasses the lower layers 1 and 2. The physical interface at the physical layer is defined as X.21. Other physical interfaces at layer 1 which are supported by X.25 are EIA-232-D, V.24/V.28, RS-449, V.35, and X.21 bis. Within the context of the X.25 standard, the different physical interfaces are defined as X.21 bis (i.e., X.21 bis is equivalent to them). The physical layer provides only an electrical path for sending out the packets and, hence, does not have much control over the network X.25 standard.

One of the main objectives of packet-switched networks is to use short packets so that routing decisions can be made quickly while the packets (to be sent) are still in the main memory. The user's data field in the original ARPANET was a maximum of 128 octets, and X.25 derives its information field size of 128 octets from the original ARPANET. Now we have packet-switched networks which can extend their information fields up to 4096 octets. This maximum size of information field is relatively shorter and offers a very short time for sending packets from a node to another, thus reducing the processing delay at the nodes.

The X.25 provides an interface between DTE and DCE but does not provide the peer-to-peer network interface up to the network layer of OSI-RM. It includes the three layers of the OSI model: physical, data link, and packet (network) layers. Each of these layers is discussed in more detail in the following paragraphs, with appropriate standards and different types of protocols.

X.25 physical layer: The X.25 recommendation supports various physical layer interfaces, but CCITT has recommended X.21 over RS-232 (DTE/DCE interface), although both X.21 and X.21 bis provide circuit switching and, as such, have similarity. The X.21 interface is basically defined for digital circuit-switched networks and may offer digital control signals at the customer's premises. It is now clear that the digital signals offered by X.21 over analog circuit-switched networks or even digital circuit-switched networks are completely overshadowed by the advent of **integrated services digital networks (ISDNs)**. The X.21 recommendations define various features of this interface and, for these reasons, it is also termed an "interface between **data terminal equipment (DTE)** and **data circuit terminating equipment (DCE)** for synchronous operation over public data network (PDN)" [CCITT]. The X.21 uses a 15-pin connector and uses only six signal pins and a ground pin. It supports both balanced and unbalanced configurations. It offers data rates of 9.6 Kbps (unbalanced) and 64 Kbps (balanced). The data rate offered by RS-232 is 19.2 Kbps.

X.25 data link layer: The data link layer of the X.25 recommendation deals with the data exchange procedures between DTE and DCE. One of the main functions of this layer is to ensure error-free, reliable exchange of data between network users. The services offered by this layer (i.e., data link) include link establishment/de-establishment, error control, flow control, and synchronization. In asynchronous transmission, as explained earlier, every character is attached with start and stop bits and a parity bit. These bits provide synchronization and error control. The flow control is achieved by using Request to Send/Clear to Send signals in RS-232 (also termed "out-of-band") and Pause-and-Resume protocol X-ON/X-OFF defined within the data stream.

X.25 is essentially a data link control (DLC) layer, which is a subset of HDLC. This layer supports two types of procedures for the link layer: **single link procedure (SLP)** and **multi-link procedure (MLP)**. The SLP typically uses LAP-B. The link layer MLP is similar to the transmission group layer of IBM's System Network Architecture (SNA).[26] In this link layer, the standard MLP defines transmission over multiple links to offer higher throughput. A typical SLP frame contains the following fields: address, control type, MLP header, data, CRC, and sync. The MLP header of two octets uses 12 bits for bit sequence, and along with data, defines the data for SLP.

The character-oriented synchronous protocol is a refined version of the asynchronous protocol and has been used by IBM in its binary synchronous communication (BSC or BISYNC) protocol. In this protocol, two synchronous characters (Syn) are attached at the beginning of every character for synchronous protocol, while the identification of the frame is achieved by using control characters (start of header, start of text, end of text). Error control is achieved by 16-bit CRC or FCS algorithms. In the event of an error frame, the sender re-transmits the frame to the receiver. Flow control is generally based on the start-and-wait technique and is achieved by using an acknowledgment (Ack) for error-free frames and a negative acknowledgment (Nak) for frames containing error. Ack0 and Ack1 are usually used alternately for detecting sequence errors, which are caused by either duplicated or lost frames.

High data link control (HDLC): ISO defines a standard protocol as high-level data link control (HDLC) which is a modified version of asynchronous and bisynchronous protocols and is mainly based on IBM's **synchronous data link control** (SDLC), a data link protocol which replaced IBM's BSC. Both SDLC and HDLC protocols are bit-oriented protocols, while BSC is character-oriented. The BSC is based on stop-and-wait concepts and runs over half-duplex. The SDLC and HDLC are based on Go-Back-N or selective-repeat concepts and run over full-duplex ARQ protocols. Other protocols based on bit orientation are ANSI's **advanced data communication control procedures (ADCCP)** and **burroughs data link control (DLC)**.

The bit-oriented protocols define variable length for the frame; also, the number of framing and control characters required in character-oriented protocols are not required here. This is because each bit in an eight-bit character may be used independently for some control operation. In order to compare these protocols, a brief description of each of these is given below. For details of the data link protocols, see Chapter 11 (data link layer). A typical frame structure in bit-oriented protocols looks like that shown in Figure 12.5.

The flag field consists of a unique bit pattern of **01111110** and is used as a reference field to provide synchronization. The HDLC frame is enclosed by flags on each side. The flag field sends the bits continuously over the line in the idle condition to maintain synchronization between the sender and receiver. When the receiver receives the frame at the receiving node, after it detects the flag field, it looks at the address field. This address does not represent the DTE's address, but it comes out of the data link protocol being used (HDLC, SDLC, etc.). The size of the frame between two flags must be at least four bytes long; otherwise, it will be considered an invalid frame and discarded. In some cases, only one flag may be used to define the boundary of the frame; i.e., it indicates the end of one frame and the beginning of a second frame.

The X.25 uses HDLC in a point-to-point mode, and the HDLC must provide distinction between the link layer protocol **command** and **response** frames. The address field of HDLC, when used with X.25, offers two values for these two link layer features: **01** (00000001) and **03** (00000011). In the link access procedure-balanced (LAP-B) version of the X.25 standard, 01 represents the frames containing commands from DTE to DCE and responses from DCE to DTE. Similarly, 03 represents the frame containing commands from DCE to DTE and responses from DTE to DCE. The control field defines the frame

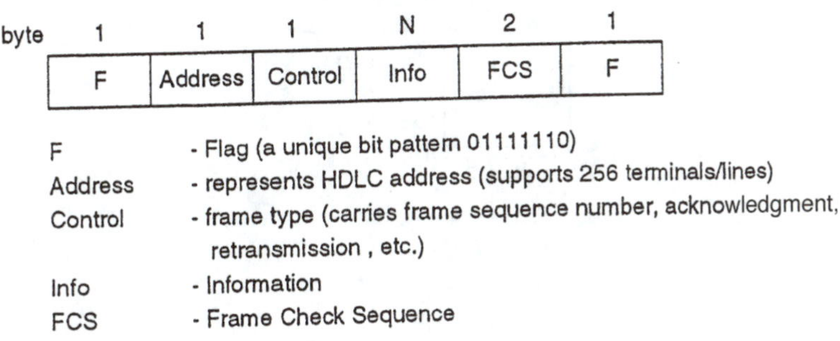

Figure 12.5 Typical frame format in bit-oriented protocol.

type and is used to carry frame sequence numbers, frame acknowledgments, re-transmission requests, error-recovery options, and other relevant control information.

The transfer of command and response frames always takes place between primary and secondary nodes, respectively; i.e., the frame transmitted by the primary is considered a command frame, while the frame coming out from the secondary is considered a response frame. The secondary nodes define the following modes of operations: **normal response**, **asynchronous response**, **asynchronous balance**, **normal disconnected**, and **initialization**. These nodes and modes of operations have been discussed in Section 11.8, but in order to provide continuity, a brief description of these nodes and modes of operations along with frame format of HDLC are explained below. The details on HDLC can be found in Chapter 11.

In **asynchronous response mode**, the secondary node sends a response frame without receiving any poll command frame or other command frame from the primary. In **asynchronous balanced mode**, there is no primary–secondary relationship between nodes. Any node can initiate a transmission. This mode is used in LAP-B (used in LANs). In **normal disconnected mode**, the secondary node is logically disconnected from the link, and this node can send only a control frame to change its status before it can send or receive any data packets. In **initialization mode**, the secondary nodes are usually in initialization mode and another command U (unnumbered) frame is needed to change its status to some other mode, e.g., normal response mode for participating in the data communication.

The primary–secondary configuration typically uses a tree structure topology in such a way that the station management acts like a primary node on some channels and like a secondary on other channels. The primary node is usually responsible for initiating error-recovery procedures. In balanced mode, routing is defined via switching nodes (communication processors) and usually controls and manages the data flow through the network. It also allows the sharing of an equal level of link authority by a number of user's packets within the network.

HDLC control frame: The supervisory frame (S-frame) is used to provide flow control, request of re-transmission, acknowledgment I-frames, etc. A control field of the HDLC frame consists of the following subfields, as depicted in Figure 12.6.

- P/F — poll/final bit
- RR — receive ready
- RNR — receive not ready
- REJ — reject

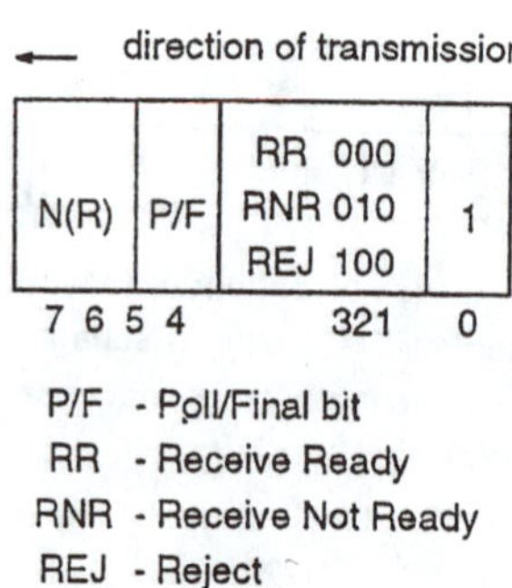

Figure 12.6 Control field of HDLC control frame.

The three sub-control commands are RR, RNR, and REJ. The RR command is used by either DTE or DCE for receiving I-frames. The command RNR indicates a busy condition or the fact that the node is not ready to receive I-frames. The REJ command makes a request for the re-transmission of the I-frame. For example, when an I-command is used without a P/F bit, we may use I, RR, RNR, and REJ responses, but when I is used with a polling bit (P), all of the responses will be associated with the final bit (F). For details of the HDLC frame, see Chapter 11.

As we know, most physical circuits pose the problem of undesired error caused by electromagnetic interference, radio signal interference, interference from power lines, thunderstorm, lightning, sunshots, etc. Telephone lines are susceptible to noise from a variety of sources. The HDLC protocol uses a 16-bit frame check sequence (FCS) to eliminate some of the errors caused by these sources — in particular, all single-error bursts. The protocol detects bursts up to 18 bits long, offering a very low error rate of less than 1 in 10^{15} bits.

Link access protocol-balanced (LAP-B) procedure: The data link access procedure defined by LAP-B is used between DTE and DCE over a single physical link. (LAP-D is a subset of the asynchronous balanced mode of HDLC). Other data link protocols such as BISYNC (binary synchronous control) have also been used by some vendors in the link layer of X.25.

The LAP-B assigns I (information field) for X.25 packets. If there are multiple physical links between DTE and DCE, then each link will use a LAP-B, which requires the use of the ISO **multi-link procedure (MLP)** on top of the link layer as one of the sublayers.

The LAP-B version of X.25 uses three types of frames: Information (I), Supervisory (S), and Unnumbered (U). The I-frames use sequence numbers as they carry the data across the network. Typically, the user information (user's data) is defined within the I-frame. The I-frame format is shown in Figure 12.7(a).

The information field of the N-byte further defines two subfields: packet header and user data defining the data packet. This data packet is defined by the network layer protocol. A pair of sequence numbers is assigned to I-frames for each direction of transmission (send and receive) and offers an ordered exchange of frames across the network between nodes. The sequence number is usually based on a modulo 8 window supporting sequence numbers of zero to seven. The control field of the I-frame is shown in Figure 12.7(b).

Bit stuffing or zero insertion: On the receiving side, the receiver computes its own FCS using the same polynomial generator and compares it with the received FCS. If the computed FCS is the same as that received, it indicates an error-free frame; otherwise, it

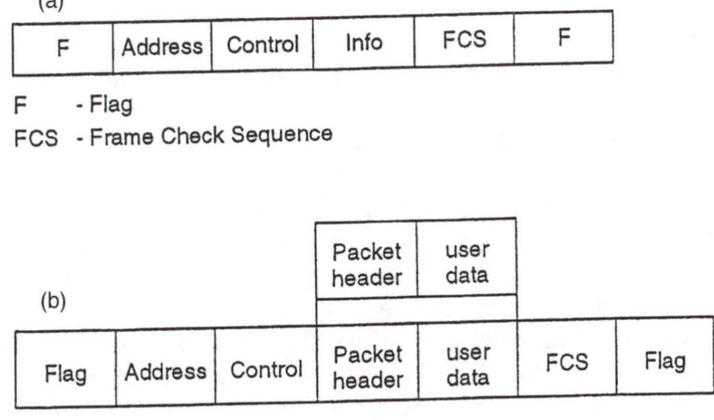

Figure 12.7 (a) Information (I) frame format. (b) Control field of I-frame format.

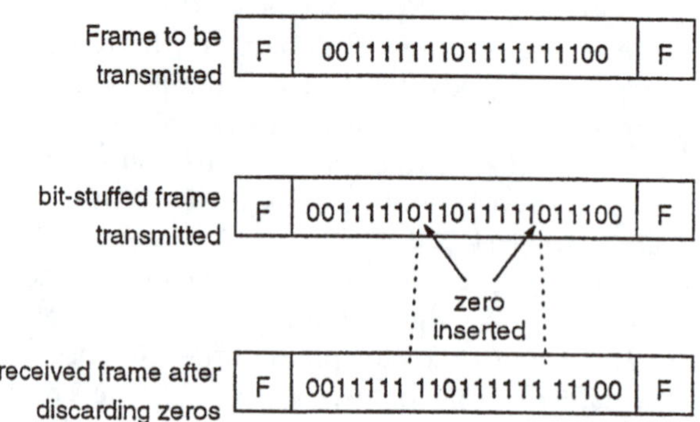

Figure 12.8 Bit stuffing or zero insertion.

discards the frame and requests the sending side to re-transmit the same frame. The **bit stuffing** or **zero insertion** technique (providing data transparency) is used to avoid situations where the data accidentally or coincidentally has the same bit pattern as that of the flag (01111110). In the bit stuffing or zero insertion technique, the transmitting DCE or DTE will insert a zero after every five continuous 1s in the frame (including all fields, i.e., address, control, FCS). The flag fields remain unchanged. The receiving DCE or DTE will remove the inserted zero after every five consecutive 1s in the frame to extract the original user information data from the received frame. The steps required in bit stuffing are shown in Figure 12.8.

The bit stuffing or zero insertion technique is automatic and is usually available as a built-in option in transmitter and receiver hardware units.

Multi-link procedure (MLP): In 1984, ISO and CCITT introduced a **multi-link procedure (MLP)** facility in the data link layer of X.25. The MLP, in fact, is considered an added function within the data link layer that transmits the frames (data link layer receives packets from its network layer and converts them into frames) onto multiple physical circuits or links. The packets from the network layer are received by MLP (within the data link layer) and transmitted over various single link procedures (SLPs). Each SLP is defined as a conventional LAP-B link. The MLP adds its own multi-link control field and information field within each frame (constructed by the data link layer) and provides a sequencing function over a sequenced frame (defined by the data link layer).

The MLP offers the following advantages:

- Support of load-balancing within the data link layer.
- Availability of layer bandwidth without affecting the existing links.
- Improved response against failed link.
- Reliable and protected link to the packets.

The ISO and CCITT have also added an extended feature within LAP-B of X.25 that supports a frame window size of up to to 127 (modulo = 128). This type of transmission is useful in satellite communication. This frame window may contain up to 127 frames before any acknowledgment can be received.

The **set asynchronous balanced mode (SABM)** command supports modulo 8 and can also be used with satellite communication, but with significant delay (between 220–260 ms for one direction). For voice communication over a satellite link, this delay may not be significant, but for data communication, this delay is very significant. The LAP-B uses another command for modulo 128 known as **set asynchronous balanced mode extended (SABME)**, which is usually used during the setup time.

X.25 network layer: The network layer of X.25 is the most complex and sophisticated layer within X.25, as it provides access to the network. It is also known as the *packet layer*. It offers two types of virtual circuit: **permanent virtual circuit (PVC)** and **switched virtual circuit (SVC)**. It includes the features of the network and transport layers of OSI-RM. The main function of this layer is to establish network connection, as the X.25 network layer provides connection-oriented connection. These connections are usually termed **virtual circuits** or **links**. The multiplexing of various virtual circuits over one physical connection is a very important function provided by the network layer. The network layer of X.25 uses a **statistical time-division multiplexing (STDM)** technique where the bandwidth can be allocated to various terminals or devices dynamically. Other functions of the network layer include routing, relaying, packet-sequencing, and flow control. In addition to these functions, the network layer offers services to transport and higher layers which include addressing, connection establishment, termination or de-establishment, data transfer, flow control, error recovery (detection and notification), and quality of service (QOS).

The **switched virtual circuit (SVC)** is analogous to a dial-up connection over the public switched telephone network (PSTN), while the **permanent virtual circuit (PVC)** supports point-to-point connections and is analogous to a leased line on PSTN. The difference between these two circuits lies in the way they are defined:

1. In **SVC**, we need three phases (as in any connection-oriented service): connection (call) establishment, data transfer, and connection (call) termination/de-establishment.
2. In contrast to SVC, the **PVC** needs only one phase, **data transfer**. The PVC service has to be requested in advance from PDNs and is not available in all public data networks (PDNs). Some PDNs don't support this service across X.25 or even on international gateways. It is advisable to check with the vendor about the availability of this service at the international level across the X.25 before subscription. Each of the user's packets is assigned a unique number that is transmitted over a physical circuit to and from the network (collection of various sub-network switching nodes). All packets are not assigned with unique numbers at one time but are assigned in a group depending on the performance constraints.

The **logical channel numbers (LCN)** are usually assigned across DTE/DCE interfaces, and it is unlikely that the LCNs at local and remote interfaces will be the same. The X.25 offers nearly 4096 logical channel numbers which are divided into various variable-width regions and can go up to five regions. The regions are reserved PVC, one-way, two-way incoming, and one-way outgoing. These numbers (channels and regions) are different for different networks and usually depend on a number of parameters such as link access time, available bandwidth, etc.

The logical channel is established between DTE and the network. The packets from a DTE to another DTE can be sent or received only when a virtual circuit through the packet network is established between DTEs. The LCN0 is reserved, while the PVC may use LCN1 to LCN*n* (where *n* is some predefined number). The channel number ranging from *n* to 4096 is used by SVCs in different modes. The DCE uses LCNs from *n* to some predefined channel number in two-way regions including one-way incoming numbers. In contrast to this, the DTE uses some numbers of two-way and one-way outgoing regions. Each region is defined by lowest and highest channel numbers.

The problem of mapping a logical channel into a virtual channel or circuit and the assignment of unique numbers to the packets are up to the network manager (administrator) within the specific limits and requirements of X.25. There are a number of ways of establishing a user's session between DTEs and the packet networks. X.25 defines five categories of packets, and the packets in each category have different primitives attached to them which, in turn, depend on whether the packet is coming out from DTE or DCE. For example, the packet **Call Request** going from DTE to DCE becomes Incoming Call when it is coming out from DCE to DTE. The other packets are **Call Setup/Clearing, Data/Interrupt**, **Flow Control**, **Reset**, **Restart**, and **Diagnostic**. The above part is clearly shown in the categories of packets. In the packet of call setup/clearing category, the call setup packets have different primitives when they are initiated from DTE to DCE than when they are initiated from DCE to DTE. The call setup packets have primitives such as call request, call accepted, clear request, and DTE clear confirmation, while the clearing packets have incoming call connected, clear indication, and DCE clear confirmation.

12.6.4 Data communication in X.25 networks

The exchange of packets across the network needs some kind of flow control which will maintain transmission and receiving rates of packets between the DTEs of two nodes. This can be obtained by using the appropriate allocation of resources managed by the data link layer protocol. Flow control is required for packets in both directions. The network protocol should be able to establish many virtual circuits with different stations simultaneously, as packet-switching networks offer multiplexing service. The discussion of these packets is partially derived from References 2–4, 6, 7, 29 and 30. For more details, please refer to these.

As many as 4100 logical channels can be supported by X.25, which means that 4100 user sessions can be established on these logical channels. The logical channel is established between the DTE and the network. The packets from a DTE to another DTE can be sent or received only when a virtual circuit through the packet network is established between DTEs. There are a number of ways of establishing a user's session between DTEs and the packet networks: permanent virtual call/circuit (PVC), virtual call/circuit (VC), fast select call, and fast select with immediate clear.

12.6.4.1 Classes of X.25 connection circuits

Under connection-oriented service, X.25 provides two classes of virtual circuit services (connection): normal virtual call/circuit (VC) and permanent virtual call/circuit (PVC) at the packet level.

Virtual circuit (VC): In the normal virtual call/circuit, a virtual call request packet is used to establish a virtual circuit (in a dynamic mode) using call setup and call terminating/clear request packet procedures. The DTE issues a call request packet to the network with an assigned logical channel number. The receiving DTE receives this as an incoming call packet from its network node which has a logical channel number. After it decides to acknowledge, it sends a call accept packet back to the network, which in turn sends it to the sender's DTE as a call connecting packet.

Once the virtual circuit is established, the packets are transmitted (channel changes its state to data transfer) over it until a call-terminating request is issued by either DTE. The transmission of clear indication and clear confirm packets between them will terminate the virtual channel between them. This type of service is analogous to telephone dial-up line service.

Permanent virtual circuit (PVC): In a permanent virtual call/circuit, a limited number of permanent network-assigned virtual circuits are established by the network manager, thus avoiding delay in call request packet or clear request packet calls to certain destinations.

This certainly requires mutual agreement between two parties and the network manager. Once it has been agreed upon between the two parties, the manager assigns a specific channel number to the virtual circuit. A packet arriving at the network will have this specific number which will automatically route it on the virtual circuit having the same channel number. This avoids the delay in negotiating for establishing sessions between sender's and receiver's DTEs. The X.25 virtual call/circuit connection is very complex and requires the exchange of various control bits (events) before a virtual call/circuit can be established between two nodes.

As mentioned above, the network layer supports both types of connection services: connection-oriented (virtual circuit) and connectionless (datagrams). In connectionless service, no call setup and clear control signals are needed; the only requirement imposed on users is that the complete destination address be specified. The network layer does not support any kind of error control or error recovery. The earlier version of X.25 had a provision of supporting datagram service, but the latest version of X.25 does not support datagram service. The main reason for this is its inability to provide data integrity, data security, error recovery, and, of course, end-to-end communication for these parameters. However, the short packets (user data up to 128 octets) can be sent using datagram service by using **fast select call** and **fast select with immediate clear call**.

The fast select call is useful in applications where a few transactions or short-duration sessions are required with some sort of data integrity and security. The fast select with immediate clear call is useful in applications where inquiries or very short responses are required (e.g., checking validity of credit cards, account information, money transfer information, etc.).

In both of these categories of datagram services, a user has to make a request to the X.25 network manager to get permission for these calls. In these applications, the virtual circuit or permanent virtual circuit calls are very expensive, and, at the same time, pure datagram service is not reliable. Thus, X.25 provides these two calls for applications where very short packets or inquiry for a short session are required.

In addition to virtual circuit and datagram packets offered by X.25, other packets have been defined, such as data transfer packet, diagnostic packet, error-recovery packet, and re-establishment of a session packet.

Other connection circuits of X.25: The public data networks (PDNs) such as X.25 provide support for other categories of connections such as **leased line** and **dial-up** connections.

For **leased line connections**, the X.25 recommendation defines a dedicated or leased line connection between PCs and networks. This connection is also known as a *direct connection* for X.25 interface to the PDNs. A line connection (can be routed from PDNs or value-added networks — VANs) leased on a monthly basis offers a data rate of up to 19.2 Kbps over analog lines, 56 Kbps over digital lines, and 1.544 Mbps over digital carrier lines (T1 service). The X.25 protocol works in synchronous mode at the physical layer. Some PDNs provide support to the synchronous mode of leased line connection to non-X.25 devices (e.g., front-end processors, IBM cluster control, etc.).

Non-X.25 leased line connection supports both configurations (point-to-point and multi-drop) and generally requires a packet assembly/disassembly (PAD)[19] device for converting the protocols of SNA to the X.25 interface which resides within the network layer of X.25. The protocols of the X.25 interface support error control and statistical multiplexing of many packets over a single physical circuit.

The **dial-up connection** in X.25 interface over synchronous mode has been described in CCITT specification X.32 (in 1984). The X.32 standard includes most of the facilities defined in X.25 and is useful for low/medium traffic volume. The majority of requests for a packet-switched network usually comes from low-speed asynchronous PCs or terminals at data rates of 300–2400 bps. It is expected that a data rate of 9.6 Kbps will be made available universally on packet-switched networks in the future. In order to provide services to these requests, each PDN node must include dial-up modems as well as support for compatibility between modems at both sides of the dial-up connections. These **asynchronous terminals** or PCs (also known as start–stop DTE) submit their dialed requests directly onto the ports of PAD. There are a variety of dumb terminals or asynchronous terminals available in the market, and some of these terminals support seven bits with start or stop bits and parity, while others may support just eight bits without parity, start or stop bits, or any other combination. The PDN must be able to support most of these terminals.

12.6.4.2 Types of X.25 packets

There is a variety of packets defined for X.25, and these packets fall into two categories: data and control packets. The **control packets** are used during call setup, resetting and managing, flow control, error control, interrupts, and many other controlling functions within X.25 networks. The **data packets** carry the data to the destination. These packets transform the corresponding requests, responses, commands, etc., into appropriate primitives between DCEs of both sending and receiving sites. The requests are initiated from a DTE to its own DCE as a control request packet primitive at the sending node and is transmitted toward the receiving site. On the receiving site, this request packet primitive is received as another request primitive by the DCE to its own DTE. After the establishment of connections and other negotiated parameters for quality of service (QOS), the data packets are then transmitted from the sending site to the receiving site. Table 12.2 describes all the types of packet primitives used for data communications between DCEs across X.25.

Other X.25 packets: Other packets for flow control, reset, and diagnostic tests are also defined between DTEs. The diagnostic packets are mainly used for performing diagnostic tests under unusual conditions. Further, the unrecoverable errors can be detected using diagnostic codes along with these packets.

Table 12.2 Packet Primitives Used in X.25 Networks

From DTE to its own DCE	Remote DCE to its own DTE
Connection establishment[a]	
Call request	Incoming call
Call accepted	Call connected
Clear request	Clear indication
DTE clear confirmation	DCE clear confirmation
Data transfer[b]	
DTE data	DCE data
DTE interrupt	DCE interrupt
DTE interrupt confirmation	DCE interrupt confirmation
Registration[c]	
Registration request	Registration confirmation

[a] These packets are used during call sets and resetting of the connection between two nodes.

[b] These packets are used to exchange the data and interrupt request between DTEs.

[c] This request packet is used for the registration.

Two types of packets are defined in X.25 to recover from the errors. The **reset packet** is used when both transmitter and receiver are out of sync. This packet resets both transmitter and receiver to start with sequence number 0. In the event of any interrupt packet being lost during transit, the upper layers recover it by resetting only the existing virtual circuit numbered packet (VC#). The **restart packet** is used for more serious errors, and VC#0 is used to restart by clearing all the packets. The **interrupt packets** are transmitted with higher priority over the normal data packets. The interrupt packet includes a 32-octet field of user's information and is always supported by end-to-end acknowledgment. The second interrupt packets will be transmitted only when the previous packet has been acknowledged.

Fast select service packets are used for applications requiring verification or validity of any information from the central database system, for example, checking the validity of a credit card. Here, the Call.request packet with fast select facility contains 128 bytes of information (in contrast to 16 octets of data in the normal Call.request packet). This information includes a credit card number, and one exchange of packets over an established virtual circuit with fast select service facility will verify the validity of the card.

Optional packet-switched user facilities (X.2 standard): The X.25 network allows the negotiations of facilities as defined in X.2.[23] It defines international user classes of services in public data networks and ISDNs. This standard includes the following facilities:

- **Throughput:** Defines this parameter on a per-call basis and allows the users to select the number of bits to be transmitted per virtual circuit. The typical range is 75 bps to 48 Kbps.
- **Flow-control parameters:** Specify the window size and maximum packet size on a per-call basis.
- **Incoming calls barred:** Prevents any incoming calls from getting into the DTE.
- **Outgoing calls barred:** Prevents any outgoing calls from being accepted by the connected DTE.
- **One-way logical channel outgoing:** For a mix of incoming and outgoing calls, any number of VC can be chosen. For example, if we choose, for outgoing, VC#20 as the upper limit on the VC, then all the outgoing packets will have a number less than 20 and will be transmitted in descending order, i.e., VC#19, VC#18, and so on.
- **Fast select acceptance:** Allows the DCE to send fast select calls to the DTE. If the DTE does not have this facility implemented on it, it will simply block all calls from the DCE.
- **Closer user group:** Protects the DTEs from receiving any unauthorized call. If a DTE belongs to any closer group(s), the DCE will send the incoming message to it only if it belongs to the user group(s) number defined in the first byte of the X.25 packet.

X.25 Tariff structure: The X.25 tariff structure includes three components:

- Distance independent within country
- Dependent on volume
- A charge for attachment to X.25 public data network, services, and optional facilities

For an attachment with X.25 PDN, there is a connection charge which includes the cost of physical attachment to DTE with the network. The monthly charge depends on the speed of the connection and chosen user's options. A monthly charge for PVC per call, along with the options, is charged for connection. The switched VC is charged per time connection, along with the options. The monthly charge for volume of data is charged at the rate of the segment (64 bytes) being transmitted. The charges of segment transmission during the evening are expected to be cheaper than during the daytime.

12.6.4.3 X.25 network protocol data unit (NPDU)

A typical packet exchange between DTEs uses the following sequence of packet transmission:

- The sending DTE sends a call request packet to its DCE, which sends it to the remote DCE across the network.
- The remote DCE renames it an incoming packet and sends it to its receiving DTE.
- The receiving DTE sends its response via call accepted packet to its DCE, which transmits it to the sender DCE across the network.
- The sending DCE renames this packet call connected and passes it on to its DTE.

The call request packet may not be a successful packet for establishing a connection between DTEs, and a **call refusal (reject) packet** may be sent back to the sender either by the network itself or by remote DTE. In the first case (refusal by network), the local DCE informs its DTE by sending a clear indication packet, while in the latter case, the remote DTE sends a clear request packet to the sending DTE, which receives this as a clear indication packet by its DCE.

The network layer, after getting packets from the transport layer, defines a **network protocol data unit (NDPU)** and passes it on to the data link layer. The data link layer defines its own **data protocol data unit (DPDU)** and passes it on to the physical layer. The NDPU contains the destination address of the remote DTE on the network and other control information. The attached DTE receives this packet from its network and, based on control information, routing of the packet across the network through sub-network switching nodes is determined. In addition to services (e.g., connection-oriented and connectionless services) provided by the network layer, it must also perform various functions across the networks:

1. Moving of data packet to its attached DTE through the data link layer and physical layer protocols
2. Interpretation of control information supplied by DTE to determine routes
3. Transfer of data for DTE to DCE
4. Some sort of error control and recovery between DTE and DCE
5. The delivery of packet by DCE to DTE
6. Sending the packet to the transport layer at the receiving node

The network layer breaks up the packet it receives from the transport layer as TPDUs into packets of fixed size specified by X.25. The maximum length of user data can be 128 octets, which includes **fast select** service in it. Another packet, a **call request** packet (which is one of the control packets), is also constructed. The network layer sends the control packet and data packet (which includes user's data) to the link layer LAP-B, which adds a header and trailer to it and transmits frames (one frame per packet) to its DCE. Various control information such as flow control, acknowledgment, header and trailer, etc., are local to DTE and are not transmitted. Only relevant information in the control field of the network layer protocol which provides end-to-end connection service is transmitted. Thus, we can see that communication between DTE and DCE is not symmetrical.

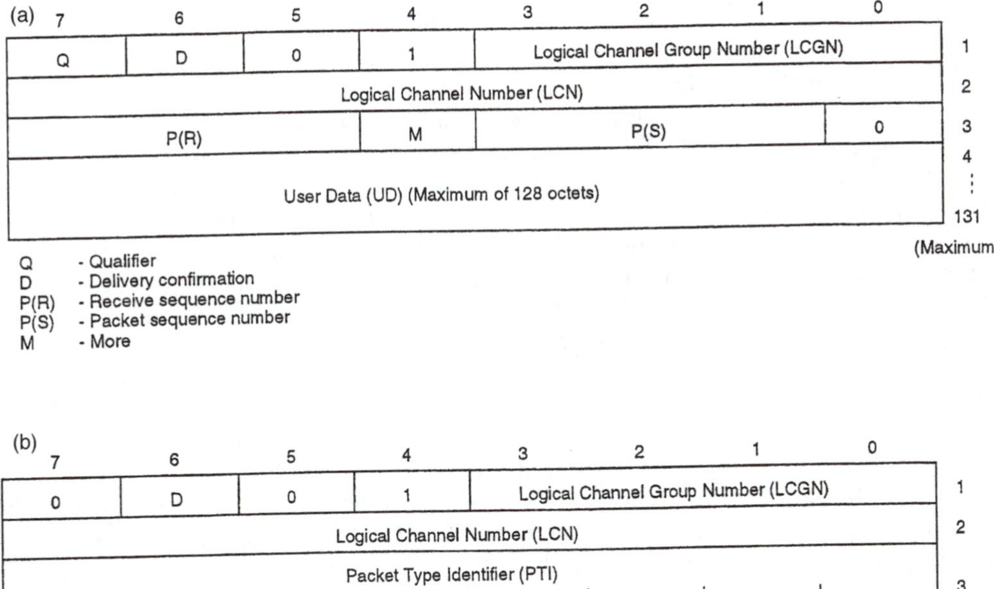

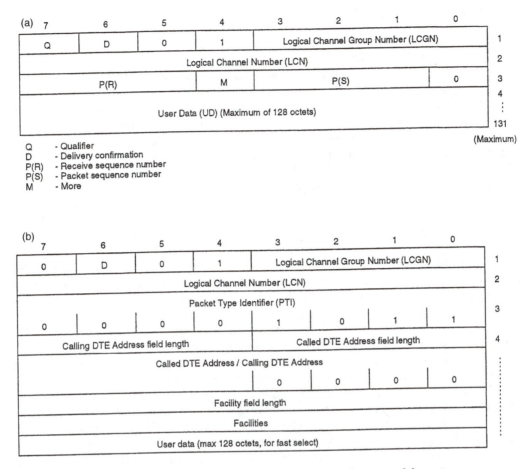

Figure 12.9 (a) X.25 data packet format. (b) X.25 call request packet control format.

X.25 data packet format: A typical X.25 data packet format is shown in Figure 12.9(a), and a typical X.25 control packet format is shown in Figure 12.9(b). Various types of packets of X.25 are shown in Table 12.2. There are also variations in the formats of the packets which are being used by these X.25 packets. The latest version of X.25 supports data packets of different sizes — 16, 32, 64, 256, 1024, and 4096 octets. The data packet which has a maximum size of 128 octets discussed below is, in fact, a standard default size of user data. The X.25 data and control packet formats are also discussed in a number of publications and presented here as derived from References 2–5, 28, and 29.

As shown in Figure 12.9(a), the maximum length of user data is 128 octets, and we are transmitting a packet of 131 octets, which means that for each data packet, we have to transmit a three-octet packet header. The first octet of the header has two fields: the first four lower bits define the format type being used for user data. The least significant bit (1) for each octet is transmitted first. Bits 5 and 6 define the sequencing scheme used for the packets. X.25 supports modulo 8 and modulo 127 sequencing schemes (i.e., 0 to 7 and 0 to 127). The D-bit (delivery confirmation bit) of the format type field defines a desire of the sender's DTE for an acknowledgment of the data it is transmitting.

The latest version of X.25 provides two types of facilities for acknowledgment. If we set D = 0, it indicates that P (R) will get acknowledgments of the receipt of the data packet from the packet network, while if we set D = 1, it indicates that P (R) will get an end-to-end acknowledgment of the data packets (i.e., the source DTE will receive acknowledgment from the destination DTE).

The eighth bit (most significant bit) is a qualifier (Q) bit, which is an optional bit available to the source DTE and used for a special type of packets, or for distinguishing user information and control. The information network will ignore a Q = 1 value and pass the packet to the destination. This will also indicate to the destination DTE that the data packet contains PAD (packet assembly/disassembly) information, while Q = 0 will indicate that data packets contain user information. The logical channel group number and logical channel number are assigned during call setup and are known as **virtual circuit assignment**.

The second octet defines the logical channel number. Depending on the length of the data packet, both the logical channel group number and logical channel number together can support data packets of up to 4095 octets. A logical channel number 0 is used for some control packets. The logical channel number assigned to various packets is valid only between DTE and DCE before the network. The network provides mapping between these two different LCNs for end-to-end communication between DTEs.

The third octet of format type field defines the sequencing of data packets. The first bit (LSB) with value 0 indicates that the packet is a data packet. Bits 2, 3, and 4 together define P(s) (packet sequence number). The fifth bit is the more data bit (M) and indicates that more data is to follow. As stated earlier, the network layer breaks the TPDU (transport protocol data unit) into various packets, and each of the packets is of the same TPDU. The M-bit is important in two situations. The first situation is where the packet is carrying a network service data unit larger than the network allows, and the second situation is where a large data/text has to be presented in a proper sequence to the destination DTE.

The remaining three bits of the third octet in the header defines the receive sequence number P (R). This header represents end-to-end communication confirmation and possesses the same function as N (R) in LAP-B. X.25 supports piggybacking for the acknowledgment, and for each acknowledgment by piggybacking, P (R) is incremented. The new value of P (R) defines the sequence number of the next frame that the destination is expecting. Both P (R) and P(s) headers are at the network layer and define the sending and receiving sequence number of the packets between DTEs. The sequence number of the frames between DTE and DCE is maintained by LAP-B. Both LAP-B and P (R) can support up to seven outstanding packets.

X.25 control packet format: This includes virtual circuit number, type of packet, and other control information. A typical X.25 control format for a call request packet is shown in Figure 12.9(b). The delivery confirmation bit (D-bit) in the first octet provides the same information as in the data packet (Figure 12.9(b)). The logical channel group number in the lower four bits of the first octet and the logical channel number in the second octet together support up to a maximum data length (in X.25) of 4095 octets and are used during call setup. The third octet defines the packet type (corresponding to different control packets) identifier. This control packet has an identifier (1011) for the call request packet. The fourth octet defines the field addresses of calling (source) and called (destination) DTE such that the lower four bits represent the address field of the called DTE while the higher four bits represent the address of calling DTE. The BCD (binary coded decimal) representation of these field addresses is shown in the fifth octet. The sixth and seventh octets provide optional services to the user and are included in the packet only when the calling DTE wants to use it and has permission to do so.

The data field octet is a maximum of 128 octets in a call request packet when the user has chosen the fast select option. It is important to note that the first three octets of the call request packet format represent the minimum size of the header that each packet must include, while the call setup process may go up to 16 octets. When the sender uses a fast select option, the call request packet does not contain user data and represents up to 16 octets. This packet is used during the call setup or establishment. On the receiving side, the receiver sends its acknowledgment on another packet known as an accepted packet (which is again a small packet).

The logical channel number plays an important role and can be used in a variety of ways for the exchange of data bi-directionally (full-duplex configuration). It may be used to identify different calls, types of circuits (permanent virtual circuits or virtual circuits), and the DTE and its packet node DCE or vice versa.

The DTE of any node can initiate and establish up to 4095 virtual circuits with the DTE of another node at any time over a single DTE-DCE link. The DTE-DCE link provides bi-directional communication (i.e., full-duplex configuration) for packets which contain a virtual circuit number (defined by the logical channel group number and logical channel number bits together). The logical channel number 0 is reserved for special types of control packets, such as restart and diagnostic packets. In order to avoid any interference between the logical channel number assignments to the packets, it has been generally agreed upon to follow the logical channel number assignment scheme as follows.

Numbers beginning with 1 are assigned to permanent virtual circuits. The next assignment of numbers to incoming virtual calls (coming out of the networks) is done by DCE. Two-way virtual circuits are assigned the numbers next, where the DCE, after getting a call request packet, assigns an unused number to the call indication packet it sends to its DTE. The DTE uses the remaining numbers to assign numbers to one-way outgoing virtual calls to its DCE using a call request packet. The above scheme allows the assignment of numbers to these types of packets in ascending order starting from permanent virtual circuits.

The logical channel number 0 is used for restart and diagnostic packets; these packets are used for recovering from errors. The initiation of a restart packet (either by DTE or DCE) will reinitiate the virtual circuit. The sequence numbers of sender and receiver will become 0, indicating that now a virtual circuit has been established. If there is any packet already on the way, it will be lost and will have to be re-sent (upper-layer protocols will decide how to recover the lost packets). A request for a restart packet (which may be caused by some error) will not only establish a new virtual circuit with number 0 but will also abolish the entire number assignment scheme. That means that the scheme has to be reused to assign numbers to various circuits/calls as discussed above. A reset packet will be initialized (either by DTE or DCE) at the event of some error (e.g., packet is lost, sequence number is incorrect, or traffic is congested).

X.25 interrupt packet format: Sometimes in the network, the traffic becomes so heavy that we get lots of errors and initialization of reset packets. This decreases throughput of the network. In order to avoid flooding in the network, the DTE transmits an interrupt packet, as shown in Figure 12.10. This packet does not go through the flow control procedure as do the data packets, as it possesses the highest priority. It carries up to 32 octets of user data and provides an end-to-end confirmation link between DTEs.

A flow-control technique based on window concepts (which has been used in various data link protocols such as HDLC, LAP-B, and SDLC) is also supported by X.25. The flow of packets between DTE and DCE is controlled by using both send and receive sequence numbers with the packets. The X.25 supports up to a maximum sequence number of 127 in the sequence number field of the packet. The flow control of packets at the DTE-DCE interface is managed by the user in the form of receive sequence numbers in **receive ready (RR)** and **receive not ready (RNR)** control packets.

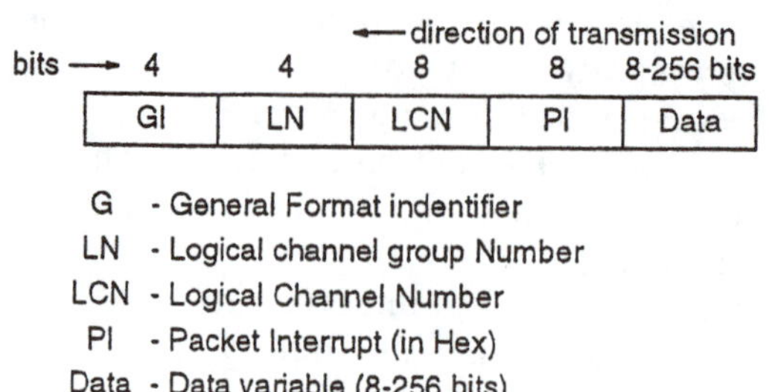

Figure 12.10 Interrupt packet.

Any sender DTE may assign a logical channel number to each of the virtual circuits it wants to establish with any number of destination. For each destination, the DTE itself assigns a logical channel number to a virtual circuit. The network manager can use a suitable scheme for the exchange of call requests between DTE and DTE to avoid any kind of interference. Each call request from the DTE is initiated with a particular logical channel number to its DCE, which considers each such request an incoming call.

The DCE recognizes the type of circuit and logical channel number (assigned by its DTE) in the packet. A new logical channel number (using the mapping technique) is assigned to the packet at its header and uses a routing algorithm to determine the path between sender and receiver. The DCE then waits for a link to the next node available on this path and sends the packet to the destination. It is important to note that X.25 does not solve the problem of routing; it has to be implemented by the user and is based on the implementation of routing technique(s).

The DCE at the receiving side receives this packet and sends an incoming call packet to its DTE with the logical channel number on the received packet. The DTE of the receiver side sends a call accepted packet with the same logical channel number as that of the incoming call packet it received from its DCE. The exchange of data between receiver DTE and DCE should avoid the interference of the logical channel number and is decided by the network manager. After the DTE sends a call accepted packet, the receiver DCE keeps its logical channel number (which is the same as that of the incoming packet) in the data transfer mode.

When the sender DCE receives a call accepted packet from the receiver DCE, it transmits a call connected packet to its DTE and keeps the assigned logical channel number (same as that of the call request packet) in data transfer mode. Similarly, the virtual circuit call can be cleared by using clear request and clear confirmation packets. The clear request packet is transmitted by the DTE to its DCE which, upon getting a response from the receiver DCE, notifies its DTE by sending a clear confirmation packet. The DCE sends a clear confirmation packet to the sender DCE only when it has received the response from its DTE on its clear indication packet.

Network addressing in X.25: The network addressing for international users over the public data network (PDN) via different hosts needs to be standardized. CCITT has defined a standard X.121[22] addressing scheme which assigns a unique address to various DTEs across the network globally. The network addresses assigned to these DTEs, in fact, are part of the international network addressing scheme defined by X.121. This standard X.121 includes up to 14 digits and divides the entire network environment into various regions or zones. Some of the regions are controlled and managed by CCITT, while others are allocated mainly to the PTTs of individual countries. The regions can be further divided into sub-regions in the cases where multiple PDNs are used, e.g., in the U.S., where there

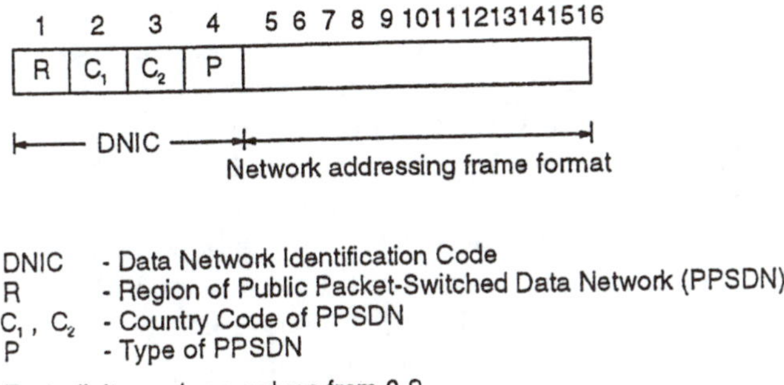

Figure 12.11 X.121 network addressing frame format.

are quite a few PDNs installed throughout the region. Every PDN defines a unique address to its own users' network addresses which conforms to the international standard addressing scheme. This international addressing scheme offers various advantages of uniformity, easy maintenance, etc., and has been widely used across the globe for the electronic mail system (X.400) over PDN and the transmission of hard copy over the telex network (fax).

The X.121 addressing scheme is valid for X.25-based networks. For non-X.25-based networks, the CCITT has defined three different recommendations for addressing: **International Telephone Numbering Plan (E.163)**, **International Telex Numbering Plan (F.69)**, and **International ISDN Numbering Plan (E.164)**.

The X.121[22] network addressing frame has the following format, as illustrated in Figure 12.11. The first four digits of the frame define the **data network identification code (DNIC)** field and it, in fact, defines a particular public packet-switched data network (PSDN) being used in that particular region or zone. The first digit typically represents the region (R), while the second and third digits (C1 and C2) are used to indicate the country of that region. The last digit (P) in the DNIC defines the type of public PSDN being used. The value of each digit may range from 0 to 9.

Following is a description of various regions, countries, and public data networks being defined within DNIC. The numbers 0 and 1 are reserved; specific regional numbers include 2 (Europe), 3 (North America), 4 (Asia), 5 (Oceanic and South East Asia), 6 (Africa), and 7 (South America). The numbers 8 and 9 are reserved for telex and telephone networks, respectively. The combination of C1 and C2 represents the addresses of various countries within the region. The following combinations have been standardized so far (the first three digits **(RC1C2)** define the **data country code (DCC)**): 208 (France), 234 (U.K.), 250 (Commonwealth of Independent States (CIS) and Northern Ireland), 302 (Canada), 311 (U.S.), 425 (Israel), 440 (Japan), 450 (Korea), 454 (Hong Kong), 502 (Malaysia), 505 (Australia), 602 (Egypt), 604 (Morocco), 724 (Brazil), and 730 (Chile).

The fourth digit (P) represents the type of public PSDN, and at this time the U.S. is the only country which uses the last two digits (C2 and P) for defining the public data network. It is interesting to note that the U.S. has more than 10 PSDNs and, further, many of the **regional Bell operating companies (RBOCs)** have defined their own PSDNs. The following is a partial list of PSDNs defined by CCITT: 3104 **(MCI)**, 3106 **(Tymnet)**, 3110 **(Telenet),** 3126 **(ADP Autonet)**, 3132 **(CompuServe)**, 3136 **(GEIS-Marknet)**, 3137 **(Infonet)**, 3140 **(SNET-ConnNet)**, 3142 **(Bell South-Pulselink)**, 3144 **(Nynex-Infopath)**, and 3149 **(Wang Pac)**. Other PSDNs include NSF, IEEE, and many others.

The remaining ten digits of the addressing frame (5–14) define the local PSDN address. This address is also known as the **national terminal number (NTN)** and is assigned by the PSDN and is not standardized. The majority of PSDNs keep the last two digits of the

NTN reserved as optional sub-addresses. These addresses are assigned by the network, and PSDNs don't process the last two digits. These addresses may represent devices, ports on a PAD, an electronic mail facility, a transaction-processing facility, or any other application which is using the X.25 host connection via the same X.25 switch. Sometimes, to differentiate between international request and request within the boundary of a PSDN, an additional single bit known as an **optional digit (Opt)** is used in front of the DNIC field of the addressing frame. If the user is within the boundary of the local PSDN, the digit is set to 0; otherwise, 1 is set for outside the boundary of the PSDN (international).

The OSI networks, when used with X.25 for large distributed LANs, are not able to use the X.121 addressing frame, as it cannot support a large number of addresses. This problem can be alleviated by using the addressing scheme of the OSI transport layer for internetworking, which can define addresses for both LANs and WANs. This address of the transport layer is known as the **network service access point (NSAP)** and supports all global addressing schemes such as the X.121, **telephone numbering scheme (E.163)**, **telex numbering plan (F.69)**, and **ISDN numbering plan (E.164).**

According to recommendations made in **Technical Office Protocol (TOP)** version 3.0 (routing and relaying description), the NSAP address should be divided into three fields: **enterprise Identification (EI**), **logical sub-network identification (LSI)**, and **address within logical sub-network (A)**. These fields handle the routing, relaying, and addressing for a globalized network environment.

The first field, **EI**, includes the addresses of hosts within the same organization or enterprise for routing of the packets. These organizations are typically government agencies, corporations, or multinational companies. The routing strategy in this field possesses the highest level of priority. The second field, **LSI**, defines the address of a specific logical sub-network within an organization enterprise. The routing strategy may be implemented as a structured and hierarchical transmittal, as the grouping within the groups of sub-networks can be defined at different levels of abstractions. The last field, address (A), defines the routing of packets from the OSI to the destination end system. This field, in fact, maps the address onto the sub-network point of attachment (SNPA) of an OSI network.

12.6.5 Packet assembly and disassembly (PAD)

As mentioned earlier, access to the public switched network is via the X.25 standard. This standard defines the lower three layers of OSI-RM, and two DTEs can communicate with each other over X.25 if these have X.25 interfaces. Here the DTE terminals are expected to possess some programming capabilities. If the terminals do not have any programming capabilities (e.g., dumb terminals), then the communication between terminals and remote host can take place through the X.25 public data network via a device known as **X.25 PAD**.[19]

The following are some of the features provided by PADs:

- PADs can support multiple packet networks which will provide permanent or switched network connections. The data link of PAD can be individually set for different networks.
- PADs can be used for point-to-point communications between terminals and host with increased throughput and fallback capabilities.
- PADs can provide higher data rates (19,200 bps) with higher throughput and improved fault tolerance (in case any data link fails).
- PADs can support various asynchronous standard speeds of 50, 75, 110, 134.5, 150, 300, 600, 1200, 1800, 2000, 2400, 3600, 4800, 7200, and 9600 bps and the speed of a 4800-bps computer port.

- PADs are used in a variety of applications. They can be used for local switching where both local and remote terminals can be connected to various local resources and other terminals. These PADs are programmable and can adapt to different data rates automatically. The adaptive speeds may range from 110 bps to 1200 bps.

PADs are usually used for display terminals but have also been used with computers, printers, teletype machines, terminal servers, etc. The available PAD devices support asynchronous, IBM bisynchronous, IBM SNA, DECnet, HP, Burroughs, and poll/select protocols. PADs do not provide any protocol conversion between different end-system communication devices. PADs are always required for an X.25 interface (with similar protocol on the network) for synchronous communication mode.

CCITT defined X.25 as a specification for a network interface and uses the protocols of the three lower layers (physical, data, and network). Many people think that X.25 is a protocol, but it is not; as just stated, it is a specification for the network interface. According to the CCITT recommendation, X.25 is an interface between DTE and DCE for terminals which are functioning in the packet mode in PDNs.

The information on the X.25 packet network is in the form of a packet/frame. A terminal (which is typically asynchronous non-buffered/buffered) device cannot be connected to X.25 directly through DCE (modems or multiplexers) and, hence, cannot access the packets from the network. The PAD can be connected between terminals and the X.25 network. The PAD gets characters from the terminals and makes a packet out of these characters in block mode. Thus, the input to PAD is a string of characters and the output is a block of characters defining a packet. This packet is transmitted to the X.25 network. On the receiving side, PAD receives the packet from the network and disassembles it into characters, and these are sent to terminals. Terminals with different speeds are supported by PAD.

CCITT has defined standards for PAD in the form of X.3, X.28, and X.29 documents.

CCITT X.3 standard: The X.3 standard defines the functions and various parameters for controlling the operations of PAD. There are 22 parameters maintained for each of the PAD devices connected to a terminal. The PAD may be connected to more than one terminal. Some of the material in this section is derived from References 2–4, 7, 29, and 30.

Responses to the following X.3 parameters are defined by their respective values. These parameters are usually addressed by the vendors in their specification manuals.

- Can the terminal escape from data transfer to PAD command mode?
- Can PAD echo characters back to the terminal?
- Can PAD send partially full packets?
- Time-out which allows the PAD to transmit partially full packet.
- Can PAD use flow control over terminal output, using X-On, X-Off?
- Can PAD use control signals to the terminal?
- Actions taken by PAD upon break signal from terminal.
- Can PAD discard DTE data sent from terminal?
- Can PAD insert control characters to prevent terminal overflow?
- Number of padding characters inserted after carriage return to the terminal.
- Can terminal use flow control over PAD using X-On, X-Off?
- Can PAD insert line feed after sending carriage return to the terminal?
- Number of padding characters inserted after line feed to the terminal.
- Can PAD support line editing (as given below)?
  - Character delete
  - Line delete

- Line display
- Terminal type for editing PAD service (line delete)
- Character that should not be echoed, even when echo-enabled
- Parity check to/from terminal
- Number of lines to be displayed at one time

CCITT X.28 standard: The X.28 standard defines the protocol between the terminal and the PAD device. Two phases of operation for X.28 have been defined. The first is the data transfer phase, in which the exchange of data takes place between the terminal and host; the second phase is the control data phase, in which the exchange of data takes place between the terminal and PAD in the form of characters. Various facilities, such as the desired baud rate between the terminal and PAD, opening of a connection with remote DTE or host, etc., may be obtained with the help of parameters defined in X.3. The PAD terminal protocol defines the following X.28 messages:

- **Stat** — A terminal request about the status of a virtual call connected to remote DTE. PAD's response is Free/Engaged.
- **Call** — Sets up a virtual call.
- **Clear** — Clears a virtual circuit. PAD's response is Clear Confirmed/Clear Zero.
- **Interrupt** — Transmit interrupt packet to DTE.
- **Reset** — Resets a virtual circuit call. The window is set to sequence number 0 at both transmitting and receiving nodes.
- **Parameter** — Specified parameters of PAD.
- **Set** — Sets or changes current PAD parameter values.

CCITT X.29 standard: The X.29 standard defines an interface and a protocol between PAD and host. The PAD–host protocol uses X.29 messages as shown below:

- **Set PAD** (to PAD) — Sets selected parameters to new values of PAD.
- **Read PAD** (to PAD) — Gets values of specified parameters.
- **Set and Read** (to PAD) — Set and Read.
- **Invitation to Clear** (to PAD) — Tells PAD to clear the virtual call.
- **Reselection** (to PAD) — Tells PAD to clear the virtual call and make a new call to selected DTE.
- **Parameter** (from PAD) — List of parameters and their values, which basically is a Response to Read command.
- **Indication of break** (to or from PAD) — Indicates break from DTE or terminal.
- **Error** (to or from PAD) — Indicates error in previous message.

These PADs are available in 4, 8, 24, and 48 RS-232 ports. One of the ports can be assigned to supervisory mode, and this port is then used as a manager which sets up the parameters for PAD operations. The manager assigns addresses (14 bits) to certain mnemonics and then sets this port as a supervisory port. For example, suppose we use the following PAD selection command (X.28):

Call, Hura, RC = M ⟨ret⟩

This command will set up a virtual circuit to the present location with name Hura, with **reversed charged (RC)**. All the parameters of the Hura site have been set up by the manager through the supervisory port. When PAD receives this command, it sends the

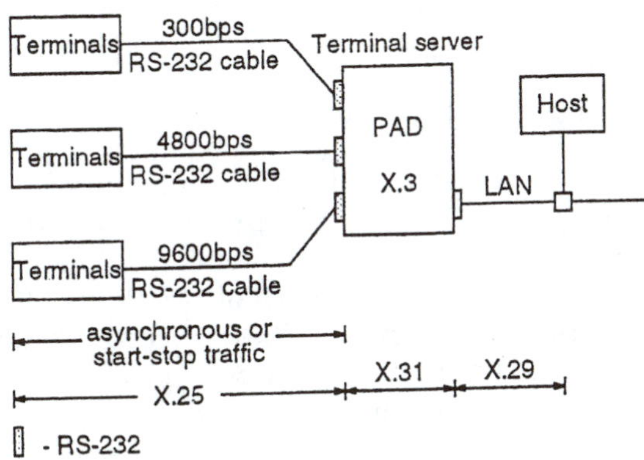

Figure 12.12 A typical PAD configuration for LAN.

Call.request packet along with the logical channel number and group number and submits it to the network. The PAD receives the X.25 packet from the network (depending on the virtual circuit carrying it) and stores it in the buffer of the designated RS-232 port. It extracts the data part and delivers it to the device connected to it via the RS-232 port.

The X.25 PADs are also available with **multi-link procedure (MLP)** features for achieving a higher throughput. A dual data link speed of up to 19.2 Kbps over full-duplex can be achieved. Each of the data link's parameters is separately programmable (e.g., window size, time-out, retry counters, or type of network). This allows each link to have different network identification codes and call addresses. For example, one port of X.25 may be connected to a remote resource via a public data network, say **Tymnet**, via one data link, while another public data network, say **Accunet**, can be connected to a second port over a second data link.

These X.25 PADs also support U.S. packet-switched networks such as **Accunet**, **CompuServe**, **Telenet**, **Tymnet**, and **Uninet**, as well as other international packet-switched networks such as **Datapac**, **DCS**, **DN1**, **PSS**, and **Transpac**.

A typical X.25 uses a generic function for DTE (PCs, terminals, etc.) which may be a network-attached, end user device or any other communication device working in the packet mode. For example, a minicomputer or any other communication device running at X.25 specification is a DTE, the front-end processor loaded with X.25 code is a DTE, etc. An alternative to these DTEs is packet assembly/disassembly (PAD), which allows the users of X.25 or non-X.25 to communicate with public or private packet-switched networks.

A typical configuration of PAD for LAN is shown in Figure 12.12. It allows the transmission of traffic from LANs or WANs in a packet/frame mode. The non-buffered asynchronous terminals cannot access the public data networks directly. These terminals can access the networks if these are connected via device PADs, which basically accept the characters from the terminals and prepare a packet by assembling character by character. This packet is transmitted over the X.25 network in a block mode. On the receiving side, this device accepts the message in block mode, de-assembles character by character, and sends the characters to respective terminals at various speeds.

Another type of DTE is a **dumb terminal**, which uses asynchronous protocols for its operations. This terminal transmits a character at a time (as opposed to a block of characters in other character- or byte-based terminals). This terminal uses start–stop bits with every ASCII seven-bit character for providing synchronization between sender and receiver. The CCITT has defined a separate non-X.25 network interface specification for this terminal (also known as start–stop DTEs). A pair of DTE/DCE is needed to use X.25, and, due to this, it has been defined as a standard interface specification by CCITT. Other possible combinations such as DTE-DTE and DCE-DCE interfaces have not been defined or standardized.

12.6.6 Accessing X.25 networks

As indicated earlier, users can access X.25 via a dial-up connection over public telephone networks or via X.25 packet-mode terminals. Users having leased lines or asynchronous mode terminals (start–stop) cannot access X.25 networks. The CCITT defined three X recommendations — X.3, X.28, and X.29 — which offer a PAD option on PSDNs to users having leased lines or asynchronous terminals. These protocols (X.3, X.28, X.29) sometimes are also known as the **Triple X Protocol** as discussed above.

The PAD includes a set of user parameters as defined by X.3 which resides in PAD. PAD supports the set of parameters for each of the asynchronous ports defined by PAD. The parameters specified for the ports of PAD can be used to define the network configurations. Usually the ports of PAD use the same number of configurations. These numbers were standardized as a set of 12 parameters in the initial version of PAD; 18 parameters were defined in the 1980 version, and more than 22 parameters are in use at this time.

12.6.6.1 PAD primitive language

X.28 protocol defines the procedures and primitive language between the dial-in terminal (start–stop) or local DTE and X.3 PAD (sitting in PSDN). It defines two types of primitive language: **PAD commands** and **PAD responses**. The primitive language requests initiated by DTE for PAD, such as call setup, are known as *PAD commands*, while requests such as clearing are known as *PAD responses*. X.29 protocol defines the procedure for controlling the functions of X.3 PAD port parameters over the network for its transmission to remote DTE (X.25 host).

The X.3 PAD parameters are classified in various groups such as data forwarding, editing, flow control, line characteristics, recalling, display formatting, and terminal display, and each of the categories contain different functions with different identification numbers.

X.28 commands allow users to communicate with PAD directly, and users can also submit their requests. The users have to dial up the local PSDN to make use of X.28 commands. The X.28 offers nine PAD commands and these commands allow the user to clear the call, read or change the X.3 parameter for any configuration, and also check the call. For each of the commands issued by the users, the PAD receives the command as a set of 12 PAD service signals, and these signals are used to define the response for DTEs. Some of the service signals include resetting, clearing, error, PAD identification, PAD service signal, status (free or busy), and a few network-dependent service signals.

According to CCITT, the X.25 defines the procedures for the exchange of control information and user data between the PAD and packet mode DTE or another PAD. The packet-mode DTE is typically an X.25. This protocol (X.29) uses X.25 at both PAD and packet-mode DTE and imposes some restrictions on the host. It supports asynchronous communication between the terminal and PAD, and the X.29 messages are defined in the data field of X.25 data packets which are sent across the network. It possesses more or less the same capabilities as X.25, especially in the areas of reading and setting various PAD X.3 parameters.

Asynchronous PADs must support all three standards of CCITT for X.25 as X.3, X.28, and X.29. These standards permit the configurability of each of the PAD ports and also provide the mapping between these terminals and remote-host computer. Further, a simple set of command language must be provided to dial-up users for setting up the data calls for transmitting the messages to the remote host computer or even the database.

PAD itself is defined by the X.3 standard that provides the assembly and disassembly of the packets into data and control fields. The X.28 standard defines an interface between the terminal and PAD. The standard X.29 defines an interface between the PAD and public switched network. The packet from a terminal goes to the public network through PAD, and the packets from the public network go to another public network(s). The standard X.75 defines an interface between various public networks such as Tymnet, Uninet, and Telenet (all U.S.), Datapac (Canada), PSS (U.K.), and so on. For all of these networks, the standard X.25 may provide a unified interface. The public network acts as a DTE to a second net's DCE. The CCITT proposed three possible configurations. In the first configuration, an X.25 DTE will be connected to a public network that in turn communicates with another X.25 DTE via a public network. In the second configuration, the DTE will interact with an X.25 LAN, which will send the packet to an X.25 DTE via a public network. In the third configuration, the end X.25 DTE communicates with public networks that communicate with each other through private networks. The public networks can be accessed via either public access ports or private ports. The public port is usually dial-up or wide-area telephone service, while the private ports can also be dial-up or leased lines.

12.6.6.2 *Data switching exchange (DSE)*

Other synchronous DTEs that work in the packet mode are directly connected to the X.25 nodes (composed of the data switching exchange (DSE) and the DCE of the packet-switched network) with leased-line connections. Because there is no standard specification defined for DCE-DCE or DTE-DTE interfaces, the transmission of packets to the network needs an intermediate node or exchange known as a **data switching exchange (DSE)** or **packet-switching exchange (PSE)**. This node can be connected to DCEs and is usually viewed as the end-switching node in the network.

A typical DTE/DCE interface used with X.25 is defined by a number of packet-switched network nodes connected together, where each node consists of DCE and DSE. Each node may be connected to the DTE directly or via modem. Further, the DTEs support a number of terminals. The X.25 interface provides physical connectivity between different DTEs and the network and is supported by its protocol. This protocol offers a different set of conversions for its three layers (physical, data, and network), while the user's network and relevant equipment share a compatible set of interfaces and signaling for their respective lower layers.

The protocols of these lower layers are implemented from the bottom layer (physical) to the network layer individually to establish and implement a X.25 circuit or link. The DTEs are connected to the network via X.25 interfaces. If DTEs are connected to the network via modem, the **carrier detect signal (CDS)** of the modem is always ON (indicating that the physical circuit is active and can transfer a bit stream). The data link layer protocol deals with error control and allows the DTE/DCE pair to communicate. The network layer protocol defines the set of procedures for setting up calls so that packets can be transmitted across the network.

In the above typical configuration, we have considered both DSE and DCE in one node, while in actual practice, the DSEs are connected together or with other DSEs in a mesh topology in such a way that each node defines two circuits to provide a link or circuit redundancy in the network. The DSE nodes may be connected by digital links (56 Kbps) or T1 carrier (1.544 Mbps).

12.6.6.3 *X.32 protocol (synchronous transmission/terminal)*

As mentioned above, the majority of dial-up access to PSDNs support asynchronous transmission to make the service of X.25 available to different types of users. The CCITT recommended the **X.32** dial-up protocol synchronous transmission which, in fact, redefines

X.25 for operating on a dial-up connection. It allows the users to access X.25 services at a cheaper rate than leased lines. The X.32 synchronous transmission protocol has several advantages over asynchronous dial-up protocols, e.g., the establishment of multiple sessions over single dial-up link, less protocol overhead, error-recovery facilities (due to the fact that it uses LAP-B), etc. It allows users to use PSTN directly to establish connection (via dialing) into the DCE or packet-switched node, bypassing PAD. X.25 is being used as a DTE/DCE interface by X.32, and the problem of providing an addressing scheme is usually mutually agreed upon between users, who may use various facilities/options available in X.25 and X.32 recommendations.

X.25 offers various services at different levels. For example, at level one (physical layer of OSI-RM), the standard X.21 provides operations for establishing connections (connection setup) and termination or clearance of the connection. At level two (data link layer of OSI-RM), a standard such as HDLC offers error control, flow control, and various operations for data transfer over a logical channel connection between DTE and DCE. This logical connection is again established by this standard. At level three (network layer of OSI-RM), X.25 supports all the services offered by the lower-level layers. It is worth mentioning here that many countries have no access to X.25 and, even if it is available, the X.25 interface sometimes seems to be very complex for simple DTEs (teletype writer) or even DTEs supporting multiple virtual circuits.

A variety of DTEs are available from different vendors, and these devices define or support various options provided by all three layers of the X.25 protocol. An attempt has been made to define a universal type of DTE which must not only provide various options offered by X.25 but also support different types of networks. It is important to note that most of the existing networks such as Telenet (U.S.), DDX (Japan), Datapac (Canada), Transpac (France), and many others do not support each other; i.e., these are totally incompatible with each other. Except for one or a few services, these networks have altogether different formats and implementation schemes. One of the compatible services is virtual circuit, which supports packets 128 bytes long. This size of packet is supported by all the existing networks.

Thus, the universal DTE must support a variety of options provided by networks, including error control and many other default control information supports. It must also be flexible and adaptable to new changes made during internetworking. To be more specific, the universal DTE must support various performance parameters such as response time, throughput, data speed of transmission media, size of packets, etc., at the physical layer level. The majority of public networks offer a data access rate of 2.4 Kbps, 4.8 Kbps, and 9.6 Kbps.

At data link layer level, the universal DTE must support LAP-B, as it has been defined as a common standard by CCITT and ISO standards organizations. An option for symmetric versions of implementation is also available on LAP-B which must include support for asymmetric versions of LAP-B and set default values for different parameters. The symmetric version of LAP-B inherently defines DTE-DCE and DTE-DTE compatibility.

At the network layer, the packets generated by simple DTE do not support any error control but, instead, support only network-supported functions. Similarly, for complex DTEs (supporting multiple virtual circuits), error control and other options are negotiated, and they avoid the use of re-transmission for controlling flow in the network.

12.6.7 Network service protocol primitives and parameters

Various connection-oriented and connectionless service primitives offered by the OSI network layer, along with the fields used in different control packets, are discussed in this subsection. These are defined in various standards such as ISO 8473 for connection-oriented service,[14] ISO 8473/DAD1 for connectionless mode network service,[15] and ISO 8473/DAD3 offering provision for ISO 8473 over subnets which offer data link service.[16]

12.6.7.1 Connection-oriented service primitives

The network service primitives of the OSI network layer are **N-CONNECT.request**, **N-CONNECT.indication**, **N-CONNECT.response**, and **N-connect.confirmation**, and each of these uses the following parameters in the fields: calling address, called address, Ack-required, Exp-required, QOS, and user data, where calling and called addresses are represented by network service access point (NSAP) addresses.

The **Ack-required** parameter is defined by a flag of Boolean type for acknowledgments. **Exp-required** is defined by another flag for selecting the expedited data mode. In this case, packets will be given some sort of priority. The **QOS** parameter defines the quality of service requested, while user data represents user's data in bytes (zero or more). The sender user may send a few bytes of data to the destination user (if he/she wishes) during the establishment of connection between them.

The field corresponding to Ack-required and Exp-required has to be negotiated and mutually agreed upon through the flag status. The QOS defines the sender's desire for various features such as throughput, delay, error rate, costs, etc., which are provided by the network layers during the establishment of connection. These features are negotiated by both sender and destination nodes back and forth. If a network cannot provide the minimum values of these parameters for QOS, then the connection establishment process is terminated. Typically, the minimum values of the parameters are the default values for them.

The N-CONNECT.request primitive is initialized by the calling (source) address for establishing connection, and all the requested fields are defined appropriately for negotiation with the destination user. The network layer sends an N-CONNECT.indication primitive to the destination user. If the network layer cannot provide the minimum features defined in the appropriate field, it sends an N-DISCONNECT.indication primitive back to the sender; thus, connection establishment fails. The network layer looks at the user data (if any) and the status of its own model and available resources. It then sends its acceptance for the establishment of connection in an N-CONNECT.response primitive or sends rejection in another primitive, N-DISCONNECT.request. It also responds to various features desired by the sender via setting flags, etc., in its response. The N-DISCONNECT.request primitive also includes the reason for rejection. The sender network, upon receiving the primitive from the destination network user, confirms the establishment of the connection for negotiated features by sending an N-CONNECT.confirmation primitive to the sender.

After the establishment of connection, the sender user uses the following primitives for sending the data to the destination user which, in turn, invokes **indication primitives** after receiving data through **request primitives:**

- N-data.request (user-data.confirmation.request)
- N-data.indication (user-data.confirmation.request)

For features such as acknowledgment and expedited data, the following primitives are used:

- N-Data-Acknowledge.request
- N-Data-Acknowledge.indication
- N-Expedited-data.request (user data)
- N-Expedited-data.indication (user data)
- N-Expedited-data.indication (user data)

It is important to note that the acknowledgment primitives do not define the segment number of the packet; hence, the sender site has to maintain its own mechanism for keeping record of the reference numbers of received packets at the receiving site. The network layer here tries to provide a better QOS to its user and is not responsible for any error-recovery or error-free service. These services are provided by other layers such as the data link and transport layers.

Other primitives which provide services for natural calamities, crashing of the system, broken links, etc., are given below:

- N-Reset.request (reason)
- N-Reset.indication (reason, calling/called address)
- N-Reset.response
- N-Reset.confirm

12.6.7.2 Connectionless service primitives

The primitives N-unitdata.request and N-unitdata.indication are used to send and receive data (up to 65,512 bytes) and contain the following fields: sender address, destination address, QOS, and user data. These primitives don't establish any connection between the sender (NSAP) and destination (NSAP) user; they simply send/receive the data back and forth. The QOS is the only feature which is supported in this primitive service. The N-facility.request (QOS) and N-facility.indication (destination address, QOS, reason) primitives define some statistics on the packets delivered to the destination and are used between the sender user and network layer. Finally, the primitive N-report.indication reports problems associated with the destination user to the source user network service.

As discussed earlier, the **calling (source)** and **called (destination)** addresses used in connection-oriented network service primitives are network service access points (NSAPs). Similarly, the sender and destination addresses in connectionless network service primitives are also NSAPs. These addresses provide endpoints of communication across the network and define the ports/sockets within the OSI model. Various services offered by network layers are available to users only through these service access points (SAPs) or ports or sockets. The sub-network/subnet addressing also needs to be defined for both sender and destination sites. There may be many NSAPs (endpoints) connected to a particular subnet and there may also be many sub-networks connected to a particular NSAP (endpoints). ISO 7498/3 discusses all the limitations, properties, etc., of the NSAP addressing scheme so that a uniform internetworking addressing scheme can be controlled by one authority.

12.6.7.3 Global network addressing

A global network addressing domain has been defined which includes all the addresses of the OSI models. This domain is divided hierarchically into a number of network addressing domains. All NSAPs belong to a network addressing domain. No matter how we assign the addresses to any domain or who is controlling a domain, all the network addressing domains defined in a hierarchical way are administered and controlled by one authority, as defined in the ISO 8348/DAD2 document.

The following six subdomains of the global network addressing domain have been defined by the ISO document. Four domains have been defined for the public telecommunication network and are controlled by CCITT. Each category of public network (packet-switched, telex, telephone, and ISDN) is supported by a domain. The fifth domain, known as the ISO geographic domain, is assigned for an individual country where representatives of each country within the ISO committee allocate the addresses to its users.

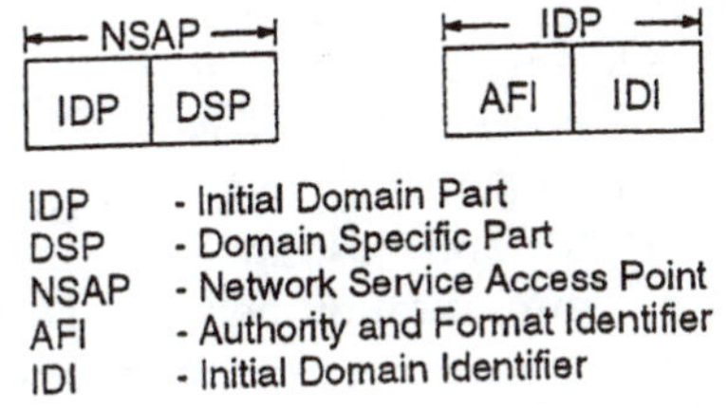

Figure 12.13 NSAP address fields and IDP field.

The sixth domain is defined for various international organizations such as NATO and is known as the **international organization domain (IOD)**.

The NSAP defines a uniform name and addressing scheme to the transport layer. The OSI network addressing scheme has a well-defined system and semantics, but the actual implementation of this scheme depends upon a number of factors such as services and functions of the network layer and transport layers (if any) and the standard protocols used by them. The network service primitives define source and destination addresses from the NSAP addressing format/structure.

The NSAP address is defined in terms of two fields: initial domain part (IDP) and domain specific part (DSP), as shown in Figure 12.13. The field IDP, in turn, is defined in terms of **authority and format identifier (AFI)** and **initial domain identifier (IDI)**. The AFI defines the structure of IDI and the type of address defined in DSP. The DSP contains address information (in coded form) for defining end system addresses or packet network addresses, numbers of various data communication systems such as telephone number, ISDN number, etc., and many other numbers. These numbers are in binary form and also in packet decimal form and range from 10 to 90. Currently, only a few of the numbers in this range are being used; the remaining numbers have been reserved for future data communication systems.

The IDI defines the domain or sub-network address for which the DSP field has been considered. It contains up to 14 decimal digits and defines the actual sub-network address. The decimal digit is taken from the value assigned to AFI. For example, if DSP represents a telex system, then IDI will represent the initial code of the country. The NSAP address scheme is of variable length of up to 20 bytes (40 decimal digits).

The NSAP address, in fact, represents the port number or socket number defined in the network through which it obtains the data from the transport layer and also supplies data through the same port to the transport layer.

12.7 *Internetworking and packet-switched data network (PSDN) protocols*

Internetworking different packet-switched networks (public or private) has become an important issue in recent years. The private packet-switched networks usually request gateway connections to PDNs. The gateway may provide support for multiple physical circuits and a large number of simultaneous call requests. For internetworking public and private packet-switched networks, the X.25 has been the most widely used and acceptable interface for small applications. This interface cannot handle a heavy volume of traffic. The CCITT defined another standard interface, X.75,[21] which supports the existing X.25 and handles a large volume of traffic. The X.75 resides on another node known as **signaling terminal equipment (STE)**. This STE node interface provides communication between different packet-switching networks via the X.75 standard protocol.

The problem of interconnecting various networks via a standard interface for global communication across them has taken a major step toward solution with the introduction of X.25. The standard interface defined by CCITT is X.75,[21] which operates at the lower three layers of OSI-RM (physical, data link, and network layers). The X.25 typically provides connections between PSDN and host and offers communication between DTE and DCE (with no support for communication between DCE and DCE). The reliability and other performance parameters are also not specified and imposed, in spite of the use of multiple sessions over a single link. Further, the X.25 does not support high-volume traffic on the Internet (discussed in Chapter 17 on Internet).

Most of these deficiencies identified in X.25 are well taken care of in defining the X.75 standard, which is really an enhanced version of X.25. A gate node **(STE)** is defined within each PSDN and a multiple-link protocol allows the transmission of a virtual call over multiple physical links. The STEs can be linked by either a single link or multiple links, and in both cases communication across the PSDNs takes place via STEs. The local DTE still communicates with its local DCE via X.25, but the standard interface for the network is defined via STEs for communication between them. The X.75 call request packet has an additional field of network utility over the X.3 call request packet format.

12.7.1 Internetworking

Data communication networks have become an integral part of any organization. These networks allow users to share data, resources, etc., and allow them to communicate with other through various services such as e-mail, file transfer, remote access of programs and databases, remote login, etc. These services may not be available on the same computer connected to a particular network, but these services can be accessed if that network is connected to another network which provides those services. Thus, a user on any network can communicate with another user on a different network and both can share the resources, programs, databases, etc., of the other. The physical linking of the networks promotes an internetworking of sub-networks or subnets. The subnet has a completely autonomous configuration which includes three layers — physical, data link, and network — and the protocol for each layer for providing different services to the users. The sub-networks may be public networks (X.25, Telenet, Datapac, etc.), privately owned networks, or even local area networks. In OSI-RM, the services offered by the network layer are routed through subnets, with connection transparency to lighter layers.

There are many internetworking devices, e.g., repeater, bridge, router, and gateway. A brief description of each of these is presented here. For more details, please read Chapter 7 on Local Area Networks.

A **repeater** accepts a data signal at its input and regenerates the signal via some kind of error correction. The regenerated signal is then re-transmitted. This device works at layer 1 of OSI-RM and can be bidirectional for bidirectional cables. In the case of token ring networks, it is a unidirectional device. It is very inexpensive.

A **router** is used at layer 3 of OSI-RM and connects different types of networks that have the same network layer protocol. It accepts the packet from one network, finds the best route (based on a routing algorithm), and uses the functions of the data link layer defined within the MAC address of the bridge (another internetworking device). Routers are used in LAN/WAN interconnectivity and provide different categories of ports. Usually, they support a few ports of each category. The port for a LAN is typically 100 Base-T, but it may have ports of other LANs as well. These ports are available in the form of a card. In addition to providing ports for LANs and WANs, routers also provide interfaces to various serial devices such as CSU/DSU. The router maintains a routing table giving information regarding the routes a packet will take going from its LAN to WAN and finally to the destination LAN. The decision regarding the routing is based on the routing algorithm being used by the router. There are a number of protocols that provide routing of the packets through the internetwork, including interior gateway routing protocol, exterior gateway protocol, border gateway protocol, intermediate system to intermediate system, and so on.

The **socket** defines an address through which the data transfer between PC and network takes place. There may be a few sockets within the interface unit. The received packet is partitioned into two fields: information and control. The message goes to the correct process in the PC's memory. The socket must be unique. The CCITT has defined the socket address of 14 bits. The first four digits represent the segment of a particular network. The remaining 10 bits are used to define the network station and unique socket within it.

A **gateway** is used to provide interconnectivity between LANs and other LANs, or packet-switched networks, or private networks, databases, videotex services, or any other services. The gateways provide compatibility of protocols, mapping between different formats, and easy data communication across them. The PC will send the packet to the gateway, which now becomes our address, which will send it to another network. In this way, the gateway basically becomes another node between networks that provides compatibility between the operations of these two different networks. It provides an interface between two different networks having different protocols. It operates at layer 7 of OSI-RM. The networks having different protocols can be interconnected and the gateway provides mapping between their protocols. Typically, our PC may be connected through a gateway to a mainframe or another computer having different protocols so that the packets can be sent to it and the mainframe resources can be used. These are expensive devices.

Gateways connected between the Internet and the destination network (target network) receive Internet e-mail and forward it to the destination. As noted earlier, the message is broken into IP packets. The gateway receives it and uses TCP to reconstruct the IP packet into the complete message. It then provides mapping for the message into the protocol being used at the destination. A variety of tools exist for managing the TCP/IP suite of protocols, such as simple network management protocol (SNMP). Network management protocols have been introduced by Cascade, Cabletron, Fore, IBM, NEC America, Hughes, Network Security Corporation, Siemens, and many others. Every gateway connected to the Internet defines and maintains a database known as a management information base (MIB), which contains information regarding the operations of gateways, e.g., operating system used in gateways, TCP, UDP, ARP mapping, IP, statistics such as the number of datagrams received and forwarded, the number of routing errors, datagram fragmentation, reassembly, and IP routing table.

LAN environment: A typical LAN environment that exists today in universities, corporations, organizations and government agencies uses FDII as a backbone network which connects different types of LANs such as Ethernet and token. These LANs are connected to FDDI via a 10-Mbps link for Ethernet and a 4- or 16-Mbps link for a ring network. Various servers are usually connected to FDDI via 10-Mbps links. The FDDI is also connected to routers via a high-speed link at 50 Mbps or above. The routers are connected to the Internet or other routers via different links ranging from T1 (1.544 to 100 Mbps), providing access to the outside world. ATM can be used for the existing LAN environment in a number of ways. In one environment, the FDDI can be replaced by an ATM switch, and some of the servers may be connected via high-speed links. The FDDI can also be reconfigured as an access network for different LANs. A second environment may define the ATM hub (ATM-based) where ATM switches are connected together via OC3 links and can be used as a backbone network in place of FDDI. Here the ATM switches will be providing interfaces (ATM adapters) and accesses to different types of LANs. These ATM switches will also provide interfaces to routers via links of DS3 or above for access to the outside world. In video-based applications, the ATM can be connected directly to workstations or terminals via optical fiber, thus giving a data rate of 100 Mbps with the options of transmitting video, voice, and data. A number of servers dealing with these types of traffic (video, voice, and data) can also be directly connected to ATM via ARM adapters.

WAN environment: As we have seen, WANs are used for larger distances more often than LANs and are slightly more complex. LANs use their own links to provide services, while WANs use the public carrier or a private line. LANs and WANs can communicate with each other through an interconnecting device known as a channel service unit/data service unit (CSU/DSU). This device provides mapping between the LAN protocol and network digital protocol corresponding to T1 or any other link in the same way that the bridges provide two types of LANs at the MAC sublayer. The LANs are connected to WANs via CSU/DSU devices which provide mapping between the protocols of LANs with that of the public carrier (such as T1 or any other link) or private lines.

The majority of networks are typically public networks and are required to provide services to the different agencies. In the U.S., these networks fall into the following classes. The local exchange carriers (LEC) provide local-access services (e.g., U.S. West), while interexchange carriers (IEC) provide long-distance services (e.g., AT&T, MCI, Sprint). Regional Bell operating companies (RBOCS) operate the LECs. The IEC also provides interstate or inter-local access transport area (LATA) service. In most countries, government-controlled post, telegraph, and telephone (PTT) agencies provide these services.

ATM-based environment: ATM is used in different classes of public networks. It can be used to replace the end office and intermediate office by connecting them with DS1/DS3 links. As we know, the subscriber's phone is connected to the end office via a twisted pair of wires which must be replaced by coax/fiber. The public network system and CATV can use ATM for providing video-on-demand services. This may require the use of compression techniques such as asymmetrical digital subscriber loop (ADSL) that uses the MPEG2 coding method.

For the WAN environment, the ATM hub can be defined by connecting ATM switches via OC3 links, and then ATM switches of the hub can provide interconnectivity with LANs through an ATM LAN connected via OC3 links. The ATM switches from the hub can be connected to WANs via OC3/DS3 links for access to the outside world. A typical WAN environment includes various TDM devices connected together via T1 carrier lines defining the TDM hub. The TDMs are connected to various routers located at different locations miles apart. The routers are used as access points for the LANs (as discussed above). There are a number of ways of providing WAN interconnectivity. In one method, each TDM may be replaced by an ATM switch, and these switches are connected via either T1 links or T3 (DS3) links (45 Mbps). The ATM switches transmit cells between them for the traffic coming out of the routers. The ATM hub supports different types of traffic (video, data, and voice). Here the ATM hub defines the MAN/WAN backbone network. In another method of WAN interconnectivity, the ATM switch may be connected via OC3 links and also can be directly connected to various routers and video services through video servers. It is interesting to note that with OC3 links, the optical signals now carry the information, which may allow us to use the public services available on ATM-based backbone public networks. These networks are connected to ATM switches via OC3 links.

Various networks are linked by Internet links, while different subnets are linked by gateway computers. A protocol for this computer network has to deal with various issues and problems of internetworking, including different sizes and formats of the packets, different addressing schemes, different types of services (connection-oriented, connectionless), different performance security criteria, etc. In addition to these technical differences, other nontechnical differences such as billing, maintenance, and diagnostic test criteria, are to be supported by the protocol. There will still be many unsolved issues and problems.

Various standards have been proposed to solve these problems and issues and are being used in various internetworking methods. The standards/protocols accept the packet from one subnet, transform its format for all three layers of another subnet, and transmit it to the subnet. They also maintain a unique routing table which automatically formats the addressing, size of packets, routing through the sub-network, etc. Other incompatibility problems such as support for a particular connection, performance, security, throughput, etc., are also resolved by the protocols.

12.7.2 Internetworking interface

The problem of internetworking can be solved in different ways. In one way, all the homogeneous subnets (having same architecture, functions, and sets of identical protocols for peer-to-peer communication) may be connected together. This type of configuration is dependent on a specific type of subnet and is known as subnet-specific; it uses identical protocols at each layer. CCITT X.75 provides internetworking for X.25 public packet networks. This scheme puts a restriction on the type of subnet which should be used. Only one specific type of subnet can be used. Since all the subnets are the same and use the same protocol, it provides a uniform scheme for all issues and problems and further reduces administrative burden.

The existing packet-switched network X.25 provides interfaces to routers, which are connected to routers, bridges, and LANs. The interface offers a dial-up connection via a 56-Kbps link from LANs. The function of the X.25 protocol is to provide the functions of the lower three layers of OS-RM. Since the transmission media are not reliable, the X.25 must support error control and error detection. Further, the packet going through three layers requires processing time, and, hence, the network is very slow. There are a number of ways an ATM can be used in X.25 networks, which are typically used for data communication. In one X.25 switched network environment, each of the X.25 switches is replaced by a frame relay (connection-oriented services). This will increase the speed from 56 Kbps to T1 (1.544 Mbps) speed. For connectionless services, switched multimegabit data service (SMDS) switches offering higher speed than T1 are used; they can offer T3 (45 Mbps) speed and provide public broadband services. The underlying frame relay may become an access point to the SMDS network. In another X.25 environment, ATM may be used as a backbone network offering virtual circuits with a much higher bandwidth between T1 and OC 12 (622 Mbps). We can use frame relay and SMDS as access networks to ATM. We can transmit voice data and video over an ATM backbone but with circuit switching. We can also use the complete B-ISDN network using B-ISDN protocols and interfaces (ATM/SONET).

One of the main advantages of X.25 packet networks is that users can use one network for communicating with each other. For sharing data and resources, there needs to be a connection between two different X.25 networks. This connection will allow users to send their data packets back and forth and also use the resources and other services of the X.25 network. The CCITT X.75 provides a logical connection between different X.25 networks and internetworking or gateway computers for users.

X.75 includes all the features offered by X.25. It uses call request packets and other control packets. The subnet X.75 is defined by three layers — physical, link, and packet — and resides in the network layer over the X.25 protocol. The various layers of X.75 are implemented in the same way as those in X.25: the physical layer uses V.35 (V series standards), while the link layer uses HDLC subnet and LAP-B. Link speeds of 64 Kbit/s, 48 Kbit/s, 56 Kbit/s, 1.544 Mb/s, and others are being supported by X.3 for signaling. In X.75 documentation, data circuit terminating equipment (DCE) is known as **signaling terminal equipment (STE)** and performs the same functions as DCE in X.25.

The X.75 and X.25 packet formats are identical, except for minor differences. The address field of X.75 defines international data numbers (X.121). In addition to this, it also has a separate field of network utilities which can be requested by an X.25 user's packet in its facility field.

The DTE of a network issues a call request to its DCE, which it sends to its signaling terminal equipment (STE) of X.75. The STE assigns a new logical channel number and transmits it to the subnet defined by X.75. This subnet provides an interface to all the requests or other control packets through STEs. The STE of the X.75 sub-network assigns a new logical channel number to the request call control packet and forwards it to the STE of the destination network. Upon receiving the control packet from the STE, the DCE assigns a new channel number to it which will also be the logical channel number for any of the response packets coming out of its DTE. In this way, the mapping of the logical channel number between various devices across the network is implemented. The response from a remote DTE will have the same logical channel number as was assigned by the source DTE in its call request packet.

12.7.3 *Modes of internetworking configurations*

The X.25 supports different categories of user facilities. Some of these facilities are provided by the network and defined in the address field of X.75, while some of the facilities may be mutually agreed upon between users during a certain period of time. These facilities are requested in a call request packet during the establishment of a **virtual circuit/call.** These facilities are defined in the X.2 document. Some of the facilities are essential in the sense that the network has to provide them to the users, while others are optional and may be requested.

The internetworking can be implemented for both connection-oriented and connecitonless configurations, and various aspects of their implementations are discussed below.

Connection-oriented configuration: In this configuration, a logical connection (virtual circuit) between two DTEs of the same sub-network can be established. Thus, internetworking can be defined for this configuration if each sub-network supports this type of connection-oriented configuration between respective DTEs. The sub-networks are connected through an **interworking module (IM)** in such a way that the IM may support more than one sub-network (if desired). Each IM basically defines a DTE for that sub-network. The number of IMs connected to any sub-network represents the number of DTEs supported by that network. Each IM is capable of providing more than one virtual circuit/call. In this configuration a logical connection is established through the network (across the sub-networks between source and destination DTEs). The logical connection between DTEs provides a unique path for the transfer of data packets and is made up of a logical connection between the source DTE and IM, between IMs across the sub-networks, and finally between the IWU and DTE across the destination sub-network.

Connectionless configuration: In this configuration, internetworking is again implemented by connecting various sub-networks through IMs. The operations in this configuration are datagram operations provided on a packet-switched network. The network protocol data unit (NPDU) initiated by the network layer at the source DTE is considered a separate and independent datagram (containing all control information, addresses, etc.)

and is transmitted on different paths. The decision as to which datagram data unit should be transmitted is typically made by the source DTE. In other words, data units corresponding to NPDUs may be travelling via different IMs. When the data unit reaches a particular IM, the decision for selecting the next **internetworking unit (IWU)** is made by that particular IM, which may choose different IMs for the data units until these data units reach a destination DTE.

12.7.4 Protocol structure for architecture configurations

As mentioned in the previous section, the internetworking can be implemented using either connection-oriented (virtual circuit) or connectionless (datagram) configurations. Each configuration has its own protocol structure which can be classified as a subnet-based protocol or subnet-independent protocol. The subnet-based protocol is a logical connection in homogeneous subnets for a virtual circuit. On the other hand, subnet-independent protocols provide internetworking for both homogeneous and heterogeneous subnets and reside in the host computer and gateways. The datagram protocols for OSI connectionless configuration include ISO 8473 and DoD's Internet protocols (IP).[25] These protocols can be used on both LANs and WANs. The reliability, data security, etc. in these protocols are usually supported by protocols of the fourth layer of OSI-RM (transport layer).

In a typical situation where a computer is connected to the Internet and wishes to send and receive data over it, it needs a set of protocols such as Telnet, TCP, IP, and protocols for LANs or other protocols for direct access of the Internet. Telnet (a program) breaks the data into a number of packets. This must run on both our PC and the host. The TCP breaks the application data (already broken into Telnet packets) into TCP packets where each packet has a header with the address of the host and information regarding assembly/disassembly and error control. The IP again breaks the TCP packets into IP packets, with a header with address information, routing information, and TCP information and data. These IP packets are picked up by subnets (LANs, backbone FDDI, etc.), which break the IP packet further into the appropriate frame format and add their own address information.

It is possible that the IP packet may go through a number of subnets (like the backbone FDDI to the Ethernet of the main division to the Ethernet of the local department/lab and so on). Each subnet adds its own header. In the case where users are not connected via subnets, they can dial access to the Internet by using point-to-point protocol (PPP) or serial line internet protocol (SLIP). The link packets (defined by subnets) are transmitted over the transmission media (analog telephone lines, twisted pair, coaxial, optical fiber, RS-232, Ethernet, etc.). The receiving-side host will have the same set of protocols, which will unpack the packets at each layer using appropriate protocols.

Virtual circuit configuration: As pointed out earlier, the sub-networks are connected through IWUs, which also serve as DTEs. If we want to send data from a source DTE to a destination DTE through a packet-switched network, the lower three layers of OSI-RM provide a logical channel across subnets, and the protocols for these layers collectively are defined in X.25. Thus, for any message from application layer up to transport layer, we simply add the header of each layer and pass it on to the next layer. These layers (4 to 7) are not supported by X.25.

The network layer of DTE receives a packet from its transport layer at the network service access points (NSAPs), which are defined by ports or sockets or even hardware addresses. The network layer adds its header (defined in X.25) and passes it on to the data link layer of OSI-RM, which has X.25 protocol, i.e., LAP-B. The header defines virtual circuit numbers between DTEs and IWUs, logical numbers, and a few other options. The link layer adds both header and trailer to the packet from the network layer and defines these as frames.

Table 12.3 OSI-RM Network Service Primitives and X.25 Service Primitive Equivalents for Connection-Oriented Services

Network service primitive	X.25 service primitive and action
Connection Establishment Phase	
N-CONNECT.request	Send call request
N-CONNECT.indication	Call request arrives
N-CONNECT.response	Send call accepted
N-CONNECT.confirmation	Call accepted arrives
Data Transfer Phase	
N-DATA.request	Send data packet
N-DATA.indication	Data packet arrives
N-Expedited-DATA.request	Send interrupt
N-Expedited-DATA.indication	Interrupt arrives
Connection Termination or De-establishment Phase	
N-DISCONNECT.request	Send clear request
N-DISCONNECT.indication	Clear request arrives
N-RESET.request	Send reset request
N-RESET.indication	Reset request arrives
N-RESET.response	None
N-RESET.confirmation	None

These frames are transmitted over the physical link between DTE and DCE (modem or multiplexer). The packet is sent across the X.25 network which has mapped the logical channel number and will receive the packet from others with this number only. The packet will be received by another DCE across the X.25 and then by IM (which is like DTE). The IWU receives the packet at the packet layer (which has the X.25 protocol for the network). This packet is then transmitted through its lower layers to the destination LAN or DTE of another host. If it is going to the destination LAN, then IM X.25 of the IWU must support both sublayers of the data link layer (media access control (MAC) and logical link control (LLC)). But if the destination is a host with DTE, it will go through the lower layers all the way up to the application layer.

Network service primitives are used by the network layer in OSI-RM. But if we have X.25 sitting in for the lower three layers, then we have to define an equivalent action/operation in X.25 which performs the same function as that of the network service primitives. The equivalence between them is given in Table 12.3.

A request call packet is issued from the source DTE containing the destination DTE address parameters containing the decision about which IM is to be routed for virtual connection. The source DCE extracts the address parameters of the destination DTE and establishes a virtual circuit with its DCE. Once the virtual circuit/connection is established between two DCEs, the destination DCE sends an incoming call packet to the IM (for which the routing decision has already been made). This IM will then make a decision about the next IM or, if there is no other IM, it will establish a virtual circuit with the destination DCE; it will receive an incoming call packet which will be indicated to its DTE or LAN for relaying the packet to a specific address defined in the call request packet. Otherwise, this IM will decide the route through other IMs until the packet reaches the destination DTE.

The network layer of ARPANET has been defined and modified several times, and in the early 1980s, after many iterations, the network layer adopted a protocol known as **Internet protocol (IP)** for internetworking. The network layer protocol supports connectionless services, while the X.25 network layer supports connection-oriented services. The

IP is used with a transport layer protocol of ARPANET which is known as a **transport control protocol (TCP)**. The TCP provides reliable transmission of datagrams which are going through various networks. The host inserts its transport header and IP header with the packets. At the data link layer protocol, these packets are encapsulated by the header and trailer of that host. The network address is defined in the header and trailer of a frame coming out from the host.

The network may be connected between hosts and gateways. The gateways include the same network layer formats as those of the hosts, with the only difference being in the data link layer, where the gateways define all the frames which travel from host to network or network to gateway or host. The combination of IP and transport header of the host or gateway is known as an Internet packet.

The combination of TCP and IP has been adopted for internetworking and offers various services in different applications. This combination can be implemented on different types of networks and is a very popular protocol for data communication interfacing. More discussions of TCP/IP and various procedures or protocols defined within this communication interface can be found in Chapters 12, 13, and 17.

Datagram configuration: Another technique of implementing internetworking is through Internet protocol (IP) used in connectionless configuration. There are two standard protocols for IP: ISO 8473 IP (ISO 8473/DAD1)[15,16] and DoD's IP.[25] These protocol standards don't depend on subnets and provide connectionless configuration or datagram service between homogeneous or heterogeneous subnets. The protocols reside in host computers and gateways and support both LANs and WANs. The advantages of connectionless configurations include robustness, flexibility, and the ability of the protocol to connect different types of subnets. In the event of any error or fault, the network does not assume responsibility; hence, users have to define error control information. Due to the nature of the configuration, reliability and connectionless service at the transport layer have to be supported by the transport protocol.

In an internetworking environment, we have hosts, DTEs, DCEs, IMs, LANs, etc., and an addressing scheme becomes crucial to data communication. According to the ISO 8473 document, the addressing of destination should follow the addressing scheme of X.25, which defines the addresses of both subnets and hosts/LANs.

12.7.5 Internetworking protocols

The preceding section discussed various service primitives offered by the network layer to its transport layer. The following section discusses various internetworking protocols along with their applications for providing interoperability across different types of networks via different types of terminals. Although some of the standards such as X.25 were discussed earlier, our main emphasis in those sections was the architecture of X.25, various standards for each of the layers, and different ways to access the X.25 and other public networks. Now we will be looking at the interconnectivity issues within X.25 protocols over public networks and discuss how internetworking protocols can be used for designing and developing our own application protocols for interconnection between different LANs.

12.7.5.1 CCITT X.25 standard protocol

X.25 serves as a standard for transfer of data across the public packet networks. It provides an interface between a computer DTE and public packet network. The physical level of X.25 uses the CCITT's X.21 physical layer standard and offers a full-duplex, point-to-point synchronous transmission between the DTE and public packet network. The link level of

X.25 controls the exchange of user's data between DTE and X.25. The top level of X.25, packet level, provides the user's data interpretation to packets and accepts virtual circuit services from ISO's network layer. In general, the public packet networks for internetworking are administered by a government agency for **Post, Telegraph and Telephone (PTT)**, which usually provides uniform and consistent services within a country. The situation in the U.S. is different, as PTT is not controlled by any government agency and, hence, standards are required to provide internetworking to various vendor products. The X.25 network should not be considered as the one which provides end-to-end protocol across the network.

If we want to send data across an X.25 network, it has to be in X.25 format in one way or another. If DTE cannot produce an X.25 format packet, then this packet has to be converted into an X.25 format packet using software/devices. There are different ways of using an X.25 network for internetworking.

In one method, we can develop our own X.25 interface software package which can reside in the host to which terminals are connected. The user's data can be sent from any of the terminals. This scheme can also provide end-to-end error control between the sender and destination nodes across the network. This, of course, is very expensive and requires extra overhead in terms of additional equipment and X.25 interface software development time.

In a second method, we can lease a vendor network where users will not get an end-to-end error control facility and will experience a slower response than in the first model, but the cost is significantly reduced. Some vendor networks support asynchronous data of DTE.

In a third method, one host DTE can be directly connected to X.25 via a CCITT packet assembly and disassembly (PAD) device which will make X.25-format packets from the asynchronous data of DTE. PAD does not provide an end-to-end error control facility and introduces considerable time delay. CCITT defined three standards with PAD standards:

- X.3 defines the operations of PADs (i.e., how the packets are assembled and disassembled, and control information).
- X.25 defines an interface between DTE and PAD and converts asynchronous data (which the majority of microcomputers and terminals produce) into the form which can be packetized by DTE.
- X.28 standard provides an interface between terminals/microcomputers.

Thus, these X.3, X.28, and X.29 standards together define the PAD standard. The X.25 provides one DTE/DCE connection and can transmit packets from the microcomputer/terminal to the public packet network.

CCITT developed international standard network access protocols for layers 1, 2, and 3 so that users from different countries and also from incompatible networks can communicate with each other. X.25 is a connection-oriented packet-switched network which defines the interface between host (data terminal equipment, DTE) and the carrier's data circuit-terminating equipment (DCE). It conforms to the network services defined by ISO 8348.[10-13] The switching nodes (IMPs) in the network are defined as data switching equipment (DSE). The details on X.25 data and control packet formats, call request packet format, virtual circuit control, etc., were given in earlier sections.

The DECnet can use either the VAX PSI standard or VAX PSI access gateway software products to communicate with X.25 packet-switched data network. The VAX PSI gateway is installed in all systems which are using DECnet and it allows multiple direct access to an X.25 public switched network from a single VAX system. Other non-DEC systems can also be accessed indirectly by DECnet. The VAX PSI access gateway software, on the other hand, provides multiple direct access to X.25 from multiple systems which are running under DECnet. VAX machines using the DEC X.25 gateway can be used as DTE or DCE, which is a very important feature of a gateway, as it allows communication between a VAX machine using a DEC X.25 software product and a machine using a non-DEC X.25 software product. VAX PSI provides multiple lines to be connected to a network(s) and also direct access to a multiple-packet network.

12.7.5.2 CCITT X.75 standard protocol

The CCITT also defined another standard, X.75, which can connect various public packet networks around the world. For example, public packet networks of the U.S. — Tymnet, Uninet and Telenet — are connected together with X.75 and also to public networks of other countries such as Datapac of Canada, PSS of the U.K., and so on. Tymnet and Telenet are connected to Datapac and PSS via X.75. Similar to X.25, the X.75 standard also defines three protocols: packet layer protocol (PLP), data link protocol (DLP), and physical layer protocol (PHY). X.75 is defined between signaling terminal exchange (STE), which is a special type of DCE used for interconnecting various networks.

The STE supports both single link procedure (SLP) and multi-link procedure (MLP) at the data link of X.75. The MLP considers the various links as one link of a higher data rate for transmitting the frames. This means that the transmission of a frame will be carried out over any available link without caring about the virtual circuit with logical channel identifiers defined within the frame. The MLC function is invoked to define a new sequence number for the frame, and then these frames are sent to PLP. As indicated above, the X.25 is connected between DTEs via DCEs and, as such, both DCEs are involved during the establishment of the connection and are included in the circuit and as the PLP of X.25 exchanges more packets. On the other hand, the PLP of X.75 handles fewer packets as it provides a connection between STEs. Due to this, the X.75 frame format will be simpler than that of the X.25 frame format as discussed above.

X.75 can interconnect only the public packet networks and does not provide any internetworking to other private networks (LANs) such as SNA (IBM), DECnet (DEC), and others. This problem was considered by CCITT, and its study group VII provided its recommendations for use of X.25 for the interfacing of any type of private network. In the recommendation, it is mentioned that a uniform internetworking across different types of public packet networks or private networks can be obtained by considering public networks as the second network's (either public or private) DCE. The X.25 also provides a permanent virtual circuit/call (PVC) between two nodes. The PVC offers the quickest possible error-free data exchange between two nodes, and a few other features. This has been considered an optional feature of X.25 (by the 1980 and 1984 versions).

As mentioned above, the public packet networks used in the U.S. are not controlled by any government-based PTT agency, but are controlled by various commercially available network vendors. Of these network vendors, Telenet (GTE-Telenet) of Vienna, Virginia and Tymnet of San Jose, California, provide very popular networks and the two support more than 75% of the total traffic. Tymnet is connected to Datapac (of Canada).

Tymnet consists of five types of nodes which are, in fact, intelligent communication processors responsible for accepting input traffic and routing it over appropriate links up to the destination. Two nodes, Tymcom and Tymsat, are used with the host and are installed within the premises of an organization. The Tymcom node accepts data from both synchronous and asynchronous devices at high speed, while Tymsat supports low-speed data out of asynchronous devices. The remaining three nodes offer leasing facilities where these nodes (Engine, Mini-Engine, and Micro-Engine) can be leased and installed either on the premises of the owner or within the Tymnet network. These nodes provide the same services as other nodes (discussed above in the Tymnet network). The packets produced by these nodes are in Tymnet format, not X.25 format. The various services and functions of the Tymnet node include control of electronic messages, network monitoring, protocol, speed and code conversion, etc. One of the main differences between various engine nodes lies in the capacity of the packet for both asynchronous and synchronous devices.

The public packet networks can be accessed via public ports (sockets) and private ports. The public ports can be obtained either by using public dial-up service or wide-area telephone service, and the private ports can be obtained by using dial-up service or leased lines. The number of ports (both public and private) vary from one public packet network to another.

12.7.5.3 ISO 8473 internet protocol (IP)

Internet software usually follows a layered architecture composed of three layers. At the bottom, it gets connectionless packet delivery service, while at the middle layer, it receives reliable transport service. At the top of this architecture is the Internet application program. Thus, we can say that the fundamental Internet service may be defined as reliable, best-effort, connectionless packet delivery. The Internet protocol basically defines an unreliable connectionless delivery means for transmitting the basic unit of data format defined by it. For the transfer of data packets, it uses a set of rules which handles the processing of packets and also error control. The discussion on formats and protocol data units defined by this standard has been partially derived from References 2–4, 6, 7, 29, and 30.

ISO IP resides on the top layer of ISO-RM as a sublayer and provides its services along with network layer services to the transport layer. If we consider the same example of sending data from sender DTE to destination DTE on the host or LAN along subnets through **internetworking modules (IMs)**, the IP sublayer receives the data from the transport layer and, after adding its header, sends an **internet protocol data unit (IPDU)** to the packet sublayer (another sublayer of the network layer). The header in IPDU defines the global sub-network address. The IPDU is then treated by X.25 in the same way as discussed in the preceding section. That is, the X.25 packet containing the addresses of the destination host or LAN and of IM are sent to a designated IM (the decision as to which IM will be sent out is already made by the X.25 packet layer).

The X.25 packet goes through a packet-switched network to the designated IM, which looks into the IP header to determine whether this packet belongs to it or if it has to be sent to another IM. In the first case, it further extracts any control information, while in the latter case, it needs to define the routing for the X.25 packet to a farther IM.

The control information contained in the IP header also defines the **quality of transport layer services (QTLS)** at the source DTE. The datagram service is defined on a connectionless configuration, various parameters such as error control, reliability, cost transmit time, re-transmission of the packet in the case of lost or damaged packets received, etc. While requesting for QTLS, the transport layer protocol allows users to choose a particular sub-network which offers the best parameters for the requested quality (QTLS).

The maximum size of user data in an IP packet is 64,512 octets. IM plays an important role in making routing decisions and error message control information. If the packet received at IM belongs to itself, then the packet will be transmitted to the host connected to it at the address contained in the X.25 protocol, but if the packet does not belong to it, it will make the routing decision (based on minimum hops, cost, and reliable paths) and add the address of the next IM through which the packet has to rerouted until it reaches the destination DTE. If in any circumstance the IM does not find a destination address, it will inform the source DTE via an error message.

If the packet received from the host is longer than the one supported by the sub-network, then ISO 8473 breaks the packet into smaller-sized packets, and each packet is sent as a separate X.25 packet. Similarly, IM may also break the packet it receives from a packet-switching network into smaller-sized packets and reroute them into the next IMs until they reach the destination DTE, where all these smaller-sized packets are buffered and the original PDU is reassembled. These assembled PDUs are then sent to the transport layer by the IP sublayer for arranging them into the original block of data and also for error control. In both cases, a software module for converting the IP format into X.25 format is required, which should conform to ISO 8473/DAD1. This software module will reside in the host and IMs and is used to manage the transfer of long datagram IP packets through the *M*-bit of X.25.

Each node and IM maintain a routing table which contains information regarding various possible routes between two nodes. When a sender DTE assigns an address of IM, it looks into the table and assigns the address of that subnet. The routes could be of different lengths (hops) between nodes. In other words, if two subnets N1 and N2 are connected by IM1, then all the nodes of N1 and N2 are said to be connected directly and can communicate with each other directly without any hop. If we add another sub-network, say N3, and connect N3 with N2 through IM2, then the nodes of N3 will require one hop to communicate with any node of N1. All the possible routes, along with the hops required through different types of sub-networks and addresses for IP packets, are defined in the route table.

As we know, a connectionless configuration does not support error control or flow control (to some extent). This is also valid in internetworking IP facilities in the sense that, if an IP packet is not accepted by IM, then it is expected that it will send this report to the source DTE, but in reality, it does not happen, as it is not the responsibility of the IM protocol. The source transport layer and higher-layer protocols may take action only after they are notified through their network layer protocol. The reasons for discarding IP packets at IM are traffic congestion, FCS error, etc. In some cases, the source address may have been damaged and, as such, even if the IM protocol determines the error, it will not be possible to notify the source.

Connectionless protocol format: The Internet protocol (IP) datagram is defined as a basic unit for carrying the user's data across the Internet. The generic format of an IP datagram includes two fields: header and data. The source and destination addresses are defined in Internet format. The IP datagram has a frame header and datagram fields where the datagram field further has two fields, a datagram header and user's data. ISO has developed a standard for connectionless configuration internetworking known as ISO IP 8473/DAD1 and DAD 3. It provides datagram service and internetworking facilities. It also offers a variety of service functions to users, who can use the primitives defined in the standard for performing those operations. The functions offered by ISO IP 8473 are usually arranged in three categories:

1. In the first category, the functions are mandatory or mandatory and selectable, which must support any type of implementation of the protocol.
2. The secondary category offers optional/discard functions to the user and these functions can become like the functions of the mandatory category (if requested). Otherwise, if a particular subnet or IM protocol does not support the IP packets, then these will be discarded.
3. In the last category, the functions may be requested as an option and if the subnet or IP protocol supports it, then it is fine — otherwise, it is just ignored.

For sending a datagram packet, the source DTE transport layer uses IP's primitives N-UNITDAT.request and N-UNITDATA.indication, which include the source and destination addresses, and control field (for error control, quality of network service protocols, etc.), as shown in Figure 12.14.

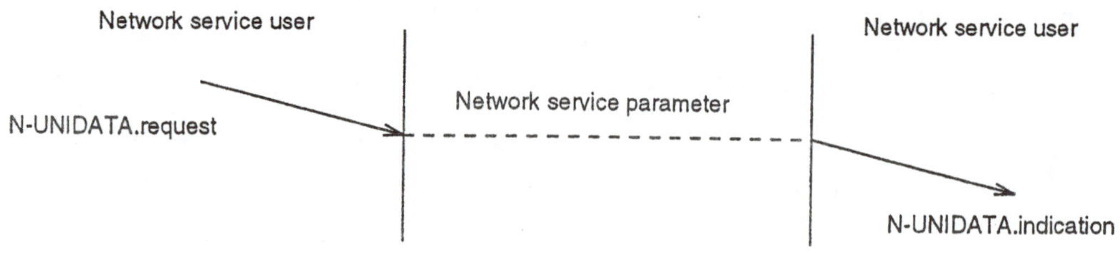

Figure 12.14 Sending a datagram packet using IP's primitives.

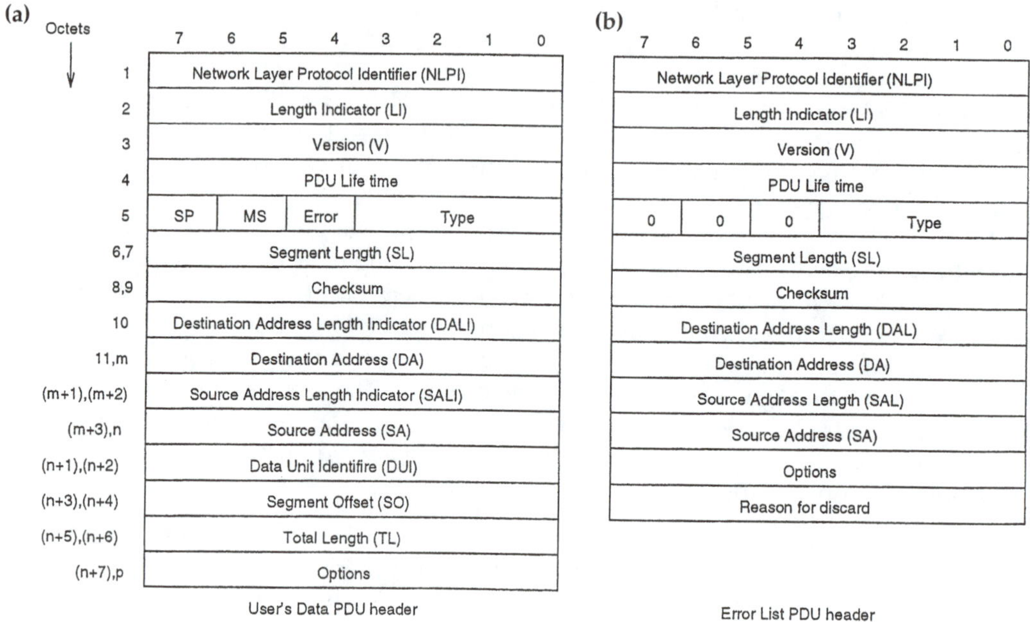

Figure 12.15 (a) User's data PDU header. (b) Error list PDU header.

Protocol data unit (PDU): ISO 8473 has defined two types of PDUs: **user's data PDU** and **error list PDU**. Each type of PDU contains an IP header which is further divided into a number of segments.

The **user's data PDU header** consists of four parts: fixed part, address part, segmentation part, and options part (Figure 12.15(a)). The fixed part is defined by octets 1 through 8, 9; the address part by 10 through $(m + 3)$, n; the segmentation part by $(n + 1)$, $(n + 2)$ through $(n + 5)$, $(n + 6)$; and option part by $n + 7$, p.

The fields of each part are discussed in the following[2,3,29,30]:

- **Fixed part:** The network layer protocol identifier in the first octet indicates whether Internet protocol ISO 8473 is required or not. If both source and destination nodes are connected to the same sub-network, then there is no internetworking and hence the Internet protocol is not required.
- **Length indicator:** Indicates the length of the header in octets and can have a maximum value of $2^8 = 256$.
- **Version:** The third octet defines the version of the protocol being used, which in fact represents the changes in header or semantic of the protocol.

- **PDU life time:** The fourth octet defines lifetime in units of 500 ms. The source node defines this value and includes it in the IP packet. Each time the IP packet goes through IM, the value (defined by the source) is decreased after every 500 ms of delay of a hop (from source to IN including processing time). During the remaining time of IP datagram the lost packets are purged from the system.
- **Flags:** The fifth octet represents a number of flags. The segment permitted (SP) flag of one bit indicates that the gateway can segment the datagram so that it can accepted by the network and make the user information field smaller than the PDU length. The more segment (MS) flag of one bit informs the destination ISO 8473 that the segmentation has been carried out and more segments are following. The error flag of one bit indicates the action to be taken in the event of occurrence of error. It suggests whether to discard the datagram or whether an error list is required. The type field represents the type of PDU (user data or error list).
- **Segment length:** It defines the length of the user's data PDU, including the header, in two octets.
- **PDU checksum:** It uses a two-octet (eight and ninth octets) checksum algorithm (LRC-link or other standard algorithm) on the header of the packet for calculating the result of the algorithm. In order to avoid any changes in the header field, the checksum algorithm is executed on it at each gateway.
- **Address part:** Both destination and source fields are of variable lengths so that each of these fields is preceded by another field — the destination address length indicator and source address length indicator, respectively. These indicator fields define the length of the address in octets being used. The maximum length of the address is 20 octets.
- **Segment part:** As indicated above, if SP is 1 in fixed part, the gateway can segment the datagram with a view to making the user's information field smaller than the PDU. Under this situation (when SP = 1), a two-octet data unit identifier assigns a unique sequence number to the PDU destination and also indicates the duration for which the PDU is present in the Internet. A two-octet segment offset defines the locations of segments (in terms of a 64-bit unit) present in the data field and this location is represented with respect to the start of the data field in the original PDU. Finally, a two-octet total length field defines the total length of the original PDU (including header and data) and is present in each segment of the PDU.
- **Option part:** Defines various option fields for providing services such as security, source routing preferences, quality of service, priority, and precedence. The datagram options are usually used for network testing and debugging. The option code octet format includes three fields: copy, option class, and option number. The option classes defined by the Internet datagram are class 0 (datagram or network control), 1 (reserved for future use), 2 (debugging and measurement), and 3 (reserved for future use).

The **error list PDU header** consists of three parts: fixed part, address part, and options part, as shown in Figure 12.15(b). The error list PDU header has more or less the same format as that of the user's data PDU header. The fixed part is exactly the same except that the status of flags indicates that the error list cannot be segmented. The address part is exactly the same. Here the location of the error list is defined in the source address field, while the location of the discarded PDU (due to the error list) is defined in the destination address field. Due to the status of the flag SP, the error list header does not have any segment part. The option part basically indicates the options that are defined in the option part of the user's data PDU. If a particular option is not defined in the user's data PDU, then that option will not be indicated in the header of the error list PDU. Depending on the type of PDU, the headers are defined; i.e., if the PDU is a user's data PDU, then the header will be followed by the user data field. If the PDU is an error list PDU, the header will be followed by the reason for the discard field and the header of the discarded PDU.

12.7.5.4 DCA's DoD internet protocol (IP)

Another protocol which can provide internetworking for homogeneous and heterogeneous computers for transfer of data across a network using datagram service is Internet protocol (IP).[25] This protocol was issued by the Defense Communications Agency (DCA) for the U.S. Department of Defense (DoD). The IP addresses for the Internet are assigned by the Internet Network Information Center registration service (InterNIC) using 32 bits. IP is an internetworking protocol, while the transmission control protocol (TCP) is a transport protocol. TCP/IP was defined as a LAN protocol and has continued the evolution of the Internet, which started with the advent of ARPANET and MILNET. The TCP/IP suite of protocols offers end-to-end data transport (TCP) at layer 4 and routing (IP) at layer 3 of OSI-RM.

The Internet has become very popular in every walk of life and a user-friendly, multimedia browser application accessing the World Wide Web has made it very easy and useful for retrieving information and allowing us to create our own home or Web page and store needed information. The Internet has attracted tremendous and ever-growing interest in the business, corporate, scientific, and other communities. It has helped create a new kind of commerce in which customers may order products and make payments electronically. The TCP/IP suite protocols provide necessary communications across connected networks. TCP offers a connection-oriented, end-to-end, reliable, and sequenced delivery of data, as well as priority, segmentation and reassembly, and adaptive flow-control services. The user datagram protocol (UDP), on the other hand, provides connectionless, unreliable, and unacknowledged datagram service. A real-time protocol defined at the transport layer provides real-time applications that deal with transmission of voice and multimedia data.

The Internet and TCP/IP together provide connectivity between computers at two remote locations anywhere in the world and allow the transfer of information between those computers over the Internet. The architecture of the Internet or TCP/IP is defined in terms of layers as process/application, host-to-host transport, and Internet and network access. The TCP/IP includes all the needed application software and is also known as Internet protocol. The top layer, process/application, includes the functions of application, presentation, and a part of the session layer of OSI-RM. It provides the following services: FTP, telnet, domain name service (DNS), simple mail transfer protocol (SMTP), and routing information protocol. The next layer, host-to-host, offers the same functions as that of transport and part of the session layer of OSI-RM. It offers two types of service: connection-oriented and connectionless. The TCP provides end-to-end error control for the entire frame and hence reliable communication (connection-oriented). The UDP, on the other hand, provides error control only for its header and hence provides unreliable service (connectionless). The UDP is faster than TCP and is used in small data transfers or applications where errors can be tolerated.

The Internet layer provides the same function as that of the network layer of OSI-RM. This layer includes three protocols: Internet protocol (IP), address resolution protocol (ARP), and Internet control message protocol (ICMP). The IP does not provide any control other than header checking via checksum. Network access is the bottom layer and includes the functions of data link, physical, and a part of the network layer of OSI-RM. It defines two sublayers as 1A and 1B. 1A is equivalent to the physical layer, while 1B is equivalent to the data link layer and includes X.25, FDDI, and 802.x LAN protocols.

IP is a datagram protocol that is resilient to network failure, but it does offer the sequenced delivery of data. It offers end-to-end reliable data communication across the Internet when used with TCP. The function of ICMP at the Internet layer is to provide a ping (echo packet) to determine the round-trip propagation delay between source and IP-address host. Routers use this protocol for sending error and control messages to other routers. This layer provides multicasting data communication using Internet group multicast protocol. It also uses ARP, which provides a direct interface to the data link layer of the underlying LAN. This protocol provides a mapping between the MAC address (physical) and the IP address. It uses some routing protocols. The Internet layer supports a variety of data link layers (Ethernet, token ring, FDDI, frame relay, SMDS, ATM, etc). The data link layer, in turn, supports a variety of transmission media such as twisted pair, TDM, fiber, etc.

The transport layer with UDP and TCP supports a variety of applications such as file transfer protocol (FTP; a secured server login, directory manipulation, and file transfers), Telnet (remote login facility), hypertext transfer protocol (HTTP; for World Wide Web), simple network management protocol (SNMP; for network configurations, setting, testing, data retrieval, alarm and other statistical reports, etc.), remote procedure call (RPC), network file server (NFS), trivial FTP (TFTP; reduces the implementation complexity), and many others. The domain name system (DNS) offers a centralized name service that can use either UDP or TCP.

IP datagram services use a connectionless mode of operation. These services are easier to implement and define in a multi-vendor environment. In order to handle IP datagrams effectively, another protocol known as Internet control message protocol (ICMP) has been defined. All three protocols — UDP, TCP, and ICMP — interface directly with IP. The user datagram protocol (UDP), transmission control protocol (TCP), Internet control message protocol (ICMP), and routing control protocols are the protocols of the transport layer and interact with the Internet protocol (IP) of the network layer.

The IP also includes another module protocol which converts the IP address into the medium address. It allows the host to maintain the address in the table. For transmitting the information, the source node broadcasts an ARP request that includes the destination IP address. The destination host, after identifying its address, sends its NIC hardware address. It also includes the host IP address and hardware address in its routing table. The reverse ARP mechanism is used when the host broadcasts its own hardware address. The server sends the host's IP address to it. Another facility within ARP is proxy ARP, where the host needs to know the destination address. The router connected to the destination host responds by sending its own address. In this way, that router becomes proxy for the remote host.

As said earlier, IP may be used to communicate with both LANs and WANs. The first version of IP was aimed at virtual circuit service on one network. The later versions of IP not only included datagram services but also expanded its use on different subnets for greater internetworking. In the case of datagram-based subnets, the responsibility of reliability, delivery of sequenced packets, and even error recovery from lost packets were laid on the transport layer protocol. Thus, the transport layer protocol has to include many services in addition to its own services defined in OSI-RM and even has to support any incompatibility between the network and data link layer. In other words, the use of DoD's IP requires a very powerful transport layer protocol. The following section will discuss in brief the datagram structure and operation of IP.

A new version of IP has been developed to meet ever-growing demands of Internet applications and is known as **Internet protocol v6 (version 6)**. The first three versions of IP have been replaced by IPv4 (the one in use currently). Version 5 has been defined for connection-oriented Internet-level protocols. The Internet Engineering Task Force (IETF) has finalized the changes in IP addressing as the next-generation IP (IPng) and published

RFC 1752, recommending the IP next-generation protocol. It basically uses 128 bits instead of 32 bits and includes features such as dynamic assignment of IP addresses, addressing flexibility, resource allocation support, security capabilities, etc. Another protocol, ICMP, has also been upgraded to ICMPv6 to include features such as larger size of header (576 octets), error messages, etc.

Most of the international Internet service providers use ATM for switching the traffic. Two standards defined by the Internet Engineering Task Force (IETF) as **real-time protocol** and **resource reservation protocol (RSVP)** support voice communication over the Internet with confirmed quality of service. The use of these protocols over the Internet with IP has led to some packetization delay caused by IP frames which are relatively much larger than the frame of an ATM cell (53 bytes).

Quality of service (QOS): As mentioned earlier, IP does not provide QOS, except that the IPv4 header defines a field to indicate the relative priority (seldom used). The RSVP allows users to make a request for a level of quality of service and bandwidth for their applications from IP networks. In fact, RSVP is a signaling protocol and does not offer any quality of service by itself, but it allows the sender to indicate the needed quality of service and bandwidth using an IPv6 packet header. It also allows the receiver to request the needed quality of service and bandwidth from a multicast sender. Microsoft's WinSock2.0 specifies the level of quality of service in a LAN environment.

As the number of applications requesting quality of service increases, the size of the routing table also increases; at the same time, it cannot deny the RSVP-requested connections, as the underlying network offers connectionless services. In contrast to this, ATM provides end-to-end quality-of-service connection over a connection-oriented network. There are a number of vendors that are involved in both ATM and RSVP products. For example, 3Com's PACE technology allows users to indicate the priority over Ethernet networks. Cisco routers running the Internetwork Operating System allow the users to make a request for needed bandwidth using RSVP for its applications. Bay Networks and Lucent Technologies introduced private network-to-network or node interfaces that support ATM quality-of-service end-to-end connections for voice, video, and data traffic.

The routing strategies for IP routing are defined in RFC 1058 in a routing protocol known as **routing information protocol (RIP)-IP** that is based on the hop count concept. This is the most widely used routing technique for gateways. It was designed at the University of California at Berkeley. It uses a physical network broadcasting technique for routing. At Xerox Corporation's Palo Alto Research Center (PARC), the protocol was derived as Xerox NS routing information protocol (RIP). It was defined, implemented, and distributed without any formal standard. This protocol must support three parts of errors (fixed, address, and option): it does not detect routing looping, it uses a hop count of 15 (as a upper limit), and the protocol creates a slow convergence or count to infinity situations.

Routing: The IP determines whether the packet should be sent directly to the local network or another router based on the address of the destination. Each time the packet is sent, the router looks at the routing table and forwards it accordingly. The routing tables are updated either manually or automatically by another router. The routing decision is based on a number of addresses such as destination network, sub-networks, hosts, routers, number of hops, etc. In this technique, the entire routing tables are available at the routers. Another routing technique known as the OSPF standard was defined in 1990 which is based on the clustering concept. The routers in each of the clusters know about the routers and links of other clusters. The entire routing table is not required at the clusters; the changes regarding clusters and links need to be sent to the routers.

The basic operation for sending the datagram packet across subnets remains the same as was discussed in ISO 8473 PDU. Here the IP defines an IP datagram (containing the user's data) and transmits it across the subnet. If the destination host is connected to the same subnet, the IP datagram packet will be transmitted to the destination node. But if the destination host is not on the same subnet, the packet will be sent to the gateway. The gateway has the outing strategies for all the subnets. The gateway connected to a subnet looks at the subnet routing table and, based on some criteria, sends the packet either to the subnet (which has the destination) host or to the other subnets and so on until the packet is delivered to the destination host or terminal.

As indicated above, the transport layer provides a large number of services to the network. In IP terminology, we have only four layers for the network, as shown in Figure 12.16, as opposed to seven layers in OSI-RM.

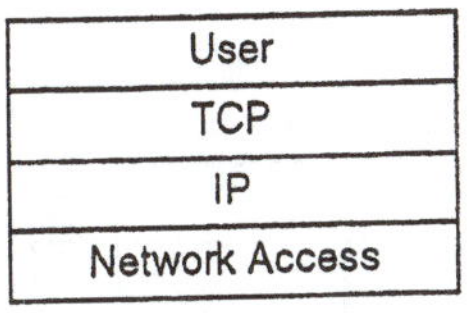

User
TCP
IP
Network Access

Figure 12.16 IP layered architecture.

IP supports both connection-oriented and connectionless services and does not want to know much about the network. Services such as reliability, error control, and flow control are provided by its higher layer, the transmission control protocol (TCP), which resides in the third layer on top of the IP. The host user's data in IP becomes the top layer of the network which offers various services to the users. The connectionless service via the subnet-independent protocol IP requires that each host and gateway should have same application protocol and should implement the same IP.

The TCP combined with IP provides an entire communication protocol and is referred to as TCP/IP. The TCP defines formats for data and acknowledgment and offers an orderly arrival of the packets at the destination node. It also supports multicasting and broadcasting and distinguishes between the addresses and provides error control (recovery). It allows negotiations on the mode of data transfer, status of various stages during transmission, etc. However, it offers a totally system-dependent implementation and does not offer any application interface, thus giving flexibility to the users to define their own application interface.

The TCP resides above IP and, conceptually, the Internet is composed of three layers: the network layer at the bottom, the Internet protocol (IP) in the middle, and at the top of IP is the TCP offering two types of application support — reliable stream TCP and **user datagram protocol (UDP)**. The TCP offers a connection-oriented service, while UDP offers a connectionless service. The TCP allows multiple application programs to communicate simultaneously and offers de-multiplexing for incoming TCP traffic for respective application programs. It is connection-oriented and requires two endpoints for meaningful data communication. The TCP considers the data stream a sequence of octets or bytes that it will divide into segments. For efficient transmission and flow control, it uses a sliding-window scheme (at the byte level), where multiple packets are sent before the arrival of acknowledgment, and handles the flow control for end-to-end communication. The window size can be changed over a period of time. It supports full duplex and allows simultaneous transfer of data in both directions.

The TCP establishes a TCP connection under the assumption that the existing packet delivery system of the network is unreliable. It offers a three-way **handshaking procedure** for establishing a TCP connection. An event is caused by a Send Syn sequential = n primitive and is transmitted over to the other side as a network message. The primitive (Receive Syn segment) on the receiving side causes another event and sends the reply as another primitive (Send Syn sequential = $n1$) and an acknowledgment Ack = n +1. The transmitting side sends another primitive (Send Ack $n1$ + 1) as an acknowledgment to the receiving side, confirming the establishment of the TCP connection.

The user of the sender host requests its TCP for sending user's data to the receiving host. The TCP, after adding its header to this data, passes it on to its IP layer. IP adds its own header, defines an IP datagram, and presents it to the **network access protocol (NAP)**.

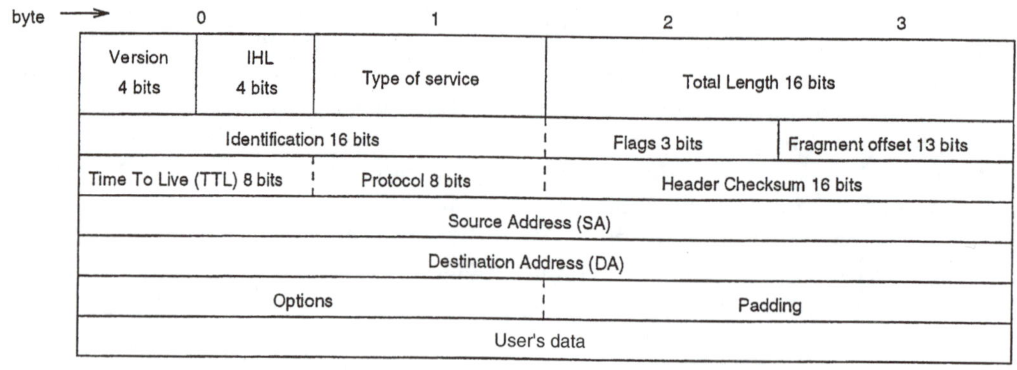

Figure 12.17 IP datagram header.

The IP accepts packets from both TCP (offering connection-oriented services) and UDP (offering connectionless services). The frame formats and various primitives defined for UDP are discussed below. The network access protocol, after attaching its header with the IP datagram, transmits it to the gateway via a sub-network. The gateway is defined by two layer protocols: the network access protocol (NAP) and IP. The NAP of the gateway passes this packet to its IP. The IP looks at the subnet-routing table and, depending on chosen criteria such as cost, length of routes, number of hops, etc., reroutes the IP datagram to the appropriate gateway via another sub-network.

If both hosts are connected to the same sub-network, then the IP datagram will be delivered to the sender host and then to the receiver host directly without going through the gateway, but if the hosts are on different sub-networks (which are connected via gateways), then the IP datagram will be routed over gateways across the sub-networks. Further, the gateways have to fragment the IP datagram packets they receive, as sub-networks have different sizes of packets. In general, the gateway IP constructs a few fragments out of the data portion of the datagram and provides an IP header with each of the fragments. The fragments are defined by identifying the boundaries in octets on the data field of the IP datagram packet which, in turn, depends on the sub-network which is going to receive it.

All the fragments constructed by the IP are sent independently until the destination IP gateway has reassembled these fragments and presented them to the NAP of the receiving host. It is important to note that these fragments may be fragmented each time they pass through a gateway. Each fragment contains a destination address, source address, security, and other facilities defined by the protocol at the source host.

IP datagram header format: TCP and IP communicate with each other through a set of IP primitives: SEND and DELIVER. The TCP of the source host will use the SEND primitive to indicate source address, destination address, type of service, data length, etc., while the DELIVER primitive will be used by the IP of the destination host to deliver the user's data (containing the TCP header) to the TCP. The DoD IP datagram header format shown in Figure 12.17 has been derived and modified appropriately.[29]

The various fields defined in an IP datagram header are discussed below.

Version: Indicates the version of IP being used by the datagram. The header format may change if the protocol evolves.

Internet header length (IHL): Indicates the length of the IP header in 32-bit words. The minimum value of an IP header is five.

Type of service (TOS): Defines the IP parameters which provide quality of service such as precedence, reliability, delay, and throughput.

Total length (TL): Indicates the total length of the fragment (obtained by fragmenting the data field of the packet IP received from its TCP) including the header and data.

Identification (I): Defines the identification of a datagram and will be the same throughout until the datagram is in the Internet. Each of the fragments of a datagram is also associated with this identification.

Flags (F): Indicates whether a datagram is to be fragmented and that more fragments in sequence are following. The first bit of a more flag (1 bit) indicates fragmentation and reassembly, while the second bit will indicate no fragmentation. At present, the third bit is unused.

Fragment offset (FO): Defines the location of a fragment's data with respect to the beginning of data (from the original datagram) in terms of 64-bit units. In other words, the fragment contains a data field of 64 bits or a multiple of 64 bits.

Time to live (TTL): Defines the time limit for a datagram to be in transit. It is measured in one-second increments, which is the time a gateway usually consumes for processing the datagram. Each time a datagram goes through a gateway, the TTL is decreased by one second until it reaches the destination host and can have a value between 0 and 255. If at any time during transit the TTL becomes zero, that datagram will be discarded.

Protocol: Indicates the next higher-level protocol (at the destination) which will receive the datagram.

Header checksum (HC): Defines 16 bits (which is 1's complement of the 1's complement of the sum of all 16-bit words) in the header and is calculated using the above rule. It does not include the checksum.

Source address: A fixed 32-bit address field that defines the subnet address and the address of the node within the subnet. The first eight bits are used to define the subnet, while the remaining 24 bits are used for the node address.

Destination address: Also a fixed 32-bit address field, defined similarly for the destination node.

IP address structure: Both IP source and destination addresses of a host represent a unique 32-bit Internet address which is used in all data communication with that host. The IP address frame format defines two fields as network identifier (netid) and host identifier (hostid). It usually defines three classes of IP frame formats of addresses: **Class A** (0 bits in the MSB, seven bits for netid, and the remaining 24 bits for hostid); **Class B** (first two bits are "10," followed by netid (14 bits) and hostid (16 bits)); and **Class C** (first three bits are "110," followed by netid (21 bits) and hostid (8 bits)). For multicast, the first 4 bits are given by "1110," while the remaining 28 bits define a multicast address. A frame format defining the first four bits as "1111" and the remaining 28 bits undefined is reserved at present.

Similarly, the IP subnet frame format defines two fields known as Internet routing and local address. The Internet routing field further defines two sub-fields as Internet and netid. The hostid becomes a part of the local address field. This frame also defines 32 bits and offers the same formats of 8, 16, and 24 bits (Classes A, B, and C) for the Internet address, while the remaining bits in each class are defined as the hostid address. The Class B IP frame format address has been modified to represent the first two bits as "10," followed by a netid of 14 bits. The hostid is divided into two sub-fields as Internetid (eight bits) and hostid (eight bits). The network address defines a hostid with all 0's, while the broadcast address has a hostid with all bits 1. These IP addresses define the connection and do not represent the host; i.e., if a host is moving from one network to another, its

Internet address must also change. If we are using Class C, and if the number of hosts is more than 255, we have to change our present class to Class B.

The Class A frame address defines seven bits for netid and 24 bits for hostid and, hence, it can support up to 2^{24} hosts which can be connected to the network. Class B defines 16 bits for hostid and can support up to 2^{16} hosts. Class C defines eight bits for hosts supporting up to $2^8 = 256$ hosts. At the same time, it defines 21 bits for netid. This means that this class can define 2^{21} networks, and each network supports only 256 hosts. A typical example of Class C is a single-site LAN, while ARPANET falls into the category of Class A. These IP frame format addresses are typically divided into four fields. Each field represents a byte; these fields are separated from each other by a dot symbol, and each decimal number can take values from 0 to 255. The byte representation in each field is expressed in its equivalent decimal representation as shown below

00001000 00000000 00000001 000000010 = 10.0.1.2 (Class A)

Here, netid for ARPANET defined in the first field is 10, while the other bytes of the remaining three fields define hostid.

10000001 0000010 00000011 00000001 = 129.2.3.1 (Class B)

Here, the netid is given by the first two fields as 129.2, while the hostid is defined by the remaining two fields as 3.1.

11000000 00000001 00000011 00001111 = 192.1.3.15 (Class C)

Here, the netid is defined by the first three fields as 192.1.3, while the hostid is defined by the last field as 15. It is interesting to note that if the last field has all 1s, it will represent a broadcast configuration and hence all the hosts having a netid of 192.1.3 will receive the packets simultaneously. The addressing for interconnection of LANs follows the same concept. As we know, when we interconnect different LANs, bridges are used at the MAC level while interconnection of subnets is defined at the network layer via routers. The routers expect each LAN to have its own netid, and if a site has a large number of LANs, it is expected that the routers will participate in the routing strategies of the overall Internet routing function. Thus, the LANs defined at a particular site may not get their own netid, but instead the site may be assigned with a netid and then each LAN at the site is defined as part of a hostid.

Option: This field is of variable length and defines the options between subnets over which a datagram is to be sent. The option may be to include a security time stamp and other time-related parameters and is mutually agreed upon between them. This option is originated at the source host. It defines various option fields for providing services such as security, source routing preferences, quality of service, priority, and precedence. The datagram options are usually used for network testing and debugging. The option code octet format includes three fields: copy, option class, and option number. The option classes defined by an Internet datagram are class 0 (datagram or network control), 1 (reserved for future use), 2 (debugging and measurement), and 3 (reserved for future use).

Padding: This field is of variable length and is used to fill the extra bits so that the Internet header is 32 bits long.

Data: This field includes the user's data and is a multiple of one byte (eight bits) in length. The maximum length of the header and data is 65,535 octets.

Version 4 bits	Header Length 4 bits
Type of service 8 bits	
Total Length 16 bits	
Identifier 16 bits	
Flag 3 bits	Fragment offset 13 bits
Time to Live 8 bits	
Header Checksum 16 bits	
Source Address 32 bits	
Destination Address 32 bits	
Options and Padding (variable)	
User Data (variable)	

Figure 12.18 Internet protocol data unit (IP diagram).

The services of the network are requested by the SEND primitive, while the DELIVER primitive is used to inform the destination end user process about the received data packets. The IP offers the required interface with a variety of networks, and the IP network interface has limited functions to perform. This leaves IP to perform routing functions most of the time. The IP DU (IP diagram) format is shown in Figure 12.18.

All the IP addresses are assigned by a central authority known as the Internet Assigned Numbers Authority (IANA). The network administrators can assign a unique address within any class of IP address and, as such, the allocation of a unique address by this authority seems to be inefficient. A new method using the concept of classless interdomain routing (CIDR) was introduced in 1993. It does not use the class system and is dependent on the concept of variable length subnet masked. Various protocols supporting this concept include OSPF, RIP II, integrated IS-IS, and E-IGRP.

The next version of IP, known as IPv6 (or Internet protocol–next generation: IPng), is based on CIDR. CIDR defines Class A address space reserved for expansion (64.0.0.0 through 126.0.0.0). The IEFT issued RFC 1752, a recommendation for IPng, in July 1994. It supports all applications of Ipv4 such as datagram service, FTP, e-mails, X-Windows, Gopher, Web, and others. It defines flow label fields that support RSVP for confirmed bandwidth and quality of service.

A number of IP-switching architectures based on ATM have been proposed and, in fact, IFTE has formed a Multiprotocol Label Switching group to look into these architectures with a view to defining a standard architecture protocol. A number of vendors have proposed IP-switching architecture protocols, such as Ipsilon, Toshiba's cell switch router, and Cisco's tag switching. For more details, readers are advised to read the References for Future Reading and also review appropriate Web sites listed at the end of this chapter.

IP addressing implementation for UNIX networks: In order to understand IP addressing and other aspects, we consider the following general guidelines for the implementation and network setup procedures required for the IP-addressing scheme as used in UNIX systems. These steps may not be valid in all network setups, and, as such, readers are advised to get more details about these procedures from their computer centers. Further, it seems quite reasonable that similar procedures will be used for implementing IP addressing in other operating systems.

To run TCP/IP, every network must be assigned with a IP number (netid) and every host is assigned with an IP address (hostid). The netid is obtained from the **network information center (NIC)**, while the hostid is assigned by the local computer center administrator. As indicated above, an IP address consists of 32 bits which are divided into four fields of eight bits each. Each IP address, in fact, corresponds to administrative divisions defined by appropriate domains, and this does not define how the networks are connected. These domains are hierarchical, with fields separated by a period symbol. The entire domain representation is unique around the world. The NIC administers the top level by assigning users to appropriate domains. Some of the top-level domains in the U.S. are:

.com — Commercial companies
.edu — Educational institutions
.gov — Government agencies

For more details, see Section 17.3 of Chapter 17.

Each host must know its domain, and this is usually defined at installation time and must be stored in a file /etc/defaultdomain. For diskless machines, this is done in a file which is in the directory /export/root/client/etc while the user is logged in on the server. The TCP/IP is usually maintained in the following network file databases: hosts (host names and IP addresses), Ethers (host names and their Ethernet addresses), networks (network names and their numbers), protocols (IP protocol names and their numbers), services (IP service names and their port numbers), netmasks (network names and their netmasks).

The hosts database is usually in the /etc/hosts file, which gives the list of all the machines in the local domain and is assigned with its IP address defined as host name nickname followed by #comments. For example, we may have a machine with the following address: 192.2.345.2 Hura # This is a comment.

The ethers database is in the /etc/ethers file and all the files are arranged in alphabetical order.

The networks database is in the /etc/networks file and uses the following address format: networkname networknumber nickname #comments. An example of such an address is: Eng 192.3.2 # engineering.

The protocols database is in the file /etc/protocols. This is not maintained by the user but it will display the list of protocols such as Internet control message protocol, Internet group multicast protocol, gateway-gateway protocol, user datagram protocol, etc.

The services database is in the /etc/services file and contains various TCP/IP network services available to the users. It gives a list of the services, the port number, and also the transport protocol layer that handles those services, and this again is not maintained by the user.

The TCP/IP network can be divided into subnets by using subnet masks. A network mask determines which bit in the IP address will represent the subnet number, and the implementation at any site within the organization remains transparent to users outside the organization. The routing within the organization will have the format as a pair of #network and #subnets, while outside the organization, only the #network address will be used. The netmasks database is in /etc/netmasks, and it contains the default netmask of the system. To set up a netmask, this file needs to be edited and it may look like

132.24.1.34   255.255.255.111

where the first address represents the actual network address and the second one is masked. To change to a subnet format, the following procedure is used: identify the topology of the network, assign subnet and host addresses, edit the /etc/netmasks file, and edit /etc/hosts on all hosts to change the host addresses and reboot the machine.

The TCP/IP has gained corporate acceptance for UNIX environments. It provides a very useful communication environment, as it provides higher-layer services such as FTP, SMTP, terminal emulation application, and Telnet. It is available on a wide category of platforms and from many vendors. Many people think that this can be used only for UNIX platforms. Since it operates above the data link and physical layers of OSI-RM, it can be used for any type of LAN, as well as wide-area communication links such as ATM, SDLC, frame relay, SMDS, and so on. In the TCP/IP environment, all the computers are known as hosts, and each host usually runs UNIX and contains internetworking software.

12.7.5.5 User datagram protocol

We have discussed TCP, a connection-oriented protocol for the transport layer. The ARPANET has also defined a connectionless transport layer protocol known as the **user datagram protocol (UDP)**. It allows a program on one machine to transmit datagrams to a program(s) on another machine(s) and can also receive acknowledgments. Each application needs to get a UDP protocol port number from its operating system. Thus, the applications communicate with each other via the protocol port number. This scheme offers various advantages, as the protocols are independent of specific systems, and it allows communication between applications running on heterogeneous computers. The sender can identify a particular receiver on a single machine using UDP, as each UDP message contains the destination and source protocol port numbers. It offers unreliable delivery of messages and depends on the underlying Internet protocol (IP) to provide transportation to the messages from one machine to another. It does not use any acknowledgment, and, as such, the UDP messages may be lost or duplicated and also may arrive out of order at the destination.

The conceptual architecture of network software may have a physical layer (Ethernet) as the transportation mechanism at the bottom level over which IP offers routing and network protocols. The UDP sits on top of the IP over which we use the user program. The client–server approach is used to define the interfaces for application programs. The UDP port number is assigned by the operating system to application programs, and these application programs interact with each other via these port numbers (system-independent). The application programs can send datagrams back and forth using these port numbers. The client–server concept is used in most communications hardware in one way or another. A single-server application may receive UDP messages from many clients.

The server cannot specify IP address and, as such, cannot specify a UDP port on another machine, as it needs that machine to send the message to it. Alternatively, it defines only a local port number. Any message coming out from any client to a server must specify the client's UDP port and also the server's port number. The server receives the message from the client and uses its port number as the destination port number for sending its reply. Thus, the interface for many-to-one communication must allow the server to specify information about the destination each time it tries to send datagrams by extracting the port number of the client and using it as a destination port number. In this case, the destination IP address or UDP protocol port number cannot be assigned on a permanent basis.

As we have seen, in OSI transport protocols for conectionless services, two primitives, T-UNITDAT.request and T-UNITDATA.indication, allow users to send a message across the network before establishing any transport connection between source and destination transport users. Similarly, the UDP does not require the establishment of any transport connection for sending any data over the network. Due to the connectionless services, the network does not give any guarantee of delivery or sequencing. The ARPANET UDP format includes a UDP header which defines source and destination port addresses.

12.7.6 Internetworked LAN protocols

The following subsections describe the internetworking between various nonstandard LANs and also between these LANs and X.25.

12.7.6.1 Internetworking with DECnet

In previous sections, we have discussed various internetworking protocols for standard LANs (IEEE 802). In this section, we will discuss various internetworking protocols defined for nonstandard (or proprietary) LANs and also the protocols which can provide interoperability between standard and nonstandard LANs.

DECnet network layer: In order to understand internetworking between SNA[26] and DECnet,[8,9] we will briefly discuss the functions of their layers equivalent to the network layer of OSI-RM.

The network layer of DEC's network architecture is defined as DNA and provides the routing of packets over communication links to various switching nodes (IMPs). The protocol for the DNA network routing layer supports connectionless service to the transport layer and, as such, routes the datagram packets to destination nodes. The DNA subnet layer protocol does not support end-to-end service directly, but this service is usually provided by higher layers (end communication layer). This communication layer in DEC's network architecture (DECnet) is analogous to the transport layer in OSI-RM. The detailed discussion on these internetworking protocols has been partially derived from References 5–7.

The DNA network routing layer does not provide all the services as defined for the network layer in the OSI model, but most of the services offered are comparable to those offered by the OSI model for its network layer.

Under DECnet architecture, the specific services offered by the network routing layer (DNA) include routing, traffic congestion control, and packet hop counter management control. When the packets are sent to IMPs, each IMP contains information regarding the length and cost of the path length on the line for each destination in the network. The routing technique in DNA is based on distributed and dynamic routing and provides the most cost-effective route between a source IMP and destination IMP. Further, the routing tables at each IMP are regularly updated in the event of change in the topology of the network. Although the actual cost of the network routing layer to some extent depends on its implementation, under DECnet architecture terminology, the cost of the link is defined as inversely proportional to the link capacity. The link capacity is the most important parameter for determining the cost, although other parameters, such as the type of traffic, transit delay, response time, error rate, and error control, are also considered.

DECnet connectionless header: The connectionless network routing layer (DNA) header is discussed below and is shown in Figure 12.19.

The first, third, and eighth bit of the routing flag are typically 0 bits. The second bit defines the version of the protocol being used by the packet. The fifth bit is a return request,

Routing Flag	Destination Address	Source Address	Hop Counter Flag

Figure 12.19 Connectionless network routing layer header (DNA).

and it is set by the source node. The packet will be returned to the source in the event of undeliverability to the destination node. The third bit (returning bit) indicates the returning of the packet, and this bit is set. The sixth bit is a choke bit which indicates the choking condition of the line. In other words, the packet queue length and utilization of a particular line has exceeded the upper limit. Each source maintains a list of a number of packets which have been sent to each of the destinations. Upon receiving a choke bit (set), it waits until the choke bit is cleared. The destination and source fields define the destination and source addresses, respectively. The hop counter flag indicates the number of hops a packet has undertaken. The first two bits are zero, while the remaining six bits provide the maximum limit of hops (i.e., 63) that a packet can take before it arrives at a destination node. Each time a packet is received by it on a switching node, the counter is incremented (initially it is at zero value). After it reaches the upper limit and a packet has not yet arrived at its destination, it will be discarded. The following brief discussion of internetworking between proprietary LANs is derived from References 4–7.

DECnet protocols: The **network virtual terminal protocol (NVTP)** defined by DEC provides a common higher-layer protocol for providing communication supporting different operating systems across the network. This protocol offers a variety of peer-to-peer services across the network and is divided into two modules (sublayers): **command terminal protocol module (CTPM)** and **terminal communication protocol module (TCPM)**.

In the **command terminal module** (sublayer), the functions such as recognition of ASCII codes, configuring the characteristics of terminal devices, managing and controlling input and output buffers, support of input and output printouts, CRT and video transmission, etc., are being performed. The **terminal communication module** (sublayer) is concerned mainly with management and control of session connection between application program module and terminal and provides an end-to-end logical connection between host system (local) and logical terminal (remote). The users can establish a telephone-like session via terminals by using utilities Phone or Talk. The information can be stored and retrieved when required, and this facility depends on the way of implementation of the functions Phone or Talk. Users may also use a remote node from their local terminal and use the services at the remote node as if it were a local host.

If a DECnet user uses a standard LAN such as Ethernet, the terminals need to be connected to a terminal server. In this way, any terminal can be logged on to any computer connected to the LAN. The NVT protocol may be used at the application layer, but it does not require all of the functions of it. Instead, another protocol known as **local area transport protocol (LATP)** has been defined. This protocol resides on a terminal server. The LATP port driver is attached to each host machine. Any terminal may be connected to any host machine within an Ethernet LAN via CTPM. A terminal may interact with any host outside the Ethernet by converting LATP to CTPM.

Internetworking with X.25: The DECnet may be internetworked with a public X.25 network via two components of gateways: VAX PSI and VAX PSI access. The first component, VAX PSI, offers two modes of configuration: single-mode and multi-mode. In the single-mode configuration, a DECnet allows its node to provide direct access to one or more than one X.25 packet-switched public network. In this configuration, only one VAX system is required for establishing direct access to the X.25 network. On the other hand, in multi-mode, the single VAX system offers direct access to one or more than one X.25 network and indirect access to other systems. All of these systems must run DECnet VAX PSI protocols.

The VAX PSI software provides an interface between the DECnet node (VAX) and X.25 packet-switched public network which, in turn, must operate in multi-mode configuration. The DEC system allows the DECnet node to include both components — VAX PSI and VAX

PSI software access. It is interesting to note that both the VAX and X.25 public network run the DECnet system. The DEC–X.25 gateway offers support to both terminals and modems. The VAX PSI allows the VAX machine to define connections to more than one X.25 network and also offers more than one physical channel between VAX and X.25 networks.

DECnet operations with X.25: As mentioned in Chapter 11, there are two virtual circuits — namely, switched and permanent — which are provided by an X.25 network. Similarly, a DECnet X.25 network supports these circuits and is managed by the network management layer. The routing layer of DEC DNA gets the commands for the network management layer for initializing the circuits of the X.25 network. Within the routing layer, a sublayer known as DECnet routing initialization, after getting commands for the network management layer, includes a database containing DTE addresses, size of packets, call status, call tries, etc. of the X.25 calls. Based on the information in the header, the databases also provide mapping for an X.25 packet into appropriate X.25 permanent or switched circuits; this is performed between the routing layer and routing initialization sublayer (within the routing layer). A typical configuration for internetworking DECnet LANs with other DECnets or non-DECnets over X.25 is shown in Figure 12.20. Various protocols used during data communication through the DECnet model are shown in Figure 12.20. It is worth mentioning here that this interconnection configuration is similar to one we have discussed earlier for LAN interconnection. The interconnection usually defines the data communication through various layers and protocols between two LANs or a LAN and a WAN through connecting devices (bridges, gateways, etc.).

The routing initialization sublayer, in turn, includes its parameters into the header for all outgoing X.25 call packets. The routing initialization sublayer looks at the addresses of DTE in the X.25 call packet for accepting/rejecting the packet. The other sublayer of the routing layer routing control sublayer is mainly concerned with the control part of routing. The routing initialization sublayer also supports connectionless services where the DNA datagrams are segmented and reassembled as packets. It also performs transmission error checking of any other side of the link. Various events occurring during the connection of X.25, such as start, reset, clear, status, etc., are being controlled by routing initialization sublayer.

The connection between the local DECnet node and remote node via gateway node provides a relationship between the various facilities/services and layers of DECnet systems.

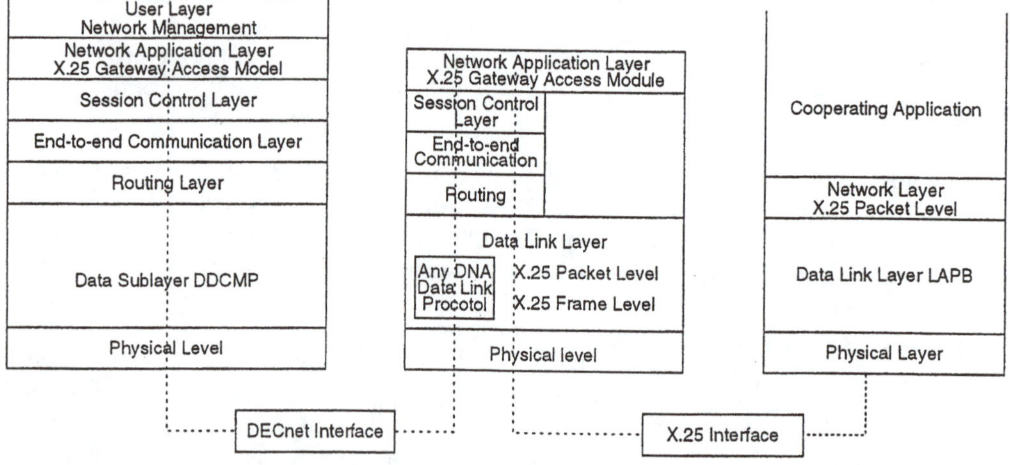

Figure 12.20 Local gateway remote.

For example, in a simple configuration, a request from a local node going through all seven layers of DEC SNA is received by DECnet via its router, which sends it to the gateway node. The call packet goes all the way to the network application layer, where it interacts with the X.25 application server module. The packet from the gateway nodes goes through all the layers (similar to DEC DNA layers), with the only difference being that now the packet goes via the X.25 protocol of the packet and frame levels. When the packet is received by the gateway, it goes through the DNA data link protocol (different types). The X.25 call packet from the gateway node then goes to the appropriate DCE of the X.25 network. The packet from the X.25 network finally arrives at the destination remote node at the cooperating program (top layer). The data link layer of the remote uses LAP-B, while the X.25 network uses the protocol of the X.25 packet level. There are two types of interfaces defined for this configuration: an interface between local and gateway nodes known as *DECnet interface* and another interface between gateway and remote nodes known as *X.25 interface*.

12.7.6.2 Internetworking with IBM SNA

Network layer of SNA: In the terminology of SNA architecture, the network layer is termed as the path control layer. The path control layer performs the same services and functions as those defined by OSI for the transport layer. It uses a static directory containing routes between each source and each destination of the network which is obtained from a session routing table or can be prepared by the network managers. The main function of the path control layer is then to select the route from this table dynamically.

The SNA network architecture consists of nodes which, in turn, define the domains of the network. There are two classes of nodes: one class defines a sub-area set of nodes which includes the mainframe and contains a routing decision capability, while the other class of nodes defines peripheral nodes that include controllers, terminals, and so on. These nodes do not have any decision capabilities. The sub-area nodes are connected by a transmission group of lines. Each sub-area node contains one or more terminals; hence, the address of any node will contain two parts of the fields: one for the sub-area node and one for the terminals or controllers or any peripheral devices.

A logical connection between two endpoints termed virtual route (VR) is defined when the session between the nodes is established. The virtual route provides the logical connection of all links and sub-area nodes between those endpoints. The users can also define a particular route and a particular level of quality (class of service) which is assigned each time a session is established, or the user can assign it each time he logs on the terminal/host, and that class of service will be available for the entire duration of the session between endpoints, i.e., if a session for a particular session has been established (virtual route). Once a virtual route is defined (it implies that the session is also established), it is mapped onto a table to define the explicit route (ER). The ER is basically a physical connection of transmission links where the transmission may correspond to land or radio links between endpoints. The maximum number of links between two sub-area nodes can be eight for each class of service. It is important to note that the routing in SNA architecture is based on the session; i.e., if a session for a particular session has been established (virtual route), multiple sessions (of similar or different type) can use the same virtual route for communication between endpoints.

Classes of network services: The various classes of services available are remote job entry, file transfer, interactive transaction, etc.

SNA also defines a transmission group (TG) for virtual routes, as there may be multiple links between sub-areas. The TG, in general, consists of identical lines and may be a leased satellite channel, parallel land line, analog line and digital channels of some defined data rates, etc. In each group, we may have multiple channels to offer greater bandwidth, and there is only one transmission queue explicit in each. Each time a packet (containing a route and destination address) comes to a sub-area, the routing decision table is searched and accordingly puts in the transmission queue of appropriate TG. The decision as to which TG is to be used is defined by the data link layer. When this packet arrives at another sub-area, the same process is performed, and this process is continued until the packet is delivered to the destination node.

It is an interesting feature of the SNA routing strategy that the sequencing of the packets is done at each sub-area node. In other words, the sub-area nodes have no knowledge about the virtual route between endpoints — they have knowledge only about their neighboring sub-area nodes. This unique feature of SNA routing does not require additional overhead should the network topology change. This is because the explicit route consists of the set of routes between two consecutive sub-area nodes, and the only change needed in the case of change in topology (addition or deletion of sub-area nodes) is the routing between neighboring sub-area nodes in the routing table. The routing between other sub-area nodes remains unchanged. The routing table defines the destination sub-area node's address and explicit route.

SNA protocols: The network addressable unit (NAU) is application software which allows users to access the network. The NAU, in fact, defines the address through which users can interact with each other and is considered a socket or port (similar to that in UNIX operating systems) of the network. The LUs allow the application programs to communicate with each other, and also the user processes can access the devices connected to the network. On the other hand, the PUs are associated with each node, and the network activates/deactivates the nodes; i.e., these nodes are connected to the network online or off-line. The addresses of devices connected to the network are defined by these PUs, which do not have any idea about the processes which are going to use them except that it allows these processes to use them via PUs.

The centralized manager node or system services control unit (SSCU) is usually implemented by hosts (PU type 5). LUs, PUs, and SSCUs, in general, are defined as subsets of NAU. The SSCP carries knowledge about other nodes of types 1, 2, and 4 and also controls the operations on these nodes, connected to the host. In the terminology of IBM, each domain of the SNA network consists of SSCPs, PUs, LUs, and other resources/devices, and the control and management of these devices are provided by the SSCP. The SNA network may operate in more than one domain.

Structure of LU6.2: The upper layer protocol widely used in SNA networks is LU6.2, which offers communication between various application programs. The SNA not only provides communication between different types of devices, but it also optimizes the performance of these devices or application programs. Various services provided to the end users by SNA include file transfer, sharing of application programs, devices, etc., and with the advent of high-speed intelligent and dumb terminals, powerful workstations, PCs, etc., the LU6.2 may not be able to provide services to the end users. The new version of the LU6.2 is the LU6.2 APPC (Advanced Program-to-Program Communication) for the SNA network. It interacts with the application program interface (API) to offer application-to-application communication and is usually called an application transaction program (ATP). The ATPs interact with LU6.2 to provide API, and then the LU session is established between LU6.2. The LU6.2 only provides an interface between different application programs and, as such, expects the transformation of syntax programming languages, commands, etc., to be taken care of by the application layer.

There are different categories of verbs used in commands for various operations, e.g., program-to-program, end user high-level languages, control functions, synchronization and coordination of various control activities, etc.

Program-to-program communication is a basic service provided by LU6.2 and is used by the LU service transaction program (STP). There are many vendor products defined under STP with different versions and options. The conversion of formats of programming languages into a standard data format at the sending node and reconversion of the standard data format back into programming languages at the receiving node are provided as a category for allowing users to use any high-level programming language.

The LU6.2 communication requires the transmission of primitives such as Allocate followed by a send.data primitive to the receiving side. The receiving node receives a receive.and.wait primitive. After receiving the data, the receiving node can send the data by using a send.data primitive. The sending node will receive the primitive receive.and.wait. After having received the data, the sending node may then send a Deallocate primitive to the receiving node, which sends back the confirmed primitive to the sending node.

12.8 Network routing switching techniques: Basic concepts

One of the most important functions defined for the network layer (OSI-RM) is the routing of the user's data packets or messages from a sender node to a destination node. The routing technique may deliver the packet or message to the destination node directly (if both nodes are on the same network, i.e., no hop) or indirectly (going through various subnets by making a sequence of hops). Further, in the latter case, the problem of choosing a suitable algorithm for the shortest route and data structure defined is a crucial and important parameter of the network protocol design. The decision as to which algorithm routing technique should be considered for the network depends on the type of network services required; for example, in the case of datagram service, the decision about selecting a route is defined after the message/packet has arrived at that subnet, while in the case of virtual circuit service, the decision about a particular route is decided before the establishment of virtual circuits. The virtual circuits will remain active as long as the session is maintained.

12.8.1 Routing switching configurations

Message switching: This technique for data communication was introduced during the 1960s and 1970s. This technique is based on messages which are defined as complete **data units (DUs)**. Each data unit contains the destination address and the routes (sequence of switches). A message (data unit) arrives at a switch (dedicated computers with disk storage) known as an **interface message processor (IMP)**, which checks the address. If it belongs to its own switch, it will send it to the appropriate destination. If it does not belong to it, it will look for a free trunk to the next switch (already selected using a routing algorithm during the establishment of the connection between them) and transmit the message to it. Since the unit is complete, the errors may also be checked by the switch and stored on its disk. A network which uses this type of switching is also known as a store-and-forward network.

The message-switching technique supports the priority defined for messages which can be routed over high-speed communication links/media. Messages of short duration usually get high priority. The lower-priority messages are stored temporarily on the disk during peak hours and later are transmitted during the times when the data unit traffic is not heavy. Electronic mail (e-mail), bank transactions, and bulletin boards are a few examples of communications which use message-switching techniques for their transmission. These messages are of variable length and, hence, require a disk at the switches to buffer them. Since the messages are stored on the disk, this technique is slow and is not good for interactive traffic or real-time traffic applications. The message circuit networks are more vulnerable and can fail if the switch(es) fail/are lost. This problem can be overcome by using duplicate (duplexed) switches. Switches are considered to be a bottleneck in message switching. Further, the communication line utilization here is poor and less efficient.

Packet-switching: In the early 1970s, a new data communication switching technique was developed and quickly adopted by many organizations and industries. The packet-switching technique overcomes all the problems with message switching. In packet-switching, there is a limit on the size of the packet and, further, these packets are stored temporarily in main memory instead of on a disk. These two features help in providing a better utilization of the link, higher throughput, and reduced delay at the switching nodes. This method of switching also avoids the main problem experienced in message switching: that the failure of one node causes failure of the network.

The packet-switched network is composed of many switching nodes (i.e., IMPs) and lines connecting the nodes. Thus, the packets can have different routing in the network. The nodes receive the packets (voice and data) and decide a route(s) for each packet through the network. Since a packet is going over nodes, each node receives the packet, stores it in main memory temporarily, decides the route, and transmits the packet to another node, which in turn performs the same set of operations, until the packet is delivered to the destination node. The packets are transmitted and received in both directions between the end users; it will be useful if we can interleave messages from different terminals or hosts into one channel by using a multiplexing technique. Multiplexing can be implemented on either a time or frequency basis; hence, time division multiplexing (TDM) and frequency division multiplexing (FDM) may be used. Here, TDM can be used to interleave messages from different terminals/hosts on the basis of time slots. The channel so defined can be assigned a port or socket in the network through which multiplexed packets can be received or transmitted. All these operations are transparent to the user and a user thinks that this particular part is assigned to him/her only. This mechanism not only helps in increasing the link utilization but also offers load balancing for different types of communication configurations (symmetrical or asymmetrical).

A packet-switched network supports the transmission and routing of packets of connection-oriented and connectionless configurations. In the connection-oriented configuration, a logical connection or virtual circuit between the destination and source is established. The nodes (including source and destination) of a packet-switched network use the primitives of the network layer for establishing the connections: call request packet and call accept packet. Once the connection is established, all the packets from the source node are transmitted over this virtual circuit and received in the same sequence. Each packet will now contain the data (user's data and control information) and the established virtual circuit pointer/sequence. The routing for the packets is decided once and will remain active until all the packets are transmitted. Obviously, initial setup time causes some delay. In the connectionless configuration, each packet or datagram is handled separately. Since there is no pre-established logical connection, each packet may be routed over a different route depending on a number of parameters, such as queue length, shortest packet, at each node. Due to this, it is possible that datagrams arrive at different times and also out of sequence. This configuration does not need any setup time for the transmission of datagrams.

12.8.2 Routing algorithms

One of the main functions of the network layer is to provide routing for the packet from one node to another in the network. Each node in the network uses a routing algorithm in such a way that a link to its adjacent node should provide a logical connection composed of established nodes up to the destination node. The selection of various parameters associated with the link to be established depends on the network designers. The parameters include cost of the link, capacity of the link, number of packets to be transmitted on the link, format of the packets, etc. If all the nodes are on the same network, then the packets can be transmitted directly to any node within the network or can be broadcast; i.e., these packets require only one hop for arriving at all the nodes connected to the network. Each of the nodes connected is considered a destination node in the network. If the nodes are on different networks, then the packets require multiple hops, the number of which, in turn, depends on how many networks are involved between source and destination. In any case, finding the routing algorithm that will provide the "best" routes and its associated data structures becomes one of the most important issues in the design of a network. Further, the routing algorithm should support both types of configuration operations: connection-oriented and connectionless. In the connection-oriented configuration, the entire logical connection (or route) between source and destination nodes is established (from the routing table) during the call setup. After the establishment of logical connections only, the packets are sent on this connection. The connectionless configuration requires each node to make a routing decision after the packet arrives at it. Some of the material herein is partially derived from References 2–7, 29, 30, and 36–39.

The routing algorithm for any network is affected by various parameters such as transit time, transit cost, throughput, connectivity, topology, efficiency, quality of service, etc. During the design of a routing algorithm, an obvious question is which parameters need to be optimized. In a combination of certain parameters, we may have to make a compromise, since maximizing one parameter may minimize other parameters and vice versa.

There are different types of routing algorithms available, but here we will group them into two categories: static routing (nonadaptive) and dynamic routing (adaptive). In each category, the routing algorithms may be implemented differently, as the requirements of users are different and vary continuously.

Static (nonadaptive) routing algorithms: The static routing algorithms maintain a table listing all possible routes between any two nodes. This table is constructed during the time when all the nodes of the network are off (unoperational). Before the network can be used for packet transfer, the table is stored on all nodes (IMPs) of the network. The static (nonadaptive) routing algorithms don't depend on the topology of the network, the type of configuration (connection-oriented or connectionless), or the size and type of packets.

The static algorithms can further be classified as shortest-path routing and bifurcated or multipath routing. In **shortest-path routing**, a shortest path is calculated between each connected pair of switching nodes (i.e., IMPs). The shortest-path routing is hoped to provide least-cost routing between connected switching nodes and minimum delay between end users. The **bifurcated routing** method calculates the number of routes between switching nodes in contrast to one route in the shortest path routing. The multiple routes may help in improving the efficiency and load capability of various links in the network. The main objective of this routing method is to provide minimum network delay.

Static routing algorithms can be defined for various parameters such as minimum transit time, actual cost, etc., and accordingly a table is constructed before the network is initialized. In other words, if we want to minimize the transit time, then a routing table is constructed defining all routes between each pair of switching nodes by assigning the transit time rate of each link in the network. The routing algorithm will then be used to calculate the least transition time between a source and destination and will define the

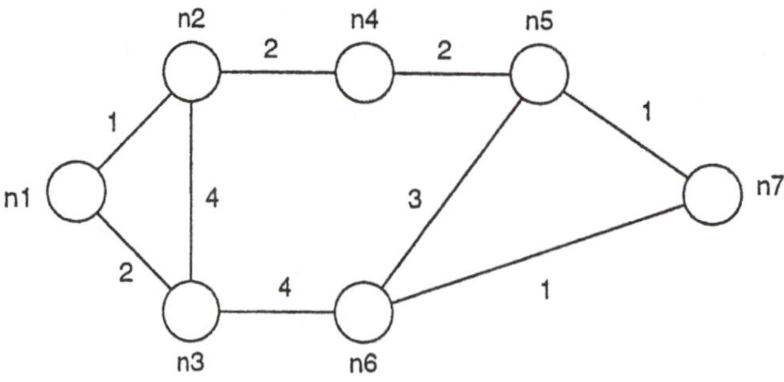

Figure 12.21 Network example.

transition time of each node with every possible neighboring node which will lead toward the destination node. The table not only contains one route with the least transit time, but it also contains all possible routes within the range of the least transit time, so that an alternate route may be used in the event of failure or traffic congestion. For other parameters, we can construct a similar table. In general, we define all the parameters by a generic word, *least-cost*, where the cost parameter can be substituted by any parameter. Once the routing table is constructed, it is loaded on each of the switching nodes (IMPs), and the entries of the table are not changed thereafter.

Shortest-path routing: There are many forms of shortest-path routing algorithms which have been developed. The one proposed by Dijkstra[27] seems to be the most popular. The following is a discussion on Dijkstra's technique for constructing a routing table. These steps are also discussed in other publications.[2-4] We assume that the cost associated with each link present in the network is known, where *cost* could be any parameter (as discussed above). The aim of the routing algorithm is to calculate all the possible routes between the source and destination with a minimum cost.

Each link (defined between nodes, say, i and j) is assigned with its cost weight. If there is a direct connection between i and j, then we consider this cost weight; otherwise, we substitute infinity. Consider a network as shown in Figure 12.21. The cost weights are assigned along the lines between the nodes. These cost weights are considered for calculating the static (or directory) routing of the network.

Let $c\ (ni, nj)$ be the cost of a link between nodes ni and nj.

$$C\left(n1,\ n2\right)=1,\ C\left(n1,\ n3\right)=2,\ C\left(n2,\ n3\right)=4,\ C\left(n2,\ n4\right)=3$$

$$C\left(n3,\ n6\right)=4,\ C\left(n5,\ n6\right)=3,\ C\left(n6,\ n7\right)=1,\ C\left(n5,\ n4\right)=1$$

The cost weight for other possible links between each pair of nodes will be infinity.

Let $D\ (nk)$ represent the distance between two nodes, i.e., the sum of the link cost weights on a given path from node ni to nk, assuming node ni is a source node. We start from the source node and set $N[ni]$ for each node nk not in N, and we set $D\ (nk)$ to $C\ (ni,\ nk)$. Then we determine a node nl, not in N, for which $D\ (nk)$ is a minimum, and we add nk to N.

Since in our example we have seven nodes, we enter the paths from source n into a tabular form as shown below:

$$N[n1] = \text{direct path (cost weight for the direct path)}$$

$$D(n2) = n1, \textit{ to } n2\ (1)$$

$$D(n3) = n1, \textit{ to } n3\ (2)$$

$$D(n4) = \text{no path } (00)$$

$$D(n5) = \text{no path } (00)$$

$$D(n6) = \text{no path } (00)$$

$$D(7) = \text{no path } (00)$$

$$N\ [ni]\ D\ (n2) = (n1, \textit{ to } n2)\ (1)$$

$$N\ [n2]\ D\ (n3) = (n1, \textit{ to } n3)\ (2)$$

$$D(n4) = (n1, m2, n4)\ (4)$$

$$D(n5) = (n1, n2, n4)\ (00)$$

$$D(n6) = (n1, n2, n4)\ (00)$$

$$D(n7) = (n1, n2, n4)\ (00)$$

Since *D* (*n*3) gives minimum distance compared to other nodes, we do not include *n*3 in set *N*, and we repeat the operation to get an updated table.

$$N\ [n1]\ D(n2) = n1 - n2\ (1)$$

$$N\ [n1, n2]\ D(n3) = n1 - n3\ (2)$$

$$N\ [n1, n2, n3]\ D(n4) = n1 - n2 - n4\ (4)$$

$$D(n5) = -(00)$$

$$D(n6) = (n1, n3, n6)\ (6)$$

$$D(n7) = -(00)$$

Since *D* (*n*4) gives minimum distance compared to other nodes, we include *n*4 in *N* and repeat the operation to get the following updated table.

A node with minimum cost weight is included in set *N*; i.e., now *N* = *N* [1, 2].

The distance *D*(*nk*) will be updated for all nodes still not in *N* with a minimum of either the current distance *D*(*nk*) or the current distance to the added node added with the distance from the node included to node *nk*. In other words, for each step until all nodes are included in the set *N*, we perform the following operation:

$$D(nk) - \min[D\ (nk), D\ (ni) + C\ (ni, nk)]$$

We compute the distance for all the nodes using the above operation and update the table as shown below. For example, we show how to compute the distance for $n3$.

$$D(n3) - \min\left[D\ (n3),\ D\ (n2) + C\ (n2,\ n3)\right]$$

$$D\ (n3) - \min\left[2,\ (1+4)\right]$$

$$D\ (n3) - \min\left[2,\ 5\right]$$

$$D\ (n3) - 2$$

Similarly, for $n4$, the distance is given by

$$D(n4) = \min\left[D\ (n4),\ D\ (n2) + C\ (n2,\ n4)\right]$$

$$= \min\ \left[4,\ (1+3)\right]$$

$$= 4$$

For other nodes, since there is no direct path between them and n1, the cost will be 00.

$$D(4) = \min\ \left[D(4),\ D\ (3) + C\ (3,4)\right]\ \ \left[00,\ (2+1)\right] = 3$$

$$D(5) = \min\ \left(D5,\ 1,25\right)\ \ \left[00,\ (1+3)\right]$$

$$N\left[n1\right] \qquad D(n2) = n1,\ n2\ (1)$$

$$N\left[n1,\ n2\right] \qquad D(n3) = n1,\ n3\ (2)$$

$$N\left[n1,\ n2,\ n3\right] \qquad D(n4) = n1,\ n2,\ n4(4)$$

$$N\left[n1,\ n2,\ n3,\ n4\right]\ \ D(n5) = n1,\ n2,\ n4,\ n5\ (\ )$$

Dynamic (adaptive) routing algorithms: In the dynamic (or adaptive) routing network, the routing algorithms provide different routing strategies depending on the topology conditions of the network, types of packets, types of errors, types of configuration, etc. The routing tables are updated constantly at regular intervals by a maintenance control packet which keeps a record of congestion of traffic, change in the network topology, delay on particular node(s), etc.

The static (or nonadaptive) routing algorithms don't adapt to any changes in either parameters or the network topology. Although the static routing algorithms can be made adaptive (as discussed in the previous section), this is accomplished at the expense of extra overhead and, further, the algorithms are, in general, not efficient. On the other hand, the dynamic routing algorithms constantly and continually update the routing table at regular intervals. In other words, the switching nodes have a decision capability which causes them to make an appropriate decision of routing based on parameters such as traffic, costs, delay, etc. The network maintenance control packet sends routing-related problems such as failure of a particular node, delay in a particular route, the traffic queue length at a particular node, etc., to all the switching nodes of the network. The transmission of routing-related information by the maintenance control packet can be implemented using either a centralized controlling mechanism or a distributed controlling mechanism.

In a centralized control routing scheme, this packet gathers information from all subnets and provides this information to one central node of the network. This node executes the routing algorithm using the information for the maintenance control packet and broadcasts a new routing table to all the nodes of the network. In the centralized control implementation, all the information from the maintenance control packet is gathered by one dedicated or centralized switching node. This information is used as data for the algorithm which is executed by the centralized node, which then broadcasts an updated or revised routing table to all the switching nodes of the network. The main feature of this technique is that all the nodes have the same updated routing table at the same time. Of course, this mechanism has a disadvantage, as there is only one centralized node and the whole network can go down if this node crashes.

In a distributed control routing scheme, each node takes information from its neighboring nodes which, in turn, gather information from the maintenance control packet. In other words, the maintenance control packet does not broadcast its information about various subnets to all the nodes but instead sends it to a few selected nodes. This packet routing obviously has the disadvantage that the nodes don't get an updated routing table at the time it is updated, and therefore the routing selected by a node may not be optimal since it has not adapted to the changes yet. The greatest advantage with these algorithms is that they help the network to gather information regarding congestion, errors, delay in a particular route, etc.; adapt to these changes; and modify the routing tables stored at various nodes under the changing conditions of the above parameters.

The distributed controlling mechanism also adapts very quickly to changes in the network topology. Based on parameters such as delay, sudden changes in topology may not necessarily give the best routing to the nodes, but the control packet immediately adapts to this sudden change and provides information to the nodes for passing on to neighboring nodes to execute the algorithm for necessary action.

The technique works well with smaller networks. As the size of a network grows (i.e., the number of switching nodes (IMP) increases), the technique may not support all the nodes using a neighborhood concept. In such a case, the adaptive routing uses the hierarchical concept where the entire network is divided into different IMP locations. Each of the IMPs of the location has the routing table for all the nodes present in the location and has no information about the structure, topology, type of network routing technique, etc., of other locations. This scheme is known as a two-level hierarchical routing technique and may be suited to small networks.

We may have to use a multi-level hierarchical routing technique for longer/huge networks; i.e., the entire network can be divided into regions, regions can further be divided into subregions, which in turn can be divided into smaller subregions, and so on. Each region will have a dedicated IMP which has a detailed internal structure of other regions. Further, that IMP can also recognize if the packet does not belong to that network. In that case it will send it to a common dedicated IMP which will be able to recognize the network type to which the packet belongs and also whether the packet is for the domestic network or for the network of some other country. If the packet does not belong to any of the domestic networks, then it will be sent to a similar dedicated IMP (destination country) which will perform the same operations and functions as the dedicated IMP of the source country.

This routing controlling mechanism is very useful in adapting to traffic congestion problems such as failure of nodes or routes, any malfunctioning, etc. One of the main drawbacks with this technique is that the updated tables maintained by each of the nodes in the network are not the actual updated tables at the time, due to the time lag between sending and receiving the routing problems from the maintenance control packet. An efficient adaptive routing algorithm for network management can be found.[32]

Flood search (directory-less) routing algorithm: In this algorithm, the packets are transmitted on all available lines with a hope that the destination node will receive it. The packets which are received are the ones which are not transmitted. This algorithm obviously sends the duplicate packets but at the same time utilizes the lines fully. Each packet transmitted is associated with a counter-value (equal to the number of hops), and each time it arrives at any switching node, its value is decreased. In the situation where the source node has no information about the number of hops between it and the destination node, the worst-case value (maximum number of hops possible in the subnet) is considered.

This algorithm also offers the shortest path between each pair of switching nodes. This algorithm may be useful in applications where the messages have to broadcast or the databases at each switching node need to be updated constantly. This algorithm provides a form of broadcast routing.

The transmission of duplicate packets causes flooding problems, which have been solved in a variety of ways, and a modified version of the algorithm itself has also been proposed, known as the selective flooding routing (SFR) algorithm. In this version of the routing algorithm, all packets are not transmitted; instead, only selected packets are transmitted, and the selection process can be either the lines connected to the switching nodes followed by routing the packets toward the destination node or sending only to a chosen switching node(s). In each case, some sort of decision capability is associated with the switching nodes which uses logically maintained databases for recording duplicate packets, thus reducing the flooding effect.

The U.S. Defense Communication Agency (DCA) uses a directory-less or flood search routing strategy for packet transfer in its Defense Switched Network (DSN). CCITT has proposed a standard signaling system no. 7 (SS7) which can be used for providing all the possible routes between source and destination, and it is left to the user to make a request for a particular route. This standard is also used in integrated services digital networks (ISDNs).

Directory routing algorithm: This is another form of static routing where routing information is defined in a directory routing of the network. The routing information is computed using various parameters such as least cost, least and best routes, maximum throughput, minimum delay, etc. The routing information for a given network topology remains constant, and hence all the decisions regarding routes for all nodes or some selected nodes are predefined. The directory may be updated in the event of changes in the network topology, or an adaptive method may be used to keep the directory routing updated for the traffic and length of queue at various switching elements. Depending on these conditions, the directory routing may use different routes for the packets, and further, these packets may arrive at the destination at different times and out of sequence. This scheme is not efficient, due to extra overhead for maintaining directory routing information.

Broadcast routing algorithm: In some applications, the source node has to send the same packet to all the nodes connected to the network at the same time. This can be achieved by a variety of implementations. Each implementation provides a broadcast or multi-destination or flood search routing. Each implementation has advantages and drawbacks. For example, if the implementation is based on the fact that the same packet is to be transmitted to all the destinations, this requires the listing of all destinations with the source and does not use the resources (communication lines, etc.) effectively. In the second implementation, we may use a flood search routing algorithm for broadcast routing. As discussed in the previous section, this routing algorithm sends duplicate packets defining a flooding situation which can be reduced at the cost of extra overhead.

In the implementation based on a multiple-destination routing algorithm, the packet is sent by the source node to all the destination nodes as a separate addressed packet on different lines. This implementation uses the communication lines effectively but has no control over the cost of different packets being sent on the same line, i.e., using the same route. Other implementations based on spanning tree, reverse path forwarding, and many other algorithms are also being used for broadcast routing, and the selection of a particular implementation is based purely on the application under consideration. These applications may be in distributed systems, office automation, bulletin board news, private networks, etc.

12.9 Network congestion control

Congestion is a condition in a network that occurs when there are too many packets in the network. The situation is similar to a highway that is congested due to too many vehicles. In network congestion, the number of packets in the network increases beyond the network capacity, the network is choked, and the performance (e.g., network throughput) degrades significantly. Some of the material herein is derived from References 33–35 and 40.

Congestion in networks can be caused in a number of ways. A typical congestion-creating scenario is as follows: due to statistical fluctuations or a burst of packets for a destination, an output link at a switching node in the network becomes overloaded, and the buffer at the node becomes full. This causes the node to drop packets, causing time-outs and re-transmission of those packets. Note that a node cannot discard a packet until it has been acknowledged by the next node on the route. This, in turn, has two consequences: first, the nodes retain packets for a longer period and second, it generates more traffic in the network. A direct consequence is that some additional switching nodes (specifically the ones feeding into the already loaded node) become overloaded. This phenomenon feeds on itself and quickly spreads until the entire networks gets overloaded; considerable resources are wasted on re-transmissions, packets move very slowly in the network, and the network throughput falls sharply.

Network congestion is handled in two ways. First, we can design a network such that congestion never occurs in it. This technique is called *congestion prevention*. The other approach is to allow congestion to occur, periodically monitor the network state for congestion, and take corrective measures to recover from the congestion. This is called the *congestion control technique*.

Congestion control requires detection of a congestion condition in the network. This can be done using an explicit or an implicit mechanism. In an explicit mechanism, the information about a congested network state is transferred in control packets. In an implicit mechanism, a node infers that congestion has taken place in the network by using some heuristic means. For example, if the round-trip delays to a node are very long, this implies congestion on the path.

Congestion occurs due to statistical fluctuations, where the load in a part of the network temporarily exceeds the capacity of that part of the network. If this overloading is not eliminated, it may rapidly spread to the entire network. Handling of this overloading by increasing the network resources (such as link bandwidth, buffer space at nodes, etc.) is generally not possible, and reducing the load by coercing is commonly employed to curb congestion.

One method of handling congestion, called **preallocation of buffers**, works when virtual circuits are being used within the subnet. At the time of a virtual circuit setup, a call-request packet reserves enough buffers at the nodes for the ensuing data packets. If an insufficient number of buffers are available at a node, the call-request packet is rerouted or returned with a busy signal. Once a path from source to destination has been charted and reserved, the data packets can be sent. Clearly, the problem of congestion is eliminated.

In another method, called **load shedding**, switching nodes or routers that are overloaded with packets start dropping packets to relieve the congestion. However, random packet dropping may aggravate the problem. For example, if a packet is dropped from all messages, the receivers may require the senders to transmit all of those messages again (depending on the underlying protocol). Generally, it is a good strategy to drop all or as many messages as possible from the same message. In case of real-time traffic (where some packet loss can be tolerated for a timely message delivery), it may be judicious to discard older messages. Important considerations are when to start taking the corrective measures and how much load should be shedded.

The **isarithmic congestion control** method attempts to tackle the problem by setting an upper limit on the number of packets allowed to be present at any given time in the network. This is done by introducing a system of permits within the subnet. If an IMP wishes to transmit a packet, it must first obtain and destroy one of the permits circulating within the system. Once the packet has been delivered to its destination, the receiving IMP regenerates the permit back into the subnet.

The isarithmic congestion control method has several problems. First, the method does nothing about the fact that individual IMPs can become congested. Even though there are a limited number of packets allowed in the subnet, all of these could conceivably be going to the same IMP, causing congestion. Also, the procedure for dealing with the permits themselves is not very good. Providing a fair way of distributing the permits throughout the subnet and allowing for a situation where a permit is accidently destroyed are both very difficult to do efficiently and effectively.

Another method of handling congestion is called **packet choking**; when a switching node or a router finds itself overloaded, it sends a choke packet to the senders of all arriving packets. A choke packet contains the destination of the arriving packet. When the sender receives a choke packet, it reduces its rate of packet transmission to the desination in the choke packet by some percentage (say 50%) for a fixed time interval. This reduced traffic helps the congested node or router to clear the backlog of load.

12.10 Value-added networks (VANs)

Value-added networks (VANs) have their origin in ARPANET, a nationwide network. The VANs were defined by the FCC in 1973 as a mechanism by which leased lines from the carrier were obtained, the processing facilities were added to it, and then services were offered to end users. This concept of VAN is regulated as a common carrier and is based on an open entry policy allowing potential public network operators to apply for FCC approval. It is important to note that VANs don't have any free data-processing or data communication services but instead offer the services to end users on the open market at a reasonable price. The first VAN, created by **Telenet Communication Corporation (TCC)**, was approved by the FCC in 1974, and following this, other vendors also defined VANs like Tymshare with Tymnet in the U.S., public packet-switched data networks (PSDN) in European countries, and so on. VANs are usually not cost-effective for leased lines or dial-up network lines dealing with high volumes of traffic, but they are suitable for low-volume traffic at low data rates. At present, there are two organizations which offer value-added carrier services: TCC and Tymnet Corporation[30]

One of the main services offered by VANs is network connectivity, i.e., providing interconnection between terminals and hosts for interactive exchange of data. Further, the interconnection between two host computers may offer services or applications such as file transfer, remote resource request services, and other remote-related services. For example, if a user wishes to access a file on a remote machine, dialing up VAN's local public dial-in asynchronous port enables terminal access. This technique of terminal access is cheaper than direct terminal access to X.25. Once the asynchronous dial-up terminal is connected to the VAN's port, the PAD (resident at PSDN) converts this asynchronous data stream into the X.25 packet format. It also offers dedicated or direct point-to-point links to the host and other resources.

The VAN offers a variety of services to users who send data from their PCs by dialing up the network at 1.2 Kbps, which can be connected to a host computer connected to the network by 64-Kbps lines. For internetworking via VAN, there are various issues to be addressed, as the terminals and hosts may have different speeds, codes, and protocols. The VAN offers conversion for each issue. The speed conversion between a PC's low data rate to the high data rate of a host (connected to the network by appropriate links) is provided by protocols of VANs. The speed conversion services offered by VANs are available in different forms. For example, the speed-conversion between the access-line speed of terminals and the connecting link of a host with the network is usually handled by an internal network flow-control service and other related procedures at regular intervals to ensure the conversions.

The protocol conversion has been a major issue for the transfer of data across various types of computers and X.25 hosts. It must support synchronous hosts such as mainframes, synchronous terminals, asynchronous dial-up terminals, and many other terminals and hosts. Various VAN-based hardware schemes are available which offer protocol conversion within the network.

Consider the following situation. We have to send data from an asynchronous terminal to synchronous terminals (say IBM 3270). In one approach, we may take the help of SNA's SDLC connection to the network if the host is IBM. It does not require any special software for a PC, and even dumb terminals can be used. A protocol converter at the PSDN handles the protocol-conversion process. The SDLC runs on a front-end processor (FEP) connected to the host. Each of the asynchronous terminals is connected to the PAD at the PSDN via asynchronous dial-up connections and includes emulation software. The dumb terminals are simply connected to the PAD without having any software on them. The host contains 3270 application software.

In another approach, we may run a virtual telecommunication access method (VTAM) on the host while an X.25 to SNA/SDLC protocol conversion procedure is executed on IBM's **NCP packet-switch interface (NPSI)** FEP. This offers various advantages such as end-to-end error recovery (by SDLC), improved 3270 emulation at the PC, etc.

Finally, code conversion deals with the codes being used in the terminals and hosts. The majority of computer vendors use seven-bit American Standard Code for Information Interchange (ASCII), which incidentally has also been adopted as the International Alphabet Number 5 (IA5) standard coding scheme by CCITT. IBM has defined its own eight-bit Extended Binary Coded Decimal Interchange Code (EBCDIC) for its own machines. The protocol conversion between the protocols supporting ASCII or EBCDIC is usually performed by a code-conversion process which is a part of protocol conversion.

12.10.1 VAN services

VANs support a variety of protocols such as X.25 SNA/SDLC, bisynchronous (BISYNC or BSC), leased lines (based on 3270 BSC protocol), file transfer, and remote job entry (RJE). The network provides protocol and code conversions between the underlying protocol and X.25. The PSDN provides support to multi-host switching, various dynamic routing, multiple dial-up over a single host/FEP port, private synchronous dial-up, leased lines for RJE nodes, etc.

A variety of PSDNs (such as Tymnet and Telenet) have also used the concept of VANs in providing transport services to users. PSDNs like GEIS and CompuServe are also providing transport services for network-based and database applications. Some data-based and network-based applications over PSDNs based on VANs include financial, travel agencies, bulletin boards, news, weather forecast, insurance, law, medicine, music, etc., and these are widely used in the market. Some of their supporters/users include Dow Jones, Nexis, Airline Guide Services, Dun and Bradstreet, Information Services, Videotex, Mintel Video Service running over Transpac PSDN (France), consumer-oriented Prodigy Videotex (IBM and Sears), etc. The majority of these services are available to the users at very cheap rates — as low as $5–$8 per minute for any type of inquiry on VANs.

Other VAN services include direct long-distance dialing (DDD) using public switched networks; wide area telephone service (WATS) offering dial-up and inward or outward calling services; telex/TWX offering low-speed switched network service for transmission of a message between teleprinter terminals; Dataphone 50 offering high-speed 50-Kbps switched data services; private leased channels (analog) offering transmission of voice, digital data, and video; and digital services for transmission of digital data rates of 2.4, 4.8, 9.6, and 56 Kbps.

AT&T offers dataphone digital service (DDS) to many cities with a very low bit error rate of 10^{-9} to 10^{-12}. Domestic satellite services for point-to-point configurations are offered by Western Union through its satellite Westar, by RCA's satellite SATCOM, by American Satellite Corporation (ASC) via RCA's and Western's satellites, and by AT&T–GTE through leased channels from COMSAT.

Electronic mail (e-mail) and **electronic data interchange (EDI) services:** E-mail and EDI are the most important transport services being offered by VANs and PSDNs, and in fact the world market has already poured billions of dollars into this area. As such, we can now see that these services are available to users from every part of the world at a very reasonable price. The widely accepted CCITT X.400 Message Handling System (MHS) recommendation for both VANs and PSDNs has made possible global communication across the world via wide area networks.

The VANs have dominated the European market for X.400 electronic facilities such as copy, reply, return, priority, broadcasting, saving, etc. Somehow, VANs have not become as popular in the U.S. as in European and other countries. E-mail services offered by VANs provide the following facilities: creation and storage of PC word-processing texts and spreadsheets, sending and receiving of telex, accepting e-mail in ASCII for its transmission as images or facsimile (fax) over switched telephone networks, transmission and storage of digital signatures, voice-annotated messages, and integrated text, images, audio, and video documents as one mail message.

EDI service is slightly different from e-mail service in the sense that it accepts documents (mainly from business-oriented applications) which may be purchase orders, delivery orders, invoices, etc., converts them into a predefined form (standardized) with appropriate fields, and then transmits them in this new format over the network. This method of transmitting documents in a standardized format not only saves data-processing efforts and time but also improves significantly the cost of transmission and accuracy in handling data communication. The EDI service is available to users around the clock and includes storing, distribution, and many other network-based services. The EDI service packages for PCs and mainframes supporting the transport services of VANs are available on the market at very reasonable prices.

As stated above, the Tymnet and Telenet VANs offer various dial-in services. These VANs also offer dial-out services on public switched networks and private asynchronous terminals where users can submit their request via dial-up, leased-lines terminals, host to the user, devices connected to a public switched telephone network, or the asynchronous terminals. In other words, a user can submit a request from his asynchronous terminal (via asynchronous modem) or host to PSDN (X.25 leased lines) which will be transmitted to remote PSTN and then to appropriate asynchronous terminals (via asynchronous modem) or host (connected by X.25 leased lines to PSDN). Thus, VANs provide both dial-in and dial-out services on PSDNs and PSTNs to various terminals, hosts, devices, etc. These services are available at different asynchronous speeds of 300, 1200, and 2400 bps on modems which are typically Hayes-compatible.

12.10.2 VAN connections

VANs offer a variety of connections between terminals, PCs, etc., and the host across the network. These connections are usually defined at main nodes (which are located in large cities) for access to the network. The U.S. has one of the largest VANs covering the continent via nearly 750 access locations throughout, and these locations are defined based on various parameters such as the volume of data, location of the place, population, etc. It is expected to assign higher priority to the large cities (with large population and high traffic of data).

The public asynchronous dial-up ports being supported by modems include 300 (Bell 103), 1200 (Bell 212A compatible), and 2400 (V.22 bis compatible) and are Hayes-compatible.

The CCITT modems include V.32, V.32 bis, and V.29. A new asynchronous dial-up modem offering 9.6 Kbps was introduced in 1988. The above modems have, in one way or another, become standards for the industry and are widely used for data communication over VANs.

For sending high volumes of data, private or dedicated asynchronous dial-up services are available on most of the PSDNs. We expect to get optimal use of the network with improved performances for file transfer and also the amount of data sent. Since these are dedicated lines, no setup time or dial-up connection time is requested, as it can be left indefinitely.

Dial-up connections have a few limitations of speed, low reliability, and higher data error rates. These problems are alleviated in the leased-line connections. Due to this, the dedicated leased lines (asynchronous) are preferred over dial-up connections between hosts with synchronous device attachments to the PSDNs. The dedicated leased lines over the PSTN can be purchased at a nominal installation fee and monthly charges which depend on the requested line speed, protocols, configurations, optional features, terminals being used, etc.

These VAN-based leased lines offer speeds of 1.2 Kbps to 19.2 Kbps for analog transmission access links and 9.6 Kbps to 50 Kbps for digital transmission access links. The 64-Kbps data rate is used in European countries for digital access links. Obviously, the 1.2-Kbps asynchronous line is the cheapest dedicated or leased-line circuit available for low-speed applications, and asynchronous protocols running on these circuits support one session at a time. In asynchronous leased lines, only one session/connection can be established on the physical circuit or link at a time, and only a limited number of speeds are available on these links.

The synchronous dedicated (leased) lines offer a variety of speeds, such as 1.2, 2.4, 4.8, 9.6, 14.4, 19.2, and 56 Kbps with an X.25 connection. Further, these lines can be used for multiple sessions/connections simultaneously for data transmission. In other words, multiple users can share a synchronous leased line, offering a cost-effective communication system between them. For example, Tymnet offers 16 sessions/connections over one 64.8-Kbps X.25 line and 48 sessions/connections over a 14.4-Kbps X.25 synchronous leased line. These synchronous lines support X.25 synchronous protocols.

There are non-X.25 synchronous protocols which are also supported by many VANs and offer services to users over leased lines. IBM's SNA and BISYNC networks support X.25 and can use the transport services offered by VANs without changing their hardware or software configuration. For example, FEP using SNA/SDLC protocol over leased-line connections can communicate with VAN, and the SDLC can be used on the leased lines to support 3270-type terminals for both point-to-point and multi-point configurations. VANs have also provided support for SDLC connections to IBM mainframes which are being used in many stores, such as Series/1 Systems 36 and 38. VANs offer line speeds of 4.8 or even 9.6 Kbps over these SDLC connections. Private services can also be obtained from SDLC dial-up lines at lower line speeds of 2.4 or 4.8 Kbps. These services are also available with BISYNC protocols. For example, Tymnet uses BSC 3270 to provide connection to either SNA or another host (based on ASCII).

The X.25 access connection usually requires the leased line attachment unit (AU) to the PSDN, and these leased lines are quite expensive. The CCITT X.32 recommendation offers the same services and capabilities as dial-up connections. It has become widely useful and acceptable and is usually termed the X.25 dial-up standard in the network industry. This standard provides access to a variety of connections of PSDNs, as discussed above (e.g., asynchronous, SNA/SDLC, BISYNC 3270, and other host applications).

A synchronous V.32 modem operating at 9.6 Kbps offers the same services with the same performance as leased lines offering lower speeds of 1.2, 2.4, and 4.8 Kbps, which are widely accepted.

The internetworking between various networks for VANs can be implemented in a number of ways, of which the following two have received wide acceptance from various PDN vendors and government agencies.

The PSDN is connected to an **international record carrier (IRC)** which provides X.75 connections to Telenet and Tymnet. In order to use the international packet-switched services of VANs, one needs to make a request to the U.S., VAN, and the remote or foreign PSDN. The local PTT of the country controls the services or connections offered by VANs. It has been seen that these foreign nodes defined by IRC have technical contacts with local PTTs and can obtain these services of VANs via NUIS. This reduces the request and ordering process, thus offering a straight channel to users, particularly those of multinational companies.

Other classes of VANs (Infonet and GEIS) have defined their own networks directly for foreign countries without using IRC transit services. In other words, these VANs have defined their own international services by creating their nodes in various countries (particularly the countries of high demand). For example, GEIS has created about 35 foreign nodes, while Infonet has created its own packet-switching nodes (dedicated) in about 20 foreign countries. These nodes of VANs can be accessed by their respective PTT/IRC X.75 connections. This approach is very useful for users who don't want to go through the respective PTTs — instead, they can deal directly with a VAN.

12.10.3 Popular network vendors

There are quite a number of vendors for private packet networks, and it has been observed that these vendors often don't offer cost-effective protocols and products across the networks. In other words, these vendors have given little consideration or effort to improving the performance of the networks. These networks can broadly be divided into two categories:

low-sized and **large-sized networks**. The size of the network depends on parameters such as type of transmissions, backbone network, transmission media, number of terminals, and number of nodes in the networks.

Low-sized networks: The low-sized networks typically support around 500 terminals, about 10 nodes (excluding PADs), low-speed analog leased lines, etc. A few vendors offering low-sized networks include Amnet, Dynatech, Memotec, Digital Communication Associates, EDA Instruments, General Datacom, and Protocol Computer. A brief description of some of the popular low-sized networks is given below.

- **Amnet:** This network uses a standard PC/AT hardware platform and includes a multi-port board for PADs. Each port is associated with a microprocessor and its own memory and offers a data rate of 256 Kbps. It supports a variety of PAD protocols such as Async, SNA/SDLC, 3270 BISYNC, etc. The ports can be configured to provide an interface to any software-protocol module.
- **Memotec:** This vendor provides a variety of network products, in particular PADs and switches. One very popular and widely used PAD is SP 8400, which is installed on firmware and does not use expensive leased lines. It supports a variety of PAD protocols such as SNA/SDLC, BISYNC 3270, 2780/3780, Burroughs poll/select, inquiry/acknowledgment (Inq/Ack) of HP, etc. It can be connected to different types of terminals such as POS terminals, cash registers, etc. It supports a trunk speed of 256 Kbps and is capable of transmitting 1800 packets per second.
- **Micom:** This network offers low-cost PADs and switches which can be used on general-purpose hardware platforms. The unit is known as a Micom Box, which offers different configurations of PADs such as Async PAD, Sync PAD, packet-switching models, etc., and can be changed by the switch. The interface and installation specification procedures are defined in a user-friendly environment. Various other features offered by these PADs include address conversion, head balancing, and different diagnostic tools.

Large-sized networks: These networks support thousands of terminals, powerful nodes, and high-speed digital backbones for 56-Kbps to T1 (1.544 Mbps) data rates. In general, the trend for large-sized networks has been toward reducing the cost per port by considering powerful machines for network nodes. The following vendors provide these services and products: BBN, Hughes, Northern Telecom, Telenet, Tymnet, and Siemens. Of these vendors, Siemens is providing the services and public networks to the Regional Bell operating companies. In the following we provide brief descriptions of some of these networks.

- **BBN:** This vendor had a long experience with ARPANET and is now concentrating on large-sized private packet networks. It has developed a number of products, e.g., X.25 FEP for IBM, SNA network, software products for SNA networks such as routing network management, etc. It has provided networks to client agencies such as MasterCard, International U.S. Treasury, and many laboratories.
- **Northern Telecom:** The DPN-100 product line has offered different types of access and network switches which provide support to over 30,000 access times with the capability of handling over 30,000 packets per second. The trunks connected to these switches are supported for data rates of T1 (1.544 Mbps) and 2.048 Mbps (Europe). The DPN-100 switches, along with other equipment, are being used in Canada's Datapac, Swiss PTT, Telecom Australia, and many other organizations. Their use in private networks such as the American Airlines SABRE system, U.S.

Federal Reserve System, and Society for Worldwide Interbank Financial Telecommunication (SWIFT) has made this company a very useful vendor in this area. Further, these switches have been used in X.25 to ISDN internetworking.

- **Telenet:** This is a public network introduced by Sprint Corporation (U.S.), and lately it is being driven in the private packet network market. This is evident from the fact that the company now has over 130 private networks worldwide. The Telenet TP3/II-3325 PAD can provide support of up to five different protocols under a multi-vendor communication environment. It has also taken an interest in CCITT's X.25 and X.400 recommendations.[31]
- **Tymnet:** The private networks offered by Tymnet support a variety of packet-switches which can be chosen from Tymnet engines. Each switch configuration can be loaded and managed for a PC running local engine monitor (LEM) software of Tymnet. Publishing companies, electronic industries, and some banks are the customers of these private networks, and so far at least 45 private Tymnet networks have been installed worldwide.

Other vendors in the market include Amdahl (Texas) and Hewlett-Packard (California). Even IBM and DEC are offering software and hardware products for X.25-compatible networks. IBM offers this service in its SNA X.25 Interconnection DECnet X.25 postal software for respective backbone networks.

Some useful Web sites

More information on the network layer can be found at *http://ntrg.cs.tcd.ie/4ba2/network*, *http://ntrg.cs.tcd.ie/4ba2/network*, *http://cio.cisco.com/warp/public/535/2.html*, *ganges.cs.tcd.ie/4ba2/network/overview.html*, *www.stockton.edu/~stk6458/concepts/index.htm*, and *http://skhuang.iis.sinica.edu.tw/course/network/network2/index.htm*.

X.25 switched networks, protocols, components, and other related products for internetworking can be found at *http://www.home.ubalt.edu*, *http://www.it.kth.se*, and *http://www.techweb.com*. A knowledge of the database index and quick searching can be seen at *http://www.shiva.com.sg*, and *http://www.wanresources.com/x25cell.html*.

TCP/IP can be found at *http://www.tcp-ip.nu* and *http://www.cisco.com*.

The Internet protocol next generation can be found at *http://playground.sun.com*, *http://www.protocol.com*, and *http://news.cnet.com*.

IP switching architecture protocols such as Ipsilon, Toshiba's cell switch router, and Cisco's tag switching are found at *http://www.ipsilon.com*, *http://www.cisco.com*, and *http://www.toshiba.com*.

Value-added networks, their applications, and services can be seen at *http://www.edi.wales.org*, *http://www.nafta.net*, and *http://www.spokane.net*, and related tools are found at *http://www.amicus.ca/legallinks*. Various value-added network vendors are *http://wwwglobalknowledge.com*, *http://www2.vinylvendors.com*, *http://www.iwin.nws.noaa.gov*, and *http://www.analysis.co.uk*.

References

1. ISO 7498, "Open Systems Interconnection — Basic Reference Definition, Information Processing Time," ISO, 1986.
2. Tanenbaum, A.S., *Computer Networks,* 2nd ed., Prentice-Hall, 1989.
3. Stallings, W., *Handbook of Computer Communications Standards,* vol. 1, 2nd ed., Howard W. Sams and Co., 1990.
4. Black, U., *Data Networks: Concepts, Theory and Practice,* Prentice-Hall, 1989.

5. Schlar, S.K., *Inside X.25: A Manager's Guide,* McGraw-Hill, 1990.
6. Meijer, A. and Peeters, P., *Computer Network Architecture,* Computer Science Press, 1982.
7. Kauffles, E.J., *Practical LANs Analyzed,* Ellis Horwood Ltd., 1989.
8. DECnet Digital Network Architecture, General Description, Digital Equipment Corp., May 1982.
9. DNA Digital Data Communication Message Protocol (DDCMP), Functional Specification, Digital Equipment Corp., 1982.
10. ISO 8348, "Data Communication — Network Service Definition Information Processing Systems," ISO, 1986.
11. ISO 8348/DAD 1, "Data Communication — Network Service Definition Addendum 1: Connectionless Mode Transmission, Information Processing Systems," ISO, 1986.
12. ISO 8348/DAD 2, "Data Communication — Network Service Definition Addendum 2; Covering Network Layer Addressing, Information Processing Systems," ISO, 1986.
13. ISO 8348/DAD 3, "Additional Features of Network Service, Addendum 3," ISO, 1986.
14. ISO 8473, "Data Communications — Protocol for Providing the Connectionless Mode Network Service, Information Processing Systems, ISO, 1986.
15. ISO 8473/DAD 1, "Data Communications — Protocol for Providing the Connectionless Mode Network Service, Addendum 1: Provision of the Underlying Service Assumed by ISO 8473, Information Processing Systems," ISO, 1986.
16. ISO 8473/DAD 3, "Data Communications — Protocol for Providing the Connectionless Mode Network Service, Addendum 1: Provision of the Underlying Service Assumed by ISO 8473 over Subnetworks which Provide the OSI Data Link Service, Information Processing Systems," ISO, 1986.
17. ISO 8208, "X.25 Packet Level Protocol for Data Terminal Equipment," ISO, 1986.
18. ISO PDTR 9575, "OSI Routing Framework," ISO, 1986.
19. CCITT X.3, "Packet Assembly/Disassembly (PAD) in a Public Data Network," CCITT, 1984.
20. CCITT X.25, "Interface between Data Terminal Equipment (DTE) and Data Circuit-Terminating Equipment (DCE) for Terminals Operating in Packet Mode and Connected to Public Data Network by Dedicated Circuit," CCITT, 1984.
21. CCITT X.75, "Terminal and Transit Call Control Procedures and Data Transfer System on International Circuit between Packet-Switched Networks," CCITT, 1984.
22. CCITT X.121, "International Numbering Plan for Public Data Network," CCITT, 1984.
23. CCITT X.2, "International User Classes of Service in Public Data Networks and ISDNs."
24. CCITT X.213, "Network Service Definition for Open Systems Interconnection for CCITT Applications."
25. Defense Communications Agency, *DDN Protocol Handbook,* vol. 2, *DARPA Internet Protocols,* U.S. DoD, Dec. 1995.
25. "IBM Systems Network Architecture: Technical Overview," IBM Corp., Research Triangle Park, NC, No. GC30-3073, 1985.
27. Dijkstra, E.W., "A note on two problems in connection with graphs," *Numerical Mathematics,* 1, 260-271, 1959.
28. Halsall, F., *Data Communications, Computer Networks, and Open Systems,* Addison-Wesley, 1992.
29. Markeley, R.W., *Data Communications and Operability,* Prentice-Hall, 1990.
29. Loomis, M.E.S., *Data Communication,* Prentice-Hall, 1983.
31. U.S. DoD, MIL-STD-1782, TELNET Protocol, Military Standard, U.S. DoD, May 1984.
32. Srikantan, T. and Hura, G.S., "An efficient adaptive routing algorithm for a network management system," *Computer Communications,* 20, 988–998, 1997.
33. Jacobson, V., "Congestion avoidance and control," ACM SIGCOMM'88, 1988.
34. Ramakrishnan, K. and Jain, R., "A binary feedback scheme for congestion avoidance in computer networks with a connectionless network layer," ACM SIGCOMM'88, 1988.
35. Chou, W. and Gerla, M., "A unified flow and congestion control model for packet networks," Proceedings of the 3rd International Computer Communication Conf., 1976.
36. Schwartz, M. and Stern, T.E., "Routing techniques used in computer communications networks," *IEEE Trans. on Communications,* April 1980.
37. Gerla, M. and Kleinrock, L., "Flow control: A comparative survey," *IEEE Trans. on Communications,* COM-28, April 1980.

38. Ahuja, V., "Routing and flow control in SNA," *IBM Systems Journal*, 18, 1979.
39. Kleinrock, L. and Kamoun, F., "Hierchical routing for large networks," *Computer Networks*, January 1977.
40. Yang, C. and Reddy, Z., "A taxonomy of congestion control algorithms in packet switching networks," *IEEE Network*, 1995.

References for further reading:

1. Newman, T.L. and Misnshall, G., "Flowlabelled IP: Connectionless ATM under IP," Networld+ Interop Presentation, April 1996.
2. Ipsilon Networks, "IP switching: The intelligence of routing, the performance of switching," Feb. 1996.
3. Katsube, Y., Nagami, K., and Matsuzawa, S., "RFC 2098: Toshiba's router architecture extension for ATM," Overview, *IEFT*, Feb. 1997.
4. Rekhter, Y., Davie, B., Katz, D., Rosen, E., and Swallow, G., "RFC 2105: Cisco Systems' tag switching architecture overview," *IEFT*, Feb. 1997.
5. Freeman, R., *Telecommunication System Engineering*, Wiley, 1996.
6. Bertsekas, D. and Gallager, R., *Data Networks*, Prentice-Hall, 1992.
7. Spohn, D., *Data Network Design*, McGraw-Hill, 1994.
8. Bradner, S. and Mankin, A., *IPng: Internet Protocol for Next Generation*, Addison-Wesley, 1996.
9. Comer, D., *Internetworking with TCP/IP*, vol. 1, *Principles, protocols and architecture*, Prentice-Hall, 1995.
10. Miller, S., *Ipv6: The Next Generation Internet Protocol*, Digital Press, 1998.
11. Jain, R., "Congestion control in computer networks: Issues and trends," *IEEE Network Magazine*, May 1990.
12. Yang, C. and Reddy, A., "A taxonomy for congestion control algorithms in packet switching networks," *IEEE Network*, July/Aug., 1995.
13. Perlman, R., *Interconnections: Bridges and Routers*, Addison-Wesley, 1992.
14. Zhang, L., Deering, S., Estrin, D., Shenker, S., and Zappala, D., "RSVP: A new resource reservation protocol," *IEEE Network*, Sept. 1993.
15. Stevens, W., *TCP/IP Illustrated*, vol. 1, Addison-Wesley, 1994.
16. Slattery, T., Urton, B.B., and Burton, W., *Advanced IP Routing with CISCO Networks*, McGraw-Hill, 1998.
17. Huffman, D.A., "A method for construction of minimum redundancy codes," *Proceedings of IRE 40*, 1098-1101, Sept. 1952.

chapter thirteen

Transport layer

"Each year it takes less time to cross the country and more time to get to work."

Mary Waldrip

"The transport layer determines how to use the network layer to provide a virtual error-free, point to point connection so that host A can send messages to host B and they will arrive un-corrupted and in the correct order. It establishes and dissolves connections between hosts. It is used by the session layer."

The Free Online Dictionary of Computing
Denis Howe, Editor (*http://foldoc.doc.ic.ac.uk/*)

13.1 Introduction

All three lower layers (physical, data, and network) of OSI-RM provide an environment for transmitting user data from source to destination stations across the network. The transmission of user data includes various functions and services such as routing provided by the network layer with every packet received from the transport layer, framing, error control (offered by the data link layer), and sending of frames (provided by the physical layer). After going through these functions, it is then transmitted over transmission media. The user data, as it travels to the network layer, is termed a **packet**, and after the packet is delivered to the data link layer, it is termed a **frame.** The main difference between a packet and a frame is that the packet contains only a header from each of the layers, while the frame includes both header and trailer. The transmission of frames from the lower three layers does not provide any facility regarding error recovery, error-free delivery of frames at the destination, action on lost frames either on local LAN or across the subnets, or any other hardware failure.

As indicated earlier, the subnet interface, routing table, etc., including the lower three layers in both host IMPs (subnet part of OSI-RM) are defined as the **subnet**, while the transport layer and other upper layers along with their protocols reside only in the **user's host**. It offers a transparent transfer of data between session layer entities and provides a reliable and cost-effective means of transport between them. The session layer entities are not concerned about how the data transfer over established transport is achieved.

The transport layer accepts data from the session layer, splits it into smaller units if required, passes these to the network layer, and ensures that the pieces all arrive correctly at the other end. Furthermore, all this must be done cost effectively and in a way that isolates the session layer from the changes in the hardware technology. The routing and relaying of the messages are also not important to the transport layer, as these are provided by the network layer. The network layer provides network-independent services to the application-oriented layers above it. It accepts available network services and provides the session layer with data transportation facilities at a required quality of services in an optimal and cost-effective way. In other words, this layer offers a bridge to a gap (with respect to QOS) between what the network layer offers and what the session layer wants. The IEEE 802 standard document has left the higher-layer protocols (above LLC) undefined and gives a lot of flexibility to the users to define their own networks. This is convenient with LANs, as the design of the LAN model is based on a bottom-up approach. The design process actually goes from the physical layer to the media access layer, which distinguishes it from that of WANs. The LLC sublayer is basically considered the uppermost top and formal layer in LAN mode, and this is this common LLC sublayer which brings various LAN technologies together at one common interface (LLC).

13.2 *Layered protocols and interfacing*

The communication control procedures are a set of rules and procedures for controlling and managing the interaction between users via their computers, terminals, or workstations across the network. The control procedures are called by different names in the industry, such as protocols, line control procedures, discipline procedures, and data link control procedures. The protocols provide a communication logical link between two distant users across the network, and the handshaking procedures provide the call establishment within the protocols.

Each node (PCs, terminals, etc.) connected to a LAN must contain identical protocol layers for data communication. Peer protocol layers exchange messages which have commonly understood formats, and such messages are known as **protocol data units (PDUs)**. It also corresponds to the unit of data in m-peer-to-peer protocol. The peer transport entities communicate with each other using the services provided by the network layer. The services of this layer are provided to the session layer at a point known as a **transport service access point (TSAP)**. For interface between the terminal and session layers, the transport primitives are defined. The **transport service data unit (TSDU)** associated with its primitives is delivered to the session layer and, similarly, the session layer uses its primitives to interact with the transport layer. The session layer sends its session protocol data unit (SPDU) to the transport layer. The SPDU consists of protocol, control information, and session user data. The session control information is exchanged between session layer entities.

Each layer defines active elements as entities, and entities in the same layer of different machines connected via network are known as **peer entities.** The entities in a layer are known as service providers, which implement services used by their higher layer, known as the user. The **transport service access point (TSAP)** is a unique hardware or port or socket address (software) through which the upper layer accesses the services offered by its lower layer. Also, the lower layer fetches the message from its higher layers through these SAPs. The TSAP defines its own address at the layer boundary and is prefixed defining the appropriate layers. The **TSDU** is sent as message information across the network to peer entities of all the layers. The transport layer protocol data unit (TPDU) contains user's data and transport control information for peer-to-peer entities. The SDU and control information (addressing, error control, etc.) known as the **Interface Data Unit (IDU)** is message information sent by an upper-layer entity to its lower-layer entity via SAPs. The IDU contains transport interface data and session-to-transport entities' interface control information. For sending an SDU, a layer entity partitions the message information

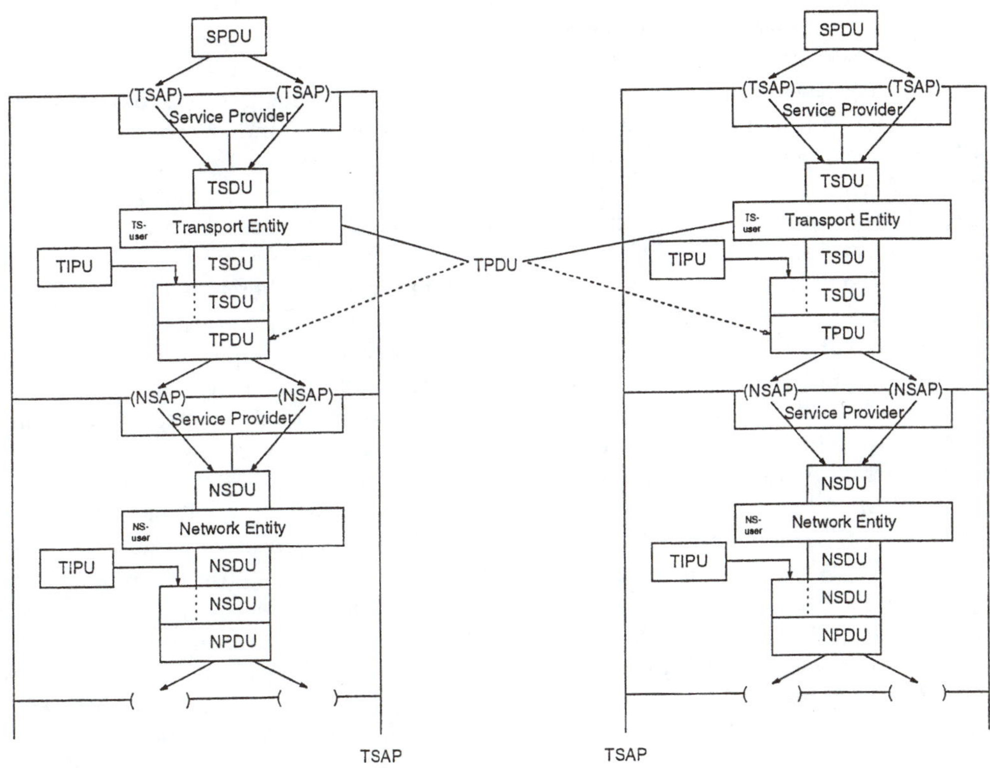

TIPU - Transport Interface Protocol Unit
TSAP - Transport Service Access Point
NSAP - Network Service Access Point
TS-user - Transport Service-user
TSDU - Transport Service Data Unit
TPDU - Transport Protocol Data Unit

Figure 13.1 Relationship between entities, TPDUs, and TSDUs.

into smaller-sized messages. Each message is associated with control information (IDU) defined by the appropriate layer as a header and sent as a **PDU**, also known as a **packet**. We use herein terms defined by ISO. The relationship between entities, TPDUs, and TSDUs is shown in Figure 13.1.

13.3 ISO 8072 transport layer services and functions

The OSI-RM poses some problems with higher layers (session, presentation, and application), as these layers are very complex and it is not transparently clear how to use them. The standards available offer very little help in this direction. The most common user requirements in using communication at present seem to be around file transfer, remote login, and remote job submission. In LANs, there are many applications which are derived from these mail requirements and are taking advantage of the high speed and reliability of LAN models to build coupled systems for dedicated applications. In such systems, the user requirements may include file transfer but will consist of interprocess communication. The processes or tasks running either in the same machine or in different computers need to pass information or activate the processes of one another. Such activation of processes (local or remote) is achieved by message passing or by sharing information in shared memory.

Looking at the functionalities and services of various layers of OSI-RM, a typical combination of network and transport layers usually provides internetworking and true end-to-end services to the users. A combination of session, presentation, and application offers a full user interface to the network, providing a synchronized dialogue (session), standardized user data representation (via presentation, common syntax), and common user semantics (application) between user application programs or processes. The lower four layers usually provide a transport service of required and acceptable **quality of service (QOS)**. It is generally felt among network users and professionals that the upper three layers (session, presentation, and application) should be considered one layer. Some efforts in this direction have already started with the notion of pass-through services, whereby all the services of the session layer are made available to the application layer (as opposed to the presentation layer in existing OSI documents). In addition to these efforts, the upper layer should have embedded primitives for request and response categories of these layers.

We will briefly discuss the objectives and various issues of the transport layer, and after explaining these basic concepts, we will identify services and functions offered by the transport layer (ISO 8072[6]) and connectionless services.[7]

One of the main objectives of the transport layer is to provide a reliable, end-to-end communication channel between source and destination user processes which are on different hosts. This type of communication channel is, in fact, between user processes and not between the user's host machines. The packets which move over different subnets may arrive at their destination out of order or may be lost during their transmission. Further, a link between the local host and interface message processor (IMP) or across IMPs may be broken. The protocol of the transport layer must provide error recovery, re-transmission through different types of subnets, and reasonable quality of service on different layers of networks. In other words, the protocol of the transport layer must provide a transport connection between user's processes with a reasonable quality of services on different types of networks. Some of the material herein has been derived partially from References 2–5.

13.3.1 *Transport QOS parameters*

The transport connection is defined between the session layer entities which are identified by the transport addresses. It encompasses different types of services which are negotiated at the time of establishing the transport connection. The types of services depend on different parameters such as throughput, transit delay, error rate, etc., and these collectively are defined as parameters in quality of service (QOS) at the global level. These parameters of QOS are typically used for all types of traffics coming out of the session entities. Once the expected or desired QOS is negotiated, the data transfer takes place over the transport connection in accordance with the mutually agreed upon QOS. In situations where the QOS is not maintained for some reason, the transport connection is terminated and the session layer entities are informed about this.

The parameters defined by the transport layer for QOS are transport connection establishment delay, transport connection establishment failure probability, transport connection release delay, transport connection failure probability, transport connection protection, transport connection priority, throughput, transit delay, and resilience of the transport connection. The following discussion of these is derived from References 2 and 3.

- **Connection establishment delay:** This delay is defined as the time difference between the time a request by a session entity for transport connection is issued and the time the connection request is confirmed (after it receives a Connection.confirm primitive) by the user of the transport service. This difference of time includes the processing time by both transport entities.

- **Connection establishment failure probability:** For this parameter an upper limit on the establishment delay (based on a number of parameters) is defined. This parameter then gives us the failure probability that the connection was not established within the limit. This failure may be due to many reasons, such as broken link, network traffic congestion, etc.
- **Throughput:** This parameter usually defines the number of bits, bytes, blocks, or packets transmitted over a period of time. This parameter may be different for different directions of transmission. It has been observed that the actual throughput is always less than that specified by the network due to reasons such as mismatching of interfaces, speeds, and capacity, as well as the fact that, for different options, control information increases the size of packets, which reduces the throughput. Throughput is usually defined as the average and maximum values for low-cost connection.
- **Transit delay:** This parameter gives the time difference between the time the packet is transmitted from the source transport user to the destination user, and this parameter includes the transmission and propagation delay and processing time at each end. This parameter must be defined separately for both directions of transmission.
- **Residual error rate:** This parameter is the ratio of the number of packets lost to the total number of packets transmitted over a period of time. This difference is caused by errors occurring in the network layer, and based on the ISO function of the transport layer, this error rate should be zero, as the higher layers usually provide layer protocol transparency and this error must hide the errors of the network.
- **Transfer failure probability:** As we said earlier, the transport connection is established only after the necessary negotiations for various parameters of QOS (transit delay, residual error rate, etc.) have been agreed upon, and it is defined as the ratio of unsuccessfully transmitted packets to the total number of packets transmitted over an established transport connection. This parameter gives us information regarding how many times the packets were not transferred within agreed QOS over the transport connection.
- **Connection release time:** This parameter is the difference between the time when a disconnect or release primitive is issued and the time the connection actually is terminated. The termination of connection is achieved via disconnect primitives and includes the processing times at both ends.
- **Connection release failure probability:** This parameter is also based on some upper time limit for termination of connection and is given by the probability that the connection was not terminated over a period of time for a number of tries through primitives. In other words, it defines the number of unsuccessful attempts when a connection was not terminated during the release time limit.
- **Protection:** This parameter allows the user to provide protection to transmitted data from being accessed by unauthorized users over the transport connection. Unauthorized users must not be allowed to modify the data.
- **Priority:** This parameter allows the users to define priority for their services so that, in case of network congestion or any other malfunctioning, the transport layer will give preference to high-priority services on its next or other transport connections.
- **Resilience:** This parameter allows the transport layer to terminate the connection by itself, and this usually happens in typical situations due to unmanageable traffic, traffic congestion, or some internal malfunctioning. It is usually defined as the probability of a transport connection termination being initiated by the transport service provider.

For establishing a transport connection, the transport layer receives a request from its session layer entity (SS-user) which defines its transport address. The transport layer fetches this request from its address and looks at the network address for a transport entity (TS-user) which offers a service to the correspondent session layer entity. The transport layer offers an end-to-end transport entity connection; it does not require any intermediate relay between the end-to-end transport entities. In other words, the transport layer, after receiving a request from the session layer, defines an end transport address. The end transport layer entities are derived from the network address onto which the transport layer maps from its transport address. The network address may support more than one transport address, and hence more transport entities are supported by the same network address which defines network entities. The mapping functions between transport and network addresses take place within the transport entities.

The mapping between transport connections and network connections uses multiplexing for efficient utilization of the network connection. The data transfer taking place over this established transport connection goes through three phases of connection-oriented service: connection establishment, data transfer, and connection termination or release. These phases are used with different functionality here during the data transfer.

As mentioned above, the transport connection is established between session entities, and the transport layer determines the network address corresponding to this transport entity which will provide services to the session entity. For this request, a network connection is established which offers a match in terms of cost and quality of services requested with the session entities. At this point, it must be decided whether multiplexing is to be used over the network connection. The size of the transport layer protocol data unit is determined using the mutually agreed upon parameters of QOS. The transport address is mapped onto the network address and a unique identification is assigned to various transport connections which are defined for similar TSAPs. The data transfer takes place over the established transport connection, after which this connection is terminated or released with the attached parameters such as reason for termination, release of identifiers of transport connections, etc.

The above was a discussion of general concepts, features, and objectives of the transport layer. We now discuss the functions and services of the transport layer separately below. It must be reiterated that the functions provided by the transport layer are the objectives of the layer, and some of the services provided by the network layer while the services of the transport layer are defined for its higher layer (session). Connection-oriented and connectionless services are detailed in References 6 and 7.

13.3.2 ISO transport layer services

1. Define the transport connection between user processes and establish it between session entities.
2. Transfer data between them over the established transport connection.
3. Close or terminate or release the transport connection.

13.3.3 ISO transport layer functions

1. Translate or map transport addresses and connections into equivalent network addresses and connections.
2. Define and release the transport connections between user processes and session layer entities.
3. Provide end-to-end sequence control for each of the transport connections.

4. Provide end-to-end error detection and error recovery, flow control for each connection, segmenting for controlling, blocking and concatenation, quality of service, expedited transport service data unit transfer, and other administrative and layer management functions (a subset of OSI layer management).

The transport layer protocol establishes a transport connection between user processes after the session entities between them have been defined. The services offered by the transport layer are then concerned with the definition of connection, controlling, and maintenance of the connection, and finally, after the data is transferred, release of the connection. The session entities (on different end host systems) support more than one transport connection between the user processes, providing different types of services for the users to choose from. Once the transport connection is defined between user processes, the synchronization of data flow between them is defined by the session layer. Further, the session layer protocol controls the transport connection and offers error recovery anywhere between the host and IMPs or across the IMPs.

A basic function of this layer is to perform a multiplexing function to achieve the necessary QOS at optimum cost. This means that several network connections may be required to provide the required rate of service for a single transport layer user, or one network connection may be sufficient to support more than one transport connection. Basically, this layer offers connection-oriented services, but ISO has defined connectionless protocols which are built on either connection-oriented or connectionless network services (ISO DP 8602).

The transport addresses are, in fact, transport service access points (TSAPs), while the network addresses are network service access points (NSAPs). The transport layer maps TSAPs into NSAPs where NSAP defines the unique end system address for each of the subnets. This address is transparent to the transport layer and is defined as a global network address. Once the connection between session entities is established, it is recognized by TSAP, which then may define more than one TSAP address between transport entities. The transport layer protocol translates such TSAP addresses into NSAPs for different network addresses between network entities.

13.4 Transport layer service control protocol

In order to understand the concepts and working principles of transport layer protocols, we first discuss various issues with different types of connections defined by the transport layer (connection-oriented and connectionless). The connection-oriented service has three phases: connection establishment, data transfer, and connection termination or release. The connection-oriented service is also known as **virtual circuit (VC)** service and establishes the logical transport connection before data transfer can take place. The transport connection is established between session layer entities, and more than one transport connection can be defined for any network connection. After the data transfer is complete, the transport connection is terminated or released.

In connectionless service, there is no establishment of transport connection, and it requires the data units to contain all the necessary information (such as destination address, source address, and other control information). This service is also known as *datagram service*, as here the datagrams containing the necessary control information are transmitted on the basis of stop-and-forward. ISO has developed standard documents for the transport layer known as ISO 8072 (connection-oriented transport services),[6] ISO 8072/DAD1 (connectionless transport services),[7] ISO 8073 (transport protocol for connection-oriented services),[8] and ISO 8073/DAD2 (transport protocol for connectionless services).[9]

13.4.1 Transport connection functions

When the transport layer gets a message from the session layer entity (through primitives), the transport layer package (entity) performs the following functions:

1. It interprets the parameters of QOS (requested by users).
2. It derives a set of transport packages and also the type of underlying network.
3. It matches the requested QOS parameters with those of the available transport package and selects the network service that provides the minimal cost. Meanwhile, it will also decide whether to multiplex multiple transport connection and act accordingly.
4. Based on functions 1–3 above, it defines an optimum size of TPDU.

Once the transport connection is established, the data can be transmitted in any mode (half-duplex or full-duplex) created by subnets. Flow control, segmentation, sequencing operations, etc., may be performed by the transport protocol for efficient utilization of the network, although the network layer protocol may also perform these operations.

As mentioned earlier, error detection is generally provided by the data link layer of OSI-RM, and CCITT has defined standards for error detection known as **checksum and cyclic redundancy check (CRC)**. These standards are very popular in the data link layer (or second layer from the bottom) of various OSI-based networks. In situations where the data link layer does not support checksum or does not indicate an occurrence of error to higher layers (in particular, to the transport layer), the CCITT checksum standard can be used in the transport layer. The cyclic redundancy check (CRC) is more effective in detecting errors, but the software implementation of CCITT checksum makes it more efficient than CRC. The CCITT standard documentation X.224[12] describes the algorithms for defining checksum parameters, as discussed below. The algorithm in the standards document is defined in such a way that the amount of processing required for each TPDU is minimal. The algorithm defines two checksum octets as *X* and *Y*, which are given by

$$X = -A + B \quad Y = B - 2A$$

where *A* and *B* are calculated using the following:

$$B = \sum_{i=1}^{L} a_i \ (\text{modulo } 256)$$

$$A = \sum_{i=1}^{L} (L + 1 - i)a_i \ (\text{modulo } 256)$$

where *i* represents the position of an octet within the TPDU; a_i represents the value of an octet in the *i*th position, and *L* defines the length of the TPDU in octets.

The two checksums are calculated using the following steps:

- Initial values of *A* and *B* are cleared and set to 0.
- The values of *A* and *B* are calculated by adding the values of *i* from 1 to *L*.
- At each value of *L*, the value of the octet is added to *A* and the new value of *A* is added to *B*.
- The values of *X* and *Y* are calculated using the above expressions.

These two checksum octets are calculated and inserted into TPDU, and this appended TPDU is transmitted. On the receiving side, the same algorithm is used with similar steps for calculating the checksum octets on the received TPDU. If either of the values of *A* and *B* is zero, it indicates the absence of error. If both *A* and *B* are non-zero, it indicates the occurrence of error, and the received TPDU is discarded. The values of *A* and *B* are defined within the range of 0–255, which does not offer any overflow, and carry is also ignored. Further, the checksum octets are calculated using one's complement.

The quality of service (datagram or virtual circuit) offered by the network/subnet layer greatly affects the protocol design and selection for the transport layer. If the subnet offers connectionless service (datagram), then the transport layer protocol (at the user's host) is very complex, as it has to perform all the tasks for error recovery, sequencing of user data, appropriate action in the event of a lost packet during transmission or duplication, etc. But if the subnet offers connection-oriented services (virtual circuit service), the transport layer protocol is very simple, as the above-mentioned tasks will be performed by subnets/networks (hosts and IMPs).

The ISO 8072 standard for connection-oriented transport services is the same as those defined by CCITT X.214 and X.225[11,12] for PSTNs and T70[13] for PSDNs. One of the main objectives of the transport layer is to provide end-to-end transport for a given quality of service with minimum cost and efficient utilization of the network service and network connection. The data transfer takes place over an end-to-end transport connection so defined and allows no restrictions on the format, information coding, or content of the user's data. The standard includes a set of parameters to be used for negotiating during the establishment of transport connection, as defined in the QOS (discussed above); these include transport connection establishment delay, transport connection establishment failure probability, transport connection release delay, transport connection failure probability, transport connection protection, transport connection priority, throughput, transit delay, and resilience of the transport connection.

13.4.2 Connection-oriented services

Various facilities under end-to-end communication are very important, as the transport layer protocol provides a transparent interface between the host and subnet; i.e., the upper layers (host part in OSI-RM) and users have no knowledge about the functions and services offered by the lower layers (subnet part in OSI-RM). The users are not concerned with how their various operations are implemented on the networks which provide these services to them. The transport layer service may define the services offered by lower layers, where upper layers have no information about any technical aspects of communication taking place in the lower layers. If the services offered by subnets change, the transport protocols have to be changed accordingly without affecting higher-layer protocols.

The transport layer addresses are defined by various **TSAPs**, or **ports** or **sockets**, which are handled by operating systems used at the hosts. The socket concept was initially used in the UNIX operating system and represents OSI TSAP. The socket is the endpoint between the connections from the operating system, and the processes from users can also be connected or attached for communication. In other words, the messages corresponding to control and data are going through these sockets. The sockets can be defined for connection-oriented or connectionless services. The operating system usually defines the endpoints and then the socket system call defines a socket (data structure within the operating system), and a buffer can be allocated to each of these sockets. The buffer is used to store incoming messages. The listen system call allocates this space, and the socket defined by listen becomes an inherent endpoint, and since it uses listen call, it is always in listening mode waiting for a request to arrive. Each of the sockets (corresponding to TSAPs) is assigned a name using another system call *blind*. A few popular system calls used in the UNIX operating system include accept, connect, shutdown, send, receive, select, etc.

The operating system in the hosts provides a library of system call primitives which may be used to define a communication channel between transport entities. Each transport address (or port/socket defined by the operating system) defines a network number, destination host number, and port/socket assigned by the host. The number of bits for each of these fields is different depending on the transport layer protocol standards being used.

The services offered by the transport layer (connection-oriented, reliable, and efficient) are transparent to the lower layers and should not depend on the types and characteristics of the networks. Before any data can be transferred between users, they must both agree on the type of service, network connections, etc., by transmitting the request packet back and forth between them. The transport layer knows the type of service being offered by its network layer. Based on the user's request for network connection and the quality of service (QOS) desired, the transport layer selects appropriate library subroutines in the transport protocols which satisfy the user's QOS. The variables of QOS are then used by the network layer to access the set of procedures in the protocols.

Standard protocol X.25 (CCITT) for connection-oriented services: The X.25 standard[14] provides connection-oriented service and defines a virtual circuit interface. In reality, X.25 does not support connection-oriented service. In order to use X.25 and expect a good quality of service, higher layers, in particular a transport layer, must include error recovery and a reliable communication channel and must contain these facilities in their respective protocols. The network addresses are used by the transport layer to define the transport connection. Each of the network addresses may either use a separate IMP line number or share an IMP line by multiplexing different transport connections (defined in different ways by different X.25s) into a single virtual circuit. These virtual circuits can then be multiplexed into one physical circuit. The difference between the two types of multiplexing of transport connections lies in the type of final virtual circuits which transfer the user data onto the transmission link/medium. In the first type (based on the use of a separate IMP), we are multiplexing multiple transport connections into various virtual circuits at the transport layer, while in latter type (based on sharing of IMP), we are multiplexing various virtual circuits into one physical circuit at the network layer. On the receiving side, the header of the frame indicates the network number (defining the virtual circuit at a given NSAP) and it is passed up to the transport layer.

The difference between connection-oriented and connectionless networks is influenced by the QOS offered by subnets. ISO 8073 has defined a standard connection-oriented transport protocol (virtual circuit service) which offers five classes of service at the transport layer, and users may choose any class of service (depending on the QOS) of the network being used.

Connectionless service (datagram) across the subnets can be obtained by using the U.S. Department of Defense's (DoD) Internet protocol (IP).[19] The DoD's transmission control protocol (TCP)[20] can be used to obtain an end-to-end, reliable communication channel which will be provided by the host machine.

CCITT has also defined its own transport layer which provides reliable, end-to-end services to the users and also to the upper layers. The upper layers have no information about the type of network used and how the communication is taking place across them. Various functions of the transport layer (CCITT) are discussed in X.224, while X.214 discusses how the services of the transport layer can be accessed by the session layer.

13.4.3 Types of transport service primitives (ISO and CCITT)

The services of the transport layer are implemented by transport protocols and offered to its higher layer (session). The transport layer protocols also support error control and flow control and seem to resemble data link layer protocols. These two protocols (data link and transport layer) are not the same, but due to some common features between them, they are sometimes compared with each other. There are many differences between them. The data link layer defines a frame by attaching a header and trailer around the packet, and error control and flow control are provided at the frame level, while the transport layer attaches a header and these services are available at the packet level. The data link provides communication between two IMPs over a physical channel, while the transport layer protocol offers services over the lower three layers (subnet composed of IMPs). The data link layer does not expect IMPs to know which IMP they are going to communicate with, while the transport layer protocol expects an explicit address. The network services of transport protocols are generally divided into three types: A, B, and C. The quality of service offered by a network configuration (connection-oriented or connectionless), in general, affects the end users and obviously the transport layer. As such, the following types of networks have been recommended by both ISO and CCITT.[2-5]

Type A: The network connection of this type supports an acceptable error rate and an acceptable rate of signaled failures or errors (defining the quality). The transport layer will not provide error recovery or any resequencing services, as it assumes that no packet will be lost. This type of network is supposed to provide a very small error rate (small number of lost or duplicate packets), and the use of the primitive N-Reset is very rare. Some LANs may offer this service and hence are known as Type A LANs. The transport protocols in this type of network are very simple and easy to implement.

Type B: This type of network service assumes a small error rate, and also the loss of packet is minimal. The loss or duplication of packets is usually recovered by data link and network protocols. The errors in this type of network are caused by the loss of packets or re-transmitted packets (after receiving Nak by the sender). The errors which are handled by network entities are known as error rate, while if transport entities are used for error recovery, the error is known as a *signaled failure*. The network connection of this type supports an acceptable error rate and an acceptable rate of signaled failure (e.g., disconnect or reset, such as in X.25). Error recovery has to be supported by the transport layer. In this type of network, the network layer uses the N-Reset primitive quite often to make sure that the network maintains an acceptable error rate and signaled failure rate. In this type of network, the transport protocols are more complex than in type A because the N-Reset primitive is used quite often. The public networks (X.25) are usually considered Type B networks.

Type C: The network connection of this type supports an error rate not acceptable to the user (transport layer), yielding unreliable service. The transport layer has to support both error recovery and resequencing. Connectionless (datagram) networks, packet radio networks, many other internetworks, and LANs fall into this category of networks. The transport protocol is generally very complex in this type of network, as it has to deal with all the situations that are present with data link layer protocols. It offers network connection with a residual error rate not acceptable to transport service users.

These networks usually provide a reasonable level of service and offer various configurations. For example, Type C networks support transmission in connectionless mode, while Type A networks support reliable network service. Type B networks may use X.25 protocol.

The transport layer services define various parameters, which in turn describe the type of services required. Each parameter (defined in QOS) is associated with a set of parameters (or predicates) which define the actions and events to be offered by the networks. These parameters associated with primitives are used by the users for mutually agreeing on the quality of service requested during the establishment of connection between them. During this process of negotiation, users may keep interacting with other users by going to various levels until a mutual agreement is established. In the event of no negotiations, the default services offered by the network will be used. After mutual agreement on the QOS, the user data is sent through its transport layer up to the physical layer, then to the transmission media, and finally through the physical to the transport layer of the end user.

13.4.4 *Classes of ISO 8073 transport protocol*

As indicated earlier, the transport layer defines transport connection between user processes on hosts over which data is transferred. After the data has been transmitted, the transport layer releases the transport connection. For all of its operations (establishing a transport connection, data transfer, and release of transport connection), the transport layer uses the commands of its session layer. Each user process is assigned a TSAP or ports (in DoD's terminology) through which the data can be transmitted as transport protocol data units (TPDUs) or segments. The QOS, based on various parameters such as throughput, error rate, reliability, transit delay, response time, data rate, etc., is defined in ISO 8073.[8]

We have discussed three types of networks, and the transport layer knows the characteristics of these types and can select the following five classes of standard protocol packages to meet the requirements of the user's requested QOS. The three types of network services and three types of networks may represent different situations for different applications, and hence the transport layer protocols will be different for each of these situations. For worst-case network services, the complexity of the transport layer protocol increases. For these reasons, the ISO defined five classes of transport protocols — Class 0, 1, 2, 3, and 4 — representing these different situations with increasing level of functionality. These classes are usually grouped as (0, 1) and (2, 3, 4).

Class 0 (simple class): The protocol of this class offers a very simple transport connection-oriented package during defining and releasing of the network connection. This class of package may detect errors but does not correct the errors; i.e., it does not support error recovery. Each network connection establishes a transport connection, and it is assumed that the network connection does not introduce any errors. In the event of error, if the network layer detects the error and does not inform the transport layer, then the network may correct it and the network connection remains active. But if the transport layer has been signaled, then the transport layer releases the connection and informs the user about it. The transport layer protocol does not support any sequencing or flow control, and it assumes that these are provided by the underlying layer protocol (network). However, the establishment and release of the transport connection is provided by this protocol. It is useful for Type A networks and also for X.25 networks, and it is defined as a telex standard by CCITT.

Class 1 (basic error-recovery class): This class package supports the numbering of TPDUs, segmenting of data, and acknowledgment. This class of protocol recovers from the N-Reset primitive to the network connection for which a transport connection has been restarted. The network connection will restart from the sites of transport entities (after resynchronization). This is achieved by keeping a record of the sequence number during resynchronization, and as such no packets are lost. It also does not support any error control or flow control and assumes that the underlying network layer protocol provides these. If an X.25 reset packet is sent, then it provides automatic recovery from the following situations: (1) network disconnect, (2) resynchronization, and (3) reassignment of the session (network connection). It offers expedited data transfer in the event of prioritized data where the transport layer protocol requests the network layer for moving that data ahead of other data in the network queue. The sequencing of TPDUs helps for acknowledgment (Ack) or negative acknowledgment (Nak) and for error recovery. This class of package supports Type B networks. It only allows recovery from N-Reset primitives with no support for error control and flow control.

Classes 2 through 4 are discussed below in increasing order of functionality with respect to Class 1.

Class 2 (multiplexing class): This class package offers the multiplexing of a number of transport connections into one network connection. It is designed to be used with reliable networks (similar to Class 0). It allows multiplexing of a number of transport connections over the same network connection. It provides flow control of an individual transport connection which pipelines all the data TPDUs, thus avoiding any traffic jam at DTE sites. The flow-control algorithm may use either the concept of window (no piggybacking) or the concept of credit allocation. In the former (window), a separate acknowledgment is required for each packet, while in the latter (credit allocation), only one acknowledgment for the packets in the credit is required. Another application of this class of protocol is in airline reservation terminals. Here each terminal defines a transport connection with a remote computer and there may be a few dedicated network connections defined between terminals and the remote computer. All of these transport connections may be multiplexed over the network connections, thus reducing the cost of the network. It does not support error detection and correction. If an X.25 reset or clear packet is transmitted, the session (network connection) between user processes will be released (or disconnected) and the users will be informed about the release of the multiplexed transport connections. A Type A network with high reliability may find this class of package useful.

Class 3 (error recovery and multiplexing class): In addition to classes 1 and 2, this class of package provides the multiplexing and flow control of Class 1 and error recovery from network failure (disconnect and reset) of Class 2. It supports both multiplexing and recovery from the N-Reset primitive. The recovery from errors is signaled by the network layer (as the network layer has clear and reset capability services). The only difference between Class 3 and Class 1 for error recovery is that in Class 3, the users are not informed about the network disconnect. Instead, Class 3 stores the user data until it receives acknowledgment from the transport layer. The Class 3 package uses the same technique for flow control as the data link layer for "timed" packets with acknowledgment; i.e., the packets requiring acknowledgments are timed out, and if it does not receive the acknowledgment during that time, it re-transmits the same packet, and the error-recovery technique may be used in conjunction with it. Interestingly, this type of flow control is not part of Class 3, but it supports it; however, users have to request it. Type B networks may use a Class 3 package. It uses explicit flow control.

Class 4 (error detection and recovery class): This class offers the same flow-control mechanism used in Classes 2 and 3 and supports detection and correction (i.e., recovery) of errors due to (1) lost packets, (2) duplication, and (3) packets delivered out of sequence at the receiving site. The package provides expedited data and also acknowledgment from the receiver when it receives a valid packet (TPDU or segment). It also provides the same flow control as that of Class 3; i.e., in the event of a lost or damaged packet, the same packet is re-transmitted after a timed-out interval. Type C networks usually use this package. The Class 4 package is generally used for a low-quality network and hence is best suited for connectionless networks. It is capable of handling situations such as lost packets, duplicate packets, N-Reset, and many other situations which network layer protocols do not handle. Obviously, the Class 4 transport layer is very complex and takes a lot of burden from the network layer protocols.

This protocol is also known as TP4 and possesses some similarities to the transmission control protocol (TCP). The transport protocols of both TCP and TP4 are based on reliable, connection-oriented, end-to-end communication and assume that the underlying network layer protocols do not provide error control and flow control so that the transport protocols have to deal with those controls. These protocols have some differences which are discussed in the following section.

The choice of a particular class of protocols depends on the application and is usually negotiated during the establishment of connection. The sender makes a request for a class of protocol or more than one class of protocol, and it is up to the receiver to accept a particular protocol. In the case where the receiver does not respond for any class of protocols, the request for transport connection is rejected or discarded.

13.5 ISO 8073 transport layer protocol primitives and parameters

Various transport layer primitives ISO 8073[8] and ISO 8073/DAD2[9] are shown in Table 13.1. These primitives are defined for session layers through which the session layer can request various transport services.

13.5.1 ISO connection-oriented transport protocol primitives and parameters

For connection establishment, the local session layer or upper layer sends a T-CONNECT.request primitive to the transport layer entity at the TSAP to indicate its intention of sending the data. The parameters defined in this primitive are used to generate a transport connection request and TPDU. The transport layer sends a primitive T-CONNECT.indication containing the parameters for QOS to the distant session layer. The distant session layer entity then sends a primitive T-CONNECT.response back to the originator, which includes the parameters for the confirmed TPDU. The TPDU goes through transport and network layers, which define equivalent parameters in the primitive T-CONNECT.confirm. This primitive provides an acknowledgment for the establishment of connection to the local session layer. Figure 13.2(a–d) shows the exchange of primitives between two transport entities for different operations.

After establishment of the connection, data is transferred from the local session layer entity by using T-DATA.request. The distant session layer entity receives this data at its TSAP as a T-DATA.indication primitive. The expedited data can also be transmitted using T-EXPEDITED.DATA.request and T-EXPEDITED.DATA.indication, respectively. The Expedited category of data has priority over the normal data, and this data is transferred over the transport connection before the normal data. For sending this type of data, the primitives of expedited data are used which indicate the category of data.

At the end of the data transfer session, the connection can be de-established (disconnected) by either side's transport layer entity by sending a T-DISCONNECT.request primitive.

Table 13.1 Transport Layer Protocol Primitives and Parameters

Primitives	Parameters
	Connection-oriented
Transport connection establishment between two session layer entities	
T-CONNECT.request	Destination address, source address, expedited data facility, QOS, transport service user data (TSDU)
T-CONNECT.indication	Destination and source addresses, expedited data facility, QOS, TSDU
T-CONNECT.response	Responding address, QOS, expedited data facility, QOS, TSDU
T-CONNECT.confirm	Responding address, QOS, expedited data facility, TSDU
Data transfer between two transport entities	
T-DATA.request	TSDU
T-DATA.indication	TSDU
T-EXPEDITED.DATA.request	TSDU
T-EXPEDITED.DATA.indication	TSDU
Release or termination of transport connection	
T-DISCONNECT.request	TSDU
T-DISCONNECT.indication	TSDU
	Connectionless
T-UNITDATA.request	Destination and source addresses, QOS, user data
T-UNITDATA.indication	Destination and source addresses, QOS, user data

The transport layer entity of either side will terminate the transport connection by sending a T-DISCONNECT.indication primitive. A reason parameter is defined in the T-DISCONNECT.indication primitive, which offers the reason for disconnection. There may be several reasons for the disconnection, such as no mutual agreement on the QOS, no resources available on either side, unexpected termination of the connection, no data transfer, and others. For connectionless services, the connectionless service primitives as discussed above are used.

For connectionless services, there are two primitives known as T-UNITDATA.request and T-UNITDATA.indication, and these primitives allow users to send data before establishing any transport connection. The primitives include destination and source addresses along with the parameters of QOS. The source node transport layer entity sends a primitive T-UNITDATA.request to the distant transport layer entity, which receives the primitive as T-UNITDATA.indication, and there is no acknowledgment for the source transport entity.

In the above discussion, the transport layer protocol does not provide the acknowledgment of data (TPDU). The acknowledgment (Ack) of TPDU can be implemented either by requesting Ack for each TPDU received by the distant session layer or by requesting only one Ack at the end of the last TPDU received. These facilities are available in the transport layer protocol classes above.

13.5.2 Transport protocol data unit (TPDU)

The transport layer protocol consists of ten types of transport protocol data units (TPDUs), where each TPDU performs a specific function. The structure of these functions is the same and includes four fields, as described below and shown in Figure 13.3.

The **length indicator field** defines the length of the TPDU header in octets. The maximum length of this field is 254 octets. The **fixed part of the control header field**

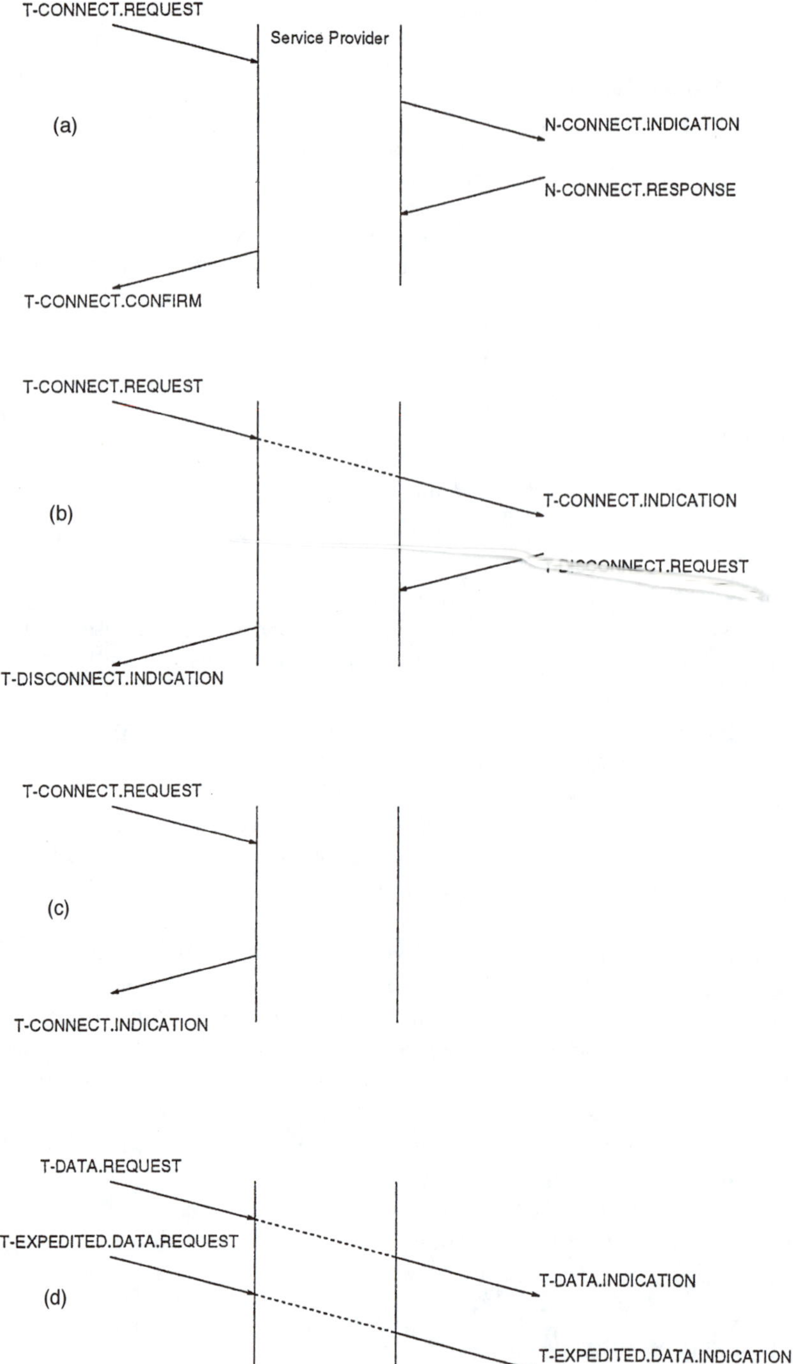

Figure 13.2 Exchange of primitives between two transport entities. (a) Connection establishment. (b) Destination transport user rejects connection request. (c) Transport layer rejects connection. (d) Normal and expedited data flow.

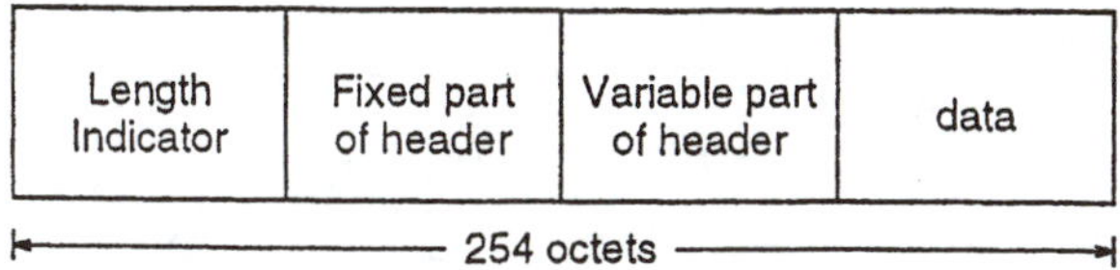

Figure 13.3 Structure of TPDUs.

defines the type of TPDU parameters which are used frequently. The **variable part of the control header field** defines the type of TPDU parameters which are used less frequently. Finally, the **data field** defines the user data coming from the application layer of OSI-RM. The majority of public networks use connection-oriented OSI transport service (ISO 8072) and OSI transport protocols (ISO 8073). See Section 13.4.4 to review the five classes of ISO 8073 transport protocol (classes 0–4).

Types of TPDUs: There are ten TPDUs that are used to exchange control information between transport entities over network connections and carry transport protocol information and transport user's data.[2-4] Each of these TPDUs uses the usual primitives to indicate an appropriate exchange of information between entities. These TPDUs include

Connection request (CR): Available in all classes; used to establish connection (used by source at source reference).
Connection confirm (CC): Available in all classes; used by destination for connection identification; adds destination reference.
Disconnect request (DR): Available in all classes.
Disconnect confirm (DC): Not available in class 0.
Data (D): Available in all classes.
Expedited data (ED)
Acknowledgment (ACK)
Expedited acknowledgment (EACK)
Reject (REJ)
TPDU error (TPDUER): Available in all classes.

To establish a transport connection, the source transport entity sends a request primitive TPDU (CR) and peer replies with connection confirm (CC) primitive. The termination of transport connection is issued by either of the users (source or destination) by using DISCONNECT.request, and the peer replies with DISCONNECT.indication. There are two types of data transfer TPDUs: DATA and EXPEDITED.DATA. Similarly, there are two types of acknowledgments: DATA.ACKNOWLEDGMENT and EXPEDITED.ACKNOWLEDGMENT. The REJ and EACK TPDUs are concerned mainly with error handling.

The control field header (both fixed part and variable part) is used for negotiations on QOS (between two end users) and includes parameters such as flow control, establishment and termination of connection, etc. The fixed part of the control field header defines a specific type of TPDU used. As mentioned above, each of these TPDUs has a fixed part and variable header field length indicator and user data field.

For understanding the workings of the TPDUs, we consider the formats of only three TPDUs (CC, DR, and DC), as shown in Figure 13.4(a–c). The formats of other TPDUs are similar and not discussed here, but they can be found in References 2 and 3.

The length indicator (LI) of one octet represents the length of the header. The fixed part of the control field header defines connection confirm (CC) in the four higher-order

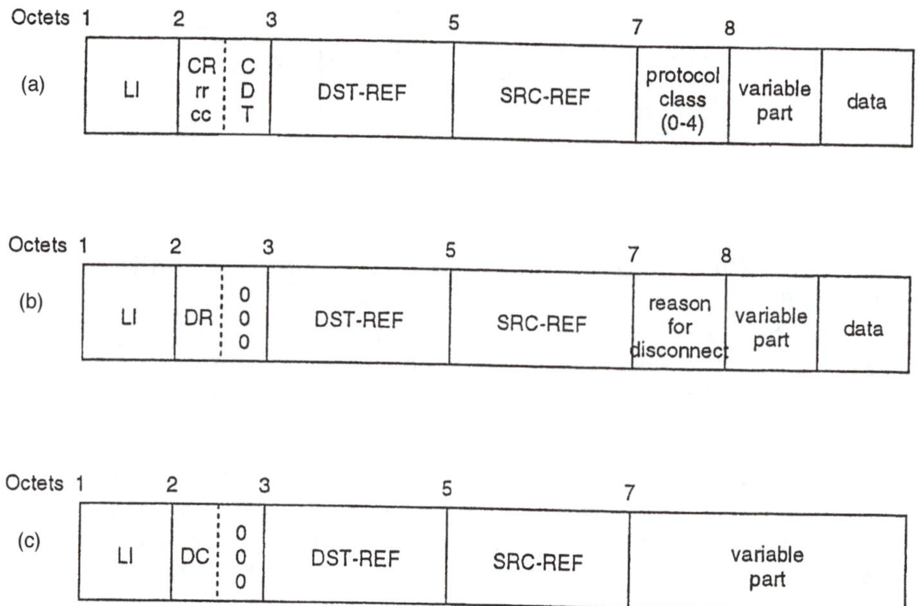

Figure 13.4 (a) Connect request (CR) or connect confirm (CC) TPDU. (b) Disconnect request (DR) TPDU. (c) Disconnect confirm (DC) TPDU.

bits, while the lower four bits are used to define credit allocation code (CAC). The destination reference (DST-REF) of two octets is initially set to 0s during connection establishment (CR) and is used by the peer transport entity to define a unique transport connection. The source reference (SRC-REF) of two octets is used by the local transport layer entity to define a unique transport connection which issues connection request. These addresses (SRC-REF and DST-REF) are used for data transfer between two transport entities. The variable part of the control request packet defines the complete source and destination address and is used in the setup connection request TPDU. These addresses (DST-REF and SRC-REF) are, in fact, the transport entities at TSAPs.

The next octet in the fixed part of CR defines various transport protocol classes (classes 0–4; see Section 13.4.4) may contain other facilities such as the flow control mechanism, etc., and is of four bits. An option field of four bits is used to define explicit flow control in different combinations, as a seven-bit sequence number and four-bit credit, or as an extended 31-bit sequence number and 16-bit credit. It also allows the use of explicit flow control in a Class 2 protocol. The variable part of the CR header defines various parameters of QOS and also includes the parameters in T-CONNECT.request which are translated into an N-CONNECT primitive along with those parameters. The parameters defined in the N-CONNECT.request primitive are used by the lower-layer protocol to define PDU.

For de-establishing the transport connection, disconnect request (DR) and disconnect confirm (DC) TPDUs are used (Figure 13.4(b,c)). The structure of these two TPDUs is the same as that of CR and CRC TPDUs. The four higher-order bits of the first octet of the fixed header contain DR, while the remaining four lower-order bits are initially set to 0s. The last octet of the fixed part contains the reason for the disconnection request. The variable part of the DR TPDU defines the parameters of various options and is usually specified by users. The user data field contains various control field information of the upper layers. The maximum size of the user field is 64 octets. The DC TPDU may include

Figure 13.5 (a) Data (DT) TPDU. (b) Extended (ED) TPDU.

checksum in the variable part of the header, while other fields contain the same information as that of DR except that DC TPDU does not have a data field, and the four higher bits of the first octet of fixed part defines DC.

The data TPDU and extended data TPDU of the transport layer (Figure 13.5(a)), as defined by ISO 8073, have the same structure as discussed above for other TPDUs and are influenced by a number of options such as classes 0–4, extended facility with classes 2, 3, and 4, facility of expedited TPDUs for classes 1–4, etc. The four higher bits of the first octet of the fixed part define the data TPDU (DT) (Figure 13.5(b)) or expedited data (ED), and the remaining four lower bits are set to 0s. The DST-REF field of two octets defines the destination TSAP of the remote site. The last octet of the fixed part contains two parts: the four higher bits define EOT bits (set to 1111), indicating the last data sequence transport sequence data unit (TSDU). The remaining four lower bits define the sequence of the TPDU as (TPDU-NR) for each of the TPDUs sent. The number of octets for the normal structure of the TPDU is five octets, while for the extended structure, it may be between five and eight octets. The DST-REF defines the transport layer entity. The variable part defines the two-octet checksum along with parameters depending on the protocol classes used. Finally, the data field contains user data, and the maximum size of this is 16 octets, particularly in the case of extended data (ED) TPDU.

The variable part of the header defines additional parameters to be used in CR or CC TPDUs during connection establishment. Some of these parameters are calling (source) TSAP ID, called (destination) TSAP ID, TPDU size (from 128 octets to 8192 octets), version number (of protocols), checksum, acknowledgment time, throughput, residual error rate, transit delay, flow-control information, and invalid TPDU.[3]

13.5.3 Design considerations of transport layer protocol primitive services

As discussed earlier, the type A (defined by ISO and CCITT) network connection supports an error-free and acceptable rate of signaled failure or error (defining quality of service). The transport layer will provide error recovery or any re-sequencing services, as it assumes that no packet will be lost. The design of the transport layer is greatly influenced by the type of services of the network layer, e.g., size of packets, reliability, etc., and different types of transport layers can be developed and used. Some of the services can be assigned to the network layer to make the design of the transport layer less complex. For example, the X.25 protocol offers reliable service. Here, the network layer services include processing of messages of variable length with optimal reliability and also the delivery of the packets in sequence to the destination. In such a case, the transport layer will provide services such as addressing, flow control, and connection establishment/de-establishment. An essential function of the transport layer is multiplexing, which allows multiple users to share the same transport layer; users are identified by TSAPs assigned to them by the transport layer protocol.

As described earlier, there are three types of network services defined by ISO, and in most of these networks, an **error** is considered a major issue in the design of the protocol. The error is defined for either loss of packets or duplication of the packets. The protocols provide support for error control in a variety of ways. In one case, the network service allows error detection, while in another case, it will also correct the errors. In these two cases, errors are reported to the transport entities, and this is considered a network service to the transport. In a special case where the network service can detect the error but does not provide any recovery, the transport protocol has to be very complex to handle the situation and is known as **signaled failure.**

In the following sections, we discuss *reliable* and *unreliable network services*. Within each type of service, there are different issues to be handled by the transport layer protocol. We will also discuss how errors are handled by the protocols these two types of service.

Connection-oriented (reliable network) service: In reliable service, the network is assumed to be reliable such that it accepts messages of variable length and will provide reliable delivery of the packet. Further, to make the protocol more practical, different situations may be considered during the design of protocols. For example, we may start with sequencing or non-sequencing of packets to be accepted by the network service, limit on the size of TPDU for network service, and other network performance parameters such as multiplexing, flow control, support for block-oriented and stream-oriented traffic, etc. Each of these issues will be discussed below.

When user A wants to communicate with distant user B, user A's transport entity establishes a connection (based on either connection-oriented or connectionless services) with B's transport entity. A unique TSAP is assigned to each user such that the user can access the transport services via this TSAP. The transport layer transmits this TSAP to B's transport entity. This information is defined in a transport layer header and passed to the network service access point (NSAP). The network layer establishes routing and transmission of the packets for transmission to the data link layer, then to the physical layer, and finally on to the transmission media. The main problem with addressing is how to know B's transport layer address. This problem can be solved by considering static and dynamic strategies (discussed in Chapter 12).

As stated above, the transport layer protocol also performs the multiplexing operation for any network service in such a way that in one direction, multiple connections can be implemented on one connection, while in the other direction, the multiplexed connections are de-multiplexed from a single connection into multiple connections. The multiplexing of the transport connection in both directions improves the throughput of the network.

If the network layer does not offer sequencing services for TPDUs arriving at the destination site, the problem becomes more complex. The transport layer not only has to support sequencing for connection-oriented services (in particular flow control), but it also has to control the data flow in the transport entity. In other words, the transport entity will be forced to send the packets in a sequence to its network layer. Further, the transport entity is now required to keep a record of control and data TPDUs. During connection establishment and termination, the control TPDUs between transport entities establish and de-establish connections using primitives, and it is possible that the acknowledgment and flow-control TPDUs regarding the availability of resources, buffers, etc., which are transmitted may have been lost, or there can be a mismatch in different types of TPDU.

This means that the local transport entity (after having sent TPDUs) might get information (resources available, buffers, etc.) at different instants of time, which will cause a major synchronization problem. The solution to this problem revolves around sequencing of TPDUs; otherwise, some of the packets may be lost during connection establishment and termination. In other words, these packets have to be sequenced, and a record of this sequence has to be maintained until the packets are received by the receiving node.

If the size of the TPDU packet is limited, the transport entity will process packets of fixed size. In this situation, the main concern lies in identifying the two types of TPDU data packets formats: stream-oriented and block-oriented. In the stream-oriented data format, the transport entity accepts data from users as a continuous stream of bits and delivers to the transport entity of the distant site without knowing anything about the limits and delimiters of the packets. In the block-oriented format of a data packet, a block of data (also known as a transport data service unit, TDSU) is transmitted. If the size of the TDSU exceeds the maximum size of the TPDU, the transport entity will break the TDSU into a fixed size before the data is transmitted. On the receiving site, the TDSUs coming in out of sequence are reassembled by the transport entity before the data is given to the users. Here, we have assumed that the TDSUs are arriving out of sequence. This problem can be solved by assigning sequence numbers to the blocks of TPDUs (of maximum size) within the TDSU. This will increase overhead, as various TPDUs are already assigned with a sequence order by the sequencing function of the transport entity.

Depending on the application under consideration, any category of reliable network services with sequencing or without sequencing of TPDUs, a connection-oriented service for all types of networks, i.e., types A, B, and C (already discussed earlier) can be selected. For example, for voice, electronic mail and bulletin boards, share-market data, reservation systems, etc., we may use non-sequencing with a limited size of packet, while sequencing with an unlimited-size transport protocol may be used in X.25 networks. Although reliability and error control for a connection-oriented network are provided by the network layer service protocol, it is possible that due to network failure, the network connection may be reset or cleared. This will result in the loss of TPDUs. The network layer entity will inform only that transport layer entity whose TPDU is lost. The error recovery from this situation has to be dealt with by the transport entity, which can use the sequencing scheme to detect it. Usually, the transport entity does not expect an acknowledgment from the distant transport entity, but in reality, as soon as the network connection is cleared or reset, the network layer entity informs its local transport entity about this event. The transport entity sends a control TPDU to the distant transport TPDU about the failure of the network connection and also the last TPDU it received from or sent to the distant transport entity.

Unacknowledged connectionless (unreliable network) service: The above discussion on the types of networks and various problems such as sequencing, size of TPDU, reliability, etc., was based on the assumption that the network layer provides acceptable flow control and error rates. This assumption means that the network layer by and large provides a reliable service to its transport layer. If the network layer does not provide reliable services to its transport layer, then the transport layer protocol must handle the situations related to unreliable services, such as re-transmission of the lost packets and lost acknowledgments, receiving of duplicate packets at the receiving site, error recovery from hardware failure, and related situations such as flow control and connection establishment and termination (as discussed in connection-oriented services). Some of these issues are discussed in Chapter 11 (data link layer).

The transport layer protocols are based on various concepts such as windows, credit allocation, etc. The discussion on window-based protocols can be found in Chapter 11. In the credit allocation technique, the receiving site sends a credit allocation number to the sending site. If the receiving and sending sites are not synchronized in terms of credit allocation at the receiving site, some of the packets may be lost. The solution to this problem is to assign a sequential number to each of the credit-allocation-numbered packets sent by the receiver. For each of the problems, a timer may be used to provide proper functions of the transport layer, e.g., re-transmission timer, reconnection timer, window timer, inactivity timer, etc.

The transport layer protocols support both types of configurations: connection-oriented and connectionless, and standards for both of these have been defined. The ISO 8073 and CCITT X.224 are the standard protocols for connection-oriented configurations, while ISO 8602 defines a standard protocol for connectionless configuration within the transport layer protocols. The ISO 8073 offers various services and functions which are used by both higher-layer protocols (user's requests) and network layer protocols. As detailed in Section 13.4.4, five classes (0-4) have been defined by ISO for the transport layer protocol: Class 0 (simple class), Class 1 (basic error recovery class), Class 2 (multiplexing class), Class 3 (error recovery and multiplexing class), and Class 4 (error detection and recovery class). A detailed discussion on these classes can be found in the preceding section.

13.5.4 Transport operational method

There are three main operational steps in the transport layer protocol. The set of primitives and parameters for this standard have already been discussed. The three main steps for transport operation are

- Connection establishment
- Data transfer
- Connection termination

Various design issues and implementation details of each of these steps will be discussed in more detail for both configurations — connection-oriented and connectionless — in the following sections.

13.5.4.1 Design issues in connection-oriented configuration

Connection establishment: In the first step, transport connection is established between two user processes which requires the use of a primitive by the session layer at the source site (which wants to transmit the data message information). The mutual agreement on services, options, classes, size of TPDUs, etc. is contained in the CC field of the TPDU. The called site may use either CC or DR for its response in the TPDU. Various parameters used in the control packet are shown in Table 13.1. The parameters of CC at the source site defines the characteristics of requested options, classes, etc., while the parameters of CC or DR of the destination site represents acceptance or rejection of the parameters defined by the source site. Once both sites have mutually agreed on the parameters of QOS, a transport connection is established between those user processes (transport entities) corresponding to the transmitting and receiving sites. At the same time, the transport entity assigns a unique service access point (SAP) address to the user as a TSAP, so that multiple users with unique TSAPs can be multiplexed. This information is also included in CC and CR TPDUs and is passed on to the network entity.[2–5]

The network SAP receives a packet from the transport entity which includes the address or name of the system on which the transport user is connected. This address allows the network layer to identify the destination address of the system on which the transport entity is connected. This address (transport) is needed only for the local network entity NSAP and is not sent in TPDU. Once the connection is established, each transport connection is assigned an identifier which will be used for all TPDUs between those transport protocol entities. By doing so, we can multiplex more than one transport connection (by their appropriate entities) on a single network connection.

During the establishment of the transport connection, the source transport entity must provide the address of the destination transport entity to which a connection establishment request must be attached. The transport layer offers its services to session layers at its TSAP, and this TSAP is the address to which the primitives can be attached. There is one-to-one mapping between the TSAP and NSAP. The source transport user issues a request primitive, T-CONNECTION.request, by attaching to the local TSAP. This address of TSAP is, in fact, a socket and is usually supported by the operating system. It also specifies the destination TSAP. The transport entity on the local machine selects the NSAP of the destination machine and establishes a network connection.

The transport entities of source and destination machines interact with each other over this network connection. Once the network connection is established, the transport entity of the local machine establishes a transport connection between the source and destination TSAPs. When the destination machine sends a T-CONNECT.indication primitive, the transport connection is established. The TSAP address is usually a hierarchical address where a number of fields are defined within the address (X.121). With this addressing scheme, it is possible to locate the TSAP of any machine by defining the TSAP as a concatenated NSAP address and a port such that whenever the transport entity provides a TSAP of the destination machine, it uses the NSAP address defined within the TSAP to identify the remote transport entity.

Data transfer: Once the connection is established and both sites (source and destination) have agreed on the quality of services, the data transfer in a normal mode takes place using the primitive DT TPDU. The data unit may be sent separately (if the size of the packet is within the maximum size of the TPDU), or the transport entity will segment the TDSU into a size within the maximum limit if the total size of the packet, i.e., TDSU and DT header, is more than the maximum size of TPDUs. The end of data is defined by an EOT bit sent in the TPDU. Each DT unit packet is assigned a number in a sequential order. In addition to this, we use the allocation method for flow control, where initial credit is described in CC and CR TPDUs. After the initial credit is described, an acknowledgment may be used to allocate credit, thus reducing overhead of the transport layer protocol. For acknowledgments, we use a separate packet; further, the acknowledgment is never sent via DT packet (i.e., no piggybacking). The choice of whether acknowledgments are required for each TSDU or TPDU or only for a block of TPDU is left to the transport protocol entity.

The TPDUs for expedited data are expedited data (ED) and expedited acknowledgment (EA), and these are numbered sequentially in the same way as in normal data transfer, except that the user process at the source site will transmit ED TPDU only after getting an acknowledgment via ED TPDU from the destination site. For the transport layer protocol of classes 0 through 4, we go even further in defining a queue of DTs for each transport connection, and then only ED is sent. The next ED will be sent only after receiving EA from the destination site. The Class 4 protocol has another issue to handle, and that is the assumption that the packet that first arrives or is given to the transport entity is not valid.

Connection termination: The connection can be terminated or released by either of the transport users by sending a T-DISCONNECT.request primitive. The transport layer responds to this primitive by sending another primitive, T-DISCONNECT.indication, on the other side, releasing the connection. There are a number of situations during the termination of connection in which data may be lost, because these primitives can be used by either transport user and at any time. Further, it is also likely that the transport layer itself may terminate the transport connection.

Another way of terminating the transport connection is via the TPDUs disconnect request (DR) and disconnect confirm (DC). After the transport entity receives a request from its user to terminate the connection, it issues a TPDU DR and discards all the pending DTs. The transport entity which receives the DR issues another TPDU DC to the other site and at the same time informs its user. It also starts a timer (to recover the DC if it is lost). The DC arrives at the source transport user, which issues an ACK TPDU and deletes the connection. When this ACK TPDU arrives on the destination transport user side, the transport layer deletes the connection. It is worth mentioning here that these primitives delete the transport connection, which basically deletes the information about the connection from their respective tables. The actions by these primitives are different from the T-DISCONNECT.request and T-DISCONNECT.indication primitives in the sense that here the connection is only deleted and not terminated. Further, now we are dealing with TPDUs as opposed to primitives. The use of TPDUs can be accomplished in a variety of ways using the handshaking concept, and it has been observed that three-way handshaking has been adequate to handle most of these situations.

13.5.4.2 Design issues in connectionless configuration

A transport protocol standard for this configuration has been defined by ISO 8602,[10] while various services of the transport layer for it are defined in ISO 8072/DAD1.[7] There are two primitives for connectionless service of the transport layer: T-UNITDATA.request and T-UNITDATA.indication.

As discussed earlier, each source and destination address of the transport user process is defined by its respective TSAP, which is passed on to the network layer entity. Based on the transport user address, the network layer identifies the system on which the user is connected. We notice that both transport layer service primitives have the same structure as that of network service primitives.

Both primitives (T-UNITDATA.request and T-UNITDATA.indication) of connectionless transport protocol service include the same parameters as in other primitives, such as source and destination addresses, quality of service (QOS), and user TS-user data. In this configuration, connection establishment is not required, and the data transfer takes place directly using the above primitives. The quality of service depends on the parameters, e.g., transit delay, protection, priority, error probability, etc. These parameters, in fact, were defined in the network service for connectionless configurations.

TPDU frame format: In order to provide flow control, delivery of packets in order, confirmed delivery of packets, and error control, the transport protocol provides formats for the above-mentioned primitives as a UNITDATA (UD) TPDU. Different protocol standards may have different TPDU formats, but, in general, all the protocol standards include the following fields, as shown in Figure 13.6.

Length indicator (LI): Defines the length of TPDU in octets.
Transport protocol data unit TPDU): Defines the user data (UD) packet being used for transmission.

Each of the addresses (source and destination TSAP) includes the following parameters:

1. **Parameter code:** Defines the code of the parameters used in TPDU (i.e., checksum, source TPDU-ID, destination TPDU-ID, and user data (UD)).
2. **Parameter length (PL):** The length in octets of the next parameter present in the field.

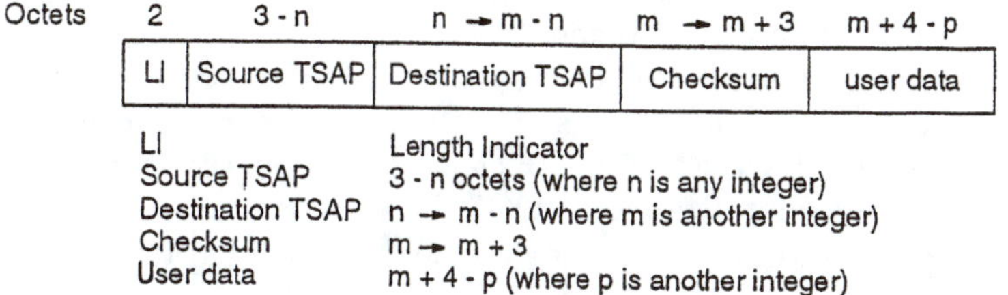

Figure 13.6 Transport protocol standards fields.

Other parameters in different fields were discussed earlier and have the same meanings and functions here in UNITDATA TPDU format.

A T-UNITDATA.request is requested by the session layer to its transport layer. The transport layer passes the source and destination addresses (obtained from its user process) into appropriate NSAPs and defines the source and destination TSAPs in TPDUs. These NSAPs, along with TSAPs, are given to the network layer service primitives.

Various parameters can be defined in T-UNITDATA.request for checksum errors. An ISO checksum algorithm may be used (16-bit checksum in each of the PDUs). This algorithm allows the source site to determine its value and place it in each of the PDUs transmitted. On the destination site, the checksum algorithm is used in the entire PDU, and if the result is the same as that of the transmitted value in PDU, then there is no error; otherwise, error is present.

Each T-UNITDATA.request contains a TS-USER-DATA parameter and is defined as a TSAP. The user data field in a TPDU is, in fact, a TSDU, and it is the TPDU which is sent to the network service and transport entity of the destination site. The destination site receives the TPDU, calculates the checksum value, and matches it with the one included in the TPDU. If the checksum is correct, then the transport layer entity sends T-UNITDATA.indication to the transport user. The transport user is defined in the destination TSAP-ID field of TPDU. If the checksum is not valid, the TPDUs will be discarded by the transport entity.

13.5.5 *Types of transformational services*

The network layer of OSI-RM offers services of both types of configurations (connection-oriented and connectionless) to its transport layer. In turn, the transport layer supporting these two types of configurations offers this as a service to its higher layers (user process). The transport layer protocol offers the translation of transport layer entities into network services. The translation of the transport layer can offer four different classes of transformation services defined as standards, and they are discussed next.

Connectionless transport and network (ISO 8602)[10]: This standard provides mapping between TPDU and NPDU using the primitive N-UNITDATA.request. This primitive is invoked after the transport layer sends its request to the network layer, which defines NSAPs for both source and destination nodes. The transport layer defines the parameters of QOS in its primitive T-UNIDATA.request. This primitive, issued by the transport layer entity, is transmitted by the network PDU by considering the TPDU as user's data. On the receiving side, the transport layer entity receives this in another form of primitive (N-UNITDATA.indication). It is worth mentioning here that these are the only two primitives required in this standard protocol.

Connection-oriented transport and network (ISO 8073): In this standard, all the primitives defined for connection-oriented network services are used. As usual, different types of TPDUs are considered user's data in the network primitives which are used during all three phases of connection establishment. Before any TPDU can be sent to the network layer, the first step requires the establishment of a transport connection over the network connection by the transport entity. If there already exists a network connection, the transport entity may use it; otherwise, it has to create a transport connection which matches the requirement parameters defined for QOS. The existing network connection will be used by the transport entity only when it meets the requirements of the requested transport connection. The request for establishing a transport connection is known as T-CONNECT.request. The transport connection is invoked by the CR TPDU, which will be sent to the network layer as its user's data part. Once the transport connection is established, other TPDUs such as connection confirm (CC), disconnect request (DR), and TPDU error (TPDUER) including CR are transmitted as user's data in the N-DATA primitives. These TPDUs are exchanged over the transport connection (so established). The N-DATA primitives are used in most exchanges and are different for different classes of protocols. Similarly, the termination or release of the transport connection is defined by DISCONNECT.request (DR) used as user's data in the N-DATA primitive. The transport connection is terminated, which also releases network connections. The network connection may or may not be terminated at this point and can be used for future communication.

Connectionless transport and connection-oriented network (ISO 8602)[10]: As discussed above, the transport connection is defined over a network connection, and it is possible that the transport layer wants to send a TPDU using connectionless transport but the network connection does not support it. In this case, the network connection must be defined separately for transmitting those TPDUs using connectionless services. As before, the primitive for sending the data is defined by a transport entity known as a T-UNITDATA.request primitive which is transmitted as a user's part in the network request primitive N-CONNECT.request. If the receiving side does not support connectionless service, it does not define any TPDU to handle this. In this case, the network layer sends another primitive, N-DISCONNECT.indication, back to the transmitting side. This indicates the non-availability of connectionless services on the receiving side. But if the receiving side does support connectionless services, then the receiving side will define a UNITDATA TPDU and is considered user's data in the network primitive (N-DATA.request). The UNITDATA TPDU will be transported over the network connection which has already been established by these two sides. The TPDUs are received as the user's part in the network primitives. The termination of the network connection is recognized by appropriate primitives. The transport understands this primitive and sends its response as N-RESET.response for the primitive it receives over the network connection as N-RESET.request.

Connection-oriented transport and connectionless network (ISO 8073/DAD2)[9]: This standard allows only the Class 4 connection-oriented protocol to use the connectionless services offered by the network layer. The standard primitives of connectionless service are used by the network layer as N-UNITDATA.request and N-UNITDATA.indication. As usual, the TPDUs which are used during connection establishment are considered user's data in the network primitives.

In this chapter, we discuss services, formats, primitives, and parameters for both services (connection-oriented and connectionless) of the standards defined by ISO and CCITT. Other standards organizations have been involved in defining standards for the transport layer, including ECMA[16,17] and U.K. Department of Trade and Industry.[18] Some of these standards, as well as ISO and CCITT standards, can also be found in Reference 15. Some of the material discussed above is derived partially from References 2–5.

13.6 Transmission control protocol (TCP)

The U.S. Department of Defense (DoD)[20] funded the Defense Advanced Research Projects Agency (DARPA) for designing a highly distributed system which could provide connectivity to four supercomputer centers at San Diego, Cornell, Pittsburgh, and Illinois. In 1979, the Internet Control and Configuration Board (ICCB) was formed to look into protocols and standards for communication across the distributed system. Under the guidance of ICCB, the communication protocol suite TCP/IP was introduced and implemented. The TCP/IP was implemented over the ARPANET in 1980. In 1983, ARPANET was split into two separate nets: ARPANET and MILNET. In 1985, the U.S. National Science Foundation (NSF) used TCP/IP for communication across the supercomputers, and many other networks connected together and, thus, the Internet was created. The Internet included over 3000 different types of networks (LANs, regional networks, etc.) and over 200,000 computers in 1990. By 1993, the Internet had grown to the extent that it included over 2 million computers. The number of users on the Internet was also growing exponentially. Today, over 200 million users are using the Internet for different applications. The Internet and TCP/IP together provide connectivity between computers at two remote locations anywhere in the world and allow the transfer of information between those computers over the Internet. TCP provides the following services: FTP, Telnet, domain name service (DNS), simple mail transfer protocol (SMTP), and routing information protocol. The IP does not provide any control other than header checking via checksum.

The TCP/IP protocol was defined to be used in ARPANET, industry, unreliable subnets, many vendor networks, various universities, and research and development (R&D) divisions. In the original version of ARPANET, the subnet was supposed to offer virtual circuit services (reliable) and the transport layer was named appropriately the network control protocol (NCP). TCP was designed to work on unreliable subnets (Type C networks). It was mainly concerned with the transmission of TPDUs over the networks. It offers services similar to those of ISO 8073 Class 4 protocols and also CCITT transport protocols, and it offers reliable and sequenced-order packets at the destination site. This means that it provides error recovery for lost or damaged packets, duplicated packets, and also non-sequenced packets. The error-recovery facility includes procedures for sequencing the data TPDUs (expressed in octets), appropriate algorithm for checksum for error detection, and methods for acknowledgment, re-transmission, and avoidance of duplicate packets. The TCP protocol resides in the transport layer under upper-layer protocols. The Internet Protocol (IP)[19] resides in the network layer. Using both TCP and IP together, users can transmit large amounts of data (e.g., large files) over the network reliably and efficiently. The TCP offers support for connection-oriented service, while another protocol within TCP/IP, user datagram protocol (UDP), supports connectionless service.

Features of TCP: The top layer, process/application, of TCP/IP includes the functions of the application, presentation, and a part of the session layers of OSI-RM. It provides the following services: FTP, Telnet, domain name service (DNS), simple mail transfer protocol (SMTP), and routing information protocol. The next layer, host-to-host, offers the same functions as the transport and part of the session layers of OSI-RM. It offers two types of services — connection-oriented and connectionless. The TCP provides end-to-end error control for the entire frame and hence reliable communication (connection-oriented). The user datagram protocol (UDP), on other hand, provides error control only for its header and hence provides unreliable service (connectionless). The UDP is faster than TCP and is used in small data transfers or applications where errors can be tolerated.

The TCP/IP offers a communication protocol independent of underlying networks (LANs, WANs, or any other interconnected network). It assumes that the network layer is using IP and that the protocols of the transport layer (TCP) and the network layer (IP) interact with each other via a defined set of primitives. It provides data and acknowledgment formats, procedures to ensure the orderly arrival of packets, initialization of data stream transfer, and indication for the completion of data transfer. This protocol does not specify application interfaces, and, further, it offers system-dependent implementation, thus offering flexibility to the users. It offers users two important options: data stream push and urgent data signaling.

TCP/IP offers full-duplex configuration. Some of the TCP/IP-based products include terminal emulation using Telnet, which allows a dumb terminal to act like 3270 (basically used to access a remote IBM host), simple mail transfer protocol (SMTP), FTP, and Unix (Berkeley and AT&T). It offers ports similar to service access points and usually defined within a socket address field of two octets of the network address. TCP/IP has become very popular and has been included as a built-in module in most operating systems such as Unix, Windows, etc. It has its own shell of about 25 Kb of code that makes it very popular on PCs.

The first option allows the formation of TPDU only after it has received enough data for its transmission and also includes data identified up to the push flag boundary. The TCP user can also request the TCP to transmit all outstanding data using this flag.

In the second option, the protocol data units of the transport layer are sent as datagram packets by IP and are also known as *IP datagrams*. The IP does not define any type of underlying network. In fact, it does not care for the underlying networks. In contrast to this, while using ISO, the set of protocols for the network and transport layers is network-dependent. This means that the protocols used by the network layer define the type of underlying network being used. In other words, the network layer protocol offers a connection-oriented interface for connection-oriented networks such as X.25, while it offers a connectionless interface for connectionless networks which use ISO-IP protocol.

Difference between TCP and ISO 8073: There is a conceptual difference between TCP/IP and ISO protocols in terms of layered architecture. In the case of TCP/IP, we do not have any layer over the TCP and, as such, it interfaces directly with the application, while in the case of ISO protocols, the protocol of the transport layer has to provide services to its higher layers, as well.

The TCP defines logically distinct processes similar to those of ISO 8073 between which the data is transferred, and it offers full-duplex line configuration between them. The TCP transport entity accepts the message, breaks it into a fixed-size packet (64 Kb), and transmits each packet as a datagram. A sequence field of 32 bits is used to provide a sequence number to every byte transmitted by the transport layer of the TCP. The TCP offers services to the users through various application protocols being used at the top of the TCP; these services are known as *reliable stream transport services* and are similar to those provided by the Class 4 version of the transport ISO protocol. While using the ISO set of protocols, a logical connection is made with peer application protocols running on a different remote host. This logical connection offers a full-duplex operation, allowing end users to interact with each other via protocol data units. The establishment of logical connections between end hosts requires the service primitives of the application protocols through presentation and session layers.

A request for connection establishment is issued from the local transport entity to the distant transport entity using the service primitives as described above. After the connection is established, the data transfer takes place between them. At the end of this session, connection termination can be issued by either one of them. The **higher-layer protocols (HLPs)** issue a request for connection and assign a specific transport layer through which connection is to be established. They can also request a specific transport layer to wait for a connection request that it is expecting from the distant transport layer. This scheme is very useful for allowing remote users to access a database, electronic bulletin board, or any other shareable files/programs. In this way, both ISO 8073 Class 4 and TCP define three-handshake protocols. During the data transfer, the sequencing of all data TPDUs is accomplished by both TCP and ISO 8073 Class 4 protocols.

Services: The TCP protocols usually offer an error-free environment; i.e., they make sure that no packet is lost or duplicated and also that the sequence of arrival of packets is maintained. All the packets corresponding to the user's data or control packets are treated as a sequence of message primitives. These primitives are request, indication, response, and confirm primitives. These primitives are usually grouped into request and response packets. All of the data or control message is treated as units known as *segments* (in the terminology of TCP/IP; similar to *data units* defined in ISO protocols). The segments are transmitted by the TCP protocol and may include the messages from one or more users, which in turn depend on the size of the user message. Since the TCP connection offers full-duplex operation, the segment in the form of a data stream can be transmitted in both directions simultaneously, and TCP makes sure that these streams of segments are delivered to their respective destinations reliably.

The user data (UD) from higher-layer protocols (HLP) is given to the transport entity in the form of a stream-oriented (as opposed to block-oriented) CCITT TPDU. The TCP defines the data packets for stream-oriented data it receives from HLP in the user's data packet format containing user data and control information. Each of the data segments (expressed in octets) defined by the TCP is numbered sequentially and acknowledged appropriately. These data segments are, in fact, given to the IP (network layer), which transmits them to the distant TCP user process (the transport layer entity). When a packet arrives on the destination side, this packet is stored in a memory buffer assigned with the application. These packets are delivered to the destination node when this buffer is full. TCP supports priority for small messages, and these messages are transmitted by bypassing the normal flow control. Similar options are available with ISO protocols, and the packets with priority are transmitted via expedited data service primitives.

The packets containing user data and control information are divided into segments of fixed size. These segments are numbered sequentially, and the acknowledgments are also received by these numbers. The interface between TCP and user process is defined by a set of primitive calls, including various parameters. For sending/receiving the data, commands such as OPEN, CLOSE, SEND, etc., similar to system calls in operating systems, are used. The segments are passed to the IP, which transmits them via subnets to the distant TCP user process. Each transmitted octet is assigned a sequence number which forces the distant TCP user process to send an acknowledgment for a block of octets, which implies that all the preceding octets have been received. A complete list of TCP service primitives, along with associated parameters, is given in Table 13.2.

If an acknowledgment is not received within the "time-out" limit, the data is retransmitted. The segment number of the receiving site is used to check duplicates and also arranges the packets received out of order. The checksum can be used to detect errors at the receiving site. The TCP also supports flow control and addressing of multiple processes within a single host simultaneously. The TCP/IP assigns a socket for each destination, and the socket has a 16-bit port and 32-bit internal address. For each connection, a pair of sockets is defined (for both source and destination). The security range allows users to specify the range of security required. The push flag and urgent flag represent the data transfer immediately (within normal flow control) or urgently (outside normal flow control), respectively. Unspecified-Passive-Open and Full-Passive-Open primitives are to be used by the server to indicate when it is ready to accept the request for connection.

Table 13.2 Parameters Associated with TCP Service Primitives

Primitive	Parameter
Unspecified-Passive-Open.request	Source port, time-out, time-out action, precedence, security range
Full-Passive-Open.request	Source port, destination port, destination address, time-out, time-out action, precedence, security range
Active-Open.request	Source port, destination port, destination address, time-out, time-out action, precedence, security range
Active-Open-Data.request	Source port, destination port, destination address, data, data length, push flag, urgent flag, time-out, time-out action, precedence, security range
Open-Id.response (local)	Local connection name, source port, destination
Open-Success.confirm	Local connection name
Open-Failure.confirm	Local connection name
Send.request	Local connection name, data, data length, push flag, urgent flag, time-out, time-out action
Deliver.indication	Local connection name, data, data length, urgent flag
Allocate.request	Local connection name, data length
Close.request	Local connection name
Closing.indication	Local connection name
Terminate.confirm	Local connection name, reason code
Abort.request	Local connection name
Status.request	Local connection name
Status-Response.response (local)	Local connection name, source port, source address, destination port, destination address, connection state, receive window, send windows, waiting Ack, waiting receipt, urgent, precedence, security, time-out
Error.indication	Local connection name, reason code

Multiple connections can be established over a single host at the same time. An application of this feature can be seen in the public bulletin board service, where it may have a fixed socket(s) through which users can access the bulletin board services simultaneously.

13.6.1 TCP header format

The architecture of the Internet or TCP/IP is defined in terms of layers such as process/application, host-to-host transport, and Internet and network access. The TCP/IP includes all the needed application software and is also known as Internet protocol. The top layer, process/application, includes the functions of the application, presentation, and a part of the session layers of OSI-RM. The next layer, host-to-host, offers the same functions as the transport and part of the session layers of OSI-RM It offers two types of services: connection-oriented and connectionless. The TCP provides end-to-end error control for the entire frame and hence reliable communication (connection-oriented). The user datagram protocol (UDP), on other hand, provides error control only for its header and, hence provides unreliable service (connectionless). The UDP is faster than TCP and is used in small data transfers or applications where errors can be tolerated. The Internet layer provides the same function as the network layer of OSI-RM. This layer includes three protocols: Internet protocol (IP), address resolution protocol (ARP), and Internet control message protocol (ICMP). The network access is the bottom layer and includes the functions of the data link, physical, and a part of the network layers of OSI-RM. It has

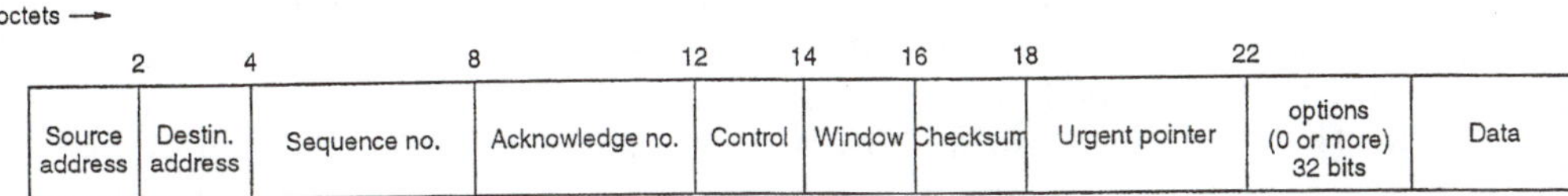

Figure 13.7 TCP header.

two sublayers known as 1A and 1B. 1A is equivalent to the physical layer, while 1B is equivalent to the data link layer and includes X.25, FDDI, and 802.x LAN protocols.

The TCP resides above the IP and allows multiple application programs to communicate concurrently; it also de-multiplexes incoming TCP traffic among application programs. The ports are assigned a small integer to identify the destination, and these port numbers are local to the machine. The destination of TCP traffic includes the destination host Internet address and TCP port number on the host machine. The TCP modules communicate with each other for providing communication between different application programs. The application program on one end machine uses a **passive** open primitive to indicate its desire to accept connections, and the port number is assigned to one end of the connection by the operating system. The application program on the other machine uses an **active** open primitive request to establish connection.

Different standards have been defined for TCP headers, and it is beyond the scope of this book to discuss all the header formats here, but in order to get a detailed understanding of its various fields, we consider the MIL-STD-1778[20] TCP TPDU header format below. This format is derived from References 3 and 4 and can be seen in graphic form in Figure 13.7.

SA — Source address (two bytes)
DA — Destination address (two bytes)
SN — Sequence number (four bytes)
AN — Acknowledgment number (four bytes; typically a piggyback acknowledgment)
Control (two bytes)
Window (two bytes)
Checksum (four bytes)
Urgent pointer (four bytes)
Options (0 or more than 32 bytes)
Data (variable)

The **source and destination addresses** are, in fact, port addresses of the transport layer (endpoints of the connection, i.e., TSAPs). The allocation of ports is usually handled by the host. In the **sequence number** field of 32 bits, we assign the sequence number to the first octet of data in the segment. The **acknowledgment number** field of 32 bits uses a piggybacking acknowledgment scheme and provides the value of the next octet sequence number that the sender of the segment is expecting from the receiving site. The minimum length of a TCP header is 20 bytes. The control field of 16 bits contains the following small fields:

- **DO (data offset):** This field contains three bits and indicates the number of 32-bit words that are defined in the TCP header. The beginning of the data can be determined.
- **XXX:** This field contains six bits and at present is reserved for future use.

- **URG:** This field contains one bit and indicates (when set) the use of the urgent pointer field. This defines a byte offset from the current sequence number at which the urgent data can be found.
- **ACK:** This field contains one bit and indicates an acknowledgment. The set bit (ACK = 1) indicates that piggybacking is not used. This can also distinguish between CONNECTION.request and CONNECTION.indication primitives.
- **PSH:** This field contains one bit and indicates whether the PUSH function is used. The PUSH function is a request which the sending user process can make to the local TCP for transmitting all the data it has. The PUSH bit indicates to the destination TCP to deliver the data to the receiving user process.
- **RST:** This field contains one bit and is used to reset the transport connection due to malfunctioning of the host, crashing of the host, etc.
- **SYN:** This field contains one bit and is used to establish transport connection. The connection request defines SYN = 1.
- **FIN:** This field contains one bit and indicates that the user is done with sending of the data and that no more data will be sent, and that it wants to terminate the transport connection.
- **Window field:** This field is of 16 bits and defines the size of the sequence number. The beginning of the data is defined in the acknowledgment number field which the sender will be accepting from the distant user. The flow-control problem in TCP is handled by a variable-size sliding window and it indicates how many bytes can be transmitted after the acknowledgment. It is nothing to do with the number of TPDUs.
- **Checksum:** This field of 16 bits detects error in both the header and data. The checksum offers a very highly reliable transport connection, and the checksum algorithm typically adds up the entire data and expresses it in terms of 16-bit words. The sum so obtained is then complemented using the 1s complement technique.
- **Urgent Pointer:** This field of 16 bits indicates the location (offset) from the current location number where the receiving user can find the urgent data.
- **Options:** This field is of variable length and is reserved for miscellaneous parameters. This may be used for requesting enough buffer storage to store necessary control information needed during connection setup.

13.6.2 TCP protocol operation

The user process and HLP interact with TCP using primitives or system calls in the same way that the operating system does for the user process to access files. The HLPs have a set of primitives which are used to issue commands to TCP, while the TCP also has its own set of primitives used for sending responses to the HLP for those commands. These primitives are defined for two modes — **passive** and **active** — as discussed earlier in Table 13.2. These modes allow the user process to open a transport connection. In the passive mode open command, a user process informs its TCP about the open requests it is expecting to get from the distant TCP entity (or sockets), while in the active mode open command, the user process requests its TCP to establish a connection with the distant TCP entity (socket).

A user process issues a primitive request UNSPECIFIED-PASSIVE-OPEN by passing parameters such as source entity (port or socket), HLP time-outs, and other optional time-out parameters, e.g., time-out action, precedence, and security. The destination port and address are defined in the primitive FULL-PASSIVE-OPEN, which includes parameters in addition to the above-mentioned primitives such as destination port and address. The primitives ACTIVE-OPEN and ACTIVE-OPEN-WITH-DATA include the common parameters source and destination ports, destination address, HLP time-out, and other time-out parameters with a difference that the latter primitive includes additional control fields such as data length, push flag, urgent flag, and the data. The SEND primitive includes control fields such as local connecting name, data length, push flag, urgent flag, HLP time-out, time-out action, and the data. The primitives ALLOCATE, CLOSE ABORT, and STATUS include the local connection name, and ALLOCATE also includes the data length control field.

The above set of primitives is issued by the user process from the HLP to TCP. The following set of primitives are used by the TCP to send its response back to its HLP. The OPEN-ID includes the local connection name, both source and destination ports, and destination address. The OPEN-FAILURE, OPEN-SUCCESS, CLOSING, TERMINATE, and ERROR parameters all include the local connection name, and the TERMINATE and ERROR parameters also define the description. The primitive DELIVER includes the local connection name, data length, urgent flag, and data. The primitive STATUS-RESPONSE includes parameters such as local connection name, both source and destination ports and addresses, connection state, receive and send window parameters, and time-out parameters.

TCP congestion: The congestion problem in TCP has been considered for a long time, and various mechanisms have been proposed to solve this problem. The flow-control technique (based on the credit-based control mechanism) has been used to control the flow of traffic between source and destination nodes. In this mechanism, the destination node will accept the number of packets without causing any overflow of its buffer. The same flow-control mechanism finds its use in controlling the congestion between source and destination over the Internet. The TCP flow-control mechanism has to identify the node in the network that causes the congestion and then control the traffic flow to that node. In order to handle the TCP congestion problem, Request for Comments (RFC) 793 was published. Since its publication, a number of techniques to control the TCP congestion problem have been proposed. Since TCP offers connection-oriented services, it uses a timer for re-transmission of the lost or errored packets. The value on the timer should practically be more than or equal to at least twice the round-trip propagation delay. The propagation delay may change with changes in Internet conditions.

There are a number of algorithms available that can be used to estimate the round-trip propagation time for setting the value of the timer for a reasonable performance of the network, including Jacobson's algorithm and Karn's algorithm. The RFC 793 provides an equation for estimating the current round-trip propagation time in the networks. Most of the algorithms use this equation to estimate the delay for different situations and modify it accordingly.

The TCP congestion problem can also be affected by window-based protocols by considering the window size for sending and receiving the packets. It is quite obvious that if window size is large, we can send a large number of TCP packets before acknowledgment. The TCP connection, once established, can send packets of the entire window size to the Internet. This will certainly create a congestion problem, if every TCP connection has sent all the packets of up to the maximum size of the window. This problem has been solved in a variety of ways. One of them is *slow start*. In this technique (proposed by Jacobson), the TCP connection allows the transmission of a limited number of packets to the Internet. It starts with one packet and waits for the acknowledgment. Each time an acknowledgment is received, it increases its window size by one. The next time, it waits for two acknowledgments. After receiving these two acknowledgments, it increases the window size by two, so that next time it can send four packets, and so on. According to RFC 1323, the TCP window size has gone from 64 Kbps to 1 Gb. The philosophy behind the X.25 protocol was to define a very complex set of procedures so that the packets could be delivered reliably between nodes. The simple design of switching protocols and digital communications (replacing analog communication) together have provided us with error-free physical layer communication. This has resulted in simple network protocols such as frame relay, SMDS, and ATM over high-performance digital fiber-optic transmission that gives extremely low error rate.

Many carriers have already started using high-performance digital communication technology by providing services to customers for home applications over fiber. A variety of traffic (data, voice, video) is being handled by ATM interfaces and networks. There exist a number of standards such as LAN emulation, multiprotocol over ATM, and many others.

13.6.3 User datagram protocol (UDP)

We have discussed a connection-oriented protocol for the transport layer known as TCP. The ARPANET has also defined a connectionless transport layer protocol known as **user datagram protocol (UDP)**. It offers unreliable service and, in contrast to TCP, it does not offer error control (no guarantee for the loss or duplication of packets). It also does not provide ordered delivery of packets. Even with these drawbacks, the complexity of the protocol is not reduced and the overhead remains the same as that of TCP. It allows a program on one machine to transmit datagrams to a program(s) on another machine(s) and can also receive acknowledgments. Each application needs to get a UDP protocol port number from its operating system. Thus, the applications communicate with each other via the protocol port number. This scheme offers various advantages, as the protocols are independent of specific systems, and it allows communication between applications running on heterogeneous computers. The sender can identify a particular receiver on a single machine using UDP, as each UDP message contains the destination and source protocol port numbers. It offers unreliable delivery of messages and depends on the underlying Internet protocol (IP) to provide transportation to the messages from one machine to another. The UDP is used mainly by Simple Mail Transfer Protocol (SMTP) that is used in various LAN internetworking devices such as bridges, routers, hubs, etc. It is used for multicast or broadcast services, while TCP is not suitable for these applications.

The conceptual architecture of network software may have a physical layer (Ethernet) as the transportation mechanism at the bottom level over which IP offers routing and network protocols. The UDP sits on top of the IP, over which we use the user program. The client–server approach is being used to define the interfaces for application programs. The UDP port number is assigned by the operating system to application programs, and these application programs interact with each other via these port numbers (system-independent). The application programs can send datagrams back and forth using these port numbers. The client–server concept is used in most machines in one way or the other.

A single server application may receive UDP messages from many clients. The server cannot specify an IP address and cannot specify a UDP port on another machine, as it needs that machine to send the message to it. Alternatively, it defines only the local port number. Any message coming out from any client to the server must specify the client's UDP port and also the server's port number. The server receives the message from the client and uses its port number as the destination port number for sending its reply. Thus, the interface for many-to-one communication must allow the server to specify information about the destination each time it tries to send datagrams by extracting the port number of the client and using it as a destination port number. In this case, the destination IP address or UDP protocol port number cannot be assigned on a permanent basis.

The IP address is defined by two fields known as network ID (netid) and host ID (hostid) and is used to transmit the datagrams to the destination host. The network

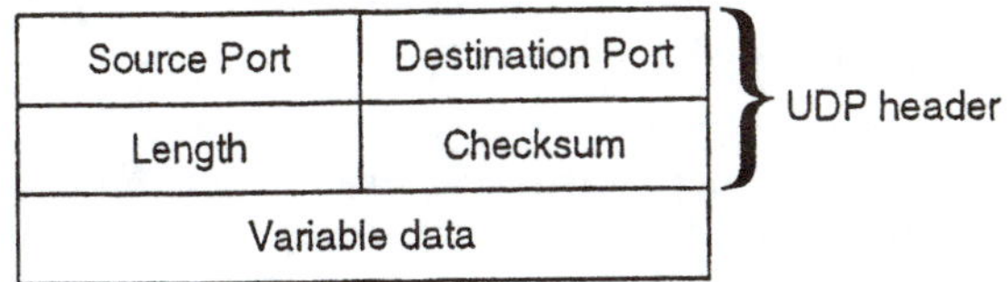

Figure 13.8 ARPANET UDP format.

interface supports different types of networks and is independent of underlying networks. The network interface defines the network (LAN, WAN, or any other network) frame header and is associated with the trailer for error control using CRC. The protocol field defined in an IP datagram header is interfaced with the protocol used in the destination host. The checksum computation is performed on the entire UDP segment. It also includes the header field, as in TCP. This header is a *pseudo-header* prefixed to the actual header before the computation of the checksum. In the event of error, the segment is discarded. The checksum in UDP is optional. If it is not being used, then it is usually set to zero.

The IP checksum is used only on an IP header and is not used on IP data. The IP user data field includes the UDP header and UDP user's data, and no checksum computation is performed on the user's data if the UDP does not use error checksum. Similar to TCP, the IP also supports multiple applications. The user host submits a request as a protocol data unit from the application processes, which will be delivered to either the TCP or UDP with the associated protocol port address in the respective PDU header (TCP or UDP).

As we have seen, in OSI transport protocols for connectionless services, we are using two primitives — T-UNITDATA.request and T-UNITDATA.indication — which allow users to send a message across the network before establishing any transport connection between source and destination transport users. Similarly, the UDP does not require the establishment of any transport connection for sending data over the network. Due to the connectionless services, the network does not give any guarantee of delivery or sequencing. The ARPANET UDP format includes a UDP header which defines source and destination port addresses, as shown in Figure 13.8.

13.6.4 *Differences between TCP and Class 4 protocols*

As mentioned above, the TP4 and TCP transport layer protocols have some common features — both offer reliable connection-oriented services for end-to-end communications and assume that the underlying network layer protocols are unreliable as they do not provide error control and flow control. These protocols have some differences, which are briefly explained herein. Some of this material is derived from References 2–5 and other publications.

The TCP protocol defines one TPDU (minimum size of TCP header is 20 bytes), while TP4 uses 10 TPDUs (discussed above). In the case of a connection collision (two requests for connection simultaneously), the TP4 establishes two connections over full duplex, while TCP establishes only one connection over full duplex with two transfers proceeding simultaneously over each direction. The TCP uses 32 bits for TSAP, while no exact format for TSAP is defined in TP4.

The TP4 offers various parameters to be defined for QOS (between source and destination transport user) before establishing transport connection, while TCP does not support any QOS parameters, but limited QOS parameters such as speed and reliability can be specified via the QOS field in its underlying Internet protocol (IP) layer.

The TCP uses a piggybacking scheme for acknowledgment, while the TP4 does not use this scheme but instead allows the transmission of two acknowledgment TPDUs (DT and ACK) in one network packet. The acknowledgment scheme in TCP always specifies the number of the next byte the receiver is expecting, and as such it is cumulative. With this scheme, the acknowledgments are easy to generate and are unambiguous, and lost acknowledgments do not always necessarily require re-transmission. The only problem with this scheme is that the sender does not know about successful transmission and the TCP protocol may be considered inefficient. It uses an adaptive routing algorithm, and each time a segment is transmitted, it starts a timer and waits for acknowledgment. It also keeps a record of what time a segment is transmitted and what time its acknowledgment arrives (round-trip delay). It reduces its transmission rate to solve the congestion problem and detects this problem automatically by observing round-trip delay.

In terms of the window method, the receiving transport user can reduce the size of the window over a period of time. Each acknowledgment contains window information (e.g., how many bytes the receiver is ready to accept, the response of the sender for increase or decrease of window sizes, etc.). For the increased size of a window, the sender also increases its window size and sends octets that have not been acknowledged, while in the case of a decreased window, it also decreases its window size and stops sending octets beyond the boundary size. The TCP does not offer any solution for arrival of packets in the wrong order due to this facility (mismatch between the window sizes of two sites), while the TP4 uses a sequencing scheme to handle this problem.

The main difference between these protocols for data transfer is in terms of the interpretation of messages. The TCP considers this message a continuous stream of bytes (also known as segments) without any explicit message framing or boundary. It uses a sliding-window scheme (operates at the byte level) for efficient transmission and flow control, and it can send multiple packets before the arrival of acknowledgment. Further, it offers end-to-end flow control. The TP4 defines the message in terms of an ordered sequence of TSDUs. In reality, the TCP defines a system procedure push which provides a boundary for the messages.

Both protocols define some kind of priority for the data. In TP4, regular and expedited data can be multiplexed and the expedited data will get priority. In TCP, this is achieved by defining an urgent field to indicate the priority for data (TPDU).

During the termination of connection, the TP4 protocol uses DR TPDU, which may be sent after data TPDUs. In the case of loss of data TPDUs, recovery is not allowed in TP4, while in TCP, three-way handshaking is used to avoid the loss of packets. The three-way handshaking guarantees that both sides are ready to transfer the data and also agree on initial sequence numbers. The sender side sends a primitive, SEnd Syn sequence = n, as a network message. The receiving side receives this as a Receive Syn segment and uses another primitive, SEnd Syn sequence = m, with Ack = $n + 1$. The sender receives this as a Receive Syn + Ack segment and sends a primitive confirming the sequence and acknowledgment as SEnd Ack $m + 1$. The receiving side receives this as another primitive, the Receive Ack segment, confirming the initial agreement for the connection. The connection request can be issued by either end side or it can be issued from both ends simultaneously.

The OSI transport layer protocols are used in other networks such as Technical Office Protocol (TOP) and Manufacturing Automation Protocol (MAP), which use ISO 8473 network layer protocols, and the TP4 (connection-oriented, reliable for end-to-end communication) has to be used (as ISO 8473 offers connectionless services). The use of TP4 must be negotiated before communication. This can be accomplished by requesting the use of TP4 in CC TPDU and, after getting a response via CC TPDU only, the connection can be established; otherwise, the connection request will be discarded. The option field must be implemented in most cases; however, the seven-bit option is also possible. The implementation must also support expedited data transfer capabilities, and if these options are not acceptable to the destination transport entity, the transport connection must be rejected, as it will create confusion to the higher layers which use these option fields. The MAP and TOP networks must support the checksum algorithm where the transport entity checks each TPDU via its algorithm, as it provides protection of loss of data during the transmission. The checksum algorithm can be executed either before sending or after receiving each TPDU.

It is not necessary to use transport layer protocols in all networks, and in fact it can be avoided if the underlying layer protocols can provide reliable connection between two machines to which users can log in. In this case, the two transport users may or may not use any transport layer, and if they want to use one, they may mutually agree on any type of transport layer protocols. For example, USENET does not use any transport layer, and there are many machines which use only TCP/IP, X.25, or even UNIX to UNIX copy (UUCP) (see below).

USENET offers the following three services: **file transfer**, **news**, and **electronic mail**. These services are available over dial-up telephone lines, and accessing any service requires the establishment of connection with a remote machine. A typical sequence of operation is defined for these services. For example, when a user wants to send an e-mail, the mail program on the local machine defines the mail, attaches this mail with a file transfer command, and sends it to the destination machine as a mail file in its mailbox. The file transfer program is invoked (either regularly or any time the network manager wants), and it looks at the spooling directory to see if there is any pending request for file transfer. If it finds requests, it tries to contact the remote by looking at a system file which contains all the relevant information about that remote system (e.g., contact number, line speed, how to log in, line protocols being used) and how to interface it (via X.25, dial-up line, or any other communication interface). This file contains all the information required to provide connection with that remote machine and log-in facility. After the connection with the remote machine has been established, the local machine is assigned a special user name and is treated as a slave process by the remote machine. The data transfer considers the packets of eight-bit bytes under sliding windows.

The UNIX-based systems also allow the transfer of files between them by using a system call known as UNIX-to-UNIX copy (UUCP). This system call allows automatic file transfer between distant machines via modems and automatic telephone dialer systems.

A number of application interfaces to TCP and UDP have been developed, e.g., FTP (secure server log in, file and directory manipulations), Telnet (remote terminal log in), hypertext transfer protocol (HTTP) supporting Web services, simple network management protocol for managing the network and maintaining various statistical reports, network file transfer, and remote procedure call for interacting with IP networks. Address resolution protocol (ARP) provides a link to the data link layer of a LAN. Its main function is to provide mapping between MAC and IP addresses. The next-generation IP protocol, known as IPng, is documented in RFC 1752 (issued by IETF in 1994). It contains 128 bits and provides multicasting very easily. It allows specification of quality of service, provides authentication and data integrity, and supports all the routing strategies of IPv4 (OSPF, RIP, etc.).

TCP/IP vendors: A number of vendors support TCP/IP in various applications; e.g., NetWare version 3.1 uses TCP/IP for server–host applications, Novell SNMP, network file systems, DOS- and Windows-based workstations, AppleTalk on Macintoshes, etc. Novell has developed the following protocols that are used by NetWare: Internetwork packet exchange (IPX), sequenced packet exchange (SPX), NetBIOS emulation, and NetWare core protocol (NCP). IPX supports half-duplex, connectionless service and is very fast, as it does not require an acknowledgment. The packet size of IPX has only 512 bytes and, as a result, the efficiency becomes poor on WANs and also the routing is not efficient. These two problems have been reduced by the introduction of burst mode protocol and NetWare link service protocol (Novell). The SPX (Novell) offers connection-oriented services and uses IPX as a protocol for routing the packets. It is interesting to note that SPX is similar to TCP in that it uses IP for routing the packets. The function of NCP is to provide the set of procedures for server–workstation communication. It also uses IPX for routing the packets.

Novell's NetBIOS emulation uses IPX for carrying the traffic. NetBIOS is, in fact, an interface between application and network. The local program can access the network without changing the environment and does realize that the communication is being implemented externally. Due to its proprietary nature, NetBIOS was originally implemented on a ROM chip, but its software version is now available. Some vendors (like Phoenix) have also implemented NetBIOS on a ROM chip. Other equivalent interfaces are APIs. The NetBIOS protocol does not provide any support for internetworking, and as such NetBIOS frames have to have headers like TCP/IP, IPX, or an equivalent of these for its routing. If it is used in LANs, then it must have a header in its front corresponding to LANs (IEEE 802.2 LLC and IEEE 802.3 or other LAN MACs for bridging). NetBIOS is very popular as an interface, and in fact it has been used in a number of applications such as IBM's OS/2 LAN server and Microsoft's LAN manager.

13.7 Internet protocol (IP)

The U.S. Department of Defense (DoD) defined a protocol for internetworking to be used with TCP known as the Internet protocol (IP).[19] The development of this protocol took place during the DARPA internetworking research project. It offers connectionless services to the user processes and does not require any connection establishment between them, thus reducing connection setup time. The structure of IP is somewhat similar to those structures discussed in connection-oriented protocols, and it does not restrict any site in having both types of configurations (connectionless and connection-oriented services). As the protocol supports connectionless service, it constructs a datagram for each packet it receives from the transport layer entity.

IP uses 32 bits to define an IP address. The IP addressing scheme has three classes known as A, B, and C. Class A starts with a "0" bit in the most significant position. It uses seven bits for the network, while the remaining 24 bits are used for local addresses. Class B defines the first two bits as "10" at most significant bits of the network address. It uses 14 bits (plus these two bits) as the network address, while the remaining 16 bits are used for the local address. Class C defines the first three bits as "110" at the most significant position, leaving 21 bits for the network address and eight bits for the local address. In Class A, the local address of 24 bits represents over 16 million nodes connected to the Internet. In Class B, 14 bits of the network address represent over 16,000 networks and 16 bits are used to represent the number of nodes. The Class C IP address is the most common class of the IP addressing scheme. It represents over 4 million networks through 21 bits of network bits and 255 nodes via an eight-bit local address.

The Internet layer of TCP/IP provides the same function as that of the network layer of OSI-RM. This layer includes three protocols: Internet protocol (IP), address resolution protocol (ARP), and Internet control message protocol (ICMP). The IP does not provide any control other than header checking via checksum. The network access is the bottom layer, and it includes the functions of the data link, physical, and a part of the network layers of OSI-RM. It defines two sublayers as 1A and 1B. 1A is equivalent to the physical layer, while 1B is equivalent to the data link layer and includes X.25, FDDI, and 802.x LAN protocols.

The header of an IP datagram defines global addresses of distant sites. Different networks are connected via gateways; the header and IP datagram will be sent to appropriate IP gateways. The gateway looks into the control information field and the datagram header which defines the route within the networks for datagram traffic. If the original packet does not belong to the network of the connected gateways, that network will send it to another IP gateway, and it will keep being routed via different gateways until the packet is delivered to the gateway to the network that contains the destination address.

The routing decision by an IP gateway is a very important factor during the implementation of a protocol, as it allows a check at each IP gateway if the destination site(s) is connected to it. If it contains data, then it will be delivered to the destination site. If the destination site is not present in any of the networks connected to it, it will find an optimal route of gateways. The packet will be routed over gateways and networks until it is delivered to the final destination site. In each routing decision, the objective is always to reduce the number of hops, or simple paths of length one, where a simple path is a direct link between two nodes. Routing tables containing information regarding the shortest routes, minimum-cost routes, alternate routes, etc. are maintained at each site and also at each gateway. The decision tables may be static or dynamic, depending on the algorithms used to develop them.

IP is a connectionless protocol and uses all the functions (e.g., routing, segmentation, and reassembly) defined by ISO 8073. Further, datagrams may be lost during transmission for any other reason, e.g., insufficient buffer space, hardware failure, link failure, violation of other functions, etc. In order to avoid the loss of datagrams, the transport layer protocols are expected to provide error-control strategies. Due to some of the above-mentioned reasons, both standards (TCP and IP) defined by DoD have become very popular and are being used together around the world. The entire address with TCP/IP includes the Internet-wide IP address of the host, and it also includes an additional protocol port address. The first field, netid, defines the network address, while hostid defines the host address.

The decision routing tables stored at each site and gateway help the gateways to identify the sequence of the next gateways over which TPDU should be transmitted. The decision routing tables, in general, adopt the changes in the event of failure of any site, link, or gateway. The neighboring gateways for a broken link or gateway transmit timed-out packets to all other gateways. IP datagram packets are segmented into a mutually agreed upon packet size (within the maximum size limit) NPDU, where each of the NPDUs is assigned a unique identifier. The field length indicator defines the relative address of PDUs with respect to the IP datagram and is placed in the NPDU. Two primitives, SEND and DELIVER, are used in IP for providing communication between end user processes. The services of networks are requested by SEND, while the DELIVER primitive is used to inform the destination end user process about the received data packets. The IP offers interface with a variety of networks and has limited functions to perform; much of the time, IP is busy in routing functions. The IP PDU data and control formats are shown in Figure 13.9(a,b). Details about IP, its formats, and different services offered by it have already been discussed in Chapter 12. Details of the IP frame format, header, and classes can be found in Section 12.7.5.

IP is a datagram protocol that is resilient to network failure, but it does offer sequenced delivery of the data. It offers end-to-end reliable data communication across the Internet when used with TCP. The function of ICMP at the Internet layer is to provide a ping (echo packet) to determine the round-trip propagation delay between source and IP-address

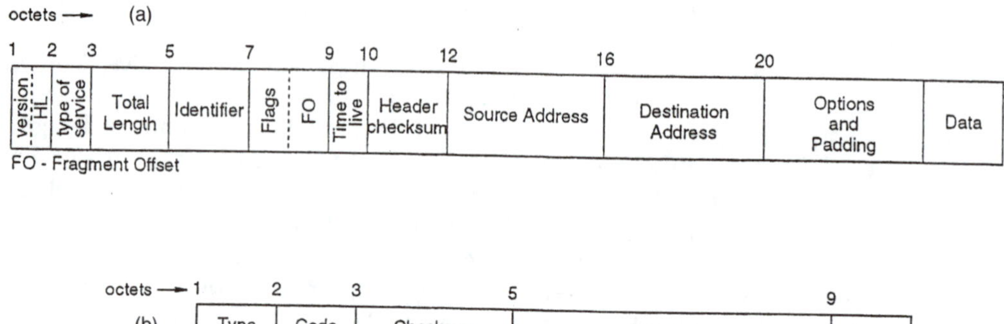

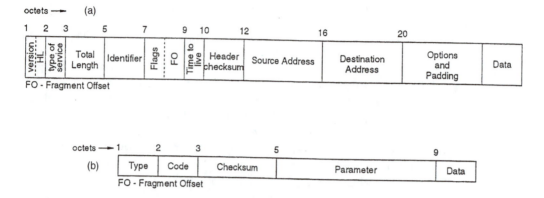

Figure 13.9 (a) IP data unit format (IP diagram). (b) IP control unit format.

host. Routers use this protocol for sending error and control messages to other routers. This layer provides multicasting data communication using Internet group multicast protocol. It also uses address resolution protocol (ARP) that provides a direct interface to the data link layer of the underlying LAN. This protocol provides mapping between a MAC address (physical) and an IP address. It uses some routing protocols. The Internet layer supports a variety of data link layers (Ethernet, token ring, FDDI, frame relay, SMDS, ATM, etc.). The transport layer with UDP and TCP supports a variety of applications such as file transfer protocol (FTP; a secured server log in, directory manipulation, and file transfers), Telnet (remote log-in facility), hypertext transfer protocol (HTTP, for the World Wide Web), simple network management protocol (SNMP, for network configurations, setting, testing, data retrieval, alarm and other statistical reports, etc.), remote procedure call (RPC), network file server (NFS), trivial FTP (TFTP; reduces the implementation complexity); and many others. IPv4 is version 4 of IP. The next version of IP, known as IPv6 (Internet protocol — next generation: IPng), is based on CIDR. CIDR defines a Class A address space reserved for expansion (64.0.0.0 through 126.0.0.0). The IEFT issued RFC 1752, a recommendation for IPng, in July 1994. It supports all applications of Ipv4 such as datagram service, FTP, e-mails, X-Windows, Gopher, Web, and others. It defines flow label fields that support resource reservation protocol (RSVP) for confirmed bandwidth and quality of service.

Some useful Web sites

Some Web sites that are relevant to the transport layer are *http://www-mac.uni-inc.msk.ru/tech1/1994/osi/transp.htm*, *http://www.scit.wlv.ac.uk/~jphb/comms/std.osirm4.html*, *http://www.csc.vill.edu/mnt/a/cassel/html/4900/transport.html*, and *http://www.maths.bath.ac.uk/~pjw/NOTES/networks/section2_7_4.html*.

More information on TCP/IP can be found at *http://www.cisco.com/warp/public/535/4.html*, *http://pclt.cis.yale.edu/pclt/COMM/TCPIP.HTM*, *http://www.rvs.uni-hannover.de/people/voeckler/tune/EN/tune.html*, and *http://ipprimer.windsorcs.com/index.cfm*.

More information on UDP can be found at *http://www.freesoft.org/CIE/RFC/1122/72.htm* and *http://www.udp.org.bz/*.

Information on transport layer security (TLS) can be found at *http://www.graco.c.u-tokyo.ac.jp/~nishi/security/ssl.html*.

References

1. ISO 7498, "Open Systems Interconnection-Basic Reference Model, Information Processing Systems," ISO, 1986.
2. Tanenbaum, A. S., *Computer Networks*, 2nd ed., Prentice-Hall, 1989.
3. Stallings, W., *Handbook of Computer Communications Standards*, vol. 2, 2nd ed., Howard W. Sams and Co., 1990.
4. Markeley, R.W., *Data Communications and Interoperability*, Prentice-Hall, 1989.
5. Black, U., *Data Networks: Concepts, Theory and Practice*, Prentice-Hall, 1989.
6. ISO 8072, "Open System Interconnection: Transport Service Definition, Information Processing System," 1986.
7. ISO 8072/DAD1, "Open System Interconnection: Transport Service Definition Addendum 1: Connectionless-Mode Transmission, Information Processing Systems," ISO, 1986.
8. ISO 8073, "Open System Interconnection System: Connection-Oriented Transport Protocol Specification, Information Processing Systems," ISO, 1986.
9. ISO 8073/DAD 2, "Open System Interconnection System: Connection-Oriented Transport, Addendum 2, Operation of Class 4 over Connectionless Network Service Protocol Specification, Information Processing Systems," ISO, 1986.
10. ISO 8602, "Protocol for Providing the Connectionless Mode Transport Service," ISO, 1986.
11. CCITT X.214, "Transport Service Definition for Open Systems Interconnection for CCITT Applications," 1984.
12. CCITT X.224, "Transport Protocol Specification for Open Systems Interconnection for CCITT Applications," 1984.
13. CCITT T.70, "Network Independent Basic Transport Service for Telematic Services," 1984.
14. CCITT Recommendation X.25, "Interface between Data Terminal Equipment (DTE) and Data Circuit-Terminating Equipment (DCE) for Terminals Operating in the Packet Mode and Connected to Public Data Networks by Dedicated Circuit," 1984.
15. Kinghtson, K.G,. "The transport layer standardization," *Proc. IEEE*, 71(12), Dec. 1983.
16. ECMA, "Transport Protocol Standard ECMA-72, 2nd ed., European Computer Manufacturers Associations, Sept. 1982.
17. ECMA, "Local Area Network Layer One to Four: Architecture and Protocol," Report TR/1`4, ECMA, Sept. 1982.
18. U.K. Department of Trade and Industry, "Intercept Recommendations for OSI Transport Layer, Focus: Standardization for IT, Technical Guide TG 102/1," March 1984.
19. U.K. Department of Trade and Industry, "Intercept Recommendations for OSI Network Layer, Focus: Standardization for IT, Technical Guide," March 1984.
20. U.S. DoD, MIL-STD-1778, "Transmission Control Protocol, Military Standard, U.S. Department of Defense, Aug. 1983.
21. U.S. DoD, MIL-STD-1782, "Telnet Protocol," Military Standard, U.S. Department of Defense, May 1984.

References for further reading

1. Slattery, T., Burton, B., and Burton, W., *Advanced IP Routing with CISCO Networks*, McGraw-Hill, 1998.
2. Iren, S., Amer, P.D., and Conrad, P., "The transport layer: Tutorial and survey," *ACM Computing Surveys*, 31(4), Dec. 1999.
3. Sanghi, D. and Agrawala, A., "DTP: An efficient transport protocol," in *Computers Networks, Architectures and Applications* (Eds. Raghavan, Bochmann, and Pujolle), North Holland, pp. 171-180, 1993.
4. Sanders, R. and Weaver, A., "The Xpress transfer protocol (XTP) — A tutorial," *Computer Communications Review*, Oct. 1990.
5. Watson, R., "The Delta-T transport protocol: Features and experience," *14th IEEE Conference on Local Computer Networks*, Oct. 1989.
6. Williamson, C. and Cheriton, D., "An overview of the VMTP transport protocol," *14th IEEE Conference on Local Computer Networks*, Oct. 1989.

chapter fourteen

Session layer

"You can't build a reputation on what you are going to do."

Unknown

"The third highest protocol layer (layer 5) in the OSI seven layer model. The session layer uses the transport layer to establish a connection between processes on different hosts. It handles security and creation of the session. It is used by the presentation layer."

The Free Online Dictionary of Computing
Denis Howe, Editor (*http://foldoc.doc.ic.ac.uk/*)

14.1 Introduction

The session layer provides a session link between two end users' processes (i.e., application programs). The most important services of the session layer include defining organized synchronization data transfer between end users and controlling data transfer provided to the users over the transport connection (fourth layer of OSI-RM). At the same time, the session layer establishes a logical link between presentation layer entities (the sixth layer of OSI-RM). The operations of establishing and terminating a session connection between end users are performed in a very systematic manner through structured interaction between end users over the session connection (also known as *dialogue*). The services and functions of the session layer are defined in the OSI document ISO 8326.[8,9]

The services of the session layer include various procedures (available in vendors' products and also in the specifications of standards organizations) which will, upon execution in a sequence, organize and synchronize the data transfer between end users. The session connection is well organized and well structured, especially during the release operations, as the session connection is released only when all the **session service data units (SSDUs)** are delivered to the destination, irrespective of the mode of operations (half-duplex, full-duplex, or two-way alternate). This feature of session layer service allows the user to define the number of SSDUs to be transmitted or discarded, which will take place only when the users inform the session service to do so. At the same time, this feature leads to a limitation on the session buffer space requirements.

14.2 Layered protocols and interfacing

The communication control procedures are a set of rules for controlling and managing the interaction between users via their computers, terminals, and workstations across the network. The control procedures are known by different names in the industry, such as protocols, line control procedures, discipline procedures, data link control procedures, etc. The protocols provide a communication logical link between two distant users across the network and the handshaking procedures provide the call establishment within the protocols.

14.2.1 *Layered protocols*

Each node (PCs, terminals, etc.) connected to the LAN must contain identical protocol layers for data communication. Peer protocol layers exchange messages which have commonly understood formats, and such messages are known as **protocol data units (PDUs)**. The PDU also corresponds to the unit of data in layer peer-to-peer protocol. The peer session entities communicate with each other using the services provided by the transport layer. The services of this layer are provided to the presentation layer at a point known as the **session service access point (SSAP)**. This is the point through which its services can be accessed by the presentation layer. For description of the interface between the session and presentation layers, the session layer primitives are defined. The **session service data unit (SSDU)** associated with its primitives is delivered to the presentation layer, and similarly, the session layer uses its primitives to interact with the presentation layer. The presentation layer sends its presentation protocol data unit (PPDU) to the session layer. The PPDU consists of protocol, control information, and presentation user data. The presentation control information is exchanged between session layer entities.

14.2.2 *Interfacing*

Protocols are defined in terms of rules (through semantics and syntax) describing the conversion of services into an appropriate set of program segments. Each layer has an **entity process** that is different on different computers, and it is this entity (peer) process through which various layers communicate with each other using the protocols. Examples of entity processes are file transfers, application programs, electronic facilities, etc. These processes can be used by computers, terminals, personal computers (PCs), workstations, etc.

Each layer has active elements known as entities, and entities in the same layer of different machines connected via a network are known as **peer entities.** The entities in a layer are known as service providers and the implementation of services used by the higher layer is known as a user. The **session service access point (SSAP)** is a unique hardware or port or socket address (software) through which the upper layer accesses the services offered by its lower layer. Also, the lower layer fetches the message from its higher layers through these SAPs. The SSAP defines its own address at the layer boundary and is prefixed defining the appropriate layers. The **session service data unit (SSDU)** is defined as message information across the network to peer entities of all the layers. The SDU and control information (addressing, error control, etc.) are known as an **interface data unit (IDU)**, which is basically message information sent by the presentation layer entity to its entity via SSAPs. For sending an SDU, a layer entity partitions the message information into smaller-sized messages. Each message is associated with control information defined by the appropriate layer as a header and sent as a **protocol data unit (PDU)**. This PDU is termed a **packet**. We will be using the terms defined by ISO in this book. Some of these terms are also discussed in other publications.[2-5,7] The relationship between SSDU, SPDU, and session entity is shown in Figure 14.1.

14.3 *ISO 8326 session layer services and functions*

The purpose of the session layer is to provide a service environment where the presentation entities interact in such a way that the dialogue message and data transfer between them

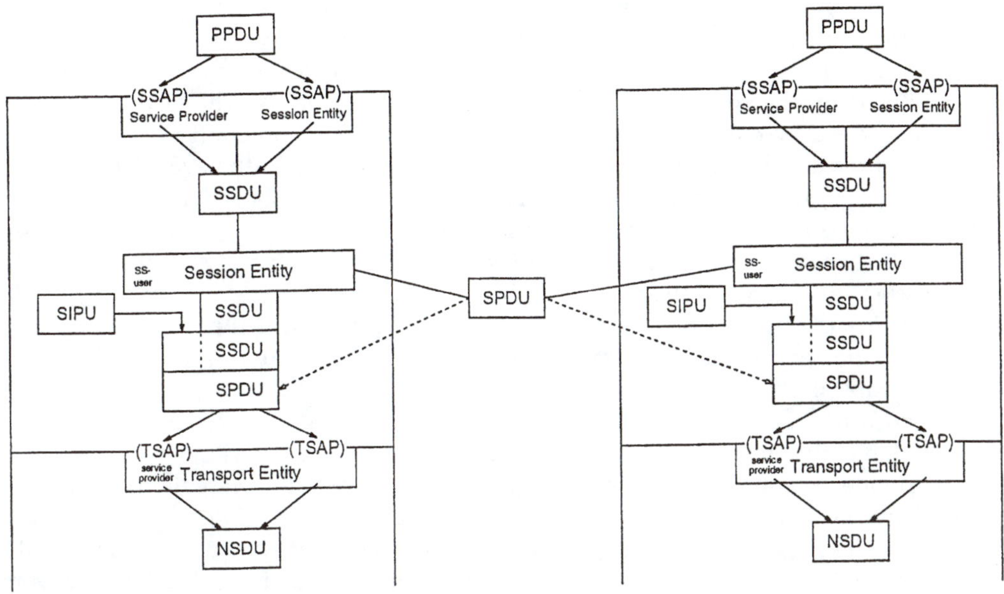

Figure 14.1 Relationship between SSDU, SPDU, and session entity.

are synchronized. The service environment, in fact, is a session connection between presentation entities which supports orderly exchange of data between presentation entities. This is achieved by mapping a session connection into a transport connection and using it for establishing a session connection. For further information, please refer to References 2–9. Some of the material herein has been partially derived from these references.

The presentation layer entity makes a request for data transfer to the session layer at its **session service access point (SSAP)**, which establishes a session connection. Once this session connection is established, the presentation layer uses the services of the session layer at this point during the session connection. These session layer services are used to offer orderly exchange of dialogue messages between presentation entities over this session connection and also maintain the status of the dialogue message between presentation entities even if there is a loss of data by the transport layer. A presentation entity may have more than one session connection at any time, and more than one session connection (consecutive) can be defined between two presentation entities. The source presentation entity may define the address of the destination presentation entity by the session address which is mapped to the transport address on a one-to-one correspondence basis. It has been observed that we can define more than one session address for a transport address.

14.3.1 ISO 8326 session layer services

ISO session layer services:

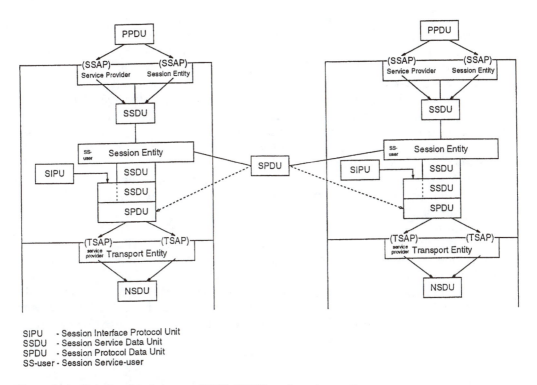

- Provides organization and interaction between user processes through dialogue (session connection establishment)
- Identifies validity of end user's right to make a request for connection and also use an application program of the network

- Defines synchronization points to offer synchronization and error recovery in data transfer between session entities. In reality, error recovery is not provided by the session entity; it offers the synchronization points, which are then used as services offered by the session layer to the transport layer entity. The error recovery is performed by the application layer entity between session entities.
- Controls and monitors transport connection, such that in the event of its failure, re-synchronization and re-establishment of transport connection can be defined without any interruption in the session service.
- Defines a set of procedures, rules for establishing a logical session for user applications between end users.
- Defines the time slots for end users to send the data over the session connection such that, at the end of its transmission, it informs the other user about its turn to transmit the data.
- Expedites data exchange (expedites handling of transfer of SSDU by any presentation entity over the session connection).
- Provides session connection synchronization (allows the presentation entities to define synchronization points and reset the session connections).
- Provides exception (session malfunctioning, etc.) notification to presentation entities.
- Provides quarantine service (presentation entity may define SSDU as quarantine, which may be discarded on sender's request after these have been trasnmitted to the destination).

14.3.2 ISO 8326 session layer functions

Session layer functions include mapping and transformation of session connections into transport connections, session connection, flow control, expedited data transfer, session connection recovery, session connection release, and session layer management (a subset of OSI layer management).

14.3.3 Modes of operation (ISO session layer services)

Session layer services operate on three types of modes of operations (for the data transfer and managing of dialogue between end users): **full-duplex**, **half-duplex**, and **two-way alternate**.

In the full-duplex mode, end users first mutually agree on parameters such as quality of service, session requirements, token types, session service (SS) user data, and initial synchronization points during the session connection establishment. The standard primitives such as S-Data.request and S-Data.indication are used to transmit data. After the session connection establishment, both end users start transmitting the data back and forth over the session simultaneously, thus reducing any kind of controlling or managing of session connections. It is possible that the higher-layer protocols may not support a full-duplex mode of operation, but that simply indicates that only half-duplex modes of operations are supported by them.

In half-duplex mode, the end user data flows in only one direction. The destination site is responsible only for accepting the data from the service user and storing it for its user, and it is known as the **destination server**. It uses the primitives S-Data.request and S-Data.indication. In order to provide proper synchronization between sender and receiver, the primitives S-Token-Give.request and S-Token-Give.indication are used to transmit the data in one direction at a time by either one of the sites. The token data is used to decide which site will be transmitting and which site will hold the token transmit first. After finishing its transmission, it issues a primitive S-Token-Give.request to its peer to transmit its message.

In two-way alternate mode, the end users define their turns to send the data using primitives. Once an end user has completed its turn of sending data, it informs the other end user of its turn to send the data. This type of mode of operation is useful for the transfer of very short data (i.e., inquiry/response application). In this mode, the end user informs the session entity about its intent of sending the data and, after it gets its turn, sends the data and informs the session entity about the completion of its turn. The session entity (service site) transmits all SSDUs to the session entity of the destination site along with its turn to transmit the data. The destination site session entity sends all the SSDUs to its users and informs them about its turn to send the data.

Various operations such as organized synchronization, connection establishments required for transfer of data between two user processes, etc., are performed by the session layer on behalf of the user processes. The session connection between session entities and the transport connection between transport entities for structured and synchronized transfer of data between user processes are also provided by the session layer.

14.4 ISO 8326 session layer service operation

The session layer services have been defined in ISO 8326 standards, and similar standards for the services of the session layer have been defined by CCITT X.215/X.225[12,13] to be used with PSDNs. The CCITT has also defined another standard known as T62 to be used with PSTNs.[14] According to these standards (or documents), the session layer must provide an organized and synchronized environment for data transfer between user processes (session entities). The services provided by the session layer to the user are described as session service user (SS-user), which uses session protocol primitives (dialogue) to establish session entities between user processes (SS-user). Some of these services are described in more detail in the following sections, particularly, session service dialogue, token services, different types of synchronization points, and session protocol primitives and parameters. Some of these service operations are detailed in References 2–5 and 7.

14.4.1 Session service dialogue

The session service provided to the SS-user by session protocol primitives for data transfer includes the following steps:

1. After a user process (SS-user) sends a primitive for a connection request, the session connection between session entities (or SS-users) is established.
2. The data transfer takes place between these two SS-users over an established session connection (session entities) in an organized and synchronized manner. At the end of the data transfer, the session connection is released in a systematic and orderly way.
3. The mutual agreement for the use of tokens to define an organized and synchronized environment for data transfer, release of connection, support of different modes (half-duplex, full-duplex, and two-way alternate), etc., is established before the data transfer can take place.
4. There is mutual agreement on synchronization points within a dialogue (session entities), and if there is any error, the process of resuming the dialogue from agreed upon synchronization points is also predefined or settled.
5. The dialogue is interrupted and is resumed later on at a prearranged synchronization point.

Based on OSI-RM documents, the lower layers are required to provide reliable service to the higher layers, and the higher layers (4 through 7), in general, are not responsible for error-control operations. The session layer, however, does not provide error control but offers reliable service in delivering all session service data units (SSDUs) before a release of the session connection. It is interesting to note that if lower layers are releasing their connections, the session service data units will be discarded.

14.4.2 Token service

The session protocol offers services to the SS-user through tokens. The tokens give the SS-user an exclusive right to use the services. Only limited services can be involved with tokens. The tokens are exchanged between SS-users by using session protocol primitives, and once a token is under the control of a user, the user can request token(s) from another user or it can pass the token to it. Each token obtained or passed over provides a specific service. If tokens are not available during the established session, then the services corresponding to those tokens are not available to SS-users.

The session layer has four types of tokens: **data token**, **release token**, **synchronize-minor token**, and **major/activity token**. The data token is used in the half-duplex mode of operation. The primitive S-Token-Give.request is used to pass the token to its peer entity. Parameters are defined to request a particular type of token by issuing a primitive S-Token-Please.request. The release token indicates the initiation of orderly release of the token. The synchronize-minor token allows the insertion of minor synchronization points, while the major/activity token defines major synchronization operations. The primitive S-Control-Give is used to surrender all the tokens immediately.

14.4.3 Activity management

This important feature of the session layer allows the user to break his/her message into a number of independent logical units known as **activities**. The session connection can transfer the data stream between two computers where the data stream represents a sequence of messages. Each message must be separated from the others, and this is usually obtained by using delimiter control characters. These control characters (start of text, end of text, etc.) provide a boundary between different messages. The control characters may also become a regular data character of the data message, and as such the boundary of the data messages is corrupted. This problem is solved by using a technique known as *stuffing*. These control characters are further supported in separating the data messages by considering each data message transfer an independent activity in such a way that each activity is invoked by using a request primitive to indicate the start of the data message and another primitive to indicate the end of the data message.

The primitive to transmit a data message is S-Activity-Start.request, which is received by the destination machine as S-Activity-Start.indication, which indicates the start of a data message following the primitive. At the end of each data message transfer (activity), another primitive, S-Activity-End.request, is issued by the sender to indicate the end of the data message and is received by the destination as S-Activity-End.indication. The session layer does not define the activity. It is up to the user to define his/her own activity and use the primitives defined by the session layer. The main responsibility of the session layer is to provide these primitives in such a way that when a sender machine uses the S-Activity.request primitive, the receiver machine immediately receives the indication about this via an S-Activity.indication primitive. The session layer will execute these primitives and is not concerned with what they contain, their use, or even their semantics.

There are many other primitives that perform various activities (discussed below), and all these activies are controlled and managed by an activity manager. The activities must be negotiated and agreed upon by both users before they can be used. In order to avoid the use of the same activity by two users simultaneously, the activity manager defines a token for activities (as is done with synchronization points). Any node wishing to use any activity must possess the token first by using the activity token primitive. After the execution of an activity primitive, the token is passed on to other users, after which a request for other tokens (for synchronization and data message) can be made separately. Any node wishing to send a data message must possess three different types of tokens before an appropriate operation can be performed. For example, using any activity primitive requires an activity token. After this, a node needs a synchronization token (major or minor) to define and insert these points into the message. Finally, the possession of a data token is required to send the data messages.

There is some relationship between the operations of activity and the synchronization point in the sense that whenever an activity is invoked, it internally resets the value of the synchronization point to a serial number of 1 and initializes the major synchronization point. This point is also the starting point of the activity and cannot be changed to any value lower than this value or before this value even by a re-synchronization primitive; i.e., this initial point of the activity cannot become any synchronization point in the previous activity. Once an activity with a major synchronization point of initial number 1 has been started, other synchronization points (major or minor) can easily be defined afterward within an activity.

14.4.4 Synchronization points

Tokens can invoke certain session services such as data transfer procedures, release of certain user dialogues, supporting synchronization services, and supporting operations across different synchronization services. Each of these services can be invoked by tokens using session layer protocol primitives, and once these services are under the control of an SS-user, the user gets an explicit right for re-synchronization at the end of each dialogue and can now define minor synchronization points within a dialogue. These synchronization points can be used by tokens to synchronize with data transfer.

Consider a situation where the session layer receives an SPDU and sends its acknowledgment. After sending acknowledgment, if this message (SPDU) is likely to be lost due to any error either at the devices or malfunctioning, it is impossible to recover it, as the sender does not have this message and is not going to send it again while the transport connection remains active during the transmission of SPDU and its acknowledgment can be sent. In order to solve this problem, the session layer divides the messages into blocks and inserts synchronization points between blocks. In case of any error or malfunction, the synchronization point will reset its position to the previous value and continue from there. In this way, the entire message is not lost, but a few blocks may have been lost. For achieving this, another process re-synchronization is defined where the sender session layer must keep this message for a longer duration. The session entity has nothing to do with this.

The synchronization points are useful in coordinating and managing data transfer between session entities. The synchronization points are just like checkpoints/resetting in a data file transfer, or a label in a data file, or any operation in the text data file related to its transfer. The synchronization points can be included in the message by the users, and each synchronization point is assigned a number (between 0 and 999,999). The use and indication of synchronization is defined with the help of primitives such that whenever a user is requesting a synchronization point, this is immediately received as an indication primitive by another (destination) user. Similarly, when a user issues a request for re-synchronization, it will be received by another user as an indication primitive. The synchronization points define the boundaries or isolation of dialogues and can also be used for error recovery. Some of the material herein is derived partially from References 2–4 and 7.

Error recovery is not one of the services of the session layer, but the session layer determines the synchronization point by going back to the previous synchronization point from where the user can recover by using error-recovery procedures. Further, the session service data unit (SSDU) is not saved by the session layer once it is sent, and the session layer does not perform any operation on error recovery, as this is the function of the application layer. The session layer simply provides a means or framework where these defined synchronization and re-synchronization signals are transmitted across the network. The saving of these signals and their re-transmission are handled by layers above the session layer.

In order to provide synchronization between session entities, two types of synchronization points with their own set of primitives are defined: **major synchronization** and **minor synchronization**. These tokens provide different characteristics for different points and are also different from the one used for controlling data flow across the network. These tokens are restored to the locations corresponding to where they were used as synchronization points by using a re-synchronization primitive.

Major synchronization points: The major synchronization point defines the structuring of data transfer in terms of the sequence of dialogue units. Each of the dialogue units defines its lower and upper boundaries such that the data contained within it is separated from other data at the boundary points. After the source SS-user has defined synchronization points, the user has limited session services until acknowledgment/confirmation of the synchronization point by the destination SS-user is received by it. Each major synchronization point needs to be acknowledged. The dialogue unit offers atomic communication within it and is not accessible by the data outside the unit. Error recovery requires storing of the synchronization points information for which acknowledgment/confirmation is awaited, and as soon as it gets acknowledgment/confirmation, it purges the synchronization points information stored at the beginning of sending the dialogue unit. This concept of major synchronization points through dialogue units is very useful and is a highly application-based service function offered by the session layer.

Each major synchronization point inserted into the data stream is explicitly confirmed (acknowledged), and the user can insert points only after getting a token. The possession of major synchronization is required to use any primitive for inserting points into the data stream. Each primitive specifies the serial number of points it wants to set or wants to go back to that number.

Minor synchronization points: Minor synchronization points are defined for structuring data transfer within dialogue units. These points offer more flexibility than synchronization points in error recovery. In the case of major synchronization points, the next dialogue unit can be sent only when it has received acknowledgment/confirmation of the previous dialogue unit; thus, the waiting time may be a problem. However, in the case of minor synchronization points, the user may define one or more synchronization points within dialogue units. In other words, these points do not get acknowledgment or confirmation. Here the SS-user does not have to wait for acknowledgment/confirmation, and the user can start sending the data through dialogue units. Again, the primitive must get possession of a minor token to insert points within the data stream and, once the token is captured, it may define a serial number for the points it wants to use or wants to go back. It is possible to re-synchronize any minor synchronization points within dialogue units or re-synchronize to the beginning of a dialogue unit at the major synchronization points which are saved for error recovery via re-synchronization primitives.

An acknowledgment/confirmation of a dialogue unit confirms the receipt of all previous minor synchronization points. The user can decide between major and minor synchronization

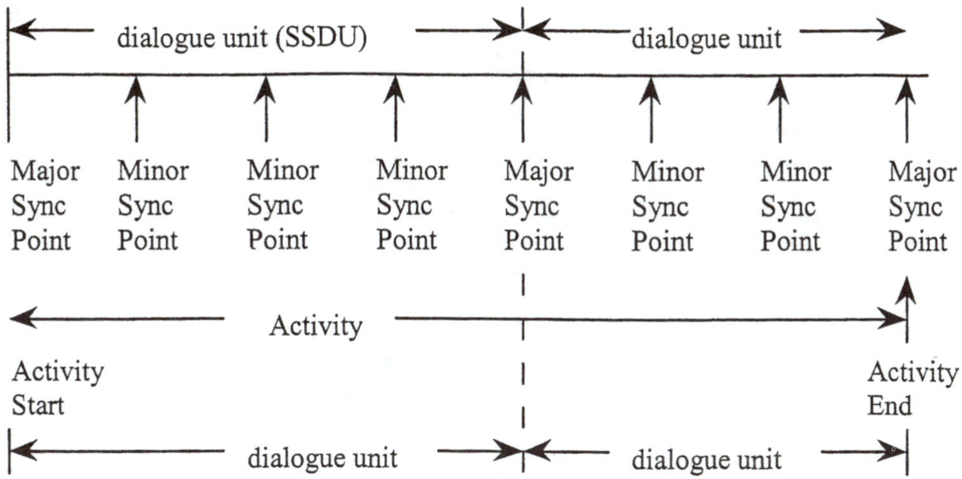

Figure 14.2 Structured dialogue unit, activity, and various synchronization points.

points for a dialogue unit which, in turn, depends on backup information for recovery, speed, waiting time, and the number of checkpoints to be used during the transfer of data. A combination of logically related operations corresponding to dialogue units can be defined and is considered active dialogue. This dialogue contains related operations which are performed on a text file or data. It can also be interrupted and resumed as other regular dialogue units. More than one active dialogue unit can be invoked during the established session connection. An active dialogue unit can perform re-synchronization, as various dialogue units together define an active dialogue unit which can be transmitted over different session connections. The active dialogue unit can be interrupted and reinvoked during another session connection if it can be acknowledged/confirmed or needs to be re-synchronized with the dialogue unit at its beginning, as shown in Figure 14.2.

14.5 ISO 8326 session layer service primitives

All the services provided by the session layer are grouped into three phases: session connection establishment, data transfer phase, and connection release phase, as shown in Table 14.1. The primitives corresponding to request and confirmation are handled by the session layer entity, while the primitives for indication and response are handled by a lower-layer entity. There are 58 primitives for connection-oriented session services defined by ISO 8326[8] and two for connectionless session layer services defined by ISO 8326/DAD3.[9]

14.5.1 Session connection establishment phase

This is an initial step for the establishment of a session connection between two session users (transport connection is already established), mutual agreement over the parameters for quality of service, initial synchronization point number, initial assignment of tokens, and use of session services (options). Other options to be negotiated during session connection establishment include activity management, exception reporting, etc. A list of primitives for connection-oriented services is given in Table 14.1.

Table 14.1 Session Layer Service Primitives and Description

Primitive	Description
Connection-oriented service primitives	
Establishment of connection	
S-CONNECT.request	Establish a connection of service, requirements, serial number, token, data
S-CONNECT.indication	Establish a connection SSAP, quality of service, requirements, serial number, token, data
S-CONNECT.response	Establish a connection SSAP, quality of service, requirements, serial number, token, data
S-CONNECT.confirm	Establish a connection SSAP, quality of service, requirements, serial number, token, data
Release of connection	
S-RELEASE.request	Release or termination of connection
S-RELEASE.indication	Release or termination of connection
S-RELEASE.response	Release or termination of connection
S-RELEASE.confirm	Release or termination of connection
S-U-ABORT.request	Initiated by user for abrupt termination or release of connection
S-U-ABORT.indication	Initiated by user for abrupt termination or release of connection
S-P-ABORT.indication	Initiated by service provider for abrupt termination or release of connection
Data transfer	
S-DATA.request	Normal data transfer
S-DATA.indication	Normal data transfer
S-EXPEDITED-DATA.request	Expedited data transfer
S-EXPEDITED-DATA.indication	Expedited data transfer
S-TYPED-DATA.request	Out-of-band data transfer
S-TYPED-DATA.indication	Out-of-band data transfer
S-CAPABILITY-DATA.request	Control information data transfer
S-CAPABILITY-DATA.indication	Control information data transfer
Token management	
S-TOKEN-GIVE.request	Give a token to its peer
S-TOKEN-GIVE.indication	Give a token to its peer
S-TOKEN-PLEASE.request	Request for a token from its peer
S-TOKEN-PLEASE.indication	Request for a token from its peer
S-CONTROL-GIVE.request	Give all tokens to its peer
S-CONTROL-GIVE.indication	Give all tokens to its peer
Activity management	
S-ACTIVITY-START.request	Request for starting of an activity
S-ACTIVITY-START.indication	Request for starting of an activity
S-ACTIVITY-RESUME.request	Request for restarting of suspended activity
S-ACTIVITY-RESUME.indication	Request for restarting of suspended activity
S-ACTIVITY-INTERRUPT.request	Interrupt or suspend an activity
S-ACTIVITY-INTERRUPT.indication	Interrupt or suspend an activity
S-ACTIVITY-INTERRUPT.response	Interrupt or suspend an activity
S-ACTIVITY-INTERRUPT.confirm	Interrupt or suspend an activity
S-ACTIVITY-DISCARD.request	Discard or abandon an activity
S-ACTIVITY-DISCARD.indication	Discard or abandon an activity

Table 14.1 (continued) Session Layer Service Primitives and Description

Primitive	Description
S-ACTIVITY-DISCARD.response	Discard or abandon an activity
S-ACTIVITY-DISCARD.confirm	Discard or abandon an activity
S-ACTIVITY-END.request	End of an activity
S-ACTIVITY-END.indication	End of an activity
S-ACTIVITY-END.response	End of an activity
S-ACTIVITY-END.confirm	End of an activity
Synchronization	
S-SYNC-MINOR.request	Request for inserting a minor sync point
S-SYNC-MINOR.indication	Request for inserting a minor sync point
S-SYNC-MINOR.response	Request for inserting a minor sync point
S-SYNC-MINOR.confirm	Request for inserting a minor sync point
S-SYNC-MAJOR.request	Request for inserting a major sync point
S-SYNC-MAJOR.indication	Request for inserting a major sync point
S-SYNC-MAJOR.response	Request for inserting a major sync point
S-SYNC-MAJOR.confirm	Request for inserting a major sync point
S-RESYNCHRONIZE-MINOR.request	Go back to a previous sync point
S-RESYNCHRONIZE-MINOR.indication	Go back to a previous sync point
S-RESYNCHRONIZE-MINOR.response	Go back to a previous sync point
S-RESYNCHRONIZE-MINOR.confirm	Go back to a previous sync point
Exception reporting	
S-P-EXCEPTION-REPORT.indication	Report by provider about exception
S-U-EXCEPTION-REPORT.request	Report by user about exception
S-U-EXCEPTION-REPORT.indication	Report by user about exception
Connectionless service primitives	
S-UNITDATA.request	Connectionless data transfer
S-UNITDATA.indication	Connectionless data transfer

Source: This table is partially derived from References 2–4 and 7.

14.5.2 Data transfer phase

This phase includes operations such as user's request for tokens, transfer of data, release of tokens, synchronization between SS-users for flow control and error reporting, etc. Various related services are grouped into appropriate groups. The following is a list of different types of data being defined by the session layer services. Each type of data is handled by its established set of class primitives.

Normal data transfer (NDT): Offers normal transfer of data units (SSDUs) between two SS-users over a session connection session and supports full-duplex and half-duplex operations. Data transfer needs tokens. The primitive of class S-DATA is used for this type of data.

Expedited data transfer (EDT): Offers expedited data transfer SSDUs (of up to 14 octets) over a session connection, with no constraints on tokens and flow control. This type of data is assigned with a priority over normal data and is transmitted before the normal data. This type of data finds its applications in short inquiries, messages, etc.

Typed data transfer (TDT): This data is just like regular data and can be sent without possessing a token. This data is received at the destination via another primitive as S-TYPED-DATA.indication. There is no need to assign tokens for sending the SSDUs; data can be sent in half-duplex mode. Typically, this type of data is used in control messages or for some other use, as decided by the session layer service. This type of data is defined with a view to providing an out-of-band data stream to its higher layers for control information such as network maintenance, system maintenance and management, etc. It is worth mentioning here that the transport layer does not recognize this type of data.

Capability data exchange (CDE): This type of data is usually used for controlling options of the session layer to be exchanged between users. It sends data SSDUs (up to 512 octets) when activity services are available. It may also be used for controlling various activities which are exchanged during session connection establishment, and many other uses of this type of data can be defined by the session layer. This type of data is also acknowledged and uses its own set of primitives.

14.5.3 *Synchronization-related services*

This type of service defines various synchronization points to be used to provide synchronization between users. All the synchronization primitives require possession of a token as a prerequisite condition for using the appropriate primitive. The synchronization points are assigned with a serial number between 0 and 999,999.

Major synchronization point: This is a kind of acknowledgment that allows the user to define major synchronization points which structure the data transfer into a series of dialogue units (SSDUs). These points provide a clear-cut boundary of data flow before and after them. Each point is assigned a serial number by a primitive that can be set to any value or revert to a previous value. The data will be transmitted only after it receives the confirmation of receipt of the major synchronization point. It requires the possession of a major synchronization token to use its allocated primitives. All primitives are confirmed.

Minor synchronization point: This structures the data transfer within dialogue units. It allows the user to define minor synchronization points with the dialogue units (SSDUs) when it is transmitted. No confirmation is required for each minor synchronization point (in contrast to each major synchronization point), but users may request explicit confirmation of the points received by the user at the receiving side. Any time the sending site receives a confirmation, it means that all the previous minor synchronization points have been received. These points are more flexible than major synchronization points and can be re-synchronized within the dialogue. Here also, each of the minor points is assigned a serial number by its primitive and can be set or go back to the previous value. All primitives are confirmed.

Re-synchronization: Re-synchronization allows the user to re-establish or set a session connection to a previous synchronization point. It sets a new value of the synchronization point, but this point cannot go beyond the first major synchronization point in the previous synchronization point value. It does not consider a minor synchronization point, but as soon as it sees the major synchronization point, it sets the value to that point and, hence, the session connection is restored or re-established or reset at that point. It discards all undelivered data.

14.5.4 *Exception reporting–related services*

These types of services basically deal with the reporting of any error or unaccepted situation. The reporting of errors can be initiated by either the user or the service provider (session layer itself).

Provider-initiated exception reporting (PIER): In this service, the SS-provider, using P-EXCEPTION reporting service, informs the SS-users about the error problems (not covered by other services), session protocol errors, and any exceptional situation. The SS-users cannot initiate any other service until the action by the SS-user has cleared the error-exception reporting information. With this service, the SS-provider discards any outstanding or undelivered data.

User-initiated exception reporting (UIER): The SS-users can report an error condition or exception situation. The U-EXCEPTION service is used by the SS-user for error reporting, which may be caused when the data token is assigned to the other user after the first user has sent the data to it.

14.5.5 Activity management–related services (ARS)

The activity service manages various activities within the session layer services. There are five categories of functions which are managed by various activities, as discussed below.

Activity start (AS): This activity indicates the start of a new activity which has been entered. The synchronization point value is initially set to 1. This activity service can only be used when there is no activity service already existing.

Activity resume (AR): This activity resumes or restarts and re-enters a previously interrupted service.

Activity interrupt (AI): This activity causes an abnormal termination of current activity so that all the computational work and states of the activity will not be lost and can be resumed later.

Activity discard (AD): This service is similar to activity interrupt (AI), with a difference in the resumption of the interrupted activity. Here the computational work and state of interrupted activity is canceled; hence, when it is resumed, it has a new state and is entered afresh.

Activity end (AE): This activity causes a permanent termination or end of the activity and sets a new major synchronization point.

14.5.6 Session connection release phase services

The services in this phase cause the termination of session connection between SS-users and can be achieved in the following three ways:

Orderly release (OR): This service allows the SS-users to mutually agree on the session connection release after the data has been delivered and a complete accounting of the data has been considered. The parameters associated with primitives provide a dialogue between SS-users in the form of an inquiry/result parameter and also inform them as to whether the connection release is accepted or not.

User-initiated abort (UIA): This service allows SS-users (using the U-ABORT primitive) to release the connection immediately. All the outstanding services will be canceled and all undelivered dialogue units (SSDUs) will be discarded.

Provider-initiated abort (PIA): The SS-provider (using the S-ABORT primitive) releases the session connection immediately due to reasons within the SS-provider domain. All the undelivered dialogue units will be discarded.

14.5.7 Connectionless service primitives

The session layer service contains two primitives for connectionless service where the datagrams with necessary control information are transmitted over the network (see Table 14.1).

14.6 ISO 8327 session layer protocol primitives and parameters

The preceding section discussed various session layer service primitives used for connection-oriented and connectionless services. This section is an extension of the previous section, and these primitives are now discussed in more detail along with associated parameters and other design issues, as described in Table 14.2.

14.6.1 Connection-oriented services

The session layer services are divided into three main phases: (1) session connection establishment, (2) data transfer, and (3) session connection release. The following sections present a detailed discussion of various session protocol primitives and parameters for each of these phases.

Session connection establishment phase: In this phase, various primitives with associated parameters and tokens are used to establish connection between two session service users (SS-users). The mutual agreement between two SS-users is defined using these parameters and tokens during this phase. The parameters may include messages or attributes associated with them, e.g., request, indication, response, and confirm (Figure 14.3(a)). The protocol using these parameters along with attributes is transparent to session users. Depending on the parameters defined within services, the resultant action is known to the session users.

In this phase (session connection establishment), a user process uses the S-CONNECT.request primitive with various parameters. These parameters are used for mutual agreement on various issues required to support the session connection. The parameters associated with this primitive are discussed below:

1. **Session connection identifier:** Defines a unique identifier value for the requested connection.
2. **Calling (source) and called (destination) session service access points (SSAPs):** The service access points at the session layer are used to define the addresses of SS-users. These addresses may also contain common reference points or any other reference points, if so desired.
3. After the SS-user sends a request, the destination site may either accept or reject the request, depending on the status of the called SS-user. If the called SS-user is busy or the SS-provider is busy, the called SS-user is inactive, or the called SSAP is incorrect, then the called SS-user will reject the request by sending a primitive containing the parameters to their effect.
4. **Quality of service (QOS):** The QOS parameter defines the services which are provided by the SS-provider to SS-users. These parameters are mutually agreed upon between SS-users, after which a session connection is established. These parameters are, in fact, part of the connection establishment and provide information about the features and characteristics of an established session connection to SS-users. The various parameters associated with QOS include delay and failure probability during connection establishment, throughput, transit delay, residual

Table 14.2 Session Layer Service Primitives and Parameters

Primitive	Parameters
S-CONNECT.request	Identifier, calling SSAP, quality of service, requirements, serial number, token, data
S-CONNECT.indication	Identifier, calling SSAP, called SSAP, quality of service, requirements, serial number, token, data
S-CONNECT.response	Identifier, calling SSAP, called SSAP, quality of service, requirements, serial number, token, data
S-CONNECT.confirm	Identifier, calling SSAP, called SSAP, quality of service, requirements, serial number, token, data
S-DATA.request	Data
S-DATA.indication	Data
S-EXPEDITED-DATA.request	Data
S-EXPEDITED-DATA.indication	Data
S-TYPED-DATA.request	Data
S-TYPED-DATA.indication	Data
S-CAPABILITY-DATA.request	Data
S-CAPABILITY-DATA.indication	Data
S-TOKEN-GIVE.request	Tokens
S-TOKEN-GIVE.indication	Tokens
S-TOKEN-PLEASE.request	(Token, data)
S-TOKEN-PLEASE.indication	(Token, data)
S-CONTROL-GIVE.request	
S-CONTROL-GIVE.indication	
S-SYNC-MINOR.request	Type, serial number, data
S-SYNC-MINOR.indication	Type, serial number, data
S-SYNC-MINOR.response	Serial number, data
S-SYNC-MINOR.confirm	Serial number, data
S-SYNC-MAJOR.request	Serial number, data
S-SYNC-MAJOR.indication	Serial number, data
S-SYNC-MAJOR.response	Data
S-SYNC-MAJOR.confirm	Data
S-RESYNCHRONIZE-MINOR.request	Type, serial number, tokens, data
S-RESYNCHRONIZE-MINOR.indication	Type, serial number, tokens, data
S-RESYNCHRONIZE-MINOR.response	Serial number, tokens, data
S-RESYNCHRONIZE-MINOR.confirm	Serial number, tokens, data
S-P-EXCEPTION-REPORT.indication	Reason
S-U-EXCEPTION-REPORT.request	Reason, data
S-U-EXCEPTION-REPORT.indication	Reason, data
S-ACTIVITY-START.request	Activity, ID, data
S-ACTIVITY-START.indication	Activity, ID, data
S-ACTIVITY-RESUME.request	Activity, ID, old activity, serial number, old SC ID, data
S-ACTIVITY-RESUME.indication	Activity, ID, old activity, serial number, old SC ID, data
S-ACTIVITY-INTERRUPT.request	Reason
S-ACTIVITY-INTERRUPT.indication	Reason
S-ACTIVITY-INTERRUPT.response	
S-ACTIVITY-INTERRUPT.confirm	
S-ACTIVITY-DISCARD.request	Reason
S-ACTIVITY-DISCARD.indication	Reason
S-ACTIVITY-DISCARD.response	
S-ACTIVITY-DISCARD.confirm	
S-ACTIVITY-END.request	Serial number, data
S-ACTIVITY-END.indication	Serial number, data
S-ACTIVITY-END.response	Data

Table 14.2 (continued) Session Layer Service Primitives and Parameters

Primitive	Parameters
S-ACTIVITY-END.confirm	Data
S-RELEASE.request	Data
S-RELEASE.indication	Data
S-RELEASE.response	Result, data
S-RELEASE.confirm	Result, data
S-U-ABORT.request	Data
S-U-ABORT.indication	Data
S-P-ABORT.indication	Reason

Source: This table is partially derived from References 3 and 7.

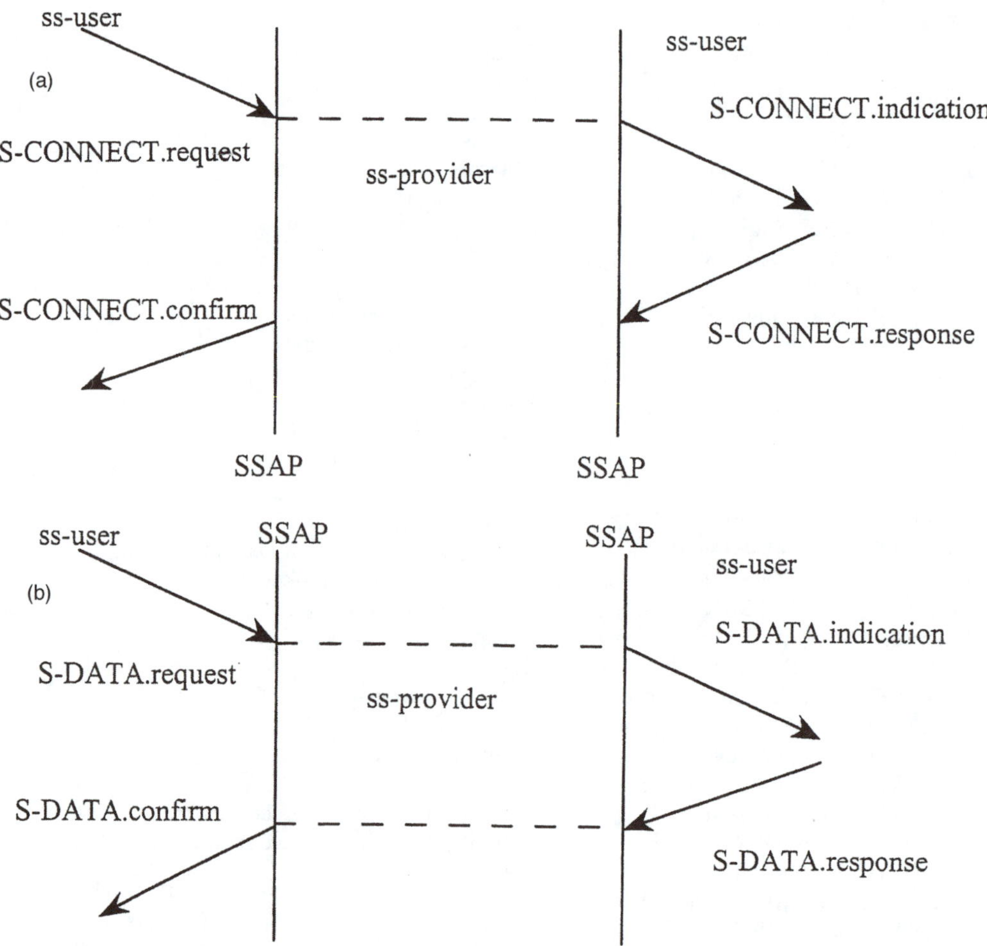

Figure 14.3 (a) Session connection primitives. (b) Session data transfer primitives.

error rate, delay in session connection relay, protection, priority in session connection, etc. The mutual agreement on QOS parameters is established during the connection establishment phase.

5. **Session requirements:** The session services are functional units and may be requested by SS-users. The entries in session requirement parameters represent the use of these functional units (see list of functional units below). The functional units are used to define related services offered by the session layer in logical groups. These functional units are used by the SS-user to mutually agree upon requirement parameters required for the session connection establishment with the destination SS-user and also will be used by higher layers. CCITT X.215 offers 12 functional unit services, and each of the functional units is associated with the actual service offered by it. The functional units are discussed in References 4 and 5.
6. **Initial synchronization point number:** An initial number is proposed in the range of 0 to 999,999 when synchronization services are used. This value is assigned to the first major or minor synchronization point and re-synchronization functional unit, whichever is defined.
7. **Token:** This parameter defines the initial assignment of tokens to the SS-user side to which the tokens are initially assigned, and it contains the right to access the services of the SS-user side.
8. **SS-user data parameter:** Defines the user data and range up to 512 octets.

The mutual agreement on the QOS parameters is established during the connection establishment phase. These parameters are used with appropriate primitives between calling and called SS-users. For example, a calling SS-user, in an S-CONNECT.request primitive, may define session connection protection and priority as a Boolean variable, error rate and delay as the minimum acceptable value desired, etc. The primitive S-CONNECT.indication for every parameter for mutual agreement is used by the provider to define the agreed upon state of any parameter(s), value(s), or desired state(s). In the primitive S-CONNECT.response, all the mutually agreed upon parameters are conveyed to the called SS-user. The called SS-user will respond to these agreed upon parameters by communicating with the SS-provider, which then uses an S-CONNECT.confirm primitive to inform the calling SS-user about the response from the called SS-user.

Once these parameters have been mutually agreed upon between SS-users during connection establishment, next comes the agreement on session requirements parameters for session connection establishment. Here again, the functional units which have been agreed upon in attributes (indication and response) are the ones supported by the SS-provider and hence will be used for session connection establishment. Similarly, for the initial assignment of token parameters (for all available tokens), attributes (request, indication, response, and confirm) will define the value or number selected (in the range 0–999,999) by either the calling or called SS-user.

A list of functional units, along with various parameters associated with them, is given below:

- **Non-negotiable services (Kernel):** Session connection, normal data transfer, orderly release, U-Abort, and P-Abort.
- **Two-way alternate communication** of data by getting tokens between SS-users by turn, give tokens, please tokens.
- **Two-way simultaneous communication** of data between SS-users, typed data transfer between SS-users with no token restrictions.
- **Expedited data transfer** without token with resetting of flow control.
- **Capability (limited)** data exchange without operating within an activity.
- **Negotiated release** (orderly) by passing tokens back and forth between SS-users, orderly release, give tokens, please tokens.

- **Invoke major synchronization** points, minor synchronization point, give token, please tokens.
- **Re-establish connection** and reset connection, re-synchronize informing of exceptional problems by SS-user or SS-provider, provider exception reporting, user exception reporting.
- **Define** different functions within an activity.
- **Activity start,** activity resume, activity interrupt, activity discard, activity end, give tokens, please tokens, give control.

For the connection request from the calling user, the called user can accept the session connection by

1. Accepting the parameters and quality of service parameter proposed by the calling user,
2. Accepting the connection request for a few parameters and proposing an acceptable parameter to the calling user, or
3. Accepting the connection request and proposing unreasonable parameters not acceptable to the calling user.

In the first two situations, the connection is confirmed and hence is established after the users mutually agree on the parameters. In the third situation, the calling user will issue an abort SPDU after it invokes a confirmed primitive response using the session protocol entity from the SS-provider, which will then notify the called user.

The confirmed primitive issued by the SS-provider may also be used to either establish a connection between users or reject the connection request due to reasons such as non-support of QOS parameters, non-support of establishing transport connection, and inability to provide required communication. Since the users and SS-provider interact locally, no transmission of SPDUs is required.

Data transfer phase: During this phase, the data transfer and dialogue structuring are defined through primitives between established SS-users (Figure 14.3(b)). The main aim of the session service provider is to accept various data units from the calling SS-user and deliver them to the called SS-user based on mutually agreed upon services and parameters during the session establishment phase. The primitives provide various operations on data units for different services such as normal rate, expedited rate, segmentation, synchronization, re-synchronization, error handling and notification, token services, etc. The tokens are used to offer various equivalent modes of operation (half-duplex, full-duplex, two-way alternate) for the application layer user process.

The session layer protocol primitives perform operations such as segmentation (if it is required); then the session layer protocol will segment the data units at the calling site and reassemble them at the called SS-user site. The normal rate of operation is controlled by primitives for token control and flow control limitations, while the expedited rate of operation is free of any token control and flow control limitations. The parameters used with primitives for data transfer and dialogue structuring are S-DATA.request and S-DATA.indication being used by SS-users. The other options provided are the transmission of data units as typed data transfer, data exchange service, which are dependent on the limitations of token control and flow control.

Once the session connection between end users is established, a set of SPDUs for data transfer is transmitted for the exchange of data between them by issuing appropriate primitives. The user data obtained from the presentation layer contains user data (from its previous layer, i.e., application layer) and the presentation layer header. The session layer adds its own header to this user data and passes it on to the transport layer. The user data is defined as a service data unit (SDU) and will be passed on to the transport layer by the session layer as SSDUs. Since the session layer protocol contains transparent implementation of transport layer services and structuring of data units, the SPDU will try to match the size of the SSDU with that of the transport service data unit (TSDU). If the size of the SSDU exceeds the size of the TSDU, the session layer will segment the user's data and transmit it using SPDUs during the data transfer phase. If the SSDU size matches that of the TSDU, then only one data SPDU is required to be transmitted, which will carry a session header, SSDU, and other parameters defined in the primitives which invoke the transmission procedure using session protocol entity.

The information regarding whether the size of both SSDUs and TSDUs match is defined by parameters of connect and accept SPDUs. This interrupts the session protocol entity which will invoke the segmentation procedure. This procedure will segment the user's data and transmit it in more than one data transfer SPDU. The data transfer SPDUs also include a parameter which defines the last SPDU in that group of segments. The last segment of data will be sent on a single SPDU. On the receiving site, the session protocol entity will store the segments sent via SPDUs in a buffer until it receives the last SPDU for the data group. Once it receives it, it reassembles the data and sends it to the user using an S-DATA.indication primitive.

The expedited data transfer service is provided by the prepare SPDU. The prepare SPDU, when sent, informs the session entity of the receiving site about the expedited data transfer SPDUs. These SPDUs may include parameters such as synchronized major acknowledgment, re-synchronize activity interrupts, activity discard, and activity end. These parameters inform the session entity of the receiving site to get ready to receive expedited data transfer SPDUs indicating the priority of the incoming data transfer SPDUs. In some cases, if the ordinary data transfer SPDUs either have arrived or are expected to arrive, they may be discarded in favor of expedited data transfer SPDUs.

Use of a token allows the user to obtain control of certain services offered by the SS-provider. Only four tokens involve services of the SS-provider: the **data token**, **release token**, **synchronization-major token**, and **synchronization-minor token**. These tokens were described in brief earlier, and we now discuss them in more detail, along with their functions.

Data token: Each user gets a token, sends the data, and passes the token to another, which, after sending its data, passes the token to other users and so on in a controlled manner. The users communicate with each other via half-duplex connection.

Release token: This token will release the connection major synchronization token and minor synchronization token (i.e., it defines two types of tokens of synchronization points). These synchronization tokens are used to define synchronization-major and synchronization-minor points, respectively. Each of the synchronization points is associated with a unique serial number that distinguishes between synchronization-major and synchronization-minor points. These serial numbers are assigned to these points by the session service provider in such a way that error control and flow control for data transfer can be handled and maintained. These numbers are then used in a request primitive for synchronization points. These numbers are assigned in equispace sequence for consecutive synchronization-major and -minor points which are helpful in providing re-synchronization (due to loss of primitives or any reason) by the session service provider.

The difference between the major synchronization primitives and minor synchronization primitives is that, in the case of the latter, a calling user may use a S-SYNC.MINOR.request primitive which requests an acknowledgment for every synchronization-minor point it receives before it can send the data. In the case of the synchronization-major primitive, for each of the issued primitive S-SYNC-MAJOR.requests, the calling user it must receive S-SYNC-MAJOR.confirmation from the called user before it can send the data. This means that the major synchronization points are always acknowledged and hence use more powerful primitives than minor synchronization points.

As explained earlier, the half-duplex connection is offered to the SS-users, and due to the type of data transfer and structured dialogue, the token service may be used to request tokens using token primitives **please token**, **give token**, and **give control token**. The please token service allows the SS-user to make a request for tokens, while the give token service allows the SS-user to return the tokens. An SS-user may pass on tokens to another SS-user using a give control token service.

The synchronization-major and synchronization-minor points facilitate re-synchronization by the session service provider, and once a calling SS-user has issued an S-SYNC-MAJOR.request primitive, it cannot send the data until it receives an S-SYNC-MAJOR.confirmation from the called SS-user. However, during this time the calling user may issue requests for token primitives and other non-synchronization activities. Similarly, the called SS-user, after receiving an S-SYNC-MAJOR.indication from the SS-provider, cannot issue any synchronization primitives until it issues an S-SYNC-MAJOR.response to the SS-provider. During this time, no data transfer takes place.

During the connection establishment phase, the use of tokens between SS-users is also mutually agreed upon, and the tokens, after the established connection, will be used by SS-users accordingly. The tokens may have different states (CCITT X.215) and, depending on the states, the tokens can be used by the SS-users. In other words, if a token is available, an SS-user assigned to this has complete control of it. If the token is unavailable, this means that the token is being used for either data transfer or synchronization. In each case, the SS-user does not get complete control of the token service.

The parameters used in the S-RESYNCHRONIZE primitive are used to define the type of value assigned to the serial number to be used by the SS-provider. Depending on the value assigned, the following options are offered:

- **Abandon:** Re-synchronizes the session connection for any number value of number equal to or greater than the next synchronization point number. No error recovery is required; the connection is not disconnected/terminated; only the present dialogue is terminated.
- **Restart:** This option is used mainly for error recovery to the previous mutually agreed upon synchronization point of the dialogue such that the number of this synchronization point should always be greater than the previous synchronization-major point number which is acknowledged and hence confirmed. After the error recovery, the SS-user may either re-transmit the same SSDU or perform any other recovery function.
- **Set:** This option does not provide any error recovery; it terminates the current dialogue but retains the connection. Any number value can be assigned to the synchronization point.

These options — abandon, restart, and set — use the re-synchronization primitive (S-RESYNCHRONIZE) with parameter attributes request, indication, response, and confirm and can be implemented using different semantics by SS-users.

Session connection release phase: The session connection can be released by either SS-user by issuing an S-RELEASE.request. This primitive offers orderly release of the session connection, which includes mutually agreed upon parameters indicating the completion of data transfer, accounting of data transfer between them, and, of course, acceptance

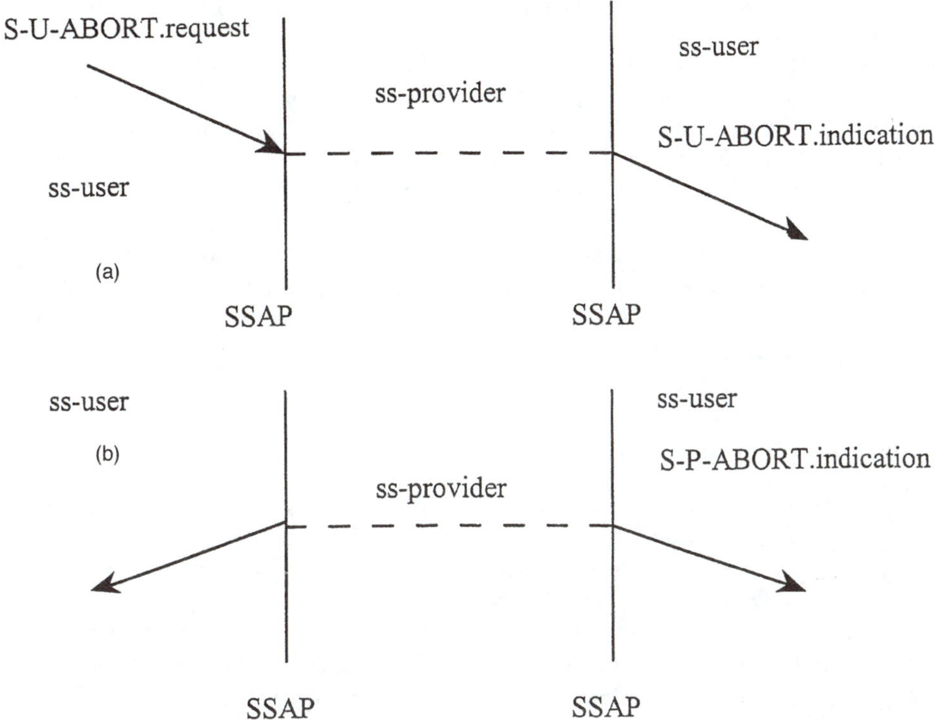

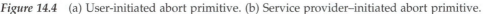

Figure 14.4 (a) User-initiated abort primitive. (b) Service provider–initiated abort primitive.

or rejection of request for the release of session connection. The primitives S-RELEASE.response and S-RELEASE.confirm define the results of acceptance or rejection of session connection in the parameters associated with them.

The SS-users can also use the primitive U-ABORT to release the session connection immediately, and all the outstanding data units will be discarded (Figure 14.4(a,b)). The session connection can also be released by the SS-provider through the primitive P-ABORT, and the session connection is released immediately as in the primitive U-ABORT. All the outstanding data is discarded. The only difference between the U-ABORT and S-ABORT service primitives is that, in the case of the S-ABORT primitive, the reason parameter for the abort primitive is issued by the SS-provider. The orderly release primitives provide a mutually agreed upon continuation or release of the session connection between SS-users such that the use of these primitives with agreed upon parameters will prevent any loss of data units.

Error recovery and notification: Error recovery and error notification are among the important services offered by the session layer service protocol. In the preceding paragraphs, we discussed the primitives which are used to provide error recovery to the SS-users, through the SS-provider. The error notification service is also provided by the SS-provider to its SS-users. The primitives P-EXCEPTION and U-EXCEPTION are used to inform the SS-users about the error problems and possible conditions and error notification options. The SS-provider invokes a P-EXCEPTION service primitive to inform the SS-users about the error conditions. The U-EXCEPTION service may be invoked by SS-users to get notification/information on error conditions and problems.

Once the SS-user has been informed about error conditions by the SS-provider, the SS-user must first take appropriate action to remove the error before it can issue a primitive for any other service. The SS-user may remove an error by issuing primitives for re-synchronization service, abort service, or give token service. The SS-provider rejects all outstanding data units after issuing a P-EXCEPTION service primitive. Various activity services are defined for session layer protocol data units (SSDUs) which provide interaction between the SS-user and SS-provider regarding the error conditions, reason for errors, etc. The U-EXCEPTION service primitive gets the notification from the SS-provider and may also request the reason by passing reason parameters to the SS-provider in the request primitive.

The request primitive is provided by interrupt and discard activity services, and the reason parameter may be requested by indicating various possibilities for error conditions such as the SS-user's not handling data properly, local SS-user error, sequence error, data token, non-specific error, unrecoverable procedural errors, etc. The SS-provider passes on the reason parameter values to indicate the reason for the errors.

14.6.2 Connectionless session services

The session layer also supports connectionless services provided by the transport layer. The ISO (DIS 9584)[11] standard defines the connectionless session protocol. The connectionless session protocol standard includes two session service primitives (connectionless) known as S-UNIT-DATA.request and S-UNIT-DATA.indication. The parameters defining quality of service are sent by the session layer entity to the transport layer entity only; they are not sent to the SS-user of the receiving site. The connectionless service standard uses the same parameters of QOS as the connection-oriented standard (ISO 8373). The S-UNIT-DATA.request primitive includes the calling session address (calling SSAP), called session address (SSAP), quality of service, and SS-user data, while S-UNIT-DATA.indication includes all parameters except quality of service.

The connectionless session service does not provide orderly delivery, flow control, or guaranteed delivery of the message to the destination address. It does not require the establishment of a session connection with the destination SS-user; it requires only one SPDU as a unit data (UD) SPDU. For the connection release, no other SPDU is needed.

14.7 ISO 8327 session layer protocol data unit (SPDU)

The service primitives with parameters are used to offer layer services and are passed from one layer to another layer. The primitives used in session layers are the ones we have discussed in the preceding paragraphs, and the functional units along with the associated session services are given in Table 14.1 and in the list in Section 14.6.1.

The session layer services do not provide confirmed and reliable services; instead, one can expect these services to be provided by the transport layer services. Even if this is not true, the structuring of data transfer is defined with the help of token control where these token-based primitive parameters continuously inform the SS-users about the transmission of data between them. Different types of token primitive parameters (discussed above) define different types of states and the status of the data. In other words, for these two services, the session layer protocol must have transparent implementation of transport layer services. At the same time, session layer primitives defining QOS parameters are transferred directly to the transport layer. These parameters were discussed and explained earlier.

14.7.1 Functions of SPDUs

The session protocol data unit (SPDU) defines a mapped function of various session service primitives such that it represents a primitive (with appropriate parameters) issued by the calling SS-user and the response (with results in parameters) by the called SS-user. In other words, one SPDU will represent a set of inquiry/response-type primitives. The SPDU is invoked by the session protocol entity and is transmitted carrying a primitive and its parameters to the called site for passing to the session protocol entity. This entity then delivers this information to the response-type primitive.

Let us look at the session connection establishment phase, which is considered to be the most sophisticated phase, as the mutual agreement on a number of parameter requirements, etc., is established and confirmed before the establishment of connection. On the calling site, a request for connection, S-CONNECT.request, is issued. This request invokes a procedure known as Connect SPDU using the user's session layer protocol entity. During this time of invoking, all the parameters of S-CONNECT.request are copied into the SPDU along with a command to transmit to the other user's site. The parameters contained in the SPDU typically include the connection ID, the proposed initial serial number value for synchronization points, token-control information (the initial site selected for assignment of token), requirements parameters (defining functional units), calling and called SSAP, the result parameter (indicating acceptance or rejection of connection request), reason parameters, and parameters for quality of service (already agreed on by selecting options of the session layer protocols).

The SPDU containing the parameters is transmitted to the other site, where it invokes a procedure known as Receive SPDU using the other side's session layer protocol entity. This entity delivers the parameters and connection request to its primitive S-CONNECT indication. If the called user accepts the connection request and mutually agrees, then the response primitive with the results of the parameters transmits an Accept SPDU using its session protocol entity, which invokes the primitive S-CONNECT.confirm. Once the calling site has received this primitive, the session connection is established. If the called user does not want to accept the connection request (due to its busy state, unavailability of a particular application, non-agreement with the parameters, or any other reason), it notifies its session protocol entity and sends its response primitive to it. This primitive includes the reason parameter for the rejection of the session connection request. The session protocol entity transmits a Refuse SPDU to the calling site. It also includes the result of parameters (refusal of connection request) and the reason for the refusal. Various session layer primitives along with associated parameters are shown in Table 14.2.

14.7.2 Types of SPDUs

As mentioned earlier, there are a large number of service primitives provided by the session layer services, and the implementation of all these service primitives will make a protocol very complex and unmanageable. As such, ISO has defined four basic subsets of services (discussed below).[10] For each of the service primitives, the protocol defines a session protocol data unit (SPDU) which is transmitted when this particular service primitive within the protocol is requested. There are some SPDUs for which response SPDUs are generated. A list of SPDUs and corresponding response SPDUs is given in Table 14.3. Here the service primitive initiated from the initiator invokes an SPDU, and if it has a response, another SPDU will be invoked to give the response. Except for two (out of 21 SPDUs), all the SPDUs are invoked by user processes.

Each of these SPDUs is transmitted by the session entity whenever a primitive of the **.request** type is issued. For example, when a user process issues an S-CONNECT.request primitive, a corresponding routine in the session layer protocol is invoked and the session entity prepares a Connect SPDU. This SPDU is transmitted toward the destination. When this SPDU arrives at the destination machine, the session enitity of that node defines another primitive, S-CONNECT.indication. The user process at the destination may either

Table 14.3 SPDUs and Corresponding Response SPDUs

Primitive	Request/response SPDU
Connect	Session connection initiated (request SPDU)
Accept	Session connection established (response SPDU)
Refuse	Session connection request rejected (response SPDU)
Finish	Order release initiated (request SPDU)
Disconnect	Order release acknowledged (response SPDU)
Not Finished	Order release rejected (response SPDU)
Abort	Session connection release abnormally aborted (request SPDU)
Abort Accept	Abort primitive acknowledged (response SPDU)
Data Transfer	Normal data transferred (request SPDU)
Expedited	Expedited data transferred (request SPDU)
Typed Data	Typed data transferred (request SPDU)
Capability Data	Capability data transferred (request SPDU)
Capability Data Ack	Capability data acknowledged (response SPDU)
Give Tokens	Tokens transferred (request SPDU)
Please Tokens	Tokens assignment requested (request SPDU)
Give Token Confirmation	All tokens transferred (request SPDU)
Give Token Ack	All tokens acknowledged (response SPDU)
Minor Sync Point	Minor sync point defined (request SPDU)
Minor Sync Ack	Minor sync point acknowledged (response SPDU)
Major Sync Point	Major sync point defined (request SPDU)
Major Sync Ack	Major sync point acknowledged (response SPDU)
Re-synchronize	Re-synchronization requested (request SPDU)
Re-synchronize Ack	Re-synchronization acknowledged (response SPDU)
Prepare	Arrival of SPDU notified (system SPDU)
Exception Report	Protocol error detected (request SPDU)
Exception Date	Error state of protocol is retained (request SPDU)
Activity Start	Arrival of signal (request SPDU)
Activity Resume	Resumption of signal (request SPDU)
Activity Interrupt	Interrupt activated (request SPDU)
Activity Interrupt Ack	Interrupt acknowledged (response SPDU)
Activity Discard	Activity canceled (request SPDU)
Activity Discard Ack	Canceled activity acknowledged (response SPDU)
Activity End	Signal activity ended (request SPDU)
Activity End Ack	Ending of signal activity acknowledged (response SPDU)

Source: Partially derived from References 2, 3, and 7.

accept this request or reject it. In either case, another primitive, S-CONNECT.response, is invoked with proper parameters to indicate clearly whether it accepts it or rejects it. Depending on the decision, the appropriate SPDU (either Accept or Refuse) will be transmitted back to the sender. Similarly, for other request SPDUs, a response SPDU (if it exists) will be received from the destination node.

14.7.3 Session service subset (SSS)

The session standard protocol provides a variety of services, and if we start implementing all the types of services of the session layer, it will make the protocol complex and it will also occupy a relatively large memory space. No application will require all of the services, and hence those services will simply make the implementation of services an unnecessarily complex protocol. In order to avoid this problem, both CCITT and ISO have jointly recommended the partitioning of session layer services of the standards into four subgroups:

Table 14.4 Session Layer Service Subgroups

Service	Kernel	BCS	BSS	BAS
Session Connection	Y	Y	Y	Y
Normal Data Transfer	Y	Y	Y	W
Expedited Data Transfer	N	N	N	N
Typed Data Transfer	N	N	Y	Y
Capability Data Transfer/Exchange	N	N	N	Y
U-Abort (user-initiated)	Y	Y	Y	Y
P-Abort (provider-initiated)	Y	Y	Y	Y
Orderly Release	Y	Y	Z	Y
Give Token	N	Y	Y	Y
Please Token	N	Y	Y	Y
Major Synchronization Point	N	N	Y	N
Minor Synchronization Point	N	N	Y	Y
Re-synchronize	N	N	Y	N
Provider-Initiated Exception Reporting	N	N	N	Y
User-Initiated Exception Reporting	N	N	N	Y
Activity Start	N	N	N	Y
Activity Resume	N	N	N	Y
Activity Interruption	N	N	N	Y
Activity Discard	N	N	N	Y
Activity End	N	N	N	Y
Give Control	N	N	N	Y

Note: Y = Yes, N = No, W = Half-duplex, Z = Negotiated release option available.
Source: Partially derived from References 2 and 3.

kernel (K), basic combined subset (BCS), basic synchronization subset (BSS), and **basic activity subset (BAS)** (see Table 14.4).[2,3]

The **kernel subgroup** defines a minimum implementation of the services which must be provided in any implementation. It invokes only those session layer functions which are required to provide the session connection establishment phase, and hence it must provide the transparent use of transport connections and transport protocol entities. It does not include any of the options or features provided by the session layer services.

The half-duplex connection and associated functional units of duplex operations are defined in the **basic combined subset (BCS)** and can be selected by the user. This subgroup finds applications in user terminal–host communication, host-to-host communication, etc. In each of these applications, a user terminal or host first requests the token and, after sending the data, passes the token on to another host which can then send the data back to the terminal/host.

The **basic synchronized subset (BSS)** defines the functional units for synchronization and connection release dialogue between users for mutual agreement on the quality of service, services required, etc., by passing parameters in SPDUs back and forth. This subgroup is used in many applications depending on the type of services required, such as connection-oriented services for file transfer, minimum quality of services required for a particular application, and so on.

The **basic activity subset (BAS)** is a structured subgroup of most of the services and functional units defined in session layer standards. The BAS includes an activity management functional unit and does not support various functional units such as full-duplex operations, minor synchronization, re-synchronization, and negotiated release. The relationship between session layer services, functional units, and these four subgroups is shown in Tables 14.4 and 14.5.

Table 14.5 Session Layer Functional Unit Subgroups

Functional unit	BCS	BSS	BAS
Kernel	Y	Y	Y
Half-duplex	Y	Y	Y
Full-duplex (Duplex)	Y	Y	N
Typed Data	N	N	Y
Expedited Data	N	N	N
Capability Data Transfer/Exchange	N	N	Y
Negotiated Release	N	Y	N
Major Synchronization Point	N	Y	N
Minor Synchronization Point	N	Y	Y
Re-synchronize	N	Y	N
Exception	N	N	Y
Activity Management	N	N	Y

Source: Partially derived from References 2 and 3.

14.7.4 ISO 8327 session layer protocol data unit (SPDU) structure

The session layer protocol ISO 8327 provides the above-mentioned operations on behalf of user processes and interacts with a reliable connection-oriented transport service. The session protocol provides a direct transformation of session connection into transport connection in the sense that multiplexing of session connections over a single transport connection and vice versa can be defined. This helps in identifying the boundaries for different types of services, functions, and procedures of protocols such that some of the responsibilities can be defined for the protocols. In other words, the common functions and services can be assigned to either the session protocol or transport protocol to reduce the burden on higher-layer protocols, which can be used for future enhanced services and functions to be provided to the presentation layer. In particular, the flow-control service can be assigned to the transport layer protocol.

As discussed in the last chapter, the transport layer protocols offer different classes of service (Class 0 through Class 4). A suitable class transport protocol for compatible network connection can be considered, thus reducing the burden on the session layer protocol.

Services and functions provided by the session connection and transport connection should be clearly identified, as each of the connections may be multiplexed, which would lower processing efforts and response time, provide information regarding the states of different connections, maximize the throughput of the system, etc. According to the OSI document, a transport connection can be multiplexed in a sequence over a single session connection, and a transport connection can support various session connections in sequence. The latter form of multiplexing has been considered for future work, while the former one has been accepted, and based on it, vendors' products and standards have been defined and implemented in the protocols.

The protocol does not discriminate between the type of file text (short or long), but short text, e.g., inquiry/response type, should not be multiplexed with long file texts, as it would provide a long delay to short file text service.

ISO DIS 8326 defined various session services and functions such as the establishment of session connections, organized and synchronized data transfer between user processes, mutual agreement process for session entities (dialogue) using tokens, release of session connections, etc. The ISO DIS 8327 document defines functions and protocol data units of the session layer such that services offered by the transport layer to the service requested by the session layer can be matched and interfaced. CCITT has also defined identical and

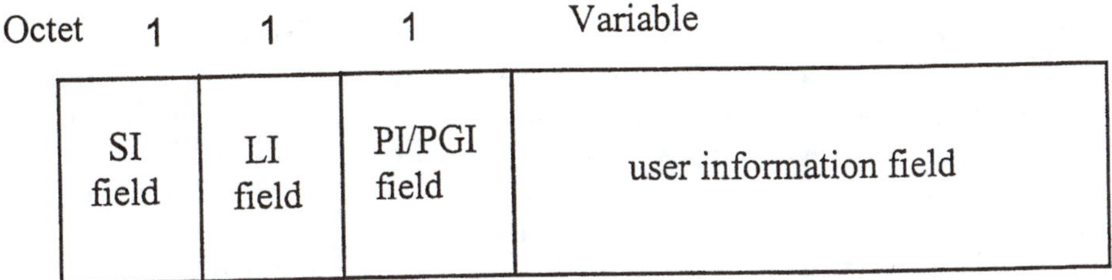

Figure 14.5 SPDU structure.

comparable standard forms CCITT X.215 and X.225, which correspond to ISO DIS 8326 and ISO DIS 8327, respectively.

General format of ISO 8327 SPDU: As we have seen in the previous section, the session protocol data units (SPDUs) provide various session layer services and are defined as routines within session layer protocols. Each SPDU carries a number of parameters which define the services offered by the protocol. Figure 14.5 shows the description of all the SPDUs used for both normal and expedited data transfer.

ISO standards have defined a flexible structure of SPDU due to the fact that the length of the parameters are different for different types of services, and therefore all these can be defined in SPDUs. The structure of SPDUs for different services generally includes four main fields: SPDU identifier (session identifier: SI), length indicator (LI), parameter identifier or parameter group identifier (PI or PGI), and user information. Some of the material herein is partially derived from References 2 and 7.

The type of SPDU (out of 34 SPDUs) being used is defined by a one-octet SI field (*CONNECT, DATA, TOKEN, etc.). The LI field (one octet) defines the length of the header indicating parameters used in the SPDU defined in the PI/PGI field. The value of the LI field varies from 0 to 255, indicating how many bytes of parameters are following it. In the case of more than 254 bytes of parameters, it typically takes the value of 255 and includes two extra bytes immediately after it. These additional two bytes give a maximum length of up to 65,535 bytes of parameters. The last field (user information) is optional and may be used (if required) to accommodate the user's data.

In the PI/PGI field, the parameters can be defined in a variety of ways, depending on the mutual agreement between SS-users. The parameters defined in the PI field may be described in terms of a parameter identifier, a length indicator, and the parameter value. In the PGI field, the parameters may be described in terms of a parameter group identifier, a length indicator, and the parameter value. The parameter values indicate the session layer services provided by the SS-provider and SS-users. Each group starts with a PGI followed by an LI which gives the length of the group.

There are many session layer primitives whose SPDUs carry parameters (e.g., destination address, synchronization point serial number, etc.). These parameters must be encoded within the format. There are a number of ways these parameters can be encoded. These parameters are usually grouped into two categories, as parameter identifier (PI) and parameter group identifier (PGI) (as discussed above). The session enitity defines an SPDU which becomes the user's data for the transport layer. The session layer protocol can combine all these SPDUs into one message and send it to the transport layer. This technique requires fewer invocations of transport primitives. This process is known as **concatenation.** It can be achieved by one transport header for a number of SPDUs, which are separated from each other by synchronization point values. The SPDUs may be multiplexed and defined in one TPDU such that one TPDU may contain several SPDUs.

The ISO 8327 standards of the session protocol supports 34 types of SPDUs. Each SPDU provides a particular option in session services and has a header of variable length which defines different parameters during various phases (connection establishment, data transfer, and connection release) of the session layer protocols. The number of octets of SPDUs is different for different phases. For example, the data transfer SPDU has a header of 30 octets in length, one octet for each SI, LI, and parameter (beginning or last SPDU of that data group) field. The user information field containing user data (a file, record, or any other type of information) is defined within a length of an integral number of octets. The user may choose any option in the types of services (provided by ISO 8327) by choosing appropriate SPDUs and the parameters associated with them.

Connection-oriented SPDU: The ISO session protocol provides an interface between the services offered by the transport layer and the services required by the session layer, and the session layer protocol also offers a transparent transport connection. In other words, we can say that the session layer provides the following two services:

1. Establishment and control of a transport connection between two users, with a minimum quality of service properties/characteristics.
2. Reliable data transfer for normal and expedited user's data. The session layer protocol provides a variety of services (SPDUs) on top of the services of the transport layer. These services are responsible for providing a structured and manageable environment for data transfer and exchange of data between users.

ISO 8327 and CCITT X.225 session protocol standards have been defined for these two services. The ISO 8327 includes 34 SPDUs which interact with 10 TPDUs such that the transport layer protocol will use an appropriate mechanism to provide a reliable data transfer environment to the session layer protocol. The session layer protocol will offer various options and services to the users. Therefore, the mapping of various transport service primitives into TPDUs seems to be more involved and complex than in the session layer protocol.

The interface or matching between transport connection and session connection is provided by the session protocol entity. The session connection should be established only after the transport connection has been established. The transport connection is responsible for providing connection to the normal and expedited data transfer. If it supports normal data transfer, then the primitives for normal data transfer service (T-DATA. primitives) are used to send the Abort and Accept SPDUs, and if the transport layer supports expedited data transfer services, then primitives for expedited data transfer (T-EXPEDITED-DATA. primitives) are used to send Expedited Data and Prepare SPDUs. The T-DATA. primitives are used in other SPDUs during the connection establishment and connection release phases.

For release of the connection, each layer protocol entity provides an option to terminate the connection. For example, if the session between SS-users is to be terminated, the session protocol entity which initially established the transport connection is the one which may release the transport connection by issuing an appropriate S-CONNECT. primitive. If it does issue this primitive, the transport connection remains active and may be used by the session layer entity to support a new session connection request. It is interesting to note that the transport connection will offer the same quality of services that it offered to the session layer entity which initially established this transport connection.

Connectionless SPDU: The session layer also supports connectionless services provided by the transport layer. The ISO (DIS 9584) standard defines a connectionless session protocol. The connectionless session protocol standard defines two session service primitives (connectionless) known as S-UNIT-DATA.request and S-UNIT-DATA.indication. The parameters defining quality of service are sent by the session layer entity to the transport layer entity only; they are not sent to the SS-user of the receiving site. The connectionless service standard uses the same parameters of QOS as the connection-oriented standard (ISO 8373). The S-UNIT-DATA.request primitive includes the calling session address (calling SSAP), the called session address (SSAP), quality of service, and SS-user data, while S-UNIT-DATA.indication includes all parameters except quality of service.

The connectionless session service does not provide orderly delivery, flow control, or guaranteed delivery of the message to the destination address. It does not require the establishment of a session connection with the destination SS-user; it requires only one SPDU as a unit data (UD) SPDU. For the connection release, no other SPDU is needed.

The structure of the unit data (UD) SPDU includes the following fields:

- The session identifier (SI) field defines the type of SPDU. Here SDPU is of type UD.
- The length indicator (LI) defines the length of parameters.
- The parameter identifier (PI) defines a particular parameter for which source and destination SSAP identifiers indicate the version number (id).
- The parameter value (PV) defines the value of the parameter being used and describes the version number for ISO standards as 1.
- The user information field includes the user data provided by the SS-user.

The connection establishment phase requires the same steps as those discussed earlier in ISO 8372 standards. For each S-UNIT-DATA.request primitive from the SS-user, an SPDU contains the parameters of the primitive, as well as calling and called SSAPs. The SPDU containing parameters and SSAPs is sent to the transport layer addresses, i.e., the transport service access points (TSAPs).

The TSAPs are determined from the corresponding session service access points (SSAPs). The SPDU is transmitted to the SS-user of the receiving site. The SS-provider sends an S-UNIT-DATA.indication primitive enclosed in an SPDU to the SS-user, which delivers the data to the users. Thus, the information field of the SPDU contains SS-user data parameters which indicate that the segmentation of data by the session layer is not possible. If the size of the SPDU does not match the TPSU, the data cannot be sent using the connectionless service. The chance of this situation is very rare, however, as this service is basically used for electronic mail, bulletin board services, or smaller user data transaction.

Some useful Web sites

Information on the session layer can be found at the Web sites *http://ic.net/~epn/session.htm* and *http://www.echelon.ca/handyman/nettech/session/sess.htm*.

References

1. ISO 7498, "Open System Interconnection — Basic Reference Definition, Information Processing Time," ISO, 1986.
2. Tanenbaum, A.S., *Computer Networks*, 2nd ed., Prentice-Hall, 1989.
3. Stallings, W., *Handbook of Computer Communications Standards*, 2nd ed., Howard W. Sams and Co., 1990.
4. Markeley, R. W., *Data Communications and Interoperability*, Prentice-Hall, 1990.

5. Henshall, J. and Shaw, A., "OSI Explained End-to-End Computer Communication Standards," Chichester, England, Ellis Horwood, 1988. (OSI session layer services and functional units are discussed in Chapter 6).
6. Emmons, W.F. and Chandler, A.S., "OSI session layer; services and protocols," *Proc. IEEE,* 71, 1397-1400, Dec. 1983.
7. Black, U., "Data Networks, Concepts, Theory and Practice," Prentice-Hall, 1989.
8. ISO 8326, "Open System Interconnection: Basic Connection-Oriented Session Service Definition, Information Processing Systems," ISO, 1986.
9. ISO 8326/DAD3, "Basic Connectionless-Mode Session Service."
10. ISO 8327, "Open System Interconnection: Basic Connection-Oriented Session Protocol Specification, Information Processing Systems," ISO, 1984.
11. ISO 9548, "Session Connectionless Protocol to Provide the Connectionless-Mode Session Service."
12. CCITT X.215, "Session Service Definition for Open Systems Interconnection for CCITT Applications."
13. CCITT X.225, "Session Protocol Specification for Open Systems Interconnection for CCITT Applications."
14. CCITT T.62, "Control Procedure for Telex and Group IV Facsimile Services."
15. Blaze, M. and Bellovin, S., "Session-layer encryption," *Fifth USENIX Security Symposium,* Salt Lake City, Utah, June 1995.

chapter fifteen

Presentation layer

"There's too much said for the sake of argument and too little said for the sake of agreement."

Cullen Hightower

"The second highest layer (layer 6) in the OSI seven layer model. Performs functions such as text compression, code or format conversion to try to smooth out differences between hosts. Allows incompatible processes in the application layer to communicate via the session layer."

The Free Online Dictionary of Computing
Denis Howe, Editor (*http://foldoc.doc.ic.ac.uk/*)

15.1 Introduction

The presentation layer offers an appropriate format to the data it receives from the application layer. The data from the application layer is end user application service and must be expressed in the same format as that of the receiving site. In addition to format, the presentation layer of both the sending and receiving sites should provide the same coding scheme, syntax, and semantics so that the encoded data can be transmitted to the peer application layer entity of the receiving site.

The lower layers (session to physical of OSI-RM) are, in general, responsible for establishing sessions and connections; controlling and managing the transfer of data, error control, and flow control; and offering these services to the user processes (application processes). These services are essential to establish a reliable network connection between user processes. The lower layers have no information about the format, code, or syntax used for user data and also are not concerned with services such as encryption and decryption of private or useful data, compression and decompression of text (i.e., file transfer), terminal handling, etc., to be provided by the presentation layer. None of these services adds any value to the data transfer services of the lower layers. A major problem with the upper layers (session, presentation, and application) of OSI-RM has been that these layers are very complex, and it is not transparently clear how to use them to build various applications, in particular for distributed applications. Further, the standards offer very little help in this direction.

The most common user requirements in using communication at present are for file transfer and remote log-in applications. In LANs, there are many other applications for these requirements, as these take advantage of the high speed and reliability of LANs to build a coupled system for dedicated applications. In such systems, the user requirements may include file transfers but will also consist of some kind of interprocess communication, as processes running either on the same machine or on different machines need to pass information or activate these processes. This type of interprocess communication for activation of the process of another machine is achieved by message passing or by placing information in shared memory.

Looking at the functions and services of different layers of OSI-RM,[1] it seems clear that a combination of network and transport layers usually offers end-to-end services and an internetworking environment. If we consider all the lower layers (transport to physical), we can say that the combined function of these four lower layers is to provide a transport service of required quality of service (QOS). Similarly, different combinations of higher layers (session, presentation, and application) offer a user interface to the network including synchronized dialogue (session), standardized user data representation (presentation), and common user semantics (application). Due to the fact that the higher layers are complex and also their usage is not clearly defined, it is felt among the network professionals and practitioners that the upper three layers should be considered one layer. A notion of pass-through is being advocated where the services of the session layer are also made available to the application layer (as opposed to the presentation layer as defined in the OSI document). Further, the request and response protocol data units for all of these layers should be defined as embedded data units.[2-6]

The encryption and decryption of data are very useful services for providing multiplexing of data transfer on an established transport connection between user processes. The services of the session and presentation layers have a very strong relationship through layer interfaces, as the session layer services have to know about the format type, coding technique of the presentation protocol data unit (PPDU), encryption algorithm for the data on text (or data) compression, etc., which are being provided by the presentation layer. Further, the detailed services of the session layer should be accessible to the application layer entities.

15.2 Layered protocols and interfacing

In OSI-RM, it is assumed that the lower layer always provides a set of services to its higher layers. Specific functions and services are provided through protocols in each layer and are implemented as a set of program segments on computer hardware by both the sender and receiver. This approach of defining/developing a layered protocol or functional layering of any communication process not only establishes a specific application of the communication process but also provides interoperability between vendor-customized heterogeneous systems. These layers are, in fact, the foundation of many standards and products in the industry today. The layered model is different from layered protocols in the sense that the model defines a logic framework on which protocols (that the sender and receiver agree to use) can be used to perform a specific application on it.

15.2.1 Layered protocols

Peer protocol layers exchange messages which have commonly understood and agreed upon formats, and such messages are known as **protocol data units (PDUs)**. PDUs also correspond to the unit of data in the layer peer-to-peer protocol. The peer presentation entities communicate with each other using the services provided by the session layer via a **session service access point (SSAP)**. The services of the presentation layer are provided to the application layer at a point known as the **presentation service access point (PSAP)**. This is the point through which its services can be accessed by the application layer. For a description of the interface between presentation and application layers, the presentation primitives are defined. The presentation service data unit (PSDU) associated with its primitives is delivered to the application layer and, similarly, the session layer uses its primitives to interact with the presentation layer. The application layer sends its application protocol data unit (APDU) to the presentation layer via the PSAP. The APDU consists of a protocol, control information, and application user data. The application control information is exchanged between session layer entities.

The ISO has defined the following terms used for data communication through OSI-RM. Each layer has active elements known as *entities*, and entities in the same layer of different machines connected via a network are known as *peer entities*. The entities in a layer are known as the service provider, and the implementation of services used by the higher layer is known as the service user. The **transport service access point (TSAP)** is defined by a unique hardware or port or socket address (software) through which the upper layer accesses the services offered by its lower layer. Also, the lower layer fetches the message from its higher layers through these SAPs. The PSAP defines its own address at the layer boundary. The **presentation service data unit (PSDU)** is defined as message information across the network to peer entities of all the layers. The PSDU and control information (addressing, error control, etc.) together are known as the **interface data unit (IDU)**, which is basically message information sent by an upper-layer entity to its lower-layer entity via SAPs. For sending a PSDU, a layer entity partitions the message information into smaller-sized messages. Each message is associated with control information defined by the appropriate layer as a header and sent as a protocol data unit (PDU). This PDU is also known as a *packet*. We use herein the terms defined by ISO. The relationship between the PSDU, presentation entity, and PPDU is shown in Figure 15.1.

15.2.2 Interfacing

Each layer passes message information (data and control) starting from the top layer (application) to its lower layer until the message information reaches the lowest layer (physical), which provides actual data communication by transmitting a bit stream over transmission media. An interface is defined between layers which supports the primitive operations and services the lower layer offers to its higher layer. The network architecture is then derived as a set of protocols and interfaces.

The functions and services of OSI-RM are a collection of processes at various levels. At each level, we have one layer for a particular process(es). Each layer in turn represents a specific function(s) and service(s), and these can be different for different types of networks. However, each of the layers offers certain services to its higher layers. The implementation of these services is totally hidden from the higher layers. The services offered by the layers are provided through protocols and are implemented by both source and destination nodes on their computers. Protocols are defined in terms of rules (through semantics and syntax) describing the conversion of services into an appropriate set of program segments. Each layer has an **entity process** that is different on different computers, and it is this entity (peer) process through which various layers communicate with each other using the protocols. Examples of entity processes are file transfer, application programs, electronic facilities, etc. These processes can be used by computers, terminals, personal computers (PCs), workstations, etc.

15.3 ISO 8822 presentation layer services and functions

The main objective of the presentation layer is to provide services for representing and formatting data into a suitable description of data structure which has no concerns about

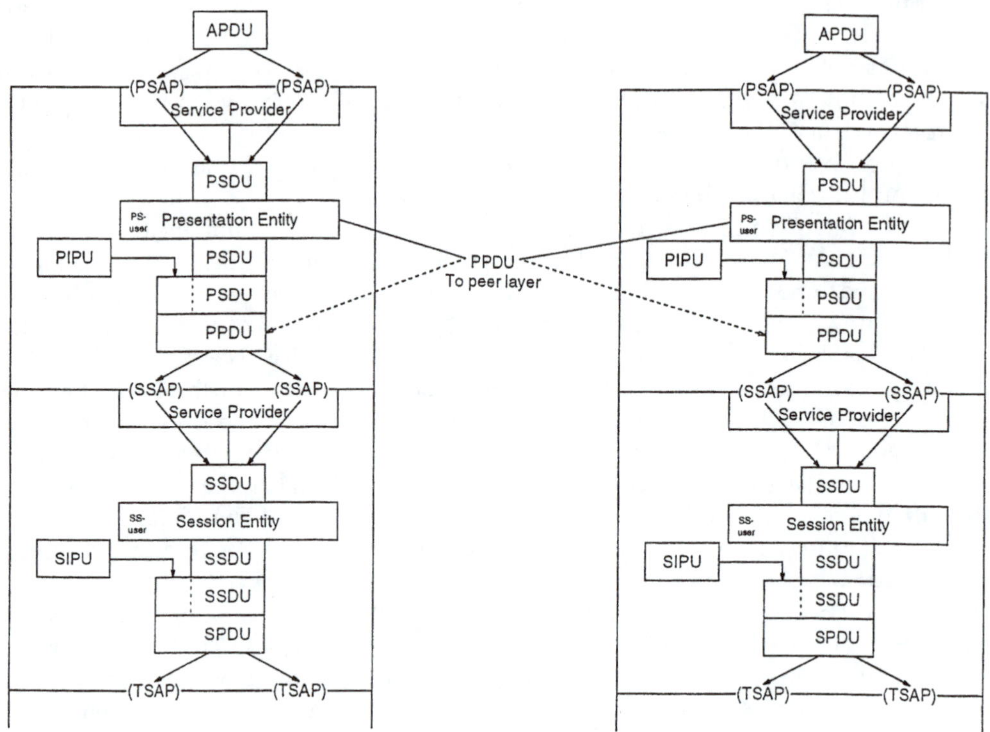

Figure 15.1 Relationship between the PSDU, the presentation entity, and the PPDU.

the semantics of the data. This form of representation of data is used by application entities with each other and is also used in their dialogue messages. Through mutual negotiations, all the message data sent by user processes is accepted by the presentation layer protocol, which performs functions such as encryption, compacting of messages and data, etc., and transmits it to the receiving site by sending it through its lower layers. The presentation layer of the receiving site decrypts and opens (or expands) the message data before it is given to the user process.

The dialogue message carries information regarding data syntax and presentation common syntax (data structure) between the presentation layer entities, and the dialogue message is requested by the application layer entity. This dialogue message also deals with negotiation over the type and location of the mapping between application syntax and presentation common syntax.

The presentation layer offers both syntax (for presentation layer entities) and semantics (for application layer entities) for defining a presentation form of data which offers both internal and external attributes of the document, respectively (ISO document [7]). The presentation layer does not offer any multiplexing or splitting of the messages but provides a one-to-one correspondence between the presentation address and session address.

The structure of message data can be implemented in a variety of ways depending on the application, and vendors have developed different methods for representing data fields and types and order of fields in the data structure (syntax of protocol data units) that are available on the market. The representation of formatting or syntaxing of the data in the presentation layer is always expressed in a common representation scheme for transmitting the encrypted data between two user processes. The common syntax representation is defined for presentation layer entities, while transmitting and receiving sites may have their own local representation or syntaxing of data (between their respective application entities).

It is important to mention here that the presentation layer is primarily concerned with the syntax part (internal attributes) of the data, while the semantic part (external attributes) of the data is mainly defined for application layer entities. The presentation layer offers a common syntax representation to the data transferred between application entities. The application layer entities may use any syntax representation, but this common syntax representation scheme offered by the presentation layer maps different forms of syntaxes into a common syntax required to provide data communication through OSI-RM between application entities.

The common representation (or syntax of data) between presentation entities is known as **transfer syntax**, while the data representation at application entities is known as **transmitting syntax** and **receiving syntax**. These syntaxes may be the same or different, depending on the applications and requirements. Each presentation layer entity selects the best syntax between it and the user process syntax (based on the requirements defined by application layer entities during the initial phase of connection establishment) and then negotiates with this selected syntax. After mutual agreement, this syntax will be used in the presentation layer entity of the receiving site. In other words, both the presentation layer entities decide, through mutual agreement or negotiations, a transfer syntax which provides a mapping or transformation between application entities and the users' syntaxes. The negotiation takes place through dialogue messages between the presentation layer entities after the presentation layer receives a request from the application layer entity (of the sender's side). The selected transfer syntax must also handle other services such as compression, terminal handling, file transfer, etc.

The above was a discussion of the features of the presentation layer. The services and functions of this layer are detailed below:

ISO presentation layer services: Data syntax mapping or transformation, mapping of presentation syntax, selection of data syntax, selection of presentation syntax.

ISO presentation layer functions: Session connection establishment request, negotiation of data syntax and presentation syntax, mapping or transformation of data syntax and presentation syntax, session connection termination request, presentation layer management (a subset of OSI layer management).

15.4 ISO 8822 presentation layer service primitives

Based on OSI-RM terminology, the services of the presentation layer are used by application layers, while it uses the services of the session layer. These services are available through various service primitives which basically represent procedures. These procedures have to be invoked by choosing a particular primitive which defines a set of parameters that can be used during establishment of a session between presentation entities (or presentation service users (PS-users). The interaction between session and presentation layers is provided via these service primitives. Let's consider a very general case of establishing sessions between session and presentation layer entities. Here, the PS-user is considered a presentation entity, while the PS-provider is considered a presentation

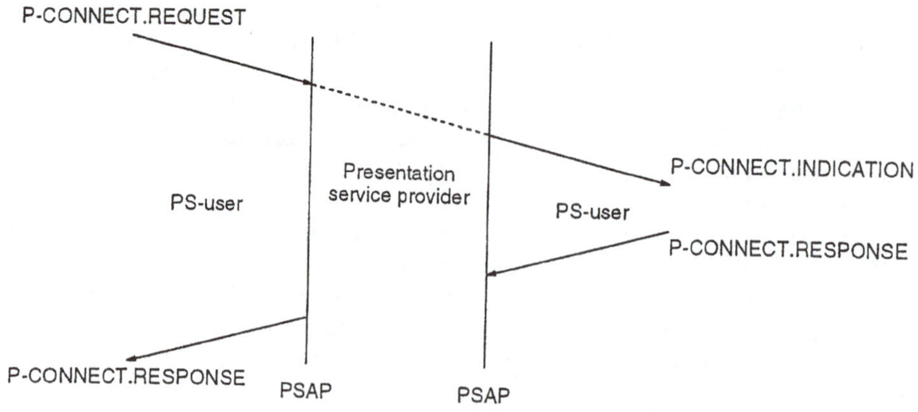

Figure 15.2 Function of presentation service provider.

service provider (presentation protocol data units) which provides a session between primitives of these two layers, as shown in Figure 15.2.

When an SS-user sends an S-CONNECT.request primitive, it invokes the procedure to transmit a CONNECT SPDU (at the SS-provider) toward a distant site SS-user. This primitive invokes the procedure S-CONNECT.indication on the receiving site, which in turn invokes another procedure to transmit another SPDU as an Accept or Reject SPDU back toward the SS-provider. This invoked procedure transmits an Accept or Reject SPDU toward the calling site. If confirmed service is required, two SPDUs are exchanged between users, while if unconfirmed service is required, only one SPDU can be transmitted and received between users. The services of any layer are implemented by its PDU; therefore, when a layer transmits its PDU to a lower layer, it becomes a part of the user data to be transmitted to the lower layer. In this case, it is the session layer which is providing its service through the SPDU to the transport layer and so on. This SSDU will be accepted as an SSDU by the session layer.

The presentation layer protocols have to deal with the services provided by the session layer, and at the same time it has to offer services to the application layer. In general, the OSI service primitives of the presentation layer are the same as those of the session layer, and as such all the presentation layer service primitives go over to the session layer. The application entity establishes a session using a P-CONNECT.request primitive, which simply invokes the presentation layer entity to issue an S-CONNECT.request primitive. The user data initialized by the application layer is sent to the presentation layer, which provides the digital logic 0 and 1 form of the user data expressed in octets and also converts various parameters associated with service data primitives into digital form. The digital form of the data is then grouped into presentation service data units (PSDUs) which are sent to the protocol entities within the presentation layer. These are also known as the semantics of the data. The lower layers (session through data link) are responsible for processing the service primitives (with which various parameters are associated) which are expressed in a sequence of bytes or octets. Table 15.1 shows the presentation layer service primitives for connection-oriented and connectionless services (ISO 8822).[8]

The user data coming out of the application layer is basically a structured set of messages and could be any application-oriented service offered by the application layer to the users, e.g., text file, database, personal file, or information regarding a visual display of some video-based services. These services are provided by the application layer in any

Table 15.1 Presentation Layer Service Primitives for Connection-Oriented and Connectionless Services

Primitive	Description
Connection-oriented service primitives	
Establishment of connection[a]	
P-CONNECT.request	Establish a presentation connection of service requirements, serial number, token, data
P-CONNECT.indication	Establish a presentation connection SSAP, quality of service, requirements, serial number, token, data
P-CONNECT.response	Establish a presentation connection SSAP, quality of service, requirements, serial number, token, data
P-CONNECT.confirm	Establish a presentation connection SSAP, quality of service, requirements, serial number, token, data
Release of connection	
P-RELEASE.request (data)	Release or termination of connection
P-RELEASE.indication (data)	Release or termination of connection
P-RELEASE.response (result, data)	Release or termination of connection
P-RELEASE.confirm (result data)	Release or termination of connection
P-U-ABORT.request (data)	Initiated by user for abrupt termination or release of connection
P-U-ABORT.indication (data)	Initiated by user for abrupt termination or release of connection
P-P-ABORT.indication (reason)	Initiated by service provider for abrupt termination or release of connection
Data transfer[b]	
P-DATA.request	Normal data transfer
P-DATA.indication	Normal data transfer
P-EXPEDITED-DATA.request	Expedited data transfer
P-EXPEDITED-DATA.indication	Expedited data transfer
P-TYPED-DATA.request	Out-of-band data transfer
P-TYPED-DATA.indication	Out-of-band data transfer
P-CAPABILITY-DATA.request	Control information data transfer
P-CAPABILITY-DATA.indication	Control information data transfer
Token management	
P-TOKEN-GIVE.request (tokens)	Give a token to its peer
P-TOKEN-GIVE.indication (tokens)	Give a token to its peer
P-TOKEN-PLEASE.request (token, data)	Request for a token from its peer
P-TOKEN-PLEASE.indication (token, data)	Request for a token from its peer
P-CONTROL-GIVE.request	Give all tokens to its peer
P-CONTROL-GIVE.indication	Give all tokens to its peer
Activity management[c]	
P-ACTIVITY-START.request (AI, data)	Request for starting an activity
P-ACTIVITY-START.indication (AI, data)	Request for starting an activity
P-ACTIVITY-RESUME.request	Request for restarting of suspended activity
P-ACTIVITY-RESUME.indication	Request for restarting of suspended activity
P-ACTIVITY-INTERRUPT.request (reason)	Interrupt or suspend an activity
P-ACTIVITY-INTERRUPT.indication (reason)	Interrupt or suspend an activity
P-ACTIVITY-INTERRUPT.response	Interrupt or suspend an activity
P-ACTIVITY-INTERRUPT.confirm	Interrupt or suspend an activity
P-ACTIVITY-DISCARD.request (reason)	Discard or abandon an activity
P-ACTIVITY-DISCARD.indication (reason)	Discard or abandon an activity

Table 15.1 (continued) Presentation Layer Service Primitives for Connection-Oriented and Connectionless Services

Primitive	Description
P-ACTIVITY-DISCARD.response	Discard or abandon an activity
P-ACTIVITY-DISCARD.confirm	Discard or abandon an activity
P-ACTIVITY-END.request (SN, data)	End of an activity
P-ACTIVITY-END.indication (SN, data)	End of an activity
P-ACTIVITY-END.response (data)	End of an activity
P-ACTIVITY-END.confirm (data)	End of an activity
Synchronization [d]	
P-SYNC-MINOR.request (T, SN, D)	Request for inserting a minor sync point
P-SYNC-MINOR.indication (T, SN, D)	Request for inserting a minor sync point
P-SYNC-MINOR.response (SN, D)	Request for inserting a minor sync point
P-SYNC-MINOR.confirm (SN, D)	Request for inserting a minor sync point
P-SYNC-MAJOR.request (SN, D)	Request for inserting a major sync point
P-SYNC-MAJOR.indication (SN, D)	Request for inserting a major sync point
P-SYNC-MAJOR.response (D)	Request for inserting a major sync point
P-SYNC-MAJOR.confirm (D)	Request for inserting a major sync point
P-RESYNCHRONIZE-MINOR.request	Go back to a previous sync point
P-RESYNCHRONIZE-MINOR.indication	Go back to a previous sync point
P-RESYNCHRONIZE-MINOR.response	Go back to a previous sync point
P-RESYNCHRONIZE-MINOR.confirm	Go back to a previous sync point
Exception reporting	
P-P-EXCEPTION-REPORT.indication	Report by provider about exception
P-U-EXCEPTION-REPORT.request (R, D)	Report by user about exception
P-U-EXCEPTION-REPORT.indication (R, D)	Report by user about exception, where R represents reason while D represents data
Context management [e]	
P-ALTER-CONTEXT.request	Change the context
P-ALTER-CONTEXT.indication	Change the context
P-ALTER-CONTEXT.response	Change the context
P-ALTER-CONTEXT.confirm	Change the context
Connectionless service primitives	
P-UNITDATA.request	Connectionless data transfer
P-UNITDATA.indication	Connectionless data transfer

[a] Parameters include source and destination presentation addresses, context definition list (containing identifiers for each item), default context name (automatically used), quality of service, presentation requirement (context management and context restoration functional units), mode (normal value is CCITT X.410), session requirements, serial number of token, session connection identifier, and data. The primitives' corresponding attributes indication and confirm include the result of the context definition (entry for each item either accept or reject) and default context (agreed or refused).

[b] These primitives use only one parameter (data). The expedited data is transmitted or processed with priority over normal data, and typically this type of data is used in signaling or interrupt situations.

[c] AI represents an activity ID, SN represents serial number. The primitives (P-ACTIVITY-RESUME with attributes request and indication) include the following parameters: activity ID, old activity ID, serial number, old serial number, and data.

[d] SN represents serial number, T represents type, and D represents data. The primitives RESYNCHRONIZE include the following parameters: type primitives with attributes request and indication only, serial number, tokens, data, and context identification list (primitives with attributes indication and confirm only).

[e] Parameters include context addition list, context deletion list, data, context addition list result (primitives with attributes indication, response, and confirm), and context deletion list result (primitives with attributes indication, response, and confirm).

Source: Partially derived from References 2–5.

form it likes without knowing the contents or value of the data within the messages. The users, of course, must know the contents or meaning of the data within the message. The contents, value, and meaning of the data within messages are known collectively as the **semantics of the data**. The data message is converted into digital form by the presentation layer via some representation scheme. Again, the presentation layer does not know the contents of data within messages; it simply converts data messages into digital form. The conversion of the data message into digital form defines the **syntax of the data**.

Both application and presentation layers are transparent to each other in the sense that each one knows about the semantics and syntaxes of data being used. The representation scheme of the application layer has complete knowledge about the syntax technique used by the presentation layer, and the syntax technique of the presentation layer has complete knowledge of the type and value of data within the message and is not concerned about the meaning of the data being used.

15.4.1 Functional units

Various presentation layer services (ISO 8823) are grouped into different functional units (similar to ones defined in session layer services) which are used between presentation service users: **PS-users (presentation entities)** for agreeing on presentation contexts (both transfer syntax and abstract syntaxes) and session services to be used during the establishment of presentation connection. The functional units are used during the presentation connection phase through which PS-users can negotiate on transfer syntax, transformations, conversion techniques, session service primitives, etc. The presentation layer services include three groups of functional units: **presentation kernel, context management,** and **context restoration**. Each of these units is discussed below in detail.

Presentation kernel functional unit: The presentation kernel functional unit offers basically all the essential services required for negotiating presentation context and the session services selected by PS-users. These services are available to PS-users who use the same representation scheme and transfer syntax for the data transfer and exchanges; i.e., both PS-users use the same presentation contexts (abstract syntax and transfer syntax) between application layer protocol entities. The transfer syntax converts the message data into digital form and allows the PS-users to use session services. Under this mutual agreement for the kernel functional unit, the presentation layer's request and response primitives for the connection are passed on to the session layer through the presentation service provider.

The session connection indication and confirmation primitives are sent by the session layer protocol to the application protocol entities. In this case, no conversion or transformation is performed by the presentation layer protocols. If both PS-users are not using the same representation (abstract syntax), then the presentation layer protocols have to provide conversion at both sites between respective entities. Both PS-users use the connection primitives to mutually agree on both abstract syntax and transfer syntaxes during the establishment of presentation connection between them.

If the presentation kernel functional unit is used, then both PS-users can use only those services which are provided by the PS-provider, provided no negotiation about the presentation context has taken place, i.e., the presentation context field corresponding to the mutually agreed upon context field is empty.

Context management functional unit: The context management functional unit allows users to change the present context during the establishment phase of presentation connection between PS-users. The presentation context may be either added with more options of presentation context or deleted, depending on the user's requirements and also whether it is offered by the presentation provider.

Context restoration functional unit: The context restoration functional unit provides communication between the context management functional unit and the session synchronization functional unit. This functional unit determines the state of the mutually agreed upon presentation context in the case of the re-synchronization functional unit's being used by the session layer.

The re-synchronization functional unit is composed of three options: **abandon, restart,** and **set**. The presentation services are provided in terms of parameters and primitives similar to the ones used in session layer services. These options in the re-synchronization functional unit are required to re-synchronize the presentation connection between PS-users in the event of failure or any error.

The **abandon** option allows the user to release the dialogue between PS-users without releasing the mutually agreed upon current context between them. If a request for a new connection is made, then the same context set will be reused. In the **restart** option, the dialogue between PS-users is shifted back to the previous synchronization-major point. The mutually agreed upon context point is also set to the value to which it had been set for the previous synchronization-major point. The **set** option depends on the serial number of the context set. If the serial number falls in either the abandon or restart option range, then the actions as discussed for these options are taken; otherwise, the mutually agreed upon context set for the known synchronization-major point is reset; i.e., the context set (mutually agreed upon) is restored to the value agreed on during the presentation connection phase. A detailed discussion of other implementation issues is discussed later in the chapter.

15.4.2 Presentation connection establishment service

ISO 8822 and ISO 8823[7,8] define the presentation context set (both abstract syntax and transfer syntax) at the synchronization-major points in the session layer operations. PS-users can change the presentation context via mutual agreement between them at synchronization-major points within the session operations. PS-users may select any transfer syntax (i.e., Huffman coding, run-length code, ASCII, etc. — to be discussed in the following sections) with the abstract syntax.

A presentation connection establishment requires P-CONNECT primitives with additional parameters compared to ones used in the session connection S-CONNECT primitive. The establishment of a presentation connection provides a step-by-step session connection. All the primitives along with parameters of the session connection will be supported by the currently established presentation connection. The parameters common to both presentation connection and session connection primitives are, in fact, used by the presentation layer to define mutual agreement and are then sent to the session layer entities, e.g., quality of service (QOS) and session requirements. Although these parameters have been defined in session connection primitives, they can be defined by the presentation layer, too. These two parameters, along with other parameters associated with session connection primitives, are then used to establish a session connection. The additional parameters associated with presentation connection primitives which are not present in session connection primitives are discussed below.

Context definition list: This list includes both abstract syntax and transfer syntax defined by PS-users. Each component of this list has an identifier which represents the types of syntax considered. Thus, for a given or known abstract syntax (defining the syntax used by the application layer), an appropriate transfer syntax is defined or selected by the PS-provider. Once these two syntaxes are defined, the initial presentation context list is established and will be used during the presentation connection.

Context definition result list: For each component in the context definition list, a parameter in the Boolean form is used to indicate acceptance or rejection of that component in the list.

Default context name: In order to transmit expedited data (which has a priority), the default context is used to avoid any delay in defining and mutually agreeing to the components of the context definition list. The PS-user sends a default context name to the PS-provider to indicate the abstract syntax of its application layer which is supported by the default context. This context will remain active until the expedited data has been transmitted.

Default context result list: It contains a Boolean type of parameter corresponding to the result in each of the components of the default context name.

Presentation requirements: This parameter defines two functional units of the presentation layer, context management and context restoration (discussed earlier), and can be selected by the user.

Mode: This parameter includes two modes of operation of the P-CONNECT primitive: normal and X.410 (1984). In the normal mode, all the parameters discussed above are used, while in X.410 P-CONNECT.request primitives, the parameters of context, default, and presentation are not used, and also there are some restrictions on a few user data presentation service primitives.

15.4.3 Presentation connection release

For the release of the presentation connection, similar primitives (as used in the session layer) are used. For example, the S-Release service available at the session layer can access the P-Release service to release the connection. Similarly, P-U-Abort and P-P-Abort services are sent to the session layer entities. If the P-U-Abort service is used for the release of the connection, a context identifier list offers the type of contexts used in the data transfer primitives. These service primitives are passed down to the session layer entities.

15.4.4 Context managers

The primitives defined in the context management functional units allow users to change or delete the contents of the defined context list. The user on the source site may request addition or deletion of parameters in the context list. The presentation service PS-provider passes the proposed addition or changes of the parameters along with the context list (which it can support) to the user of the receiving site. After the PS-provider gets a response from the receiving site user, it defines the confirmed services along with the response of the receiving site regarding the sender user's proposed additions or deletions in the define context list to the user of the sending site. Thus, the defined context list (supported by the PS-provider) and mutually agreed upon options for addition or deletion into the defined context list is decided by the context management primitives before sending any data.

15.4.5 Dialogue control primitives

These primitives allow users to use the corresponding session layer service primitives to mutually agree on the context list by transmitting to the distant user. The services of the session layer during dialogue control are not lost.

15.4.6 Information transfer primitives

The primitives of information transfer allow users to transmit either data or expedited data to the distant site with which the sending site has established a presentation connection and has also mutually agreed on the context list or defined context list. In the case of expedited data, users have no choice of defining parameters for presentation context; instead, the default context list will be used between the application layer entities of these two sites. The reason for using the default context list is that expedited data has higher priority over normal data, and users may not agree on either the defined context list (supported by the PS-provider) or any addition or deletion options within the defined context list. Thus, we always use the default context list for the expedited data.

For normal data, the defined context list, including mutually agreed upon options and also supported by the PS-provider, is established before the data can be transmitted. The transmission of normal data includes various parameters associated with it, e.g., typed data, capability data, or any other data parameters. The normal data can also be transmitted with the default context list (where both PS-users agree to the parameters defined within the defined context list primitive), provided that the defined context list is empty.

From the previous sections, it has become clear that the main purpose of the ISO 8823 presentation protocol is to support all the services of the OSI session protocol through the syntax (abstract or transfer) transformation process into suitable binary form. The PS-users may change abstract and transfer syntaxes (i.e., presentation context) by simply defining new synchronization-major and -minor points or by using tokens. The users have flexibility in defining any type of abstract syntax, transfer syntax, or any other algorithm for various activities (of both session and presentation layer protocols) between application layer entities. In other words, ISO 8823 does not force users to use a particular type of transfer syntax, abstract syntax, security algorithms, etc.

15.5 ISO 8823 presentation layer protocol

ISO 8823[8] has defined a standard ISO 8824[9] and ISO 8824.4[10] for transfer syntax used in the presentation layer protocol and known as **abstract syntax notation one (ASN.1)**.[9] It describes the data types and data values. Other standards defined by ISO are 8825[11] and ISO 8825.5,[12] which provide a set of rules for representing ambiguous data into a model of unambiguous description of data within the message at the lowest level, i.e., the bit level. These standards are known as **basic encoding rules (BER)**. Standards ISO 8824 and 8825 together are known as ASN.1.[9,10]

The CCITT has also developed standards for presentation layer protocols known as X.208 and X.209, where X.208 contains similar procedures and rules defined in ANS.1 and X.209 contains the same set of encoding rules as in BER. The encoding rules in X.209 or BER are the ones defined for ANS.1.

15.5.1 Features of presentation layer protocol

According to ISO and CCITT recommendation documents, the conversion between semantics and syntax of data within message data is provided by the presentation layer and employs the following steps:

- The data type and data value of message data are defined by abstract syntax (ANS.1) provided by the application layer.
- The data type specifies a class of message, e.g., numeric, Boolean, etc.
- The data value specifies an instance of data type, e.g., number or part of text file.

The abstract syntax service is the first service in the application layer which expresses the data in a form without having any knowledge of a particular representation scheme. This is a very useful feature of presentation layer services for the following reasons:

1. The data type allows the machine to interpret the data which can process the data values.
2. The abstract syntax defines data types which are found in programming languages, e.g., Pascal, C, and Ada.
3. X.208 uses a grammar **Backus-Naur Form (BNF)** for defining data type and values, and it also has the same data type definitions found in the languages Pascal, C, and Ada.
4. The abstract syntax (defined either in ANS.1 or X.208) defines the **application protocol data units (APDU)** which interact with the presentation layer entity. This interaction between abstract syntax (offered by the application layer) and presentation layer entities is totally dependent on the implementations.
5. The presentation layer defines a transfer syntax which provides an interaction between the application layer and session layer service in such a way that the abstract syntax can be converted into a digital form of data within the message data.

The transfer syntax is mutually agreed upon between user processes and is provided between their presentation entities. Each presentation layer protocol entity knows about the abstract syntax of the user connected to it and has a list of transfer syntaxes that are available for encoding the data. After both presentation layer protocol entities have agreed on transfer syntax, the conversion between abstract syntax to digital form automatically takes place for data transfer. The combination of both abstract syntax and mutually agreed upon transfer syntax is sometimes also known as *presentation context*.

As discussed earlier, the services and functions of the presentation layer include data encryption and decryption, data compression and expansion, text file, file transfer, etc. The transfer syntax offers services with different attributes/features; i.e., it supports data encryption, data compression, etc. Each of these features defines a transfer syntax, each of which will have all the required services, with the only difference being in their features and implementation. In other words, two transfer syntaxes can be defined for data encryption and data compression, and then any combination out of four possible combinations (depending on requirements such as cost, security, etc.) can be selected.

15.5.2 Data transfer service

In order to provide data transfer between two application layer protocol entities, the presentation layer offers services for mutual agreement of transfer syntax and conversion of abstract syntax into suitable digital form. The abstract syntax at both sites may be different at the local level, but the data transfer can still take place between two application entities (user processes). Once the transfer syntax is mutually agreed upon between two user processes, the transformation or conversion user's data representation is provided by transfer syntax between presentation layer protocol entities. The application layer services also interact with session layer services for providing dialogue management between them. Since the application layer does not have direct access to the session layer services, all requests for session services are sent through the presentation layer.

The presentation connection supports various abstract syntaxes (used by different users) and transfer syntaxes, which together are sometimes known as **presentation context**. The presentation provider provides communication between presentation service users (PS-users) for mutual agreement on transfer syntax. The PS-users can define their transfer syntax through the PS-provider by using presentation service primitives, or both PS-users can use the transfer syntax provided by the PS-provider. The PS-users are not negotiating; i.e., their respective field is empty. The presentation context defines data types and data values; i.e., the data type represents the presentation context of the data values and meaning of the data within the presentation context. As soon as the presentation connection is established, it may support more than one presentation context, where each of the contexts may have different user's abstract syntaxes and mutually agreed upon transfer syntaxes which are used for transmitting the user data (for which the presentation connection is established) to the destination site.

15.5.3 ISO 8823 presentation layer protocol primitives and parameters

As mentioned earlier, the presentation service primitives are similar to the session layer service primitives, and as such the presentation protocol also seems to be simple, as the presentation service primitives will simply invoke the corresponding session layer service primitives when these are requested. For each of the primitive services, a **presentation protocol data unit (PPDU)** is defined which is transmitted over the corresponding connection. Since the presentation layer service primitives are similar to the session layer service primitives, any request at the presentation layer will invoke the presentation layer service primitive, which will then be used by the presentation layer entity to invoke the corresponding session service primitive. The session layer entity will then decide whether it wants to send a PDU. If there is a request for a token, then it will send its own SPDU, while no PPDU will be transmitted. It is worth mentioning that this PPDU is sent by the session layer as an SPDU, while the PPDU is never transmitted and instead simply establishes a session between PS-users.

Presentation service protocol primitives: The services of the presentation layer are used by the application layer entities using the presentation layer service primitives as shown in Table 15.1. The ISO standard (ISO 8823[8]) for the presentation layer protocol offers the following connection establishment (session) services to application processes.

A user service request at the application layer triggers a P-CONNECT primitive for the presentation layer. The user service request contains parameters for both session layer and presentation layer primitives. The presentation layer, after getting parameters (set by the application layer) and the user data, defines a connect PPDU primitive. This connect PPDU contains two field parameters pertaining to the presentation layer and the session layer. The PPDU **(connect PPDUs and response PPDU)** primitives are defined by the application layer of OSI-RM using a service request command. The PPDU primitives (with attributes connect and request) are enclosed inside the appropriate primitives of the session layer. In other words, the connect PPDU primitive (encoded in some binary form) will be converted or enclosed inside the SPDUs. The session layer primitive connect request is used to retrieve the PPDU and also the session parameters defined by the user request primitive, and it also defines its SPDU. The connect SPDU primitive will contain the parameters defined at the application layer and the PPDU which is in encoded form sent by the presentation layer. The presentation layer also establishes abstract and transfer syntaxes. At any time all three connections — **application**, **presentation**, and **session layers** — are synchronized with each other and hence are defined simultaneously.

The conversion or enclosing of the presentation layer PPDU into an appropriate SPDU for connection and confirmation should allow the users to mutually agree on various parameters present in the three upper layers simultaneously. In the case where simultaneous connection is not possible, reviewing two connection primitives can be defined. For example, if the presentation layer connection for some reason cannot be established, this does not mean that the session connection should not be established. Similarly, if an application connection cannot be established, the presentation and session connections can be made and the establishment of the application connection can be tried at a later time.

When the presentation layer uses token service for the establishment of connection, it first uses a request to invoke P-TOKEN-GIVE.request and transmit it to the session layer. It does not send any user data parameters; i.e., no PPDU is sent. The session layer, after receiving the request, uses an S-TOKEN-GIVE primitive to invoke the transmission procedure of a Give-Token SPDU. These primitives are used to transmit or receive the parameters defined between the appropriate layer entities. The PS-provider transmits the PDU to the session entity of the receiving site. This entity invokes a primitive S-TOKEN-GIVE.indication to the presentation entity, which in turn invokes another primitive, P-TOKEN-GIVE.indication, to its presentation user. Table 15.1 shows the parameters used with the PDUs of both session and presentation layer primitives for different presentation layer services.

The presentation connection is provided by a P-CONNECT primitive, which becomes a part of the user data for the session layer SPDU when the session layer uses S-CONNECT primitives. In other words, various connection-oriented (ISO 8823) standard protocols (presentation layer protocols) are assumed to use the ISO 8327 standard (session layer protocol) and include three phases (similar to the session layer protocol) — connection establishment, data transfer, and connection release. Since the services of the session layer and other lower layers are implemented through their respective PDUs, the presentation and application layers do not use PDUs for implementing their services. Out of all presentation service primitives, only one primitive, P-ALTER.CONTEXT, has a different primitive name (S-TYPED-DATA), while other primitives have exactly the same names with the only difference at the prefix (P for presentation and S for session).[2-6]

Once the session and presentation connections have been established, the application layer connection can be established. As can be seen, this type of scheme of establishing all three connections simultaneously wastes time in connection establishment and connection release phases. It requires an additional transfer of PDU between users. Further, the application layer of the transmitting site cannot mutually agree or negotiate with the application request primitive of the receiving site's application layer service since some of the parameters of the application layer entity primitive can be used only when presentation and session connections have been established.

Presentation protocol data unit (PPDU): The functions of presentation layer protocols are implemented via PPDUs (based on ISO standard protocol definitions). Each PPDU has various fields including data header, control information, and user data. The user data includes the data either from the application layer or from the session layer, depending on whether it is transmitted or received. Various ISO 8823 PPDUs are shown in Table 15.2. It is interesting to note that some of the parameters are the same in the PDUs of presentation and session layers, including a strong mutual dependency between session and presentation layers. The PDUs of any layer, in fact, carry out the commands of layer entities; i.e., both service primitives and PDUs of any layer have a direct relationship between them on a one-to-one basis, as discussed below.

1. The presentation protocol data units (PPDUs) allow the transfer of data and control information between presentation entities using the codes shown in Table 15.2. The transfer of data and control is obtained through structure and encoding of PPDUs.
2. The presentation layer entities allow for the selection of procedures for referencing to functional units.
3. A one-to-one mapping is provided between the services of both presentation and session layers and the presentation layer protocol. PPDUs are converted and encoded by S-CONNECTION primitives within the user data field of connection-oriented SPDUs. These PPDUs are also discussed in References 2, 3, and 5.

Table 15.2 Presentation Protocol Data Units

PPDU	Description
Presentation connection (session) establishment	
Connect Presentation.request (CP)	Establish a session
Connect Presentation.indication (CP)	Establish a session
Connect Presentation Accept.response (CPA)	Accept the session
Connect Presentation Accept.confirm (CPA)	Accept the session
Connect Presentation Reject.response (CPR)	Reject the session
Connect Presentation Reject.confirm (CPR)	Reject the session
Termination of session	
Abnormal Release-U.request (ARU)	Abnormal termination of session (user)
Abnormal Release-U.indication (ARU)	Abnormal termination of session
Abnormal Release-P.indication (ARP)	Abnormal termination of session (provider)
Data transfer	
Transfer Data.request (TD)	Normal data transfer
Transfer Data.indication (TD)	Normal data transfer
Transfer Expedited.request (TE)	Expedited data transfer
Transfer Expedited.indication (TE)	Expedited data transfer
Transfer Typed Data.request (TTD)	Typed data transfer
Transfer Typed Data.indication (TTD)	Typed data transfer
Transfer Capability.request (TC)	Capability data transfer
Transfer Capability.indication (TC)	Capability data transfer
Transfer Capability Confirm.response (TCC)	Capability data acknowledgment
Transfer Capability Confirm.indication (TCC)	Capability data acknowledgment
Context control	
Alter Context.request (AC)	Requesting a context
Alter Context.indication (AC)	Requesting a context
Alter Context Acknowledgment.response (ACA)	Acknowledgment of the context
Alter Context Acknowledgment.confirm (ACA)	Acknowledgment of the context
Resynchronize.request (RS)	Request for re-synchronization
Resynchronize.indication (RS)	Request for re-synchronization
Resynchronize Acknowledgment.response (RSA)	Acknowledgment for re-synchronization
Resynchronize Acknowledgment.confirm (RSA)	Acknowledgment for re-synchronization

As discussed in the previous chapter, one of the primary functions of the session layer is to offer orderly release of connection between SS-users (session layer entities). The presentation layer has no PPDU defined for the orderly release of connection service. As soon as the session connection is released by the SS-user, the presentation connection is automatically released. However, the presentation layer user may abort connection by initiating an abort primitive for the entities of the presentation layer. The kernel functional unit provides an Abnormal Release User (ARU) service PPDU, which aborts the connection, and the user may send some user-defined data in the user data field of the ARU PPDU. This field may also contain a context list if the user wants to send. The following discussion is partially derived from References 2, 3, and 5.

On the other hand, if the PS-provider wants to abort a connection between PS-users, it uses an Abnormal Release Provider (ARP) PPDU functional unit. This functional unit initiates the ARP PPDU from the presentation layer entities, and the PS-provider can initiate one of these presentation entities. Here, no user is involved for aborting the presentation connection. Since one of the presentation layer entities is initiating the abort PPDU under the control of the PS-provider, the initiating presentation entity has to provide the reason for the abortion of connection. The PS-provider initiates one of the presentation entities to abort connection because the PS-provider receives either an

1. Unrecognized PPDU from the PS-user (presentation entities)
2. Unexpected/undefined PPDU from the PS-user
3. Unexpected/undefined session service parameter
4. Unrecognized/undefined PPDU parameters

The PS-provider recognizes the PPDU through abort data parameters, and this PPDU, after being recognized as a PDU (containing the abort parameter), is used to initiate the abort procedure. This procedure sends an ARP to both the users at their respective presentation layer entities, which in turn send a primitive P-P-ABORT.indication to their users. As pointed out earlier, the presentation layer does not have any PPDU for aborting the connection; the ARP PPDU is enclosed in the user information field of user data and is sent to the session layer. It is at the session layer that the SS-provider requests to abort session connection by issuing an S-S-ABORT.request to its session entities, which in turn issue an S-S-ABORT.indication to SS-users. In other words, it is the session layer user who initiates the abortion of connection between SS-users and informs the PS-provider. The PS-provider then indicates the request to abort connection by sending an ARP PPDU to presentation layer users.

In some situations, the presentation connection can be aborted simultaneously with the session service provider's request for aborting the session connection between SS-users; i.e., when the SS-provider sends an S-P-ABORT.request to the session layer entities of its users, it also sends an S-P-ABORT.indication to both the SS-users and presentation layer simultaneously. The presentation entity then sends a P-P-ABORT.indication to its users and aborts the connection between presentation users at the same time that the connection between SS-users is aborted.

The P-ALTER-CONTEXT service primitive in the presentation layer offers a confirmed service and defines two types of PPDUs: Alter Context (AC) and Alter Context Acknowledgment (ACA). These PPDUs are used only between PS-users and are independent of session connection. Therefore, there are no corresponding session SPDUs, and the session layer considers the PPDU typed data which is operated upon by the S-TYPED-DATA primitive. This primitive (as discussed in previously) is not controlled by data token control management.

Once the presentation connection between PS-users is established, the data transfer takes place. All the different types of data (normal, typed, expedited, and capability) supported by the session layer are also supported by the presentation layer, and as such appropriate PPDUs are defined (Table 15.3). As we can see in Table 15.2, for each of these services, all of the PPDUs have the same parameters (i.e., user data at the presentation layer).

The AC PPDU is a very useful primitive of the presentation layer protocol. The presentation protocol assumes that the session layer protocol ISO 8323 is being used and accordingly offers similar phases of services (through PPDUs), such as **connection establishment**, **data transfer**, and **release of connection (connection termination)**.

The AC PPDU allows users to change the currently defined presentation context using the following steps. The sending site uses an AC PPDU for negotiating the availability of new context definitions in the mutually agreed upon defined context set. The application layer of the sending site sends a primitive, P-ALTER-CONTEXT.request, to its presentation layer entity, which in turn sends an AC PPDU to the presentation layer entity of a distant site. The AC PPDU contains the requested information and proposed parameters for the presentation context. The sending site sends the abstract syntax information and other related information on conversion, compression of data, etc., to the distant presentation entity. The distant presentation entity responds by sending an Alter-Context-Acknowledgment (ACA)

Table 15.3 Equivalent Presentation and Session Service Primitives and Their PPDUs

Presentation Service Primitive	Equivalent Session Service Primitive	PPDU
P-CONNECT	S-CONNECT	CP, CPA, and CPR
P-U-ABORT	S-U-ABORT	ARU
P-P-ABORT	S-P-ABORT	ARP
P-ALTER.CONTEXT	S-TYPED-DATA	AC, ACA
P-TYPES-DATA	S-TYPED-DATA	TTD
P-DATA	S-DATA	TD
P-EXPEDITED-DATA	S-EXPEDITED-DATA	TE
P-CAPABILITY-DATA	S-CAPABILITY-DATA	TC, TCC
P-RESYNCHRONIZE	S-RESYNCHRONIZE	RS, RSA

PPDU, and it also may send information regarding any new context definitions it supports and that are not proposed by the sending site. Once both presentation entities have mutually agreed on the context and its parameters, the sending site presentation entity informs the application layer of the P-ALTER-CONTEXT.confirm primitive.

Presentation layer service primitives and protocol data units: As we mentioned earlier, the service primitives of the presentation and session layers are similar to each other, and there is a one-to-one correspondence between them. For example, the presentation layer service primitive P-CONNECT has an equivalent primitive in the session layer known as S-CONNECT. For each of the service primitives, a PDU is defined by the protocol of the layer. Similarly, the presentation layer defines PPDUs for each of its service primitives. The PPDUs simply establish a session between presentation and session entities, and the PPDUs are transmitted only by the session layer entity. There are certain presentation (and session layer) service primitives for which we have no presentation protocol data units, and others for which we have one or more. Table 15.3 presents these presentation service primitives along with their equivalent session service primitives which have PPDUs.

As discussed above, the presentation layer defines PPDUs for its service primitives and does not necessarily define a PPDU for each service primitive (see Table 15.3). The PPDUs are usually divided into different groups (depending on their functions and operations). Various functional units of the presentation layer, along with different PPDUs required for essential or mandatory services, are shown in Figure 15.3. The mutual agreement between two PS-users is provided through the context list and context result list for transfer syntax (supported by the PS-provider) and abstract syntax requested by the PS-user. Both abstract and transfer syntaxes considered are the ones which have been agreed upon by both PS-users. Once these are defined, the functional units, along with associated PPDUs, are also known and are used to define the requirements of the presentation layer. The user data obtained from the application layer is converted into suitable and appropriate PPDUs and is sent to the session layer.

15.6 Presentation context, representation, and notation standards

Earlier in this chapter, we discussed the services and functions of the presentation layer. The presentation layer defines the description of data structure and representation in such a way that it provides conversion of different types of data and translates the representation into binary form. This layer does not provide the measuring (semantics) of data but instead facilitates the conversion of application data to suitable binary form at the sending site and conversion of the binary form of data to the appropriate application form. In both cases, the conversion/representation is obtained through abstract syntax.

Functional Unit (mandatory) defines the following PPDUs:

Connect Presentation (CP)
Connect Presentation Accept (CPA)
Connect Presentation Reject (CPR)
Abnormal Release User (ARU)
Abnormal Release Provider (ARP)
Normal Data Transfer (TD)
Expedited Data Transfer (TE)
Typed Data Transfer (TTD)
Capability Data Transfer (TC)
Capability Data Acknowledgment (TCC)

Context Management Functional Unit (optional) defines the following PPDUs:

Alter Context (DC)
Alter Context Acknowledgment (DCA)
Resynchronize (RS)
Resynchronize Acknowledgment (RSA)

Context Restoration Functional Unit (optional):

Nil

Figure 15.3 Presentation layer functional units and their PPDUs.

Another conversion/representation known as **transfer syntax** takes place between presentation layer entities which should support the application layer functions at both sites. The representation of data types and encoding rules at the local site is obtained by abstract syntax, while the conversion between two sites is obtained by transfer syntax. Both international organizations ISO and CCITT have defined a standard notation for the presentation of the data types, values, and transfer syntax which will be used by the application layer protocol. The ISO 8824 standard specification (abstract syntax) is known as **Abstract Syntax Notation One (ASN.1)**.[9] The representation of data at the bit level is described by a set of rules discussed in the standards ISO 8825 (**Basic Encoding Rules (BER)**) and ISO 8825.2.[11,12] Standards ASN.1 and BER together are known as ASN.1. ASN.1 thus defines an abstract syntax for different data types and their values, while BER defines the actual representation of these data types.

The CCITT defines standards X.216 and X.226[15] for the presentation layer, and these are collectively discussed in CCITT recommendation X.409.[14] X.208 defines the rules used in ASN.1, while X.209 is primarily concerned with basic coding rules for ASN.1. The standard notation of data types and values in the CCITT X.208 standard is the same as in the ANS.1 standard, and the representation of data types in X.209 is the same as that of BER for ASN.1 in ISO and is known as standard representation. Thus, CCITT defines standards for standard notation and standard representation (encoding), while ISO defines them as abstract syntax and representation standards (encoding). ISO ASN.1 and CCITT X.409 are compatible with each other, and the following discussion on various presentation, abstract syntax, notation, and basic encoding rules applies to both standards.

15.6.1 ISO abstract syntax ASN.1 (ISO 8824, ISO 8824.2, or CCITT standard notation)

The main issue with both standards is how to define the information exchanged between users, and this is based on the concept of data type, i.e., defining the information in terms of data type and its value. The same concept is used in programming languages, where the data structure of the information can be defined in terms of data types in a different way. The notation for the data type information in X.208 is based on **Backus-Naur Form (BNF)**. Many programming languages such as Pascal, C, and Ada use the data types defined by BNF, which basically describes the syntax. The value of the data type defines the instance of that type. The type may be expressed as a class of information, e.g., Boolean, numeric, integers, characters, alphabetic, etc. It offers a class of primitive types such as **integer** (arbitrary length), **Boolean**, **bit string** (number of bits), **octet string** (number of bits), **null** (no type), and **object identifier**. These primitives are considered building blocks in programming languages and are available as built-in library routines.

Some of the programming languages have a built-in data type class which can be used to represent the information data in terms of complex data structures, i.e., arrays, records, etc. The recursive technique may be used on various data type classes to generate the binary code for complex data structures of data types. The concept of data type is a very important tool for the presentation layer services, as it allows computers to know the data structures being used by sending it to computers and hence provide values of the data type for interpreting and processing the data information. Further, a tag may be used to distinguish a class between different types and values of data, i.e., record, database record libraries of job control languages (JCL), etc. Tags are also used to inform the receiving site about the contents, class, etc., of the data type. Various PPDUs dealing with abstract syntax are based on BNF. The syntax transfer represents data structure in a very simple and unambiguous way. In this scheme of defining data structure, each value transmitted typically includes fields such as identifier (type of tag), length of data field in bytes, and data field followed at the end by a content flag indicating the end of the data value type. The tag can have different data types, as discussed in the next section.

15.6.2 CCITT X.409

As mentioned earlier, CCITT X.208 offers an abstract structure representation of the information via abstract syntax, while CCITT X.409[14] defines encoding rules for determining the values of data. The abstract syntax describes the structure representation of data type and may be different between PS-users. The **transfer syntax** (X.409) standard defines the values of data type which will be sent from the application layer to machines in encoded form. Each of the syntaxes (abstract and transfer) is defined by a standard representation as a **data element** or **type-length value (TLV)**. There are three fields in the data element: **identifier**, **length**, and **information of element**.

The **identifier** is used to distinguish between different types and also contains the way to interpret the data in the information field. The **length** of data in the information field is defined in the second field of the data element. The data structures for values of data type are defined in the **information of element**. A similar data element is defined for transfer syntax. The data elements can be either a single data element (STLV) or multiple data elements (MTLVs).[2,3]

15.6.3 Classes of data or tag types

In terms of ISO and CCITT standards, the identifier field represents the data type, while the information field represents the value of data used in the data element. The identifier field in both standards ANS.1 and CCITT X.409 contains four classes of data or tag types (i.e., data information): universal, application-wide, context-specific, and private-use. The identifier field assigns two bits (toward the most significant bit (MSB) side) to represent these tag types. The binary values assigned to these types are shown in parentheses in the following descriptions.

Table 15.4 Tag Identifier Associated with Universal Class of Data or Tag Types in X.208

Tag		Type
Universal	1	Boolean
Universal	2	Integer
Universal	3	Bit string
Universal	4	Octet string
Universal	5	Null
Universal	6	Object identifier
Universal	7	Object descriptor
Universal	8	External
Universal	9	Real
Universal	10	Enumerated
Universal	11	Encrypted
Universal	12–15	Reserved for future use and addenda
Universal	16	Sequence and sequence-of-types
Universal	17	Set and set-of-types
Universal	18	Numeric string (character string)
Universal	19	Printable string (character string)
Universal	20	Teletex string (character string)
Universal	21	Videotex string (character string)
Universal	22	IA5 string (character string)
Universal	23	VTC time
Universal	24	Generalized time
Universal	25	Graphics string (character string)
Universal	26	Visible string (character string)
Universal	27	General string (character string)
Universal	28	Reserved for future use and addenda

Universal (00): The universal tag is application-independent. Tag assignments for the X.208 universal type defined by the standard given in Table 15.4.

Application-wide (01): This tag type is relevant and specific to an application and is generally defined in other standards.

Context-specific (10): This tag type is relevant and specific to a particular application but limited to a defined context within the application layer.

Private-use (11): This type of tag is defined by users for private use and not covered or defined by standard X.409.

Each type is distinguished by a tag which defines the class of type and the particular type and has two parts: a **class identifier** and a **number**. The data type and data value are defined by the syntax.

The X.409 identifier can be either **single-octet** or **multi-octet**. In a **single-octet identifier**, the last two bits (toward the MSB side) are used to define the four classes of types discussed above. One bit is reserved for the class of data type and another bit is reserved for the form of the data element. The remaining bits of the single-octet identifier can be used to define the number of octets in the identifier field.

In the case of a **multi-octet identifier**, the length field of the data element defines the length of the data information field. The information field contains the actual data information. The information field is described in multiples of an octet (i.e., eight bits) and has variable length. The data type defined in the identifier field of the data element is used to interpret the information as bit string, octet string, etc.

Table 15.5 X.208 Standard Built-In Types and Descriptions

Type	Description
Boolean	Logical data with two states: true or false.
Integer	Signed whole numbers.
Bit string	Binary data arranged in a sequence of 1s and 0s in an orderly fashion. Names can be assigned to individual bits which provide specific meaning.
Octet string	Data or text in a sequence of octets — used to represent binary data type.
Null	A single value is assigned to a simple type.
Sequence	A structure type (variable number of data elements) which can be referenced by an ordered list of various types.
Sequence of	A structure type (variable number of data elements) which can be referenced only by an ordered list of single type. The ordered list corresponds to each value in the type, making it similar to an array. The sequence can also be defined as a structured type of a fixed number of data elements which may be of more than one type. Another form of sequence of a fixed number of data elements could have optional data elements, and each of the optional elements is a distinct type. The ordered list of data elements in a sequence is very important during the encoding of a sequence into a specific representation of sequence type.
Set	Similar to sequence type; set can be referenced by data types arranged in any order.
Set of	Similar to sequence of; set of can be referenced by an unordered list of a single type.
Choice	Offers a choice to users for a data type group of alternative data types.
Selection	Allows users to select a variable whose type belongs to one of some alternatives of a previously defined choice.
Tagged	Allows users to consider a new type from the existing list of types and assign a different identifier to it.
Any	Allows users to consider data whose type is not under any constraint.
Object identifier	A unique value is assigned with an object or a group of objects and encoded as a set of integers. For example, a set of rules or abstract and transfer syntaxes can be defined by object identifiers.
Object descriptor	Provides a brief description or comments of object identifiers to the user.
Character string	A string of characters for a defined set of characters.
Enumerated	A simple type notation with a value assigned to distinct identifiers.
Real	Allows representation of a real value.
Encrypted	A type whose value is determined by encrypting another data type.
External	A type whose values are either not defined or not specified and should be defined using standard available notations.

Source: Derived in part from Reference 5.

Built-in data types: The X.208 standard also defines different built-in types and their descriptions, which are, in fact, standard notation used in various commonly used types (Table 15.5).

In **set** type, the order of data elements is not important during its encoding into any specific representation of the set type. The set also may be of a fixed number of data elements and could have optional data elements, as discussed in sequence type. The set type, in general, has one element of data type name which may be one of the defined sequence. Another element of data type may be given a name. The data type may be used in different contexts; for this reason, each data type is associated with a tag.

In order to explain how different built-in types can be represented by data elements (of three fields: identifier, length, and information), consider the following example of Boolean built-in type.

Boolean Type := BOOLEAN

Boolean Value := TRUE/FALSE

From Table 15.4, the identifier assigned to Boolean type is 1 (associated with the universal class). The coding of true and false has to be defined. If we code false with all 1s and true with a particular combination, then the built-in Boolean type for the false condition can be represented by a data element as

00000001 : 00000001 : 11111111

where the first octet represents the identifier, the second octet represents the length of the information field (in our case, it is one, as only the true or false condition is represented), and the information field denoted by the third octet represents the false condition.

15.6.4 BNR encoding rules and notation

The previous section discussed how different classes of data types (X.208) and identifiers associated with each type (X.409) can be defined. Both types that have a single data element and those that have multiple data elements (X.409) were also discussed. The first field of a data element is the identifier type which is encoded into tags (representing class and number) of the type of data value. Each field has a length of one octet. The first two bits of the identifier (or type) field define the four classes of types (universal, application-wide, context-specific, or private-use).

One bit is reserved for defining the form of the type **primitive** or **constructor**. The **primitive** element does not have any data structure defined and represents the value of the data type. The **constructor** bit, on other hand, defines the contents of information in the information field; i.e., one or more data values are completely encoded and, hence, it contains a series of data elements.

The remaining bits represent the distinct representation of one data type from another data type of the same class. If the tag has a number greater than or equal to 31, all five bits are made 11111, representing the series of octets. The first bit of each additional octet should indicate the last octet in the type field.

The length (L) of the information field defines the length (L) in octets of the data information in the field. The length field contains a single octet starting with zero if the value of L is less than 127. But if L is greater than 127, then the first octet of the length field will represent a seven-bit code which specifies the length of the information field.

Document compaction algorithms: As stated earlier, syntax provides a transformation of incoming data from the application layer into a format (binary form) suitable for transmission to the receiving peer application entity. Depending on the applications, various presentation layer services have been defined. For example, if the data is a file transfer, compaction and decompaction algorithms will be useful, while for confidential/secret data it may recognize encryption and decryption of the data on the same transport connection. The data transmitted over the transmission media can be a sequence of bits, bytes, characters, symbols, numbers, and so on, depending on the applications.

The transmission media offer a limited bandwidth and capacity, and as such the information may not be sent directly over them but instead compressed to include a small number of characters or bits which typically define the relative frequency and content of the message within a finite range. This offers two advantages: higher capacity and data rate of the link, and lower cost of the services. This is particularly true when we are transmitting the data through PSTNs, where charges are based on time and distance. If we can reduce the time to send the block or frame over the networks, it will reduce the cost for the duration of the call. Ideally, a 50% reduction in the amount of data can be expected over the PSTN for sending data across it. Since we are using modems for data communication, we may achieve similar performance using a higher-data-rate modem over a lower-data-rate modem.

The transmission of a compressed message does not lose the information, but it allows more information to be carried within the bandwidth and capacity of the link. On the receiving side, this message is decompressed to get the original message. The compression technique is based on defining a suitable encoding technique which can provide different lengths of codes for different characters or symbols, depending on their occurrence within the messages.

There exists a range of compression algorithms being used as built-ins with smart and intelligent modems. These modems provide facilities of adaptive compression algorithms, which allow the users to select any compression algorithm depending on the data being handled.

Text compression: In order to reduce the total number of bits transmitted for data information, users may choose a compaction algorithm which allows the user to compact the data into a code at the sending site. The receiving site should choose appropriate decompaction from the code. Thus, it will not only save the bandwidth of data transmitted from the sending site but also reduce the cost and enhance the throughput of the network. The compaction algorithms can be used for compacting both text of fixed-length symbols and binary data. The frames contain data characters, control characters, symbols, etc., and each of these characters is represented by codes, e.g., ASCII and EBCDIC. Each of these coding schemes defines a fixed number of bits for representing the characters, symbols, etc. If the number of bits within each representation of character (seven bits in ASCII) is changed to a lower number by distinguishing between numeric and alphabetic in the code set, the number of bits transmitted for the characters is lowered, improving the throughput of the network. This concept of reducing the number of bits in each code set by defining a set of rules, in fact, defines the compression technique.

For the compaction test, the data symbols (text) are represented such that frequently used symbols are defined by short codes, while infrequently used symbols are defined by long codes. This type of compaction algorithm (Huffman code) is used in areas such as library automation, manufacturing automation, etc. By sending a particular item to various sections, the frequent use of a particular item can be calculated. The transmission of information about items may require the section number and list of items requested per day. It is assumed that a complete and finite list (set) of items is available and that each item is assigned a code. If a particular item is not present in the list, then that particular item has to be represented in its full form and can be referenced through the name of that item.

Huffman coding: In Huffman coding, all data symbols (long or short) in a set of symbols are coded with a fixed length of code representing symbols. This type of algorithm for coding is useful in transmitting a text message or files in which each symbol is of eight-bit length and hence can be represented and transmitted as ASCII code.

Table 15.6 Percentage of Occurrence in English of Various Letters, Digraphs, and Words

Letters	Percentage	Digraphs	Percentage	Words	Percentage
E	13.0	TH	3.2	THE	4.7
T	9.0	IN	1.5	OF	4.0
D	8.2	ER	1.3	AND	3.2
A	7.8	RE	1.3	TO	2.4
N	7.3	AW	1.1	A	2.1
I	6.8	FE	1.1	IN	1.8
R	6.6	AR	1.0	THAT	1.2
S	6.5	EN	1.0	IS	1.0
H	5.8	TI	1.0	I	1.0
D	4.1	TE	1.0	IT	0.9
L	3.8	AT	0.9	FOR	0.8

The encoding of a long binary string can be obtained by another compaction algorithm known as **run-length coding**. Here it is assumed that the binary string contains either more 0s or more 1s. This type of technique is used in facsimile machines, video, digital data, digitized analog data, etc., where the data does not change rapidly and also most of the data occurs frequently; i.e., the data is repeating. The following discussion is partially derived from References 2–5, 14, and 16.

In any text, the probability of the occurrence of a particular symbol is different from that of any other symbol. Similarly, the probability of the occurrence of a particular character, word, or symbol will be different from that of others in the English language. Accordingly, the percentage of occurrence of some letters, digraphs, and words will vary, as shown in Table 15.6.

Huffman coding reduces the length of a text file. This is possible because Huffman coding assigns shorter codes to the most frequently used characters, while it assigns longer codes to the least-used characters.

The Huffman code defines the coding for various symbols in such a way that each symbol in a set should be coded to yield the minimum average bit length per symbol. The sending site must know the frequency of occurrence of each symbol being used at the receiving site. Huffman coding can be calculated using the following steps:

1. Start with all the symbols written with their probability of occurrence at the bottom. The symbols with higher probability are written on the left-hand side, and other symbols with descending probability are written from left to right. These symbols become the terminal nodes for the tree which is constructed as the algorithm executes.
2. Select two symbols with the lowest probabilities and create a new symbol by connecting it with the two new symbols selected. The probability of the new symbols is the sum of the probabilities of the two connected symbols.
3. Repeat step 2 until all the symbols have been considered except one. The probability of this symbol (which is not considered) will always be equal to one.
4. The symbol at the top of the tree provides the starting node for all the paths for symbols situated at the bottom (i.e., terminal symbols). The sequence of arcs (left and right) may have 0s if the arcs are on the left side and 1s if the arcs are on the right side for each of terminating symbols. This sequence can then be encoded in terms of the symbol nodes (starting from the top symbol to the terminating symbol) by assigning 0 to the left arc and 1 to the right arc.

After the Huffman code for all the terminal symbols has been determined, encoding can be obtained by considering one symbol at a time and entering this in the information field of the data element. These compressed symbols may not be accommodated in the octet boundary, so initially we create a table to translate the compressed symbol codes back into the original symbols and transmit it to the receiving site. The receiving site, using the table, decompresses the code into the original symbols. This is how the receiving site host gets an implementation of Huffman code to decompress or decompact the original symbols (symbol by symbol) from the Huffman code of the symbol tree transmitted.

The Huffman code algorithm described above defines the table in radices two. We can also implement the Huffman code algorithm in radices of a byte where we have to choose 256 symbols of lowest probabilities at each step until the top symbol node of probability one is achieved.

Consider an information field containing five letters A, B, C, D, and E. We have to find the Huffman code for each of the letters used (in information) with the following probability of occurrences:

A = 60%, B = 30%, C = 20%, D = 10%, E = 10%

Step 1:

6	3	2	1	1
A	B	C	D	E

Step 2: See Figure 15.4(a).
Step 3: See Figure 15.4(b).

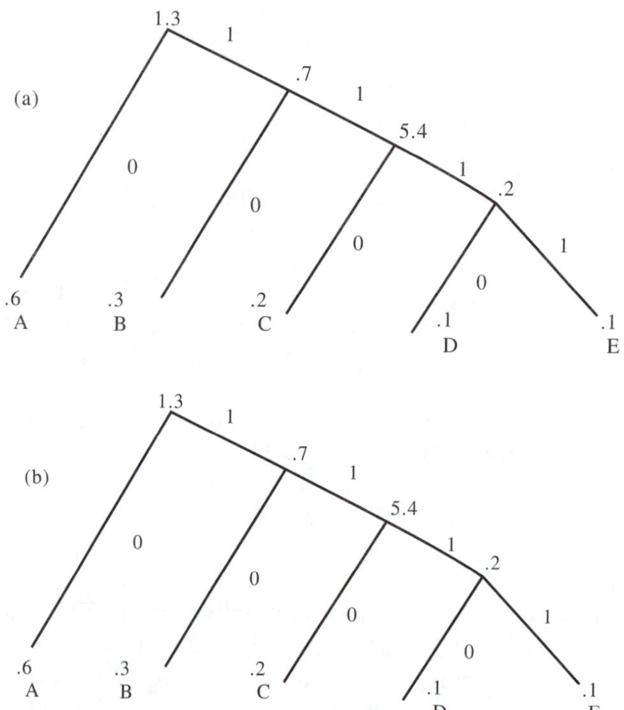

Figure 15.4 (a) Step 2. (b) Step 3.

Assigning 1 for the right side and 0 for the left side, we get

A = 0

B = 10

C = 110

D = 1110

E = 1111

The Huffman code algorithm discussed above assumes that the probability of occurrence of a new symbol created is independent of its immediate predecessor. A better scheme could consider the conditional probability (instead of probability of occurrences) for each symbol for each of its possible predecessors. This method is well suited to situations where, if we can obtain a correlation between symbols and their successors, there is a big savings in computational efforts. If this is not possible, then this becomes the main disadvantage of this method, as the size of table will be very large.

Run-length coding: Another related method of coding is run-length code, which is used to encode long binary bit strings, e.g., video, facsimile data, digitized analog signals, etc. The bit strings contain mostly 0s and the number of identical successive bits are considered or counted for the code; i.e., each n-bit symbol will indicate the number of 0 bits occurring between consecutive 1 bits. The count is coded in n-bit symbols, and only the number is transmitted. The string of identical bits is equal to or greater than $2n - 1$ (where n represents an n-bit symbol), which also represents the true distance of $2n - 1$ followed by the symbol for the remainder. In this case, the symbol consists of all 1 bits and is very useful to represent a string having long 0 runs. Consider the following bit string:

001000100000100000000100101101001

This string consists of runs of length 2, 3, 5, 8, 2, 1, 0, 1, and 2. The run-length coding of this string is given by

010 011 101 111 001 000 001 000 001 010

We observe here that run-length coding has fewer bits than the bit string.

Facsimile compression: As indicated above, one may ideally expect a reduction of 50% in the volume of data to be transmitted using a compression algorithm over the network, and it has been observed that Huffman coding usually offers close to a 50% reduction in the transmission of text files. One very useful application which offers this savings has been a scanner, which digitizes images and documents and transmits them over PSTNs using facsimile (fax) machines. The scanner digitizes each line at a rate (typical) of eight picture elements (Pels) per millimeter for black and white images by assuming logic 1 for the black mode and logic 0 for the white mode of images or documents. This data rate offers a number of bits per page equal to about two million. The time for transmission of one page may be a few minutes, depending on the speed of the modem being used.

CCITT has defined the following standards for facsimile transmission. In most of these standards, the data compression technique is used, and a compression ratio of 10:1 is very common. Many vendors have developed products offering higher compression ratios.

Group 1 (T2) — Rarely used
Group 2 (T3) — Rarely used
Group 3 (T4) — Used with PSTN and uses modulation
Group 4 (T5) — Used with ISDN and uses digital networks

15.7 Security

The goal of security is to protect the integrity and confidentiality of user's data when it is transmitted from one host to another over the network. The problem of security becomes more complicated when microwave radio transmission and satellite physical media are used to carry the data. It becomes very easy to access the data and modify it, as the transmission media are exposed to the open atmosphere and provide easy access to intruders. Some type of encryption needs to be defined to protect the data in such a way that it should become useless to unauthorized users while remaining useful to the intended receiving site/user of the network. Security is crucial for applications in the military, government agencies and contractors, banks, universities, corporations, etc.

Security against attacks from within an organization can be handled by various methods, e.g., assignment of a secret, unique password to users, regulation of access permission given to selected users, using file access methods to protect access, etc. The attacks can be made through various access points or at interconnectivity interfaces. For example, the gateway that connects different types of networks may be used for authentication. Within large corporations, organizations, or universities, LANs can be installed in different departments/divisions and not be connected via any internetworking devices such as a bridge. In this way, we can isolate various departments, but then the purpose of LANs within organizations is defeated. For e-mail services, security can be provided by a digital signature that goes at the end of a message. A unique bit pattern defines a digital signature, which cannot be forged easily. A message with a digital signature cannot be tampered with during its transmission over the packet network; the receiver also cannot tamper with the received message. The discussion herein is partially derived from References 21–24.

There are a number of tools available that can be used to provide security in our PCs that are connected to mainframes, public telephone networks, data networks, and so on. Such tools are needed to provide security against misuse of the data during its transmission and also to make sure that data transmission is authentic.

The issue of security concerns not only protection and confidentiality of the data information against unauthorized access or modification but also identification of the correct ID of both sender and receiver. The encryption of data information provides protection against misuse and can be included in the presentation layer as one of the services, just like data compression.

As we just saw, in compression we define a unique code for a given data stream which can be transmitted. On the receiving site, it is decompressed to get the original data stream to be given to the application user. Encryption also uses the same concept of transforming the data stream into encrypted code which can be decrypted on the receiving site. The encryption can be done at any layer, from physical to application, and provides security for the data information.

There are two popular methods of encryption: **data link encryption (DLE)** and **end-to-end encryption (ETEE)**. In the case of data link encryption, the encryption device is attached at the host-IMP line when the data leaves the host, and the decryption device is attached into the line just before the data enters the IMP. This type of encryption is also known as **hardware encryption (HE)** and does not affect the software. The end-to-end encryption may be defined between presentation layers or, in some cases, between application layers and can be done by either software or hardware. Both of these methods of encryption provide encryption action of their respective previous layers; i.e., the data link encryption method encrypts the physical circuits, while end-to-end encrypts a particular session and is not transparent to the software of the presentation layer. ANSI X3.105 is the only standard defined for physical and data link encryption for security of the data.

15.7.1 Data encryption

ISO and CCITT have defined standards for application layer protocols which provide the local representation syntax (based on the scheme used in the host machine) and a transfer syntax (defined through standards). The standards provide one-to-one mapping for each of the presentation connections between presentation layer connection primitives and session layer connection primitives. The local representation syntax is usually defined by the application protocol, while the transfer syntax is defined and implemented by the presentation layer protocols. Each presentation entity has to convert the local representation syntax into common syntax. Different techniques of data encryption have been defined and are used in different applications. Some of these techniques have been standardized and are discussed below.

15.7.2 Terminology and a model of cryptographic systems

Encryption can be defined as a process which converts/translates the data information (text or symbols) into another form of information (unintelligent) using a set of predefined rules of transformation. The original data information to be encrypted is known as *plaintext*. The conversion/translation of plaintext generates a new form of information (which has been encrypted) called *ciphertext* or *cryptogram*. The cryptogram is the information which is transmitted over the network. The conversion/translation of plaintext into ciphertext is obtained by a function parameterized by the operations (conversion, rules of transformation, etc.) defined in the encryption key. The encryption key is a very important component of security, and without this an unauthorized user (intruder) cannot decrypt the ciphertext. The problem of security sometimes gets complicated when an intruder copies the complete ciphertext and inserts his/her own data into the ciphertext, modifies the ciphertext, and then transmits it on the network.

The *cryptanalysis* deals with listening to the ciphertext while cryptography is concerned with designing a cipher (key) to allow the intruder to inject or modify the ciphertext. Sometimes these two processes are collectively known as cryptology. The encryption key is a short string of characters, and one potential encryption key is selected out of the string. For analog or text communication, the key may be a string of characters, but for digital communication, the key is a binary string. Depending on the application, the key may be kept secret (no additional security) or the key and encryption algorithm together may be kept secret (for additional security). The processes required for encryption of data (cryptography) are shown in Figure 15.5.

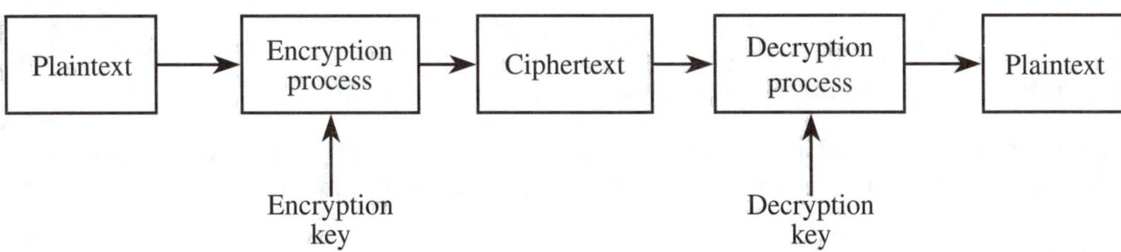

Figure 15.5 Encryption of data.

An encryption operation on message M using key Ke to produce ciphertext C is denoted as C = EKe(M), and a decryption operation of ciphertext C using Kd to produce message M is denoted as M = DKd(c).

15.7.3 Encryption cipher (code)

The encryption process deals with the design of ciphers (key) where the coding for encryption is defined. There are many different codes available for different applications (defense, data integrity for data communication, security, etc.), and these can be categorized into two main codes as substitution ciphers (or codes) and transposition ciphers (codes). Some of the codes in this subsection have been derived from References 3–5.

15.7.3.1 Substitution ciphers

The encryption algorithm based on substitution ciphers is one of the oldest methods of coding. In this method, the data (letters or a group of letters) in a message is replaced by another form of data (letters or group of letters) in a systematic way.

Caesar cipher: The Caesar cipher is the oldest method of coding and replaces each letter (lower-case) by skipping three letters (upper-case) in the alphabet. For example, a will become D, b will become E, z will become C, and so on. The plaintext is represented in lower-case letters and the ciphertext in upper-case letters. A new alphabet message will be defined to decrypt the ciphertext to get the original message.

Original alphabet

a b c d e ....x y z

Substitution code (cipher)

D E F G H ....A B C

The following example gives plaintext and ciphertext using the above substitution code (key is number 3):

Plaintext: WAIT FOR MY ORDERS
Ciphertext: ZDLW IRU PB RUGHUV

The encryption key is made up of number 3 for encrypting and decrypting the message. The key has a fixed number. An improvement of the Caesar cipher provides a variable number (instead of 3) for shifting the letters in the alphabet. This new variable number is known to both sender and receiver and remains valid during the transmission of the message between them. Thus, this new variable number is the encryption key which shifts the alphabet in a circular way. Both encryption keys (of constant number and variable number) could easily be decrypted and hence are not useful for secret documents, military operations, government contracts, and so on.

Monoalphabetic substitution cipher: An improvement over previous methods is to define a new alphabet for the original alphabet in a less systematic way, as shown below:

Original alphabet

a b c d e .... x y z

Substitution code (cipher)

Q W K O S .... P B M

Plaintext: FIRE ON MY ORDERS
Ciphertext: YIRS FA CB FRORL

This method of defining a substitution code (cipher) is also known as monoalphabetic substitution, with a key that is a 26-letter string for the entire original alphabet. Although the cryptanalyst knows the scheme of obtaining the encryption key (defining a new letter for each letter of the original alphabet), it would take a long time to find the actual key from a large number of possible combinations, so this is not a promising approach. Further, using the probability of occurrences of letters, words, diagrams, etc., in a given message, one can easily determine the key used for the transmission of the message, and hence the data message can be decrypted.

Polyalphabetic substitution cipher: In order to give more resistance to intruders, we should devise a coding method which has a revised order of substitute alphabet and also avoids the frequencies of occurrences of letters in ciphertext. This type of design concept can be implemented in a variety of ways. One method of cipher is to consider multiple cipher alphabets and use them on a rotational basis; this is known as **polyalphabetic cipher**. The Vigenere code is an example of polyalphabetic substitution cipher. It defines 26 ordered alphabets (Caesar alphabets) in the form of a square matrix. Each row starts with the same letter as defined in the original alphabets; i.e., the first row is known as A and is defined as A, B, ..., Z, the next row is known as B and is defined as B, C, ..., Y, Z, A, and so on. The last row is known as Z and is defined as Z, A, B, ..., X, Y. Each row has a string of 26 distinct characters (Figure 15.6). For the first letter of plaintext, we write down the letter from the intersection of the row and column of the square matrix.

A secret alphabetic key (known to both sender and receiver) is selected for the encryption of data. This key, in turn, selects the appropriate equivalent alphabet for each letter of the message. Consider the same example of plaintext and assume we have a key WONDERFUL. Using Figure 15.6, we derive the following:

Plaintext: FIRE ON MY ORDERS
Key: WONDERFULWONDE
Cipher: BWEHSERQZNRRVW

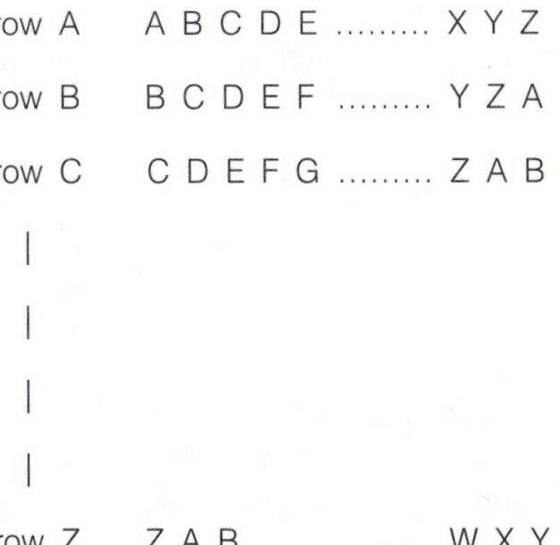

Figure 15.6 Vigenere code.

The plaintext is encrypted using the method of determining a letter at the intersection of the letter of plaintext and the letter of the key by considering one letter of plaintext at a time. It is useful to note that the letter at the intersection of rows and columns must be chosen carefully, as otherwise the decryption of code at the receiving site will not result in the message the sender intended to transmit.

In the Vigenere code method of polyalphabetic cipher, the code is constructed using Caesar alphabet in the rows of A, B, …, Z. Instead of Caesar alphabets, if monoalphabetic ciphers are used, then it is hoped that this code will be better than the previous one. The only limitation polyalphabetic code has is the construction and writing down of a 26×26 square table for the transmission of the message. Further, this 26×26 table also becomes a part of the encryption key. Although the polyalphabetic code is certainly better than monoalphabetic code, it can be decrypted by intruders.

15.7.3.2 Transposition ciphers

Previous methods of substitution code maintain the order of plaintext symbols and disguise them. In transposition codes, the letters/symbols in the message are reordered but are not disguised. The key is a word or a phrase, and such probable key should not contain any repeated letters. The main aim of the key is to assign the columns with a number, e.g., number 1 to the first column, which is under the key letter closest to the start of the alphabet, and so on. The following steps are required for this code:

1. Break the message using the probable key.
2. Rearrange the message.
3. Identify and recognize a probable word.

The plaintext is written horizontally in terms of rows and is aligned under a key word. The ciphertext is read through columns, starting with the column whose key letter is the lowest. If the length of the message is greater than that of the key, the message text is wrapped around the key. A filler is included at the end of the plaintext. The columns of characters are transmitted in the same order of alphabetic sequence of the key.

In order to understand this code, consider the same example of plaintext by assuming the key is VERTICAL.

Plaintext: FIRE ON MY ORDERS

The plaintext is expressed in terms of a row and written under the key such that it contains the same number of letters or characters as in the key, as shown below. If the last row does not have enough characters, extra characters can be added to match its number of characters with that of the key.

V	E	R	T	I	C	A	L
8	3	6	7	4	2	1	5
F	I	R	E	O	N	M	Y
O	R	D	E	R	S	R	S

This coded text is defined by transmitting the columns under the key. The columns are numbered starting with the lowest letter (of key) in the alphabet. The column having the lowest letter in the alphabet (A) is transmitted, following by column with the next-lowest letter (C), and so on. The plaintext is defined horizontally as a series of rows, while the ciphertext is defined by columns, starting with the column with the lowest letter of the key and transmitting characters of plaintext.

Plaintext: FIREONMYORDERS
Ciphertext: MRNSIRORYSRDEEFO

In both encryption methods (substitution and transposition code), computers can be used to determine the key based on the probability of occurrences of different letters or characters and evaluating all possible solutions. This becomes a major problem with both encryption methods. Somehow, we have to define the encryption technique by integrating the concepts of both techniques, i.e., developing a technique by combining the concepts of substitution and transposition codes and defining a complex algorithm instead of straightforward algorithms so that the cryptoanalyst is unable to decrypt the key.

One of the main objects of an intruder is to extract information being transmitted over the network during data transmission. The contents of information (files, electronic mail, etc.) can be accessed by intruders. We may use some kind of masking technique to encrypt the information, but this also can be analyzed and accessed by intruders. In other words, these two attacks seem to be difficult to detect, but the possibility of prevention may reduce the misuse of information. Another way of misusing data transmission is to create or modify the data stream (virus) and spread it over the networks. In this way, users will not be able to use the Internet with an acceptable quality of service and confidence. This type of attack is difficult to prevent, but it can be detected and then recovered from by using a variety of tools (encryption, anti-virus). Even if some of these attacks can be detected or prevented and recovered, the security issue remains unsolved in terms of authentication. How can we ensure that the information we have received is coming from the valid user and is genuine? The standard encryption mechanism can be used to provide authentication using available error-control protocols, time stamping to identify any delay, and so on. The information authentication can also be provided without using any encryption techniques. Different authentication algorithms have been proposed in the literature and can be combined with encryption methods to provide network security. Some of the information presented herein is partially derived from Reference 24.

15.7.4 Private-key cryptography

In private key cryptography, the same key is used for both the encryption and decryption operations. These systems follow the principle of "open design" in the sense that underlying encryption and decryption techniques (algorithms) are not kept secret. However, the value of the key, called the *private key,* used in encryption and decryption is kept secret. Private key systems use Shannon's principles of diffusion and confusion for added security.[22]

1. **Permutation:** Permutation operation permutes the bits of a word. The purpose of the permutation operation is to provide diffusion (because it spreads the correlation and dependencies among the bits of a word).
2. **Substitution:** A substitution operation replaces an m-bit input by an n-bit output. There is no simple relation between input and output. Generally, a substitution operation consists of three operations: (1) the m-bit input is converted into decimal form; (2) the decimal output is permuted (to give another decimal number); and (3) the decimal output is converted into n-bit output. The purpose of the substitution operation is to provide confusion.

We next discuss the Data Encryption Standard (DES) which was developed by IBM Corporation and has been the official standard for use by the U.S. government. A more detailed description can be found in Reference 23.

15.7.5 Data Encryption Standard (DES)

The encryption of data can be provided by any layer of OSI-RM. If the encryption is to be provided at the physical layer or data link layer, then data encryption equipment (DEE) is used between the host and IMP connections. Here, the data is encrypted by encryption hardware devices before it leaves the host, while decryption devices decrypt the data before it is received by IMP. If the encryption is used by the application layer or presentation layer or even the transport layer using host software or special hardware devices, the encryption is known as **end-to-end encryption (ETEE)**, while encryption used in the data link or physical layer is known as **data link encryption (DLE)**. End-to-end encryption involves encryption by presentation layer entities, while data link encryption provides encryption of physical connection circuits. For more details about these codes and implementation, see Reference 2.

With data link encryption, users have no control over the encryption technique being used by the data link layer. In other words, the same encryption technique will be used for all data link connections, while in the end-to-end encryption method, users have control over the encryption technique, in the sense that users can choose an encryption technique at any time. Data link encryption provides encryption to all the fields of data, including the header of the data link frame, while in end-to-end encryption, the header of the data link layer is not encrypted. The description of data structures and data conversion/translation services (operations) are provided between different user processes, different applications, and so on by the set of procedures in the standards of the presentation layer protocols defined by the application and presentation layer standard protocols. Some of standardization aspects of various cryptography techniques for higher layers are discussed in Reference 15.

15.7.5.1 Implementation of DES

The implementation of substitution and transposition codes can easily be obtained by manipulating data on hardware devices. The transposition code can be generated by using a Permutation box (P-box) which has the same number of input and output lines (Figure 15.7(a)). By interconnecting input lines with output lines in a different way, various

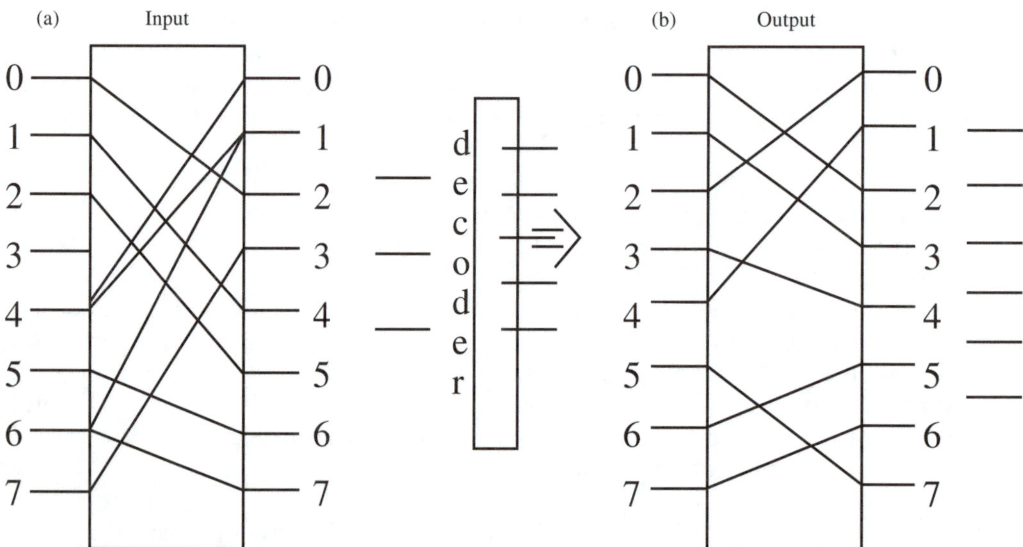

Figure 15.7 (a) Permutation box. (b) Substitution box.

types of transposition codes can be generated. The key issue in the transposition code is the internal wiring or integral connection between inputs and outputs.

The substitution codes can be generated by a substitution box (S-box). This device consists of a pair of decoders/encoders and can be used to generate various substitution codes (Figure 15.7(b)). The key issue in the generation of codes is the internal connection between the output of the decoder and the input of the encoder.

These two devices — P-box and S-box — provide basic components of the transposition and substitution codes, respectively. The algorithms for generating those codes seem to be straightforward and can easily be decrypted. Instead of defining algorithms for each code in a systematic way, we can combine these two algorithms in such a way that the algorithms for generating codes are complex and provide better protection against misuse by intruders. Such a scheme leads to the development of a product cipher algorithm, which provides much better and more secure encryption by combining a series of substitution and transposition codes. Numerous combinations can be achieved for a product cipher algorithm. The product cipher can be made powerful by defining a large number of stages for the combination of both codes.

15.7.5.2 Data Encryption Standard (DES) Algorithms

A cipher algorithm developed by IBM in 1977 was adopted by the U.S. government as a federal encryption algorithm standard for unclassified and sensitive information. The National Bureau of Standards (now known as the National Institute of Standards and Technology — NIST) published this standard in 1977 — known as the Data Encryption Standard (DES) — in hardware (due to fast speed), thus encouraging a number of vendors to implement the DES standard due to the increasingly low cost of hardware chips.[13] It uses 64 bits, out of which 56 bits are used for the key while the remaining eight bits are used for parity check. Thus, DES maps 64 bits of plaintext into 64 bits of ciphertext by passing the data through a key of 56 bits.

Data Encryption Algorithm (DEA): DEA is a national standard (U.S.) adopted by ANSI and available to both users and vendor suppliers. The encryption algorithm is known to both users and vendors, making it completely available to the public standard. Users have to define their own security schemes for the key and also for the data. The DEA provides interoperability between various users who are using different types of computers from different vendors with different security mechanisms but are using the same key.[2,3,5]

We now discuss the working of the DES algorithm (Figure 15.8). It encrypts the plaintext in 64-bit blocks, which generates a ciphertext of 64 bits. The key has 56 bits and is defined in octets, where each octet contains seven key bits and one odd parity bit for error detection. This key is used by all authorized users for both encryption and decryption of data. The algorithm is evaluated in 19 stages. The plaintext of 64 bits is converted into 64 equivalent bits using a P-box in the first stage, which is key-independent. The resultant block of 64 bits is executed by 16 iterations of a set of permutation (transposition) and substitution codes where each stage is functionally identical but provides different functions due to different parameters of permutation and substitution codes. The output of the 64-bit block at each stage depends on the key. At each stage, the key of 56 bits is used and hence gets different subsets of bits from the key. At the eighteenth stage, the leftmost 32-bit block within the 64-bit block obtained from the previous stage is exchanged with the rightmost 32 bits, and then this 64-bit block is passed to the last stage of inverse permutation (transposition), yielding 64-bit ciphertext.

The ciphertext is generated by combining permutation or transposition and substitution codes. The standard uses P-box and S-box operations on the data; i.e., the P-box permutes bits while the S-box substitutes the bits of data.

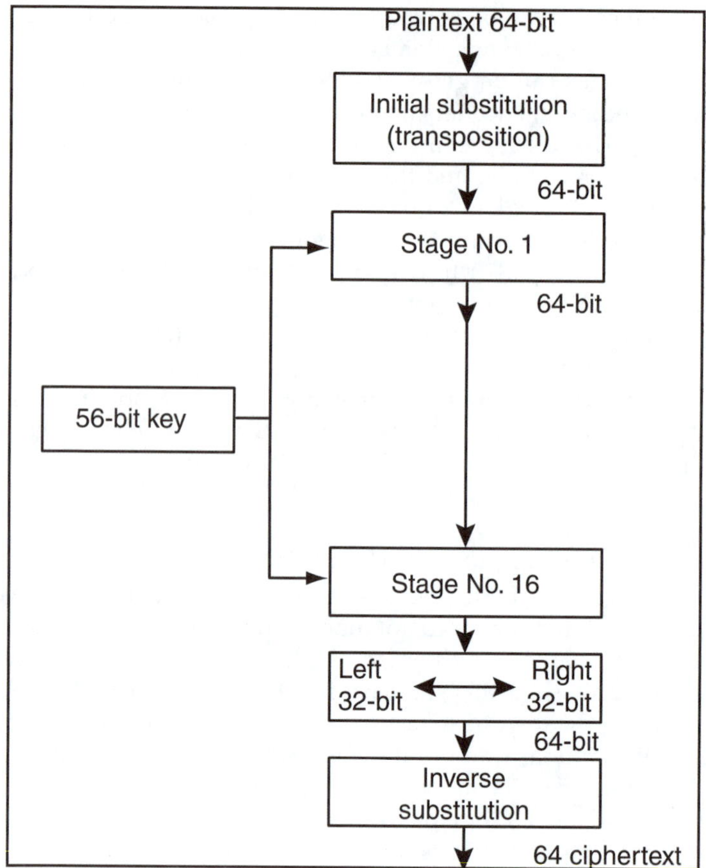

Figure 15.8 DES algorithm.

As stated above, there are 16 identical functions with different parameters and the same key, and these define 16 stages or iterations. At each iteration, a set of permutation and substitution is performed. At each stage, the 64-bit block is broken into 32-bit blocks, and each stage takes two 32-bit inputs and generates two 32-bit outputs. These inputs and outputs are defined in such a way that the left output is same as the right input, while the right output is exclusive-ORed with the left input and a function of the right input and the key for that stage. An S-box accepts a binary number and substitutes another one for it. This box in DES has six inputs and four outputs. The function is generated by considering a 48-bit number which is obtained by expanding the rightmost 32 bits using a transposition and duplication rule. Then the exclusive-OR operation is performed on the 48-bit number and the right input and key of that stage. The output of the exclusive-OR is divided into eight groups where each group consists of six bits.

Each group is put into a different S-box which produces four outputs. Thus, each of the 64 possible inputs to an S-box is mapped onto a four-bit output. The same outputs can be obtained using different input combinations, yielding a list of eight four-bit numbers. These 32 bits obtained through the blocks are then used as input to the P-boxes. The P-box performs a permutation on the position of bits in an ordered manner. This operation guarantees that the logical nature of the bits before the operation (at the input) remains the same at the output also. A different key is used at each stage, and at the beginning of the DES algorithm, the key is substituted (transposed) by 56 bits. After the beginning of the algorithm, the key is partitioned into two 28-bit blocks before the start of each stage. After the key is partitioned before any stage, the units of 28 bits are shifted left by a number of bits corresponding to the stage number. The key for the next stage is obtained by substituting the shifted key with another 56-bit transposition to it.

The input to the S-box is defined by a six-bit binary number, and the S-box substitutes this number with another number of four bits at the output. From the input of six bits, the last and first bits are used to define/indicate the row of the substitution matrix. The remaining four bits define the column. The DES algorithm, although it seems to be very complex, offers a high degree of protection and is, in fact, a monoalphabetic substitution cipher based on 64-bit characters in the plaintext with a key of 56 bits.

The level of security can be enhanced by defining a stream cipher instead of a block cipher, as discussed above. The DES algorithm based on block cipher is very difficult for cryptanalysts to decrypt, but having defined the stream cipher, it gets more difficult and too complex for them to even try it.

The DES algorithm based on hardware chips at the sender and receiver requires that the sender site be in encryption mode while the receiver site be in decryption mode, and appropriate hardware chips are used at the sites. In the DES algorithms based on stream cipher, the algorithm requires both sending and receiving sites to be in encryption mode. The encryption hardware chips at both sites have a 64-bit input shift register and output register. The plaintext is exclusive-ORed with the output register's first bit (least significant bit) only, leaving the other seven bits unchanged and unused. The resultant exclusive-OR operation (ciphertext) is transmitted to the receiving site's output register and also to its own 64-bit input shift register. The shift register is shifted left, thus removing the most significant bit which triggers the output of the chip to compute a view input value (substitution).

On the receiving site, the incoming data character (ciphertext) is exclusive-ORed with the first bit (least significant bit) of the output register (plaintext) and then shifted left by its 64-bit input shift register. If the sender and receiver sites use the same bit of output registers at the same position for exclusive-ORing the plaintext at the same time, then the plaintext received on the receiving side will be exactly the same as that at the sender site; i.e., both sending and receiving transmitters are properly synchronized.

The first bit of the output register plays an important role in making this algorithm more powerful and effective, as its value is dependent on the contents and previous information of plaintext, which will avoid the generation of repeated ciphertext for repeated plaintext. This algorithm is very useful for both computers and terminals.

15.7.5.3 Data link encryption (DLE)

ANSI X.3.105 has defined standard data equipment encryption (DEE) which can be used between DTE/DCE. This standard can be used either at the physical layer or the data link layer. At the physical layer, encryption is provided on the raw bits, while in the data link layer, the encryption is interfaced with the data link protocol. The encryption for a DTE/DCE connection can either be provided by a separate hardware unit or be integrated with DTE or DCE.

The DEE must have the same functionality, electrical and physical characteristics and parameters, signaling rate, etc., as the associated DTE/DCE (interface). In other words, the function of DEE must be transparent to DTE/DCE. The triggering pin voltages, transfer of control action for encryption and decryption, etc., must have the same parameters as DTE/DCE for the interface connector.

If the encryption is provided at the physical layer, the entire raw bit data is encrypted in synchronous mode, while in the case of asynchronous only, the raw bit data (excluding start and stop control bits) is encrypted. The data link encryption can be obtained either by encrypting the frames of the data link layer or by encrypting only the control data (defined in the information field).

Frame encryption is provided by encrypting all the bits of a frame after the flag in a bit-oriented protocol while it follows the SOH field or its equivalent field in the case of a character-oriented protocol. Although each frame should be encrypted and is supported by frame encryption, many frames may be encrypted in succession, thus avoiding the execution of the encryption algorithm for each of the frames. If the encryption is provided in the information field (I-field), the encryption algorithm is allowed to switch on and off each of the frames separately.

The DEE (based on the DES algorithm) is used between DTE and DCE for DTE/DCE combination, while it is used between computer and modem for computer-to-telephone-system configurations. In the case of computer-to-telephone configuration, the DEE is connected to the computer and telephone system via a physical layer standard RS-232 link.

Another variation of the DES-based DEE algorithm is a **cipher feedback (CF)**. In this algorithm, the input to DEE is 64 bits and the plaintext is combined with DEE-generated ciphertext. This combining of ciphertext with generated DEE, in fact, is a logical integration of the bits of both and is activated each time 64-bit data input is applied at the input of DEE. This algorithm is executed by first sending an uncrypted control variable to the receiving side DEE. This control variable provides compensation for the initially absent feedback operations. This allows us to send ciphertext attached with the control variable to the receiver site. For the occurrence of any error bit in the transmission, all the ciphers are going be affected, as each cipher is obtained by combining plaintext output and ciphertext input into its input stage. The number of bits used in feedback gives an idea of the type and extent of bit errors present in the transmission process.

When the algorithm is used with physical layer DEE, the DTE first sends a request-to-send (RTS) signal to the distant DTE through the local DEE. The receiving site responds to the sender by sending a clear-to-send signal. This will result in the establishment of a circuit on simplex or full-duplex. Once the connection is established, the control variable is transmitted. The DEE initially sends a six-octet control variable before the encrypted eight-bit characters (i.e., asynchronous mode). The encryption is then provided by DEE by encrypting eight bits between the start and stop bits of each character, asynchronously (i.e., start and stop bits are not encrypted, but the bits between them are encrypted). In the case of synchronous communication, a 64-bit control variable is sent after the establishment of connection (i.e., after request-to-send and clear-to-send procedures have been executed).

Security problems have become a serious threat to users of the Internet. There are a number of places where attacks can be made into the Internet. These points, in fact, are the access points or interconnecting points. If we are using dial-up ports, a password is a standard practice. We may buy special software that can respond to a particular tone being produced by a device. The callback to an authorized location only has also been used in some cases. The access points or taps of transmission media may also experience attacks from intruders. Data encryption (as discussed above) has been very useful in scrambling the data so that it can be unscrambled only by an authorized receiver.

15.7.5.4 Telex standards

The primary task of the presentation layer is to define syntax, and as such we have standards defined by international organizations. The syntax modifies the binary form of any application software into suitable binary form and sends it to the session layer. The presentation layer supports a variety of applications in the data communication area, e.g., electronic mail, file transfer, etc. Data communication can also be obtained by emerging technologies such as facsimile communication, digitized voice and data communication, etc. All the different types of devices/interfaces for various applications use the presentation layer protocol in accordance with joint recommendations by CCITT and ISO organizations. The presentation layer protocols, presentation context, coding binary number rules, etc., for different types of facsimile devices are defined in CCITT's T-series recommendations. The coding structure known as International Alphabet #5 is widely used for different types of communication devices and systems. It is also known as ASCII codes or IA5. These codes, in fact, accept input from the application layer and negotiate with its peer layer for the syntax representation. There are different types of syntaxes supporting applications such as teletype, videotext, facsimile, ASCII, etc.

The Telex International Alphabet #2 standard or TWX 50 or 95 character set has been redefined as Teletex Alphabet #2. In telex, the text message is transmitted and therefore it must use a terminal which can provide the characters, punctuation signs, and other equivalent key operations of a mechanical typewriter into a terminal for full-text set operations. Other interesting applications of telex include supporting office equipment with text generation and processing, transferring files, documentation and data-entering systems, saving and retrieving various types of documents, etc. Looking at the variety of telex applications in organizations, universities, and corporations, the CCITT in 1980 defined the following standards for telex systems:

F.200 — Teletex service
S.60 — Use of terminal equipment in teletex service
S.61 — Character repertoire and coded character sets for teletex
S.62 — Control procedures for teletex service
S.170 — Basic support service for teletex

Teletex systems have become very popular, and many countries have implemented teletex systems in their own infrastructures and organizations. The implementation of teletex systems in various countries and by different vendors caused the problem of interoperability and interfacing among systems. This issue was handled by CCITT, which has advised (in its recommendation) all teletex terminals to use a common standard graphic set. On the receiving side, the teletex terminal must process the capability of storing and responding to all the codes, characters, symbols, etc., defined in the teletex graphic set. This requirement was forced on users so that users from different countries could also interact with each other in their respective languages. This teletex set provides symbols, characters, etc., of 35 languages and allows users to retrieve any language symbol characteristics from the set.

Another use of presentation layer standards is to extend the presentation layer syntax to video screens. Two standards have been recommended for defining the screen representations in two communication systems: videotex and teletex devices. These two communication systems have incompatibility problems, and an effort has to be made to provide compatibility between these two systems. The CCITT standard T.101 has recommended the following three categories of syntaxes which will provide compatibility between videotex and teletex, internetworking between these systems, etc.: Data Syntax I, Data Syntax II, and Data Syntax III. The coding syntaxes, protocols, and other possible services are currently under way to approve these recommendations as standards.

15.7.6 Public-key cryptography

Private-key cryptography requires distribution of secret keys over an insecure communication network before secure communication can take place. This is called the "key distribution problem." It is a bootstrap problem; a small secret communication (over an insecure communication network) is required before any further secret communication over the network can take place. A private courier or a secure communication channel is used for the distribution of keys over the network.

Public-key cryptography solves this problem by announcing the encryption procedure E (and associated key) in the public domain. However, the decryption procedure D (and associated key) is still kept secret. The crux of public-key cryptography is the fact that it is impractical to find the decryption procedure from the knowledge of the encryption procedure. This revolutionary concept was advocated by Diffie and Hellman.[17] Encryption procedure E and decryption procedure D must satisfy the following properties:

1. For every message M, D(E(M)) = M.
2. E and D can be efficiently applied on any message M.
3. The knowledge of E does not compromise security. In other words, it is impossible to derive D from E.

Public-key cryptography allows two users to have secure communication, even though these users have not communicated before. This is because the procedure used to encrypt messages for every user is available in the public domain. If a user X wants to send a message M to another user Y, X simply uses Y's encryption procedure E_y to encrypt the message. When Y receives the encrypted message $E_y(M)$, it decrypts it using its decryption procedure D_y.

15.7.6.1 Implementation issues

Diffie and Hellman suggested that one way to implement public-key cryptography systems is to exploit the computational intractability of the inversion of one-way functions.[17] A function f is one-way if it is invertible and easy to compute; however, for almost all x in the domain of f, it is computationally infeasible to solve equation $y = f(x)$ for x. Thus, it is computationally infeasible to derive the inverse of f even though f is known. Note that given f and output y $(= f(x))$ of the function, what we want is that the computation of input x should be impossible.

Diffie and Hellman introduced the concept of "trapdoor one-way" functions.[17] A function f is referred to as a trapdoor one-way function if the inverse of f is easy to compute provided certain private "trapdoor" information is available. An example of private trapdoor information is the value of the decryption key K_d. Clearly, a trapdoor one-way function f and its inverse can be used as matching encryption and decryption procedures in public-key cryptography. Various implementations of public-key cryptography which make use of such one-way functions have been proposed. Next, we discuss a popular implementation by Rivest, Shamir, and Adleman.[18]

15.7.6.2 The Rivest–Shamir–Adleman method

In the Rivest–Shamir–Adleman (RSA) method, a binary plaintext is divided into blocks, and a block is represented by an integer between 0 and $n - 1$. This representation is necessary because the RSA method encrypts integers.

The encryption key is a pair (e, n) where e is a positive integer. A message block M (which is between 0 and $n - 1$) is encrypted by raising it to the eth power modulo n. That is, the ciphertext C corresponding to a message M is given by

$$C = M^e - 1 \left(\text{modulo } n\right)$$

Note that ciphertext C is an integer between 0 and $n - 1$. Thus, encryption does not increase the length of a plaintext.

The decryption key is a pair (d, n) where d is a positive integer. A ciphertext block C is decrypted by raising it to the dth power modulo n. That is, the plaintext M corresponding to a ciphertext C is given by

$$M = C^{d} - 1 \left(\text{modulo } n\right)$$

A user X has his/her own set of encryption key (e_x, n_x) and decryption key (d_x, n_x) where the encryption key is available in the public domain but the decryption key is secret and is known only to user X. Whenever user Y wants to send a message M to user X, Y simply uses X's encryption key (e_x, n_x) to encrypt the message. When X receives the encrypted message, it decrypts it using its decryption key (d_y, n_x).

Determination of encryption and decryption keys: Rivest, Shamir, and Adleman[18] identified the following method to determine the encryption and decryption key. First, two large prime numbers, p and q, are chosen, and n is defined as

$$n = p \times q$$

Note that p and q are chosen sufficiently large so that even though n is public, it will be practically impossible to determine p and q by factoring n. After p and q have been decided, a user can choose any large integer d as long as the chosen d is relatively prime to $(p - 1) \times (q - 1)$. That is, d should satisfy the following condition:

$$\text{GCD}\left(d, (p-1) \times (q-1)\right) = 1$$

Integer e is computed from p, q, and d such that it is the multiplicative inverse of d in modulo $(p - 1) \times (q - 1)$. That is,

$$e \times d = 1 \left(\text{modulo } (p-1) \times (q-1)\right)$$

When n, e, and d are computed in this manner, the encryption and decryption processes in the RSA method work correctly.[18] Note that every user must compute its own set of n, e, and d.

Even though n and e are made public, determination of d requires that n be factored into two primes p and q so that the product $(p - 1) \times (q - 1)$ is known. (Note that this product is needed to compute d.) The main hurdle here is that if n is a sufficiently large number, say 200 digits, factorization of n will require an enormously long time, even on the fastest computers.

An example: Assume $p = 5$ and $q = 11$. Therefore, $n = 55$ and $(p - 1) \times (q - 1) = 40$. We choose d as 23 because 23 and 40 are relatively prime (GCD(23, 40) = 1). Now we must choose e satisfying the following equation:

$$23 \times e \left(\text{modulo } 40\right) = 1$$

Note that $e = 7$ satisfies this equation.

15.7.6.3 Signing messages

To maintain the confidentiality of a message in public-key cryptography, it is encrypted with the public key and is later decrypted with the secret key. However, in public-key cryptography, a message can be first encrypted with the secret key and later decrypted with the public key. Note that by encrypting a message in this manner, a user is creating a signed message, because no one else has the capability of creating such a message. The encryption and the decryption operations in this situation are referred to as *signing* and *verifying* a message, respectively. However, if public-key cryptography is to be used for signing messages, the following condition must hold:

$$(\forall\, M)(\forall\, D_{PK})(\forall\, E_{SK}):: M = D_{PK}(E_{SK}(M))$$

where M is a message and PK and SK are, respectively, the public and secret keys.

15.8 Secure socket layer protocol

As the World Wide Web (WWW) is growing in popularity and is increasingly being used to transfer critical information such as financial transactions, credit card information, and corporate secrets through the Internet, organizations and individuals have to worry about electronic fraud. The Internet does not provide built-in security. Technologies are needed to provide a high level of security for Internet communication. The **secure socket layer (SSL)** is a technology that provides such secure Web sessions over the Internet.

The secure socket layer is an open, non-proprietary protocol designed by Netscape Communications for providing secure data communication between two applications across computer networks. SSL sits between the application protocol (such as HTTP, telnet, FTP, and NNTP) and the connection protocol (such as TCP/IP and UDP). SSL provides server authentication, message integrity, data encryption, and optional client authentication for TCP/IP connections.

When data is sent over the Internet, it is likely to travel through several intermediate computers before reaching its destination (the trusted server). While the data is in transit, any of these intermediate computers can potentially access the information and abuse it. SSL allows organizations and users to exchange information on the WWW (text, pictures, forms, etc., transmitted through the Web browsers) in an encrypted manner such that it can be decrypted only by its intended recipients, not by a third party who may be able to intercept the information while in transit. SSL ensures that intermediate computers cannot eavesdrop, copy, or damage communications.

The secure socket layer security protocol resides in the transport layer. The advantage of residing in the transport layer is that it makes SSL application-independent, allowing protocols such as HTTP (hypertext transfer protocol), FTP (file transfer protocol), and telnet to be layered on top of it transparently.

15.8.1 The design of a secure socket layer

There are two main issues to be addressed in the SSL protocol:

1. How to guarantee the confidentiality of the data in transmission
2. How to securely communicate a secret key to be used for data encryption

15.8.1.1 Confidentiality of data in transmission

SSL ensures the confidentiality of the data to be transmitted by encoding it using the technique of private-key cryptography. The sender and the receiver have a secret key and use it on an agreed upon algorithm to encode the data in the transfer such that only a person or software with the secret key can decode it. SSL uses a secret key system called RC4, developed by RSA, Inc., to encrypt data or information.

SSL comes in two strengths, 40-bit and 128-bit, which refer to the length of the "session key" generated by every encrypted transaction. The longer the key, the more difficult it is to break the encryption code. Most browsers support 40-bit SSL sessions, and the latest browsers, including Netscape Communicator 4.0, enable users to encrypt transactions in 128-bit sessions — trillions of times harder to break than 40-bit sessions.

By using the method of encryption described earlier, a hacker (an unauthorized user) will not be able to read messages that are not intended for him/her. However, a hacker can stand between two users and damage messages sent between them, although the hacker cannot read them. The hacker can just replace the message with garbled information. This is possible because a hacker may know the protocol being used. SSL prevents this by using a technique called the message authentication code (MAC). A MAC computes a secret piece of data that is added to the message. A MAC can be 40 or 128 bits, which makes it impossible to figure out what the right MAC for a message is.

15.8.1.2 Secure communication of a secret key

The use of a secret key for data encryption implies that the participants in the information exchange have a secret key and have communicated it among themselves in a secure manner. When an Internet connection is established, generally there is no prearranged secret key for information exchange among a set of participants, and techniques must be designed for securely generating a secret key and securely exchanging it among the participants.

SSL uses RSA public-key cryptography for data encryption and user authentication. Recall that public-key encryption uses a pair of asymmetric keys for encryption and decryption. Each pair of keys consists of a public key and a private key. The public key is made public by distributing it widely, whereas the private key is never distributed; it is always kept secret. Data that is encrypted with the public key can be decrypted only with the private key. Conversely, data encrypted with the private key can be decrypted only with the public key. This asymmetry property makes public-key cryptography very useful.

After the initial connection, SSL does a security handshake, which is used to start the TCP/IP connection. This handshake enables the client and the server to agree on the secret key to be used. The server announces its public key to the client. No encryption is used initially; therefore, any eavesdropper can read this key. However, now the client can transmit secret information to the server by encrypting it with the server's public key that only the server can decrypt. The client generates a shared secret (46 bytes of random data) and sends it to the server after encrypting it with the server's public key. Only the server can decrypt the message using its private key and determine the shared secret. This shared secret is used to generate a set of secret RC4 keys to encrypt and decrypt the rest of the session.

However, this is not enough. The client needs to make sure that it is communicating with the server, not with an interloper who is impersonating the server by sending out its public key. Thus, the client needs to perform a server authentication. (Authentication is the process of verifying that the user is actually who he or she claims to be.) Server authentication is accomplished by means of ISO X.509 digital certificates in conjunction with RSA public-key cryptography. A digital certificate does connection verification between a server's public key and a server's identification. These certificates are issued by trusted third parties known as the **certificate authorities (CAs)**. A certificate authority is a trusted authority responsible for issuing certificates used to identify individuals, systems, or other entities which make use of a computer network. When a user wants to use SSL as a part of a secure system, he/she first obtains a digital certificate from a CA. A certificate contains the following information:

- The certificate issuer's name
- The server's identity
- The public key of the server
- Some timestamps

When a CA issues a digital certificate to the user, it encrypts the certificate using its private key so the digital signature cannot be forged. If the user has the public key of the CA, it can decrypt the certificate to find out the server's public key. Timestamps are used to determine the expiration time of the public signature in the certificate.

A CA attaches a signature to a certificate, which is generated in the following way: a hash value of the signature is computed and it is encrypted with the CA's private key. If even a single bit is changed in the certificate for any reason (like an intruder attempting to destroy the certificate), the hash value in the certificate changes and the signature becomes invalid.

15.8.2 A typical session

When an Internet browser engages an SSL server, the server's site certificate is presented to the browser. It is the certificate authority's public key which verifies the SSL server and its public key. Immediatley after the verification process, the Internet browser automatically generates a secret session key. This key may be either 40 or 128 bits, depending on the level of security desired. This session key is then encrypted with the SSL server public key and sent to the server, where it is decrypted with the SSL server private key. At that point, both server and browser have established a secure link using the same secret session key.

Once the handshake process is completed, all transmission is encrypted using the RC4 stream encryption algorithm that contains a 40-bit key. A message encrypted with a 40-bit RC4 key takes a 64-MIPS computer a year of dedicated processor time to break. This encryption remains valid between client and server over multiple connections. Since the encryption changes from time to time, the same amount of effort must be expended to crack every message.

15.8.3 Other security protocols

Two other security protocols are in use today, namely Secure Hyper Text Transfer Protocol (S-HTTP) developed by Terisa Systems, and Private Communications Technology (PCT) developed by Microsoft. The S-HTTP protocol is used in a different manner than SSL and PCT. It operates at the application level, encrypting only the hypertext markup language (HTML) pages. The client and the server negotiate with each other to apply certain security constraints and protocols for each Web document or set of documents. Microsoft's PCT, on the other hand, attempts to correct the perceived shortcomings of SSL. Microsoft has made several functional changes to Netscape's SSL protocol but has kept the underlying data structures the same to enable compatibility between the two. Microsoft's PCT, however, is still not fully developed. Microsoft has had to include SSL in its browser, Internet Explorer, because it is still the security protocol of choice in the industry. However, PCT is expected to be included in the next version of Internet Explorer.

Various tools for network security include denial of service, fireballs, IP spoofing, privacy, S-MIME, secure electronic transaction (SET), security shareware, secure shell (SSH), secure socket layer-TLS, and many others.

Some useful Web sites

Information on cryptography can be found at *http://www-dse.doc.ic.ac.uk/~nd/netsecurity/*, *http://www.seas.upenn.edu/~dmurphy/cryptogr.htm*, and *http://www.trincoll.edu/depts/cpsc/cryptography/pgp.html*. Information on private-key cryptography can be found at *http://faqs.jmas.co.jp/FAQs/cryptography-faq/part06*, and information on public-key cryptography can be found at *http://www-path.eecs.berkeley.edu/~ee122/FALL1996/Discussion/DiscWeek12/discweek12/node4.html*.

An excellent Web site for information on network security is *http://www-dse.doc.ic.ac.uk/~nd/netsecurity/*.

Information on SSL can be found at *http://developer.netscape.com/tech/security/ssl/howitworks.html*, *http://www.rit.edu/~esp3641/ssl.html*, *http://www.safe-mail.net/site/SSL.html*, and *http://cie.motor.ru/Topics/121.htm*.

The ASN.1 consortium is located at *http: www.asn.1elibet.tm.fr*. Various computer security products include anti-virus, fireballs, intrusion detection systems, public-key infrastructures, and authentication. Some Web sites for these are at *http://ipw.internet.com* and *http://www.security.com*.

References

1. ISO 7498, "Open System Interconnection — Basic Reference Model, Information Processing Systems," ISO, 1986.
2. Tanenbaum, A. S., *Computer Networks*, 2nd ed., Prentice-Hall, 1989.
3. Stallings, W., *Handbook of Computer Communications Standards*, 2nd ed., Howard W. Sams and Co., 1990.
4. Markeley, R.W., *Data Communications and Interoperability*, Prentice-Hall, 1990.
5. Black, U., *Data Networks, Concepts, Theory and Practice*, Prentice-Hall, 1989.
6. Henshall, J. and Shaw, A., *OSI Explained — End-to-End Computer Communication Standards*, Ch. 6, Chichester, England, Ellis Horwood, 1988.
7. ISO 8822, "Connection-Oriented Presentation Service and Protocol Definitions," ISO, 1986.
8. ISO 8823, "Connection-Oriented Presentation Protocol Specifications," ISO, 1986.
9. ISO 8824, "Specification of Abstract Syntax Notation One (ASN.1)."
10. ISO 8824.2, "Specification of Abstract Syntax Notation One (ASN.1).
11. ISO 8825, "Specification of Basic Encoding Rules for Abstract Syntax Notation One (ASN.1)," ISO, 1986.
12. ISO 8825.2, "Specifications of Basic Encoding Rules for Abstract Syntax Notation One (ASN.1)," ISO, 1986.
13. National Bureau of Standards, Data Encryption Standard, Federal Information Processing Standard Publication, NBS, 1977.
14. CCITT X.409, "Message Handling Systems: Presentation Transfer Syntax and Notation."
15. CCITT X.216 and X.226.
16. Tardo, J., "Standardizing cryptographic services at OSI higher layers," *IEEE Communications*, 23, July 1985.
17. Diffie, W. and Hellman, M.E., "New directions in cryptography," *IEEE Trans. on Information Theory*, November 1976.
18. Rivest, R.L., Shamir, A., and Adleman, L., "A method for obtaining digital signatures and public key cryptosystems," *Communications of the ACM*, February 1978.
19. Popek, G.J. and Klein, C.S., "Encryption and secure computer networks," *ACM Computing Surveys*, December 1979.

20. Needham, R.M. and Schroeder, M.D., "Using encryption for authentication in large networks of computers," *Communications of the ACM,* December 1978.
21. Steiner, J.G., Neuman, C., and Schiller, J.I., "Kerberos: An authentication service for open network system," *Proceedings of the Winter USENIX Conf.,* Feb. 1988.
22. Shannon, C.E., "Communication theory of secrecy systems," *Bell Systems Journal,* October 1949.
23. Denning, D.E., *Cryptography and Data Security,* Addison-Wesley, 1982.
24. Stallings, W., *Data and Computer Communications,* 5th ed., Prentice-Hall, 1997.

chapter sixteen

Application layer

"Successful people are very lucky. Just ask any failure."

Michael Levine

"The top layer of the OSI seven layer model. This layer handles issues like network transparency, resource allocation and problem partitioning. The application layer is concerned with the user's view of the network (e.g., formatting electronic mail messages). The presentation layer provides the application layer with a familiar local representation of data independent of the format used on the network."

The Free Online Dictionary of Computing
Denis Howe, Editor (*http://foldoc.doc.ic.ac.uk/*)

16.1 Introduction

The application layer is the highest layer of OSI-RM and provides an interface between OSI-RM and the application processes which use OSI-RM to exchange the data. The application process uses application entities and application protocols for exchange of information with OSI-RM. It receives services from its lower layer, the presentation layer, which are also used during information exchange. The presentation layer is the only layer which provides services directly to the application processes. The interface points between the top three layers (application, presentation, and session) are different from those of the lower layers. The lower layers do not offer services and functions independent of each other; i.e., the services and functions are supported by many layers. In the case of higher levels, in many applications, the presentation layer may not provide any services to the dialogue unit primitives between application and session layers. For example, the token primitives P-TOKEN-PLEASE and S-TOKEN-PLEASE are used between application and session layers, and when these are passed through presentation entities, the presentation layer simply passes them to the session layer (this is due to the fact that most of the service primitives of the presentation layer are similar to those of the session layer, as discussed in previous chapters).

The application layer provides communication between open systems and provides functions which have not been defined by lower layers. It is concerned mainly with system management and application layer management activities. The application layer contains user programs for different applications, and the user programs are also known as applications. There are many applications, but of these, two applications — file transfer and remote file access — seem to be essential for computer networks. The remote file access is somewhat similar to file transfer, except that in the case of remote file access, part of the files are either read or written (as in diskless workstations), as opposed to the entire file (file transfer). Regarding access of a file, both applications have the same technique and are managed by a file server.

ISO has developed a set of application protocols, and these are application-specific protocols. Each protocol offers a specific application service and can also provide services to TCP/IP application protocols. These application protocols of ISO include virtual terminal (VT), file transfer access and management (FTAM), message-oriented text interchange standard (MOTIS), job transfer and manipulation (JTM), remote database access (RDA), draft international standard (DIS), manufacturing message service (MMS), etc. Some of these application protocols have similar services provided by the TCP/IP set of protocols.

Each of the above-mentioned protocols uses a client–server computing model for implementation. In client–server implementation, we have a client application (which here corresponds to application protocols such as virtual terminal, FTAM, etc.) and a server application program. These two application programs may run either on the same computer or on different computers. Details of the client–server model and various aspects for implementing applications using a client–server model may be found in Chapter 20 on client-server.

In the implementation, the client application protocol runs on one computer and interacts with server application protocols via Structured Query Language (SQL). The server application protocol, after running the application, sends the results back to the client. The client VT application protocol does not offer interfaces for the terminals (scroll mode, page mode, form mode, etc.). The interface to these terminals is provided by the local application program (user element or user agent). Since there are different types of terminals available, the VT application protocol allows users to negotiate over the types of terminals being used, and both users have to agree on the **virtual terminal environment (VTE)** which is used by the application protocols. The negotiations between users are usually provided by a token (a special pattern of control characters) and use different types of token primitives for different types of transmissions (asynchronous and synchronous). Some of the material herein is partially derived from References 1–3, 19, and 20.

The file transfer has become one of the founding blocks for the application layer, as other applications within and outside OSI are implementable over it and can be used on OSI and non-OSI networks. The application layer protocols allow users to access or transfer any file from any machine to any other machine, and a standard interface has been offered by a **virtual filestore (VF)**. Another application which is basically a subset of file transfer is electronic mail (e-mail), which has become an integral part of many organizations and universities. The e-mail systems offer a human interface (for reading, editing, etc.) and transportation for the mail (managing mail lists, mailing to remote destinations, etc.).

The e-mail facility was standardized by CCITT in the X.400 series of recommendations and was called a **message handling system (MHS)**. This standard is called a **message-oriented text interchange standard (MOTIS)** by ISO, and the X.400 standard was modified by CCITT in 1988 with a view to merging with MOTIS (initially as an application in the application layer). Other applications include directory systems, which can be used to search for information such as network addresses and services offered by the networks.

Remote job entry (RJE) is another interesting application which allows a user working on one machine to submit a job on another machine. A typical use of this application is connection between PCs connected to a mainframe machine where PCs can submit jobs to the mainframe and the mainframe is responsible for collecting all the data files, programs, and statements needed for job control (which are available on different machines) at one point. These applications are concerned with text messages. The application layer also provides application of telematics services, which in general fall into two categories: teletex and videotex. Teletex provides services to users based on one-way communication (broadcasting through TV), while videotex provides interaction between users and database systems where users can access the data from databases in interactive mode.

Manufacturing message service (MMS) provides services related to automated manufacturing-related messages to different divisions involved in different aspects of processes within the manufacturing process. An application protocol may send manufacturing-based messages to other application process protocols like numerical machine tools, programmable machines and controllers, etc.

The **remote database access (RDA)** offers users access to a remote database management system. The discussion of some of these application protocols has been partially derived from References 3, 4, and 19.

16.2 Layered protocols and interfacing

The layers of OSI-RM are the foundation of many standards and products in the industry today. The layered model is different from layered protocols in the sense that the model defines a logic framework on which protocols (that the sender and receiver agree to use) can be used to perform a specific application.

16.2.1 Layered protocols

The services of the presentation layer are provided to the application layer at a point known as the **presentation service access point (PSAP)**. For interface between the application and presentation layers, the application layer primitives are defined. The **application service data unit (ASDU)** associated with its primitives is delivered to the presentation layer and, similarly, the presentation layer uses its primitives to interact with the application layer. The application layer sends its **application protocol data unit (APDU)** to the presentation layer. The APDU consists of the protocol, control information, and application process. The application control information is exchanged between application layer entities.

16.2.2 Interfacing

The service access point (SAP) is a unique hardware or port or socket address (software) through which the upper layer accesses the services offered by its lower layer. The lower layer also fetches the message from its higher layers through these SAPs. The PSAP defines its own address at the layer boundary. Similarly, other layers define their own SAPs with the appropriate prefix delineating the corresponding layers.

The application service data unit (ASDU) is defined as message information across the network to peer entities of all the layers. The SDU and control information (addressing, error control, etc.) together are known as an interface data unit (IDU), which is basically message information sent by an upper-layer entity to its lower-layer entity via SAPs. For sending an SDU, a layer entity partitions the message information into smaller-sized messages. Each message is associated with control information defined by the appropriate layer as a header and sent as a protocol data unit (PDU). The PDU is also known as a packet. The relationship between APDUs, ASDUs, and application entities is shown in Figure 16.1.

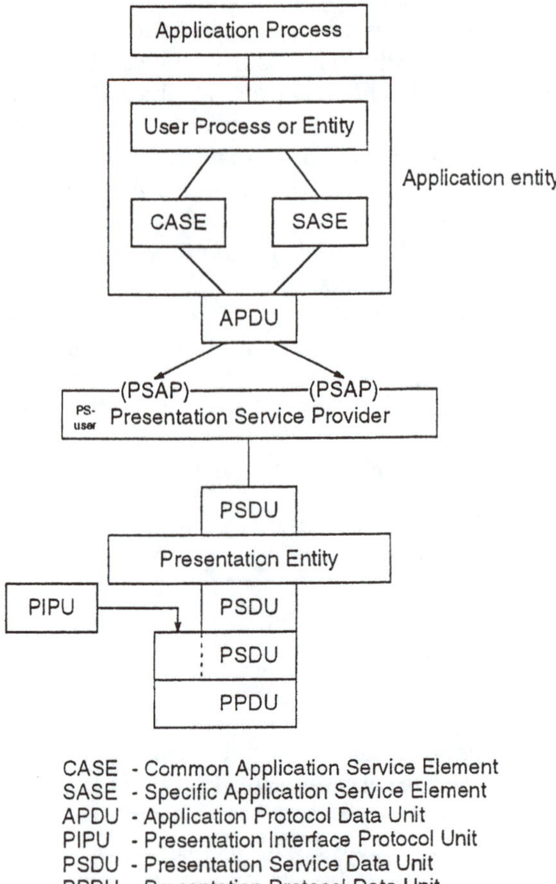

Figure 16.1 Application layer structure.

16.3 ISO application layer services and functions

In the previous section, we discussed some of the main purposes and objectives of the application layer. Now we will discuss the functions and services separately as described by ISO. As indicated earlier, there is significant overlapping of the services and functions offered by different layers; however, OSI-RM remains popular among users, and more and more vendors are entering the market to produce OSI-RM–based products.

The application layer, being at the top of OSI-RM, does not provide any services to its higher layer (as there is no layer over it!), and as such it does not have any service access point (SAP). The services provided by it are usually provided by OSI management.

ISO application layer services: Application layer services include identification of the destination address, identification of its communication establishment authority, cost of the features of communication (privacy, resource allocation strategies, etc.), quality of service (determined by parameters such as link error rate, response time, their costs, etc.), synchronization between cooperating application processes, constraints on data syntax, control procedures for data integrity, control procedures for error control and error recovery, etc. This list is not complete, as these are some of the services provided by OSI management. Other services can be defined by users, depending on the applications.

ISO application layer functions: The application layer includes all of those functions required for communication between open-end systems which have not been defined by lower layers, and similar functions can be grouped together. Thus, any application process will look into these groups for its use, and that group will contain the application entities. The application entities include application protocols and services offered by the presentation layer to the application layer. The application process accesses the services of OSI through these entities, and we can consider application entities to be service providers to the application processes.

The application layer also includes various aspects of management of OSI, e.g., application management (for application processes), application management and entity (application entity invokes application management functions), resource management (of OSI and the resource status across the network), layer management (protocol deals partially with activation of respective functions such as error control, etc., and the remaining part is usually defined as a subset of this management), etc.

The user process can access all the services provided by the application layer using application layer protocols, and since the application layer is the highest layer in OSI-RM, it does not have any application service access point (ASAP). Hence, all the services of the application layer are provided to application processes which are not a part of OSI-RM. The application layer provides an interface between the application services and OSI-RM. An application service request primitive is sent from the application layer to the presentation layer. The primitive has a set of parameters, some of which are defined for the application layer, while the rest are sent to the presentation layer. When the presentation layer sends a presentation service request primitive to the session layer, it passes on the parameters received from the application layer and the parameters defined by the presentation layer.

The application process is an element within OSI-RM which performs the information processing for a particular application (OSI document) and lies outside the OSI-RM architecture. It can also be defined as an end user program which implements the services of the application layer. These services are accessed by user processes by using application layer protocols.

16.4 Application layer architecture (structure)

The application layer of OSI-RM consists of a number of entities which allow the transfer of information between application layer protocols and presentation services. The entities are also known as **service elements (SEs)** and define application processes within OSI-RM. The entities provide both standard and unique services to the network users. The application processes access the services offered by the application layer through these entities or service elements. Thus, the entities are responsible for providing services to one or more application processes. The entities define the services offered by the layer and also correspond to the service provider. Since the application layer does not have any layer over it, these services are used by the application processes. The implementation of these entities by the appropriate application process is defined as an application service provider (AS-provider). The discussion presented below is partially derived from References 1–3 and 19.

The application entity can be divided into two groups known as **user element (UE)** and **application service element (ASE)** (needed to support that particular application). The former represents a specific application which is represented by an **application process (AP)** (outside the OSI-RM) to the UE and is responsible for the actual application services of OSI-RM. In other words, the UE is a part of AP which provides a means for accessing OSI services, as shown in Figure 16.2. The AP represents an end user program.

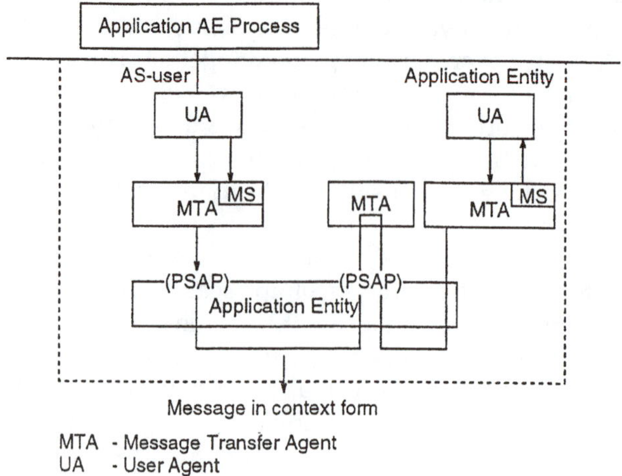

Figure 16.2 Data flow in OSI-RM.

It provides an interface for application service elements which are defined as functions in OSI-RM. The application service elements, on the other hand, are responsible for providing the essential services of OSI-RM.

The application service element identifies the functions provided by the application layer services. The application layer contains various application entities which employ protocols and presentation services to exchange the information. These application entities provide the means for the application process to access the OSI environment and offer useful services to one or more application processes. The functions offered by the application service entities are typically not defined but are dependent on the applications under consideration. According to the OSI document, the functions are grouped into a collection of functions which defines a specific task of the application layer. This group of functions includes the application service elements which provide an interface for accessing OSI services and also includes other service capabilities of accessing a particular (specific) or common set of applications.

Typically, the group of functions has three components. The first is the **user element**, which is a part of the application process specifically concerned with accessing of OSI services. It also offers interface with application service elements of the functions which provide the OSI services. The second component is the **common application service element (CASE)**, which has capabilities that are generally used in a variety of applications. The third component is the **specific application service element (SASE)**, which has specific capabilities that are required to satisfy particular needs of specific applications. The SASE and CASE components will be discussed in more detail in the following section.

16.4.1 *Specific application service element (SASE)*

An application entity operates through one presentation layer service access point (PSAP) address with the presentation layer. It provides data exchanges between peer-to-peer protocols. The application entity usually contains one user element and a set of ASEs. The set of ASEs can be defined in a variety of ways to support a variety of applications.

It is the responsibility of users to implement UE, which provides accessing of the services of the application service element, needed to exchange the data at the application entity. The user element has no direct access to the services of the presentation layer but can access the application service elements by invoking its services within the application entity. Some examples of this application entity include **file transfer, job transfer, message exchange**, and **remote terminal access**. Both ISO and CCITT have defined a series of standards for these applications.

The application entities may call each other and also can access the services of the presentation layer. There is a unique PSAP for each application entity defined in the application layer such that for each PSAP, there is a unique SSAP. However, if an application process is using more than one application layer protocol during a single session, then the application entity supporting this application process must support a single user element (UE) and have capabilities for supporting a variety of applications for a particular application, i.e., multiple specific application service elements (SASEs) and common application service elements (CASEs) kernels.

16.4.2 Common application service elements (CASE)

The CASE category of elements includes Abstract Syntax Notation standard ASN.1 (discussed in the previous chapter) and the service elements. This set of elements provides a logical connection between two SASE entities before defining the specific application function and is considered another example of CASE which is obtained through the application function known as **association control service elements (ACSEs)**. This function includes other parameters in the functions **association control** and **commitment, concurrency and recovery (CCR)**. It is interesting to note that ACSEs define various aspects of connection-oriented services which later on are used by lower layers to provide connection-oriented services to the users. Standards include ISO DP 8649/1 and DP 8650/1, which are discussed in the following section. It is worth mentioning here that these elements (CASE or SASE) should not be considered sublayers of the application layer (similar to LLC and MAC in the data link layer). The CASE is available as a peer with SASE elements and offers a set of functions that would be present in each specific application service if not provided separately. Within an application entity, application service elements may call each other and use the services of the presentation layer.

16.5 Application layer standards and protocols

The standards for both categories of elements (CASEs and SASEs) have been defined by ISO and CCITT, and a variety of vendor products are available on the market. The standards for common application service elements (CASEs) are defined by ISO in DP 8649/1 and DP 8650/1. The specific application service elements (SASEs) include the following application service applications: file transfer (FT); job transfer and manipulation (JTM); virtual terminal (VT); message exchange and remote terminal access (MERTA); file transfer, access, and management (FTAM); electronic mail (e-mail); and telematics.

ISO has defined protocols for two SASEs known as message handling service (MHS — CCITT's X.400–X.430) and FTAM.[18] MHS and FTAM are compatible with the standards defined for equivalent DoD protocols.

16.5.1 Association control service element (ACSE)

The association control service element (ACSE) is an indication of establishment of application connection between two applications entities (an essential requirement for all layers) over which communication can take place. This application connection so established is also known as **association**; thus, the establishment of association provides a one-to-one mapping between the association and presentation layer connection, which in turn defines a session layer connection. These connections (presentation and session) are defined over a unique **presentation service access point (PSAP)** and session service access points (SSAPs). Once an association between two application entities is established or defined, the application service elements allow the association to exchange the information of application protocols for mutual agreement over the services. At the same time, this association may be used by other protocols (corresponding to various applications) over the same application connection.

For each ACSE, there is a one-to-one correspondence with a presentation layer service primitive. It is interesting to note that these ACSEs should have used the presentation layer primitives (as done with other layers within OSI) directly, but this concept of association is basically used to define the primitives of the application layer and also may be used for future enhancement in the functionality of the application layer. The various OSI primitives defined for ACSE include A-ASSOCIATE, A-RELEASE, A-ABORT, and A-P-ABORT. Details about these primitives along with parameters are provided later in this chapter.

ISO-DIS 8649 and ISO-DIS 8650 standards and protocols: In order to use a variety of applications supported by the application layer, a set of services defined as standard ISO-DIS 8649 and protocol ISO DIS 8650 is used as an essential service. This service provides connection establishment, maintenance of connection, and termination of connection. This set, in fact, is part of the common application service elements (CASEs). An application connection (connection-oriented service) needs to be established for the exchange of messages between peer application protocols via application entities. This type of connection is also known as *association control* (AC). There are two types of association controls: application association control and application context control.

The **application association control (AAC)** defines a relationship between two application entities which allows them to mutually agree on the procedures (supported by the AS-provider) and also the semantics of the data to be used on the presentation connection. The AS-provider represents implementation of application services within the application entity for application processes. The presentation connection is defined by a relationship between presentation entities which provides the means over which abstract data values can be transferred and also delineates a relationship or function (application association) between application entities within a defined application context.

The **application context control (ACC)**, on the other hand, defines a mutually agreed upon function and relationship (application association) between application entities for the transfer of data across different open systems. This mutually agreed upon relationship is valid for the duration of time during which these entities are transferring the user data back and forth. For different functions, different entities of the application layer are used; thus, only those entities of the application layer service which have been mutually agreed upon and employed for this application provide a valid application association for a defined application context. This relationship between application entities is automatically canceled or aborted after the task of transfer of data between them is completed.

ISO-DIS 8650 protocol primitives and parameters: Various ISO application control service primitives along with the parameters associated with them (DIS 8649) are discussed below. This set of services is, in fact, a collection of service elements for connection establishment, maintenance of connection, and termination of connection, and is also known as common application service elements (CASEs). It offers services to user elements and specific application service elements (SASEs). A list of parameters is associated with every service primitive, and a few parameters are implemented by the application association services while the remaining parameters are implemented by the presentation and session layer services. An application association is provided by the association service primitive A-ASSOCIATE, which provides a direct link between application association and presentation connections using different attributes (request, indication, response, and confirm) between application service users via an application service provider (AS-provider). This service is supported by its equivalent counter-primitive P-CONNECT (of the presentation layer) and includes the parameters defined in the P-CONNECT primitive; it also provides a direct mapping for the parameters used by session layer services. The following are various application service primitives along with associated parameters in parentheses.[2]

A-ASSOCIATE.request (mode*, application context name, calling AP title, calling AE qualifier, calling AP invocation identifier, calling AE invocation identifier, invocation identifier, called AE invocation identifier, user information, calling presentation address*, called presentation address*, presentation context definition list*, default presentation context name*, quality of service**, presentation requirements*, session requirements**, serial number**, token assignment**, session connection identifier*)
A-RELEASE.request (reason, user information)
A-RELEASE.indication (reason, user information)
A-RELEASE.response (reason, user information, result)
A-RELEASE.confirm (reason, user information, result)
A-ABORT.request (user information)
A-ABORT.indication (abort source, user information)
A-P-ABORT.indication (reason)

The A-ASSOCIATE.request primitive is originated by the sending user and sent to the application service provider, which gets a response parameter either from the receiving user via the association control service element (ACSE) or by the presentation service in the A-ASSOCIATE service indicating the states accept or reject. In the case of accept, the A-ASSOCIATE service will create an application association corresponding to presentation and session connections. The parameters used in these three created connections are the ones which have been mutually agreed upon by the application service user for each of the connections. For each connection, the result parameter maintains a list of parameters which indicates the address of the source (in A-ASSOCIATE service) that created the three types of users: ACSE service user-receiving, ACSE service provider of the application layer, and presentation service provider of the presentation layer. In the case of reject, the A-ASSOCIATE service primitive request is discarded.

Various ISO service primitives used in ACSE are A-ASSOCIATE (for establishing connection), A-RELEASE (for releasing or terminating connection), and A-ABORT and A-P-ABORT (abort initiated by user and service provider, respectively). The A-P-ABORT.indication primitive is implemented by the presentation service.

The parameters defining the application context control (ACC) are used by sending and receiving application service users to mutually agree on the application context via assigning names to it. These parameters are not defined in the standards and hence can be defined and agreed upon by users via defining the names of the application contexts. These names, e.g., **ASN.1 type object identifier**, are usually defined or used by the standards organizations as a subset of the application layer service protocols. It is important to note here that these names for application context are dependent on the application under consideration. Various terms used in the application layer protocol standards can be found in DIS 9545. These terms can be found in application association control protocol ANS.1 under the heading External, which clearly indicates that this standard (ANS.1) does not include both syntaxes (abstract and transfer) defined by the presentation layer service protocols within application context control.

Diagnostic parameters of ACSE primitives: The application service provider (or ACSE service provider) has the following diagnostic parameters for association control service primitives with confirm and response attributes:

- No reason
- No support for application context
- Calling AP title unrecognizable
- Calling AE qualifier unrecognizable
- Calling AP invocation identifier unrecognizable
- Called AP title unrecognizable
- Called AE qualifier unrecognizable
- Called AP invocation identifier unrecognizable
- Called AE invocation identifier unrecognizable

The termination of application association along with the presentation and session connections is caused by the primitives A-ABORT and A-P-ABORT. The parameters defined in A-P-ABORT are initiated by the presentation service provider, which terminates the presentation connection using a P-P-ABORT service primitive. The mapping of a P-P-ABORT service primitive into an A-P-ABORT service primitive terminates all three connections (application association and presentation and session connections) simultaneously. The primitive A-ABORT is used by either the association control service (application service) provider or other service users (presentation or session layers). The parameters associated with these primitives define the source which initiates the primitives.

Application protocol data unit (APDU) primitives: The protocol for application control services is defined in DIS 8650. Various application protocol data units (APDUs), similar to protocol data units for the different layers of OSI-RM, along with associated parameters are given in Table 16.1. The parameters defined on the APDUs are the ones which are actually sent to the presentation layer as presentation user data and which are defined within the user application process as APDUs. The presentation layer includes its own parameters in the PPDUs and sends them to the session layer. The structure or format of the APDUs is the same as that of PPDUs and also defines the abstract syntax.

Associate APDU (linking function): Associate APDUs can also be thought of as a linking function between the primitives of the application and presentation layers; hence, they are also known as *linking functions*. As mentioned earlier, some of the parameters in the application layer service primitives belong to the application layer, while other parameters belong to the presentation and session layers. In other words, the session layer service

Table 16.1 Associate Application (Association Control) Protocol Units and Parameters

APDU	Parameters
A-ASSOCIATE-REQUEST (AARQ)	Protocol version, application context name, calling AP title, calling AE qualifier, calling AP invocation identifier, calling AE invocation identifier, called AP title, called AE qualifier, called AP invocation identifier, called AE invocation identifier
A-ASSOCIATE-RESPONSE (AARS)	Protocol version, application context name, responding AP title, responding AE qualifier, responding AP invocation identifier, responding AE invocation identifier, result, result source-diagnostic, implementation information, user information
A-RELEASE-REQUEST	Reason, user information
A-RELEASE-RESPONSE	Reason, user information
A-ABORT	Abort source, user information

Source: Partially derived from References 1–4.

Table 16.2 Associate APDU Link Functions between Presentation and Application Layers

Associate APDU (provides linking function)	Presentation layer service primitive	Application layer service primitive
A-ASSOCIATE-REQUEST	P-CONNECT.request	A-ASSOCIATE.request
	P-CONNECT.indication	A-ASSOCIATE.indication
A-ASSOCIATE-RESPONSE	P-CONNECT.response	A-ASSOCIATE.response
	P-CONNECT.confirm	A-ASSOCIATE.confirm
A-ABORT	P-U-ABORT.request	A-ABORT.request
	P-U-ABORT.indication	A-ABORT.indication
No linking function	P-P-ABORT.indication	A-P-ABORT.indication
A-RELEASE.request	P-RELEASE.request	A-RELEASE.request
	P-RELEASE.indication	A-RELEASE.indication
A-RELEASE.response	P-RELEASE.response	A-RELEASE.response
	P-RELEASE.confirm	A-RELEASE.confirm

primitives will define and negotiate over a few parameters, e.g., quality of service, session requirements, token assignments, etc., which will be used by the application layer service primitives and similarly by the presentation layer. The associate APDUs provide a direct relationship or linking function between the application and presentation layers over those parameters defined by the lower-layer service primitives. Table 16.2 describes the linking functions between the presentation and application layer service primitives along with appropriate attributes.

Flow of application processes: The flow of various application processes through the protocol data units from application layer to session layer is discussed in the following paragraphs.

An application process initiated by the sending user sends a service request to the A-ASSOCIATE.request primitive of the application layer. This primitive triggers a transforming function of the application layer and a P-CONNECT.request primitive of the presentation layer. The transforming process creates presentation service user (PS-user) data by including all the application-related parameters and user information defined by the application layer. The associate APDU links the primitives of the application and presentation layers. The application-related parameters are defined in the application header field, while application user information is defined in the user data field.

The presentation-related parameters are passed directly by the application layer primitive to the P-CONNECT.request primitive of the presentation layer. The presentation layer also includes its parameters in this primitive. These parameters are then passed to the presentation header field, and at the same time session-related parameters are passed to the session layer S-CONNECT.request primitive. The user data defined by the application layer is transformed to the user data field. The linking function is defined between the presentation layer service primitive (P-CONNECT.request) and the session layer service primitive (S-CONNECT.request) by a CONNECT.request PPDU. This protocol data unit provides encoded user data provided to it by the associate application protocol data unit (APDU).

The PPDU defines a user data–encoded packet containing a presentation header and user data fields and sends it to the session layer. The encoded packet becomes the user data for the session layer, and the session header field contains the parameters defined in the S-CONNECT.request primitive. The session-related parameters are defined in the session header field; the application-related parameters are defined in the application header and are included in the user data field of the presentation layer. The presentation layer service primitive parameters are defined in the presentation header and are included in the user data field of the session layer. Finally, the session-related parameters are defined in the session header and are included in the user data field of the transport layer. The encoded user data packet from the presentation layer is then accepted by the user data field of the session layer through a CONNECT.request SPDU.

16.5.2 *Message-handling system (MHS)*

An electronic mail (e-mail) facility has become a useful service that enables people around the world to communicate. Users can send their typed text or images stored in a file or drawings to a remote destination. The CCITT X.400 series document contains procedures for handling electronic mail facilities and includes X.400–X.430 standards. In 1988, CCITT modified this standard to merge it with MOTIS (message-oriented text interchange standard), an original version of e-mail protocol defined for the application layer by ISO 10021. This standard X.400/MOTIS allows users to send their mail (or message) to the host connected in the network. The e-mail application offers plenty of implementation styles and user experiences. The hosts connected to the network are known as message transfer agents (MTAs), which, along with the user agent (UA), are a component of the message-handling system (MHS). The MTA accepts the messages and processes them on a store-and-forward basis until they are delivered to the destination host. Some of the material herein is partially derived from References 1–3, 6–8, 19, and 20.

The MHS can be characterized as follows:

- Contains standard message header format similar to an office memo
- Is based on the store-and-forward concept and uses it for delivery of messages to multiple destinations
- Provides inherent conversion of message contents and formats to allow message transfer between different types of terminals (dissimilar terminal types)
- Provides submission and delivery time stamping
- Offers notification of delivery or nondelivery status
- Offers standard service access control which offers a global connectivity for data communication across the network between users anywhere around the world

The e-mail service provided by X.400/MOTIS services of standard MHS documents is becoming very useful and important for LANs and WANs. E-mail can be provided by using a mail program (MP) or mail system (MS) which allows the users to send mail messages back and forth. It offers various features such as composition, transfer, reporting, conversion, formatting, and disposition.

- **Composition:** This aspect deals with creation of messages and replies to received mails. The mail program offers different options within the mailing system which can be used for receiving, replying, creating, editing, etc.
- **Transfer:** This aspect of e-mail deals with transmission of a mail message from one machine to another machine over the network. This aspect needs an interface to either the ACSE or presentation layer entity and is responsible for establishing a connection, moving the mail message, and then releasing the connection. The mail program provides these services to the users transparently so that the user knows how these are defined and implemented on the local user's or destination user's machines.
- **Reporting:** This aspect is concerned with reporting on the status of a mail message (i.e., is it delivered or not?).

- **Conversion:** Due to a number of incompatible devices connected to the network, this aspect provides conversion between different terminals and other devices (performing the same functions, such as printers). Also, the compression technique is used for sending a high bandwidth of data over a network of limited bandwidth and capacity, and then the decompression technique is used on the receiving side to extract the original mail message; conversion of format, pixels, and many other parameters is necessary for this process.
- **Formatting:** This aspect of e-mail is concerned with formatting of a mail message in different formats on the receiving-side terminals. The mail program must support different formatting systems and could be better if it could display the mail formatted in a particular format by invoking that particular formatting program. The current mail systems allow the user to use editors to modify the mail and offer limited formatting capabilities. UNIX-based systems offer standard Troff system calls for editing.
- **Disposition:** This aspect of the mail program deals with various options the user has after receiving the mail message, e.g., reading, deleting, saving, forwarding, etc. Further, the saved messages must be retrievable with the options of forwarding them, processing them, etc.

The majority of the mail programs provide a mailbox for users to store incoming messages. They also offer a number of options such as creation of a mailbox, deletion of a mailbox, insertion and deletion of messages into and from the mailbox, etc. The programs also allow users to send mail messages to either a group of people (distribution list) or everybody on the network (broadcasting) by including the e-mail addresses of these people. Other facilities available with mail programs are "carbon copies," high-priority mail, secret mail, alternate recipients (if the boss is not there, then his secretary may receive it), etc. The mail can be replied to using the received mail message by using an appropriate command. The mail programs or systems typically include three types of mail messages: user message, reply message, and test message. The user message is the mail message sent out by a user, and the reply message is a system-generated message sent out to the sender to indicate the status of the message (delivered or not). The test message is usually used to determine if the destination is reachable and informs the sender, who can then determine the route of the message.

The computers are connected via a network which simply provides a communication link between them. Each computer is required to manage its own resources and can have its own operating system. The network protocols provide services for sharing of resources, file transfer between these computers over the network, sharing of databases and programs, etc. In e-mail, the text message is exchanged between users. The e-mail service is provided by the mail program, which sends mail to the destination network and is stored on the destination network until it is read by the destination users. After the user reads the mail, other services such as answering, forwarding, saving, replying, distributing, and broadcasting are provided by the same mail program.

The **mail program (MP)** or **mail system (MS)** consists of two components: **user agent (UA)** and **mail or message transfer agent (MTA)**. The UA is a program which offers an interface to the mail program. It allows users to compose, send, and receive mail and also to manipulate the mailboxes. It prepares mail from the typed text, images, or any other form of message, and allows users to read the mail. The MTA, on other hand, is concerned with the delivery of mail to the destination. It accepts the mail from the UA and transmits it to the destination node. The UA typically runs on a PC, while the MTA runs on a mainframe or dedicated workstation. Since PCs offer limited storage, all the incoming mail is stored in the user's mailbox by the MTA. The PC user logs on and reads the messages from this mailbox. Until the user reads the mail, these messages will remain in the mailbox. The interface between the UA and MTA has been standardized by OSI, and protocols are available. The MTA defines a mailbox for every user, and the interface between the UA and mailbox has also been standardized.

The structure of the MHS also consists of two components, UA and MTA, as mentioned in the X.400 series of standard documents. The functions of these components are discussed briefly below and are explained in detail in the following section.

Functions of UA:

- To provide a communication interface with the user for sending and reading mail.
- To provide editing facilities for mail to the users; the editors are different for different systems (e.g., UNIX has vi, VAX/VMS has VMS editor, etc.).
- To provide other features such as saving and forwarding of mail, printing of mail, replying to mail, etc.
- To provide transparent implementation of the mail system to users.

Functions of MTA:

- To receive pre-formatted mail messages from users.
- To decide/determine the routing (defining the sequence of the networks to be used for forwarding the mail). This routing, in turn, depends on the formatting used in the addressing scheme of the mail program.
- To re-format the message (if needed) to conform to standards.
- To receive messages from different networks via monitoring and controlling procedures of mail systems.
- To check if the mail belongs to its valid user. If it finds the message for a valid user, it passes it on to the local mail delivery program which, after accepting the mail, delivers it to the user's mailbox (a file of fixed block of memory assigned to each user and usually managed by the computer center of the organization or university).
- If the mail message does not belong to a local user, it considers this mail message an incoming message, re-formats it, and re-transmits it to another network, and this process is continued until the mail message reaches the destination network.

The MTAs accept messages from users, process the messages on the basis of store-and-forward, and eventually deliver them to the destination users. The collection of MTAs is known as a message transfer system (MTS).

X.400 Recommendation Document: The IEEE 802 committee has mainly concentrated on the physical layer and data link layer of OSI-RM. It defines two sublayers of the data link layer as logical link control (LLC) and media access control (MAC). The higher-layer protocols are usually considered above the LLC sublayers. The higher-layer protocols are left to the client and users to define with their own options and requirements, and as such they offer flexibility in configuring the network framework to suit different applications. This is convenient with LANs, as the LAN models are organized in a bottom-up approach starting from the physical layer to media access control. The LLC sublayer is the uppermost formal layer in the LAN mode. It is worth mentioning here that the LLC layer is common to different types of LANs, and it brings together various LAN technologies to a common interface. One of the interesting concepts introduced in distributed systems is **remote procedure call (RPC)**. Both CCITT and ECMA are cooperating on **remote transaction services (RTS)** as a part of X.400. ISO's term for RPC is **remote operation service (ROS)**.

The CCITT X.400[18] document offers the following recommendations:

X.400 System Model Service Elements
Naming and Addressing
Overview of Layer Architecture
X.401 Basic Service Elements and Optional User Facilities (defines service elements on interpersonal messaging service)
X.408 Encoded Information Type Conversion Rules (for example telex, IA 5 text, teletex, G3 fax, text interchange format 0, videotex, voice, etc.)
X.409 Presentation Layer Transfer Syntax and Notation
X.410 Remote Operations and Reliable Transfer Service Server (RPC facilities)
X.411 Message Transfer Layer
X.420 Interpersonal Messaging User Agent Layer
X.430 Access Protocol for Teletex Terminals

The standard X.408 provides **telematics** services, which are typically terminal-based services and have been used in a variety of applications such as teletex, videotex, public banks, etc. The telematic service is classified broadly into two categories as **teletex** and **videotex**. The teletex service is basically a one-way communication service which transmits data via a TV broadcasting system to a user terminal. The entire page remains on the screen until either it is changed by the service or the user makes a request for the next page. The videotex offers two-way communication between the user terminal and the database where the information is stored. Users access the data of the database (located at the remote location) over the network. Both services allow users to access information stored in a database, with the only difference being that in teletex, the users can only see the message on the screen, while videotex allows users to view the data interactively. The telematic service provides support for text, graphics, images, pictures, data, etc. This service is used in a variety of applications, e.g., news, sports, tele-shopping, tele-negotiation, tele-advertisement, electronic mail, telemetry, financial information retrieval, banking transactions and many other applications of the information retrieval category.

X.400 electronic mail services are used in both LANs and WANs and are becoming standardized through the X.400 series of standard documents. The method used to transfer mail messages from one site to another differs between different computer systems and networks but offers electronic mail services across the network.

The remote procedure call (RPC) is a very important building block in distributed systems and is becoming very useful in these applications. This facility is designed to activate remotely located software or resources just like a local software or resource where the user does not feel any difference in running the software or using the remote resources. For the user, this facility gives a feeling that the software is running on the local machine and also that the local resource is being used for applications. The RPC transaction consists of a call message being sent from the sender to the receiver computer. The requested action is carried out at the remote machine, and the reply to it is sent back to the sender computer. For further details, please see References 4–9, 19, and 20.

Both sender and receiver interact with each other via message-passing primitives, and the idea behind the RPC protocol is to provide syntax and semantics of the local procedure call for RPC so that the users can define the remote procedure calls for distributed systems. The RPC has been included in CCITT recommendation X.410 (Remote Operations Protocol) within X.400. This facility of RPC has been used in a number of distributed applications, and a protocol has been developed in the **Xerox Network System (XNS)**. The RPC can be considered an **interprocess communication (IPC)** where both sender and receiver, located on either the same machine or different machines, interact with each other for data information exchange. Due to the high performance requirement for protocols for LANs, the RPC protocol is usually included on top of the data transport mechanism of LAN. The underlying protocols of the LANs could be either datagram or virtual circuit services (i.e., they support both connection-oriented and connectionless services).

Structure of MHS: These recommendations are used by the MHS as a layered model providing application-independent system services within the application layers of OSI-RM. As mentioned earlier, the MHS consists of two components: user agent (UA) and message transfer agent (MTA). The MTAs are connected either to other MTAs or to UAs and are usually considered hosts connected to the networks. Each registered user of the MHS is assigned a UA. The UA facilities provide interpersonal messaging services. The MTA on which UA is built, in fact, provides a general store-and-forward information delivery service, as shown in Figure 16.3(a). The UA receives a request as user data from a user's terminal and defines an envelope known as a message envelope (ME) to the MTA. The MTA may interact with other MTAs or get requests from different UAs. The ME is defined in terms of two fields, the message header and message body. This ME is also known as a UA protocol data unit (UAPDU), as shown in Figure 16.3(b).

For the sake of understanding, and also based on CCITT recommendations, these will be considered similar to lower layers one (physical) and two (data link layer). The second

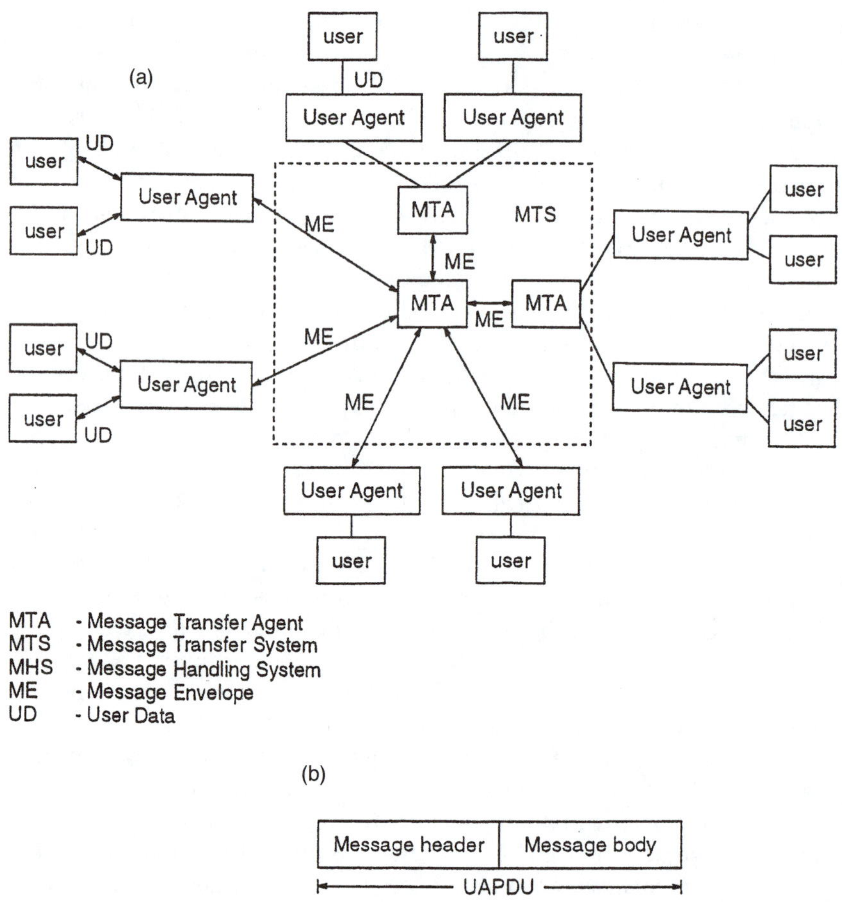

Figure 16.3 (a) Message-handling service. (b) Message envelope frame format.

layer (host) contains the protocol (X.411) which provides the transfer services to the envelopes (containing the messages) and also manages the envelopes. It does not know the contents of an envelope. An envelope provides a means of exchanging instructions.

The application layer includes various protocols of different applications, including the message-handling protocols. As said earlier, ACSE allows the application layer of OSI-RM to characterize the connection-oriented connection features for lower layers which establish connections between two different systems for the reliable transfer of messages between them. The addresses of the network and the interfaces defined by the MHS protocol and contained in enveloped messages support the network layer routing to map to the network node or an X.25 interface address. In general, the network to which a user's computers are connected is assumed to be X.25 for MHS applications with a transport service of class zero (see Chapter 13 on the transport layer). A variety of service elements is supported by the MHS. The user agent (UA) needs to obtain a service element from the MTS to provide a specific class of service to the users.

User agent (UA): The first-layer user agent (UA) contains functions for managing the contents of message envelopes. A UA defines the general functions in the MHS and may be a user or a computer process. It provides an interface between the user and MHS and is also known as the **user agent entity (UAE)** of the first component user agent. A sending user uses the general function described by the UA and generates the message. This UA, in fact, is an application layer entity which provides an interface between the user and message transfer system. The message generated by the sending user is received by the user agent entity (UAE) which interacts with the MTA's **message transfer layer (MTL)**. The **message transfer agent entity (MTAE)** (which supports the message transfer layer function via a specific protocol) transfers the user message to the receiving-side user(s), where it is received by the submission and delivery entity (which provides MTL services to the user agent entity and interacts with the message transfer agent entity in a specific protocol) of the message transfer layer.

The collection of UA layers and message transfer layers provides an application (corresponding to a specific function) known as a message handling system (MHS). In other words, the MHS is composed of two layers: UA (layer 1) and MTA (layer 2). The UA offers different types of interfaces within a mail program. It provides interface between the user terminal and mail program, where it manages interaction with the terminals and then offers interface with the message transfer system (MTS) for delivering or accepting mail. Finally, it interacts with a mailbox or mail store defined by the MTA for every user. The mail program invokes the UA and then offers a set of commands to the user that control composing the mail message, sending the mail message, receiving the mail message, and accessing the mailbox. The mail program checks the mailbox and displays the list of incoming mail messages on the screen. The display list of mail messages usually shows different types of fields maintained as a user profile (a file defining various display formats). These fields typically show the status of a mail message such as new, old but unread, answered and forwarded, etc.

Other fields include the length of the mail message, message ID, subject, carbon copy, and others, followed by the text message. At the end of the message, another control command to indicate the end of message is used. The commands are usually associated with parameters. Various commands used with a mail system after the display of mail messages from a mailbox include display header (number), display current header, type message on screen (number), send a message, forward, answer (number), delete (number), move messages to another mailbox (number), read a new mailbox (number), go to next message and display, go back to previous message, go to a specific message (number), and exit the mail system.

Format of mail message envelope: The system may receive mail from different sites or users via different networks. The mail message envelope frame format generally consists of two components, **message header** and **message body**, as shown in Figure 16.3(b). This message envelope frame format is also known as a UA protocol data unit (UAPDU). The message header includes various information concerning the message arranged as a series of lines or header fields. Each header field has a name which indicates the meaning of the line header. A general message header format is shown below:

From: G.S.Hura < >
Date: Thursday, 21 Dec. 1994, 16:10:05 GMT
Message-ID: < 1234.678.@ cs. abc.edu>
To: Destination user's address
CC: Copy to
Subject: ...

The message header contains the names of both sender and destination in the fields **From** and **To**. The CC defines the names of recipients who will get a carbon copy of the message. The Message-ID field simply contains a unique identifier which refers to this message and is produced automatically by the mail transfer agent (MTA), who delivers it when a message is sent. It is used to trace the message or delete duplicates or as a search string in a file of messages. The subject field contains a string which may be used by the user agent (UA) to provide a summary of the message.

The message body includes the actual typed text in any format and is simply a sequence of lines of text followed by a control character to indicate the end of the message body.

For each message generated by users, the UA submits a message-handling system (MHS) which includes the contents of the original message defined by the user and a submission envelope which includes the instructions to the local MTAs and control information needed for the MTS to provide its services. The envelope has a header (containing addresses, subjects, etc.) and body (containing the contents of the original message). The local MTA will provide different user programs which may be on either the same computer or a different one.

Message transfer agents (MTA): The message transfer system (MTS) is responsible for relaying the mail message from sender to receiver. In this application, the application layer association is always existing and the users can send mail messages at any time. The mail messages will be stored in the appropriate mailboxes. The sending UA submits the submission envelope to the MTA. This envelope is converted to a relaying envelope, which in turn submits it to other MTAs for transmitting the message until the envelope is passed on to the appropriate UA of the receiving side. This relaying of the envelope in the MHS is used basically for transmission of messages between MTAs.

At the receiving side, a delivery envelope is provided with a UA and is transmitted between the MTA and UA of the receiving side. The delivery envelope contains the original message and control information to be used by the MTA and UA for delivery of the message to the receiving user. The MTA can receive messages from either the MTA (via presentation layer entities) or UA. If it finds that a mail message is from the UA, it checks the syntax and sends it back to the sender if it finds any error (with reason). If it finds this UA to be valid, it will assign a message identifier and time stamp on it and considers this message as if it came from the MTA.

After it has received the mail message, the MTA checks for the address of the local mailbox of the receiver. If it finds the receiver to be on some other network (i.e., it is not local), it will send it to another MTA. For a local mailbox, it will store the message in the local user's mailbox. The mail messages can be of different types, such as ASCII, analog facsimile, telex, digital voice, images, drawings, etc. In some cases, the MTA may provide conversion from one form to another. The addressing in MTAs is a very important issue and is discussed below.

The addressing scheme used in a mail system is based on a domain-oriented addressing scheme. We briefly discuss a simple address scheme which is built in a hierarchical form. The details of addressing schemes are discussed in Chapter 17. In the address *g.hura@cs.abc.edu*, the *g.hura* field represents the user-id, the *cs* field represents the name of the department, the *abc* field represents the name of the university where *g.hura* is working, and the *edu* field indicates the nature of the organization (*edu* is commonly used for universities in the U.S.). This field may have different attributes, such as *com* (company such as AT&T, IBM, etc.) and *org* (organizations such as IEEE, ACM, etc.), and may also represent the country (Ca represents Canada, Au represents Australia, etc.).

Message-oriented text interchange standard (MOTIS): MOTIS is an electronic mail system developed by ISO and is a complete mail system as opposed to a protocol. It offers the same services as those of SMTP of TCP/IP. Due to its working concepts and services offered, it is sometimes also known as the message-handling system (MHS), which is based on CCITT's X.400 standard. A typical configuration for the X.400 standard includes a request from the user via the application's user agent to the **submit/transmit service element (TRSE)**, which is sent to the **message store (MS)** via peer protocol (STSE). The message transfer agent (MTA) uses another application's process known as the **message transfer service element (MTSE)** and sends it to the other user's computer after changing its format back to that of the sender's user. That is, it sends it to the MTA via its associated MTSE followed by the MS via STSE. In other words, the users submit their requests to their UAs while MTAs interact with each other to provide proper mapping between them. Each of these elements used with UAs and MTAs is defined by respective protocols. For more details about e-mail, readers are referred to References 6–8.

16.5.3 File transfer access and management (FTAM): ISO 8571 standard

File management is the main process for the transfer of file data between hosts which handles the management, transfer of files, communication link, administration, etc. The file management specification known as **file transfer access and management (FTAM)** and detailed in **ISO 8571** provides interaction between users working on different types of hosts and using different techniques of data formatting, accessing, storing, etc. It has been a big problem to various communication and data-processing organizations to provide file management including file transfer and many other operations for their data-processing applications due to non-availability of a standard file format and to management issues. Here also, the client program corresponding to the application process will access and manage a remote file server. This file server is implemented by another application process and may be running on either the same computer (on which the client program is running) or a different computer. The primitives used by the client program to access and manage the file store are usually machine-dependent and, hence, can be considered local to the client program. The mapping function is implemented by various associated primitives.

The file management function is also known as data management or data administration or file management and administration. It should be re-emphasized here that the application service elements (either of CASE or SASE) are not responsible for providing any specific services but are mainly concerned with providing a means as primitives by which those services offered by the application process or program can be accessed by the users.

There is similarity between the working of MTA (discussed above) and FTAM in the sense that both provide relaying of messages between sender and receiver, but the fundamental difference lies in the way these two services are implemented. In the case of MTA, the users can send mail messages at any time, as the application layer association is always there, while in the case of FTAM, there is no access to a particular file if the file server is down. The ISO[10] FTAM specification is mainly concerned with file transfer and management and is categorized into four distinct groups providing various operations on file transfer:

ISO 8571/1 — File transfer problems (concepts of FTAM and its introduction)
ISO 8571/2 — File attributes, names, contents, accessing codes, etc. in FTAM (discusses vocabulary and terms used)
ISO 8571/3 — Interface between entities for file transfer (describes application entities between which transfer takes place)
ISO 8571/4 — Rules, conventions, activities defined for interfacing (for ISO 8571/3)

Each of these standards will be discussed in detail in the following sections.

ISO 8571/1 FTAM: The application users have to define a data format, data-accessing method, data storing, etc., on the logged-on hosts, and these will be different for different applications. The file server design is mainly concerned with concurrency control, file replication, and other implementation issues. The file servers or virtual file store (VFS) basically deal with file structure, file attributes, and various operations of a file. For providing interaction between these types of hosts for the transfer of files between them, we have to define a common framework/model which provides an acceptable and compromised method for file data formatting, file accessing, and file management. This framework provides a mapping between the various characteristics of files into a standard file which is known as a file store. The application users using FTAM protocol have to have an acceptable nonexistent file server known as a virtual file store (VFS), which converts the characteristics of different host servers into a standard file store. The VFSs are defined by a file's characteristics, formats, attributes, types, etc. The main advantage of the VFS is that it does not require the details of data structures, data formats, etc., and is only responsible for establishing communication with distant hosts across the network. For more information on virtual file stores, see References 4 and 5.

The set theory concept has been very successful in defining and describing the overall relationship between the elements of database management systems at various levels. The same concept is also extended in the protocol for defining and describing the structure or format of the data file through sets and subsets. The data file is defined by a set, while the description of data is defined by a subset. The set assigns attributes to the data elements defining the relationship between them for a data file. This relationship does not establish the structure or formats and also does not provide any information regarding the accessing of the data file. The data file can be equated with that of a database which also uses the set to map the relationship between various components into a database. The FTAM file service primitives used between distant host file servers and the VFS have the same structures and are used on virtual file stores only.

Structure of files (ISO 8571/2): The file servers define the file structure in various forms, and each server provides different models for file structures. There are three models for file structures: file server, flat file, and hierarchical file. The *file server* is an unstructured collection of data and does not have any internal structure and hence does not support any operations on these files. A *flat file* is an ordered sequence of records. Each record is assigned a label or position in the model. This label or position is used as an address to access this record, and the records may have different sizes and lengths. The third model, a *hierarchical file*, uses tree structure to define the model of a file where each node of the tree is assigned a label and also a data record. This model of file is more popular than others and allows user access either by giving the label name or by relative position with respect to a labeled node. There are a number of efficient schemes available to access these files. Some nodes are assigned a label only, while others are assigned a record.

The structure of files and the file-accessing methods are usually expressed in terms of subsets, and users may choose a particular subset from a large number of subsets to meet the requirement of different schemes of accessing, modifying, and storing the data files. The implementation of these requirements becomes very complex and also requires extra memory at each host connected to the network. This problem has been overcome by the ISO FTAM specification standard which allows the hosts (using different rules, conventions, and structures of data file) to use the procedures of the FTAM protocol for connecting the hosts at the presentation layer (by passing appropriate parameters in the application service primitives to the presentation layer).

All of the files in any of the models are associated with a set of attributes (ISO). These attributes generally describe the characteristics of files stored in file servers. Typical ISO file server or file store attributes include file name, allowed **operations** (by creator, e.g., read only, no permission to insert, replace, extend or delete, etc.), access control (who can access the file and also the access technique), **file available**, date and time of file creation, date and time of last file modifications, date and time of last file read (also provides security), owner, identity of last modifier, **contents type** (record structure), maximum future size, etc.

ISO file server operations which can be performed on the files stored in the file server include create file, delete file, select file (for attribute management), de-select file (cancel the current selection), open file (for reading or updating), close file, read attribute (of a file), change (or modify) attribute of the file, locate (record), read (data from the file), insert (new data into the file), replace (overwrite existing data), extend (append data to some record), and erase (delete a record). All these operations are self-explanatory and will not be discussed further.

Once the connection between different hosts at the presentation layer is established through the procedures of FTAM protocol, the control and management of the file service dialogue between hosts can be provided again by another set of procedures of FTAM at the session layer. Thus, the ISO 8571 FTAM protocol is a very useful standard for providing file transfer between hosts having different structures and different accessing methods of data files.

The data file characteristics, e.g., name, attributes, and so on, offered by different file stores are the real file stores and are transformed into virtual file stores which are handled by FTAM file service primitives. Any time a user wants to transfer a file from one location to another location, an appropriate pre-transfer activity is initiated by the sender through FTAM protocol. Similarly, the receiver accesses the data file sent by the sender through an appropriate file-accessing activity of the FTAM protocol.

FTAM data unit (ISO 8571/3): The data elements of the data file in a virtual file store are called data units (DUs). The DUs are connected in a hierarchical structure based on which a suitable file access technique can be defined. This structure is known as **file access data units (FADUs)**. The DUs are transmitted through the hierarchical structure under the control of commands used in FADUs. By associating attributes with DUs, various behavioral characteristics of the files can be described. The FADUs, in fact, are the DUs which define the contents of the file. The behavioral characteristics of the file can be divided into four main classes: minimum and common properties for all files, properties for file storage, properties for access control, and properties for private or personal files (non-FTAM operations).

A *group of minimum and common properties of all files* includes the properties of single file name, description attributes (structural behavioral of the file, e.g., hierarchical, sequential, accounting-based record, history record of the file, any constraints or restrictions on accessing technique, read, update, deletion, operations, etc.), addresses of application entities between which communication will take place, addresses of application users, etc.

The *group of properties for file storage* deals with the characteristics of past and current operations performed on the files. In the past operations, the properties include data and time for any changes in the operations and attributes, address of sender, reader modifier, attributes, and changing the size of the file. The current operations include the address of the receiver, requirements for file storage, file access control, any synchronization primitive (such as locks), the address of the sender, etc. The FTAM protocol is initiated by the sender of the file transfer application. The attributes associated with files define a sequence of operations to be performed on the data file and also define the logical structure and the size of the data stored in the file.

The *group of properties for access control* is mainly concerned with attributes on access and security control, e.g., access right, encryption procedures and integrity for data files, any other legal actions, etc.

The *group of properties for private or personal files* is non-FTAM standard and hence does not use any attributes for virtual file store operations. All non-FTAM service primitives are mutually defined and agreed upon between the users of a private group connected to the network. Some of the material discussed above is partially derived from References 1–4, 19, and 20.

FTAM service primitives (ISO 8751/4): A list of ISO 8571 FTAM service primitives which perform specific functions on the DUs of a file is shown in Table 16.3. These services are managed and controlled by primitives. A particular file service process is invoked for a specific primitive, and each primitive is associated with application parameters. The file service primitives may be initiated by either the application service provider or the application service user and are required to define the file activity. The FTAM protocol offers different types of file activities such as open/close of a file, read and update a file, data file transfer addition or deletion in a file, etc., and is supported by appropriate file service primitives.

Most FTAM primitives shown in Table 16.3 are self-explanatory and are simple to understand. However, some of the primitives are more complex, not in terms of attached parameters but in terms of various operations involved in one primitive. For example, the implementation of the primitive F-INITIALIZE for establishing a connection with a file server or file store requires invoking A-ASSOCIATE, which then calls the primitive of the presentation layer for connection known as P-CONNECT. This primitive, in turn, calls the S-CONNECT primitive, which further calls T-CONNECT, and finally the primitive N-CONNECT provides the connection between the file store and the user.

Common management information service element (CMISE): The ISO has developed a management protocol known as common management information service element (CMISE) which represents only the management entity of OSI-RM and is termed a system management application entity (SMAE). As the name implies, this protocol is responsible for managing various aspects within a networking environment which are defined at remote computers in a variety of ways. It handles the following aspects within networking:

Table 16.3 ISO 8571 FTAM Service Primitives and Parameters

Primitive	Parameters
	Application connection
F-INITIALIZE (establishes a connection with the file server)	Calling and called address, responding address, diagnostic, service type and class, functional units, attribute groups, rollback availability, communication QOS, presentation context, initiator's address, current account, file store password, checkpoint window, initiator's request and confirmed service
F-TERMINATE (releases a connection with the file server)	Charging parameters, initiator's requested confirmed service
F-U-ABORT	Diagnostic requested by either provider or initiator, a (user-initiated) confirmed service
F-P-ABORT	Diagnostic, provider's requested unconfirmed service (provider initiates)
	File selection
F-SELECT (for manipulation or transfer)	Diagnostic, file name, attributes, access control, access passwords, current access structure type, current account, initiator's requested unconfirmed service
F-DESELECT (terminate selection)	Diagnostic, charging, initiator's requested confirmed service
F-CREATE (a file)	Diagnostic, file name attributes, access control, access passwords, concurrency control, commitment control, oversize, current account, delete passwords, activity identifier, recovery mode, initiator's requested confirmed service
F-DELETE (destroy a file)	Diagnostic, file name attributes, access control, access passwords, concurrency control, commitment control, oversize, current account, delete passwords, activity identifier, recovery mode, initiator's requested confirmed service
	File access
F-OPEN (for reading or writing)	Diagnostic, processing mode, presentation context name, concurrency control, commitment control, activity identifier, recovery mode
F-CLOSE (close an open file)	Diagnostic, commitment control
	Data transfer
R-READ	Read one or more file access data units (FADUs)
F-WRITE	Write one or more FADUs
F-DATA	Data arrival (F-DATA.indication)
F-DATA-END	End of FADU (F-DATA-END.indication)
F-TRANSFER-END	End of transfer
R-READ-ATTRIBUTE	Read file attributes
F-CHANGE-ATTRIBUTE	Diagnostic, attributes (modify operations)
	Miscellaneous file store management
F-LOCATE (to a specific FADU)	Diagnostic, FADU identity, concurrency control
F-ERASE (destroy an FADU)	Diagnostic, FADU identity, concurrency control
F-BEGIN-GROUP	Mark start of an atomic action
F-END-GROUP	Mark end of an atomic action
F-RECOVER (re-create after a failure)	Diagnostic, activity identifier, bulk transfer number, access control access passwords, recovery points
F-CANCEL	Abrupt termination of data transfer
F-CHECK	Set a checkpoint
F-RESTART	Go back to a previous checkpoint

protocol management, bridge management, router management, etc. Additional functions are provided by other application service elements, and the protocol using these facilities is termed the system management application service element (SMASE).

We do not have to provide additional setup for interaction between network management and local management application processes. Each host includes the SMASE, CMISE, and ACSEs over the presentation layer. At the top of this stack of protocols is the UA, which provides services to the users. The internetworking environment is provided by ISO-IP, while the stack of protocols corresponding to SMAE provides an OSI management environment. The internetworking can be defined for a number of networks such as LANs, public or private data networks, WANs, X.25 networks, ISDNs, etc. The CMISE has a set of service primitives which are used to provide interaction between network and local management applications. These primitives include M-INITIALIZE, M-TERMINATE, M-ABORT, M-EVENT-REPORT, M-GET, M-SET, M-ACTION, M-CREATE, M-DELETE, and M-CANCEL-GET.

Manufacturing message service (MMS): The MMS is an application service element which offers services between various application processes used within an automated manufacturing environment. In this environment, various computer-based processes may be allowed to exchange messages among themselves. These application processes may include programmable logic controllers, industrial robots, and other controllers within a manufacturing environment. In a normal operation, at a high level, the FTAM protocol may be used to communicate with a particular application process which uses MMS to transmit automated equipment associated with it and sends commands for its manufacturing or assembling (depending on the application). Each application process has its own set of primitives for exchanging the messages. Basically, the sequence of operations may follow a selection of a particular application process, selection of a particular part and transport of it over to some other application process, and issuance of appropriate commands. Readers are referred to Reference 4 for more details.

16.5.4 Virtual terminal service (VTS)

The terminal is a very important component in any computer application, and it is unfortunate that standard terminals have not been defined properly and also have not been accepted widely. This has resulted in a number of terminals being used in different applications. Most of the terminals accept a certain character sequence (escape sequence) which allows different operations on terminals (moving cursor, inserting and deleting characters or lines, entering or leaving video mode, etc.). The terminals used in different applications do not use the same character sequence (even ANSI's standard character sequence is not accepted widely), and as such the terminal (in particular, the input keyboard) is always a problem. Further, the screen (output device) does not provide any compatibility with other output devices for the screen editor. OSI has developed a concept of a virtual terminal (VT) to solve this incompatibility problem in terminals and screen editors. It represents an abstract data structure for the terminal, which defines various states of a real terminal. Users can access this data structure and accordingly change the existing data structure of their keyboard to offer an acceptable display on the screen, thus making the users feel that they are working on their real terminals. Some of the material herein is partially derived from References 1–4, 9, 19, and 20.

Virtual terminal service (VTS) primitives: OSI has defined **virtual terminal service (VTS)** to support the virtual terminal concept for both synchronous and asynchronous terminals. OSI has defined the following parameters for virtual terminals which the user selects to provide mapping between terminals to get the required display on their screens:

Table 16.4 VTS Primitives and Descriptions

Primitive	Description
VT-ASSOCIATE	Establish a connection
VT-RELEASE	Release a connection
VT-U-ABORT	User initiates this primitive
VT-P-ABORT	Provider initiates this primitive
VT-START-NEG	Start of parameter negotiation process
VT-END-NEG	End of parameter negotiation process
VT-NEG-INVITE	Invite the peer to propose a parameter value
VT-NEG-OFFER	Propose a parameter value
VT-NEG-ACCEPT	Accept a proposal parameter value
VT-NEG-REJECT	Reject a proposal parameter value
VT-DATA	Data operations
VT-DELIVER	Deliver buffered data
VT-ACK-RECEIPT	Acknowledgment of VT-DELIVER
VT-GIVE-TOKEN	Pass the token in synchronous mode
VT-REQUEST-TOKEN	Request the token in synchronous mode

mode (synchronous, asynchronous); number of dimensions (1, 2, or 3); character sets allowed; fonts allowed; emphasis types allowed; foreground colors allowed; background colors allowed; maximum *X*, *Y*, and *Z* coordinates; and addressing constraints.

In order to implement various operations on virtual terminals (such as addressing, text, attributes for the three coordinates, erase operations on all three coordinates, etc.), OSI has defined the VTS primitives shown in Table 16.4.

Types of terminals: The terminals are available in three main classes: **scroll**, **page**, and **form**. The **scroll class of terminals** is very simple and does not offer any editing facility. When any key on the keyboard is hit, the terminal displays that character and it is transmitted over the line. At the end of the line, the old line goes upward (scrolls) and a new line starts and is displayed. The hard copy terminals and some CRT terminals are of this class. The scroll class of terminals does not offer any processing power, but different options may be defined differently such as line length, echo, line feed, carriage return, different kinds of tabs (horizontal or vertical), etc.

The **page class of terminals** typically represents CRT terminals which are used with our systems. This class offers 25 lines, with each line consisting of 80 characters. The cursor can be used to select a particular area for its display. The problems with this class of terminals include page length, cursor addressing, etc., in addition to the problems of the scroll class of terminals. Due to these problems, the main difficulty in using these terminals lies with the development of common screen editors and other application software for display devices. The Berkeley UNIX system has defined a virtual terminal which defines all the commands in different types of page classes of available terminals. When using this program, the invoked editor inquires about the terminal type and gets all the relevant information about it from the database (created by the virtual terminal) known as **terminal capability (TERMCAP)**. A set of character sequences for this terminal is obtained from this database.

The **form class of terminals** is more complex than the previous classes, as it offers processing power and it also has built-in microprocessors which allow local editing and offer macros and many other facilities (depends on applications). These terminals are used in banking, financial institutions, airline reservations, etc. They allow the user to display a standard form on the screen (read only) and the necessary information can be input via keyboard onto different fields. A modified version of display-oriented terminals was developed which shares the data structure. Here virtual terminal software defines an abstract data representation of the display image which can be read and updated by the terminal and also the application (display-oriented) on some other host connected to it by the network. The changes made in abstract data representation are taken care of by virtual terminal software.

The user sends a virtual terminal command from his/her terminal (with built-in microprocessor) to a distant host over the network using **virtual terminal protocol (VTP)**. The host user reads an updated version of the data structure using a different set of application commands. Now the application program (host computer) can also modify this data structure and its display. In this way, both have replicated copies of the data structures. There is a possibility that both terminal user and application program may try to modify the data structure at the same time, and this problem is solved by defining a token which will be held by either the host computer or terminal. This mode of operation for virtual terminal software is based on synchronous operation, where a single data structure or dialogue is used between terminal user and host computer.

In an asynchronous mode of operation, each end has a data structure for input and another one for output. When any user is making changes on its output data structure, the input data structure of the other end user is changed. Further, it does not change the input data structure, and the output data structure of other end remains unchanged. In other words, each copy of the structure at any end has one reader and one writer, and conflicts between similar operations (read or write) simultaneously will not happen. It requires extra storage for keeping the data structures at two ends.

Virtual terminal support protocol (VTP): We use mail services to receive mail from different types of terminals, and obviously it is too cumbersome to remember all the different types of terminals with their different characteristics. In order to access different types of networks and also different types of computers and terminals, we need to use a virtual terminal model protocol which can be easily configured to different types of computers, networks, and terminals. The X.25 networks support asynchronous terminals and other devices, and interfaces need to be defined to connect these devices to X.25 networks. A standard offering protocol conversion and another standard offering interface to asynchronous devices such as **packet assembly and disassembly (PAD)** will be used by asynchronous devices and terminals to access X.25 networks. This includes interface between terminal (scroll) and RS-232 and communication protocol to interact with the network. CCITT[17] defined a standard PAD interface as a combination of three interfaces — X.3, X.28, and X.29. The standard interface X.3 defines PAD parameters, X.28 defines an interface between terminal (scroll class) and PAD, and X.29 defines an interface for a PAD-based computer (DTE).

CCITT defined X.25 as a specification for a network interface that uses the protocols of the three lower layers (physical, data, and network). Many people think that X.25 is a protocol, but it is a specification for the network interface. According to CCITT recommendation, X.25 is an interface between data terminal equipment (DTE) and data circuit-terminating equipment (DCE) for terminals which are functioning in the packet mode in a public data network (PDN). A typical X.25 uses a generic function for the DTE (PCs, terminals, etc.) which may be network-attached, an end user device, or any other communication device working in the packet mode. For example, a minicomputer or any other communication device running an X.25 specification is a DTE, the front-end processor loaded with X.25 code is a DTE, etc. An alternative to these DTEs is PAD, which allows users of X.25 or non-X.25 to communicate with public or private packet-switched networks.

CCITT recommendations of PADs: The CCITT has three recommendations known as X.3, X.28, and X.29 interfaces which will provide connection of asynchronous terminals with X.25 networks. These three interfaces are collectively known as 3X protocols or PADs. PAD offers protocol conversion for a user device (DTE) to a public or private network and protocol conversion on the receiving side. It allows different terminals to communicate with each other and provides transparent service to the users. The PADs support asynchronous devices (using X.3, X.28, and X.29 standards), and many vendors have developed PAD services to other link control protocols (synchronous) like IBM's BSC and ISO's SDLC. We have already discussed the PADs in greater detail. We present the main features and protocols below just to get a feel for a comparision with other application protocols of this layer discussed here. The following section describes these standards in brief. Details of PADs can be found in Chapters 3 (signal transmission basics) and 12 (network layer).

X.3: Defines the functions and various parameters for controlling the operations of PAD. There are typically 22 parameters maintained for each of the PADs connected to a terminal. A PAD may be connected to more then one terminal.

X.28: Defines the protocol between the terminal and PAD device. Two phases of operation for X.28 have been defined. The first one is *data transfer phase*, in which the exchange of data takes place between the terminal and host, while the second phase is *control data phase*, in which the exchange of data takes place between the terminal and PAD in the form of characters. Various facilities such as desired baud rate between terminal and PAD, opening of a connection with remote DTE or host, etc., may be obtained with the help of parameters defined in X.3.

X.29: Defines the interface and a protocol between PAD and host.

16.5.5 Job transfer and manipulation (JTM)

In a typical data-processing and management environment, users submit their processing-based jobs via different means such as punched cards, **job control language (JCL)** statements, etc. Various operations performed on these jobs include file retrieval, file manipulation, file accessing and manipulation, file updating, etc. In the old days, the processing task was defined on punched cards and the job control language (JCL). Today, we can submit the processing tasks from our terminals by executing JCL statements. These JCL statements already stored on the disk select a particular processing task for its execution. The JCL distinguishes between the source code and the data which are defined for a particular processing task. It also supports other functions such as output, disposition of programs, files, listings, etc.

ISO defined a standard ISO 8831 for job transfer and manipulation (JTM), which provides interface between open systems, whereby the users can submit the tasks from any open system, and also allows the execution of this task on different open systems connected via networks. Here again we would like reiterate that similar to other application protocols (discussed above), the JTM does not provide any services, but it offers an environment whereby the client application may request operations related to job specifications to be transferred between the end users. The users submit the request through an associated user entity (UE) or user agent (UA). The users must know the options and services provided by the JCL of the remote open system where the task is to be executed. The JTM does not mention anything about the JCL which has to be taken care of by the users. The primary functions of the JTM are to provide the users with the ability to run their jobs or tasks on any open system, provide transfer of task-related data between them, and keep complete control of various activities involved in the execution and data transfer

Table 16.5 Job Transfer and Manipulation Primitives and Descriptions

Primitive	Description
J-INITIATE	Create and initiate a work specification (user)
J-GIVE	Ask source to give JTM document
J-DISPOSE	Pass a document to sink or executor
J-ENQUIRE	Ask source or executor for document name
J-SPAWN	Create a sub-job
J-MESSAGE	Report progress of work
J-END-SIGNAL	Indicate completion of JTM work
J-STATUS	Report status of work
J-HOLD	Suspend the work (temporarily)
J-RELEASE	Continue the suspended work
J-KILL	Abort work
J-STOP	End work

between open systems. It defines service parameters to invoke the JTM procedures and also the services of the presentation layer. It uses the services offered by the presentation layer through CASE and CCR protocols.

The protocol for JTM provides different types of interfaces. The first interface is between the user and the system through which the users submit their jobs. The second interface is inherently within the protocol which tries to find out the locations of the files required to run this task and is known as *source interface*. The third interface is again internal between these locations and a common storage for them and is known as the *sink interface*. Finally, the protocol, having loaded all the required files to run the submitted task, actually executes the task and is known as the *executor*. It is important to note that these interfaces may be located either on the same computer or on different computers. The main functions of the JTM protocol are to assign the required work specifications for each of the interfaces through primitives and accept the primitives from them as the results or replies. The OSI has defined the JTM primitives shown in Table 16.5.

16.5.6 Directory services

The primary philosophy behind the directory system defined by OSI is to search for names (of people, organizations, universities, corporate offices, etc.) based on attributes. The directory system is based on the concept of hierarchical structure (used in addressing systems X.121 and also telephone addressing). The directory system is defined in terms of fields starting from the root. At the root level, the attribute corresponds to the name of the country. From the root level, we go down to the first level, where the category of the requested search is assigned with another set of attributes. From this level, we go down to a lower level and assign another type of attribute, and this process of defining a directory system is continued with proper attributes at each level until the user's ID is reached. The attributes assigned at various levels of the directory system may have a type, interpretation, value, etc. They can also represent an integer, string, or any other data type.

The directory system is used in a number of applications such as mailing lists, aliases, searches for organizations, geographic entities, etc. Earlier we explained the address domain system for e-mail. The directory system uses the same addressing domain system in its implementation. In order to understand the directory system and its structure for maintaining files, we use the same example: *g.hura@cs.abc.edu*. The root level is defined by the highest field (*edu*) which indicates a university in the U.S. It may represent a university in another country at a lower level, as in *edu.Au*. Here the highest level (root) is the country (Australia), while *edu* is at the lower level representing a university. The next field, *abc*, represents the name of the university, and the next, *cs*, represents the name of the department (Computer Science). The lowest field represents the user's ID (*g.hura*).

CCITT X.500 has also been used by ISO and, in fact, was developed jointly by these two organizations. This standard includes various aspects of directory functions, and it is also known as the directory system. It maintains a directory information tree and follows the scheme discussed above for its implementation. It offers different services to users via an application process or program known as a **directory user agent (DUA)** in the same way that other application protocols have defined their respective user agents.

The DUA provides interface between users' requests coming from the terminal working under some operating systems and the stack of protocols used. The request is received by the directory service agent which, after retrieving (from directory information service) and processing, sends it back to the user. There are different types of services available from the directory system, but in general, these services are classified as *directory query* and *directory updating*. For each of these classes, a separate set of primitives is defined. For example, the directory query service includes the primitives read, compare, list, search, etc., while the directory updating service includes the primitives add-entry, remove-entry, modify-entry, etc.

16.5.7 Transmission control protocol/internet protocol (TCP/IP)

The TCP/IP has also defined standards for the application layer which typically provide services for electronic mail (e-mail) and file transfer. These standards include Telnet, file transfer protocol (FTP), simple mail transfer protocol (SMTP), and simple network management protocol (SNMP). The TCP/IP[12,13] defines a set of procedures which provide integration with the required application-specific protocol. The implementation of TCP/IP procedures is based on the client–server computing model, where the client interacts with the server for any application. For example, a client application program running on one computer may request a file transfer. Another server application program for file transfer may be running on another computer. The TCP/IP provides network communication interaction between them such that both client and server application programs may interact with each other irrespective of whether client and server are running either on the same or on different computers located a distance apart. The TCP/IP assigns a port to each of these application programs. The user's request is transmitted over the TCP/IP communication interface in the form of requests and replies between client and server. The application protocols sitting at the client and server provide necessary mapping to different applications. As a result of this, we may expect TCP/IP to be more complex than its counterpart ISO set of application protocols.

In a typical situation where a computer is connected to the Internet and wishes to send and receive data over it, the computer needs a set of protocols such as Telnet, TCP, IP, and protocols for LANs or other protocols for direct access of the Internet. Telnet (a program) breaks the data into a number of packets. This must run on both our PC and host. The TCP breaks the application data (already broken into Telnet packets) into TCP packets where each packet has a header with the address of the host and information regarding assembly/de-assembly and error control. The IP again breaks the TCP packets into IP packets of header with address information, routing information, and TCP information and data. These IP packets are picked up by subnets (LANs, backbone FDDI, etc.), which break the IP packets further into the appropriate frame formats and add their own address information. It is possible that the IP packets may go through a number of subnets (such as the backbone FDDI to the Ethernet of the main division to the Ethernet of the local department/lab and so on). Each subnet adds its own header. In the case where users are not connected via subnets, they can dial access to the Internet by using point-to-point (PPP) protocol or serial line Internet protocol (SLIP). The link packets defined by subnets are transmitted over the transmission media (analog telephone lines, twisted pair, coaxial, optical fiber, RS-232, Ethernet, etc.). The receiving-side host will have the same set of protocols, which will unpack the packets at each layer using appropriate protocols.

The user datagram protocol (UDP), transmission control protocol (TCP), Internet control message protocol (ICMP), and routing control protocol are the protocols of the transport layer, and these interact with the Internet protocol (IP) of the network layer. TCP offers a connection-oriented, end-to-end, reliable, and sequenced delivery of data, as well as priority, segmentation and reassembly, and adaptive flow-control services. UDP, on the other hand, provides connectionless, unreliable, and unacknowledged datagram service. Real-time protocol defined at the transport layer provides real-time applications that deal with transmission of voice and multimedia data.

IP is a datagram protocol that is resilient to network failure and offers the sequenced delivery of the data. It offers end-to-end reliable data communication across the Internet when used with TCP. The function of ICMP at the Internet layer is to provide a ping (echo packet) to determine the round-trip propagation delay between the source and IP-addressed host. Routers use this protocol for sending error and control messages to other routers. This layer provides multicasting data communication using the Internet group multicast protocol. It also uses address resolution protocol (ARP), which provides a direct interface to the data link layer of the underlying LAN. This protocol provides mapping between the MAC address (physical) and IP address and uses some routing protocols. The Internet layer supports a variety of data link layers (Ethernet, token ring, FDDI, frame relay, SMDS, ATM, etc.). The data link layer, in turn, supports a variety of transmission media such as twisted pair, TDM, fiber, etc.

The transport layer with UDP and TCP supports a variety of applications such as file transfer protocol (FTP; a secured server log-in, directory manipulation, and file transfer), Telnet (remote log-in facility), hypertext transfer protocol (HTTP; for the World Wide Web), simple network management protocol (SNMP; for network configurations, setting, testing, data retrieval, alarm and other statistical reports, etc.), remote procedure call (RPC), network file server (NFS), trivial FTP (TFTP; reduces the implementation complexity), and many others. The domain name system (DNS) offers a centralized name service that can use either UDP or TCP.

As indicated above, the ISO application protocol defines a virtual device along with a set of primitives, and as such the client and server have to use these primitives for communicating with each other. Further, the virtual device offers a set of services, and if these services are changed, then mapping for additional services must be provided at the device level. As we described earlier, each application protocol defines mapping software known as a user element (UE) or user agent (UA). These protocols are used over a standard TCP/IP communication protocol environment. In the remainder of this subsection, we present a brief introduction to each of these application layer protocols. These protocols, along with their applications and other services, are discussed in more detail in Chapters 13 (transport), 12 (network), and 17 (Internet).

Telnet: This protocol was developed by the ARPANET project as **virtual terminal protocol (VTP)** to be used by scroll (asynchronous) terminals at initial stages, but now it can also be used with other types of terminals.[16] It is based on the concept of **network virtual protocol (NVP)**, which basically defines a generic and common terminal for various applications and user processes. This terminal is not the same as the ones we use, but it is a virtual terminal defined to make network access easier for users. Without Telnet, users must know the behavioral characteristics of a remote terminal before transmitting any message to it and also to access its services. With Telnet, users do not have to send messages in the format of remote terminals.

The protocol VTP provides an interface not only between terminal devices but also between their application user processes. The VTP defines two simplex configurations between two nodes, one for each direction. It defines identical data structure at both ends in the sense that, whenever a key is hit on a terminal keyboard, it transmits a stream of eight-bit bytes. The terminal conversion process converts the terminal output (typically ASCII) into a suitable form acceptable to the network. A list of 15 commands has been defined to use Telnet. Each command is attached to an IAC character (interpret as command) at the beginning and indicates that the following character is a command. Some of the commands include end of command, mark, break, start of command, go ahead (synchronous), erase line, erase character, are you there, do, don't, etc., and are assigned a code number.

Telnet works over the standard communication protocol TCP/IP and, hence, is connected over the TCP. All the data and commands used by the users at the terminals are accepted by the local operating system, which passes them on to the local Telnet application protocol. The TCP offers a reliable stream service for Telnet application protocols which can transmit the user's data to the Telnet server application protocol sitting at a remote computer. The TCP/IP encodes the user's message in **network virtual terminal (NVT)** form, as opposed to different forms used in ISO protocols for different applications. In the ISO protocol, we have a virtual file server, virtual mail server, etc. In contrast to this, the TCP/IP provides one common form for the user's message in NVT form.

The TCP connection provides an interface between application user processes and NTP, and hence after the establishment of TCP connection, the data transfer over it can take place through the NTP across the network. This virtual terminal works in only one direction, and supports half-duplex (two-way alternate) line configuration. It uses standard ASCII information coding, which provides seven-bit character representation, with the eighth bit used as a parity bit for error detection. It also includes a pritner and keyboard. It supports both types of terminals (synchronous and asynchronous) and offers mapping between these terminals and the NVT. It uses a simple mail command to negotiate for the data transfer. This command includes various options to be chosen by the users. The user will choose an option and, after the execution of that option, it will get positive or negative acknowledgment (Ack or Nak). It also allows remote users to start executing the options, and execution of chosen options will return the Ack or Nak.

Some of the options offered by Telnet include binary transmission, reconnection, suppress, go ahead, status, output line width, remote controlled transmission and echo, echo, output line feed disposition, and many others. Each of these options is assigned a unique number (which can be found in the vendor's practical handbook/manual).

File transfer protocol: FTP was developed by DoD in 1970[14] as a communication transfer protocol for transfer of files among the DoD Internet users. The Internet is a collection of different types of computer networks and offers global connectivity for data communications between them. FTP works over a standard TCP/IP communication protocol. It defines a set of commands which are used by users to access the FTP for transferring a file(s) from any location to any other location across the Internet. It deals with four types of files: image, ASCII, EBCDIC, and logical byte.

The image file is transmitted on the basis of bit-by-bit (no changes). ASCII is a standard format for text exchanges not applicable in IBM mainframe, and EBCDIC is used for IBM-based machines. The logical files are in fact binary files and can be defined in different sizes of bytes (from eight bits). The files can be transferred in three modes using FTP: stream mode (ordinary file), compressed mode (higher bandwidth and capacity application), and block mode (paged files where a header is associated with each page). Both compressed and block modes support error recovery by using checkpoints.

In general, the commands of FTP provide facilities for sending and receiving files, manipulation of directories, defining parameters, defining modes of transfers, etc. The FTP commands can be either control or service (data) commands. These commands are exchanged through either a control TCP connection or a service (data) TCP connection. The user uses control commands along with parameters to request access to the remote host over the TCP control connection and, after mutual agreement over negotiated parameters, the FTP connection is established. Now the file transfer (downloading or uploading of any file) can take place between two sites over the data TCP connection. The control commands are managed by a file server which receives the request from the user and responds to the user directly by establishing TCP data connection.

There is another file transfer protocol known as **NETwork BLock Transfer (NETBLT)** which transfers large files and does not use TCP or any higher-layer protocols; instead, it runs over IP. This protocol requests a transport connection and negotiates on parameters (buffer size, packet size, etc.) during the connection establishment. The sender then passes the entire buffer to this protocol NETBLT, which defines packets out of this entire buffer and transmits them to the receiver side. The size of the packet is the one which has been negotiated. The receiver will send acknowledgment about the packets it did not receive and hence only those packets are re-transmitted. During this time, the sender can start sending the packets of the second buffer.

The service or data commands and control commands are associated with a list of parameters which check the validity of users, status of file transfer, and options in file transfer (reading, writing, saving, printing, etc.). For example, a *user* command is used for making a request to FTP a file. The file server checks the validity of the users and permissible operations on file via the parameters selected. In some file servers, a number of parameters are required, such as password, account number, and others, to check the validity of users. For general catalogue applications, the file server may not require any of these parameters and may allow users to access the information or file transfer operations by anonymous FTP. Once the validity has been checked, the FTP assigns a port to the user data on a host which may not be the one on which the user is working. That port (assigned on a host), irrespective of location, will provide a service access point through which data transfer takes place.

A partial list of FTP commands is given in Table 16.6, which shows the two classes of commands — control commands and service or data commands.

The file transfer applications over LANs require a less complex interface than that defined in TCP's FTP. It follows the same sequence of operations in FTP, i.e., submission of a request by a client application, transmission of this request across the network over TCP connection, processing by server application protocol, and returning of replies to the client over the same TCP connection. In order to support file transfer across the LANs, another version of FTP has been developed, known as **trivial file transfer protocol (TFTP)**, which uses the **user datagram protocol (UDP)** services as opposed to TCP. For error control, suitable error-handling techniques based on automatic repeat request (ARQ) versions can be used. In most of the applications, the simple version of the ARQ one-bit window will be sufficient. It has defined four types of PDUs as read, write, data block, and acknowledgment. Since this protocol is also based on client–server, it also supports different clients simultaneously.

Simple mail transfer protocol (SMTP): The primary use of ARPANET was to provide electronic mail, and the format of this electronic mail is defined in the RFC-822 (Request for Comment) document. The standard protocol that supports e-mail services (which can

Table 16.6 Partial List of FTP Commands

Control commands	
User name	USER
Password	PASS
Account	ACC
Reinitialize	REIN
Log out	QUIT
Data port	PORT
Transfer mode	MODE
Structure of file	STRU
Service commands	
Retrieve	RETR
Store	STOR
Allocate	ALLO
Restart	REST
Abort	ABOR
Delete	DELE
List	LIST
Status	STAT
Help	HELP
No operation	NOOP

be transmitted over the Internet between hosts) is known as simple mail transfer protocol (SMTP).[15] RFC-822 was originally designed as a protocol for sending ASCII text lines only, and no other characters can be transmitted. The RFC-822 provides the format of a mail message by defining a header (also in ASCII) as a part of the message which can be changed by anyone using appropriate editors. MOTIS, on the other hand, defines different fields for the header and the message itself, and also the header is defined in binary form (using ASN.1) so that it becomes difficult for anyone to change it.

The majority of the mail systems use SMTP. A typical configuration of this protocol for handling mail messages between two hosts follows a layered approach where, at the top level, the user's request is accepted by the lower layer of the local mail system through its operating system. It then is interfaced with the SMTP client application protocol which transmits it over the TCP connection across the network. The local mail system interacts with the local user, while the SMTP application protocol interacts with the local mail system.

At each host there are two modules of SMTP — transmitting and receiving — which are invoked during the TCP connection establishment. The transmitting module is invoked whenever a node wishes to send mail by sending some control command and a test message, while on the receiving side, the receiving module is invoked when it receives the test message from the transmitting node. The receiving node will send some kind of acknowledgment back to the sender whether it can receive the message or not. After the sender receives this acknowledgment (control command) from the receiver, it sends an acknowledgment indication of the receiving node. The commands used to handle mail transfer are usually encoded in ASCII using either a three-digit number or text commands.

When the transport connection is established (after the receiving node or host has been identified), the data command is transmitted. After connection has been established, the sender sends a *mail* command, which will indicate the sender's address to the receiver, who can use it to communicate any error messages. The *data* command indicates to the receiving node that following this command is the text message (identified by the start of the text control character). The end of the text message is identified by sending an *end of text* control character.

The *quit* or some other control command is used to terminate the transport connection, after which both the transmitting and receiving modules go back to their original states. The message travels over a number of intermediate hosts using the store-and-forward concept until it reaches the final destination.

The above discussion for mail transfer between two hosts using TCP/IP assumes that both hosts use the SMTP protocol. If the mail protocols are different, then the hosts can still interact with each other via gateway protocols known as *mail gateways*. The gateway will accept mail from one host and convert it into another form, and on the other side it is converted back into the mail system being used. For example, the SMTP will be changed to MOTIS form (ISO mail protocol) and transmitted.

Network management: Network management is very important for any type of network (LANs, WANs, etc.). In many organizations, the top-level management may not have sufficient and relevant hands-on experience with all the details of network implementation, and as such they as well as the users look at the network as a transparent network environment. The management issues include not only the efficient use of various resources of the network, but also the ability of the user to use it conveniently and effectively. J.V. Solman[6] has identified five categories of network management: problem determination, performance, change, configuration, and operations.

Problem determination deals mainly with the determination of failure in the network. It does not address the reason for the failure. Based on that information, appropriate vendors and service providers can be contacted. **Performance management** deals with response time and availability. For different types of networks, this may be defined differently; e.g., in CSMA/CD LANs, this may be defined as the number of collisions. **Problem** or **change management** is primarily concerned with reporting, tracking, and some kind of collective agreement regarding various components of the network. **Configuration management** is concerned with creation and maintenance of a database containing an inventory of various components of the network. This also includes various characteristics (physical and logical) of these components. **Operations management** provides various attributes of operations such as linking new devices, documentation, file transfers, remote facilities, etc.

For documentation management, three levels of documentation are network, user interface, and user. Network documentation should include a complete description of the network. This includes routes of the cable, various devices, locations and characteristics of primary nodes, technical specifications of each of the components, ports, nodes, etc. User interface management documentation includes the codes of all application software, manuals, troubleshooting guides, software packages to check the current network configurations, etc. Performance management deals with diagnostic testing equipment that can monitor the performance network.

One of the main objectives of LANs and other networks is to provide easy and convenient access to various resources and applications within the organizations with some degree of security. The security can be provided at different levels. For example, for the least amount of security, we may have terminal access, account number and password, passwords for files, access rights for files (provided by the operating system), and hard-wired terminals. A higher amount of security can be provided through encryption, different types of public keys, key distribution keys, and other software modules as a part of communication protocols.

The network management standard protocols have been defined for both LANs and WANs. In WANs, the switching nodes are interconnected and carry the packets from a source to a destination node. Many different types of WANS are used in different countries. The WANs are usually controlled and maintained by the telephone companies. Some of the popular WANs includes Catenet (USA); British Telecom's Packet Switch Stream (PSS), and Joint Academic NETwork (JANET). The network management in the case of WANs mainly deals with accounting and billing, as the users are getting these services from the same carriers. A number of software tools can be used to manage the networks, congestion control, charges for packet transmission, and many other related functions. Network management protocols have been introduced by Cascade, Cabletron, Fore, IBM, NEC America, Hughes, Network Security Corporation, Siemens, and many other companies.

LANs also need to be managed properly for effective and efficient utilization of the resources. The ISO has identified five categories of network management within the OSI management framework: configuration, fault, performance, security, and accounting.

Configuration management is concerned with capabilities for remote initialization, reset, software distribution and installation, cabling (broken cable, poor grounding, intermittent problems, noisy equipment, etc.), naming and addressing control duplicate addresses, duplicated network interface cards, etc., and design and planning. **Fault management** provides error control that handles error detection, error correction, and error reporting; alarming of potential faults; and hardware and software repair. **Performance management** is responsible for modeling and performance evaluation of LANs, collection of traffic statistics, and set/reset counter functions. **Security management** mainly deals with access control, monitoring of data integrity, violation detection, and all the security-related issues. **Accounting management** is primarily responsible for billing of services, distribution of costs, capacity planning (based on performance evaluation parameters and statistics), software license agreements, and accounting-related issues.

The network management of LANs within the OSI framework deals with the development of protocols for transfer of information over the network. The standard must be easy to use, organize, and control the communications. It must be flexible to adapt in a dynamic environment and provide security and fault tolerance. It performs three specific tasks: processing, storage, and data communication. Some of the material in this subsection is derived from Reference 25.

Simple network management protocol (SNMP): SNMP is based on IETF's network management protocol and is a very popular management protocol for data communication networks. This protocol is based on the concept of object-oriented methodology and requires very small memory (RAM) of about 50–70 Kb in any internetworking device (bridge, router, gateway, etc.). Internetworking devices containing SNMP need to be polled for invoking management protocol. SNMP supports both connection-oriented (TCP) and connectionless (UDP) transport layer protocols, both of which work with IP (network layer protocol). The SNMP is independent of vendor, and as such there are various vendors manufacturing this protocol, which offers a set of functions and services.

The condition of any physical device is usually defined by a management information–based (MIB) object within SNMP. The earlier version of MIB included eight groups of objects with a total of over 100 objects. Two more object groups were added in MIB II. The transmission group supports different types of LANs such as Ethernet, T1, and token ring, while the other group is responsible for monitoring activity. The condition of devices is defined by variable data that can be retrieved by five simple command messages: get, get next, set, response, and trap. The object-variable values of devices are sent from a device to the manager by another software module known as an *agent*. A proxy agent defines an interface between a non-standard device and an SNMP management station.

This protocol does not offer any service to the users but instead offers the networking framework within which the users using different mail protocols may exchange messages across the network. It manages all the protocols being used within various hosts. It is responsible for handling the following aspects within networking: fault management, name management, security management, accounting management, addressing management, router management, etc. The exchange of management messages between the central network management and local management does not require any special connection or configurations, but it uses the same setup for initiating any request from central management (or local management) to invoke the network-dependent protocols and transmit the request to the internetwork. A set of PDUs and primitives has been defined for exchange of management messages and includes **Get.request**, **Get-Next.request**, **Set.request**, **Get.response**, and **Trap**. The first three are mainly for the central network manager, while the remaining two are for local management application protocols.

16.5.8 Commitment, concurrency, and recovery (CCR) protocols

The ISO application service and protocols are defined by DIS (Draft International Standard) 9804 and ISO 9805 standards. This is one of the common application service elements which has been defined to be used in a variety of applications and offers coordination between multiple users interacting with each other simultaneously. Association control defines a set of services which are common to a variety of applications basically used for establishing connection, maintaining it, and finally terminating it after the data transfer. As such, the protocols for association control are very complex, and their services for connection are required by all applications based on connection-oriented services. The only exception where these services are not used is the connectionless applications. The CCR standard protocol is defined in a limited way but is being defined as common application service elements. The CCR specification is used where the database activities are to be maintained over multiple sites and allows a number of operations such as data transfer, access to the database, updating of databases, etc. The primary objective of CCR specification is to provide consistency across the database for updating and defining correct and accurate database values in data fields. Some of the material presented herein is derived from References 1–4, 19, and 20.

In normal operation, the protocols provide interaction between sender and receiver in such a way that the sender sends a request (action) to the receiver, which performs some action on it. This action can have different states such as acknowledgment, refusal, and no response. The sender will send the data or next request only after it receives an acknowledgment from the receiver. Since the sender and receiver are interacting synchronously with each other, any mismatch in the reply or action from either side will put them out of synchronization, and hence the communication channel between them will corrupt the data communication. This problem is serious in many time-critical applications such as updating of databases, time-based data communications, etc. Further, in the event of any failure of link between them due to hardware or malfunctioning, or if there are many activities defined by different systems within the communication protocols, the network is likely to collapse, and the restarting of the networks will restart the process of communication, offering a highly inconsistent behavior. These problems are dealt with differently in centralized and distributed database systems.

In order to reduce these problems (preventing two nodes from accessing and updating the same data simultaneously), a new concept has been introduced to define various performance terms such as commitment, recovery, rollback, and concurrency. In the following subsections we briefly explain each of these concepts.

Commitment (failure) and recovery: When a sender sends a request to the receiver, the receiver performs different actions on it depending on the state of the protocol at the receiving side. It is expected that the communication protocol must handle all the situations (worst case). With this concept, when the sender sends a request to the receiver, the receiver can perform three actions: acknowledgment, refusal, or no response over the received request. In the event that it sends an acknowledgment, we associate a primitive with it in the form of **Commit to Perform** assigned for process. This acknowledgment of Commit to Perform will be responded to by the sender in another primitive, **Commit**, sent back to the receiver. The receiver uses another primitive, **Done**, for this acknowledgment. In this way, the receiver is now committed to perform that process or action (requested). This type of communication between sender and receiver is based on the master–slave concept or primary–secondary model. This new concept of commitment will also offer some problems which must be addressed. We will discuss some of the possible worst-case situations below.

When the sender sends a request, it may crash before getting any response from the receiver. In this concept, this request is stored on its disk and it will send the same request with the same assigned ID in the event of restarting of the system.

When the sender sends a request, it is acknowledged by the receiver by sending Commit to Perform. Let's assume that this acknowledgment is lost during its transmission due to malfunctioning or link failure. The sender, not having received any response from the receiver after a predefined time, will re-transmit the same request with the assigned ID to the receiver. The receiver considers this re-transmitted request a duplicate request and will send another Commit to Perform primitive, thus maintaining consistency between sender and receiver in the sense that when the system restarts, the sender starts from the same point by sending the request with the same ID.

The sender sends a request to the receiver and both have agreed to **Commitment**. The receiver finishes its process and sends a Done primitive to the sender. If this primitive is lost during the transmission due to malfunctioning, link failure, or failure of the application entity, the sender sends the same Commit primitive to the receiver. The sender gets a primitive **Unknown ID**, which means to the sender that the receiver has completed its process (thus offering recovery from the malfunction or any other failure).

The receiver usually keeps a record of request ID on its disk. The receiver has received a request from the sender and before it can send Commit to Perform, it crashes. The sender times out and sends another request to the receiver after not getting any response from the receiver during the predefined time. When the system is restarted, the receiver may send the Commit to Perform primitive back to the sender by attaching the same ID (recovery).

Similarly, consider a situation where the receiver, after receiving a request from the sender, sends a primitive Commit to Perform and keeps a record on the disk, but it then crashes before it can get a Commit primitive from the sender. In this case also, the sender, after getting no response from the receiver to its Commit primitive, will re-transmit the same primitive to the receiver. The receiver will recognize this (as it has recorded it in its disk) and send its acknowledgment to the Commit primitive.

In this concept, the sender or receiver has started processing the respective processes, and in the event of failure (as discussed above), necessary action for recovery is taken. If both sender and receiver are coordinating properly, then this concept basically becomes the ordinary protocol (as discussed in Chapter 11, Data Link Layer) for error detection and correction strategies.

Rollback: This concept is derived from recovery, with a minor difference. Here, after the receiver has sent its primitive Commit to Perform, the sender may send either the Commit primitive to it or another primitive, Rollback. This Rollback primitive indicates to the receiver to delete its earlier request and no action will be performed by the receiver. It is important to note that this concept allows the sender to rollback the process which the receiver has committed to perform before the receiver starts working on it. This concept is useful in applications where it is possible to request to the receivers without having the process or application ready to be processed, and the sender, after getting a Commit to perform primitives from a number of receiving nodes, may rollback by deleting that process to be processed by the receiver, as this process is not yet ready to be executed or the sender has realized a mistake.

This concept may not offer advantages for applications involving two nodes. But if a number of nodes are controlled by a sender node in a tree-like structure, then the sender issues requests to various receiving nodes. If it gets responses from those nodes in the form of Commit to Perform primitives, it does not have any problem in sending its Commit primitives to them over the tree. It is interesting to note that, in this tree-like structure, the sender sends requests to its connected nodes, which in turn send requests to nodes which are connected to them; thus, these nodes become sender nodes for them (based on atomic action). The sender will send Rollback to its connected node(s) which has received refusal from their lower connected nodes, and this process is continued until the primitive Rollback from the sender reaches the lowest-level node(s) which has refused. In this tree-like structure, we have more than one application entity active at any time, and each node represents an application entity, while the branch between them represents an association of the application entities.

When implementing atomic action using the two-phase technique, the sender first sends a request to all its receiver nodes, and the receiving nodes which can perform the assigned processes will lock this data so that no requests from other sender nodes can interfere with this request. The receiving nodes send a Commit to Perform primitive back to the sender. The sender node, after getting this primitive, sends a Commit primitive to each receiver node to indicate that it can start the process. During the Commit primitive exchange, if any receiver node indicates the failure to the sender, the sender rollbacks and aborts the entire action and informs all the receiving nodes to unlock the request and restore their initial states. In this case, no action will be performed at all.

Concurrency: This concept makes the recovery concept more realistic and useful. In the case of commitment (as we discussed above), the sender sends a request to its connected receiving nodes and receives a Commit to Perform primitive from them (depending on their states). When the sender has sent its Commit primitive to a particular receiver, this basically indicates that the application entity of the sender has authorized to the application entity of the receiver to perform that requested process. In this case, the application entity of the receiver has started processing the request and must be able to access the resources, and no other application entity must deny access to this application entity for using the resources. Further, there must not be any interference from any application entities during the processing phase of the particular application entity which has been awarded permission to start.

The above-mentioned concepts are being used in standard CCR services defined by the ISO DIS 9804 standard document and offer the following features:

- **Atomic action (transaction) tree:** This is useful feature for representing involvement of more than two applications in an atomic action (i.e., indivisible) such that application entities can be allocated to nodes of a tree-like structure with appropriate association between them by branches until we reach the leaf nodes where actions are performed. The atomic action is basically a collection of messages and operations which may be used to finish the assigned task, or it can rollback to the original state indicating that no action has been performed. Each node except the root may behave as both a receiver and sender in receiving the request and sending a request to its connected nodes. The atomic action scheme is usually implemented using a two-phase commit technique.

- **Failure:** The CCR service makes sure that secure data is not lost due to failure of any application entity during transmission. The CCR service and secure data primitives provide a successful restarting of atomic action, and the lost information is recovered.
- **Concurrency control:** Each node representing an application entity is capable of making changes in local data (atomic action) and offers a concurrency control facility at the local level by using a locking mechanism.
- **Restart and rollback:** The state of any node (application entity) can be restored to the initial state of any commitment after getting rollback and restart primitives.

The application layer defines CCR common application service elements (CASEs) and CCR users in such a way that both sender and receiver nodes are allowed to interact with each other via application entities.

CCR service primitives and parameters: A list of primitives and parameters of CCR service is given below.

C-Begin.request (atomic action identifier, branch identifier, atomic action timer, user data)
C-Begin.indication (atomic action identifier, branch identifier, atomic action timer, user data)
C-Prepare.request (user data)
C-Prepare.indication (user Data)
C-Ready.request (user data)
C-Ready.indication (user data)
C-Refuse.request (user data)
C-Refuse.indication (user data)
C-Commit.request
C-Commit.indication
C-Commit.response
C-Commit.confirm
C-Rollback.request
C-Rollback.indication
C-Rollback.response
C-Rollback.confirm
C-Restart.request (resumption point, atomic action identifier, branch identifier, restart timer, user data)
C-Restart.indication (resumption point, atomic action identifier, branch identifier, restart timer, user data)
C-Restart.response (resumption point, user data)
C-Restart.confirm (resumption point, user data)

For example, the primitive C-Begin.request is issued by the sender to begin atomic action. This primitive is mapped into the primitive of the presentation layer P-Sync-Major. The primitive C-Ready.indication sends a reply when it is ready to Commit to Perform and also to Rollback if requested. It is mapped to the P-Typed-Data primitive defined by the presentation layer. The C-Rollback primitive is issued by the sender to indicate its decision to revert and is expecting the receiver to restore the initial state before Commit to Perform. It is mapped to P-Resynchronize. The C-Restart primitive can be issued by either the sender or receiver and it indicates to initialize all the procedures for recovery. It is mapped to the P-Resynchronize (Restart) primitive of the presentation layer.

The following is a list of CCR primitives used in two-phase commit operation.

C-Begin (sender initiates and an atomic action begins)
C-Prepare (sender indicates the end of phase I, prepares to commit)
C-Ready (receiver indicates its desire to do the assigned task)
C-Refuse (receiver indicates its inability to do the assigned task)
C-Commit (sender commits the action to receiver)
C-Rollback (sender aborts the action)
C-Restart (either sender or receiver indicates the occurrence of a crash)

16.6 Application layer protocols of proprietary LANs

We have seen that the layers of the proprietary LANs are not the same as those defined in OSI-RM and also do not conform to OSI-RM, but the architecture of these LANs depends on the layered approach. The functionality of these LANs includes the entire functionality of OSI-RM, and there seems to be a one-to-one correspondence between their layered architecture and standard OSI-RM LANs. For more information on these proprietary LANs, see References 20–22.

16.6.1 IBM's SNA application layer protocol

One of the proprietary LANs is IBM's System Network Architecture (SNA). The upper layer protocol widely used in SNA network is LU6.2 which offers communication between various application programs. The SNA not only provides communication between different types of devices but also optimizes the performance of these devices or application programs.

Various services provided to the end users by SNA include file transfer and sharing of application programs, devices, etc. With the advent of high-speed and intelligent dumb terminals, powerful workstations, PCs, etc., LU6.2 may not be able to provide the services to the end users. A new version of LU6.2 was developed known as LU6.2 APPC (Advanced Program-to-Program Communication) for SNA networks. It interacts with the application program interface (API) to offer application-to-application communication and is usually known as an application transaction program (ATP). The ATPs interact with LU6.2 to provide API, and then an LU session is established between them.

The LU6.2 provides only an interface between different application programs and expects the transformation of syntax programming languages, commands, etc., to be taken care of by the application layer.

These features are included in APPC. In addition to these features, LU6.2 offers a variety of options/facilities which are available and implementable at higher layers and supported by lower layers, thus providing communication between two vendors if mutual agreement on the facilities at higher layers can be accomplished. The API enables total transparency for providing communication between various applications and distinguishes the functions to be performed by APPC and ATP. The application programs are not aware of APPC being used underneath. Although the conversion between syntax, verbs (used during communication), and commands is provided by the application layer, LU6.2 has to support the conversion process offered by the application layer. It is interesting to note that, during an established session, conversion of various parameters can be performed in a sequence and can be used for communication of messages of any duration (short or long), depending on the mutual agreement.

There are different categories of verbs used in commands for various operations, e.g., program-to-program, end user high-level languages, control functions, synchronization and coordination of various control activities, etc.

The program-to-program communication is a basic service provided by LU6.2 and is used by the LU service transaction program (STP). There are many vendor products defined under STP with different versions and options.

The conversion of formats of programming languages into a standard data format at the sending node and re-conversion of the standard data format back into programming languages at the receiving node are provided as categories for allowing users to use any high-level programming languages.

The control functions define synchronization points which are used to check the communication process, monitor the communication process in the event of any failure or interrupts, etc. Finally, synchronizing and monitoring the activities of various control functions during the communication process need to be handled and are usually performed by a category of command verbs.

The LU6.2 communication requires the transmission of primitives such as Allocate followed by a Send.data primitive to the receiving side. The receiving node receives .receive and .wait primitives. After receiving the data, the receiving node can send the data by using a Send.data primitive. The sending node will receive the primitives .receive and .wait. After having received the data, the sending node may then send a Deallocate primitive to the receiving node, which sends back the confirmed primitive to the sending node.

LU6.2 services: In order to get a feel for the types of verbs along with their meanings, we consider the following command verbs of the categories of basic and conversion modes of LU6.2 services.

The **Allocate** verb has the following attributes attached LU, ATP, Mode, and Sync. This command verb defines a new LU and assigns a name with which it is used to establish communication in ATP. The type of mode may be defined as batch, interactive, time-sharing, etc., while the Sync is defined as levels for acknowledgment.

Other command verbs of this category include Deallocate, Confirm, Flush, Backout, etc. There are a few command verbs which have primitives associated with them. For example, the send.data command verb transfers to the data output buffer and returns the control back to the application programs. The general data stream (GDS) under LU6.2 defines the format for the data to be transferred. The maximum length of a data stream in GDS can be up to approximately 32 Kb. Other command verbs with primitives are Receive.wait, Send.error, Prepare.to.receive, Request.to.send, Get.attributes, Post.on.receipt, Receive.immediate, etc.

16.6.2 DEC's DECnet application layer protocol

The application layer of DECnet or DNA is known as the network application layer, and typically it does not have any similarity with the functions and services of the application layer of OSI-RM. As we mentioned earlier, this proprietary LAN follows the layered approach and is comparable to OSI-RM but does not follow the OSI-RM standards. The network application layer of DECnet provides the functions for data access, terminal access, and various communication services to the users. DECnet has defined the following communication protocols[21]:

- **Network Virtual Terminal Protocol (NVTP):** Provides interface to the terminals with DECnet host systems
- **Data Access Protocol (DAP):** Provides support for remote file access and file transfer
- **Loopback Mirror Protocol (LMP):** Provides support for transmission and reception of one message

- **X.25 Gateway Access Protocol (XGAP):** Provides support for communicating with X.25 networks
- **SNA Gateway Access Protocol (SGAP):** Provides support for communicating with SNA networks

This protocol resides in the network's application layer and provides an interface between the **file access listener (FAL)** and the subroutine doing the accessing. The users cannot access DAP directly but can be involved by external subroutines to use it. It controls network configuration, data flow, transportation, formal conversions, and monitoring of the programs. As stated above, this protocol is mainly concerned with managing and controlling remote file access, file transfer, etc. The main function of DAP is to perform different types of operations on files, e.g., reading and writing of files onto I/O devices, deletion and renaming of files, directory of remote files, execution of files, support of different data structures of files (sequential, random, indexed file organizations) and their access techniques (sequential random and indexed), etc.

The user makes a request for a remote file via I/O commands which are transformed into DAP messages. These messages are transmitted over a logical connection established between the terminal and remote server. The interpretation of DAP messages into actual I/O operations on a file is carried out by the server. The server sends an acknowledgment indicating the status of the file back to the user. Initially, when the message is sent from the user to server, it is required that this message contain information about the files (file system organization, size of the file for the buffer, record size, etc.). DAP is composed of very complex software handling the situations of varying size of files, data record size, data structures of files, databases, etc.

The DEC system includes the following protocols for remote file access and file transfer. These protocols are based on the concept of DAP.

1. **File access listener (FAL):** This utility targets the program of a file access operation, accepts the user's request, and processes it on the node where FAL is residing. This protocol resides at the remote server node and responds to user I/O commands for establishing a listening session between them.
2. **Network file transfer (NFT):** This protocol resides at the top layer (user layer of DNA) and maps the DAP-speaking process into an appropriate DAP function. It offers file transfer and various operations on the files.
3. **Record management services (RMS):** The majority of the operating systems of DEC use this scheme for their file management systems. The DAP messages are transmitted over a logical connection by the file system (RMS), and in some cases, it also includes remote node address and access control message. The DAP message includes the user request for remote file, remote node address, and possibly access control information. This message is transmitted to the remote server node where FAL, after receiving the message, performs the operations.
4. **Network file access routines (NFAR):** This protocol is a set of routines which is a part of the user process (when making an I/O request). These subroutines allow the user to access remote files and are executed under the control of FAL. These subroutines reside at the application layer of DNA at the server node.
5. **VAX/VMS command language interpreter:** Various commands for file access and file transfer are interfaced with RMS for allowing users to access remote files and perform operations on them; NTF is included in VAX/VMS.
6. **Network management modules (NMM):** These modules are used to provide services like down-line loading of remote files, transferring of up-line dumps for storage, etc.

Some useful Web sites

Information on the application layer can be found at *http://ntrg.cs.tcd.ie/4ba2/application/*, *http://www.csc.vill.edu/faculty/schragge/html/CSC8560/chap4lec.htm*, and *http://www.cs.wpi.edu/~cs4514/b98/week8-appl/week8-appl.html*.

The SNMP forum is found at *http://www.snmplink.org*.

There are a number of emulation and simulation tools available. The emulation tools are available at *http://www.lognet-systems.com*, *http://is2.antd.nist.gov*, and *http://ctd.grc.nasa.gov*. The simulation tools are available at *http://www.cnet.com*, being used at *http://www.cs.uwa.edu.au*, Opnet modeler at *http://wwwopnet.com*, ns-2 at *http://www.mash.cs.Berkeley.edu*, X-sim (X-kernel protocol) at *http://www.cs.arizona.edu*, and COMNET III high-fidelity network simulation at *http://www.caciasl.com*.

References

1. Tanenbaum, A.S., *Computer Networks*, 2nd ed., Prentice-Hall, 1989.
2. Stallings, W., *Handbook of Computer Communications Standards*, 2nd ed., Howard W. Sams and Co., 1990.
3. Markeley, R.W., *Data Communications and Interoperability*, Prentice-Hall, 1990.
4. Henshall, J. and Shaw, A., *OSI Explained: End to End Computer Communication Standards*, Chichester, England, Ellis Horwood, 1988.
5. Linningtin, P.F., "The virtual filestore concept, *Computer Networks*, 8, 1984.
6. Solman, J.V., "Design of public electronic mail system," *IEEE Network Magazine*, 1, Oct. 1987.
7. Huffman, A.J., "E-mail — Glue to office automation," *IEEE Network Magazine*, 1, Oct. 1987.
8. Hutchison, G. and Desmond, C.L., "Electronic data exchange," *IEEE Network Magazine*, 1, Oct. 1987.
9. Gilmore, B., "A user view of virtual terminal standardization," *Computer Networks and ISDN Systems*, 13, 1987.
10. ISO/DIS 8571, "Open System Interconnection: File Transfer Access and Management, Information Processing Systems," ISO, 1986.
11. ISO/DP 8850, "Open System Interconnection: Specification of Protocol for Common Application Service Elements, Information Processing Systems," ISO, 1984.
12. MIL-STD-1777, "Internet Protocol, Military Standard," U.S. Department of Defense, Aug. 1983.
13. MIL-STD-1778, "Transmission Control Protocol, Military Standard," U.S. Department of Defense, Aug. 1983.
14. MIL-STD-1780, "File Transfer Protocol," U.S. Department of Defense, May 1984.
15. MIL-STD-1781, "Simple Mail Transfer Protocol, Military Standard," U.S. Department of Defense, May 1984.
16. MIL-STD-1782, "Telnet Protocol, Military Standard," U.S. Department of Defense, May 1984.
17. CCITT Recommendation X.3, "Packet Assembly/Disassembly (PAD) in a Public Data Network," CCITT, 1984.
18. CCITT Recommendation X.400-X.430, "Data Communication Network: Message Handling System," CCITT, 1984.
19. Black, U., *Data Networks, Concepts, Theory and Practice*, Prentice-Hall, 1989.
20. Kauffles, E.J., *Practical LANs Analyzed*, Ellis Horwood, 1989.
21. DEC, "DECnet Digital Network Architecture, General Description," Digital Equipment Corp., May 1982.
22. DEC, "DNA Digital Data Communication Message Protocol (DDCMP), Functional Specification," Digital Equipment Corp., 1982.
23. ANSI 90, "Directory for Information Systems, X3.172," American National Standards Institute, 1990.

24. Nabielsky, J. and Skelton, A.P., "A Virtual Terminal Management Model," RFC 782, for Defense Communications Agency, WWMCCS ADP Directorate, Reston, VA, (*http://www.faqs.org/rfcs/rfc782.html*).
25. Freeman, R.B., "Net management choices: Side stream or mainstream," *Data Communications*, Aug. 1982.

chapter seventeen

Internet: Services and connections

"If it doesn't absorb you, if it isn't any fun, don't do it."

D. H. Lawrence

17.1 Introduction

The U.S. Department of Defense (DoD) over the last 24 years has developed a tool for the scientific elite: an amorphous computer network known as the Internet which now allows more than 20 million users to exchange electronic mail and share other resources. The Internet has become very popular in every walk of life, and user-friendly multimedia browser applications for accessing the World Wide Web have made it very easy and useful for retrieving information as well as creating a home page and storing needed information. It has attracted tremendous and ever-growing interest in the business, corporate, scientific, and other communities. It has enabled a new type of commerce which allows users to order products and make payments electronically. The number of Internet users is increasing at a rate of over 20% per month.

The Internet is a collection of different types of computer networks and offers global connectivity for data communications between them. The communication between these computer networks is defined by a common language through a standard protocol, transfer control protocol/Internet protocol (TCP/IP). ARPANET (sponsored by DoD's Applied Research Project Agency) was the first experimental network, designed to provide support for military research. Although the Internet is derived from ARPANET, it now includes a variety of networks such as NSFNET, BITNET, NEARNET, SPAN, CSNET, and many others. The workstations with Berkeley UNIX were introduced about 10 years ago and have Internet protocol (IP) networking facilities.

The local area networks (LANs) were also introduced more or less at the same time and were used to provide the services of ARPANET to the users connected to them. At the same time, other networks were also introduced and defined. One the most useful and important networks was NSFNET, developed by the National Science Foundation (NSF), a U.S. federal agency, to provide connectivity for supercomputers. Users can access the supercomputer facilities via the NSFNET network. It is based mainly on the ARPANET concept of connecting various supercomputer centers (about five) with every university in the U.S. directly over 56-Kbps telephone lines. Various regional networks were also defined across the country which provide interconnection with various universities in those regions. These regional networks are then interconnected with the supercomputer center over NSFNET. The NSF allows educational institutions to make the services available to their users on campus, thus providing Internet services to those who don't have access to the Internet. The Internet, until the time NSFNET was introduced, was available mainly to researchers, government employees, and computer professionals, but the introduction of NSFNET has offered Internet services to various university communities; it is under consideration to provide Internet services to the secondary school communities, as well. It has been shown that more than a million users use the Internet daily for different services.

The National Science Foundation in the U.S. provides and maintains an NSFNET backbone that connects supercomputers across the country. This also provides communication links for long-haul communication between various research and academic institutions across the U.S. and around the world. These communication links are usually maintained by commercial companies such as AT&T, MCI, etc., and are leased to different Internet service providers. The Internet started with four supercomputer centers in San Diego, Cornell, Pittsburgh, and Illinois Universities. Now many other resources have been added to the Internet, including University of Washington Information Navigator (WIN), University of British Columbia, Portland Area Library Services, Stanford Linear Accelerator, NASA Ames Research Center, University of Colorado Weather Underground, Minnesota Supercomputer, Cleveland Free-net, Space Physics Analysis Network Information Center, SUNY at Buffalo, and University of Delaware OCEANIC.

The global Internet (connecting the wide area networks and resources of other countries) provides connectivity with all the countries around the world. One of the supercomputer centers in the U.S. is connected to entities around the world including Canada; Universidad de Guadalajara, Mexico; Puerto Rico; Venezuela; Ecuador; Univeridate de Sao Paulo, Brazil; Argentina; Univers Netherlands; United Kingdom; Ireland; Free University of Brussels; Belgium; France; CERN: European Laboratory for Particle Physics; Italy; Spain; Greece; Bulgaria; Magpie BBS; Croatia; University of Ljubljana, Slovenia; Crez Technical University, Poland; Austria; Germany; Hungary; Iceland; Finland; Sweden; Hebrew University of Jerusalem; Kuwait; Education and Research Net (ERNET), India; Malaysia; National University of Singapore; Thailand; Hong Kong; Taiwan; Japan; Australia; New Zealand; Rhodes University Computing Center, South Africa; United States Research Station; and many others.

In the U.S., many of the networks connected to the Internet are developed for various research and educational applications of different federal agencies. Each federal agency has its own network for its respective applications which likely creates overlap and wastage. Recently, legislation has been passed to define or create a **National Research and Education Network (NREN)** which will be a part of the Internet. This will provide research and educational applications to various federal agencies (such as NSF, NASA, DoD, etc.) There are quite a number of Internet connection and service providers, and the requested services are routed via appropriate routers depending on the type of service requested. For example, the research and educational application services are routed via a router "RE" (research or education) of NREN, while communication application services are routed via "C" routers.

Due to the emerging use and importance of ARPANET, it was decided in 1983 that ARPANET should be divided into two types of networks (based on packet-switching) — MILNET and ARPANET — and together these two networks are known as the **Internet**. The **Defense Communication Agency (DCA)** decided to impose the use of TCP/IP on ARPANET and MILNET via packet-switching software and defined a separate network number and gateway for each of these networks. The number of computer networks connected to the Internet also expanded from 100 in 1983 to more than 2220 by 1990. The information regarding the number of computer networks connected to the Internet is mainly maintained by the **DDN Network Information Center (DDN NIC)**. Both ARPANET and MILNET also expanded as the Internet.

The backbone network includes backbone links (typically T3 links of 45 Mbps) and routers connected to regional networks and supercomputers via optical fiber or other media of higher data rate. The function of a router is to transmit the packets between different types of networks (LANs, MANs, regional networks, proprietary networks, or any other networks) and also between networks and backbones. The links connecting to various routers carry the data at a very high data rate. To get a feel for this, a CD-ROM usually contains over 550 Mb of information. A T-1 link (DS-1), offering a date rate of 1.544 Mbps, will transfer the entire data from a CD-ROM to anywhere in less than about six minutes. A T-3 link, on the other hand, offers a data rate of 45 Mbps and will take less than about 13 seconds to transfer the data. In the same way, the time taken to transmit the data of a CD-ROM in other media can be calculated.

Many backbone networks were also introduced during this time which were connected to the Internet, e.g., CSNET in 1984 (interaction between computer and engineering professionals and researchers who were not connected to either ARPANET and MILNET) and NSFNET in 1986 (to interconnect supercomputer centers for their use by faculty members of universities). NSFNET has now become a backbone network for the U.S. national research network and uses TCP/IP. The main feature of NSFNET is that it connects a number of low-level networks which are then connected to universities and other commercial enterprises. We can say that the NSFNET itself is just like a mini-Internet.

The TCP/IP protocols provide necessary communications across connected networks. The user datagram protocol (UDP), transmission control protocol (TCP), Internet control message protocol (ICMP), and routing control protocols are the protocols of the transport layer and interact with the Internet protocol (IP) of the network layer. TCP offers a connection-oriented, end-to-end, reliable, and sequenced delivery of data, as well as priority, segmentation and reassembly, and adaptive flow-control services. UDP, on the other hand, provides connectionless, unreliable, and unacknowledged datagram service. Real-time protocol defined at the transport layer provides a real-time application that deals with transmission of voice and multimedia data.

IP is a datagram protocol that is resilient to network failure and offers sequenced delivery of the data. It offers end-to-end, reliable data communication across the Internet when used with TCP. The function of ICMP at the Internet layer is to provide a ping (echo packet) to determine the round-trip propagation delay between the source and IP-addressed host. Routers use this protocol for sending error and control messages to other routers. This layer provides multicasting data communication using the Internet Group Multicast Protocol. It also uses the Address Resolution Protocol (ARP) that provides a direct interface to the data link layer of the underlying LAN. This protocol provides a mapping between the MAC address (physical) and the IP address. It uses some routing protocols. The Internet layer supports a variety of data link layers (Ethernet, token ring, FDDI, frame relay, SMDS, ATM, etc.). The data link layer, in turn, supports a variety of transmission media such as twisted pair, TDM, fiber, etc.

The transport layer with UDP and TCP supports a variety of applications such as file transfer protocol (FTP; a secured server log-in, directory manipulation, and file transfer), Telnet (remote log-in facility), Hypertext Transfer Protocol (HTTP; for the World Wide Web), Simple Network Management Protocol (SNMP; for network configurations, setting, testing, data retrieval, alarm and other statistical reports, etc.), remote procedure call (RPC), network file server (NFS), trivial FTP (TFTP; reduces the implementation complexity), and many others. The domain name system (DNS) offers a centralized name service that can use either UDP or TCP.

The Internet must provide security to users, and there are a number of ways by which it is possible to break into ASCII terminal ports via interactive terminal port file, transfer port, e-mail port, and many other application ports.

In order to handle this problem of breaking in and misusing the systems connected to the Internet, an organizational committee known as the **Computer Emergency Response Team (CERT)** has been set up which interacts with manufacturer vendors and discusses these problems along with possible solutions. This committee may be contacted at *cert@cert.sei.cmu.edu.*

To some extent security problems can be avoided by choosing a proper password. In fact, according to CERT, nearly 80% of break-ins are caused by a poor password choice. A proper password should have at least six characters (upper-case, lower-case, or any number) and should not be a word or sequence of letters from the keyboard, as these can easily be interpreted by intruders.

One of the most widely used encryption programs both in the U.S. and outside is **PGP (Pretty Good Privacy)**. This program is available free of charge. The National Security Agency (NSA) has authority to read our e-mail!

There are a number of ways users can check whether their computers have been broken into, and service commands can be used to check (depending on the operating system used). For example, if we are working on a UNIX-based PC, we can use the following sequence:

- Look at your password file and watch for unusual entries (type: /etc/password)
- Look at the tasks running with BSD-UNIX (type: ps aux — or ps-el with system V version)
- Look at the extended directory listing (type: ls-la).

Applications of the Internet: The Internet has offered tremendous potential for various applications which can be shared by users around the world. The Internet is growing continously at a rapid speed and there are now 3.2 million computer hosts on the network spread in 46,000 domains.[1-8,10] In one year alone (1994), over 1 million computer host users started accessing Internet services. The U.S. alone has 2 million computer hosts and this number is growing at a rate of nearly 40% compared to last year. A collection of papers on how Internet services are being used in various companies and businesses can be found in Reference 11.

Internet facilities are provided through a variety of services, and these services can be accessed in different ways. The services include electronic mail, online conversation, information retrieval, bulletin boards, Gopher, games, newsgroups, etc. These services can be accessed in some of the following ways. If the user has an account with an online service (e.g., CompuServe, MCI Mail, or any similar mail system), then these services can be accessed electronically (electronic mail system) on the Internet. If the user uses a bulletin board system (BBS), the services can be accessed via e-mail over the Internet. UNIX-based networks use the same networking conventions as that of the Internet.

17.2 The Internet

A variety of information is available on the Internet, including directories, statistical databases, university catalogues, library online catalogues, weather charts, census data, and many other categories of information. In the following subsections, we discuss all of the services provided to the users by the Internet.[1-5,9-13]

17.2.1 Definition

The Internet is an interconnection of a variety of networks (federal networks, international or national networks, campus networks, and other networks), all of which use **Internal**

Protocol (IP). The Internet has also become popular in non-IP-based networks, and some connecting devices such as gateways provide interconnection between the Internet and networks such as DECnet, Bitnet, UUnet, etc. These devices provide exchange of e-mail services across the network. The ARPANET, defined in 1969, eventually became a tool offering services such as remote log-in, file transfer, electronic mail, and sharing of information. Now the Internet offers connections to more than 5,500 backbone regional and local computer networks of more than 55 countries. Due to heavy use and its ever-growing popularity, the Internet offers services to more than 20 million users around the world. A host of directories, information servers, navigational tools, and other application programs have been defined and many more tools and application programs are under development.

The Internet is a forum for users to share worldwide information resources. The resources are incomprehensibly vast, and many of us really don't understand the Internet fully. We can define the Internet as a computer network which provides information services, resource-sharing services, etc. to a large number of users worldwide. We can also consider the Internet a "global information library" which allows users to participate in group discussions, search for any information, start any discussion for others, and so on.

The Department of Defense (DoD) funded the Defense Advanced Research Projects Agency (DARPA) for designing a highly distributed system which should provide connectivity to four supercomputer centers at San Diego, Cornell, Pittsburgh, and Illinois Universities. In 1979, the **Internet Control and Configuration Board (ICCB)** was formed to look into protocols and standards for communication across the distributed system. Under the guidance of ICCB, the communication protocol suite TCP/IP was introduced and implemented. The TCP/IP was implented over the ARPANET in 1980. In 1983, the ARPANET was split into two separate nets known as ARPANET and MILNET.

In 1985, the U.S. National Science Foundation (NSF) used TCP/IP for communication across the supercomputers and many other networks connected together and introduced this as the **Internet**. The Internet included over 3000 different types of networks (LANs, regional networks, etc.) and over 200,000 computers in 1990. By 1993, the Internet had grown to the extent that it had over 2 million computers. The number of users on the Internet was also growing exponentially. As of today, over 20 million users are using the Internet for different applications.

The NSF has installed a very high-speed backbone network service **(vBNS)** that is based on ATM and uses OC3 links for the five supercomputer centers located at San Diego, Pittsburgh, Cornell, Illinois, and Austin. This backbone network was implemented by MCI in 1995. These supercomputer centers are connected via OC-3 and provide Virtual Path connections between them over MCI's commercial HyperStream ATM service. The vBNS now uses an OC12 trunking system, and a number of universities and organizations have been added for higher-bandwidth applications. Each of the supercomputer centers is connected to a LightStream ATM switch that is connected to Cisco routers. The supercomputer centers are connected by OC12/STS12c links, while most of the routers are connected via DS3links.

The Internet and TCP/IP together provide connectivity between computers at two remote locations anywhere in the world and allow the transfer of information between those computers over the Internet. The architecture of the Internet or TCP/IP is defined in terms of layers as process/application, host-to-host transport, Internet, and network access. The TCP/IP includes all the needed application software and is also known as *Internet protocol*. The top layer, **process/application**, includes the functions of the application, presentation, and a part of the session layer of OSI-RM. It provides the following services: FTP, Telnet, domain name system (DNS), simple mail transfer protocol (SMTP), and routing information protocol. The next layer, **host-to-host**, offers the same functions as the transport and part of the session layer of OSI-RM. It offers two type of services — connection-oriented and connectionless.

The TCP provides end-to-end error control for the entire frame and, hence, reliable communication (connection-oriented). The user datagram protocol (UDP), on the other hand, provides error control only for its header and hence provides unreliable service (connectionless). The UDP is faster than TCP and is used in small data transfers or applications where errors can be tolerated.

The **Internet layer** provides the same functions as that of the network layer of OSI-RM. This layer includes three protocols: Internet protocol (IP), Address resolution protocol (ARP), and Internet control message protocol (ICMP). The IP does not provide any control other than header checking via checksum. The **network access** is the bottom layer and includes the functions of the data link, physical, and a part of the network layer of OSI-RM. It defines two sublayers as 1A and 1B. 1A is equivalent to the physical layer, while 1B is equivalent to the data link layer and includes X.25, FDDI, and 802.x LAN protocols.

In very typical and standard (to some extent) configurations, the user's PC or terminal is connected to LANs which are connected to a high-speed LAN link or backbone LAN (e.g., FDDI LAN) to define a campus-wide wide area network (WAN). Most WANs are connected via leased telephone lines, although other media such as satellite links are available. On a campus (of universities, companies, research organizations, government/private sectors, etc.), the LANs are connected by routers to define WANs, and WANs may be connected by routers to define larger campus-wide WANs. The high-speed backbone LAN link is connected to the outside world (i.e., Internet).

17.2.2 Accessing Internet services

The Internet provides wider area connection over the telephone system. Within the Internet, each computer connected to the LAN or LANs connected to WANs is considered an Internet host. A user's PC or workstation may access the servers (offering different services) via the Internet directly over different links. Each link defines an Internet line connection. The PCs or workstations may also use the services via access systems connecting users' computers to the Internet over leased lines. The access systems allow the users to use the services offered by the Internet, and there are many Internet connection and service providers available (these will be discussed in more detail later).

The transmission link between a subscriber's phone system and the end/central office supports only a voice-grade channel occupying a frequency range of 300 Hz to 3.4 KHz and offering a very low data rate of 56 Kbps. Internet traffic needs much higher rates, as it includes lots of graphics and also video, which may take a longer time to download on our PCs. The links between the Internet and end offices provide data rates of as high as 2.5 Gbps. Although modems offering 56 Kbps and even higher rates have been available on the market since 1996, the bottleneck remains the twisted-pair cable connected between the user's phone and the end office.

High-speed Internet traffic: A number of new technologies are being considered to provide a high rate for Internet traffic. Some of these include ISDN, cable modems, and asymmetric data subscriber line (ADSL). Telco has suggested the removal of loading coils from twisted-pair cable as a means of increasing the data rate at the expense of reduced bandwidth. Cable modems provide very large bandwidths (due to the cable medium). They can be used on FDM, as is being done in the case of community access television (CATV).

The television program, in fact, provides simplex channel configuration where the traffic flows in one direction. In the case of cable modems, the traffic will flow in both directions, giving full-duplex channel configuration. It is expected that with the reconfigurations of CATV into full-duplex mode, data rates of 35 Mbps (downward) and 10 Mbps (upward) can be achieved. IEEE 802.14 has developed standards and specifications for the **multimedia cable network system (MCNS)**. Europe uses DAVIC and DVB standards.

A number of manufacturers are involved in the development of components for cable modems, including Analog Devices, Inc. (chipset), Hybrid Networks, Inc. (wireless modems), and Wantweb, Inc. (high-speed Internet access in microwave band). It is hoped that data can be transmitted in different frequency bands, e.g., downward (from cable to Internet service provider to user) in the band of 42–850 MHz and upward (from user to cable to Internet service provider) in the band of 5–42 MHz. The U.S. uses the band of 5–42 MHz for upward, while Europe uses a band of 5–65 MHz.

The ADSL modem, on the other hand, provides much higher data rates than ISDN. Here also, Telco suggests a higher data rate by removing the coils from unshielded twisted-pair cable (being used between the subscriber's set and end office). ADSL modems offer a higher data rate for downward but a lower data rate for upward communication. The telephone link is used for both data communication and the telephone system at the same time. The modem is connected between the twisted pair and end office. On the receiving side, it uses a plain old telephone service (POTS) splitter, which provides two different paths on the same link between an ADSL modem and the user's PC and between a splitter and telephone system.

Hughes Electronics, Inc., offers high-speed Internet access over its Hughes' Galaxy IV geosynchronous satellite system. The downward (from satellite to user's PC) data rate is 400 Kbps. The user has to go through the Hughes Satellite control center via PSTN (as is done in a normal course for Internet access). The request for any file or program is sent to the satellite, which then relays it to the satellite receiver. The files are then downloaded from the receiver onto the user's PC. The downward communication uses a frequency band of 11.7 to 12.2 GHz. The data communication in this frequency band is transmitted using TDM technique.

There are a number of vendors that are developing modem analyzers. The HP 79000 ADSL test station is being introduced for testing ADSL standard ANSI T1.433 and POTS standards ITU T0.133. This analyzer tests both local modem (end office ADSL transceiver ATU-C) and remote end ADSL transceiver modem ATU-R.

As we know, there are three operating systems defined for digital VAX: VMS, Ultrix, and OSF/1. The Ultrix and OSF/1 are basically versions of UNIX. There is no single version of TCP/IP for VMS. About five versions of TCP/IP exist and all of these versions offer a different set of services. The VMS offers its own network known as DECnet and defines a file management system (RMS). The VMS uses DEC's standard mail system or another compatible mail system or even an All-in-1 office automation mail system.

Mac users, on the other hand, communicate with each other via a built-in mail system, AppleTalk. The Mac Internet software for TCP/IP at the lower level is known as MacTCP, and the official product is known as TCP/IP connection for Macintosh. The two most important protocols defined for MACs are the Internet's TCP/IP and Apple's AppleTalk, and three types of hardware schemes have been defined as LocalTalk (or Phonenet), Ethernet (any version), and token ring. The Mac can handle AppleTalk on LocalTalk, Ethernet, or token ring and also TCP/IP on Ethernet or token.

Internet users access information from the network, where the information can be sent either by people or by the computer itself. The Internet allows users to access a variety of resources (typically stored information); e.g., a user can access information about space technology (past, present, and future), holy scriptures, court proceedings, musical lyrics, and much other stored information. Other popular, useful, and widely available services offered by the Internet are electronic mail (e-mail), bulletin boards, file transfer, remote log-in, shareable databases, programs and files, indexed programs, and so on. The Internet allows users to connect their computers either directly (to access these resources corresponding to the above-mentioned services) via application protocols such as Telnet, FTP, etc., or via an access system (defining dial-up connection). For access to limited applications, an account on the Internet may allow the users to access the services via some terminal emulator which can be dialed in from the user's computer. After the user has logged on to the computer, the user can retrieve files or programs. Each of these services is available as a remote service to the users on the Internet connection.

According to Global Reach (a San Francisco–based firm), English is the native language for only half the online population. The researchers at Angus Reid Group in Vancouver, British Columbia, Canada estimate that the U.S. online population recently dropped below 50% of the world's total for the first time. The Internet Engineering Task Force (IETF), a standards committee composed largely of U.S. citizens, is looking at alphabets with accent marks and non-Roman characters. It is estimated that by 2005, most of the Internet will be in languages other than English. At this time, the number of online users using English is about 72 million users, while it is expected to rise to 265 million by 2005. Within the same period, non-English, Chinese, Spanish, and Japanese will go from 45, 2, 2, and 9 million to 740, 300, 100, and 70 million users, respectively. More information can be found at Global Reach's Web site, *http://glreach.com/globstats*.

17.2.3 Implementation of Internet services

Users share resources over LANs and WANs using two independent application programs running on different machines known as client and server programs. The client–server computing model is an application of distributed computing systems where one application program (client) interacts with another application program (server). The client basically is concerned with user interface and is required to provide the following functions: translating the user's query request into appropriate commands, sending the command to the server, waiting for the response from the server, translating the response into the user's interface on the screen, and presenting the results to the users. The server, on the other hand, accepts the user's query (expressed in SQL), processes it, and returns the results to the client using SQL commands.

The server program provides services for the resources, while the client program allows the users to use them. For any time-sharing application, the client–server implementation concept is used, and we may have different servers for different applications, such as file server, disk server, print server, newsgroup server, Gopher server, etc. In each of these applications, the server program offers the services to the request coming out from the client program. The client program also provides the interface to our machine (i.e., it interprets the keystrokes, displays the menu (via keyboard or mouse), and allows users to issue command requests from their keyboards). The client–server computing model offers various advantages; we discuss the main ones here (for more details, see Chapter 20). It offers a flexible interface development which is independent of the server hosting the data. This means that the users may work on Macintosh, define their own user interface, and access the data from a server which has its user interface on the mainframe. Different types of systems and user interfaces are supported by a central server.

The development of a user interface in the client–server computing model is the responsibility of the users, and the servers can be used for processing queries and sending the results back to the clients. The computing and storage capabilities of mainframes are enormous, so they are typically used as servers, but for limited and less-demanding applications, a PC or Macintosh can also be used as a server. One of the main features of a client–server computing model is that here any application which is to be developed is considered a task, and this task is running over the client and server in a cooperative way.

The X-Window-based machine under the UNIX system offers various advantages when used for accessing the Internet. X-Window supports graphical user interfaces (GUIs) which allow users to use computers via a keyboard or mouse or any other pointing device. X-Window offers a standard set of tools for designers who develop graphical applications and interfaces for interacting with these applications. In X-Window (or X-system), the X-server maintains all GUIs, and the user interfaces are well defined and standardized. The window manager is an application program which provides the "look and feel" of the interfaces and runs on top of X. Two widely used window managers are **Motif** (Open System Foundation) and **Open Look** (Sun). The X-system provides different application programs which move on different windows and are seen on the screen of the X-display. These programs run on either a local machine or a remote machine. The X-client program requests services for managing the user interfaces to the server program. The X-client may run on a different computer from the X-server. Typically, the X-server runs on the user's machine which is connected to our machine via network. The client–server configuration on the X-system provides various advantages, such as running of client programs on different machines; availability of many client programs on the Internet to copy, read, use, transfer, etc.; and use of various resource service package tools (such as Gopher, Archie, WAIS, and many others) as clients. Further, the X-system offers powerful user interfaces to these clients (in contrast to character-oriented clients) in non X-systems.

17.2.4 Internet services: An overview

As stated above, the Internet offers a variety of services to users, and most services in turn offer a large number of resources. All of these services are implemented using client–server architecture, in which two separate programs known as client and server run on different computers. These two programs interact with each other and establish connections between two nodes for time-sharing of resources over the Internet. Usually, the server program runs over a remote mainframe or workstation and provides services to users, while the client program makes use of these services. For each service, a server program is defined which can be accessed by a number of users simultaneously by running appropriate service application programs on their computers. For example, a *Gopher* client program running on a user's computer will make a request for a Gopher service which is being accessed by another Gopher program running on a remote mainframe. Another application program (a combination of over 100 protocols) which provides reliable communication between the nodes is TCP/IP. This combination also is based on client–server implementation. The client–server architecture offers various advantages, such as reduction of communication costs, faster response, simultaneous accessing of stored information by users, etc.

The Internet is heavily used for information sharing. Searching for particular information may be accomplished by a number of available tools (browsers or net surfers). These tools offer a listing of sources or URLs from which the information can be accessed (read, downloaded, printed, etc.). *Archie* uses client–server-based application software such as **file transfer protocol (FTP)** that looks at the databases to get the file listings and stores them into the database servers. A **wide area information server (WAIS)** provides the databases with keywords. Here we have to provide the keywords for searching a particular file/program/information.

Gopher was introduced by the University of Minnesota and is easier to use than FTP. It offers a user-friendly menu to enable searching for files/resources and provides connection to those sites easily. *Veronica* acts as a search engine for WAIS. WWW has become an integral part of the Internet, and a number of Web sites have been created by different vendors with information about their products, catalogs, and related information. Online service is another step toward getting information about a product, including pictures, specifications, data sheets, etc., instantaneously. With the advent of e-commerce, it is possible to perform online shopping, along with payment using credit cards. A variety of encryption codes provide protection against the misuse of credit cards.

Commercial online services are available to users for various applications including Internet e-mail service. Some of the online service providers are America Online, Delphi (smaller online system patterned after CompuServe; offers FTP and e-mail service), Prodigy (a centralized, commercial online service), CompuServe (oldest and very extensive online service), Genie (smaller commercial online service by General Electric), Fidonet (an international network composed of local computer bulletin board systems), MCImail (electronic mail network, usually used by corporations), Applelink (an e-mail message network by Apple), UUCP (similar to Fidonet; operates between computers running the UNIX system and uses UNIX-to-UNIX Copy Protocol — UUCP), and many others.

There are six main Internet services over which other options/facilities can be provided: **electronic mail (e-mail)**, **Telnet**, **Usenet**, **FTP**, **anonymous FTP**, and **newsgroups**. Although other books may describe different main services, we consider the above-mentioned services as the main ones and others as either extension options or build-ons. The following subsections discuss these services in brief. Details regarding the use of these services and the options/facilities available with each of them are given in Section 17.4.

Electronic mail (e-mail) service: E-mail service deals with the transmission and receiving of messages between nodes across the global networks (defined by interconnecting public network hosts and local networks). It offers a variety of options and facilities such as read, write, reply, save, broadcating, etc. This resource service is available on mail systems such as CompuServe, MCImail, and many other systems which have connections with the Internet. It allows users to send and receive messages (any type of data, e.g., text files, graphical pictures, programs, images, video images, pictures, etc.), electronic magazines, announcements, etc. The Internet mail system is the backbone for the network mailing service and, similarly, the mail systems of CompuServe and MCImail are the backbone for e-mail resources.

Telnet service: Telnet provides a session link (logical connection) between a user's computer system (PC, terminal, workstation) and a remote computer system (mainframe, workstation, server) across the Internet. After the session link is established, other services such as remote log-in, sharing of resources remotely, and many other services can be obtained on the local computer system, giving full transparency to users, who may feel as if these are local services available on their own computers. With a valid log-in and password, the user can use resources of the remote computer (connected to the Internet) using Telnet. For those users who do not have a valid log-in and password to access the remote computer, a public account has been created on the Internet for anyone to use the public services (such as weather, culture, announcements, news). It is important to know that this service protocol is one of over 100 service protocols defined by TCP/IP. It offers access to public services, libraries, and databases at a remote computer (servers).

Usenet (Users Network): The University of North Carolina started Usenet in 1979 to create an electronic bulletin board which can post articles and news messages, notices, etc., and also allow users to read them. In the beginning, only two universities (North Carolina and Duke) had Usenet, but now a large number of universities use Usenet, providing news and information to a large number of users. Each Usenet site has its own news administrator who manages the system.

Usenet is a system that collects messages from various newsgroups on a particular topic. The newsgroup is nothing but a database of messages on different topics. There are many newsgroup servers available, and these can be downloaded on a user's PC using reader software. There are two types of newsgroups: moderated and unmoderated. In the former, news on a particular topic is mailed to moderators who will store it on a server containing messages on that topic. In the case of unmoderated newsgroups, messages go straight to the newsgroup server. The Usenet hierarchy newsgroup is based on the concept of **domain name system (DNS)**, where different domains or group names starting from the right side are defined for major names followed by subnames, and so on. The newsgroups are organized into the same hierarchy by defining the major topic, followed by subtopic, and so on.

This resource service provides a platform for a group discussion on articles of different areas/categories across countries over the Internet. It is derived from the Users Network and does not define an actual network but represents a resource service to users worldwide. The Usenet discussion groups are available at various Internet nodes and are usually managed by individual system administrators locally. Each newsgroup is serviced by a server, and within each groups various subgroups are defined. The majority of the server's newsgroups are obtained from Usenet, a set of newsgroups available to a user globally, free of charge. The user, in fact, uses a program known as a *news reader,* which accesses various servers (such as news, program, database, etc.) and requests the item files in that category. The server collects this information from various sources such as Usenet, local news sources, mail reflectors, Clarinet, depending on the type of request. It allows users to read news, messages of newsgroups or bulletin boards, etc. For additional information, see References 1–5, 9, 12, and 13.

There are about 80,000 Usenet sites, and a very large number of users use Usenet. Not all of the 80,000 Usenet sites are on the Internet. The amount of data received by a news server can be measured in terms of the number of characters. The number of characters received by the news servers is very high (tens of millions) and is defined by different Usenet sites and users separately.

The information in a newsgroup is held for a predefined time and then discarded. Usenet includes the following seven newsgroups:

- **Comp:** This group contains information about computers and related topics.
- **News:** This group is mainly concerned with the news network and is composed of two main subgroups: news.newsusers.questions and news.announce.newsusers.
- **Rec:** This group deals with recreational activities such as penpals, hobbies, arts, etc.
- **Sci:** Information regarding science research and applications is available in this group.
- **Soc:** Social and other political or regular news is discussed in this group.
- **Talk:** Religion issues and controversial, unresolved, or sensitive topics are exchanged in this group.
- **Misc:** This group discusses any topic which has not been covered by previous groups.

File Transfer Protocol (FTP): FTP allows users to transfer files from one computer to another across the Internet. The files from a remote computer can be copied on a user's computer using FTP (downloading) and the files from the user's computer can be copied onto a remote computer using FTP (uploading). It is also possible to transfer files from one remote computer to another remote computer, and this can be done on the local computer. It is important to note that this service protocol is one of more than 100 service protocols defined by TCP/IP. It allows returning of files from the public domain, moving of files between various Internet nodes across the Internet, etc.

Different versions of files are available over the Internet, and we need to have an appropriate software package to access them; e.g., a PostScript file format defines the format of documents, formatting information used in desktop publishing. Software to open and print the PostScript files is needed on the user's PC. In order to print the files, the special high-resolution PostScript device driver for typesetting the documents is needed. An executable program is defined in binary language, and the source code for the program is written in terms of instructions using programming language (low-level or high-level). The source code is usually defined within ASCII file format and needs to be compiled into program files. The images are defined in GIF format, which can be downloaded quickly. For accessing the image GIF reader to read binary GIF files, the PC must have a video display card which translates the data into signals that displays the image on the monitor.

In order to save the bandwidth of the network and for faster downloading of a very large file, a compression technique is used to compress the file into a compact smaller size. It is then decompressed on the receiving side. An extension of the compression technique is added at the end of the file. There are a variety of compression techniques available: PkZip, ARJ, LHArc, Pak Zoo, BinHex, etc., and for MS-DOS; MacB, SelfExtract, Stuffit for Macintosh; and Gzip and unixcompress for UNIX systems. The UNIX command *Tar* combines a group of files and is defined as one file with an extension of .tar. The compression technique extension is added at the end of the files.

Anonymous FTP: As mentioned above, the FTP service allows the transfer of files from one computer (local or remote) to another (local or computer). Another application program service based on FTP is anonymous FTP (AFTP), which allows organizations to create certain files as public files. These public files can be accessed by any user free of charge using "anonymous" for user ID or login, and no password is required. Files can only be downloaded; users may not upload any of their files. This service is a very important feature of the Internet, as it can be used as a **distribution center** for created programs or files on the Internet. Many applications, such as discussion groups, technical discussions, electronic documentation, etc., are available on the Internet worldwide through AFTP for wide circulation throughout the world.

News group: Services for searching/browsing any information or catalog can be obtained using a variety of newsgroups (Gopher, News, WAIS, Archie, etc.). Each of these newsgroups is considered a client application program and corresponds to a software package tool. A particular client program of any tool displays a menu and, based on the selection from the menu, the client program connects the user computer to the appropriate server without regard to the location of the server as long as it is connected within that client program over the Internet. Gopher service offers a unique and uniform interface across the network. The Internet resources and various textual information can be accessed by it. The majority of the Gophers can be connected to each other, and users can also switch from one Gopher to another with the same interface.

It is important to note here that the Internet is a collection of networks and hence should be considered a network. Usenet, on the other hand, represents a **Users Network (Usenet)** and is not a network but is a collection of discussion groups which can be shared on the Internet. The Internet can be considered a carrier for information worldwide, while Usenet is one of the services offered by it. The Internet is not the only platform for providing Usenet services; others, such as UUCP on the UNIX-based system, also provide access to Usenet without going to the Internet. The Internet is faster and better, with the only disadvantage being high cost, which depends on geographic location. For example, in the U.S., most of these services are either free or very cheap. Although communication costs are getting lower, in some countries these services may still be expensive or even unavailable.

17.2.5 Other options/facilities/services on the Internet

As we said earlier, these six main services provide a platform for other options/facilities or even services to build upon it, and various Internet resources can be accessed via these services. A brief description of available resources is given below.

Wide Area Information Service (WAIS): This is another important service for searching the information and is based on indexed browsing. It allows the user to access a large number of databases and, after receiving keywords from the users, it searches for matches in the entire file or document and presents them to the user in menu form.

Archie: In the preceding sections, we discussed three categories of data traffic on the Internet as interactive terminal sessions (Telnet), file transfer (using FTP), and e-mail. A number of application programs have been developed to access these different types of data over the Internet. In order to access information from all of these types of data servers, an application program called *Archie* was developed at McGill University, Canada and is known as the Archie Server Listing Service tool. This tool is composed of two application programs: one provides various Internet FTP archive sites in a local server and the second allows users to identify the owner of a database.

The list of archive sites is updated regularly every month. It operates under both UNIX and VMS operating systems. A number of anonymous FTP servers are connected to the Internet, and these often contain very large numbers of files. The main function of the Archie server is to identify the location of the anonymous server which contains the particular file a user wants to access. Once this information is available to the user, the file can be downloaded onto his/her computer system using anonymous FTP, irrespective of the location of the server. Archie can be accessed by Telnet or remote log-in to *quiche.cs.mcgill.ca* (132.206.2.3), and one then looks at the directory *ftp/archie/listings* in the file *archie.mcgill.ca*. Archie also allows users to search archival databases and query the databases via strings. The result will be the size of the program and any additional information required. Three commands have been defined: **Site** (lists all anonymous FTP sites of databases), **List** (lists the sites which are tracked by Archie) and **Servers** (lists other Archie servers around the world).

Finger: As we know, every user gets a user ID or log-in name for accessing the Internet. This user ID or log-in name is not the complete name of the user, and this may sometimes cause confusion over widely used ID or login names. Various other interesting facilities/options such as mailing addresses, phone numbers, fax numbers, messages for selected groups, etc., are provided by the Finger resource service and can be accessed by the Finger service application program. This service can also identify computers which are active. In addition to these resources, the Finger application program service is also available as a public service for specific information over the Internet.

Other facilities available on the Internet include finding a user on a specific machine (type finger), the e-mail address of a user (type finger <e-mail address>), and access to the directory of a DDN network which contains the list of people working for the Internet and doing research work (type Whois). This list of DDN network can also be accessed by Telnet: *Telnet nic.ddn.mil* or e-mail: *main service@ nic.ddn.mil*. Similarly, the directory list of Usenet newsgroups can be accessed by an MIT server, which contains the list of newsgroups which are hosting news worldwide. To obtain this list, contact *mail.server@pit-manager.mit.edu* and then send Usenet-address/search-string.

As described above, the majority of client application programs corresponding to services are provided to the users as client programs. The client program requests a particular resource, and it is provided by the server program. Both client and server programs may run on different machines. The Finger client program allows users to get the following information about any user: userid, name, if currently logged in, last time logged in, phone number, office location, etc. A list of users currently using the Internet host and the ability to communicate with other Internet hosts are other services offered by the Finger program. To use the service, the Finger program must be run over the user's machine: Finger <userid>.

Another way of using Finger service is to indicate the address only. The finger server at the remote host will display the names of all users who are logged in on finger@address. This service can also be used to find out about a computer connected on the same local area network as our computer: Finger@computer. The command Finger will tell only about the user's own computer.

Fred: The X.500 directory system follows a top-to-bottom tree structure for its hierarchy. Various domains are defined at various levels. At the top, the world root directory may provide access to various countries. Each country further provides various network servers, which in turn, provide access to individual users. For example, the U.S. may offer news servers such as PSI, ANS, etc. For accessing the X-500 directory system (an OSI standard), another application program, Fred, is defined by NYSERnet and PSInet, and it is said to provide a better and simpler, friendly interface.

The server containing Fred is available at *wp.psi.com* and *wp2.psi.com*. For accessing Fred, we have to Telnet one of the above machines, and once we are connected, we have to use the users to search for a particular person in an organization by giving some clues or keywords about the name of the organization, person, employment, etc. Fred will return the list of persons/organizations which closely match the given clues. Fred can also be accessed by e-mail: mail whitepages@wp.psi.com.

Talk service: Talk service is a very useful resource on Internet Talk (client application program) being offered to users. Two users may be thousand of miles apart but can send and receive messages back and forth simultaneously (as if they were sitting in front of each other and talking). The two users communicate with each other by typing messages at the same time, and this does not affect each other's typing of messages. The Internet Talk program partitions the screen into two sectors, and the user can see the typed messages from the remote user on the top (or lower) sector while the local user's message will be displayed on the lower (or top) sector.

Internet relay chat: This is an extended version of Internet Talk service in the sense that here more than two users can participate in the conversation. It may be thought of as a telephone conference call among users sitting miles apart and talking to each other. It may be used either for a public discussion on various topics (open to all users) or for selected people in conference on any topic of their choice.

Veronica or Jughead: As mentioned above, there are a large number of Gophers connected to the Internet, and the Gophers may have different menus. Veronica is an application program which keeps a record of these different menus offered by different Gophers. Jughead is an indexing facility that indexes a particular set of Gopher servers. The application program (Veronica or Jughead) searches menus that contain keywords we specify for the Gophers. To access Veronica, we have to look for any Gopher server and may use the following command: search titles in Gopherspace using Veronica. The menu will give us a list of items to be selected. It acquires data similar to Archie. This client application offers a customized menu where, when a user chooses a particular menu, it is automatically connected to that Gopher (which offers that particular menu). This includes the information of all available Gophers.

World Wide Web (WWW): This application software was developed as the WWW project at the European Laboratory for Particle Physics, Geneva, Switzerland (CERN) in 1989 for defining a distributed hypertext and hypermedia system. It is based on **hypertext**, which allows users to select a particular word which may be linked to other documents and, in this way, a big logical link of the word data across the documents can be defined. The concept of hypertext offers the advantage that we can get more information about a particular topic or subject just by clicking on it. When that particular topic or subject is chosen, we can read the entire information contained therein. The documents can also be linked to other documents which may be on the same topic or subject but with a different author or source. If we look at articles published in technical journals, we usually find footnotes. With the hypertext system, we can select a footnote number in the body of an article and be transported to an appropriate citation in the Notes section. Further, the citation can be linked to the cited article and so on, and we can create an infinite link of citations which can also be backtracked in the same sequence in which it was formed initially.

The **hypertext transfer protocol (HTTP)** was originally defined to provide communication between Web servers and clients. This application program ran under Next computers and its use was limited to these computers only. In the meantime, the Gopher application program, which is available on different platforms, was defined and became more popular than the HTTP protocol.

In 1993, the Web came back with a better set of protocols which ran over a variety of platforms. This new version of the Web was developed at the National Center for Supercomputing Applications and defined as **Web client** and server application programs. The server application program (httpd) is available on UNIX-based systems and Next computers as well, and the servers can be used at different sites. The client application program (NCSA Mosaic for the X-window system) supports graphics, WAIS, Gopher, FTP, and other accessing techniques.

Due to the overwhelming response for the Web, IEEE has created a discussion group at *Web-d@ieee.org*. Interested readers may send their queries there.

The terminal-based client (Lynx) was developed at the University of Kansas and allows the VT100-based terminals to access the Web. With the successful introduction of Macintosh amd Microsoft Windows versions of Mosaic, the Web has bounced back into the Internet arena. Now we have four popular client applications to access the Web: NCSA Mosaic for Microsoft Windows, Lynx, NCSA Mosaic for X-Window systems, and MacWeb.

The Web offers access to the services of Gopher, WIAS, FTP sites, and Usenet. It also allows users to access information available in hytelnet, hyper-g, techinfo, texinfo, man pages, hypertext document, etc. It is important to know that WWW is the only client application tool which requires only two commands (**Follow a link** and **Perform a search**), compared to a number of commands used in other client application tools. This service is set up to help users find information they need. The cross-linking of Web services allows many entries to reach the same specific area with different starting points in different documents. The hypertext concept used in this service allows us to produce articles which can be made available on compact disk (CD) or over the network via the Web. The articles available on CD can be read on our PCs using CD-ROMs. In order to access the same information over the network, users can access it over an Internet access connection and by using client software like Mosaic on their PCs. The retrieving and searching of information on CD can be done either under Microsoft Windows or MS-DOS. The CD installation requires roughly three Mb of memory and allows two versions for retrieving the information from CD: Microsoft Windows and MS-DOS. This software includes 144 bytes for error detection and error correction code for every one Kb of data. This addition of codes and header information is known as *premastering*, and it remains the same on CD whether it is of computer or of music.

On a traditional tape (audio cassette tape), the data is read magnetically. A CD recorder writes data along a spiral track made up of tiny pits and lands which change the reflectivity of the disk's surface. The CD player uses a laser beam to follow the track and read the data encoded in the transition between these lands and pits. The silvery alumunium plastic discs can last for years and are immune to magnetic fields and also to sunlight (to some extent). The disc can be played thousands of times without losing quality and contents of the disc. The only drawback of this system is that it cannot be rewritten.

In a typical networking environment, every academic institution, organization, and company has its own Web site where all the relevant information about it is maintained. For example, the University of Idaho has maintained a Web site which provides general information, admission procedures, various academic departments, degree and course work needed, faculty members' resumes and links to their personal home pages, course schedules, Engineering Outreach course schedules, various campus branches, links to other universities, a list of other search engines, etc. For convenience, students can register for a course, withdraw from a course, view course grades, and utilize many other categories of information. Microsoft's Internet Explorer and Netscape's Communicator are the leading browsers which allow users to surf the Web (search for particular information), while search engines such as Yahoo!, AltaVista, Lycos, and many others allow the users to find sites that have the desired information. The information stored on a Web site uses a programming language known as Hypertext Markup Language (HTML). Another language, Java, is also becoming popular. These two programming languages support different types of information, such as video, data, images, graphics, etc. There are a number of Web editors that can be used to construct a home page. A number of search engines also assign an e-mail address, such as *userid@yahoo.com*. Further, they may assign limited memory on their servers for a customized home page.

White Pages directories (WPD): This client application program runs on special-purpose servers which allow users to search for the name of an Internet user, resource, etc. At this point, we don't have any central source for names, addresses, and other information of Internet users. The servers have limited White Pages directory service, and if we can find somebody's name from it, various services such as e-mail, Talk, Finger, and other services can be performed on the user's name. This service may be considered similar to the telephone directory (Yellow Pages). There are a large number of WPDs on specialized servers available to choose from.

The directory system consists of names, addresses, and also resources such as databases of organizations, corporations, government agencies, etc. The **directory server agent (DSA)** maintains the White Pages. There is another program known as the directory client agent which keeps a record of all DSA sites and also allows the user to search the resources maintained by the DSA. Telneting to the Whois database or X.500 DCA can find the name and address of a person. The Whois system keeps a record of names and addresses.

Electronic magazines (EM): This is yet another interesting client application program service offered to Internet users. The electronically published magazines contain articles as text files which can be accessed on the Internet. These magazines can be of either specialized or general categories. This service saves a lot of paper and offers only the relevant articles without any advertisement, thus reducing transmission time over the links (in contrast to facsimile). It also saves time in reading the main articles of interest, and it also allows users to select the type of articles they want.

Mailing list: This is a very useful and interesting client application program service offered by the Internet. Here, a mailing list of people interested in a particular area/topic may be defined. This list will receive messages pertaining to that topic. Similarly, a mailing list can be defined for professional organizations (IEEE, ACM, etc.). In fact, IEEE has already defined a mailing list of all the section offices based on the regions (it has divided the world into ten regions). This mailing list may be used to relay messages pertaining to professional society activities to all of its members.

Bulletin board system (BBS): This client application program service is another Internet service which allows users to share information, messages, and files over the Internet. We can select items from a series of menus defined on the Internet. There are quite a number of BBSs in the world. Usually the BBS is managed and controlled by a single organization/person. Most of these systems can also be accessed by Telnet. A new BBS can be established in any organization for providing courses and information; one such example is the IEEE Student Branch at the University of Cincinnati.[7]

Games: There are a large number of client application program games which can be played on the Internet (e.g., chess, bridge, game of diplomacy, etc.). Users can also play these games on their computers by downloading them (by using anonymous FTP) and executing them on their computers. The games are usually self-explanatory, and various commands are displayed through various menus/buttons.

Multiple user dimension (MUD): MUD is a client application program service which offers virtual reality (hot topics with added potential in the area of medicine,3-D images, space technology, creation of a village, house design, and many other applications). It allows users to use the virtual reality of a variety of applications such as a puzzle, create strange places with new buildings, bridges, houses, people, etc. of a particular MUD. To access MUD, we have to Telnet a MUD server and then select the virtual reality of a particular MUD.

17.2.6 Interconnection with TCP/IP

As discussed above, the Internet is a collection of computer networks and offers Internet services to users worldwide. The Internet uses a pair of protocols, transport control protocol (TCP) and Internet protocol (IP), for connecting various computers and networks. TCP/IP is, in fact, a large number of procedures (or protocols) which provide various services to users. These procedures actually define application programs such as Telnet, FTP, etc.

TCP/IP is used to provide the following services: transport between two end users, flow control, and routing for the messages being transmitted across the networks and computers connected to the Internet. It is interesting to note that TCP offers services similar to those of the transport layer, while IP offers the services of the network layer of OSI-RM. This combination of TCP/IP includes over 100 protocols which are used to provide interconnection between computers and networks. We have earlier discussed two of its popular protocols, Telnet and FTP. The TCP partitions the message into packets which are assigned the sequence number and address of the destination. It also includes error-control information for error recovery and error detection for the packets (during the transmission). The IP accepts the packets from the TCP, determines the route(s), and then transports them over the network to the remote destination. On the receiving side, the TCP receives these packets and checks for errors. If it finds any error in the packet, that packet needs to be re-transmitted by the transmitting site. If it does not find any error in the received packet, it will use the sequence number attached to the packet to construct the original message from the received packets.

The partitioning of messages into packets performed by TCP offers the following advantages:

- The same link can be used by the Internet to transmit messages of various users.
- Flexibility can be offered by the Internet in terms of routing and allowing the re-transmission of the packets containing errors only over the link until the packets are received error-free, which in turn depends on the type of protocols being used.

In a typical situation where our computer is connected to the Internet and we wish to send and receive data over it, a set of protocols is needed such as Telnet, TCP, IP, and protocols for LANs or other protocols for direct access of the Internet. Telnet breaks the data into a number of packets. This must run on both our PC and the host. TCP breaks the application data (already broken into Telnet packets) into TCP packets where each packet has a header with the address of the host, information regarding assembly/deassembly, and error control. The IP breaks the TCP packets into IP packets with a header with address information, routing information, and TCP information and data. These IP packets are picked up by subnets (LANs, backbone FDDIs, etc.), which break the IP packets further into appropriate the frame format and add their own address information.

It is possible that the IP packet may go through a number of subnets (e.g., from backbone FDDI to Ethernet of main division to the Ethernet of local department/lab and so on). Each subnet adds its own header. In the case where users are not connected via subnets, they can dial access to the Internet by using **point-to-point (PPP) protocol** or **serial line Internet protocol (SLIP)**. The link packets defined by subnets are transmitted over the transmission media (analog telephone lines, twisted pair, coaxial, optical fiber, RS-232, Ethernet, etc.). The receiving-side host will have the same set of protocols, which will unpack the packets at each layer using appropriate protocols.

A PC can be connected to a host via modem using client–server software, where it becomes a client that can use the graphical interface provided by the host. A PC can also run a terminal emulation program and allow itself to be a remote terminal to the host. *VT-100* is the most popular terminal to emulate, and this configuration provides very limited Internet services (ones provided by the host). A *dumb terminal* can also be used for Internet access by connecting it directly to the host. Its applications are usually found in libraries, government offices, and other organizations.

As said earlier, a PC can dial access to the Internet via PPP or SLIP protocols. These protocols, in fact, create a complete connection over telephone lines using high-speed modems. The PPP makes sure that the packets arrive undamaged and is more reliable than SLIP. In the event of a damaged or corrupted packet, it re-sends the packet. PCs and UNIX workstations can use either PPP or SLIP.

The Telnet facility can be used to access different types of servers such as file, database, Archie, etc. Once a user has logged onto any server using Telnet, the files of the server can be accessed. The Archie server searches all the FTP sites and keeps a record of all the information about the files. In order to access (download, view, print, etc.) the files from an FTP site, a user may either use the FTP program to be connected to that site or send an e-mail requesting the particular file. Archie can also be used to search other archives or Archie servers.

Bandwidth requirements: In order to provide connectivity with the Internet from our systems, there are a number of choices and requirements. Of these requirements, one of the most important is the bandwidth requirement. The bandwidth requirement further depends on a number of factors such as amount of data, number of users requesting services simultaneously, etc. For a normal e-mail, Usenet news, and use of FTP for small files via Web or Gopher, we may easily request a 56-Kbps connection line to the Internet and it may serve a few users (50–60) on a corporate LAN. The normal use of these services (without heavy use of graphics) may provide reasonable performance for this number of users. With a 56-Kbps line connection, a file of size one MB will take approximately three to four minutes for transfer over the Internet, provided there is one user working.

With increasing demands for bandwidth requirements and number of Internet hosts, many popular servers are using T1 connections for themselves. The 56-Kbps line connection is slowly being replaced by ISDN and frame relay link connections. One of the main problems with 56 Kbps is that this is not scalable. The ISDN offers a 128-Kbps (two B channels) bandwidth for a network connection. The telephone companies usually charge anywhere between US$30.00 to US$500.00 per month for ISDN connection. The ISDN connections may support up to 23 data channels, giving us nearly the same bandwidth that one could get from a T1 carrier. A frame relay connection is usually provided by a third-party network provider to provide connection between users and the Internet service provider. These frame relay connections are available in 64 Kbps and can go up to 512 Kbps. T1 connection (also known as DS-1) offers 24 channels each of 64 Kbps, offering a total bandwidth of 1.544 Mbps.

17.2.7 Internet standards organizations

Various activities and the future direction of Internet technology are handled by the **Internet Society (ISOC)**, which is a voluntary organization. The organization is composed of technical experts and other management personnel to discuss various standards, and it is controlled by the **Internet Architecture Board (IAB)**. It handles issues such as 32-bit address scheme, assignment of addressing, standards, etc.

Another voluntary organization, the Internet Engineering Task Force (IETF), has been created to provide a forum for users to give their opinions and suggestions. Based on these opinions, different working groups are formed to look into those areas and can make recommendations to IETF or even IAB.

The Internet has been an international network for a long time and offers services to many countries, particularly the allies of the U.S. and various military bases. The Internet provides services to more than 40 countries, and this number is increasing rapidly. With the ability of the Internet to communicate with OSI-RM-based protocols, the Internet will expand exponentially to many other countries. As stated above, the telephone line connection to the Internet is a major link for various applications, and the quality of the telephone system plays an important role in access to Internet services. The overall Internet hardware architecture consists of corporate networks connected to regional or national service providers via routers.

Service providers basically have collections of routers connected to each other over telephone lines and also to local area networks via routers. The LAN used can be one of the standard LANs (Ethernet, token ring, token bus, etc.). The Internet protocol takes care of Internet addressing and defines an IP envelope for each packet containing the source and destination addresses. Each address is of four fields, each containing a maximum of 256 decimal numbers. The IP envelope or packet may have the following format (details about the format are discussed Section 17.3):

From: 123.456.7.8
To: 987.654.32.10

The TCP makes a packet 1–1550 characters, and each packet is connected to an IP packet with an address as shown:

$$\text{TCP (1–1550 characters)} + \text{IP packet or envelope}$$

17.2.8 Internet connections

Section 17.2.6 describes interconnection with TCP/IP. Some of the connections to access the Internet are described below:

1. Large institutions, corporations, or organizations can request dedicated network access to the Internet and can get all the services offered by the Internet. A service provider assigns a leased dedicated telephone line at requested speed with a router sitting on the dedicated computer. This router receives messages from all the computers connected to this connection and sends them to the appropriate destination. Similarly, all the messages coming out from different users of that place are received by the router and are sent to the appropriate terminals or computers within that place. The computers are connected to the dedicated connection via LAN along with the router.
2. An Internet connection may be bought from the Internet administration at an inexpensive rate (a few thousand dollars), and monthly charges of a few thousand dollars are paid as the usage and maintenance fee of the Internet.
3. There are two methods of providing interconnection between Internet nodes (PCs, workstations, Macs, terminals) and the Internet. The first method provides Internet connections to PCs, Macs, and workstations, while the second method is basically for terminals or even other computer systems. The computer nodes (PC, Mac, or workstation) are connected directly to LANs or campus WANs, which are connected to the Internet, and each of these nodes (PC, Mac, or workstation) is known as an Internet host with its own e-mail address. The terminals can also be used to access Internet services by connecting to one of these hosts (PC, Mac, or workstation).
4. The computer nodes or terminals may be connected to LANs or WANs via either hard-wired connections or dial-up connections (over telephone lines). The dial-up connection requires a modem between computer and telephone lines, and modems are available with different data speeds. The modem converts digital signals into analog signals, and these are transmitted over telephone lines. On the receiving side, the modulated signal is demodulated to get the original signal. Modems with data speeds of 1.2, 2.4, 9.6, 14.4 Kbps and higher are available on the market. A modem with a data speed of at least 14.4 Kbps should be chosen.

   Modems are available either as a separate device or as an adapter card which is built into the computer. It seems reasonable to expect at least three times higher the data speed if compression techniques are used with modems. The computers can access a remote computer host over the Internet. In general, these Internet hosts are configured to provide connection to the terminal systems. In other words, these hosts can be accessed by terminals via modems over the telephone lines. On a PC, Mac, or workstation, a **terminal emulation program** must be run on the user's computer. This program enables the user's computer to communicate with a remote host, which treats this as a terminal. The standard choice of terminals has been VT-100 (now VT-102 is available) developed by Digital Equipment Corporation (DEC). It is common to see the following message on our computer regarding the terminal after we log in.

   Prompt sign > VT 100

If it is not seen, we can find out the type of terminal by looking at the configuration of our computer.

5. In the situation where a user's computer is being used as an Internet host for obtaining a full Internet connection over telephone lines, we have to install an **Internet application program (IAP)** known as **point-to-point protocol (PPP)**. The TCP/IP facilities can be used by the user's computer after the telephone connection between the user's computer and the Internet host has been established. Now the computer can read the Internet host and can be identified by its electronic address. Another Internet application program which allows the computer to read an Internet host is the **serial line Internet protocol (SLIP)**.

   There are a few versions of Internet application programs (PPP and SLIP) available which use high-speed modems (9600 baud) and run over the telephone lines. Although a 2400-baud modem can also be used, it will be very, very slow. The ideal modems would be V.32 bis or V.42 bis (CCITT). These application programs are much cheaper than dedicated telephone lines (connection) and can allow users to dial in to the network to access the Internet, thus making the telephone lines available to other users. These application programs allow users to connect their computers (home or work) to a LAN, which in turn is connected to the Internet, thus enabling full access to the Internet services. The users can also be connected to service providers such as **UUNET (UNIX–UNIX Network)** or **Performance Systems International (PSI)**, which usually charge a nominal fee (a couple of hundred dollars) per minute for offering SLIP or PPP services. Further, the service providers are providing these services at cheaper rates via toll-free numbers (i.e., 800) or local access numbers in important cities.

   These application programs (PPP and SLIP) can run on a user's computer which may be connected to the Internet host via either a dial-up line (with modem) or a dedicated line. In the case of dial-up lines, the number (usually assigned by the computer center for setting up the host) needs to be dialed for connection, while in the case of dedicated lines, the connection is made permanently. The dedicated telephone line (acquired by any organization or university) using a PPP application program will provide Internet access to all the computers connected to the network (line). This way, Internet access can be achieved by all the computers over a single line as opposed to multiple lines (expensive).

6. A PC or terminal may access a remote computer or a list of resources using application programs (Telnet) and retrieve files or records from the directory/database of a remote computer via FTP onto the user's computer across the Internet. This type of configuration is known as an *Internet connection.*

7. Another way of accessing these remote services is via an access system which offers a dial-up Internet connection. In this configuration, we can read or write electronic mail or read any file, but the electronic mail cannot be sent from a PC at home or at work until it is connected to some remote access point via modem. The files can be read and saved at the remote system but cannot be saved on our computer directly. If we want to save the file on our computer, we have to first save the file on the disk of the remote system; then the communication program resident on our desktop or PC will move the files onto our disk or memory. Similarly, the fetched files from the Internet public domain can be moved from the access computer system to our PCs. If we have an access right for CompuServe or Bitnet (Because It's a Network — a public network), we can send electronic mail to the Internet, read Internet bulletin boards (also known as news), and access much other information. If we are running UNIX on our PC, we may use the UUCP command to use electronic mail and news.

17.2.8.1 Internet connection and service providers

The majority of organizations and companies and a large number of universities of different countries around the world have access to the Internet, and Internet services are available to a large number of users. Although most users have Internet connections either from the university community or from an organization, there are other ways for users to access these services. Connection to the Internet can be obtained on a public network known as **Freenet**, which is a typical community-based library and e-mail computing system to access Internet services. There are quite a number of Freenets in the U.S., Canada, and European countries. Access to the Internet can be requested from these **Internet service providers (ISPs)**, which offer a variety of Internet services with some constraints on the time of use. These Freenets are controlled by the service providers. Some of the material herein is partially derived from References 1, 2, 9, and 12.

It should be noted that these Freenets offer very limited services such as mail, Usenet discussion groups, etc. They do not provide full Internet access, and many services such as Gopher, FTP, and others are not available, as they do not provide connection to the remote Internet. In order to manage these Freenets (belonging to the service providers), a nonprofit organization has been set up called the **National Public Telecomputing Network (NPTN)**. This is a coordinated body for Freenet organizations worldwide and offers help on starting a Freenet. The NPTN maintains a list of existing Freenets and organizing committees on anonymous FTP host **NPTN.org** in the directory *pub/info.nptn*.

These nets are usually connected to a university or organization computer service and can exchange the e-mail messages with Internet Usenet. It is important to note here that the Internet users from universities, organizations, or companies can also access these Freenets via Telnet if they so desire. Some of the service providers include UUNET, PSI, MCI, etc.

Various services of the Internet can be accessed by different service providers, and there is a big market of service providers around the world. The service providers can be classified as regional, national, and international, and they offer services within their premises (based on their classification). A partial list of international service providers is given below.

Service provider	Type of services
Advanced Network and Services (ANS) (*Maloff@nis.ans.net*)	Dedicated (1.5–45 Mbps), all T series: T1 (1.544 Mbps), T2 (6 Mbps), T3 (45 Mbps)
Performance Systems International (PSI) (*info@psi.com*)	PSI link dial-up, SLIP/PPP, UUCP dedicated (9.6 Kbps to 1.5 Mbps)
Sprint Link	Dedicated (9.6 Kbps to 1.5 Mbps)
UUNET	SLIP/PPP, UUCP, dial-up, dedicated (9.6 Kbps to 1.5 Mbps)

In addition to those international service providers, there are some national service providers that also provide services at the international level, and some of these are listed below.

Service provider	Type of services
AARNET (Australia; *aarnet@aarnet.edu.au*)	Dedicated (9.6 Kbps to 2 Mbps), SLIP/PPP
a2i Communications (U.S.; *infor@rakul.net*)	Dial-up
Demon Internet Services (U.K.; *internet@demon.co.uk*)	Dial-up, SLIP/PPP
EUnet (Europe; *glen@eu.net*)	
Vknet (U.K. countries; *Postmaster@uknet.ac.uk*)	Dedicated, dial-up, UUCP
PACCOM (Pacific Rim countries, Hawaii; *Torban@hawaii.edu*)	Dedicated (6 Kbps to 1.5 Mbps)
The World (U.S.)	Dial-up
The Well (Access via XVII.25 and dial-up; *info@well.sf.ca.u*)	Dial-up

The cheapest way of accessing the Internet would be to request an account on some computer host which is connected to the network (offering dedicated access to the Internet). This technique allows you to log on to this computer (considered a remote) and use the Internet services which are available to the network connected to the Internet by the dedicated connection. The users share the connection with other users of the Internet and also access Internet services at a very low rate. The above-mentioned service providers allow users to access Internet services via their dedicated connection(s). In order to access Internet services, the user needs a modem and **terminal emulation package** to access the network. The users are limited to the services allowed by the service provider, which are usually very cheap (around $20.00 per month). One of the service providers, PSI, has announced a new dial-up service software package called PSI-link free of charge to users. This software package allows the PC running DOS to connect to PSI's system to use the following services of Internet: e-mail, bulletin boards, news, file transfer services, etc.

The UNIX-based systems can allow users to access Internet e-mail and Usenet news via the UNIX service UUCP. This service transfers the data onto telephone lines allowing the users to first read the mail on their system. The PC uses UUCP for dialing a remote system and then provides the transfer of news and mail services on it at regular intervals. The UUCP basically requires a UNIX-based PC or terminal and modem only. No additional software or hardware is needed to access Internet services on UNIX-based systems.

So far, we have discussed the ways to access Internet services in a somewhat restricted or limited manner. The users can access only a few services of the Internet, and most of the networking services like Bitnet and CompuServe usually include gateways, which in fact allow the users to exchange e-mail with the system on the Internet. Some gateways allow reading of Internet bulletin board (Usenet news) services only. The retrieving of a file involves fetching the file and then mailing it to the users automatically on the user's system.

The accessing of the Internet requires a telephone connection which may be either the ordinary voice link or digital link. These telephone connections offer different speeds and data rates and support different software packages for connection establishment. For example, ordinary standard telephone lines offer a speed of up to 19.2 Kbps and support SLIP or PPP or dial-up connections, while dedicated leased lines offer speeds of 56–64 Kbps. The digital links (T-series: T1, T2, T3) offer much higher data rates and are used as dedicated links carrying high volumes of data. The data rates of T1, T2, and T3 are 1.544 Mbps, 6 Mbps, and 45 Mbps, respectively. Out of these three, T1 is widely used and accepted (although it is not a standard), while T3 is generally used for large companies, organizations, and universities.

17.2.8.2 Internet mail system (IMS)

The Internet mail system is a very important and useful Internet resource which provides transportation for various types of information, documents, programs, databases, queries, publications, personal and official letters, etc. A standard program has been defined as a part of TCP/IP which can accept the messages, format them, and transmit them to the destination across the network — the **simple mail transfer protocol (SMTP)** and **multipurpose Internet mail extention (MIME)**. SMTP offers basic electronic mail services while MIME standards add multimedia and graphic capabilities. Protocols is a part of the system's transport system (TCP) and cannot be accessed by users. The user uses another application program known as a user agent (UA) which provides interface with the Internet mail system. This user agent allows the user to read the message, create new messages, reply to messages, save messages, delete messages, and many other options. The transport agent (TA) program within TCP/IP is responsible for providing orderly transportation of the messages. The TA program is known as **Daemon** in the UNIX operating system.

There are different UAs used on the Internet. The UNIX mail program is the most common UA in the UNIX environment. Other UAs which have been defined and are used in appropriate platforms include Elm (full-screen program), Pine (simple program), Message Handler (MH; a single-purpose mailing program), **Mush** (Mail User's Shell), and **Emac** (a complete UNIX environment for editing the text).

As stated above, the LANs are connected to form WANs which are connected to the Internet. The LANs provide shareable services to users connected to them via PCs, Macs, workstations, or terminals. Each LAN has to run one of the UAs for the mailing system, e.g., Novell's Message Handling Services (MHS), Pegasus (P-mail), and Post Office Protocols (POP). The P-mail runs on both PCs and Macs and is very popular among Internet users. The POP runs on PCs and Macs. This protocol gets the messages on PCs or Macs for the users to read from. This protocol offers various advantages, such as complete control over stored messages, use of graphical user interface, etc. Some of the POP-based UAs include Eudora (Mac-based), NuPOP, and POPMain/PC.

17.3 Internet addressing

As stated above, the Internet is defined by interconnection of different networks, and these networks are connected by gateways. Within the Internet, the corporate networks or LANs (Ethernet, token ring, or even telephone lines) are usually connected by routers. The routers are, in fact, dedicated computers which connect the networks and provide routing for the messages among the users based on the **store-and-forward** concept. In this scheme, the message is received by the node processor and it stores it in its buffer and searches for a free link to another node processor on its way to the destination. It sends the message to that node processor, which again searches for a free link, and this process of storing the message and then forwarding continues until the message reaches the destination node.

The **Internet protocol (IP)** networks carry packets of size ranging from 1 to 1500 characters in envelopes, and these may arrive out of order at the destination. The IP envelope includes source and destination addresses. For providing reliable transportation, another protocol at the transport layer — transmission control protocol (TCP) — is used in conjunction with IP. This protocol breaks the data packets into smaller-sized packets and defines envelopes for each of the packets. One of the main problems with TCP is that it does not deliver the packets in the same order that they were transmitted and also does not address the issue of lost packets. Further, set-up time for TCP itself is time-consuming, and it introduces delay and overhead which affects the efficiency of the transmission. Some of these problems can be alleviated if we use another standard protocol (in conjunction with IP) known as **user datagram protocol (UDP)**. This protocol is useful for short messages and is very inexpensive. The packet envelopes are placed inside IP envelopes bearing the source and destination addresses. A checksum is used with TCP envelopes to detect any errors in the information data.

The Internet domain addresses are converted into IP addresses by an application program, Domain Name System (DNS), offered by TCP/IP. Internet addresses are usually different for different networks (e.g., UNIX-leased UUCP, CompuServe, MCImail, Fidonet, Bitnet). Each of these networks uses different commands for e-mail services. The UUCP network uses the **UNIX-to-UNIX copy (UUCP)** program, which is available on all UNIX-based systems. This program allows users to copy files from one UNIX system to another. The UUCP application program is a collection of programs and possesses fewer facilities and capabilities than TCP/IP.

At this point, we feel that the Internet and DNS offer advantages over UUCP in regard to Internet addressing where the user has to specify the address of destinations and the mail system automatically determines the routes for the messages.

A friendly environment can be created using a software package to access various Internet application services such as remote log-in, file transfer, electronic mail, etc. These applications are available via appropriate application programs sitting on top of the TCP or UDP and IP and are usually proprietary application programs offered by vendors. The **Network Information Center (NIC)** maintains a host file with addresses and distributes it to every machine connected to the network. The distribution of host files to a large number of users has become a problem, and the addressing scheme needs to handle long file names by using the **domain name system (DNS)**.

17.3.1 Domain name system (DNS)

The DNS defines subsets of names and attaches them to different groups at different levels. Each level is termed a *domain*, and domains are separated by periods. Each DNS consists of four domains. As indicated earlier, each Internet host has an electronic address, and this address is also known as the **Internet address (IA)**. This address includes the user ID and the name of the host computer, and these two are separated by the symbol "@" (pronounced "at"), for example, *g.hura@ieee.org*. Here the userid is *g.hura*, while the computer host is *ieee.org*. The computer host part after the symbol @ defines the domain. The combination of userid and computer host addresses will be unique and is very unlikely to correspond to more than one Internet user anywhere in the world.

The domain defined for the host address may have a number of sub-domains which are separated by a dot (period): *ieee.org*, *abc.cs.edu*. The sub-domain (separated by a period) defines the hierarchy within the host computer, and each sub-domain represents a particular type of computer, organization, departments within an organization, different schools within a university, etc. The sub-domains are always read from right to left.

The rightmost (top-level) domain may represent either the organizational entity (educational institution, commercial entity, or professional entity) or geographic entity (the name of the country). The entities associated with either organization or geography are usually abbreviated and always represent the top level (rightmost position in the Internet address). Table 17.1 is a partial list of organizational entities and Table 17.2 is a partial list of geographic entities for sub-domains which are being used in the Internet addressing scheme.

For typing the addresses, the Internet does not make any distinction between upper-case or lower-case letters. An Internet address typically has the following standard format:

userid@domain

The domain may have two (minimum), three, or four sub-domains, depending on the Internet host location and the format of the domain.

Table 17.1 Organizational Entities for Sub-domains in Internet Addresses

Domain	Represents
.com	Commercial organizations (i.e., AT&T, IBM, HP, etc.)
.edu	Educational organizations (institutions, universities, etc.)
.gov	Government organizations (federal agencies, state agencies, etc.)
.mil	Military agencies (Navy, Army, Air Force divisions of U.S.)
.int	International organizations (i.e., NATO, UNO, European Commission (EC), etc.)
.net	Networking organizations and resources
.org	Nonprofit professional organizations (i.e., IEEE, ACM, IEE, etc.)

Table 17.2 Geographic Entities for Sub-domains in Internet Addresses

Domain	Represents
.at	Austria
.au	Australia
.ca	Canada
.ch	Switzerland (Cantons of Helvetia)
.de	Germany (Deutschland)
.dk	Denmark
.nl	Netherlands
.es	Spain (España)
.fr	France
.gr	Greece
.ie	Republic of Ireland
.in	India
.jp	Japan
.nz	New Zealand
.uk/.gb	United Kingdom
.us	U.S.

To the left of the top-level domain (i.e., organizational), the next sub-domain represents the name of the university or organization. For example, if we use .edu as the top-level domain, the next sub-domain may represent the name of the university. But if a geographic entity (e.g., name of country) is used in the top-level domain, then the next sub-domain level represents the organizational entity, such as academic (.ac), company (.co), or even .edu.

The geographic entity is usually defined by a two-letter abbreviation for the country (Table 17.2). It is interesting to note here that Great Britain uses two different geographic entities: .gb for England, Scotland, and Wales, and .uk for Northern Ireland, England, Scotland, and Wales.

Consider the following Internet address:

g.hura@ABC.cs.edu

Here *g.hura* is the userid. The first domain represents the name of the host with the assigned IP address. The name of the computer is *.cs* for the computer science department, which represents the department where the computer is situated. This computer is a part of the University of ABC. *ABC* represents the national group of institutions (*.edu*). The domain *.edu* indicates the computers in all U.S. universities, the sub-domain *cs.edu* indicates all the computers at the University of ABC, and so on. The domain *ABC.edu* can have a different group name as, say, *MN.edu*; what it needs is to get a new name in the worldwide database. Some Internet addresses may have only two sub-domains.

Sometimes we see a percentage sign (%) in an Internet address. In such an address, the % sign is always on the left side of the symbol @, as shown below:

g.hura%Moon@ABC.edu

Here, the computer is defined by sub-domain *ABC.edu*, while the computer is *Moon* but is separated from the userid by the % symbol.

UUCP addressing: The UUCP does not provide remote log-in and is slower than TCP/IP, but it is a standard feature with all UNIX-based systems. It is available free of charge and supports both dial-up and hardwired connections. In the UUCP addressing scheme, the user has to define the entire route for the mail, and various computers through which the mail travels down to the destination are separated by a exclamation mark (!), as shown in the Internet address below. Here each name separated by the ! symbol represents the computers which are defined in the sequence for data communication.

g.hura!A!B!C

Similarly, if another domain such as *ABC.CS.edu* wants to add a new computer in the CS department, it assigns a name to it, e.g., *Moon.ABC.CS.edu*, and adds it to the network. The rules and procedures of this DNS guarantee no duplication of machine names. The universities in the U.S. and in many other countries have their last domain as *.edu*. But in some cases, we may add the geographic entity also at the top level before the organizational entity, e.g., *Moon.ABC.CS.edu.ca* (Canada). Similar to this, other organizations are identified by the respective domains.

17.3.2 Domain name system (DNS) servers

Most of the Internet uses the domain system for addressing. For this scheme, the computer looks at the local DNS server for the address. If the address lies in the local server of the database or a particular address has been requested, the local server knows the address. If it knows the address, then it certainly knows how to find this address and it informs the requesting user accordingly. The software looks at the root server which knows the rightmost domain of the address part (for example, *.edu*, *.com*, *.mil*, etc.). The root server, after finding this domain, searches the server for the address of the second domain (moving from right to left). For example, the CS server will be contacted by the root server (which finds the top-level domain as *.edu*). This server of the second domain (after checking .CS) then looks for the address of third domain server (e.g., ABC). Finally, this server (of the third domain) contacts the address of the last domain (fourth) server of the host, which corresponds to the destination of the application program. This host address (e.g., ABC) routes it to the appropriate users connected to it.

It is important and interesting to note that New Zealand and Great Britain are the only countries which have defined the domain part of the Internet address in reverse order. The mailing system accepts the following two formats of the IP addresses in these countries:

hura@uk.ac.oxford.comp or
hura@comp.oxford.ac.uk.

17.3.3 Classes of Internet addressing

We see that the Internet addresses are composed of names of users, organizations, countries, departments, etc. in different domains. But underneath the Internet, these addresses are converted into numbers rather than names. The Internet address expressed in number form is also known as the Internet protocol (IP) address, which is expressed in a domain format. Each host on the Internet is defined by a 32-bit Internet address. Each host is represented by a pair of addresses as **"netid"** and **"hostid."** The netid on the Internet corresponds to the network while the hostid corresponds to the Internet host connected to the network. The addressing scheme for the Internet is divided into four fields, with each field containing a number less than 256. Each field is separated by a period symbol. The Internet addresses can be divided into four categories: Class I, Class II, Class III, and Class IV. In each class, we have four fields, each of one byte or eight bits.

In **Class I**, the first bit of the first byte is 0, while the remaining seven bits define the address of the netid. The remaining three fields (24 bits) together define the hostid. There are a few networks of this category, and these can support up to 2^{16} hosts. The network bits indicate the number of networks. Class I supports 126 networks, where bits 0 and 127 are reserved for all 0s and 1s addresses.

In **Class II**, the first two bits of the first byte are 1 0. The remaining bits of the first byte and the entire second byte together correspond to the netid, while the remaining two bytes (16 bits) correspond to the hostid. There are intermediate networks which use this scheme supporting up to 2^{16} hosts.

In Class III, the first two bits of first byte are defined as 1 1. The remaining bits of the first byte and the other two bytes together define the netid, while the fourth byte corresponds to the hostid. These networks support up to only 2^8 hosts. The number of networks represented are between 192-223.

In **Class IV**, there are no bits for networks, but it can represent the number of networks between 224 and 254, while no bits are assigned for the hosts.

All the IP addresses are assigned by a central authority known as the Internet Assigned Numbers Authority (IANA). The network administrators can assign a unique address within any class of IP address, and the allocation of unique addresses by this authority seems to be inefficient. A new method using the concept of *classless interdomain routing* (CIDR) was introduced in 1993. It does not use the class system and is dependent on the concept of variable length subnet mask. Various protocols supporting this concept include OSPF, RIP II, Integrated IS-IS, and E-IGRP.

The next version of IP, known as IPv6 (**internet protocol-next generation: IPng)** is based on CIDR. CIDR defines Class A address space reserved for expansion (64.0.0.0 through 126.0.0.0). The IFET issued RFC 1752, a recommendation for IPng, in July 1994. It supports all applications of IPv4 such as datagram service, FTP, e-mails, X-Windows, Gopher, Web, and others. It defines flow label fields that support resource reservation protocol (RSVP) for confirmed bandwidth and quality of service. IPv4 is version 4 of IP.

The Internet address of *ieee.org* may have an IP address of 112.108.12.5. Each field, in fact, represents its equivalent binary number expressed in eight bits. The above address basically corresponds to:

01110000. 01001100. 00001100. 00000101

It is worth mentioning here that the IP address does not correspond to the Internet address. The standard formats for the Internet address and IP address are given below:

Internet address: Host address
IP address: Host IP address

The following two addresses of the same userid are valid:

g.hura@ABC.CS.edu (Internet address)
g.hura@132.45.78.2 (IP address)

We specify the host address, while DNS converts or maps this address onto the IP address.

17.3.4 Mailing to other networks

As stated above, the Internet is a collection of networks, and these networks are connected by gateways. Each of the networks has defined a different Internet-like address format to send the messages across the Internet. The address format of the mail gateway of a particular network (connecting to the Internet) needs to be converted by the users. The addressing schemes being used in different mailing systems supported by different vendors for some commercially available networks include CompuServe, MCImail, Fidonet, Bitnet, etc.

UNIX-UNIX copy (UUCP): The Internet domain addresses are converted into IP addresses by an application program, domain name system (DNS), offered by TCP/IP. Internet address schemes are usually different for different networks (e.g., UNIX-based UUCP, CompuServe, MCImail, Fidonet, Bitnet). Each of these networks uses different commands for electronic mail services. The UUCP network uses the UNIX-to-UNIX copy (UUCP) program and is available on all UNIX-based systems. This program allows the users to copy files from one UNIX system to another. The application program UUCP is a collection of programs and possesses fewer facilities and capabilities than TCP/IP.

At this point, we feel that the Internet and DNS offer advantages over UUCP in regard to Internet addressing, where the user has to specify the address of the destination and the mail system automatically determines the routes for the messages.

CompuServe: CompuServe defines the addresses as a sequence of digits into two domains which are separated by a comma. Consider the following Internet-like address defined by 1234.567. If we have to send e-mail to this address, we have to type the following address

mail > **1234.567@ compuserve.com**

It is important to note that in a CompuServe account address, the comma is replaced by a period symbol.

MCImail: An MCImail address is somewhat similar to the Internet, except that the userid may be either a sequence of digits or an actual name. For example, the mail will reach the same user with following two different addresses:

mail > **12431@mcimail.com**
mail > **ghura@mcimail.com**

Fidonet: Fidonet provides connection to PCs around the world through telephone lines and hence is a worldwide network of PCs. The domain address for this network is given by *fidonet.org*. In this network, the computer address is defined as three sub-domains which are separated by different symbols.

Consider the following address:

2:142/867

The first number (from the left) represents the zone number (2), while the second number (usually a pair of numbers) after the colon represents the net number (142) followed by the node number (867), which is separated from the second number by a slash. The general format of the addressing scheme here is given below:

zone:netnumber/node number
nnet.zzone:fidonet.org/fnode

If we are sending messages from the Internet to this address, then it is defined in reverse order, as shown below:

fnode.nnet.zzone.fidonet.org

The following address will be used to send e-mail to say, g.hura:

mail > **g.hura@f867.n142.z1.fidonet.org.**

Bitnet: Bitnet, whose name is an abbreviation of **Because It's Time Network**, is a collection of networks in the U.S., Canada, Mexico, and Europe. This is another worldwide network which connects computers from different universities, organizations, and industries of nearly 50 countries. This network is different from the Internet. Bitnet serves a large number of users over IBM mainframe machines which are located at various Bitnet sites. IBM mainframe computers were connected by this network in 1981. Bitnet is, in fact, now used for educational institutions and is managed by a nonprofit consortium of educational institutions known as **Educom**. This network is also used by various federal agencies which use duplicate data over the network. In order to avoid these duplications and other management issues, a unified **Corporation for Research and Educational Networking (CREN)** was formed to administer Bitnet.

Bitnet mailing list: It is interesting to note that Bitnet has different names in different countries — e.g., Netnorth in Canada, European Academic Research Network (EARN) in Europe, and other names in Asian countries.

It defines a pseudo-domain for Internet addressing which requires mailing software on our computer system which can recognize the address in this domain, rewrite the address, and send a message to the computer, which eventually sends the message to a particular network. The mailing address software needs to send the following message to the computer if the local mail software does not recognize the pseudo-domain address defined in Bitnet:

Bitnet/Internet gateway

If we have to send mail to say, g.hura, on Bitnet, the address looks like the following:

g.hura@computervm.bitnet

Usually in Bitnet addresses, the last two letters in the second domain are **vm**, which represents the standard operating system of **Virtual Machine (VM)** for the IBM mainframe. This indicates that in Bitnet, the mail program runs on IBM under the VM operating system.

Bitnet addressing is based on the UUCP path notation (i.e., includes the sequence of computers separated by an explanation sequence). These computers provide mapping for the commands between Bitnet and Usenet. The information about these gateways, Bitnet-oriented newsgroups, Bitnet mailing lists, and names of gateways can be obtained from the newsgroup hosts **bit.admin** or **news.answers**. These can also be obtained by **anonymous FTP, rtfm.mit.edu**, and directory **/pub/Usenet/news.answers/bit**.

Bitnet uses either of the following formats for its addressing scheme:

gateway!computer.bitnet!userid
userid%computer.bitnet@gateway

The Bitnet standard address usually follows the following format:

g.hura@host.bitnet
g.hura%host

As we discussed earlier, the Internet services are available over the network by a set of protocols and procedures known as TCP/IP. Bitnet, on the other hand, uses a collection of IBM protocols known as **remote spooling communication subsystems (RSCS)** and **network job entry (NJE)**.

The majority of Bitnet lists can be retrieved via either Usenet or mail. All the Bitnet lists start with **bit.listser** in the Bitnet hierarchy. The exchange of Bitnet lists between Usenet and Bitnet is provided by specialized computer Bitnet/Usenet gateways.

There are different kinds of mailing lists available, and each mailing list corresponds to a set of people in a particular area/topic. Such a list is usually known as a list-of-lists. The following is a partial list of the mailing lists, and we also describe how these lists can be accessed.

1. **SRI list-of-lists:** SRI International maintains a list of public lists offered by the Internet and Bitnet. It also provides a brief description of these lists along with the methods of subscribing to this list. This list can be downloaded by anonymous FTP, as follows:
   ftp.nisc.sri.com
   login:
   the file interest-groups.z in the directory/netinfo which describes the list of Bitnet and Internet mailing lists.
2. **Dartmouth list-of-lists:** This is another large list-of-lists (similar to SRI list-to-lists) managed and controlled by Dartmouth University. This list also keeps a record of both Bitnet and Internet mailing lists. The list can be accessed by anonymous FTP: *dartcms1.dartmouth.edu*.
3. **WAIS:** Both of the above services provide a complete list of mailing lists. Sometimes, we are interested for the mailing list of a particular area. This service allows users to search for mailing lists of a particular area/topic. We have to search for the mailing list source in the WAIS database.
4. **Internet list-of-lists:** As the name implies, this service provides a list of public Internet mailing lists. This list can be obtained by various Usenet newsgroups, such as news.lists, news.answers, or anonymous FTP, such as rtfm.mit.edu. Then look at the directory/pub/Usenet/news.answers/mail/mailing-lists for files part1, part2, etc.
5. **Academic mailing lists:** These lists include the names of various scholarly and academic subjects. They provide very useful information about various disciplines, majors, and other descriptions of the subjects and can be used by both instructors and students. They can be obtained by anonymous FTP, such as *Ksuvxa.Kent.edu*. Download the files acadlist.readme from the library. The various parts of the file can be seen as acadlist.file1, acadlist.file2, and so on.

17.4 Detailed descriptions of Internet services

The Internet is being used heavily for information sharing. A number of tools (browsers or net surfers) are available to assist in searching for particular information. These tools offer a listing of sources or uniform resource locators (URLs) from which the information can be accessed (read, downloaded, printed, etc.). Some useful services are mentioned here again for ready reference.

Archie uses client–server-based application software such as file transfer protocol (FTP) that looks at the databases to get the file listings and stores them in the database servers. The wide area information server (WAIS) provides the databases with keywords. Here we have to provide the keywords for searching a particular file/program/information.

Gopher was introduced by the University of Minnesota and is friendlier than FTP. It offers a user-friendly menu to enable searching for files/resources and provides connection to those sites easily. Veronica acts as a search engine for WAIS.

WWW has become an integral part of the Internet, and a number of Web sites have been created by different vendors with information about their products, catalogs, and other related information. Online service is another step toward getting information about a product, including pictures, specifications, data sheets, etc., instantaneously. With the advent of e-commerce, it is possible to perform online shopping, along with online payment using credit cards. A variety of encryption codes provide protection against the misuse of credit cards.

Microsoft's Internet Explorer and Netscape's Communicator are the leading browsers which allow users to surf the Web (searching for particular information), while search engines such as Yahoo!, AltaVista, Lycos, and many others allow users to find the sites that have the desired information. The information stores on a Web site use a programming language known as Hypertext Markup Language (HTML) that we have to use whenever we visit a particular Web site. Another language, Java, is also becoming popular. These two programming languages support different types of information such as video, data, images, graphics, etc. The following subsections discuss useful Internet services in detail.

17.4.1 Electronic mail (e-mail)

This service operates on telephone lines at data speeds of less than 64 Kbps, and the delivery time for e-mail includes two time components: the time a network takes to deliver the e-mail to the user's PC or terminal, and the time the user takes to read the message on the computer. E-mail service is asynchronous in nature and requires an address of the particular user on the computer. The Internet address and the domains are defined by an Internet administrator, while the e-mail address uses the symbol @ (at) to separate the user and host addresses. For example, the e-mail address *hura@ABC.CS.edu* indicates the userid as hura, who is logged in to a computer host with a domain of *ABC.CS.edu*.

There are three regions of a mail message: header, body, and signature (optional). The body includes the actual text of the message. The signature includes the user's name, address, phone number, fax, e-mail, etc. The header includes a few lines of information before the message. In the following, we describe the meaning of various lines in the header.

The first line of the header indicates the source:

From: <email address> Day Date Time Year

The lines below this line represent various systems through which the message has gone and may come in different order. The number of lines depends on the particular system's configuration and also the mail program being used. We see the following lines with appropriate headings: Data line, From line, Message line, and X-mailer line. After these lines, we will see these lines: To line, Subject line, CC line, and Status line. A typical complete header may look like this:

```
From: <email address> Day Date Time Year
Received: from <address>
   id . . to, day date
Received: from <e-mail> by
   id . . Day, date, time, year
send mail/2.0 SUN
   Day PST for <e-mail>
Received: by <name of the host>
   id . . day, date, time, year
```

Received: tsi.com by <email>
id . . day, date, time, year
sendmail/SUN via SMTP
Date: Day Date PST
From: <email address>
Message: ID <@tsi>
X-Mailer: Mail User's Shell (7.2.14 date)
To:
Subject:
CC:
Status: RO

The Received lines show the path, time, dates, and the application programs being used at each step. The different user agents used in the mailing systems are also mentioned in these lines. If we look at the Date line, it gives the day, date, and time. This time represents the time that the message was received by that particular location. The Internet uses **Greenwich Mean Time (GMT**) as a standard time. The **Universal Time (UT)** is also used. Other time standards available are listed below:

UT: Universal Time (similar to GMT)
GMT: Greenwich Mean Time
EST: Eastern Standard Time
EDT: Eastern Daylight Time
CST: Central Standard Time
CDT: Central Daylight Time
MDT: Mountain Daylight Time
PST: Pacific Standard Time
PDT: Pacific Daylight Time

The Message_id line assigns a unique identification tag. The X-mailer line indicates what user agent program is being used by the sender. For example, in Mail User's Shell (MUSH), the CC line allows the user to send a copy to himself/herself (carbon copy). The user may hit return if it is not required. The last line of the header indicates the status of the mail (R — reading the message, RO — message is already Read and now it is an Old message).

The Internet mail system allows users to send a text file (containing characters, numbers in documents), ASCII file (using ASCII code), or a binary file (containing a picture with no characters in it). The simple mail transfer protocol (SMTP) sends text files, while another protocol known as multipurpose Internet mail extension (MIME) is used for binary files. Typically, the binary data is stored in a file which is attached to a regular text file. Support of MIME will send this as a text file over the Internet. We may also record our voice as a binary file, which can then be played back.

Sometimes, the e-mail address may have a different interpretation on a different machine, and we have to make it look like an Internet address. To make this point clear, we take the following Internet service providers and look at the address scheme being used by them carefully.

- The **Bitnet** standard address usually follows the following format:
  g.hura@host.bitnet
  g.hura%host
  Note that in order to compare different types of mail services, we discuss Bitnet here, although it is not being used nowadays.

- The **CompuServe** address is divided into two parts separated by a comma, but this is changed to a period and is used in the left-hand side of the address. For example, *g.hura* would be represented as *g.hura@CompuServe.com*.
- The **MCImail** addressing scheme defines both address and user's name. The gateway name is written on the right-hand side, while the given name is written on the left-hand side, as shown below:
  first name-last name@mcimail.com
- The **UUCP** addressing scheme has different forms, such as *!uunet!host!name* and *Name@host.UUCP*. In the first scheme, the UUCP path is given. This indicates that the mail is for the system vunet which will send it to the host. We have to change this to *Name%host@gateway machine*.

Various options in electronic mail: As mentioned in the preceding sections, e-mail offers various options and facilities. In the following section, we will discuss some of these facilities/options. The mail program can be executed under different operating systems and uses different commands and prompt symbols. To start electronic mail service, we type

prompt sign> mail (optional address showing)

For empty address strings, mail goes into command mode. We may get the following message:

prompt sign> mail
You have five messages, or
prompt sign> mail
Blank or No mail message

For a mail message, we may get the list of messages by typing **directory** (or in some cases, it gives the list of messages in mail command mode itself):

Mail> directory

1. (e-mail address) day and time, subject
2. (e-mail address) day and time, subject
3. . . .

In the UNIX system, it may give the following message:

"/user/spool/mail/hura"; 5 messages, 1 New

1. . . .
2. . . .

U &

Each of the messages can be read either by typing **r** or by hitting the return key. After each message is read, it goes to the directory command mode.

Mail can be sent by typing:

Mail> send (address) or (address list)

After hitting the return key, the subject command will appear, followed by CC (carbon copy to self). Once the message typing is finished, users may type their names and exit from the mail mode by using **Ctrl Z** or some other command (depending on the operating system).

For replying to any of the messages, one can type **R**, which will show the sender's address. The reply can be typed and will be transmitted back to the source address.

Other facilities available with e-mail are listed below:

- **Aliasing:** This allows users to define their nicknames instead of the complete address. Whenever the user wants to send a copy of a message to himself, only the nickname needs to be typed. Also, a group of selected addresses can be defined, and each time mail is sent out, it will be received by each of the addresses in this group.
- **Forwarding:** This facility allows the user to forward the received mail to a given address.
- **Attachment:** A text file can also be mailed to the destination address via e-mail. We can insert a copy of the text file into our message to be sent. The text file can also be received by the destination address by using FTP.
- **Mailing list:** A list of a group of users can be created. Each time a message is sent, it will automatically be sent to all these users simultaneously.
- **Carbon copy (CC):** The user can send a copy of the message to himself/herself by typing his/her nickname after the command **CC**.
- **Signature file:** This file is usually used to carry information about the user such as full name, postal address, phone numbers, fax numbers, and e-mail addresses. Consider the following:
  G.S. HURA
  e-mail address
  work address
  phone and fax number
  Quotations
  <. .>

The maximum size of a signature file is usually 20 lines, but this may vary for different systems. In general, none of the operating systems supports the signature files, but a signature file can be created by the users and can be inserted at the end of the message using appropriate commands of the underlying operating system (UNIX, VAX/VMS, etc.).

Mailing in UNIX and VAX/VMS environment: In order to give a brief introduction of how the facilities of e-mail can be used in VAX/VMS and UNIX, we consider the following sequence of events which occur in the mailing program. For details of the commands, please refer to the appropriate manuals.

Facilities	VMS	UNIX
1. Starting mail	Mail	% Mail user/speed/mail/: 2 messages 1 new 1 2 U(unread) > N(New)
2. Reading message	Mail> Read, or Mail> Dir, or Mail> Read message number	
3. Print	Mail> Print Queue = Queue-name	& p message & p 1–2 & p 3

Facilities	VMS	UNIX
4. Sending message	Mail> Send To: user name (IN% < >) subject	% mail address-list-or and mail address-list subject
5. Replying to a message	Mail> Reply	% r message # (original and other recipients), or % r message # (original sender only)
6. Filing message/folder	Mail> File hura	& s hura
7. Forwarding message	Mail> set forward <address>	Create forward file and place forwarding address here (-r or -m)
8. Editing	Mail> edit <file name>	% v (for vi editor) % setnev Editor <name of editor> and then use e
9. Mailing list	Mail> alias	Create a file.matrixVII, add a line to define alias Alias nickname <address> Alias staff <address>
10. Getting help	Mail> help	% help or ?
11. Exiting	Mail> Ctrl z or exit	% quit

The electronic mail service also allows the transfer of files (binary form) from WordPerfect (disk dumps) to any destination node on the Internet. Both sender and receiver must have a utility which can convert their binary files into ASCII and back. UNIX has a utility known as **"uuencode."** Attached to the uuencode utility, there is a specification defined by the multipurpose Internet mail extension (MIME), which automatically sends objects other than those present in e-mail messages.

FTP is another way of transferring a file across the network. One of the main concerns or problems with FTP may be its limited reach, while e-mail has a wider reach around the world across the network. It should be simple to convert binary files into ASCII format, and most of the operating systems have utilities to do this conversion (e.g., the UNIX command "uuencode," as discussed above). There are mailing lists which we can subscribe to by sending our requests. Some of the mailing list types include listserv, Internet, etc., and we have to type the subscription address followed by the subscription message; e.g., the Internet list type has the following subscription address:

list-request@hostname

When we send a message to a particular destination node, the message may come back with the following series of messages, depending on the system we are using.

Mailer-Daemon @<address> date Returned mail, and unknown host,

or

Postmaster @. .date Returned mail, and unknown host.

When we read this message, we get the following description:

from: Mailer-Daemon @:< address>, day, date, time, year
Date:
From:
To:
Subject: returned mail: unknown host
Transcript of session
number (TCP) number (unknown host)

Unsent message follows:
Received from:
Received by:
Date:
From:
Message-Id:
To:
Subject:

17.4.2 Telnet

One of the important services offered by the Internet is Telnet, which offers remote log-in on an Internet host. The program Telnet runs on our computer (local), and the Internet provides connection between the local and another computer (remote). After the establishment of a connection between local and remote computers, various facilities and resources of the remote host can be accessed by using Telnet commands from our local computer. Telnet is both a protocol and a program that allows users to perform remote log-in and is the standard TCP/IP remote log-in protocol.

There are two methods of establishing a connection with a remote host using Telnet. In the first method, we have to type the Internet address of the remote host.

```
Telnet <address of host> (Internet address)
login:
   Password:
          or
Telnet sequence of digits 123.23.26.2 (IP address)
```

After this, we will get messages of progress, status, etc.

```
_ Trying-----
_(connected to <add>
   Telnet>
   login:
   password:
_Host is unreachable.
```

In the second method, instead of giving the address of the host along with Telnet, we can type the following commands in sequence:

```
Telnet
Telnet>     open     <host address>
```

Sometimes, the hosts within the Internet providing Telnet access to public services require users to also type the port number. The port number on a host also indicates the type of services requested. For example, the University of Michigan maintains a host (port number 3000) to provide weather information of the U.S. and Canada as a public service. This information can be accessed by typing

```
Telnet> downwing.sprl.umich.edu 3000
```

or

```
Telnet> open downwind.sprl.umich. edu 3000
```

This particular port number runs on a special server program which displays weather reports and supports nearly 100 users simultaneously.

Similarly, port number 23 has been assigned to the host for establishing Internet connection:

Telnet <address> 23

This host (by default) listens to the connection program for Telnet connection.

A list of all such hosts is maintained by the **Internet Assigned Number Authority (IANA)**. The hosts usually provide support to quite a number of users simultaneously. Typical Internet commands include close display, mode, open, quit, send, status, toggle, etc. A special key combination of **ctrl** plus] pauses the Telnet connection and displays a Telnet prompt symbol. (An analogous feature in UNIX is provided by the Job Control command.)

As mentioned earlier, all the services are based on client–server implementation, where the client application program from our computer makes a request to the server program for the requested service on the remote server. This application program allows users to log in on a remote system on the Internet. The remote system can be a computer located either locally or remotely on the geographic map. The keyboard of the user's computer system also seems to be connected to that computer (local and remote) and can be used to type commands. The users can execute any interactive client program (access the library or access the resources of the remote system) on the servers.

The commands for using Telnet will remain more or less similar on different types of computers (UNIX-based, DOS, VAX/VMS, or Macintosh). The only difference will be some details or a prompt sign during the running of Telnet. The following is a generic sequence of action which is seen on the screen during the execution of Telnet:

Prompt sign> Telnet < remote- computer- name>
_ Trying
_ Message regarding connection establishment to remote computer
_ Operating system information
_ Login: hura (to log in to remote)
_ Password: <…> (not visible on screen)
_ Message regarding last log-in information
_ Message regarding last interactive log-in information
_ request for list/directory
_ Mail <NEWS>
_ other appropriate commands
_ logout
Prompt sign>

When we type **Telnet**, the client program creates a TCP network connection with the server. After the connection is established, the client accepts the input from the keyboard, re-formats it, and sends it to the server. The server executes these commands and sends its output, which is re-formatted into the form acceptable to the client for its display on the client's computer. The server runs the server program to provide the services, and it has different names in different domains (e.g., in UNIX, it is known as *daemons* which means the execution of system tasks in the background).

A list of commands used in Telnet can be displayed by going to command mode, for example,

Telnet>. help or ?

This will show the list of commands such as close, display, open, quit, set, status, send, echo, carriage returns, etc. The services on the Internet are usually available on a particular port number (assigned to the server). When the client program runs, it must tell the specific address of the remote server computer along with the port number (from which services can be obtained). Some ports are standard ports for some specific application services on the Internet. Other applications can be implemented and assigned with what we call a *nonstandard port* for the server. These nonstandard ports have to be documented and are displayed on the screen with the port number. The standard ports are always considered default ports.

Telnet also runs on IBM mainframe computer hosts and provides services to the users in two modes: **live mode** and **3270 mode**. In live mode, the terminal connected to the mainframe sends characters to it one at a time. In order to see the character being sent, we can set **echo** to see the characters on our screen. After this, Telnet will allow the users to use the services of the mainframe.

In 3270 mode, IBM uses a full-screen terminal (known as 3270) as a proprietary. To use 3270 mode application, we need a terminal emulator (3270) that will make most of the terminals behave like a 3270 terminal. In general, IBM mainframes have a built-in terminal emulator (tn3270). PCs and workstations can also be connected to a mainframe by running the 3270 emulator program on them.

17.4.3 Usenet newsgroup

The Internet allows users to read and participate in discussions on a wide variety of topics via discussion groups known as network news or bulletin boards (available on CompuServe, on private dial-up systems, and also from service providers).

The newsgroups can be accessed by using a news reader program installed on our PCs. There are different news organization systems available, e.g., Clarinet, Usenet, etc. Each of these news organizations implements the news systems via some servers which are connected to other newsgroup organizations via news servers.

The network news is organized into the main newsgroups and each group offers a structured and orderly menu-based discussion on various topics. Different types of commands are used to participate in these newsgroups (e.g., **"nn"** and **"on"** in UNIX, **News** in VAX/VMS systems, and so on). The newsgroup is organized in a hierarchical structure, with the main group broadcasting on the topic followed by various subgroups underneath until it reaches to the lowest subgroups. Each group may have a parent and subgroup and the name of the group separates the parent from its subgroups by a period symbol. Thus, a particular group may have the following Internet-like address:

soc.culture.country name

This group for participation is "social" with a category of "culture" for a particular country name. Each group is assigned a unique Internet address. All of these newsgroups provided by different types of servers are arranged in alphabetical order. Within each group, subgroups corresponding to country and culture are defined. The popular ones are misc.jobs, misc.sports, misc.travel, etc.

Usenet is also defined as a collection of discussion groups. These discussion groups cover different topics such as engineering disciplines, biology, science fiction, philosophy, and many other topics. These discussion groups (typically over 5000) are accessed by millions of users across the Internet and around the world. Some of the discussion groups are at the local level, some are at the national level, and over 2500 discussion groups are at the international level. Any user who has access to the Internet can access any of the discussion groups. Those who don't have access to an Internet computer can also access these discussion groups by paying a minimal fee to Freenet (community-based network). The discussion groups can be created as required, and the topic of general interest can also be defined.

Usenet is also known as **news** or **network news** (netnews). Within each newsgroup, each of the contributions is known as an **article** or a **posting**. Whenever we submit an article in a newsgroup, we post that article in the newsgroup.

A client application program, the **news reader**, provides an interface between the newsgroup and our computer and allows users to read the newsgroups. There are a number of news reader programs available, and these can be selected depending on the topic or group we are interested in. The news reader programs, in general, are very complex and offer a large list of commands. The UNIX-based systems offer four very popular news readers: nn, rn, tin, and trn. The news readers offer the options of moving from one article to another, saving an article, replying to an article, creation of an article, composition of read and created articles, etc.

The newsgroups are usually created locally and can be accessed by the servers. The local newsgroups have different subjects and can pass the information across various servers. The local groups can either have their own selected users to access them or the access can be made public by sending them to the public servers. In other words, the local and remote servers exchange the newsgroups and are, in general, managed by their respective server managers. Each server chooses a unique name which avoids any conflict between other users. Usenet is composed of various newsgroup servers where each server may correspond to a particular category or topic area. Each category will have subjects, subtitles, etc. in a hierarchical way. Some of the common newsgroups are discussed below (see also Table 17.3).

- **alt:** This group (defined as alt.gopher) allows the creation of an official newsgroup without going through any bureaucratic setup (time consuming).
- **bionet:** This includes groups of people interested in biological sciences.
- **bit:** This group offers a platform for the discussion (listserv) of Bitnet.
- **ieee:** This group offers services and information regarding technical activities, memberships, life insurance, and other benefits to its members throughout the world.
- **vmsnet:** This group discusses technical information about DEC's VAX/VMS operating system and its LAN, DECnet.

In addition to these, there are other newsgroups available: biz (business), k12 (kindergarten to high school teachers and students), de (technical and social discussion on German), fi (technical and social discussion on Japanese), etc. For more information about these newsgroups, see References 1 and 2.

The network news also handles various commercial information services offered by different groups (one such group is Clarinet, known as United Press International (UPI), typically used by corporations, organizational campuses, etc.). It is important to note that all these groups gather a very high volume of data, and a typical server may support more than 1600 newsgroups and handle more than 10 Mbytes/day. Thus, the newsgroups across a typical server address heavy data traffic over the network. Commonly used keywords with distribution lists are World (by default), att (for AT&T, U.S.), Can (Canada), Eunet (European sites), NA (North America), US (USA), IL (Illinois), NY (New York), FL (Florida), and many others corresponding to states within the U.S.

It is possible to send replies to an individual of a newsgroup via e-mail; users may refer to the manual of the operating system for appropriate commands. The reply by e-mail will be sent only to the person who has sent that particular news (file) from the newsgroup, and other subscribers of the newsgroup will not get this reply.

Structure of a news article: A news article in any newsgroup is typically composed of three components: header, body, and optional signature (already discussed above in the section on e-mail). The header defines the subjects of the news and includes various lines, each defining an attribute such as path, newsgroup, subject, keywords, data, distribution, organizations, etc. For understanding the articles posted in a newsgroup, the newsgroup usually defines a set of acronyms.

Numbering scheme: The numbering scheme for the articles submitted in any newsgroup follows the sequential pattern of assigning the number in ascending order. The articles usually remain in the newsgroup for a few days, and after that these are replaced by new articles. The news reader program maintains a record regarding which articles have been read in a newsgroup by using a file *.newsrc*. As discussed above, UNIX provides four types of news readers: **rn**, **trn**, **nn**, and **tin**. For scanning a large number of articles quickly, use *nn*. For reading news page by page, use *rn*. For reading an article back and forth, use *trn*. For reading the entire news group, use *tin*. For posting an article, UNIX offers two of its popular editors: **vi** and **Emac**. The X-window system also offers the news readers **xrn** and **xvnews**.

The decwrl computer receives nearly 26,500 news messages a day containing about 56.5 Mb (56.6 million characters) of data. There are a large number of Usenet newsgroups, and many new groups are initiated regularly. Not all of these groups use Usenet sites. Each newsgroup has two components in its name address, and these are separated by a period. The first component indicates the name of the newsgroup, while the second component indicates the topic/area within that newsgroup. All the questions regarding Usenet are directed to

news.newsusers.questions

If we choose the newsgroup **recreation**, it gives a list of various topics/areas within it:

rec.arts.startrek.tandam
rec.arts.startrek.reviews
rec.arts.sf.movies
rec.boats

and so on.

Usenet newsgroup hierarchy: Table 17.3 describes the popular and available Usenet newsgroup hierarchies. Out of these newsgroup hierarchies, the mainstream newsgroups available on the Usenet news server include comp, news, rec, sci, soc, talk, while other newsgroup hierarchies represent the alternate Usenet newsgroup hierarchy and are optional. The number of newsgroups within each mainstream and alternate newsgroup hierarchy is as follows: comp (460), news (22), rec (273), sci (71), soc (87), talk (20), alt (586), bionet (41), bit (192), biz (32), ieee (12), info (39), and vmsnet (32). The total number of newsgroups seems to be increasing every day, but as it stands, today we have about 5600 newsgroups, about 2000 from mainstream and alternate newsgroups worldwide plus 3350 from cultural organizations and other local groups plus 250 clarinet newsgroups.

It is interesting to note that not all the sites carry the entire list of newsgroup hierarchies, but the particular newsgroup hierarchy is designated to a particular Usenet site(s).

Table 17.3 Popular Usenet Newsgroups

Name	Topics/Areas
alt	Alternate newsgroups on a variety of topics
arpa	Deals with ARPA activities
bionet	Deals with biology
bit	From Bitnet mailing services
biz	Deals with business, marketing, computers
clari	Articles, news
comp	Computers
ddn	Defense Data Network
gnu	Free software foundation and electronics engineers
ieee	Institute of Electrical and Electronic Engineering (a professional society)
info	From University of Illinois mailing services, many topics/areas
k12	Kindergarten through high school
news	About Usenet
rec	Recreation, hobbies, arts, etc.
sci	Science of different disciplines
talk	Discussions on controversial topics
U3b	AT&T 3B computers
vmsnet	DEC VAX/VMS and DECnet-based computer systems

The most popular newsgroup, which contains information for new Usenet users, seems to be[1,2]

news.announbce.newusers

Other newsgroups are usually devoted to local, national, or a particular group of people of different cultures with different languages. Four of these are well-known for cultural services to the people in those regions: de (German [Deutsch]), jp (Japanese), aus (Australia), and relcom (Russian). The majority of newsgroup hierarchies are in English but can use other native languages, too. For example, the jp newsgroup hierarchy uses non-English characters and is based on the Kanji alphabet. The topics/areas along with articles in these newsgroups use the Kanji alphabet.

Clarinet: Usenet does not carry any news information but uses the news service offered by Clarinet. Clarinet offers its own hierarchy in the same way as the newsgroups. The news service of Clarinet is offered by a number of vendors (private companies). This service can be obtained by paying a very minimal fee to these companies. The Clarinet hierarchy includes a large number of newsgroups (typically about 250) which are available at both the international and national or local levels on the topics of appropriate interests.

The Clarinet client program allows users to read the news by using a news reader program available on Clarinet. The Clarinet program allows the reading of news only and has no facility for posting (in contrast to Usenet). The following is a partial list of Clarinet newsgroups available to users. For accessing newsgroups, users may type the following for getting samples of, say, feature newsgroups:

feature.clarinet.sample

A few other newsgroups in Clarinet are given below:

clari.biz.economy.world
clari.canada.politics
clari.feature.movies

clari.news.books
clari.news.sex
clari.sports.tennis

Using Usenet: While accessing the newsgroup on Usenet, most of the frequently asked questions are answered in a documented list known as Frequently Asked Questions (FAQs) available with the newsgroup. This list can be accessed by a number of methods, e.g., retrieving FAQ after reading a newsgroup, reading *news.answers*, or anonymous FTP the FAQ list.

As mentioned above, the Usenet is a large distributed bulletin board system with a very large number of discussion groups or newsgroups. Each newsgroup has its own FAQ file that must be read before posting messages, as it describes the newsgroup in detail. If we want to submit an application for a job, we must send it via e-mail or send an ASCII or Postscript file. We should not post job-wanted messages in these newsgroups, as there are other groups which handle these, and a partial list is given below:

misc.job.offered: many companies post job vacancies
biz.job.offered: group for general job postings
misc.jobs.offered.entry: list of entry-level jobs
misc.jobs.contract: contract work
misc.jobs.resumes: for posting resumes; frequently scanned by employers
misc.jobs.wanted: for posting job-wanted messages
misc.jobs.mis: newsgroup for miscellaneous job-related issues, e.g., experience, how to write a cover letter

In addition to these newsgroups, there are job openings based on geographic regions. Some of these are given below. For more listings, see References 1–5, 9, and 12.

cle.jobs: job listings for the Cleveland area
in.jobs: job listings in Indiana
dc.jobs: job listings in Washington, D.C. area

Many universities have also created their own newsgroups on jobs, e.g., *usjobs.resume* (U.S.), *us.jobs.offered* (U.S.), *can.jobs* (canada), *aus.jobs* (Australia), etc. Others include *vmsnet.employment* (VAX/VMS-related jobs), *bionet.jobs* (for scientific jobs), etc.

17.4.4 File transfer protocol (FTP)

FTP (file transfer protocol) allows users to move a file(s) from one computer to another irrespective of their physical locations, connections, and the operating systems they are using. The only requirement of these computers is that they must use FTP and must have access to the Internet. FTP allows users to look into various databases and services offered by the Internet. Since files can be structured in a variety of ways (binary, ASCII, compressed or uncompressed, textual, images, etc.), FTP handles a variety of these structures and is a fairly complex protocol.

The FTPmail server also allows retrieval of files anywhere on the Internet via anonymous FTP. The previous two types of servers allow the retrieval of files located at a known or one location on the Internet, while FTPmail allows retrieval of files from any location across the network. For moving a file, type

FTP: remote-machine-name

In some operating systems, userids and passwords may be requested. After checking the validity of the user, the user is placed into FTP command mode. The FTP commands may be used to perform different operations on the files. The following is a partial list of commands of FTP:

FTP> get source-file destination file
FTP> put source-file destination file
FTP> mget.txt (multiple file with text)
FTP> put read txt
FTP> directory, directory name local-file-name
FTP> is directory-name local-file-name

The list of anonymous FTP sites can be obtained by typing

User-Info.Notes.Conference

In order to optimize the storage space and transfer time, most of the public access files are stored in a compressed format. These files are identified by file name extensions.

The **get** and **put** commands are used for transferring a file from one location to another. The same commands with slight modifications like **mget** and **mput** are used to move multiple files from one location to another. A list of FTP commands can be seen by typing **help**. Some of the widely used commands include

ASCII, Binary — for transferring appropriate files
dir or ls file destination — files of remote machine
mget, mpt file-list — transfer of multiple files
open, close, quit
delete file (user), name (user name)

Some of the popular FTP and Bitnet servers include

ftpmail@decwrl.dec.com (U.S.)
bitftp@plearn.edu.pl (Poland)
bitftp@vm.gmd.de (Germany)

As we discussed earlier, most of the services provided by the Internet are, in fact, client programs sitting on top of TCP/IP. The FTP client application program connects to another program server on a remote computer. The commands from the client program are received by the server program which, after executing them, performs the action (e.g., for transferring a file, it will send a file from the remote host to the user's computer).

The client program on the user's computer will receive the file and store it in the directory. We may consider the transfer of files from the remote host to our computer and from our computer to the remote host as downloading and uploading of the files, respectively. The FTP program, when executed on our machine by typing **FTP** and the address of the remote computer, provides FTP connection between them. We may have to type our log-in name and password to access remote services. If we are not a valid user, then obviously we cannot access the files remotely from our computer.

17.4.5 Anonymous FTP

The previous section discussed how files can be transferred in different modes (single or multiple copies) and how these files can be accessed by computers connected to the Internet via FTP. The implementation of these facilities of file transfer sometimes becomes too complex and difficult to provide access right to read, copy, print, and other options to everyone. This requires a list of valid users along with their passwords. This problem of checking the validity of user and password can be avoided in another service offered by FTP for file access: **anonymous FTP (AFTP)**. The user of anonymous FTP can get the files and copy them (with some restrictions) onto his/her computer memory but cannot create or install new files or perform any modifications on existing files. The users, in fact, use an anonymous login name, and then FTP will accept any string of characters and users can access the files. The anonymous FTP may impose restrictions on the time interval during which it is available to the users. The access of files or copying of the files on our system shows sequences of messages on the screen, and at the end it informs us about completion of transfer.

Useful information is stored in files at different locations controlled by local computers throughout the world. Many of the files can be accessed free over the Internet. These files can be accessed and also transferred on our systems from a remote computer using AFTP. In order to find a particular file, the anonymous FTP host contains another client application service known as **Archie**, which searches the FTP database of all the hosts that contain the requested file.

One of the main advantage of anonymous FTP is that it can be used to distribute files and programs (software) on the Internet to users worldwide free of charge and is available to users who have access to the Internet and also to those who want to set up on an Internet host for accessing Usenet discussion groups. For accessing Usenet discussion groups, we must have Usenet application software installed on our system. Interestingly, the information about this software can be obtained by anonymous FTP. The following lines may appear while using FTP commands:

FTP <host> ---> downloading of files from <host>
connected to < > 200 FTP server (version date, time, year) ready
Name
Password
Anonymous: Guest login OK, send e-mail as password
Password
FTP>

Anonymous FTP can be obtained on the FTP mail server by

mail ftpmail@decwrl.dec.com

Subject: Request for any newsgroup

The accessing of AFTP requires users to know the address of the site and follows the following sequence of steps:

Prompt sign FTP < host address)(type host address at prompt sign)
Connected to <host address> (user is connected to FTP site)
120.NIC.DDN.. Server Process.... Sun 12 Aug.1994 (site shows the time of user log-in)
Name: <host address: ghura> : anonymous (user log-in as anonymous)
Password:<host address> : ------- (uses *guest* for password)
220 Anonymous user OK, send real ident as password
230 User anonymous logged in at Sun 12 Aug, 1994 12:56-PDT,
job 22 (acknowledges the user log-in and notes the time)
ftp> (ftp prompt sign)

The following is a partial list of other FTPmail servers available:

FTPmail@grasp insa-lyon.fr (France)
FTPmail@decwrl.dec-can (California, U.S.)
BitFTP@vmgmd.de (Germany)
FTPmail@ieunet.ie (Ireland)

Each of these FTP mail servers has a unique IP address.

As we mentioned earlier, most files are stored in a compressed format, which saves not only memory but also the transmission time over the network. It has been shown that the compression technique can reduce the size of a binary file quite significantly (anywhere between 40% and 75%). The compression technique programs are treated by most of the operating systems as **utilities**. The compressed binary files are still treated as binary files, and hence anonymous FTP can transfer them just like regular binary files (of very small size). When the files are transferred to the destination user's computer, these have to be decompressed (a reverse process of compression to extract the original file).

The compression technique offers various advantages (discussed above), but the decompression technique offers disadvantages in terms of availability of nonstandard decompression techniques. The following are the commonly available compression and decompression techniques and are usually identified by some suffix attached to the file names.

Compression/decompression technique	Suffix with file name
Compress/Decompress	(.z)
Pack/Unpack	(.z) 200210/200210 (.z00)
Pack it/Unpit	(.pit)
PkZIP/Unzip41	(.ZIP)

The percentage of required compression can be defined by the users. For example, in UNIX, the following sequence of commands is executed for compression techniques (with required or requested percentage of compression):

% <file name>
% compress-v file name <file name>
compression: y % replaced with file name .z
% <file name. z>

The file name .z is y% smaller than the original file. Similarly, decompression will bring the original file of the same size.

% Uncompress file name .z
% <file name>

It is interesting to note that anonymous FTP does not transfer the entire directory but instead transfers multiple files at a time. This method of transferring multiple files (a reasonable number of files may be 20–30) may be a suitable or effective way of handling the files. But more than this number of files in the directory may not be convenient to transfer using FTP.

In the situation where software tools may be stored in a given FTP tool, the number of files in the FTP tool may be very large. Instead of transferring the large number of files corresponding to the FTP tool, these files can be combined using **tar**, and this file is compressed and transferred as a .z file. The decompression and distribution of files are performed on the target machine.

17.4.6 Archie

In the previous sections, we have seen how news files from various newsgroups can be transferred, read, etc., by Internet users. The servers keep a record of all these files and newsgroups and allow users to retrieve them via anonymous FTP and read them using commands. In order to identify the locations of files, program data, and text files (which are available on servers) on the Internet, we use an application client program known as **Archie** which contains the indexes of over 1250 servers and contains more than 2.2 million files across the Internet on various servers. The user inserts some keywords (defining the category of files), and the Archie program compares these keywords and returns the actual names of the files along with the location of the server where they are found.

The main philosophy behind Archie is the same as that of anonymous FTP, with the only difference being that anonymous FTP searches the directory in the server using its *ls* (list) command. It gets the listing of all the files in that directory of the server. Thus, using the anonymous FTP *ls* command, the directories in various servers along with the listings of files in each directory can be searched. *Archie*, on the other hand, accepts the keywords from the users, scans all the merged directories of servers, and sends the file names which match the given keywords along with the location of the server which contains those files.

Archie can be considered a "card catalog" for the largest library in the world. The Archie client program sends a request to the Archie server program. The Archie application program on the Archie server then locates the anonymous FTP host and also the directory which contains that file. It also indicates the directory path for the file or directory to the users on their computers via the Archie client program.

To access Archie, we can Telnet a remote host and enter **Archie** as the userid to log in. Archie will search the database and display the results. Information about programs, data files, documents, hosts, etc. is described by another facility within the Archie program known as *whatis*. Information regarding Archie can be obtained via e-mail by typing

info@bunyip.com

There are a number of ways to access the Archie application program, and these are discussed below in brief:

1. Type **Telnet**
   login name: Archie
   Password: hit return
   In UNIX, the command looks like
   % Telnet archie.nov.net
   login: Archie
   Once in Archie command control, use the command for searching the file name by name, descriptive index, etc.
   Archie> show search
2. The Archie application cannot be accessed via UUCP (UNIX-based system) or Bitnet. Archie can be accessed via e-mail or FTP mail (in particular, the networks which support FTP). The accessing of file names typically requires a significant time for look-up. Various commands are available to access Archie via e-mail

interfaces, including path e-mail address, compress, servers, prog regexp (searches for regular expression), list regexp, quit, etc.

3. The Archie commands can be installed on our PC. It is loaded in the system. Thus, the program allows the user to use Archie for searching using the following command:

   % Archie modifiers *string*

   where *string* defines the search string. Various modifier controls have been defined to search and return the file names along with the location of the server.

The Archie program also runs under the X-system using XArchie. By typing **XArchie**, the X-system pulls down the menu, which can be selected by mouse, and then a particular item's Search Term box can be selected.

There are a large number of Archie servers available worldwide, and a partial list of those servers is given below. Again, each of these servers has a unique IP address defined as four fields of digits separated by a period symbol (e.g., 126.23.45.2).

Archie.au (Australia)
Archie.uguam.ca (Canada)
Archie.doc.ic.ac.uk (England)
Archie.funet.fi (Finland)
Archie.th-darmstadt.de (Germany)
Archie.luth.se (Sweden)
Archie.sura.net (Maryland)
Archie.rutgers.edu (NJ)

To use an Archie server, follow the sequence:

Telnet
Telnet>
login: Archie
Password: hit return

Archie can also be accessed via mail. We have to first find the Archie server and then access it via mail:

%mail Archie@Archie.rutgers.edu
Subject:
Set mail < give your path address >
Search: . . .

17.4.7 Talk

Another interesting service on the Internet is **Talk**, which allows two users to communicate with each other by typing and then watching the messages that the remote user is sending to the local user without affecting each other's typing of messages. In this way, the users can "talk" to each other. An application program for talk service is known as *Talk Daemon*, which runs on the background but is responsible for establishing connections between two users. The talk program also partitions the server into two regions where the upper region displays what we type while the lower region displays what the other user is typing. This service can be used only if both users have an account on the Internet and also they must be logged in on different computers on the Internet at the time of the session. To use this service, type

talk <user-id>

We may also use this service to talk to another user connected to the same computer by typing this user-id only:

talk hura

Similar to talk service, there is another service known as **Internet relay chat (IRC)**, which is available on the Internet. This service allows more than two users to have a chat conversation over the Internet. When the IRC client program on our machine makes a request for IRC service to the IRC server, a channel with some name beginning with a symbol (#) is created. But if there already exists one channel, it will be allocated to the requester (client user). The IRC server program establishes connection with the client computer and allows the users to use IRC commands. If we want to participate in other conversations, too, we can use appropriate IRC commands on our computer. This command will connect the IRC server to another IRC server, and we find that all the IRC servers will eventually be connected with others, thus allowing users to participate in a large number of conversations over different channels. We can leave the conversation group at any time and the channel will be assigned to other users.

Some existing channels are very specific to a particular conversation group, while some channels are assigned for discussing anything users would like to talk about. The IRC also reserves the lower portion of the screen for typing your message which will be transmitted when you wish to do so. The upper portion of the screen will show messages from remote users. It is important to note that IRC allows and encourages users to use a unique nickname (up to nine characters) during the conversation, and thus the anonymity of users is maintained. The users can define their nicknames and store them as valid names in the IRC database maintained by the IRC server.

Information about IRC and a list of commands, along with tutorials, can be obtained using the following anonymous FTP:

Anonymous FTP nic.funet.fi
Look at the directory /pub/unix/irc/doc
Anonymous FTP cs.bu.ed (/irc/support)
Anonymous FTP coombs.anu.
(/pub/irc/doc)
edu.au

These hosts will download the IRC primer (file) in these directories. Further, many versions of the files are available as txt (text file); ps (postscript file); pdf (portable document format); jpg, gif, and bmp (graphical images); doc (word document); etc. Look for files with names tutorial (tutorial 1-----) in directory /irc/support. The tutorials can be obtained by accessing the host.

17.4.8 Wide area information service (WAIS)

We have discussed application program tools for searching an index of data (such as Archie, Whois). Another application program tool based on indexing is wide area information service (WAIS), which also provides Internet services (searching an indexed data file) indexing. This service was developed jointly by Apple, Thinking Machine, and Dow Jones. It works with the collection of data programs, newsgroups, or databases on the Internet. Here, data is defined as an index, and the result (from the index) of any query is a document. The index is a searchable page and returns the document with links. It is a distributed text searching system and is based on ANSI's recommendation Z39.50.

To access WAIS, the WAIS client application program needs to be installed. This program is available on a number of platforms such as Mac, DOS, X-Windows, UNIX, and so on. Now we are ready to access this service:

Telnet> quake.think.com or nnsc.nsf.net
Telnet> quake.think.com
login: Wais
Password: hit return

This service can also be accessed via Gopher or the World Wide Web (WWW). The menu will show information about Gopher servers explicitly. The access to Gopher servers will be performed by WAIS by selecting WAIS-based servers and information lists which will search for the resources via index interfacing of Gopher. The available forms of WAIS are SWAIS (character-oriented interface for UNIX), XWAIS (X-Window version), etc.

WAIS searches for a requested collection (via keywords) and displays an index on the screen. For example, in the category of a list of books, one may ask to get the books which have chapters on LANs, OSI models, etc.

We interact with the WAIS client program on our system, which sends the commands to the WAIS server program and also displays information and commands to choose from. As usual, there are a large number of WAIS servers available and these can be accessed by the WAIS client program. The results are sent back and displayed on our machine. Information about WAIS can be obtained from the Usenet discussion groups comp.info-system.wais (also includes FAQ files) and alt.wais.

The FAQ WAIS list can also be downloaded from

Telnet> rtfm.mit.edu

Then look at the directory /pub/Usenet/news.answers/wais-faq.

A partial list of public WAIS clients (use Telnet) available is given below:

- Info.funet.fi (Finland)
- Swais.cwis.uci.edu (California, U.S.)
- quake.think.com (Massachusetts, U.S.)
- sunsite.unc.edu (North Carolina, U.S.)

17.4.9 World Wide Web (WWW)

A relatively new information service on the Internet is known as the World Wide Web (WWW) or, in short, the **Web**. This tool is based on the concept of *hypertext*, which also is an emerging concept in information technology. In hypertext, information is given in a special format in which a word can be selected and expanded at any time. By expanding, we obtain other meaning or information from the selected word, as it may be linked to other documents (text, files, images, etc.).

Hypertext is a technique of data which may contain a variable or a word to provide a link to other data. Similarly, hypertext in the context of a document is a document which contains requested words which might provide links with other documents, and so on. The hypertext document is read by a program browser, while the linkage between documents is identified by a program navigator. This program was developed at CERN Research Center in Switzerland. The hypertext concept may be new in the application program tool for information searching, but it has been used as a hypercard program on Mac and also on IBM RS/6000 (in Info Explorer).

Execution of a browser program allows users to access the Web. This program reads the documents and also fetches the documents from other sources. It also allows users to access files by FTP, Internet News Protocol (NNTP), Gopher, etc. Further, a browser program allows the searching of documents and databases if appropriate servers offer searching capabilities. As we mentioned earlier, the hypertext documents displayed on the Web offer pointers to other texts which are used transparently by the browser program. In other words, this program allows the users to select a pointer and will display the text that has been pointed to.

This application tool is based on a line-oriented searching technique and can be used with traditional terminals. The WWW-based online-oriented searching/browsing seems to be simpler to use than similar application program tools available (e.g., ViolaWWW (based on X-Windows), NEXT UNIX, Workstation, Mac, and PC). The information regarding its support for different computers can be obtained by typing **anonymous FTP >info.cern.ch** and looking at the directory pub/www/bin.

To access the Web or WWW, use the commands

Telnet> info.cern.ch
Login: www
Password: hit return

To access a line-oriented browser, we may use one of the following commands:

Telnet info.cern.ch

or

www

There are two different ways of using browsers: we can Telnet the browser or we can use our own browser. It is always advisable to use our own browser. However, the browsers which are available via Telnet are given below:

- http://info.cern.ch/hypertext/www.FAQ.Bootstrap.html
- www.cern.ch (no password required)
- www.cc.ukans.edu (a full screen browser, Lynx, requires a VT100 terminal; install Lynx on your terminal)
- www.njit (or Telnet 128.235.163.2)
- www.huji.ac.il (a dual-language Hebrew/English database, line-mode browser, Israel)
- sun.uakom.cs (slow link, Slovakia)
- info.funet.fi (Telnet 128.214.6.102; offers various browsers including Lynx)
- fserv.kfki.hu (Hungary, slow link)

It is worth mentioning here that in most of the browsers, we have to type www for login and no password is required.

There are different cateogories of browsers available which can be used on different machines under different operating systems. A complete list of browsers along with source and executable forms is available on

http://info.cern.ch/hypertext/www/clients.html

It is important to note that these browers require communication protocols such as SLIP, PPP, and TCP/IP.

The **uniform resource locator (URL)** is a very important component of the Web. It is used to identify both protocols used by and the location of Internet resources. Typically, the URL is defined by the following format:

Protocol://host/path/file

Here, *Protocol* defines the class of Internet resource and these resources are Gopher, WAIS, FTP, Telnet, http, file, mailto, etc. The *host* corresponds to the name of the IP (Internet protocol) address of the remote computer (e.g., 122.34.13.26 or www.cs.ncsa.edu). The *path* corresponds to a directory or subdirectory on a remote computer. Finally, the *file* represents the file we want to access.

With the help of URLs (as defined above) and Web browsers, it is possible for users to access any resources of the Internet. For example, the following URL is used to establish an FTP session:

FTP://ftp.cs.ncsa.edu/pub/stacks/alawon/alawon-f100

In this URL, we are requesting an ftp for ftp.cs.ncsa.edu and log-in as anonymous. After this, we are changing the directory to/pub/stacks/alawon and accessing the file with name alawon-f100. The important thing to remember here is that the translation of TCP/IP into URL is provided by the Web browser and we only have to know how to create a URL for the FTP session. Some of the material herein is derived from References 1–3, 5, 9, 12, and 13.

MSDOS browser: Lynx is a basic Web browser which is basically defined for DOS computers or dumb terminals running under UNIX or VMS operating systems. This browser does not support images or audio data but supports the mailto URL. The mailto URL is used for the simple mail transfer protocol (SMTP). It is worth mentioning here that SMTP is a standard mail service used by the Internet. When a user selects a mailto URL, the user sees a form to be completed. The resulting text from this form is then delivered to the destination computer or user specified by the URL over the Internet. DosLynx is an excellent text-based browser for use on DOS systems. This browser can be obtained by anonymous FTP:

ftp2.cc.ukans.edu

Then, look at the directory pub/www/Doslynx.

Macintosh browsers: MacWeb is a browser for a Macintosh. This application browser was designed at Microelectronics and Computer Technology (MCC) and is distributed over Enterprise Integration Network (EINet). The MacWeb requires System 7 and MacTCP (operating system defined by Apple which allows Macintosh computers to understand TCP/IP). For using MacWeb, an application program **StuffIt Expander (SIT)** is used which translates and decompresses files. This is a very useful utility program, as AFTP always gets the compressed files. MacWeb is very fast and offers a customized user interface, supports automatic creation of HTML documents, and supports MacWAIS. It does not allow users to select or copy text directly from the screen, and the displayed text, after being saved, loses all its original formatting.

The following available browsers work on different platforms:

- **Mosaic for Mac** can be obtained by anonymous FTP
  ftp.ncsa.uiuc.edu, look at the directory Mac/Mosaic
- **Samba** from CERN can be obtained by anonymous FTP
  info.cern.ch, look at the directory /ftp/pub/www/bin for file **mac**

- **MacWeb** can be obtained by anonymous FTP
  ftp.einet.net, look at the directory einet/mac/macweb
- **Amiga:** The browser for AmigaOS is based on NCSA's Mosaic and can be obtained by anonymous FTP
  maXVII.physics.sunysb.edu, look at the directory /pub/amosaic

NeXTStep browsers: The earlier version of browser for NeXt computers was defined as hypertext transfer protocol (HTTP) which runs under NeXTStep operating systems. The NeXt system runs on X-based browsers. The OmniWeb browser for NeXTStep can be obtained by anonymous FTP

ftp.omnigroup.com, look at the directory /pub/software/directory

Another browser for NeXTStep is CERN's browser-editor, which can be obtained by anonymous FTP

info.cern.ch, look at the directory /pub/www.src

NCSA Mosiac browser for X-Window system: This browser, coupled with NCSA's Web server (httpd), offers copy and display options. It also supports WAIS and URLs. It requires a relatively powerful computer and direct access to a UNIX or VMS machine running the X-Window system. The Macintosh Windows computer (based on MacX or Microsoft Windows) needs HummingBird Communications's Exceed/W which can run X-Window terminal communication sessions.

X/DECWINDOWS browsers: This browser runs under graphica UNIX and VMS. A variety of browsers are available and a partial list of those browsers is given below:

NCSA Mosaic for X (UNIX browser using X11/Motif)
NCSA Mosaic for VMS (uses X11?DECWINDOWS/Motif)
tkwww browser/Editor for X11 (UNIX browser/Editor for X11)
Midaswww browser (UNIX/X browser from Tony Johnson)
Viola for X (two versions: UNIX/X with Motif and a second without Motif)
Chimera (UNIX/X browser using Atena; does not require Motif)

Text-mode UNIX and VMS browsers

Line Mode Browser
Lynx full-screen browser
Tom Fine's perlwww (tty-based browser)
Dudu Rashty's full-screen client based on VMS's SMG screen
Emac w3-mode (W3 mode for emacs)

Batch-mode browsers

url_get browser can be obtained by anonymous FTP: fto.cc.utexas, look at the directory /file/pub/zippy/url_get.tar.z.edu.

Web server software: For offering Web services, the Web server application program executes after getting a request from the client application software. A list of some of the popular server application software is described below. By the way, these application programs are also known as Web server software and run on a variety of platforms (Macintosh, UNIX, VMS, Microsoft Windows systems, etc.).

- **MacHTTP:** This server requires System 7 to support features like AppleScript. It runs on Macintosh II-type machines (e.g., Macintosh IIci, Centris, Quadra, etc.). It requires MacTCP and does not run on Macintosh Plus, SE, and PoserBook 100.
- **NCSA httpd:** This server software runs under UNIX and defines source code and binary form to support different forms of UNIXVII. It is widely supported by the Internet to include a number of postings and allows access to other applications like Gopher, WAIS, or a list server. Many of its common gateway interface (CGI) scripts are written in Perl language, commonly used on UNIX. The main feature of this interface is that it runs in the background and returns only the results. It includes the ability to display the current time or the number of users who have accessed the server. The CGI can be written in any language: C, Perl, AppleScript, Visual Basic, or DCL. The scripts are processed by the interface and the results are sent back to the Web servers (usually in the HTML form).
- **CERN httpd:** The NCSA httpd cannot be used on a VMS system, while the CERN httpd is defined for VMS systems. The CERN httpd server is available in binary form and in source code form.
- **NCSA httpd for Windows:** This server software supports all the standard features like forms, CGI scripts, graphics, and access control. It is relatively slower and requires a lot of memory and CPU power and a WinSock-compatible TCP/IP driver.

The documents obtained from Web servers are usually formatted, and the formatting can be in different languages. The hypertext markup language (HTML) is the language used to format documents in Web servers, and the HTML standard can be accessed from the CERN server. HTML files are simple ASCII files containing rudimentary tags which describe the format of the documents. To get an idea of how the tags are used, consider the following document formatted by HTML:

```
<HTML>
<Head>
<Title>Good Morning HTML document</Title>
</Head>
<Body>
Hello, How are you??
</Body>
</HTML>
```

The tags are always used as a pair. The <HTML> and </HTML> tags define the document as an HTML document, the <Head> and </Head> tags define the main text of the document, while the <Title> and </Title> tags define the title of the document. The <Body> and </Body> tags represent the location of the formatted text. The backward slash sign indicates the completion of logical formatting. In order to format the document, various other tags such as <P> (paragraphing),
 (line breakers), etc., can be used within the body of an HTML document. Many tools have been defined to create HTML documents, such as Simple HTML Editor (SHE), which requires Macintosh and Hyper-Card Player; HTML Assistance, which is a Windows-based HTML editor; and Converter, which converts word processor file format into HTML document format. The word processor documents could be in any form, e.g., Microsoft Word, WordPerfect, RTF, etc.

Dynamic HTML has been defined to encompass many new technologies and concepts that can be complex and are often at odds with Web site construction. Dynamic HTML is a combination of HTML and JavaScript and includes several built-in browser features in fourth generation browsers (Netscape version 4.0 and Internet Explorer version 4.0) that enable a web page to be more dynamic. The dynamic feature indicates the ability of browser to alter a Web page's look and style after the document has been loaded. The DHTML used in IE-4's is far more powerful and versatile than NS-4's.

Extensible Markup Language (XML) is a document description language, much like HTML, used to construct Web pages. It is more versatile than HTML. As said above, HTML uses a defined series of tags to create Web pages and these tags provide instructions to the software reading them on how to present the information. XML also includes a series of tags that describe the components of the document. It allows the users to define the tags and is considered an advanced version of HTML. Due to these features, it offers a tremendous power to describe and structure the nature of the information presented in a document. It does not allow the description of presentation style of data. An associated language — Extensible Style Language (ESL) — must be used.

More and more companies are entering the Web. The Web address of a company is usually determined by using the format **http://www.ABC.com/**, where ABC is the name of the company. For example, IBM may have the Web address http://www.ibm.com/. This addressing scheme is not applicable to all companies.

17.4.10 Mosaic

Mosaic is one of the finest browsers offered by the Web. It offers a multimedia interface to the Internet. The hypermedium is a superset of hypertext; i.e., it allows a linking of media by a pointer to any other media. Here also, the browser program offers a transparent pointer which, when selected, will display images or sounds or animation that have been pointed to by the pointer. As it presents hypertext and supports multimedia, it is also known as a hypermedia application software tool. Mosaic can be accessed on time-sharing, by 9600 baud modem, or on X-system (UNIX). All the versions of Mosaic are available at

ftp.ncsa.uiuc.edu

It was developed at the National Center for Supercomputing Applications (NCSA) at the University of Illinois, Champaign-Urbana, U.S.

Another source for navigation or browsing is by online magazine and is known as Global Network Navigator (GNN). Other options provided by Mosaic are searching, saving, and printing a file. The different file document formats supported by it include plain text, formatted text, PostScript, and HTML.

Mosaic also allows the users to use hotlist and window history. A hotlist offers a permanent list of documents that can be accessed with one menu. Any item on the hotlist can also be added. The window history gives a list of windows which have been seen or used by the users via the navigate menu. One of the interesting features of Mosaic is the ability to introduce a message with Web documents (annotations).

When accessing the Web using WWW or Mosaic, a list of searchable indexes appears on the screen, and these searchable indexes can be from other application tools such as Veronica, WAIS, or any other look-up tools. It also allows users to access servers of different application tools such as FTP, Telnet, WAIS servers, WAIS directory, Gopher, and newsgroups.

The hypertext documents used in WWW and Mosaic are becoming very useful and are popular application software tools for offering information delivery systems on the Internet. The information delivery systems offer a variety of applications in different disciplines. The WWW clients program (browser) allows users to access a variety of services from the system (e.g., WWW server program, Usenet accessing of news, Gopher, Telnet session, and many other services). The WWW client program connects the system to the WWW server, which offers hypertext documents. As usual, there are many WWW servers available over the Internet and each WWW server provides specific services in a specific topic(s) or area(s). The WWW searches both text files and indexes (for searching).

The WWW client or browser program is the key component in the Web tool, as it is available as a graphical user interface browser, character-based browser, etc. As such, there exist public WWW browsers for accessing Internet services. The WWW client programs can run on a variety of systems, such as X-Window, PCs (with Windows), Mac, VMS, and UNIX-based systems. A partial list of public WWW browsers is given below:

- info.funet.fi (Finland)
- info.corn.ch (Switzerland)
- www.njit.edu (New Jersey, USA)
- fatty.law.cornell.edu (USA)

As usual, we have to Telnet one of these hosts and enter the WWW at the userid.

CERN is the European Particle Physics Laboratory in Geneva, Switzerland. The home page may have information about WWW, CERN, subjects, help, etc. Each of these may have different areas. For example, other subjects will offer indexes of academic information. This will show a number of subjects to choose from (e.g., aeronautics, computing, etc.). There is a draft standard for specifying an object on the Internet known as the uniform resource locator (URL). The general format of this shown below:

telnet://dra.com

The first part of the URL, before the colon, defines the access method, while the second part of the URL, after the colon, defines a specific access method. The two slashes after the colon define the machine name.

17.4.11 Gopher

In the previous sections, we have discussed a variety of client application programs which allow users to search for files (binary or ASCII) and directories, read files and directories, transfer files and directories, search for various services/servers available, list directories of people working on different available networks, etc. Most of these application programs, in general, allow the users to access these services by typing appropriate commands and do not provide user-friendliness. A few new application program tools which offer a more friendly interfacing environment to the users to access Internet services have been implemented. These tools allow users to search for a variety of online resources which can be selected from the menus. These also allow users to access those selected resources from their local computer systems.

One very popular application program tool is *Gopher*. As we mentioned earlier, Gopher is a client communication application developed at the University of Minnesota and offers users access to more than 5000 Gopher servers worldwide. It allows users to search for various services (just like searching in catalogs), and the requested item is automatically communicated to them. The library of these services may be located anywhere, and it does not affect access so long as it is available on Gopher.

Gopher is also known as **Internet Gopher (IG)**. It gives a list of source organizations or universities which are providing various newsgroup services, libraries of resources books, records, and many other resources. These services are available by choosing an appropriate domain from the menus which will offer the category of services being offered. Note that users don't have to type the domain name, IP address, or other relevant information to access Internet services. The Gopher application program has already established connection with these sources (servers) and the selected request will be available to users online.

The Gopher services are, in general, not structured or standardized, and as a result, we have Gopher services available from different servers in very different formats. The accessing of the files could be achieved via Telnet, FTP, or e-mail — it does not affect the Gopher — and, in fact, users don't have to perform these operations, as the accessing by any of these means is performed by Gopher itself. The Gopher application program is available on many client computers such as UNIX, Mac, IBM/PC, X-Window, VAX/VMS, and many others. A list of Gopher clients can be obtained in a variety of ways (e.g., anonymous FTP, Archie, etc.). The Gopher services can also be obtained on two public Gopher clients by typing

consultant.micro.umu.edu

or

gopher.ninc.edu

These two Gopher clients will help users install the Gopher application program on their client computers. We can also install the Gopher client application program (similar to other application program tools we have discussed earlier):

```
>Telnet
>login: Gopher
>Password: hit return
```

The main attraction behind Gopher is the ability to select a resource from a given menu. The Gopher will display that selected resource, which may be a text file or another menu. The selected text file (irrespective of its location) will be displayed on the screen, and the list of resources offered by the selected menu will be displayed on the screen. The connection to a remote machine for displaying the resources to the screen is totally transparent to the users. To keep users informed about the progress, the following messages will appear on the screen:

_connecting
_retrying
_please wait

In response to a request, Gopher will give the names of organizations (containing the client), names of universities, and topics of general interests such as weather and forecasts, software and data information, information about other Gopher servers, etc. Once the user chooses any one of these, it will further give a list of detailed information available under that selected category. The selected information may yield a set of relevant information, and so on, until the user accesses the files for the appropriate selected field/information.

The Gopher client application program is installed on the user's computer which, after executing, displays the menus. The Gopher client program, in fact, requests the Gopher server program for the connection, retrieval, etc. The Gopher client program also allows users to make a request for downloading a file or connecting to Telnet or any other server, and the connection for these requests is provided by the Gopher server program. As expected, there are a large number of Gopher servers available, and these servers provide services for accessing a variety of information available with them. These Gopher servers are managed and maintained by companies, universities, and many other agencies, and each server contains information useful to users either worldwide or locally. The worldwide information is available to all Internet users, while local services are available to those local users (e.g., a particular university has a Gopher server for its employees, or a company has bylaws of the company, insurance, and other benefits information for its employees, etc.).

Gopher offers two types of entries in the menu: directory and resource. Each directory is represented as a menu shown on the screen via graphical interface. These interfaces are very user-friendly and easy to learn and use.

The Gopher's FTP facility allows users to transfer files from anonymous FTP servers to their computer running the Gopher client. This gopher client may not be on the user's computer but instead may be running on some other computer The client running on the user's computer enables transfer of files from servers to that computer. Gopher allows users to search the resources and then the resources can be accessed by Telnet. In other words, using Gopher, we are connected to online resources, various organizations, universities, directories, and many other services. At the end of each session, a marker <TEL> is attached to indicate that this resource belongs to Telnet. If we try to use this resource from Gopher, we will get an error message, but if we use Telnet, we will be able to access it. Once Telnet starts, the Gopher loses its control, but it regains it when Telnet is finished.

The following is a partial list of public Gopher client application programs available to Internet users worldwide. Each of these Gopher servers will have a unique IP address containing four regions of digits separated by a period symbol.

info.anu.edu.au (Australia)
gopher.denet.dk (Denmark)
gopher.brad.ac.uk (England)
gopher.th-darmstadt.de (Germany)
gopher.ncc.go.jp (Japan)
gopher.uv.es (Spain)
gopher.sunet.se (Sweden)
infopath.ucse.edu (Canada and U.S.)
gopher.uiuc.edu (Illinois)
gopher.msu.edu (Michigan)
gopher.unc.edu (North Carolina)
gopher.ohiolink.edu (Ohio)
gopher.virginia.edu (Virginia)

When we run a Gopher client program on our machine, it displays a menu showing different servers. We can select any of the servers to get more information about that particular Gopher. If we don't see that Gopher server in the menu, we can get information about it by accessing Usenet newsgroups reserved for Gopher servers only: *comp.infosystems.gopher* or *alt.gopher*. These two newsgroups will provide all the relevant information about all the Gopher servers of the Internet. A list of frequently asked questions (FAQs) about Gopher can be obtained by

Anonymous FTP > rtfm.mit.edu

Then look into the directory /pub/Usenet/news.answers. Now download the file with name *gopher.faq*. Similarly, this directory also contains the FAQs about other client application programs.

Internet Gopher client information: A typical Gopher menu may show a number of resources along with its category or type, as shown below:

Name of Company or University of Gopher Server

1. About the university
2. Search online
3. Other Gopher servers
4. Other Gopher servers around the world
5. Directory <CSO>
6. A particular resource
7. Catalog
8. <Picture> <TEL>
9. Fun and games

Press ? for help, q to quit, u to go up a menu.

In most of the Gopher servers, we come across the following symbols very frequently: slash (/), <CSO>, <TEL>. The slash symbol indicates that this resource item represents another menu of other resources. A very useful symbol, Computing Services Office <CSO>, indicates that a facility is available at that server which tells about the people of that organization or university. Since <CSO> is attached to each server, we have CSO name servers on the Internet. The CSO name server is a collection of programs which basically works as a White Pages directory and can allow users to find someone on the Internet.

The symbol <TEL> is a different category of the resource and requires a request for Telnet connection. If we choose this resource, the Gopher server will initiate a Telnet connection to the remote host. It is important to note that now the Gopher server becomes the source host (on behalf of the user) which has initiated the Telnet connection request. After accessing that server on Telnet, the Gopher server will bring us back to the main menu (from which we selected that resource). In the Gopher servers, when a Telnet connection is initiated, we may see some warning messages with some help commands.

The symbol <PICTURE> indicates a binary file which contains a picture; the symbol <> indicates that the binary file contains a sound signal.

Gopherspace is a centralized catalog which contains all the menus and types of resources that are available. It defines resources of different sites for application software tools, e.g., Gopher, FTO, Telnet, etc. Another resource service which can provide the above information on Gopher is **Veronica**, which runs within the Gopher server program. It searches the Gopher space for that menu containing the requested resource. For example, if we want to get information on modern arts, it is certain that this resource is in some menu of Gopher and is in Gopherspace. The resource service Veronica will search for the users. It can be accessed using the menu item for: *search titles in Gopherspace using Veronica.* Sometimes Veronica can be found under the Gopher server. After we have selected the Veronica menu from Gopher, we may get the following messages on our screens:

Search Gopherspace using Veronica at <host>.
Search Gopherspace using Veronica at <host>.
Search Gopherspace for Gopher directory <host>.
FAQ: Frequently Asked Questions about Veronica.

The Veronica application program within Gopher searches for various menus of Gopherspace. This service is a kind of public access and searching time in some applications may be significant. All of the requests for searching Gopherspace for all the menus go through the Gopher servers supporting Veronica across the Internet.

In certain situations where we would like to keep a record of Gopherspace menus of a particular region, we may use another application service known as **Jughead** instead of Veronica. The Jughead application tool is similar to Veronica, with the only difference being that Jughead searches for a small designated area of Gopherspace. The servers within that area (company or university) will search the database for all menu items within the limit of Gopherspace specified. A Jughead server maintains this database of all the menus of items. The user can confine the search within the limited Gopher space. The search time is faster than for Veronica, as it will go only to the Jughead Gopherspace server rather than the Veronica server and then to the server of that Gopherspace. Further, it does not have any limit as to how many items can be searched (in contrast to Veronica). From Gopher, we can select Jughead in the same way as Veronica for our search.

There are a large number of Jughead servers around the world, and each Jughead provides searching for a specified Gopherspace within a small area. Jughead seems to be faster, allows the users to make a request for more items, and will look into the Gopherspace Jughead server within that area. It also allows users to establish connections with either other Gopherspace servers or the Internet servers worldwide.

It is important to note that none of these tools (anonymous FTP, Gopher, Archie, and others) or even the Internet as a whole is controlled or managed by one organization. Instead, the Internet provides an open forum for users to make use of these application tools to access information. Internet users can access services via these tools and also switch to other newsgroups via Telnet, Usenet, etc.

A unique service offered by the Internet is to search for an e-mail address or somebody's name. There are a variety of application program tools available on the Internet for this type of service, and we have servers for each of the tools. A brief summary of each of the tools is given below.

White Pages Directory (WPD) services: As we noted earlier, a White Pages Directory (WPD) client program gives information about any Internet user in the following form:

- User's name
- e-mail address
- Postal address
- Department/Division
- Telephone number
- Fax number

and so on. The Internet offers a variety of WPD application tools, and each of the tools offers information about the user in a different format. Some of the tools have already been discussed above but will be described here in brief.

1. **Gopher:** This application tool allows users to use a CSO name server or some other category of the White Pages Directory supported by it. If we know the organization or university or where the individual works, then we can obtain the menu item for the CSO or something similar to it from the Gopher server of that organization or university.
2. **Whois Server:** This is another way of finding out someone's name and e-mail address on the Internet. There are quite a number of Whois servers on the Internet. This list can be obtained by anonymous FTP. One can use the Archie search for

looking at the list of Whois servers. A widely accepted and very useful server which contains the list of Whois servers is *whois.internic.net*.

3. **Knowbot:** This is another application program tool which searches for the requested information. This tool is an automated robot-like program which uses some kind of predicate/knowledge to provide searching. Information about this service is available at port number 185 and can be obtained by Telnet:

   Telnet> nri.reston.va.us 185.

4. **Fred and X.500 directories:** CCITT defined an international standard XVII.500 as a White Pages Directory for maintaining a list of users. As it stands today, the XVII.500 has not been fully accepted worldwide and, instead, we have seen a variety of application tools on the market which are not standards but are being used effectively and efficiently for accessing information, WPDs, etc. An application program, *Fred*, gives a list of all of these software tools, which can be obtained by

   Telnet> wp.psi.com

   or

   wpz.psi.com

   The host *rtfm.mit.edu* contains information about Usenet archives and also copies of all FAQ.

5. **Netfind:** Netfind is another application program which can look for a computer which has knowledge about the user we are trying to find. The requirement to use this program is that we must know to which computer that user is attached. The program will find the name, address, and also Finger information. Netfind runs only on UNIX-based systems.

   Netfind > hura place
   Netfind> hura name of university

   The Netfind services can be obtained by Netfind servers which are available as public access servers. A partial list of such servers is given below. These Netfind servers have unique IP addresses.

   Archie.au (Australia)
   macs.ee.mcgill.ca (Canada)
   monolith.cc.ic.ac.uk (England)
   bruno.cs.colorado.edu (U.S.)
   netfind.oc.com (U.S.)
   ds.intermc.net (U.S.)

WWW (Web) vs. Gopher: WWW and Gopher are similar in that they are application program tools which provide Internet services to users. However, there are some fundamental differences in the way the services are offered and also in how the searching/browsing technique is being used and implemented. WWW is based on hypertext, which offers a variety of links, but all of the links for a selected word may correspond to the same subject with different pointers within the article. Gopher presents only the list of resources or the directory but has no idea about contents of the files or resources it is accessing, and hence no different meaning to the selected word with linkage to other documents can be derived. The data is either a menu, a document, an index, or a Telnet connection.

In WWW, everything is defined as a hypertext document which may be searchable. WWW uses only two commands — **follow a link** and **perform a search** — as opposed to a number of commands required in Gopher, depending on the interfaces and the type of resources we are using on our computer. WWW can represent the Gopher (where a menu is a list of links), while a Gopher can be considered a hypertext document without links; the searches are the same and Telnets are also the same.

WWW is very flexible, provides more information than any of the existing application program tools, and offers cross-referencing facilities. It also allows users to read Usenet and does not distinguish between public data and personal data. WWW allows users to search for a WAIS directory of servers, interface to FTP resources and Telnet resources, read White Pages, etc.

Both Gopher and WWW servers are clearly competing with each other and providing a variety of services in a different format. An obvious question could be asked: which server should be used? This requires a detailed comparison of Gopher and WWW servers.

The Web servers are more efficient than Gopher servers, as most of the sharable information processing is performed by the client application software. Web clients use HTML language and the Web servers offer services either via menus or without menus. We can add more texts and abstracts to the hypertext links so that the users can evaluate their choices in different ways. Also, Web servers are easier to maintain since the single file defined in HTML can be used with public domain editors or database programs. In contrast, Gopher servers are doing most of the information processing and offer a standard menu system to access the services; further, the hypertext linking of all files can be accessed by Gopher servers. The Web servers can be obtained from Usenet newsgroups: *comp.infosystems.www*, *comp.infosystems.www.users*, and *comp.infosystems.www.misc*.

The unofficial newspaper of the Web is known as *What's New With NCSA Mosaic* and can be obtained at

http://WWW.ncsa.uiuc.edu/SDG/Software/Mosaic/Docs/whats-new.html

The subject catalogue of the Web can be obtained from the WWW Virtual Library maintained by CERN at

http://info.cern.ch/hypertext/DataSources/bySubject/Overview.html

Some useful Web sites

For Global Reach, refer to *http://glreach.com/globstats*. For information on the European Internet use *http://www.proactiveinternational.com* and *http: www.icann.org*.

Internet protocol of the next generation can be found at *http://playground.sun.com*, *http://www.protocol.com*, and *http: news.cnet.com*.

Information TCP/IP is found at *http://www.tcp-ip.nu* and *http://www.cisco.com*. Other useful issues of history, deployment, socket programming, services, and applications of TCP/IP are available at *http://www.cisco.co*, *http://www.users.uniserve.co*, *http://www2.hursley.ibm.com*, *http://www.tcpip-gmbh.de*, and *http://www.sunworld.com*.

A variety of Internet client applications have been developed and can be found at any Internet Web site. Some of the clients include Chat, Directory, Encryption and Privacy, File Sharing, File System, Filtering, FTP, Hotline, Mail, Media, News, Printing, Push, Search, Telnet, Time, Usenet, Utilities, Videoconferencing, VRML, Web-phone, WWW, and others.

Various resources, books, Web sites, newsgroups, etc., can be found at *http://www.private.org*.

Visit *www.geocities.com* for DHTML and *www.tdan.com* for XML.

References

1. Krol, E., *The Whole Internet: User's Guide and Catalogue,* O'Reilly and Associates, Inc., 1994.
2. Hahn, H. and Stout, R., *The Internet: Complete Reference,* Osbourne McGraw-Hill.
3. Butler, M., *How to Use the Internet,* Ziff-Davis Press, 1994.
4. Fraase, M., *The Windows Internet Tour Guide,* Ventana Press, 1994.
5. Godin, S. and McBride, J.S., *The 1994 Internet White Pages,* IDG Books, 1994.
6. Alden, B., "Traveling the information highway," *The Institute (IEEE News),* supplement to *IEEE Spectrum,* Sept. 1994.
7. Alden, B., Traveling the information highway, *The Institute (IEEE News),* supplement to *IEEE Spectrum,* Sept. 1994.
8. *Proc. First International Conf. on World Wide Web,* CERN, Geneva, Switzerland, May 24-27, 1994. [Further information can be obtained at m.haccon@elsner.nl and papers are available on WWW and can be accessed at *http://w.elsevier.nl/.*]
9. Levine, J.R. and Baroudi, C., *The Internet for Dummies,* IDG Books Worldwide, 1993.
10. *Internet World,* 5(5), July/Aug.,1994.
11. Cronin, M.J., *Doing Business on the Internet,* Van Nostrand Reinhold, 1994.
12. Hura, G.S., special issue editor, "Internet: State-of-the-art," *Computer Communications,* 20(16), Jan. 1998.
13. Hura, G.S., "The Internet: Global information superhighway for the future," *Computer Communications,* 20(6), 1998.

Additional reading

Huitema, C., *Routing in the Internet,* Prentice Hall, 1995.

Perlman, R., *Interconnections: Bridges and Routers,* Addison-Wesley, 1992.

Shenker, S., "Fundamental design issues for the future Internet," *IEEE Journal on Selected Areas in Communication,* Sept. 1995.

Part V

High-speed networking and internetworking

chapter eighteen

Integrated digital network (IDN) technology

"You have to be first, best, or different."

Loretta Lynn

18.1 Introduction

Digital technology has become an integral part of new public telephone networks, as it is cheap and offers a variety of services with better quality of voice transmission over the networks. The **integrated services digital network (ISDN)**, based on the concept of a complete digital network, has proved that digital technology will not only provide fast and reliable services but will also help in introducing other technologies such as **multimedia**, **virtual reality**, **concurrent engineering**, etc. The integration of different types of signals (audio, video, text, graphics, images, and many other application types) over high-speed networks is becoming an essential service of communication networks.

The main philosophy behind **integrated digital technology (IDT)** for integrated digital networks is based on digital switching and transmission. The digital switching technique is based on **digital time division multiplexing (DTDM)**, while **synchronous TDM (STDM)** is used for carrier transmission between switching nodes. It must provide a standard interface for ISDN and non-ISDN services and provide access to these services on demand. As such, it is expected to provide multiple capabilities and offer global connectivity to enable users to access the services of other networks. The full-duplex transmission of digitized voice signal is generally used between the subscriber and switching nodes. For control information regarding routing and other control-related parameters, **common channel signaling (CCS)** over public packet-switched telecommunication networks is being used.

The integration of these techniques (switching and transmission) were tried in an **integrated digital network (IDN)**. In the conventional public telephone networks, TDM is used to multiplex the modulated voice signals, and these are transmitted over the communication link using frequency division multiplexing (FDM). The modulated signals pass through various intermediate switching nodes before arriving at the destination. At each switching node, the signals are de-modulated and de-multiplexed. The process of de-modulation and de-multiplexing followed by modulation and multiplexing for further transmission certainly is expensive and affects the quality of signals (due to the introduction of noise at the switching nodes). But if we can digitize the voice signal (no modulation) at the transmitting node itself and, using TDM, transmit the digitized signals, then it will improve the quality of services considerably and offer a higher speed of transmission. These signals will be going through various intermediate switching nodes without going through a transformation process, since modulation is not required at the transmitting side.

The digitization of an analog signal is obtained by the **pulse coded modulation (PCM)** technique. The digital transmission offers different advantages (such as lower costs, higher speed and capacity, and improved quality of services) and has been used in different networks. An initial version of PCM systems was defined mainly for increasing the capacity of existing cables over a short distance.

Digital switching has also evolved in parallel with digital transmission, and the main philosophy behind this type of switching is the use of solid-state cross-points over the existing reed-relay cross-bar switching systems. The digital switching systems can support many networks used in organizations of different sizes and can support thousands of lines.

The CCITT has recommended two encoding standards for PCM: 24 channels operating at 1.544 Mbps (North America) and 30 channels operating at 2.048 Mbps (Europe). Both of these standards have different characteristics and cannot internetwork with each other directly, and both have different digital hierarchies.

One example of switched networks (based on integrated communication networks) is *circuit-switching networks*, where the nodes of the network use time-division switching instead of analog-division multiplexing. The switching of the signals is performed on digital switches (in contrast to analog switches). Thus, both transmission and switching of signals are done in digital form, and integration of these two aspects of communication can be achieved easily.

The intermediate switching nodes in a circuit-switched network can be either a digital switch (time-division) or an analog switch (space-division). The former switch receives the TDM channels (streams of bits), and these are routed on a different circuit (already established) by multiplexing the channels of the data. These multiplexed channels are transmitted to another switching node(s) until the data reaches the destination node. In contrast to this, in the analog networks, the FDM channels are received by an analog space-division switch. The switch provides a switching connection between each input line to each of its output lines, and this is possible only if both inputs and outputs have the same frequency.

Since FDM channels have different frequency bands, these have to be first transformed (de-modulation and de-multiplexing) to have the voice-grade frequency band of 4 KHz. Once the voice-grade channels of 4 KHz are obtained, the analog switch will provide a logical connection between input and output accordingly. At the output again, the channels of data are modulated and de-modulated on a different circuit depending on which circuit has been assigned to a particular channel. On each circuit, the assigned channels are multiplexed and de-modulated and transmitted to appropriate switching nodes until the channel of data arrives at the destination node.

The integration of digital transmission and switching in IDNs offers various advantages over analog networks, such as lower cost (due to VLSI chips), longer distance transmission with smaller error rate (due to the use of repeaters in contrast to amplifiers), higher bandwidth and capacity (due to the use of a digital time-division switch), security and privacy (due to encryption techniques which can be applied to the digital signal), and, of course, integration of different types of signals (data, graphics, audio, video, images, etc.). Digital transmission is also used in trunks carrying multiple channels of both voice and data multiplexed via synchronous TDMs. It has also been used in subscriber's loops for full-duplex operations if the digitized voice signals can be derived.

Normal service in digital networks requires two twisted pairs of lines/wires between the subscriber and the local office so that the transmission of signals between them can take place on different pairs in opposite directions. The service on a single twisted pair can be obtained by using time compression, multiplexing, and echo cancellation.

Currently, users who are connected to conventional telephone lines by PCM **coder/decoder (codec)** can use customer-to-network signaling on the premises by digital transmission, and an ISDN can be defined.

In time-compression multiplexing, the channels of data are sent in one direction at a time, and at the end of transmission, the mode of transmission in the other direction is changed. The data compression techniques are used to accommodate a large number of digital data elements in the allotted time to achieve a high data rate. The channels of data on the other side are then expanded in full form (with the same data rate as on the transmitting side) from the bursty type of data it receives. A central timing mechanism provides timing slots for transmission of channels of data by either of the stations. Since the channels of data are moving in opposite directions at the same time in the form of bursty (nondeterministic and random information from digital devices) data, the actual data rate of the circuits must be greater than twice the data rate required by the local subscriber node.

In the case of echo cancellation, the transmission of digital signals takes place in both directions over a pair of lines simultaneously within the same bandwidth. A switch known as a *hybrid* allows the transmission in both directions, and any mismatch at any point of the interface will cause a reflection of signals (or echo).

18.2 Evolution of digital transmission

In the late 1960s, the digital interoffice transmission and stored program with controlled electronic switches was introduced in the local office for voice communication (a step toward ISDN technology). The main reason behind this was to define a digital carrier. The tandem or poll offices use **analog-to-digital (A/D)** and **digital-to-analog (D/A)** converters between them for transmission of analog signals. The toll or tandem office on the other side, using an A/D converter, transmits the signal over a four-wire digital carrier. The digital signal is then converted back into analog signal and transmitted from the local office to the subscriber's telephone set in analog form.

The digital carrier offers 24 channels, and the noise is totally eliminated in the transmission. The local office is located physically nearer to the subscriber's telephone set, while the toll office or central office (telephone exchange) is situated at a central location connecting all the local offices. The local office and toll office are connected by four-wire digital carrier offering A/D signal conversion.

During the 1970s, digital toll and tandem switching was introduced. In this configuration, the signal from the local office is transmitted over four-wire digital carrier directly to the toll/tandem office. The digital switching further transmits the signal over four-wire digital carrier toward the local office, where the analog signal is recovered from the received signal. The toll/tandem and local offices provide digital services, and hence D/A conversion before the toll/tandem office is avoided. In other words, only one conversion at the source local office (A/D) and one conversion at the destination local office (D/A) are required. Due to digital transmission and switching, the noise is reduced. The major evolution toward ISDN during this period was, in fact, the development of the **digital toll office** — the toll office uses a digital switch.

In the early 1980s, the **public switched digital service's (PSDS)** switched 56-Kbps service allowed the transmission of both analog and digital signals from the local office to the toll/tandem office over four-wire digital carrier as a digital signal which is transmitted to the other side's local office. At this office, both analog and digital signals are received separately. In other words, both local loop and local control offices use digital transmission and switching. The digital local offices further reduce the cost by converting the analog signal to digital signal at the local office and transmitting it over the toll/tandem office until it reaches the local office on the other side, where the digital signal is converted back into analog signal from the subscriber's telephone system.

In the late 1980s, a new configuration of the public telephone network was introduced: the **integrated services digital network (ISDN)** and the use of a two-wire digital carrier between the subscriber's telephone set and the local office. In this configuration, the telephone set includes a built-in A/D converter, converts the analog signal there itself, and transmits it over a two-wire digital carrier to the toll/tandem office as a digital signal. The toll office then transmits it to the local office on the other side as a digital signal, where the subscriber's telephone set recovers the original analog signal from it. Here, ISDN is used to provide an end-to-end digital connection which offers network service access to integrated two-wire carriers. These integrated two-wire carriers carry the digital signal over local loop connection (between the subscriber's set and the local office).

According to **CCITT SG XVII**,[1] the ISDN is defined as "a network evolved from telephony of **Integrated Digital Network (IDN)** that provides end-to-end digital connectivity to support a wide variety of services, to which users have access via a limited set of multi-purpose standard interfaces." ISDN is a network which provides various services such as voice, data, and images, and also supports various other networks. It supports voice and data services over the same channel via switched or non-switched connections. It offers digital end-to-end connectivity for existing networks.

The amount of data and voice communication to be transmitted over the network has been increasing with different data rates, particularly on the higher side. The requirements for data services are increasing every year at a much higher rate (about 25%) than voice communication services (about 10%). Due to the integration of data, voice, and multimedia applications, it is quite evident that voice communication service data rates are going up due to a number of new requirements, and it will not be surprising that this rate may become at least equal to the data rates (if not higher). This is mainly due to the fact that recent years have seen the introduction of many new telecommunication services; computer technologies; higher-performance, high-speed, and digital switches; more efficient access to databases and specialized services; etc. A single high-performance and high-speed unified network which can provide different types of services (data, audio, video, images, etc.) is currently under development.

The main features of any digital communication system dealing with integrated voice and data communication are usually defined by its switching and transmission. The transmission is based on digital communication, and the first such system was introduced in 1962 by AT&T — the T1 carrier. The first large-scale switch, **4 ESS**, was developed by Western Electric in 1976. Here also, both switching and digital transmission concepts were integrated into ISDNs.

18.3 Digital transmission services

The digital transmission service is transparent to users and independent of the underlying ISDN which provides the circuit or virtual circuit for transmission, and as such users have the freedom to define their own application protocols. It supports both voice (telephone) and non-voice (digital data) communications and hence supports various integrated applications. One of the main requirements with ISDN is that it should offer leased (non-switching) and switched (circuit and packet) services independent of the data being transmitted. This will allow users to get both data and voice services at a reasonable cost with access to a single network.

The standard data rate for digitized voice communication was defined as 64 Kbps. At that time this data rate was sufficient to support various applications for circuit switching.

The digital communication and switching techniques may offer a data rate lower than this (e.g., 32 Kbps). These data rates (64 or 32 Kbps) may not be sufficient for many digital data applications. Interface standards are helpful in selecting the equipment, type of service required (from various patterns of traffic), response time, etc., without changing the existing equipment. Users also get advantages from ISDN, as this network has provided many opportunities for different vendors to develop products for both digital switching and digital transmission and, as a result, manufacturers can spend more time on developing new applications. The set of protocols used in ISDN also follows the architecture approach in defining them, due to advantages already explained in the preceding section and chapters.

Digital subscriber line: This new technology appears to be very hot in the communication industry. The xDSL defines a family of technologies that is based on the DSL concept. The variable "x" indicates upstream or downstream data rates. The upstream data rate is provided to the user for sending data to the service provider, while the downstream data rate is provided to the service provider for sending data to the user. Different DSLs with data rates and applications as defined by the ATM forum are given below:

- Digital subscriber line (DSL) — 160 Kbps, ISDN voice services
- High data rate digital subscriber line (HDSL/DS1) — 1.544 Mbps, T1 carrier
- High data rate digital subscriber line (HDSL/E1) — 2.048 Mbps, E1 carrier
- Single-line digital subscriber line (SDSL/DS1) — 1.544 Mbps, T1 carrier
- Single-line digital subscriber line (SDSL/E1) — 2.048 Mbps, E1 carrier
- Asymmetric digital subscriber line (ADSL) — 16–640 Kbps (upstream), 1.5–9 Mbps (downstream), video on demand, LAN access
- Very high data rate digital subscriber line (VDSL) — 1.5–2.3 Mbps (upstream), 13–52 Mbps (downstream), high-quality video, high-performance LAN

A number of new technologies are being considered to provide high data rates for Internet traffic. Some of these include ISDN, cable modems, and asymmetric data subscriber lines (ADSL). Telco has suggested the removal of loading coils from twisted pair as a means of increasing the data rate at the expense of reduced bandwidth. A cable modem provides a very large bandwidth (due to the cable medium). An ADSL modem, on the other hand, provides much higher data rates than ISDN. Here also, Telco suggests a higher data rate by removing the coils from unshielded twisted-pair cable (being used between the subscriber's set and the end office). An ADSL modem offers a higher data rate for downward and a lower data rate for upward communication. The telephone link is being used for both data communication and the telephone system at the same time. The modem is connected between twisted pair and the end office. On the receiving side, it uses a plain old telephone service (POTS) splitter which provides two different paths on the same link between the ADSL modem and the user's PC and between the splitter and the telephone system.

With digital technology, most of the networks are shifting toward digital transmission over the existing cables. The digital links so defined are supporting the addition of overhead bytes to the data to be transmitted. This payload is a general term given to a frame (similar to the one defined in LANs). The frames used in LANs carry a lot of protocol-specific information such as synchronization, error control, flow control, etc. In the case of ATM networks, the data is enclosed in these frames (if available). For more details, see References 29 and 30.

ISDN offers a variety of services of **CCITT I.210**[3] such as voice and data applications, and these applications can be grouped into two specific application areas: **facsimile and**

telex. With **facsimile,** any text (handwritten or printed) can be transmitted over the telephone network in the analog form standards. Due to the advantages of digital technology, new standards for digital transmission have been defined. With the help of digital facsimile standards, a page of text can be transmitted over the network at a data rate of 64 Kbps in less than five seconds. This time is very significant if we send the same text using analog facsimile standards (about five minutes for one page with a 4800-bps modem). The **telex** services offered by ISDN include **telex**, **teletex**, **videotex**, and **leased circuits**. All of the services except images are supported by ISDN at 64 Kbps and are also known as *teleservices*. The teleservices are described in CCITT I.240.[4] For high data rates of more than 64 Kbps, various services may include music, high-speed computer communication networks for teletex, video phone, TV conferencing, and many other multimedia and broadband applications.

Teletex allows users to interact with each other through the terminals and perform different types of operations over the messages. The operations could be creating and closing files and editing, transmitting and receiving, printing, etc., of the messages where messages are created and stored in the form of files. It is becoming a new widely acceptable international communication service which can be provided by different communication devices such as PCs, word processors, etc., and can be transmitted over telephone networks, packet-switched networks, ISDN, and other upcoming networks. This new service will slowly replace the existing telex machines (bulky, slow, and unreliable). It offers a data rate of 2400 bps (a default) as opposed to 50 bps (maximum) in the case of telex. With the advent of graphic tools and visual programming, it is possible to include various features on graphics and other controls. Due to a wide variety of software and hardware products and their incompatibility, **CCITT T.60** has defined a standard which not only allows different terminals and PCs to interact but also offers the data rate and character set characteristics of document images being supported by telex, etc.

Facsimile and teletex services are collectively known as **tele-ISDN** services and are supported by existing OSI-based standards. CCITT has also defined standards which support ISDN teleservices (both facsimile and teletex). These services defined by **CCITT T.70** are used over packet-switched data networks (PSDNs), circuit-switched data networks (CSDNs), and also public switched telephone networks (PSTNs).

PSDNs use the X.25 standard, and the teletex messages, documents, and images, and facsimile documents and images, are transmitted in an X.25 packet format in sequence. The CSDNs using digital transmission over a digital circuit use two steps of operation for the communication: **call establishment** and **data transfer**. The analog telephone lines of PSDNs support half-duplex and full-duplex configurations over a modem interface.

In contrast to facsimile, in **videotex** an interactive two-way communication of the message data between users is defined via terminals. The transmission of one page can take place in around one second at a data rate of 9.6 Kbps. This application service is defined as a system which combines both textual and graphic information, displays, and terminals (sometimes equipped with a television receiver). The control selection panel is very simple and is attached to the receiver terminals. The advantages offered by videotex include low cost, readily available display devices (television receiver), and the medium.

According to **CCITT I.200,**[5] videotex as a communication system provides two services: **broadcast** or **one-way videotex** (also known as **teletex**) and **interactive videotex** (also known as **videotext**). This technology has found applications in travel and insurance industries, as well as home banking (particularly in the U.K.). In France, videotex has captured the domestic market mainly in video games and personal services. Videotex is used in retailing, electronic shopping, and a few other areas in the U.S. It is also used in corporations, mainly for providing inexpensive, easy-to-use access to computing facilities.

For using this service in general, users have to buy a modified version of a TV or an adapter to access the pages of information. In most countries (e.g., U.K., France, Finland), the intent is to provide videotex service to users over the existing PSTN.

ISDNs support the following services which represent various classes of voice, data, text, pictures, images, etc. **Voice communication** (voice-grade of 4 KHz) requires a telephone set, and the leased circuits of a public network may be used to provide communication media. The digitalization of voice at 64 Kbps can also be transmitted over the telephone network. The transmission of data over ISDN can be either via circuit-switching or packet-switching, and it also supports leased circuits, telemetry, etc. The transmission of integrated voice and data takes place over a single transport connection, and users can have all the services on it. As such, the users need only a single access to the ISDN link, and all the services can be obtained on it, even though each of these services may have different requirements in terms of volume of data traffic, response time desired, interface type, etc. For more details, see References 2, 16, 29, and 30.

The data link layer protocol LAP-B (defined in X.75) is useful for full-duplex, while LAP-B protocol is used for half-duplex.

ISDN supports **teleservices** in addition to the different networks discussed above. The teleservices can also be considered a standard protocol defined by **CCITT** as the **X.400**[6] Message Handling System (MHS). The teleservices can be considered electronic mail, but they do not offer all the functions of e-mail services, e.g., creation, sending, receiving, and storing.

The electronic mail service is being offered as an essential service by many vendors in their networks version which simply allows users to send their mail on the system of the same vendor. In order to send e-mail over the systems of different vendors, we need to adopt the CCITT standard **X.400**. This standard does not describe the services for users but instead offers the services to the scheme of sending the message across the network and can be considered a platform for developing the **user interface (UI)**. This user interface is located inside the message-handling capabilities and is not available to users directly. The X.400 standard defines protocols for eight layers (**CCITT X.400–430)**,[2] describing two services offered to application and presentation layers of OSI-RM, while the services of the remaining layers, in general, are defined by specific protocol services.

A global network service known as **International Business Service (IBS)** defined by IBM provides services to more than 52 countries. This network is being used by various videotex services and also offers services, such as validation of traveler's checks by U.S. banks, inventory management in factories across countries, etc. PC-based videotex terminals offer greater functionality and require a special modem to access the services. The future extension of applications in this technology seems to be in the direction of use of different media like satellite, CD-ROM, and display on low-cost terminals.

18.4 Switching techniques

A computer network may be either a set of terminals connected to a computer(s) or an interconnection between computers. In general, computer networks include computers, terminals, transmission links, hosts, etc. This interconnection between computers allows the computers to share data and interact with others. Sometimes a computer network so defined is also known as a **shared network**. In shared networks (e.g., telephone networks, packet-switched networks, etc.), switching for routing the data packets provides the following:

1. The routing information between source and destination across the network
2. The link utilization techniques which can transmit the traffic equally (depending on the capacity of the links) on all links

These switching functions can be performed on interconnected computers by the following three switching techniques:

1. Circuit switching
2. Message switching
3. Packet-switching

18.4.1 Circuit switching

A circuit-switched network is a collection of circuit-switching nodes where each node consists of various components such as network interface, control unit, and digital switch on one side, while different types of devices (digital and analog) are connected on the other side. The network interface offers an interface to digital devices and phones and also to analog phones. The interface must be capable of converting analog signals into digital signals.

The circuit-switching technique is used in voice communications (e.g., public telephone networks, private branch exchange (PBX), etc.) and data communications (e.g., interface between terminals and computers via hardware switching devices, etc.). Different versions of circuit switching have been defined: **space-division**, **time-division**, **time-slot interchange**, and **time-multiplexed switching**. The following discussion of these switching techniques is derived partially from References 2, 16, and 30.

Space-division switching establishes a distinct and unique path via a switch on an interconnection matrix configuration. It is useful for both analog and digital signal switching.

The concept of synchronous **time-division multiplexing (TDM)** is used in the TDM bus (topology of LAN) circuit-switching technique where a time slot is created on the bus, and during this time slot only, the circuit/channel is established. The switching technique is useful in PBX and other small data switches.

The long-distance carrier systems have, in general, agreed to use TDM in their digital networks which transmit voice signals over high-bandwidth (capacity) transmission media such as optical fiber, coaxial cable, satellite, etc. The digital carrier system uses the standard TDM which supports 24 multiplexed channels. Each channel contains eight bits and a framing bit for $24 \times 8 + 1 = 193$ bits. The 193rd bit is used for synchronization between sending and receiving sites. Each channel contains one word of digitized voice data using PCM at a rate of 8000 samples per second. For a frame length of 193 bits, a data rate of $8000 \times 193 = 1.544$ Mbps is available. This digital system (DS) frame control also provides digital data service.

Voice communication: For voice communication, in five of every six frames PCM samples of eight bits are used, while the sixth frame uses a PCM sample of seven bits and the eighth bit is used for control, routing information, etc., and for establishment/de-establishment of the connection. For data communication, the 24th channel is basically used for signaling and synchronization, while 23 channels are used for data information. In each channel, the first seven bits contain the data, while the eighth bit represents a control bit indicating the type of information (data or control) contained in that frame. If seven bits in each channel define the data information, the data rate is reduced to 56 Kbps (8000×7). Using multiplexing, the lower data rate can be achieved such that more sub-channels can be multiplexed within this bandwidth.

For example, if we take an additional bit from each channel, then each channel will have a data rate of 48 Kbps (8000×6). This also indicates the bandwidth of the channel. As such, sub-channels of different data rates can be multiplexed within the data rate limit of this channel.

Table 18.1 Bell Systems Digital Signal Number Scheme and Data Rates

Digital signal scheme/carrier number	Number of channels	Data rate (Mbps)
DS-1/T1	24	1.5
DS-2/T1-1C	48	3.15
DS-3/T2	96	6.312
DS-3/T3	672	44.736
DS-4/T4	4032	274.176

Table 18.2 CCITT TDM Carrier Standards and Data Rates

Number of channels/level number	Data rate (Mbps)
30/C1	2.048
120/C2	8.448
480/C3	34.368
1920/C4	139.264
7680/C5	565.148

Let us consider a seven-bit channel which has a data rate of 56 Kbps. In this channel, we can multiplex five sub-channels of 9.6 Kbps, 13 sub-channels of 4.8 Kbps, 26 sub-channels of 2.4 Kbps, and so on. For combined voice and data communication, all 24 channels are used, and no bit is reserved or used for control or signaling or synchronization. Different digital signal numbers for the TDM carrier standard have been defined and are currently being used.

The digital signal number scheme defined by Bell Systems, and also known as North American standards, ranges from level 1 to 4 with numbers as shown in Table 18.1. CCITT has also defined its own set of standards for TDM carriers which uses all eight bits in the channels (Table 18.2). For details of these ranges and standard framing and hierarchy, see References 2, 7, and 8.

The multiplexing among the TDM carrier standards is also possible in the same way as in the case of TDM channels; e.g., T1-C may accommodate two T1s, T4 may accommodate 168 T1s or 84 T1-Cs or 42 T2s, and so on.

Routing strategies: The routing strategies in circuit-switching networks usually require the establishment of a complete path or circuit in the network between two stations before voice or data communication can take place. The establishment of a path is defined in terms of various switches, trunks, capacity, etc. One of the main objectives of routing strategy would certainly be to increase the throughput of the network by minimizing the number of switches, trunks, etc., and optimizing the utilization of the capacity of the trunk during heavy traffic loads. With a minimum amount of switches and trunks, it is expected that the cost will go down subject to the condition that the traffic is handled properly during heavy traffic.

Public telephone and telecommunication carriers offer both local and long distance services and, in general, the local services are offered by **Bell Operating Companies (BOCs)** which operate under telephone carriers (e.g., AT&T in the U.S.), while the long-distance services are provided by various telephone carriers such as AT&T and MCI in the U.S. The public telephone network has a very generic architecture which includes stations (telephone sets), an interface (between telephone set and network), switching nodes, and the links (usually known as **trunks**) connecting these nodes. The trunk carries multiple channels and is usually connected to the switching nodes by a link known as a **trunk link**. It carries both voice and data signals, and each trunk carries multiple channels using synchronous time-division multiplexers. The telephone set consists of a transmitter and receiver for sending/receiving the analog or digital signals to/from the networks.

The telephone set is connected by a pair of wires (generally, twisted pair) to the switching node(s) of the network and is usually known as **local loop** or **subscriber loop**. It uses digital transmission. It carries digitized voice signals over a full-duplex configuration. The local loop usually covers a distance of a few miles. The end loop also connects the telephone sets to the switching nodes located at a common place known as the **end office**. This end office offers services to a few thousand users. These end offices are not connected directly but rather are connected via intermediate switching offices, which offer routing to various requests. The switching centers are usually connected by high-speed links known as **trunk links**. These trunk links support multiplexing to carry a large number of voice-frequency channels/circuits.

To get an idea of the architecture of a public circuit-switched network, we take as an example the U.S., where the currently used public circuit-switched network (PCSN) architecture has ten regional centers which are connected to 67 sectional centers. Each sectional center is connected to **primary centers** (about 230), which in turn are connected to **toll centers** (about 1300). The toll centers are further connected to end offices (about 19,000), which provide services to more than 150×10^6 users. For more details, see Chapter 6. Various routing approaches used in circuit-switching telephone networks include **direct**, **alternate hierarchical** (includes additional trunks), **dynamic nonhierarchical** (selection of a route takes place dynamically during the establishment of connections), etc.

The establishment and de-establishment of connections, maintenance and control of connections, etc., are performed by control signals which are exchanged among users via switches of different hierarchical architecture of the network. The control signals are used to help users make a request call and to help the network manager manage and control the connection and other aspects of network management.

Signaling techniques: Control signaling provides a means for managing and controlling various activities in the networks. It is exchanged between various switching elements and also between switches and networks. Every aspect of network behavior is controlled and managed by this signaling, also known as **signaling functions**.

Various signaling techniques being used in circuit-switching networks are **in-channel in-band**, **in-channel out-of-band**, and **common channel**.[2,9,30] These techniques have been discussed in detail in Chapter 6, and we summarize them in brief here.

With circuit switching, the popular control signaling has been **in-channel**. In this technique, the same physical channel is used to carry both control and data signals. It certainly offers advantages in terms of reduced transmission setup for signaling. This in-channel signaling can be implemented in two different ways: **in-band** and **out-of-band**.

In **in-channel in-band** signaling, the control signals (defining different types of tones for different functions in the telephone systems) are sent in the same frequency bandwidth as voice-grade signals, and it has been observed that this technique becomes ineffective and useless, as the tones used for control signaling may get mixed with the signals of voice-grade frequency and cause a considerable problem, especially with the public telephone centers. Also, in in-channel in-band signaling, the control signaling uses the same physical channel. In this way, both control and data signals can have the same properties and share the same channel and transmission setup. Both signals can travel with the same behavior of the network anywhere within the network at any location. Further, if any conversions (analog-to-digital or digital-to-analog) are used in the networks, they will be used by both signals in the same way over the same path.

The **in-channel out-of-band** signaling technique uses the control signals of frequency outside the frequency range of a voice-grade band (a small signaling band within 4 KHz), thus offering continuous monitoring of the connection between stations. The main problem with both in-band and out-of-band signaling is that in both techniques, the voice and control signals are sent on the same circuit. In out-of-band in-channel signaling, a separate narrow frequency band is defined within voice grade (4 KHz) for control signals and can be transmitted without even sending data signals. In this way, the monitoring of the channel is continuously performed. Due to lower frequency for control signals, the transmission speed of signaling is lower, and further, the logic circuit to separate these signals is complex.

Both of these methods of in-channel signaling suffer from the fact that the circuit-switched networks always introduce delay and the frequency bands for control signals are also very low. A new signaling technique known as **common channel signaling (CCS)** has been introduced which alleviates these drawbacks. The main philosophy behind this is to define two separate circuits for control and data signals. **CCS** requires one set of procedures for using a group of trunks, and control information is sent on a control channel (consisting of control switches) and need not go through the voice-channel processing procedure. In other words, the control information does not require the voice-channel equipment. Due to this feature, the interface between control and data information is minimized, and also the misuse of control information is avoided (in contrast to in-band or out-of-band techniques). The details of CCS can be found in Chapter 5 and Section 19.9.5 of Chapter 19.

The CCS scheme is better than in-channel signaling and, in fact, all the public telephone networks including integrated digital networks are now using this technique, as it provides a unified and general common channel signaling system for all standards.

The broadband systems use two channels for transmitting the data. One channel, known as an **outbound** channel (54–400 MHz), is used for data leaving the **headend** toward the network, while the other channel, an **inbound** channel (5–30 MHz), is used for carrying the data from the network toward the headend. The headend can be considered a repeater which provides frequency conversion of signals. It is an analog translation device which accepts the data defined in inbound band channels and broadcasts it to outbound band channels. All the nodes send their data on inbound band channels and listen for their addresses on outbound band channels. There is a small band known as a **guard** which separates these two bands. These two bands define a number of channels in their respective bands. Each channel can carry different types of data, e.g., voice, video, data, etc., at different rates. No multiplexing of voice mail signals with video signals will be carried out. It works on bus and tree topologies and is suitable for multi-drop operations.

Signaling System 7 (CCITT standard): Signaling System 7 (SS7) provides a common channel signaling to provide support to a variety of integrated digital networks. It is also known as inter- and intra-network protocol and offers more flexibility than the traditional in-channel signaling scheme discussed above. The CCITT has recommended this standard SS7 as an interface to translate the existing in-channel signaling into a **common channel signaling (CCS)** scheme. SS7 is a very useful interface, as it supports different types of digital circuit-switched networks and also is recommended for use in upcoming integrated digital networks (ISDN, B-ISDN, etc.). It offers all the controlling functions within ISDN and other integrated digital networks. Some of the material herein is derived from References 10–12 and 30.

When used with digital networks, SS7 utilizes a 64-Kbps digital channel and provides a reliable signaling scheme to route data and control packets in a point-to-point configuration. The control packets contain control information such as setup, maintenance, connection establishment and termination, controlling of information transfer, etc. These control packets are transmitted across the circuit-switched network using a packet-switched control signaling scheme. Although the underlying network is a circuit-switched network, the controlling and accessing of network services are performed using the packet-switching technique.

SS7 offers signaling procedures to various services of ISDN which typically include user-to-user signaling, call forwarding, calling number ID services, automatic callback, call hold, and many others. Each of these procedures is defined in appropriate documents. More discussion on SS7 can be found in Chapter 19. The CCITT has recommended the following Signaling System 7 standards:

- **Q.701 to Q.710** standards define various aspects of message transfer part (MTP) services which are similar to the X.25 packet-switching network. The MTP offers a connectionless transport system for reliable transfer of signaling messages.
- **Q.71X (Q.711 to Q.714)** standards define the signaling connection control part (SCCT), which offers OSI-RM network layer functions. These functions of OSI-RM include addressing, connection control, etc., which are not included in SCCP.
- **Q.72X (Q.721 to Q.725)** standards define the telephone user part (TUP), which allows the use of MTP's transport services for circuit-switched signals and also provides control for both types of signals (digital and analog).
- **Q.76X (Q.761 to Q.766)** standards define the ISDN user part, which also uses the MTP's transport services and SCCP's network layer services for ISDN.
- **Q.79X (Q.791 and Q.795)** standards basically deal with the various operations, administration, and maintenance (OAM) issues for defining the standards of SS7. SS7 offers the services and functions which can be used in both packet- or circuit-switched networks. The main function of SS7 is to provide interfaces and control functions to the applications being offered by integrated networks (circuit-switched and packet-switched). It does not possess any data transfer facilities. It uses the functions of the network layers of OSI-RM in defining routing, addressing, and other performance-related parameters of the network.

In conclusion, circuit switching in communication networks is typically responsible for providing efficient, reliable, and high-quality voice communication. Due to high demand for digital switching, a dimension of the integrated digital network has emerged, and we hope that the new techniques in digital switching (like TDM of data or digitized voice, synchronous TDM, etc.) will certainly reduce the cost and improve the quality of the signal received.[10-12]

18.4.2 Packet-switching

As mentioned earlier, circuit-switched networks are basically designed for voice communication over short and long distances, and the main traffic in the network is voice. For each call request, a dedicated circuit is established during the duration of the conversation, and utilization of resources during this time may be very high. There is some delay during the establishment of connection between two stations, which may become a bottleneck if we want to send data over the circuit-switched networks. Further, the data can be sent at a fixed data rate which might restrict the use of circuit-switched networks for high-speed terminals or computers to transmit the data over the network. During the establishment of connection, the communication link is not utilized, resulting in a wastage of resources.

In principle, packet-switching is based on the concept of message switching, with the following differences:

1. Packets are parts of messages and include control bits (for detecting transmission errors).
2. Networks break the messages into blocks (or packets), while in message switching, this is performed by the users.
3. Due to very small storage time of packets in the waiting queue at any node, users experience bidirectional transmission of the packets in real time.

In message switching, a message is divided into blocks, and the first block contains the control information (routing across the switching nodes of the network). Since there is no limit on the size of a block, blocks can be of any size, and this requires the switching node to have enough buffer to store these blocks. Further, it reduces the response time and throughput of the networks, as the access of these blocks from secondary devices requires significant time.

In packet-switching, the message is divided into blocks (or packets) of fixed size and, further, each packet has its own control information regarding the routing, etc., to be transmitted across the network. This means that the packets may follow different routes across the networks between source and destination and also may arrive at the destination in a different sequence. The receiver, after receiving the packets out of sequence, has to arrange the packets in the same order that they were transmitted from the source.

Routing strategies: As discussed above, in packet-switched networks, the data information is transmitted as packets of fixed size (typically 700–950 octets or bytes). Each packet contains two fields: user data and control information. The control information includes routing, destination addresses, and a few parameters pertaining to the quality of service, etc. The following steps are required for delivering a message data from a source to a destination node. The messages are defined in terms of packets of fixed size. The packets are routed via different switching nodes or stations until they reach the destination node. At each node, the packets are stored, and free links to the next node(s) are searched.

In fixed routing, the routes for the packets are predefined and fixed and, hence, cannot be changed without a change in the conditions. This technique may be useful for a fixed network, which offers stable operations under all conditions. In the flooding routing technique, a transmitted packet is replicated, offering a high degree of reliability. The routing decision in the adaptive, isolated, distributed, and centralized techniques, in general, is determined at each node based on some prerequisite preferences, the knowledge of interconnectivity of the network, delay conditions, and other parameters. The main difference between distributed and centralized routing strategies is that in the former, the routing decision is made at each node based on the information available there, while in the latter, the routing decision is made at a central controller based on the information it gets from various switching nodes.

In the event of a free available link, the packets are sent over it to that node (connected to this free link), and this process is repeated until the packets are received by the destination. As is evident from these steps, the packets belonging to the same message may arrive at the destination at different instants of time and also may follow different routes. It is the responsibility of the destination host to deliver the message in proper order by extracting it from the received packets.

The switching technique in a packet-switching network can be implemented either by a **datagram-switching** approach (for connectionless services) or by a **virtual circuit–switching** approach (for connection-oriented services). In the former approach, each packet contains all the required information and is transmitted independently, while in the latter approach, a predefined route is identified and established before any data can be transmitted. No setup time is required in datagram switching, and as such it is useful for short data messages. Further, the performance and reliability can be improved by bypassing congested traffic routes and network failures. In other words, we can say that the datagram-switching technique does not require setup time and hence is more suitable to connectionless applications. The virtual-switching technique, on the other hand, supports connection-oriented applications, and the routing is predefined and decided during the setting up of the connection. The main disadvantage with this technique is that if any node in the predefined or pre-decided route crashes, the virtual circuits using that node are lost (in contrast to circuit switching).

In general, most of the available packet-switched networks use virtual-circuit switching for data communication. Some private packet-switched networks use the datagram switching technique. As users, we really are not concerned with the switching technique as long as our needs and requirements are met reliably and efficiently. As a network manager, the switching technique may become an important issue, as it may affect parameters such as performance, cost, reliability, packet size, efficiency, network management, etc.

18.4.3 Features of circuit- and packet-switching techniques

To provide continuity and easy understanding, we review below the features of circuit and packet-switching. A more detailed discussion can be found in Section 3.8 (Chapter 3).

Circuit-switching features:

- A constant delay for each packet
- Establishment of connection for call requests
- Transmission of data information over a dedicated circuit or link or channel
- No need to establish a connection, as it already exists
- Little or no delay at the switching nodes, and a small transmission delay
- After the exchange of data information is complete, the connection is terminated by either one of the nodes (source or destination)
- Useful and fast for interactive communication
- Fixed data rate and bandwidth
- No extra bits for synchronization or any other purpose after the establishment of a dedicated circuit connection

The circuit-switching technique is very useful in applications of an integrated environment of voice and data, where a communication set of procedures and protocols can be used. The circuit-switched network uses dial-up lines, and billing charges depend on data rate and distance and duration of the connection. This is very useful for small traffic (low volume), but for heavy traffic (large volume), leased-line, semi-permanent, or even dedicated circuits may be more cost-effective options, as the billing charges depend on the data rate, distance at a fixed cost per month, and so on. These circuits (leased lines or dedicated) may be leased from the telephone carrier companies, satellite carrier companies, microwave, or any other transmission media companies.

Packet-switching features:

1. Virtual-circuit (connection-oriented)
    - A delay at each switching node for a call request due to waiting time in the queues of the nodes
    - Establishment of a virtual circuit on the link one at a time until the final destination node
    - Transmission of packets over established virtual link or circuit

- No dedicated link or circuit
- The total delay during the establishment of connection may be more than a constant delay during the setup phase in circuit switching; transmission delay for packets; some sort of speed and code conversion
- The delay at each node may vary with the number of packets it has on its queue
- Reliable and fast for interactive communication
- Variable data rate and hence more bandwidth to the packets in dynamic mode

2. Datagram (connectionless)
    - No setup phase for establishing a connection
    - Routing of each datagram separately and independently
    - Delay at each node (same as in virtual circuit and packet-switching)
    - Useful for small-sized packets (small delay)
    - Significant delay introduced at each switching node, as the packets have to be accepted, stored, re-routed, etc.
    - Not suitable for large-sized packets
    - No dedicated link or circuit
    - Fast for interactive communication
    - Speed and code conversion
    - Transmission delay, call setup delay

The packet-switching technique is used in integrated voice and data with simultaneous code and speed conversion options, efficient utilization of switches, and communication/link/trunk, with the only drawback being delay. This switching technique is very useful in public data networks, value-added networks (for resource sharing and utility services for terminals and computers), private packet-switched networks, and of course, packet-switched data networks (with very low delay). This switching offers two types of communication circuits: virtual circuit (connection-oriented) and datagram circuits (connectionless).

18.4.4 Fast packet-switching

The integration of data and voice over high-speed networks has led to the development of high-speed switching techniques, and one of these techniques is **fast packet-switching (FPS)**. This switching technique has been very successful in ISDN technology. The packet-switched networks provide various services to high-speed wide area networks, e.g., data rate conversion, transmission of busy traffic, and flexibility. These services, in fact, can be considered options (to the users) provided by fast packet-switched networks. FPS has given a new direction to the development of high-speed networks based on ISDNs such as B-ISDN and ATM-based.[13-16,30]

Packet-switched networks, however, suffer from drawbacks such as delay at the switching nodes due to variable size and fixed capacity of the links. Due to this drawback at the switching nodes and also due to the high-speed requirements, this new concept of FPS has been introduced. As a result, the fast packet-switching (retaining the main services offered by the packet-switching technique and reducing its drawback) is becoming popular and offers services in the following two modes: **virtual circuit** and **end-to-end error control**.

The main concept behind fast packet-switching is that the error- and flow-control protocols are not used for every link connection but instead are used for the end-to-end connection only.

In the case of voice communication, the small packets are sent over a high-speed communication channel connecting fast packet nodes. In spite of the high transmission speed, the delay at the switching nodes may become significant in most cases, as the queue delay gets updated/accumulated over the predefined virtual-circuit route. In order to provide minimum delay, voice packets may be assigned a higher priority over data packets. This may offer smaller delay and still maintain a constant data rate. In the event an error is detected, the packet is discarded, and the use of error control is avoided at each node. Instead, the error-detection and -correction techniques may be used together to ensure that the packets (voice or data) may be transmitted at a constant data rate. In the case of data communication, the packet having the error may not be discarded, and an end-to-end strategy for error control and flow control may (sometimes, but not always) be used. This scheme may work well with the data packets, as we can assign a different level of priority to it with respect to voice packets. The fast packet-switching concept is used in both voice and data communications.

18.5 Public networks

In light of emerging directions and trends of high-speed and bandwidth requirements within multimedia applications, a public network can be used to provide specific telephone/telecommunication network applications and services. These public networks do not offer larger bandwidth and speed and are usually referred to as service-dependent networks. Some of the public networks available are summarized below.

- **Plain old telephone service (POTS):** Telephone services are provided on analog dial-up telephone lines and transported through PSTNs. The dial-up lines use different interfaces to support voice and data communication. Leased lines (analog or digital) are also provided to the users by the telephone network carrier.
- **Telex:** The messages (characters, symbols, alphanumeric, etc.) can be transmitted over a telex network at a very low speed of 300 bps. Usually a five-bit code (Baudot code) is used to code the messages for transmission. This service can also be added to POTS. Similarly, the facsimile service can be performed either as public packet network carrier service or private packet network service. It can also be used on a private circuit-switched network installed within an organization.
- **Public and private data networks:** The data signal can be transmitted over a public **packet-switched data network (PSDN)** using standard public network X.25 (CCITT) protocol (defining the lower three layers of OSI-RM) or a **circuit-switched data network (CSDN)** based on X.21 protocol. However, the circuit-switched networks are not very popular and are being used only for limited applications in very few countries.
- **Community antenna or cable TV (CATV):** Television signals can be broadcast in different ways on various transmission media, e.g., radio wave, microwave, satellite link, and coaxial cable using CATV. The satellite link is also used for telephone conversation, faxes, etc.

Private networks usually are local within organizations or universities and allow users to share resources and mutually exchange data. The popular local networks are IEEE 802 LANs: Ethernet (IEEE 802.3), token ring (IEEE 802.5), token bus (IEEE 802.4), FDDI, etc.

Each of the above-mentioned public or private networks is defined to provide a specific service, and usually this service will not be provided by other networks. Only in very restricted and narrow applications can the same service be provided by different networks using additional appropriate protocols.

In general, for any public high-speed network, there is an ever-increasing demand for greater capacity and a high-speed data rate under the global connectivity environment. There are a number of developments taking place with new technologies, and these are playing a very important role in providing various services to users from one network at a reasonable cost. For higher speeds, usually the transmission and switching criteria are considered to be crucial. On the basis of transmission, we have two main classes of subcriteria as line coding **(asymmetric digital subscriber line — ADSL, and high bit rate digital subscriber loop — HDSL)**, while another is based on multiplexing and data bit rate. There have been two main approaches for this subcriteria: **synchronous** and **asynchronous** modes of transmission. Asynchronous transmission is now defined over DS3, while the synchronous mode of transmission defines a **synchronous digital hierarchy (SDH)** and **synchronous optical network (SONET)**.

Various types of digital subscriber lines are described in Section 18.3. ADSL supports high-speed transmission of digital signals. A number of links have been introduced for providing high data rates, including high data rate digital subscriber line (HDSL; Bellcore), single-line digital subscriber line (SDSL), very high data rate digital subscriber line (VDSL), and asymmetric digital subscriber line (ADSL). Each of these links uses a different signaling technique and offers a different mode of operations. ADSL and VDSL support an asymmetric mode of operations, while the other two support a symmetric mode of operations. These links are used for a limited distance of about one to six km. HDSL and SDSL support T1 and E1 data rates, while ASDL and VHSL support different dates rates for upstream and downstream. For more details, see References 29 and 30, and also visit the Web sites of ADSL, SONET, and SONET interoperability forums.

SONET: The original name of SONET was synchronous digital hierarchy (SDH), as SONET is a synchronous system. The signals represented in electrical form are known as synchronous transport signal (STS). An equivalent optical representation is known as an optical signal. The SDH and synchronous transport module (STM) are popular in Europe. The STM is defined within SDH. SONET, STS, and OC are popular in the U.S. The SONET hierarchy is shown below:

SDH	SONET	Data rate (Mbps)
—	STS-1/OC-1	51.84
STM-1	STS-3/OC-3	155.52
STM-3	STS-9/OC-9	466.56
STM-4	STS-12/OC-12	622.08
STM-6	STS-18/OC-18	933.12
STM-8	STS-24/OC-24	1244.16 (1.244 Gbps)
STM-12	STS-36/OC-36	1866.24 (1.866 Gbps)
STM-16	STS-48/OC-48	2488.32 (2.488 Gbps)

In this hierarchy, each row represents a level of multiplexing. For example, if we want to move from STM-4 to STM-5, the number of multiplexed data signals in STM-5 will be five times the number of STM-4 signals.

On the switching side, we have seen two main classes of switching techniques: **circuit** and **packet**. Circuit switching is used in switched 56 Kbps, ISDN (giving BRI and PRI rates), and voice communication networks. Packet-switching, on the other hand, can be defined as either a fixed or a variable length. The fixed length of packets used in switching and transmission has been used in **asynchronous transfer mode (ATM)** and **switched multi-megabit digital service (SMDS)**, while variable-sized packets have been used in **frame relay** and **X.25 networks**.

18.6 Integrated services digital network (ISDN)

A single network using digital switching and transmission must use an efficient coding technique and VLSI chips for describing various functions and services which will help in reducing the bandwidth of the link. Further, it must allow the network to configure to any new services for their adaptation and transmission over the network. Some of the common resources can be shared by different services (in contrast to the use of these resources by different networks).

The first attempt to provide two services (data and voice) in a single network was the development of an integrated services digital network (ISDN), which transmits both voice and data over the same medium. These two types of signals can be transmitted over the network (which uses both circuit and packet-switching) in a limited way, and the data rate of the services depends on the **basic rate interface (BRI)** or **primary rate interface (PRI)**. No other service (e.g., TV broadcasting, video, graphics, etc.) can be offered by this network due to limited bandwidth.

It is hoped that having all the services provided by one network will make design, maintenance, etc., much more convenient and easy compared to maintaining different types of networks for providing individual services. ISDN uses the existing telephone networks and also can use satellite links and third-party packet-switched networks. These networks are different from the existing telephone service providers. The services of ISDN can also be obtained over packet-switched networks via X.25. It uses a standard 64-Kbps channel for circuit switching carrying voice and data signals.

The data rate of 64 Kbps may be lower or higher for certain applications, but it is considered an acceptable standard which provides excellent voice reproduction on the receiving side. A lower data rate of 32 Kbps (due to digital technology) has been introduced which can accommodate more channels and more data rates over the channel. Data rates of 64 Kbps or 32 Kbps offer restricted services due to lower bandwidth; for this reason, ISDN is also known as **narrowband ISDN (N-ISDN)**.

The development of ISDN is the first step for integrating low-speed services such as audio, data, facsimile, etc., simultaneously over the medium, and a similar concept seems to be applicable with any of the upcoming high-speed networks where the communication link offers a high data rate or bandwidth. ISDNs support both switching (circuit and packet) and non-switching connections for various applications, and each new service to be supported by ISDN has to be defined with a 64-Kbps bandwidth connection. The implementation of ISDN can be carried out in a variety of ways based on the type of applications at a specific place, as it follows layered protocol structure for accessing ISDN. This offers various advantages, as the standards already defined for OSI-RM may be used with ISDN, e.g., X.25 to access packet-switching services and LAP-D which is derived/based on LAP-B protocol (subset of high-level data link (HDLC) protocols). The standard protocols for each layer can be developed independently to meet the services and functions defined by the standards organization. Various services, maintenance, and other network management functions have to be developed by using AI techniques, as it has to perform a different combination of functions before the data information can be transmitted. Some of the material herein is derived partially from References 7–16, 22, 23, 29, and 30.

In the evolution from the current telephone telecommunication computer networks toward the integration of voice and text, many new technological directions have recently been implemented. ISDN became the first network to allow the integration of voice and data (text) using a few existing services in the network. These services conform with the recommendations of standards organizations, in particular CCITT's recommendation specifying the required interface and transmission scheme for integrated signals. All of the available telephone telecommunication computer networks, public and private, are generally characterized by suitable applications. The principles of ISDN, evolution of ISDN, and definitions and services of ISDN have been defined in **CCITT I.200**.[5]

18.6.1 ISDN services

The I.200 series of the CCITT recommendation is the most useful, important, and valuable standard, as it defines all the services offered by ISDN. These services are categorized as **bearer services**, **teleservices (telematic)**, and **supplementary services**.

Bearer services: The bearer services category deals with the transmission of audio, data, video, etc., and these services are usually known as services of the lower layers of OSI-RM. These services define the function of the network (lower layers 1–3). The higher layers (4–5 of OSI-RM) offer functions and transportation of data and are provided by teleservices. These services define the functions of terminals and are offered by the network.

The CCITT I.221[17] recommendation has accepted 12 bearer services of ISDN along with the attributes. The first five services offer a 64-Kbps data rate for data and audio. ISDN provides 64 Kbps (standard data rate) for digitized voice, but the present technology in coding may require 32 Kbps to digitize and transport the voice signal. The unrestricted option with bearer service indicates that the message data will be transferred without modification/alteration. Further, the eight-KHz structured option indicates that a structure (typically of eight bits) is always maintained between the customers, and as such each piece of information includes eight KHz of timing information. Different bearer services, along with different modes of operations (circuit or packet), can be found in CCITT I.230, I.231, and I.232.

Teleservices or telematics: The teleservices defined by CCITT I.240[4] recommendations typically deal with file transfer operations and are being offered over the basic bearer services (defined by the attributes of the lower layers 1–3) by the upper layers (4–7). Various services of this category include telephony, teletex, telefax, videotex, telex, etc. In the telephone services, the speech/voice communication is defined by a 3.4-KHz digital signal. The user information is provided over the B-channel, while the D-channel contains signaling information. Teletex and telefax services provide end-to-end communication for text and facsimile, respectively. The attributes are assigned to higher layers, and different protocols are defined for these services, e.g., I.200 for teletex and facsimile group 4 for telefax. A list of services supported by ISDN is described in CCITT I.241.[4]

Supplementary services: Supplementary services are provided in conjunction with bearer services and teleservices for a specific function of switching the mode of operation, e.g., reversing of switching mode, creation of collect messages in the teleservices of the message-handling system, etc. Each of the services offered by ISDN is associated with various attributes to define the services from which users can compare and select a particular service for their requirements. Some of the attributes are communication and connection configuration, information transfer rate, topology, access channel rate, quality of service connection, performance, etc. The services offered by ISDN are usually defined by the attributes of communication, while the connection configurations are defined by the attributes of connections. Definitions of supplementary services, call-offering and call-completion services, and charging and additional information about these services are available in CCITT I.250 to I.257.

Broadband ISDN and future applications: The ISDN switches are designed for a 64-Kbps voice channel. With the rapid progress in speed coding, image processing, and use of various data-compression techniques, different interfaces for low data rates of 32 Kbps (using adaptive PCM) and 13 Kbps (mobile network) are available, and the existing switches have to be replaced for efficient utilization of resources at these lower rates. The future network must provide all the existing services as well as accommodate future services, and hence it must offer minimum bandwidth for full utilization of the resources (link, etc.). This is what has been considered in **broadband ISDN (B-ISDN)**, and most of the new upcoming high-speed networks are expected to offer higher data rates and bandwidths.

ISDNs provide the services over existing public telephone networks, although other services based on circuit switching and packet-switching of data networks are also offered by ISDN. The development of ISDN, in fact, evolved around telephone networks, and as such many of the services to be provided by the networks are based on voice communication, with additional services to be provided on top of those basic services; ISDN services will be implemented through **digital technology** (digital transmission, digital switching, digital multiplexing, etc.). The packet-switched services will be offered via X.25, and it is hoped that interfaces for high-speed networks, fast packet-switched networks, and other multimedia applications will be developed and interfaced with ISDNs.

With digital technology, most of the networks are shifting toward digital transmission over the existing cables. The digital links so defined are supporting the addition of overhead bytes to the data to be transmitted. This payload is a general term given to a frame (similar to the one defined in LANs). The frames used in LANs carry much protocol-specific information such as synchronization, error control, flow control, etc. In the case of ATM networks, the data is enclosed in these frames (if available). The frames are generated at a rate of 8000 frames per second, and digital links offer a very high data rate. The commonly used digital link in the U.S. is the Synchronous Optical Network (SONET). The basic rate is defined as STS-1, which offers 51.84 Mbps, while STS-3 offers a data rate of 155.52 Mbps.

Presently, ISDN is using existing analog technology and equipment (based on modems) to provide some of the services, but with the operation of digital subscriber loops, digital exchange, and the availability of standards, it will provide all the services to users around the world. One such application of ISDN which has been popular is **videoconferencing**.

The CCITT recommended the following specifications for various services of ISDN:

> The bearer services of circuit-switched are transparent at 64 Kbps. The teleservices include telephone at 64 Kbps, facsimile at 64 Kbps (Group IV), teletex at 64 Kbps, and mixed mode teletex/fascimile at 64 Kbps. Further, the teleservices at 64 Kbps include telephony at seven KHz, audio conference at 64 Kbps, videotex alpha-geometric at 64 Kbps, image transmission and data communication at 64 Kbps, etc. The adapters recommended for non-ISDN services include the X.21 bis adapter and asynchronous V.24 terminal adapter.

18.6.2 Pipeline services

ISDN supports the concept of pipeline services where all the services are arranged in long **"pipes,"** as shown in Figure 18.1. The following services can be integrated, pipelined, or multiplexed over ISDN, which carries the integrated or multiplexed signal (data, voice) to the destination site. It accepts the signals from the following communicating devices: terminals, facsimile, PCs, LAN (supporting many terminals), digital PBX (connecting many subscribers' telephone sets), mainframe computers, and private networks.

The advent of ISDN offered various telecommunication services, and the revenue generated from these services is growing fast. In the mid-1980s, nearly 90% of the services were from voice communications and the remaining were from data communication; in the early 1990s, the utilization of voice communication services was reduced to nearly

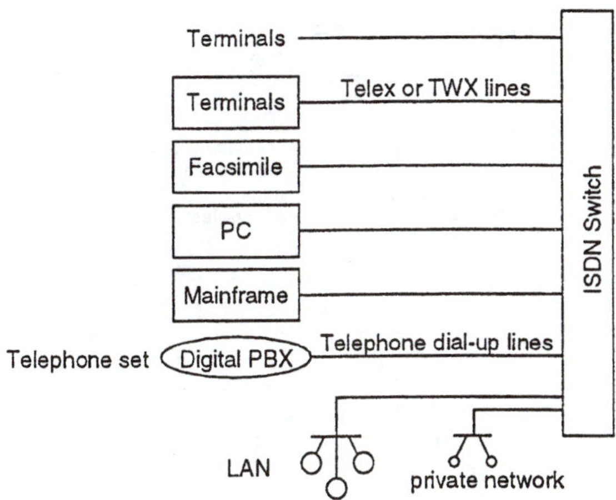

Figure 18.1 ISDN pipeline services.

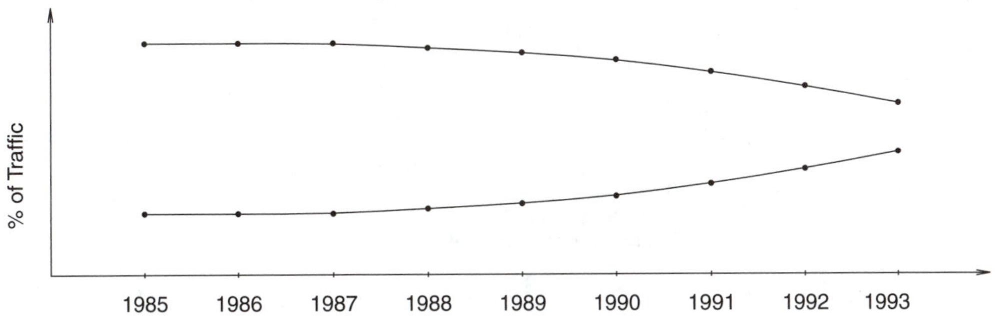

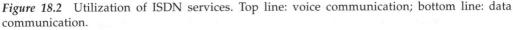

Figure 18.2 Utilization of ISDN services. Top line: voice communication; bottom line: data communication.

80%, while the data communication services were increased to 20% (Figure 18.2). It is expected that with multimedia technology, the upcoming high-speed networks not only will provide high data rates but also will provide and support more video, voice, and data communication services.

The integrated access provided by ISDN is a single two-wire loop which allows the customer to access different types of networks. For example, in a non-ISDN framework, each of the networks has separate access and interface with the switch (Figure 18.3(a)). ISDN offers one common access and interface for all networks over a single two-wire loop, as shown in Figure 18.3(b).

The services offered by ISDN are cheaper compared to individual network services or a combination of services. For example, in a conventional network, if we want to transmit data, voice, and packet, we require PSDS, PPSN, and message-handling service (MHS), while with ISDN, only a 2B+D-channel combination is required. The cost of this channel in some cases may be less than half than that of the existing combination of PSDS, PPSN, and MHS. The channel combination of 2B+D can also serve applications with different requirements such as two voices, one data; two data, one packet; two voice, one packet, etc.

From the user's point of view, ISDN should provide a single device which can be plugged in to an ISDN line and, after confirming the authenticity of the users, can offer services on the basis of call-by-call service, e.g., B-channel services. With the use of a 5 ESS

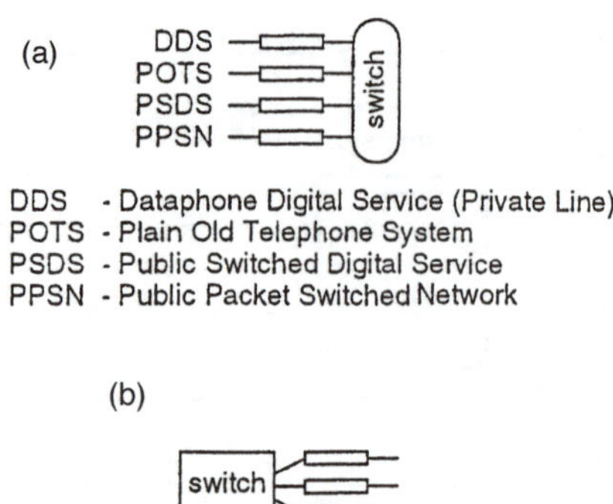

Figure 18.3 (a) Non-ISDN interfacing framework. (b) ISDN interfacing framework.

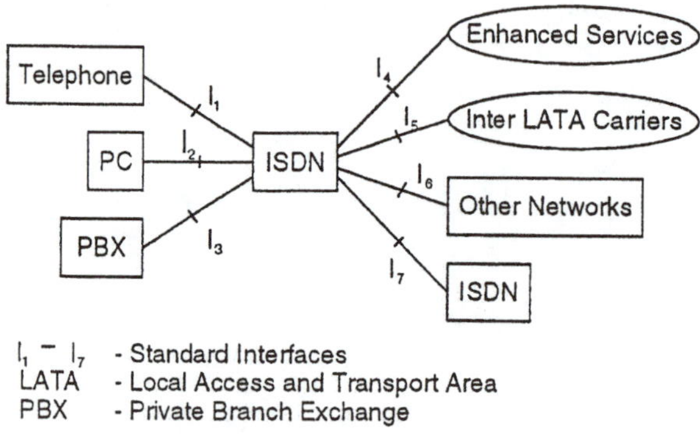

Figure 18.4 Standard interfaces of ISDN.

switching device, ISDN becomes a **telecommunication service provider (TSP)**. It offers access and interface to the subscriber's telephone set, PCs, PBXs, inter-local access and transport area (LATA) carriers, and other ISDNs, and it supports enhanced services (Figure 18.4). The standard protocols and services and various performance characteristics are already defined and currently being used.

The standard protocols for ISDN are a layered model composed of different layers for each layer of ISDN architecture. This layered type of ISDN protocols offers the same advantages as the layered architecture reference model, OSI-RM (discussed earlier). These protocols define an interface for communication between ISDN users and networks and also between two ISDN users. The protocol architecture may not follow the standard OSI-RM architecture, but the concept of defining the protocols at each layer is certainly defined with a view that the entire functionality of ISDN is considered. The exact relationship between OSI-RM and ISDN architecture is not yet standardized.

18.6.3 ISDN reference model

The reference model of ISDN is shown in Figure 18.5(a,b). It was approved by **CCITT I.411**[18] as a **user/network interface (UNI)** and is composed of functional groups and reference points. The following discussion is partially derived from References 2, 10, 12, 13, 16, 22, and 30.

Terminal equipment 1 (TE1) devices use ISDN and support standard ISDN interfaces. Various interfaces which are supported by TE1 include digital telephone, integrated

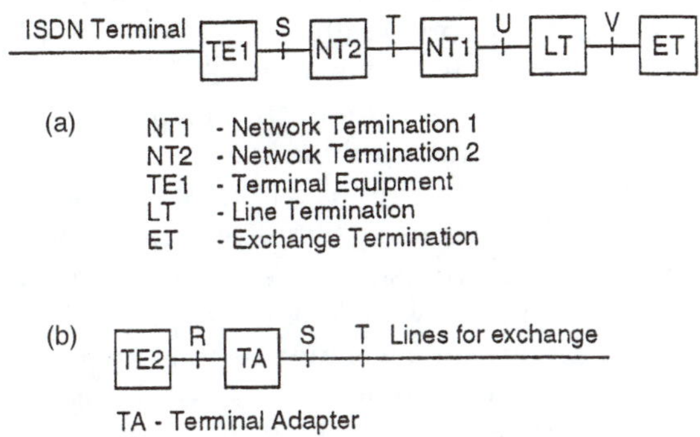

Figure 18.5 (a) ISDN reference model for ISDN terminal. (b) ISDN reference model for non-ISDN terminal.

voice/data terminals, digital devices for telex, facsimile, etc. TE2, on the other hand, supports all non-ISDN devices such as terminals with RS-232 and hosts with X.25. Each of these devices uses another device known as a terminal adapter (TA) to provide interface to the ISDN interface.

TE1 provides a channel or link (over twisted-pair four-wire) between the terminal and the ISDN. The terminal is the same as that of the DTE we discussed earlier (Chapter 10). TE1 is an ISDN terminal capable of connecting directly to the ISDN user/network interface (CCITT T.430).[19] Three channels are supported over the link via time-division multiplexing (TDM). The ISDN interface is defined by CCITT in terms of bandwidth and channel configuration.

There are two **bearer (B)** channels of 64 Kbps each and one **delta (D)** channel of 16 Kbps. The basic access rate (BAR) interface is defined by 2B+D. The TE1 offers a basic rate of 2B+D over a four-wire link.

The basic rate interface offers a 144-Kbps basic data rate. ISDN supports eight TE1 where each shares the same basic rate interface of the 2B+D channel. The same basic rate interface can also be used by a non-ISDN terminal via a terminal adapter (TA). The TA is interfaced with TE2 via an R-series interface and offers services to non-ISDN devices, terminals, etc. The TE2 offers connecting slots to physical device interfaces (RS-232) or V-series interfaces, which are then interfaced with TA to get ISDN services from non-ISDN communication devices. The CCITT I.420[20] standard is applicable to both S and T reference points to enable the connection of identical TE1 and TE2 to both NT1 and NT2 functions (see below).

The reference points between various devices usually describe the characteristics of ISDN signals and provide a logical interface between them. These points are defined by CCITT, as shown in Figure 18.5(a,b).

The devices TE1, TE2, and TA are connected to ISDN devices (for providing basic access rate), and there are two types of devices, **NT1** and **NT2**, where **NT** stands for **network termination**. An NT1 device is primarily used for functions associated with the user's premises (ISDN) and loop connection interface, whereas NT2 offers a standard for multi-way interfaces. These devices are connected at some reference points (defined by CCITT recommendations). These points offer a logical interface between devices or functional modules.

The NT1 is a customer premise device, connects the four-wire subscriber wires to local loop two-wires, and supports eight terminals. It offers similar functions to those of the physical layer of OSI-RM, i.e., synchronization, physical interface, signaling, and timing. Specifically, the functions offered include physical and electrical termination of ISDN as it defines a boundary of the network. The loop connection functions include line maintenance, loop-back testing, etc. It supports multiple channels (2B+D) at the physical layer and multiple devices in multi-drop arrangement where various devices for a particular application in the multi-drop configuration may send signals over the common connection by identifying the device and submitting it there itself.

The NT1 accepts input from ISDN terminals (digital telephone offering digital voice and customer–network signaling over the line, multi-functional terminals offering digital voice, data, telemetry message, or even slow-speed data along with customer–network signaling). The interface with non-ISDN terminals is defined via terminal adapter (TA). The TA offers data communications and customer–network signaling for circuit-switched terminals and also supports low-speed data of packet-switched terminals. The existing X-series interfaces such as X.21 (circuit-switched) and X.25 (packet-switched) are also interfaced via appropriate TAs. These interfaces with NT are defined on customer premises and are also known as **customer premises equipment** or **computer (CPE** or **CPC)**.

NT2 is, again, a CPE but is more intelligent than NT1. It offers the combined functions and services of the data link and network layers of OSI-RM and typically is used with PBXs and other non-ISDN terminals. It allows multiple devices attached to ISDN, e.g., digital PBXs, LANs, terminal controllers, etc., to transmit information across the ISDN. The NT2, in fact, represents devices such as LANs, terminal controllers, etc. The combined functions of both NT1 and NT2 have been defined in another device known as **NT12,** which includes all the functions of NT1 and NT2 in one device. This device can be owned by many service providers and, in fact, these service providers offer services of both NT1 and NT2 to various users.

Each of the devices inside the box represents a functional module (Figure 18.5(a,b)). The reference point S offers the **basic access rate (BAR)** or **basic rate interface** (2B+D) to NT1 or NT2. The T point represents the reference point on the customer side of the NT1 device. The S and T reference points provide interface on four-wire twisted pairs up to a length of one km. These points also support point-to-point or multi-point configuration over a limited distance of less than 200 m. The B channel can support a wide range of standardized communication services, while the D channel is mainly used for outband signaling but can also be used to transport packet-switched data services.

The multiplexing of 23 B+D channels onto one line is known as the **primary access rate (PAR)** or **primary rate interface** and is defined by an interface structure of 23 B channels (each B is 64 Kbps) and one D channel (16 Kbps), giving an interface rate of 1.544 Mbps (Bell's T1). The NT devices/interfaces typically offer communication for networks, while TE devices/interfaces define the communication for users. The PRI provides users with a primary rate of 2.048 Mbps within the **European Conference of Posts and Telecommunications (CEPT)** networks and 1.544 Mbps within North America. At a 2.048-Mbps primary rate, the interface structure is defined by 30 B channels of 64 Kbps and one D channel of 64 Kbps. This interface rate is useful for PBX installations. The electrical specifications of the 30-channel PCM standard for Europe is defined by **CCITT G.703.**[21]

18.6.4 ISDN interfaces

The following is a description of the ISDN channel structure:

D channel: D16 (16 Kbps for BRI)
  Q.931 signaling
  Support for X.31, X.25, LAPD packet user data
D channel: D 64 (64 Kbps for PRI)
  X.931 signaling for N channels (B)

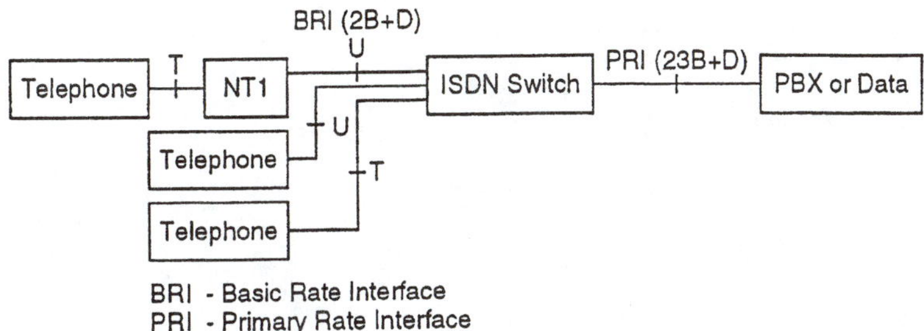

Figure 18.6 ISDN interface.

B channel: 64 Kbps
Support for voice (3.4 KHz), 56-Kbps or 64-Kbps circuit-switched data, and X.31/X.25 packet-switched data
H0 channel: 384 Kbps
H11 channel: 1.536 Mbps

A typical user/network ISDN interface may be defined as shown in Figure 18.6. Here the interface T is defined for use inside the network and uses four wires. The U interface, on the other hand, uses two support wires with **2 binary 1 quaternary (2B1Q) line code,** a standard defined by ANSI T1.601.

The **BRI** offers 2B+D service at a data bit rate of 64 Kbps + 64 Kbps + 16 Kbps = 144 Kbps. The channels are controlled by Q.931 signaling protocol in the packet D channel. The services on demand include voice and data from the first B channel; the second B channel may offer data, video, images, and X.31 packet data, while the D channel basically provides support to Q.931 signaling protocols but can also support the X.31 packet data.

The **PRI** offers 23B+D service at a data bit rate of 23 × 64 + 64 = 1.544 Mbps (U.S.) and 30 B+D = 30 × 64 + 64 = 2.56 Mbps (Europe). The B channels are controlled by Q.931 signaling protocol in the packet D channel. Here all the B channels are typically used for different types of services such as voice, data, video, images, etc., while the D channel provides support to Q.931 signaling protocol only. The PRI interface, on a call-by-call basis, provides all B channels, while the D64 channel provides Q.931 signaling.

18.6.5 ISDN architecture application

A standard ISDN architecture for low bandwidth (home application) is shown in Figure 18.7. The standard architecture or reference model of ISDN is composed of functional groups and reference points. The functional group defines an arrangement of physical

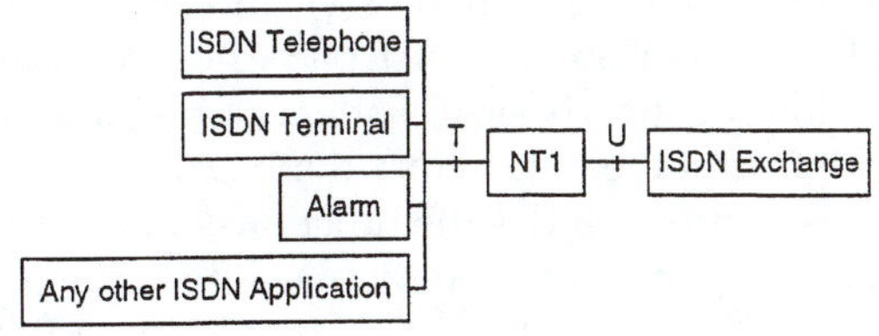

Figure 18.7 Standard ISDN architecture for low bandwidth (home application). (Derived from Reference 22.)

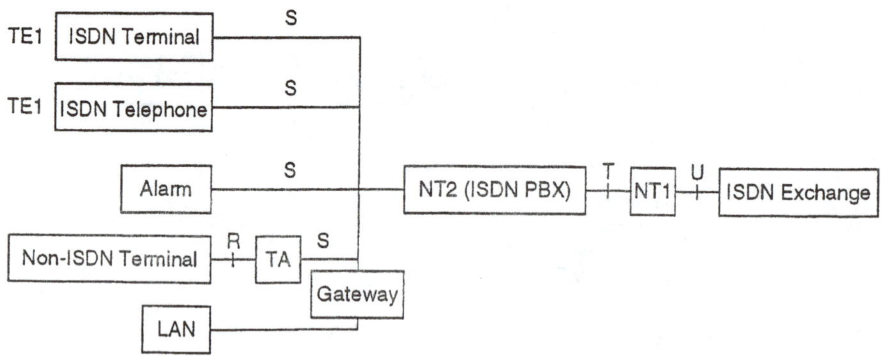

Figure 18.8 Standard ISDN architecture for ISDN and non-ISDN applications.

equipment or combination of equipment, e.g., NT1, NT2, NT12, and TE. Reference points are used to separate these functional groups.

The NT1 functional group offers functions for physical and electrical behavior of the physical layer usually controlled by the ISDN service provider. It also defines the boundary of the network. The line functions such as loop-back testing are also provided by NT1. It supports multiple channels (2B+D) at the physical level and also multiple devices in a multi-drop mode (e.g., PCs, ISDN terminals, telephones, alarms, etc.). All of these devices may be connected to NT1 via a multi-drop mode.

In Figure 18.7, the user's premises equipment includes all the ISDN-based devices up to point T. If we include NT1 also, then this boundary is known as the **user's carrier office (UCO)**. The ISDN exchange falls in the boundary of the carrier's office; thus, the U point defines the boundary between the user's carrier office and the carrier's office.

Figure 18.8 shows a standard architecture for ISDN and non-ISDN applications.

18.6.6 ISDN physical configurations

The CCITT has defined a number of possible configurations of ISDN applications at various reference points (Figure 18.9). These configurations are different for different locations of reference points, as the reference points may be either part of the user's carrier office or it may be a separate reference point in the user's premises. The user's premises may have an ISDN interface at S and T points, or combined S and T where NT1 and NT2 are combined in one device, usually referred to as NT12. In the above configurations, the reference points R and T are not considered part of the carrier's office. The NT2 functional group performs switching and concentration and includes the functions up to layer 3. Some of the popular NT2 devices are digital PBX, terminal controller, and LANs. Consider a configuration where there is a private network which uses circuit switching at various sites and each site may include a PBX. This PBX acts as a circuit switch of the host computer, which acts as a packet-switch. The concentration function basically offers access to the multiple devices attached with a digital PBX to transmit the data across the ISDN.

The NT12 functions group includes the functions of NT1 and NT2. In most countries, ISDN service provides NT12 to users who can use it to access various services offered by ISDN.

Figure 18.9(e,f) presents the configuration (defined by CCITT) where the R and S points are part of the user's office and interface with NT1 for both ISDN and non-ISDN devices is defined at T points. This configuration is used in packet-switched networks where the host computer offers the services of packet-switching to the terminals in ISDN.

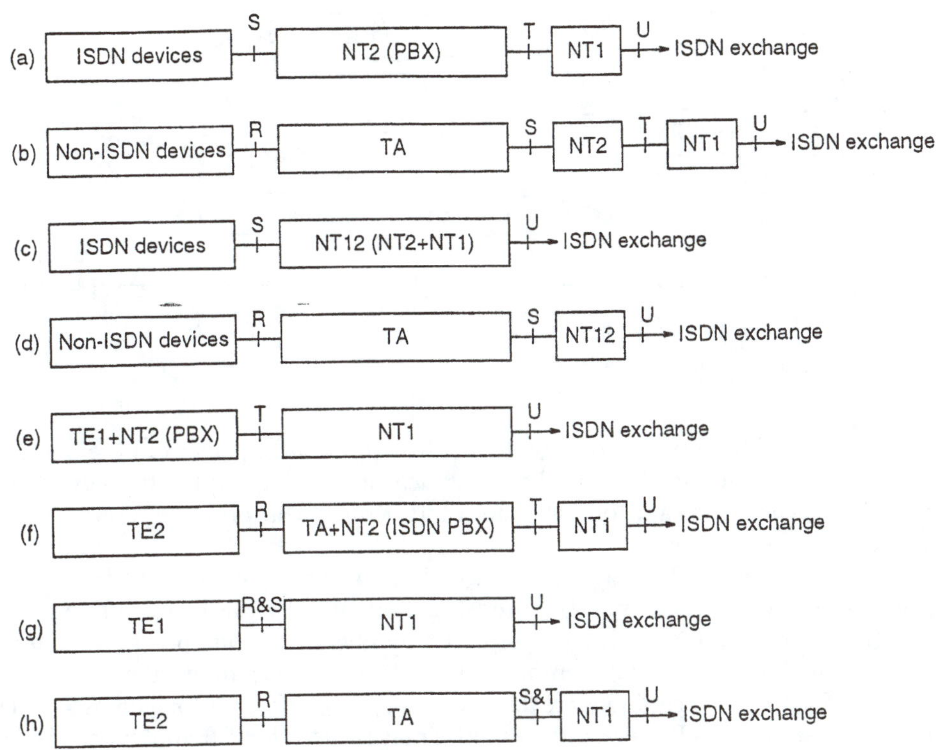

Figure 18.9 ISDN physical configurations.

In situations where both S and T points are at the same location (point), the configuration can be defined as shown in Figure 18.9(g,h). The configuration may be used to connect an ISDN telephone or any other ISDN devices directly to the NT1 or into PBX or LAN, and it offers portability for ISDN interface compatibility.

These configurations may be implemented in different ways for different applications by implementing different functions and features. For example, NT2 may be a multi-drop distribution PABX, LAN, etc., while the combination of NT1 and NT2 may include LAN, multi-drop distribution PABX, etc. These configurations define different implementations which depend on the location of reference points. Similarly, for other configurations, different implementations and functions can be derived depending on the location of reference points.

CCITT has also defined a few configurations whereby users can have more than one physical interface at a single point. The ISDN terminals and other devices can be connected through a multi-drop line or through a multi-port NT1. All TE1s may be connected to NT2 via S reference points. Multiple connections can also be provided between TE1 and NT2 and offer a configuration similar to PBX or LAN.

18.6.7 Standard ISDN architecture

The standard architecture of ISDN proposed by CCITT I.310[23] is shown in Figure 18.10. This architecture defines a digital local loop, connections, and interfacing for various services and support of these services.

A common interface to the network must provide support to all the services of the telephone, PC terminals, video terminals, PBXs, non-ISDN devices, etc. The protocols are

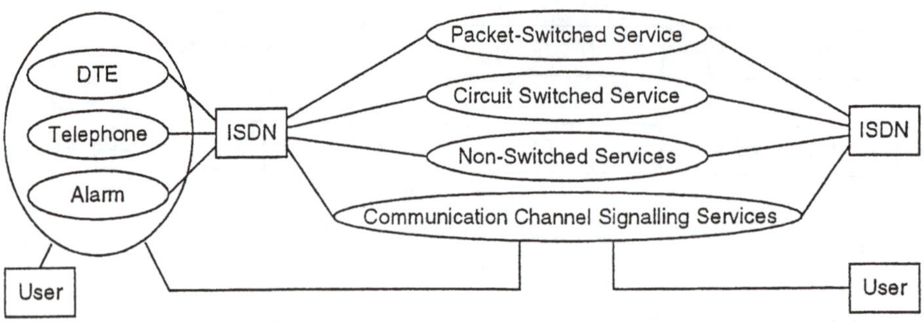

Figure 18.10 Standard ISDN architecture.

defined for each of the services for moving control and data signals from the devices to the networks. The interface offers a basic service of three multiplexed channels (two channels of 64 Kbps each and one channel of 16 Kbps). The primary service includes multiple channels of 64 Kbps.

The local loop provides a physical connection between the device and ISDN central office and is usually a twisted pair of wires. It is hoped that with the evolution of high-speed networks, this may eventually be replaced by fiber optics offering a data rate of at least 100 Mbps. For each of the communication devices, a separate interface, physical connector, and accessing techniques are defined/used, and different protocols are available for voice and data signals. Further, different lines may be used for different types of signals (analog and digital).

At present, more than 800 million telephone users are making use of telephone networks for conversation (voice communication) services, and the telephone networks use different types of transmission media: coaxial lines, twisted pair, satellite link, etc., at different locations within the architecture of the network. With the advent of digital technology, these services are extended to other types of communication services such as data video communication, graphic communication, music communication, and upcoming multimedia applications.

The local loop still uses a telephone twisted pair of lines, and it is expected that digital technology will provide further digital connectivity to end user terminals via appropriate standards and interfaces. Various interfaces shown in Figure 18.10 also need to be replaced by one common interface which will provide user–network signaling.

This communication interface is also known in some books as an *ISDN switch*. The ISDN architecture offers these services at two different layers: at lower layers (layers 1–3 of OSI-RM), it provides access and interface to all of the communication devices and various networks (circuit-switched, packet-switched), while at higher layers (layers 4–7 of OSI-RM), it offers application services such as teletex, facsimile, and other application services. These services may be provided either by a separate network or within ISDN.

The main purpose of communication channel signaling protocol is to provide control and management of various calls/requests for establishing logical connections between users so that the services can be accessed by the users. Some of the issues within the CCS protocols are still under consideration and have not yet been standardized (by CCITT). It has been widely accepted that these functions of CCS within lower layers will be defined within ISDN. In some cases, it has been defined within lower layers on separate packet-switched networks. Some of the material discussed above is derived from References 2, 9, 10, 16, and 30.

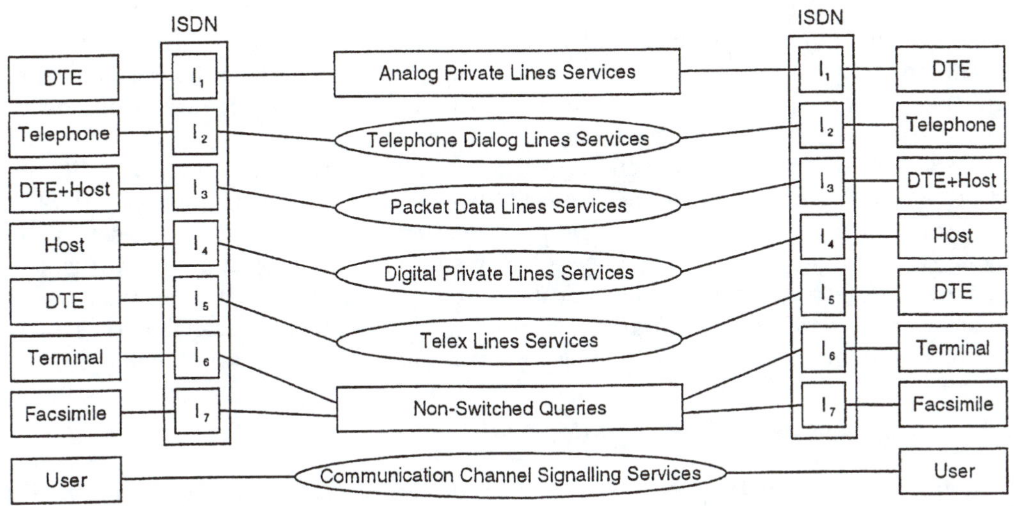

Figure 18.11 ISDN communication channel signaling standard for user–network interfaces (UNIs).

18.6.8 CCITT ISDN communication channel signaling standard for user–network interface (UNI)

For the user–network interfaces, the layering concept used in OSI-RM is being used. In the case of OSI-RM, each layer has a specific service(s) and function(s) and offers these to higher layers at specific points via user processes or entities. These points offer interfaces between layers so that services can be exchanged using the control and data frames. In the same way, various functional groups are defined in ISDN for various reference points (defined by CCITT). The interfaces at these reference points offer compatibility to different vendor equipment and applications. At the same time, these also support the specific functions for each of the reference points. The I.411[18] recommendations for a communication channel signaling (CCS) standard for ISDN user–network interfaces are shown in Figure 18.11.

This standard is basically for ISDN operations and services. This corresponds to the top layer of the call control communication (CCC) frame. This frame accepts the packets from higher layers and defines a CCC frame by defining an I.451[24] call control message at the top layer of the CCC frame. The I.451 passes this packet to its LAP-D (I.441), which defines an appropriate frame by attaching a header and trailer and passing it on to the physical layer. The physical layer supports both BRI (I.430) and PRI (I.431).[25] This message frame is transmitted over the D channel and specifies the procedures and other control messages to establish connection over the B channel. Both B and D channels share the same physical layer supporting BRI and PRI.

The control information for establishing connections, controlling calls, termination of calls, etc., is exchanged between the users and network over the D channel. The format of standard I.451 is defined in terms of the message fields protocol discriminator, call reference, and message synchronization. A fourth message field may be mandatory containing additional information. Each message field contains a set of messages and each message defines its state as a **command** or **response**.

The ideal and conceptual architecture of ISDN may include a set of appropriate ISDN interfaces for various signals coming out from telephones, DTEs, PBXs, gateways (connected to LANs), home safety controllers, etc. The subscriber end loop and ISDN channel/link will carry integrated signals to the ISDN central office. Again from the central office, the signals are going out to different networks via appropriate interfaces over digital links. The signals from the central office are transmitted to different networks (for onward transmission), e.g., circuit-switched, packet-switched, private networks, LANs, and other networks. The digital links between the central office and various networks offer different bit rates (and capacity) and can be requested by the user at the local interface of ISDN. The bit rates or capacity requirements for different applications are different; e.g., a telephone set or single terminal may have a lower data rate, while multiple terminals, telephone sets, DTEs, and other service applications may require a higher data rate and capacity. Either way, the services to these different applications can be made available to users via appropriate interface and control signals and multiplexed onto the same digital link.

18.7 Layered model of ISDN

As indicated above, ISDN offers all the services to end users through layers of OSI-RM. The concept of layering as used in OSI-RM has proven to be useful and has also been used in ISDN structure. Two layers (**bearer** and **teleservice**) together define the complete functionality of OSI-RM. The lower three layers (subnet) of OSI-RM are denoted as the bearer layer, while the top four layers (host) of OSI-RM are denoted as the teleservice layer. The teleservice layer offers services such as telephone, telefax, teletex, videotex, etc., and these are well supported by the bearer layer. The entities and user processes of the layers together provide end-to-end connection for these applications. Typical services offered by the bearer layer include almost all the services offered by each of the lower layers of OSI-RM (physical, data link, and network layers). The services offered by these layers are discussed below briefly for ready reference.

- The **physical layer** offers the transmission of bit stream, multiplexing of channel, connection establishment, and establishment of the physical layer.
- The **data link layer** offers services such as flow control, error control, synchronization, framing, and connection establishment and de-establishment.
- The **network layer** offers addressing, congestion control, connection establishment and de-establishment, multiplexing of network connection, and routing of the data packet.

The services offered by the upper four layers of OSI-RM are collectively provided by the teleservice layer of ISDN. The services offered by the top four layers of OSI-RM are discussed below briefly for ready reference.

- The **transportation layer** offers flow control, error control, segmentation, establishment and de-establishment, and multiplexing of transport connection.
- The **session layer** deals with session management, mapping between session and transport, establishment and de-establishment, and synchronization of session layer connection.
- The **presentation layer** is mainly concerned with encryption/decryption and various techniques for data compression and expansion.
- The top layer (**application**) offers various functions of teleservice to the lower layers.

The ISDN protocol layered model is shown in Figure 18.12. The standard layered protocol, common channel signaling (CCITT I.451),[24] is shown in Figure 18.13. The protocols

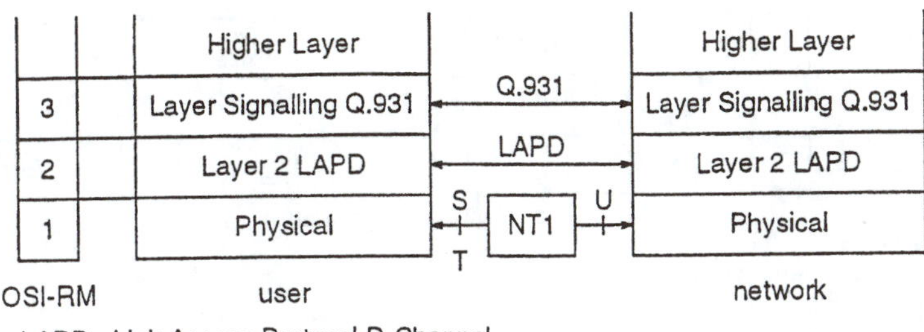

Figure 18.12 ISDN layered protocol model.

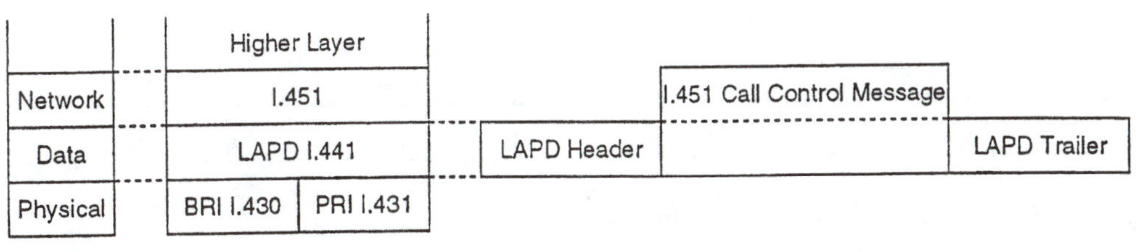

Figure 18.13 Standard network common channel signaling, CCITT I.451.

provide interaction between the ISDN user and network and also between one ISDN and another ISDN. They define procedures for establishing connection on the B channel that shares the same physical interface to ISDN as the D channel. This connection establishment requires three phases: establishing, controlling, and terminating call requests.

The main objective of the Q.931 protocol is to provide a means for establishing, maintaining, and terminating the network connection across the ISDN between application entities. In particular, the Q.931 protocol offers the following functions: routing and relaying, network connection and multiplexing, error control, sequencing, congestion control, etc. It defines a different set of messages for each of the phases; e.g., the call-establishment message includes alerting, connect, call processing, setup, etc., while the call-clearing message may include disconnect, release, and release complete. Other messages include status, status enquiry, etc. Some of the material presented herein is derived partially from References 2, 9, 10, 16, and 30.

The **link access protocol** or **procedure-D channel (LAP-D)** layer is equivalent to a data link layer of OSI-RM and offers transportation for data transfer over the D channel between layer 3 entities. It offers the following functions: establishment of multiple data connections, framing (delimiting, transparency), sequence control, error control, and flow control. It supports two types of services: acknowledged service with support for error recovery and flow control, and unacknowledged service without support for either error control or flow control.

18.7.1 Standard rate interface format

In the following paragraphs, we discuss the frame formats for various rate interfaces. The format for **basic rate interface (BRI)** is shown in Figure 18.14. The frames between terminal

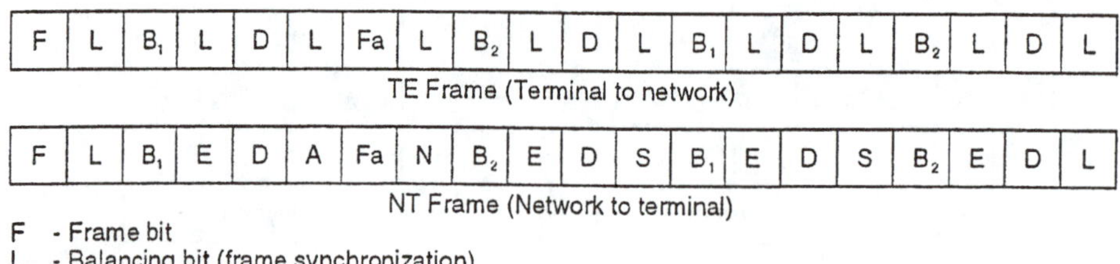

Figure 18.14 Format of basic rate frame.

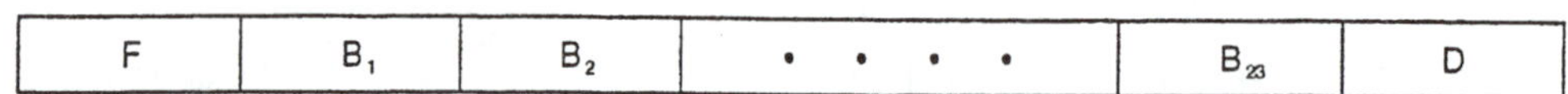

Figure 18.15 Format of primary rate frame.

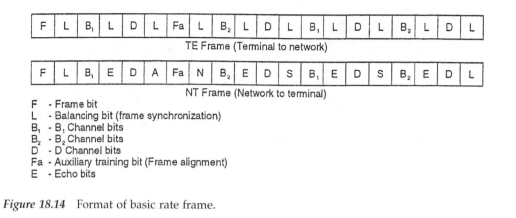

to network and network to terminal have the same number of bits (48), are similar, and are transmitted every 250 ms.

The format of the **primary rate interface** frame **(PRI)** (T1 or 1.544-Mbps frame) consists of 23 B channels and one D channel, as shown in Figure 18.15. This frame is popular in the U.S., Canada, and Japan. Similarly, the frame (2.048 Mbps) consisting of 30 B channels and one D channel is popular in Europe. F is a framing bit and represents 001001 for a length of 24 frames with its value in every fourth frame. The format of the primary rate frame is similar to this, except for the value of F, which is equal to 0011011 in the position of frames 2 to 8 in channel time slot 0 of every other frame.

18.7.2 Link access protocol-D channel (LAP-D) frame

The data link layer protocols of OSI-RM provide framing synchronization, error control, flow control, and connection of data link layers. The standard protocols for the data link layer are SDLC, HDLC, and LAP-B (ISDN). These protocols have been already discussed in Chapter 11 (data link layer). As indicated in that chapter, the majority of the new protocols are a subset of HDLC, and LAP-B is also a subset of HDLC. The LAP-D frame includes user information and protocol information and transfers the user's data it receives from the network layer. It offers two types of services to ISDN users — **unacknowledged** and **acknowledged service** to the user's data of the network layer.

The unacknowledged service basically transfers the frame containing user's data and some control information. As the name implies, it does not support acknowledgment. It also does not support error control and flow control. The acknowledged service, on the other hand, is very popular and commonly used. It offers services similar to those offered by HDLC and LAP-B. In this service, the logical connection is established between two LAP-Ds before the data transfer can take place (as happens in connection-oriented services).

The D channel of the ISDN rate or access interface is used mainly for signaling and offers connectivity to all the connected devices to communicate with each other through it. The protocol for this service is LAP-D, which requires a full-duplex bit transport channel and supports any bit rate of transmission and both point-to-point and broadcast configurations. The format of the LAP-D frame is shown in Figure 18.16.

F	SAPI	TEI	CONTROL	I	FCS	Flag

F - A typical bit pattern of (01111110) for synchronization.
SAPI - Service Access Point Identifier. It also includes a bit representing command/response category and also an extension bit for addressing.

Figure 18.16 Format of LAP-D frame.

The **flag (F)** represents a typical bit pattern of 01111110 (for synchronization). The **service access point identifier (SAPI)** includes a bit representing the command/response category and an extension bit for addressing. The user sends commands by setting this bit to 0 and sends responses by setting it to 1. The network responding to the user's request uses the opposite polarity of the bits for commands and responses. The **terminal endpoint identifier (TEI)** defines the number of terminals in the extension bit for addressing. Both the SAPI and TEI have an EI bit which offers more bits for addressing. The **information (I)** field represents information. The **frame check sequencing (FCS)** field has the same functions as discussed in HDLC and LAP-B protocols.

The LAP-D protocol also offers various modes of operations (similar to HDLC), such as numbered, supervisory, information transfer frame, etc. The control field distinguishes between the formats of these modes. In LAP-D, flow control, error control, etc., are usually not supported and a faster data transfer may be expected. These options are provided (in HDLC) by the **unacknowledged information frame (UIF)**.

18.7.3 Structuring of ISDN channels

As mentioned above, the transmission structure of the ISDN access link can be defined by the B channel (64 Kbps), D channel (16 or 64 Kbps), and H channel (384 Kbps, 1536 Kbps, 1920 Kbps).

The B channel carries the digital data, PCM-encoded digital voice, combination of low-rate traffic (digital data and digitized voice in a fraction of 64 Kbps), 301-KHz audio, 56-Kbps or 64-Kbps circuit-switched data, X.31/X.25 packet-switched data, etc.

The B channel supports the following connections:

1. **Circuit-switched:** For the user's request for call, a connection is established with another network. After the connection is established, the call request is transmitted.
2. **Packet-switched:** Here the user is connected to a packet-switched node and data is transmitted over the network via X.25 between users.
3. **Leased-line:** Here the dedicated connection between users is established and it requires a prior arrangement.

The D channel offers two functions: signaling to control a circuit-switched call on the B channel at the user's interface, and using the D channel for low-speed (100 Mbps) telemetry services when no signaling is waiting. It offers 16 Kbps for BRI and supports Q.931 signaling. It supports X.31/X.25 LAP-D packet-user's data. It also offers 64 Kbps for PRI and supports Q.931 protocol for signaling for N channels.

The H channel is basically used for transmitting the user's information at higher rates, e.g., high-speed trunk. This channel also allows different sub-channels for applications such as fast facsimile, video, high-speed data, high-quality audio, multiple information streams at lower rates, etc. The H0 channel defines 384 Kbps, while H11 defines 1.536 Mbps.

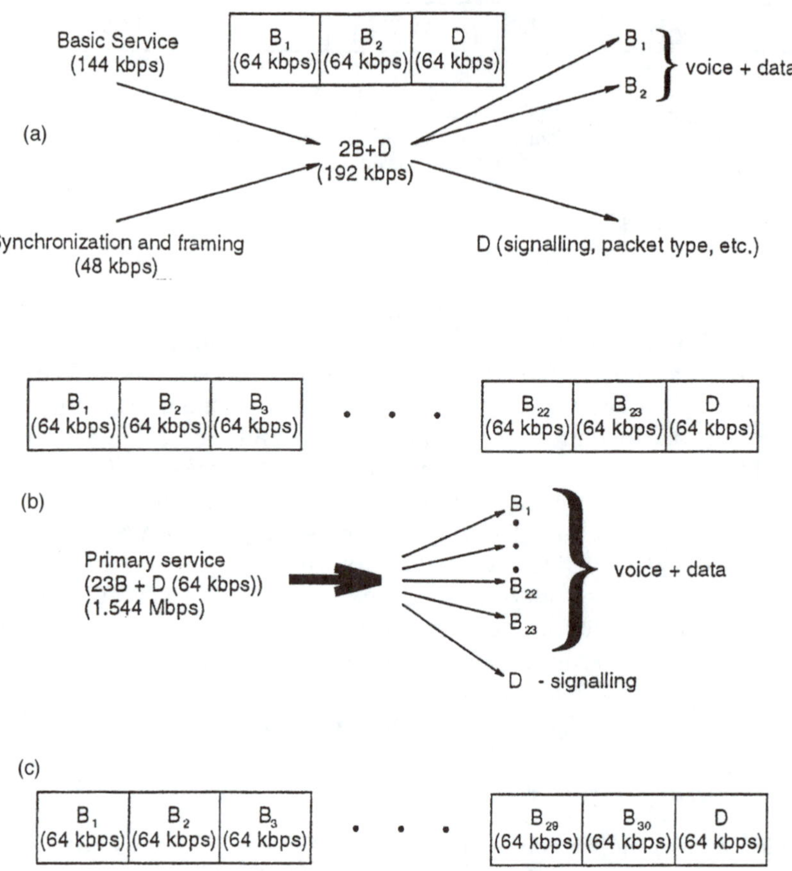

Figure 18.17 Structure of standard ISDN interfaces. (a) Basic rate interface (BRI). (b) Primary rate interface (PRI) (AT&T's T1, CCITT's recommendation for U.S., Canada, and Japan). (c) Primary rate interface (PRI), 2.08 Mbps, for Europe.

As indicated above, ISDN offers two types of access interfaces: basic access interface or basic rate interface and primary access interface or primary rate interface. The recommendations for these interfaces are shown in Figure 18.17(a–c).

Basic rate interface (BRI): The **BRI** is the most common data rate offered by ISDN and is defined by 2B and 1D channels, giving rise to a total of 144 Kbps. There is certain control information, framing control, etc., making the data rate 192 Kbps. All three channels (2 B and 1 D) of the basic rate interface are multiplexed using TDM and, also, each of the B channels can further be divided into sub-channels of eight, 16, or 32 Kbps. The users have complete control over the use of both B channels and can use them in any way. The user information is transmitted over the B channel, and supports a variety of applications, e.g., voice, data, and broadband voice within the range of 64 Kbps.

In the D channel, control and signaling information is transmitted, but it may be used for sending the user's information in some cases. In contrast, the B channel does not carry any control or signaling information. The signaling information for different types of signals is declared type signals and these signals are then statistically multiplexed for transmission. Other channels defined in ISDN are E and H channels, and they are defined for faster speeds. The E channel is of 64 Kbps and is used to carry control information and signaling information for the circuit-switched network. The H channel offers a data

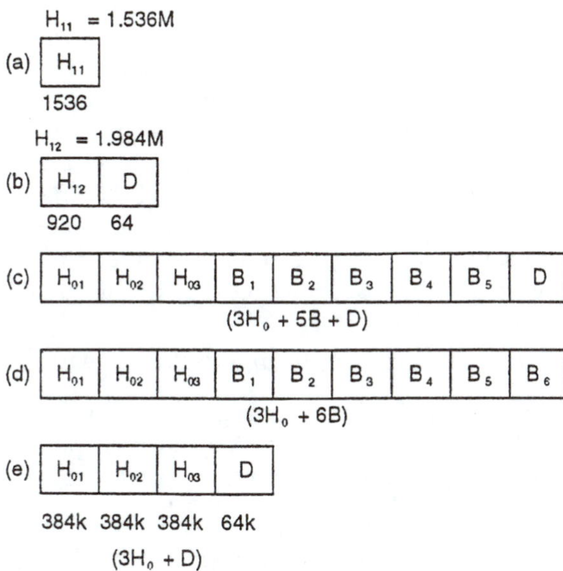

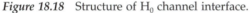

Figure 18.18 Structure of H_0 channel interface.

rate which is a multiple of the B channel data rate and is classified as H0, H11, and H12 for 384 Kbps, 1536 Kbps, and 1920 Kbps, respectively. The structure of these channels is shown in Figure 18.18(a–e).

Users have complete control over the B channels, which can be used in any combination for various applications. In the case of high data rate applications, more primary access interfaces are required. A D channel of any one of the primary rate interfaces may be used for signaling and other control information, while other D channels may be used for voice data in conjunction with B channels.

Primary rate interface (PRI): The **PRI** supports H channels (H0, H11, H12). The structure of these interfaces is shown in Figure 18.18(a).

The primary rate interfaces of 3 H0+D and 4 H0+D provide a 1.544-Mbps data rate, while the 5 H0+D interface supports a 2.0048-Mbps data rate. The primary rate H11 and H12 channel interface structure is shown in Figure 18.18(b).

As stated above, the B channel carries user information (voice data) while the D channel carries signaling information. In order to use the entire B channel for voice data, we need to have devices to support that data rate, but the existing devices (terminals, PCs) operate at a much lower data rate (maybe a few Kbps; e.g., RS-232 can offer a maximum of 19.2 Kbps), and the proper utilization of the channel is not obtained. However, with the advent of ISDN cards and other high-speed interfaces, it is hoped that the devices will offer the data rate of B channel interfaces. Further, the B channel can be used efficiently if we can multiplex various applications (heading for the same destination) on one B channel rather than sending them on different B channels.

The multiplexing of different applications for sub-channels will be the main concept for circuit-switched connections. The multiplexing of various applications in the sub-channels is usually defined in terms of 8, 16, or 32 Kbps coming out of different interleaved applications.

There is a possibility that the applications may not be representable in the order of 8, 16, or 32 Kbps, and in these cases, the applications have to be expressed in terms of a bit stream of 8, 16, or 32 Kbps before multiplexing. The basic and primary rate interfaces usually operate in the frequency range of over 200 KHz, while the voice signal (20 Hz to 4 KHz) is transmitted over a local loop which supports this frequency range. The local loop usually uses unshielded copper wires and usually covers a distance of a few kilometers.

For low frequency (voice), the attenuation may not be a problem, but for high-frequency signals (ISDN), the attenuation may become significant. The **T1D1.3 Working Group** has made recommendations for local and long-distance basic access signaling that reduces this problem. According to the recommendation, the **2 binary, 1 quaternary (2B1Q)** technique for signaling offers a higher data rate. In this technique, two binary bits are used to represent each change in the signal, and a data rate of twice the signaling speed can be achieved. For more details, see References 2, 9, 10, 16, and 30.

18.7.4 Realization of functional groups and reference points

CCITT recommendations have defined functional groups and reference points in the ISDN architecture such that the reference points define interface points between functional groups in which each functional group represents a specific process.

Reference points: The following section discusses in brief the significance of each of these reference points, along with the services and functions offered by them. Other standard user–network ISDN interfaces are shown in Figure 18.19(a–c). The reference points provide interfaces between various groupings.

The U reference point provides interface between NT1 and the transmission line. It offers interface to two-wire NT1 equipment and defines a boundary between NT1 and the line termination equipment. It operates at full duplex with channel splitting, which is used for echo cancellation. This point is not specified in the CCITT recommendation for ISDN but is specified for the U.S. network. According to the U.S. **Federal Communications Commission's (FCC)** recommendation, as shown in Figure 18.19(c), this point is specified on the central office (CO) side of NT1. The standard structure of the ISDN user–network interface is shown in Figure 18.19(a).

In Figure 18.19(a), the reference point T over four-wire twisted pairs (for a typical length of 1100 m for a point-to-point link or 140 m for multi-point configurations) is specified on the user side of NT1. In Figure 18.19(b), the reference point U is specified on the central office (CO) of the NT1 side and supports a two-wire configuration on full duplex.

It is interesting to note here that NT1 may be considered a DTE (defined in OSI-RM) and is known as **network channel terminating equipment (NCTE)**. The reference point S

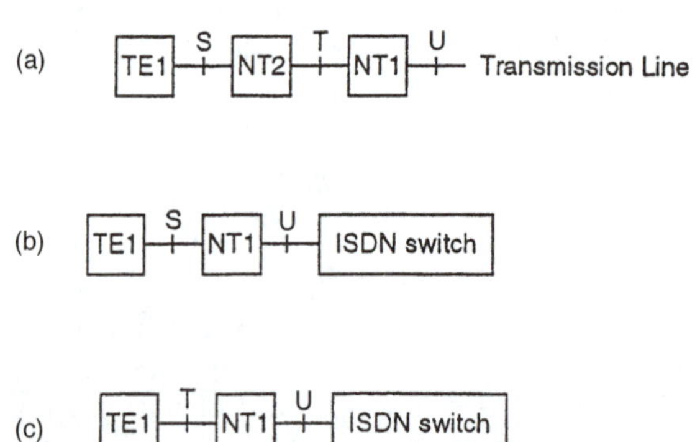

Figure 18.19 Structure of standard user–network interface (UNI) (CCITT I.411). (a) Standard ISDN user–network interface. (b) CCITT's recommendation for user–network interface. (c) FCC's recommendation for user–network interface.

offers basic rate interface (2B+D) to NT1 or NT2 devices, while T defines the interface of NT1 on the customer side. It may be considered similar to the S point.

The reference point T (terminal) defines a boundary between the network provider devices and the user's devices. At this point, the user's device communicates with the ISDN. The reference point **system (S)** offers an interface for different types of ISDN terminals and defines a boundary between user terminal devices and the network communication function offered by ISDN. The reference point **rate (R)** defines the interfaces for non-ISDN terminals with non-ISDN-compatible and other adapter devices. These non-ISDN devices typically follow X or V series CCITT recommendations. The full-duplex user's operation for the data signal over transmission lines is defined by the reference point **user (U)**.

Functional groups: The following is a description of various functional groups used in the reference model of ISDN.

Network termination 1 and 2 (NT1 and NT2) supports the functions which are required to change the format of characteristics of signals being transmitted over the line into a format which is independent of any switching or transmission techniques. These functions are offered by TE1 at the T point. The terminal equipment 1 (TE1), in fact, is a terminal which offers a direct connection between TE1 and user–network interface (I.430).

The NT1 can be considered the lowest layer of the ISDN and is similar to the lowest layer of OSI-RM (physical layer). This layer provides electrical and physical characteristics. In the case of ISDN, the electrical and physical characteristics of connections are provided at the user's premises and are controlled by the ISDN. The transmission line maintenance functions and services are also provided by NT1, and it supports multiple basic rate interface (2B+D). The NT1 supports multiple devices where various application devices (e.g., telephone, PCs, etc.) may be attached to a single NT1 via a multi-drop line configuration. All the channels of these applications are multiplexed using synchronous TDM and are transmitted at the physical level as a bit stream.

In contrast, NT2 includes services and functions of the subnet (all three lower layers of OSI-RM) and is a more intelligent device than NT1. Since NT2 represents a subnet, it could specify LANs, digital PBXs, etc. The NT2 device also allows the integration of multiple-device LANs, digital PBXs, etc., to transmit data over an ISDN, and this process is usually called **concentration** in ISDN terminology. It offers multiple isolated S reference points where each point is identical to the basic access T reference point. It also multiplexes the services and functions offered by layer 2 and 3 protocols, switching to the data packets, etc.

The NT1 is usually located on the user's premises and offers connection to ISDN exchange in the carrier's office. All the signals are multiplexed and transmitted to NT1 via the T point. The user's premises equipment is connected to the carrier office (several miles away from the premises) via a twisted pair of telephone lines which already exists between the carrier office and telephone set. The NT1 has a connector in it to which a cable can be connected. The cable can connect between 8 and 10 ISDN devices, and the signals from these devices can be transmitted over it at the T in a digital bit piping fashion. In addition to the connector, the NT1 also includes the electronic circuit which manages various administrative tasks such as loop-back testing, addressing, contention resolution, etc.

The **terminal equipment** 1 and 2 (TE1 and TE2) are the devices which use the services of ISDN. TE1 supports the standard ISDN interface, which typically includes services for devices such as digital telephone, integrated voice/data terminals, digital facsimile, etc. TE2, on the other hand, supports the standard non-ISDN interface which typically includes physical DTE/DCE interface, RS-232, etc. These devices cannot be connected directly to an ISDN interface but instead have to go through another device known as a **terminal adapter (TA)**, which provides a rate adaptation function between reference points R and S/T. These functions basically accept the data rate of less than 64 Kbps and translate it into 64 Kbps. In the case of circuit-switched networks, various rates of 8, 16, or 32 Kbps or less than 32 Kbps and 64 Kbps over the B channel are supported. These data rates are identified during the call setup which is being requested over the D channel via common-channel signaling (CCS). If the data rates of two users are not the same, then the request for the setup call is not processed.

In the packet-switched network, a user does not have direct connection with another user, but instead, the user uses the circuit connection to one of the nodes of the packet-switched network. This packet-switched network may be connected to another packet-switched network, and the communication between the users of these packet-switched networks takes place via public packet-switched network X.25. The LAP-B frames carry X.25 packets, and if the data rate is less than the data rate of the B channel, it will use a terminal adapter which offers buffers to these packets. These packets are then transmitted over the B channel at 64 Kbps.

The process of **bit stuffing** or **character stuffing** is used to provide the indifference of speed of the packets accepted and transmitted with some stuffed octets which are eventually discarded at the receiving side. This concept of stuffing removes any distinction between the data rate of packets, and the D channel offers the same signaling to both packets of different data rates.

As indicated earlier, the process of digital pipelining carries the signals from the digital telephone terminal or digital facsimile without regard for the type of devices it is dealing with. It only supports the transmission of bits in a piped fashion in both directions. It can be considered as multiplexed (using TDM) independent channels where the channels are carrying signals from different devices. Two standards have been defined by CCITT for digital piping that cover low-bandwidth and high-bandwidth applications. The low-bandwidth applications typically include home applications such as alarms, terminals, etc., while the high-bandwidth applications typically include business applications such as LANs, non-ISDN devices, and terminals.

The higher-bandwidth applications, where the signals are coming from ISDN and non-ISDN devices, may be multiplexed through NT2 (or PBX). This offers a real interface to each of the ISDN and non-ISDN devices, with the only constraint being that these devices should belong to one organization, as PBX or NT2 can handle only a few channels without the use of the carrier's ISDN exchange. In an existing PBX for telephone channels, "9" is typically dialed to get off the premises, while in the ISDN exchange, only a few fixed numbers (4 or 5 digits) are dialed for communication within the organization. NT1 defines the boundary of the network, while NT2 defines the user's PBX. Similarly, TE1 defines ISDN terminals or devices, while TE2 defines non-ISDN devices such as RS-232.

The reference points T and S offer the access points to ISDN for various B channel services, while reference point inputs of TE1 and TE2 offer teleservices. TE1 offers access to ISDN standard terminals, while TE2 offers access to non-ISDN standard terminals, and the services of both terminals are obtained via terminal adapter. The B channel services of 64 Kbps (circuit-switched connection) can be offered at either T or S points.

The new ISDN terminals offer many interesting features, such as **feature phone** (includes directories, recent number redial, call timing and costing, voice answering, hands-free dialing, calling number display, logging of calling and called numbers, etc.); high-speed desk-to-desk file/screen transfer of spreadsheets, documents, graphs, pie charts, data or program files, telephone directories, etc.; high-speed desk-to-database connection to view data and mainframe applications; and advanced messaging (with unattended operation, multiple addresses, automatic retry, copy lists, etc.). The ISDN terminals with these advanced features are very useful in a variety of applications such as management workstations, messaging within large corporations, computer terminals (offers low-cost, high-speed access), bureau access terminals, etc.

Video coding/decoding (codecs) have been introduced which support 56 Kbps for NTSC frames and 64 Kbps for PAL frames with color versions. The resolution of these codecs usually ranges from 128 × 128 picture elements (pixels) offering an acceptable quality of transmission to very high-quality 256 × 256 pixels. The 56/64-Kbps codecs include features such as detection of changes in the picture, graphics mode for high resolution, built-in audio codec (for transmitting both video and audio over the same channel of 56/64 Kbps), etc.

18.7.5 Addressing and numbering scheme in ISDN

For global connectivity over the public telephone network, every ISDN user is assigned a unique number which allows the network to establish the physical connections and initialize the required functions for charging (billing information). The CCITT E.163 defines the addressing scheme, and it uses 12-decimal digital. The original I.163 included only telephones, while later on it was enhanced to larger digits to include other ISDN devices (e.g., ISDN terminals). The numbering scheme for the ISDN is discussed in CCITT I.330[26] recommendations and, in general, follows a similar scheme used in telephone numbering, for example, only the ISDN number for inter-ISDN communication, the country code followed by the ISDN number, the decimal number representation, etc. The ISDN number defines the ISDN network in some cases as the reference point T. This number is used by the ISDN network to route the call.

An addressing scheme, on the other hand, includes the ISDN number (as described above) and any additional information regarding the addressing. The ISDN number includes the information of routing the call, billing information, etc., and in some cases, reference point S. A unique number may be assigned to each NT2 which may have multiple terminals connected to it (at the S point) with separate ISDN addresses. The reference point T between NT2 and NT1 defines the ISDN number. Another way of defining the ISDN number is to assign the ISDN number to the D channel which provides signaling to a number of users where each user may have a unique ISDN address. For non-ISDN devices, the T reference point for NT1 defines multiple ISDN numbers for each of the devices.

The addressing schemes follow more or less the same pattern as that of national or international telephone calls. It includes the country code (1 to 3 digits), nation code (variable length), and ISDN number area code address which includes the sub-address (typically 40 digits). These definitions can be found in I.163 recommendations. A unique ISDN number for international communication includes the country code and the national user number, along with the national destination code, which together define the national ISDN number. The maximum number of digits for the international ISDN number is 15, while the ISDN address is typically 55 digits, which incidentally also includes sub-addresses. The sub-addresses along with the international ISDN number define the ISDN address.

The numbering scheme in internetworking includes the number of the public network, ISDN number, existing public packet telephone network, and existing public data networks (e.g., X.25, telex network, etc.). Each of the networks has a different scheme of numbering, and it becomes quite difficult to define a unique numbering scheme for internetworking. The CCITT has defined a variety of standards, and it appears at this point that the majority of the standards suffer from the same common problem of incompatibility. In spite of this incompatibility, it seems that the E.163 and ISDN standards (CCITT I.330)[26] are compatible and E.163 also seems to be compatible with the public data network (X.121).

ISO has also defined an international numbering scheme for its OSI-RM known as the **Authority and Format Identifier (AFI)**. It supports public networks (like packet-switched networks), telex, PSTNs, and ISDNs, and assigns a unique number to the individual country.

18.7.6 Internetworking with X.25 and LANs

The transmission link between the subscriber's phone system and the end/central office supports only a voice-grade channel occupying a frequency range of 300 Hz to 3.4 KHz, offering a very low data rate of 56 Kbps. Internet traffic needs much higher rates, as it includes complex graphics and also video, which may take a longer time to download on PCs. The links between the Internet and end offices provide data rates as high as 2.5 Gbps. Although modems offering 56 Kbps and even higher rates have been available on the market since 1996, the bottleneck remains the twisted pair connected between the user's phone and the end office. A number of new technologies are being considered to provide a high rate for Internet traffic. Some of these include ISDNs, cable modems, and asymmetric data subscriber lines (ADSLs). Telco has suggested the removal of loading coils from twisted pair as a means of increasing the data rate at the expense of reduced bandwidth. The cable modem provides very large bandwidth (due to the cable medium). It will be used on FDM, as is being done in the case of **community access television (CATV)**. For more information on DSL, see Reference 29.

The television program, in fact, provides simplex channel configuration where the traffic flows in one direction. In the case of cable modems, the traffic will flow in both directions, giving a full-duplex channel configuration. It is expected that with the reconfigurations of CATV into full-duplex mode, data rates of 35 Mbps (downward) and 10 Mbps (upward) can be achieved. The standards and their specifications for the multimedia cable network system (MCNS) and IEEE 802.14 have developed modem. Europe uses DAVIC and DVB standards.

It is hoped that data can be transmitted in different frequency bands, e.g., downward (from cable to Internet service provider to user) in the band of 42–850 MHz and upward (from user to cable to Internet service provider) in the band of 5–42 MHz. The U.S. uses a band of 5–42 MHz for upward transmission, while Europe uses a band of 5–65 MHz.

The ADSL modem provides much higher data rates than ISDN. Here also, Telco suggests a higher data rate by removing the coils from unshielded twisted-pair cable (being used between the subscriber's set and the end office). The ADSL modem offers a higher data rate for downward and a lower data rate for upward communication. The telephone link is being used for both data communication and the telephone system at the same time. The modem is connected between the twisted pair and the end office. On the receiving side, it uses a plain old telephone service (POTS) splitter which provides two different paths on the same link between the ADSL modem and the user's PC and between the splitter and the telephone system.

As mentioned above, the basic rate interface offers a data bit rate of 144 Kbps (two B channels and one D channel of 16 Kbps). The D channel uses a LAP-D protocol offering both signaling and data packets to be statistically multiplexed between the ISDN exchange and customer terminals. The requirements for ISDN are based on the need to offer access from the X.25 terminals to existing X.25 packet data networks. The CCITT I.461 (X.31)[27] defined a support for X.25 terminals. ISDN is based mainly on circuit-switched 64-Kbps connectivity, and the digital exchange may not support packet-switching.

The CCITT I.462[28] recommended two categories of packet-switched communication I series terminals. In the first category, a transparent circuit-switched connection is provided from the user terminal to a port of packet-switched networks. The B channel offers access to packet calls on a physical 64-Kbps semi-permanent or switched B channel. In this category, the integration for internetworking is minimal.

In the second category, greater maximum integration is defined. Here ISDN offers packet-switching within it where both B and D channel access is supported with a packet handler offering support for processing packet calls, standards for X.25, and also setting

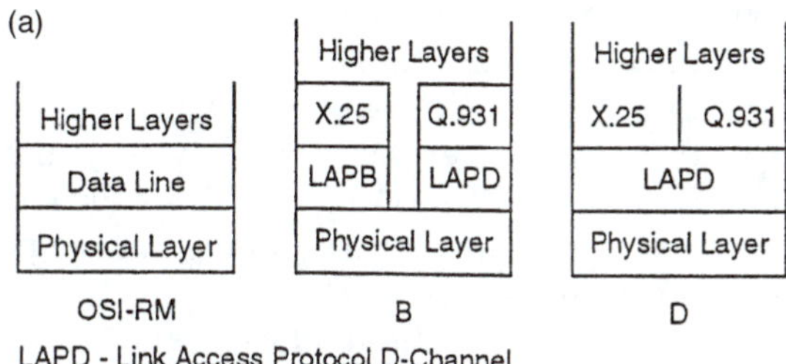

Figure 18.20 ISDN architecture and internetworking.

of functions for adaptation. The call setup for each channel first uses ISDN access to LAP-D signaling procedures on D channels, and then the control phase of virtual circuits is established using X.25 procedures over the B channel. D channel access defines permanent access with no additional phase required for its establishment. It requires only X.25 procedures to establish a call into and across the packet networks.

The reference points between ISDN-based applications via ISDN terminals for internetworking are shown in Figure 18.20. In order to provide an internetworking environment between non-ISDN public networks and ISDN networks using an **internetworking unit access point (IUAP)**, a variety of reference points are defined in addition to the existing S and T reference points (available with ISDN-compatible user equipment); these reference points are K, L, M, N, and P. Each of these reference points has a specific function and provides interface between ISDN and existing telephone networks, dedicated networks, specialized networks, or even other ISDNs. The internetworking between two ISDNs does not pose a problem if both ISDNs offer the same bearer and teleservices. If the ISDNs do not offer the same bearer and teleservices, the internetworking can be obtained by first negotiating over the acceptable services (control phase) followed by the establishment of connection between them (user phase).

The **internetworking between ISDN and PTSN** is already in place, and the implementation of the digital transmission switching mechanism and signaling techniques has already been defined. The only part of the digital transmission segment which is left out or experiences low progress is the digital local loop transmission. Some of the material herein is derived partially from References 2, 9, 10, 16, and 30.

The **internetworking between LANs and X.25** is a very complex process and, in fact, requires the interface protocol to consider the following fundamental issues:

1. The X.25 is defined for DCE and the user has to define its own DTE protocol which is compatible to X.25.
2. X.25 offers asynchronic operations for DTE-to-DCE as opposed to synchronic operations for DTE-to-DTE on LANs.
3. X.25 supports point-to-point configuration as opposed to broadcasting by LANs.
4. Many X.25 DCEs offer a large number of virtual calls to their attached DTEs, as opposed to one in the case of LAN DCE.
5. The LLC (a sublayer of the data link layer) used in the IEEE 802–based LANs offers connectionless services, while the data link layer protocol LAP-B in ISDN offers connection-oriented and reliable service to X.25.

In spite of these fundamental issues during the internetworking between X.25 and LANs, an ISO 8881 protocol offers a variety of configurations for internetworking.

The communication between a LAN (containing standard protocols at various layers of OSI-RM such as user application, X.25, LLC/MAC, etc.) and WAN (X.25) can be defined by inserting an internetworking unit (IWU) between them. The IWU consists of the same LLC/MAC and X.25 protocol up to layer 3 on the LAN side, and an X.25, LAP-B, and X.25 **packet level protocol (PLP)** on the WAN side. In other words, the IWU offers a function similar to gateways to provide internetworking between LANs and WANs. The ISO 8881 contains the various procedures for coupling the left-side protocols of LANs with the right-side protocols of WANs without affecting the lower layers. The X.25 looks at LANs as its DTE and offers point-to-point configuration between them (ISO 8208).

The LANs can also be connected via two IWUs providing two gateway functions. The first IWU is between the LAN and X.25 on one side, while the second IWU is between X.25 and the LAN on the other side.

The LANs can also be connected directly via one IWU which supports the emulation of a DCE on the terminal side and a DTE on the network side. The LANs may be internetworked directly via two IWUs. The IWU, when used with LANs and WANs, provides mapping for a logical channel number across each side of the packet level protocol (PLP) levels. This is exactly what X.25 offers to LANs on each side by mapping the logic channel numbers. It performs the function of an X.25 network in the sense that it receives the call setup and other control packets from either side of the gateway (i.e., LAN). It offers various operations such as editing, mapping of formats of one packet type to another, and many others.

The **X.25 DTE can communicate with the X.25 network via ISDN** node, which is composed of the basic rate interface and appropriate software for controlling the packets (packet handler). The interface between the ISDN node and X.25 is defined by the standard X.75. The basic access rate allows access either by B channel or D channel. The access via D channel is more common and is available in most of the vendor products, as it provides the transmission of data over LAP-D. In some situations, only B channels are used to transmit the packets to X.25 via ISDN node, which consists of the NT1 interface and packet handler. All the packets originated from DTEs will travel over the ISDN node to the X.25 network. The ISDN node interfaces with the X.25 network via the X.75 internetworking protocol.

The internetworking (or interworking) of various networks is defined in the CCITT X.300 internetworking recommendations standard. It discusses the usefulness and importance of X-series recommendations. Various aspects of internetworking for different configurations, cases, and signaling interfaces are also discussed.

As discussed above, different types of networks such as LANs, MANs, and FDDIs can be interconnected via different devices that can be used at different layers of OSI-RM. These devices are repeaters, bridges, routers, gateways, etc. Of these, the routers seem to be most popular, as they offer connectivity to a variety of networks. The routers operate at the network layer and support both types of services: connection-oriented and connectionless. A network interface card (NIC) for LANs and a WAN board interface for the router network are used for interconnectivity. The router networks look at the address in the datagrams and route them over the network from one hop to another until they reach the destination. The ATM network, on the other hand, is basically a connection-oriented network. The router-based networks do not require any specific protocol or user; they are very general and can be used for LANs, WANs, and so on. The router-based networks resemble the old version of X.25, which usually provides connection-oriented network service. If we look at the layered architecture of X.25, we find that it has a physical layer (X.21), link access protocol-balanced (LAP-B) layer, high-level data link layer (HDLC), and network layer (packet layer protocol). Thus, routers typically provide connectionless services, while switches (like ATM) provide connection-oriented services. The transport layer provides end-to-end communication between nodes, while other interfaces are user–network interfaces (UNI) and network-to-network (network node) interfaces (NNI).

One of the main differences between private and public sector WANs is the option for end-user application development facilities. The public WAN is basically a public carrier and does not support any end-user application development facilities in its specifications. On the other hand, the private sector WANs are more concerned with the end-user application developments than any internetwork functionalities. The public sector WANs are used in a variety of areas, e.g., telex, teletex, public-switched telephone networks (PSTN), telefax, public data networks (both circuit-switched and packet-switched), ISDN, B-ISDN, and upcoming high-speed networks.

These public sector WANs provide services to different organizations over leased lines, public links, or dial-up connections; e.g., a bank may use leased lines with public circuits and services to define its own network for electronic funds transfer (EFT), department stores may use their own network of overall management for inventory, revenue collection, etc. Some organizations offer additional services on top of services being offered by public WAN via value-added network service (VANS). These additional services are provided to customers who already receive services from public networks.

Digital, Intel, and Xerox (DIX) is the best known LAN in the marketplace and is based on the principles of the earlier ALOHA radio network developed at Palo Alto Research Center under Metacaffe and Boggs. In 1980, DIX defined a new specification of Ethernet.

The CCITT recommendation X.31 of the ISDN provides two types of internetworking with ISDN where the X.25 node is considered an ISDN node and hence internetworked with ISDN. There are two types of internetworking supported by internetworked ISDN.

In the **first type of internetworking**, each X.25 node supporting a B channel of ISDN is interfaced with NT1 of the ISDN. The ISDN interface node is connected to the X.25 network on the other side and uses X.75 protocol, providing internetworking between X.25 DTE nodes and the X.25 network via ISDN node. Here, both X.25 DTEs communicate with the X.25 network via ISDN and support only the B channel. The transfer of call request packets between the X.25 DTE and the network is transparent to the users. It is interesting to note that for all request call packets, the normal route will be same; i.e., the packets initiated from the DTE will go all the way to the X.25 network via ISDN node even if the X.25 DTE is searching a packet of another X.25 DTE on the same side supporting the same B channel.

In the **second type of internetworking**, the X.25 DTE supports both B and D channels, which are interfaced with NT1 of the ISDN node. The ISDN node also includes a packet handler which scans the packets received by ISDN via NT1 and, by looking at the options requested by the packets, the packets are transmitted over the channels. The options supported by the packet call are access to the X.25 network via either B or D channels of ISDN. As pointed out earlier, the B channel supports multiple terminals, and multiple terminals can access the network via the B channel in a multiplex mode. The D channel is basically is used for signaling and supports the establishment of logical channel connections of the D channel. Also, various packets of X.25 such as call setup, call request, call disconnect, and actual data are transmitted over the D channel on LAP-D link.

The X.300 internetworking standards were defined by CCITT in 1988 as a collection of various internetworking documents ranging from X.300 to X.370. This recommendation includes various configurations for internetworking, discussing the aspects for individual configurations, different configurations, and signaling interfaces. For example, if an ISDN is to be internetworked with another ISDN, it uses X.320 protocol with signaling interface provided by X.75. It includes configurations for internetworking different types of networks such as circuit-switched data networks, private networks, telex networks, telephone networks, etc. The sequence of the protocol and standards to be used for a particular configuration is also defined.

18.8 ISDN protocols

ISDN offers a variety of applications, e.g., multi-point connections, multimedia and multi-based protocols, etc. These applications are not supported directly by OSI-RM, but if we look at the overall functionality of OSI-RM and ISDN, it seems quite clear that both are compatible with each other and support each other's applications. The network-to-network connection is offered by the lower three layers (1–3), while end-to-end user connection is offered by the remaining layers (4–7) of OSI-RM. The ISDN is not concerned at all with the higher layers, as the users have to be provided with this connection for the transfer of information data. The standards I.430 and I.431 define the physical layer interface supporting both basic and primary data access interfaces. These standards support both channels, as the channels B and D are multiplexed and transmitted over the same physical media.

18.8.1 Layered model for ISDN protocol

Figure 18.21(a–c) shows the layered model of both channels B and D and their counterpart layers in OSI-RM. As we can see, different data link layer protocols are defined for B and D channels, known as link access protocol-B (X.25) and link access protocol-D (CCITT I.441), respectively. The functions of the network layer of OSI-RM are translated in terms

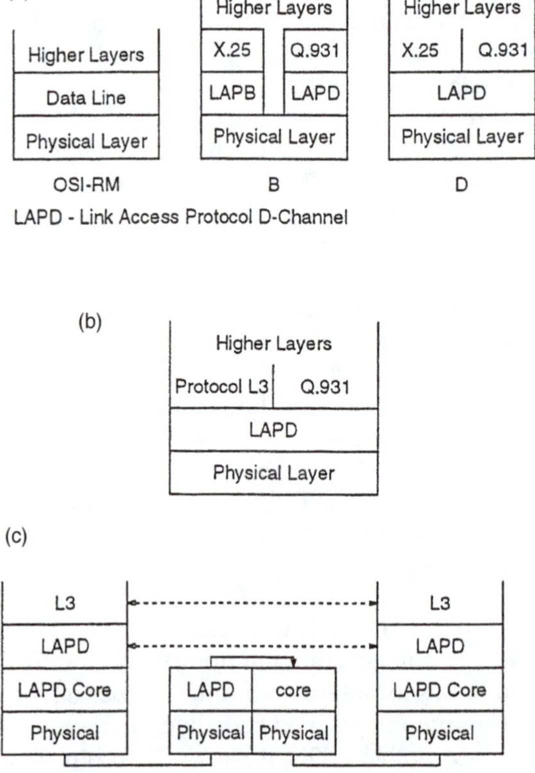

Figure 18.21 Layered model for ISDN protocols. (a) Layered model of B channel. (b) Layered model of D channel. (c) OSI-RM for B and D channels.

of call control signaling protocol (I.451), interface with the X.25 network at packet level, and telemetry interface. As such, LAP-D supports three applications: control signaling, packet-switched network, and telemetry.

Control signaling is mainly concerned with establishing, maintaining, and terminating connections for B channels. The end-to-end connection is established by the D channel layer of user signaling, which is equivalent to the top layers of OSI-RM. The B channel, on the other hand, can be used for circuit-switching and packet-switching services. In summary, we can say that the B channel supports circuit-switched and packet-switched services, while the D channel supports circuit-switched services only.

In the circuit-switched services, for a request call, the B channel is initiated by either NT1 or NT2 (depending on whether ISDN services or non-ISDN services are used). The D channel services use a three-layer network access protocol which interacts with the local exchange (at the D channel controller). The local exchange switches interact with CCITT's signaling transfer point which, in turn, interacts with SS7 on the other side up to the destination ISDN node. In this way, the complete circuit is established via these switches. Some of the material herein is derived partially from References 2, 9, 10, 16, 29, and 30.

Similarly, for packet-switched services, the D channel provides a local channel connection to packet-switched options in the exchange. This option offers a packet link between the other options over the transit exchange. Here access to the packet-switched network is defined over a high-speed B channel of 64 Kbps. The local user interface is required basically for providing and maintaining a transparent connection between the user and packet-switched node. Layers 2 and 3 of X.25 are used to provide communication between these entities.

The D channel can also be used to support packet-switched services at layer 3 (same as for virtual call in the B channel). The I.450 and I.451 recommendations support all three types of connections, e.g., circuit-switching, packet-switching and user-to-user (equivalent to higher layers of OSI-RM). The network connection at the ISDN user–network interface is established, managed, or cleared by a set of procedures defined in the I.450 and I.451 recommendations. It defines various types of messages for establishing connections, transferring the data, and finally de-establishing the connection. A list of the message commands for establishing connections is given below:

CONNECT
CONNECT ACKNOWLEDGE
CALL PROCEEDING
SETUP
SETUP ACKNOWLEDGE
ALERTING

Similarly, the set of message commands for de-establishing connection is as follows:

DETACH
DETACH ACKNOWLEDGE
DISCONNECT
RELEASE
RELEASE ACKNOWLEDGE

The message commands for information such as status, options, etc., are as follows:

RESUME
RESUME ACKNOWLEDGE

RESUME REJECT
SUSPEND
SUSPEND ACKNOWLEDGE
SUSPEND REJECT
USER INFORMATION

In addition to these message commands, there are some miscellaneous messages which may be used for determining options, congestion, register, information, etc. The options and other parameters in these messages include ISDN number, source and destination addresses, etc.

A variety of facilities such as X.25 services, call completion, reverse charge acceptance, etc., are supported by ISDN and are controlled by various messages such as CAN (cancel), FAC (facility), etc. Most of the message commands are self-explanatory. Some of the message commands are shown below:

CANcel — A request to discontinue
CANcel ACKnowledge — A request to discontinue with acknowledgment (confirmation)
FACility — Initialization of access to a network
REGister — Initialization of a register of facility
REGister ACKnowledge — Initialization of a register with acknowledge

The standards and protocols for various layers, interfacing, and accessing in ISDN are defined mainly by CCITT, although many other standards organizations may have been involved in the standardization process.

18.8.2 CCITT I-series recommendations

The I-series of recommendations of ISDN was defined in 1984. The CCITT, in fact, started working toward ISDN in 1968 with the aim of introducing digital technology in public telephone networks. As a result of these efforts, a digital system and integration of services came into existence in the time period of 1968–1976. The digitalization of analog signals by pulse-coded modulation (PCM) was defined as a milestone toward digital networking. During 1976–1980, the integrated digital network (IDN), along with the integration of services, was discussed in detail. In the following two sessions, various issues and aspects of ISDN, the transmission of voice and data over the same digital switching and transmission (digital paths), and many other questions were addressed.

The first standard of ISDN was defined in 1980 as the G.705 CCITT recommendation, while various issues and concepts of ISDN were discussed during this time. During 1980–1984, various standards of ISDN ranging from I.100 to I.605 addressing services (I.200–I.257), aspects and functions of ISDN (I.310–I.340), user–network interfaces (I.410–I.470), internetworking interfaces (I.500–I.560), and maintenance services (I.601–I.605) were defined and introduced. These sets of standards were adapted by manufacturers to develop ISDN products and ISDN-based services. During 1984–1988, the work on the I-series continued, and some of the recommendations were added or modified during this term. The I-series recommendations were sufficient to implement ISDN in the early 1990s.

The I.200 series of the CCITT recommendations is the most useful, important, and valuable standard, as it defines all the services offered by ISDN. These services are categorized as **bearer services**, **teleservices**, and **supplementary services**. See Section 18.6.1 for details.

18.8.3 Teletex services (CCITT standard)

CCITT has also recommended attributes for higher layers for teletex service and family group 4, mixed mode (CCITT I.200), with the same arrangement for channels as in teletex or telefax; i.e., user information is transmitted over the B channel, while the D channel is basically used for signaling. The videotex service offered by ISDN is similar to the existing videotex services. The telex services offer interactive communication for text. The user information is usually transmitted by either circuit-switched or packet-switched mode-bearer channel services, and as usual, the signaling is provided over the D channel.

The transmission of documents, messages, images, etc., can take place via computer networks, but each one has a different technique for transmission. For example, the teletex and electronic mail techniques are used to transmit messages composed of only characters (also known as texts). Image signals for visual images can be transmitted over the computer networks, while images, texts, graphics, symbols, etc., can be created by using bit-map representation, which is then transmitted over the network as a facsimile.

The facsimile technique of transmitting messages, information, or documents is better than the existing technique of transmission (videotex, teletex, electronic mail), as here different formats of messages (texts, pictures, images, handwritten documents, or any other documents) can be transmitted. It saves the setup time, as the scanner accepts the message as input and transmits it. There is no need to enter this message into the system via keyboard. Although facsimile communication is not a new technology (it has existed since the eighteenth century), the medium is different, and now with the use of the public telephone network and availability of widely accepted standards, facsimile transmission has become very cheap and fast and offers better quality.

Since the cost of machines (hardware) has gone down, digital technology has further been given a boost in offering services such as storage of messages on disk, tape, etc., manipulation by computers, improved security, etc. Also, there is no need to use paper, as the entire message (containing images, texts, drawing charts, etc.) can be created by graphic packages and can be transmitted as an electronic mail message.

18.8.4 Facsimile service classes (CCITT standard)

CCITT has defined four classes of facsimile standards to be used on public telephone networks.

Group I: This standard depends on the low-speed analog transmission scheme using frequency modulation with various levels of gray color in the black and white range. It allows the transmission of documents of ISO a4 size at the speed of four lines/minute.

Group II: This standard uses the compression technique and offers twice the speed of that of Group I with the same resolution. The phase modulation technique is used for the transmission and supporting gray colors in the black and white range.

Group III: This standard offers the transmission of analog/digital signals over an analog telephone network via modem. It uses a digital encoding technique and some sort of filtering before modulation. The filtering identifies a redundant set of information from the signal to be transmitted. This standard is considered the first digital facsimile standard and offers higher speed than Group II as well as black and white values within the samples.

Group IV: With the acceptance of the Group III standard for digital transmission, this standard also supports black and white values. It is used over digital networks, supports error-free reception, and offers a data rate of 64 Kbps. Due to high speeds, the transmission time is in seconds rather than minutes.

A typical configuration for the facsimile standard over the public network includes the following components:

1. **Preprocessor:** This component takes input from the scanner and converts it into different white or black values. The scanner is an analog device which scans over the input document image.
2. **Compressor:** This component takes the input from the preprocessor and compresses the black and white values in the form of a bit matrix. This is a reduced form of the initial document or picture and does not include any redundant information.
3. **Storage:** This is used for storing control information regarding error control, address, link control, etc., in addition to the compressed data.
4. **Communication interface:** This interface provides a link between the compressed data (stored in storage) and different types of networks, i.e., packet-switched and circuit-switched networks and ISDNs.
5. **Decompressor:** This device regenerates the original document/image from the data signal received at the receiving side in the form of bit-matrix representation.
6. **Post processor:** The decompressed bit-matrix representation of the original document/image is further processed for improving its quality via interpretation or any other technique (based on resolution of the devices under consideration).

18.8.5 Compression techniques

For the transmission of document images on a digital facsimile machine, the compression technique is very useful, as it allows more data messages to be transmitted in less time. For example, sending one page containing 200 picture elements/inch (where a picture element is basically the smallest line element of black/white value) may generate about 3 million bits. If we are sending this page via ISDN at a data rate of 64 Kbps, it would take close to one minute to transmit one page. The use of compression technique will not only reduce the transmission time but will also transmit more pages using the same data rate.

There are mainly two categories of compression techniques: **information-processing** and **approximation**. In the former method, the document/image is scanned and converted into digital form, while in latter method, the original document/image is approximated using efficient approximate technique. The CCITT has approved two algorithms as standards — modified Huffman and modified READ (MR) — and both of these algorithms are based on the information-processing technique. The facsimile standard Group III uses the MR algorithm, while Group IV uses MH. Group III also offers MR as an option for the transmission of documents/images.

18.8.6 Character and pattern telephone access information network (CAPTAIN)

An international videotex internetworking protocol known as VI-protocol offers a border across videotex service. The **character and pattern telephone access information network (CAPTAIN)** system is the first and most standard videotex system in Japan. With the advent of ISDN and digital networks, the **digital videotex communication network (DVCN)** is also becoming popular, and we hope that many small-sized private CAPTAINs based on digital technology will make a tremendous impact in the information market. Some of the applications of CAPTAIN in Japan include user-car auction service, medical and information service for doctors and pharmacists, travel information and hotel reservations, education information service, stock dealing and stock quotation services, electronic encyclopedia services, and many other services.

The advent of ISDN has opened a new door to advanced telecommunication functions that are needed in an information society. Since DVCNs are based on public-switched telephone networks (PSTNs), the digitization of both PSTNs and DVCNs has to be accomplished at the same time, as it will be based on ISDN digital network technology. Digital CAPTAIN systems offer high-quality audio signals and full-color photographs. They offer a transmission speed of 64 Kbps. The digital terminals can transmit the control data (information frame data created by digital terminals) to information centers through digital CAPTAIN networks using a high-speed path of transmission at a rate of 64 Kbps. The types of messages which can be transmitted include alphanumeric data, graphic information consisting of photographs, geometric data, hand-down messages, etc.

The CAPTAIN consists of three components: video communication network, information center, and user's terminals. The video communication network is a public telephone network and includes a videotex communication processing (VCP) unit, PSTN, and transmission lines connecting various equipment. The VCP is a very important component of a videotex communication network, as it offers functions such as protocol conversion, code/pattern conversion, etc., based on the characteristics of the user's terminals. It is a Data Syntax I of CCITT T.101 and is based on hybrid coding structure in the sense that it uses both modes: pattern and coding for transmission of images and joint transmission of characters and graphics, respectively. It supports different types of graphics for videotex, e.g., geometric graphic, music graphic, photo graphic, etc. The standard resolution is 204×248 (vertical × horizontal). The high resolution typically is defined by 408×496.

The private CAPTAIN network provides videotex services to users in the local area. The user terminals are connected to information centers via existing communication facilities (like PSTNs, PABXs, etc.) and leased circuits. The information stored in private CAPTAINs can be accessed by a limited number of users. The private CAPTAINs can be connected to public CAPTAINs to provide services to users across countries as these networks use these services.

Different versions of files are available over the Internet, and we need to have an appropriate software package to access them; e.g., a PostScript format of file defines the format of documents and formatting information used in desktop publishing. Software to open and print the PostScript files is needed on our PC. In order to print the files, the special high-resolution PostScript device driver for typesetting the documents is needed. An executable program is defined in binary language, and the source code for the program is written in terms of instructions using programming language (low-level or high-level). The source code is usually defined within an ASCII file format and needs to be compiled into a program file. The images are defined in GIF format, which can be downloaded quickly. For accessing the image GIF reader to read binary GIF files, the PC must have a video display card which translates the data into signals that display the image on the monitor.

In order to save the bandwidth of the network and enable faster downloading of very large files, a compression technique is used to compress the file into a compact size. It is then decompressed on the receiving side. An extension of compression technique is added at the end of the file. A variety of compression techniques are available, e.g., PkZip, ARJ, LHArc, Pak Zoo, BinHex, etc. for MS-DOS; MacB, SelfExtract, and Stuffit for Macintosh; and Gzip and unixcompress for UNIX systems. The UNIX command Tar combines a group of files and is defined as one file with an extension of .tar. The compression technique extension is added at the end of files.

18.9 Broadband ISDN: Why?

According to CCITT, broadband service is defined as the one which requires transmission channels capable of supporting rates greater than 1.5 Mbps, the primary rate in ISDN, or

T-1 or DS-1 in digital terminology. Many people think that B-ISDN is a new technology, but it is in fact a *framework* that supports different types of technologies such as ATM, SONET, frame relay, SMDS, and many other intelligent networks. The data going into a switched network is routed using these different technologies. Some of the applications needing higher data rates and computing power at a lower cost include videoconferencing, graphics, video telephone, imaging, high-definition television, multimedia, visualization, LAN connectivity for higher data rates, computer-aided design tools, etc.

According to ITU-T, the ISDN offers services below 64 Kbps, and any services above 64 Kbps are known as broadband communication services. As a result of this recommendation, the traditional ISDN has been renamed N-ISDN, while B-ISDN provides the services above 64 Kbps and below 1.544 Mbps. It offers complete integrated services of low rate bursty; high rate of real time; low-speed traffic such as data, voice, telemetry, and faxes; and high-speed traffic such as videoconferencing, high-definition television (HDTV), high-speed data communications, etc.

The main idea behind defining new ISDN service (broadband) is to develop a transmission channel which can support data rates greater than the primary data rate, in contrast to narrow-band ISDN. This new service will, in particular, support video and images which require data rates much higher than can be handled by N-ISDN, thus supporting various applications of multimedia technology. The optical fiber transmission, VLSI chips for high-speed, low-error for switching, transmission, high-quality video monitors are some of the latest advances which have contributed significantly to the development of new services (B-ISDN) such as video and images over the existing services offered by N-ISDN. B-ISDN (CCITT) is defined as a service which requires a transmission channel capable of supporting rates higher than the primary rate in addition to the basic services of N-ISDN. Various switching techniques have been proposed lately, but asynchronous transfer mode (ATM) seems to be an acceptable mode for the transmission of integrated voice and data over B-ISDNs. For more details on broadband services, see Reference 30.

Broadband that is a logical extension of the old ISDN includes ATM and offers higher data rates than T1 (1.544 Mbps, used in North America) and E-1 (2.048 Mbps, used in Europe). Since an ATM switch has a very open functionality, it can be used at any layer of OSI-RM. The ATM can be used as a MAC sublayer of OSI-RM, thus avoiding the use of LAN MAC. This means that there is direct mapping between the LLC sublayer and ATM, and all of the packets carrying LLC information are enclosed in ATM cells. Thus, the entire ATM replaces the physical layer and MAC sublayers of OSI-RM. This is precisely what has been implemented in SMDS. Further, the LAN Emulation Service (LES) group within the ATM Forum is working on the standardization process for IEEE LANs. The main advantage of this is that now ATM has become another LAN like Ethernet and token ring, but it loses its features such as flexible bandwidth, multimedia applications, etc., compared to routers for interconnectivity.

As indicated above, the ISDN is a telephone network which uses digital voice lines for both internal network and local loops. It is interesting to note that the telephone network with analog voice lines for local loop also supports the transmission of data and voice together, with the only difference being in speed and time for data and voice signals. In other words, both of these signal types (data and voice) can be transmitted with the same speed at the same time. With ISDN, two standard types of services are defined as basic rate interface (BRI) and primary rate interface (PRI).

The basic services include two 64-Kbps channels (B channels) and one 16-Kbps channel (D channel). Each 64-Kbps channel can be used for ordinary voice communications (data transfer). The 16 Kbps is typically used for internal signaling of the telephone network and for connections, network management, etc., of both channels of 64 Kbps of the network. It can also be used for low data-rate applications such as alarm, security system, etc.

The B channels are useful in setting up a direct connection to some destination by using ISDN as a circuit-switched network and then allowing the users to use it as an access line into a node of a packet-switched network. But if we try to transmit a high-resolution graphic image of 10^9 bits over a 64-Kbps access line, we have to wait four to five hours. This limits the application of N-ISDN in the areas where we require very high data-rate images and video services.

Due to these reasons and due to the fact that optical fiber has been introduced and accepted as an alternative for existing twisted pairs (used in local loop), it seems clear that higher data rate services cannot be offered by N-ISDN.

Some useful Websites

Information on ISDN is available at *http://www.3com.com*, *http://www.ccg4isdn.com*, the California ISDN group is at *http://www.ciug.org:8080*, and an ISDN tutorial is at *http://www.ralphb.net*.

Modems, wireless LAN products, HomeLAN, broadband access cables, ADSL, and analog modems are available at *http://www.zoomtel.com*. A very useful site describing the cable modem glossary can be found at *http://vidotron.ab.ca* and cable modems in general can be found at *http://www.cable-modems.org*. The ADSL forum is at *http://www.adsl.com*, and some of the vendors manufacturing or providing digital subscriber line services can be found at *http://www.alcatel.com*, *http://www.3com.com*, *http://www.zyxel.com*, *http://www.virata.com*, and many others.

Routers, ATM products, and other network products can be found at *http://www.networking.ibm.com*. X.25 switched networks, protocols, components, and various other related products for internetworking can be found at *http://www.home.ubalt.edu*, *http://www.it.kth.se*, and *http://www.techweb.com* A knowledge database index and quick searching can be found at *http://www.shiva.com.sg*.

Optical standards can be found at *http://www.siemon.com*, *http://www.bicsi.com*, and *http://www.ansi.gov.* A standard glossary about standards organizations is maintained at *http://www.itp.colorado.edu*. The SONET interoperability forum site can be found at *http://www.atis.org*.

References

1. CCITT Study Group XVII, Digital Networks including ISDN, 1990.
2. Stallings, *ISDN: An Introduction,* Macmillan, 1989.
3. CCITT I.210, "Principles of Telecommunications Services Supported by an ISDN and the Means to Describe Them," Blue Book, Fascicles, III.7, Geneva, 1991.
4. CCITT I.240, "Definitions of Teleservices," Blue Book, Fascicles, III.7, Geneva, 1991.
5. CCITT I.200, "Guidance to I.200 Series of Recommendations," Blue Book, Fascicles, III.7, Geneva, 1991.
6. CCITT X.400, "Message Handling System."
7. Ruffalo, D., "Understanding T1 basics: Primer offers picture of networking future," *Data Communications,* March 1987.
8. Bell Telephone Laboratories, Transmission Systems for Communications, Murray Hill, NJ, 1982.
9. Bellamy, J., *Digital Telephony,* John Wiley, 1982.
10. Freeman, R., *Telecommunications Engineering,* John Wiley, 1985.
11. Roehr, W., "Inside SS No. 7: a detailed look at ISDN's signaling system plan," *Data Communications,* Oct. 1985.
12. Schlanger, G., "An overview of signaling system no. 7," *IEEE Journal on Selected Areas in Communications,* May 1986.

13. Turner, J., "Design of an integrated services packet network," *IEEE Journal on Selected Areas in Communication,* Nov. 1986.
14. Rahnnema, M., "Smart trunk scheduling strategies for future integrated services packet switched networks," *IEEE Communication Magazine,* Feb. 1988.
15. Bauwens, J. and Prycker, M., "Broadband experiment using asynchronous time division techniques," *Electrical Communications,* No. 1, 1987.
16. Handel, R. and Huber, S., *Integrated Broadband Network: An Introduction to ATM-Based Networks,* Addison-Wesley, 1993.
17. CCITT I.221, "Common Specific Characteristics of Services," Blue Book, Fascicles, III.7, Geneva, 1991.
18. CCITT I.411, "ISDN User–Network Interfaces — Reference Configurations," Blue Book, Fascicle III.8, Geneva, 1991.
19. CCITT I.430, "Basic User–Network Interface Layer 1 Specifications," Blue Book, Fascicle III.8, Geneva, 1991.
20. CCITT I.420, "Basic User–Network Interface," Blue Book, Fascicle III.7, Geneva, 1991.
21. CCITT Revised Draft Recommendation G.703, "Physical/Electrical Characteristics of Hierarchical Digital Interfaces," CCITT SG XVII, Report, June 1992.
22. Turner, J., "Design of an integrated services packet network," *IEEE Journal on Selected Areas in Communication,* Nov. 1986.
23. CCITT I.310, "ISDN Network Functional Principles," Blue Book, Fascicle III.7, Geneva, 1991.
24. CCITT I.451, "ISDN User–Network Interface Layer 3 Specifications," Blue Book, Fascicles III.7, Geneva, 1991.
25. CCITT I.431, "Primary Rate User Interface — Layer 1 Specification," 3 Blue Book, Fascicle, Geneva, 1991.
26. CCITT I.330, "ISDN Numbering Schemes," Blue Book, Fascicles III.7, Geneva, 1991.
27. CCITT I.461, "Support of X.21 and X.21 bis Based DTEs by an ISDN," Blue Book, Fascicles III.7, Geneva, 1991.
28. CCITT I.462, "Support of Packet Mode Terminal Equipment by an ISDN," Blue Book, Fascicle III.7, Geneva, 1991.
29. Maxwell, K., "Asymmetric digital subscriber line: Interim technology for the next forty years," *IEEE Communications Magazine,* Oct. 1996.
30. Kumar, B., *Broadband Communications,* McGraw-Hill, 1994.

Additional readings

1. Black, U., *ISDN and SS7: Architectures for Digital Signaling Network,* Prentice-Hall, 1997.
2. Stallings, W., *ISDN and Broadband ISDN with Frame Relay and ATM,* Prentice-Hall, 1999.
3. CCITT I.120, "Integrated Services Digital Networks," Blue Book, Fascicle III.7, Geneva, 1989.
4. CCITT I.320, "ISDN Protocols Reference Model," Blue Book, Fascicle III.8, Geneva, 1991.
5. CCITT I.412, "ISDN User–Network Interfaces — Interface Structures and Access Capabilities," Blue Book, Fascicle III.8, Geneva, 1991.
6. CCITT I.432, "B-ISDN User–Network Interface — Physical Layer Specification," Blue Book, Fascicle III.8, Geneva, 1991.
7. CCITT Q.761, "Functional Description of ISDN User Part of Signaling System No. 7," Blue Book, Fascicle VI.8, Geneva, 1989.
8. CCITT Q.921, "ISDN User–Network Interface Data Link Layer Specifications," Blue Book, Fascicle VI.10, Geneva, 1989.
9. CCITT Q.930, "ISDN User–Network Interface Layer 3-General Aspects," Blue Book, Fascicle VI.11, Geneva, 1989.
10. CCITT Q.931, "ISDN User–Network Interface Layer 3 Specification for Basic Call Control," Blue Book, Fascicle, VI.11, Geneva, 1989.
11. CCITT X.1, "International User Classes of Services in Public Data Networks and Integrated Services Digital Networks (ISDNs)," Blue Book, Fascicle VIII.2 Geneva, 1989.

chapter nineteen

High-speed networks

"Let the wind blow through your hair while you still have some."

Dave Weinbaum

19.1 Introduction

The idea of introducing a single network became evident when integration of voice and data was provided by a digital network known as an integrated services digital network (ISDN). In most countries, the existing public switched telephone networks are being converted or replaced for full digital operations. This means that new networks will be based on all-digital transmission and switching, which is expected to offer fast connection setup for voice communication across the network anywhere around the world. This new interface will handle data communications, offering both data and voice communications simultaneously. The ISDN became the first network to transmit both voice and data over the same medium or channel. Although N-ISDN allows the integration of voice and data outside the network and offers two types of interfaces (**basic rate interface** (popular in the U.S.) and **primary rate interface** (popular in Europe), inside the N-ISDN we can see two types of networks being defined. These networks are based on different switching techniques and are known as the **circuit-switched network (CSN)** and **packet-switched network (PSN)**.

ISDN switches are available for voice channel communication (based on circuit switching), but recent hardware technologies in different areas have introduced a number of lower-bit-rate chips, e.g., adaptive differential PCM chips of 32 Kbps, chips of 13 Kbps to be used with mobile networks, etc. This concept of defining a single service-independent network is picking up the market, and it is expected that in the future, this type of single, unified, and universal network must provide the existing services and also must support future services of both voice and data communication efficiently.

Due to bandwidth limitations of ISDN's basic rate interface of 2B+D, where the B channel offers 64 Kbps and the D channel offers 16 Kbps, TV signals (typically of 6-MHz bandwidth) cannot be transmitted over it. For transmitting TV signals, a separate TV network may be required.

The introduction of the ISDN and its design, prototyping, and implementation have take more time than expected, and the concepts of integration of voice and data signals, digital networking, etc., have become a kind of symbol. The advantages of the ISDN are enormous, but due to ever-evolving technologies and greater demand for capacity, bandwidth, and higher bit data rates, the ISDN has not been accepted widely. In order to extend the concept of ISDN's integration of voice and data, a new concept of defining a unified network known as broadband ISDN (B-ISDN) that will be geared toward a revolutionized ISDN has emerged to provide broadband services (data, voice, images, video, etc.) to users at a reasonable cost over a single network. It is hoped that this will meet the demands of various multimedia application requirements. The B-ISDN offers not only larger bandwidth or capacity but also higher bit data rates of hundreds of megabytes per second. Thus, a single network offering a variety of services must provide the following features: high bandwidth, bandwidth on demand, low latency, fault tolerance, multiple service access interfaces, support of integrated applications, etc. Although Chapter 18 discussed these issues in greater detail, we will address each of these features in brief in the following sections for understanding the philososphy behind various broadband (high-speed) networks.

One very interesting and powerful application of N-ISDN is **videoconferencing**, and, in fact, this has become a standard service of N-ISDN. In this service, lectures or seminars can be transmitted over TV screens in any part of the world. N-ISDN may not be able to introduce other services such as point-to-point, file transfer, facsimile, multi-point, etc., but in pure videoconferencing, there is no exchange of files, facsimiles, etc., and hence it offers two-way communication for both point-to-point and multi-point configurations between the users. Different conference places can be connected to receive point-to-point service. Videoconferencing can also be implemented for multi-point configuration, where a small number of users can communicate with each other and discuss common documents or display a user's live pictures either one at a time or all together on the screen. All of these options are usually handled by video service providers.

Looking at the existing networks which provide different types of services, we can generally say that the existing networks are service-dependent and will require additional resources and enhancement to offer services not already provided to the users. These networks are offering different bit data rates, and as such very little flexibility may be expected from them to cater to the services of audio, video, and images even with compression and new coding techniques. To provide support services by the existing networks may not be that difficult compared to services such as high-definition TV (HDTV), multimedia applications, virtual reality, and many other upcoming applications, as in these applications, the requirements of bit rates are unknown.

Some of the drawbacks (such as service dependency and inflexibility) seem to be overcome by a new category of integrated networks (a modified version of N-ISDN) — **broadband ISDN (B-ISDN)**. This service-independent network transports different types of services (audio, video, voice, images, etc.) requiring transmission channels supporting more than primary rate interface (PRI) defined by N-ISDN, and allows users to share the resources of various networks for different services. It is more flexible and efficient in the optimal allocation of resources among the users. An **asynchronous mode transfer (ATM)** has been considered a transfer mode solution (dealing with transmission and switching) for implementation of B-ISDN. Further, with the advances in coding algorithms, transfer modes, compression techniques, transmission strategies over fiber optics, and other relevant technologies, it is hoped that this will be able to accommodate all the **teleservices** (telex, teletex, videotex, facsimile, etc.) of future applications with less expense, as only one network is needed to offer these services.

According to **CCITT I.113**,[24] the B-ISDN is defined as "a network offering a service or system which requires channels capable of supporting the data rates of greater than Primary rate." B-ISDN also includes ISDN's 64-Kbps capabilities but offers them at data rates of 1.5 Mbps or 2 Mbps. The upper limit of the data rate at present seems to be 100 Mbps. The channel rates of 32 to 34 Mbps, 45 to 70 Mbps, and 135 to 139 Mbps are designated as H2, H3, and H4, respectively. These data bit rates (including a 140-Mbps network interface rate) were aimed toward the data rates of **pleisochronous hierarchy (CCITT recommendations**

G.702, i.e., H channels).[19] Pleisochronous hierarchy is defined as a set of data bit rates and multiplexing schemes for multiplexing different types of high data rates. It multiplexes not only synchronous data rates but also ISDN's 64 Kbps into high bit rate signals. For this reason, B-ISDNs are suitable for both business and home application services. It supports data bit rates (fixed and variable), data, voice (sound), still and moving picture or image transmission, and multimedia applications (including data, picture, and voice services together in one application).

19.2 *Applications of integrated networks*

The integration of voice and data has played a very important role in both home and business application services. In the area of home services, we have seen the introduction of the analog **cable antenna TV (CATV)** with separate networks for **telephone** (audio) and **videotex** signals. The CATV network offers services such as pay TV, games, etc. The future of the digitization process in the home application services is already on cards, and we will soon see digital processing and digital transmission of video, video telephone sets, etc. (although initial attempts have already shown some of the applications and service available in a very limited and restricted way). The low-speed videophone will be available in the telephone networks which use ISDN. The digital TV distribution and information of switched video services will offer the services of videophone and videotex as an integrated broadband communication network on **high-definition TV (HDTV)**. The evolution of television has seen the following technologies: black and white TV (B/W TV), color TV, cable TV (CATV), videoconferencing, stereo TV, video on demand, satellite news broadcasting, and now high-definition TV (HDTV).

In the area of business application services, videoconferencing has already been established, as this service saves traveling time and cost for attending conferences or business meetings. It has become a very important and useful telecommunication tool, as it offers a high-quality picture. Due to its high data bit rate capabilities, it offers provision for allocating the data rate for interconnecting different types of networks. We have seen two types of services at the initial stages: **digital data network (DDN)** and **telephone network (TN)**. Telephone communications have evolved, and some of the technologies that have found their use in telephone networks include digitalization of voice, voice conferencing, voice storage, HiFi telephony, text-to-voice, and voice-to-text, and now there is a big market for multimedia applications. The network offers access to data via modem.

The integration of voice and data has given a new direction to data communication over ISDN which allows the transmission of data, telex, fax, low-speed image retrieval, low-quality videophone, etc. High bit data rates, leased lines, and video leased lines have also been introduced. ISDN also offers to home applications some of the services of business applications, such as low-quality videophone and low-speed image transmission.

The merger of ISDN and leased lines of high-volume data and video signals resulted in services such as videoconferencing, high-quality integrated domain for document and video retrieval, CAD/CAM, CASE tools, high-speed data transmission, high-speed services over switched networks, multimedia applications, etc. The multimedia applications may be defined as integrated systems which send a combination of real- and non-real-time transfer of time-based and non-time-based information. The traditional data communication applications supported by networks (LANs, high-speed networks, ISDNs, B-ISDNs, etc.) are mainly non-time-based transfer of data with other real-time (distributed computing) or non-real-time (interactive computing) aspects of computing. Multimedia supports two types of information: time-based (video, audio, animation, etc.) and non-time-based (graphics, images, texts, etc.). Similarly, the delivery requirements for these two types of information are different and are real-time- and non-real-time-based. The real-time-based information is required instantaneously in applications such as image browsing, while non-real-time-based information may be required at a later time, e.g., video mail.

Real-time delivery requirements can further be provided for both time-based (videoconferencing, video on demand, etc.) and non-time-based (image browsing, interactive computing, etc.) information. Similarly, the non-real-time delivery requirement can be defined for time-based (e.g., video mail) and non-time-based (e-mail, file transfer, etc.) information. The information can be either periodic or bursty type traffic. The periodic type of traffic occurs at regular intervals as in time-based information; e.g., 64-Kbps PCM voice grade generates samples at 125-μs intervals with each sample of eight bits. For uncompressed full motion **North American Television Standards Committee (NTSC)** video, the video frames are generated at a regular interval of 1/30th of a second, or 30 frames per second. When this signal is compressed, the frames are generated at a regular interval of 1/30th of a second for NTSC or 1/25th of a second for a **phase alternating line (PAL)** frame. The bursty traffic is characterized by messages of variable length generated at random times and separated by some random duration of time when no data is being sent.

With the digital communication technology being adopted very rapidly in both home and business applications, and services such as telephone and videotex and high-speed services over a switched network at high data rates with high quality of videophone and high-speed video, etc., it is quite convincing that a single service-independent network will become a reality very soon, and efforts in this direction have already proven so. Wireless communication in mobile telephony and now LAN is another technology which is capturing the market rapidly. The various technologies in the area of mobile telephones include paging, cellular phone, digital cellular, cordless phone, portable phone, paging at the national and international level between ground-to-ground and ground-to-air communications, etc.

19.3 Evolution of high-speed and high-bandwidth networks

The telephone system has become an essential tool of communication in business, homes, organizations, and anywhere else. It allows the exchange of information transfer between different places between users around the world. This communication system is very simple to use and handle. The underlying transmissions are different and offer faster response during conversations. This communication system has become very efficient and useful due to added features such as global availability and connectivity, fast response (minimal delay), and user friendliness. Integrated broadband network concepts and vendor products for these networks are becoming a reality to replace the existing old telephone systems and are considered an extension of the first-ever digital network 64-Kbps-based ISDN.

The ISDN is based on the concepts of digital telephone networks. It is a complete digital network offering integrated services over the same communication link. All the information (data, voice, video, images, etc.) is transmitted and switched as digital signals over the networks, which offer end-to-end communication between nodes. Thus, we can consider ISDN basically a new development for plain telephony systems, as the basic service offered by ISDN still remains voice communication. Due to the fact that the upcoming services require high bandwidth and higher data rates, the ISDN is not suitable for these applications. The broadband-ISDN (B-ISDN) has been defined to add new features such as high data bit rates for fast data or moving picture transmission and high-quality video applications with greater flexibility.

B-ISDN follows the same principle as 64-Kbps ISDN and utilizes a limited set of connections. It offers a variety of services requiring higher bandwidth and requiring transmission channels of supporting the bit data rate of more than primary rate interface (PRI). It supports multi-purpose user–network interfaces. The **CCITT I.121**[25] presents an overview of B-ISDN capabilities:

> B-ISDN supports switched, semi-permanent, and permanent, point-to-point and point-to-multi-point, connections and provides on demand reserved and permanent services. The connections in B-ISDN support both circuit mode and packet mode services of mono- and/or multi-media type and of a connectionless or connection-oriented nature and in a bidirectional or unidirectional configuration.

A B-ISDN implementation will depend on ATM (CCITT I.121). It will include intelligent capabilities for providing advanced service characteristics, supporting powerful operations and maintenance, and providing various tools, network control, and management. We hope that the new technologies will allow users to access a high bandwidth at high data rates on WANs, e.g., ISDNs, synchronous optical networks (SONETs)[18] used in the U.S., synchronous digital hierarchy (SDH)[20] at the international level, the currently proposed asynchronous transfer mode (ATM), frame relay, switched multi-megabit data service (SMDS),[17] and broadband ISDNs (B-ISDNs).

Of these technologies, the one that will allow users to access higher bandwidths of public networks depends on many criteria. In general, however, **transmission** and **switching** are considered the criteria for assessing the technology which fits the application best. Of course, now with artificial intelligence (AI) technology, intelligence may also become a third criterion for assessing a particular technology for a particular application. Some of the important technologies which have received much attention from network professionals are discussed in brief here.[1-10,46-52] In no way should the following list be considered a complete list of the technology advances for networking.

High-performance transmission media: Various advances in areas such as microprocessor-based distributed computing systems, diskless workstations, parallel workstations, HDTV, high-speed switching, and internetworking are a few examples of the technologies which are leading toward high-performance and high-speed communication networks. We have seen ever-growing interest and great advances in various technologies which, when integrated, will provide a unified high-performance and high-speed network. This uniform network will provide a variety of services ranging from video and images to multimedia applications across the network for a globalized environment at very high speeds. Some advances are geared toward transmission media, e.g., **fiber optic transmission media (FOTM)**, offering data rates in Gbps. Other advances within fiber optics include **laser diode**, **optical amplifiers**, **optical transmission with very low attenuation**, etc. Network providers have already started manufacturing and providing the network products based on a high-performance transmission-media-based communication network.

Optical transmission is the main force for **B-ISDN**, as it offers very high bandwidth and data rates. The optical transmission is typically used for a longer distance and covers a greater distance for repeaters. The diameter of the fiber is very small, and as such it offers low weight and volume. It is immune to electromagnetic fields and offers low transmission error probability. There is no cross-talk between fibers and it is very difficult to tap. The basic B-ISDN interface type across a broadband user–network interface offers 150 Mbps, while a second type of interface offers 600 Mbps. Various standards have also been defined and accepted, and large corporations, universities, and research and development laboratories have already started pushing the **B-ISDN** interfaces, protocols, and reference models as a WAN at both national and international levels. This WAN will provide high speeds for voice, data, video, images, etc. applications in an open-communication, multi-vendor environment for global connectivity. High-performance and high-speed networks for voice, data, images, video communication, and multimedia applications have revolutionized the market in the area of data communication, and this trend certainly demands still higher data rates, integrated services, and high-performance communication networks with unlimited services available to users worldwide.

Signaling control: The signaling scheme plays an important role in routing strategies for providing end-to-point digital signal call requests and introduces an intelligence into the network protocols. The existing network protocols such as TCP/IP are using the same kind of intelligence in the sense that the services offered by the network are compatible with the Internet for certain applications such as e-mail, file-transfer services, information retrieval, bulletin board services, etc. The intelligence here is based on the fact that users are provided with the services by network layer protocols. For applications such as e-mail with voice, video, and others, the higher-layer protocols must define and offer the services to its lower layers. This will offer interoperability between various networks.

Currently, these services (better known as **multimedia applications**) are under consideration and are not available, as the intelligent network protocols have not been standardized yet or a forum has not yet agreed on the technology it is going to fit into the existing public networks. Attempts are being made in the U.S. and European countries to introduce intelligence into the network services; for example, in Europe, Deutsche Bundesport Telecommunication (Bonn), France Telecommunication (Paris), and AT&T are working together toward the development of intelligent networks which can support multimedia applications.

AT&T has defined an out-of-band signaling scheme known as **Signaling System number 7 (SS7)** (also known as **Common Signaling System 7**) which provides a standard link for establishing a signaling connection between switches and call setup request information. The digitization technique used in video, image, and audio communications with compression techniques is another major breakthrough leading toward high-performance and high-speed networks. These provide signaling systems which can handle and control the user's data separately using SS7, which offers a signaling scheme based on the out-of-band concept. The D channel defined in ISDN for signaling also defines a separate channel for signaling, although this may be used for data communication.

Transmission and switching: In terms of transmission, various issues which are discussed include encoding techniques, the network's data rates, multiplexing schemes used, transmission media, etc. In terms of switching, we are mainly concerned with routing strategies across the network — either packet-switching or circuit switching.

As discussed earlier, in packet-switching the full bandwidth is allocated to the users during communication, while in circuit switching the delay time (required during the setup transmission, propagation, etc.) is minimal. On the intelligence criteria side, we manage or control the flow of data through the network.

Synchronous optical network (SONET) and synchronous digital hierarchy[18,20] (SDH): SONET is Bellcore's Synchronous Optical Network. In the past few years, the telecommunications service carrier companies have been busy replacing the existing copper wires used in the network with an **optical fiber (OF)** medium, with an intention of using the new upcoming technologies to allow users to access a larger bandwidth of WAN and send the data at very high speeds. The fiber-optic cable certainly offers a virtually unlimited bandwidth, but it cannot be used with slower-speed channels until the standards for data rates, integration techniques, etc., are fully defined, in particular, SONET and SDH.

The data transmission rates up to 2.4 Gbits for optical standards such as **DS-3** (44.736 Mbps) and **E3** (34 Mbps) are already available and can be extended up to 10 Gbits for SONET and SDH. The standards defined for SONET and SDH have some similarities but are altogether different, and interestingly the standards for these two networks have been defined by different standards organizations; e.g., the specification of SONET is defined by the **American National Standards Institute (ANSI)**, while the European data rate and application of SDH are defined by the **European Telecommunication Standards Institute (ETSI)**.

CCITT recommendation G.707[20] has also defined standards for SONET and SDH, but these are a superset of the ANSI and ETSI standards. The specification and features of SONET and SDH have some similarities, and usually they support lower-speed (T1/E1 or T3/E3) signals which are mapped and multiplexed onto optical data hierarchy. These networks will allow carriers to build fully interoperable networks where different vendors (using switched networks) will be able to communicate and exchange network management information. This is possible, subject to the approval of standards, and at present the standards provide services only for switching and multiplexing. Due to these features and specifications of SONET and SDH for transmitting data, these networks can be used with any technologies, including asynchronous transfer mode (ATM) cells (CCITT recommendation I.150[26] specification for ATM defines how ATM cells can be mapped onto SONET and SDH channels and transmitted). It describes all the techniques for establishing VPCs and, to be accurate, there are three techniques defined. The establishment of VPC can be on a semipermanent basis and can be customer-controlled or network-controlled.

In the public switched telephone network (PSTN), there are two main components which are essential for voice communication and data communications and which can be designed and developed separately: **switching** and **transmission**. The switching component defined by the **network node interface (NNI)** provides interface between various elements within the network, while the transmission component defined by the **user network interface (UNI)**[34] provides interface between the user and network with different variations. The infrastructure so defined by PSTNs can also be used with other non-PSTNs, e.g., private circuits, leased lines, telex, etc. The introduction of pulse-coded modulation (PCM) allows the analog signals to be digitized and transported in a digital mode for various obvious reasons (advantages offered by digital transmission over analog transmission were discussed earlier). The voice-grade signals are sampled at a rate greater than twice its maximum frequency (Nyquist theorem). Each sample is measured by assigning eight bits within the channel, and the resultant signal is then transmitted. The set of these samples derived from different channels is transmitted sequentially over the transmission medium using time-division multiplexing (TDM).

Encoding: Various encoding techniques for existing twisted-pair wires are **high-speed digital subscriber line (HDSL)**, **asynchronous digital subscriber line (ADSL)**, and **very high digital subscriber line (VHDSL)**. HSDL offers full-duplex T1 services over two twisted pairs of copper cabling (this will avoid the use of repeaters). ADSL supports 1.544 Mbps in one direction over a single twisted pair and provides a low-speed data channel of typically 16 Kbps in another direction. A very high bit rate digital subscriber line (VHDSL) offers services at three to six Mbps over two twisted pairs.

This new technology (ADSL) is very important and useful in the communication industry. The xDSL defines a family of technologies that is based on the DSL concept. The variable "x" indicates upstream or downstream data rates. The upstream data rate is provided to the user for sending the data to the service provider, while the downstream data rate is provided to the service provider for sending data to the user.

ADSL supports high-speed transmission of digital signals. There are a number of links that have been introduced for providing high data rates. Some of these links include high data rate digital subscriber line (Bellcore), single line digital subscriber line, very high data rate digital subscriber line, and asymmetric digital subscriber line. Each of these links uses different signaling techniques and offers a different mode of operations. ADSL and VDSL support asymmetric while the other two support a symmetric mode of operations. These links are used for a limited distance of about one to six km. HDSL and SDSL support T1 and E1 data rates, while ASDL and VDSL support different dates rates for upstream and downstream.

Different DSLs with data rates and applications as defined by the ATM forum are given below:

- Digital subscriber line (DSL) — 160 Kbps, ISDN voice services
- High data rate digital subscriber line (HDSL/DS1) — 1.544 Mbps, T1 carrier
- High data rate digital subscriber line (HDSL/E1) — 2.048 Mbps, E1 carrier
- Single line digital subscriber line (SDSL/DS1) — 1.544 Mbps, T1 carrier
- Single line digital subscriber line (SDSL/E1) — 2.048 Mbps, E1 carrier
- Asymmetric digital subscriber line (ADSL) — 16–640 Kbps (upstream), 1.5–9 Mbps (downstream), video on demand, LAN access
- Very high data rate digital subscriber line (VDSL) — 1.5–2.3 Mbps (upstream), 13–52 Mbps (downstream), high-quality video, high-performance LAN

19.3.1 Pleisochronous digital hierarchy (PDH)

There is a general consensus worldwide for a sampling rate of eight Kbps (eight KHz) and a channel rate of 64 Kbps (8 KHz × 8 samples) as a minimum requirement for switching. The transport layer standards, however, are different for different countries (Europe, North America, etc.). Japan has collaborations with North America for the standards, except that it uses higher rates. A set of standards describing these aspects is known as **pleisochronous digital hierarchy (PDH)**. The digital technology in networks has given a new direction for enhancing the data rate of transmission and switching components. Various levels of PDH are discussed in CCITT recommendation G.702.

1984 was the year for a major shift of control in the U.S. from AT&T to seven regional **Bell Operating Companies (BOCs)**, and these companies started working as **local exchange carriers (LECs)** and one long-haul carrier. For LECs, regulated restrictions were imposed, while the long-haul carrier was made open in the market. The LECs were further divided into **Local Access and Transit Areas (LATAs)**. The inter-LATA transport service is to be provided by long-haul carrier. Each of the LECs was required to request/assign a **point of presence (POP)** for network traffic over the long-haul network. These changes in the U.S. telecommunication infrastructure have given a boost to the development of a high-cost, standardized transport interface delivered by multiple vendors on the open market.

The bifurcation between the local access and long-haul access mechanisms has given a large market to private networks (based on digital leased lines) at the primary rate. This end-to-end service cannot be provided by LECs, and they need to be bypassed by the long haul carrier which provides this service to corporate users. These kinds of infrastructure changes in the U.S. have spread to other parts of the world, in particular Europe, where the **European Economic Community (EEC)** was set up to look into the liberalization of telecommunication in Europe. The technical framework of base standards in Europe

Table 19.1 CCITT G.702 Recommendations for PDH

Level	North America (Mbps)	Europe (Mbps)	Japan (Mbps)
1	1.544 (DS1)	2.048	1.544
2	6.312	8.448	6.312
3	44.736	34.368	32.064
4	—	139.264	97.728

for the telecommunication network is usually based on recommendations of the **Committee European de Post et Telegraph (CEPT)**. ITU-T advised various groups G.707, G.708 and G.709 to define an SDH standard which was based on ANSI's SONET. Europe defined a basic data rate of 2.048 Mbps as E1, which is equivalent to the basic rate of 1.544-Mbps T1 or DS1 of North America.

SONET is a set of interfaces for an **operating telephone company (OTC)** optical network and is proposed by Bell Communication Research (Bellcore). Due to high data rates of fiber optics, there was a need for defining a set of digital signal interfaces. These interfaces will support a base rate for SONET of 50 Mbps, DS-3 electrical signal data rate of 44.736 Mbps, and transport of broadband 75-Mbps applications. The three recommendations — G.707,[20] G.709,[21] and G.709[22] — are the SONET recommendations, and all of these digital hierarchies or systems support ATM.

CCITT recommendation G.702 defined the asynchronous pleisochronous digital hierarchies (PDH) as shown in Table 19.1. It is clear from the table that the European signal hierarchy has no level near 50 Mbps, and therefore CEPT wanted a new synchronization hierarchy to have a base signal near 150 Mbps to transport its 139.264 Mbps.

19.3.2 SONET signal hierarchy (SSH)[20]

The first level of SONET signal hierarchy is the synchronous transport signal level 1 (STS-1).[20] It offers a bit rate of 51.84 Mbps and is assumed to be synchronous with an appropriate network synchronization source. The STS-1 frame structure is shown in Figure 19.1.

In Figure 19.1, the B field represents an eight-bit type. This frame represents 90 bytes, with each byte representing a field. There are nine rows of such frames in one STS-1 frame. This STS-1 frame consisting of nine rows of 90 byte-long frames is transmitted in 125 µs. It offers a data rate of 64 Kbps, as these frames are encoded using one byte per 125 µs. The first three column fields represent the line overhead bytes. The remaining 87 columns and 9 rows are used to carry the STS-1 **synchronous payload envelope (SPE)**. The SPE carries the SONET payload, which includes nine bytes of path overhead. The STS-1 can carry a clear channel DS-3 (44.736 Mbps) or a variety of lower-rate signals such as DS1, DS1C, and DS2. The physical interface for the STS-1 signal is via optical carrier level 1 (OC-1) and provides electrical-to-optical conversion at the layer. OC-1 is the lowest-level optical signal used in SONET and network interfaces. The STS-1 signal carries a payload pointer in its line overhead.

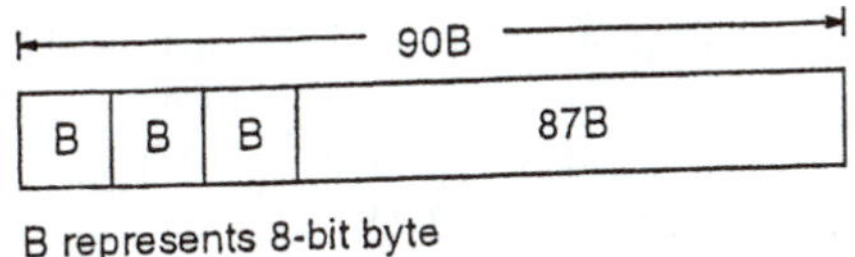

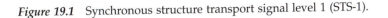

Figure 19.1 Synchronous structure transport signal level 1 (STS-1).

If the standards of the above networks (SONET and SDH) are fully defined and developed, then the users can immediately get access to larger bandwidth of WANs. One very useful feature of SONET and SDH is that these networks use synchronous multiplexing, which is better suited to handle channel **drops** and **inserts**. If we have asynchronous multiplexed circuits, then in order to extract a low-speed channel, a user has to demultiplex the high-speed signals, extract the low-speed channel, and again use the multiplexing technique to multiplex the signals (requiring many conversions). With synchronous multiplexing, the user knows exactly where and how the speed channel can be found out, thus making dropping and inserting a channel much easier and hence enabling the carriers to route low-speed channels through their networks.

19.3.3 SONET/SDH networks

Fiber optics has been used as a medium of transmission in all broadband communication protocols. Specifications and interconnection of different network systems through it has been considered by two international standards organizations, which have defined two standards: **synchronous optical network** (SONET, used mainly in North America) and **synchronous digital hierarchy** (SDH, used mainly in Europe).

The SONET/SDH network installation requires the following three types of equipment: **intelligent switches**, **digital cross-connect switches**, and **multiplexers**. The intelligent switches are the most problematic type of equipment which most carriers are currently dealing with. This switch includes an optical interface and is situated at the central office. The switch must be capable of rerouting the cells automatically in the event of link failure as well as performing other functions. The intelligent switch may be based on ATM technology, and we hope that very soon these switches will be standardized and used in the networks.

Both SONET and SDH support point-to-point and ring link configurations. In point-to-point configurations, SONET or SDH multiplexes over the channel, while ring configurations use digital cross-connect switches (DCCSs) with SONET or SDH interfaces. Both SONET and SDH standards provide direct synchronous multiplexing to different types of traffic over their higher data rates. There is no need for converting the traffic signals into any format before these can be multiplexed, and as such SONET/SDH networks can be interconnected directly to any network. Both SONET and SDH can be used in different types of networks, such as long-haul, LANs, and loop carriers. CATV networks also use these standards for video transmission. One of the unique features of SONET/SDH is their ability to provide advanced network management and maintenance, and this is due to the fact that these standards allocate about 5% of their bandwidth for these functions.

The following are the characteristics of point-to-point and ring configurations. **Point-to-point configuration** is simple and needs only one SONET port. It lacks fault tolerance, has limited flexibility, and does not support drops and inserts. **Ring configuration**, on the other hand, offers multiple DCCSs in concatenated point-to-point configurations and has two or more ports in the ring. It offers rerouting in the event of a failed ring and supports drops and inserts.

The SONET/SDH protocol architecture consists of path, line, section, and photonic layers. The photonic layer is at the bottom, while the path layer is at the top of the architecture. The **path layer** receives the video, voice, and data and provides transport of services to DS1 and DS3 between terminating equipment. The **line layer** provides reliable transport, synchronization, and multiplexing for the path layer. The **section layer** provides a few functions similar to those of the data link layer of OSI-RM, such as framing, error monitoring, communication, scrambling, etc. The **photonic layer** offers optical transmission at very high data rates. This layer provides mapping between the electrical signal and optical signals such that STS electrical frames can be converted into OC3 optical frames. The frame format of SONET and SDH may have some differences, but the working of SONET and SDH beyond STS-3 is exactly the same.

Digital communication is primarily based on a sampling rate of 8 KHz (based on PCM) and it defines eight bits in each PCM-coded sample. This type of sampling gives a frame rate for digital transmission of 125 μs (eight bits per channel), providing a basic channel rate of 64 Kbps. A number of basic rate PCM-coded channels can be multiplexed into a primary rate digital signal (DS1). Different numbers of channels have been defined for this rate, e.g., 24 channels in 1.544 Mbps (North America) and 30 channels in 2.048 Mbps (Europe). The least significant bit in every sixth channel sample is used to provide channel signaling and is known as channel associated signaling (CAS) in North America, while in Europe, the CAS offers an additional time slot 16 (TS-16) of the 2.048-Mbps frame. Similarly, the frame synchronization and transmission aspects in these primary rates are implemented in a different manner. For example, one bit per frame, or eight Kbps in 1.544 Mbps rate, and one octet per frame or 64 Kbps in 2.048 Mbps, are used. The path layer error control in both the rates was enhanced using cyclic redundancy check (CRC) and data channel capacity. **Common channel signaling (CCS)** has been defined by 64 Kbps in both the formats. The standards are defined as TS-16 in 2.04 Mbps and TS-24 in 1.544 Mbps. The discussion herein is partially derived from References 46–54.

The multiplexing technique used in pleisochronous digital hierarchy (PDH) usually increases the data rate of the tributary signals in such a way that the multiplexed signals provide a channel rate which matches the synchronous data rate derived from the new aggregate data rate. The aggregate rate in a PDH multiplexer must be greater than the sum of tributary data rates to cover up the multiplexing overhead. The aggregate format is defined by choosing an appropriate frame rate such that part of it is reserved for frame alignment and operation and administration overheads.

The STS-1 payload pointer is a unique feature of SONET, as it offers multiplexing synchronization in a pleisochronous environment so that higher rates can be obtained. There are two methods of multiplexing payloads into higher rate signals: bit stuffing and fixed location.

The first method, **bit stuffing**, increases the bit rate of the signal to match the available payload capacity in a higher data rate signal. At specified locations, a bit stuffing indicator with respect to the data frame is defined; it also indicates whether the stuffing bits correspond to real or dummy data in the higher data rate frame. Every frame will be followed by the bit stuffing indicators, and at the end of the indicators, we have a bit which indicates the status of data (real or dummy). This technique of bit stuffing for multiplexing payloads into higher data rates is used in multiplexing four DS1 signals into DS2 signals and multiplexing seven DS2 signals into DS3 signals (asynchronous). It is also used in accommodating a large number of asynchronous payloads of different frequency ranges. On the receiving side, the multiplexed payload signals need to be de-stuffed, and this sometimes becomes a difficult task.

The second method, **fixed location**, translates the signals into higher data rate signals. The digital switches are very useful in providing synchronization at the network level, and the fixed location can be easily mapped to define specific bit positions in the higher data rate synchronous signal. These mapped locations can carry the low data rate synchronous signals. The main advantage of this method of multiplexing is that the fixed specified bit position in higher data rate synchronous signals always carries the signal from the seam tributary payload. This method offers easy access and does not require any stuffing/de-stuffing. The DS3 (Syntran) signal uses this method for mapping of DS1 signals. The payload pointers offer easy access to the synchronous payload and at the same time avoid the use of 125-μs buffers (to provide phase synchronization) and associated

Table 19.2 ANSI Standards for Optical Carrier Levels

Level	Line rate (Mbps)
OC1	51.84
OC3	155.52
OC9	466.56
OC12	622.08
OC18	933.12
OC24	1244.16
OC36	1866.24 (1.244 Gbps)
OC48	2488.32 (2.488 Gbps)
OC192	9600 (9.6 Gbps)

interfaces. The STS-level signal is converted to an optical carrier-level signal. The SONET standard is set at a 51.84-Mbps base signal for a new multiplexing hierarchy known as synchronous transport electrical signal 1 (STS1), and its equivalent is optical carrier level 1 (OC1). This is equivalent to the 1 of the DS3 and 28 of DS1. The levels of optical carrier — along with the line rates — have been defined by the American National Standards Institute and are shown in Table 19.2.

For different data rates, both SONET and SDH offer different frames; e.g., at data rates of 50, 150, and 622 Mbps and 2.4 Gbps (equivalent to OC1, OC3, OC12, and OC48), SONET frames are STS-1, STS-3, STS-12, and STS-28. On the other hand, SDH supports only three data rates of 150 and 622 Mbps and 2.4 Gbps with frame formats of STM-1, STM-4, and STM-16, respectively.

19.4 Evolution of switching techniques

A switch in networking can be considered the heart of the data communication (similar to an engine in a car). Depending on the type of switching (packet or circuit), we have two categories of switches associated with switching equipment. Appropriate switches have to be used when a network (be it public or private) based on that switching technique is selected, although plug-in switches are available which can be used interchangeably on the networks. When the switches are used with different types of networks such as frame relay, X.25, switched multimegabit data service (SMDS),[9,10,17] synchronous optical network (SONET),[7,16,18] and synchronous digital hierarchy (SDH),[20] different interfaces have to be provided by the switching equipment. But with the introduction of ATM, the concept of switching (packet or circuit) may disappear, as it supports both voice and data in addition to video and other information and also integrates both switching techniques for its switching and transmission facilities.

The introduction of ISDN and its standardization in 1984 (CCITT COM XVIII-228, Geneva 1984) and more comprehensive specification in 1988 (**CCITT I.121**)[25] somehow have created some serious problems at the initial stages due to the slow pace and lack of interest in the U.S. As it was about to emerge, the LANs took the job of linking desktop terminals; also, different vendors in the U.S. were not interested in the interoperability of ISDN products. As a result of this, the ISDN technology became more popular in European countries and other countries which are typically governed by respective PTTs, and it could not get enough support from vendors within the U.S. This technology was popular in countries such as France, Japan, Germany, and the U.K. without looking at the multi-vendor or interoperability aspects of the internetworking. In the U.S., instead of ISDN, AT&T's switched S6 services which offer dial-up 56-Kbps circuits became accepted. ISDN technology in the U.S. currently is gearing toward teleconferencing, remote access to computers, videoconferencing at a low rate of 128 Kbps, and a few more applications requiring low data rates.

In order to appreciate and understand the main concepts behind the high-speed and high-bandwidth networks, the following subsections will discuss the evolution of switching techniques along with their roles and application in telecommunication services. For details on various switching techniques, see Chapters 3 and 18.

19.4.1 Circuit switching

In the classical approach, a circuit or link or logical connection is defined for the entire duration of the connection. In circuit switching, a complete physical connection is established between two sites and then data is transmitted from the source site to destination site. It supports end-to-end connection between the computer nodes or sites. An initial setup time (for establishing end-to-end connection) in some applications, e.g., voice and video, communications may become a serious problem. During the established connection, the entire bandwidth is allotted to the users. Unused bandwidth will be wasted and cannot be used by other users or even by the network. Circuit switching offers minimum delay and guarantees the delivery of the packets in order. It is still being used in N-ISDN, where it offers a fixed data rate and fixed-capacity network for the entire duration of the connection. The message data is transmitted at different bit data rates; e.g., for a 64-Kbps data rate, eight bits per 125 μs are transmitted, while for an eight-Mbps data rate, the number of bits is 1000 per 125 μs.

Switching supports different types of multiplexing techniques (space, time). At each node, for the input data message, an assigned link along with the time slot is maintained. For each circuit or link which is multiplexed, the same time slot with the frame is used for the duration of the connection. Similarly, a table describing the relationship between incoming and outgoing links along with time slots at each switching node is maintained. The time slots are fixed for both sides of switching for the duration of the connection. Further, the time slot for a given data rate is fixed, and hence this technique dose not provide any flexibility. For example, if we consider a 64-Kbps data rate channel, it activates the time slot for the duration of eight bits/125 μs.

CCITT uses the term *circuit switching* for describing features of link transmission, multiplexing, switching, and interfacing aspects in computer networks. A synchronous and fixed time division multiplexing can be used to mix a number of data rates and multiple users within the primary data rate range. But this multiplexing technique may not be suitable for multiplexing different higher data rates for different applications. In other words, there is no common interface which can provide services to these applications, and as such this switching in B-ISDN does not offer any flexibility. Furthermore, the existing switches used in N-ISDN can handle only 64 Kbps, while in B-ISDN we need to use a different switch which can handle a high data rate. The synchronous TDM technique will make the transmission more complex and difficult, in particular for the channel speeds of H2 (30–45 Mbps) and H4 (120–140 Mbps). As a result of these drawbacks, this switching technique is suitable for fixed rate bit stream–based applications of integrated voice and data communications, but it is not suitable for variable data rate applications such as burtsy traffic.

Since the channels are defined as the time slots, the channels may not be appropriate for all broadband services due to their different data rates. In order to support a variety of services, a high bit data rate is usually chosen. The main problem with this technique is that even the small data rate applications will use the channel for the entire duration, which is a wastage of resources. For these reasons, the classical circuit-switching technique is not suitable for broadband services, which require variable high data rates for different services.

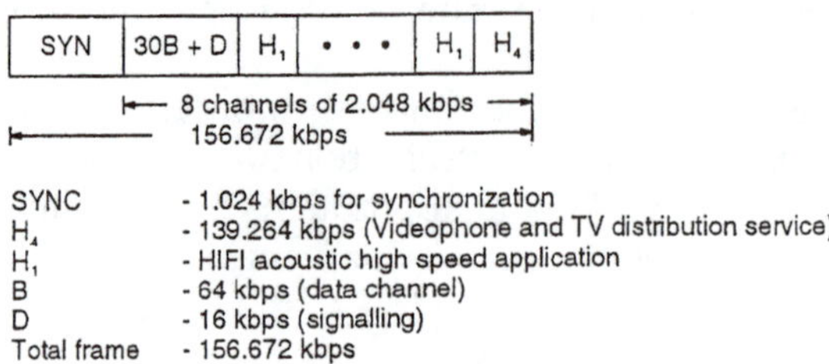

Figure 19.2 Basic frame of MRCS.

19.4.2 Multi-rate circuit switching (MRCS)

The above-mentioned drawback of resource wastage in circuit switching has been overcome in another switching technique known as **multi-rate circuit switching (MRCS)**, which is a modified version of circuit switching. This switching uses synchronous time-division multiplexing which provides flexibility in terms of multiple synchronous circuits/links. This type of flexibility is certainly expensive due to synchronous multiplexing, but it supports a variety of applications of integrated data and voice communication which usually are described as fixed or constant data rates of the bit stream. Since it is mainly based on the concept of circuit switching, it also does not support variable data rate applications (bursty traffic).

In MRCS systems, each connection is defined as a multiple of the basic channel rate. Any connection can have n basic channels, where n is an integer greater than one. The standard recommendation H.261 offers a bit rate of $n \times 64$ Kbps for videophone on N-ISDN. The maximum value of n is 30. Since there exist more than one connection, each connection needs to be synchronized separately during the connection. Further, in this scheme, the selection of the minimum data rate becomes a problem, since the services require a wide range of data rates. For example, services such as telemetry may require a very low data rate (as low as 1 Kbps), while services such as high-definition television (HDTV) may require a very high rate (as high as more than 120 Mbps). If we choose a minimum basic rate of 1 Kbps, then services with higher data rates will require a multiple number of 1-Kbps channels. But if we take a high data rate as the basic rate, then there is a wastage of channel bandwidth for the services of lower data rates. For these reasons, the MRCS-switching scheme is proposed, where the basic frame time slot is divided into different time slot lengths, as shown in Figure 19.2.

A channel H4 of data rate 139.264 Kbps is multiplexed with eight channels of 2.048 Kbps each. The H4 offers videophone and TV distribution services, while the H1 channel is for HI-FI acoustic high-speed data and some other medium-speed services. There are 30 B channels, each of 64 Kbps, and one D channel of 16 Kbps (for signaling).

The SYNC of the 1.024-Kbps channel is used to provide the synchronization between the transmitter and receiver for the boundary of the frame. Each channel is defined by a switch. Thus, the information from the user is multiplexed/de-multiplexed (depending on the application) to the different channels, which basically are connected to appropriate switches. Control and other maintenance management for all of the channels and switches are provided by a common set of procedures. One of the drawbacks with this switching is that the resources are not used efficiently. For example, if any application has used all H1 channels, then it has to wait until the previous channels are free; it will not get an H4 channel.

19.4.3 Fast circuit switching

This switching technique is a modified version of circuit switching in which the resources are allocated during the transmission of message information and are released immediately after the transmission is completed. This technique is useful in applications requiring bursty traffic (applications requiring variable and high data rates). This technique is a form of connectionless packet-switching in the sense that, during the setup of connection time, a request for destination bandwidth for connection is stored inside the switching system, which allocates a header for this information. The header has information about the source and other request control information from it. The resources will be allocated immediately after the source starts sending the information. One of the major drawbacks with this technique is its inability to allocate enough resources to unexpected bursty traffic.

It will be interesting to see the combinations of fast circuit switching (FCS) and MRCS, as they will provide not only different channels but also support for bursty traffic applications due to available channels of different bandwidth/data rates. Due to its complexity, this combination of FCS and MRCS has not become popular and useful, and this technique has not been considered for high-speed transmission of message information data.

19.4.4 Packet-switching

In packet-switching, a fixed-size packet (defined by the network) is transmitted across the network. All the packets are stored in main memory (instead of on a disk as in message switching) and, as a result, the accessing time of the packets is reduced. Due to this, the throughput offered by this type of network is very high. In packet-switched networks, a circuit, connection, or logical link is not established for transmitting the packets in advance. After the packets have been transmitted, whereby they use the entire bandwidth of the link, the bandwidth is released and can be used by other packets. In other words, packet-switching offers a larger bandwidth to the packets.

There is a problem of congestion, i.e., storing of these packets, and it is likely that due to limited space, some of the packets may be lost. Also, the packets may not be delivered in sequence, as they are arriving at the destination node from different routes at different times. This may pose a problem for voice communication, as this application requires a dedicated circuit to be established before voice communication (as done in circuit switching) or a circuit which offers minimal or no delay. As we have seen, circuit switching allocates the entire bandwidth to the application (voice communication) during the entire duration of the connection. Unfortunately, there is under-utilization of resources, particularly the circuit/link (between terminal and host) and, further, both the transmitting and receiving nodes must have suitable interfaces to support the fixed data rate being used for the application (no flexibility).

In packet-switching, the packets define the user's information header and other control information such as flow control, error control, etc., and are transmitted over the packet-switched networks. The packets have variable lengths, and as such the requirement for buffer size and the probability of delay become major problems with packet-switched networks, even in low data rate of transmission applications. In spite of these drawbacks, this technique is very efficient and useful for such applications, as we can see in X.25. The high data rates in the packet-switched networks (such as X.25) are usually not supported, due to the fact that the protocols used in these networks are very complex. Also, the packet-switched networks do not offer delivery of the packets in the same order in which they were transmitted at the sending end; thus, these networks are not suitable for applications of voice communications or applications requiring a fixed data rate for the bit stream.

CCITT has defined the X.25 packet-switching standard over the B and D channels of N-ISDNs. This form of switching is sometimes known as **X.25 packet-switching**. The packet-switching for N-ISDN over the X.25 network is now being defined via different switching techniques such as **frame relaying** and **frame switching**. These two switching techniques definitely have less functionality compared to X.25 switching.

Frame-relaying switching technique: In this technique, the end-to-end terminals transmit/receive the re-transmission of the user data frame for error recovery. The error detection (provided by protocols) uses CRC algorithms and, after the detection, the errored frame is discarded. The flow control is also provided between end-to-end users at the frame level, and no multiplexing for flow control is performed at the packet level. The packet-switching supports error control, which requires extra overhead. Due to high-speed telecommunication networks, it is becoming quite obvious that the underlying transport system will be providing very low bit error rates and high data rates for data communications. This will reduce the overhead required by the network for error control, and as such we may expect data rates as high as 2 Mbps (compared to 64 Kbps in traditional circuit-switched ISDNs).

Frame-switching technique: This technique provides both error and flow control at the frame level between end-to-end terminals. This allows the re-transmission of a frame using any link-level protocols that can perform this function at the data link layer level. This technique does not provide any de-multiplexing at the packet level (as is done in X.25). Various feasibility studies have revealed that the X.25 packet-switching protocol offers around a 2-Mbps data rate, while frame switching offers data rates somewhere between 5 and 9 Mbps. The frame-relay switching protocol may offer a data rate of up to 145 Mbps. The frame-relay switching technique does not support error control (e.g., ARQ protocols), flow control, or multiplexing of logical channels. It offers more or less all the services offered by X.25 such as frame boundary flap, bit stuffing, CRC checking, error control (ARQ), flow control, etc.

Point-to-point network: The point-to-point network is usually defined as a set of point-to-point links interconnected by switching nodes. A computer node accesses a point-to-point network either directly through the switching node or through intermediate access point-to-point nodes which accept the traffic from different nodes and are multiplexed. The multiplexed data information is then sent to the node. Examples of point-to-point networks are the Internet (switching nodes are defined for the routers) and ATM-based networks (where fast packet-switching nodes are defined for data information).

The following sections concentrate on each of the high-speed networks with a view to understanding the concepts, standards, frame formats, and their applications. The following high-speed networks will be discussed:

- Fiber-distributed data interface (FDDI)
- Fiber-distributed data interface II (FDDI-II)
- Switched multimegabit data service (SMDS)
- Frame relay
- Broadband ISDN (B-ISDN)
- Asynchronous transfer mode (ATM)

19.5 *Fiber-distributed data interface (FDDI)*

The existing LANs have limitations on speed and capacity and hence cannot be used for interconnecting host computers for sending the data at a very high speed. The first version of FDDI[15] was primarily intended for data applications. Although FDDI can handle circuit switching, it is mainly based on the packet-switching approach for data communication. A detailed discussion of its use in interconnectivity, the MAC layer, and related protocols can be found in Section 8.2.9 (Chapter 8). In order to provide continuity, we are summarizing some of the information here with a view to building the foundation for discussion of high-speed networks.

The existing systems (public networks, premises wiring systems) have offered transmission speeds of 1–2 Mbps and are now even offering transmission speeds of 32–45 Mbps. The LANs (IEEE 802 standards) are offering data rates of 1 to 16 Mbps within campus or organizational premises. The fiber-distributed data interface (FDDI) LANs offer a data rate of 100 Mbps. This can be used to implement a metropolitan area network (MAN) defined in the IEEE 802.6 standard.

FDDI uses ring topology, which offers various advantages. Ring topology does not impose any restrictions on the number of stations or length of links. In the total length of a MAN, the ring topologies generally offer high performance under heavy conditions. Further, the performance of ring topology can be improved by using two rings, such that in case of a failed link or node, it can be reconfigured to bypass the failure and still offer the same response without affecting the topology or the nodes. The ring topology offers point-to-point communication. A review paper on FDDI is available; see Reference 2.

An enhanced version of FDDI known as FDDI-II offers integrated services of packet-switching, and plans are underway for handling isochronous services (voice and video over circuit switching). The FDDI-II ring supports high-performance processors, high-performance workstations, mass storage systems, distributed database systems, a switchable backbone LAN for a number of functions, etc. One of the main features of the FDDI-II ring network is its ability to support isochronous service offered by wide-band channel (WBC), which describes circuit-switched connections. A total bandwidth of 98.304 Mbps is available for a variety of applications. The isochronous bandwidth (WBC) may be allocated to different sub-channels supporting different virtual circuit services such as video, voice, image, control data, real-time data, etc., within the bandwidth of one or more WBCs.

If the FDDI-II ring network is working in a packet-switched mode, it may well be used as a backbone for interconnection of devices (e.g., gateways) to the public data network. The maximum length of a ring is about 200 km, and it supports up to 1000 users. The maximum distance between two nodes is 2 km. It supports packet-switched traffic only, but it is being developed for handling isochronous traffic in the future (voice, video).

19.5.1 FDDI local area networks

There are three fast Ethernet LANs — 100Base T, 100 Base VG (voice grade) or AnyLAN, and fiber-distributed data interface (FDDI) LAN. The 100 Base T can use unshielded twisted pair (categories 3, 4, and 5) and optical fiber. It is based on CSMA/CD, as it uses the same MAC sublayer, and also is compatible with 10 Base 10. 100 Base VG, or AnyLAN, was introduced by Hewlett-Packard and uses twisted pair (categories 3, 4, and 5) and optical fiber. It supports the transmission of audio, data, and video. FDDI is based on the token concept and allows many frames to travel simultaneously over the media. It uses dual ring and hence offers fault tolerance. The American National Standards Institute (ANSI) introduced it; it uses copper twisted pair and coax. It supports synchronous traffic (audio, video) and asynchronous traffic (data). It uses a timed-token protocol that provides fairness to all the nodes to access the token ring. The FDDI is based on fiber optics and covers a distance of nearly 150 miles. It uses two dual-counter rotating rings and uses a token-passing MAC frame (IEEE 802.5 token ring standard). It offers a data rate of 100 Mbps and supports both synchronous and asynchronous data services by allocating the required bandwidth dynamically.

FDDI is being used as a backbone LAN in many corporations and universities. It is a 100-Mbps LAN which uses optical fiber as a transmission medium. The FDDI protocol is based on the token ring media access technique. Initially, FDDI was developed by **Accredited Standards Committee (ASC) X3T9** as an input/output interface standard. With subsequent changes and advances, the FDDI LAN was standardized by **IEEE 802**, offering a data rate up to 100 Mbps. The IEEE 802 committee advocates this LAN as a packet-switched data network, and FDDI also followed their recommendations for its architecture.

The token ring access method of the IEEE 802.5 token ring LAN became the main ingredient of the FDDI LAN. Due to their use of packet-switching and token ring access, the FDDI LANs have become by and large backbone LANs (*de facto* industry standard) and also follow-on standard LANs of IEEE 802. Other standards organization such as OSI provided needed help in defining separate standards for different layers. The backbone LAN is expected to provide a high access data rate and high bandwidth to different types of nodes, e.g., LANs, hosts, PCs, terminals, etc., within campus premises.

With the advent of high-performance video workstations, high-speed FDDI LANs have become front-end, high-performance LANs for a variety of applications. FDDI connects different types of LANs being used in various departments or divisions. These LANs usually offer data rates of 4 and 16 Mbps (token ring LAN) and 10 Mbps (Ethernet) and, when connected with FDDI, a data rate of 100 Mbps is achieved. The typical average throughput with a reasonable number of LANs can be as high as 80 Mbps. Although they can carry isochronous (audio) messages, they still use the public telephone network for carrying audio over PBXs.

The FDDI LAN supports both types of transmissions: **synchronous** and **asynchronous**. Synchronous frames are transmitted whenever an agreed upon bandwidth is available, e.g., video or audio messages. On the other hand, asynchronous frames are transmitted even if the bandwidth is not available, e.g., datagram-based messages. For each type of transmission, the node has to capture the token as soon as it receives either synchronous or asynchronous frames. FDDI also is used in baud-end input/output interfaces. The diversified application areas of FDDI have resulted in the development of different sets of standards, and different services from FDDI have been derived for different applications. The FDDI networks offer minimum acceptable guaranteed response time (useful for time-dependent applications) and high link utilization (possibly under heavy traffic) and, in general, depend on a number of parameters like transmission time, propagation time, data rates, number of nodes connected, length of the ring topology, etc.

Improved and increased services can be offered by integrating circuit switching with packet-switching in a traditional packet-switched FDDI. This resulted in the development of FDDI-II, which supports digital voice, video data, sensor control, and data required in any typical real-time applications (e.g., mission-critical, time-dependent, etc.). The material herein is derived partially from References 1-10 and 46-54.

Two types of services are provided by LANs, including FDDI LANs: connection-oriented and connectionless. In **connection-oriented service**, the following three phases are defined:

- Connection establishment
- Data transfer
- Connection de-establishment (termination)

In the first phase, the logical connection is established before data can be transmitted. The logical connection is usually initiated by the source node and includes a sequence of switching elements between source and destination nodes which is used for forwarding the data frames. This logical connection is also known as a **virtual circuit** or **link** or **circuit**. The virtual circuit allocates the entire bandwidth to the data message being transmitted from the source to the destination and is available during the connection. In the second phase, after the circuit has been defined, the data transfer takes place through **virtual packets**. In the third phase, either of the nodes (source or destination) can make a request for the termination of the connection between them.

In contrast to this, in **connectionless service**, the messages include the source and destination addresses and are usually sent separately over different circuits. The message is broken into packets of fixed size known as **datagram** packets.

As mentioned earlier, the IEEE 802 committee's main objective was to standardize LANs and, as a result, we have seen five standard LANs: IEEE 802.3, Ethernet (CSMA/CD), IEEE 802.4 token bus, IEEE 802.5 token ring, and IEEE 802.6 metropolitan area network (MAN). These standard LANs offer data rates in the 1–20 Mbps range and can be considered low-speed LANs. The IEEE 802 committee was also looking at high-speed LANs which can be used as backbone LANs. The backbone LAN must possess performance parameters such as high data bandwidth, security, high performance, reduced size, reduced weight, etc. The significant improvement in the price and performance of optical fiber (including transmitters and receivers) has made the all-fiber LAN one of the leading candidates for a high-speed backbone LAN. Since the FDDI design is optimized for the use of optical fiber as a transmission medium, it has also become a leader in optical fiber LAN technology. FDDI standards conform to the OSI model and satisfy the requirements of the broad, high-speed LAN marketplace.

FDDI uses a ring topology which offers features such as reliability, availability, serviceability, and maintainability even in the presence of a broken link or failed node. The ring topology offers simple interconnection of physical hardware at the interface level, ease of dynamic configuration capabilities when the network requirements change, no limit on the length of ring link or number of stations, etc. The optical fiber provides enough bandwidth for serial transmission, which offers added features in terms of reduced size, cost, and complexity of the hardware circuits required by the network.

FDDI supports two types of nodes: dual-attachment and single-attachment. A dedicated node providing connection with fault-tolerance known as a concentrator is also used. Realizing the need to send video, images, audio, and data over the network, FDDI-II was developed to meet the requirements at a higher data rate.

The dual counter-rotating ring is composed of two rings, primary and secondary. The primary ring usually carries the data and the secondary ring is used as a standby ring. In the event of a fault on the primary ring, the secondary ring takes over and restores the functionality of the ring by isolating the fault; now the network is running over a single ring. After the fault is repaired, the network operates with dual rings. If more than one fault occurs in the network, then only those nodes that are still connected by the ring can communicate with each other. Various states of FDDI ring operations can be defined as ring initialization, connection establishment, ring maintenance, and steady state. The steady state represents the existence of operational FDDI where the nodes connected to the ring can exchange data using a timed-token protocol.

The protocols of FDDI offers various options such as addition or deletion of any stations without affecting the traffic, removal of a failed station or fiber link, initialization, failure isolation, recovery, reconfiguration, load balancing, bounded access delay, fair allocation of available bandwidth, low response time, etc. It seems quite clear that ring topology is a leading candidate to meet the requirements of high-performance networks offering data rates between 20 and 500 Mbps. The high-performance networks at these data rates are characterized by high connectivity, large extents, etc.

FDDI layered reference model: Figure 19.3 shows a reference model of an FDDI station.[15,16] The layers of FDDI conform to both IEEE 802 and OSI-RM. Some of the material presented herein has been derived from References 2, 4, 11, 12, and 42.

The layered structure of FDDI IEEE 802 has defined a layered structure for the lower two layers of OSI-RM. This layered structure is used in the LAN and MAN standards of IEEE 802. The LLC sublayer (IEEE 802.2) is independent of the topology and medium

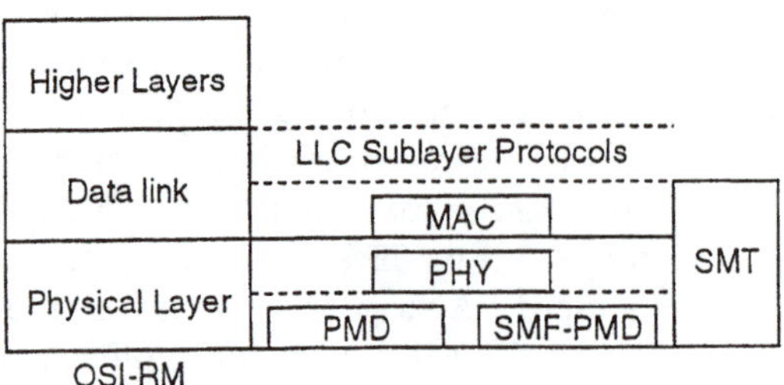

SMF - Single-Mode Fiber
PMD - Physical layer Medium Dependent
PHY - Physical layer
SMT - Station Management

Figure 19.3 Reference model of FDDI station.

access technique used by the MAC sublayer. The LLC supports both connection-oriented and connectionless services, and the protocols for these services are standards defined by IEEE. The MAC sublayer protocols are different for different LANs and are available as MAC of IEEE 802.3 (Ethernet), IEEE 802.4 (token bus), and IEEE 802.5 (token ring). An overview of FDDI and FDDI-II and a discussion of various attachments can be found in Reference 2.

The physical layer is divided into two sublayers: **physical (PHY)** and **physical medium dependent (PMD)**. The PMD sublayer protocols depend on the medium and at this time, two versions of PMD are available: basic **multimode fiber (MMF-PMD)** and **single-mode fiber** (**SMF-PMD**).[39] In MMF-PMD, a 1325-nm optical window is used for generating light by LEDs. This mode is used for limited distance, and the connection is established between nodes via dual-fiber cable. In the SMF-PMD version, a single mode fiber and later a diode transmitter (in contrast to LEDs in multimode) are used. This extends the length of links up to 100 km.

The physical layer offers transmission media connectors, optical bypassing, driver–receiver interface requirements, encoding and decoding, and clocking for defining data bit rate and framing. The framing so defined by clocking is transmitted either over the media or to its higher layers. The MAC sublayer defines the flow of data over the ring and offers access to the medium using a token which is moving around the ring. The incorporated token-passing protocols control the transmission over the network and also offer error recovery. The MAC defines a packet frame (containing header, trailer, addressing, and CRC), while the MAC token offers features such as priority, reservation, etc. The LLC (IEEE 802.2) is independent of LANs, and hence different LLC protocols can also be used in FDDI.

A typical combination of SMF and PMD offers a data rate of 100 Mbps over a link of up to 100 km. A frame can have a maximum of 4500 octets, and multiple frames can be transmitted on the same accessing of the network. Each station defines two physical connections which are connected by 100-km duplex cable.

The **station management (SMT)** layer describes the network management application process and also the control required for internal configuration and operations within a station in the FDDI ring. It is part of the OSI network management standards. It interfaces with the bottom three sublayers of FDDI, including error detection and fault isolation. It also acts as a means of providing a connection between nodes and its higher layers. It is important to note that the FDDI generally follows the layer-based protocol architecture over the LLC sublayers and upward.

The **media access control (MAC)**[38] layer describes the access technique to the media, addressing, data checking, and data framing, and is the lower sublayer of the data link layer. The physical layer protocol PHY deals with encoding/decoding, clocking, and transmission of frames, and it corresponds to the upper sublayer of the physical layer. The PMD is mainly concerned with power levels and behavior of the optical transmitter and receiver, requirements for optical signal interfacing, acceptable bit error rate, requirements of optical cable plants, etc., and corresponds to the lower sublayer of the physical layer.

The physical sublayer has a data transmission speed of 100 Mbps. The FDDI frames (defined for FDDI packets) can have a maximum length of up to 4500 octets. The PHY sublayer defines recovery times which are defined by a standard default FDDI configuration of 500 stations and 100 km (length of optical fiber) working in a full-duplex mode of operation. The FDDI does not define any cycle time.

The FDDI-II (enhanced version of FDDI) (to be discussed below) defines a cycle time frame of 125 μs. This cycle frame allocates a maximum of 16 isochronous channels, each of 6.144 Mbps, giving a total bandwidth of 96.344 Mbps. Each channel of 6.14 Mbps is equivalent to four times T1 (1.544 Mbps in the U.S.), with each offering 1.536 Mb, and three times 2.048 Mbps (European countries). Even if all 16 isochronous channels are used, a token channel (residual) of 768 Kbps is still available. The allocation/de-allocation of isochronous channels is done dynamically. The residual token channel of 768 Kbps may be useful for packet data. The MAC sublayer field consists of the control fields (preamble address) and user's data to be transported to the LLC sublayer.

If we look at the model, we notice that the FDDI model has two sublayers of physical layers — PHY and PMD or SMF-PMD — while the FDDI MAC provides services to the IEEE 802.2 LLC sublayer. Other LLC sublayers are also supported by this MAC.

The FDDI trunk ring represents a pair of counter-rotating rings. Any station can be connected either directly to the token ring or indirectly via a **concentrator**. As mentioned above, there are two types of attachment configurations defined for this type of LAN — **single attachment unit (SAU)** and **dual attachment unit (DAU)** — and these are discussed below.

FDDI attachment configuration: A single attachment unit has one PHY and one MAC and does not connect the station to the main FDDI ring. This station is indirectly connected to the ring via a concentrator. Figure 19.4(a–c) describes the internal configuration of various attachments.

In Figure 19.4(a), M represents MAC while P represents PHY + PMD (both sublayers of the physical layer). It represents the configuration of a single attachment. The single attachment station is connected to the FDDI ring via another attachment (concentrators), shown in Figure 19.4(b).

A dual attachment unit (DAU) node has two PHY entities and may have one or more MAC entity; one PHY entity may be in the counter-rotating ring or both may be on the same ring. A dual attachment node also possesses the capability of bypassing it via an optical bypass switch, which removes it from the ring if a station is disabled by SMT or if there is a power failure (shutdown). Different combinations of a dual attachment configuration are shown in Figure 19.4(c). It is important to note that the first combination, (i), in Figure 19.4(c) must connect one of its PHY connections to the FDDI ring via concentrator, while the second connection may be connected to some other attachment.

As stated above, the protocols of token ring remove the failed station or broken link without affecting the network via different mechanisms. One of these mechanisms is counter-rotating ring topology, as shown in Figure 19.5, which offers a normal flow of data around the ring topology composed of single or dual attachment unit nodes. The

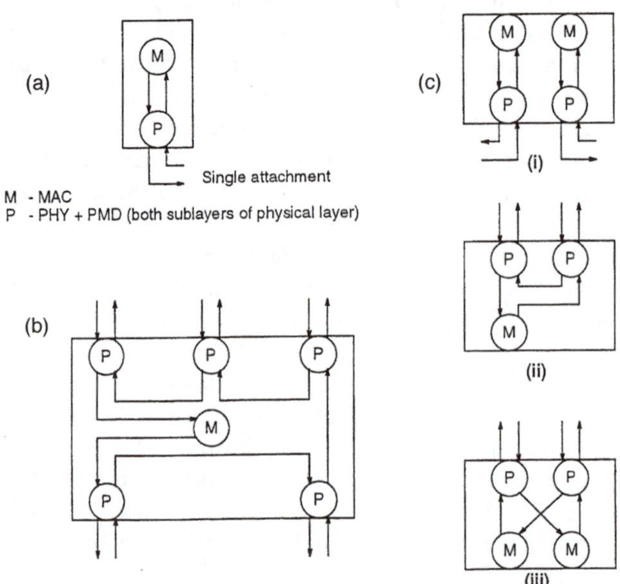

Figure 19.4 FDDI attachment configurations. (a) Internal configuration of attachments. (b) Trunk ring or dual attachment station. (c) Possible combinations of dual attachment configuration.

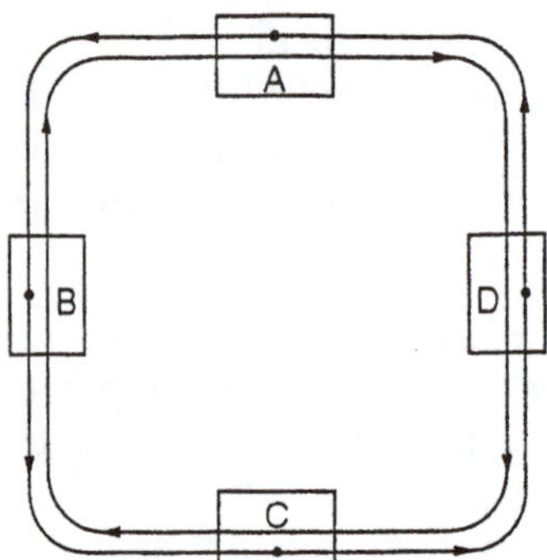

A, B, C, D correspond to nodes (stations or concentrates) with MAC attachment and PHY

Figure 19.5 Normal flow of data in counter-rotating ring topology.

counter-rotating topology includes two physical rings connecting stations and concentrators, and it allows the data to flow in one direction while the second ring carries the data in the opposite direction. The FDDI protocol reconfigures the internal ring architecture in such a way that in the event of a failed station or broken link, it will still work and offer the same performance. The diagram shown in Figure 19.6(a,b) illustrates the flow of data under the condition of a failed node and a failed link, respectively. The reconfiguration capability of the protocol does not affect the normal flow of traffic in the network.

The bypassing of a failed node or link is reconfigured by the protocols internally, as shown in Figure 19.6(a,b). The dots (.) in the figures represent MAC attachment along with PHY, and in the event of a failed node or link, this attachment connection is reconfigured internally at the neighboring stations of the failed link or node.

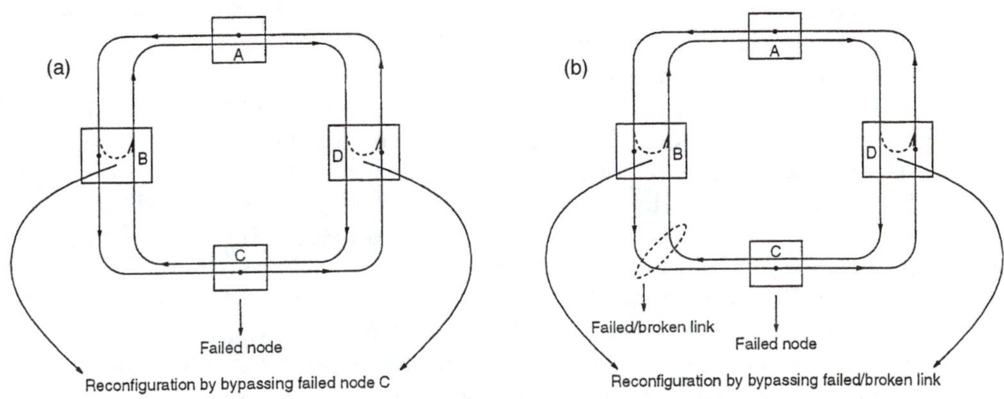

Figure 19.6 (a) Reconfiguration for failed node. (b) Reconfiguration for failed/broken link.

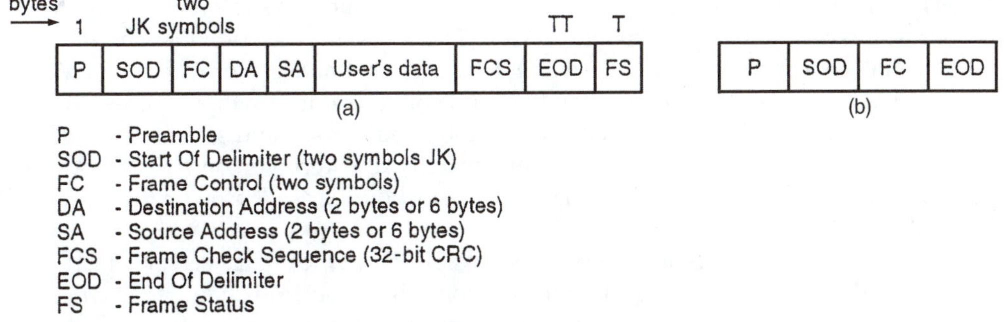

Figure 19.7 (a) FDDI frame format. (b) FDDI token frame format.

In some cases, as mentioned above, the stations may use an optical bypass switch which is activated in the event of a failed node or link, and both the transmitter and receiver of the failed node are bypassed. The neighboring nodes activate the failed node.

FDDI frame and token frame formats: The frames which carry the user's data around an FDDI ring are of variable length. Within a frame, a token frame of small size is defined (similar to the token frame of three bytes defined in the IEEE 802.5 token ring network). Figure 19.7(a) shows a typical frame format.

The **preamble (P)** field includes 16 IDLE symbols, is used for providing synchronization between source and destination nodes, and is sent with every transmission. The **start of delimiter (SOD)** field includes a two-symbol sequence (JK) which is different from the ones used in the preamble field. These symbols are non-data symbols. It also defines the boundaries for the symbols to be used in the user's data field. The **frame control (FC)** field is a two-symbol field which defines the type of frame being transmitted. It distinguishes between synchronous and asynchronous frames, 16-bit or 48-bit address fields, and LLC or SMT frames. There are two types of tokens supported in an FDDI network: **restricted** and **nonrestricted**. The restricted token is usually used for a special category of services, or in some cases, the node activates itself and the connections to the node are removed by protocols without affecting the network traffic.

Concentrators may also be used to remove the failed node, as the nodes connected via concentrator are controlled and managed by the concentrators. The reconfiguration of the FDDI ring network is automatic and eliminates the failed node or link without affecting the normal flow of traffic. It is possible to lose some frames during reconfiguration, and those frames need to be re-transmitted.

Two symbols — TT — are defined in the **end of delimiter (EOD)** and are attached with three fields P, SD, and FC to define a token, as shown in Figure 19.7(b).

The **DA** and **SA** fields represent destination and source addresses, respectively, and may be either 16 bits or 48 bits long, depending on the value selected in the FC field. The DA field can define individual, group, or broadcast addresses. The **FCS (frame check sequence)** field uses a 32-bit cyclic redundancy check (CRC) error-control polynomial protocol (used in IEEE 802 LANs). The user's data field consists of the symbols of the user's data and does not use symbols which are not defined in FCS. The EOD field uses one delimiter symbol (T). The last field is the **frame status (FS)** field, which indicates the status of the frame (set, receiving by destination, and copy of user's data in its buffer). The three states (set, error detection, and copy) of FS are defined by three different symbols to provide the above-mentioned conditions and are updated by the node itself after the necessary action has taken place.

Physical layer functions: The physical layer is partitioned into two sublayers: physical (PHY) and physical layer medium dependent (PMD). The PHY sublayer provides the protocols, while the link connection (optical hardware) between FDDI nodes is provided by PMD.

The MAC receives the data from the LLC and converts it into symbols. These symbols are then translated by the transmitter into a five-bit code group and sent as encoded data to the transmission medium. On the receiving side, the receiver decodes the original symbols from the received data, identifies the symbol boundaries (based on the SOD field), and finally sends the data to its MAC.

The PHY sublayer to some extent is responsible for avoiding any error which might be introduced or inserted during the transmission. This problem is handled (by PHY) by using a bit clock for each node in such a way that the total number of bits in the FDDI should remain the same length. The receiver uses a variable-frequency clock (phased locked loop oscillator), while the transmitter uses a local-frequency clock for bit clocking. The difference between the clocks can be defined in the P field of the frame to maintain synchronization between the transmitter and receiver.

The scheduling and data transfer are handled by the MAC sublayer in the frame. The MAC receives a frame to transmit from LLC or SMT and tries to get a token frame (which is rotating around the ring). The token frame defines various features such as availability of frame, priority requirements, and other options within it. Once the token is obtained by a node, its MAC removes it from the ring, attaches the user's data to it, and transmits it back onto the ring. After the transmission, MAC issues a new token (indicating the availability of the medium) back into the ring.

When a node receives a frame from another node, it compares the destination address (DA) with its MAC's address. If these addresses match, then an error is checked. If the frame is error-free, it copies it onto the local buffer and MAC informs its LLC or SMT about the receipt of the frame. If these addresses do not match, it simply forwards it to the next node connected in the ring topology.

As mentioned above, the FS field defines three states (set, error detection, copy), and the symbols in this field are updated depending on the status of the received frame. After the frame is copied, the frame is sent back to the transmitter with appropriate symbols, which is an indication to the transmitter for acknowledgment and also provides the status of the received message. The MAC sublayer of the transmitting node removes the frame(s) from the ring. MAC recognizes the frame by looking at its own address SA field in the frame.

Various control operations used in an FDDI frame include **quiet**, **idle**, **halt**, **reset**, **set**, and **violation**, and each of these operations is represented by different symbols with different coding (of five bits) or decimal numbers (of two digits).

The PHY sublayer has been standardized by ANSI X3.148[15] in a 1988 recommendation and by ISO 9314-1 in 1989. The MAC sublayer standard was published in 1987 as ANSI X3.139[38] in its 1987 recommendations and by ISO 9314-2 in 1989. PMD has been standardized as ANSI X3.166[39] in its 1989 recommendations and by ISO DIS 9313-3 in 1989.

MAC layer and token functions: A **timed token rotation (TTR**) protocol is used by the MAC sublayer to manage access control, priority, and bandwidth for the nodes. The TTR allows the node to measure the time that has elapsed since a token was last received. It also allows the nodes to request a bandwidth, priority for asynchronous frames, response time for synchronous frames, etc. It makes sure that no station can keep the token forever. This protocol assigns priority to voice and video communication to avoid any delay and also assigns the remaining bandwidth to packet data applications where the delay is not a major problem. The TTR protocol measures the time that has elapsed for non-time data since a token was last received. If this time exceeds a predefined time limit, the token is allowed to go around the ring and no data is transmitted.

Each station defines a **target token rotation time (TTRT)** which may be the same for all stations. Each station also includes a **token rotation timer (TRT)** and **token hold timer (THT)**. The TRT defines the time since the token was last received at that station, while the THT defines the allowed time any station can keep the token. Token time represents the physical ring latency and is determined mainly by propagation time over a ring at zero load condition. For optimal ring efficiency, the TTRT should be high, and this will make response time very high. For voice and video communications, TTRT should by very small. The disadvantage of this is that the ring efficiency is reduced, as a token is moving around the ring most of the time.

A node can make a request for minimum time transmission given by the TTRT. The FDDI ring network supports both synchronous and asynchronous frames. The token is obtained whenever MAC has synchronous frames. If the time since a token was last received by MAC has not exceeded the minimum time defined by TTRT, the MAC will obtain a token and transmit the asynchronous frames. The FDDI utilization may be defined as the ratio of (TTRT-Token) and TTRT ((TTRT-Token)/TTRT) where TTRT represents the lowest value requested by any node and **Token** represents the time for a token to go around the ring under no load conditions. This time includes transmission time, propagation time, and the data rate offered by the network.

The value of TTRT is useful for considering the required link utilization. For example, a low value of TTRT may offer reasonable guaranteed response time which may find an FDDI network useful in time-dependent applications (integrated services of digitized voice and data, etc.). On the other hand, a larger value of TTRT may offer higher link utilization under heavy traffic. The token time depends on the number of nodes connected to the ring, the length of the ring, and its data rate, and it may have a very low value. Since a token ring network offers theoretically 100% utilization under heavy traffic, we may expect a link-utilization value very close to 100% (maybe over 99.5% or 99.6%) under heavy traffic.

Station management layer functions: Station management (SMT) defines the application process within the network management at the local node and includes various control information for proper operation of an FDDI station in an FDDI ring network. Various options/facilities available at nodes — e.g., initialization, performance monitoring, error control, flow control, maintenance activation, data transfer after getting a token, etc. — are controlled by SMT protocols. It also interacts with the entities of other SMT nodes to define overall control and management of the network. The typical entities of SMT include addressing, allocation of bandwidth, configuration, etc. Within SMT, the establishment of a physical connection between adjacent nodes is established by the **connection management (CMT)** library function. For establishing a physical connection between two nodes, CMT makes use of low-level symbols of protocols, e.g., quiet, halt, and idle. The physical connection, in turn, helps CMT to define logical connections between PHY and MAC within the station. All these logical connections within sublayers of a node or nodes use the established physical connections between nodes for data transfer and are consistent within each node, as the logical connections within nodes defines the behavior of those nodes. Further, this type of flexibility of logical connections for describing different functions of the nodes supports a wide variety of topologies and makes it useful for a variety of applications.

The following is a list of ANSI/ISO standards that have been defined.

- Hybrid ring control (HRC) — ANSI X3.186, ISO 9314-5
- Physical layer medium dependent (PMD) — ANSI X3.166, ISO 9314-3
- Single mode fiber physical layer medium dependent (SMF-PMD) — ANSI X3.184, ISO 9314-4
- SONET physical layer mapping (SPM) — ANSI T1.105
- Token ring layer protocol (PHY) — ANSI X3.148, ISO 9314-1 PHY-2
- Token ring medium access control (MAC) — ANSIX3.139, ISO 9314.2 MAC-2

19.5.2 FDDI-II LANs

An enhanced-compatible version of FDDI is FDDI-II, which offers a bandwidth of 100 Mbps. Since FDDI-II supports both packet-switching and circuit switching for data transfer, the entire bandwidth can be allocated to either a packet-switched network or a circuit-switched data network. With packet-switched service, FDDI packets (also known as FDDI frames) carry the user's data across the network, are of variable length, and contain various control symbols, addresses, and delimiters for identifying the fields within the frame. Contrary to this, circuit switching does not need an address but instead establishes a connection upon request and requires knowledge of the time slot and prior mutual agreement with the destination user node.

FDDI-II concepts: In FDDI, a cyclic clock of a pair of symbols is used to provide synchronization, while in FDDI-II, a timing marked clocking known as **basic system reference frequency (BSRF)** is used which is a 125-μs clock used by the public network.

The data communication in the circuit-switched mode goes through three phases: connection establishment, data transfer, and connection de-establishment. For digital voice data, a data rate or bandwidth of 64 Kbps is supported, while for other data (video, images, and other applications requiring high bandwidth), data rates of tens of megabits per second are provided.

The data communication in the packet-switched mode follows a different approach in FDDI-II. Here the packets (containing the user's information) arrive at a random time and are of different length. The packets following these characteristics are sometimes known as **asynchronous packets or traffic**. In contrast to this, if we get the packets at regular intervals with a predefined number and size, the network traffic may be termed **synchronous packets or traffic**.

There is a third category of network traffic known as **isochronous traffic**. In this category of traffic, the packets are well defined in size, and their arrival during time slots, the use of the entire bandwidth, and other features are well known in advance. All of these events are well defined in a precise sequence. The digital samples (of voice, video, images, etc.) have to be synchronized with clock information accurately. On the receiving side, the original samples are extracted from the received samples, which are again synchronized accurately with the sampling clock. The bit clocks are used to regenerate the original signal from the bits within the samples. It is expected that these cycle clocks are synchronized accurately and precisely to minimize the distortion during the original signal reconstruction process. Further, the delay in the received signals with respect to transmitted signals must be minimized. For these reasons, the isochronous data traffic is usually transmitted in the circuit-switching mode. The FDDI node inserts a delay in the FDDI rings for isochronous data in such a way that the rings, after some time, provide their length in exact multiples of 125 per second. The delay inserted by the node does not cause any delay in the packet traffic.

FDDI-II layered reference model: The FDDI-II layered architecture model is shown in Figure 19.8. FDDI does not support isochronous services as required by MAN. In FDDI-II, the synchronous operation typically works on packet-switching but can also operate on circuit emulation (used in ATM networks). This emulation accepts all continuous bit rate information (generated by **continuous bit rate — CBR**), which is defined in frames and is then transmitted. As we know, switching and multiplexing operations are always performed by higher layers (data link and network layers and above), while transmission functions are performed by lower layers (e.g., physical). For the packet-switched mode, the processing of these operations (switching and multiplexing) will be done at a packet level. In packet-switched networks, during the packet transfer mode (based on packet-switching, frame switching, frame relaying, fast packet-switching, ATM, etc.), errors may be introduced in both packets and bits (being transmitted) during transmission.

In the original version of FDDI, the optical fiber included a light-emitting diode (LED) which transmitted the signals at a wavelength of 1325 nm over a multimode fiber. A dual-fiber cable containing a polarized duplex connector was used to connect the nodes. The discussion in this section is derived partially from References 2, 11, 12, and 41.

As discussed above, FDDI uses ring topology (as discussed in IEEE 802.5 token ring LAN), which offers various advantages. The ring topology does not impose any restrictions on the number of status or length of links. In the total length of MAN, generally the ring

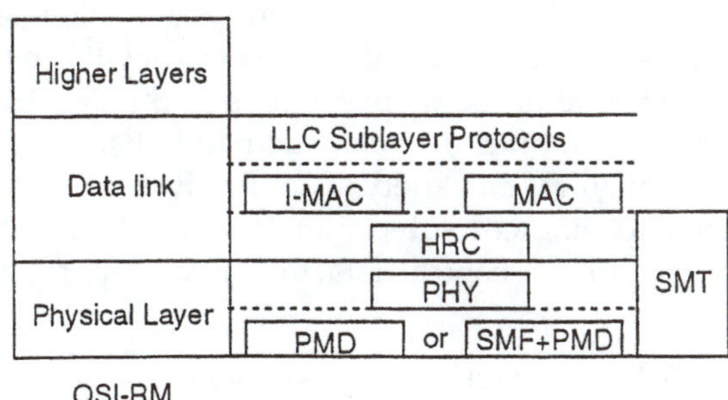

SMF - Single-Mode Fiber
PMD - Physical layer Medium Dependent
PHY - Physical layer
SMT - Station Management
IMAC - Isochronous MAC

Figure 19.8 FDDI-II layered reference model.

topologies offer high performance under heavy traffic conditions. Further, the performance of ring topology can be improved by using two rings such that in case of a failed link or node, it can be reconfigured to bypass the failure and still offer the same response without affecting the topology of the network and also the failed nodes. A counteracting arrangement composed of two rings has been proposed and used in ring topology. The ring topology offers point-to-point communication. The FDDI IEEE 802 has defined a layered structure for the lower two layers of OSI-RM. This layered structure is used in LAN and MAN standards of IEEE 802.

Difference between architecture models of FDDI and FDDI-II: The architecture model of FDDI-II is an enhanced version of FDDI, with minor differences which are discussed here.

In the FDDI layered architecture model, the physical layer is divided into two sublayers: physical (PHY) and physical medium dependent (PMD). The PMD sublayer protocols depend on the medium and, at this time, two versions of PMD are available — basic PMD (multimode fiber) and SMF-PMD (single mode fiber). The main function of the physical layer is to define the data bit rate and framing. The MAC sublayer is mainly concerned with accessing of the medium using a token which is moving around the ring. The token ring MAC frame also includes addressing, data checking, and data framing, while the token offers features such as priority. The LLC (IEEE 802.2), is independent of the LAN, and hence different LLC protocols can also be used in FDDI.

The data link layer of FDDI-II includes a set of two sublayers: MAC or **isochronous MAC (I-MAC)** and **hybrid ring control (HRC)**. The top sublayer of the data link layer uses LLC (IEEE 802.1) or any other LLC sublayer protocol. The I-MAC sublayer is an additional feature in the FDDI network which supports isochronous network traffic data. Further, MAC has to deal with the continuous transmission and receipt of data (synchronous, asynchronous, and isochronous) as the packets are interleaved. The HRC multiplexes the data from the MAC and I-MAC sublayers. A bandwidth of 100 Mbps may be assigned to packets (of the packet-switched mode) totally, or a part of it can be assigned to packets (of the circuit-switched mode). The **continuous bit rate (CBR)** is recovered from the received frame by removing the network jitter. The enhanced version of FDDI ensures upward compatible operation between FDDI and FDDI-II.

The FDDI-II ring network provides added applications over existing FDDI ring networks. Some of these added applications are highlighted below briefly.

The FDDI-II ring supports high-performance processors, high-performance workstations, mass storage systems, distributed database systems, switchable backbone LANs for a number of functions, etc. One of the main features of the FDDI-II ring network is its support for isochronous service offered by a **wide-band channel (WBC)**. A total bandwidth of 98.304 Mbps is available for a variety of applications. The isochronous bandwidth (WBC) may be allocated to different sub-channels supporting different virtual circuit services such as video, voice, image, control data, real-time data, etc., within the bandwidth of one or several WBCs.

Wide-band channel (WBC): The circuit-switched connection in FDDI-II is described in terms of a wide-band channel (WBC) number. For the user's request, a connection is defined as *A* bits beginning at byte *B*, after a predefined number on WBC. The FDDI-II has 16 WBCs which may be allocated to either packet-switched or circuit-switched data packets. The WBC offers a data rate of 8 Kbps to each connection and data rates up to 6.144 Mbps, allowing as many as 643 connections each of a multiple of 8 Kbps. The maximum number of WBCs in FDDI-II is 16, and multiple WBCs can be used for higher data rates (if required). The data rate of 6.144 Mbps is nearly four times higher than that of a T1 network popular in North America and three times higher than the 2.14-Mbps network popular in European countries. The 16 channels each of 6.144 Mbps over full-duplex operation gives a total bandwidth of 98.304 Mbps available from FDDI.

The basic time reference in FDDI-II is 125 μs (as used in PCM). Each 125-μs cycle consists of a preamble for synchronizing an 8-KHz clock, a cycle header, and 16 WBCs with a total data rate of 6.144 Mbps. A residual frame of 768 Kbps can be used for packet data. It consists of 12 bytes generated every 125 μs. This packet channel is known as a **packet data group (PDG)** and is interleaved with all 16 WBCs.

Each WBC can operate either in isochronous (circuit-switched) mode or in packet mode. Within each WBC, the bandwidth can be allocated by the node in terms of a multiple of 8 Kbps, including 64, 384, 1536, and 2048 Kbps.

FDDI-II also defines two types of tokens — **restricted** and **unrestricted** (as defined in FDDI) — and supports different levels of priority. The level of priority can be assigned either in isochronous or in packet mode. The highest priority level is used for isochronous mode and is associated with WBC for a circuit-switching option. This level of priority typically is provided to continuous bit rate (CBR) services and transmitted without getting a token. The next level of priority is for synchronous packet traffic data. The delay in this mode does not exceed twice the target token rotation time (TTRT). This time is defined by the lowest value requested by any node. The packets are transmitted after the token is held. The third level of priority is assigned to packet traffic with no delay. Any type of token may be held for transmitting the packet. The restricted tokens are usually used for negotiations for making sure that the bandwidth limit of the node does not exceed the available limit. The lowest level of priority is assigned again to the packet data where the nodes use an unrestricted token for transmitting them.

If an FDDI-II ring network is working in a packet-switched mode, it may well be used as a backbone for interconnection devices (e.g., gateways) to a public data network. The existing LANs have limitations on speed and hence cannot be used for interconnecting host computers for sending the data at very high speeds. The first version of FDDI was primarily intended for data applications; although it supports circuit switching, it is mainly based on the packet-switching approach for data communication. FDDI-II offers integrated services of packet-switching and isochronous services (circuit switching).

Physical medium–dependent (PMD) layer functions: In multimode fiber (MMF) PMD, a 1325-nm optical window is used for generating light by LEDs. This mode is used for a limited distance, and the connection is established between nodes via dual-fiber cable. In a single-mode fiber (SMF-PMD)[39] version, a single-mode fiber and laser diode transmitter (in contrast to LEDs in multimode) are used. This extends the length of the links up to 100 km. The physical sublayer defines the data transmission speed of 100 Mbps. The FDDI frames (known for FDDI packets) can have a maximum length of 4500 octets. The PHY sublayer defines recovery times which are defined by a standard default FDDI configuration of 500 stations and 100 km (length of optical fiber) working in a duplex mode of operation. The FDDI does not define any cycle time.

FDDI-II defines a cycle time frame of 125 μs. This cycle frame allocates a maximum of 16 isochronous channels (WBC), each of 6.144 Mbps, giving a total bandwidth of 96.344 Mbps. Even if all 16 isochronous channels are used, a token channel (residual) of 768 Kbps is again available. The allocation/de-allocation of isochronous channels is done dynamically. The residual token channel of 768 Kbps may be useful for packet data. The MAC sublayer field consists of control fields (preamble address) and user's data to be transported for the LLC sublayer.

Isochronous traffic service is supported by WBC's 8-Kbps bandwidth as virtual service, and this service can be obtained by various nodes which are assigned with one or more WBCs (depending on the application). If a WBC is assigned to a node, the MAC

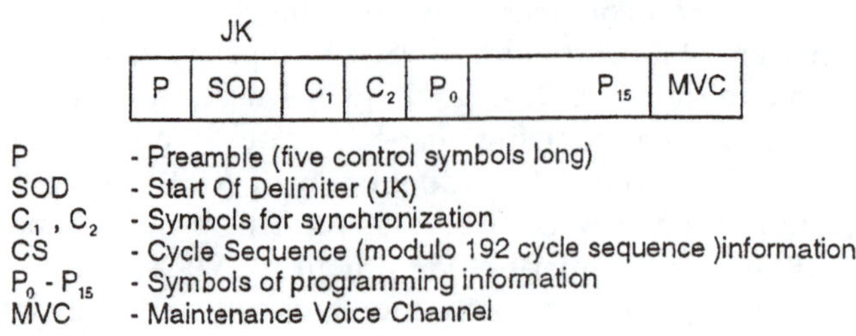

Figure 19.9 Cycle header frame format.

sublayer at the node can offer a channel of 8-Kbps bandwidth or channels of multiples of 8 Kbps. In other words, the B channels of 16, 32, and 64 Kbps and other popular channels of 384, 1536, 1920, and 2048 Kbps are also supported, thus offering a mixture of any of the channels that are multiples of 8 Kbps. For more bandwidth, WBCs may be assigned to the node, which in turn defines the channels within the allocated bandwidth.

For data communication through FDDI-II, irrespective of mode (packet-switching or circuit-switching), the first step is the initialization of a ring which generates a token and puts the network in the basic mode of transmission. After generation of the token, the FDDI ring is switched to the hybrid mode of operation by a node which has been designated as the cycle clock controller (after going through negotiation, etc.), after which the node is allocated the requested synchronous bandwidth to support various operations on both circuit-switched and packet-switched packets. Now the MAC sublayer of this node can support any type of packet (synchronous, asynchronous, and isochronous). The node generates a cycle of 8 KHz (125-μs cycle time) and introduces it into the ring such that the ring is synchronized to a multiple of 8-Kbps bandwidth.

Cycle header frame format: The cycle header frame format is shown in Figure 19.9. The **preamble** field is of five control symbols and is used to provide synchronization between source and destination nodes. The **start of Delimiter (SOD)** field allows the setting of the mode to either basic mode (FDDI) or hybrid mode (FDDI-II). The SOD will define the basic mode by symbols JK and the hybrid mode by symbols IL. These symbols are non-data symbols. The node designated as cycle master uses symbols C_1 and C_2 to transfer the information about various WBCs defined by p_0 to p_{15} fields. The **cycle sequence (CS)** field of one byte provides a 192-modulo cycle sequence count. The fields p_0 to p_{15} are the symbols of programming information corresponding to the status of WBCs, i.e., whether they have been allocated to synch and async packets of the packet-switching mode or isochronous packets of the circuit-switching mode. The last field, **maintenance voice channel (MVC)**, of one byte, offers a voice channel of 6 Kbps for maintenance of frames.

The PHY sublayer has been standardized by ANSI X3.148 in a 1988 recommendation. The MAC sublayer standard was published as ANSI X3.138 in a 1987 recommendation. ISO has also standardized MAC and PHY sublayer standards as 150 9314-3 and 150 9314-1 recommendations, respectively. The structure of FDDI and FDDI-II standards is shown in Figure 19.10. The relationship between the FDDI reference model and OSI-RM is shown in Figure 19.11.

19.6 *Switched multimegabit data service (SMDS)*

Local area networks (LANs) have played a vital role in making multimegabit data communication widely available within the premises of universities, organizational campuses, etc. LANs offer effectiveness in high-speed environments by offering capabilities of sharing

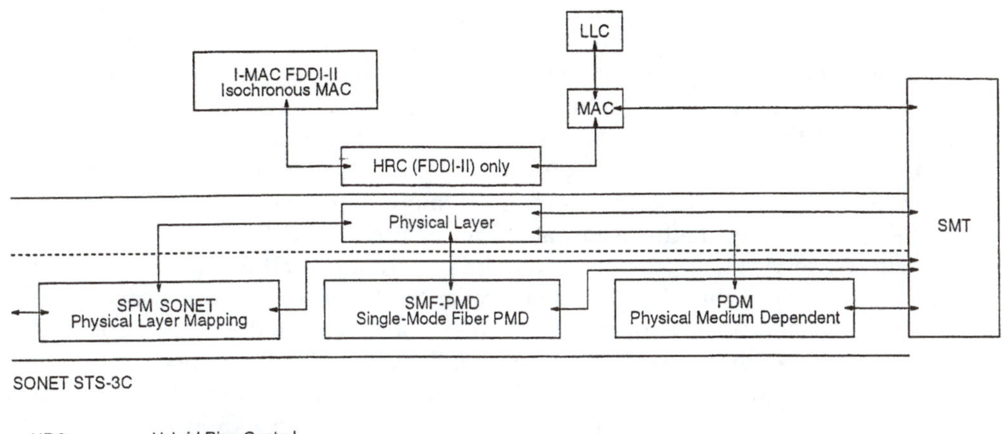

Figure 19.10 Structure of FDDI and FDDI-II standards.

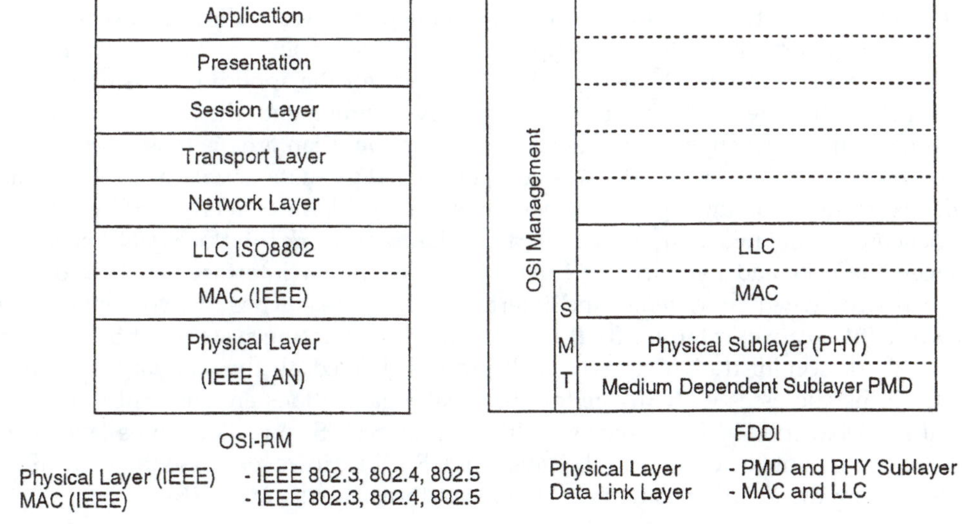

Figure 19.11 FDDI reference model vs. OSI-RM.

resources among multiple users with a high degree of reliability and security. Further, the high data capacity and rates of LANs with low per-port interconnection cost has made them an important application in the development of distributed processing. A high-speed LAN connecting low-cost workstations provides a single system which can allow users to share files, executable software, databases, and many other resources such as printers and parallel machines, high-performance supercomputers, etc. The material presented herein is partially derived from References 9, 10, 17, 46, 47, and 50.

Due to these new trends of using high-speed and high-performance networks for distributed integrated applications with higher bandwidth, various new technologies have evolved lately. The main objective of these technologies is to define a single high-speed network which can meet the high bandwidth requirements and offer wide area and global connectivity in the data communication business so that a wide variety of services can be provided to users at a global level at low cost in one uniform network. One of the services meeting some of these requirements is known as **switched multimegabit data service (SMDS).**[10] It is a high-performance, high-speed, packet-switched, connectionless data communication service which can be integrated into the local environment (LAN). This is Bellcore's[10,17] proposed high-speed data services provider (DS1–DS3) connectionless packet-switched service that provides LAN-like performance and features over a metropolitan area. This new technology has some similarities to that of frame relay and is based on fast packet-switching. It supports LAN interconnectivity and the protocol for the transfer of data across the network. Bellcore introduced the SMDS network for providing connectionless (datagram) services, and it is developed for local, intra-LATA, and WAN services. The Regional Bell Operating Companies (RBOCs) are providing SMDS as public broadband services. Bellcore has defined a number of interfaces. The intent of SMDS is to provide a public connectionless packet-switched service to users so that they can use variable-length data units up to a maximum length of 9188 octets. It is technology-independent and will be available to users via MAN for a variety of services. It will be one of the very first switched broadband services which will be fully supported by B-ISDN.

SMDS is a public interconnection and hence can be interconnected with LANs and host computers within a **local access and transport area (LATA)**. During integration with the local environment, the existing hardware and software is not changed. Since it provides a connectionless service, the datagram packets are transmitted and handled without establishing any prior network and logical connection between two nodes. The existing high-speed LANs (token ring, FDDI, etc.) support various operations and features which have also been defined in SMDS. It therefore becomes easy for the applications using LANs to use SMDS without restructuring the existing LAN environment. The SMDS can be used in a broadband ISDN environment defined by a **metropolitan area network (MAN)** and multi-service broadband networks. It is now considered a generic service and can be made available on all upcoming high-speed networks, e.g., ATM, frame relay, B-ISDN, etc.

Bellcore has defined a number of interfaces. **User–network interface (UNI)** is defined between the user and network, an **inter-switching system interface (ISSI)** is defined between two switching systems, and interfaces between switches of two networks are known as **inter-carrier interface (ICI)** and **interface as operating system (OSS)** for billing and network administration services. Bellcore has defined all the standards of SMDS. Based on the successes with SMDS, international organizations are engaged in defining global B-ISDN services. The protocol architecture of SMDS consists of three layers, and these layers do not correspond to the layers of OSI-RM. However, these three layers are implementing the functions offered by the lower three layers of OSI-RM.

19.6.1 Features of SMDS

The features of SMDS include a high degree of reliability, quality, availability, serviceability, and maintainability, and these features usually depend on the interface card used to access the network services. The SMDS can be considered a public interconnection service which will allow users to define their virtual private networks by using the addressing features of SMDS. The services offered usually depend on the capabilities of the network that provides SMDS and also on the underlying operating system which supports the interfaces to the network.

The SMDS consists of a metropolitan area network (MAN) switching system which typically offers switching to the data packets going through the network. This switching

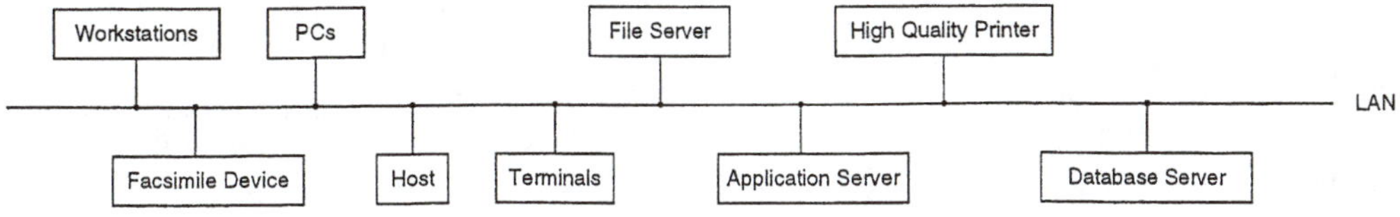

Figure 19.12 Typical networking environment.

can be based on either token ring or another type of switching (e.g., Batcher–Banyan). The switching elements based on this type of switching are usually connected by MAN switching trunk lines. The higher-layer protocols provide a common and single service to all the users, irrespective of the switching used in their networks. The users interact with the network offering SMDS via **subscriber network interface (SNI)**, which supports a DS-3 link (44.736 Mbps). The SNI was defined by Bellcore[10,16,17] as an interface (for providing services to the users) which complies with IEEE 802.6. The optical facilities are used on data communication networks offering SMDS to provide conversion for electrical signals into optical at the physical layer.

The SNI provides a maximum bandwidth in the DS-3 range rather than SONET's STS-3 and offers data capabilities. It does not support isochronous traffic. The SNI is typically attached to access facilities and are available to the users over fiber-based transmission to a **MAN switching system (MSS)**. User gets their own interfaces and cannot share the data with others, and hence it provides security and privacy in services to the users.

A typical networking environment offering multimegabit data services (MDS) may be defined as a framework composed of existing LANs to which various devices (workstations, peripheral devices, intelligent peripherals, hosts, etc.) can be connected, as shown in Figure 19.12.

This type of networking environment needs some changes in hardware and software. It requires training and investment, and it also introduces compatibility for multi-vendor applications, interoperability, etc. All of these parameters and issues have to be considered during the design of multimegabit data communication networks. High-speed data communication between various devices provides/supports the performance of functions either at the local or remote host in a transparent mode without affecting the performance of the network services.

High-speed networking applications are rather restricted on the local environment (due to speed limit) and may not even support high-performance applications (e.g., distributed processing, client–server implementation, time-dependent applications, image processing, multimedia applications, etc.), but they may provide a minimum acceptable level of confidence or performance of the network. However, applications such as e-mail, access to news, remote login, access to a remote computer, Gophers, various client applications for information sharing, etc., can be supported easily by low-speed networking. The local environment LANs with data rates of 1 to 20 Mbps may not support video, voice, facsimile, and image communication at high speed, but inter-local communication across LANs can certainly be achieved at very high data rates using private fiber-optic networks, satellite-based transmission services, or even leased DS-3 (44.736 Mbps).

19.6.2 SMDS services

The SMDS services are provided to large corporations, universities, and public platforms, particularly in metropolitan areas. Typical applications using these services include desktop publishing, word processing, spreadsheets, databases, multimegabit communication to interconnect high-speed LANs, etc. They may also be useful as logical private network services within premises for internal communication or to interconnect to organizations (high-performance, supercomputer center) or even a public switched service. Further, SMDS is supported by a network which uses the switching system used in MANs. SMDS can now access a broadband network and will be part of the multiservice features of any high-speed network.

SMDS supports the network security that can check the transmission of valid frames. It is a very useful feature of SMDS, as it can connect the public backbone network via a single entry point and limit the accesses to the resources. It offers services to end users and network users. Some of the end user services include address validation, group addressing, address screening, access classes (4 M, 10 M, 16 M, 25 M, and 34 M on DS-3), etc. Some of the network user services include operations, administration and maintenance, billing information, etc.

It is important to note that one of the main objectives of SMDS was to keep the cost of the network service to a minimum so that new or existing **customer's premises equipment (CPE)** should be able to access this service easily without changing the overall configuration of the underlying network. This objective has made the functionality, features, and performance of SMDSs in a sense independent of network high-speed architecture, or any new technologies within switching transmission of data communication. In other words, SMDS is well suited to meet the requirements of users for higher productivity and extension of data communication capabilities from LAN applications to enhanced wide-area and broadband services. Based on this, we may use SMDS as a private logical network service for selected and specified interfaces, as a public switched service where it may be shared among all users, or even as a MAN service. In any of these applications, the interesting thing to realize is that the amount of changes to the existing CPE is minimal, and the broadband services are easily available to the users.

The protocol architecture of SMDS consists of three layers, and these layers do not correspond to the layers of OSI-RM. However, these three layers implement the functions offered by the lower three layers of OSI-RM. The lowest layer is defined in the SMDS interface protocol (SIP) specification. It contains two protocols — physical layer convergence protocol (PLCP) and transmission system. The transmission system defines digital carrier systems supporting DS-1 and DS-3 with a hope for supporting SONET STS3c in the future. The presence of PLCP offers compatibility with the IEEE 802.6 MAN standard. The layer-two protocol provides access to MANs. The top-layer protocol accepts the data from higher layers, routers, gateways, or bridges. For more details on SONET, see References 7, 12, and 16.

19.6.3 End-to-end communication

A typical configuration for end-to-end communication over a customer's communication architecture[9,10,47] connected by bridges across SMDS is shown in Figure 19.13. The SMDS can also allow the broadcasting of messages to selected users. In this configuration, a screening function is performed on source and destination addresses.

In Figure 19.13, both the LAN and network offering SMDS are considered subnets, and each provides transport service (connectionless) to the higher layers (IP layers). The user has to access the network using the SMDS interface protocol (SIP) and then can use higher-layer communication protocols (sitting on top of SMDS). The protocols currently used in CPEs support connectionless services. The connectionless services provided by SIP can easily be integrated into protocols of CPE and hence can use the communication protocols directly. The SIPs of both end hosts are connected to the network providing SMDS via a DS-3 transmission link (44.736 Mbps). The SIP allows the users to access the

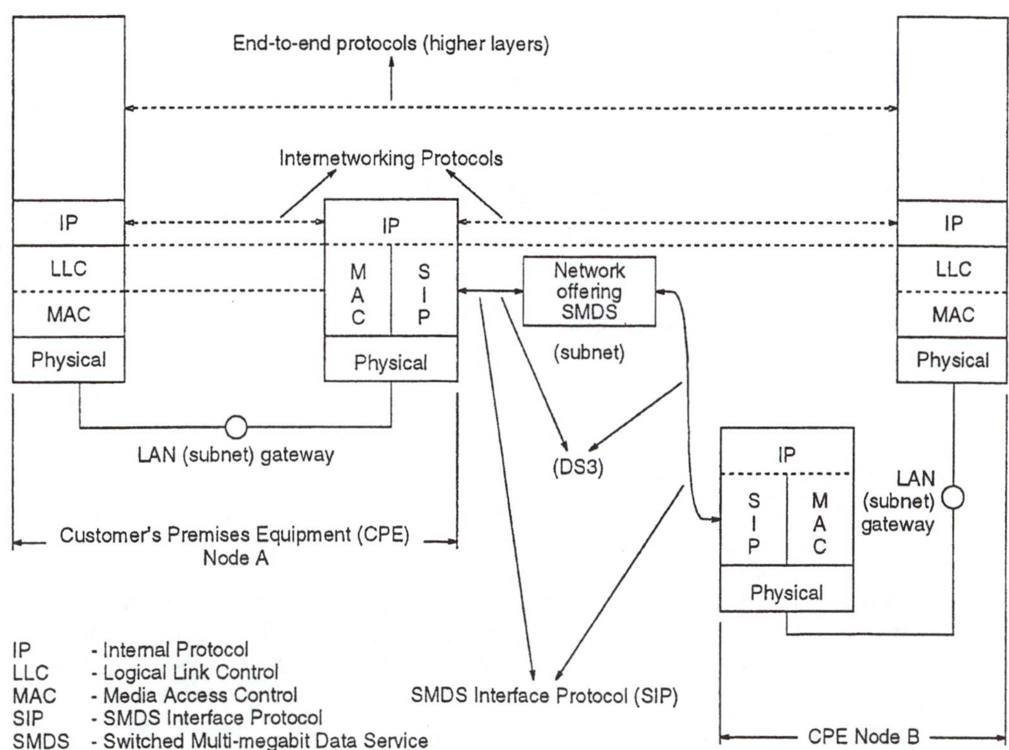

Figure 19.13 End-to-end communication vs. customer's communication architecture.

services provided by the network which support SMDS at SNI. It is basically a connectionless three-layer protocol (similar to Internet protocol — IP) and offers similar functions such as framing, addressing, and physical transport (typically a fiber link). Since it offers connectionless services, it does not support error control and flow control or even error detection, and these must be provided by higher layers, as is done in TCP/IP protocol.

The SMDS customer interface uses a protocol which is compatible with the network interface board and implements a driver similar to a LAN interface driver. Here the customer premises equipment acts as an IP gateway between the LAN and network providing SMDS. The details of these connections are defined in a **customer interface driver (CID)**, as shown in Figure 19.14.

In a typical configuration, the message originating from a terminal is received by a router. This router is connected to the terminal via a 100-Mbps fiber-optic channel. These are connected to SMDS CSU/DSU nodes via a **high-speed serial interface (HSSI)** running at 34 Mbps. The CSU/DSU nodes are connected to public SMDS networks via 45-Mbps or DS-3 speed using **SMDS interface protocols (SIPs)**. The messages from the routers are usually datagrams of variable length. The CSU/DSU node converts the frames into cells with 48 bytes of data and 5 bytes of header. These cells are transmitted to the SMDS cells at a rate of 45 Mbps. The SMDS switches extract five-byte headers from the cells, provide error control (CRC check), and identify the type of segment, i.e., beginning of message (BOM), cells of message (COM), and end of message (EOM).

The customer interface includes the physical medium and SIP. The physical medium is typically a DS-3 transmission link (44.736 Mbps), while the framing, error control, addressing, flow control, etc., are provided by SIP. Some of these functions provided by SIP, known as **SMDS service data units (SSDUs)**, are discussed below.

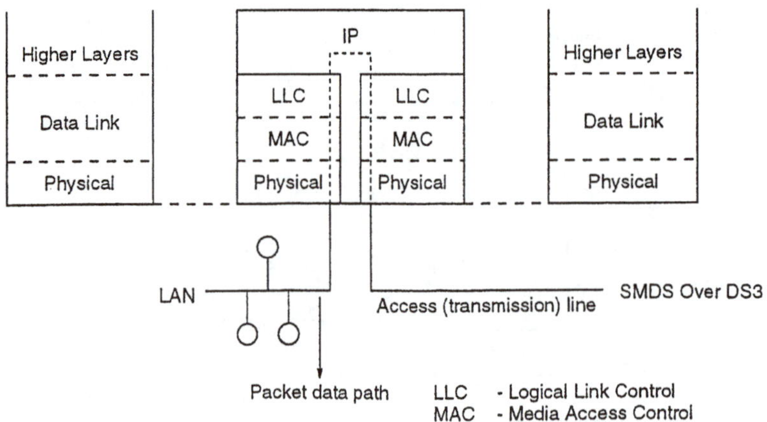

Figure 19.14 Customer interface driver.

19.6.4 SMDS service data unit (SSDU)

As stated above, the SMDS provides transparent transport for control information associated with SMDS, known as SMDS **service data units (SSDUs)**. The control information typically defines addresses, error-control information, and user data. Since it is based on connectionless service, the SDUs are transmitted as datagrams and may arrive at the destination out of order. The existing customer premises equipment (CPE) handles the situations of lost, out-of-order, or duplicate SDUs to provide end-to-end communication for SMDS.

SSDUs contain 8191 bytes, and the larger-sized SDUs will require less processing time with minimum delay. Due to this, the SMDS can also be used as a bridge between two IEEE 802.4 token bus networks. This is because the MAC frame of IEEE 802.4 is a larger frame than found in other LANs and is easily encapsulated in the SDU of SMDS. This avoids the segmenting and reassembly of MAC frame transmission over SMDS.

SMDS utilizes the end-to-end capabilities provided by the majority of protocols (TCP/IP, SNA, DECnet, etc.), and no additional mechanism is required for flow control. The rate of data transfer needs to be controlled between users to the network in both directions and is usually handled by **access class (AC)** and **credit managers (CM)** algorithms.

The AC algorithm defines the parameters associated with the CPE which allows the transfer of SDUs between the user and network. The CM algorithm defines predefined parameter values for a credit window. These parameters offer different types of traffic such as data rates and access class in the network.

The error control (recovery and reporting) in high-speed LANs is performed by higher-layer protocols. The same is true while using SMDS, in that SMDS does not detect and report errors found on SDUs; instead, the higher-layer protocols perform these operations on SDUs.

The addressing scheme in SMDS follows CCITT recommendations E.164 (defined for the ISDN numbering scheme) for defining the address of the SMDS user interface. The addressing scheme in SMDS includes features such as multiple addresses at a single interface, group addressing (a set of customer interfaces), source address validation, mobility of the SMDS address (no reconfiguration required for relocated user; i.e., the same SMDS address may be used), etc. The addressing scheme allows users to define logical connection configurations which are similar to those defined in virtual private networks.

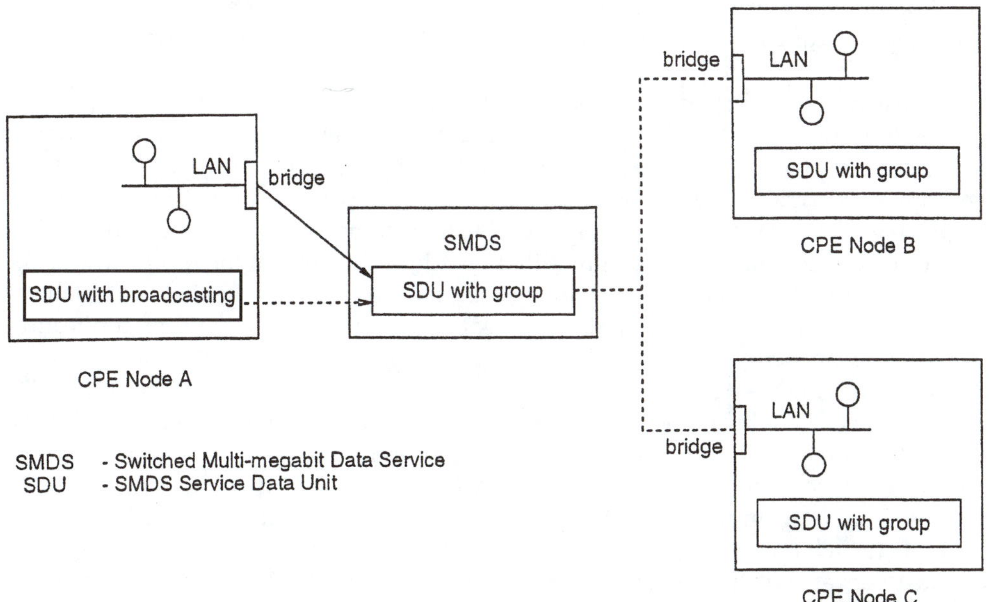

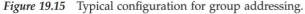

Figure 19.15 Typical configuration for group addressing.

With one transmission, the users can transmit the data packet to multiple addresses. A group address contains a predefined number of network addresses. The network offers mapping between group addresses and actual network addresses.

The network also provides screening to both source and destination addresses in the sense that the user will receive the packets from a selected address. The source address is screened before the packets are sent to the destination. Similarly, the destination addresses are screened before packets are sent across the internal SMDS network. A typical configuration for group addressing (a class of broadcasting) can be seen in Figure 19.15. Here, SMDS provides interconnections between customer premises having different LANs — A, B, and C. The transmitting bridge device (say A) will send an SDU to the group address which corresponds to the bridges B and C. The bridges receive the SDU and it is transmitted on the connected LAN for broadcasting. In this way, a message from any LAN can be broadcast either to its own users or to the users of other LANs.

19.7 Frame relay

The idea of sending voice and data packets over a packet-switched network has provided a new era in data communication. This concept seems to be cost-effective, as we have seen the development of various hardware and software protocols which have made the packetization of voice (or sending voice as packets) feasible. The transmission of voice and data packets over transmission media simultaneously defines a vision of digital networks. The new concept of fast packet-switching over X.25 has led to this new network protocol known as frame relay (jointly supported by ANSI and ITU-T). It works on a connection-oriented concept and supports the following data rates: 56 Kbps, T1 (1.544 Mbps in North America and Canada), E1 (2.048 Mbps for Europe), and ISDN's basic rate of 64 Kbps and its multiple rates.

The establishment of a connection is initiated by a call setup frame and, if it is accepted, a connect frame is sent by the destination node. The network sends an acknowledgment to the frame relay user about the establishment of connection. Other frames used in a frame relay network include call proceeding, progress, disconnect, release, and release complete.

The fast packet-switched network is another digital network which supports integrated services of data and voice packets. The fast packet-switch is based on the concept of a T1 multiplexer, and the protocol executes at the physical layer only. The conventional T1 multiplexing carrier system generates 24 channels of 64 Kbps and uses the time-division multiplexing (TDM) technique. In this multiplexing technique, a well-defined time slot is allocated to each channel with a bandwidth of 64 Kbps. This technique results in wastage of bandwidth to other channels, and not enough bandwidth in a channel is available to handle the larger bandwidth required in applications such as bursty traffic or applications with variable and high data rates. The material herein is partially derived from References 1–13 and 46–54.

The fast packet-switched network uses a different approach of multiplexing. Here the entire T1 data stream is partitioned into discrete, fixed-length packets and the entire frame is carried from one node to another. In standard T1 multiplexing, eight bits of information per channel are transmitted; as such, the number of bits required for one frame is given by 193 bits [24 (channel) × 8 (bits/channel) + 1 (for framing)].

In fast packet-switched networks, the address field fast packet is used to determine the routing across the network. The control field offers priority for the type of traffic. As stated above, error recovery is performed on the user data by a higher-layer protocol. The priority facility within the control field allows the assignment of minimal delay to packets carrying voice messages.

The frame relay digital network supports both circuit- and packet-switching techniques for sending voice and data over the same physical link or circuit. Traditionally, voice signals were transmitted over analog telephone networks, which offer a transmission line speed of about 15 Kbps without analog impairments (alteration, echo, distortion, etc.). At higher speeds, the analog impairments may become a problem. The frame delay digital network offers two frame modes of services associated with ISDNs: frame-relay and frame-switching. Both of these frame modes use the same signaling procedures. Frame relay offers minimum overhead, while frame switching requires the network to perform error control and flow control. Further, the frame relay–based network performs multiplexing and routing at the data link layer, as opposed to the packet and link layers together performing these functions in X.25 networks. Frame relay networks are used in information systems, client–server, CAD/CAM, graphics, multimedia, and other applications requiring bursty-type traffic. The frame relay network offers two types of connection within packet-switching: **permanent virtual circuit (PVC)** and **switched virtual circuit (SVC)**.

The analog telephone network uses circuit switching and, as a result, there is a wastage of resources and limited bandwidth. The resources (switches, physical connection) are allocated to the user's request and remain busy during the conversation. At the same time, for voice communication, circuit switching offers no delay, while packet-switching may present a delay — sometimes a significant delay — and we can notice this delay during a conversation with a friend (in particular during overseas telephone conversations).

The drawbacks of analog telephone networks (limited bandwidth and wastage of resources) have been overcome in the packet-switched network. In this network, the analog speed is converted into digital signals (the most popular technique in pulse-coded modulation — PCM). The PCM voice signal still requires a circuit-switched connection for its transmission. The PCM voice signal does not have any noise and is considered another form of digital data. The PCM uses analog-to-digital conversion to sample the speech at a rate of 2000 samples per second (or 8000 Hz) with a pulse duration of 125 μs. Each sample is assigned an eight-bit binary value which represents an amplitude (or height) of analog wave form. In order to encode the speech signal, the number of bits per second required is eight (bits per sample) × 8000 (samples/sec) = 64,000 bits/sec (64 Kbps).

The packet-switched networks need extra bits to provide redundancy for error control, and they require extra processing at various switching nodes, which adds up to overhead. With the advances in telecommunication networks offering high data rates, the error rate is very small. This low-error data rate reduces the overhead required for error control and offers higher data rates, as now the switching nodes are not processing fully for error control. The advantages of higher data rates and lower error rates of underlying networks are enormous, as now the data can be transmitted at a very high data rate. An overview of frame relay and other congestion control issues can be found in References 40 and 41.

19.7.1 Frame relay services

As discussed earlier, the frame-relay switching technique offers higher data rates using these advantages, and as a result the overhead required in frame relay–based networks is very small. This switching technique offers a variable size of frame and also does not process for error control at any switching nodes, as this is handled by the network itself. Due to this feature of frame relay, a data rate as high as 2 Mbps can be achieved (as opposed to 64 Kbps in traditional packet-switched ISDNs). Further, frame relay allows multiple calls to different destinations to be handled at the same time by the data link layer. Thus, once a virtual circuit (path) has been established using the D channel, a unique identifier is assigned to this circuit. This path is used for all subsequent frames. This path identifier has a local significance, and the frames travel over different links associated with virtual paths. When the frame handler receives frames during the data transfer phase, it checks the data link connection identifier and combines it with all incoming numbers to determine the corresponding outgoing link and data link connection identifier. It assigns a new data link connection identifier, and the frames are forwarded to the appropriate outgoing link. This maintains the order of arrival of frames, and the routing also becomes faster. The only disadvantage with this switching is the congestion during heavy network traffic over the outgoing links.

The **asynchronous transfer mode (ATM)** of transmission offers advantages in terms of reliability and quality of services (due to digital transmission, switching, and technology). The ATM-based and frame relay–based networks offer higher data rates than packet-switched X.25 networks. All of these networks have a common switching technique — packet-switching — and each one of these allows the multiplexing of multiple logical connections over a single physical interface. The ATM-based network defines the switching via cell relay (derived from frame relay). Frame and cell relay have some common features, e.g., no support for link-by-link error control and flow control, support of error control by higher-layer protocols, minimal support at lower layers for error control, etc. Due to these features, the overhead in each of these switching techniques is minimal, and higher data rates are achieved from these networks.

Frame relay defines a variable size of the data frame and supports data rates up to 2 Mbps. In cell relay, the overhead for transmitting the data frame is further improved by defining a small size of the data frame (53 bytes). With this reduced size, the data rate as high as 500 to 1000 Mbps may be achieved from cell relay–based ATM networks.

The data (derived from packet-switching) and packetized voice data (derived from circuit switching) are multiplexed (using TDM) and transmitted over the same physical link. The packetization of voice has been a driving force behind the acceptance of integrated services digital networks (ISDNs). The advent of ISDN has given a boost to a variety of voice-based applications (toll-free numbers, conferencing, call forwarding, etc.). ISDN offers two channels: **bearer (B)** and **delta (D)**. The B channel has a bandwidth of 64 Kbps and can be configured either for circuit-switching or packet-switching connection. One of the B channels is always used for carrying digital voice (or PCM voice) in the circuit-switched mode. The remaining speed portion of the B channel can be used for either circuit-switched or packet-switched connection. Although 64 Kbps may look like a low speed dial-up configuration, it can accommodate as many as six high-speed modems of 9.6 Kbps.

The B channel line speed of 64 Kbps is used in high-bandwidth applications where time is a critical factor. The time of transmission over the B channel will be faster than over the analog line via modem for same size of data. Here we assume a 9.6-Kbps modem, but if we have a 2.4-Kbps modem, the time will be significant. It may be even worse if we use a synchronous modem (requiring at least 10 bits for each character in asynchronous as opposed to eight bits in synchronous) offering a speed of 2.4 Kbps.

The D channel, on the other hand, is for signaling purposes and operates in packet-switched mode at a speed of 16 Kbps. It can also be used for carrying non-signaling data (packetized voice or data) and uses the protocols (packet-switched) which have similarities with X.25 protocols. This makes the D channel useful for various applications of telemetry.

19.7.2 Frame relay protocol

Various nodes and links of packet-switched networks, in general, offer the following functions:

- Packet sequencing
- Congestion control
- Acknowledgment
- Error recovery
- Re-transmission

The protocols used at packet-switching nodes support these functions in one way or another and, as a result, the packet-switch at the node is always busy during the performance of these functions and no time for its main function of moving the packets is left. These functions are to be performed within the network layer. It was felt that burdening the switch to perform these functions instead of doing its switching of the packet and other processing functions should be avoided. As a result of this, the **frame relay (FR)** protocol, which is based on LAP-D (data-link protocol for the D channel used in ISDN) and derived from X.25, was developed. As we know, X.25 does not use any packet layer (analogous to the network layer of OSI-RM), and the functions of layer 3 and above can be performed by the end system. This end system may be a high-performance MIMS system which executes most of these functions (outside the network), leaving the switch to perform more switching functions than the normal and routine functions of layer 3 of OSI-RM.

The frame relay network has a two-layer protocol: physical and MAC. The terminal connected to a LAN sends a packet to the router (working at the network layer). The router sends the packet to the frame relay network, which passes it on to the router connected to the destination terminal. The function of the router is to use the lower three layers of OSI-RM to determine the route and send the packet to the frame relay network. The frame relay, using MAC, forwards the frame to the appropriate router. The physical layer of the frame relay performs the same functions as those of OSI-RM. The function of the MAC is to provide error control through the FCS field.

The frame relay offers an interface between the user and frame relay via the **frame relay interface (FRI)** protocol. This protocol is being used between the router and the frame relay network. The frame relay, after receiving a frame from the incoming router, checks for any error. If it finds an error, it will discard the frame; otherwise, it will forward

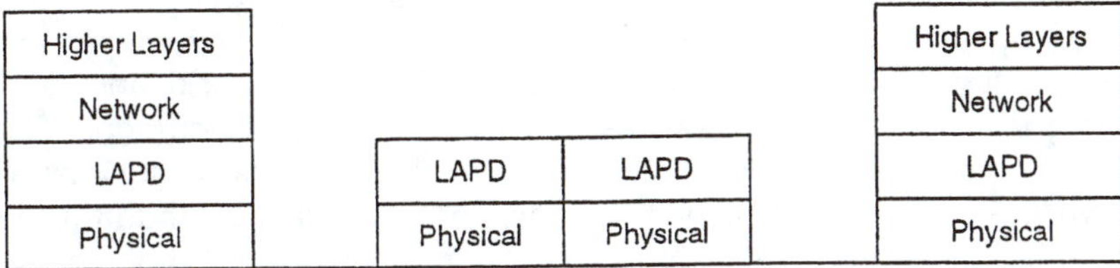

Figure 19.16 Frame relay.

it to the outgoing router. The frame relay network does not perform any error recovery; this is performed by the **customer premises equipment (CPE)** at the destination node.

A typical connection of frame relay between end systems is shown in Figure 19.16. The frame relay (FR) layered protocol architecture uses the physical layer standards CCITT I.430 or I.431, depending on the interface being used (basic rate interface or primary rate interface). The data link layer standard protocol is defined as Q.922, which is derived from **CCITT I.441/Q.931**.[41b] The I.441/Q.931 standard is basically an enhanced version of the LAP-D protocol used in X.25. It is important to note that the data link layer of X.25 also uses LAP-B, and the selection or choice between LAP-D and LAP-B depends on whether the D or B channel is used. The standard Q.922 offers a set of core functions which typically include framing, synchronization, multiplexing/de-multiplexing, error detection, and other functions defined for the data link layer of OSI-RM. The protocol architecture defines two planes as the C (control) plane and U (user) plane.

The C plane uses the I.451/Q.931 protocol for its network layer, while the Q.922 is used as a data link protocol. This protocol provides reliable data link control service and also error control and flow control. The Q.922 provides transportation to I.451/Q.931 packets. The U plane allows users to define terminal functions which are interfaced with core functions of Q.922. The user may use either Q.922 protocol for terminal functions or any other compatible protocols. Users can also define additional functions for the terminal if they so wish, and these depend on congestion and other performance control functions. Similarly, the additional functions may be considered on top of the core functions offered by Q.922. The physical layer remains the same (I.430 or I.431) for C plane or U plane architecture.

The **CCITT Q.922** standard has recently been defined, and it includes all the core functions defined in I.441/Q.931 and offers additional functions not present in Q.931. The core functions of Q.922 include frame delimiting, data transparency, frame multiplexing and de-multiplexing, checking of frame bits before zero bit insertion, checking the size of frame, error control, and flow control. These are also core functions of I.441/Q.931. In the protocol architecture, it seems quite clear that the processing involved in these layers is minimal and the data frames are transmitted over the network. The intermediate switching nodes do not provide any support for error control except that they discard the frame if it has an error. Error recovery will be provided by higher layers.

The frame relay protocol deals with some of the functions of X.25, which include error checking, recognizing frames, generating flags, checking of valid/invalid frames, translating addresses, etc. The intermediate switching nodes will have a minimal error-checking facility, and a bad frame is discarded right away. The higher-layer protocols at end systems detect only missing, lost, or out-of-sequence frames. The X.25 logical channel multiplexing is done by layer 2 (LAP-D). A virtual circuit may be selected from the data link control identifier (DLCI), a 13-bit address field. This DLCI, in fact, replaces the X.25 logical channel number. There seems to be a one-to-one correspondence between frame relay and X.25 architecture.

For ready reference, we mention the protocols used for the three layers of X.25: physical layer protocol (for the physical layer), protocol LAP-B or LAP-D (data link layer;

depends on whether it is using the B or D channel), and X.25 protocol (network layer). All three layer protocols are implemented by the interface and network. The architecture of frame relay does not define any protocol for the third layer but defines protocols for the remaining layers as core functions as defined in Q.922 and additional functions in physical layer protocol I.430 or I.431 (depending on primary or basic rate interface). Here, the lower two layers are implemented by interface and network, while additional functions at the data link layer or network layer have to be implemented by the interface and are not the responsibility of the network.

In this frame relay network architecture, the user frames are processed without providing any support for error control or flow control, and if any frame contains errors, it will simply be discarded. In other words, error control has to be provided by the higher layers. As we have noted, frame relay does not define any protocol at the third layer; however, this does not mean that it does not provide the functions associated with this layer (compared to X.25). In fact, many of the functions of X.25 are also provided by frame relay, including flag generation, flag recognition, transparency, FCS generation, discarding of invalid frames, multiplexing of logical channels, and address translation. The remaining functions provided by X.25 are not included in frame relay. Since frame relay is involved with a minimum number of functions, the processing time is considerably reduced compared to X.25, and a higher data rate is offered by frame relay.

The frame relay switches easily handle the data rates of T1 (1.544), and these switches are basically the routers. The node routes the frame as soon its address header is received by the frame relay switch, and the delay may be reduced considerably. The bandwidth in frame relay is allocated to various requests dynamically and hence is useful for applications such as higher bandwidth, bursty traffic, etc. Further, the frame relay switch can enhance the number of connections in the backbone network. The connections can be made with other frame relay switching nodes, LANs, bridges, routers, ISDN switches, and many other application nodes. It is interesting to note that frame relay will allow protocols of different vendors to interact and communicate with each other over the same backbone network. Users may get an opportunity to select a particular carrier for the frame relay or use the frame relay switch over the existing backbone network.

The frame relay switch offers frame-oriented datagram service and is derived from the X.25 packet-switching data network protocol. It allows the end system to perform error recovery, lowering the burden on the switch at layer 3 protocols. The frame relay switch operates at a 1.544-Mbps or even higher data rate over the backbone network, as opposed to the 64-Kbps data rate we get from the X.25 network. The frame relay frame format is similar to other data link layer protocols with some minor differences in the fields. It is composed of the following fields (shown with the number of octets for each of the fields): flag (one octet), address (two to four octets), user's data (variable), FCS (two octets), and again a flag (one octet). The address field may have two, three, or four octets and defines different subfields for each of the formats. The subfields for these formats typically deal with congestion notifications, data link connection identification, status of the frame (command/response), etc. The frame relay does not define any control frame. No in-band signaling is used; instead, a logical connection is defined to transport the user data. As mentioned above, the frame relay does not support error control or flow control, since the transmission of frames (via Q.922) does not use any sequence scheme.

The following standards for a frame relay network have been defined:

- Service description — ANSI T.606, ITU-T (I.233)
- Core aspects — T1.618, ITU-T (Q.922)
- Signaling — ANSI (T1.617), IYTU-T (Q.933)

The Frame Relay Forum has asked the ANSI TIS1.2 group to look into network-to-network interfaces (NNI) that define the interconnectivity of different types of networks with frame relay networks. Frame relay switches within a frame relay network share with each other through this interface.

Frame relay networks are used in information systems, client–server, CAD/CAM, graphics, multimedia, and other applications requiring a bursty type of traffic. As mentioned above, the frame relay network offers two types of connections within packet-switching: permanent virtual-circuit (PVC) and switched virtual-circuit (SVC).

Some of the vendors who are developing PVC services include AT&T, MCI, Wiltel, Sprint, CompuServe, etc. It is interesting to note that PVC services are under consideration and are not available. The establishment of connection is initiated by the call setup frame, and if it is accepted, a connect frame is sent by the destination node. The network sends an acknowledgment to the frame relay user about the establishment of connection. Other frames used in the frame relay network include call proceeding, progress, disconnect, release, and release complete.

19.8 Broadband integrated services digital network (B-ISDN)

Due to the growing demand for high-performance and high-speed networks, national and international standards organizations and network manufacturers and providers have already started defining B-ISDN interfaces, protocols, reference points, layered architecture, switching, multiplexing, and transmission techniques. The combined efforts seem to be directed toward defining B-ISDN as a unified, universal wide area network which can support high-speed, voice, data, video, and image communication with low latency and error rate. According to ITU-T, the ISDN offers services below 64 Kbps, and any services above 64 Kbps are known as *broadband communication services*. As a result of this recommendation, the traditional ISDN has been renamed N-ISDN, while B-ISDN based on ISDN provides services above 64 Kbps and below 1.544 Mbps. It offers complete integrated services such as low bursty rate; high rate of real-time, low-speed traffic such as data, voice, telemetry, and faxes; and high-speed traffic such as videoconferencing, high-definition television (HDTV), and high-speed data communications.

The network should serve for both interactive and distributed communication at data rates of over 140 Mbps (for video services). It must also offer both connection-oriented and connectionless services and support different connection configurations (point-to-point, multicast). In addition to standard connection configurations, other configurations defining a parallel connection among multiple users and multiple connections should also be supported by B-ISDN, as these configurations are very common and useful in the new technology of multimedia communication. Finally, B-ISDN must be able to provide high-capacity facilities with transparent transport connection. This is very useful in the digital signal-processing applications where the digital bit streams may be compressed/decompressed to carry very large amounts of data. The material herein is partially derived from References 1–13, 46, and 52–54.

19.8.1 Requirements for B-ISDN

In applications of interactive and distributed system developments, the data rate requirements may be much higher than 150 Mbps (at least 650 Mbps). B-ISDN offers services for high-resolution video applications which require data rates of more than 150 Mbps. The existing telephone networks using circuit switching over coaxial cable or even optical fiber may not be able to support these high data rates, and this limitation seems to be pushing toward the fast packet-switching technique over optical fiber with a different user network interface protocol such as asynchronous transfer mode (ATM). Although frame relay and cell relay (ATM) have some common features, frame relay can offer data rates of only up to 2 Mbps, while cell relay (small size of fixed cell of 53 bytes) offers data rates as high as in the gigabits-per-second range.

The narrow band channels of N-ISDN are B, H0, H11, and H12 (CCITT) and are used for integrated data and voice applications such as teleconferencing, home applications (lower data rates), business applications (higher data rates), etc., using two interfaces (BRI and PRI). The broadband services to be provided by B-ISDN include video and high-speed data rate applications. The channels proposed for broadband services are **H2 (30–45 Mbps)**, **H3 (60–70 Mbps)** and **H4 (120–140 Mbps)**. There is no general consensus on any single H2 data rate globally, but variations of this have been accepted as standard broadband channels, e.g., H21 and H22 within H2.

North America (U.S.) has proposed an H22 rate of 44.160 Mbps over existing DS-3 transmission links (44.736 Mbps) supporting the T1 (1.544 Mbps) hierarchy defined in the U.S. Europe has proposed H21 (32.768 Mbps) supporting the CEPT hierarchy (2.048 Mbps) within 34.368 Mbps. Nippon Telegraph and Telephone (NTT) has defined its own H2 as 44 Mbps. For more information on these channels and other services, see References 1, 4, 5, 8, 11, 42, and 46.

19.8.2 Transfer modes

It is believed that the B-ISDN network will actually offer a high-speed, high-performance, high-capacity user interface and universal open standard interface which will meet all the above-mentioned requirements and that it will create a multi-vendor environment for global connectivity.

The high-speed network interfaces have to concentrate on three very important and critical issues during their development: **switching, transmission**, and **multiplexing**. In the literature, the network professionals have given a common name to switching and multiplexing: **transfer mode (TM)**. In some books, however, the names switching and multiplexing may be used separately.

CCITT Study Group XVIII focuses on defining standards for B-ISDN and has defined a transfer mode (TM) for switching and multiplexing operations combined. The CCITT Study Group XVIII task group[25] on ISDN broadband aspects (BBTC) chose asynchronous transfer mode (ATM) as the basis for B-ISDN. According to the recommendations of this group, ATM is a high-bandwidth, low-delay mode based on a packet-like switching and multiplexing technique. Although it is based mainly on connection-oriented services, connectionless services are also supported by it. It is important to note that in ATM interface, the word "asynchronous" has nothing to do with asynchronous transmission. The standard functional architecture for information transfer across the B-ISDN consists of three layers: **transmission** (bottom layer), **transfer mode** (middle layer), and the top layers which provide the **service modules**.

The internal structure channel plays an important role during data communication and depends on the underlying applications (video, audio, voice, etc.). One suggested solution has been to consider a structure where the payload should be divided into fixed-size segments. Each segment carries an embedded narrow-band interface structure (nB+D) where n can be 2, 23, or 30. At the same time, the broadband interface can be defined by appending additional segments for higher data-rate channels. Each segment may be divided into sub-channels to carry more than one channel over the time slots. Each segment has a predefined capacity to carry the channel. For example, the H4 channel will be carried by segment C4, H2 by C2, and so on. In this way, end-to-end performance is provided by the network providers.

The transfer mode is important to both switching and transmission aspects of the network interface. There exist two types of transfer modes: synchronous transfer mode (STM) and asynchronous transfer mode (ATM).

Synchronous transfer mode (STM): In STM, the network uses **synchronous time-division multiplexing (STDM)** and circuit-switching technique for data communication. This mode is not being used in any high-speed network, as it was considered an interim technique of switching and multiplexing during the evolving of B-ISDN. It offers a fixed bit rate, very few virtual connections per channel, and small information delay. It is suitable for applications of data rates of over 35 Mbps and does not lose any information blocks during the transmission. The standards of this mode are available. A variety of applications over a single user–network interface (UNI) is supported by ISDN.

Typically, ISDN defines various channels within a framed interface. A frame is a periodically predefined or preassigned set of time slots. The delimitation and other control functions within a frame are obtained by specific time slots within the frame, while the remaining slots can carry the data. Out of the time slots assigned for data (payload), a few can be assigned to a specific bearer service for the duration of the call, and hence the channel is known as a **bearer (B) channel**. On the other hand, the D channel is also a channel with time slots but is not assigned to any particular bearer service. It carries signaling and other services of the lower layers by the service access point (SAP). It offers higher bandwidth and lower information delay.

The set of channels present in any UNI is defined by the STM interface structure and can be defined either as fixed or variable. The *fixed-sized internal structure* defines the channels with permanent assigned slots, e.g., the basic rate interface (BRI) of ISDN (2B+D). The *variable internal structure* channels will have a variable number of time slots, e.g., the primary rate interface (PRI) with bit-map channel within Q.931 (to support a dynamic combination of B, H0, and H1 channels). The fixed internal structure supports various combinations of 2B+D to specific network switching, and therefore the sharing of capacity among different services is limited. Further, it is attached to a particular set of service requirements and does not provide any compatibility features if used in different countries by different users, as service requirements may change. Although flexibility seems to be the reasonable choice, multiplexing and other service functions make the structure more complex and difficult to use and implement for internetworking.

In the STM frame, each channel is assigned a time slot. Each sequence of channels (predefined) is preceded by a control header or flag of a few bytes (for the framing signal) and followed by a control header. The sequence of frames encapsulated between the control headers is known as a **periodic frame**. Each piece of data is specified by its position within a periodic frame. Each frame gets a fixed bit rate as the bit rate channels (B, H2) are already defined, and hence offer no flexibility (very rigid structure of transmission). This switching offers a limited selection of the combination of channels at the corresponding interfaces, as these interfaces are defined for a predefined data rate channel (B or H). The switching offers an interface for each of the B and H2 channels, and hence switching must handle the selection of either B or H simultaneously based on the requested data rate of the application at appropriate time slots.

The main drawback with circuit switching or the STM-based network is that it does not offer any flexibility of requesting higher data rates. On the other hand, while it is possible to define a high-speed channel by combining several low-speed channels by using multi-channel switching, the data rate obtained from one high-speed channel is lower than in an ATM-based network (which offers a bit rate on demand).

Asynchronous transfer mode (ATM): ATM resembles (to some extent) the conventional packet-switching technique and provides communication with a bit rate that is tailored to actual need, including time-variant data bit rates for real-time applications. ATM uses asynchronous time-division multiplexing (also known as address multiplexing or cell interleaving) and fast packet-switching technique. The fast packet-switching in reference to ATM has also been referred to as **cell** or **block** switching. The **asynchronous time-division multiplexing (ATDM)** allows the dynamic allocation of bandwidth to the slots. The CCITT has considered this transfer mode for broadband ISDN (B-ISDN) networks. This mode offers a variable data rate, integration of services at all levels, and support of various virtual circuits per channel.

ATM does not assign any periodic time slots to a channel. Here the packet (containing information) is defined as a block prefixed with a header. The header contains an identification number (or label) of a channel. These numbered (or labeled) channels are multiplexed onto a stream of blocks. The multiplexing technique used in ATM is based on dynamic allocation of bandwidth on demand with a fine degree of granularity. It offers a unique interface which can support a variety of services. The same network interface can be used for low data bit rate and high data bit rate connections and supports both stream and bursty data traffic in each category.

The ATM switching mode (also known as cross-connect mode) performs the following two functions: translation of **virtual channel identifier (VCI)** and **virtual path identifier (VPI)** and transport of the call request from the input line of the switch onto the dedicated output line. In ATM switching, the switching element is a basic unit of the switch fabric. The switching element is usually defined as a set of lines, an interconnection network, and a set of output lines. These input and output lines are controlled by respective controllers as the input controller and output controller, respectively. These controllers are connected to the interconnection network. All the incoming request cells are handled by the input controller (it selects the request, defines respective links and connections over the transmission path, provides synchronization between the internal clock and the arrival time of the requests, etc.).

Other functions of ATM switching include management of input and output buffers and their locations in interconnection networks, memory management, implementation of interconnection networks, optimal routing for request cells, performance measures for ATM switching systems, etc. The ATM switching system uses either single-stage network configuration or multi-stage configuration. In the former, the shuffle exchange and extended switching matrix are examples, while the latter uses single-packet network (known as Banyan), multi-path, etc.

ATM-based networks usually consist of a single link-by-link cell transfer capability which is common to all services. It provides end-to-end support for ATM cells which encapsulates the service-specific adaptation functions defined by the higher layers. It allows a grouping of several virtual channels into one virtual path, and various virtual paths define a transmission path. ATM-based networks have a few problems, notably variable information delay and the possibility of losing packets. Each channel in an STM-based network contains time slots and offers services to both narrow-band and broadband applications.

In ATM, both user–network and network–node interfaces are provided in a better way than in STM. Here the specific periodic time slots are not assigned within the channel, but the information is always transmitted as blocks. Each block has two fields: **header** and **information**. The header identifies a logical channel established for multiplexing and also routing by high-speed packet-switching, while the information field contains the user's data. This virtual channel associated with any ATM cell header may occur at any position within the cell.

An ATM interface structure supports a variety of **bearer** services with different information transfer rates. ATM virtual circuits allow ISDN to be more flexible in changing the service requirements, as the capacity can be allocated dynamically on demand. This feature of ATM interface allows users to define dynamically the labeled interface structures to meet the requirements of the user. If the service requirements are changed, then the existing physical interface structure and also network access technique need not be changed and will be done by allocating capacity dynamically within the channels. This makes the ATM interface more suitable for bursty traffic (requiring high and variable data rates, also known as **variable bit rate — VBR**) and **continuous bit rate (CBR)** services and avoids the need for separate packet- and circuit-switching fabrics. It also carries real-time isochronous services (based on circuit-switching emulation) provided it offers sufficient switching and transport bandwidth. It also supports bursty traffic in a statistical mode by assigning lower priority to it. In other words, ATM interface is well suited for narrowband and broadband services and supports both isochronous and inisochronous services.

For isochronous application services, STM interface may be more efficient than ATM interface, due to higher bandwidth and low information delay, while ATM interface may be more efficient for inisochronous application services. It is interesting to note that both isochronous and inisochronous services do not require the entire bandwidth during any time in a continuous fashion and, as a result of this, the resources can be shared statistically by other services.

ATM switching suffers from the following problems:

- Loss of cell
- Cell transmission delay
- Voice echo and tariffs
- Cell delay variation
- Lower quality of service

It is important to mention here that the transfer mode simply refers to switching and transmission information in the network. If we see the ATM operation in the context of multiplexing transmission, the ATM switching allows the cells allocated to the same ATM connection to exhibit different network requirements (or offer an irregular pattern of data information), as the cells are defined or filled with the information based on demand by the users.

We have discussed two transfer modes (STM and ATM) and compared some of their capabilities and drawbacks. The partitioning of bandwidth of an interface and allocation of bandwidth to the user services are the fundamental steps common to both STM and ATM. We will discuss in brief how these fundamental steps are used in both STM and ATM in the following subsections. These two modes of operations are also discussed in References 3, 5, 8, and 42.

19.8.3 STM connection for B-ISDN

A typical STM-based network may offer an access service to the users, while the channels required for the services are multiplexed using a synchronous time-division multiplexing system which, in fact, offers the STM connection, as depicted in Figure 19.17. Here, the network terminal (NT) multiplexes the channels of various applications. The STM (working as a local switch) is connected to sub-networks which are either circuit-switched or packet-switched. The circuit-switched sub-network offers a switching connection for 64 Kbps, 2 Mbps, 135 Mbps, or any other data rates.

Limitations of STM connections: STM supersedes the frequency division multiplexing (FDM) frame. The time slots within the STM structure or frame are allocated to a service during the call (setup phase). The position of time slots within the frame defines the STM

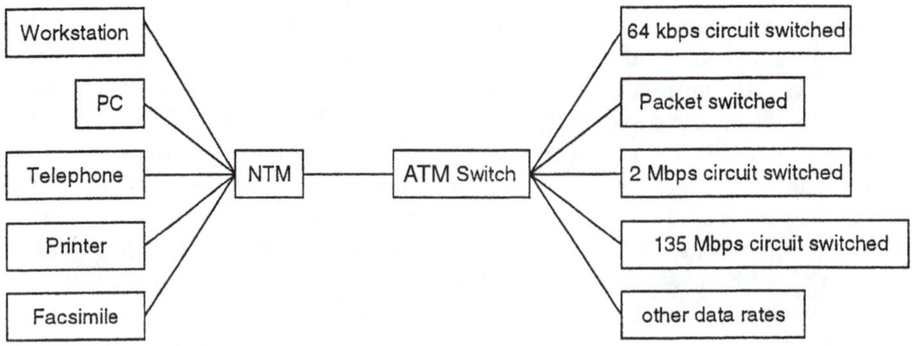

Figure 19.17 STM-based B-ISDN network.

Table 19.3 Data Rates of Various STM Channels

Channel	Data Rate
B	64 Kbps, common to U.S. and Europe
H0	384 Kbps, common to U.S. and Europe
H1	1.920 Kbps (U.S.), 1.536 Kbps (Europe)
H2	37.768 Mbps (U.S.), 43–45 Mbps (Europe)
H4	132.032–138.240 Mbps, common to U.S. and Europe

channel identification number which is replaced by the frequency range of that slot. This is based on the concept of FDM which is typically used in analog communication systems. On the other hand, the STM-based network is a digital communication system and is defined as a digital transmission hierarchy (as defined in ISDN with 2B+D as a basic rate interface, where *n*B+D is the primary rate interface and *n* represents the number of channels and has a value of 2, 23, or 30). The primary rate interface has different data rates for different countries (North America uses $n = 23$ while Europe uses $n = 30$). The B channel is a 64-Kbps bearer channel while the D channel is 16-Kbps in the basic rate interface (BRI) and 64-Kbps in the primary rate interface. The B channel carries pulse-coded modulation (PCM) voice. The STM internal structure frame supports the following bearer channel (B channel) rates which are extensions of (*n*B+D) structures (Table 19.3).

The STM internal structure interface is a fixed structure and does not offer flexibility. This lack of flexibility in the number of channels for the basic rate may not be significant (due to its limited capacity) but may become significant in the case of primary rate interfaces.

The assignment of time slots to a channel within a frame for switched services may be flexible, where each channel defines one or more slots per frame. Although the partitioning of time slots to a channel may be useful for the basic rate (to limited capacity), it may become a serious problem if we have a large number of time slots, as the time slots and the coordination between them need to be mapped into user and network interfaces. For example, a 150-Mbps interface allows about 2000 usable eight-bit time slots (the remaining bits are used for control) with an ability to supporting more slots in the higher rate interface. The switching system is further affected by multiple-rate STM structure, as

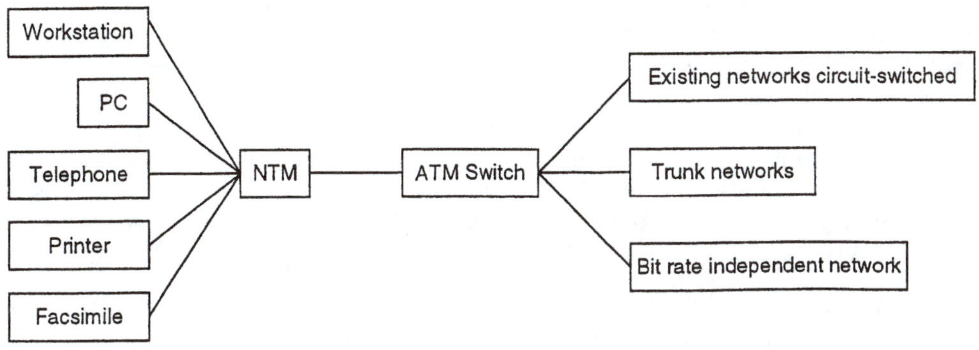

Figure 19.18 ATM-based B-ISDN network.

it involves management, maintenance, etc. Thus, it is clear from the above discussion that STM-based networks are suitable for fixed-rate services but are not suitable for bursty services (one of the main services supported by ISDN), as the STM channels do not support these services efficiently.

19.8.4 ATM connection for B-ISDN

A unified network (with a fully integrated environment) for information transfer at all network levels which can handle information of virtually all types of services over the same switching and transmission setup is known as an ATM-based B-ISDN network, as shown in Figure 19.18.

The CCITT has discussed procedures for managing small fixed-length blocks in periodic time slots in its recommendations on ATM. The fixed-length block is known as a **cell** in ATM terminology. A new network and network node interface have been standardized.

Although there was a general consensus to use ATM as a target transfer mode solution for implementing B-ISDN, some critical issues, such as the underlying transmission system for the user–network interface (UNI) which interconnects a user's system to public B-ISDN, and other ATM parameters, needed to be standardized to make B-ISDN a unified international standard network. Further, it has been agreed to internetwork voice communication with the existing networks. Also, the parameters of an ATM network may be useful in high-quality video and high-speed applications.

The introduction of a new optical transmission system known as a **Synchronous Optical Network (SONET)** has given yet another push to ATM, which can be used by any digital transmission system (mentioned in SONET CCITT recommendations G.707, G.708, and G.709).[20,22] Some of the ATM services can be carried over DS3 transmission media (44.736 Mbps) with a particular type of interface and will be carried over another medium using another interface. Thus, it looks quite natural and obvious to have a single standard UNI[34] which can provide all the services of ATM-based B-ISDN, without changing the internal structure of the ATM frame. In other words, if we have to use SONET, we may have to embed the ATM internal structure into the SONET frame, and do similarly for other transmission media, or we can use one interface with all of its bandwidth defined within the ATM internal structure, thus limiting the ATM for restricted applications. These matters have been left unresolved and are mentioned in CCITT's I.121 recommendations.[25]

One of the required features of B-ISDN is its high-level service capabilities and low-level transmission capabilities, and none of the existing available structures seems to fulfill these requirements except ATM, which to some extent offers lower-level service capability with higher-level transmission capabilities. The UNI must totally structure the payload capacity of the interface (i.e., the entire transmission capacity except a small portion which is used for operating the interface) into the ATM cell. Further, the UNI does not define any preassigned identification of cells for any specific applications. In other words, the assignment of cells is defined on demand at the interface. There are two interface bit rates defined for ATM-based networks: 150 Mbps and 600 Mbps.

Features of ATM connections: Some of the limitations (as discussed above) of STM structure interface have been eliminated in the ATM structure interface, as it supports bursty services more efficiently than STM-based networks and also supports **continuous bit rate (CBR)** services without affecting the normal performance of the network. One of the fundamental concepts of an ATM-based network is that it adopts the procedures for supporting various services with higher grades of performance as opposed to the standard store-and-forward concept used in traditional networks. Various user network functions are supported by different layers of ATM. The lowest layer (analogous to the physical layer of OSI-RM) provides functions such as regeneration of digital signals, assembly and disassembly of data frames (byte stream), and signaling control. The signaling control is usually defined for endpoints for each of the end user's data information as well as the cell header control information function.

The ATM layer defines two types of connections as **virtual circuit connection** (analogous to virtual circuit in X.25) and **virtual path connection** (group of logical virtual connections). It also provides support for fixed-size cells in full-duplex configurations for either signaling control or data exchange across the network. In other words, both control and data cells are defined within the fixed size of the cells. The routing and other network management functions are associated with each cell. The second type of connection is known as a virtual path connection (VPC), which is typically defined for a group of logical virtual connections which share a common path. The virtual paths are established and used by different groups of virtual connections which share the same logical connection to the network. Each path carries a different set of virtual connections.

The broadband ATM switching system supports a very high speed (about 80 Gbps) and gets the message data from interfaces at about 50 Mbps. As said earlier, ATM supports statistical multiplexing. The ATM layer of B-ISDN is mainly concerned with switching and offers the following functions: generic flow control, cell header, payload type identification, and loss priority indication. The virtual path identifier (VPI) and virtual channel identifier (VCI) and cell are multiplexing.

The data communication over the ATM network requires the establishment of a physical path over coax-based DS-3 or fiber-based SONET (OC3) or any other higher medium. This physical path allows the definition of virtual paths. The number of virtual paths depends on the number of bits used on the ATM header and is defined in the VPI field. A typical value of the number of bits at the user interface is 12. Similarly, the virtual path can accommodate a number of virtual channels, which are defined in the VCI field of the ATM header.

The ATM adaptation layer (AAL) in B-ISDN protocol provides a mapping between higher-layer protocol data units to a fixed size of cells and passes it to the ATM layer. That adds the header with appropriate VPI and VCI to the cell. The AAL layer protocol consists of two sublayer protocols — **convergence (CL)** and **segmentation and reassembly (SAR)**. The convergence sublayer provides a service access point (SAP) to each of the services offered by AAL. There are five classes of AAL layers: AAL1 (constant bit rate or circuit emulation), AAL2 (variable bit rate video and voice), AAL3 (connection-oriented services), AAL4 (connectionless services), and AAL5 (high-speed data communication). The SAR sublayer maps the incoming messages (protocol data units from the higher layers) into cells of fixed size (48 bytes). On the receiving side, the reverse process consists of extracting data messages from cells, reassembling into the message, and passing SAR PDU to the higher layers. The physical layer consists of two sublayers — physical medium and transmission convergence. The ITU-T has recommended fiber optics as a transmission medium for B-ISDN and has defined the physical interfaces of SONET/SDA for fiber and pleisochronous digital hierarchy (PDH) for coaxial. Details can be found in Section 18.9 (Chapter 18).

The ATM structure usable bandwidth is partitioned into fixed-size units known as **cells**, and no periodic time slots are assigned to a channel (as is done in STM structure). Each channel consists of two fields: **header** and **information**. The size of a cell is 53 bytes (CCITT), out of which five bytes are reserved for the header while the remaining 48 bytes are used for user's information. These cells are allocated to the services on request or demand. The header defines a **virtual channel identifier (VCI)** as opposed to time slot positions for identification of the channels in the STM-based network. In other words, in STM-based networks, time slot positions are used for sequencing the channels being transmitted.

The ATM interface structure defines a set of labels instead of slot position (as in STM) in the channel. These labeled channels may be used for both fixed and variable data rates. Since the service rate and information transfer rates are different, the mapping of time slots to the channels during call establishment is not required. Further, the ATM multiplexers and switches don't depend on the requested bit rates, and a variety of services can be provided by ATM without de-establishing multiple positioned time slots (or channel). This avoids the use of multiple overlay rate-dependent circuit channels or fabrics and adds more network integration.

ATM is suitable for both bursty (variable bit rate) and continuous bit rate (CBR) services. The CBR services are supported by circuit-switching emulation (ATM has to offer sufficient switching and bandwidth). The bandwidth allocation takes place during call establishment, and signaling procedures, once established, provide end-to-end communication between the users.

The ATM forum has defined some additional interfaces as T1 (1.544 Mbps), E1 (2.048 Mbps), T3 (45 Mbps), E3 (34 Mbps), E4 (140 Mbps), SONET (STS3c/SDH STM1 — 155 Mbps), and FDDI PMD (100 Mbps). For each of these physical media protocols, a separate convergence sublayer protocol has been defined similar to the MAC protocol in OSI-RM.

Functions of ATM connection: In an ATM network, cells (53 bytes) carry the information and are transmitted across the network. The layered protocol architecture of an ATM network must support the mapping of information into cells, transmission of cells, loss of cells and cell recovery, etc. Five functions of the protocol architecture for handling cells have been identified:

1. Routing of cells; must follow connectionless mode of operation.
2. Maintaining ATM connections to support different types of services, e.g., multimedia, video, audio, data, graphics, etc.
3. Segmenting and reassembling or mapping of information into cells at the sender site, and of cells into information at the receiving site. The traditional routing is based on wait-and-forward philosophy, where after every hop of the ATM switch, it determines the route for forwarding the cell.
4. Supporting different types of transmission media for cell transmission.
5. Transmitting all of the bits of cells serially, as is being done in the existing networks.

All these functions of ATM cells have to be implemented in one way or another. Different standards organizations have defined different numbers of layers to implement all of these functions. For example, ITU-TSS has defined three layers — physical layer, ATM layer, and ATM adaptation layer (AAL) — which implement the lower five functions. A separate layer realizes the connectionless function. Bellcore has defined the layers for switched multimegabit data service (SMDS) which is based on distributed queue dual bus (DQDB) protocol. This supports the LAN connectionless services for cells. The cells in SMDS consists of three ATM layers as SMDS interface protocol (SIP). For connectionless service, the top function, routing, must support connectionless network service (CLNS) at the top layer of ATM networks.

ATM interconnectivity with LANs: For interconnectivity with LANs or WANs, different types of ATM hubs can be defined, such as router-based, ATM-based, or a combination of these. In the router-based hub, all the routers are connected together and ATM provides interfaces between different LANs and routers. The ATM contains interfaces for different types of LANs and also interfaces with routers, and these become the access points for LANs. The routers so connected form a router-based hub. In an ATM-based hub, all the ATM switches are connected together and the routers provide interfaces between different LANs and ATM. Here the router defines interfaces to various LANs and these become the access points for LANs. The combination of these two types of hubs can be obtained by combining these two classes of hub architecture, which is usually used in large corporations.

ATM interconnectivity with WANs: WAN interconnectivity is slightly more complex than that of a LAN. The WAN, as said earlier, is used for greater distances than LANs. The LANs use their own link to provide services, while WANs use the public carrier or a private line. The LANs and WANs can communicate with each other through an interconnecting device known as the channel service unit/data service unit (CSU/DSU). This device provides a mapping between the LAN protocol and network digital protocol corresponding to T1 or any other link in the same way that the bridges provide two types of LANs at the MAC sublayer. The function of a router is to provide interconnectivity between LANs located within a building or at different locations (via WAN). The LANs are connected to WANs via CSU/DSU devices which provide mapping between the protocols of LANs and that of public carrier (such as T1 or any other link) or private lines. The protocol used on public carrier or private lines use TDM technique to accommodate the data from different sources. Unfortunately, the existing carrier supports data traffic only. Other types of traffic such as video or voice can also be handled by a TDM device via codec and PBX.

For WAN interconnectivity, the ATM hub can be defined by connecting ATM switches via OC3 links, and then ATM switches of the hub can provide interconnectivity with LANs through an ATM LAN connected via OC3 links. The ATM switches from the hub can be connected to WANs via OC3/DS-3 links for access to the outside world. In another form of WAN interconnectivity, the ATM switch may be connected via OC3 links and also can be directly connected to various routers and video services through video servers. It is interesting to note that with OC3 links, the optical signals now carry the information which may allow us to use the public services available on an ATM-based backbone public network. These networks are connected to ATM switches via OC3 links.

Limitations of ATM connections: ATM usually operates in a statistical mode to handle bursty traffic and is assigned with a lower priority. It is quite clear that ATM does not provide efficient bandwidth utilization switching and transmission techniques and offers a significant delay in continuous bit rate for various services, and we may consider ATM-based networks to be inefficient for these services. But there are advantages offered by ATM, particularly during the transmission of bursty data and other features such as resource utilization, performance, etc. In spite of its limitations, the ATM-based network will be used for time-dependent and bursty traffic application services. Further, for the bursty data, it may not require the continuous allocation of its maximum bandwidth, and hence more services may share the resources.

19.8.5 Services of B-ISDN

B-ISDN must support a variety of services with different network requirements and data rates. Further, it must support bursty traffic and take care of delay and loss of information in the case of sensitive applications.

Although it is difficult to estimate the network requirements of different applications, attempts made in this direction define certain bandwidth requirements of some of these applications. These estimates are not accurate but at least provide some idea of the network requirements. The connection-oriented services typically require data rates in the 1.5-Mbps to 130-Mbps range with a support of bursty traffic of 1–50. The connectionless services require the same bandwidth with a support of 1 bursty traffic. Document transfer/retrieval offering 1–20 bursty traffic requires 1.5–45 Mbps. Videoconferencing and video telephony offer 1–5 bursty traffic in the range of 1.5–130 Mbps. TV distribution typically requires 30–130 Mbps, while HDTV requires 130-Mbps bandwidth.

Different categories of service class can be defined for B-ISDN services, depending on the applications. As we mentioned earlier, B-ISDN offers services for both home and business applications. For small business, the services are available within a small area (like a small room), provided it has the necessary internal transmission and switching equipment capabilities in the room. For larger businesses, e.g., factory, organization, etc., the interactive services (e.g., telephone, videoconferencing, and high-speed data communication) may be defined for users (around 100) and must have internal switching and transmission capabilities. A larger organization may require a network to cover a distance of more than the LAN's limit; in such a case, distributed services may be obtained. For a larger geographic distance, we may have to use MANs and other wide area networks.

B-ISDN service has been defined as a service which needs a transmission channel capable of offering data rates higher than the primary data rates offered by ISDN (CCITT's recommendation I.211).[27] The need to define B-ISDN was to provide services to broadband applications such as images, audio, video, and many other multimedia-based applications requiring higher bandwidth and higher data rates. The last decade has seen tremendous changes in certain technologies and advances which can provide support to define a unified universal high-performance and high-speed network. The advances are taking place in the following areas and many others:

- Optical-fiber transmission media offering a very high data rate of 100 Mbps, to be used for user's lines and network terminals
- Introduction of asynchronous transfer mode switches
- High quality of video monitor cameras
- High-definition television (HDTV) at low costs

With these technologies and advances combined, the idea of defining a universal network which can provide a variety of services such as sharing culture and social lives without visiting an individual country, educational information sharing, retrieval and sharing of a variety of information, multimedia-based applications, and many others seems highly likely.

The services offered by B-ISDN are based on the category of information which is being transmitted. The following are the categories of information along with their applications (these services and applications are also detailed in References 1, 3, 8, 11, 12, 27, and 42).

Data transmission: High-speed digital information services are high-speed data transfer, LAN-to-MAN connectivity, computer-to-computer connectivity, video information transfer, still images, scanned images, drawings, documents, CAD/CAM, etc. High-volume file-transfer services include data file transfer, telemetry, and low-bandwidth applications such as alarms, security systems, etc.

Document transmission: High-speed telex services are user-to-user transfer of text, images, drawings, still pictures, scanned images, documents, etc. High-resolution image and document communication services offer the transfer of images, medical images, remote games, and mixed documents (text, images, drawings, etc.).

Speech transmission: Multiple speech or voice program services offer the transmission of multiple programs containing speech, voice, and sound signals.

Speech and moving pictures transmission: Integrated services of speech and moving pictures are provided in a number of ways, e.g., video telephone, video-scanned images and documents (teleservices such as tele-education, tele-shopping, etc.), videoconferencing (tele-education, tele-advertising, tele-business conference or meetings, etc.), video surveillance (offering security for traffic, buildings, banks, etc.), video/audio information services (TV signals, video/audio dialogue session, etc.), video mail service (e-mail box and fax services), document mail services (mixed documents including text, graphics, still and moving pictures and images, and voice annotation transfer), etc.

Moving pictures and images retrieval: B-ISDN allows the retrieval of videotex services (moving pictures, tele-software, tele-shopping, etc.), video retrieval services of high-resolution images, data retrieval services, video services (TV distribution services such as North American Television Standards Committee [NTSC], Phase Alternation Line [PAL], and Systems en Conleur avec Memoire [SECAM]) for TV programs, high-definition and enhanced-definition TV programs, high-quality TV services such as pay TV, pay-per-view, pay-per-channel, etc.

The available bandwidth in demand for ATM interface in B-ISDN ranges from 155 to 620 Mbps and even higher. Some of the application services offered by B-ISDN (CCITT I.121 recommendation on broadband aspects of ISDN) include videotex, high-resolution image retrieval, electronic newspapers, broadband teleconference, high-speed color telefax, video telephony, video mail, document retrieval, HDTV, multilingual TV (MLTV), high-quality audio distribution, tele-software, video surveillance, etc. High-performance and high-speed networks are providing integrated services and can be categorized as high-speed data service networks and video service networks.

LANs have already played a important and vital role for information and resource sharing among the users connected to them. PCs, workstations, sharable resources, and terminals may be connected to LANs. Different LANs offer different data rates (e.g., IEEE 802.3 Ethernet offers a data rate of 10 Mbps, IEEE 802.5 IBM token ring LAN operates at 4 Mbps and 16 Mbps, and FDDI LAN operates at 100 Mbps). Higher-speed LANs for connecting PCs, diskless workstations, and CAD/CAM terminals are in big demand nowadays. These high-speed LANs may be used as front-end networks which provide interconnection between PCs and host computers and may be used as back-end networks providing interconnection across the multiple hosts. In other words, high-speed LANs can be used both as front-end and back-end products for sharing of resources, information, etc. The high-speed networking across PCs, workstations, and host computers is limited to a local distance of a maximum of 10 km of a campus, building, or corporation premises.

Hardware devices such as bridges, routers, and gateways may be used to enhance the interconnectivity of LANs at a greater distance, but these devices connect these high-speed LANs to public or private networks. The public network (wide area network) offers a channel speed of 56 Kbps or 64 Kbps. B-ISDN provides interconnection of LANs to wide areas at data rates equal to or greater than today's LAN speed. Further, B-ISDN uses a standard numbering scheme which extends the LANs over a wide distance (thousands of miles), making it a worldwide type of network. Various applications which can be performed over B-ISDN worldwide include exchange of large files, file back-ups, display of animated graphics, digital signal processing, support of diskless workstations, etc.

19.8.5.1 Classes of B-ISDN services

The services offered by B-ISDN (as recommended by **CCITT I.211**[27]) are classified as **interactive** and **distributive**. Other recommendations discussing some aspects of these services include CCITT I.121,[25] CCITT I.311,[28] and CCITT I.113.[24]

The **interactive** class of services includes conversational services and services of two-way communication (e.g., messaging, retrieval, etc.) and does not include control signaling information. Typical applications of conversational services may include moving pictures of video and sound (video telephony, multi-point and point-to-point videoconferencing, video/audio transmission), data (file transfer, digital information, etc.), and documents (high-speed text transmission, e.g., telefax). The messaging services may include video mail services, document mail services, etc. Finally, the retrieval services may include broadband videotex (data, text, graphics, etc.), video retrieval, document retrieval, etc.

The **distributive** class of services includes any service of one-way communication, e.g., broadcast or individual presentation control services. It does not include control information. The services are for applications such as high-quality TV distribution services in different modes (e.g., phase alternating line (PAL), North American Television Standards Committee (NTSC), and Systeme en Couleur avec Memoire (SECAM)), pay-TV (pay-per-view, pay-per-channel), documentation retrieval services, high-speed digital information, and video information.

Video telephony includes a video transmitter and receive/display facilities so that both voice and line pictures can be received during the dial-up call. This may be used in office-oriented services like sales, consulting, etc. **Videoconferencing** offers a point-to-point and multi-point communication environment, and many other services can also be used in conjunction with this service, e.g., facsimile, document transfer, etc. **Video surveillance** is usually unidirectional and transmits the information (video image) to the user's computer system. The **video/audio information** transmission service offers more or less the same facilities as those of video telephony. The **data services** offered may include file transfer, program or file downloading, data for LAN connections, etc.

Videotex service is an interactive system which provides support to a variety of retrieval-based applications. It contains one videotex computer which accesses different types of databases, including public databases and vendor-supplied services. These services are provided on video terminals, PCs, telephone sets, etc., by transmitting them across the public switched telephone networks. The broadband videotex is an enhanced version of the videotex system where the user can select sound, high-resolution images of TV, short video clips, text, graphics, etc.

Another **teleservice** based on user presentation control is teletex, which is a simple one-way system that uses the unallocated bandwidth of a TV signal. The transmitter sends a fixed number of text messages which are shown on TV sets (after decoding and storing temporarily). The decoder reads pages from incoming signals, stores them, and displays them on the screen with the help of a keypad. The broadband version of teletex is known as *cable text*, which uses the entire digital broadband channel (as opposed to the unallocated

Table 19.4 CCITT I.35B Service Quality Categories for Video Applications

Service quality category	Data rate (Mbps)	Application
A	92–220	HDTV
B	30/45–145	Digital components coding
C (analog broadcast television)	20–45	Digital coding of PAL, NTSC, SECAM for distribution
D	0.384–1.92	Reduced spatial resolution and movement portrayal
E	0.064	Highly reduced spatial resolution and movement portrayal

Table 19.5 CCITT I.121 Channels for N-ISDN

Channel	Data rate	Application
D	16 or 64 Kbps	Control signaling, packet-switched
B	64 Kbps	Circuit- and packet-switched data, voice, facsimile
H0	354 Kbps	Data, voice, facsimile, compressed video
H11	1.536 Mbps	PBX access, compressed video, high-speed data
H12	1.920 Mbps	PBX access, compressed video, high-speed data, broadband
H2	30–45 Mbps	Full-motion video for conference, video telephone, video messaging
H2	44 Mbps	Nippon Telegraph and Telephone (NTT)
H21	32.768 Mbps	Europe
H22	44.160 Mbps	North America
H3	60–70 Mbps	Not identified
H4	120–140 Mbps	Bulk data transfer of text, facsimile, enhanced video information

part of the bandwidth of an analog TV channel). The cable text includes text, images, and video and audio components in the continuous transmission of the text.

19.8.5.2 Quality of services for video applications

CCITT I.35B[23] has defined five categories of service quality for video applications. Each category is defined for a specific application service along with the allocated data rate (Table 19.4).

CCITT I.121[25] has defined the channels for narrow-band ISDN (N-ISDN) shown in Table 19.5. A bandwidth of H2 was further partitioned into two groups, H21 and H22. The H21 bandwidth of 32.768 has been accepted by Europe, and this channel can be carried within the 34.368 signal of 2.048-based CEPT hierarchy. The H22 group defines a bandwidth of 44.160 and has been adopted by North America. This channel can be carried over the existing DS3 (44.736) facilities of 1.544-based hierarchy of North America. Nippon Telegraph and Telephone has accepted the H2 channel of 44 Mbps as its standard channel.

The CCITT I.327[29] recommendation of B-ISDN defines the capabilities of information transfer and signaling to be used in the architecture of B-ISDN. The information and signaling are defined for different interfaces with different capabilities: broadband and 64-Kbps ISDN capabilities and signaling for user-to-network, inter-exchange, and user-to-user applications. The signaling information for a connection is communicated using different pointers or identifiers (out-of-band signaling). The broadband information transfer is obtained by ATM switching, which offers call sequence integrity; i.e., if a call request assigned with a virtual channel has been sent, then another call request assigned with the same virtual channel will never supersede the previous call request. In other words, the switching of ATM follows the connection-oriented concept.

B-ISDN uses the out-of-band signaling scheme (also used in 64-Kbps ISDN). In ISDN, the physical signaling is defined by a D channel of 64 Kbps. As mentioned earlier, the

ISDN is based on a digitized telephone network which is characterized as a 64-Kbps channel. The voice-grade frequency of 3.4 KHz is used in voice communication, and when this voice signal is sampled at the rate of eight samples with a frequency of 8 KHz, it offers a data bit rate of 64 Kbps (8 KHz × 8 bits). The ISDN is basically a circuit-switched network but can also offer access to packet-switched services. It offers two main interfaces — basic rate interface (BRI) and primary rate interface (PRI). Similarly, it provides two types of physical signaling — D16 (16 Kbps for basic access) and D64 (64 Kbps for primary access).

In B-ISDN, the logically signaling channels for users are distinguished by a virtual channel (VC)–based signaling scheme, and it supports 64-Kbps applications and other new broadband services. Although the existing signaling scheme Q.931 (CCITT I.451 recommendation) can also be used in B-ISDN, it does not fully support the multimedia applications. This means that any signaling scheme we choose for B-ISDN must support, establish, maintain, and release the ATM's **virtual channel connection (VCC)** and **virtual path connection (VPC)** for information transfer and negotiation for traffic requirements of connections. Further, it must not only support internetworking for ISDN devices but must also support non-ISDN services.

In B-ISDN, the signaling scheme is based on the out-of-band signaling technique and is defined as a dedicated signaling virtual channel (SVC). Different types of SVCs which have been proposed for use in B-ISDN are meta-signaling channel (bidirectional), broadcasting SVC (unidirectional), multicasting SVC (unidirectional), and point-to-point SVC (bidirectional).

19.8.5.3 B-ISDN architecture model

A typical broadband architecture integrating all services is shown in Figure 19.19. The reference points S, S(B), and S(D) correspond to standard 64-Kbps ISDN, broadband, and asymmetrical interfaces based on STM internal structure–based networks (useful for TV signals), respectively. The network termination (NT) provides either direct connection to the distribution of module STM at line termination or indirect connection via remote unit over glass fiber. The local switch also receives the signals from NT over existing copper wires.

The B-ISDN switch at the local end includes three modules — **STM distribution, broadband ATM**, and **dialogue module** — and a narrow-band STM ISDN switch. The N-ISDN STM switches can also be connected by Signaling System No. 7 (SS7) for providing signaling within the trunk network at **signaling points (SP)**. A glass fiber connects the broadband NT to the local exchange, providing a star topology for accessing the network. The local exchange includes a local B-ISDN switch which defines modules for broadband dialog and distribution services. The B-ISDN trunk network is defined as an overlay network consisting of a few transit switches and offers optimum adaption to all different broadband data services.

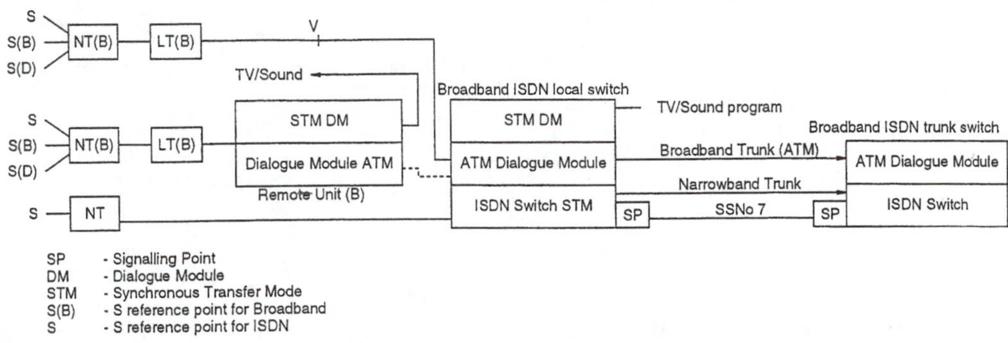

Figure 19.19 Broadband ISDN architecture. (Partially derived from References 8 and 48.)

The broadband switches provide access to continuous bitstream and bursty data, and one of the possible configurations for supporting the services could be the use of different switches for different types of data, as shown in Figure 19.19. Since standard digital switches on a 64-Kbps basis are available, the lines using continuous information protocol (based on STM) can be connected directly to a circuit-switched interface such that it supports basic primary rate access and analog signals on different lines. The broadband information can be transmitted over broadband access to the ATM switch. Both STM and ATM switches have a common control circuit. Each switch is interfaced with respective trunk lines. Since ATM and STM switches have different multiplexing and switching techniques, an interworking interface provides functions of blocking, segmentation, and reassembly between the switches.

The ATM switch receives all ATM channels over the same access line and hence allows multiple channels with different bit rates to be switched in a single switching network. Another configuration offering access to both services would be a common ATM switch for both types of data, as shown in Figure 19.20(a,b).

An ATM packetizing interface converts all continuous bitstreams into cells at the input of the ATM switch. The interworking interface and ATM packetizing interface introduce delay whenever ATM switches are used in an ATM trunk, but if we use interfaces without using an ATM trunk, this delay may be significant in some applications. A typical B-ISDN user–network interface (UNI)[34] is shown in Figure 19.21.

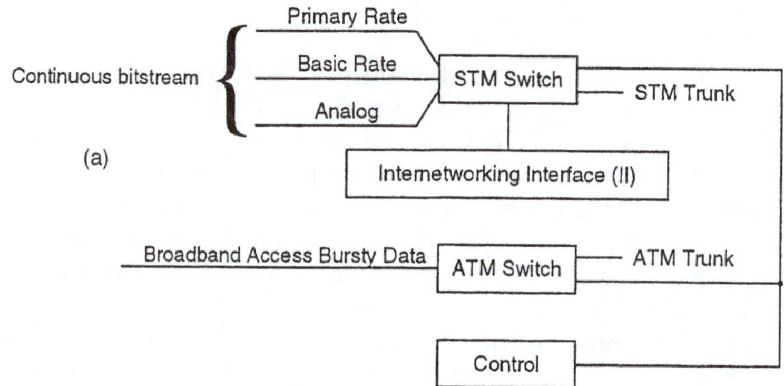

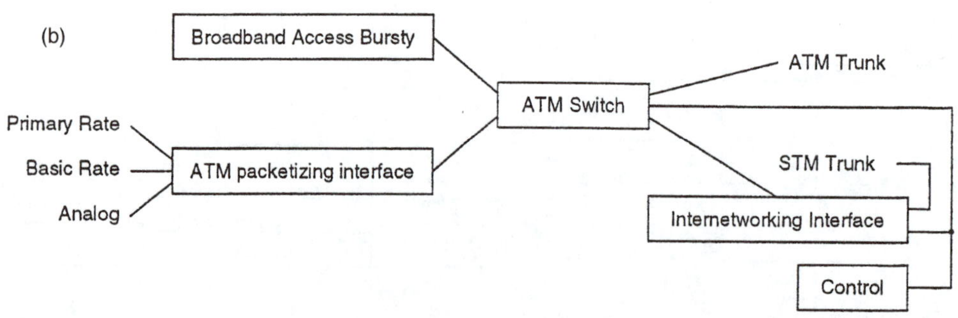

Figure 19.20 Common ATM switches. (a) Separate access type. (b) ATM access type. (Partially derived from Reference 8.)

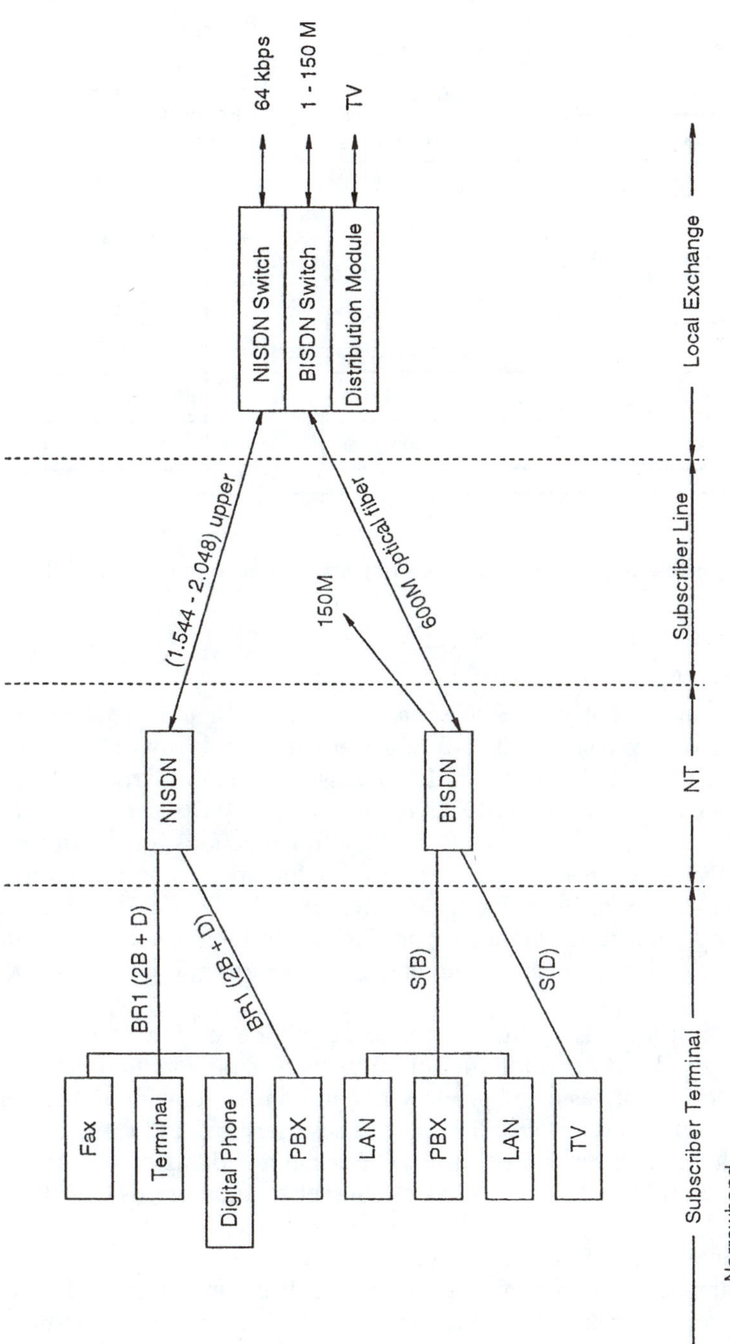

Figure 19.21 Broadband ISDN user–network interface.

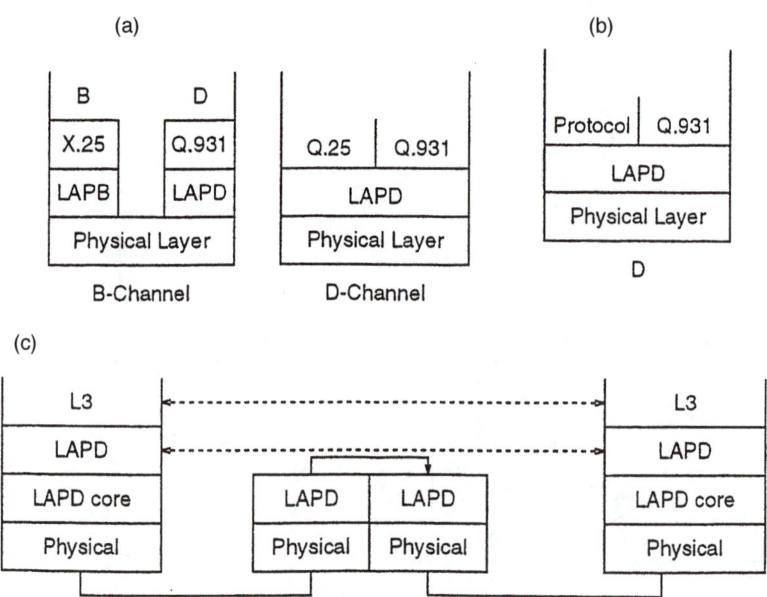

Figure 19.22 (a) X.25 access to ISDN. (b) Functions of bearer channel frame. (c) LAP-D connection to ISDN.

19.8.5.4 B-ISDN protocol model

The layered architecture model of N-ISDN is based on OSI-RM and has defined standards for the lower three layers (physical, data link, and network), as shown in Figure 19.22(a–c) for both B and D channels (CCITT I.413).[34] The discussion herein is partially derived from References 12 and 48. In Figure 19.22(a), X.25 offers access to ISDN in a packet-switching mode. In Figure 19.22(b), all the bearer channel control functions are supported by the control plane; i.e., Q.931 possesses complete control, while the protocol at level 3 will be responsible for user data transfer. In Figure 19.22(c), the LAP-D offers a connection on a link-by-link basis and offers transparency for bit, CRP computation, and frame delimiting. Other functions are performed on an end-to-end basis including flow control and packet re-transmission.

The ISDN physical layer is defined at reference point S or T to the users. Various functions of the physical layer include the full-duplex transmission mode in both B and D channels, multiplexing for basic and primary rate interfaces, activation/deactivation of the physical layer, and encoding of digital data. These services and functions for basic and primary rate interfaces are defined in CCITT recommendations I.430 and I.431.[43,44] CCITT recommendation I.430 defines the basic user–network interface (2B+D) channel at 192 Kbps.

The basic rate interfaces (2B+D) defining a total data rate of 144 Kbps are multiplexed over a 192-Kbps interface at an S or T reference point. The remaining bandwidth (192 – 144 = 48 Kbps) is used for framing and synchronization. The basic rate interface defines a frame 48 bits long and operates at a rate of one frame after every 250 μs.

CCITT's recommendation I.431 defines the primary user–network interface at the S or T reference point. Usually this interface defines point-to-point configuration at the T reference point for multiple TES which are being controlled by PBX for synchronous TDM access to ISDN. It defines two rates: 1.544 and 2.048 Mbps. The B-ISDN protocol reference model will have a physical layer at the bottom, ATM layer, ASTM adaptation layer (AAL), control and user sublayers (planes), and management layer at the top of the model. The control sublayer or plane is responsible for setting up the connection, maintaining the signaling information, and other relevant control signals. The user sublayer (as the name suggests) is responsible for the transfer of bits (defined within cells) across the network.

The interface at 1.544 Mbps is defined for North America over DS-1 structure (based on T1). One frame contains 193 bits, with a repetitive one frame at 125 μs. Each frame contains 24 channels or time slots, with each slot of eight-bit length. A framing bit precedes the frame which is used mainly for synchronization and other management purposes. A frame repeats at a rate of every 125 μs or 8000 samples per second, giving rise to a data rate of 1.544 Mbps. Each channel offers 64-Kbps service. The typical **primary rate interface (PRI)** in the U.S. consists of 23 B channels and 1 D channel. Interface at 2.048 Mbps is primarily defined for European countries. Each frame consists of 32 eight-bit time slots, and the bitstream is defined into repetitive 256-bit frames. Out of 32 frames, the first frame is used for framing and synchronization while the remaining 31 are used for user's data. Here, a frame repeats at a rate of one frame every 125 μs (8000 samples per second). There are 32 channels, and each channel offers 64-Kbps service. This transmission rate supports 30 B channels and 1 D channel.

The standard protocol reference model of B-ISDN defines three planes: management, user, and control. The management plane is at the top of the model and mainly deals with layer management functions and plane management, resources, and parameters defined within protocol entities. The user plane provides functions such as flow control and error control. The control or signaling plane, at the bottom, provides carious call and connection control functions. The physical and ATM layers offer functions similar to those of the user and control planes. The B-ISDN uses different types of cells, where a cell is a frame of fixed length (ITU-T). The cells used in B-ISDN include idle, valid, invalid, assigned, and unassigned. These cells are exchanged between the B-ISDN physical and ATM protocol layers. The physical layer functions in the B-ISDN protocol are composed of two sublayers — physical medium and transmission convergence. The physical medium, as the name implies, is concerned with the functions of transmission media. The transmission media are responsible for bit transmission of electrical to optical signals, and can be optical, coax, or even wireless media. The ITU-T recommends optical fiber for the B-ISDDN (SONET/SDH). The transmission convergence sublayer is mainly concerned with cell rate decoupling, header sequence generation and verification, cell delineation, transmission frame adaptation, and transmission frame generation and recovery.

The broadband ATM switching system supports a very high speed (about 80 Gbps) and gets the message data from interfaces at about 50 Mbps. As said earlier, ATM supports statistical multiplexing. The ATM layer of B-ISDN is mainly concerned with switching and offers the following functions: generic flow control, cell header, payload type identification, and loss priority indication. The virtual path identifier (VPI) and virtual channel identifier (VCI) and cell are multiplexing. Data communication over the ATM network requires the establishment of a physical path over coax-based DS-3, fiber-based SONET, (OC3), or any other higher medium. This physical path allows the definition of virtual paths. The number of virtual paths depends on the number of bits used in the ATM header and is defined in the VPI field. A typical value of number of bits at the user interface is 12. Similarly, the virtual path can accommodate a number of virtual channels, which are defined in the VCI field of the ATM header.

19.8.5.5 Physical configuration of B-ISDN

CCITT I.413[34] defined different classes of physical configurations of B-ISDN which basically give the implementation or realization of a user–network interface (UNI).

In the star configuration, each B-ISDN terminal equipment (TE) is directly connected to network termination 2 (NT2) via a separate dedicated line. The NT2 is interfaced with NT1 over an access line. The implementation of NT2 can be defined in different ways, i.e., centralized or distributed. The distributed implementation may be defined using LAN-like topology (bus or ring) where the terminals are connected to common media via special medium adapters. In this configuration, modification (addition or deletion of terminals) requires extra cost and larger multiplexers.

The B-ISDN terminals can also be connected directly with NT2 via a common shared medium (e.g., dual-bus). All the TEs are connected to a common medium, which is connected to NT1 over an access line and also through NT2 (if it exists). Here, the terminals must include MAC function and are usually supported by a generic flow control (GFC) protocol residing at the ATM layer. This protocol offers orderly and fair access of terminals to a shared medium by supervising the cell stream. Each terminal is assigned requested capacity on a per-cell basis. This physical configuration is easy to implement and modify (i.e., addition or deletion of any terminals).

A combination of star and bus configurations is known as *starred bus*, where a group of TEs are connected via a common medium which is connected to NT2 via different links. The NT2 is then connected to NT1 via an access line. Here NT2 is also acting as a multiplexer.

According to CCITT I.413,[34] the following functions of NT2 have been defined:

- Adaptation function for different interfaces, media, and topologies
- Multiplexing/de-multiplexing/concentration of traffic
- Buffering of ATM cells
- Resource allocation
- Signaling protocol handling
- Interface handling
- Switching of internal communication

The NT2 (B) defining NT2 for broadband covers actual implementation and may be nonexistent (null NT2 (B)) if interface definitions allow direct connection of terminals with NT1 (B). It uses layer 1 connection (wires) and provides connection and/or multiplexing functions, or it can be a full-blown switch (PBX).

B-ISDN is a fiber-based network which offers user-interface bandwidths of 155.520 Mbps and may support 622.080 Mbps (under consideration). The interface bandwidth of 155.520 Mbps offers a transfer capability of 149.760 Mbps (SDH). The 155.520-Mbps interface bandwidth is based on SDH and each frame has nine rows and 270 columns. With a repeating frequency of 8 KHz, the bandwidth becomes $9 \times 270 \times 8 =$ 155.520 Mbps. The 622.080-Mbps bandwidth is four times that of 149.760 and, defined as STM-4, offers $4 \times 149.760 = 599.040$ Mbps. The remaining bandwidth basically is defined for section, path, and other control overheads. The architecture of B-ISDN must support all the services of 64-Kbps ISDN (both basic and primary data rate accesses: BRI and PRI).

Flexible multiplexing facilities are provided by asynchronous transfer mode (ATM), which provides broadband width of various B-ISDN interfaces to a large number of data, voice, and video communications. It has been standardized that B-ISDN networks will transport ATM multiplexed information over fiber optics, which adheres to Bellcore's Synchronous Optical Network (SONET).

The ATM interface multiplexes and transmits the B channels of 64-Kbps ISDN using cell interleaving. The broadband channel so defined (multiplexed B channels) can be used as either H channels with predefined fixed bit rates (using STM connection) or virtual channels with variable data rates (using ATM connections).

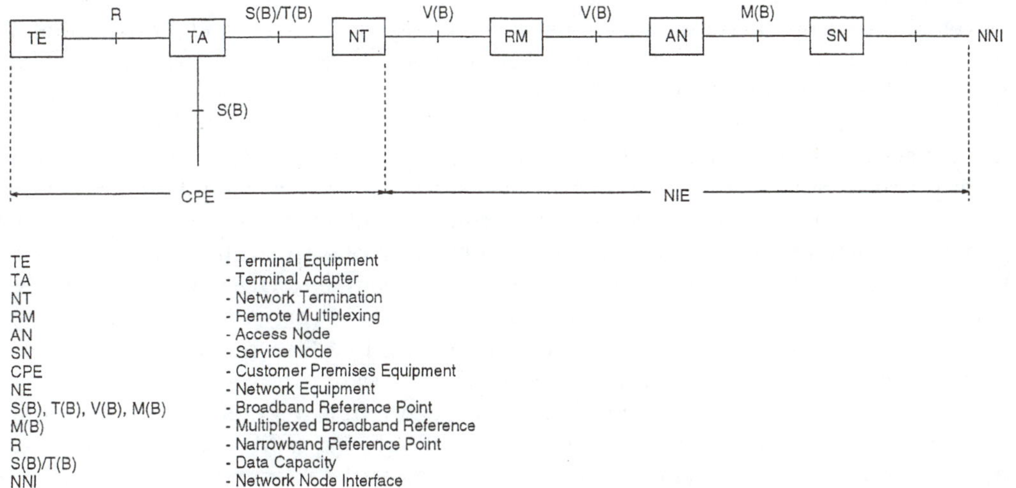

Figure 19.23 Reference model of B-ISDN. (Partially derived from References 29 and 48.)

The SONET-based UNI can support both transfer modes (STM and ATM) using a cell-structured payload. The STM connection can be established by allocating one or more ATM cells to a channel in every frame (using deterministic multiplexing), while ATM can be established by dynamically using the cells for variable bit rate requirements of virtual circuits (based on statistical multiplexing).

A reference architecture model of B-ISDN is shown in Figure 19.23. It also shows various functional modules defined in the architecture of B-ISDN with standard reference points indicated between them. When we compare this architecture with that of narrow-band ISDN, we notice that this is an extension of narrow-band ISDN standard reference point architecture. In order to distinguish between narrow-band and broadband ISDNs, we are using (B) for broadband points. Each functional module represents a specific function and can be implemented in different ways, but the reference points between functional modules provide the interface between them and are fixed.

The B-ISDN architecture can be partitioned into two segments: **customer premises equipment (CPE)** and **network equipment (NE)**. The first segment, CPE, basically consists of three functional modules: **terminal equipment (TE)**, **terminal adapter (TA)**, and **network termination (NT)**. The remaining functional modules, namely remote multiplexing (RM), access node (AN), and service node (SN), are defined by second-segment network equipment (NE).

In **customer premises equipment**, the TE module provides an interface to a variety of devices, such as LANs, hosts, PCs, workstations, N-ISDNs, terminals, video terminals, etc. The TA module includes and provides functions of media access control (MAC), segment assembly/reassembly (SAR), address checking, etc. The function of a TA is to support the terminal devices which don't support B-ISDN interfaces, providing the connection for these devices with B-ISDN architecture. It will provide support to MAC protocols of LANs, segmentation and reassembly of ATM cells, ATM signaling, and label processing (address checking). It may also provide analog-to-digital signal conversation in the case of voice communication and may provide compensation for any delay introduced during voice communication.

Both TE and TA have a reference point **R** between them. The communication devices which have B-ISDN interfaces are directly attached to NT, while the devices which don't have B-ISDN interfaces need an additional connector, a TA. The TA adopts the native mode protocols of the TE at reference point R. The NT module is used mainly for media conversion, with optional multiplexing and switching (or transfer mode) capabilities. A typical media conversion would be from single-mode fiber (SMF; very common in public networks) to multi-mode fiber (MMF; very common for cabling within a short distance or premises). Another useful function of NT is to multiplex several end user terminals onto a standard reference point V(B) interface. The switching between terminals on customer premises is also performed by NT. The users may have different terminal devices connected to a broadband network which may run different protocols (e.g., IEEE 802 LANs with appropriate protocols, N-ISDN basic and primary rate application protocols, hosts under different operating systems, etc.). The reference point interface between TA and NT is defined as a ratio of S(B) and T(B) and is based on ATM connection. The ATM offers the same data capacity at S(B)/T(B) as that at the U(B) interface. The ratio of the S to T interface corresponds to the ATM interface of NT.

Each module of the **network equipment** segment offers a specific function and the fiber connection terminals at NT. The remote multiplexing (RM) module multiplexes several user interfaces onto a single fiber and sends data to the access node via the reference point U(B). This reference point is on both sides of remote multiplexing, indicating the same interface with NT and the access node.

The access node (AN) module multiplexes (statistically) various user's services onto a small number of high-speed fibers which carry the data from the access node to service node. The access node may also include video switches and a distributed metropolitan area network (MAN). The interface between the access mode and service mode is defined by a reference point interface M(B), which is also known as a multiplexed broadband interface (MBI). Thus, the access node performs the functions of multiplexing, concentration, and switching, while the service node (SN) performs the functions of switching and most of the service logic and control point for **operations, administration, and maintenance (OA and M)**. The CCITT recommendation M.610[36] defines OA and M aspects as "the combination of all technical and corresponding administrative actions including supervision actions intended to retain an item in, or restore it to, a state in which it can perform required functions." This recommendation defines the following functions for the OA and M:

- Performance monitoring
- Defect and failure detection
- System protection
- Failure or performance information
- Fault localization

The reference points U(B) and M(B) have some common functions such as multiplexing but differ in a few aspects. The reference point M(B) also carries signaling channels between the access node and service node, while the U(B) interface carries only information data. The NNI interface carries the data from the service node onto the network via the transmission medium. Maintenance terminology and definitions are defined in CCITT M.60.[37]

19.8.5.6 Functional model of B-ISDN

The control of B-ISDN is based on common channel signaling (CCS), and the use of Signaling System No. 7 (SS7) within the network enhances the support for expanded capabilities at a higher speed. Similarly, the user–network control signaling protocol used is an enhanced version of I.451/Q.931.[45] The B-ISDN must support 64-Kbps ISDN (both circuit- and packet-switching), while at the UNI, these capabilities will be provided with the connection-oriented ATM facility.

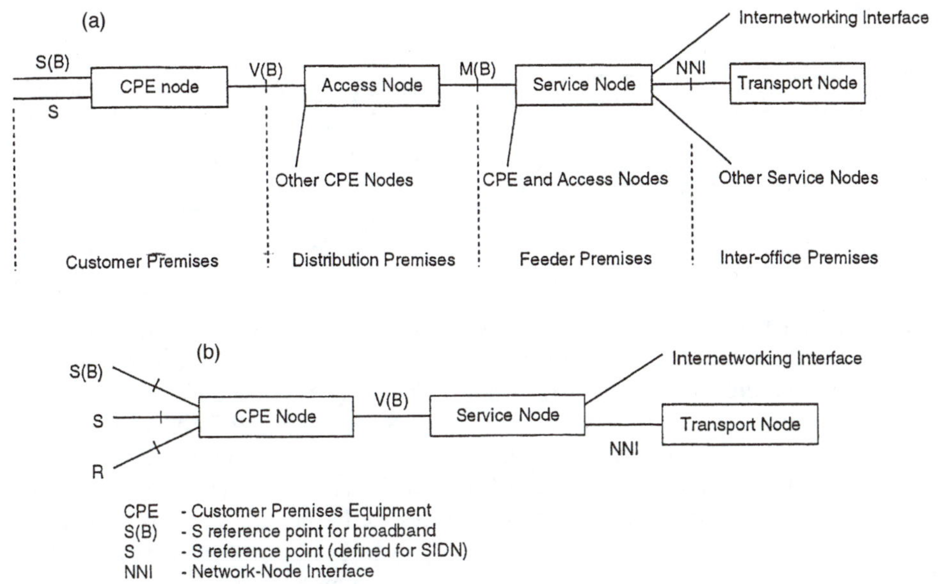

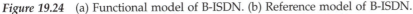

Figure 19.24 (a) Functional model of B-ISDN. (b) Reference model of B-ISDN.

A more detailed functional reference model of B-ISDN is shown in Figure 19.24(a,b). Here, the three modules of the customer premises segment are combined together in one module known as **customer premises equipment (CPE)**. This module is one essential block in B-ISDN and is also known as a **customer premises network (CPN)**. This will have reference point interfaces [S, S(B)] at the input. The S interface is already standardized to connect terminals of 64-Kbps ISDN as defined in Reference 48. It may be used in cell-structured broadband U(B) interface (using SONET frame), S(B) optimized for broadband service, and has ATM structure. Here S is derived from ISDN architecture, while S(B) corresponds to broadband. The module interacts with the access node for ATM-based B-ISDN information (using SONET), which goes through reference interface points U(B), M(B), and network node interface (NNI). The access node may also get information from other B-ISDN services, while the service node may also get information from other access nodes or even CP nodes and offer outputs to other service nodes.

It is also possible to have a direct transfer of ATM-based B-ISDN information to the service node by bypassing the access node. The service node carries the information either to interworking interfaces, transport nodes, or service nodes. The input to the CPE node for direct interaction with the service node corresponds to S and R interfaces of ISDN and S(B) of B-ISDN. The input reference points and CPE nodes belong to the customer premises. The reference point U(B) between the CPE node and access node defines distribution premises, while the reference point M(B) between the access node and service node defines the feeding of information premises. From the service node to other interfaces or nodes via NNI reference point basically defines interoffice premises.

The CCITT I.413[34] has defined two types of physical configurations for CPE or CPN. In the first type of topology, all the terminals are directly connected to NT2 by respective dedicated lines. Each terminal is connected by a separate line. In the second topology, the terminals are connected to a common shared medium (e.g., dual-bus), which in turn is connected to an NT device.

We can have a service node and transport node (optional) common to both configurations as shown in Figure 19.24(a) and Figure 19.24(b).

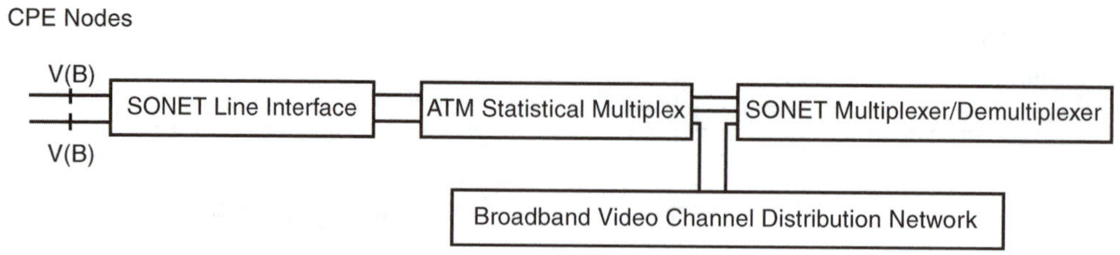

CPE Nodes

V(B) —— SONET Line Interface —— ATM Statistical Multiplex —— SONET Multiplexer/Demultiplexer
V(B)
Broadband Video Channel Distribution Network

Figure 19.25 SONET access node architecture.

The customer premises architecture usually performs switching at the **network termination/terminal adapter (NT/TA)** interface and uses IEEE 802.6 multi-point bus configuration. This configuration uses IEEE 802.6 protocols for the switching and shared-media access control sublayers. The access node accepts the SONET/ATM information at U(B) from the CPE nodes and provides SONET line interface before these are statistically multiplexed by the ATM statistical multiplexers. The broadband video channel distribution network provides multiplexed video channels to the SONET multiplexer/de-multiplexer along with ATM multiplexed services. The SONET multiplexer/de-multiplexer sends the ATM-based information to the service node over SONET. Access node architecture is shown in Figure 19.25.

The service node, along with associated logic circuits and processing units, provides the main functions of B-ISDN: switching and processing. These functions of B-ISDN provide support for wide-area connectionless services, connection-oriented narrow-band and broadband data services, video and image distribution, switched video services, and other ATM-based services. The service node receives the ATM-based information from the CPE nodes connected directly at U(B) and information from the indirectly connected CPE nodes at M(B) reference interface at the SONET data rates of 1.2 Gbps (OC-24), 1.8 Gbps (OC-36), or 2.4 Gbps (OC-48) on interoffice links. This node contains video channel switch (handling video TV channels), narrow-band switch (handling ISDN services), and ATM switch (handling B-ISDN services). Other SONET optical interface rates and formats can be found in Reference 16.

The narrow-band services are sent over narrow-band networks via appropriate interworking interfaces (e.g., DS1 and DS2 circuit networks). The ATM-based services can also be sent over narrow-band networks in addition to their normal transmission over transport node via trunk interface and SONET multiplexer/de-multiplexer. Finally, the video channel services are sent over video-headend interface. The narrow-band services include 64-Kbps and X.25 packets, while broadband services include ATM, video channel, etc. Interworking interfaces for the narrow-band network include DS1 and DS3 circuit networks. The interface between service nodes (directly or via transport) is usually defined by reference interface NNI.

19.8.5.7 Physical layer interface standards

SONET provides transmission (encoding, decoding) and multiplexing for backbone networks at a very high data rate range of 51.84 Mbps to 2.4 Gbps. The physical layer of SONET is given by the STS/OC family. The **STS (synchronous transport signal)** defines an interface to electrical signals, while the OC defines the **optical carrier** based on light. Various standards for the physical layer interface are listed in Table 19.6.

Europe uses the standards shown in Table 19.7 for the physical layer interface (of SONET). CCITT G.707 and G.708 have defined standards for the physical layer interface (of SONET) based on synchronous digital hierarchy (SDH), as given below:

Table 19.6 Physical Layer Interface Standards

Standard	Data rate
STS-1 and OC-1	51.840 Mbps
STS-3 and OC-3	155.52 Mbps
STS-12 and OC-12	622.08 Mbps
STS-24 and OC-24	1244.16 Mbps (1.244 Gbps)
STS-48 and OC-48	2488.32 Mbps (2.488 Gbps)

Table 19.7 European Standards for Physical Layer Interface

Standard	Data rate
E0	64 Kbps
E1	2.048 Mbps
E2	8.448 Mbps
E3	36.368 Mbps
E4	139.264 Mbps

SDH1 — 155.52 Mbps (same as OC-3)
SDH4 — 622.08 Mbps (same as OC-12)
SDH16 — 2488.32 Mbps (same as OC-48)

Some NNIs include ATM-based, STS-3 (155 Mbps), and STS-12 (620 Mbps) SONET framing. The NNIs are multiplexed to offer very high data rates of SONET's OC-24, OC-36, and OC-48 over interoffice links. Further details on these hierarchies and interface rates and formats can be found in CCITT G.707, CCITT G.798, and CCITT G.709[20-22] and ANSI standard T1.105-1988 SONET.[16] Articles on SONET standards can be found in References 7 and 53–54.

Table 19.8 lists all the access interfaces (UNI) and trunk access interfaces (NNI) along with their speeds. Although some of these interfaces have been described earlier, we show them here in tabular form for ready reference.

Table 19.8 Transmission Speeds of UNIs and NNIs

Interface	Speed of transmission (Mbps)
User–Network Interfaces (also known as access interfaces)	
DS1 ATM UNI	1.544
DS3 ATM UNI	45
STS-3c ATM UNI and STM-1 optical interface	155.520 (supported on OC-3/SDH)
DS1/DS3 circuit emulation (SAC)	1.544/45
NTSC video codec interface	3–12
DS1/E1 SMDS SNI	1.544/2.048
DS3/E3 SMDS SNI	45 and 34 (Class 1–5)
DS1/E1 frame relay	1.544/2.048
Network–Node Interfaces (trunk interfaces)	
DS3 ATM NNI	45
STS-1 ATM NNI	51.84 (supported on OC-1 optical interface)
STS-3c ATM NNI	155.52 (supported on OC-3/SDH STM-1 optical interface)
DS3 ICI (Bellcore for SMDS)	45
DS1 frame relay NNI	1.544

Broadband ISDN is a logical extension of ISDN and is also known as an **integrated broadband communication network (IBCN)**. Due to the introduction of a large number of transport services with different requirements, there is a need to define a uniform global network which can support different types of services. Further, the global networks are also using high-speed switching and multiplexing (due to semiconductor and optical technology advances). As a result of these advances, both in flexible network and conceptual system design, we have seen a new multiplexing and switching technique known as asynchronous transfer mode (ATM) in the communication network. One of the main objectives of these advances has been to define a universal global network which can provide a flexibility to reconfigure to any new architecture to meet the requirements and support a variety of services (voice, data, video, etc.).

The existing networks offer different services, and specific services are available on specific networks (based on the application). Each of the networks uses a different concept for switching, multiplexing, and transmission media. For example, the telex network is based on the store-and-forward principle and is used for transmitting telex messages. The computer data can be transmitted over either a packet-switched data network (PSDN), e.g., X.25, or a circuit-switched data network (CSDN), e.g., X.21. Telephonic conversation takes place across a public switched telephone network (PSTN), which is based on circuit switching. Television transmission takes place over different TV networks, e.g., cable television (CATV) and broadcasting from ground antenna as well as satellite (direct broadcast satellite).

Each network operates on a different concept and supports different services. But defining a global universal network will offer flexibility in adapting to changes or adding new services, efficient use of resources, service independence, experience, etc. Although it is a very difficult task, if it can be implemented it will be a big breakthrough in the existing computer communication network technology.

The existing networks for different services are not utilizing their resources efficiently, and users are usually confined to using a specific network for a specific service. For example, ISDN allows integration of voice and data only for a small bandwidth of about 2 Mbps, and hence it is not suitable for TV signals (requiring a 6-Mbps bandwidth). Even within ISDN, two types of switching are used, and as such we have two overlay networks; integrated service over an ISDN network is primarily for providing access to the users and not the services. The transmission media or transport mechanisms used in these networks are also suited to specific services. For example, transmitting voice data over an X.25 network will be very inefficient (due to jitter noise, transmission delay, etc.), but at the same time, the data can be transmitted over PSTN via modems.

19.8.5.8 Classes of services for future networks

To provide integrated services, the CCITT I.211[27] has identified two classes of services for future networks: **teleservices** and **bearer services**. The teleservices are usually defined by higher-level functions which are performed by communication devices (terminals, interface, etc.). When we compare the teleservices with OSI-RM to find out which layer provides a particular service, we find that teleservices use all the layers of OSI from the transport and its higher layers. Various teleservices include voice communication, telex communication, telefax communication, computer conferencing, videoconferencing, broadcast video, audio, and many other services.

The bearer services deal mainly with various parameters used for transmission of data. These parameters define different types of techniques/methods used for that particular set of parameters. Some of the parameters include data transfer switching (circuit, packet, asymmetry, symmetry, etc.) access control strategies (access control protocols, channel rate access, etc.), and a set of general parameters defining the quality of service (QOS).

One very critical function for the design of a network to adapt new services is the *semantic transparency*, which offers a guarantee for correct delivery of the packet to the destination. The data link protocols usually provide error control to achieve an acceptable QOS of the network, which has been defined as the probability of errors in the message by CCITT in its recommendations. For example, Q513 for ISDN discusses this issue of error control.

In the existing packet-switched networks, the acceptable end-to-end QOS in the presence of errors is performed by data link protocols such as high-level data link protocol (HDLC) and link access protocols (LAP) for the B and D channels. These protocols include frame delimiting, bit transparency, error recovery, error detection, and other features which together provide semantic transparency within the network.

Cyclic redundancy check (CRC) and other error-detection protocols are available which, along with re-transmission, provide error control for the packets. In this scheme, the error control is provided for end-to-end communication on every established link and in the literature is known as **full error control (FEC)**. This FEC is required due to poor quality of transmission media and inefficient implementation of switching and multiplexing.

With the advent of ISDN for narrow-band services, the quality of transmission switching was slightly improved, and this caused less error within the network. As a result of this, functions such as delimiting, bit transparency, and error checking were required only on a link-by-link basis (similar to core functions), while error recovery (detection and re-transmission) was required on an end-to-end basis (similar to optional). Using a layered approach, we can say that these functions (core and optional) together will provide full error control (FEC), while core functions will provide only **limited error control (LEC)**. This concept of defining core and limited functions within the same layer is also known as *frame relaying*. Here we can consider optional and core functions to be part of the data link layer protocol.

In B-ISDN, the core functions are moved to the physical layer (edge of the network) using cells (instead of packets), thus making the network more nearly error-free. This concept of moving these functions to the edges of the network is known as ATM. This reduces the performing functions for error control in ATM networks. A typical layered architecture protocol model of B-ISDN is shown in Figure 19.26. The conceptual issues associated with ATM and layered architecture ATM networks are derived partially from References 48–54.

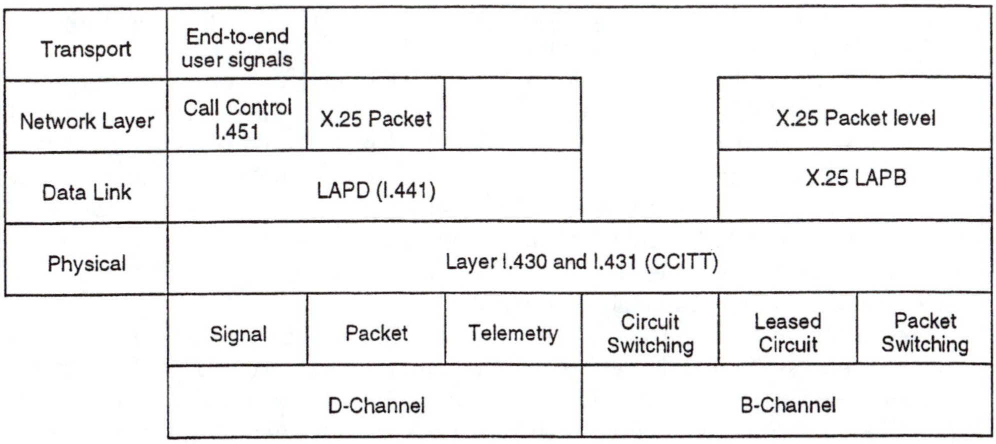

Transport	End-to-end user signals					
Network Layer	Call Control I.451	X.25 Packet			X.25 Packet level	
Data Link	LAPD (I.441)				X.25 LAPB	
Physical	Layer I.430 and I.431 (CCITT)					
	Signal	Packet	Telemetry	Circuit Switching	Leased Circuit	Packet Switching
	D-Channel			B-Channel		

Figure 19.26 Layered architecture and protocols of B-ISDN.

In X.25, FEC requires loading a complex protocol at each node, while in the case of LEC, frame relaying requires a less complex protocol than that of X.25 at the switching nodes. In ATM, the nodes have a very simple protocol and hence support a higher bit rate (600 Mbps).

19.8.5.9 Link access protocol D (LAP-D) channel

The LAP-D channel corresponds to layer 2 and defines a logical link between peer layer 3 entities for exchanging information. Various functions provided by LAP-D include establishing multiple data link connections between nodes, frame delimiting, alignment, sequencing control, error control, flow control, and error recovery (although a major part of this function is also supported by higher layers). The LAP-D protocol is used for all the traffic over the D channel.

The LAP-D protocol mainly provides link control functions, e.g., flow control, error detection, and error control. LAP-D provides services to the network layers and supports multiple entities of layer 3 (e.g., X.25 and I.451). It provides two types of network services: **unacknowledged information transfer service** and **acknowledged information transfer service**. The former service provides support for point-to-point and broadcast connection with no support for flow control or error recovery and no guarantee of delivery of packets, but all services are performed without acknowledgment. The latter service is similar to the one offered by LAP-B (link access protocol-balanced) and HDLC and provides error recovery and flow control, since it is based on a connection-oriented connection which includes three phases: connection establishment, data transfer, and connection de-establishment. A logical connection is established based on the call request. During the connection establishment phase, both users exchange information frames with acknowledgment before the data can actually be transferred over it.

The unacknowledged service provides data transfer of data frames for layer 3 (equivalent to the network layer of OSI-RM) in a numbered fashion (as it requires an acknowledgment). The protocols for this type of service are not expected to provide any error control or flow control, as it will be provided by the higher layers. If any data frame has an error, it will simply be discarded. The acknowledged service, on the other hand, provides data transfer of data frames in a numbered fashion (as it requires an acknowledgment). The protocols for this type of service provide error control and flow control. Both of these services provide services to layer 3 and have the same frame format as that of LAP-D (discussed below). A detailed discussion of these versions of LAP protocols was presented in Section 11.8 (Chapter 11). We describe these versions of LAP protocols here in brief, as these are used in all high-speed networks and we will be referring to them frequently, with more emphasis on their use in ATM-based high-speed networks.

The LAP-B protocol was defined for X.25 and HDLC, and later on another version of LAP was known as LAP-D, which provides two types of functions corresponding to the two services supported by it. The first function within the protocol, unacknowledged, does not provide any error control or flow control. The error-detection control function discards the damaged frames. The second function, acknowledged, includes error control and flow control, and the numbered frames are transmitted and acknowledged. The acknowledged function enables the LAP-D to establish multiple logical LAP-D connections (similar to multiple virtual circuits in X.25).

A typical frame format of LAP-D is shown in Figure 19.27. As usual, the flag field provides delimitation for the frame and includes a unique pattern of 01111110 at both ends of the frame. The flag field also provides synchronization between the sender and receiver and provides bit stuffing when used with Signal System No. 7. The address field defines two sub-field addresses within it: **terminal endpoint identifier (TEI)** and **service access point identifier (SAPI)**. The TEI address corresponds to the user–device interface which

Flag	Address	Control	Information	FCS	Flag

Figure 19.27 LAP-D frame format.

is shared by a number of users at the subscriber's site. The SAPI address corresponds to an address of the device interface which provides different services over it defined by the interface (typically a TEI), e.g., call control, packet-switched communication facilities for X.25 level 3 and I.451, and exchange of magnet information of layer 2. TEI and SAPI together are used for establishing a logical connection at layer 3, where each SAPI has a unique layer 3 entity for a given device interface addressed by TEI.

The control field defines three types of frames: **information (I)**, **supervisory (S)**, and **unnumbered (U)**. The frame check sequence (FCS) field defines a code used for error detection. This code is obtained by performing an algorithm (corresponding to a chosen protocol) on the entire frame (excluding the flag field). A standard code used in the FCS field is the CRC-CCITT code of 16 bits.

LAP-D uses seven-bit sequence number fields and piggybacking for the acknowledged function. The maximum window size supported is 127 and uses the Go-Back-N ARQ version for error detection and correction. The address field uses 16 bits, and bit stuffing is used in the flag field for synchronization.

Other protocols include SS7 signaling link level, LAP-B, and X.25 level 3. Most of these protocols use piggybacking, window sizes of 7 or 127, CRC-CCITT codes, and the Go-Back-N version of ARQ protocol for error detection and correction. X.25 level 3 is the only protocol which does not use stuffing and the CRC code, while the SS7 signal link level does not define any address field in the frame.

The frames defined in the control field (I, S, and U) are used for providing acknowledgment functions (of LAP-D) between TE and the network over the D channel. The acknowledged function requires three phases: connection establishment, data transfer, and termination/de-establishment of connection. In contrast to this, the unacknowledged function does not provide any error control or flow control and as such does not use any Ack frame. The user's data is sent in the user's information field via an appropriate LAP-D entity.

IEEE 802.9 defines an architecture for D channel signaling protocol which typically accepts the data frame from the DTE (IEEE 802.9 IV) controlled by the station management layer. This data is sent to LAP-D by the network layer Q.931 D channel protocol (standard call control for IEEE 802.9). The LAP-D, after defining the data frame, transmits it onto the physical layer over the D channel established between the DTE and its peer node. The Q.931 provides the following connection functions: establishment, maintenance, and termination. It defines a network connection across ISDN for accessing both BRI and PRI between communicating application devices.

Other functions offered by Q.931 include routing and relaying, user-to-network and network-to-user exchange of information, error detection, error recovery, sequencing, congestion control, etc. The following is a list of messages exchanged between the user and the network during the data communication.

Call establishment messages
- .Alerting
- .Call proceeding
- .Connect
- .Connected acknowledge
- .Setup

Call data transfer phase
.User's data transfer
Call clearing messages
.Disconnect
.Release
.Release complete
Miscellaneous messages
.Status
.Status enquiry

The connection establishment over the B channel is obtained by a set of procedures (control, signaling, etc.) also termed the D channel layer 3 interface in CCITT's recommendations I.450 and I.451. This recommendation defines a common interface to ISDN which is being used by both B and D channels. Further, end-to-end control and signaling are provided by I.450 and I.451 over the D channel.

The ISDN is used as a data transporter within OSI-RM, providing user-to-user communication where ISDN services are mainly used by the protocols to provide user-to-user protocols between the sender and receiver. The users access the physical layer recommended protocols of both the basic rate interface and primary rate interface data link layer LAP-D and I.451 call control protocol. The network uses a set of protocols which are transparent to the user and provide various services to the user. This set of protocols is also known as Signaling System No. 7 and provides a common channel signaling function with ISDN.

19.9 Asynchronous transfer mode (ATM) operation

For telecommunication and communication networks, the transmission scheme is the most important component and should be able to provide support for switching, signaling, multiplexing, and various transmission and communication configurations, etc., in the network. As we know, the current trend in communication technology is toward digital signal transmission due to the obvious reasons of speed, brightness, higher data rate, lower error rate, reliability, etc. The networking infrastructure of existing network service providers usually offers the services for narrow-band needs. The predominant services are offered at 56 Kbps or 64 Kbps, with new switched services up to T1 carriers (1.544 Mbps). Within this infrastructure, the applications requiring voice, data, and videoconferencing are based on circuit switching and channelization of the services. These services are available on a variety of overlay networks, e.g., public switched analog networks, private-line analog networks, private-line digital data service networks, dedicated high-speed T3 digital networks, and narrow-band ISDN (N-ISDN), and now we see a big push for high-speed networks.

Digital technology: With digital technology, most of the networks are shifting toward digital transmission over the existing cables. The digital links so defined are supporting the addition of overhead bytes to the data to be transmitted. This payload is a general term given to a frame (similar to the one defined in LANs). The frames used in LANs carry a lot of protocol-specific information such as synchronization, error control, flow control, etc. In the case of ATM networks, the data is enclosed in these frames (if available). The frames are generated at a rate of 8000 frames per second, and digital links offer a very high data rate. A commonly used digital link in the U.S. is the Synchronous Optical Network (SONET). The basic rate is defined as STS-1, which offers 51.84 Mbps, while STS-3 offers a data rate of 155.52 Mbps.

The high-speed networks based on high-speed fiber optics include switched multi-megabit data service (SMDS), frame relay, ATM, B-ISDN, etc. With the advent of ATM and its acceptance in B-ISDN high-speed networks, various applications are being developed or are under consideration, including distributed multimedia (sales, engineering, support organizations, etc., using multimedia applications over ATM), distance learning (video broadcasting of training sessions will be cheaper than existing satellite technology), video documentation (new CD-ROM technology documenting a large volume of video for sharing over high-speed and high-bandwidth ATM networks), medical imaging (transmission, sorting, and sharing data from medical imaging devices via a high-performance network within premises which may be extended for WANs based on ATM networks), etc.

Both CCITT and ANSI have adopted ATM protocols and architecture. ANSI is responsible for extending ATM protocols within the U.S., while CCITT is responsible for those within Europe. In 1991, an ATM forum composed of Cisco, NTI, U.S. Sprint, and Adaptive was formed (within the U.S.) with a view to using ATM products and services for connectivity to different types of networks. With a remarkable growth in membership and other activities, many European and Japanese companies have also joined the forum. In 1993, three documents were published: ATM user–network interface (UNI) specification, version 3.0; ATM data exchange interface (DXI) specification, version 1.0; and ATM broadband ISDN intercarrier interface (B-ICI) specification, version 1.0.

The transmission of different types of signals due to data, voice, video, images, etc., depends on the speed of the network. Each of these signals has different requirements for the bandwidths, speed, and mode of transmission. If we can combine the concepts corresponding to the advantages of the two commonly used switching techniques, circuit and packet-switching, then we can define a new environment of supporting various applications at high speeds (based on digital signal transmission) on networks of different sizes and possibly eliminating any distinction between the categories of networks (local area networks, metropolitan area networks, or wide area networks). This new environment has been used in many of the high-speed networks. Here, the concepts of packet-switching (larger bandwidth) and circuit switching (minimum delay) are integrated and used together.

In ATM networks, the packets containing information are transmitted asynchronously, based on fast packet-switching and delay features defined in circuit switching. The research center of Alcatel Bell in Antwerp, Belgium in 1984 introduced the concept of ATM. CNET and AT&T also published initial ATM-related documents. ITU has accepted ATM as a switching mechanism or transfer mode for B-ISDN. It offers both continuous data rates and variable data rates for any type of network traffic.

Features: ATM offers flexibility, efficient use of resources, dynamic allocation of bandwidth (bandwidth on demand), and a simple universal network with reduced transport costs. The header has very little function to perform, and processing of the header does not cause any delay. The network traffic from different sources with different data rates is mapped into fixed-size cells. These cells are then packed into a large transmission pipe using statistical multiplexing. This avoids the wastage of bandwidth. The transmission pipe includes voice, data, and video and hence supports variable data rate traffic. The ATM switch is informed only of the incoming port and outgoing port for the cells. The decision to standardize the size of ATM cells to 53 bytes was made by ITU-T in June 1989 in Geneva. The packet-switching X.25 supports packet re-transmission, frame delimitation, and error checking, while ATM does not support any of these functions.

19.9.1 Switched networks

There are four main types of switched networks which provide a variety of services to users: **circuit-switched**, **message-switched**, **packet-switched**, and **now fast packet-switched networks**. Each of the switched networks uses a different scheme of switching and multiplexing and is used in specific applications. For example, circuit-switched networks are useful for telephone conversation and data communication (at a very low speed). Based on applications, there are two categories of circuit-switched networks: **public telephone switched network (PSTN)** and **telex circuit-switched network (TCSN)**.

- **Circuit-switched networks** provide connection-oriented services and operate in three phases: connection establishment, data transfer, and connection de-establishment/termination. A typical circuit-switched telephone network may have switches as well as switched and leased lines from a private branch exchange (PBX). The switched and leased lines provide dedicated and permanent connections with the PBX, while other PBXs are connected by PBX switches. In contrast to this network, the packet-switched telephone networks allow PCs to be connected to the switching nodes of data networks via modems. The switching nodes of data networks are connected to each other based on predefined topology.
- **Message-switched networks** use the concept of **store-and-forward (SAF)**, where the incoming messages are stored temporarily in the buffer of the switching node. The node searches for a free link and, as soon as it finds one, forwards the message to another switching node. This node performs the same operations as the previous node (based on store-and-forward), and this process continues until the message is delivered to the destination node. This type of switching is very useful in the networks of teleprinters. The size of the message is variable.
- **Packet-switched networks** also use the concept of store-and-forward (as in message-switched networks), with the only difference being that here the messages are defined in the forms of packets of fixed size. The first experimental network introduced (ARPANET) is, in fact, based on packet-switching, and other wide area networks including X.25 are also based on packet-switching.
- **Fast packet-switched networks** are a hybrid approach of using both circuit switching and packet-switching in an attempt to take advantage of the low delay and fixed bandwidth offered by circuit switching and the effective utilization of resources offered by packet-switching. The fast packet-switched (FPS) networks provide support for the integrated services of data, audio, video, images, etc., onto one common physical channel. This switching is basically introduced for making packet-switching capable of supporting all switching within it. It includes various switching techniques and has minimal functions. This concept has been incorporated in various techniques with different names in different countries.

The FPS technique is very popular in the U.S., while **asynchronous time division (ATD)** was originally used in CNET and later considered by Europe. The standards organization CCITT adopted a new name for this technique: **asynchronous transfer mode (ATM)**. The fast packet-switching technique offers a higher data rate than packet-switching, while ATD and ATM techniques use asynchronous operations between sender and receiver clocks and the difference between clock speeds is minimized by inserting "Dummy" packets which do not carry any user information. All of these techniques (FPS, ATD, ATM) offer a uniform transport environment to various services irrespective of their characteristics, behaviors, requirements, etc.

With the introduction of the high-speed transmission medium, optical fiber (offering a data rate of 100 Mbps), a new trend of defining the transfer modes (combination of switching and multiplexing), was initiated. This resulted in two classes of transfer modes: **synchronous transfer mode (STM)** and **asynchronous transfer mode (ATM)**, as discussed above. The ATM transfer mode has been adopted by B-ISDN networks to provide integrated services (data, audio, video, images) to the users. The STM is a conventional time-division multiplexing scheme where each multiplexed channel is distinguished by its time position.

Asynchronous transfer mode cell (ATM) switched network: ATM switching defines a transfer mode where fixed- and variable-length information units with headers are multiplexed onto a transmission medium. Each multiplexed channel is distinguished by the attached header. The ATM switching that each channel does not have any limit on its speed other than the limit defined by the transmission medium. It does not support error control and flow control, and we do not have to synchronize the terminals with that of the network for data communications. The ATM switching provides translation of VCI and VPI at the switching node or cross-connect node (ATM switching node). It also provides cell transportation between the set of input lines into dedicated output lines. The physical resources are used on demand and are available only when the data frame is to transmitted. For these reasons, ATM offers higher data rates and also accommodates both circuit and packet-switching onto one transmission medium.

ATM offers two types of services: **permanent virtual circuit (PVC)** and **switched virtual circuit (SVC)**. The PVC is equivalent to a private line or LAN, while SVC is equivalent to voice communication over a switched telephone network. PVC, as the name implies, offers a dedicated circuit or line for any type of services based on the bandwidth-on-demand concept. There is no need to establish the circuit or line. In contrast, SVC service needs to be requested by dialing up the number of the server. Once the call is established, 64 Kbps (provided by telephone network) will be allocated to the user by default.

ATM cells: The ATM network does not distinguish between the types of networks and provides very high data speeds of 100 Mbps and greater to various types of application signals such as data, voice, video, images, etc. The ATM integrated switching technique is generally known as a **cell-based switching technique** and offers connection-oriented services to these applications. It offers a single link-by-link cell transfer capability which is common to all applications. The higher-layer information is mapped into the ATM cell to provide a service-specific adaptation function on an end-to-end basis. The main feature of cell-based networks is the packetization/de-packetization of a continuous bitstream into/from ATM cells or segmentation/reassembly of larger blocks of user information into/from ATM cells of fixed size over a connection-oriented configuration. It also supports the grouping of several virtual channels into a virtual path, which in turn is grouped into a transmission path.

A fixed-size cell length of 53 bytes is defined, out of which five bytes are used for the header (per CCITT's recommendations). In ATM switching, the switching functions are carried out by routing ATM multiplexed data frames, which depends on the header of the individual data frame. This is exactly what is done in packet-switching, where the packets are routed based on the control information in its header field. ATM supports both types of data frames or traffic (continuous and bursty).

In order to send the information (data, video, graphics, etc.), we need only ATM cells. The number of available cells depends on the number of frames per second and the data rate of the link. If we use an STS-1 link, then we may have about 15 ATM cells, as each cell has an overhead of five bytes. The frames are generated at a rate of 8000 frames per second, and digital links offer a very high data rate. The commonly used digital link in the U.S. is the Synchronous Optical Network (SONET). The basic rate is defined as STS-1, which offers 51.84 Mbps, while STS-3 offers a data rate of 155.52 Mbps. The number of frames per second is 8000. The number of data bits per cell is $48 \times 8 = 384$, while the number of header bits per cell is $5 \times 8 = 40$. The number of cells for a given rate of 51.84 Mbps is 15. Similarly, for STS-3, the number of cells per frame is about 44, and so on.

When ATM switching is used for continuous switching, it offers flexibility by supporting various data rates, e.g., B channel — 64 Kbps, H0 channel — 384 Kbps, and H1 channel — 1.536 Mbps or 1.920 Mbps, for continuous traffic in B-ISDN. It offers unified ATM multiplexed channels which can carry different types of data or traffic, and the ATM switch simply deals with the individual channel speeds; i.e., it deals with the header only.

Other channel speeds can be selected from the same configuration. The ATM switch offers a very small switching delay, and the memory requirements are also small. In the case of bursty traffic, ATM offers packet-switching, which is different from that used in X.25 networks. ATM packet-switching offers high-speed simple data transfer protocols, fixed length configurations, and high-speed capacity derived from hardware switching.

The network sets up a virtual circuit (as is done in circuit switching) before the transmission of data and, once it is established, sends all the cells (in order) across the network. Note that this is exactly what is done in frame relay networks, where frames are transmitted over the virtual networks. The frame relay network uses the packet-switching technique and offers high bandwidth, but delay (setup time, transmission of signals, propagation delay, etc.) sometimes becomes a serious problem. The frame relay network defines fixed-size cells by their switches, but these cells interact with routers using frame relays.

These networks based on ATM, ATD, or fast packet-switching provide no flow control or error control, and also no dynamic actions are defined for the lost packets. This is in contrast to ARQ protocols which re-transmit the lost or error-containing packets. However, preventive actions for error control are supported by these networks. This does not become a major drawback, as these networks work in connection-oriented mode. The establishment of connection during the setup phase is defined (based on the request received by the networks) before the user data can be transmitted. After this phase, the user data will be transmitted only when the resources are available. After the user data transfer is over, the resources are released. This type of connection-oriented mode also provides an optional quality of service (QOS) and less probability of packet loss.

The probability of losing a packet during transmission is very low (typically 10^{-8} to 10^{-12}), and also the allocation of resources is made available with an acceptable probability. For the fast processing of the packets within the network, the ATM header offers very limited functions (as opposed to a variety of functions being provided by the header of the packets in LANs, N-ISDN, etc.). The functions which are supported by the ATM header include the identification and selection of the virtual connections and the multiplexing of these virtual connections over a link. The small size of the information field reduces the size of the buffer and also helps in reducing the delay. Due to these features, ATM networks support a data rate of 150–600 Mbps.

ATM cell transmission: The transmission of ATM cells through ATM networks goes through various ATM-based components, e.g., ATM multiplexers and cross-connect nodes. Obviously, we have to provide synchronization for proper communication for the ATM cells. The transfer of ATM cells follows the following sequence of functions: generation of cells, transmission of cells, multiplexing, switching nodes, and switching of cells. The generation of cells deals mainly with the mapping between non-ATM information and ATM information. If the B-ISDN terminals are used, the information is inherently mapped onto ATM cell formats and hence no addition mapping is required. The network terminating 2 (NT2) functional group supports both non-ATM and ATM interfaces, and mapping is done at NT2. The customer premises equipment (CPE) supplies information to NT2 at S or R and gives it to NT1 at T.

The STM multiplexer multiplexes different signals originating from B-ISDN terminals onto a single access line of $N \times 155.520$ Mbps, where N represents the number of inputs of the multiplexer. The STM multiplexer does not process the signal payload carrying idle cells. On the other hand, the ATM multiplexer checks the idle cells and provides cell concentration by offering a high access rate at output as 155.520 Mbps, which is shared by the users depending on their applications. Thus, it removes any idle cell (erroneous, incorrect, invalid) and multiplexes only valid cells into one STM-1 frame.

The VPs are translated by VP switch, which provides a mapping between an incoming VP and a dedicated outgoing VP. The **ATM switch** (also known as a cross-connect node) concentrates on ATM traffic, since idle cells are not transmitted. It separates the user's traffic which is designated for the local switch from the one to be transmitted over a fixed route through the network. The ATM cells are transported over a variety of transmission systems. For a user–network interface, the reference points S(B) and T(B) offer two different options for transmission: one is based on synchronous digital hierarchy (SDH) and the other is based on pure multiplexing of cells.

Pleisochronous digital hierarchy (PDH) provides data bit rates of 2 Mbps, 34 Mbps, and 140 Mbps (popular in Europe), and 1.5 Mbps and 45 Mbps (popular in North America). These data rates and hierarchies were discussed earlier. The ATM cell transport must offer synchronization at ATM multiplexers or the ATM switch, and it carries bit timing and cell timing control signals. The synchronizer at the ATM switch adapts cell timing on all incoming signals (including STM-based audio and video) and can adjust to different statuses of cells (idle cell, stuffing, etc.). The clock timing control information allows the use of the existing clock distribution network from 64-Kbps ISDN. It is interesting to note here that there is no need to have the transmission link synchronized.

Bandwidth on demand: ATM offers a different bandwidth to the users (based on their demand) on the same digital link. The cells are not assigned to one user; instead, they are assigned to the data to be transmitted. Multimedia-based applications such as HDTV with Motion Picture Experts Group (MPEG) 3 compression may need about 30 Mbps. Based on this information, the number of cells will be allocated to this application. Since frames are generated at the rate of 8000 per second, it does not matter that much, even if a user gets one frame, as he/she will be getting another frame after 0.125 second. As said earlier, since ATM supports data, video, audio, graphics, and images, there might be a voice packetization delay which is equal to the number of voice channels (48) in the cell (DS-0 voice channel generates one byte in 125 μs (8000 per second). Thus, the delay for 48 bytes will be $48 \times 125 = 6$ ms. The round-trip delay for voice communication will be at least 12 ms. In addition to this delay, there will be propagation delay, which may cause a significant delay for voice communication. This delay is still acceptable over the bandwidth of a 64-Kbps voice channel.

With these switching techniques and advances in other areas — e.g., coding algorithms, VLSI technologies, data transmission switching, etc. — it seems quite obvious that high-speed networks based on ATM or any other fast switching techniques must support all services of smaller bandwidths and optional sharing of the resources among all the services, and they must define a single unified and universal network for global connectivity. We need to design this single network in such a way that the controlling and maintenance are managed efficiently and more simply due to the economies of scale. It is expected that the advantages and features of a single network providing a variety of services at a lower rate will benefit telecommunication companies, vendors, customers, operators, etc.

19.9.2 Evolution of ATM-based networks

When we compare the ATM-based networks with other high-speed networks, we find that ATM is hoped to dominate the network market in both private and public networks of this decade. In fact, various vendors of routers and **data service units (DSUs)** for

ATM-based networks have already defined their products, and it is hoped that very soon users will start receiving the services of these networks in both the private and public sectors. The private network based on ATM has already taken a big lead over public networks. In spite of these features and the vision of an ATM-based network, no network standards are defined yet, no public services are defined yet or available, and no standards are set yet for signaling, setup function, and many other performance parameters such as network management, flow control, congestion management, billing management, etc. We can only hope that ATM-based networks will at least not have the same fate as that of integrated services digital networks (ISDNs).

In the literature, we notice that the ATM-based networks are mentioned in conjunction with other networks such as B-ISDNs, SONET, SMDS, and frame relay networks. In the following subsections, we discuss the relationship between ATM-based networks and other individual networks. For more details on ATM networking and various aspects of LAN interconnectivity, see Reference 53.

When the ATM-based network is interfaced with B-ISDN or is compared with it, ATM offers fixed-length cells of 53 bytes, and these cells are switched through an ATM switch. Some of the issues related to services and network management with ATM-based networks are still unresolved. On the other hand, B-ISDN also supports data, voice, video, etc., and transports the data from one location to another, but there are still some issues in B-ISDN that remain unresolved. The B-ISDN is based on fiber transmission, and the asynchronous transfer mode offers new switching and networking concepts to offer new services leaning toward multimedia applications, including video communication. It also allows integration of intelligent network management and networks offering efficient communication management, etc.

The existence of B-ISDN is based on the fact that the services (requiring high bandwidths and higher data bit rates) of existing networks can be accessed for this unified network. Other networks which are integrated with B-ISDN include public data networks, analog telephone networks, 64-Kbps ISDN, TV distribution networks, etc. It offers only interactive types of services at 1.5 Mbps or 2 Mbps. Access to other networks (e.g., 64-Kbps ISDN) will be provided by appropriate interfaces as defined by CCITT I.430 (basic access interfaces).[43]

The B-ISDN services from its terminal equipment (TE) B at 155.520 Mbps (CCITT I.432) can be integrated with 64-Kbps ISDN originating from the TE (of ISDN) at 192 Kbps (CCITT I.430)[43] via an internetworking unit (IWU). The integrated access environment for ATM-based networks can be obtained at the network terminating (NT) B functional group, which accepts the signals from B-ISDN terminals (CCITT I.432) over one link and other signals over another link. The ATM cell packetization is performed at NT(B), and integrated signals are transmitted over 155.520 Mbps to the local exchange. The local exchange separates these two signals (de-packetization), and they are sent over different links. Similarly, the TV distribution networks may be integrated with B-ISDN in either switched or non-switched forms. The TV programs from TV program providers are fed into a local exchange. The exchange provides administrative services such as handling TV programs, maintenance, operations, and other administrative controls. The selection of a particular TV program is available on an access line by the functional group NT over the TV screen and is done by signaling the channel and procedures. Alternatively, the TV programs may be fed directly to the access line and users can select the TV programs by their own TV switch.

A **switched multimegabit data service (SMDS)**–based network basically defines a set of services for metropolitan area networks (MANs) based on the IEEE 802.6 standard and offers network-to-network communication. This is being defined by Bellcore (Livingston, NJ). This service transports data only (as opposed to data, voice, and video in

B-ISDN). The service offered by this network is based on the cell concept and supports a connectionless environment. Here also, the ATM cells and switching concepts are used with SMDS for transporting the data for network-to-network communication.

Synchronous Optical Networks (SONETs) define the transmission mode of signals (synchronous), data rates of optical media, and also the technique of using the bandwidth of optical circuits and, in general, use the envelope technique. Regarding the relationship between ATM and SONET, the ATM cells can be transported via these envelopes by inserting the cells inside the envelopes.

The use of ATM in public networks has already taken a lead over the public networks, and CCITT has defined a 53-byte cell composed of a five-byte header (10% overhead in each cell). The standards for signaling and call setup are still under active consideration. Without these standards, ATM-based networks do not offer any communication between nodes. Further, the currently defined ATM provides SONET data speeds of 100 Mbps. The lowest speed defined for an ATM-based network is SONET OC-3 (115 Mbps). The industry products, services, and standards for Bell's T1 and T3 rates (1.54 and 45 Mbps, respectively) lean mainly toward private networks.

Broadband ISDN includes ATM and offers higher data rates than T1 (1.544 Mbps used in North America) and E-1 (2.048 Mbps used in Europe). ATM offers the following features:

1. No error checking on cell data; it is done at the endpoints of the ATM network. In fact, AAL includes cyclic redundancy check (CRC).
2. During traffic congestion, ATM discards cells. Cells in ATM are usually attached with a cell loss priority (CLP) bit. By setting CLP = 1, the cells may be discarded.
3. ATM carries audio, video, and multimedia. So far no standards have been defined for sending video and audio together on ATM networks. The available networks use the sampling of voice at a regular continuous bit rate of 64 Kbps. The voice with any volume (low, high, or none) will generate a bit pattern of fixed bit-rate application. Video, on the other hand, usually uses compression/decompression techniques before the transmission. The compression techniques may reduce the number of bits by half and even more of a variable-bit-rate application. Thus, sending both voices of video and video over ATM does not offer any better solutions than existing analog transmission over cables or wireless communications.
4. ATM is more efficient in terms of greater flexibility of bandwidth and higher data rates.
5. It is expected that ATM will continue to be available, as it is offering many services today.
6. There is a difference between SMDS and ATM. Bellcore developed SMDS as a way of implementing all the layers of ATM, but its own layered architecture does not correspond to the layers of ATM. SMDS provides connectionless services, while ATM needs a connectionless network access protocol on top of its model. The SMDS services can be provided over ATM networks, and the user interface layer is above the connection-oriented service layer of ATM. The ITU-T and European connectionless broadband data services are working together to merge SMDS with the ATM network.

The building and installation of private networks with ATM switches will start as soon as the industry forum provides a lower-speed definition of the ATM switch, as this can be used as a backbone switch for a private network. This network will offer switching and routing of cells across the networks. The data cell traveling over the network may come from routers, DSUs, other switches, video codecs, PBXs, and other devices. The ATM interfaces to these devices to format the incoming data into ATM's 53-byte data cell and then transmit it to the backbone are being developed by various companies.

Frame relay is another public network specially defined for ISDN that has played a very important role in providing low overhead and supporting bursty heavy traffic services and, in fact, has already become a major service offered by value-added networks (VANs) in the U.S. and many European countries. For details about the services and connections, readers are referred to Section 12.9 (Chapter 12).

The advantages of frame relay over X.25 lie in minimal processing for error control, flow control, and error recovery at each node of the network. In other words, we can presume that this will provide minimum delay and hopefully higher throughput. This network supports the access of E1. Frame relay can be used with SMDS's lowest data rate T1 (defined by Bellcore) and also with SMDS's data rate of 64 Kbps (defined by MCI). These two types of public networks (frame relay and SMDS) may be competing in the market, but it seems that frame relay will be suited to interactive data transfer, and also its low overhead will be another important advantage over SMDS.

19.9.2.1 Why ATM for B-ISDN networks?

Although ATM uses packet-switching techniques, it uses fixed-length cells and very simple protocols, as opposed to narrow-band networks (X.25) which have variable-sized packets and very complex protocols. Due to the fixed size of the cells, the ATM network can transmit the data of different services (data, voice, video, etc.) at fiber rates (as high as 622 Mbps). An excellent review paper on ATM discusses some of these issues in more detail.[13]

ATM combines the concepts of packet and circuit switching and offers fully integrated support of voice, video, images, and data through a fixed size of cells. This has now been adopted by international standards organizations such as CCITT, ANSI, and the ATM Forum. Further, the ATM technology has been accepted by network vendors, telecommunication service providers, and computer vendors. SONET has also helped the ATM technology by offering increased reliability with ring topology and offering high-bandwidth digital video applications.

ATM offers support to both types of network traffic — **constant bit rate (CBR)** and **variable bit rate (VBR)** — and the distinction between them is maintained by assigning labels to every cell transmitted. These labels in the cells are used to identify the virtual paths (VPs) and virtual channels (VCs) such that each VP carries those VCs within it assigned by labels corresponding to it. The virtual paths and virtual channels (circuit capacity) are based on the type of interface of ATM being used. General guidelines for accessing these interfaces offer a minimum of 256 VPs and VP/VC termination for DS-1 rate interface. The number of VPs and VP/VC termination for DS-3 and STS-3c (synchronous transport signal; SONET nomenclature) and STM-1 (synchronous transport module; CCITT nomenclature) rate interfaces is 1400 and 4096, respectively. These numbers represent minimum numbers and should not be considered the actual numbers. These points will be made clearer later in this section.

The traditional time-division multiplexing technique can provide connection at low data rates of 64 Kbps and 384 Kbps, but with ATM, continuous bitstream circuit connections are established through a technique known as *circuit emulation*. The incoming bits from virtual circuits are analyzed or encapsulated in cells (of fixed size) and are then transmitted thorough a B-ISDN network (based on ATM) to the destination. The original message is extracted from the cells received. The established connection can offer a maximum bit rate of up to the capacity of the user–network interface. Due to its fixed-rate connection capabilities, ATM offers a greater flexibility in network synchronization. It provides transport to asynchronous DS-1 signals via ATM virtual circuits to the circuits which are not synchronized with the network clock. This feature of ATM makes it very useful to use a B-ISDN network where different clocks are being used in different parts of the network and offers a uniform interconnection between them for a variety of services without much delay in the transmission. It provides integrated services of voice, data, video, multimedia services, etc.

The B-ISDN based on ATM is basically connection-oriented, but it also supports connectionless and isochronous services. This is very important, as most LANs and MANs use connectionless protocols and this must be supported by B-ISDN. The B-ISDN offers this service (connectionless) on top of ATM, and there are two ways this type of connectionless service can be supported (CCITT I.211).[27]

In the first approach, the connectionless service is provided indirectly via existing connection-oriented services by defining the ATM transport layer connection within B-ISDN. These ATM connections are usually permanent and reserved and hence are available on demand. The connectionless service protocols are transparent to B-ISDN as connectionless services and corresponding AAL functions are implemented outside the B-ISDN; hence, the connectionless service protocols are independent of other protocols being used in B-ISDN.

In the second approach, the connectionless service function may be defined within or outside B-ISDN and the cell includes the routing information. Based on this information, the cell is routed to the respective destination and is offered on top of the ATM layer.

With digital technology, most of the networks are shifting toward digital transmission over the existing cables. The digital links so defined are supporting the addition of overhead bytes to the data to be transmitted. This payload is a general term given to a frame (similar to the one defined in LANs). The frames used in LANs carry a lot of protocol-specific information such as synchronization, error control, flow control, etc. In the case of ATM networks, the data is enclosed in these frames (if available). The frames are generated at a rate of 8000 frames per second and the digital links offer a very high data rate. The commonly used digital link in the U.S. is SONET. The basic rate is defined as STS-1, which offers 51.84 Mbps, while STS-3 offers a data rate of 155.52 Mbps. The ATM cell contains 53 bytes, of which five bytes are reserved for a header and the remaining 48 bytes carry the data. In order to send the information (data, video, graphics, etc.), we need only ATM cells. The number of available cells depends on the number of frames per second and the data rate of the link. If we use an STS-1 link, then we may have about 15 ATM cells, as each cell has an overhead of five bytes. It is calculated as follows: number of frames per second = 8000, number of data bits per cell = $48 \times 8 = 384$, number of header bits per cell = $5 \times 8 = 40$. The number of cells per frame = cells $\times 8000 \times 380 \times 40 = 51.84$ Mbps = 15.

If we use STS-3, the number of cells per frame will be about 44, and so on. ATM thus offers a different bandwidth to the users (based on their demand) on the same digital link. The cells are not assigned to one user; instead, they are assigned to the data to be transmitted. For more details on ATM networking, see References 53 and 54.

19.9.2.2 ATM-based LANs

Private networks based on ATM switches can also be defined by ATM-based local area networks. These LANs configure in the same way that other LANs do; e.g., in Ethernet, token ring, token bus, and FDDI, the nodes (PCs or workstations) share the resources by sharing the LAN bandwidth. Each LAN has a different scheme for sharing bandwidth and, in general, in all of these LANs, the LAN bandwidth is shared among the nodes connected to it. In ATM-based LANs, the ATM switch sets up a point-to-point ATM communication between the nodes and once an ATM call is established between nodes, the full bandwidth of 100 Mbps can be used by them. In the case of FDDI LANs, we also get 100-Mbps bandwidth, but it is shared between all the users using the LAN. In ATM-based LANs, each user gets the full bandwidth of 100 Mbps (after ATM connection is established). The FDDI LAN supports only data, while the ATM-based LAN supports data, voice, any combination of signals.

ATM-based LANs may have ATM LAN switches with 16 to 68 connections, internal ATM interface cards for nodes (PCs or workstations), and a fiber-optic transmission medium. The data rate for ATM LANs is around 100 Mbps, which seems to be widely accepted among the members of the ATM Forum and may become a *de facto* standard. Each ATM interface card provides a direct link between it and the switch and provides a 100-Mbps full-duplex link between the nodes.

Implementation: B-ISDN includes ATM and offers higher data rates than T1 (1.544 Mbps, used in North America) and E-1 (2.048 Mbps, used in Europe). Since an ATM switch has a very open functionality, it can be used at any layer of OSI-RM. The ATM can be used as a MAC sublayer of OSI-RM, thus avoiding the use of LAN MAC. This means that there is direct mapping between the LLC sublayer and ATM, and all the packets carrying LLC information are enclosed in ATM cells. Thus, the entire ATM replaces the physical layer and MAC sublayers of OSI-RM. This is precisely what has been implemented in SMDS. Further, the LAN Emulation Service (LES) group within the ATM Forum is working on the standardization process for IEEE LANs. The main advantage of this is that now ATM has become another LAN like Ethernet and token ring, but it loses its features such as flexible bandwidth, multimedia applications, etc., compared to routers for interconnectivity.

ATM can be implemented at the data link layer, whereby the IP will be enclosed in ATM cells, thus avoiding the use of any data link protocols. The Internet Engineering Task Force (IETF) is working on this concept. The network layer defines a unique network address; e.g., in LANs it defines a network interface card (NIC) address, in WANs it defines an HDLC identifier, and so on. TCP/IP needs to provide a mapping between network and link addresses and is performed by address resolution protocol (ARP). This is based on broadcasting the request for the address and will not work with ATM. The IETF is proposing the modification of IP and has proposed INARP as a modification to TCP/IP, which is still under consideration.

The ATM Forum has proposed another specification known as application program interface (API) to make use of ATM at the transport layer. For all the layers above transport, API may provide support for the functions and applications. It offers various advantages, including access to ATM for multimedia, video, and data applications and, of course, ATM networks. Disadvantages include no support for variable size of packets and the need for major changes in the existing interfaces at the transport layer. If we look at the B-ISDN model, we find that ATM is used at the transport layer, while B-ISDN provides the implementation of the upper layers of OSI-RM. Vendors using ATM at different layers may not be able to communicate. But if two vendors can use ATM at the same layer, then the advantages offered by ATM are enormous (higher and more flexible bandwidth, higher data rates, support for multimedia, video, graphics, and data applications).

The variable size of cells, or packets or even frames, is used efficiently in data communication, as it does not require any synchronization between bit patterns on the receiving side. But in the case of graphics, multimedia, video, and audio applications, the variable size of cells does not work, and hence ATM does not provide any real benefit for their transmission. The standards and specifications for flow control, error control, and congestion control for WAN-based ATM networks, and also transport layer implementation, are still under consideration before ATM can become a viable network like X.25, TCP/IP.

The ATM Forum has designed two categories of performance measurement — accuracy and dependability for errors and speed for delay. The first performance measurement (accuracy and dependability) includes the following parameters: cell error ratio, cell loss ratio, cell misinsertion ratio, and severely errored cell block ratio. The speed performance measurement includes the following parameters: cell transfer delay, mean cell transfer delay, and cell delay variation. Every network experiences a bit of delay during transmission. ATM offers a very low error bit rate compared to other networks because of the small size of the cell and the fact that ATM is designed to run over fiber in both public and private networks. A single bit error causes the re-transmission of the packets. If we transmit fewer bytes during re-transmission, it will improve the utilization of the network. For example, if there is a single bit error in LANs, the entire frame (typically 1600 bytes in Ethernet, 18,000 bytes in a 16-Mbps ring network) has to be sent, a wastage of network bandwidth.

Problems with ATM-based implementation: In spite of these features and the potential of ATM switches, standardization of various areas like signaling, congestion, flow control, etc., may become a significant problem in the deployment of ATM-based networks. However, once these are defined, ATM-based networks can easily be used for interoperability. Network managers and users are trying to get a larger bandwidth for their communication on wide area networks. This limited bandwidth of WAN is now becoming slightly less restricted, whereby users can request and get a larger bandwidth of the WAN. This is because users are now using distributed networking concepts to access a large amount of databases and programs which require larger bandwidth. The public networks cannot provide larger bandwidth to the users and, at the same time, the bandwidth requirements of the users are expanding due to new technologies, e.g., graphics, multimedia, icon-driven interfaces, availability of corporate desktop computers with 50 to 100 Mbps running on Windows NT, etc.

For the transmission of data, it appears that the packet-switched public network will be more useful in the future than the circuit-switched network, although at present circuit-switched public networks are well-suited for transmission of voice since they have very small delay. With the new trend of allowing larger bandwidth and large volumes of data in WANs, it is obvious that the packet-switched network is more useful, and hopefully it will dominate the public networks. If this happens, then it also appears that ATM switches will be used in these networks. Currently the packet-switched public networks such as X.25 define packets of variable length. In ATM switches, we define a small fixed-length cell of 53 bytes (of which five bytes defines the header of the cell, while the remaining 48 bytes are for the data). The small cells of fixed length can be transmitted very quickly, while the minimal delay (provided by circuit-switched networks) allows the switches to handle services of any type of information — voice, video, or even data.

Another type of technology which offers the same features (as discussed above) is SMDS (defined by Bellcore and also based on IEEE 802.6 MAN). It provides cell-based services, and various aspects/issues like network management, flow control, and billing management have already been standardized (in contrast to ATM-based networks). This network, in fact, provides specific services and defines a packet format on top of the cell-switched packet format. In other words, the user's data packet encapsulates an SMDS packet which, in turn, includes 53-byte cells. These cells can be transmitted at high speeds due to the cell switches (ATM). The ATM has been defined to run over fiber, particularly on SONET (SDH for ITU). This technology offers gigabits over fiber. ATM can also be used over other media like unshielded twisted pair, coaxial cable, fiber optics, and wireless channels. The design of ATM for fiber provides a much better bit-error rate compared to other media and also guarantees error-free communication. In most cases, it may offer a non-bursty single-bit error.

Implementation with public network: The X.25 packet-switched public network now supports access speeds up to 2 Mbps defined by CCITT (higher than 64 Kbps) and, in fact, after the approval of this recommendation, a large number of X.25 public networks now support 2 Mbps.

The majority of the networks are typically public networks and are required to provide services to the different services. In the U.S., these networks fall into the following classes. The local exchange carriers (LECs) provide local-access services (e.g., U.S. West), while interexchange carriers (IECs) provide long-distance services (e.g., AT&T, MCI, Sprint, etc.). Regional Bell Operating Companies (RBOCs) operate the LECs. The IEC also provides interstate or inter-local access transport areas (LATAs). In most countries, government-controlled Post, Telegraph, and Telephone (PTT) provides these services.

ATM is used in different classes of public networks. For example, it can be used to replace the end office and intermediate office by connecting them with DS-1/DS-3 links. As we know, the subscriber's phone is connected to the end office via a twisted pair of wires which must be replaced by coax/fiber. The public network system and CATV can use ATM for providing video-on-demand services. This may require the use of compression techniques such as asymmetrical digital subscriber loop (ADSL) that uses the MPEG2 coding method.

The existing packet-switched X.25 network provide interfaces to routers, which are connected to routers, bridges, and LANs. The interface offers a dial-up connection via a 56-Kbps link from LANs. The functions of the X.25 protocol is to provide the functions of the lower three layers of OSI-RM. Since the transmission media are not reliable, the X.25 must support error control and error detection. Further, the packet in going through three layers requires processing time, and hence the network is very slow. There are a number of ways an ATM can be used in X.25 networks, which are typically used for data communication. In one X.25 switched network environment, each of the X.25 switches is replaced by frame relay (connection-oriented services). This will increase the speed from 56 Kbps to T1 (1.544-Mbps) speed. For connectionless services, switched multimegabit data service (SMDS) switches offering higher speeds than T1 are used. SMDS can offer T3 (45-Mbps) speed and provides public broadband services. The underlying frame relay may become the access point to the SMDS network. In another X.25 environment, ATM may be used as a backbone network offering virtual circuits with much higher bandwidth between T1 and OC12 (622 Mbps). We can use frame relay and SMDS as access networks to ATM. We can transmit voice data and video over the ATM backbone but with circuit switching. We can also use the complete B-ISDN network using B-ISDN protocols and interfaces (ATM/SONET).

The packet-switching X.25 supports packet re-transmission, frame delimitation, and error checking, while ATM does not support any of these functions. A typical configuration of ATM for WAN interconnectivity includes the ATM public network connected to various ATM hubs via OC3 UNI. The ATM public network comprises a number of ATM hubs interconnected via NNI. As mentioned above, the header of an ATM cell defines routing through two fields known as the VPI and VCI. The VPI is similar to TDM in circuit switching, while the VPI uses ATM as a traffic concentrator (cross-connect). The ATM cross-connect replaces the existing digital cross-connects. A number of carriers have provided ATM PVC services, including AT&T, MCI, Sprint, Wiltel, and others. The ATM service environment using ATM offers the following features: transmission pipe composed of different types of traffic (video, data, audio, multimedia) via statistical multiplexing, bandwidth on demand that is available and allocated dynamically, a single universal network (no need for separate networks like data, voice, video, or multimedia), greater economy and reduction of overhead and administrative costs, and high-speed access to the network starting at DS-1 or T1 (1.544 Mbps).

19.9.2.3 Flexibility of ATM

ATM offers different types of flexibilities such as time, semantics, bandwidth, reconfigurations to new or modified architecture, etc. Most of the packet-switched networks offer flexibility in terms of bandwidth and also offer very efficient resource usage strategies. The initial version of the X.25 public network (based on packet-switching) offered limited support to services possessing time constraints or requiring higher bandwidth and data rates with acceptable QOS. As a result of these requirements and additional requirements for a variety of services to be offered by a unified, flexible, and universal network, the concept of packet-switching used in the upcoming networks evolved. Due to this, we have seen different types of packet-switching techniques being introduced and used in high-speed networks.

Since an ATM switch has very open functionality, it can be used at any layer of OSI-RM. This puts a limitation on interconnectivity, and vendors using ATM at different layers may not be able to communicate. But if two vendors can use the ATM at the same layer, then the advantages offered by ATM are enormous (higher and more flexible bandwidth, higher data rates, and support for multimedia, video, graphics, and data applications). A number of working groups within the ATM Forum are working on defining the protocols of LAN inteconnectivity at different layers of OSI-RM, as discussed above.

ATM can be implemented at the data link layer whereby the IP will be enclosed in ATM cells, thus avoiding the use of any data link protocols. The Internet Engineering Task Force (IETF) is working on this concept. The network layer defines a unique network address; e.g., in LANs, it defines a network interface card (NIC) address, in WANs, it defines an HDLC identifier, and so on. TCP/IP needs to provide a mapping between the network and link addresses, and this is performed by address resolution protocol (ARP). This is based on broadcasting the request for the address and will not work with ATM. The IETF is proposing the modification of IP and has proposed INARP as a modification to TCP/IP, which is still under consideration.

The ATM Forum has proposed another specification known as application program interface (API) to make use of ATM at the transport layer. For all the layers above transport, API may provide support for the functions and applications. It offers various advantages, including access to ATM for multimedia, video, and data applications and, of course ATM networks. Disadvantages include no support for variable size of packets and the need for major changes in the existing interfaces at the transport layer. If we look at the B-ISDN model, we find that ATM is used at the transport layer, while B-ISDN provides the implementation of the upper layers of OSI-RM.

The packet-switched network must offer a reliable transport and correct delivery of message data packets to the destination. In an early version of X.25 public packet-switched networks, error control (for providing end-to-end acceptable quality of service) was performed at each layer, and this requirement was provided by data link layer protocols, e.g., high-level data link control (HDLC) and link access protocol (LAP-B and LAP-D). These protocols provide functions such as frame delimiting, error detection (using cyclic redundancy check procedure), frame re-transmission, etc. Here it is assumed that the errors are caused during switching and transmission and hence are considered part of the network services.

The QOS provided by transmission and new switching techniques used in ISDN and similar integrated networks reduced the functions of data link layer protocols. In these networks, the data link protocols now provide functions like frame delimiting and error detection for link-to-link configuration, while other functions like error recovery are provided for end-to-end configuration only. In this way, the switching nodes are less complex and will provide a higher speed of transmission. The packet-switching was modified so that two sublayers of data link protocols were defined as (1) a lower sublayer providing the core functions of the data link layer and (2) an upper sublayer offering optional functions. This type of switching is sometimes known as frame relaying. In this switching, both sublayers are still with the data link layer.

The switching technique used in B-ISDN further takes these functions from the data link layer to the physical layer, where again the functions are provided by two sublayers of the physical layer as core and optional functional sublayers. Moving the error-control functions down to the physical layer will certainly make the switching nodes less complex, and these provide minimal functions. The net effect of these efforts will make switching nodes available for transmitting the message data at a faster speed. This is exactly what is done in ATM-based networks where the ATM switching nodes (also known as cell nodes) are less complex (compared to X.25 nodes, frame relaying nodes) and offer data rates as high as 622 Mbps.

ATM has the flexibility to reconfigure to a new architecture to meet the requirements of the user, and it defines a virtual path of trunk groups between B-ISDN exchanges. A cell header is appended to include an additional field of the **virtual path identifier (VPI)**, which defines a cell label to be seen by intermediate switches. In other words, the intermediate switches look at VPI, which is translated at each transit switching node. Different virtual paths denoting a call request are identified by other fields of the cell header. This avoids the use of intermediate switching nodes for establishing a connection between call requests, thus saving a lot of processing time per call. Any particular trunk on the virtual path will use the bandwidth only if it is carrying the data, or else the bandwidth can be used by other trunks.

The ATM network is used in LANs for a higher bandwidth and data rate, so that the entire ATM protocol has moved from WAN groups like ITU-T, Bellcore, and others toward the possible LAN solutions in every organization. Looking at the possible LAN solutions based on ATM, IEEE is moving toward defining 100-Mbps LANs (Ethernet, token ring). The existing 10-Mbps Ethernet will be developed as switched Ethernet, which will offer the entire bandwidth of 10 Mbps at each node rather than sharing the bandwidth in the traditional working of Ethernet. FDDI is already running over fiber at a data rate of 100 Mbps. FDDI provides features similar to those of ATM in terms of data support and time-based applications such as video and voice (isochronous channels); in some applications, it may be cheaper than ATM. Other alternatives to ATM for possible LAN solutions may include 100-Mbps switched Ethernet, FDDI-II as a LAN, IEEE 802.6 as a MAN, packet transfer mode as a WAN, and SMDS and frame relays as public network services.

As discussed above, different types of networks (LANs, MANs, FDDIs) can be interconnected via different devices that can be used at different layers of OSI-RM. These devices are repeaters, bridges, routers, gateways, etc. Of these devices, the routers seem to be most popular, as they offer connectivity to a variety of networks. The routers operate at the network layer and support both connection-oriented and connectionless services. A network interface card (NIC) for LANs or a WAN board interface for a router network is used for interconnectivity. The router networks look at the address in the datagrams and route them over the network from one hop to another until they reach the destination. An ATM network, however, is basically a connection-oriented network. The router-based networks do not require any specific protocol or user, are very general, and can be used for LANs, WANs, and so on. The router-based networks resemble the old version of X.25, which usually provides connection-oriented network service. If we look at the layered architecture of X.25, we find it has a physical layer (X.21), link access protocol-balanced (LAP-B) and high-level data link layer (HDLC), and network layer (packet layer protocol). Based on the above discussions, we can say that routers provide typical connectionless services, while switches (like ATM) provide connection-oriented services. The transport layer provides end-to-end communication between nodes, while other interfaces are user–network interfaces (UNI) and network-to-network (network node) interfaces (NNI).

Packet-switching, frame relaying, and cell switching defined for ATM offer functions in decreasing order for error control and also are being moved from the traditional data link layer to the physical layer. As a result, the switching nodes are getting less and less complex and hence will offer higher data rates. It seems clear at this time that cell or ATM switching nodes possess minimal functions for error control and are suitable for continuous bitstream data such as voice at 64 Kbps.

19.9.2.4 Issues in ATM-based broadband services

One of the main issues in the ATM-based B-ISDN has been whether this architecture supports circuit-switched services in pure packet-switching form. The parameters dealing with this issue includes grade of service, compatibility, etc., and to some extent, these parameters are the issues which have been discussed in circuit-switching systems. This issue can be handled in different ways, e.g., emulation of circuit-switched services, introduction of new circuit-switched services as a fundamental service of B-ISDN at the network level (transport layer mode-transparent), etc. In other words, a high-quality circuit-switched service must be available from B-ISDN without any constraints on the underlying switching and multiplexing techniques being used in the networks (ATM, STM, etc.). The B-ISDN must provide a quality of services to constant traffic sources (PCM voice, fixed-rate video, etc.), and according to CCITT I.121[25] recommendations, the continuous bitstream-oriented services must be handled by either STM or ATM deterministic transfer mode. The material herein is partially derived from References 1–12 and 46–52.

Issue 1: The ATM-based network offers two layered functions similar to a circuit-based network. In the existing circuit-based network, the first functional layer assigns a time slot or link to each incoming call into an outgoing time slot link based on the control memory. Each link is associated by the slot position as the channel identification. The second functional layer changes the contents of the control memory (based on the outcome of the call setup process execution), while in ATM-based networks, the time slots are replaced by ATM cells with virtual channel identification (VCI) numbers. The switching function used in ATM-based networks also consists of two functional layers as described above for circuit-based networks and performs the same functions as in circuit-switched networks. In order to provide a circuit-switched grade of service, periodic bandwidth requirements must be met. The parameters are defined during the call setup procedure and an appropriate algorithm is used to control the number of outgoing calls (based on the bandwidth requirement).

For any application (e.g., video, images, etc.) based on the circuit-switching concept, the bandwidth requirement will be used to provide the routing to ATM cells from circuit-switched networks to video sub-networks (handling video services). The video service may be provided by any network, and the changes needed in the underlying network are transparent for the access protocols.

Issue 2: The network-level performance measures for critical switching services include throughput, transit delay, transfer failure probability, residual error rate, etc. Obviously, the throughput measures need to be negotiated during the call establishment (a part of bandwidth management). The ATM mechanism defines its own performance measures which are equivalent to these measures of the circuit-switched network. Each of these measures or parameters needs to be mapped onto the ATM measures. The equivalent measures to these measures of the circuit-switched networks are cell loss rate, undelivered or wrongly delivered cells, delay, and improper sequencing of cells.

Many activities are going on to define proper and appropriate equivalence and mapping techniques between these performance measures of circuit-based and ATM-based networks. It seems reasonable that the circuit-switched services will provide a very simple interface between an ATM network and an existing PSTN. Since the majority of problems of handling voice communication in ATM networks due to packetization delay and variable

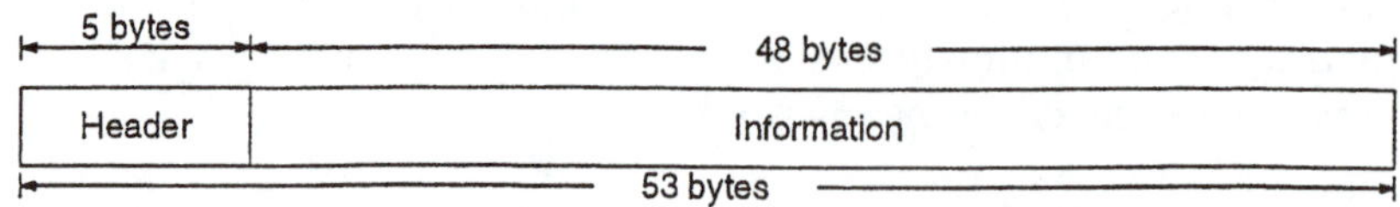

Figure 19.28 ATM cell.

transmission delay, ATM offers a partial filling of voice cells and relays in the circuit-switched grade of service to make sure that the voice packets arrive in the proper sequence without significant delay.

As stated above, the B-ISDN networks provide a variety of broadband ISDN services, and integration of service varies from user to user and varies over time. This requires both the multiplexing and switching (or transfer mode) to be more flexible, as together they control and manage the bandwidth of the B-ISDN network. In contrast to the traditional approach of multiplexing (time-division multiplexing), the B-ISDN network uses a new multiplexing structure, ATM. In this scheme of multiplexing, the bitstream of data is divided into a number of fixed length or size of 53-byte cells (CCITT recommendation). Each cell has two fields: header and information. The header field contains network control information, while the information field contains the user data, as shown in Figure 19.28.

Issue 3: In synchronous time-division multiplexing, the position of time slots within a frame defines the identification of a channel for a call request. This type of multiplexing sometimes is known as *position multiplexing*. In contrast to this, in asynchronous multiplexing (used in ATM), the cells are labeled inside the header for identifying the channel for a call request, and due to this, this method of multiplexing is sometimes known as *cell* or *label multiplexing*. There is a mapping table at the multiplexing and switching processing nodes which translate these incoming links to the appropriate outgoing links or labels, defining a connection or virtual circuit between the nodes.

ATM handles continuous bitstream services (arrival of cells at a regular rate from a source) and bursty traffic (e.g., data information and variable rate–coded video signals, PCM voice signals, etc.) at a statistically varying rate. The word "asynchronous" in ATM, in fact, represents the non-deterministic nature of multiplexed channels. The headers are checked to identify the contents of a cell. The ATM multiplexed information can also be transmitted over synchronous transmission networks (based on synchronous mode transfer) such as SONET.

19.9.3 ATM layered protocol architecture

As stated above, the concept behind three popular designs in high-speed networks (fast packet-switching, asynchronous time-division (ATD), and asynchronous transfer mode (ATM)) is to provide very high-speed switching to the message information, but advantages offered by ATM-based networks, such as connection-oriented mode of operation, support of different services (voice, data, video), no support for error control and flow control on a link-to-link basis but rather in an end-to-end configuration, etc., make ATM more popular than other switched networks. The standards organization CCITT technical group SGXVIII defined standard ATM architecture and other technical details. This technical group had to make a compromise between various integrated services (voice, data, video) and high-speed data services, as different countries have different requirements. In June 1992, CCITT approved the following I.113 recommendations.[24]

- Broadband Aspects of ISDN I.121
- B-ISDN ATM Functional Characteristics I.150
- B-ISDN Services Aspects I.211
- B-ISDN General Network Aspects I.311
- B-ISDN Protocol Reference Model and Its Applications I.321
- B-ISDN Network Functional Architecture I.327
- B-ISDN ATM Layer Specification I.361
- B-ISDN ATM Adaptation Layer (AAL) Functional Description I.361
- B-ISDN ATM Adaptation Layer (AAL) Specification I.363
- B-ISDN User–Network Interface I.413
- B-ISDN User–Network Interface Physical Layer Specification I.432
- OAM Principles of B-ISDN Access I.610

An internal architecture of ATM still has not been standardized and is still in the research and evaluation phase of various proposals. In the ATM-based network, we use both circuit- and packet-switching techniques in one way or another, except that now the packet-switching technique provides support for high-speed networks. For some of the previous concepts like routing and performance, very large-scale integration (VLSI) has been used in the data connection quite extensively. Somehow this is not useful in the ATM networks to address some unresolved issues such as traffic congestion, signaling, cost, and support of incremental deployment for applications, although a few ATM switches have been deployed to meet the bandwidth requirements. It is important that these switches must also support services of future technologies in addition to the present demand of network traffic (nearly 2 million LANs are being used to connect over 21 million PCs, terminals, workstations, etc., worldwide and support a variety of LAN/WAN interconnections, scalability, efficient utilization of resources, etc.).

Since the fast switching or cell technique is being used in the ATM network, it is sometimes known as a cell- or fast-switched network. A cell of fixed size (with a five-byte header and 48 bytes of user data) is defined for transmitting any : data (audio, video, text, etc.) and control message information. The header includes information regarding the virtual channel to which the packet belongs. The layered architecture as defined by ISO has become very popular and is now an integral part of describing any communication system. It offers various advantages and features (discussed in Chapter 1). The OSI-RM has been widely accepted as a framework of LANs for various protocols to provide connection within the model. CCITT has defined a similar layered architecture concept for the ATM B-ISDN network in its recommendation I.321. At present, only the lower layers have been discussed and explained.

The ATM protocol architecture defines two layers of B-ISDN. Messages from the higher layers are converted into a fixed size of 48 bytes (cell). The ATM layer adds its header of five bytes to each cell. The header includes routing information. There are six fields of the header: generic flow control (GFC), virtual channel identifier (VCI), virtual path identifier (VPI), payload type (PT), cell loss priority (CLP), and header error control (HEC). Out of these fields, VPI and VCI carry routing information. VPI can be considered similar to TDM in circuit switching. The header also defines the type of interface being used as user–network interface (UNI) or network–node interface (NNI). A detailed discussion of these can be found in the following sections.

The relationship between ATM B-ISDN layered architecture and OSI-RM has not been standardized or widely accepted (in contrast to the relationship between OSI-RM and other networks like SNA, DECnet, DoD, etc.). The ATM network supports variable and fixed data rates of transmission, both connection-oriented and connectionless services, multiplexing, etc., and hence is a suitable platform for multimedia applications to implement

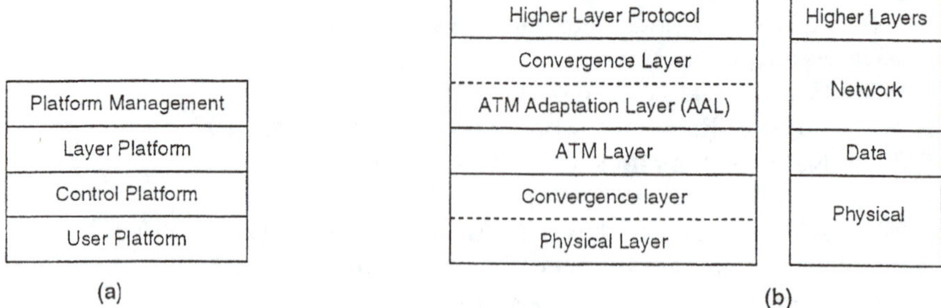

Figure 19.29 (a) ATM platform layered configuration. (b) ATM layered architecture.

the client–server computing paradigm, which was accepted as one of the most effective and efficient paradigms for distributed cooperative processing and multimedia applications in the 1990s.

A typical ATM B-ISDN–platform layered configuration model is shown in Figure 19.29(a). There are three main platforms (similar to the N-ISDN protocol reference architecture model) — user, control, and management. Each of the platforms interacts with each other in the same way as the layers within OSI-RM and offer specific functions and services. The user platform is concerned with the transportation of information, while the control platform includes the control information for signaling, timing, etc. The management platform deals with the execution of various operational functions and maintenance of the network.

On top of these platforms, one might define platform management which controls and manages these platforms. Each platform is composed of layers which interact with each other through parameters and primitives associated with the respective platforms. As stated above, the complete relationship between layers of different platforms and OSI-RM has not yet been defined by CCITT, but the layer interfacing and protocols within layers of the platform perform various functions in the same way that they do in OSI-RM. A typical ATM layered architecture is shown in Figure 19.29(b).

The user platform contains a physical layer (equivalent to the physical layer of OSI-RM) and usually performs the functions on bytes or bits. This layer ensures the correct transmission and receiving of bits on the media. It must reconstruct the proper bit timing of the signal received at the receiver.

19.9.3.1 Physical layer

The main function of this layer is the extension of a path between networks which assembles and disassembles the payload (continuous or variable byte streams) of the transmission system. The payload typically is defined as a message carrying user information along with the necessary transmission control information. The physical layer is usually a medium-dependent layer and typically defines the electrical behavior of carrier signals and the physical behavior of the transmission media. It is divided into two sublayers: **physical medium (PM)** and **transmission convergence (TC)**.

The **PM sublayer** is at the lowest layer of the model and offers physical-dependent functions like bit transmission, line coding, electrical/optical conversion facility, and support of optical fiber and other media (e.g., coaxial cable). The **TC sublayer** offers transmission frame adaptation for cell flow, cell delineation (defining the cell boundary), header error control (HEC) for detecting one-bit and some multiple-bit errors, cell rate decoupling (for controlling adaptation of the rate of ATM cell transmission between two nodes), and generation and recovery of the transmission frame.

The ATM layer is independent of the physical layer and transmission media used to transport ATM cells. The multiplexing/de-multiplexing founding of cells of different connections is performed to define a single cell stream where each connection is different for different control and user data and is identified by VCI or VPI values. The cell header is added after it receives a cell from the adaptation layer and is removed from the cell which it receives from the convergence or physical layer. Finally, the flow-control scheme may be implemented on the user–network interface and is supported by bits in the header.

The physical layer used in B-ISDN is the Synchronous Optical Network (SONET) and offers data rates of 155.520 Mbps and 622.080 Mbps. The physical layer has a provision offered by another layer sitting on its top or a part of this layer which allows the transmission of ATM cells over other physical layer transmission media and is defined as the *convergence layer*. This layer translates the ATM cell stream into bytes/bits to be transported over the transmission media.

The TC layer offers the first function of adaptation to the constructed cell, and the cell must be able to adapt to a transmission frame. It also identifies the cell boundary to provide proper delineation at the receiver. The technique used for cell delineation is based on the header error check (HEC) algorithm. The convergence layer also performs the generation and verification of the ATM cell algorithm, a mechanism for correction/detection of the error. Finally, it exchanges operations and maintenance (OAM) information with the layers of the magnet platform.

The ITU-T has recommended fiber optics as a transmission medium for B-ISDN and has defined the physical interfaces for SONET/SDA for fiber and pleisochronous digital hierarchy (PDH) for coaxial. The ATM forum has defined some additional interfaces as T1 (1.544 Mbps), E1 (2.048 Mbps), T3 (45 Mbps), E3 (34 Mbps), E4 (140 Mbps), SONET (STS3c/SDH STM1; 155 Mbps), and FDDI PMD (100 Mbps). For each of these physical media protocols, a separate convergence sublayer protocol has been defined similar to the MAC protocol in OSI-RM.

19.9.3.2 ATM layer

The ATM layer of the B-ISDN protocol performs various functions on the cell header of five bytes, and this layer is sometimes also known as the *ATM cell layer*. The functions include **cell routing, cell multiplexing, cell switching, construction of ATM cell**, etc. The functions and specifications of the ATM layer have been recommended by CCITT I.150[26] and I.361,[31] respectively. It defines **virtual channel (VC)** as a unidirectional transport means for ATM cells associated with a common and unique identifier known as a **virtual channel identifier (VCI)**, which is part of the ATM cell header.

The VCs with assigned identifiers (i.e., VCIs) are encapsulated within another logical unidirectional transport of ATM cells defined as the **virtual path (VP)**. A virtual path is also assigned a unique identifier known as a **virtual path identifier (VPI)**. A VP includes VCs which are associated with it, which in turn depend on the bandwidth requirements and requested data bit rates. The **transmission path (TP)** is defined by several virtual paths, and each path carries several virtual channels.

ATM switching node and/or cross-connect node of the interconnection network provide translation for VCIs and VPIs. The cell header is added (after the generation process) after it receives a cell from the adaptation layer and is removed (extraction process) from the cell when it receives a cell from the convergence or physical layer. Finally, the flow-control scheme, generic flow control (GFC), may be implemented on the user–network interface (UNI) and is supported by bits in the header. This supports the control for ATM traffic flow.

As mentioned above, within the ATM layer two links corresponding to VCs and VPs are defined. The **virtual channel link (VCL)** provides a unidirectional transport mechanism for ATM cells between two points. At one point, the VCI value is assigned, while at the second point, the assigned VCI value is either translated or removed. Similarly, a **virtual path link (VPL)** offers a unidirectional transport mechanism between two points; at one point, the VPI value is assigned, while its value is translated or removed from the cell at another point.

The VPC actually terminates at these points. By concatenating VCLs, a **virtual channel connection (VCC)** is defined and a **virtual path connection (VPC)** is defined by concatenating VPIs. A VPC consists of a VCC which is defined as concatenated VPLs. Each VPC consists of concatenated VPIs defining a transmission path within the ATM layer. Each VPL is implemented as a unique path in the transmission path.

The values of VCI and VPI defined within VCC and VPC are translated into VC/VP switching entities. The switches used for transmitting VCs and VPs provide the necessary switching for them. The VP switches define termination of VPL, and it translates incoming VPIs to corresponding outgoing VPIs based on the distribution of VPC. The assigned identification number to VC (VCI) usually remains unchanged.

19.9.3.3 ATM adaptation layer (AAL)

The next layer is the ATM adaptation layer (AAL), which provides interface for adaptation between the higher layers and the ATM layer. The functions and specifications of this layer have been recommended by CCITT I.363. This adaptation is performed on both signaling and user information coming from the higher layer into an ATM cell of fixed size. This layer is defined between the ATM and higher layers. The main function of this layer is to provide adaptation to the requirements of its higher layers for services provided by ATM (lower layer). The user data of the higher layers is defined as **protocol data units (PDUs)** which are mapped onto the information field of ATM (48 bytes). The remaining five bytes of a standard 53-byte cell correspond to the ATM cell header.

As indicated earlier, the address scheme is defined in two subfields as the virtual path identifier (VPI) and the virtual channel identifier (VCI), and a similar scheme is used for addressing of an ATM cell. The VPI defines the physical path, which is in turn used by a predefined set of VCIs. The header of the cell is always protected by one byte of **cyclic redundancy checksum (CRC)**.

The AAL is typically divided into two sublayers (as defined by CCITT I.362)[46]: **segmentation and reassembly (SAR)** and **convergence sublayer (CS)**. The SAR sublayer segments the higher layer's PDUs into a fixed size of ATM cells, and on the receiving side, it reassembles the cells into the higher layer's information (in terms of PDUs). The protocol for this sublayer breaks the information into 48-byte cells.

The upper sublayer is service-dependent and performs functions such as message identification and clock time recovery at the AAL service access point (SAP). There is no SAP between the SAR and CS sublayers.

Since the SAR is performing segmentation and reassembly operations, there is an extra field in the information field which distinguishes between them and also indicates the states of the cells (completely occupied, partially occupied, etc.). All of these sublayer functions of AAL are supported by AAL entities which exchange information between peer AAL entities.

CCITT I.321 has defined primitives (for data transfer between two adjacent layers) for the physical and ATM layers. In the B-ISDN layered architecture model, there are two primitives for each of these layers. For the physical layer, the primitives are **Ph-DATA.request** and **Ph-DATA.indication**.

The first primitive, when used by the ATM layer, indicates the request of transporting the service data unit (SDU), which is associated with the request to its peer entity layer.

Similarly, the second primitive, when received by the physical layer, indicates to its ATM layer the availability of an SDU it has received from its peer entity.

In the same way, the ATM and AAL layers communicate with each other via the ATM primitives **ATM-DATA.request** and **ATM-DATA.indication**. The AAL layer uses the ATM-DATA.request primitive for its lower-layer ATM to request transport of the attached SDU attached to its peer. The second primitive has the same meaning as described earlier for the physical layer primitive.

Adaptation layer protocols: As discussed above, the B-ISDN interface accepts information from different sources and converts, translates, or adapts it on the ATM cells for transportation through the B-ISDN network. The information from different sources may have different requirements for bandwidth, data rates, throughput, type of data generated from the source, etc. A 64-Kbps PCM voice source generates a continuous bitstream. LANs will generate variable lengths of frames (or packets), while in some applications, the ATM concept may be used for higher throughput.

An ATM cell must be capable of handling or adapting all these types of signals so that they can be transmitted over B-ISDN networks. These services, in general, can be divided in two categories: continuous bitstream and burtsy information. The continuous bitstream category includes PCM voice and video application sources, while bursty information includes data from LANs, subnets, etc. As such, ATM defines two adapter functions, one for each category of information (continuous bitstream and bursty), as shown in Figure 19.30.

For bursty data, the cell error (end-to-end) rate must be very small. The adaptation functions defined within ATM can be considered the ATM adaptation layer (AAL) on top of the ATM layer. The AAL may receive different types of data from different sources, e.g., signaling, OA and M, or user information from the session or lower layers of OSI-RM. A layered model for protocols handling these situations is shown in Figure 19.31.

These protocols are used by different types of terminals, and ATM-based networks must adapt these terminals in such a way that the incompatible protocols and systems may be handled by the ATM network in a uniform end system environment. Since ATM

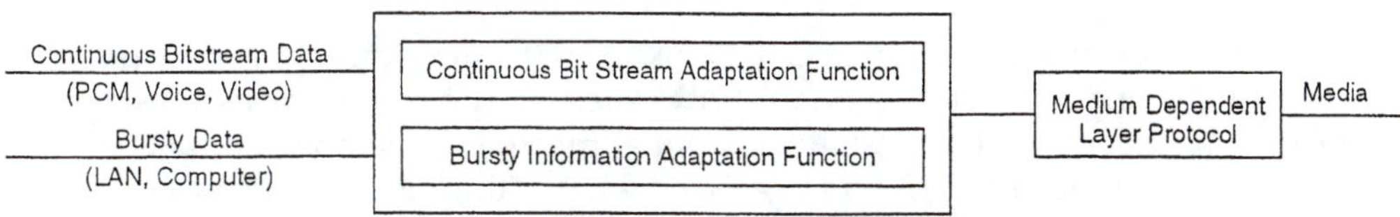

Figure 19.30 End system protocol architecture.

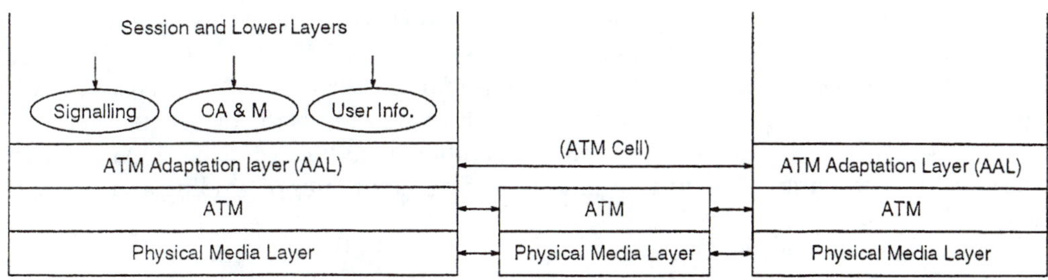

Figure 19.31 Layered ATM protocol architecture.

defines cells of fixed size, one of the most important functions of the adaptation layer is segmentation and reassembly, which basically segments all the incoming data of different services into cells of fixed sizes. On the receiving side, the original information is extracted from the cells to route to the appropriate users/devices.

The AAL protocol offers various functions which meet the requirements of user's applications and also provide an end-to-end incompatibility environment. Since the ATM layer provides a fixed-size cell, the AAL must provide segmentation of all the incoming data SDUs and also provide reassembly of ATM cells to form SDUs. The AAL protocols must detect any errors in SDUs and also correct the errors (if requested), thus providing error recovery. ATM networks usually handle the problem of congestion control by discarding ATM cells, and the technique may not provide any relationship between discarded cells and the next cells (even if both cells belong to the same SDU). The AAL protocol must handle this problem of loss of discarded cells for its recovery.

Core functions of AAL: Various services supported by the adaptation layer protocol are usually selected from the parameters of QOS, and a particular service depends on the set of parameters chosen and the service user. For example, if the service user is LLC, the user has a selected parameter of QOS for LLC, and this user service needs only segmentation and reassembly by the AAL protocol (no error recovery, as it is supported by LLC itself). But if the user service is X.25 PLP (packet-level protocol), then the AAL protocol must provide these functions: segmentation, reassembly, error detection, and error correction.

In the case of user services other than LLC or X.25 PLP, all the functions of AAL are required for user services. For any of the service users, the AAL protocol defines a list of core functions which will be used, and other functions will be provided on top of these core functions, depending on the type of service user selected.

A typical set of core functions of AAL protocols is given below:

- Segmentation of SDUs to fit into an integral number of ATM cells.
- Definition of the delimiting function via the ATM cell and also the last octet within the cell which defines SDUs.

The above list may not be complete, but looking at the functions of the AAL protocol, these are required in nearly all the services. Error detection and correction, re-transmission, and related functions may be considered optional functions. The optional functions are usually defined by different SDUs and have to be selected during multiplexing of different applications into one ATM connection.

The above discussion on the functions of AAL protocols is mainly for the bursty data coming out of LANs, hosts, etc. The functions of AAL protocol for a continuous-rate stream are different in nature, although some of the core functions will be the same. Since the bitstream is continuous in nature, for the circuit-switched emulation technique used in ATM, incoming bits are accumulated into ATM cells. The ATM cells are transmitted over the B-ISDN network and the original bitstream is extracted from the ATM cells. The AAL protocol must handle cases of lost cells due to transmission error or queue overflows, delay caused by transmission in the cell, delay jitter introduced by the ATM network, incorrect delivery of the cell by virtual circuits, and different rates of sending and receiving the cell.

One of the main reasons for the loss of a cell is different rates of sending and receiving of ATM cells, and this happens when the source and destination clocks are not synchronized. Usually, the same network clock may be used for the synchronization. In applications where this is not possible, the source adapter logic circuit (informing about the phase

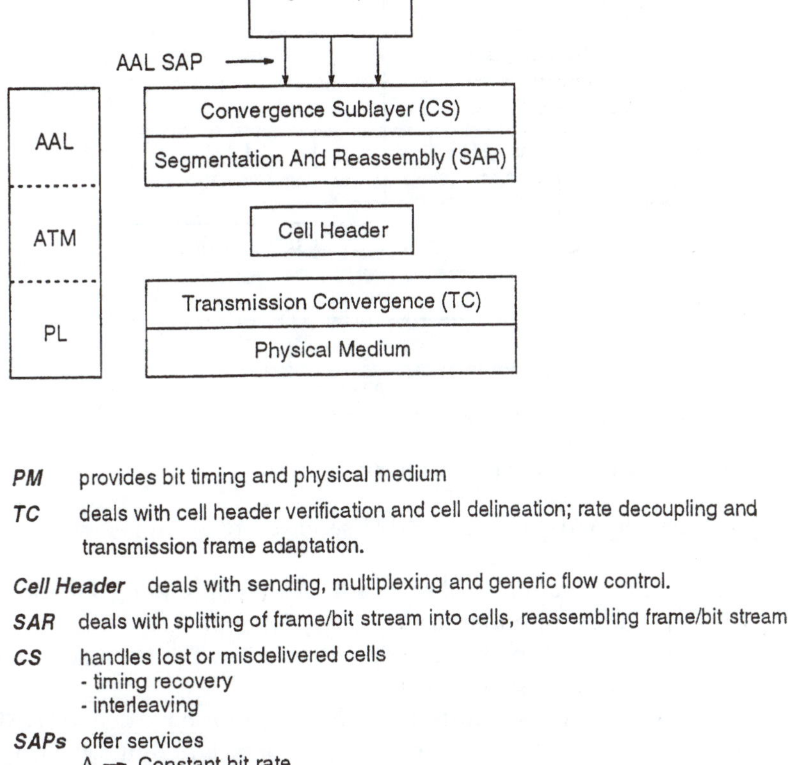
Higher Layers
AAL SAP
Convergence Sublayer (CS)
Segmentation And Reassembly (SAR)
AAL
ATM
Cell Header
Transmission Convergence (TC)
PL
Physical Medium
PM provides bit timing and physical medium
TC deals with cell header verification and cell delineation; rate decoupling and transmission frame adaptation.
Cell Header deals with sending, multiplexing and generic flow control.
SAR deals with splitting of frame/bit stream into cells, reassembling frame/bit stream
CS handles lost or misdelivered cells
- timing recovery
- interleaving
SAPs offer services
A → Constant bit rate
B → Variable
C → Connection-oriented
D → Connectionless
AAL - ATM Adaptation Layer
PL - Physical Layer

Figure 19.32 Protocol reference model.

difference between the source and network clocks) is transmitted. The destination adopter logic circuit obtains the information from the cell received. The protocol reference model of the ATM-based network follows the layered protocol reference model of OSI, as shown in Figure 19.32. The flow of information under the control of the user and control planes activity for data transfer is shown in Figure 19.33.

Classes of ATM services: CCITT has defined four classes of ATM services:

- **Class 1 or A:** Circuit emulation (transport of 2-Mbps or 45-Mbps) signals, constant bit rate video. This class of service includes the applications possessing a continuous bit rate (e.g., pulse code modulation telephone system).
- **Class 2 or B:** This class of service includes the applications of compressed video and audio in which variable data rate non-data information is transmitted.
- **Class 3 or C:** This class of service includes applications of connection-oriented data transmission.
- **Class 4 or D:** This class of service includes applications of connectionless data transmission.

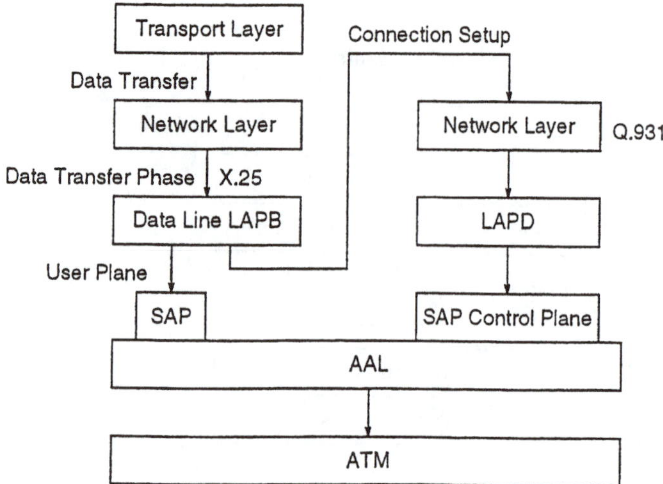

Figure 19.33 User and control planes for data transfer.

Various protocols for AAL to support ATM services are given below:

- **AAL 1:** This protocol supports Class 1 or A ATM application (continuous bit rate).
- **AAL 2:** This protocol supports Class 2 or B ATM application (variable bit rate).
- **AAL 3/4:** This protocol supports Class 3 or C and 4 or D applications (connection-oriented and connectionless services).
- **AAL 5:** Due to the complex nature of AAL 3/4, vendors have defined AAL 5, which provides limited function in the area of error control. It supports error detection only without error recovery and has lower processing and bandwidth requirements. It is also known as the **simple and efficient adaptation layer (SEAL)**.

These classes of services and protocols of AAL are partially derived from References 3, 5, 6, 8, 13, and 42.

19.9.4 *Structure of ATM cell*

CCITT's recommendation offers one channel for ISDN over which different types of signals (data, video, and voice) are sent, and this channel is defined as broadband user–network interface (UNI). This channel carries small fixed-length packets known as cells. The routing and other control information are defined in a header of five bytes (of 53-byte cells). The switching equipment (including switching techniques, multiplexing, modulation, etc.) must provide either packet-switching (public networks, X.25, frame relay, etc.) or circuit switching for routing the packets across the network. In narrow-band ISDN, the circuit-switching technique offering several digital channels supports a UNI over which different types of information (data, video, voice, etc.) can be transmitted. Although a clear distinction can be seen between the workings of N-ISDN and B-ISDN, the CCITT views both ISDNs as the same ISDNs with minor differences in accessing of the data.

The ATM information has two fields — header and user data — as shown in Figure 19.34(a), and Figure 19.34(b) shows the direction of transmission of ATM cells. The routing and other control information (no explicit addressing) are defined in the header, while the user data includes data, voice, video or any combination of the information. Since the cells are smaller and of fixed size (in contrast to variable size in X.25 and frame

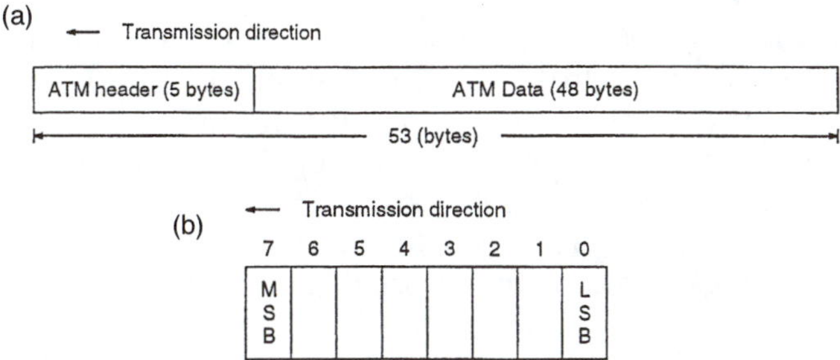

Figure 19.34 ATM cell structure. (a) Header and user data fields. (b) Transmission direction.

relay public networks), it is not justified to explicitly define the address in the cell. The switching equipment, after receiving the incoming call, looks at the routing label on its input port. It looks at its routing table and determines the output port for the call and associated label. It assigns this new label to the call for that output port. This new label will then be used by either switching equipment, and the same process is repeated by the destination.

The routing table is usually defined in advance (based on the topology and many other performance parameters or performance-related parameters of the network). The allocation of routing can be assigned with the cells either in advance or dynamically as the cells travel across the network. In both cases, the connection is established before the cells can be transported, as the ATM-based network offers connection-oriented service in contrast to connectionless service in many other public networks.

The connection setup is similar to the one established in N-ISDN for 64 Kbps. It can be set up either during the request call or dynamically (using Q.931 signaling protocol). Since an ATM-based network supports connection-oriented services, it delivers all the cells in the same order that they were received. There are two types of connections supported by the ATM-based network — point-to-point and point-to-multi-point. In each connection, the entire bandwidth is available to the users.

19.9.4.1 ATM cell transmission

The establishment of connection is defined in two steps:

1. In the first step, a virtual channel (VC) connection or link is established between two switching points, and usually this is defined in routing information contained in the ATM header. The concatenation of the virtual path identifier (VPI) defines the routing information in the header of the cell. The virtual channel links are defined in the look-up tables which basically define end-to-end connection, as shown in Figure 19.35.

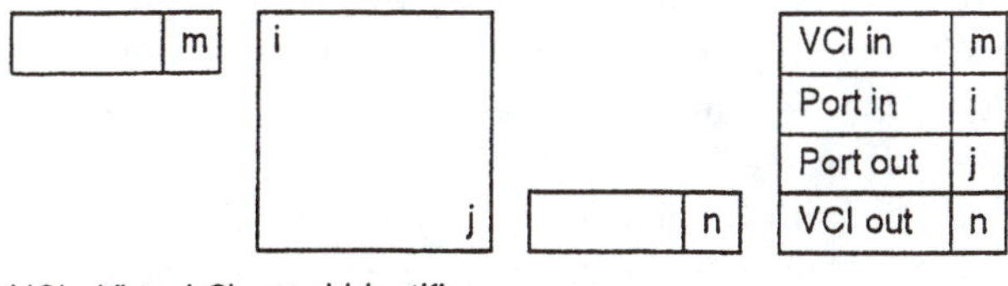

Figure 19.35 Label switching.

2. A virtual channel identifier (VCI) will be assigned a value for the calling node, which will write and send the information, while on the receiving side, a different value will be assigned to that VCI for reading the incoming information. During this time, the look-up tables are set up in the network. Similarly, different values will be assigned to respective VCIs for sending information in the opposite direction. In a virtual channel connection, the traffic-related parameters may also be associated with, for example, the quality of service parameters relative to the lost cell. These parameters, defined by associating attributes, are different for different directions of the transmission of information.

After the establishment of connection, the ATM cells are transmitted over VPs. The transmission of an ATM cell requires the transmission of a header in ascending order; i.e., the first octet is the **most significant octet (MSO)** of the header and is sent first. Within the octets, the bits are transmitted in the descending order of octets, starting with the **most significant bit (MSB)** within the octet, toward the **least significant bit (LSB)**.

The functions and services offered by the physical layer are described in CCITT recommendation I.432 and are used with different transmissions at different interfaces. The more popular and preferred medium is optical fiber, but other media such as coaxial cable have also been considered. Two types of interfaces for the physical layer have been defined for the transmission of ATM cells (the details of ATM cell coding, etc., can found in CCITT recommendation I.361).

19.9.4.2 Functions of ATM cell header

ATM networks support the connection-oriented mode, and control information such as source and destination addresses and sequence numbers of frames are not required. A virtual connection is usually assigned with a unique identifier number which basically finds its use logically per link during the virtual connection. As stated above, the ATM network provides a high quality of transmission services with a very low error rate, and usually the error control and flow control are not provided but can be offered to the user on demand in an end-to-end configuration. By the way, the X.25 provides error control and flow control (automatic repeat request protocols) as the basis of link-to-link configuration.

The ATM header also supports the priorities which are defined on the basis of logical connections and, obviously, it may not help in optional resource allocation strategies within network management but may help in the situation of traffic flooding. In this situation, we may want to lose message cells of a low-priority connection/channel. Two types of priorities have been defined in the ATM network header: **time** and **semantic**. In the former, some message cells will be in the network for a longer time and hence those connections will have more time transparency than the others. In the latter, some of the cells may be defined with a high probability of being lost, and as a result, we will have higher loss ratios for both the cells and their connections, giving rise to semantic transparency.

A few bits are used for the maintenance of the overall network and performance management of the ATM connection. Typically 0–2 bits are used for maintenance in the ATM network header. Other features of the ATM header include multiple access, header error protection, etc. Each of the features requires extra bits in the header. The material herein is partially derived from References 1–13 and 46–54.

The other function performed by the header is the identification and selection of virtual connection by unique identifiers. The header field is composed of two subfields: the virtual channel identifier (VCI) and the virtual path identifier (VPI). VCI defines a logical connection which will be allocated dynamically, while VPI defines the allocation of connection statically. As mentioned earlier, for each of the VCIs and VPIs, a virtual channel link (VCL) and virtual path link (VPL) are defined which basically provide the transport between the points where assigning and translation or removal of the assigned values take place. Further, the concatenation of VCLs is defined as a virtual channel connection (VCC), and a virtual path connection (VPC) is defined by concatenated VPLs.

19.9.4.3 Virtual channel connection/virtual path connection (VCC/VPC)

The VCC/VPC are defined between various points in the architecture of B-ISDN, such as user-to-user, user-to-network, and network-to-network. All the ATM cells associated with individual VCC/VPC are transferred over the same route established within the network irrespective of the applications, bandwidth requirements, and data bit rate requested. The VCC/VPC offer different functions at each of the interfaces mentioned above.

VCC is concerned with carrying user data and signaling at the user-to-user interface and allows access to the local connection (user-to-network signaling). The VCC performs network traffic management and routing at the network-to-network interface.

VPC offers a transmission pipe at the user-to-user interface and traffic transfer from the user to the network elements at the user-to-network interface, and finally it transfers the traffic (based on the route predefined by the VCC) at the network-to-network interface.

The VCI subfield can provide multiple channels over the links. The links may have different data rates, and an appropriate number of bits needs to be defined in the VCI subfield. Typically, for narrow-band networks, VCI supports a few kilobytes per second, but for links like fiber optics which offer hundreds of megabytes per second, thousands of channels have to be defined. This requires more bits in VCI, and a typical size of a VCI subfield may go up to as high as 15–17 bits.

The identifier numbers assigned to VCIs are used during the connection setup phase and are released after the data transfer. These numbers will be reused by other connections. Each connection is typically characterized by the number which is assigned by connection-oriented mode of the ATM networks. The multiple services are supported by multiple VCI identifier numbers for various applications, e.g., video, telephone, TV, etc. Within each of these applications, various signals (audio and video signaling) will be transported over different VCI connections. The ATM network provides the combining/separating of these signals at the transmission and receiving nodes, respectively.

The VPI subfield provides support for a large number of connections simultaneously. This requires defining a virtual path, circuit, or network which is a kind of semi-permanent link, thus allowing development of an efficient and simple resource allocation strategy within the network management. It is interesting to note that these logical unique identifier numbers in VCI are translated into a node (ATM) number and support up to 18 bits for connection. The VPI subfield, on the other hand, typically defines 8–12 bits to support 4096 virtual paths, and a virtual path, in turn, defines a very large number of channels (typically up to 64,000 virtual channels).

Considering all of the functions and services offered by the ATM header, a typical value of bytes to support these ranges from 2 to 7 bytes. The number of bits required for each of the functions and features is as follows: VCI (8–16 bits), VPI (8–12 bits), priorities (0–4 bits), maintenance (0–2 bits), other features such as multiple access error protection together (0–16 bits), and reserved bits (0–6 bits). The CCITT has defined a total of 40 bits for these functions and features. CCITT uses only one bit for priority. One bit is reserved while the other functions are implemented/realized by more or less the same number of bits as defined in the ATC cell header.

The design of ATM over fiber uses the concept of multiplexing to take it further for higher link utilization. In ATM, during the user connection, the cells per second are defined in terms of average and peak bit rate. This allows a number of ATM connections that can be multiplexed onto a single UNI. This is possible with multiple input and output ports attached to ATM switches. For any ATM network node, the capacity of an output link should be greater than the average bit rate (in cells per second) of all the input links and also should be less than the peak bit rate of all the inputs. The peak and average cell rates have to be specified before virtual channel connections (VCC) can be established. If we choose a lower value for average and peak cell rates, a large buffer is needed to store the cells, and we may lose some cells. Further, the storage of cells in the buffer may introduce some delay as the size of the buffer increases. On the other hand, if we take the average and peak values on a higher side, the ATM may not be able to utilize the bandwidth effectively, and hence there will be a wastage of bandwidth.

As stated above, in both B-ISDN and ATM protocol reference models (based on ITU-T), three planes — user, control, and management — have been defined within the layer and management layers. The ATM Forum has developed specifications for managing ATM networks based on simple network management protocol (SNMP) defined for the Internet. The interim local management interface (ILMI) works with ATM SNMP.

ATM in a LAN environment: The ATM for a LAN environment offers very high bandwidth (on demand) and much better performance with very high data rates for data, video, and voice. ATM can be used as a backbone, and routers associated with it offer interfaces to different LANs such as HSSI, DSI, DXI, and so on. Different types of ATM hubs can be defined, such as router-based, ATM-based, or a combination of these. In the router-based hub, all the routers are connected together and ATM provides interfaces between different LANs and routers. The ATM contains interfaces for different types of LANs and also interfaces with routers and hence becomes the access points for LANs. The routers so connected form a router-based hub. In an ATM-based hub, all the ATM switches are connected together, and the routers provide interfaces between different LANs and the ATM. Here the router defines interfaces to various LANs and hence becomes the access points for LANs. The combination of these two types of hubs can be obtained by combining these two classes of hub architecture and is usually used in large corporations. For more details, see References 12, 13, and 46–54.

A typical configuration of using ATM for LAN interconnectivity includes the ATM LAN hub to which terminal switch ATM adapters are connected. The hub is also connected to various routers. The ATM hub provides an interface to the outside of the premises of an organization. The ATM hub offers OC3 (150-Mbps) interface and supports different types of LANs, e.g., Ethernet, token ring, FDDI, and adaptation interfaces. The adaptation interface is responsible for the adaptation function (transmitting ATM cells) of ATM.

ATM in a WAN environment: WAN interconnectivity is slightly more complex than that of LANs. The WAN, as said earlier, is used for greater distances than LANs. The LANs use their own link to provide services, while WANs use the public carrier or a private line. The LANs and WANs can communicate with each other through an interconnecting device known as a channel service unit/data service unit (CSU/DSU), which provides mapping between the protocols of LANs with those of the public carrier (such as T1 or any other link) or private lines. The protocols on the public carrier or private lines use the TDM technique to accommodate the data from different sources. Unfortunately, the existing carrier supports data traffic only. Other types of traffic (e.g., video or voice) can also be handled by the TDM device via codec and PBX. For WAN interconnectivity, the ATM hub can be defined by connecting ATM switches via OC3 links, and then the ATM switches of the hub can provide interconnectivity with the LANs through the ATM LAN connected via OC3 links. The ATM switches from the hub can be connected to WANs via OC3/DS-3 links for access to the outside world.

A typical configuration of ATM for WAN interconnectivity includes the ATM public network connected to various ATM hubs via OC3 UNI. The ATM public network is composed of a number of ATM hubs interconnected via NNI. As mentioned above, the header of an ATM cell defines routing through two fields as VPI and VCI. The VPI is similar to TDM in circuit switching, while VPI uses ATM as a traffic concentrator (cross-connect). The ATM cross-connect replaces the existing digital cross-connects. A number of carriers have provided ATM PVC services, including AT&T, MCI, Sprint, Wiltel, and others. The ATM service environment using ATM offers the following features: transmission pipe composed of different types of traffic (video, data, audio, multimedia) via statistical multiplexing, bandwidth on demand that is available and allocated dynamically, a single universal network (no need for separate networks like data, voice, video, or multimedia), greater economy and reduction of overhead and administrative costs, and high-speed access to the network starting at DS-1 or T1 (1.544 Mbps).

Another WAN environment based on ATM includes various TDM devices connected together via T1 carrier lines defining a TDM hub. The TDMs are connected to various routers located at different locations miles apart. The routers are used as access points for the LANs (as discussed above). There are a number of ways of providing WAN interconnectivity. In one WAN interconnectivity method, each TDM may be replaced by an ATM switch, and these switches are connected via either T1 links or T3 (DS3) links (45 Mbps). The ATM switches transmit cells between them for the traffic coming out of the routers. The ATM hub supports different types of traffic (video, data, and voice). Here the ATM hub defines a MAN/WAN backbone network. In another WAN interconnectivity method, the ATM switch may be connected via OC3 links and also may be directly connected to various routers and video services through video servers. It is interesting to note that with OC3 links, the optical signals carry the information which may allow us to use the public services available on the ATM-based backbone public network. These networks are connected to ATM switches via OC3 links.

Gigbit testbeds: Research in the area of broadband networking testbeds offering gigabits is being sponsored by the National Science Foundation (NSF) and Advanced Research Projects Agency (ARPA). The work on testbeds started in 1989 and is managed by Corporation National Research Initiative (CNRI). Some of the gigabit networking testbeds include Blanca, Aurora, Nectar, Project Zeus, CASA, VISTA, MAGIC, BAGNET, ACRON, BBN, COMDisco, and so on. For more details on these testbeds, interested readers may read an excellent book on broadband communications written by Balaji Kumar.[53] Other countries such as Canada, Europe, Germany, England, France, and a few Asian countries have also proposed ATM/broadband testbeds. Two intercontinental ATM/broadband projects have also been conducted and are operational. One project is between AT &T and KDD of Japan which extends more than 9000 miles and consists of AT&T GCNS — 2000 ATM switches at Holmdel, NJ and Shinjuku, Japan. It offers a data rate of 45 Mbps (T3). Various applications over this network include data, videoconferencing, voice, and multimedia. The second network is installed between New York and Paris. This network provides connectionless broadband data service over ATM virtual paths. This offers high-speed interconnectivity for Ethernet LANs located in these two places. It uses Thompac 2G ATM switches.

19.9.4.4 ATM cell header format

Since two types of interfaces have been defined — **user–network interface (UNI)** and **network–node interface (NNI)**, there are two ATM cell header formats defined. This subsection discusses the two formats for these interfaces.

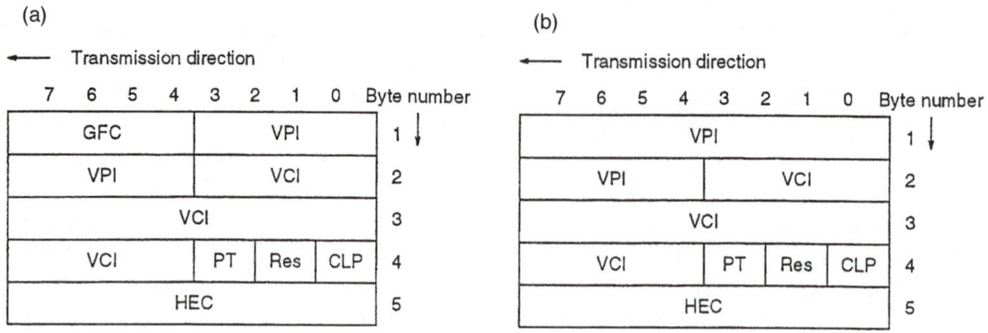

Figure 19.36 Header of ATM cell. Header format is different at (a) user–network interface (UNI) and (b) network–node interface (NNI).

ATM cell header format for UNI: The ATM cell header format for UNI is shown in Figure 19.36(a).

In the first byte of the header, the first four bits are reserved for generic flow control (GFC). It is important to note that many of the details of GFC like functionality, coding, etc., have not been standardized. The lower four bits define VPI. The second byte defines four bits each for VPI and VCI, while the third byte is reserved for VCI. The fourth byte has the higher four bits reserved for VCI while the lower bits are defined for the payload type (PT), reserved bit (RES), and cell loss priority (CRP). The last byte defines the header error control (HEC). If we see the format, we find that the routing field is defined by two subfields, such as user data, maintenance control, error control, etc.

The additional bits in the PT field may be useful for inserting special cells per virtual connection, and these cells are treated like normal cells but include special maintenance or error control information. Two header bits are used for this field. The cells which carry the user data information are usually assigned with an identifier of PT00. Since we are using two bits, the combination corresponding to PT00 is used for cells while the other combinations used for user information and network information are yet to be standardized. The payload of user information cells carry user's information and functions of the adaptation layer; i.e., the payload information includes the information required by the network for its operation and maintenance. The cells are usually inserted at a specific location within the network at switching nodes, multiplexer nodes, or end-to-end nodes.

The first bit, LSB, is the fourth byte which represents cell loss priority (CLP), which defines the priority associated with the cell. If CLP = 0, it means that a high priority is assigned, while if CLP = 1, it indicates that the cell may be discarded. The **reserved (RES)** bit allows for future enhancement in the cell header and at present is left unused. As usual, the default value of this bit is 0. The fifth byte of the header defines the **header error control (HEC)** of eight bits. It provides error control for the entire cell header. These eight bits provide error control for both single-bit or multiple-bit errors. The error control uses the similar concept of defining polynomial generates (as discussed in FCS of data link protocol). The value of HEC is calculated by multiplying the header bits which computes the polynomial generator by 8. The resultant bit pattern is divided by the polynomial generator $x^8 + x^2 + x + 1$ (standard for HEC). The remainder of this division process appears as the header bytes of the HEC field. On the receiving side, the reverse process is performed, and if the difference between the receiver and computed value (at the receiving side) is 0, it indicates an error-free condition; otherwise, it indicates an occurrence of error.

ATM cell header format for NNI: The ATM cell header format for the other interface of the physical layer, the network–node interface (NNI), is shown in Figure 19.36(b). In this format, the GFC field is replaced by VPI and the VPI field now includes 12 bits (in contrast to eight bits in the UNI ATM cell header format). The physical layer defines one bit (CLP) in the fourth byte of the ATM cell header which distinguishes between the types of cells which can be transmitted to the ATM layer. There are some cells which are received by the physical layer and are not passed to ATM, while there are some which are passed on to ATM. CLP = 0 indicates the cells to be forwarded to the ATM layer.

19.9.4.5 Types of ATM cells

The cell is a very important component for B-ISDN and accordingly has been defined as a block of fixed length. It is identified by a label at the ATM layer of the B-ISDN protocol architecture model. The CCITT I.321 has defined the following six types of cells and also includes the details of these cells:

- Unassigned/assigned cells
- Idle cells
- Valid/invalid cells
- Meta-signaling cells
- General broadcasting signaling cells
- Physical layer OAM cells

The unassigned cells don't carry user data and don't require any action at the receiver ATM node, as these cells have not been assigned any value. However, they require a cell destination function to be performed on them by the convergence layer. After cell rate decoupling, these cells are discarded. Another name for unassigned cells can also be found in CCITT's recommendation — *idle and empty cells*. The unassigned cells are transmitted to the ATM layer while the idle cells are not, and thus idle are available only at the physical layer. If the sender does not have any information to send, the unassigned cells are transmitted to allow asynchronous operations to be performed at both the transmitting and receiving nodes.

Valid cells are defined as cells whose header does not have any error or has not been modified by the cell header error control (HEC) verification process. The HEC is capable of correcting single-bit errors and it also may detect certain types of multiple-bit errors. The cell delineation field is used to identify the boundaries of cells being transmitted. The cell delineation method requires the identification of a co-relationship between the bits of the ATM header (to be protected) and the appropriate control bits defined in the HEC field of the header. This is achieved by using an associated polynomial generator (already defined).

The meta-signaling cells are used (as the name implies) for signaling of VCI and other resources between the users. The general broadcast signaling cells, on the other hand, broadcast the information to all connected terminals at UNI. Finally, the physical layer OAM cells include the maintenance information required for the physical layer.

It is worth mentioning here that the valid/invalid cells, meta-signaling cells, and OAM cells are defined for the physical layer, while assigned/unassigned cells are defined for the ATM layer.

19.9.4.6 ATM header parameters

CCITT I.121[25] recommended two data rate interfaces for B-ISDN — 150 Mbps and 620 Mbps. The ATM-like network is not only restricted to B-ISDN interface, but the

IEEE 802.6 MAC layer from the MAN known as distributed queue dual-bus (DQDB) can also be used with ATM. The DQDB will be its physical interface using DS3 (44.736 Mbps) which conforms to ATM with the intention of migrating to B-ISDN. Two committees are set up to look into the format of ATM headers. One committee appointed by ANSI, known as TISI, has the main function to develop positions on ISDN standards. The other committee from Europe, known as the European Telecommunication Standards Institute (ETSI), coordinates telecommunications in Europe.

As discussed above, the ATM cell header includes control information and virtual circuit identification labels. The header field of a cell is processed at each switching node, and it does not contain information for end-to-end transmission. The following is a list of functions which have been identified and agreed upon for the ATM cell header.

- **Access control (AC):** Sharing of a single access link by multiple terminals (eight bits).
- **Virtual circuit indicator (VCI):** Association of cell with virtual circuit (20 bits).
- **Operations, administration, and management (OA and M):** Testing of operations of virtual circuit without interfering with end user's use of circuit.
- **Priority:** Assigned for a cell on virtual circuit.
- **Header check sequence (HCS):** Error detection and correction on cell headers only.

The U.S., Japan, and Australia want a cell header of 5 bytes and information field of 64 bytes, while Europe wants 3 and 32 bytes for header and information, respectively. The CCITT has recommended an ATM cell of 53 bytes, of which the header is defined by five bytes while the remaining bytes are used by the information field.

It was a difficult task for standards organizations to arrive at an acceptable size of cell, header, and information along with ATM header parameters. Various issues considered during the standardization process of ATM cells are listed below:

- The same size of header can be used for a larger size of cell, offering more bandwidth.
- Wastage of bandwidth in large cells during the transmission, e.g., an acknowledgment.
- The rate of processing of headers at switches and terminals is reduced for large cells (due to long inter-arrival time between cells).
- Reduction in variance of delay of the network in shorter cells.
- Reduction of delay jitter with shorter cells for a given link utilization.
- Reduction in packetization delay with shorter cells.

19.9.4.7 Performance parameters of ATM cells

The variable size of cells, packets, or even frames is used efficiently in data communication, as it does not require any synchronization between bit patterns on the receiving side. But in the case of graphics, multimedia, video, and audio applications, the variable size of cells does not work, and hence ATM does not provide any real benefit for their transmission. The standards and specifications for flow control, error control, and congestion control for WAN-based ATM networks and also transport layer implementation are still under consideration before ATM can become a viable network like X.25 and TCP/IP.

The ATM Forum has designed two categories of performance measurement — accuracy and dependability for errors and speed for delay. Various parameters for the first performance measurement (accuracy and dependability) include the following: cell error ratio, cell loss ratio, cell misinsertion ratio, and severely errored cell block ratio. The speed performance measurement includes the following parameters: cell transfer delay, mean cell transfer delay, and cell delay variation. Every network experiences a bit delay during transmission. ATM offers a very low error bit rate compared to other networks because of the small size of the cell and the fact that ATM is designed to run over fiber in both public and private networks. A single bit error causes the re-transmission of the packets. If we transmit fewer bytes during re-transmission, it will improve the utilization of the network. For example, if there is a single bit error in LANs, the entire frame (typically 1600 bytes in Ethernet, 18,000 bytes in a 16-Mbps ring network) has to be sent, a wastage of network bandwidth.

The CCITT I.35[23] recommendation identified the following problems to be discussed during the performance measurement of ATM-based networks:

- Successfully delivered cell
- Errored cell and severely errored cell
- Lost cell
- Inserted cell

We have discussed above the drawbacks of ATM-based networks, and most of these drawbacks are based around the cell. These unresolved problems include loss of cells, cell transmission delay, cell delay variation, voice echo and tariffs, etc. In order to define performance measures of ATM-based networks, the following performance parameters for ATM cells have been defined:

- **Cell loss ratio (CLR):** This parameter is defined as a ratio of the total number of lost cells to the total number of cells transmitted. The total number of cells transmitted includes the cells which are lost during transmission and the number of cells successfully delivered at the destination nodes.
- **Cell insertion ratio (CIR):** This parameter is defined as a ratio of the number of cells inserted into the transmission path over a period of time which in turn is expressed by the connection second.
- **Cell error rate (CER):** This parameter is defined as a ratio of the number of cells with error (also known as errored cells) to the number of cells successfully delivered at the destination nodes.
- **Cell transfer delay:** This parameter consists of two types of time parameters: **mean cell transfer delay (MCTD)** and **cell delay variation (CDV)**. The MCTD is given by the arithmetic average of a specified number of cells' transmission delay. The CDV is expressed as the difference of a single observation of cell transfer delay and the mean cell transfer delay over the same VCC.
- **Cell transfer capacity:** This parameter defines the maximum possible number of cells delivered successfully over a specific ATM connection over a unit of time. This parameter is defined for each of the ATM connections and is averaged over a unit of time for the transmission path.

19.9.4.8 ATM packetization delay

The ATM-based network defines two types of connections for virtual channel and virtual path as VCC and VPC. These connections are defined by concatenating various VC and VP links established within the ATM layer. The functions required to transmit the cells over these connections in an ATM-based network include generation of cells, transmission of cells, multiplexing and concentration of cells, cross connection of cells, and switching of cells. These functions are defined within the ATM layers for the transmission of cells through ATM networks. The generation of cells is carried out by a routine known as

packetizer or **ATM-izer** in B-ISDN terminals. This routine maps the PDUs of higher layers into a fixed-size information field of 48 octets and control transmission information in the cell header. This mapping of the user's information into the cells is provided at the B-ISDN terminals.

The packetization is needed to provide mapping between non-ATM information and ATM cells. It may accept data from STM or non-ATM devices, and in both cases, it will map the information into cells. Usually this type of mapping or conversion is performed at network terminating 2 (NT2) of B-ISDN which provides interface to both ATM and non-ATM data packets. Similarly, a de-packetization is needed on the receiving side whenever we have to convert the ATM data into non-ATM devices. The use of packetization (conversion of non-ATM into ATM) and de-packetization (conversion of ATM into non-ATM) will cause delay known as **packetization/de-packetization delay**.

Some of the main objectives during the design of ATM-based networks should be to reduce the number of conversions, use more echo cancellation devices (compared to telephone networks), and utilize the bandwidth of the cell information field (48 octets) optimally. The packetization delay may be reduced by half if the cell information field is half-filled.

In the ATM switching connection, delay is caused by a number of conversions (ATM to non-ATM) and is usually known as packetization delay. The ATM cell offers 48 bytes for the information field, and using 8-KHz frequency for sampling, the delay at the cell is given by 48 bytes/8 KHz = 6 ms. This delay will be caused each time a signal is coming from 64-Kbps ISDN. Similarly, on the receiving side, the same value of delay will be caused for de-packetization. Another delay is caused for STM to ATM conversion and vice versa within one connection. Each type of conversion usually causes a delay of one ms.

CCITT I.311[28] further defined a set of traffic control operation functions for defining a desired performance of ATM-based networks and these operations are given below:

- Connection admission control function
- Usage parameter function
- Priority function
- Congestion control function

19.9.4.9 Vendors of ATM products

There is a large list of various products of ATM and many vendors are engaged in manufacturing those products. Some of the products include chips, interfaces (UNI, NNI, etc.), adapters, switches for LAN interconnectivity, switches for hubs (router-based, ATM switch-based), backbone switches, switches for accessing broadband services, ATM backbone networks, and test equipment to test the products for their compliance with standards. The material presented herein is partially derived from References 53 and 54.

AT&T has already announced a switching system for its network system's ATM cell-switching products — GCNS-2000. This switch is defined for global network service providers so that broadband applications can be implemented. It is based on a CCITT recommendation for ATM technology in future broadband applications which employ voice, images, and data over a common network. The network uses the ATM cell-relay concept for the transfer mode (switching and multiplexing). It offers a switching capacity up to 20 Gbps. It uses eight optical inputs to the switch fabric, each running at 2.5 Gbps. These high-speed inputs carry multiplexed inputs of many lower-speed ports with access speeds from 1.5 to 155 Mbps. The optical lines are also used as 2.5-Gbps output lines as ATM cells are routed through a switch and then de-multiplexed for routing to an individual destination port.

The components needed to create an ATM-based LAN environment include chips, adapters, and LAN switches. The function of an ATM chip is to implement the entire broadband protocol architecture and support various interface speeds by mapping cells into those payloads. There are chips available for ATM adapters, ATM access switching, backbone switches, etc. Some of the vendors manufacturing hardware chips for ATM networks include Applied Micro Circuit Corporation, Base2 Systems, National Semiconductor, QPSX, Saturn Synoptics, Chip Development, Texas Instruments, Transwitch, and Vitesse Semiconductor.

The ATM adapter interface card is similar to a network card and is plugged into a workstation or PC bus. The function of the adapter is to provide mapping between the message and cells and route them through the ATM network. The appropriate information regarding routing is defined by the VCI and VPI fields of the header and is performed by the adapter. Various vendors manufacturing the adapters are Adaptive and Fore Systems (for workstations like SUN, Silicon Graphics, DEC, Next, and HP). These adapters offer a data rate of at least 100 Mbps. Other vendors for adapters for lower data rates of about 50 Mbps are Newbridge for unshielded twisted pair for SBUS, NUBUS, Microchannel, ISA, EISA, etc.

LAN switches connect the LANs with the ATM hub and use star topology for different types of LANs such as Ethernet (16 Mbps), FDDI (100 Mbps), TAXI (100 Mbps), and workstations (45-Mbps UNI and 155-Mbps SONET OC3). The vendors are Adaptive and Fore Systems. Fore Systems, Inc., has also defined ATM switch interfaces, ATM switch SONET/SDH interfaces, ATM video adapters, network management software, ATM SBus adapters for Sun SPARCstations, etc.

The UNI or access interface (AI) supports T1, T3, and SONET/SDH formats of ATM. The NNI or trunk interface (TI) supports T3 and SONET/SDH. Various components of GCNS-2000 include switch fabric interface to fabric and UNIs and NNIs. It supports ATM, SMDS, and frame relay interfaces.

The ATM switch interface offers a line rate of 44.736 Mbps, supporting DS-3 private or public networks, can be connected to the DS-3 multiplexer and ATM DSU, etc. The ATM switch SONET/SDH offers a data rate of 155.520 Mbps and can connect SONET/SDH private or public networks over OC3c.

The ATM adapter provides optimized onboard cell-processing functions including segmentation and reassembly (SAR). This offers services to the ATM API interface via ATM API module which accepts ATM signaling. It also supports the TCP/IP for developing application software via its IP and both transport layer protocols (TCP and UDP) and uses the ATM signaling.

The ATM hub or ATM backbone network provides LAN, WAN, and switched X.25 interconnectivity and supports various interconnecting devices such as bridges and routers. The hubs are connected via fiber-optic cables so that a direct connection between them and high-speed FDDI can easily be defined. Within a hub, every component such as LAN switches will have assigned ports through which switching between different switches can be implemented. With the introduction of ATM, the ATM switches will replace the existing low-speed switches and still provide interfaces with the existing LANs. The ATM hubs offer a bandwidth of 2 to more than 10 Gbps. The interface cards offer data rates of 1.5 to 155 Mbps. Some of the vendors manufacturing ATM hub switches are Newbridge, VIVID, 3Com, DEC, Fibercom, Ungermann-Baa, Adaptive/Net, Northern Telecom, and Fore Systems. The details of each of the vendors can be found in Reference 53.

There are a number of vendors manufacturing the ATM producers for public networks, including AT&T, DSC Communications, TRW, Northern Telecom, Alcatel, and others. HP, ADTech, and other vendors are involved in manufacturing the testing equipment for ATM/broadband products.

19.9.5 Signaling system number 7 (SS7)

Signaling System No. 7 (SS7) provides common channel signaling to provide support to a variety of integrated digital networks. Within the network itself, a separate network known as a signaling network, which includes a stack of all seven layer protocols, is used. These protocols together provide common signaling to different types of services, which is known as common channel signaling system No. 7 (CCSS7 or SS7). It is more useful and flexible than the traditional in-channel signaling. CCITT approved a recommendation of changing in-channel signaling to common channel signaling in 1980 as an SS7 standard interface. This standard interface supports different types of digital circuit-switched networks, in particular integrated digital networks such as ISDN and now B-ISDN. It offers all the controlling functions within the ISDN architecture. It utilizes a 64-Kbps digital channel and provides a reliable signaling scheme to route information (control and user's data) for point-to-point configuration. For more details, readers are referred to References 51–53.

The control information packets perform various functions such as setup, maintenance, connection establishment and termination, controlling of information transfer, etc. These control information packets are sent across the circuit-switched network using a packet-switched control signaling scheme. Although the underlying network is a circuit-switched network, the control and use of the network are performed using a packet-switching technique.

The protocol architecture of the SS7 standard interface follows the layered approach (similar to OSI-RM), as shown in Figure 19.37. The protocol architecture is defined by four layers. The lower three layers together provide reliable connectionless services to the SS7 network (circuit-switched) for routing all the messages through it. These layers together are also known as the message transfer part (MTP), analogous to the subnet defined in

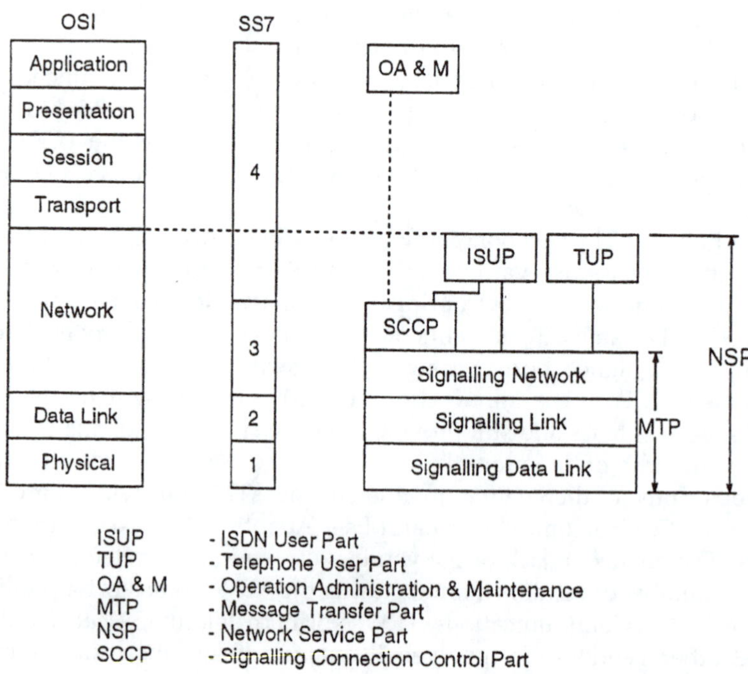

Figure 19.37 SS7 protocol architecture.

OSI-RM (the subnet is defined by the lower three layers of OSI-RM: physical, data link, and network layer). The lower three layers may be described briefly, as follows:

- **Layer 1:** The **signaling data link layer** provides electrical and physical support to various signaling links (e.g., signaling point, link signal transfer point, etc.). This layer is analogous to the physical layer of OSI-RM.
- **Layer 2:** The **signaling link layer** provides services similar to those of the data link layer (OSI-RM) and is expected to provide reliable logic connection and delivery of user's data across the signaling data link in a sequential order. The protocol of this layer is defined in CCITT I.440/441, which is same as Q.920/921 and is also known as LAP-D. The main function of this protocol is to transfer the messages (call setup) of layer 3. As we discussed earlier in Chapter 11, the LAP-D defines two sub-address fields of its address — the service access point identifier (SAPI) and terminal endpoint identifier (TEI).
- **Layer 3:** The **signaling network layer** is mainly concerned with routing of data across the signal transfer point links between the sender and receiver controls.

These three layers of MTP together provide a few functions of the layer of the subnet or subset (OSI-RM). The functions not supported by MTP include addressing and connection-oriented services. The layer 3 protocol is defined in I.450/451, which is the same as that of Q.931. This protocol is primarily concerned with sequencing of messages or packets which are exchanged over the D channel for setting up a call connection. The types of messages used by this protocol for different phases include alert, connect, connect ack, setup, and others (call establishment); user info, and others (data transfer); and disconnect, release, release complete, and others (call termination).

In 1984, CCITT introduced a new feature in SS7 which resides in the fourth layer of SS7 — signaling connection control part (SCCP). The MTP and SCCP together within layer 4 of SS7 define a network service part (NSP). The SCCP includes those functions of the network layer of OSI-RM which have not been defined in the signaling network and provides a variety of services to the users. The NSP is a user's information delivery system, while the remaining part of layer 4 defines the actual user's data. The telephone user part (TUP) is another feature within layer 4 which is invoked by a user's telephone request call. The functions provided by the TUP are the ones used in a typical circuit-switched network. The ISDN request calls are handled by the ISDN user part (ISUP), which provides control signaling to those ISDN request calls. The operations, applications, and maintenance (OA & M) part is primarily concerned with network management functions and parameters associated with operations and maintenance.

Some useful Web sites

Optical standards can be found at *http://www.siemon.com*, *http://www.bicsi.com*, and *http://www.ansi.gov*. A standards organization glossary is maintained at *http://www.itp.colorado.edu*.

The SONET interoperability forum site can be found at *http://www.atis.org*.

FDDI LAN standards are available at *http://www.server2.padova.ccr.it*.

Frame relay and its applications, services, and other implementation discussion are found at *http://www.frforum.com*, *http://wwwcomputerhelp.net*, *http://www.mot.com*, *http://www.3com.com*, *http://wwwdata.com*, *http://www.tek-tips.com*, *http://www.sangoma.com*, and *http://wwwodints.com*. Some of the frame relay vendors can be found at *http://www.internetwk.com*, *http://www.networkcomputing.com*, and *http://www.analysis.co.uk*. Frame relay standards are available at *http://cell-relay.indiana.edu* and at the Frame Relay Forum.

SMDS, protocols, vendors, and other related documents are available at *http://www.support.baynetwork.com*, *http://www.wanresources.com*, *http://wwwlads.com*, *http://www.dca.net*, *http://www.analysis.co.uk*, *http://www.tile.net*, *http://www.cell-relay.indiana.edu*, and *http://www.zdwebopedia.com*.

B-ISDN reference model, protocols, products, standards, and other related documents on B-ISDN are available at *http://www.bisdn.et*, *http://www.nhse.npac.syr.edu*, *http://www.protocols.com*, *http://www.bitcentral.com*, *http://www.fh.telekom-leipzig.de*, *http://www.techweb.com*, *http://www.internetwk.com*, and *http://www.atmforum.com*.

ATM protocols, products, news applications, and related information are available at *http://www.atmforum.com*, *http://www.atm.com.pl*, *http://www.telenetworks.com*, *http://www.encyclopedia.com*, *http://wwwsulu.lerc.nasa.gov*, *http://www.cables-and-networks.com*, *http://www.webopedia.com*, *http://www.atmdigets.com*, *http://www.att.com*, *http://www.net.com*, *http://www.chernie.fu-berlin.de*, *http://www.arl.mil*, *http://www.techfest.com*, *http://www.ora.com*, *http://www.atm25.com* (links to other ATM reference sites), and many others.

References

1. Kuehn, P.J., "From ISDN to IBCN (integrated broadband communication network," *Proc. World Computer Congress IFIP'89 San Francisco*, 1989.
2. Ross, F., An overview of FDDI: The fiber distributed data interface," *IEEE Journal on Selected Areas in Communications*, 7(7), Sept. 1989.
3. Lutz, K.A., "Considerations on ATM switching techniques," *International Journal of Digital and Analog Cabled Systems*, 1(4), Oct. 1988.
4. Byrne, W.R., Kafka, H.J., Luderer, G.W.R., Nelson, B.L., and Clapp, G.H., "Evolution of MAN to broadband ISDN," *Proc. XIII International Switching Symposium Stockholm*, vol. 2, 1990.
5. Minzer, S.E., "Broadband ISDN and asynchronous transfer mode (ATM)," *IEEE Communication Magazine*, Sept. 1989.
6. Decina, M., "Open issues regarding the universal application of ATM for multiplexing and switches in the BISDN," *Proc. IEEE, ICC'91*, 39.4.1–39.4.7, 1991.
7. Ballart, R. and Ching, Y.C., "SONET: Now it's the standard optical network," *IEEE Communication Magazine*," March 1989.
8. Schaffer, B., "Synchronous and asynchronous transfer modes in the future broadband ISDN," *ICC'88*, Toronto, Canada, 47.6.1–47.6.7, 1988.
9. Hemrich, C.F. et al., "Switched multi-megabit service and early availability via MAN technology," *IEEE Communication Magazine*, April 1988.
10. Bellcore, "Switched multi-megabit data service," *Bellcore Digest of Technical Information*, May 1988.
11. Ahmadi, H. and Denzel, W.E., "A survey of modern high performance switching techniques," *IEEE Journal on Selected Areas in Communication*, 7(7), Sept. 1989.
12. De Prycker, M., "Evolution from ISDN to BISDN: A logical step toward ATM," *Computer Communication*, June 1989.
13. Le Boudec, J.Y., "The asynchronous transfer mode: A tutorial," *Computer News and ISDN System*, 24, 1992.
14. Nikolaidis, I. and Onvural, R.O., "Bibliography on performance issues in ATM networks," *ACM SIGCOM Computer Communication Review*, 1992.
15. ISO 9314-1,-2,-3, "Fiber Distributed Data Interface (FDDI)" American National Standards Association, NY.
16. ANSI standard T1.105-1988, "SONET Optical Interface Rates and Formats," 1988.
17. Bellcore Technical Advisory TA-TSY-000772, Generic system requirements in support of switched Multi-Megabit Data service, Issue 3, Oct. 1989.
18. "More Broadband for Bell South," *Communications Weekly International*, 1990.
19. Revised Draft Recommendation G.703, "Physical/Electrical Characteristics of Hierarchical Digital Interfaces," CCITT SG XVIII, report, June 1992.
20. Revised Draft Recommendation G.707, "Synchronous Digital Hierarchy Bit Rates," CCITT SG VIII, report, June 1992.

21. Revised Draft Recommendation G.708, "Network Node Interface for Synchronous Digital Hierarchy," CCITT SG VIII, June 1992.
22. Revised Draft Recommendation, G.709, "Synchronous Multiplexing Structure," CCITT SG VIII, June 1992.
23. CCITT Draft Recommendation I.35B, "Broadband ISDN Performance," Com XVIII-TD 31, Matsuyama, 1990.
24. CCITT Recommendation I.113, "Vocabulary of Terms of Broadband Aspects of ISDN," Geneva, 1991.
25. CCITT Recommendation I.121, "Broadband aspects of ISDN," Geneva, 1991.
26. CCITT Recommendation I.150, "B-ISDN ATM Functional Characteristics," Geneva, 1991.
26. CCITT Recommendation I.211, "B-ISDN Service Aspects," Geneva, 1991.
28. CCITT Recommendation I.311, "B-ISDN General Network Aspects," Geneva 1991.
29. CCITT Recommendation I.327, "B-ISDN Functional Architecture," Geneva, 1991.
30. CCITT Recommendation I.321, "B-ISDN Protocol Reference Model and Its Application," CCITT SG XVIII, June 1990.
31. CCITT Recommendations I.361, "B-ISDN ATM Layer Specifications," Geneva, 1991.
32. CCITT Recommendation I.362, "B-ISDN ATM Adaptation Layer (AAL) Functional Description," Geneva, 1991.
33. CCITT Recommendation I.363, "B-ISDN ATM Adaptation Specification," Geneva, 1991.
34. CCITT Recommendation I.413, "B-ISDN User–Network Interface," Geneva, 1991.
35. CCITT Recommendation I.432, "B-ISDN User–Network Interface Specification," CCITT SG XVIII June 1990.
36. CCITT Recommendation I.610, "OAM Principles of the B-ISDN Access," Geneva, 1991.
37. CCITT Recommendation M.60, "Maintenance Terminology and Definition," Blue Book, Fascicle IV.1, Geneva, 1989.
38. American National Standards Association, "FDDI Token Ring Media Access Control MAC," ANSI X3.139, 1989.
39. ANSI Draft Proposal, FDDI Token Ring Single Mode Fiber Physical Layer Media Dependent-SMF-PMD," ASC X3T.9.5, April 1989.
40. Grosman, D., "An overview of frame relay technology," *Proc. Tenth Phoenix Conf. on Computers and Communications,* March 1991.
41. Marsden, P., "Internetworking IEEE 802/FDDI LANs via ISDN frame relay bearer service," *Proc. IEEE,* Feb. 1991.
41b. CCITT I.451/Q.931, "Layer User–Network Interface," 1991.
42. Handel, R. and Huber, S., *Integrated Broadband Network: An Introduction to ATM-Based Networks,* Addison-Wesley, 1993.
43. CCITT Recommendation I.430, "Basic User–Network Interface Layer 1 Specifications," Blue Book, Fascicle III.8, Geneva, 1991.
44. CCITT recommendation I.431, "Primary Rate User Interface — Layer 1 Specification, 3," Blue Book, Fascicle, Geneva, 1991.
45. CCITT Recommendation I.451/Q.931, "ISDN User–Network Interface Layer 3 Specification.
46. de Prycker, M., *Asynchronous Transfer Mode,* Ellis Horwood, 1993.
47. Sher, P.J.S., Jonathan, B.S., and Yun, K., "Service concept of switched multimegabit data service," *IEEE Globcom,'88,* Nov./Dec. 1988, 12.6.1-12.6.6, 1988.
48. Byrne, W.R., Papanicolaou, A., and Ranson, M. N., "World-wide standardization of broadband ISDN," *Int. Conf. on Digital and Analog Cable Systems,* 1989.
49. Ekhundh, B., Gard, I., and Leijonhufvud, G., Layered architecture for ATM network," *IEEE Globe Communication'88,* Nov./Dec. 1988.
50. Gechter, J. and O'Reilly, P., "Conceptual issues for ATM," *IEEE Network,* 3(1), 1989.
51. Donohoe, D., Johannessen, G., and Stone, R., "Realization of Signaling System No. 7 network for AT&T," *IEEE Journal on Selected Areas in Communication,* Nov. 1986.
52. Phelan, J., "Signaling System 7," *Telecommunications,* Sept. 1986.
53. Kumar, B., *Broadband Communications,* McGraw-Hill, 1994.
54. Goralski, W.J., *Introduction to ATM Networking,* McGraw-Hill, 1995.

Part VI

Client–server LAN implementation

chapter twenty

Client-server computing architecture

> "There's nothing so dangerous for manipulators as people who think for themselves."
>
> Meg Greenfield

20.1 Introduction

There has been a tremendous push for worldwide competition in recent years, and financial and economic sectors are gearing toward globalization for balancing the traditional centralized corporate control. This trend has forced the business community to adopt new techniques for increasing productivity at lower operating costs. This has given birth to a new concept of *re-engineering,* where the corporate-wide work-flow processes are redesigned instead of simply automating the processes within their organizations. The use of emerging technologies is being used to fulfill the performance and productivity targets and goals of corporate re-engineering. Large corporations usually make decisions for any future expansion or strategy based on the data stored in central databases or files residing on mainframes or mini-computers. The off-loading of processing and manipulation of data from these expensive mini-computers and mainframes should be accomplished on cheaper workstations that offer access to the host machine. This is exactly what is proposed in client–server computing. A *client* is an application program which runs using local computing resources and at the same time can make a request for a database or other network service from another remote application residing on the *server.* The software offering interaction between clients and servers is usually termed "middleware." The client is typically a PC workstation connected via a network to more powerful PCs, workstations, or even a mainframe or minicomputers usually known as servers. These are capable of handling requests from more than one client simultaneously.

The architecture of any system must possess the following attributes for solving any business network–based applications: maintainability, flexibility, structure, modularity, scalibility, adaptability, portability, interoperability, etc. These attributes help in defining a distributed computing architecture, flexible architecture (useful for continuous system evolution), and application to fulfill the performance and productivity goals of corporate emerging technologies (in particular the ones based on re-engineering). A flexible architecture which has been used in various business applications is generally known as **host-centered architecture**. This architecture typically allows users to enter their business-processing requests at dumb terminals (without any processing power) or even PCs. The request is sent across the wide area network (WAN) to the host processor which, after processing the requests, sends the result/reply back to the terminals. The host processor defines different logics (presentation, business rules and procedures, and data processing and manipulation, e.g., addition, deletion, updating, etc.). Some of the material herein has been derived from References 1–3.

20.2 Distributed computing architecture (DCA)

The evolution of data processing in the last two decades has seen the introduction of minicomputers, personal computers (PCs), and workstations, which not only have reduced the size of processing devices but have also added enormous processing powers with improved price/performance, software features/functionality, interconnection, etc. Although the overall configuration of host-centered architecture has not been changed significantly, the host processor of a mainframe now is being replaced by minis, workstations, and PCs, while the dumb terminals are replaced by bit-mapped terminals or even other PCs or workstations. The communication between them for data processing is now taking place over local area networks (LANs). Client–server computing offers an optimal solution in distributing computation, data generation, and data storage resources so that the users using this computing model may expect efficient, cost-effective data processing within the organizations and also across various departments/divisions of the organization.

The implementation of distributed computing architecture, on the other hand, is based on a number of approaches, but in general, the following two approaches have received wide acceptance from a number of users/vendors in various corporations, large organizations, universities, etc.

In the first approach, the entire business network–based application (to be developed) is implemented on PCs or workstations (due to price/performance advantages) and provides local autonomy and flexible functionality along with **graphical user interface (GUI)**. This approach offers advantages, as the end user now has complete control over the network application and can manipulate it (if required) via GUI. Advantages include established technology and simple architectural and functional support of system management and control. At the same time, various disadvantages include response time (depends on the usage), poor user interface, high backlogs, and high maintenance and resource costs. The discussion of GUI and its various features and applications is presented later in the chapter.

In the second approach, the end user needs to access the corporate data which requires a higher level of integration for centralized applications and resources like PCs and workstations, with the added features of availability and performance. This approach defines a distributed computing architecture at different levels of abstraction or layers and is based on layered architecture (similar to that used in OSI-RM of LANs). This architecture is based on traditional hierarchical **master–slave computing architecture**, with the only difference being that more processing and interactive (cooperative) capabilities are added among the layers of the architecture in this approach. Layered architecture offers many advantages, e.g., structure, modularity, interoperability, etc., and has also been used in non-OSI-reference models such as IBM's System Network Architecture (SNA), DEC's DECnet, and Siemens's Transdata. Here, the LANs allow the sharing of resources and offer user-friendly GUIs and flexibility. The disadvantages include high traffic on LANs, use of PCs for executing query commands, etc.

A typical distributed computing architecture based on the layered approach is composed of three layers — **processing**, **server**, and **terminal**, as shown in Figure 20.1. The processing layer (top) usually uses the most powerful processing computer with enormous storage capabilities (e.g., mainframe) for corporate data. The server layer (middle) provides

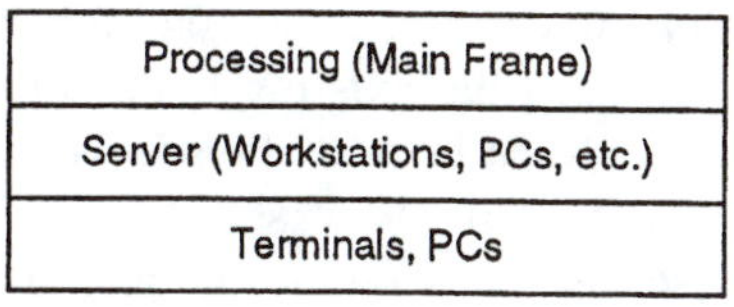

Processing (Main Frame)
Server (Workstations, PCs, etc.)
Terminals, PCs

Figure 20.1 Distributed computing architecture.

communications via LAN servers between the top layer and bottom layer (composed of PCs, workstations, terminals, etc.). The business request calls are submitted by terminals, PCs, and workstations at the bottom layer to the middle layer which provides a communication network. The request is transmitted over the network to the top layer mainframe computer which, after processing the requests, sends the results/replies back to the terminals, PCs, and workstations over the communication network.

Here, we may think of PCs and workstations at the bottom layer as the **client**, while the **server** layer functions as a server for the clients. The server (middle layer) may get requests from different clients (from the bottom layer) simultaneously and submit all of these request calls coming from different clients to the mainframe for their processing. The submission of the requests from the server to the mainframe is carried out on behalf of the clients, and hence the server can be considered a client for the mainframe, which is functioning as a server for this server layer. Thus, this layered architecture can implement multiple clients and multiple servers and other combinations for any data-processing application.

The client–server computing architecture alleviates some of problems mentioned in the above two approaches. Here, the architecture supports more openness, scalability, and better response and performance than previous approaches. The GUI tools and evolving technologies in communication networks and multi-vendor environments make this model far superior to previous ones, but this is true only when the model is implemented properly. The disadvantages include complicated solutions, new infrastructure, and development of a new methodology.

This layered architecture can be extended/defined either vertically or horizontally to adapt to new changes of the applications and usually depends on the configuration being used in the organizations. This has also effected changes in the organizations in the same way as in the layered model. In a typical organizational infrastructure, the top-level management is composed of research, manufacturing, marketing, and finance and provides necessary directions to the lower levels, horizontal updating to management, order management, and finally the client–server computing layers. A typical layered model will consist of a server at the top level which includes an OS/2 (IBM's new multi-tasking operating system) server, local database, and groupware and also a gateway to the host. The client workstations are connected to the server along with MS-DOS Windows user applications running on different machines.

For example, consider the following scenario. An organization has a central single host which is connected to different LANs within the organization via LAN server. Each of the divisions/departments of the organization uses its own LAN which is connected to PCs, workstations, printers, and other shareable devices of that particular division/department. The LAN server provides a communication network for all these PCs and workstations and sends the requests from these devices to the host computer (i.e., mainframe, workstations, etc.) at the top layer. We consider this architecture a vertical one, as it allows the updating of processing in the vertical direction without affecting the underlying architecture. At the top layer, more processing power and storage capabilities can be added while more LANs at various divisions/departments (at the bottom layer) can be added and will be serviced by the existing LAN server.

Horizontal architecture allows the addition of resources at each layer. For adding extra processing power, hosts (e.g., mainframe, workstations, etc.) may be located at different locations while different LAN servers may be located at different locations corresponding to different regions of data centers. The LAN servers are connected to the hosts via WAN. The LAN servers may be connected by different LANs at a lower level of the architecture. Thus, horizontal architecture is expanded in the horizontal direction. In general, large corporations or organizations use both of these techniques for expanding their architectures.

20.3 Components of DCA

The distributed computing architecture is very flexible and is based on the concept of distributed interactive or cooperative processing of network applications. Cooperative processing, in fact, is the driving force for any computing architecture, offering a high degree of interaction between various application components or segments within the architecture. The application components of any network business application may be organized in different ways, depending on the architecture considered. In order to understand the working of architecture for any business network application, we discuss the following logical components of network business applications which are defined in any typical DCA. The number of logical components may be different in different books or publications[1-3,15] or review articles,[7,14] but the total functionality of the entire DCA must be represented by them. The distribution of layers is not important at this point, but at a later stage it may become a serious problem and we may not get the real advantage from this model. In order to see the real benefits and advantages, we are breaking the entire business network applications into well-defined and established logic which can be designed and implemented separately.

- Presentation or user interface (UI) logic
- Business rules and procedures processing logic
- Database processing and manipulation logic
- Management function logic

20.3.1 Presentation or user interface (UI) logic

This logical component deals mainly with an interface between the end user's terminal or workstation and the required business network application. The tasks to be performed by this logic include **screen formatting, reading and writing, window management, and interfacing for input devices** (keyboard, mouse, and other pointing devices). The available interfaces offering GUIs include OS/2 Presentation Manager (IBM), Microsoft's Windows for PC DOS, X-Windows for Workstations (Microsoft), Open Look (Sun Company for UNIX), and Motif (Open System Foundation), and they have been used in various business data processing applications. Some of the material herein has been derived from References 2–5, 7, and 8.

20.3.2 Business rules and procedures processing logic

This logic accepts data from the presentation logic (i.e., screen-based) or database processing logic (i.e., database) and performs business tasks requested in the request calls on the data. It receives the request from presentation logic in the form of a **structured query**

language (SQL) statement and, based on the request, it sends it to the database logic. The business logic performs the processing on the data and sends the result back to the business logic, which can be seen on the screen at the presentation logic. The business logic interface is a part of the application and is implemented either by a third-generation language (3GL) such as COBOL and C, or a fourth-generation language (4GL).

20.3.3 Database processing and manipulation logic

This logic primarily deals with the processing and manipulation of data within the business network–based application. For the processing and management of the databases, a **database management system (DBMS)** is used in the case of data while a **relational database management system (RDBMS)** is used in the case of relational database techniques. Most of the DBMSs and RDBMSs use **structured query language (SQL)** statements for sending the queries, and their language, **data manipulation language (DML)**, is compatible with the business logic interfaces (3GL, 4GL, etc.). The actual processing on the data is performed by DBMS or RDBMS and is transparent to its higher logic layers. The application logic in the traditional host-centered architecture may reside on the same computer and be linked together in a hierarchical way to execute the entire business application. Here, the resources are limited to one platform at a central location.

20.3.4 Management function logic

This logic deals mainly with various management functions within the components of the network-based application to be implemented over DCA.

In the distributed computing architecture, we can distribute the data at different locations, and a single application can access the data from any location. This type of data processing and manipulation in a distributed environment offers various advantages such as availability, placement of the data close to the source, etc. Its main disadvantage lies with **singularity**, which affects performance, portability, etc. But if we distribute the business application processing along with the distributed data across the network, then the distributed resources (PCs, workstations, etc.) can be effectively utilized (due to their excellent price/performance characteristics). Various components of the distributed application cooperate with each other in the processing of the business logic of the application.

20.4 Client–server computing architecture

Client–server architecture, as shown in Figure 20.2, may be defined in a variety of ways, but we will consider the following definition, which by no means is a standard definition.

> Client–server architecture defines a computing environment where a network-based business application is divided into two processes **(front end** and **back end)** which run over multiple processors (client and server). These processors interact and cooperate (transparent to the user) with each other in such a way that the network application

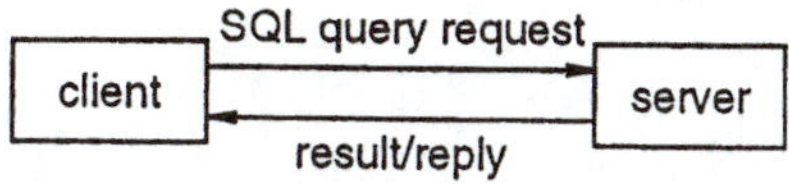

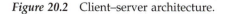

Figure 20.2 Client–server architecture.

> is completed as one unified task by them. Shareable resources are managed by the server processor, which offers services to the clients. The clients make a request to the server for data processing using a predefined language — structured query language (SQL) — over the communication network connecting clients and servers. The server, after processing the query, sends the results or reply back to the client. In some cases, the server may request another server to become its client (on behalf of the original client and so on), and the results or replies from its server are sent back to its original clients. For each of the query requests via SQL statements, only the results are sent back to the clients.

The client–server architecture contains two types of distribution within its implementation — **data** and **application** — in a highly distributed cooperative processing environment. The application logic and part of the business rules and procedure processing logic run on the client workstation, while the remaining part of the business logic and the database processing and manipulation logic runs on a high-performance and high-speed server (workstation). The actual distribution of various logics between client and server workstations depends on the application under consideration and is affected by many other performance parameters, e.g., throughput, response time, transit delay, etc.

We would like to differentiate between client–server computing architecture and file server systems, as both offer similar services but in a different implementation. Both systems offer resource sharing through networks and can access the data via a shared network.

The file server systems provide a remote disk drive which contains the files. The users can access the files on a file-by-file basis over networks. On the other hand, the client–server computing model offers full relational database services via predefined SQL and allows operations such as access of SQL, updating of record, insertion of record, deletion with full relational integrity backup/restore, and other transactions. As indicated earlier, these two application programs, client and server, interact with each other via middleware software, which defines the sequence of various operations, identification of the requests, determining which entity will process and when it will process (between client and server), etc.

Although PCs have limited processing capabilities for business processing and manipulation logic, the introduction of LANs removes this drawback and allows the PCs to request more processing power from the mainframe over the network. Thus, PCs are now not only considered terminals with limited processing power but are also used as file servers, print servers, and other peripheral device servers across LANs. The use of PCs as servers of various applications (programs, file, devices, etc.) has been adopted in a number of universities, organizations, and businesses. With this configuration, files, programs, databases, word processing applications, spreadsheets, graphics packages, computer games, printers, and many other resources can be shared by a number of users connected by LANs.

The PC servers maintain a separate queue for each of the resources. The popular servers based on PCs include **Novell NetWare**, **Microsoft MSNet-based software**, **LAN software** (e.g., **IBM PC-LAN**), etc.

20.4.1 Why Client–Server?

Due to the high processing power of workstations and information-sharing capabilities offered by LANs, LAN servers, database servers, and other application servers are becoming very useful and popular in **downsizing** and **upsizing** applications of the mainframe onto PC servers and LANs using cooperative processing environments and platforms (in particular, client–server architecture). This client–server architecture is well-supported by new **network operating systems (NOSs)**, database architectures, query languages, and graphical user interface (GUI) tools. It further offers a very important step toward the **re-engineering process**, where the corporate-wide workflow processes are redesigned totally using a different approach (open system–based multi-vendor platforms) that deviates from the traditional approach of only **automation**.

Downsizing of enterprise systems may be viewed in different ways. At the management level of any organization, we see a re-engineering concept in the infrastructure going from highly centralized and monolithic toward decentralized and more autonomous profit centers, and as a result it has been observed that this concept has enhanced the productivity at lower operating costs with reduced response time. On the other hand, the strategies for handling databases are also changing from traditional mainframes and mini-computers toward powerful PCs and workstations with improved performance.

Upsizing is mainly concerned with the replacement of LAN-based PCs with new client–server computing models. This is due to the fact that, with traditional file server systems, it becomes difficult to load the data stored at lower levels to higher-level databases, since the file server is not flexible, scalable, and reliable. On the other hand, in client–server models the server ensures the relational integrity of the client's updates and offers a single reliable data source for higher-level updates.

Now with client–server architecture, the divisional-level system will be getting more operational management functions, which in turn, offers more independence and opportunities to divisions and regions to accept new technologies and respond to the challenges at the divisional or regional level without waiting for instructions from their head offices. This environment will improve performance and productivity, as it is now also considering the local social and cultural aspects of the society to make it more global and competitive worldwide.

This re-engineering concept obviously may require a complete change in the way of thinking, designing, implementing, and marketing, but due to the availability of a variety of software tools (user interfaces, graphical user interfaces, graphic packages, network communication protocols, etc.), it is possible to integrate the existing back-office tools (financial, human resources, etc.) with front-office operational tools (spreadsheets, word-processing packages, groupware, etc.). This is exactly what has been provided in the **cooperative processing environments**, where the network-connecting hosts, minis, PCs, and workstations allow the users to share the information, available tools, and resources, and to cooperate with each other.

LANs provide a higher level of shared device processing in which PCs are connected to the network that allows them to share resources, e.g., a file on a hard disk, expensive printers, databases, programs, and graphic packages. Client–server processing architecture is also based on sharing of applications and data between users. In a typical client–server implementation, the client users share a variety of services such as files, databases, facsimile, communication services, electronic mail, library, configuration management, etc. For each of the services, a shared processor known as a server is defined. Thus, we have a file server, printer server, database server, etc. The clients make an input/output (I/O) request for a particular service to an appropriate server.

20.4.2 Client–server cooperative processing mode

In order to understand and appreciate the advantages offered by client–server architecture, we consider a particular case. For example, a file is placed in a shared file processor known as a file server. The client sends its I/O request to the file server. The database server (providing database access services) is more complex than the file server, as the **database management system (DBMS)** provides concurrent data access and data integrity. Here, the client requests retrieval of a particular piece of data or record. The data-processing logic processes the required request on the data at the database server. It also processes the requests from other client users. Thus, it allows multiple clients to access the database concurrently. In contrast to this, the file server accesses the entire file.

The shared devices are managed by servers. The servers get a request from PCs or workstations (clients) in the traditional host-centered environment. The servers offer limited processing services of shared files (read by multiple users) or printer services (printing by multiple users). A file or printer obtained by the distributed functions of I/O, print, etc., is shared by different PCs or workstations. In both distributed operations (I/O and print), the entire file is used, and as such both PCs and servers are busy until the operations are completed. Shared device-processing services over the network are available through Novell's NetWare, Microsoft's LAN Manager, and many other servers.

The client–server processing architecture is an extended version of shared device processing and, due to the large number of workstations connected to the network, the role of workstations is also changed. As a result of this, the sharing of a file or printer over a LAN is considered a fraction of the business application processing which is now distributed to servers. The servers process the requests from applications running over the client workstations. In other words, the application processing is distributed between client and server, and the requests (for a service) are always initiated and controlled partially by the clients. Further, both client and server can be configured for cooperative processing between them to execute the entire application as a single application task. Database servers (e.g., Sybase or Microsoft SQL server) are some of the distributed processing architectures which are based on the client–server processing architecture concept.

20.4.3 Client–server computing

In order to get a conceptual feel for the fundamentals of traditional and client–server-based cooperative processing computing environments, let us consider a simple application of file retrieval. In the traditional distributed cooperative architecture, the file server allows the user to send a request from his/her PC for retrieving a particular file. The file server sends the entire file to the user's PC, which then searches for the required record in the file (through the application running on it). The computer resources of the file server and the resources required by the user's PC are used or held until the search of a record in the file is complete. After retrieving the record, the file is sent back to the file database server. Now, if another record from the same file is to be retrieved, the same procedure is repeated; i.e., the same file (entire) will be sent to the user's PC which, by running the application program, searches for the required record. In this approach, the communication cost is higher due to the network traffic over the transmission lines. Further, the processing power of both client and server and other resources are under-utilized.

The client–server computing environment, on other hand, removes this main drawback of traditional architecture by using the resources efficiently and effectively. The client PC running the application sends a query for a record of a file to the database server. The database server processes this request for a database file locally and sends the result (flag, status, or value) back to the client. The entire file is never sent to the client, but instead the search program runs over the server. During this time, both client and server may be running other application programs.

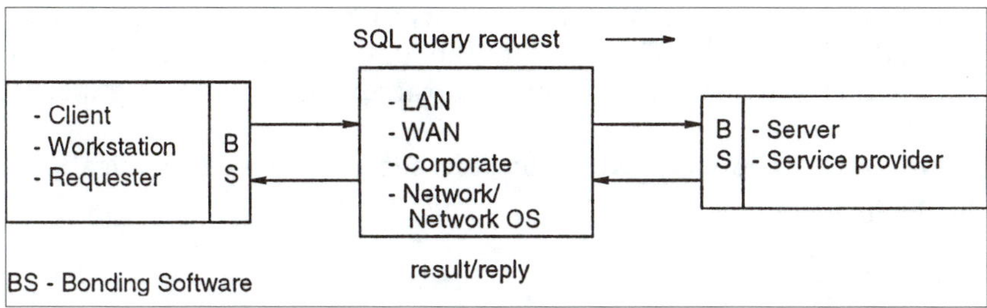

Figure 20.3 Typical client–server computing model.

In the preceding sections, we have explained the conceptual and fundamental design issues of the client–server computing model by considering a few examples. These examples illustrate the main implementation differences between traditional and client–server approaches. By no standard does this computing model define a new infrastructure or technique for implementing any application. Instead, it basically offers a new way of integrating new versions of GUI tools for improving the productivity and hopefully the efficiency of the application-development process.

A typical client–server computing model for a cooperative distributed processing environment (Figure 20.3) must possess the following features[1-3,11,14,15]:

1. A client user issues a request in a predefined language — structured query language (SQL) — for the service. The server performs the requested functions or services on the data and may also issue a request to another server (if needed) for any additional use of services or any other functions needed for completing the user's request, thus becoming a client for that server (offering peer-to-peer processing).
2. The network-based application is distributed between the client and server.
3. The server resolves the contention between various clients requesting the same service.
4. The server offers a list of services or the data which can be requested by clients.
5. The communication link between client and server is provided by LANs or WANs.
6. Peer-to-peer processing must support the database over the network in such a way that the users can move data seamlessly between multiple heterogeneous databases and heterogeneous computers.
7. Peer-to-peer processing must provide transparent cooperative processing between applications which may run over different hardware and software platforms.
8. Bonding software (BS) residing on both clients and servers interacts with the network operating system to provide a communication link or interaction between clients and servers and is transparent to the users.

20.4.4 Role of a client

The client workstation in a client–server computing model uses a variety of services provided by different servers. One of the main functions of clients (as a services consumer) is to provide a presentation function logical part to any network business application. It also presents I/O functions. It offers the following functions:

- Support for GUI.
- Support for a windowing system to establish various window sessions simultaneously.
- Support for various operating systems, e.g., Windows 3.X versions, Mac System 7 (does not support multitasking), OS/2, UNIX Power Builder for Windows, Windows NT (supporting multi-tasking), etc.

- Support for the cut-and-paste facility via dynamic data exchange (DDE) or object level embedding (OLE). These two techniques DDE and OLE usually provide a link for transfer of files or data between texts, spreadsheets, and graphics between different windows.
- Support for GUI and windowing functions such as field edits, navigation, training, data storage, etc.

20.4.5 Role of a server

The server is considered a multi-user computer. It offers shared access to server resources. Since there are many applications running on the server, it must ensure that these applications do not interfere with each other. In order to achieve independence for these applications, the operating system must support features such as shared memory, pre-emptive multi-tasking, and application isolation. Traditionally, the mainframes and minicomputers are used for storing a large volume of data and have been considered servers for various terminals and PCs which can access the network services and database services through a central host server. The client–server computing architecture downloads data from these traditional servers onto workstations for local manipulation and also allows the desktop workstations to share resources over LANs.

The Microsoft LAN Manager network operating system offers file and printer services but offers lower performance and fewer functionalities than Novell's NetWare. As we indicated above, the operating system for a server must offer features such as shared memory and pre-emptive multi-tasking, and the following operating systems do provide these features when used with LAN Manager: OS/2, UNIX, VMS, and MVS. The server does not require any special type of hardware, as its main objective is to support multiple simultaneous client requests for different types of services. Various services offered by a server include file, database, fax, image, security, network management, etc. The servers offering these services may run under different operating systems on different hardware platforms and may use different database servers.

The **file server** offers record-level services to non-data applications. The file servers provide different catalog functions for naming of files, directory systems, etc. The file services allow the users to access the virtual directories and files located on client workstations and also in the server's permanent storage. The distinction between local and remote files or directories is provided by redirection software, which is a routine in the client operating systems. The file services are usually provided at the remote server processors, but in a typical implementation, the server services (shared database, back-up, etc.) are stored on disk, tape, or optical storage devices. Due to ever-increasing demands for more bandwidth, larger optical storage devices are usually operated as servers, which minimizes the installation and maintenance of software, and various application software should always be loaded from the server for its execution on client workstations.

The **database server** initially defines space for an application within which the table and catalogued information is created and stored within the allocated space by the database engines, e.g., Ingress, Informix, Oracle, or any other database or relational databases. The earlier versions of database servers were file servers, which offer different types of interfaces, and these database servers include dBASE, Clipper, FoxPro, Paradox, and others. These database engines usually run over client machines and use the existing file servers accessing records and also for memory management. Now the trend of controlling the database servers is shifting toward application programs which use a primitive lock for creating a lock table and offer access at record levels. The records are returned to the client for filtering or any other application after the primary key has been matched. Most of the existing database servers — Sybase, IBM's Database Manager, Ingress, Oracle, and Informix — provide support for SQL for its execution on the server. Information about these database servers can be found in a number of publications.[5,11-13]

The **print server** offers a number of services, such as receiving documents, arranging them in queues, printing them, support of priority, etc. It includes a list of print driver logic software which will initialize any requested printers, and it also offers error control.

The **fax server** offers services similar to those of the print server and transmits the documents using telephone communication links during the time when charges are lower. The fax documents are compressed dynamically on the transmitting side and are decompressed on the receiving side during the printing or display. Other servers are image servers, communication servers, security, LAN management, etc.

20.4.6 Front-end and back-end processes

The client–server computing architecture typically consists of clients, processes, network, and an operating system which provides appropriate interprocess communication or transport communication between the client and server. It defines a distributed computation presentation platform where a LAN server distributes the application processing among computers. The client may be a PC or workstation and is connected to the LAN server and runs programs on it. The client PC is also known as a **service requester**, **front-end processor**, or **client workstation**. The server can also be a PC or a workstation connected to a LAN and provides services, resources, and related information to the client workstation over the network. It is known as a **service provider**, **back-end processor**, or **server workstation**.

One of the main features of client–server computing architecture is to support decision capabilities at the divisional level by automating the front-end functions. This feature can be implemented by off-loading the processing and manipulation functions on data to the mini-computers and powerful workstations and at the same time providing user access to the host data. The clients interact with high-performance relational database servers and host to perform various tasks within a cooperative processing environment.

Any business application can be divided into two processes: **front-end (client)** and **back-end (server)**. The client workstation submits a request (transaction, query, etc.) and receives a response or results from the server. In an ideal cooperative processing environment, the client workstations can off-load their time-consuming tasks onto powerful server workstations. This way of dividing the application and then allowing the application program to run on a client workstation (dealing with the presentation or user interface) is known as a **front-end process**, while the server workstation (handling the business rules and procedures, and data processing) is known as a **back-end processor** and allows the processes to run on different machines and complete the entire business application as a single task.

Both client and server computers possess intelligence and can be programmed in a variety of configurations. Both front-end (client) and back-end (server) processes are connected by a communication link and may run on either the same machine or different machines. The front-end process usually uses tools, 3GL- or 4GL-based business rules, and procedures which can be used to develop applications and define requests for data manipulation, etc., by sharing packages such as spreadsheets, graphic software packages, query tools, customized software, etc. The back-end process usually runs over a variety of platforms and its main function is to receive a request from the client, process it, and send the results back to the client. The back-end process may be a database server (PC, workstation, mainframe), OS/2 server, etc. Some of the terms and definitions used in the client–server computing model can be found in References 1-3, 5, 7, 8, 10, 14, and 15.

Functions of front-end process: The **front-end process** in a typical client–server computing architecture interacts with the user and offers the following functions:

- It defines a presentation or GUI for the users. This interface represents user queries, data retrieval, presentation of results, etc.
- Multiple clients may run concurrently, with each offering its own user interfaces (UIs) from Microsoft Windows, OS/2 Presentation Managers, etc.
- A predefined language known as **structured query language (SQL)** is used by both client and server. The language provides commands for query, security, access control, data processing, etc. Depending on the applications, the clients may or may not send any query to the server.
- Clients and servers interact over an interprocessing communication link (defined by the network communication channel), which is usually transparent to the user.
- The queries are initiated by the clients, and data processing takes place at the server. The results are sent back to the client.

Functions of back-end process: The **back-end process** may run either on the same computer as the front-end process or on a different computer and offers the following functions:

- The server offers a variety of services to multiple clients and the type of service depends on the business rules and procedures and data-processing requirements. As such, the resource utilization (particularly, processing) of the server may be different for different applications. One of the objectives of client–server implementation must be to maximize the processing power of the server for database processing (database server), image-processing server, etc. Although a file server and print server may also be used by the client, these applications require much less processing power from the servers.
- The server usually provides total transparency to the clients and users about client–server implementation, interprocess communication, and its platform (both hardware and software). Thus, clients working in any operating system (OS) environment (DOS, UNIX, etc.) must be able to interact with the server working in any operating system environment, in more or less the same way; i.e., the server must provide independence of operating system intercommunication and LANs connecting clients and servers.
- A server receives a query request from the clients and usually does not initiate a query request. In typical cases, it may send a request to another server for which it will be a client. This is done on behalf of its client user.

In essence, we can say that the server typically uses a high-performance **relational database management system (RDBMS)** and executes the functions of business rules and procedures which are not being executed by clients.

20.4.7 Evolution of PCs for the client–server model

Due to the widespread use of client–server implementation on PCs and workstations, PC vendors have started offering various features in their products at very reasonable prices which may be useful for a client–server platform. In order to get an idea of the evolving client–server computing, we consider the evolution of PCs with associated software and other specifications, updating, etc., which in one way or another provides support for client–server architecture in a variety of applications. The main philosophy behind the client–server computing model is not to propose any new environment but to enable the processing power of workstations to be off-loaded to server workstations so that the users can perform a number of applications by cooperating with server workstations.[5-7,11-13]

386-based PCs were introduced in the early 1990s and offered the following configurations: 386 SX/DX with 2 Mb RAM,VGA, 40/80 Mb hard disk, HDD, 20/33 MHz. During the same time, Intel announced the 486-chip, and soon we saw the introduction of 486-based PCs with the following configurations: 4 Mb RAM, VGA, 300 Mb hard disk, HDD, 33 MHz.

These configurations were slightly modified to offer more memory for RAM and hard disks with higher speed and, as a result, we have seen the following configurations:

- 386 DX, 4–6 Mb RAM, VGA, 80 Mb hard disk, HDD, 25 MHz
- 486 SX/DX, 4–6/8 Mb RAM, VGA/XGA, 300/600 Mb hard disk, 33/66 MHz

With the introduction of Intel's 586-chip (Pentium), we now have a 586-based PC with the following configurations, in addition to the updated versions of 486-based PCs:

- 486 SD/486 DX, 12/16 Mb RAM, VGA/XGA, 150/600 Mb hard disk, HDD, 33/50 MHz
- 586-based PCs, 20 Mb RAM, XGA, 1 Gb hard disk, HDD, 66 MHz
- 686 (Pentium II), 32 Mb RAM, 3 Gb hard drive, 300–500 MHz
- 786 (Pentium III), 64 Mb RAM, 10 Gb hard drive, 500–900 MHz
- Pentium IV, 120 Mb RAM, 20–80 Gb hard drive, 1.4–1.5 GHz

20.4.8 Advantages and disadvantages of client–server computing architecture

It is predicted that client–server computing will become the prevailing architecture in this decade, and this is well supported by the fact that a number of corporations have already implemented their future data-processing and distributed operations based on the client–server model. In order to get an idea of the state of the art of these business applications, we will present a few case studies at the end of this chapter. The client–server computing model does not propose any new model or architecture, but it simply allows users to get more processing power for developing their business network applications in a cooperative processing environment. It does not define any new infrastructure, but it uses the existing structure and new user interface tools. It integrates these new tools and the concepts of distributed architecture to define a new computing environment which will enhance productivity at much lower operating costs. The advantages and disadvantages discussed below are derived partially from References 1–3, 7, 8, 10, and 14.

Advantages of client–server computing:

1. **Improved productivity:** The applications can be distributed to front-end and back-end processes in such a way that the processing-based part of the application makes use of faster and more powerful workstations, workstation tools, and various GUIs. Further, the server workstation offers an easy-access control to information, in particular the time-dependent real-time information. The GUI offers a highly user-friendly environment for the users to develop applications by running processing-based segments of applications (spreadsheet, word processing, database server, etc.) over the server as back-end processes. A number of surveys have shown a relationship between the use of client–server GUIs and productivity. The GUI users seem to have seen an increase in their productivity (in terms of tasks completed) over users who do not use GUI tools. The new graphics-based front-end and back-end tools are also being used to improve productivity. These tools are easier to use and, also, the training costs to learn new software decreases, as most of these tools are highly user-friendly, menu-driven, and easy to use.

2. **Improved throughput:** The server offers the processing over the data which resides in the server itself, and hence the network traffic and response time are reduced considerably. This enhances the throughput and capacity of the network under heavy traffic conditions.
3. **Communication:** The clients are interacting with the server via a predefined language (SQL), and the server, after processing the local data, sends the reply or results back to the clients. This type of communication between clients and server helps in optimizing the network utilization.
4. **Performance:** Client–server computing allows the off-loading of various tasks from mini-computers and mainframes to powerful and fast workstations. In this way, the performance of distributed cooperative processing will be improved, yielding a low online response time.
5. **Reduced price:** Client–server computing allows corporations, organizations, and universities to leverage desktop computing, as the workstations are offering considerable computing power at a greatly reduced price compared to that of a mainframe.
6. **Reconfigurability:** The servers can be programmed to adapt to any new configuration of the distributed cooperative processing environment and, as such, possess intelligence.
7. **User interface:** Since clients use a **user interface (UI)** or **graphical user interface (GUI),** the desktop can also be programmed to different configurations, making it an intelligent desktop.
8. **Resource sharing:** With the help of SQL, multiple clients may request for sharing of server resources (CPU, data storage, etc.) simultaneously and efficiently, as the SQL commands are based on the message-passing concept. Different types of applications requiring different amounts of computation may run on the server at the request of the clients, and the server returns the results back to the clients.
9. **Training cost:** With the GUI, the training cost for developing new applications decreases, as the skills can easily be transferred to new applications.
10. **Time criticalness:** The time-dependent (mission-critical) applications can be implemented on PCs or workstations instead of on a mainframe.
11. **Connectivity:** Client–server computing supports connectivity and processing to both LANs and RDBMS, and as such the users can access relational database tools for different business applications. The users can access and retrieve any data across the applications, platforms (hardware and software), query facilities, etc., using the predefined SQL. Further, the processing of standard relational databases offers flexibility over the different platforms (software and hardware) in the sense that client–server architecture can be defined using any hardware and software platforms. The network connectivity provides integration of workstations and PCs into corporate networks for sharing of files, printers, program databases, etc.
12. **Off-loading:** Client–server computing offers long-term benefits in off-loading mainframe processing to server or client workstations, as this reduces the load on information system resources and supports the use of workstation platforms for providing decision support functions and processing efficiently. The workstations are suitable for computation-intensive applications, while the mainframe host is suitable for the less-computation-intensive applications but offers very large memory capabilities.
13. **Scalability:** Client–server computing supports scalability of both hardware and software, independently at both client and server levels, without affecting the original infrastructure, thus offering efficient use of data-processing resources at both the client and server levels.

14. **Availability:** Due to availability of standard front-end, back-end, and GUI tools, the interoperability between heterogeneous platforms can easily be achieved for a multi-vendor networking environment. Different processes can be linked with this multi-vendor network to communicate with other processes and also run their applications on it.
15. **Openness:** Client–server computing advocates the acceptance of an **open system** where clients and servers may run on different computers of different vendors. These computers may use different operating systems, thus creating a nonproprietary architecture environment where the available products can be used to provide economical and competitive advantages over existing cooperative systems.
16. **Software maintenance:** Software maintenance costs and effort are lower, since application development is being done by using the existing software packages and tools. Further, new technology in software engineering (e.g., object orientation) helps in reducing the maintenance cost.

Disadvantages of client–server computing: The client–server computing architecture may not be the ultimate platform for all of these applications, as it possesses the following disadvantages:

1. If the major portion of an application runs on the server, it may become a bottleneck and its resources will always be used by the clients. This may cause various problems such as resource contention, overutilization of resources, etc., without getting any real benefit, as this then becomes similar to master–slave architecture (with the mainframe working as a server).
2. There are problems with a distributed cooperative processing environment, and we may expect the same problems with this architecture, as we are using various, run-time development tools, application development tools, user interfaces, and optimization techniques for various issues at both clients and servers.
3. We don't have a complete set of compatible development tools, and also the division of applications into front-end and back-end processes needs a complex mechanism.
4. There is a lack of standard management tools and other building and testing tools.

In spite of these disadvantages, it has been established via various surveys that the client–server architecture (if implemented properly) will offer the following obvious features:

- Reduced software maintenance cost
- Portability
- Scalability
- Improved performance of existing networks
- Improved productivity
- Reduced training costs
- Reduced software development process
- LAN-based PC servers
- Higher-level network connectivity via remote procedure calls (RPCs)
- Interaction between client and server via a predefined language (SQL)
- Reduced client software time due to standard user interfaces, e.g., Microsoft Windows, OSF Motif, IBM Presentation Manager, etc.
- LAN software vendors have defined a number of low-level and high-level programming interfaces for data communication across the networks. The low-level programming interface typically provides low-level communication support, e.g., file sharing, printer sharing, etc., and usually offers high overhead (due to development and testing of various communication modules). On the other hand, the

high-level programming interface offers communication support with little effort on the part of the designers for development of various communication modules.
- Network Basic Input/Output System (NET BIOS) offers a programming interface at the session layer of the OSI-model and is supported by nearly all major LAN vendors (e.g., Novell, Microsoft, Banyan) under different operating systems, e.g., DOS, UNIX, Windows, OS/2, etc.

A number of surveys have been conducted by different vendors, and the outcomes of these surveys clearly demonstrate that many professionals are either ignorant of the client–server computing models or do not want to take a risk. According to some surveys, about 70% of the organizations consider client–server as logic processing on either the client or server, and both the client and server are processed on the same machine. About 5% like the GUI offered by it, while about 28% of the organizations think of this as PC-to-mainframe connectivity. Another survey conducted on the status of client–server offered an encouraging response in its favor. About 30% of the organizations have already developed and implemented client–server, while about 50% have started pilot program implementations of the client–server model.[12]

The survey also highlighted the following reasons for not implementing the client–server computing model:

- Lack of experience
- Cost
- Budget constraints
- Lack of applications
- Immature technologies and incompatible GUI tools
- Connectivity issues
- System management issues
- Too difficult to implement
- Security issues
- Technology breakthrough

20.5 Client–server models

Due to the recursive nature of client–server architecture, there are different client–server models defined, as described below.

- Multiple clients may access a server.
- Multiple clients may access multiple servers.
- Multiple services are provided by a server.
- Multiple servers are offering a service.
- Servers may become clients of other servers, which may become clients of other servers, and so on.
- Clients and servers are connected by networks.
- Bonding software residing on both clients and servers interacts with the network operating system to provide interaction between them, and it is transparent to users, as shown in Figure 20.4.

The client–server computing architecture has been accepted as a generic name for a cooperative processing environment, but it also has other names which are popular in industry:

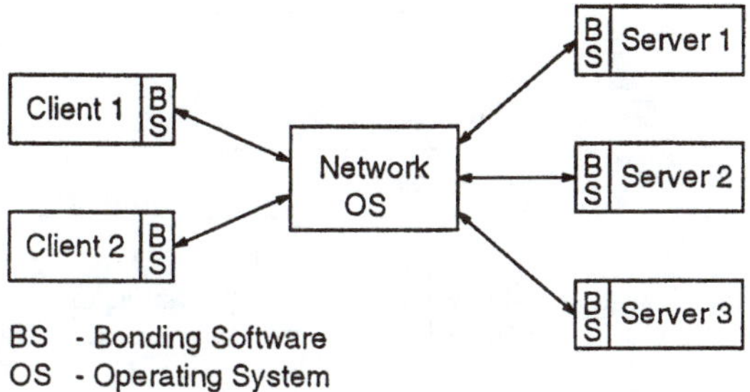

Figure 20.4 Interaction between client–server bonding software and the network operating system.

- Requester/server architecture (RSA)
- Desktop-centered computing architecture (DCCA)
- Requester/service provider architecture (RSPA)
- Workstation/server architecture (WSA)
- Network computing architecture (NCA)

Although there are many names for client–server computing architecture, the fact remains that in all of these, the network business application is divided into front-end and back-end processes so that these can run over multiple processors. These processors cooperate with each other in executing the entire application as a single task via a predefined SQL.

Clients access resources from multiple servers for executing an application. The servers may be general-purpose computers, a specific workstation performing a particular category of task(s), multi-tasking workstations, multi-tasking PCs, etc. The shared and reusable services (electronic mail, fax, graphics, peripheral sharing, communication, and other computation-based tasks) or other specific services from dedicated systems are provided by the server.

The following is a partial list of possible clients and servers:

Clients: PC (DOS-based), PC (UNIX), diskless PC, PC (OS/2), workstation (UNIX), workstation (X-Windows), mainframe.

Servers: Electronic mail, system operation, distributed database, relational database, network communication, computation-based (numeric image) peripheral sharing, AI.

The following is a partial list of standards available for client–server computing:

Online transaction processing (OLTP): This standard is generally used in outline tickets that require faster response and results.

Decision support system (DSS): This standard is basically used for responding to queries for decision making. We should have minimum DOE support, standard or structured query language (SQL), open database connectivity (ODBC), and OLE support.

20.5.1 Client–server communication via SQL

A typical interaction between the client and server via an SQL access query request follows the following steps (Figure 20.5):

1. A user (or client) makes a query request through GUI, which offers graphical presentation services.
2. The application part dealing with UI issues an SQL statement (for the request) over the network. This statement is interfaced with the network OS via the bonding software and port sitting on the client.

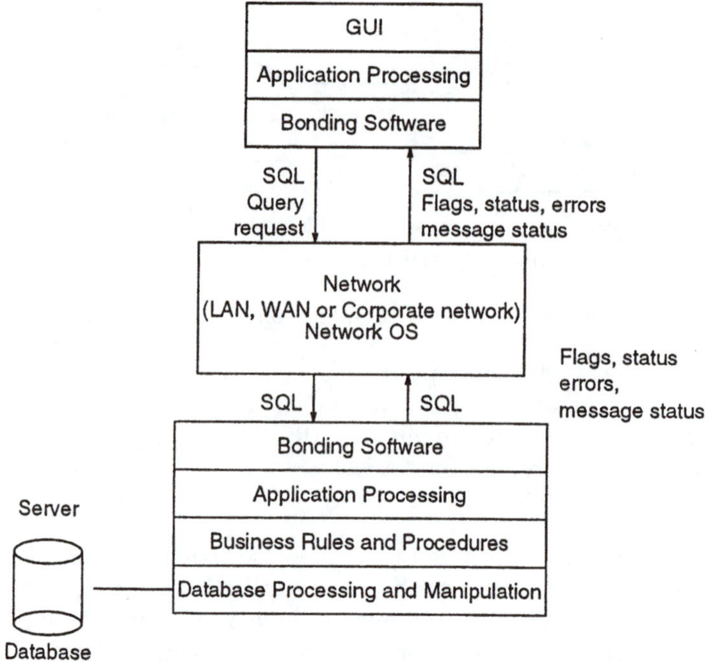

Figure 20.5 Interaction between client and server via SQL statement.

3. The server accepts the SQL statement for the network via the bonding software interface and performs the requested operation on the data, using the remaining part of the application and business rules and procedures.
4. Depending on the type of request the server has received from the client, the result or reply is sent back to the client over the network by setting value flags, status, or error message flags.
5. The user receives the result/reply back on the client, which displays the result on the screen.

20.5.2 Applications of client–server computing architecture

The client–server architecture is becoming a very useful and powerful environment to users, analysts, vendors, and many other professionals due to various advantages and benefits offered by it over traditional cooperating distributed environments by using LANs, minis, and mainframes (as discussed above). The introduction of OS/2 on LANs, various new powerful GUI tools, and RDBMS have all given a further boost to client–server computing architecture to be tried in a variety of applications. This has resulted in the development of new communication gateways, powerful and effective electronic mail systems, different groupware applications, multimedia applications, virtual reality, Internet applications, LAN management, schemes for tracking and security, connectivity of desktops, LANs and existing host-based applications, and many more applications/products.

The **graphical user interface (GUI)** is a very useful user interface in most of these applications, as it offers various features like window-based interface, icons, buttons, movable cursor, compatible pointing devices for I/O, etc. This interface provides greater friendliness to the users, supports graphics, possesses menu-driven capabilities, etc. (in contrast to characters, lines, symbols, and keyboard I/O in character-based interface for

PC- and terminal-based systems). In GUI, most of the objects, tasks, and operations are shown via familiar representational symbols which users can easily use without studying thick manuals and memorizing the combination of keys to be used for a particular task. This certainly improves the user's productivity, as now more tasks can be completed.

A **high-performance relational database** defines the information in a relational environment and is also independent of data structures used on the host-based or non-host-based systems. This allows the implementation of relational host and off-host systems on an open system platform, independent of vendors. One such example of an open system environment is a UNIX-based system using a predefined SQL statement for a database query.

Groupware-based applications based on client–server architecture are becoming very useful in utilizing processing capabilities of programmable workstations efficiently and effectively. Some interesting applications include meeting scheduling, personal agents, intelligent interrogators, etc. In the case of intelligent interrogators, groupware software on our PCs may receive various messages, i.e., text, graphical charts, images, pictures, video messages, etc., via e-mail or **electronic data interface (EDI)** within the organization. Based on the business rules and procedures defined within our application and the appropriate actions such as sorting, query initialization, etc., the interrogater will perform those operations (filtering and processing) on behalf of the user in his/her absence.

The users using electronic mail (based on client–server architecture) interact with different front ends and the same back end for the applications. In the database applications also, the front-end and back-end processes can be identified clearly, which is the basis for client–server architecture. The interprocess communication between client and server also allows the users to run only the application program on their PCs/workstations. The communication protocols can be taken to the server, thus lifting the burden of the PC to run these in its memory. A typical example is an OS/2-based communication server, which contains all the communication protocols, gateways, etc., which are executed on the server based on the user's request from the client computer, which runs only application programs.

Some of the tools for client–server applications include database administrator (DBA) tools, programmer-oriented tools, 4GL, end user query and access tools, and special tools for DBAs and system administrators (SAs). The DBA can be used to design databases or manage servers. The application development team typically uses 4GL to create a total application which is composed of interfaces, application logic, and connectivity to a particular database. The programmer-oriented tools usually require C coding. The end user query and access tools provide read-only access to server data and logical manipulation of the data. Some of the development tools include the graphical programming language Microsoft Visual Basic, Powersoft's PowerBuilder (a 4GL designed exclusively for client–server application development), Visual C (a language), and Microsoft Access (a client access tool).

The **graphical user interface (GUI)** is fast dominating the market in client–server tools such as Windows, Motif, Open Look, and Macintosh Desktop. With the emerging application of client–server architecture in the areas of database (particularly relational) communication, groupware-based system network management, e-mailing, virtual reality, and concurrent engineering, it is clear that the use of GUI tools like OSF's Motif, DEC's Open Look, OS/2 environment, Microsoft Presentation Manager, and many other upcoming tools has already made it more effective to use client–server architecture with the advantages of higher productivity, faster response time, improved scalability, greater throughput, etc.

Surveys performed by different organizations[5,6,12-14,16-18] have clearly shown a trend in the use of client–server architecture in the following ways:

- Downloading various application programs from the mainframe onto a powerful workstation (a clear indication of getting rid of mainframes).
- Re-engineering design strategies for future applications on powerful workstations.
- Modifying the currently available traditional configuration into a client–server environment. This does not require additional resources or a change in infrastructure. It requires only the adaptation of the traditional system onto the client–server architecture.

As described earlier, the application program on client–server computing architecture is divided into two processes: front-end and back-end. The front-end tools (client process) provide access to databases and help the users to develop application programs to run with database servers. The front-end products supporting database servers and other types of servers from different vendors have been defined. Some of the early front-end client products include dBase IV 1.1 (Ashton-Tate), DataEase SQL (DataEase International), XQL 2.01 (Novell), Oracle 5.1b (Oracle Corporation), and many others. Two types of interfaces are supported by front-end products: **character-based** and **graphics-based**. The communication between client and server is commonly defined by a predefined language, SQL.

Microsoft Windows OS/2 Presentation Manager provides a graphics-based interface, and SQL Windows (Computer Technologies) allows the user to develop applications using a graphics-based interface, while other front-end products typically work over a character-based interface. The following front-end client products provide direct support with their servers: **Professional Oracle, dBase IV, SQL Windows,** etc.

The tools introduced lately (for commercial application) which are being used include **1-2-3/G for OS/2, OS/2 Extended Edition (EE), Lotus's spreadsheets, servers for word processing, electronic mail, project management**, and many other tools specifically for business productivity applications.

The **back-end server (database)** products must provide links to minis and mainframes as well as support for interprocess communication via SQL predefined language. Some of these products include **Micro Software's SQL Server (Ashton-Tate), OS/2 Extended Edition Database Manager (IBM), SQL Base Server (Eupton Technologies), NetWare SQL (Novell), Oracle Server for OS/2 (Oracle Corporation)**, etc. Of these databases servers, it is seen that the OS/2 EE database server defined by IBM and strongly supported by Microsoft has taken a lead over other vendors. Further, two leading mainframe relational database management systems (DBMS), SQL/DS and DB2 (IBM), have helped the OS/2 data server to enter the market with a leading role in the area of cooperative processing and distributed database systems. A typical combination of leading products from IBM (SQL/OU, DB2, LAN server program) seems to be a leading platform for distributed database processing providing connectivity to different vendors via its OS/2 II database server.

OS/2 LAN server program: Although IBM has introduced its own OS/2 LAN server program for the network server, there are other vendors (Microsoft, 3Com, Novell, Banyan, etc.) who have developed their own operating systems for the OS/2 II database server. For example, Microsoft has developed MS OS/2 LAN Manager, which is very popular and is being considered the *de facto* industry standard for an OS/2 LAN environment. The PC Network program (1.3) allows the DOS-based workstation to interact with OS/2 servers.

Structured query language (SQL): Another technology which is also going through transformation in the area of client–server computing is the standard SQL, which is defined by the American National Standards Institute (ANSI). This organization, with SQL-92, and other developers are expanding into remote data access (RDA), remote procedure call

Table 20.1 GUI Tools for Client–Server Applications

Vendor	Product	Standard	Platform
Architech Corporation	NeWS/2	NeWS	OS/2, LAN Manager
Gupta Tech., Inc.	SQL Windows 3.0	Windows 3.x	MS-DOS, NETBOIS
MIT Software Center	X-Windows System	X-Windows	Most UNIX, Sun SPARCstations, RS/6000, DECstation/VMS, HP Apollo, Ethernet, TCP/IP
Powersoft Corporation	PowerBuilder	Windows 3.x	MS-DOS

(RPC), and call level interface (CLI) as standards for distributed database processing. Object orientation and groupware are also following the same direction as GUI and distributed databases.

Table 20.1 provides a partial list of GUI tools defined for client–server applications.

20.6 Client–server operating systems

In the preceding sections, we have discussed various aspects of the client–server computing environment and various advantages and disadvantages, along with a few simple applications. As indicated earlier, the client–server operating system offers a communication environment for clients and servers to interact with each other via a predefined SQL. We will now discuss the role of an operating system within the client–server computing model for the development of business network applications. It is important to note that clients, servers, and networks (providing a physical connection between them) may have the same or different operating systems, and both client and server will still interact with each other as if they are using the same operating systems.

20.6.1 Client and server programs

A network application in client–server architecture is divided into a number of functional groups (presentation, business rules and procedures, database processing and manipulation) and has been discussed earlier. In general, during the implementation of the client–server computing architecture, the client program deals mainly with the presentation and a small part of the database processing, while other functions are performed on the server. The main philosophy behind client–server architecture is off-loading the processing of our computer to a powerful server computer. Even in complex distributed client–server architecture, the client computer performs the functions of client–user interactions with more emphasis on the presentation function. The client can be considered a desktop workstation in a client–server model. We can also define the client as a workstation that typically is used by a single user. But when more than one user shares the same workstation for different services simultaneously, that workstation can be defined as a server.

A typical client workstation may be an Apple MAC SE, IBM PS/2 Mode 30, Compaq System Pro, Sun SPARCstation, DECstation 5000, or X-terminal. This is by no means a standard definition or a standard. The following operating systems are supported by client workstations: DOS, Windows, OS/2, Mac System 7, and UNIX. The client is connected to the server via LAN and uses services provided by the network operating system (NOS). The client workstation uses functions like word processing and can also use the services of servers via server functions supported by NOS (like file server, printer server, database server, etc.). It is interesting to note that with client–server architecture, we can access applications of minis or mainframe hosts via the client workstation, which will now serve as a terminal, in addition to the services offered by it as a workstation.

The partitioning of the network application into various functional groups and then their processing on either client or server or both depend on the configurations being defined. But in a typical client–server application, the client issues an SQL statement to the database server, which provides a reply/results back to the client. The data processing and manipulation are performed at the server processor based on the business rules and procedures defined, while the client processor may process a few calculations and formatting for the screen based on the response it receives from the server. When the clients define their queries in SQL and send them to the server, the server checks the syntax of the SQL statements, checks the access right of the client, abstracts the data type of the objects, assesses the availability of the SQL statement, interfaces with the database, etc.

The server, in fact, has the responsibility of providing data integrity and consistency and also managing various access mechanisms to the data. After checking these parameters, the server sends it to the appropriate data location for data processing and manipulation. At this time, the server chooses an optimal route for this request (it may invoke the SQL query optimizer to determine the optimal route). After the execution of these statements, the server maintains a record of this query and sends the results back to the client.

The client workstation, using the concept of *cut and paste*, can be used to create a complete document which takes inputs from different sources (spreadsheets, word processing, invoice data, application service created by client–server, etc.).

The client workstation can request a variety of services depending on the types of servers connected to the network. The server program may run either on the same processor or on a network processor. The request from the client is always defined in the same format (SQL). The request from the OS to the client workstation is accepted by NOS, which translates the request, adds extra fields into the request (depending on the application), and sends it to the destination server's operating system.

The mapping of request calls between operating systems is usually performed by a service program known as **redirection**. This program interprets the request calls from the client OS, translates them into suitable commands, and sends them to the server OS. This program accepts a variety of request calls (initialized for different resources), e.g., disk directory, files, printer, serial devices, peripheral devices, remote accessing, application program, labeled or named service, etc. This service program (redirection) itself is based on the client–server concept. The redirection program accepts the request calls from the clients. If the requests are for the local directory, files, printer, etc., it passes them to the local file system, while other request calls are sent to the respective servers via **remote procedure calls (RPC)**.

Redirection software defines an RPC for each of the requests and includes an **application programming interface (API)** call to the NOS server. The NOS server processes these requests and submits them to the appropriate server. After executing the requested queries, the server sends the results back to the client workstation. Here again, the redirection software follows the same set of procedures. It is important to note that the network server processes the requests as if they are local to it. The redirection concept was introduced by Novell for internetworking and MS-DOS, but it has now been accepted in all network operating systems, including UNIX's **Network File System (NFS)**.

20.6.2 Remote procedure call (RPC)

The RPC facilities provide a framework in which the processing load of cooperative distributed processing systems (typically based on a loosely coupled configuration) can be distributed among different types of processors (heterogeneous) which are connected to a network. Within a network business application, a local process or procedure may be executed over a remote processor, and the results are sent back to the local computer which is running that application. The users always feel that these local processes or procedures are running over the local computer. This is similar to client–server computing, where the clients request a remote call for the execution of local processes. This request call contains the arguments, services required, remote host address, etc., and is transmitted to the appropriate server (also connected to the network). The server executes this request (by interpreting the arguments passed, requested service) and sends the results back to the local computer (client) by passing flags, arguments, parameters, status, values, results, etc.). During this time, the client system is running some other application programs. The returned results come from the server by a subroutine known as **remote procedure control (RPC)** which changes the presentation logic of the network application. This change in the presentation logic is displayed on the client's computer.

The RPC offers a standard notation and primitive for the users to implement in any way such that the remote procedures can identify the calls and respond accordingly. In a typical application development process, the client requests and server responses are embedded in the RPC and are independent of the location and requested call within a cooperative processing environment. The RPC allows the clients and servers to run on different operating systems over different hardware platforms, and it will provide the necessary mapping between those environments. Many RPCs even offer mapping for different data formats, between the environments. The RPC is a subroutine within the main program to which control is passed (when it is requested) and it is executed. After the execution, it returns the results to the same point from where control was transferred from the main program to the subroutine. The only difference between this subroutine and a traditional subroutine concept in high-level languages is that here client and server programs are running on different computer systems. Further, during the execution of RPC, both client and server are still busy executing other applications without any interruption.

The RPC implementation is different for different operating system environments; e.g., Sun defines RPC as **open network computing (ONC)**, which includes the following parameters in the RPC packet: remote server address, sequence number of the procedure, name of the program, version of the program, arguments, appropriate representation for arguments. For the representation of arguments, it uses an **external data representation (XDR)** function defined within RPC. It also provides mapping of data across different types of computer systems.

The Novell Operating System provides implementation similar to Sun's ONC for RPC and is known as **NetWise**.

Apollo has taken a different approach for defining the **network computing architecture (NCA)**, an implementation for RPC. In this implementation, a set of routines known as **stubs** is defined at both client and server. These stubs are generated by the RPC compiler, which interacts with the RPC runtime library. A local call or procedure may be initiated at either client or server and it goes through the stubs, which in turn define the remote procedure calls for them. The advantages of considering an RPC runtime library is that it offers complete isolation between client and server programs, and the network protocols are transparent to the users. The RPC runtime library provides both types of services (connection-oriented and connectionless).

The database interface, as discussed above, accepts requests from clients (working on different computers under different operating systems) expressed in a predefined language, SQL. The SQL statements for required services are sent to the server, which executes these statements. The server sends the results back to the clients, which are shown to the user via presentation logic of the application.

20.6.3 Database interface

The acceptance of SQL as a standard has lowered the burden for users for defining a database interface and now the users can worry about database computing technique or aspects of the application without worrying about the database interfacing. The SQL syntax and underlying network mapping between require services into SQL statements at runtime and perform the network communication for an application. There are a large number of software tools available which offer these services as network-transparent database functions. The tools for spreadsheets, word processing, report generation, etc., also provide an SQL interface which allows the users to retrieve the information from any database server.

The acceptance of a standard database interface SQL has lowered the burden on the users, and now they have to worry about the implementation of efficient database computing techniques or aspects of the applications. The requests from the client workstation (working on different computers under different operating systems) are expressed in predefined SQL statements. These SQL statements carrying the request are sent to the database server. The database server executes these statements and, after completing the executing of all the statements, the results are sent back to the clients in the form of flags, values, status, etc. These replies can be seen on the presentation logic or user interface of the application.

When the clients define their queries in SQL and send them to the database server, the database server checks the availability of SQL statements, the syntax of SQL statements, the abstract data type or even the data type of the objects, and the interface with the database. This is because the server has the responsibility for providing data integrity and consistency, as well as managing database access controls. After checking these parameters, the server sends the SQL statements to the appropriate data locations within the database over an optimal route for data processing and manipulation of the data. At this time, the server chooses the optimal route for the request (by invoking an SQL optimizer). After the execution of these statements, the server maintains a record of this query and sends the results back to the client. The SQL syntax and underlying network operating system provide mapping between services and SQL statements at runtime and perform the network communication function for the application. There are many software tools (similar to CASE tools) available which offer these services as network-transparent database functions. The tools are for spreadsheets, word processing, report generation, etc., and also provide an SQL interface which allows users to retrieve information from any database.

20.6.4 UNIX operating system

AT&T's Bell Laboratories defined the UNIX operating system in 1969 for the mainframe, but soon after it became portable across a variety of hardware platforms (PCs, workstations). It is written in a high-level programming language, C, and is used mainly in universities. The University of California Berkeley adapted the UNIX operating system and redefined it for use in various universities, and the result of this attempt, is known as **Berkeley Software Distribution (BSD)**. Due to its openness, today we can see more than 200 versions of UNIX being used in different universities on a variety of platforms. The UNIX operating system is an AT&T proprietary operating system. Other proprietary operating systems based on UNIX have been defined by different vendors, e.g., XENIX (Microsoft Corporation), ULTRIX (DEC), and AIX (IBM).

UNIX is used on workstations and network servers as a network operating system, in a cooperative processing environment, and in a client–server architecture, and now it has been adapted to hardware architectures based on the **reduced instruction set computing (RISC) concept**, e.g., the IBM 6000 family.

The wide acceptance and adaptation of UNIX by a variety of platforms is due to the following reasons:

1. It supports a multi-vendor and open environment.
2. It runs on different platforms (mainframes, minis, PCs, workstations, database servers, etc.).
3. It supports multi-tasking, multi-user interoperability, and portability.
4. The RISC-based platform will offer improved price/performance under a UNIX environment.
5. It has been accepted and used widely for cooperative processing applications based on client–server computing architecture.
6. It is very inexpensive and easy to learn.
7. It has been accepted as a *de facto* standard operating system worldwide for various applications over networks.
8. In the U.S., it has been accepted as a standard OS for all software development processes of different federal agency–sponsored projects.

In spite of these advantages and features of the UNIX operating system, the incompatibility between multiple UNIX versions remains a major problem.

According to various surveys conducted on the use of UNIX as an operating system,[7,11,12,16-18] it is clear that UNIX-based systems provide better price/performance and network management support, and offer application portability, scalability, and interoperability. The wide acceptance of UNIX in cooperative processing environments (in particular, client–server computing architecture) has resulted in the development of standards organizations/consortiums to define standards for software development paradigms. Open System Foundation (OSF) has defined the GUI tool Motif, while UNIX International (UI) is concerned mainly with the use of UNIX globally for software development paradigms.

20.7 Graphical user interface (GUI) tools

20.7.1 Features of standard GUI tools

Adaptation of GUI tools in various business-distributed applications has become a serious challenge, since different projects require different behaviors of off-the-shelf window systems. But, in general, any GUI tools must possess the following features[1-5,7,8,11-18]:

- They must provide consistency across applications. The consistency can be defined for a number of aspects of window systems, e.g., uniform response for mouse- and menu-based commands, similar position for dialogue boxes, mouse cursor, uniform sequence of commands syntax, etc.
- They must keep users updated about the progress of processing of various commands in the form of some actions or messages on the screen.
- They must give an error message if the user chooses a wrong command at any time.
- They must support error recovery by letting users correct an error as soon as it occurs (e.g., undo, abort, cancel, etc.).
- They must provide creation of menus, dialogue boxes, prompt signatures, icons, cursors, and buttons which users are more familiar with. The help command should also be available to provide in-depth explanation or action required by the users.

- They must include familiar or popular symbols for users to choose from. These symbols (buttons or icons) should have obvious meanings, thus avoiding unnecessary memorization for the users to use the tool. Most of the available GUI tools use symbols for which meanings are easy to interpret.

20.7.2 Standard GUI tools

Although the use of standard tools doesn't necessarily guarantee a well-designed interface, it is quite clear that these tools certainly help designers spend less time on development with a low probability of any major errors. At present, windowing systems are being used or considered as widely used graphical front-end standards which can be used in-house. Each window system provides its own **application programming interface (API)** and runs over different platforms, as discussed below.

The IBM-compatible standard GUIs include Motif (OSF), Microsoft Windows (Microsoft), and OS/2 Presentation Manager (IBM). The GUI tools for workstations include Next Step, Open Look, and Motif (OSF).

The client–server architecture (if implemented properly) will reduce the development software life cycle and maintenance costs and will increase software productivity and portability. As a result, the performance of existing networks will improve considerably. Although from the standards point of view the client–server architecture is still immature and not standardized, a significant effort has been made to standardize the functionalities and interfaces and also to have compliance with existing networks for interoperability in a multi-vendor environment. The architecture develops an application program which is portable across various custom-based platforms. The interoperability, system scalability, and application portability requirements must be provided by the architecture in accordance with industry-accepted standards.

One such example for requirements is **IBM's Systems Application Architecture (SAA)** system, which was designed to provide application portability, connectivity, and consistency across its SAA-based or -supported platforms (MUS, VM, OS/2). Somehow, the SAA system does not provide any interoperability or portability between different vendors; i.e., there is no interconnection between SAA-based and non-SAA platforms for interoperability or portability. But now IBM has defined an operating system **AIX** (based on UNIX), which will comply with the Portable Operating System Interface (POSIX) standard and provide an interface across both SAA-based and non-SAA-based platforms. The POSIX operating system is defined by IEEE TCOS for the computer environment and offers consistency across different types of incompatible operating systems. The industry-accepted standards in the cooperative processing environment correspond to programming, communication networking, interfaces (between application and system services), presentation segment, system services, network management, and system services management.

Based on IBM's SAA system for interoperability, DEC has also defined its own AIA system for interoperability as an proprietary product.

20.7.3 Standards organizations of GUI tools

The standards dealing with the above-mentioned issues for interoperable application portability across multi-vendor environments have been developed by four national and international standards organizations, with each concentrating on specific aspects within a multi-vendor environment. These four organizations are ANSI, IEEE, CCITT, and ISO.

ANSI has defined the following standards (shown with the aspects being handled):

- ANSI X.3JHC (portability, scalability; written in C language)
- ANSI X.323 (portability; written in COBOL)

- ANSI X3.135 (SQL data definition, manipulation, portability, and interoperability)
- ANSI X.12 (interoperability; providing electronic data interface (EDI), remote data access)

The **IEEE** Technical Committee on Open Systems (TCOS) has defined P1003.n (Portable Operating System Interface, POSIX) defining portability, scalability, and interoperability. The POSIX for a computer environment provides a compatible interface across different types of operating systems. **ISO** has been responsible for defining an open system interconnection (OSI) reference model for networks and also for developing standards protocols for these layers. An overview of LANs, WANs, and MANs can be found in References 9, 16, and 17.

The **Application Program Interface (API)** is provided by bonding software. The bonding software capabilities for a peer-to-peer connection is defined by cooperative communication protocols (TCP-IP, OSI-RM, LU6.2, etc.), while bonding software is available as Oracle SQL-Net, Sybase open server, and others. The main feature of bonding software is that it makes complex communication implementation transparent to the users.

Industry-based standards organizations/consortia that deal with software incompatibility include **X/OPEN**, **UI**, **OSF**, **COS**, **UNIFORM**, **SQL ACCESS**, and **OMG**.

X/OPEN is a non-profit organization formed in 1984 to handle incompatibilities between various software tools. It offers a **common application environment (CAE),** which is included in the X/OPEN Portability Guide (XPG3).

The **Corporation for Open Systems (COS)** is mainly concerned with conformance testing of protocols of OSI and ISDN.

The **Object Management Group (OMG)** is another international organization whose main responsibility is to promote object orientation in the software development process. The **X-Window Consortium** was founded in 1990 (X.11.3) to discuss various issues on networking of GUIs between heterogeneous computers.

Customers come from different sectors of the professional society, e.g., U.S. military, FIPs, etc., and potential buyers such as GM using MAP, Boeing using TOP, and so on.

The U.S. military defined **TCP-IP** (1777-78) for interoperability.

Products vendors are SAA (IBM), SIA (DEC), DOS (Microsoft), NFS (AT&T), and so on.

Structured query language (SQL) is a non-procedural language for defining data, performing various operations on data, and also controlling various administrative activities. This language is independent of the structure of the database and also of how the data is stored. It supports remote data access via its built-in application program interface (API).

Interoperability is one of the main challenges of the multi-vendor environment, and it is provided by networks (LANs, MANs, WANs). It is a foundation block of cooperative processing (in particular, client–server). With the increase in network traffic for various applications, we have seen various connecting devices in internetworking such as bridges and gateways. Further, the network traffic over LANs now requires more bandwidth.

20.7.4 X-Window system and GUI tools

In the client–server computing architecture, each of the components (server, client, and network) may have a different operating system, or the client and server, or even all three components, may have the same operating system.[4] The presentation or user interface logic of any application is usually provided by the client operating system. These user interfaces are now known as **graphical user interfaces (GUIs)**. There are many GUI tools available for PCs and workstations. For PCs, the popular GUIs (offered by the underlying client OS) include **Microsoft Windows, IBM Presentation Manager, OS/2, HP NetWare,** etc.

X-Window system: The X-Window system, or X-Windows under UNIX on Workstations, has become the *de facto* industry standard for GUIs. This system is considered a standard platform by a number of organizations, including the National Institute of Science and Technology (NIST), American National Standards Institute (ANSI), Institute of Electrical and Electronics Engineering (IEEE), and X/Open Specification. It provides text and two-dimensional graphics and offers a network-transparent window system (over different remote computers; however, the users always feel as if applications are running locally). It offers a hierarchy of rectangular windows on the screen where each window can define sub-windows. These sub-windows may overlap each other, and various operations such as moving, resizing, stacking, linking, etc., can be performed on these sub-windows.

Structure of X-Window system: X-Windows defines two types of programs as **X-client** and **X-server**. The X-server is local to the user's computer, while the X-client may be executing either on the local computer or on a remote computer. The communication between X-clients and X-servers is provided by a standard **X-protocol.**[4] As stated above, in a typical client–server computing architecture, both client and server may run on the same computer or they may run on different computers. Similarly, in X-Windows, the X-client and X-server may run on different computers. If both X-client and X-server are running on the same computer, the communication between them is provided by interprocess primitives (used by X-protocols) supported by the underlying OS. But if they are running on different computers, then the normal transport provider provided by the OS is being used.

There are two types of window systems: **kernel** and **network** or **server**. The kernel-based window system is a part of the operating system or core. It supports smaller-size window systems, and debugging tools are usually poor. The entire system may collapse in the event of the occurrence of a bug, and the development, debugging, and maintenance of the window system is very slow. However, the window system is very fast. The Macintosh window and Sun window are examples of this type of window. The kernel becomes too big and memory-resident, and very little memory is available to the user programs. Further, it makes the window system small.

The network or server window system is not a part of the operating system and is considered a user-level process; as such, this window system can be quite large and complex. It supports a larger-size window system and offers easy portability to other platforms. It offers protection against any malfunction, and the entire system does not collapse due to this malfunctioning. It also allows distributed applications on the heterogeneous hardware. The main disadvantage with this window system is that it is very slow. X-Windows, NeWS, and Andrew are some of the window systems of this type.

In the X terminology, the presentation logic which resides on the client workstation is known as the X-server. The communication between the X-server and X-client conforms to the X-Window system X-protocols known as **X11 protocols**. It is important to note here that the application logic defined by the user may reside on the server but is known as the X-client. The communication between the X-client and X-server in X-Windows follows the following sequence of steps.

The X-client initiates or issues a request for communication to its X-server. After the establishment of communication, the X-client issues or requests I/O operations for display on the screen. The X-client and X-server may reside on the same computer or on different computers. If they are residing on the same computer, the network communication is provided by an **inter-process communication (IPC)** system call. But if they are residing on different computers, then the transport service of the network between them provides a communication link between them. One server may receive requests for interaction or I/O from multiple X-clients at the same time.

The X-Window system offers a standard user interface at a low level and does not allow the users to define their own user interfaces (like drawing of scroll bar, shape of the windows on the screen, new schemes for performing various operations on windows, etc.). A tool kit which can offer a higher level of abstraction than that provided by X-Windows is needed to define a customized user interface. Some of the popular tool kits are Motif (Open System Foundation), Open Look (Sun), etc.

It is very interesting to note that the X-Window system may itself become a client–server architecture for business network-based applications if users use UNIX workstations and X-Windows. The designers have to deal with non-X-Windows-based requests for implementation which are directed to the X-client. Most of the PC-based LAN software vendors support UNIX clients via an application-programming interface (API). These interfaces allow the designers to develop their client programs. We can also use X-clients on PCs, but it offers a very slow response, and memory requirements are significant. The Microsoft LAN Manager working under UNIX defines all LAN Manager APIs in UNIX using an LM/U system call.

If we want to implement our application using client–server on an X-Window system, we must be careful during the implementation, as now we have two clients and servers. The X-clients communicate with the X-server for graphical data presentation, data query, etc. Here the X-server may be running on either a local or remote computer. The client of the client–server computing architecture interacts with its server for data processing and manipulation via a database server and waits for the results/replies from it. If the underlying OS is frame is the same in all three, then the client and UNIX X-client may be supporting X-servers running on UNIX workstations or networked (LAN) DOS-based PCs, while the client may also be interacting with database servers.

The presentation or user interface logic should offer uniform and consistent interfaces across the multi-vendor environments for different applications. The program or data interface in some cases may be hidden from the users, but the presentation logic is visible to the users and designers. The presentation or user interface logic basically provides an interface between the user and the application and is also known as a GUI. Any applications (word processing, desktop publishing, different styles or texts, etc.) which require a graphic presentation require a GUI. There are a variety of GUI software tools available, but unfortunately many of the available interfaces (character-based or graphic-based) are not compatible, causing a portability problem. Some of the compatible GUI tools include Microsoft Windows, OS/2 Presentation Manager, and tools for Macintosh.

An attempt has been made to develop a standard GUI that must contain a common API which may provide compatibility with the available tools. The standard GUI must possess a number of features and requirements such as portability, flexibility, platform independence, globalization, support for development tools, interoperability, multi-vendor environment, open system, etc.

Features of X-Window systems: GUI tools define a rectangular area known as a **window** on the screen and present the information to the users within this window. Various operations (size, position, coordinates, etc.) can be performed by the users. The window also offers a number of objects which are defined on small object pictures known as **icons**. The keyboard or mouse can be used to select any of the objects by clicking on it. The new GUI tools eliminate typing of the required commands and instead provide a series of menus on which items can be selected by a mouse and another pointing device. For each selection of a menu item by a mouse, appropriate events are generated which are processed by the user interface logic in cooperation with the application logic.

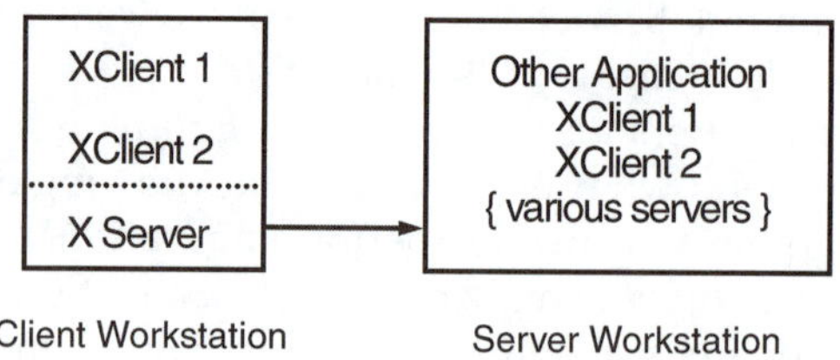

Figure 20.6 X-Window system. (Derived from References 4, 16, and 17.)

The processing of these events is distributed by the GUI between application logic and API. Thus, we can consider APIs a set of GUI-specific library routines which in fact create windows, display various graphics, offer various operations on a window, define window stretching, etc. Some of the common events generated by the I/O and system include mouse event, keyboard event, menu event, resizing event, activation/deactivation event, initialization event, termination event, creation of windows event, stacking of windows event, etc. It is important to realize that these events are implemented differently in different GUI tools.

A typical X-Window system is based on client–server architecture, as shown in Figure 20.6. In some applications, the use of the X-Window system itself offers a client–server implementation without even considering client–server computing architecture. The X-Window system offers high-performance device-independent graphics and a hierarchy of different windows (e.g., overlapping size, etc.). It allows multiple applications to be displayed on different windows simultaneously. The communication protocol for client–server interconnection offers network transparency to the users, and application logic and its presentation logic may reside on a remote workstation of which users have no knowledge. For the users, the remote display offers a feeling of local processing, local display, and other local operations.

The X-Windows applications running on one central processing unit (CPU) can allow users to see their output using either the same display or another display connected to either the same CPU or some other CPU, thus offering total network transparency to the users. The X-Windows applications are portable and the software is independent of the workstation hardware. The X-Window system supports the compilation and linking of applications with the system CPU or a combination of CPUs and operating systems.[4,15] Further, the X-Windows workstation vendors provide pre-compiled versions of XLib and other subroutine libraries which may be linked to the applications, as shown in Figure 20.7.

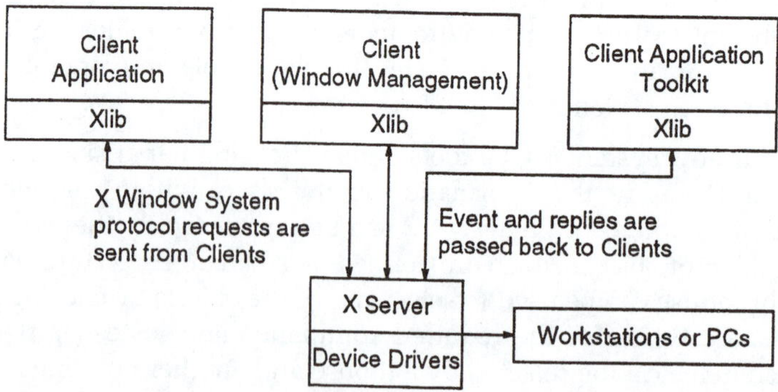

Figure 20.7 Client communication with the server via XLib calls.

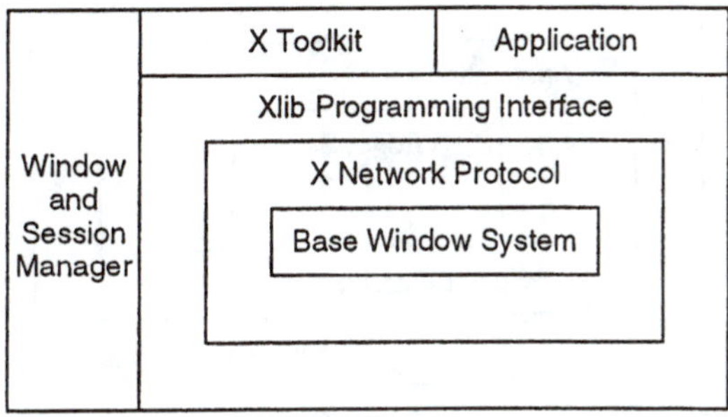

Figure 20.8 XLib application interface. (Derived from References 4 and 5.)

The X-Window system consists of a display of X-server and X-client programs. The display includes a bit-mapped screen and I/O (keyboard, mouse, or any other pointing device) interfaces. The X-server program offers interface between the X-client program and itself and also controls the display. The X-client program is considered a typical application program. Both X-client and X-server programs are linked by a communication network which is controlled by X-Windows protocol. The X-Window system defines an X-client program known as a **window manager (WM)** which interacts with network-based application client programs via the X-server and controls various operations to be performed on windows (size, shape, position, move, etc.). For initializing a request from the X-client for interacting with the X-server, a library of C routines known as an **X Library (XLib)** is defined.

The base window system interfaces with the outside world using X-Window network protocol, and it provides mechanisms and not the policies. The policies are usually implemented by window managers and toolkits. The session and window managers are treated by the X-Window system as X-Windows application software rather than privileged system software. XLib is a C-language subroutine package that allows users to let the application be interfaced to the network protocol and hence to the base window systems, as depicted in Figure 20.8.

Most of the applications use high-level X toolkits during their development. The XLib defines the lowest level of interface of the X-Window system; i.e., it offers a standard window user interface to the application program. For customizing the window (with other facilities and options not available with X-Windows), higher-level interfaced toolkits such as Motif (Open System Foundation), Open Look (Sun), DEC Windows, etc., are used. These toolkits still use the XLib for providing a low-level interface for windows. The toolkits provide high-level graphics functionality, and these are usually offered by two sets of libraries: **intrinsic** and **widget**.[4] The **intrinsic library** is concerned with the creation, deletion, addition, and linking of widgets and their management. It translates even sequence from the server into procedure calls that applications and widgets have registered. It also offers read and write operations for a widget. It allows the users to negotiate over screen real estate when a widget changes size or position and can be tied up with widgets into the user interface. It also allows the users to write new widgets for reusability and share in other programs rather than call from the XLib. The **widgets** represent abstract data objects such as scroll bars, buttons, menus, push buttons, toggle buttons, scales, etc. These objects are defined in the widget library. This defines a way of using a physical input device to enter a certain type of word (command, value, location, etc.) coupled with some form of feedback from the system to the users. It is also known as an interaction technique (IT) or user interface component. Widgets are implemented using calls to the intrinsic library and the XLib, as shown in Figure 20.9.

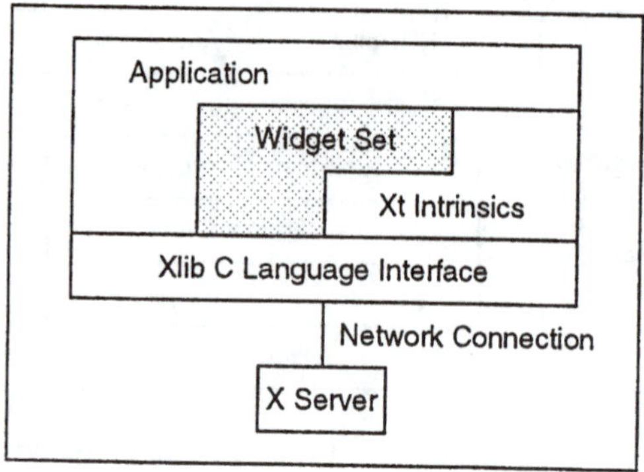

Figure 20.9 Widget implementation. (Derived from Reference 4.)

Table 20.2 X-Window System Tools

Vendor	Product	GUI style	Platform
DEC	DEC Windows	Motif	VAX/VMS, ULTRIX, Sun SPARCstations
Open System Foundation	OSF/Motif 1.2	Motif	AT&T UNIX System V, RS/6000/AIX, Unisys, Silicon Graphics
Sun Soft	NeWS[a]	SunView	PC-MS-DOS, DEC VAX/VMS, Sun-3, 4/SUNOS

[a] NeWS = Network extensible Window System.

The X-client program, in general, can run on any UNIX or non-UNIX machine. The X-server program, on the other hand, may not run on ordinary UNIX or non-UNIX-based machines. It can run only on a UNIX workstation or DOS PC which allows simultaneous users and applications to run and also provides interaction between X-clients and the underlying operating systems.

A dedicated X-terminal machine has been introduced to run X-server programs. The application logic and other presentation logic programs run on a different machine(s). Only the X-server program runs on the X-terminal. This provides easy and simple control over many administrative and security functions. Further, the X-terminals are available on the market at a much lower price than PCs or workstations.

Table 20.2 provides a partial list of tools defined for the X-Window system.

20.7.5 Toolkits

Toolkits are used for generating a complete user interface set which includes widgets, is written in C language, and interfaces with XLib. Toolkits sit on top of XLib, and include scrollbars, title bars, menus, dialogue boxes, etc. They define a widget library and provide an easy transformation among applications. Toolkits simplify the design and development of user interfaces. They save time in programming efforts, allow users to reuse the standard and uniform widgets for other applications, and offer total transparency to users by letting users adapt different widgets for the same task. In the following subsections, we present the main features of two toolkits — Motif and Open Look. For further details, see References 1-3, 8, 12, and 14–18.

20.7.5.1 Motif

As discussed earlier, the X-Window system defines a lower-level interface for windows via its X-Lib and offers a standard appearance and behavior. Further, the X-Window system is not a complete standard GUI (which must offer portability, flexibility, platform independence,

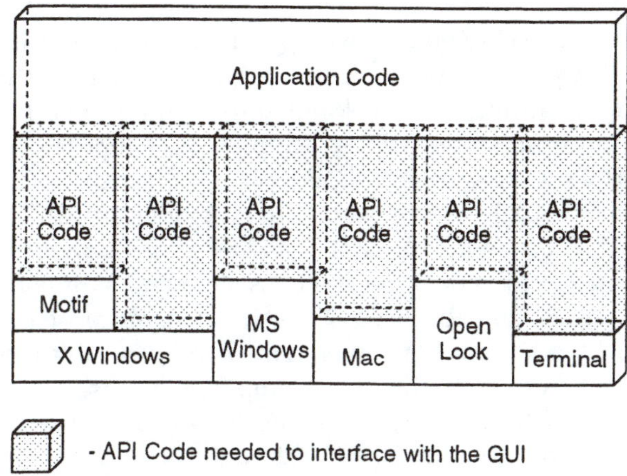

Figure 20.10 Interfaces between API and base windows.

etc.). In order to define new appearances, behavior, etc., users need a customized window system. This requires toolkits which are based on the X-Window system and offer higher-level interfaces. Some of the toolkits include Motif (Open System Foundation — OSF), Open Look (Sun), and DEC Windows.

Motif (OSF) defines a common API which is based on the DEC Windows concept. It also uses the concept used in Microsoft's Presentation Manager for behavior and three-dimensional appearances. It runs on a variety of platforms including IBM's RS/600, DEC's VAX, SUN's SPARCstations, MIP's R2000, Intel's servers (286, 386, 486, 860), and Motorola's 68020, 68030, 68040, and 88000-based systems. The interfaces between API and the base window for different tools as a standard are shown in Figure 20.10.

Motif complies with the *de facto* industry standard X-Window system. It also complies with the Inter-Client Communication Conventions Manual (ICCCM), which allows Motif to interact with other ICCCM-compliant applications and share their data and network resources. These two standards further define a common platform for developing application logic and provide interaction between them (which may have been developed on X-Window and ICCCM-based systems).

Motif offers the following set of development tools which are required for defining Motif's development environments:

- **User Interface X Toolkit:** Defines various graphical objects (widgets and gadgets).
- **User Interface Language (UIL):** Describes the visual aspects of GUI in Motif applications. Various objects for visual aspects include menus, forums, labels, push buttons, etc. The required objects are defined by the users as a text file which is compiled by the BIC compiler to produce a resource file. This resource file will be loaded at runtime during the Motif development process of application logic.

Motif contains its own **Motif Window Manager (MWM)** which is different from the window manager of X-Windows and, in fact, runs on top of X-Windows. The MWM allows various operations on windows (like moving, resizing, icons, etc.) and supports OS/2 Presentation Manager's behavior and three-dimensional appearance capabilities.

Motif is supported by the following operating systems: 4.3 BSD UNIX, MS-DOS 2.0, HP-UX version, MIPS RISC/OS, OS/2, SUNOS 3.5/4.0, ULTRIX, VMS, UNIX system V 3.2/4.0, AIX, and a few others.

20.7.5.2 Open look

Sun Microsystems and AT&T together developed this toolkit, and it is very popular in AT&T and SUN-based systems. Open Look defines three APIs (in contrast to one common in Motif) which are used during the development of applications. The first API is Sun's **News Development Environment (NDE)**, which provides support to SUN-based systems. The second API is SUN's **XView**, which supports SUN SPARC, DEC VAX systems, Intel 80386, and the Motorola 680X0-based system. This provides an API to XLib of an X-Window system. The third API is defined by AT&T as **Xt+**, which provides support for AT&T-based systems. These three APIs offer users a consistent and simple presentation logic. It is important to emphasize that the behavior and appearance of these GUI tools (MS Windows, Motif, OS/2 Presentation Manager, DEC Windows, HP's toolkit and managers, etc.) are totally different from each other, including Open Look.

Open Look complies with X-Window systems, for which it provides three APIs based on X11 protocols, as shown in Figure 20.10. Although these APIs support different platforms, portability across these platforms is very simple. These APIs have defined toolkits for developing the Open Look–based applications, as given below.

- **News Development Environment (NDE)** development toolkit (SUN Microsystems) supports an emulated PostScript interpreter for Window systems.
- **XView** development toolkit (Sun Microsystems) provides a high-level interface to XLib.
- **Xt+** development toolkit (AT&T) defines various graphical objects (widgets and gadgets).

The Open Look tool provides flexibility and scalability, and it provides support to a variety of input devices for I/O operations and display devices. Due to its X-compliance, it supports many operating systems which are supported by the X-Window system. However, following a recent announcement by SUN about discontinuing Open Look for the future market of development tools, Motif has become more useful and is available in a variety of applications.

In the previous subsections, we briefly discussed two widely used GUI tools, but there are many other GUI tools available on the market that are being used on different platforms, including Macintosh interface, Microsoft Windows version 3.0, OS/2 Presentation Manager, DEC Windows (for ULTRIX and VMS), NEXT, etc.

As described earlier, these GUI tools have different appearances and behavior, but they have a common goal of providing portability, scalability, flexibility, vendor independence, and many other features. Sometimes, it becomes too difficult to develop an application using one GUI tool and try to port on another tool-based platform. Further, we should not expect users to learn all of the interfaces (of different GUI tools). In order to avoid such situations, an attempt to define a common API has been carried out, and as a result, we have noticed the development of various APIs across the GUI tools. These APIs are known as **compatibility tools**. Oracle Corporation developed an Oracle toolkit compatibility tool which provides API to various GUI tools (Macintosh, Microsoft Windows, X-Windows, Motif, etc.). XVT, Inc. has developed the Xensible Virtual Toolkit (XVT), which defines a set of libraries and files which sit on top of GUI's XLib.

20.8 Client–server LAN implementation

The IEEE 802 Committee has defined OSI-based LAN architecture which has been widely accepted by a number of vendors. Networking via a standard architecture and set of protocols allows interconnection and interoperability between different types of computers, and also the sharing of various resources can be easily implemented.[16-18]

IBM's token ring networks (IEEE 802.5) and PC networks (IEEE 802.3) for both broadband and baseband are widely used. The IBM LAN Support Program defines interfaces (both low- and high-level) to a network. At the low level, we have a common interface and **logical link control (IEEE 802.2)**, while a high-level programming interface to the network is provided by NET BIOS.

IBM has defined a number of inter-LAN connection devices, e.g., token ring bridge program (for interconnecting similar types of LANs) and gateways for interconnecting different types of LANs. It has also defined a variety of interfaces for interconnecting host computers and WANs, and some of these are the IBM Token Ring/PC Network Interconnection Program, IBM PC 3270 Emulation Program, APPC/PC, APPC for OS/2 EE, etc.

The IBM token ring network is based on baseband transmission and offers data rates of 4 and 16 Mbps over a cable which consists of shielded or unshielded twisted-wire pairs. The MAC sublayer uses an IBM token ring network adapter card installed in PC-based LAN nodes. This network can provide connection for up to eight nodes via a multi-station access unit forming a star-based configuration. The IEEE 802.3 standard LAN (CSMA/CD) is available in two product categories: PC network–broadband and PC network–baseband. The difference between these two categories lies in the type of topology and data rate. For example, the baseband network uses a tree topology and offers a data rate of 2 Mbps, while the broadband network is based on star topology (also known as daisy chain) and offers a data rate of 2 Mbps.

As we indicated earlier, IBM defines two types of interfaces to a network, one at low level (typically at the data link layer of OSI-RM) and the other at high level (typically at the session layer of OSI-RM), and they are described in the IBM LAN Support Program. The other higher-level interface is consistent with the session layer of OSI-RM and has been defined as the NET BIOS interface. This standard protocol (based on full-duplex transmission) provides reliable data communication for connection-oriented (virtual circuit) and connectionless (datagram) services without acknowledgment. In addition to these services, other services are flow control, session services and functions, etc. Although NET BIOS interface has not been accepted as a standard interface, due to its wide acceptability and support by different operating systems, it has become the *de facto* standard of the industry. These operating systems basically emulate the NET BIOS interface.

Another higher-level programming interface defined by IBM for the session layer is APPC/PC. This interface is based on the concept of advanced program-to-program communication (APPC), which has been used as APPC protocols in IBM's SNA (System Network Architecture) Logical Unit Type 6.2 (LU 6.2). The main objective of APPC/LU 6.2 protocol is to provide peer-to-peer communication and synchronization between programs, and it can be useful for developing any application over client–server architecture. For more details, see References 1–3, 5, 8, 12, and 14–18.

Ethernet is the most popular cabling technology for local area networks. In the past, the most commonly installed Ethernet system was 10BASE-T, which provided transmission speeds of up to 10 Mbps. With high bandwidth requirements for many multimedia applications, a new LAN — 100BASE-T, also known as fast Ethernet — which could (theoretically) transmit at 100 Mbps was introduced. Fast Ethernet uses the same cabling, packet format and length, error control, and management information as 10BASE-T. Fast Ethernet is typically used for backbones, but the ever-growing bandwidth requirements have promoted its use in workstations. The gigabit LAN that can (theoretically) provide a bandwidth of a billion bits per second is becoming very popular as a backbone. Applications in the modern enterprise networks are making a heavy-use desktop, server, hub, and switch for increased bandwidth applications. Megabytes of data need to flow across intranets, as communication within enterprises will move on from text-based e-mail messages to bandwidth-intensive real-time audio, video, and voice.

An increasing number of enterprises are employing data warehousing for strategic planning, as it deals with very high volumes of data and low transmission latency. These warehouses may be composed of terabytes of data distributed over hundreds of platforms and accessed by thousands of users. This data must be updated on a regular basis to ensure that the users access real-time data for critical business reports and analyses.

Enterprise-critical applications will demand ever-greater shares of bandwidth at the desktop. As the number of users grows rapidly, enterprises will need to migrate critical portions of their networks (if not the whole network itself) to higher-bandwidth technologies. Going from megabit to gigabit Ethernet seems to be an ideal choice for backbone networks. *Enterprise* is a relatively modern term that includes any organization where computing systems replace traditional computers for higher productivity.

The switched 10-Mbps and 100-Mbps Ethernets have already captured the market, and it is hoped that the introduction of gigabit Ethernet may pose a tough competition to ATM. Nearly 80% of all desktops are using Ethernet LANs, and that high usage creates different types of LANs (fast Ethernet, gigabit Ethernet), and their support for gigabit WAN transport makes ATM the dominant switching technology for high-performance LAN connectivity. For the applications (real-time, multimedia, video, and voice) that require quality of service in LANs, ATM ensures the end-to-end quality of service communications. At this time, real-time protocol and resource reservation protocol defined within IP are offering the quality of services for Ethernet networks. The ATM provides the best LAN-to-WAN migration path. Thus, a server based on LAN can have a connection anywhere from 25-Mbps ATM to 622-Mbps ATM. The ATM has already standardized OC48 (2.5 Gbps), supporting a full-duplex mode of operations and offering much higher data rates than gigabit Ethernet (half-duplex).

20.8.1 Popular LAN vendors for client–server implementation

In order to understand popular LAN client–server implementation and the functions offered by network operating systems, we discuss a few vendor networks and network operating systems in the following subsections.

20.8.1.1 AT&T StarLAN

This local area network defined by AT&T is based on the concept of standard CSMA/CD protocol and offers a data rate of 1 Mbps. This LAN is designed for existing telephone cables (unshielded twisted-wire pairs) and supports up to ten stations on a single cable. These single cables carrying ten stations each are connected using daisy chain segments through network extensions into a star topology configuration. This topology of StarLAN usually supports up to 11 single cable segments and up to 13 units per segment; thus, it can accommodate over 1350 stations in the network topology. It supports file and printer services, e-mail services, and other services under both operating systems, DOS and UNIX. Also, it provides compatibility with IBM's PC LAN program, as it can emulate NET BIOS.

20.8.1.2 3Com

3Com Corporation is one of the leading LAN vendors which offers LAN products for a variety of PCs (IBM, Apple, and other IBM-compatible PCs). It offers a data rate of 10 Mbps for Ethernet and IEEE 802.3 CSMA/CD protocols. It also supports token ring protocols. Some of its LAN products are given below:

- 3+ Share network operating system (printer and file services)
- 3+Mail electronic mail services
- 3+Route, 3+Remote, and 3+NetConnect (for interconnecting)
- 3+ 3270 (interconnection with IBM's SNA LAN)

20.8.2 Network operating systems (NOS)

Although the client workstation provides support for OS/2 and UNIX operating systems, due to the fact that traditional OS/2 and UNIX don't support tools like spreadsheet, e-mail, and word processing, these OSs are generally not used as client OSs. Realizing this, both UNIX and OS/2 have now come up with their own versions to be considered as client OSs. The OS/2 Version 2.0 (integrated support for DOS, Windows 3.X, and X-Windows) has become more popular than traditional DOS and Windows, as OS/2 provides support for multi-tasking. Similarly, UNIX supports various PC applications such as Lotus 1-2-3, WordPerfect, dBase IV, and now has been coupled with low-cost, high-performance RISC processors. These two OSs support multi-tasking and also other PC-based tools and, as a result of this, they are becoming widely accepted and powerful client OSs in cooperative processing environments (in particular client–server, open system, and multi-vendor environments). The OS/2-based operating system for PCs is available, while the UNIX-based operating system for PCs requires a new hardware interface known as OSF11 (commercial-grade UNIX) for Intel.

Further collaborative efforts between IBM, Apple, and Motorola have resulted in a new operating system (for client) known as **Taligent**, which is based on AIX, OS/2, and Mac System 7. This operating system provides an interface to Mac System 7 and offers connectivity with AIX and OS/2 so that these services and functions can be used by the client.

The client operating system may not provide all required services for inter-process communication. These services (not provided by client OS) are provided by network operating systems (NOSs). There are a number of NOSs, e.g., Novell NetWare, Microsoft LAN Manager, IBM LAN server, Banyan Vines, etc. In the following section, we will describe these operating systems in brief.

20.8.2.1 Novell NetWare

This OS offers a LAN environment for sharing of files and printer resources and was initially defined as a LAN NOS for PCs. It allows PC users to share printers, files, software, links, peripheral devices, etc. It now supports IBM-compatible and Apple Macintosh clients, as well as IBM-compatible servers. We mentioned earlier that there may be different operating systems for clients, network, and servers, but Novell OS is common to all three and provides a complete OS environment for client–server architecture. In other words, servers do not need DOS, Windows, OS/2, Mac System 7, or even UNIX, nor do the clients and the network need different OSs. Only Novell NetWare alone can be used for client–server application. Novell defines compatible products to support other operating systems, e.g., Portable NetWare for UNIX supporting RISC-based UNIX System, IBM-compatible systems using OS/2, high-end Apple Macintosh Mac System 7, and DEC's VAX under VMS.

The NetWare architecture follows the concept of client–server and has two components: **client workstation** and **server**. The client workstation contains an application program through which request calls are initialized. The requests may also be initiated via its OS. All the requests are handled by a redirection program (a part of the client workstation) and, depending on the type of requests (local or remote), the redirection software will send it back to the client OS or to the server via RPC. Both client and server have the same redirection software. The redirection software includes communication software and a library of different protocols. The client and server are connected by a hardware connection. After the server receives the RPC, it checks the category of servers requested in it (e.g., network services, file, printer, database).

NetWare is one of the widely accepted LAN implementations providing support for connectivity on different platforms via a set of gateway protocols. It can be used with all standard IEEE LANs, Ethernet, CSMA/CD networks (AT&T StarLAN, IBM PC network), IBM token ring, and other token ring architecture networks. It provides support for a variety of services, e.g., file and printer sharing, electronic mail, remote access, and others. It emulates NET BIOS and provides interconnection between different types of OSI-based IEEE 802 LANs, NetWare bridges, proprietary networks such as IBM's SNA via NetWare SNA Gateway (over synchronous data link control (SDLC) lines), and others. It supports different operating systems, including PC-DOS and UNIX. A portable NetWare application program has been introduced to provide a direct link between NetWare and UNIX hosts. This offers a higher level of connectivity for UNIX hosts with different operating systems (DOS, Windows, OS/2, Macintosh, etc.). Each of the operating systems is well supported by NetWare.

Novell, with third-party vendor support, allows the users to use NetWare on a variety of platforms, e.g., Macintosh, Windows, DOS, OS/2, UNIX, IBM MUS, IBM VM, OS/400 VMS communication. The server processor, after receiving a request from its redirection software, processes the request on the platforms, which provide services such as memory management, scheduling file system, CPU scheduling, etc. The results/replies will be sent back to the client via the same sequence of events.

Novell has also defined products for open systems (based on openness and protocol independence). It has defined two standard interfaces, **Open Datalink Interface (ODI)** and **NetWare Stream (NWS)**, which allow other vendors to define the protocols for the NetWare environment. These interfaces are used at different layers. The open-system architecture of NetWare supports the client OS Mac System 7, OS/2, DOS, Windows, NetWare UNIX NFS, and OS/2 and UNIX OSs for servers. The server OSs are installed on the same LAN as the NetWare servers.

The open datalink interface receives a request from a client workstation and passes it on to NetWare Stream via OSI-RM, SNA (IBM) TCP/IP. The NetWare Stream allows the requests to be serviced by the selected NetWare service or server application. Each client workstation interacts with NetWare 386 via different interfaces, e.g., Mac System 7 (AFP Apple talk), OS/2 EE with LAN requester (SMB Netbem), NetWare Client (NCP IPX), or UNIX NFS Workstation (NF, TCP/IP).

20.8.2.2 Microsoft LAN Manager

This operating system is a standard OS for client–server implementation which uses OS/2 as a server OS. The LAN Manager/X version of OS is used when UNIX System V is used as a server OS. LAN Manager provides support to a variety of client OSs, e.g., DOS, Windows, and OS/2 Mac System 7, and to server OSs, e.g., NetWare, Apple Talk, UNIX, and OS/2. The client workstations can use both server OSs (NetWare and LAN Manager) simultaneously for accessing the data. The communication link protocols between client and server are also defined by LAN Manager. The NET BIOS and Named Pipes LAN communications are supported between the client workstation and OS/2 servers. As usual, these communication protocols use redirection software for determining the types of requests (local or remote). If the request is remote, the protocol provides mapping of files, printers, directories, devices, etc., from the remote workstation to client workstations. The present form of LAN Manager allows application programs, databases, and communication servers to run on the same computer on which a printer or file servers are running. This type of facility is provided to OS/2, UNIX, and Mac System 7 server operating systems and is suitable for small LANs. It is important to note that preemptive multitasking helps in maintaining the availability of other services without affecting the performance and reliability of the architecture. With the introduction of Windows NT, it looks quite obvious that LAN Manager will be well suited for applications sharing peripheral resources, since Windows NT provides all the functions of network applications to be performed on a server (database applications) and also the communication services.

20.8.2.3 IBM LAN servers

For network applications where access to more than one server is needed, an attempt by IBM to integrate its LAN server with NetWare OS has made this service possible over an integrated environment. NetWare is concerned mainly with high-performance file and print services and offers a friendly interface to users. It is easier to use on a variety of platforms. On the other hand, the LAN server is responsible for providing interconnection between different platforms. It is quite common for most of the leading organizations to use both NetWare and the IBM LAN server on an integrated platform and receive services from them simultaneously. It also offers flexibility to the users to decide which OS should be chosen for a particular service on their architecture. Both IBM and Microsoft have also defined compatible APIs for their OSs. The LAN server OS supports a variety of client OSs such as DOS, Windows, and OS/2 but it does not support Mac System 7.

20.8.2.4 Banyan Vines

This operating system is mainly defined for a distributed system to support large PC networks. It is claimed that this **distributed operating system (Vines)** defined by Banyan Systems (Westborough, MA) defines integrated services of directory, security, and network management which are handled by interconnected servers. Each server offers support to one or more PC LANs. This means that one server may provide one type of service to IBM token ring LANs while another server may offer a different type of service to other LANs (IEEE 802.3), and so on. These servers can also provide services to other LANs, as each of the these is connected to WANs. The Vines supports different operating systems, e.g., DOS, Windows, UNIX, OS/2, Mac System 7 client workstation, etc. It runs over a variety of hardware platforms like 286-, 386-, and 486-based PCs running under UNIX System V.

Vines provides connectivity to all types of LANs and supports different types of topologies used in LANs. Further, it supports a variety of communication protocols (3270 emulation, TCP/IP, X.25 switching protocols, etc.). It provides file and print services in addition to its popular **Vines Network Mail (VNM)** service.

One of the powerful features of Vines is its ability to address the resources by name, which is achieved by an **Option Streetwalk**. It also provides optional routing over WAN within a server, which can be obtained from the network operating system. This network operating system is very useful in the type of organization which offers WAN host connectivity to multiple hosts from a large number of workstations. It runs under a UNIX environment and offers file and print services in the same way as those being provided by Novell and LAN Manager.

20.9 Case studies

In order to appreciate the usefulness and importance of client–server implementation, we present a few case studies in this section. There are a large number of case studies based on the client–server technology which has been developed and implemented in the past few years, and it looks as if there will be a great demand and push for this technology in light of the re-engineering concept, which is well accepted for global competitiveness. For further information about these case studies, see References 16–18.

Prosecutor's Information Management Systems (PIMS): The Los Angeles County District Attorney's office has developed an information management system based on client–server to help 950 prosecutors handle more than 300,000 adult and juvenile cases each year. This system defines a shared network that increases the productivity of the prosecuting staff, enhances management information, and increases information sharing with other criminal justice agencies through Los Angeles County.[12]

The environment consists of several small systems with equipment and applications that do not interfere with or require duplication of data entry. The users can track the document generation on a single system, which eliminates costly and time-consuming duplicate data entry. This system provides information exchange across 40 branches and provides 950 prosecutors with current data on juvenile and adult defendants involved in the 250,000 misdemeanors, 50,000 felonies, and 25,000 juvenile cases tried each year. The system is a highly integrated network architecture and defines a distributed environment for the Prosecutor's Information Management System (PIMS).

The database table resides in DB2 on an IBM ES 9000 mainframe, while the primary application programs are accessed from local file servers. The network architecture consists of LANs connected to the county's wide area router network. The SQL statements are used to transmit the data from the mainframe to a file server and then to a workstation, where processing occurs. Information is then transmitted back to the mainframe for storage. Users are running Lotus Notes on OS/2 workstations in the office locations with servers for database, print, domain, and communications. Selected workstations can operate in a stand-alone mode when necessary to ensure that documents needed for court appearances can be printed even if the mainframe, wide area networks, and LANs are off-line.

World Cup '94 Soccer Games: Sun Microsystems, Inc., Sprint Corporation, and Electronic Data Systems Corporation joined together to develop a client–server using three multiprocessor APARCcenter 2000s in Los Angeles, Dallas, and Washington, D.C. These will anchor a network of SPARCstation 10 servers and up to 1000 SPARCclassics and LX workstations, which will be used to centrally monitor ticketing information, display scores, provide online news service feeds for journalists, and monitor track records and all related game logistics. For details, see Reference 12.

Brigham and Women's Hospital: This is one of the largest hospitals in the world. It has 6,300 employees and offers medical services to nearly half a million people each year. It has received three Nobel prizes and is one of the teaching institutions for Harvard Medical School. The nature of processing information in medical care is very complex, and there is a wide variation of processing applications required, from paycheck generation to medical research. Such a system needs a large amount of processing power, high bandwidth for movement of images, and capacity to store a large amount of clinical data over long periods of time. It must also support new applications and new platform components.

The present system uses Intel technology and defines a platform which has 2,500 Compaq Intel 486-based machines spread throughout the hospital and also serves 40 sites within a 10-mile radius. These client machines are networked to servers through a token ring network. Many of these clients are diskless. There are 120 servers in the computer room which are used for distributed data and application processing. The IBM PS2 Model 95 is used as a server. The operating system is DOS and the application software is being custom-developed by InterSystems. The hospital uses the Novell NetWare network operating system.

Some useful Web sites:

Some of the Web sites giving a list of vendors, applications, and other related information are *http://www.tucc.uab.edu*, *http://www.ougf.fi*, *http://www.analysis.co.uk*, *http://wwwsybase.com*, *http://www.faqs.com*, *http://www.cosmoline.com*, *http://www.lantimes.com*, *http://snmp.cs.utwente.nl* (about SNMP), and *http://www.w3.org* (about WWW consortia).

References

1. Berson, A., *Client-Server Architecture,* McGraw-Hill, 1992.
2. Boar, B.H., *Implementing Client-Server Computing: A Strategic Perspective,* McGraw-Hill, 1993.
3. Baker, A.B., *Intelligent Offices: Object-Oriented Multi-Media Information Management in Client–Server Architecture,* John Wiley, 1992.
4. X-Protocol Reference Manual, O'Reilly and Associates, Inc., CA.
5. Percy, T. and McNee, B., *Demystifying Client-Server,* Gartner Group, 1991.
6. *Emerging PC LAN Technologies,* Computer Technology Research Group Corp., 1989.
7. Inmon, W.H., *Developing Client-Server Applications 1991,* QED Technical Publishing Group, Boston, 1991.
8. Miller, C.A., *Client-Server: The Emerging of the Enterprise Server Platform,* Gartner Group, 1992.
9. Hura, G.S., *Computer Networks (LANs, WANs and MANs): An Overview,* Computer Eng. Handbook, McGraw-Hill, 1992.
10. Hura, G.S., "Client-server implementation," Strategic Research Group Seminar, KL, April 1994.
11. *Case Strategies,* Karen Fine Coburn, June 1992.
12. *Computerworld Client-Server Journal,* August 11, 1993.
13. *Intel Technology Performance Timed for Business,* Intel Corp., 1993.
14. Sinha, A., "Client-server computing: current technology review," *Communication of the ACM,* 35(7), July 1992.
15. Smith, P., *Client-Server Computing,* SAMS Publishing Co., 1992.
16. Chorafa, D.N., *Beyond LANs Client/Server Computing,* McGraw-Hill, 1994.
17. Guengerich, S. and Schussel, G., *Rightsizing Information Systems,* SAM Publishing, 1994.
18. Hura, G.S., "Client-server computing model: A viable re-engineering framework," *SCIMA Systems and Cybernetics in Management,* The Society of Management Science and Applied Cybernetics, 24(1–3), 1995.

Acronyms

AAL	ATM adaptation layer
AC	Access unit
AC	Acoustic coupler
ACKNUM	Acknowledged message number (in DNA)
ACM	Access control method
ACSE	Association control service elements
ADM	Adaptive delta modulation
ADSL	Asymmetric digital subscriber line
AE	Application environment (DCA)
AES	Advanced encryption standard
AIS	Alarm indication signal
AM	Amplitude modulation
AMI	Alternate mark inversion
ANSI	American National Standards Institute
ARP	Address resolution protocol
ARPA	Advanced Research Projects Agency
ARQ	Automatic repeat request
ASCII	American Standard Code for Information Interchange
ASK	Amplitude shift keying
ASN.1	Abstract Syntax Notation One
AT&T	American Telegraph and Telephone Corporation
ATM	Asynchronous transfer mode
AWG	American Wire Gauge
B-ISDN	Broadband integrated services digital network
B-NT1	B-ISDN's network termination 1
B-NT2	B-ISDN's network termination 2
B-TA	B-ISDN's terminal adaptor
B-TE1	B-ISDN's terminal equipment 1
B-TE2	B-ISDN's terminal equipment 2
B8ZS	Bipolar with 8-zeros substitution
BBN	Bolt, Barnak and Newman
BCC	Block check character
BCC	Block check code (DNA)
BDLC	Burrough's Data Lock Control (BNA)
BER	Basic Encoding Rules (ISO 8825)
BER	Bit error rate
BGP	Border gateway protocol
BIP	Bit interleaved parity
Bitnet	Because It's Time Network
BNA	Burrough's network architecture (Burroughs)

BNF	Backus-Naur form
BOC	Bell Operating Company
Bps	Bits per second
BPSK	Binary phase shift keying
BSC or BISYNC	Binary synchronous communication
BSI	British Standards Institute
CAD	Computer-aided design
CAM	Computer-aided manufacturing
CASE	Common application service element
CATV	Community antenna television
CBR	Constant bit rate
CCC	Carrier common communication
CCIS	Common channel interoffice signaling
CCITT	Consultative Committee of International Telegraph and Telephone
CCITT X.21	DTE/DCE interface for public packet-switched network (X.25)
CCITT X.25	Interface between a local DTE and its DCE
CDDI	Copper distributed data interface
CEPT	European Conference of Posts and Telecommunications
CIM	Computer integrated manufacturing
CLP	Cell loss priority
CLR	Clear service signal (X.28)
CMI	Coded mark inversion
CNA	Communication network architecture (ITT)
CO	Central office
Codec	COder/DECoder
CPE	Customer premises equipment
CPM	Customer premises network (also known as customer premises equipment — CPE)
CRC	Cyclic redundancy check
CS	Convergence sublayer
CSMA	Carrier sense multiple access (Ethernet)
CSMA/CD	Carrier sense multiple access with collision detect
CSN	Circuit-switched network
CSU	Combined service unit (used for high-speed alternative to digital transmission on digital network, asynchronous and synchronous)
CSU	Communication system user (DCA)
CSU/DSU	Channel service units/data service units
D/A	Digital-to-analog converter
DAM	Discrete (digital) analog modulation
DAP	Data access point (DNA)
DC	Direct current
DCA	Distributed communications architecture (Univac)
DCE	Data circuit-terminating equipment
DCN	Data communication network
DDCMP	Digital data communication message protocol (DEC)
DDN	Defense Data Network
DEE	Data encryption equipment
DES	Data encryption standard
DFC	Data flow control (SNA)
DIS	Draft International Standard (ISO)
DLC	Data link control

DLE	Data link escape (ASCII)
DLPDU	Data link protocol data unit
DM	Delta modulation
DM	Digital modulation
DNA	Distributed or digital network architecture (DEC)
DNA or DECnet	Digital/distributed network architecture
DNHR	Dynamic non-hierarchical routing
DNS	Domain name system
DoD	U.S. Department of Defense
DOS	Distributed operating system
DP	Draft Proposal (ISO)
DQDB	Distributed queue dual bus
DS-0	Digital signal 0
DSA	Distributed systems architecture (CII-Honeywell-Bull)
DSB-AM	Double side band-AM
DSE	Data switching equipment (CCITT)
DSE	Distributed systems environment (CII-Honeywell-Bull)
DTE	Data terminal equipment
DTP	Data transmission procedure
DXI	Data exchange interface
e-mail	Electronic mail
EBCDIC	Extended Binary Coded Decimal Interchange Code
ECMA	European Computer Manufacturers Association
EIA	Electronics Industries Association
ENQ	Enquiry (ASCII)
EPC	Even parity check
ERP	Error recovery procedure
ET	Exchange termination
ETSI	European Telecommunication Standards Institute
FCC	U.S. Federal Communications Commission
FCS	Frame check sequence
FDDI	Fiber distributed data interface
FDM	Frequency-division multiplexing
FEP	Front-end processor
FIFO	First-in–first-out
FM	Frequency modulation (analog signal)
FM	Function management (SNA)
FMD	Function management data (SNA)
FPS	Fast packet switching
FR	Frame relay
FRI	Frame relay interface
FSK	Frequency shift keying (digital signal)
FTP	File transfer protocol
GA	Global addressing
GEN	Global European Network
GFC	Generic flow control
GNT	Gigabit networking testbed
GO	Geosynchronous orbit
Go-back-N ARQ	Go-back-N ARQ protocol
GOSIP	Government OSI profile
HCS	Header check sequence

HDLC	High-level data link control
HDTV	High definition television
HDX	Half-duplex
HEC	Header error control
HLPI	Higher layer protocol identifier
HPPI	High performance parallel interface
HSLAN	High-speed local area network
HSSI	High-speed serial interface
HTTP	Hypertext transfer protocol
Hz	Hertz
IA5	International Alphabet 5
IAB	Internet Architecture Board
IBM	International Business Machines Corporation
ICL	International Computers Limited
ICMP	Internet control message protocol
ID	Identifier (SNA)
IDN	Integrated digital network
IDU	Interface data unit
IEEE	Institute of Electrical and Electronics Engineers
IETF	Internet Engineering Task Force
ILD	Injection laser diode
IMP	Interface message processor
INMAR-SAT	International Maritime Satellite Organization
Inter-LATA	Inter-local access transport area
IP	Internet protocol
IPng	Internet protocol-next generation
ISDN	Integrated services digital network
ISO	International Standards Organization
ISUP	ISDN user port
ITDM	Interleaved TDM
ITU	International Telecommunications Union (United Nations)
ITU-T	ITU-Telecommunications standard
IWU	Internetworking unit
JPEG	Joint Photographic Experts Group
JPL	Jet Propulsion Laboratory
JTM	Job transfer and manipulation
K	Kilo or 1024 bits
LAN	Local area network
LAP	Link access protocol
LAP-B	Link access protocol-balanced
LAP-D	Link access protocol-D
LAPF	Link access procedure for frame mode bearer services
LATA	Local access and transport area
LD	Line driver
LDM	Limited distance modem
LED	Light emitting diode
LFC	Local function capabilities
LHM	Long-haul modem
LLC	Logical link control
LRC	Longitudinal redundancy check
LSB	Least significant bit

LU	Logical unit
LUI	Logical unit interface
MAC	Media access control
MAN	Metropolitan area network
MAP	Manufacturing automation protocol
MCI	Microwave Communication, Incorporated
MHS	Message handling system
MIME	Multipurpose Internet mail extension
MMF	Multimode fiber
MPEG	Motion Picture Experts Group
MSB	Most significant bit
MTP	Message transfer part
MUD	Multiple user dimension
N-ISDN	Narrow-integrated services digital network
NAK	Negative acknowledgment
NNI	Network–node interface
NRZI	Non-return-to-zero inverted code
NRZL	Non-return-to-zero level code
NT	Network termination
NTD	Network television distribution
NTSC	North American Television Standards Committee
OAM	Operations and maintenance
OC1–OC192	Optical carrier (level)
OFC	Optical fiber communication
OM	Optical modulation
OSI	Open system interconnection
PABX	Private automatic branch exchange
PAD	Packet assembly and disassembly
PAL	Phase alternation line
PAM	Pulse amplitude modulation
PAR	Positive acknowledgment and retransmission
PAX	Private automatic exchange
PCM	Pulse coded modulation
PDH	Plesiochronous digital hierarchy
PDM	Pulse duration modulation
PDN	Public data network
PDU	Protocol data unit
PFM	Pulse frequency modulation
PL	Physical layer
PLCP	Physical layer convergence protocol
PLL	Phase lock loop
PLN	Private line network
PLP	Packet layer protocol
PM	Physical medium
PM	Pulse modulation
PPM	Pulse position modulation
PPP	Point-to-point protocol
PSI	Performance Systems International
PSK	Phase shift keying
PSN	Packet switched network
PSPDN	Packet-switched public data network

PSTN	Public switched telephone network
PTT	Post, Telegraph and Telephone
PU	Physical unit
PUP	PARC universal packet
PVC	Permanent virtual circuit
PWM	Pulse width modulation
QAM	Quadrature amplitude modulation
QOS	Quality of service
QPSK	Quadrature phase shift keying
QPSX	Queue packet synchronous exchange
RACE	Research and Development of Advanced Communication in Europe
RAM	Random access memory
RBHC	Regional Bell Holding Company
RBOC	Regional Bell Operating Company
RF	Radio frequency
RPOA	Recognized private operating agency
RSI	Ring station interface
RSVP	Resource reservation protocol
S/N	Signal-to-noise ratio
SAP	Service access point
SAR	Segmentation and reassembly sublayer
SASE	Specific application service element
SCCP	Signaling connection control point
SDH	Synchronous digital hierarchy
SDLC	Synchronous data link control
SDN	Software defined network
SDU	Service data unit
SECAM	Systeme en couleur avec Memoire
SHM	Short-hand modem
SIP	SMDS interface protocol
SLIP	Serial line Internet protocol
SMDS	Switched multimegabit data service
SNA	System network architecture
SNI	Specification network interface
SNI	Subscriber–network interface
SNMP	Simple network management protocol
SONET	Synchronous Optical Network
SPE	Synchronous payload envelope
SPN	Subscriber premises network
SS7	Signaling System Number 7 (CCITT)
SSB-AM	Single sideband-AM
STDM	Statistical time division multiplexing
STM	Synchronous multiplexing
STM	Synchronous transfer mode
STP	Shielded twisted pair
STS-1	Synchronous transport signal 1
SVC	Signaling virtual channel
TA	Terminal adapter
TC	Technical committee
TC	Transmission convergence sublayer
TCM	Time compression multiplexing

TCP	Transmission control protocol
TDM	Time-division multiplexing
TE	Terminal equipment
Telnet	GTE's public network
THT	Token hold time
TMS	Time multiplexed switching
TN	Telex network
TOP	Technical office protocol
TP	Telex protocol
TSI	Time slot interchange
TV	Television
Tymnet	McDonnell Douglas X.25 public network
UART	Universal asynchronous receiver and transmitter
UNI	User–network interface
URI	Universal resource locator
URL	Uniform resource locator
UTP	Unshielded twisted pair
UUI	User-to-user protocols
UUNET	UNIX–UNIX Network
VAN	Value-added network
VBR	Variable bit rate
VC	Virtual circuit/channel
VCC	Virtual channel connection
VCI	Virtual channel identifier
VCL	Virtual channel link
VLAN	Virtual LAN
VOD	Video on demand
VP	Virtual path
VPC	Virtual path connection
VPI	Virtual path identifier
VPL	Virtual path link
VRC	Vertical redundancy check
VTAM	Virtual telecommunication access method
WWW	World Wide Web

Glossary

Abstract syntax: This is used in the transmission and also known as *transfer syntax*, which is related to application-based unique syntax termed as abstract syntax and is discussed in ISO 8823 presentation layer protocol.

Abstract Syntax Notation One (ASN.1): The application layer protocol of OSI-RM uses a presentation and transfer syntax (defined by ISO and CCITT). This corresponds to specification of ISO 8824 and describes an abstract syntax for data types and values, while specification ISO 8825 defines the representation of data and is known as Basic Encoding Rules (BER).

Access: ITU-TSS X.25 is defined as a standard access method for public networks. This is a technique by which the data is accepted by public networks.

Access control method (ACM): A method for the control of access to a transmission channel.

Access protocol: A set of procedures used by an interface (described at a specified reference point) between user and network. The users will be able to access the service facilities and functions of the network with the help of the access protocol (CCITT I.112).

Access unit (AU): The distributed queue dual bus (DQDB) is a dual-bus architecture and consists of nodes. Each node is attached with an access unit (AU) and a physical attachment of AU to both of the buses. It provides access control, deposits the information into slots, and provides a physical connection to both buses for read and write operations.

Accredited Standards Committee for Information Processing Systems (ASC X3): This organization works under the guidance and control of the Computer Business Equipment Manufacturing Association (CBEMA). ASC X3 is responsible for defining the standards for the products of information processing.

ACCUNET: This is a collection or class of services offered by digital data transmission in the U.S., and represents a trademark for this class of services.

Acknowledgment (Ack): The OSI-RM data link layer of the transmitting station, after transmitting the frames expects a positive acknowledgment from the receiving station. Thus, Ack indicates that the frame is received error-free.

Acoustic coupler: A modem which contains a microphone and a speaker. It also includes interface attachment cards for communication channels and standard telephone handset and is activated by audio tones for its operation. This offers an interface through the handset rather than the transmission line as other modems do. It converts digital signals into audio tones, which are given to the mouthpiece of the handset. This is employed where a telephone set is available without wires to a telephone line, making the terminal portable.

Adapter: The CCITT recommendation has defined an X.21 adapter and asynchronous V.24 terminal adapter as services of ISDN to its users. The line adapter provides interface to low-cost terminals to share the data over the same telephone line via basic access rate interface standard. It also may be used in a videotex terminal to provide integrated voice and videotex system. The ATM adapter converts the information from computers, workstations, etc. into ATM cell format.

Adaptive delta modulation (ADM): A modification over delta modulation where step size is not kept constant. It has more quantization error, but provides a net advantage over a delta modulator. ADM operating at 32 Kbps bit rate offers performance similar to that of 64 Kbps using PCM. A variation of this modulation has been used in the Space Shuttle. It has been used in various communication networks on different transmission media, e.g., coaxial cables, optical fibers, microwave links and satellite links (radio communications), and other applications such as recording, signal processing, image processing, etc.

Address: A storage location or related location or source (destination) address in coded form. Since multiple nodes and devices use the same communication channel, each node or device or any other location will have a unique address, expressed in hexadecimal. It represents a variety of things in different applications. For example, a memory location in programming language, address of site/node/station expressed in a bit-stream or sequence of characters in computer networks, etc. It also identifies ports either via their hardware addresses (at various layers) or by software addresses created by underlying operating systems.

Address resolution protocol (ARP): This protocol extracts the hardware address of MAC from the received IP address. Based on this address, the packet will be routed to the appropriate destination host MAC address. ARP is a part of the TCP/IP communication transfer interface.

Advanced Research Projects Agency (ARPA): The former name of DARPA (Defense Advanced Research Projects Agency). This is the agency of the U.S. government that funded the ARPANET, and other DARPA Internet projects.

Algorithm: Formal and procedural steps required to execute any operation; usually defined in terms of a set of rules (represented by characters or symbols or signs). An instruction or statement typically performs a specific function in programming language and executes in a computer. The instructions for any algorithm are usually executed in some predefined, sequential manner.

Aloha: A ground-based contention packet-broadcasting network, which uses radio as a transmission channel. All of the users have equal access to the channel based on a pure random scheme; hence this type of network is also known as a *pure Aloha network*.

Aloha (slotted): In this type of network, a uniform slot time is defined on the transmission channel, where slot time is equal to frame transmission time. A non-random scheme for access of the network for independent channels is used. In an Aloha network with owner, the slots in the frame are used and owned by the user. An Aloha non-owner network selects a free slot in the frame and is reserved for that user until it is released.

American National Standards Institute (ANSI): A national coordinating agency for implementing standards in all fields in the U.S. It is a voting member of the International Standards Organization (ISO) and defines standards for data communications in OSI-RM. The basic standard defined in the area of communication is ASCII code.

American Standard Code for Information Interchange (ASCII): A 7-bit communication code used in the U.S. and defined by ANSI in ANSI X3.4. This standard represents a representative in CCITT T.50 and ISO 646. The CCITT has defined a standard from ASCII known as International Alphabet 5 (IA5). It represents 96 uppercase and lowercase characters and 32 non-displayed control characters

American Telegraph and Telephone Corporation (AT&T): A leading telephone communication carrier now working in the area of computers (after it bought National Cash Register — NCR).

American Wire Gauge (AWG): A system developed by the U.S. to specify the diameter (size) of the conductor or wire through which electric current can flow. The thinner the wire, the higher the gauge number.

Amplifier: An electronic circuit which amplifies a weak signal (analog) which is attenuated during its travel over the communication channel. Amplifiers are placed at regular distances and also amplify the noise present in the signal with the same amplification factor. In contrast, repeaters are used in the case of digital signals, where the binary logic values are regenerated. Here also the repeaters are placed at regular distances.

Amplitude: The strength of voltage or current in their waveform representation. Amplitude is usually expressed in volts or amperes, and can have both positive and negative value with respect to the y-axis.

Amplitude modulation (AM): One of the earliest forms of modulation and one of the most commonly used modulation techniques. Here the amplitude of a carrier signal is changed in accordance with the instantaneous value of lower-frequency modulating signals. The AM process generates a modulated signal which has twice the bandwidth of the modulating signal. It is obtained by multiplying a sinusoidal information signal with a constant term of the carrier signal, and this multiplication produces three sinusoidal components: carrier, lower sideband, and upper sideband, respectively.

Amplitude noise: Noise caused by a sudden change in the level of a signal; depends on the type of modulation scheme used for signal transmission. This noise is always present in amplitude modulation (AM), as in this modulation scheme, amplitude varies with the instantaneous value of the amplitude of the modulating signal. Faulty points, amplifiers, change of load and transmission line, maintenance work, etc., usually cause this noise.

Amplitude shift keying (ASK): The amplitude of an analog carrier signal varies in accordance with the discrete values of the bit stream (modulating signal), keeping frequency and phase constant. The level of amplitude can be used to represent binary logic 0s and 1s. In a modulated signal, logic 0 represents the absence of a carrier, thus giving OFF/ON keying operation; hence the name amplitude shift keying. The carrier frequency is chosen as an audio tone within the frequency range of PSTN. It can also be used to transmit digital data over fiber.

Analog data: The data represented by a physical quantity that is considered to be continuously variable and whose magnitude is made directly proportional to the data or to a suitable function of the data.

Analog modulation: This type of modulation requires the change of one of the parameters (amplitude, frequency, phase) of the carrier analog signal in accordance with the instantaneous values of the modulating analog signals. In this technique, the contents or nature of information (analog or digital) is not changed/modified. Here, the modulating signal is mixed with a carrier signal of higher frequency (a few MHz) and the resultant modulated signal typically has a bandwidth that is the same as that of the carrier signal.

Analog signaling: In the analog transmission (using modulation), the user's information is represented by continuous variation of amplitude, frequency, and phase of modulating signals with respect to a high-frequency carrier signal. The amplitude of the modulated signal gets attenuated (loses strength) as it travels over a long distance. This type of signaling used in a system is known as an *analog signaling system.*

Analog signal waveform: Speech or voice signals can be represented by analog continuous electrical signals (sinusoidal wave). The sinusoidal signal starts from zero level, goes in a positive direction, reaches a maximum level (also known as *peak*) on the positive side at 90 degrees and starts decreasing the value of signal, becomes 0 at

180 degrees and changes its direction, reaches peak value at 270 degrees (on the negative side), and finally comes to level. Sinusoidal signal represents sin function in trigonometry. Similarly, we can have cosine function for analog signal. In both representations, the signal follows a fixed pattern and is continuous, and periodic, and hence is known as an analog signal.

Analog-to-digital converter (A/D): This communication device converts an analog signal to digital signal form (binary logic values 0 and 1). The quantized samples of analog information signal obtained from a quantizer are applied to an encoder, which generates a unique binary pulse or voltage level signal (depending on the method used). Both quantizer and encoder combined are known as an analog-to-digital (A/D) converter. It accepts an input analog information signal and produces a sequence of code words where each word represents a unique train of binary bits in an arithmetic system. In other words, it produces a digitally encoded signal, which is transmitted over the communication link.

Analog transmission: This transmission scheme is responsible for transmitting analog signals (modulated) and is not concerned with the contents of the signal. For a larger distance, the signals are amplified. In the event of loss of signals, no attempt is made to recover the signals.

Angle modulation: In this technique of modulation, the phase angle of carrier signal is changed in accordance with the phase of modulating signals.

Angle of Acceptance: In the optical cable, maximum light from the cable is received at this angle.

Angular frequency: This is expressed as a product of frequency in Hertz/second and 2π and can also be expressed in radians/second.

ANSI X.34: This standard defines the 7-bit ASCII code, which is most popular communication code in the U.S.

ANSI X3.44: This standard defines a measure known as Transfer rate of information bits (TRIB) which can be used to perform comparison on the type of protocols to be used in networks of fixed and limited data transfer capabilities. In other words, we can consider this measure as one of the performance measures for selecting a protocol for the network.

ANSI X3.92: This standard is defined for security of data and defined as Data Encryption Algorithm (DEA). The concept of this standard is derived from another standard Data Encryption Standard (DES) defined by National Bureau of Standards (NBS) and is available to both users and computer product manufacturers and suppliers.

Answer record (A-record): A standard for transport layer was defined in West Germany (now United Germany) by ad hoc Working Group for Higher Communications Protocols in Feb.1979.

Antenna: It is a device, which can radiate and receive electromagnetic waves from the open space.

Application layer (AL): The top layer (seventh) in OSI-RM provides facilities to the users for using an application program independently of underlying networks. It provides an interface between network and user. It offers these facilities in common application service element (CASE) part of the layers. The services such as File Transfer Access Management (FTAM), Message Handling System (MHS), Virtual Terminal protocol (VTP) and other application-specific services to the users are provided as Specific application service elements (SASE) over CASE.

Archie: A file-searching utility which allows users to find a file in the file servers. This file can be downloaded onto the user's computer. This service is available at a large number of anonymous sites.

ARPAnet: A pioneering long-haul network sponsored by ARPA. It served as the basis for early network research and development and also for designing the Internet. ARPAnet includes packet-switching-based computers connected by leased lines.

Association control service elements (ACSE): The application layers of OSI-RM offer a variety of services to the users and are usually offered as service process entities. The services we get from these entities include electronic mail, file transfer, remote access, general-purpose services, etc. These entities are defined as specific application service elements (SASEs).

Asymmetric encryption: In this encryption technique, two different keys for encryption and decryption are used. One of the keys is a *public key* while the other one is a *private key*. This type of encryption technique is used in public key encryption protocol.

Asymmetrical digital subscriber line (ADSL): A copper-based network offering a data rate of 1.5 Mbps. It can also support video transmission if appropriate coding techniques such as MPEG or JPEG are used.

Asynchronous communication (start–stop communication): In this transmission, a start bit is attached at the beginning of each text character (of 8 bits) and represents the synchronization between transmitter and receiver clocks during the duration of the character. One or more stop bits then follow the character. Asynchronous transmission systems are inexpensive and easy to maintain but offer low speed; hence, the channel capacity is wasted. In the event of errors, only that character which is lost will be transmitted. Various input devices include keyboards, transducers, analog-to-digital (A\D) converters, sensors, etc., while the output devices are teleprinters, printers, teletypewriters, etc.

Asynchronous line drivers: Used to boost the distance over which data can be transmitted. Although a LAN can be used to transmit the data to a distance of a few hundred feet or a few miles, it has a limitation on the distance. The asynchronous line drivers can be used to transmit the data over inexpensive twisted pair (four-wire on half- or full-duplex) wires between various locations within a building or across the buildings within some complex (within a distance of a few miles).

Asynchronous mode of transmission: In this mode of transmission, every character or symbol (expressed in ASCII or EBCDIC) is preceded by one "start" bit (spacing) and followed by one or more "stop" bit (marking) and is then transmitted over the transmission channel. During the idle situation of the channel (no data is transmitted), logical 1s is constantly transmitted over the channel. The start bit indicates the receiver to activate the receiving mechanism for the reception, while the stop bit indicates the end of the present character or symbol. Due to this working concept, this mode of transmission is also known as start–stop transmission. The start and stop bits together provide synchronization between sending and receiving stations. This mode is very useful for sending text between terminals and computers asynchronously. The main advantage with this transmission is that it does not require any synchronization between sender and receiver.

Asynchronous transfer mode (ATM): This mode is defined for broadband integrated service digital networks (B-ISDNs). It supports different types of signals (data, voice, video, images, etc.). This mode of transmission and switching partitions the user's information into cells of fixed size (53 bytes). The transfer mode is basically a combination of switching and transmission of information through the network. In ATM, the user's information is encapsulated in the information field of 48 octets with 5 octets for the header. The header field contains the information about the identification of cells, labels, and the functionality of ATM cell, while the information field is available for the user's information.

ATM adaptation layer (AAL): The layer above the ATM layer is known as the ATM adaptation layer (AAL), which provides interface between higher layers (above the AAL) and ATM layers. The adaptation is performed on both signaling and user information coming from the higher layer into an ATM cell. This layer is divided into two sublayers: segmentation and reassembly sublayer (SARS) and convergence sublayer (CS).

ATM adaptation layer (AAL) type 1: This type corresponds to service type A defining constant bit ratio (CBR). This type receives the PDUs from higher layers at a constant bit rate and also delivers the PDUs to its higher layer at a constant bit rate.

AAL type 2: This type of protocol provides service type B, which defines variable bit rate (VBR) services, e.g., video, audio. The protocol supports the exchange of SDUs with variable bit rate between source and destination nodes.

AAL type 3: This type of protocol is defined for service type C, which is basically a connection-oriented service with variable bit rate. No timing relationship is supported between source and destination nodes.

AAL type 4: The protocol of this type supports service type D that is defined by connectionless services. This protocol allows one AAL user to transfer AAL-SDU to more than one AAL user by using the services provided by the ATM layer. The users have a choice from this available AAL-service access point (SAP).

ATM connection: The connection defined by ATM is in fact based on connection-oriented concepts. It also offers flexible transfer support to the services based on connectionless concepts. More than one ATM layer link are defined as an ATM connection such that each link is assigned a unique identifier. These identifiers remain the same during the duration of a connection.

ATM layer: The second layer from the bottom in the layered architecture of an ATM-based network. It performs various functions on the cell header of 5 bytes (of a total 53 bytes of an ATM cell) and as such is also known as the *ATM cell layer.* The functions of this layer are to provide routing, cell multiplexing, cell switching, construction of ATM cell, and generic flow control, and are the same for any type of physical media.

ATM performance parameters: In order to define the quality of ATM cell transfer, the following parameters have been identified by CCITT I.35B: successfully delivered cell, errored cell and severely errored cell, lost cell and inserted cell. Based on these parameters, a number of performance parameters have been defined. These parameters include cell loss ratio, cell insertion rate, cell error ratio, severely errored cell ratio, cell transfer delay, and cell transfer capacity.

ATM switch: The architecture of an ATM switch consists of switch core (performs switching of cells) and switch interface (performs input and output functions) and software to control and manage these switching operations.

Attachment unit interface (AUI): The standard specification of Ethernet defines a medium attachment unit (MAU), also known as a transceiver. The functions of the physical layer of OSI-RM are implemented at each station and the station is connected to the MAU via the attachment unit interface (AUI). This is usually provided in IEEE 802.3-based products.

Attenuation: Due to noise factors such as transmission errors, reflections due to mismatching, and other atmospheric errors, the transmitted signal loses its strength as it travels down the transmission channel. This loss of signal strength (or magnitude), usually expressed in voltage, current, or power, is known as *attenuation* and is expressed in decibels (db).

Audio frequency (AF): This frequency range includes the frequencies which can be heard by the human ear and have been assigned a range of 30 Hz to 20 KHz. For the transmission of audio signals, only a bandwidth of 4 KHz is used in voice-grade channels.

Authentication: This method is used to check whether the received message is the same as that sent by verifying the integrity of the message. There are many authentication protocols used over TCP/IP.

Automatic repeat request (ARQ): Most protocols for the data link layer support error detection and error recovery over a transmission line or link. This error-control technique provides error recovery after the error is detected. In the event of error detected by ARQ at the receiving site, the receiver requests the sending site to retransmit the protocol data unit (PDU). The combination of error detection and error recovery results in a reliable data link. Three versions of ARQ have been defined: Stop-and-wait ARQ, Go-back-N, and Go-back-selective.

i) **Stop-and-wait ARQ:** A sending station sends a frame (protocol data unit) to the destination station and waits until it receives an acknowledgment from the destination station. This means that there can be only one frame over the channel at any time.

ii) **Go-back-N ARQ:** A sending station sends more than one frame which is controlled by the upper limit of the maximum value. If an error is detected in any frame (when an acknowledgment arrives at the sending station) or the acknowledgment is lost or it is timed out, in all three cases, the sending station will retransmit the same frame until it is received error-free on the receiving side. All the frames behind the errored frames are discarded and hence have to be retransmitted.

iii) **Go-back-select ARQ:** This version of ARQ is similar to Go-back-N ARQ, with the only difference being that here the frames behind the errored frame are stored in a buffer at the receiving site (as opposed to discarding in Go-back-N ARQ) until the errored frame is received error-free.

Backbone: A high-speed connection within a network, which provides connection to shorter, usually slower circuits. In some cases, it also corresponds to a situation of creating a hub for activities.

Backbone local area network (LAN): Each building or department or a division within the organization may have its own low-capacity network connecting personal computers (PCs), workstations, resources, etc., and operating at a low-speed data rate. These low-speed networks can then be connected to a high-speed and high-capacity network. The low-speed network is referred to as a *backbone local area network*. FDDI may be used as a backbone network for applications requiring higher data rates, while, in general, the Ethernet is used as a backbone network offering 10 MBps.

Backend local area network: This network usually interconnects mainframe and storage devices within a very small area. The main feature of this network is to transfer the bulk of data (file-transfer-based applications) at a higher data rate of 50 to 100 Mbps between a small number of devices connected to it. FDDI may be used as a backend local area network.

Backus-Naur form (BNF): Used to define the syntax of the data in the presentation layer. In OSI terminology, this is used as a notation for describing the protocol data unit (PDU). The standard X.208 uses BNF for defining the data information.

Backward explicit congestion notification (BECN): This primitive is invoked in frame relay networks to inform the user about the occurrence of congestion in the network. It is the network that informs the user or source devices about the network congestion.

Balanced modulator: A device that uses a double-balanced mixing of signals.

Balanced transmission: A circuit for this type of transmission provides a separate return path for each signal transmitted in either direction. For every communication, the circuit offers two wires in the cable, hence offering a higher data rates. In contrast, in the case of unbalanced transmission, the ground is used as a return path for the signals flowing in both directions.

Bandwidth: The difference between the upper and lower boundary of frequency (of a signal). It represents the information capacity of the transmission line to transmit the signal and is measured in hertz (Hz). It also represents the capacity of the network defining the ability to carry the number of files, messages, data traffic, etc., and as such the unit of bandwidth is changed to bits per second (bps). It also represents the range of signal channel (in cycles per second) which that particular communication system can transmit.

Bandwidth balancing: This scheme is defined in a DQDB network where a node that has been queued in may not be assigned the empty QA slot. This depends on a number of factors, such as bandwidth on demand and effective sharing of QA slots.

Banyan networks: The multi-stage networks remove the drawbacks of single-stage networks and offer multiple stages, which are interconnected by different link configurations. Based on the number of reachable paths between source and destination nodes, these networks can be either single-path or multi-path multi-stage networks. In the single-path networks, there exists only one path between source and destination — these networks are known as Banyan networks. These networks offer advantages in terms of simple routing but may cause blockage.

Baseband bias distortion: The signals from the computers and terminals are square wave pulses and these pulses, experience some losses due to the parameters of the communication link, including resistance, inductive and capacitive effects, and shunt resistance. These parameters oppose the flow of signals and as such the signals are distorted, and collectively are known as bias distortion.

Baseband network: The Ethernet accepts digital signals that are fed into cable in the form of electrical impulses. A baseband network uses digital signaling and bus topology, is bi-directional, and does not use frequency division multiplexing (FDM). It also supports analog signals. One user uses the entire bandwidth at a time after the establishment of the link/circuit/session.

Baseband transmission: In this transmission, the unmodulated signals are transmitted at their original frequencies. Here each binary digit (bit) in the information is defined in terms of binary 1 and 0. The voltages representing logic 1 and 0 are directly used over the transmission lines. In other words, the entire bandwidth is allocated to the signals transmitted and also the variation of voltages is defined between these two values over the line with time as the information data is being transmitted over it.

Basic access: This ISDN standard interface was defined by the CCITT I-series recommendations as consisting of two 64-Kbps bearer (B) channels and one delta (D) channel of 16 Kbps, giving a total bandwidth of 1.92 Mpbs. This interface is also known as a basic rate interface (BRI). The ISDN is a digital network offering a channel rate of 64 Kbps.

Basic Telecommunication Access Method (BTAM): The software package known as teleprocessing access method (TPAM) provides a clear distinction between the interfacing details and the network programmers engaged in data networking. IBM defined its own software package known as Basic Telecommunication Access Method (BTAM). It supports access to bisynchronous and asynchronous terminals. Other vendors allow users to share TPAMs for their multiple applications, requiring more memory.

Batch processing: This scheme allows the tasks to be submitted for their processing in a predefined manner. Each of the tasks is submitted in a complete form and they are executed in the same order in which they were submitted.

Baud: The rate of signaling (number of discrete conditions or signal levels) per second on the circuit or the number of line signal variations per second. If a single level represents each bit over the transmission line, then this may also represent the number of bits per second over the line (data rate). When more than 1 bit is transmitted during the signaling period, the data rate and baud may not be equal, and, in fact, the bits per second will be more than the baud.

Baudout code: This code is basically defined for typewriters.

B-channel: The CCITT has defined a 64 Kbps channel which offers standardized communication services as a standard channel in both basic and primary rate interfaces. The 64 Kbps channel offers circuit-switched connectivity, as ISDN is based on the digital telephone network. The B-channel is used for accessing. In maximum integration, packet switching is handled within ISDN, and both B and D channels are supported. The packet handling function within ISDN offers processing of packet calls, X.25 functions for X.25, and also other functions needed for any rate adoption.

Because It's Time Network (Bitnet): This network is based on NJE-based international educational network. Its counterpart in Europe is EARN. This network can be accessed by FTP or anonymous FTP.

BEL: A unit of power loss or gain equal to 10 db (smallest unit).

Bessel function: A mathematical equation that is used to calculate the amplitude of harmonics of angle modulation technique.

Bi-directional amplifier: A device which amplifies analog signals of different frequencies in both directions, i.e., it amplifies signals of lower frequencies in one direction as inbound, and the signals of higher frequencies in the opposite direction as outbound. The interference between inbound and outbound signals is avoided by a guardband frequency that does not carry any signal.

Binary code: A communication code comprised of two discrete logic values representing two states (ON and OFF). The binary logic values for these two states are represented by binary bit 1 and 0. There are different classes of binary codes, e.g., BCD, ASCII, etc. In binary coded decimal (BCD), four binary bits represent the numbers in the decimal range (0–9) where each bit represents two logical states or values or combinations.

Binary synchronous communication (BSC or BISYNC): A standard protocol for the data link layer of OSI-RM defined by IBM in 1966. It offers synchronization between the sending and receiving sites of the network. It uses ASCII or EBCDIC for delineating the user data of the frame. It is also known as *character-oriented protocols*. It operates on half-duplex mode of transmission and uses a special set of characters of ASCII code for providing synchronization to frames, characters, and control frames. This protocol offers computer-to-terminal and computer-to-computer communication.

Biphase coding: This scheme provides a digital signaling for local area networks. It may have more than one transmission during bit-time and as such guarantees synchronization between transmitter and receiver. Manchester and differential Manchester coding schemes are known as biphase coding schemes.

Bisynchronous transmission: Binary synchronization (bisync) data transmission provides synchronization of characters between sender and receiver via timing signals generated by these stations. This is also known as *synchronous transmission*.

Bit: The smallest unit to represent a message in a binary system. It is also known as a binary digit and has two states: ON (logic 1) and OFF (logic 0).

Bit error rate (BER): The number of bits which occur as errors in the given number of bits. For example, a BER of 5×10^{-4} indicates that out of every 10,000 bits, 4 bits represent error bits, i.e., 4 bits are corrupted. The bit error rate can also be defined as the ratio of the total number of bits (including errored) received to the actual number of bits transmitted at the sending site over a period of time.

Bit order: In a serial transmission, the bits are sent serially bit by bit. Generally in such systems the least significant bit (right-most position of the bit in the packet/frame) is sent first. The most significant bit (left-most bit) will be the last to be received.

Bit-oriented protocol: These protocols are synchronous and support the full-duplex mode of transmission. They use a unique bit pattern (01111110) known as a *flag* for providing an initial synchronization between transmitter and receiver. They are usually independent of any coding techniques and as such allow any desired bit pattern to be defined in the data field of the frame.

Bit rate: In order to measures the performance of B-ISDN networks, CCITT recommendation I.311 defined various parameters for traffic control. The following parameters are defined for traffic source control: average bit rate, peak bit rate, burstiness, and peak duration. These parameters are useful for defining quality of service (QOS).

Bit stuffing (zero insertion): In order to distinguish between the data and control within a frame, a technique known as bit stuffing is used over a synchronous transmission line. Each block of information is enclosed between special bit pattern flags (special control bit pattern). In this technique, logic 0 is inserted after every five consecutive logic 1s at the sending node. The receiving site detects the 0 after every five consecutive 1s automatically. A similar approach is used in character-oriented protocols (BCS) and there it is known as *character stuffing*. The main idea in stuffing is to provide a mechanism of distinguishing between control and data signals within the frames being transmitted.

Bits per second (BPS): The rate of transmission of data or control bits per second over the communication channel.

Block check character (BCC): A field in the standard protocol of the data link layer (binary synchronous communication) frame format. The BCC code is calculated for the entire frame, starting with initial start of the header to the end of text (ETX) or end of transmission block (ETB). At the receiving site, the BCC for the received frame is calculated, and if it matches with the one transmitted, the receiver sends a positive acknowledgment (Ack); otherwise it sends a negative acknowledgment (Nak) for ARQ error-recovery protocols to correct the error at the transmitting site.

Block coding: Here, a logical operation is performed to introduce redundant bits into the original message such that the message defines a fixed size of the format. There are number of ways of implementing this translation process. The block coding technique is well suited to ARQ techniques and can be either constant-ratio coding or parity check coding. Two popular and widely used codes supporting parity-check coding are *geometric* and *cyclic*.

Block error-rate (BLER): This measure indicates the quality and throughput of the channel and is defined as the number of blocks (including errored) received to the actual number of blocks transmitted over a period of time. It has been used widely for determining the network topology, as the overall throughput of the channel can be expressed by BLER.

Block sum check: Used to detect the error during data transmission. It defines binary bits which are obtained using modulo-2 sum of individual characters/octets in a frame or in the information block.

Boeing Computer Services (BCS): BCS defined the local area network standard for office automation as the technical office protocol (TOP) and this standard is now controlled by SME. The standard was defined by BCS to complement General Motors' local area network known as manufacturing automation protocol (MAP) and developed in 1962. Although both TOP and MAP are similar and are based on OSI-RM, they have a different set of protocols for each of the layers, and overall, they are compatible to each other for interconnection. It uses IEEE 802.3 CSMA/CD or IEEE 802.5 protocols at lower layers and the same OSI protocols at higher layers.

Bolt, Barnak and Newman (BBN): The Defense Department's Advanced Research Projects Agency (ARPA) handed over the packet-switched network with other management and controlling functions to Bolt, Barnak and Newman (BBN) of Cambridge, MA (U.S.) to deploy the fully operational network as ARPANET. This network is still being used for military applications and at present offers support to more than 100 nodes (the first network had only four nodes).

Bridge: This device provides the interconnection between similar or different types of local area networks and is usually known as a media access control (MAC; one of the sublayers of OSI-RM) bridge, as it is connected at the MAC sublayer of the data link layer. This is a data link device which connects two similar networks where the higher layer protocols are the same on both networks. The specification for different types of LANs is defined in the IEEE 802.1 and 802.2 standards.

British Standards Institute (BSI): A standards organization which is responsible for developing standards in the U.K., and is a voting member of ISO.

Broadband integrated service digital network (B-ISDN): This service-independent network transports different types of services (audio, video, voice, images, etc.) requiring transmission channels supporting more than the primary rate interface (PRI) defined by N-ISDN, and allows users to share the resources of various networks for different services. An asynchronous transfer mode (ATM) has been considered a transfer mode solution (dealing with transmission and switching) for implementation of B-ISDN. Further, with the advances in coding algorithms, transfer modes, compression techniques, transmission strategies over fiber optics, and other relevant technologies, it is hoped that this will be able to accommodate all of the *teleservices* (telex, teletex, videotex, facsimile, etc.) of future applications with a smaller cost, as only one network is needed to offer these services.

B-ISDN network termination 1 (B-NT1): This interface is defined in ISDN as NT1 and also defined for B-ISDN as B-NT1 and offers functions similar to those of the physical layer of OSI-RM. These functions (CCITT I.413) include line transmission, transmission, and OAM functions.

B-ISDN network termination 2 (B-NT2): This interface is similar to NT2 defined in ISDN but is denoted by B-NT2 in B-ISDN. The functions of this interface are similar to those of the layer and also those of higher layers. These functions (CCITT I.413) include adaptation for different interface media and topologies, multiplexing/demultiplexing/concentration of traffic, buffering of ATM cells, resource allocation, signaling protocol handling, interface handling, and switching of internal connections.

Broadband local area network: This uses analog signaling on a single channel to offer a single transmission path for an analog signal. It uses bus or tree topologies and is highly unidirectional. It allows frequency division multiplexing (FDM) on audio, video, and data signals and supports a distance of tens of kilometers. It offers a low attenuation at low-frequency signals.

Broadband networks: The broadband signals only support analog signals, and the bandwidth is shared by a number of stations active at that time. The signals are usually transmitted at their respective frequencies and as such are not changed by the modulation. The maximum distance for its operation is typically a few kilometers.

Broadband services: The ISDN basically offers audio, video, and data services with a limited bandwidth of 64 Kbps and limited processing capabilities. The development of B-ISDN is intended to offer different types of services at higher data rates based on asynchronous transfer mode (ATM). The broadband services include the services of constant and variable bit rates and includes data, voice, and still and moving picture transmission.

Broadband transmission: In this transmission, the signal frequency is modulated (changed) to the frequency value to which it has been assigned. The entire bandwidth of the transmission is divided into a small number of frequency bands. Within each channel, the data is modulated on a single frequency. In analog transmission, each band is in the megahertz range and carries different types of signals, e.g., videotex, video teleconferencing, cable television, voice, speech, etc., while digital transmission can offer a high bit rate (megabits per second) and send moving images and pictures.

Broadcast: In this mode of transmission, all the nodes connected to the network can receive the message simultaneously over the same link. A unique bit combination of 1s is used to indicate to all of the nodes (devices) that this is a broadcast message and will be received by all connected to the network.

Brouter: An integrtaed device that includes the implementation of bridging and routing together. This device provides not only bridging but also provides routing for the packets.

Buffer: When the data packet is transmitted from one site to another, the data packets coming out from the devices with different speeds have to be stored temporarily in a storage area called a *buffer*. The buffer enables maintaining the flow-control of packets between the transmitting and receiving sites over the networks.

Bursty: A signal not regularly distributed over time and known as an irregular signal. Usually this is used to denote a type of noise, but it has also frequently been used as a term in networks, in particular in reference to traffic of packets. These packets are short and are usually not evenly distributed in time. With short messages, the time required for setting up and clearing the transmission lines may exceed the time required to transmit the data message, and further, the conversion process over the line may be underutilized, since the line is idle for a longer time.

Bus topology: One of the most common topologies to implement local area networks; provides interconnection of all digital devices over a limited distance. All the nodes (or stations) communicate through a common transmission medium over which both data and control signals can be transmitted simultaneously. Each station is assigned a unique address and we can also send a message to a particular station by using that particular address, while for broadcasting, we use a special bit pattern in the address field to send the message to all stations. The most common LAN using this topology is CSMA/CD (Ethernet).

Busy slot: In DQDB networks, once a slot has been assigned and is being used, it cannot be accessed by QA access functions.

Cable: This is the most versatile transmission medium for LANs. Broadband systems use standard 75-ohm coaxial cable for the transmission of data. There are three categories of cable used for data communication in broadband networks: trunk cable (300 MHz for 100-ft long cable, a few kilometers to tens of kilometers), distribution cable (short distance), and drop cable (short distance; also used to connect local area networks to the stations).

Call: In data communication, a link is established through a request call, which is basically defined in terms of dial-up codes. The transmission of a request call indicates a desire to establish connection. Similarly, other calls (e.g., terminate, data packet call, etc.) are used for appropriate functions.

Cambridge ring: An experimental ring system based on slotted ring was defined and installed at the Cambridge University in 1974. It consists of ring interfaces connected to each other via cables. The data containers usually are transmitted at data rate of 10 Mbps.

Carrier common communication (CCC): A telephone company which offers communication services to the telephone subscribers and is usually controlled by government agencies.

Carrier sense multiple access with collision detect (CSMA/CD): The concept of CSMA/CD allows multiple transmitters to share a common communication channel (here, it is coaxial cable). Any station wanting to send a message has to listen to the channel. If the channel is free, it will send its message on the channel; otherwise, it must wait. In the situation where another message is already on the channel, there will be a collision (causing a contention situation) between these messages. In the event of collision, both stations have to withdraw from the channel and retry transmitting the respective messages at a later time. There are a few versions of contention algorithm protocols which may be used.

Carrier signal: A constant high-frequency signal used for modulating the analog modulating signal in various modulation schemes.

Carrier system: Enables definition of a number of channels over one circuit and is obtained by modulating each channel in a different carrier frequency value.

Consultative Committee of International Telegraph and Telephone (CCITT): A treaty within the United Nations created the International Telecommunications Union in 1986, and CCITT is one of the members of this standards organization. CCITT defines and sponsors recommendations in the area of data communication networks, digital systems, switched networks, terminals, and other areas related to data communication. It has five classes of membership — some of the classes allow voting power while others do not.

CCITT 511 and 2047 bits sequence: These sequences are used to determine the quality of a channel, which is measured as the ratio of total number of bits (including errored bits) to the total number of actual bits transmitted. In this ratio, the sequence of bits plays an important role.

CCITT address signal (AS): The OSI X.213 recommendation specifies three fields for defining a network address: (1) interface to public or private data network, (2) network service access point (NSAP), and (3) network protocol control information (PCI). The third field is also known as an *address signal* (AS).

CCITT fax: The transmission of text, graphics, images, etc., can be obtained by facsimile transmission (fax) and the method of transmission uses data compression to allow users to send bulk data. T.2, T.3, T.4, and T.5 are various recommendations of CCITT for fax; the difference between these recommendations mainly lies with the speed per page.

CCITT I.113: Terms used in B-ISDN.

CCITT I.121: Principal aspects of B-ISDN.

CCITT I.150: Functions of ATM layer (cell header).

CCITT I.211: Classification and general description of B-ISDN services.

CCITT I.311: Network techniques at virtual channel and virtual path levels (B-ISDN).

CCITT I.321: Extension of I.320.

CCITT I.327: Enhancement of I.324.

CCITT I.361: ATM cell structure (53 octets) and coding for ATM cell header.

CCITT I.362: Basic principles of AAL layer.

CCITT I.363: Types of AAL protocols.

CCITT I.413: Reference configurations.

CCITT I.432: B-ISDN user–network interface.

CCITT I.450 and I.451: These standard protocols specify the procedures to establish and de-establish connection, clear the connection, and manage the connection at layer 3 of the user–network interface of ISDN. These operations are defined for different types of connections, e.g., circuit switched, packet switched, and user-to-user.

CCITT I.610: OAM principles of B-ISDN access.

CCITT Study Group (SG): There are 18 study groups working under CCITT. Each study group is assigned a specific recommendation, e.g., SG II defines the operation of telephone networks and ISDN, while SG VIII deals with the recommendation for digital networks.

CCITT V.32 trellis coded modulation (TCM): This recommendation specifies a modem which uses TCM in a 19.2 Kbps channel. The V.32 DCEs usually operate at data rates up to 9.6 Kbps on full-duplex (dial-up two-wire lines) on switched network and point-to-point leased networks and supports the synchronous mode of operation.

CCITT V-series: This series of recommendations defines a physical-level DTE/DCE interface protocol that includes an interface between DTE and DCE and also between DCE and another DCE. Interestingly, EIA-232 defines an interface between DTE and DCE only. Another standard of CCITT known as I-series recommendations defines the signaling between two DCEs. In other words, it defines a digital physical-level interface between them and has been incorporated in all Hayes modem products.

CCITT X.21: This standard was defined in 1976 and allows access to public data networks. It works on full-duplex synchronous and allows users to send and receive messages from public data networks. It is used as a physical-layer protocol for terminals which access the public packet-switched network (X.25).

CCITT X.21bis: Before the X.21 standard could be defined and used for a digital circuit-switched network, an interim standard was defined as X.21bis to be used with a public packet-switched network (X.25). It is also similar to RS-232.

CCITT X.25: Defines an interface between a local DTE and its DCE. X.25 was first published in 1976 and has been changed regularly to include new sets of protocols. It is now an integrated set of three protocols.

CCITT X.26: This standard defines electrical characteristics of RS-422 (DTE/DCE interface for balanced transmission).

CCITT X.27: This standard defines the electrical characteristics of RS-423A (DTE/DCE interface for unbalanced transmission).

CCITT X.75: Interface between terminals and public packet-switched network (X.25).

CCITT X.200: A standard defining common architecture to connect heterogeneous computers and other devices; compatible with various standards of ISO specifications.

CCITT X.350: The transmission of data, video, voice in sea ships can be obtained in a variety of ways, but most maritime users are transmitting these signals over satellite link. For communication with sea ships, an International Maritime Satellite Organization (INMARSAT) was formed to define various standards and operational facilities, and as such various operations, services, and functional facilities are governed by several CCITT standards.

CCITT X.351: Packet assembly/deassembly (PAD) facilities.

CCITT X.352: Routing strategies for internetworked maritime satellite system and public switched network.

CCITT X.353: Routing strategies for internetworked public and maritime satellite network.

Cell (or ATM cell): During data communication, two types of traffic have been defined as circuit-switched-based traffic (or data stream) and packet-switched-based traffic (bursty type). These two types of traffic have different requirements for capacity and data rates, and as such a single interface which can handle both types of traffic has been defined as a *cell*. The traffic is encapsulated within the fixed size of cell (of 53 octets), which will carry the information over any type of connection. The cell-based network is connection-oriented and allocates the bandwidth on demand.

Cell header: The ATM cell consists of 53 bytes, and out of this, 5 bytes are defined for a cell header. The bytes in the header are used for control information, multiplexing, and switching.

Cellular telephone: This system uses the FDM technique for connecting various wireless telephone connections to the existing wired switched network.

Centralized network: A common central processing node is defined within a network, which provides a link to other nodes for data and control signals.

Central office (CO): The telephone subscriber's lines are connected to the switching equipment at a place which is under the control of carrier common communication (CCC): a telephone company. It is also known as exchange, end office (EO), or local control office (LCO).

Central office vendor: A vendor supplying telephone equipment and products that conform to the central office standards. For example, Siemens supplies DQDB switches, AT&T supplies ATM switches, and so on.

Central processing unit (CPU): This is very important module of a computer that interprets and executes all of the instructions. It is nothing to do with memory or any other interfaces. It fetches instructions from memory and other peripheral devices.

CEPT: European Conference of Posts and Telecommunications (CEPT) is an international standards organization that is actively engaged in supporting digital networks and equipment. It is mainly concerned with the specification of ISDN user–network interfaces defined by CCITT I-series recommendations and it ensures that this must be implemented within Europe to comply with those recommendations. The primary rate interface (PRI) of 2.048 Mbps offers primary rates within CEPT networks.

Channel: A part of total transmission capacity of the broadband transmission process; also known as a circuit, facility, link, or path. The channel is defined over transmission media and allows the connected devices (DTEs and DCEs) to share resources, exchange data, etc. The communication channel can be defined over transmission media such as twisted pairs of lines, radio signals, satellite links, microwave links, coaxial cables, etc.

Channel bandwidth: The speech signal does not contain a unique frequency and instead consists of many harmonics. Similarly, signals corresponding to other communication circuits such as radio, television, and satellite also consist of a range of frequencies rather than one frequency. Thus, these circuits are also characterized as a bandwidth of signals which can be transmitted, e.g., a TV channel usually has a bandwidth of 6 MHz. The channel bandwidth for AM is 10 KHz, FM is 200 KHz and 4.5 MHz (video) and 6 MHz. The broadcasting bandwidth for these transmissions are AM (535–1605 KHz), FM (88–108 MHz), 54–216 MHz, 470–890 MHz).

Channel capacity: The number of bits which can be transmitted over the channel. The speed of a channel depends on many factors, including bandwidth, bit data rate, noises, modulation techniques, etc. The channel capacity indirectly depends on the bit rate, which is directly proportional to bandwidth of the channel. Although there is no direct relationship between bandwidth and capacity of channel, various frequency ranges for different applications clearly demonstrate the relationship between them. Accordingly, the greater the bandwidth, the greater the capacity of the channel.

Channel configurations: A fundamental aspect of a communication system; establishes a route or circuit (logical path) between transmitting and receiving sites over which electrical signals are transmitted. This type of connection may be defined over many transmission media, e.g., telephone lines, coaxial cable, etc. A combination of links may be defined as a communication channel over which message information can be transmitted, and is characterized by various parameters such as channel operations, transmission speed, modes of operation, transmission configurations, etc. Various modes of operations in the communication channel are simplex, half-duplex, and full-duplex.

Channel service units/data service units (CSU/DSU): A pair of channel service unit (CSU) and data service unit (DSU) is used to provide an interface between a computer and the digital switch of a digital circuit network. The DSU converts signals of DTE (computer/terminal) into bipolar digital signals.

Channel voice grade: Any information (speech, digital, analog, or any other type) in a frequency range of 300 Hz to 3.4 KHz can be transmitted over this type of channel.

Character-communication code: The American Standard Code for Information Interchange (ASCII), a 7-plus-parity-bit communication code, was introduced by the American National Standards Institute and has been accepted as the U.S. federal standard. This coding scheme is used for transmitting transparent or binary data, which is defined within the code itself. This code is used in IBM PCs.

Character impedance: The impedance offered by the transmission link to the signal flowing through it.

Character-oriented protocols: The data link layer protocols for alphanumeric communication codes are usually known as character-oriented protocols. In these protocols, two consecutive control characters defined as syn characters usually provide the synchronization.

Character stuffing: In the information field, symbols may have a resemblance with control characters, and as such we have to define a scheme which can distinguish between the characters used in two different fields, i.e., a character used in the information field should not be interpreted as a protocol control character. Character stuffing allows the sender to insert a DLE symbol before other link control characters used in information and control fields. The binary synchronous communication (BSC) defined by IBM uses a character-stuffing technique to offer transparent operations.

Check sum: A method of detecting error based on parity bits may leave some errors undetected. Checking the parity bits of an entire frame, i.e., parity bits of all the characters in the frame, can incorporate an additional check for error detection. The check sum method for error detection may be implemented in a variety of ways, e.g., 1s complement of the entire frame and others.

Ciphers: The main aim of security (provided at the presentation layer) is to provide confidentiality and integrity of the data. Encryption is a process which converts the normal form of a text message, image message, or symbols into a form which is senseless for any user to extract the message from. This conversion process uses a certain predefined set of rules and converts the original text message (often known as plaintext) into encrypted form (often known as ciphertext). On the receiving side, the decryption process delivers the original text message to the user.

Circuit-switched network (CSN): This network is based on the concept of a normal telephone network. It allows the sharing of resources among the users uniformly and as such it reduces the cost. A user makes a request for the establishment of a logical connection over the link. Once the link or circuit is established, it becomes a dedicated circuit or link for the duration of its usage. After the data is transferred, the user makes a request for the termination of connection.

Circuit-switched public data network (CSPDN): This class of network uses the circuit-switching technique for transmitting the message across the networks. This switching technique allows the transmission of both data and voice over the network. A dedicated connection is established for the duration of the message between two nodes. There is no limit on the size of the message and the duration, and as such the billing information is based on the amount of the message being transmitted.

Circuit switching: A dedicated connection is established for the duration of the message between two nodes; this type of switching is used in telephone networks and some of the upcoming digital networks. Here the messages are delivered to the destination in the same order in which they are transmitted at the transmitting node. In circuit switching, the dedicated connection has to be terminated and may have to be established again for the same nodes or different nodes. There is no limit on the size of the message and the duration, and as such the billing information is based on the amount of the message being transmitted. Both the source and destination nodes operate at the same speed.

Client: The user of the network service or the services offered by the server for any network-based application. It runs a client program, which provides a communication link between the client (PCs, workstations, etc.) and powerful workstations. The workstations offload most of the processing-intensive tasks from the client so that many users can share the services simultaneously. The server program interacts with the client program. Both programs may run on the same computer or on different computers.

Client–server architecture: This architecture allows users to move the network control access to different computing resources within the network. There are two components within this architecture — client and server. The client makes a request for any query, while the server provides the information to the client. All applications of the Internet use this concept in their implementation. The clients are typically terminals, PCs, and workstations, while the server is typically a high-performance, multi-MIPS machine.

Client–server LAN implementation: A client in fact is an application program which runs using local computing resources and at the same time can make a request for a database or other network services from another remote application residing on a server. The software offering interaction between clients and servers is usually termed *middleware*. The client is typically a PC or workstation connected via a network to more powerful PCs or workstations, or even mainframe or mini-computers. These are capable of handling requests from more than one client simultaneously.

Clock: In a synchronous system, the synchronization between sending and receiving nodes is provided by a signal known as a *clock*. The data transfer takes place after these nodes have been synchronized.

Coaxial cable (coax): A widely used transmission medium where a central core is contained within concentric layers of insulator, braided earth wire, and finally a tough outer plastic coating. It has a copper wire surrounded by a cylindrical sheath of insulation and can transmit analog signals of frequency from 300 MHz to 400 MHz. The digital signal of a 100 Mbps data rate can also be transmitted on it. Due to its larger bandwidth, it can be used to transmit data, voice, and video simultaneously. For each of the electrical transmissions (baseband and broadband), we have a separate coaxial cable.

Codec (COder/DECoder): An electronic circuit which converts voice signals to a pulse coded modulated signal for its transmission over the media. On the receiving side, the received signals are demodulated and converted back into the original voice analog signals.

Coded modulation (CM): We use a different approach of generating a coded signal for analog signal using binary coding representation schemes. In digital modulation, we convert the analog signals contained in continuous signals into a sequence of discrete codes/signals generated by code or symbol generators. This modulation process is also known as *coded modulation* (CM).

Code rate (CR): In order to determine the degree of redundant bits in any communication code, an efficient measure known as *code rate* (CR) is defined. It is a ratio of information bits to the total number of bits per symbol in the coding scheme being used, e.g., ASCII uses 8 bits per symbol. It can be used to determine the efficiency of a communication codes and it should be as small as possible.

Command: This unit converts a set of procedures into a machine code and is usually represented by a string of characters that defines an operation (or action) for the computer. The command, being the smallest unit of any program, is decoded and stored as a command that is retrieved as a machine code for the execution of the command by the computer.

Commitment, concurrency, and recovery protocol (CCR): This is one of the common application service elements which has been defined to be used in a variety of applications and offers coordination between multiple users interacting with each other simultaneously. This protocol is defined in a limited way. The CCR specification is used where the database activities are to be maintained over multiple sites and allows a number of operations such as data transfer access to data and updating of databases.

Committed information rate (CIR): In a frame relay network, the user can send the frames at CIR rate, which is in fact an average rate of frames over the network defined by the user. This rate is guaranteed and maintained by the networks. Any frame that is being being transmitted at a rate greater than CIR is likely to be discarded or will be transmitted with a lower priority.

Common application service element (CASE): Various services of a network are accessed through an application layer process entity known as CASE. This is different from the one used in software engineering known as a computer-aided software engineering (CASE) tool. The entity uses ANS.1 protocol defined by ISO 8824 and ISO 8825 and describes an abstract syntax for data types, values, and representation of data at the application layer for various services.

Common carrier: A vendor in industries that provide telecommunication services to the users. These services offer facilities of two-way conversations over a distance, transmitting messages and data communications. Such vendors include AT&T, MCI, Sprint, etc.

Common channel interoffice signaling (CCIS): A well-recognized and widely used telephone signaling scheme that transfers the messages between two stations of the telephone network. The messages include both voice and control fields where control messages in general deal with the contention establishment and de-establishment signaling and other telephone network management controls. This scheme transmits the control message for a group of trunks on one channel, while the voice message is transmitted over another channel.

Common management information protocol (CMIP): This protocol is defined for the application layer of OSI-RM to manage and control the layer management control information across OSI-RM. Each layer has its own layer management and all are controlled by the management protocol at the application layer.

Communication channel: Two communication devices are connected by transmission media (telephone lines, coaxial cables, satellite link, microwave link, etc.) which define a communication channel on different types of links. The communication channel offers a variety of services of different systems. These services are available on private networks (usually owned by a private organizations or corporations, and governed in the U.S. by the Federal Communications Commission — FCC), private leased networks (usually known as common carriers to provide leased communication services; also governed by the FCC), and public switched networks (known as communication service providers for communication services).

Communication media: Communication media provide a logic path over a physical circuit or channel or link over which data transfer can take place between two nodes. These nodes could be telephone subscribers or computer users. The movement of data takes place over different logic paths depending on the underlying transmission media. Various transmission media available are twisted pair of wires, coaxial cables, fiber optics, or wireless media such as microwave link, satellite link, waveguide, cellular, infrared, etc.

Communication port: The communication devices (computer or terminal) are connected to other hardware devices (modems, multiplexers, computer networks, etc.) via a communication port. The port may be implemented by software (sockets are created by operating systems) or by hardware (communication adapter interface, serial ports, universal asynchronous receiver/transmitter, universal synchronous/asynchronous receiver/transmitter).

Communication processors: There are basically two types of processors: data and communication. These devices include switches, multiplexers, concentrators, front-end-processors, protocols, different techniques for error control and flow control within data link layer protocols, routing strategies, etc., and are known as communication processors. The front-end processors offer input/output processing, thus reducing the burden on the main CPU.

Community antenna television (CATV): The concepts used in CATV networks are also used in data communication over local area networks. These LANs use analog technology and use a high-frequency modem to define carrier signals for the channel. It is based on the concept of broadband systems, which transmit data, voice, and video signals through cable television (CATV).

Companding: If we define, say, 8 bits for quantization, it will represent 256 levels and offer a data rate of 64 Kbps (at the rate of 8000 samples for 8 bits). This process is known as *companding*. This technique defines three segment code bits (defining eight segments), four quantization bits (giving 16 quantization levels within each segment). This technique is known as **mu** law in the U.S. and a few other countries, and is different from A law used in European countries.

Compandor: This electronic device defines functions of compression and expansion. Compression reduces the volume range of signals, while the expansion function restores the original signal. One of the main purposes of this combined function of compression and expansion is to improve the signal-to-noise ratio so that the electrical interference may not have any effect on the voice signals. It is used in satellite communication.

Computer conferencing: A group discussion or conference which uses a computer network as a means of communication.

Concentrator: This device reduces the number of physical links for connecting terminal devices to a node or even a central processing unit (CPU) in a network. The links to various terminal devices are combined together into one link or channel that is then connected to a node or CPU in a network. A concentrator provides interconnection between a large number of input lines (usually of small data rates) and a small number of output lines (usually of higher data rates). It also offers a buffer switch for both types of signals (digital or analog) with a view to reducing the number of trunks required. It works on the store-and-forward concept, where large amounts of data information can be stored and can be transmitted at a later time.

Concurrency: Executing more than one independent action or operation simultaneously. Concurrency has been used in a variety of systems, e.g., concurrent processes in operating systems, access of shareable data information at the same time by various users of databases, independent execution of processes in multi-processor systems, etc.

Conditioning: An electronic device which provides flatness to the transmission-channel response of frequency attenuation and envelope delay. In other words, it nullifies the effect of any noise introduced and offers a constant frequency response of the signal. This feature is available as a Telco service on various private-lease voice channels.

Connectionless service: This service does not require establishment of a virtual circuit. It sends packets of a message, from one node to another, where each packet contains address information and other control information. At the receiving side, these packets are reassembled into the original message. It is unreliable service. The user datagram protocol (UDP) provides this service.

Connection-oriented service: This service requires the establishment of a virtual circuit (logical channel) between the endpoints over transmission media. The data transfer takes place over this established circuit. It is a more reliable service than connectionless service. Transmission control protocol (TCP) offers this service.

Contention: A network provides resource sharing over the same channel. When more than two users try to share the resources at the same time, there will be contention, as only one user can use the channel at a time. This concept is being used in Ethernet IEEE 802.3 CSMA/CD LANs.

Convergence function: This function is defined for the DQDB protocol layer that is used to provide interfaces to higher-layer protocols of the network.

Convergence sublayer (CS): The functions offered by the AAL layer are supported by the exchange of information entities of AAL with their peer AAL entities. This layer is divided into two sublayers: segmentation and reassembly (SAR) and convergence sublayer (CS). The CS sublayer offers services which are service-based and hence provides the AAL services at the AAL service access point (SAPs).

Copper distributed data interface (CDDI): In 1980s, LANs were supporting data rates of 1 to 3 Mbps. In the 1990s, data rates of 10 to 16 Mbps were very common and standard, and with the new technologies in high-speed switching and transmission, data rates of 100 Mbps are common over twisted pair cabling (copper distributed data interface — CDDI). Due to the fiber distributed data interface (FDDI), LAN technology can be obtained over desktop transmission.

Countdown counter: This counter used in DQDB defines the number of empty slots that must pass before the access permission for a DQDB bus can be granted to a node.

CP/M: This operating system is defined for microcomputers and is known as a *control program for microcomputers.*

Cross talk: In open wire, one wire picks up the transmission of a signal on an adjacent wire, constituting interference in it. We experience this problem when we notice that our voice is either not clear or we get many other voices together while using the telephone system. Cross talk may also be created by a motor, generator, or even transmitters/receivers near the lines. Either twisting the wires or putting a separator between parallel wires can minimize this noise, which reduces the interference effect between the electromagnetic fields.

Customer premises equipment (CPE): One of the most useful features of ISDN and B-ISDN is a separate block known as customer premises equipment (CPE) which can be configured in different ways for different applications. CPE is also known as customer premises network (CPN) or subscriber premises network (SPN). The CPE is interfaced with the public networks at a reference point T.

Cyclic redundancy check (CRC): A very effective method for transmission error detection in frames of data (at the data link level). The sender determines CRC and appends the data packet at its transmitting end. On the receiving side, CRC is again determined and matched with the one received. If its value is the same as that of the transmitted, no error has occurred during the transmission; otherwise, error has occurred. Either hardware or software can obtain the implementation of CRC.

D4: This format is defined by AT&T to provide framing and synchronization for the T1 transmission system.

Database: This method keeps a record of interrelated data items, which can be processed by various applications.

Data broadcast: In order to provide reliable distribution of data, this unit provides an interface between a CPU and other asynchronous devices (terminals, printers, etc.). For each of the devices, a separate RS-232 signal is regenerated at the corresponding port. The CPU can broadcast the data to all of these devices simultaneously.

Data circuit-terminating equipment (DCE): The main function of DCEs is to provide an interface between data terminal equipment (DTE) and a communication link, and thus connecting DTEs via communication link. The DTEs convert the user data (from computer or terminal) into appropriate format for transmitting over the communication link. The DCEs can be modems or any other digital service units, which provide a comparison of incoming signal into appropriate form to be used by that communication media.

Data communication network (DCN): Data communication networks offer many advantages such as simplified and reduced wiring, reduction in operational costs, and expansion of life of obsolete and old equipment. Data communication networks can be classified as local area networks (LANs), wide area networks (WANs), or long-haul networks (LHNs).

Data compression: This technique reduces the volume of data to be transmitted within the data rate of the link, thus saving the communication costs. Further, the errors and charges for the applications are also reduced.

Data encryption equipment (DEE): The main services offered by the application layer of OSI-RM are the syntax of data and security of the data. For security, encryption techniques are used to change the format of the message (data, voice, video, images, etc.) to an unintelligible format, which is then transmitted. The encryption technique initially is defined with the services of the application layer but can be implemented at other layers also. For example, DEE can be used to provide security at the physical and data link layers, while suitable software can be used at the transport layer or even at the presentation layer for this purpose.

Data encryption standard (DES): This standard at the presentation layer defines an algorithm to be used by U.S. federal agencies to provide security to unclassified, confidential, and sensitive messages. The security of data mainly depends on the security of the key. Further, the standard provides total interoperability, whereby different vendor products can communicate with each other using the same key.

Data exchange interface (DXI): One of the interfaces defined for LANs within a router. It allows the LAN containing this protocol to be connected to a router for high-speed communications, typically over ATM networks. Other interface protocols include high-speed serial interface (HSSI), digital signal 1 (DS1), and so on.

Data file: Stores various types of information such as text, number, drawings, images, etc. The data file in most of the operating systems is considered a single unit and is stored in a data medium such as a hard disk, floppy disk, magnetic tape, etc. Each file has a unique name independent of length and is stored in a directory. The retrieval of a file requires the entire path name of the file.

Datagram: The connectionless service defines a datagram instead of a packet (as done in connection-oriented services) and it contains the entire information including data and source address and other control information. This does not require the establishment of a logical connection before the data can be transmitted. The packet in the connectionless communication is known as a *datagram*.

Data link control: The flow of data and control messages between stations over communication physical link layer controls the traffic over the link. The data link control's main responsibility is to provide error-free services between the stations. The communication link between two stations is managed via three distinct steps: connection establishment, data transfer, and connection de-establishment or termination.

Data link layer protocol: The data link layer protocol provides services to its network layer such as control, flow control, and also media access methods.

Data link protocols: One of the main functions of data link protocols is to provide synchronization to the data frame and control frame, and transmission of frames between sending and receiving sites. Depending on the schemes used to provide the synchronization, we have different classes of data link protocols, such as bit-oriented, character-oriented, and byte-oriented protocols.

i) **Digital data communication message protocol (DDCMP):** This protocol is character-count-oriented or byte-count-oriented, as the length of the user data (in octets) is defined in the header of the frame.

ii) **High-level data link control (HDLC):** ISO 3309 and ISO 4335 have defined a bit-oriented line protocol as high-level data link control (HDLC). This protocol supports both modes of operation — half-duplex and full-duplex — and both classes of line configurations — point-to-point and multi-point.

iii) **Advanced data communication control procedure (ADCCP):** This is also a bit-oriented protocol and has been defined as ANSI X.366 standard. It is similar to HDLC with minor differences.

iv) **Binary Synchronous Control (BSC):** This protocol was defined by IBM for supporting both link configurations (point-to-point and multi-point) and supports synchronous mode of transmission (see also the main binary synchronous control entry).

v) **Multilink procedures (MLP):** In manufacturing industries, the management of more than one link sometimes becomes advantageous, as a faulty link can be replaced by a back-up link quickly without causing the assembly plant to shut down. An MLP standard has been adopted by various manufacturers and is the same as other existing data link control protocols, except that here a common sequence number is used to manage window size and flow control of all of the links used.

Data management system: A system that offers a set of procedures and query programs through which a collection of interrelated data can be organized, maintained, and accessed.

Data network: The data network provides various services for transmitting information back and forth among the connected terminal equipment. The linking of terminals, routing of the information globally, management and control of data packets in the network, etc., are some of the main services offered by the data network.

Data over voice: The telephone network can transmit both voice and data together by multiplexing (using FDM) them on the same channel. Suitable filters can be used to isolate voice-grade signals (300 Hz to 4 KHz) and data signals (500 KHz to 1 MHz), whereas the data channels can further be multiplexed using TDM can reduce the cross talk at higher frequency bands.

Data packet (frame): In the data network, a packet is defined as a basic unit of information which contains a fixed number of data characters and control characters. The control characters are usually used during the transmission of a packet.

Digital or data service unit (DSU): Used for high-speed alternative to digital transmission on digital asynchronous and synchronous networks.

Data switching equipment (DSE): This switch connects various DTEs and allows each DTE to use it for communicating with other DTEs. In other words, this switch forms a networking environment and also provides alternate routes in the case of a failed link.

Data terminal equipment (DTE): It offers an interface between a computer or terminal and the transmission circuit for the data communication between end-user machines. The data usually is transferred between computers and terminals and a particular underlying application. The terminal or computer is known as *data terminal equipment* (DTE) while the communication interface is usually defined by a device known as *data circuit-terminating equipment* (DCE). The computers, terminals, workstations, mainframes, end-user applications, and even stations are known as DTEs, while modems, controllers, signal converters, etc., are examples of DCEs.

Data transmission: The transmission of packets (containing data and control information) between data circuit-terminating equipment (DTEs) across the transmission media/links/channels.

Data transmission procedures (DTPs): A set of procedures used to establish the setting up of a communication channel/link, synchronous or asynchronous transmission, serial or parallel transmission, and various configurations such as half-duplex, full-duplex, etc.

Data transparency: For distinguishing between the symbols of messages and the protocol units, the data transparency technique is defined and used in various data link layer protocols. This technique may allow the same symbols to be used in both fields: user data message and control message.

DB-25: A popular and commonly used connector in the U.S. for the RS-232-C interface standard.

D channel: The ISDN defines three types of channels as B channels (64 Kbps), D channels (16 or 64 Kbps), and H channel (384–1920 Kbps). The D channel carries signaling information for controlling and managing circuit-switched traffic on B channels at the user interface. It may also be used for low-speed packet-switched applications when no signaling information is waiting.

Decibel: This unit represents the relative strength of two signals. It is defined as log of the ratio of the signal strength. The strength could represent current, power, or voltage.

Decryption: The encryption technique maps the data or text into a format (not recognizable) known as *ciphertext*. On the receiving side, this encrypted message has to be mapped into the actual data and text format known as *plaintext*. This mapping process at the receiving side is known as *decryption*.

Defense Data Network (DDN): Based on the success with ARPANET, the U.S. Department of Defense (DoD) in 1982 initiated a project to develop the Defense Data Network (DDN). The DDN is a packet-switched network which provides security and privacy for the data integrity. It uses dynamic routing strategies whereby it manages by itself to any failure in the network without affecting the services being offered to the users. The DDN can be used as a backbone network and allows the users to access various applications on TCP/IP via public packet-switched network (X.25) and ARPANET.

Delay distortion: Lower-frequency signals travel faster than those of higher frequency if a transmitter is sending more than one frequency signal at the same time. Two types of distortions, phase delay and envelope delay, cause this distortion. The phase of transmission varies with frequency over the transmission band, giving rise to phase-delay distortion. The envelope-delay distortion is defined for the transmitted signal. This delay is usually expressed as a delay from average delay time at the central frequency of the transmission band. Thus, low-frequency signals have less delay distortion than high-frequency signals.

Delta modulation (DM): Delta modulation (DM) has a unique feature of representing a difference value of 1 bit, comparable to PCM at lower rates of 50 Kbps. At this low rate (for example, in telephone systems), for some applications in military transmission, lower quality of reception can be accepted. Further, DM may be useful in digital networks if we can digitally convert delta modulation into PCM. Differential PCM and delta modulation have some unfavorable effects on high-frequency signals.

Demodulation: The demodulation process is a reverse of modulation. It extracts the original modulating signal from the received modulated signal. Demodulation or detection is a non-linear process, although we might use linear circuits for linearizing for obvious reasons (linear circuits offer stable response and are inexpensive and less complex).

Department of Defense (DoD) standard: DoD has defined various standard protocols for application layers. For each application, the underlying TCP provides a transport connection.

DG COM VII: Deals with data communication networks.

Dial-up lines: Modems can be used over dial-up lines or private (leased) lines, and each category of modem can further be classified as using one pair or two pairs of wires. The dial-up line (a pair of wires) connecting the user's telephone set to a local/central switching office of the telephone company is known as *local loop*. One line of this pair is grounded at the switching office, while the other line carries the analog signal. The typical length of local loop is about 8 Km.

Dial-up services: The network services can be obtained in two classes: switched and leased. Switched services are offered on public switched networks and sometimes are also known as *dial-up services*. The switched (dial-up) services are controlled and managed by AT&T, Data-Phone system, and Western Union broadband channel. This service sends billing information based on the use of switched services (duration of usage).

Dibit: In the operation of phase-shift-keyed modems, if we consider the phase changes as multiples of any angle, then the information levels are identified by bits combination and the units of these levels are defined as *dibits*. For example, if the phase angle changes are multiples of 90 degrees, then four levels of information can be defined and dibits will be the units for each of them. If the phase changes are multiples of 45 degrees, then eight levels of information can be expressed with units of tribits, and so on.

Differential phase-shift-keyed modem (DPSK modem): In the operation of phase-shift modems, the logic 1s and 0s are represented by two out of phase sine wave signals, i.e., a logic 1 may be represented by a sine wave signal while logic 0 is represented by a 180 degrees out-of-phase sine wave signal (with respect to the original sine wave signal of logic 1). In the operation of differential phase-shift-keyed modems, the value of the phase is not important; instead, it is the phase changes which are important to represent logic 1s and 0s.

Differential pulse-code modulation: In PCM, we transmit the instantaneous value of a signal at the sampling time. If we transmit the difference between sample values (at two different sampling times) at each sampling time, then we get differential pulse-code modulation. The difference value is estimated by extrapolation and does not represent the instantaneous value as in PCM. If these difference values are transmitted, then by accumulating these, a waveform identical to an analog information signal can be represented at the receiving side.

Digital component: Any component that operates with digital signals (0s and 1s) is known as a *digital component*. Some examples include flip-flops, registers, microprocessors, converters, etc.

Digital cross-connect system: The T1 system may be broken into different digital signal DS0 bit rates that are used for testing and reconfiguration.

Digital distortion: This type of distortion occurs in a DTE/DCE interface if the standard defined by EIA is not properly and strictly followed. The noise is caused due to a difference in the sample transition, different values of signal at different transitions, different times to restore the decaying amplitude of a digital signal, etc. As a result of this, the transition from one logic to another takes different times and causes errors in data information. An appropriate logic error occurs at the receiving side if it takes more time than other to reproduce it. It is interesting to note that the same sources may also cause timing jitter noise.

Digital modulation (DM): This category of modulation changes the analog form of a modulating signal into digital modulated signal form. The modulated signal is characterized by bit rate and the code required representing analog signals into digital signal. For digital modulation class, the information is in discrete form, but can be used on analog channels and are treated as analog modulation with different names. The digital modulation used in switched circuits and leased circuits provides data and voice communication.

Digital network: The advent of digital switching technique to perform routing and switching functions for digitized voice and digital data has made the digital network a powerful framework for many high-speed digital networks. The digital switch uses a PCM technique and as such is based on a sampling rate of 8000 samples per second. In order to transmit 8000 samples, the switch must be capable of supporting 8000 slots per channel, at a speed of 8000 samples per second.

Digital network architecture (distributed network architecture — DNA or DECnet): Digital Equipment Corporation (DEC) defined a distributed framework for the network to communicate software, hardware, and other products. Within the framework, a dedicated computer to perform various functions of homogenous networks (connecting similar systems together). It defines a common user interface to support various applications of different LANs, public packet-switched networks, etc. It does not follow the standard OSI-RM but has defined its own layers that are compatible to OSI-based LANs.

Digital signal: Any signal representing two states (high and low) or pulses.

Digital signal 0 (DS0): There are 24 channels defined in a DS1 (T1 in North America and E1 in Europe) channel and each channel is of 56 Kbps.

Digital signal 1 (DS1): The standard 1.544 Mbps or T1 used in North America. The equivalent standard used in Europe is E1 (2.048 Mbps).

Digital signal 3 (DS3): This corresponds to 44736 Mbps.

Digital signaling: If we consider a carrier as a series of periodic pulses, then modulation using this carrier is defined as pulse modulation. In this signaling scheme, the content of the user information is important, as it is expressed in logic 1s and 0s. The digital signals are not modulated (digital form of the signal is defined only after modulation). This type of signaling used in a system is known as a digital signaling system.

Digital signal waveform: Digital devices such as computers, terminals, printers, processors, and other data communication devices generate signals for various characters, symbols, etc., using different coding techniques. The digital (or discrete) signals can have only two binary forms or logic values — 1 and 0. With the advent of very large scale integration (VLSI), we can now represent any object, e.g., symbols, numbers, images, patterns, or pictures, into discrete signals.

Digital signature: For providing integrity of data and a source, this authentication mechanism allows the users to attach a digital code known as a signature. This attached code maintains the integrity of data and source.

Digital splitter: The point-to-point and multi-point connections can also be configured on a leased line using a line splitter. Usually, a line splitter is used at the exchange to provide the connection. By using a line splitter, one assigned line of exchange can be used by more than one location at the same time. Similarly, when these terminals want to transmit the information, these are combined on that line and delivered to a computer. The digital splitter allows more than one terminal to use a single modem port for transmitting and receiving information.

Digital-to-analog (D/A) converter: On the receiving side, the combination of quantizer and decoder is known as a digital-to-analog converter. It produces a sequence of multilevel sample pulses from a digitally encoded signal received at the input of D/A from the communication link. This signal is then filtered to block any frequency components outside the baseband range. The output signal from the filter is the same as that of the input at transmitting side, except that it has now some quantization noise and other errors due to noisy communication media.

Digitization: A process of converting analog signals into digital that consists of pulses of two states only.

Digitized voice transmission: In order to prevent interference between voice-grade bandwidth of 3.4 KHz in telephone networks, a technique known as *telephone voice digitization* based on pulse-coded modulation (PCM) is performed. The PCM samples the analog signal at the rate of 8000 samples per second. In the telephone network, usually the subscriber local loop (user's set to central office) carries an analog signal, while digitization of voice takes place at a central place using a device known as a codec (coder-decoder).

Differential Manchester coding: In the case of ordinary binary coding of the messages, a loss of one or more bits during the transmission will cause sending and receiving sites to be out of synchronization. In order to avoid this problem, the differential Manchester coding technique is used. In this technique, both logics 0s and 1s are represented by 2 bits during logic duration. The logic 1 is represented by a positive signal followed by a negative signal of equal duration, while logic 0 is represented by a negative signal followed by a positive signal. The bandwidth of the transmission is halved, but the synchronization remains intact in spite of loss of any logical signals.

Dipole antenna: In this antenna, the length of the antenna is proportional to the carrier wavelength.

DIS 10022: Services offered by the physical layer.

DIS 8649: Services offered for association control service elements.

DIS 8802/2: Logical link control (LLC) sublayer of IEEE LAN model.

DIS 8886: Services offered by the data link layer within OSI-RM.

DIS 9805: Protocol specification for the commitment, concurrency, and recovery service element.

Discard eligibility bit: This bit indicates the discarding condition of a particular frame in a frame relay network when a congestion condition has occurred. This bit, when set to 1, indicates that that frame is eligible for discarding from the network.

Discrete (digital) analog modulation (DAM): In this category of modulation, the modulating signals are defined by discrete values of signal, and these discrete or binary logic values carry the information. This process of modulation basically is analog in nature, as it does not change the contents of information contained in the modulated signal (although information is in binary form). The digital devices are connected to PSTN via modem, which converts digital signals into analog signals.

Distributed network: In this network configuration, all the nodes are connected either directly or indirectly to each other via different paths. These paths may be defined through various intermediate nodes within the networks

Distributed operating systems (DOS): A collection of different types of processing elements that appear to the users of the system as a single computer system. The distributed operating system is an integrated system comprised of hardware and software and working like a time-sharing system. Here various modules of the operating system may be distributed over different processors that are considered as application programs. A small shell providing low-level I/O interfaces resides on all of the processors and invokes the necessary application programs when needed.

Distributed queue dual bus (DQDB): This has been proposed as a LAN/MAN standard, and it provides support to all three types of services (connection-oriented, connectionless, and isochronous) simultaneously. It is based on queue packet synchronous exchange (QPSX). It defines two buses which carry signals in opposite directions with multiple nodes sending simultaneously. Each bus provides a unidirectional transmission and is independent of the transmission media.

Domain: A part of the addressing scheme. Each domain may have subdomains, and each of these domains or subdomains are separated by a period symbol.

Domain name system (DNS): This addressing scheme converts the Internet names to their corresponding Internet address. Each address in DNS is a combination of four fields separated by the period symbol. Each field can have a maximum of 256 numbers, as the entire Internet address is of 32 bits. Most of the networks use Internet-like addresses, and we should check the address of the network we are accessing before we use its address, as the addressing format is different for these networks.

Double sideband-AM (DSB-AM): If the carrier signal (carrying no power) in DSB-AM can be stopped from the transmission of the modulated signal, the power wastage can be reduced, but bandwidth remains the same. This is the working principle of double-sideband suppressed carrier amplitude modulation (DSBSC-AM). Since the carrier is not transmitted, the phase reversal in the modulated signal (for transition from one level to another) becomes a problem. Double sideband amplitude modulation (DSB-AM) is a very popular transmission method used for radio broadcasting (commercial). The main advantage with suppressed carrier is that this carrier signal can be used for other control information, e.g., synchronization, timing, etc.

Download: Information (data, video, graphics, or any other form) can be downloaded on the local PC's hard drive or any other drive and can be saved.

Drop and insert TDM channel: The concept of drop and insert of TDM channels is used in applications where hosts are connected over TDM channels through intermediate hosts. The intermediate hosts typically gather data from either of the hosts and transfer them back and forth between them. Intermediate hosts as multiplexed data receive the data from any host. The intermediate hosts simply combine the data from different ports of hosts located at two different locations. This combined multiplexed data is received by the host and has to be demultiplexed there.

Dual bus: The bus topology used in DQDB where two buses are used in two types of configurations — open dual and looped dual. Every node connected can access both the buses.

Dynamic Hypertext Markup Language (DHTML): It has been defined to encompass many new technologies and concepts that can be complex and are often at odds with Web site construction. It is a combination of HTML and JavaScript. It includes several built-in browser features in fourth-generation browsers (e.g., Netscape 4.0, Internet Explorer 4.0) that enable a Web page to be more dynamic. The DHTML used in IE 4.0 is far more powerful and versatile than NS 4.0.

Dynamic Non-hierarchical Routing (NHR): In order to reduce the problems associated with hierarchical models, the non-hierarchical model for routing strategy has been adopted by many telephone companies. It allows the inclusion of a set of predefined routing paths at each office which will handle heavy traffic efficiently such that customers get fewer busy signals and are connected faster than in fixed topology.

E1: The European T1 CEPT standard for digital channel offering a data rate of 2.048 Mbps; equivalent to T1 (used in North America).

E.164 (CCITT): ISDN and PSTN use the addressing scheme defined in the CCITT E.164 recommendations. This scheme defines a variable length address up to 15 digits. It consists of a code for country followed by the ISDN subscriber number.

Echo control: This control equipment is an essential component in voice communications and in fact is an integral part of communication systems in satellite telephone systems. The echo cancellation technique is very popular in Europe, and is based on the existing technique of echo handling where the sender generates nearly the same echo signal on the line.

Electromagnetic wave: A signal wave that has a frequency between 20 Hz and 300 GHz and even more than this spectrum.

Electronic mail (e-mail): This service is handled by message handling protocol (MHP) which resides at the application layer. It allows OSI-based network users to exchange messages. This protocol works on the basis of the store-and-forward concept and uses intermediate hosts as message transfer agents (MTAs). Another protocol which provides electronic service is simple mail transfer protocol (SMTP) on DoD's Internet.

Electronics Industries Association (EIA): The first standard developed by this standards organization for the physical layer interface was RS-232 (DTE/DCE) as a 25-pin connector. This standard defines the basic functions of the physical layer (mechanical, functional, electrical, and procedural behavior) of the device. Other standard interfaces defined are RS-449, RS-422A, RS-423A, and EIA-530 for DTE and DCE. It employs serial transmission.

Empty slot: In a DQDB network, the queued request for slot access that is not being used by any node may be allocated to a node that makes a QA request.

Emulation: This is a technique based on programming concepts which allow users to design a program for another computer or application and run it on their computing systems. The computing system offers selected features to allow users to execute the programs of other systems. It also makes a programmable device invoke another device and produce the same result.

Encapsulation: This property is very useful in object-oriented programming languages and object-oriented technique, as it allows designers to provide information hiding by encapsulating the implementation details of procedures in the package. A similar concept of encapsulation is used in the layered architecture of OSI-RM to define protocols for data communication through layers. The header contains the function of appropriate layers and is usually known as *control information*. The reverse process on the receiving side is termed *decapsulation*.

Encoding: (1) The method of encoding of binary logic (0s and 1s) defines the representation in a suitable form of signal, which can be transmitted over the transmission medium under consideration. The decoding method at the receiving side should be capable of extracting the original signal from the received signal. The encoding method depends on various parameters such as cost, performance, characteristics of media, etc. (2) The performance of signal transmission depends on a number of factors such as data rate, bandwidth, signal to-noise ratio, and encoding scheme. The encoding schemes converts the data bits to signal elements. The encoding technique must provide adequate signal spectrum, error-detection facilities, clocking for synchronization, and acceptable performance in the presence of noise and other interference during transmission. There are a number of encoding techniques available, e.g., NRZ-L, NRZI, Pseudoternary, Manchester, Differential Manchester, etc.

Encryption: A process in which a text, symbol, or image file is translated into an unrecognizable form using predefined rules. The set of rules and operations depends on parameters used in this type of converter or encryption key. The key may have strings of symbols/characters or binary strings. Similarly, the *decryption* process recovers back the original text, symbol, or image file from unrecognizable text or symbol form. This process of encryption offers security for confidentiality and integrity of data, and various algorithms and keys are kept secret. Various standards for data encryption are data encryption standards (DES), data encryption equipment (DEE).

End of delimiter (EOD): This field in a MAC frame of IEEE 802.4 (token bus LAN) and IEEE 802.5 (token ring LAN) represents the end of the frame. It contains non-data symbols (I and E bits). The EOD in an FDDI frame format also contains non-data symbols to represent the end of the frame.

Enterprise network: The network defined for the entire organization that has many branches located at various locations. Networks connect all the locations.

Entropy: This measure defines the randomness in the signal received at the receiving site. The randomness in fact is for the contents of the data received. If we send a 2-bit symbol, the entropy with the received signal can be between 0 to 3 bits, and can have 4 possible states. Mathematically, it is defined as $H = -\sum_{i=1}^{n} Pi \log_2 Pi$ where, n = number of states, Si = the probability of ith symbol, and H = entropy.

Equalization: When the frequency of signal increases, the signal usually gets attenuated due to various factors such as noises, interference, etc. In order to obtain a constant signal value over the frequency range, we need to neutralize the effect of this loss of the signal due to unwanted signals present in the original signal. This is obtained by a process known as equalization where we have to design a suitable electronic circuits which provide a characteristic opposite to the variation of signals so that the signal loss can be maintained to a constant level over the frequency range. The electronic circuit, which provides this operation, is known as equalizer.

Error rate: This is defined by a ratio of the number of bits in error to the total number of bits received. Similarly, it can also be defined for characters and block of characters. The error rate of say 10^{-6} indicates that one bit out of 10^6 bits is in error. The error rate is the case of character and block of character is defined as ratio of the number of character or block of characters in error to the total number of characters or block of characters received.

Error-control: This is one of the main services offered by data link The error may be introduced by noise into digital signal. A very short transient noise may have significant effect on the number of bits transmitted. The entire data link control techniques for error-control is used by LLC and provide the following two options: i) Error-detection and ii) error-correction or error-recovery.

Error-correction or error-recovery: The most common technique for error-recovery used by the LLC sublayer on the receiving side is Automatic ReQuest repeat (ARQ). The ARQ technique includes the receipt of a positive acknowledgment from the receiving side in the case of error-free frame. If an error is detected in the frame, a negative acknowledgment is sent to the transmitting side, which retransmits all the unacknowledged frames. There are three versions of ARQ protocols as Stop-and-wait, Go-back-N, and N-Selective.

Error-detection and retransmission: The error-control strategy which detects the error and then sends a negative acknowledgment (NAK) for retransmitting the same frame as it has detected error in it. In this technique of error-control, the sender keeps on retransmitting the frame until either the frame is received error-free (indicated by acknowledgment: ACK) or time-out expired.

Error-detection: The error-detection can be obtained using either analytical or non-analytical methods and can be implemented by software and hardware chips. The MAC sublayer on the receiving side detects the error in the received frame, discards the frame and hence the frame does reach the LLC sublayer.

Ethernet: This is one of the most common and earliest local area network. Xerox using the concept of broadcasting radio network defined at University of Hawaii (ALOHA) developed it. Later on Xerox, DEC, and Intel cooperated to define Ethernet specification in 1976 and eventually was adopted by IEEE 802.3 as a national standard. It allows the multiple users to share common channel (radio frequency band or coaxial cable) and is based on Common Carrier Multiple Access/Collision Detect (CSMA/CD). It uses baseband bus topology

European Commission (EU): European Commission has formulated a Special Telecommunication Action for Regional Development (STAR) with a view to develop a telecommunication infrastructure via carriers of European countries. These carriers must adopt ETSI standards and also the standards defined by other international standards organizations. With the support of German telecom, British telecom, France telecom, telephonica of Spain and others, a Global European Network (GEN) using optical fiber transmission has been developed and is based on SDH standards. This standard supports ATM switching offering a variety of B-ISDN services. GEN is a European-wide digital network, which provides a transparent transmission speed of over 155 Mbps.

European Computer Manufacturers Association (ECMA): This organization deals with the development of standards for data processing in computer communication systems. It was instituted in 1961 and is associated with various technical committees and other groups of ISO and CCITT.

European Telecommunications Standards Institute (ETSI): ITU's four classes of services are mapped into six AAL types in ATM networks. AAL-1 is mainly used for E1 and T1 circuits. AAL-2 is for variable bit rate applications. The AAL-4 is merged with AAL-3 and is mainly used as a connectionless data transport for connectionless network services, switched multimegabit data services (Bellcore), and connectionless broadband service (ETSI).

European worldwide ATM/broadband networks: With GEN in place, various projects from different European countries have been considered with a view to offering the following services: LAN interconnection, multimedia services, videoconferencing, video transmission, circuit emulation, distance learning using videoconferencing, high-quality medical imaging, distributed group communications, advanced data visualization, and many others.

Even parity check (EPC): This method of error detection allows the detection of errors in bits, characters, or blocks of characters or bits. Here non-data bits append the actual data such that the total number of 1s (or 0s) is even in the overall data frame transmitted. In other words, the total number of 1s or 0s in the entire data frame has to be adjusted to an even number.

Exchange: All the communication services (e.g., telephone connection establishment and de-establishment, billing, and other administrative services and functions) offered by any telephone communication carrier company at a common place. The services are offered by switching equipment connected by cables and software to manipulate various operations and collectively are termed *exchange*.

Exchange termination (ET): The main function of ET is to control B channels up to eight basic access between line termination and terminal control element and also control D channel protocol. This provides protection to D channel packets against data violation.

Explicit congestion notification: In a frame relay network, a number of messages can be used to inform both source and destination nodes about the occurrence of network congestion: FECN, BECN, or CLLM.

Extended Binary Coded Decimal Interchange Code (EBCDIC): This code was developed by IBM in 1962 and is used in IBM medium and large computers. This code was published by the American National Standards Institute (ANSI) and is an 8-bit code with no support for priority. It can represent 256 symbols, but only 109 symbols are used. It also contains a few control characters not available in ASCII. In contrast, ASCII code is a 7-bit code with the 8th bit for error checking and is used on IBM PCs.

Extended super frame (ESF): AT&T defined this frame for digital networks which contains various options in it, such as signaling, diagnostics capabilities, error control, etc., and uses a simplified LAP-B.

Facsimile (fax): This allows the transmission of text and images over the existing telephone lines. Here, the text or image is scanned at the transmitter and sent over the telephone lines as a digital signal. On the receiving side, it is reconstructed and printed on the paper.

FastPacket: Packet switching technique defined by StrataCom Corp. and being used on private LANs. It uses a 192-bit packet and packetized voice.

Fast packet switching: This technique is considered an improved switching technique over the existing X.25 packet switching. It offers higher data speeds (by reducing the size of packets), and has become a standard framework for a number of high-speed connection-oriented networks such as ATM, frame relay, SMDS, B-ISDN, DQDB, and many others. B-ISDN uses fiber optics technology, low-cost computers, and other devices for a variety of applications such as LAN interconnectivity, multimedia, imaging, video, and many others.

Fault detection: In CCITT recommendations M.20 and I.610, various operations to be performed during operation and maintenance have been defined in terms of identifying the categories of faults and then defining actions for restoring the network in a required function. A continuous or periodic checking procedure or routine detects the faults caused by malfunctioning or predicted malfunctions. These routines, when are invoked, detect the faults and report to the higher layers, where the information about these is produced and appropriate indications are generated.

Fault information: The faults due to malfunctioning are detected and after their detection, information about them is sent to other management entities. In this way all three planes of the B-ISDN reference model know about the faults. It also deals with the response message to the status report request to these planes.

Fault localization: In the event of faults, if the fault information does not provide enough information to the planes or the fault information is insufficient to invoke any routines or procedures, the fault localization system test routines or procedures are invoked to determine the failed entity. This fault entity may be determined by internal or external system test routines defined by fault localization.

FDDI-II: This standard LAN is an extension of FDDI and can carry isochronous traffic (voice, data, and video).

FDDI Follow-on: ANSI has set up this committee to look into the standards for extending the data rates of FDDI up to 600 Mbps.

Federal Communication Commission (FCC): A board of commissioners responsible for regulating various aspects of communication systems within or outside the U.S. for all messages originating from the U.S.

Fiber distributed data interface (FDDI): The high-speed LANs (HSLANs) are used to provide high data rates of 100 MBps (typically over fiber) in a metropolitan area within a diameter of 100 Km. These LANs are also known as metropolitan area networks (MANs) due to the wide range of coverage. There are two popular high-speed LANs available, fiber distributed data interface (FDDI) and distributed queue dual bis (DQDB). The FDDI is based on ring topology using optic fiber medium for transmission of 100 MBps. It can be used for a distance of 100 Km and supports as many as 1000 users. This network supports only packet-switching services. The enhanced version of FDDI is known as FDDI-II, which supports isochronous services, e.g., voice, data, and video.

Fiber optics: A thin strand of glass is used as a physical medium over which a modulated light beam containing information is transmitted. It offers very high bandwidth, high data rate, and low error rate. The input signals are encoded by varying some of the characteristics of the light wave signals that are generated by laser. Fiber optics are used in FDDI LANs and are also being considered for very high-speed networks.

File transfer protocol (FTP): This protocol was defined in 1970 to allow DoD Internet users to transfer data among themselves. It uses TCP/IP services and allows users to get the services from their computers, terminals, and preferably from the user programs.

Filter: An electronic circuit network which allows the signals of certain frequencies (defined as cut-off frequency) to be either rejected or passed (accepted). Beyond or below the cut-off frequency, signals are either passed or rejected. Based on this, we have four categories of filters: low pass, high pass, bandpass, and band elimination.

Flow control: The normal flow of traffic between transmitting and receiving sites has to be controlled and maintained in a data communication system. The number of packets transmitted from the transmitting site must be received before further packets can be transmitted. In the event of failed or poorly managed flow control, packets may be lost.

Forward explit congestion notification: In frame relay networks, the network informs the destination device about the occurrence of network congestion through this message.

Fourier analysis: The most commonly used series to represent the characteristics of signals as it generates various harmonics.

Fractional T1: The transmission of a segment of at least 64 Kbps and a multiple of 64 Kbps within a T1 channel.

Frame: The data link layer (second layer from the bottom of OSI-RM) gets the packets from the network layer and adds its header and trailer around the user data. The user data now contains a header at the front and a trailer at the end of the user data packet. This user data to be transmitted to the lowest layer of OSI-RM is termed a *frame* and is transmitted over the network.

Frame check sequence (FCS): This field contains the code for error detection and is usually defined at the end of a frame. At the receiving side, the frames are checked for any error by computing a check sum sequence. In the event of any error, the frame may be retransmitted or discarded, depending on the error-control mechanism being used.

Frame relay assembler/disassembler: Unlike PAD, this device provides interface between non-frame relay protocols to standard protocols of frame realy. It in fact acts as a concentrator and protocol mapper between non-frame relay and frame relay protocols.

Frame Relay Forum: This forum consists of users, service providers, product manufacturers, and suppliers and promotes the use of frame relay networks. It uses optical-based fiber communications and intelligent customer premises equipment. It proposes specifications of various interfaces (UNI, NNI, and others) and other recommendations to ANSI and ITU-T.

Frame relay: A new technology for data communication that has been considered by ANSI and ITU-T. It is based on fast packet switching and hence offers connection-oriented services. It offers variable length of link frame and does not use any network layer. The frame relay interface supports various access speeds of 56 Kbps, T1 (1.544 Mbps), and E1 (2.048 Mbps).

Frame relay interface (FRI): This provides an interface between user and frame relay network. The protocol supports connection-oriented services over the network and supports various speeds, e.g., 56 Kbps, T1 (1.544 Mbps), E1 (2.048 Mbps), and many basic rates of ISDN.

Frequency: The rate at which a signal oscillates; inversely proportional to the time period of a cycle. Its unit is hertz or cycles per second.

Frequency division multiplexing: This method is analogous to radio transmission where each station is assigned a band of frequency. All of the stations can transmit the programs within the assigned frequency band at the same time. The receiver has to select a particular frequency band to receive the signal from that station. In FED, the entire bandwidth is divided into smaller bandwidths (subchannels) where each user has an exclusive use of the subchannel.

Frequency modulation (FM): In this modulation process, the amplitude of modulated wave is the same as that of the carrier, but the carrier frequency changes in accordance with the instantaneous value of the baseband modulating signal. It is used for high-fidelity commercial broadcasting, radio frequency transmission due its immunity to noises.

Frequency shift keying: In this modulation, the frequency of the carrier (binary values) is changed in accordance with discrete values of the signal, keeping amplitude constant. Different frequency values are used to represent the binary values (ON and OFF or 0 and 1) of the carrier. It is used in radio broadcasting transmission, LANs, and also low-speed modems.

Front-end processor: A specialized computer that performs the functions of line control, message handling, code conversion, and error control. The CPU does not have to perform these functions, thus making the system very fast. It provides asynchronous and/or synchronous ports for the system.

Full-duplex configuration: In this configuration, two stations connected by one channel can send the data back and forth at the same time. Typically for voice-grade communication, two different frequency values are used for each direction to avoid any interference between them. Other configurations include simplex (one way) and half-duplex (one way at any time with a change of direction capability).

Functional group: The ISDN reference configuration for user–network interface (UNI) defines the functional groups and reference points. The functional groups represent the interface derives, e.g., network termination 1 (NT1) and NT2. NT1 offers functions similar to those of the physical layer of OSI-RM, while NT2 offers the functions of higher layers in addition to those of the physical layer of OSI-RM. Similarly, the terminal equipment (TE) corresponds to a terminal with a standard interface. These functional groups are also defined for broadband ISDN, where these groups and reference points are attached with prefix B (i.e., broadband).

Gateway: The heterogeneous networks are defined as a collection of different types of networks (different vendors) which have different protocols at the data link and network layers. These networks with different vendor products can communicate with each other if a device which can allow them to interact with each other connects them, and that device is known as a *gateway*. A gateway computer produces internetworking connection for data transfer between these networks (public or private). The gateways maintain a routing table to handle internetworking addressing and offer compatibility between different protocols (vendors) of the network, data link, and physical layers of OSI-RM.

Generic flow control (GFC): CCITT I.413 has defined different physical configurations for customer premises equipment (CPE) or customer premises network (CPN). The terminals, which share the transmission media, have to have their own media access control function. Instead of having a different media access control, the physical configuration based on the sharing of transmission media defines a generic flow control (GFC) functional protocol. This protocol provides fair access to all the terminals and assigns appropriate capacity of the networks to cells on demand. The GFC protocol resides in the ATM layer.

Geosynchronous orbit: The orbit is at a distance of 22,282 miles from the earth, and a satellite is located at a point above the equator. The satellite at this point travels at the same speed as that of the earth on its axis. Since both orbit and earth are moving at the same speed, it makes the satellite in the orbit stationary with respect to the earth. The satellite is used for telecommunication services and requires two earth stations with huge antennas located at a distance to provide a line-of-sight communication link.

Gigabit networking testbeds: The broadband services include all three types of signal — data, audio, and video. The ATM/broadband testbed is a platform for providing these services and is appropriately known as a *gigabit testbed*, as the speed of communications is very high, in the range of gigabits. The initial work on testbeds was started in 1989 and is coordinated by Corporation for National Research Initiative (CNRI). There are a number of gigibit testbeds around the world (Aurora, CASA, VISTA, ACRON, BBEN, and many others in the U.S., Global European network in EC, Mercury in the U.K., and many others).

Global addressing: Frame relay provides services based on virtual circuits and offers two types of services — permanent and switched virtual circuits (PVC, SVC). The header of a frame of 10 bits defining a destination node is known as a *data link connection identifier* (DLCI). It also supports broadcast communication by choosing a particular type of DLCI. The global addressing makes the DLCI port-independent, and as such the packet can originate from any port and it will be delivered to the destination terminal.

Global European Network (GEN): Many carriers of European countries have decided to develop their own network based on ATM that will adopt their standards, such as ETSI. This network is defined as Global European Network, which uses SDH for transmission. The partners of this network are German Telecom, British Telecom, France Telecom, Telephonica of Spain, and STET of Italy.

Go-back-N ARQ: The error control for the data link layer protocol logical link control (LLC) sublayer of the data link layer (OSI-RM) offers various functions, e.g., error detection, error correction, and error recovery. If we use both error detection and error correction by using automatic repeat request (ARQ), then there will be reliable transfer of data. There are three versions of ARQ protocols; Go-back-N ARQ protocol provides an error-recovery connection-oriented service in the network.

Gopher: A very popular application program tool of the Internet. It allows users to search for various services (just like searching in a catalogue), and the requested services are automatically communicated to the users. This tool does not provide structured services.

Government OSI profile (GOSIP): In order to support OSI open systems and OSI-based international standards, a series of workshops was initialized by the U.S. government under the auspices of the National Institute of Standards and Technology (NIST). The workshops were designed for various organizations and vendors involved in OSI-based standards. One of the main goals of the workshops was to define a framework for implementing various interoperability requirements. This resulted in a profile of interoperable protocols and options and is termed *government OSI profile* (GOSIP).

Gray's code: In this coding scheme, the output changes by one bit position for every change in input signal in succession.

Ground loop current (GLC): This current is a major source of noise in the signaling circuits in the DTE/DCE interface of the physical layer. If the physical distance between DTE and DCE is large enough, the ground reference voltage at these locations may not be the same, and as such can cause imbalance of reference voltage. Due to this difference of reference voltages, a closed loop current is always flowing between them and is known as a *ground loop current* (GLC).

Ground wave: The electromagnetic waves of less than a few megahertz usually travel the contour of the planet.

Group address (GA): Sometimes, a data message needs to be transmitted to a group of selected terminals or PCs, connected to the network interface unit (NIU) or all the terminals or OCs which are connected to the LAN. In these situations, a group-addressing scheme is used, in which, for the selected group of terminals or PCs, a multicast-addressing scheme is used which broadcasts the data messages within the selected group of users. In contrast in broadcasting addressing scheme, all the users (PCs, terminals, etc.) connected to LAN receive the data message simultaneously. For addressing in broadcasting scheme, a unique bit pattern, usually all 1s, is used in the destination address field in the frame.

Guard band: In frequency-division multiplexing (FDM), the entire frequency bandwidth is divided into a number of channels of smaller bandwidth. For example, the voice-grade channel will have a 4-KHz frequency bandwidth. Since all the frequency channels are close together, interference between neighboring channels may affect the quality of signals received. In order to reduce/avoid interference between the channels, a band of 500 Hz on each side of the channel is considered *guard band*. This certainly reduces the bandwidth available for the channel, but without any interference.

Half-duplex transmission mode: In this mode, the signals travel only in one direction on the same transmission medium. When the transmission of a signal in one direction is over, the transmission of a signal in opposite direction can take place. This mode is used in police controller systems, air-controller systems, cellular phones, or other applications and is based on the walkie-talkie application system.

Hamming code: This code was invented by R. W. Hamming of Bell Laboratories in 1950 and is used for forward error-correction technique. It can detect multiple errors in encoded messages and also correct the errors at the receiving side. The number of redundant bits in the code determines the number of errors, which can be detected and corrected.

Handshake: This technique is used to define the type of communications two users want to establish between them. It is a kind of predetermined exchange between them before the data communication starts. There are different types of handshaking available, i.e., two-way and three-way. In two-way handshaking, the request for connection is acknowledged. In three-way handshaking, the request for a connection is acknowledged and another acknowledgment needs to be sent for the acknowledgment.

Harmonics: A frequency component whose frequency is an integer times the fundamental frequency of the signal waveform.

Hash function code: This function converts the variable size of the data information into a fixed size of blocks of data information. The code providing this type of mapping is known as *hash code*. It also provides protection and authentication to the data and is also known as *message digest*.

Headend: A starting node in a tree topology from which various topologies can be defined. From the headend we may have cables as leaves, and each of the cables may have cables at its leaves and so on. The topology is loop-free and allows multi-point configuration for the signal transmission through the topology. This topology does not require any switching or repeaters, and further, no processing is involved in the topology.

Header error control (HEC): This is one of six fields defined in an ATM cell header, and these fields are of variable size. The function of HEC is to calculate CRC on the first 4 bytes of the header for error detection and correction. Any cell found with an error will be discarded. The sequence number in this field is used to reduce the cell loss and wrong routing. The HEC does not provide any error detection or correction for the payload (data of 48 bytes in the ATM cell).

Head of bus: In DQDB networks, a specific and dedicated node that maintains the generation of empty slots and information fields within the frame.

Hertz (Hz): A unit for the measurement of frequency; also expressed as cycles per second (c/s).

Heterogeneous networks: A network which is a collection of different types of hosts, interfaces, and protocols which further are from different manufacturers. This network allows users of different hosts with different sets of communication protocols and interfaces to send messages or transfer data back and forth between them. This property sometimes is also known as *interoperability.*

High definition television (HDTV): In this type of television, the basic concept is not to increase the definition per unit area, but rather to increase the percentage of visual field contained by the image. The majority of analog and digital HDTV systems are gearing toward 100% increase in the number of horizontal and vertical pixels. This results in a factor of 2–3 improvement in the angle of vertical and horizontal fields. In 1990, the FCC announced that HDTV would be simultaneously broadcast (rather than augmented) and its preference would be full HDTV standard.

Higher layer protocol identifier (HLPI): One of 12 fields defined in an SMDS header. The HLPI is typically of 6 bits and is not processed by the network. Its main function is to provide alignment with the cell between level 3 and level 2 SMDS interface protocols.

High-level data link control (HDLC): The bit-oriented protocol for the data link layer of OSI-RM, defined by ISO. Other derivatives of the HDLC protocol are LLS (a sublayer of the data link layer), LAP-B, and LAP-D (used in ISDN).

High speed channel: The channels offering data rates of 32–34 Mbps, 45–70 Mbps, and 135–139 Mbps are denoted as H2, H3, and H4 channels, respectively, and define the data rates which fall within the range of plesiochronous digital hierarchy (PDH). The PDH mainly defines a single very high data rate signal which is multiplexed of different synchronous signals, ISDN 64 Kbps.

High-speed serial interface (HSSI): In a typical SMDS environment, the PCs are connected to routers, which carry the data to SMDS CSU/DSU. The CSU/DSU is connected to the router via HSSI offering a data rate of 34 Mbps. The HSSI carries the router datagram frames of variable length.

High-split: In the broadband system, the guard band is usually defined at about 190 MHz. A maximum number of 14 channels are defined for a mid-split system for return path of transmission.

Host: This station in the network is usually the end communicator and is assigned a unique IP address.

I.100: General information about ISDNs.

I.200: Service capabilities of ISDNs.

I.300: Overall network aspects and functions of ISDNs.

I.400: User–network interfaces (UNI) of ISDNs.

I.500: Internnetworking functions of ISDNs.

I.600: Maintenance control aspects of ISDNs.

Impedance: A communication channel offers a wave resistance known as impedance to alternating current flowing through the channel.

Implicit congestion notification: This message about network congestion is sent by higher layers such as TCP. The lower layers, e.g., network and data link, use different types of messages for informing the source and destinations nodes about network congestion.

Impulse noise: This type of noise is caused by a sudden increase in amplitude of signal to the peak value within or outside the data communication circuit. It is known by different names such as white, gaussian, random, and so on.

In-band in-channel signaling: In this technique, the control signaling uses the same physical channel and frequency within voice-grade range. In this way, both control and data signals can have the same properties and share the same channel and transmission setup. Both signals can travel with the same behavior of the network anywhere within the network at any location.

In-band signaling: Control signaling provides a means for managing and controlling various activities in the network. It is exchanged between various switching elements and also between switches and networks. Every aspect of the network behavior is controlled and managed by these signals, also known as *signaling functions*.

Information: The message data to be transmitted.

Information bits: Bits representing the information data; they have a different interpretation in different types of transmission modes. For example, in asynchronous transmission of ASCII, it may represent user information data along with some control data, while in synchronous transmission, it may be covered by control data.

Information capacity (or bandwidth): The capacity is usually expressed in terms of bits per second, which indicates the amount of information and is the same as that of the speed of DCE.

Infrared: The short-haul transmission in the frequency range of infrared (10^{12} to 10^{14} Hz) finds very interesting short-distance applications. It is comprised of optical transceivers, which can receive and transmit the signal in the infrared frequency range for a very short distance application (typically less than a mile).

Initialization vector: A code used to initialize the distant data encryption equipment (DEE) which provides encryption to a communication channel/circuit/link. The initialization of DEE is obtained by a section of code containing encrypted variables.

Institute of Electrical and Electronics Engineers (IEEE) standards: IEEE set up a committee of IEEE 802, which developed a number of local area network (LAN) standards that have been accepted as national standards. IEEE is a member of ANSI and is mainly concerned with local area networks (LANs) and many other standards. The IEEE 802 committee has defined a number of local area networks and is primarily concerned with the lower two layers (physical and data layers of OSI-RM) standards. The various standards defined by IEEE 802 are IEEE 802.1, Higher Layer Interface Standards (HLIS); IEEE 802.2, Logical Link Control Standards (LLC); IEEE 802.3, Carrier Sense Multiple Access with Collision Detection LAN (CSMA/CD); IEEE 802.4, General Motors (GM) Token Bus LAN; IEEE 802.5, IBM's Token Ring LAN; IEEE 802.6, Metropolitan Area Network (MAN) LAN; IEEE 802.7, Broadband Technical Advisory Group; IEEE 802.8, Fiber-Optics Technical Advisory Group; IEEE 802.9, Integrated Voice and Data Networks (ISDNs) Working Group; IEEE 802.10, LAN Security Working Group.

Integrated services digital networks (ISDNs): A digital network that provides services such as information retrieval, banking, and of course electronic mail. The ISDN provides direct access to digital circuits that allows it to transmit different types of signals (voice, data, facsimile, images, etc.) together.

Integrated switching: In this switching technique, various functions, e.g., management, configurations, and other related network functions, are performed by one hardware device or switch.

Integrated voice/data networks: These networks allow the integration of voice and data transmission. Although these transmissions have different characteristics and drawbacks, this concept of integration of data and voice has become very important in high-speed networks and other multimedia network applications. A variety of approaches have been adopted for integrated voice/data systems, but two switching approaches — fast packet switching and hybrid switching — seem to be more popular than other combinations.

Intelligent networks: In a typical centralized computing environment, the processors, databases, and other shareable resources are shared by the users over the link or network. The customer premises equipment (CPE) allows the users to use the network. With the advent of distributed systems, these shareable resources are located at different locations, and this offers better connectivity than a centralized system. Further, by adding intelligence with the CPE, we can define intelligent networks.

Interactive processing: In this technique, the users interact with the computers on a continuous basis and can execute programs interactively. Debugging and other operations are also done interactively.

Interchange circuits: A circuit between DTE and DCE that allows the transmission of different types of signals (data, control, and timing). These two devices (terminals and modems) are connected by an interchange circuit, which is also known as a DTE/DCE interface. It transmits/receives three types of signals: data, control, and timing. The output of DCE is connected to the communication link via a communication circuit.

Interchange circuits (balanced): In the balanced transmission, a pair of wires is used between two sites for sending and receiving messages back and forth on different lines. This offers a high data rate and low error rate. Thus, the interchange circuit for balanced transmission uses two separate lines for sending and receiving the message on/from the network, respectively.

Interchange circuits (unbalanced): In this transmission, the return path is via ground circuit, which introduces noise, lowering the data rate and increasing the error rate. Also, the distance gets reduced, e.g., a typical data rate of 20,000 bps over a distance of 50 ft can be achieved. In general, a higher data rate can be achieved for a short distance, while a lower rate can be obtained for communication over a larger distance.

Interexchange carrier (IEC): A carrier that provides inter-LATA services within the U.S. and also provides voice and data communication services around the world, e.g., AT&T, MCI, Sprint, etc.

Interface: Offers virtual connections between two layers, two equipment units, hardware and software, or even two units of software. It also defines the physical characteristics of the virtual connection in such a way that the total functionality of one layer can be mapped onto another one.

Interface bit rate: B-ISDN offers two values of data rates — 150 Mbps and 600 Mbps — across the user–network interface (UNI). The 150-Mbps bit interface rate offers the same data rates to users from either direction of data communication at the interface, i.e., user-to-network and network-to-user. This bit rate is suitable for interactive application type services, e.g., telephone, video-telephony, and database services. The 600 Mbps bit interface, on the other hand, does not provide the same data rates in both directions at the interface.

Interface data unit (IDU): The service data unit (SDU) and control information (addressing, error control, etc.) together are known as an *interface data unit* (IDU), which is basically a message information sent by an upper layer entity to its lower layer entity via SAPs.

Interface message processor (IMP): The switching system, after receiving data from the source, looks for a free transmission link between it and the switching element which is connected to the destination host. If it finds the free link, it will forward the data onto it; otherwise, it will store the data on its memory and try other route for the data. It will send the data to another switching element, which will again look for the free link until the data is delivered to the destination. The element of the switching system has been referred to by different names in the literature, such as interface message processor (IMP), packet-switch mode, intermediate system, and data exchange system. These elements provide interface between hosts and communication systems and establish a logic connection between hosts over transmission links.

Interface payload: The ISDN offers lower values of interface bit rates by primary rate interface (PRI) in the range of 1 to 2 Mbps. The advent of B-ISDN has made it possible to provide greater capacity with higher data rate services to data, audio, video, images, etc. The interface structure plays an important role for handling different types of traffic e.g., circuit-switched or packet-switched (e.g., bursty-traffic). A new concept in transfer mode has been accepted recently for encapsulating the entire payload capacity into fixed-sized small cells of 53 octets. The payload capacity of the interface includes the entire information except the control information required for controlling the operations of the interface itself.

Interim local management interface: The ATM forum has defined a user–network interface for ATM networks that has functions similar to those of SNMP.

Inter-LATA: The public telephone network in the U.S. offers local and long-distance services through local exchange carriers and interexchange carriers. The local carriers are grouped into Regional Bell Operating Companies (RBOCs) while long-distance carriers include AT&T, MCI, and Sprint. The local access transport area (LATA) defines the boundaries between these two service providers, which in fact interact with each other via inter-LATA communication.

Interleaving in TDM: The TDM can multiplex bits, bytes, characters, and packets. Based on this, there are two modes of operations in TDM. In the first mode, *bit multiplexing*, a bit from a channel is assigned to each port, which in turn represents a preassigned time slot. This mode is also defined as *bit interleaving*. In the second mode, *byte multiplexing*, a longer time slot (corresponding to a preassigned port) is assigned to a channel. This mode is also known as *byte interleaving*.

Intermodulation distortion: The mixing of two modulated signals on the receiving side creates this distortion.

Intermodulation noise: This noise is caused due to nonlinear relationship of signals at different frequencies. In the case of frequency division multiplexing, we usually keep a band of 500 Hz on each side of the voice channel as a guard band, which reduces this noise.

International Alphabet 5 (IA5): An international standard which provides binary representation to numbers, symbols, characters, etc. Two widely used binary codes are EBCDIC code (IBM) and ASCII code (ANSI).

International Business Machines (IBM): Manufacturers of large computer systems, in particular mainframes. Also a manufacturer of computer equipment.

International Maritime Satellite Organization (INMAR-SAT): The CCITT X.300 series standards define various steps for establishing an at-sea communication channel connection that creates a link from a ship to a satellite and back to the ship. Standard diameter of less than a meter is attached at the top of ship which transmits signals to and receives signals from the satellite.

International Standards Organization (ISO): A non-governmental, non-treaty, voluntary organization. Its main function is to define and develop a wide range of data communication. More than 75% members of ISO are basically representative of their respective governmental organization or agencies duly approved by the government. Usually these agencies are known as Post Telegraph and Telephone (PTT) and are controlled by their government. ISO was founded in 1946 and has so far defined and developed more than 5500 standards in a variety of areas. Technical Committee 97 (TC97) is usually responsible for coordinating various working groups.

International Telecommunications Union (ITU): This is a standards organization of United Nations that defines standards of telecommunication systems. About 170 nations around the world are members of this organization.

International Telecommunications Union–R (ITU-R): It deals with radio broadcasting.

International Telecommunications Union–T (ITU-T): It deals with telephone and data communications.

Internet: A collection of different types of computer networks that offers global connectivity for data communications between them. A common language through a standard protocol, transfer control protocol/Internet protocol (TCP/IP), facilitates communication between these computer networks. Although the introduction of the Internet is derived from ARPANET, it now includes a variety of networks such as NSFnet, Bitnet, Nearnet, SPAN, CSnet, and many other computer networks.

Internet protocol (IP): The Internet offers interconnecting capability on either connection-oriented or connectionless connections. For the connectionless services, the transport layer protocol must provide a reliable communication across the subnets, and as such we have DoD's transport layer protocol known as transmission control protocol (TCP). IP is similar to ISO 8473 and offers interconnection between various private and public wide area networks via gateways. The IP checks the destination address of a datagram and, if it belongs to the local subnet, it sends it to the destination, but if the packet does not belong to the local subnet, it will be sent to the local gateway.

Internetworking: In any business or organization (small or large), two types of communication systems are very common — voice communication and data communication. The voice communication is handled by PBX, which also handles facsimile, while data communication is handled by LANs. The existing standard LANs typically offer data rates up to 16 Mbps over a typical distance of about 10 Km. High-speed LANs (e.g., MAN, FDDI, and DQDB) offer data rates of 100 Mbps. Workstations and file servers used for interconnecting the existing LANs and also for high-speed data communication required these high-speeds LANs. Different types of subnets can be internetworked where subnets may be public (Telenet, Tymnet, Datapac) or private networks (local area networks). This type of internetworking obviously requires a variety of issues to be discussed in the design of networks, e.g., size and formats of protocols, quality of service, addressing schemes, routing strategies, performance criteria, security, etc.

Internetworking unit (IWU): Due to increasing communication requirements, interconnections must be defined between different types of LANs, between MANs and B-ISDNs and between LANs and private MANs. A hardware device known as an internetworking unit (IWU) is defined to provide the interconnections between these networks. Two networks (any) can be connected by this IWU if the distance between them is small. If the distance between the networks is large, then intermediate subnetworks interconnect the networks. The IWU is known as a *repeater* if two similar LANs are interconnected at layer 1 (physical). It is known as a *bridge* if different LANs are interconnected at layer 2 (data link), and a *router* if the networks are interconnected at layer 3 (network). If the networks are interconnected at a higher layer (normally at the transport or application layer), the IWU is known as a *gateway.*

Interoperability: This method uses a set of procedures, rules, and protocols that interconnect different types of communicating devices. These devices or protocols are from different vendors and, using interoperability, these devices or protocols can communicate with each other.

Intra-LATA: The switched multimegabit data service (SDMS) uses the fast-packet switching technique and offers LAN interconnectivity. The protocols of SMDS offer connectionless broadband services specifically useful for MANs and Regional Bell Operating Companies (RBOCs). It also offers services for local area networks, intra-LATA, and WANs.

ISDN I-series: Integrated services digital network I series recommendations.

ISO 1745: This standard defines the support services for synchronous or asynchronous modes of operation. The support for different link configurations, half-duplex and full-duplex, is also discussed.

ISO 2110: Binary synchronous control (BSC) protocol is very important and widely used for synchronous transmission and has been accepted throughout the world. IBM defined this protocol. With its widespread use, ISO defined various protocols based on BSC known as ISO 2110, ISO 1745, and ISO 2628.

ISO 2628: The error recovery and other interrupt services to support ISO 2111 and ISO 1745 are described in this standard.

ISO 3309: High-level data link control procedures (frame structure).

ISO 7498: Open system interconnection reference model (OSI-RM).

ISO 8072: Services offered by the transport layer within OSI-RM.

ISO 8072 and ISO 8073: This standard protocol for the transport layer supports all five classes of transport services and as such the users have a choice to define the quality of service. This protocol supports connection-oriented for its service primitives.

ISO 8073: Protocol of the transport layer.

ISO 8208: X.25 packet level protocol for data terminal equipment (DTE).

ISO 8326: Basic connection-oriented session services within OSI-RM.

ISO 8327: A standard protocol for the session layer responsible for defining and offering synchronization between sending and receiving stations. It defines the services and functions offered by the session layer and offers a communication channel over a reliable connection-oriented transport layer (ISO 8348).

ISO 8348: Used by ISO 8473.

ISO 8348: Defines network services and functions and supports a connection-oriented packet switching network with an X.25 network. It is, in fact, one of the very popular internetworking standards used in X.25.

ISO 8473: Offers connectionless service to the transport layer and a request for a specific quality of service offered by the network to transport layer. It supports a 512-octet user information field and defines the destination addresses in X.25 address format. This protocol is for network layer support connectionless mode and is also known as Internet protocol (IP).

ISO 8571: A standard protocol that provides the services for file transfer between the users across the network and is known as file transfer access and management (FTAM). It allows different hosts using different techniques to represent and access the data to transfer the files among them.

ISO 8602: Protocol to provide connectionless mode transport services.

ISO 8638: Internal structure and organization of the network layer.

ISO 8822: Connection-oriented services of the presentation layer within OSI-RM.

ISO 8823: A standard protocol for the presentation layer defining the format conversion of the user data for various applications offered by the application layer (top layer of OSI-RM). It offers the services of transforming the application data units into a suitable format acceptable for the transmission.

ISO 8825: Basic encoding rules for Abstract Syntax Notation One (ASN.1).

ISO 8878: Use of X.25 for providing OSI connection-oriented network service.

ISO 8886 or CCITT X.212: This standard includes primitives and parameters to be used for connection-oriented and connectionless operations.

ISO 9040 and ISO 9041: These standards define the services and protocols for virtual terminal (VT) interactive applications. The VT users share the information via conceptual communication area (CCA) which is implemented by suitable data structures, mapping, and user processes.

Isochronous traffic transmission: In this transmission, we are concerned with voice traffic only. An attempt is made to transmit voice and video along with the data over different types of networks. Traditionally, the transmission of voice takes place over the PBX through the public telephone network system using different media such as cable, satellite, microwave, and so on. With the advent of frame relay, ATM, SDMS, and B-ISDN, it is possible to transmit data, video, and voice together over the network through fiber optics at higher data rates.

Jet Propulsion Lab: A distributed computing environment has been experimented on for the CASA network that is located in California. This network was started around 1990 and it connects research centers such as Jet Propulsion Lab, Los Almos Lab, and Caltech. Some public carriers such as MCI and Pacbell are also involved in this project. The supercomputers located at these locations are connected by high-performance parallel interface (HPPI) to OC48 network (Northern Telecom).

Jitter: When the receiver receives the data signal, it tries to recover both clocking control signal and original signals from it. The deviation of clock control signal in the transmitted data message during the recovery of the clocking signal (of the master clock) from the signal causes this noise. The optical/electrical conversion takes place at the interface and it should possess a common jitter reduction function for these two types of signals, as otherwise, it will cause error in the case of misalignment for these two signals.

Job transfer and manipulation (JTM): This is one of several application service protocols being supported by the application (top-most) layer of OSI-RM. For every application, a separate interface and protocol is defined. Other application service protocols supported include FTAM, directory services (CCITT X.500), message handling system (CCITT X.400), transaction processing, and many others.

Jumping window: In ATM-based networks, the traffic control procedures have not yet been standardized, and different algorithms are used for controlling the traffic control strategies of resource allocation in the networks. These algorithms must provide an efficient traffic control mechanism and also provide reasonable quality of service. Various algorithms under consideration include leaky bucket, sliding window, jumping window, exponentially weighted moving average, etc.

Kernel: A collection of subroutines within an operating system that provide core functions such as low-level interfaces to various modules. Other modules when invoked do not have to bother about low-level interfaces.

Kilo: In the metric system, this represents 1000, while in the case of a digital computer memory module, it represents 1024.

LAN interconnection: Different types of LANs (OSI-RM or non-OSI-RM) are connected together for data communications. There are different devices which can interconnect them, e.g., repeaters, bridges, routers, etc. High-speed networks such as frame relay, SMDS, and ATM are also being used to provide LAN interconnectivity for sending different types of traffic (data, voice, video, etc.).

Laser (light amplification by stimulated emission of radiation): A device that converts electromagnetic radiation into a single frequency of amplified radiation.

Latency: In token-ring-based LANs, a token of 3 bytes circulates around the network. Any node wishing to send the data has to capture the token, append it, and then send it to the destination. In the absence of any data, the time taken by a token to make one round of the ring is called *latency time*.

Layer: The International Standards Organization (ISO) defined a layered model for LANs comprised of seven layers. Each layer offers a set of services to its higher layer and offers specific functions. Each peer-to-peer protocol (defined for each layer separately) carries information defined in terms of data units known as protocol data units (PDUs).

Layer management: Layer management within a network management protocol is concerned with the operations of layers of OSI-RM. It also allows users to maintain various statistical reports of various operations of the networks.

Layer management entity: In a DQDB network, local management of any layer is performed by the entity defined within the layer protocol.

Layer management interface: In a DQDB network, this interface is defined between layer management and network management layers.

Layered protocol: Each layer offers a set of services to its higher layer, and these services offer communication between the layers. The communication between layers takes place through a service access point (SAP).

Leased line (dedicated line): The leased line or dedicated line offers a full-time, permanent connection between two remote nodes and is billed on the monthly basis. The line is always connected and hence no dialing is required. It supports point-to-point and multi-point connection (using bridges). It provides one-loop (a pair of wires) or two-loop (two pairs of wires) modes of operation.

Least significant bit: The bit on the right-most position; has minimal binary weight in the representation.

Limited distance modems: These modems are used for short-range communications and are available both as synchronous and asynchronous versions, covering a wide range of host applications. Limited distance modem (standard V.35) is used for the transmission of a bulk of data within a local distance of a few miles (typically less than 3 miles). It supports point-to-point or multi-point configurations and offers data rates up to 64 Kbps.

Line booster: This device increases the transmission distance for asynchronous data over an RS-232 interface. As we know, RS-232 cable length goes up to 15.5 Km, while by connecting one line booster between two RS-232 cables, the signals now can travel to a distance of 30 km. One of the applications of a line booster would be to place a printer at a distance from the PC/workstation.

Line driver: A device used to increase or boost the distance for transmitting the data over transmission media. Instead of using expensive line drivers, we can use line drivers with compression technique, and there are many line drivers with this facility available on the market. These line drivers provide interface between RS-232 and short-haul modems and convert RS-232 signal levels to low-voltage and low-impedance levels. They support point-to-point configuration and also can connect more than one station to the same line via multi-point line drivers. The line drivers are available for both asynchronous and synchronous transmission.

Line extender: This device is basically an amplifier that is used to amplify the signals of a broadband system within the building.

Line interface: This communication interface device provides a communication link between computers or terminals over a particular line. The computer manufacturers usually provide these interfaces for different combinations of connection, e.g., line-terminal or line-computer and remote computers/terminals. It comforms to ASCII coding and RS-232 interface (physical layer DTE/DCE interface) and also provides interface for modems/acoustic couplers.

Line of sight: The transmission technique for the signals over 30 MHz typically uses the concept of *line of sight*, where the signals are transmitted from antennas. The height of these antennas is calculated based on the curvature of the earth. All wireless transmission media use this concept in one way or the other.

Line protocol: The set of procedures and rules defined to manage and control the transmission of data over the line or link. Again, this link can be defined over different transmission media. This set of procedures and rules is not used to provide error control or flow control, as these are typically provided by the network protocols.

Line sharer: This communication device allows a single dial-up line or leased line to be shared among multiple devices. It offers data rates of up to 64 Kbps.

Line speed: The maximum data rate that can be considered for transmitting the data over the transmission line.

Line splitter: Usually, a line splitter is used at the exchange to provide the connection. By using a line splitter, one assigned line of exchange can be used by more than one location at the same time. That assigned line transmits information from a computer on the line to various terminals connected to different exchanges. Similarly, when these terminals want to transmit information, these are combined on that line and delivered to the computer. The point-to-point and multi-point connections can also be configured on leased lines using a line splitter.

Line terminating equipment: A device used in high-speed networks that generates and terminates the optical carrier level signals. These signals are used for the performing various functions on transport overhead (generate, access, modify, terminate, etc.).

Link: This physical connection between two nodes provides the path for the transmission of data over it. The physical connection may be defined over any transmission media (open wires, coaxial cable, satellite, microwave, etc.).

Link access protocol (LAP): This standard has been derived from HDLC and uses the HDLC set asynchronous response mode (SARM) commands on an unbalanced configuration. It has been designated as VA 2, 8 and is being used to support X.25 networks. The primary and secondary nodes use SARM and unnumbered acknowledgment (UA) commands, respectively, for establishing a link between them. The extended version of that protocol is known as LAP-B.

Link access protocol-balanced (LAP-B): This standard is a subset of HDLC which controls the data exchange between DTE and public switched data networks or even private networks and establishes a point-to-point link between them. It controls the data frame exchange across the DTE/DCE interface and offers asynchronous balanced mode of operations across the interface. This standard is used in both private and public networks, including X.25. In X.25, it defines the data link protocol.

Link Access Protocol-D (LAP-D): The LAP-D frame includes user information and protocol information and in fact transfers the user's data it receives from the network layer. It offers two types of services — unacknowledged and acknowledged service — to the user's ISDN data of the network layer. The D channel of ISDN rate or access interface is used mainly for signaling and as such offers connectivity to all the connected devices to communicate with each other through it.

Local access and transport areas (LATA): Each country has its own telephone network that is managed either by the government or by a duly recognized company. The telephone companies typically offer these services for voice communications. In the U.S., the telephone network services are provided by seven Regional Bell Holding Companies (RBHCs) which control various long-distance carriers such as AT&T, MCI, Sprint, etc. The main duties of RBHCs are to install and operate about 165 local access and transport areas (LATAs), which are usually located in metropolitan areas. Inter-LATA services are provided by RBHCs and they provide equal access inter-LATA service to all long-distance carriers.

Local area network (LAN): In any organization or business environment, two types of communications are very common: voice communication and data communication. The voice communication is obtained over private branch exchange (PBX), while the data communication is obtained over computer networks. The computer networks can be classified as wide area networks and local area networks (LANs), and are in general distinguished by the distance of their coverage and topologies used.

Local bridge: An interconnecting device between two different types of LANs that operates at the data link layer of OSI-RM. It provides mapping between different types of MAC frames and protocols and is expected to have high throughput.

Local loop: Each customer (telephone subscriber) has a telephone set which connects to the Class V end office through two-wire or four-wire (copper conductor), and this connection is known as *local loop* or *subscriber loop*. The distance between the customer's telephone set (either at homes or offices) and the end office is typically a few miles.

Logical link control (LLC): OSI-RM defines seven layers for a local area network, but if we divide the data link layer of OSI-RM into two sublayers — media access control (MAC) and logical link control (LLC) — then this reference model defines eight layers and is known as *OSI-reference model for LANs* (defined by IEEE 802 Committee). The MAC sublayer mainly is concerned with the media access strategies, is different for different LANs, and supports different types of transmission media at different data rates.

Logical link control (LLC) services: The LLC sublayer offers a common interface to various MAC sublayer protocols and, in fact, is independent of the media access control strategies. The LLC protocol data unit (PDU) defines the following fields in its format: destination service access point (DSAP), source service access point (SSAP), a control field, and users' data.

Logical record: A collection of data independent of the physical locations. The same logical records may be located in different physical records.

Logical ring: Both token ring and FDDI LANs use a ring topology that provides a point-to-point link between every pair of nodes connected to the ring. Every node is connected to two of its neighboring nodes. In these networks, the token is used for data communication.

Logical unit (LU): The main aim of SNA (IBM) architecture is to provide a structural framework which can be compared with OSI-RM. The lower layers of SNA and OSI-RM seem to have common functions, but the higher layers are different. The levels in SNA are defined as *logical units* (LUs). The LU provides an interface between it and a centralized manager and also between it and other LUs of IBM networks.

Logical unit interface (LUI): The logical unit (LU), in fact, can be considered as a port through which end users communicate with each other. Although in most cases, the interface including LU is not defined specifically, it needs to be implemented as an LU, as it offers the functions of the session layer (OSI-RM).

Long-haul modems: Modems designed for long-haul transmission systems offer a frequency band of 48 KHz and use frequency division multiplexing.

Longitudinal redundancy check (LRC): The LRC and VRC are used to detect errors in rows and columns, respectively, and hence combined LRC and VRC will detect all the possible errors of any bit length. It allows the checking of parity of each character in the frame, and hence is used to detect a block of data. Combining LRC with other parity patterns may enhance the error-detection capabilities.

Looped dual bus: This topology used with DQDB uses a pair of buses for data communication. Each node connected to this topology can access both buses. This topology defines two types of access configurations — open dual bus and looped dual bus. In the looped dual bus, the node makes a loop with both buses of the network.

Lower sideband: In amplitude modulated signal, two sidebands are generated. The lower sideband is on the lower side of the carrier frequency side.

Macintosh browser: MacWeb is a browser for Macintosh. MacWeb requires System 7 and MacTCP (operating system defined by Apple which allows Macintosh computers to understand TCP/IP). For using MacWeb, an application program, Stuff It Expander (SIT), is used which translates and uncompresses files. The MacWeb is very fast and offers a customized user interface. It supports automatic creation of HTML documents and supports MacWAIS.

MAC sublayer: This second sublayer of the data link layer is concerned with frames formatting and media access control strategy services. Various IEEE 802 MAC standards available include carrier-sense multiple-access with collision detect (CSMA/CD).

Management layer: The reference model of B-ISDN consists of three sectional planes — user, control, and management. Each plane is further comprised of layers with specific functions to be performed within the plane. The management plane is mainly concerned with the management of the entire system and provides necessary control and management controlling functions to all the planes of the reference model.

Manchester encoding: A digital signaling technique which is defined by a transition at the center of a digital bit. A low-to-high transition level during the first half of the bit interval represents logic 1, while logic 0 is represented by high-to-low level during the first half of the bit interval. The transition at the center of the bit interval provides clocking and also data. This technique does not need any additional synchronization and contains no DC components. It provides error detection.

Magellan passport switches: Northern Telecom introduced three different types of ATM switches under a brand of Megellan products in 1993 — passport, gateway, and concorde. These switches support SONET/SDH. The passport switch is the smallest of all three and is used in enterprise networks. It offers a number of interfaces, e.g., T1, T3, FDDI, UNI, NNI, and others. It supports data and voice via frame relay interface. Gateway is a broadband multimedia switch supporting access to ATM networks. The concorde is being used in carrier core backbone network and offers data rates of 10 to 80 Gbps.

Manufacturing automation protocol (MAP): Developed by General Motors (automobile industry, U.S.) in 1962 for providing compatibility between various communication devices used in the manufacturing processing environment. The communication devices include terminals, resources, programmable devices, industrial robots, and other devices required within the manufacturing plant or complex. This attempt to define such a local area network was transferred to the Society of Manufacturing Engineers (SME).

Media access control (MAC) bridge: Bridges (at the data link layer), routers (network layer), and gateways (at higher layers) can interconnect LANs. A device known as a bridge connects the LANs, which define the same protocols for the physical layer and IEEE MAC sublayer. A bridge operates at layers 1 and 2 of OSI-RM and is commonly used to either segment or extend LANs running the same LLC protocols.

Media access control (MAC) services: The MAC sublayer defines a frame for each packet it receives from its LLC by attaching a destination address, length of fields, etc., and sends a primitive type MA-DATA.confirm to its LLC via the same or different MSAPs, indicating the success or failure of the operations over the request.

Medium: The physical link connecting different types of devices in a network. Each physical link has its own characteristics and offers an access technique. The data communication takes place over this link. Two types of links are wired and wireless. Wired links are copper wires, coaxial cables, and fiber optics, while wireless links include satellites, microwaves, radios, etc.

Mercury communication networks: A public network implemented in the U.K. using ATM and SMDS. The network offers broadband applications to the customers.

Mesh: A LAN topology based on the concept of a tree which connects different types of nodes (hosts, computers, etc.) operating under different environment supporting different sizes of the packets so that these different devices can exchange information. There is no set pattern for this topology. This topology was used in the Department of Defense (DoD) Internet. It usually is useful for WANs.

Message handling system: One of several application service protocols defined at the application layer of OSI-RM. The application service protocol defines an interface and sends the message from the application layer through layers of OSI-RM. On the receiving side, it goes through the layers of OSI-RM until the application layer.

Message switching: In this technique, an individual message is separately switched at each node along its route or path from source to destination. This information is contained in each message. Telegrams and data files are examples of messages. Here, the circuit is not established exclusively for the message, but instead messages are sent using the store-and-forward approach. Users, based on the capacity of the networks, divide each message into blocks of data and these blocks of messages are transmitted in sequence.

Meta signaling procedure: In order to access the services of the AAL layer, virtual channel links are defined which are used to assign values to VCI or remove values from it between two points. The concatenation of such virtual channel links is defined as a virtual channel connection (VCC). The VCC can be established in a variety of ways. Some of the techniques used for establishing VCC include metasignaling, user-to-network signaling, user-to-user signaling. The meta signaling is used for establishing a signaling VC.

Metropolitan area network (MAN): The high-speed LANs offer a data rate of over 100 MBps over a distance of 100 Km (diameter) and support more than 1000 users. The distance of over 100 Km usually is defined within a metropolitan area, and as such these high-speed LANs are also known as metropolitan area networks (MANs, IEEE 802.6). The MANs use fiber cable transmission medium. Two contra-flowing unidirectional buses, I and II, define the network topology of MAN. The stations are connected to the buses via access units (AUs), which are defined by OR-write connection over contra-flowing unidirectional buses in MAN. Each of these operations (read and write) is performed in opposite directions. It supports two types of topologies: open and looped bus.

Microcomputer: This computer system (e.g., PC) has a limited physical size, speed, and memory and typically is a single user system.

Microprocessor: This processing unit contains the logical elements for performing arithmetic and logical operations on the data signals.

Microwave communications: Works on the principle of "line of sight," in which the antennas at both transmitter and receiver should be at the same height (this means that the height should be large enough to compensate for the curvature of earth and also provide line-of-sight visibility). It carries thousands of voice channels over a long distance. It operates at a frequency range of 1.7 GHz to 15 GHz (in some books, we may find this frequency range mentioned as 250 MHz to 22 GHz). It can carry 2400 to 2700 voice channels, or one TV channel.

Microwave links: Used to transmit multiplexed voice channels, video signals (television), and data. These are very useful in television transmission, as they provide high frequency and multiplexing for different television channels. The size of antenna for television transmission depends on the wavelength of the transmission, which is defined as a reciprocal of frequency in hertz. The unit of wavelength is meters.

Minimum and maximum integration: The CCITT I.462 recommendation defines two types of architecture for handling both circuit and packet-switched connections. The ISDN is basically a digital telephone system and provides only circuit-switched connections, and it may or may not handle all the packet-switched connections. In order for ISDN to handle both connections (circuit-switched and packet-switched), the I.462 recommendation introduced the concept of two classes of integration — minimum and maximum integration.

Modem: An interface between devices which need to share data across the network; generally consists of a transmitter and receiver. It transmits data from computer/terminal connected to one side of the modem, while the other side is connected to the telephone line. It changes the characteristics of signals so that it is compatible with the telephone channel. Two main types of basic modems have been defined as asynchronous and synchronous to support appropriate transmission modes.

Modem eliminators: Some times the distance between host terminals is so small that even small distance modems or line drivers may not be used; instead, the local connection for synchronous devices running on limited distance modems (V.35) can be extended to thousands of feet apart without any line booster. The modem eliminator offers various mapped pin connections (clocking for interfaces, etc.) between synchronous host machine, terminals, or peripheral devices (e.g., printer), thus eliminating the use of a pair of modems.

Modem (null): The null modem (crossover) cable includes both modem and cable and can support a length of 15 m for local applications. The cable is usually immune to noises and is not available for higher distances than the standard 15 m.

Modem Splitter: A single modem line can be used to carry the data from multiple terminals, and a modem splitter achieves this. It offers modem line sharing among these terminals and supports physical layer interfaces RS-232/V.24.

Modulation: A communication process is an essential component for both data processing and data communication system. If we can define a process which can change the characteristics of a signal of higher frequency by instantaneous value of amplitude, frequency, and phase of baseband (modulating) signal, then this signal can be transmitted to a larger/greater distance. This process is termed *modulation*.

Modulation index: A measure used to determine the depth of modulation.

Mosaic: One of the finest browsers offered by the web (WWW). It offers a multimedia interface to the Internet. The hypermedia is a superset of hypertext, i.e., it allows a linking of media by a pointer to any other media. As it presents hypertext and supports multimedia, it is also known as a hypermedia application software tool.

Mosaic (NCSA) for X-Window system: This browser is coupled with NCSA's web server (httpd) which offers copy and display options. It also supports WAIS and URLs. It requires a relatively powerful computer and requires a direct access to a UNIX or VMS machine running the X-Window system.

Most significant bit: The bit at the right-most position; has the heaviest binary weight in the representation.

Motion Picture Experts Group (MPEG): The transmission of audio, video, and data requires a large amount of memory. The PC or workstation must have this amount of memory. For a simple video clip, the memory requirement without any compression can easily go to a few gigabits per second. Although the broadband networks offer higher data rates, this amount of data will take a very long time to transmit. A compression technique is used to compress this huge data into a small amount of data. The compression standards include Motion Picture Experts Group (MPEG) and Joint Photographic Experts Group (JPEG). With these standards, compression ratios of 10:1 to 50:1 are available.

MS-DOS browser: Lynx is a basic web browser which is defined for DOS computers or dumb terminals running under UNIX or VMS operating systems. This browser does not support images, audio, or data, but supports the mailto URL. The DOSLynx is an excellent text-based browser for use on DOS systems.

Multicast: In this technique, all the nodes connected to the network get the message from any node. This is also known as a broadcasting system. A special case is being used for addressing all the nodes. Any message from any node will be broadcasted to all the connected nodes.

Multimode fiber (MMF): Optical fiber is used in long-distance telecommunications, LANs, metropolitan trunks, and also in military applications. Two different types of light sources are used in fiber optic systems — light emitting diode (LED) and injection laser diode (ILD). It supports two modes of propagation, single and multimode. In the case of single mode, the radius of the core is reduced which will pass a single angle or few angles. This will reduce the number of reflections, as fewer angles are there. Single mode offers a better performance than multimode, where multiple propagation paths with different lengths are created. These angle signals will travel through fiber and also the signals will spread out in time.

Multiple user dimension (MUD): A very demanding client application program service over the Internet which offers virtual reality (hot topics with added potential in the area of medical, 3-D images, space technology, creation of a village, house design, and many other applications). It allows users to use the virtual reality of a variety of applications, e.g., puzzle, create strange places with new buildings, bridges, houses, people, etc. of a particular MUD.

Multiplexer products: A large number of supporting products for multiplexing are available on the market and are used with multiplexers over transmission media offering different data rates and support for physical layer interfaces.

Multiplexing: A process which combines small-capacity or low-speed subchannels into one high-capacity or high-speed channel and transmits the combined channel over the communication link or transmission media. This technique improves the link/medium utilization and also the throughput of the system.

Multi-point line: This line connects a single transmission line to more than two nodes. In the literature, it is also known as a *multi-drop line*. All of the nodes use the same transmission line to carry their respective data, and the line keeps on dropping the data to respective destination nodes.

Multipurpose Internet mail extension (MIME): The Internet mail system allows users to send a text file (containing characters, numbers in documents, etc.), ASCII file (using ASCII code), or a binary file (containing a picture with no characters in it). The simple mail transfer protocol (SMTP) sends text files, while another protocol known as multipurpose Internet mail extension (MIME) is used for binary files. Typically, the binary data is stored in a file which is attached to a regular text file. Support of MIME will send this as a text file over the Internet. We may also record our voice as a binary file, which can be played back.

Multi-stage networks: A number of ATM switch architectures are available, but in general all of these are classified into two classes: single-stage and multi-stage switching networks. In multi-stage networks, a number of intermediate stages can be defined (offering different link interconnections) between the input and output lines, and as such we have more than one available route between input and output lines.

Narrow band: The channels support a variety of data communications options and, in the literature, these options have been divided into three categories of communication: narrow band, voice band, and broadband. The narrow band communication includes the applications requiring data rates up to 300 Kbps. The voice band communication usually defines the bandwidth of voice signals in the range of 300–3.4 KHz.

Narrow band FM: In this form of angle modulation technique, the index of modulation is two or less than two.

National Television Standard Committee (NTSC): North America video signal is a standard called NTSC and is not the same as computer or RGB video. TV is an analog medium. NTSC is used in TVs, VCRs, and camcorders. In computer, video, color, and brightness are represented by digital numbers, but analog TV defines all these in voltages which are affected by length of cable, heat, connectors, video taps, etc. Most engineers now call this Never Twice the Same Color. Other standards are PAL and

SECAM. PAL is used for TV recording, while SECAM is a transmission standard. Multi-systems supporting these standards and also changing into each other are available. This is mainly used in the U.S., Japan, Taiwan, and Korea.

Negative acknowledgment (Nak): Usually a positive acknowledgment is also considered an acknowledgment (Ack). Thus, Ack indicates that the frame is received error free. In contrast, a negative acknowledgment (Nak) indicates an error in the received frame and as such suitable error control techniques have to be used. The Nak requires the retransmission of the same frame until it is received error free.

Network: The main goal of a network is to provide resource sharing to its users such that users of any computers (working under any environment) connected via a network can access the data, programs, resources (hard disks, high quality expensive laser printer, modems), peripheral devices, electronic mail (e-mail), licensed software, etc., regardless of their physical locations. The physical locations may be a few feet, miles, hundreds of miles, or even thousands of miles, but the users and computers exchange their data and programs in the same way as they do locally.

Network architecture and layered protocol: In 1977, the International Standards Organization (ISO) set up a committee to propose a common network architecture, which can be used to connect various heterogeneous computers and devices. It is also an initial step for defining international standardization of various standards and protocols to be used for different network layers.

Network layer: The third layer from the bottom of OSI-RM. This layer is responsible for proving routing of data through the network. The DoD defined a TCP/IP suite of protocol for the Internet. The network layer is realized by Internet protocol (IP), while transmission control protocol (TCP) realizes the function of the transport layer of OSI-RM.

Network management: The management of designing, planning, organizing, and maintaining the operation and status information of networks. The protocol for network management provides various statistical reports, status of devices, network resources, protocols, etc.

Network–node interface (NNI): The two interface bit rates defined for user–network interface are 155.520 Mbps and 622.080 Mbps, which are the lower side of synchronous digital hierarchy (SDH). This transmission hierarchy offers a variety of services and provides compatibility for a number of user–network interfaces with the nodes of topology of B-ISDN known as network–node interface (NNI). The NNI is usually implemented as a cross-connect or switching element and is interfaced with SHD equipment. The users access the B-ISDN through these network nodes.

Network structure: The general network structure includes hosts (computer, terminal, telephone, or any other communicating devices) and communication subnet (also referred as network node, subnet, transport system). Each host is connected to the communication subnets. The host provides various services to its users, while the subnet provides a communication environment for the transfer of data.

Network television distribution: Fiber technology seems to be ideal for accessing a variety of services from different types of networks, and emphasis is mainly toward taking the fiber cable closer to the customer's premises equipment (CPE). A number of schemes and configurations have been proposed for interactive services (e.g., TV distribution, telephony of 64 Kbps) over fiber. This can be achieved either over the same cable via multiplexing (wavelength division) or through the use of fiber for different services with a common optical/electrical conversion. After optical/electrical conversion, the TV distribution signals may be transmitted over coaxial cable, while telephony (64 Kbps) can be transmitted over two-wire copper links. B-ISDN offers internetworking with all of these networks (analog telephone networks, 64 Kbps

ISDN, TV distribution networks, etc.) to offer a variety of services to the users. B-ISDN may integrate the TV distribution networks in either a switched or non-switched manner to offer these two types of services.

Network terminating equipment: This standard defines the boundary between the functions of ISDN and the user's equipment. This is a reference point where the interface between them is defined.

Network virtual private: B-ISDN offers a variety of services at higher data rates with flexibility and reduced cost compared to MANs. The private MANs may use the services of B-ISDN by creating virtual private networks within B-ISDN. These virtual private networks offer functions and services over private or dedicated networks. They use the resources and other facilities of public networks, but the connection is provided to the users of the private network on a monthly charge basis. Thus, the services of B-ISDN are available via this virtual private network within B-ISDN.

N-ISDN: ITU-T renamed ISDN as narrowband-ISDN and defined this network as the one which provides services (data, audio, and video) requiring a bandwidth of less than or equal to 64 Kbps or up to a regular voice channel. Any service requiring a bandwidth of higher than 64 Kbps and below T1 (1.544 Mbps) is provided via broadband ISDN (B-ISDN).

Noise: Unwanted signals transmitted along with the original signals over the transmission media. These signals are introduced/generated due to many atmospheric conditions, e.g., thunderstorms, rains, etc., and also due to the equipment being used in the communication systems. A variety of noises have been defined, e.g., electrical noise, cross-talk, impulse, transient, phase jitter, harmonic distortion, echo, transmission distortion, modulation, quantization, nonlinear, digital, timing jitter, baseband bias distortion, etc.

Nonreturn to zero: This encoding technique maps the digital signal into signal elements. Here two different voltage levels are used to define two binary signals. A positive/negative constant voltage represents the binary logic 1, while no voltage represents the binary logic 0. In both representations, the voltage value does not come to zero value, and hence the name. It is easy to implement and offers a better utilization of bandwidth. Due to the presence of a DC component and lack of synchronization, this technique is not used for signal transmission applications.

Nonreturn to zero invert on ones (NRZI): This is a variation of nonreturn to zero and defines a constant voltage pulse for the bit duration. Logic 1 is represented by low-to-high or high-to-low transition at the beginning of a bit time while logic 0 is represented by no transition.

Nonreturn to zero level (NRZL): This encoding scheme uses two different voltage levels for two binary logic values. Logic 1 is represented by a negative voltage while logic 0 is represented by a positive voltage.

Nyquist bandwidth: According to Nyquist's theorem, the minimum bandwidth required to pass all the frequency components present in the input.

Nyquist's law: In analog modulations, the characteristics of carrier signal are changed by the instantaneous values of the parameters of the modulating signal. In pulse modulation, the modulating signal has to be sampled first and then the values of samples are determined and it is these values which are used with the pulse carrier signal. Nyquist's law is stated as follows: "the original signal can be obtained from the received modulated signal if the sampling rate is twice the value of the highest signal frequency in the bandwidth." This sampling rate will be just enough to recover the original signal with acceptable quality.

Octet: A block of 8 bits is defined as 1 byte or octet.

Off-loading: A method by which the processing requirements of the main CPU are shared by other communication interface devices. Here, the communication interface devices are used for a specific task, e.g., input/output operations, etc. Off-loading is very crucial and important in client–server computing, as the servers are used to off-load a bulk of processing from the client which can use the added capacity for processing other applications. The front-end processor is usually attached to the host and controls input/output operations, thus lowering the burden of the CPU. Similarly, the terminals may be used to lower the burden of concentrators.

Omninet: A baseband LAN developed by Corvus Systems, Inc. In this LAN, after the occurrence of collision, each sender node waits for a random amount of time before sensing or listening to the channel. The amount of waiting time before it could try again is known as *backoff*, and it increases exponentially as the traffic increases. As a result of this, the frequency of occurrence of collision also increases. It uses CSMA/CA variation of CSMA protocol at a data rate of 1 Mbps over a segment of 33 m serving 64 nodes.

On-line: In this configuration, the peripheral devices are directly connected to the processing module in the system. It also corresponds to the direct connection between terminal and transmission line.

On-line processing: In this technique of processing, the data is directly entered into the computer from an input device for processing, and after processing, the data is directly sent to the output device.

Open system interconnection (OSI): The ISO 7498-1984 "Information Processing System — Open System Interconnection — Basic Reference Model," developed by the American National Standards Institute, defined a reference model for open systems comprised of seven layers (OSI-RM). Each layer offers specific functions and offers services to its higher layer. Each layer defines a set of entities which provide interaction between layers.

Operation and maintenance (OAM): CCITT recommendation M.60 defined maintenance as "the combination of all technical and corresponding administration actions, including supervision actions, intended to retain an item in, or restore it to, a state in which it can perform a required function." Recommendation CCITT.20 and I.610 define the following actions for OAM: performance monitoring, defect and failure detection, system protection, failure or performance information, fault localization.

Optical carrier interface: The mapping between the functionality of narrow band and broadband communication is usually achieved by using internetworking units. These units provide mapping between asynchronous communication systems and synchronous communication systems. They provide mapping between narrow band network standards (DS3, DS1) to broadband network standards (STS levels, OC levels).

Optical carrier (level) (OC1–OC192): For uniform optical communications, two standards were defined: Synchronous Optical Network (SONET) and synchronous digital hierarchy (SDH). The former is basically for North America, while the latter is for Europe. A value of 51.84 Mbps has been defined as a base signal for SONET for a new multiplexing hierarchy known as synchronous transport signal level 1 (STS-1). The optical carrier level 1 (OC1) over fiber optical media is an optical equivalent of an electrical signal (STS-1). Different optical carrier capacities are defined as OC1/DS3 (51.84 Mbps), OC3/DS3 (155.52 Mbps), OC12/DS3 (622.08 Mbps), OC24/DS3 (1.244 Gbps), OC 28/DS3 (2.488 Gbps), OC192/DS3 (9.6 Gbps). Here we show direct mapping between OCs and DS3. Other mappings between OCs and other DSs such as DS1, DS0 can be derived.

Optical fiber communication: If we can propagate the light through a channel (free of atmospheric noises) which possesses known and stable characteristics and provides controlled reflections along the channel to the light, then it can be used to carry the entire message (without any loss) from one node to another. The channel or tube or dielectric guide known as *fiber*, which has a refractive index greater than the atmosphere light propagating through a glass fiber channel (optical fibers), can be used to carry different types of signals, e.g., voice, data, and video. Optical fibers used as a communication media in telecommunication systems offers features such as large bandwidth (50–80 MHz in multimode fibers, several gigahertz in single mode fiber) allowing high bit rate, and immunity to electromagnetic interference. It suffers from the following problems: the development of opto-electronics transducers, their optical interface with fiber, and the reliability; electrical insulation between emitter, fiber, and receiver; the problem of supplying direct current power to intermediate repeater stations; impurities in fiber; mismatching; poor alignment of connectors; operating constraints; reliability issues; coupling of light signals from one fiber to another fiber; difficulty in installing and repairing it; being expensive to replace the existing equipment for introducing and using fiber optics; lack of standardization.

Optical modulation: Optical fiber supports the following modulation schemes, known as optical modulation. (1) *Continuous:* In optics, we usually call this *intensity modulation* (IM), similar to analog modulation (AM), with the only difference being in the nature of the optical carrier signal (non-sinusoidal). (2) *Discrete:* Here, it is known as optical on–off keying (OOK) similar to shift keying with a sinusoidal carrier. The modulated signal is comprised of pulses (ON/OFF) of constant amplitude. These techniques are based on the variation of optical power according to electrical signals applied to the electro-optical transducers. It is recommended to use frequency pulse modulation (FPM) for analog transmission of a television channel by optical fiber.

Optical transmission: The use of optical fiber as a transmission medium in a network for data communication offers very high data rates in the range of gigabits per second. All the high speed LANs and integrated networks use a fiber transmission medium for data transfer. The optical transmission medium offers very low attenuation and very high bandwidth. Two of the high data rate user–network interfaces of 155.720 Mbps and 622.080 Mbps for ATM-based B-ISDN have been defined by CCITT recommendation G.707.

OSI LAN management protocol primitives and parameters: In the area of LAN management, the LANs interconnect a number of autonomous, multi-vendor, and different operating systems. The OSI committee has defined a number of management domains for future network management activities such as network management, LAN management, OSI management, and management dealing with loosely coupled interconnection and distributed computing aspects of LANs. The LAN management mainly deals with the lower layers, and IEEE considers the LAN management as system management. One of the main objectives of LAN management is to provide continuous and efficient operation of LANs.

Out-of-band signaling: CCITT I.327 defined the capabilities of architecture of B-ISDN in two components — information transfer and signaling. The broadband signaling is expected to support broadband services in addition to its support for existing 64 Kbps ISDN, multi-party connection (conferencing type environment), internetworking, etc. The signaling information is communicated via dedicated out-of-band channels known as signaling virtual channels (SVC). These channels are defined per interface and support different types of configurations, such as meta-signaling channel (bidirectional), general broadcasting (unidirectional), selective broadcasting (unidirectional), and point-to-point channels.

Overmodulation: The modulation technique in which the index value of modulation is more than one.

Packet: According to the Defense Advanced Research Projects Agency's recommendations, the user's messages or information are defined as data segments known as *packets*. The packets contain control information, the user's data, destination address, etc. In OSI's recommendations, the packets are known as protocol data units (PDUs) for each layer of OSI-RM.

Packet assembly and disassembly (PAD): The main objective of PAD is to define a packet from the characters it receives from character-oriented asynchronous terminals. These packets follow the format of the packet defined by X.25 public switched packet network (PSPN). On the receiving side, PAD, after receiving these packets, disassembles them back into character form and sends the characters to the respective terminal one-by-one. This device provides an interface between an asynchronous DTE (typically a terminal) to a public packet-switched network (typically X.25).

Packet layer protocol (PLP): There are a number of ways of providing internetworking between LANs and WANs via gateways. There are four configurationss define the possible connections for internetworking. In one of the configurations, two DEC DNAs are connected via X.25 gateway. Both LAN1 and LAN2 have peer-to-peer entities for protocols of all layers of DEC DNA (user layer, transport layer protocol, X.25 packet layer protocol (PLP), LLC/MAC). The gateway includes two tables within it. In table one, the packet goes through LLC/MAC to X.25 PLP, while the other table includes X.25 PLP, LAP-B, and X.25 PL. The gateway interface provides mapping between these two tables.

Packet switched network (PSN): This network works on the same principle as that of message-switched networks, except that here the message information is broken into packets which contain a header field along with the message information. The header field includes packet descriptor, source and destination addresses, and other control information. The packets can be routed on different routes independently.

Packet switching: In principle, packet switching is based on the concept of message switching, with differences such as the following: (1) packets are parts of messages and include control bits (for detecting transmission errors), (2) networks break the messages into blocks (or packets) while in message switching, it is performed by the users, (3) due to very small storage time of packets in the waiting queue at any node, users feel bi-directional transmission of the packets in real time, etc.

Parabolic antenna: This type of antenna has a parabolic shape.

PARC universal packet (PUP): The bridges in LAN interconnection are required to maintain the status information such as direction, receiving port, source nodes, next nodes, next ports, mapping, etc. The bridges can also accept mini-packets for a set of destination nodes by maintaining a table look-up. The transport session between internetworked LANs is achieved by an interface architecture defined by Xerox PARC, Palo Alto, CA and is known as PARC universal packet (PUP). This interface offers transport for connectionless service (datagrams). The PUP resides in the host machine, which can support different types of networks.

Parity bit: The widely used codes are ASCII (American National Standards Institute) and EBCDIC (IBM). EBCDIC is an 8-bit code. ASCII is a 7-bit code but, in general, is seen as an 8-bit code, since the vendors usually use an extra bit at the least significant side as a parity bit. This extra bit is used for error or parity checking. It can be used as an even or odd parity bit, depending on the protocol.

Path layer: This layer is the top layer of SONET/SDH protocol architecture. Its main function is to provide mapping between the services of the path overhead and STS synchronous payload envelope (SPE; format of the lower line layer). The path overhead uses pointers to define the boundary of different signals (DS1 or DS3).

Path overhead: This overhead is transported with SONET payload and mainly used for payload transport functions.

Payload pointer: This pointer indicates the starting of SONET synchronous payload envelope.

P-box: The data encryption standard (DES) is a federal encryption algorithm to be used for unclassified, sensitive information and resides at the presentation layer. It is also known as a product cipher, as it evaluates ciphertext in a sequence of transposition and substitutions. The substitution is performed by a P-box (where "P" denotes "permutation"), while substitution is performed by an S-box (where "S" denotes "substitution").

PC communication protocols: The personal computer (PC) is now being used as a very useful tool for data communications across networks. Due to availability of asynchronous serial ports and dial-up modems, the PCs have now become a useful and effective communication device for data communication.

PDTR 9575: OSI routing framework within OSI-RM.

Peak-to-peak voltage/current: The voltage/current is expressed as a sinusoidal waveform. This amplitude represents twice the peak value, as the signal represents maximum value on both the positive and negative side of the waveform.

Peer entity: OSI-RM looks at a computer network as a collection of subprocesses with a specific set of services and functions transformed into each subprocesses (corresponding to layers). Within each subprocess, the services or functions of layers are obtained through entities (segment of software or a process, hardware I/O chips, etc.). Each layer defines active elements as entities, and entities in the same layer of different machines connected via network are known as *peer entities*. The entities in a layer are known as service providers and the implementation of services used by its higher layer is known as *user*.

Peripheral devices: The devices external to the CPU that perform input–output operations and other functions are known as peripheral devices, e.g., disk, I/O devices, memories, etc.

Permanent virtual circuit (PVC): A frame relay network offers services through its two virtual circuits — permanent and switched (ANSI and ITU-T). In PVC, the connection is established permanently upon request and hence there is no need to set up the connection. The virtual circuits can carry different types of traffic.

Phase: This parameter defines the relative angle value of the signal waveform with respect to the x-axis.

Phase alternation by line (PAL): This was adopted in 1967. It is a time code and speed format used in video products. It is primarily used for TV recording. Multisystems converting this standard into other standards such as NTSC or SECAM are available. It has 625 horizontal lines making up the vertical resolution; 50 fields are displayed and interlaced per second, making for a 25 frame-per-second system. An advantage of this system is a more stable and consistent hue (tint). PAL-M is used only in Brazil. It has 525 lines, at 30 frames per second. The countries following PAL include Afghanistan, Algeria, Australia, Austria, Bangladesh, India, Indonesia, Hong Kong, Germany, and New Zealand.

Phase lock loop (PLL): In frequency modulation, the incoming DC level signals (1s and 0s) are converted into two separate AC signals that have different frequency values. When digital signals are converted into AC signals, these signals can be transmitted over the telephone lines over a longer distance. The frequency range for these signals is usually between 300 Hz and 3.4 KHz. This bandwidth is considered as a practical bandwidth for telephone networks. The higher value of frequency represents logic 1, while the lower value of frequency represents logic 0. The FSK modem contains a modulating circuit based on a voltage controlled oscillator (VCO) whose output frequency depends on the voltage at its input. The first DC level represents logic 1, while the second level represents logic 0. One of the most popular methods for FM demodulation is based on phase-locked loop (PLL), which extracts the instantaneous frequency from the modulated signal.

Phase modulation (PM): In this modulation, the phase of carrier signal varies linearly according to the instantaneous value of baseband modulating signal. The instantaneous frequency deviation is proportional to the derivative of the modulating signal.

Phase-shift keying (PSK): One of the three techniques of modulating analog signals for digital data. Here, the modulated signal contains frequency components on each side of the carrier signal frequency, thus giving a range of frequency signals. We usually term this range the *bandwidth* of the modulated signal. In phase shift keying, the phase of the carrier signal is shifted with respect to the modulating signal, and each phase shift represents a different value of data. Depending on the number of bits, the phase shifts can be defined for the modulation of the signals. Within each phase shift, we can define more than one shift within the signal. This technique is immune to noise and more efficient than frequency shift keying.

Photoelectric effect: This mechanism is used to convert an optical signal into an electrical signal. This is also known as *photovoltaic effect*.

Physical layer convergence protocol (PLCP): The IEEE 802.6 standard has discussed two lower layers of OSI-RM as physical and data link There is a direct one-to-one correspondence between the functions and protocols of the physical layers of DQDB and OSI-RM. The physical layer of DQDB defines an additional protocol known as physical layer convergence protocol that offers interfaces and speed to various media for adapting its capabilities during data communication.

Physical layer medium dependent: In FDDI LAN, the medium-based layer offers the same functions as those of the physical layer of OSI-RM.

Physical layer protocols: An interface between terminal and computer (known as data terminal equipment — DTE) and modems, multiplexers, and digital devices such as data service units (known as data circuit-terminating equipment — DCE) is usually known as an *interchange circuit*. This circuit transmits data messages, control signals, and also timing signals. The most widely used standard and connector for a DTE/DCE interface is **EIA-232-D**. Other standards include **RS-422A, RS-423A, RS-449**.

Physical layer services: Provides electrical, mechanical, procedural, and signaling functions of transmission media. At the same time, it offers services to the data link layer that include fetching the frame and sending the received frame to it. Primitives and parameters usually define the services offered by any layer in OSI-RM.

Physical unit (PU): The machine may represent terminals, hosts, controller nodes, etc., depending on the characteristics of the physical unit (PU). The PUs are the interfaces and are mainly concerned with the controlling of functions across the interface. Each node of SNA (host, communication controller, terminal, etc.) has a function of PU, which is a part of the function manager layer of that node. The PU services reside in every node of the network, as they define the function of the set of resources attached to that node.

Piggybacking: Generally, in all of the data link protocols, the frames have few bytes in the header reserved for the acknowledgment. When a frame is transmitted, the transmitter waits for the acknowledgment. The receiver, after receiving the frame, has to send the acknowledgment by informing the transmitting site of the frame number it has received. This type of arrangement requires many acknowledgment frames over the channels, and as such the throughput of the channel may be reduced. If two stations are connected and are communicating with each other, then instead of sending acknowledgment frames separately, the receiver may add a few bytes at the end of the data frame it wants to send to the receiving site. Thus, the frame going from the receiving side to the transmitting site has data for the other site and the acknowledgment of the data frames being sent from that site.

Pipelining: With pure stop-and-wait protocols, the transmitting site, after sending a frame to the receiving site, waits for an acknowledgment. Once the acknowledgment has arrived, it sends the next frame. The concept of *windows* allows transmission of multiple frames over the channels at any time. These frames are numbered and are defined within the maximum size of the windows mutually agreed upon between sender and receiver. The transmission of multiple frames in this way over the channel gives the concept of *pipelining*. The sender maintains a table of all the frames being transmitted and also the frames for which acknowledgments have been received.

Plaintext: The encryption and decryption functions are used to provide integrity of the data and source. The data to be sent is encrypted on the sending side and is decrypted on the receiving side, and is known as plaintext.

Plait: This topology is based on ring topology, and the main emphasis is given to the reliability. Here, if a central node fails, there is an alternate node which can be used for this node or any other node in the network. Similarly, if any link is failed, there is another link which will be used for this failed link. Further, if one of the concentrators goes down, it has a little effect on the network. The plait topology offers cost-effective, variable capacity and supports a number of media. It has been used in IBM token ring and FDDI LANs. One of the main drawbacks with this topology is that the latency time tends to increase with the load.

Plesiochronous digital hierarchy (PDH): The commonly used digital hierarchy for a synchronous transmission network before the advent of synchronous digital hierarchy. PDH was popular in Europe and is still being used as E1, E2, and E3 standards.

Point–to-point and point-to-multi-point switched connection: A point-to-point connection in an ATM network is defined as a collection of various ATM virtual circuits or virtual paths. It supports point-to-point connection between a parent node and a created child node. The parent node can create a number of child nodes. The child nodes cannot communicate directly; instead, they have to go through the parent node. The point-to-multi-point connection is defined between two end points where child nodes can communicate directly.

Point-to-point topology: The simplest LAN topology, which consists of a host, a communication link (transmission media), and a terminal or PC or even a workstation. Any type of terminal (batch or interactive) can be used. This topology is very simple and can easily be implemented; further, it supports different types of hosts (mainframe, minicomputer, microcomputer, etc.). Each terminal communicates with the host on the basis of point-to-point transmission.

Poll and select: This technique is basically used in a master–slave configuration of a distributed network system. The primary node in the system polls for a secondary node by asking the the secondary nodes to respond. Based on the responses from secondary, the primary selects one node to communicate with. This technique is based on the concept of *poll and select*.

Positive acknowledgment and retransmission (PAR): During error control, the receiver looks at the frame check sequence (FSC) and, if it finds any error, it discards the frame. It also has a feature of error correcting by using PAR, where the sender will retransmit the same frame again until it is received error-free. This feature is available in most data link protocols, including LAP-B, which is an OSI-oriented protocol that support bit-oriented, synchronous, and full-duplex operations.

Presentation layer: This is sixth layer from the bottom of OSI-RM. It provides interface between the application and session layers of the model. Its main function is to provide the syntax for data representation and represent the application into a form that is acceptable to the network system.

Private automatic exchange (PAX): This exchange offers services to users in an organization, and all the communication on the premises takes place via this private exchange as opposed to a public telephone exchange. In other words, this exchange offers private services to the organization and does not transmit the signals to the public network exchange.

Private automatic branch exchange (PABX): An organization has its own automatic telephone exchange network which is connected to the public telephone network exchange.

Private branch exchange (PBX): This exchange provides the connection to public telephone network systems and is usually located within a building. It may be manual or automatic. All of the users working in that building go through this exchange, which provides switching service over the lines within the building and outside the building.

Private key: One of two keys defined within the asymmetric encryption technique. This key is defined for secured communication and is known to the system administrator only.

Private line network (PLN): This network is also known as a *dedicated network* between two nodes. It offers fixed and known bandwidth in advance. Network security and independence of protocol features are facilitated when connected via a modem. It offers transparency to the data messages. A fixed monthly charge makes it more suitable for continuous, high-volume data, as resources are permanently allocated to the users. The fixed bandwidth and speed of the modem are the major drawbacks with this type of network, but high data rates may be achieved with high-speed modems (9600 baud).

Programmable sharing device (PSD): This unit can be configured to act as a port sharing unit and supports both asynchronous and synchronous devices. It can handle both leased lines and dial-up lines.

Propagation delay: The time taken by the signal to propagate from sender to receiver over the media. It depends on a number of characteristics of the transmission media.

Protocol: In OSI-RM, it is assumed that the lower layer always provides a set of services to its higher layers. Thus, the majority of the protocols that are being developed are based on the concept of the layered approach for OSI-RM and hence are known as *layered protocols*. Specific functions and services are provided through protocols in each layer and are implemented as a set of program segments on computer hardware by both sender and receiver.

Protocol data unit (PDU): Peer protocol layers exchange messages which have commonly understood formats, and such messages are known as protocol data units (PDUs). It also corresponds to the unit of data in m-peer-to-peer protocol. The peer m entities communicate with each other using the services provided by the lower layer.

Proway: The Proway LAN is required to provide high availability, high data integrity, and proper operations of the systems in the presence of electromagnetic interference and other sources of noise (due to voltage differences in ground voltages, etc.). It must provide high information transfer rate among the processes within a geographical distance. The standard supports 100 devices/equipment over a distance of 2 Km, and the devices are connected via a dual bus medium.

PS/2 LAN interface: A collection of the products defined for different types of networks (baseband, broadband, token ring, etc.). IBM baseband PC network adapter A allows the PC to interface and communicate with baseband LANs. IBM broadband PC network adapter II/A allows PCs to interface and communicate with broadband LANs. IBM token ring network adapter A offers interface for PC into IBM token ring, and the adapter supports 4 Mbps and 16 Mbps (with a 9-pin connector).

Public data network (PDN): A packet switched network that provides public data communication value address services to the public.

Public key: One of two keys defined within the asymmetric encryption technique. As the name suggests, this key is in the public domain and hence is known. When this key is used with private key, then it becomes difficult for intruders to extract data from the encrypted code.

Pulse amplitude modulation (PAM): This modulation is obtained by sampling a modulating signal through a series of periodic pulses of frequency F_c and duration t and defining the instantaneous value of amplitude of this signal for each sample. Here the value of the sample is used to change the amplitude of the pulse carrier. PAM is not useful for telecommunication, as it is sensitive to noise, interference, atmospheric conditions, cross talk (just like in AM), and propagation conditions on different routes through PSTNs.

Pulse code modulation (PCM): Nyquist's sampling theorem is used in this type of modulation, where the signal is sampled at regular intervals of time at a rate which is greater than twice the original frequency of the modulating signal. These samples will include the entire information of the modulating signal. On the receiving side, low-pass filters may be used to construct the original signal from these received modulated samples. The digital signal may be derived from digitizing the analog signal, or it may represent information already in digital form (obtained from computer or digital data device).

Pulse duration modulation (PDM): This modulation is obtained by varying the pulse duration t according to the modulating signal while keeping the amplitude constant, as shown in modulated PDM signal. Various features include immunity to noise, no amplitude linearity requirements, etc.

Pulse frequency modulation (PFM): Similar to phase modulation; the instantaneous frequency for two consecutive modulated pulses is a linear function of modulating signals as shown in modulated PFM signal. This modulation cannot be used for the construction of time division multiplexing (TDM). The features offered by this technique include immunity to noise, no amplitude linearity requirements, etc.

Pulse modulation: In this modulation scheme, a carrier is defined as a series of pulses. Here, the characteristics of a carrier signal are changed in accordance with the instantaneous value of the baseband modulating signal. The characteristics of pulses are discrete in nature, and as such the modulated signal will have the following four types of modulation schemes: pulse amplitude modulation (PAM), pulse duration modulation (PDM), pulse position modulation (PPM), and pulse frequency modulation (PFM).

Pulse position modulation (PPM): In this modulation, trailing edges of pulses represent the information and hence it is transmitted through trailing edges only, as shown in modulated PPM signal. The duration of the unmodulated pulses in this modulation represents simply a power, which is not transmitted. Thus, PPM provides an optimum transmission, as only pulses of constant amplitude with very short duration are transmitted. This requires lower power and hence the modulation circuit is simple and inexpensive. It offers features such as immunity to noise, no amplitude linearity requirements, etc.

Pulse width modulation (PWM): In this modulation method, the width of the pulse is used to represent the amplitude of the signal.

Quadrature amplitude modulation: This modulation technique is a form of digital modulation and is similar to QPSK.

Quadrature phase shift keying (QPSK): In this digital modulation scheme, the input data is represented by four states of phase corresponding to four values.

Quality of service (QOS): The networks can be used by a number of users for accessing shareable resources, programs, files, etc., over different platforms under different environments. The networks are expected to meet certain performance requirements so that both users and service providers can agree on a minimum and acceptable performance of the networks. In addition to these performance parameters, we need to define parameters for quality of service (QOS) to be obtained from the network.

Quantization: This mechanism measures the resolution of the sampled data system used in pulse-code-modulation technique.

Queue packet synchronous exchange (QPSX): The fiber distributed data interface (FDDI) and distributed queue dual bus (DQDB) networks use fiber transmission media for data communication. The main applications of these high-speed networks are LAN interconnectivity and transmission of different types of traffic such as voice (isochronous), data, images, etc. FDDI is used as a backbone in most universities and organizations, and hence can be considered a high-speed private LAN. DQDB, on the other hand, was intended to be used with IEEE 802.6 MANs, and thus can be considered a public network. DQDB is based on the concept presented in the implementation of QPSX and, realizing that, IEEE 802.6 committee renamed QPSX as DQDB.

Radian: This unit expresses the angle with which the radius arc of the circle is rotated. If a vector makes one complete rotation, it covers a distance of 2*P* radians. The circumference of a circle is thus given by 2*P* radians. It represents the unit of angular frequency.

Radio frequency: The frequency spectrum that lies between the voice-grade frequency and the lower end of the optical frequency range of 300 GHz is known as the *radio frequency band*. This band covers the entire voice grade and other forms of signal communications.

Radio waves: Electromagnetic waves carry signal information between stations. There are different forms of radio waves: ground wave, space wave, satellite wave, microwave, scattered wave, etc. Each form is transmitted at different heights with respect to the ground and is characterized by its transmission pattern, frequency range, etc.

Recognized private operating agency (RPOA): Within UNO, the International Telecommunication Union (ITU) has become a dedicated agency to provide support and guidance to three international standards organizations: IEEE, CCITT, and ISO. CCITT defined various classes of membership and individual countries are represented within CCITT via different agencies. For example, most European countries are represented by Post Telegraph and Telephone (PTT), while the U.S. is represented by the State Department (a voting member). At a lower level of membership, the U.S. also has formulated Regional Private Operating Agencies (RPOAs) within CCITT that are usually represented by Regional Bell Operating Companies (RBOCs).

Reconfigurability: This feature or behavior of the network will offer an alternate network environment in the event of failed network components (failed node, failed cable). The network is expected to provide the acceptable behavior (if not exactly the same) for any added or new requirements that occur due to these changes. The topology of the network has a great effect on the stability, reliability, and efficiency of the network.

Reference model (OSI): An international standard, open system interconnection (OSI), defined a layered network architecture for providing communication between two computers. Since this architecture has seven layers, it is also sometimes known as the *seven-layer model*. The architecture does not define any protocol or procedures but allows designers to develop protocols or procedures for each of the layers separately and independently.

Reference points: CCITT I.411 defined a standard configuration for 64 Kbps ISDN known as user–network interface (UNI) which supports integrated voice and data communications over the same channel. Two standard interfaces — basic rate interface (BRI) and primary rate interface (PRI) — have been defined. The functional groups defined in ISDN include network termination 1 and 2 (NT1 and NT2), terminal equipment

(TE), and terminal adapter (TA). These functional groups are interfaced at reference points such as terminal (T), S, U, etc.

Reflector element: This element of an antenna reflects the signal back into the active element at the receiving side.

Refraction: In this technique, the signal wave changes direction due to a bending effect.

Regional Bell Holding Company (RBHC): Each country has its own telephone network that is managed either by the government or by a duly recognized company. The telephone companies typically offer services for voice communications. In the U.S., the services of telephone network to the customers are provided by seven Regional Bell Holding Companies (RBHCs), which control various long-distance carriers such as AT&T, MCI, Sprint, etc. The main duties of RBHCs are to install and operate under about 165 local access and transport areas (LATAs), which are usually located in metropolitan areas. Inter-LATA services are provided by RBHCs, and they provide equal access inter-LATA service to all long-distance carriers.

Relay (Internet) chat: An extended version of Internet talk service in the sense that more than two users can participate in the conversation. It may be considered just like a telephone conference call among users sitting miles apart and talking to each other. It may be used either for public discussions on various topics (open to all users) or for selected people in a conference on any topic of their choice.

Reliability: A performance measure of the network where it will provide a minimum level of acceptable service over a period of time. This is affected by node or link connectivity and is directly proportional to the connectivity. Higher reliability can be achieved by considering higher node or link connectivity. The reliability is directly proportional to the redundancy; we have to make a compromise between reliability and redundancy, and this usually depends on a number of parameters such as communication traffic and cost, overhead, data rate, transmission codes, etc.

Remote bridge: An internetworking device that offers interconnectivity between remote LANs and WANs and is expected to possess high throughput.

Repeater: Electrical signals (analog or digital) lose their amplitude as they travel through the communication channel over the link or media due to the resistance offered by them. Thus, the behavior of signals gets changed, and this change in behaviors is known as *attenuation*. In order to achieve unattenuated signal over a long distance, amplifiers are used for analog signals. For digital signals, a repeater is used which regenerates the logic 1s and 0s and retransmits this logic over the media. Repeaters can be unidirectional or bi-directional, while amplifiers work only in unidirectional node.

Residual error rate: The frame relay networks are required to provide the requested quality of service (throughputs, capacity, bandwidth, etc.) during their operation. The requested services may be permanent or switched virtual circuits. There are a number of measurement parameters which can be used to ensure the quality of service and requested throughput. One of these is *residual error rate*, which is the ratio of 1 minus the total number of service data units delivered correctly to the total number of service data unit offered.

Ring: In this LAN topology, all of the nodes are connected to a circular ring link or closed loop via repeaters, defining point-to-point configuration between different pairs of nodes. Each node connected to the ring has adjacent nodes at its right and left, and as such each repeater works for two links at any time. This topology provides unidirection transmission of data, i.e., data are transmitted in one direction only around the ring. The repeaters accept the data from one node bit by bit and pass it on to the next connected node. The repeaters do not define any buffer, and as such the data can be received and transmitted at a faster speed. A bi-directional loop structure can also be defined where one loop or ring carries the data information in one direction, while the other loop or ring carries the data in the opposite direction.

Ring station interface (RSI): In the case of ring-based LANs, the frame (containing token and information) circulates around the ring until it is received by the destination node, after which it still circulates around the ring. The source node has to remove this frame from the ring and issue a new token to the ring so that another node can use the network for data transmission. The token provides a provision of assigning a priority to the stations (via the access control field defined in the token frame) which are attached with a ring station interface (RSI) or network access unit (NAU). These RSIs or NAUs perform various functions, e.g., copying the message data packets onto the buffer, making a request for token, requesting priority, changing the status of the token to indicate the acknowledgment, etc., based on the round-robin technique.

Router: The router provides interconnection between two different networks. This internetworking unit (device) is compatible to the lower three layers. Unlike bridges, it supports at least three physical links (in general, it supports other links too). A message frame transmitted over the LAN goes to all the nodes. Each node determines, by looking at the address defined in the frame, if it belongs to it. If the frame belongs to it, the router accepts this frame. The router defines the route for a frame to be transmitted to the destination.

Routing: The establishment of a virtual path over various switching nodes between source and destination nodes defines a routing for the transmission of a packet or message. For a given network topology, the virtual paths or routes are calculated between every pair of nodes within the network. These routes (either a complete set or a partial set) are stored in the routing tables at each of the nodes of the network.

RS-232-D: The most widely used physical layer (OSI-RM) standard and connector for DTE/DCE interface; approved by EIA in 1986. This conforms to other international standards such as CCITT V.24, V.28, and ISO 2110. The RS-232 standard for DTE/DCE interface works for a limited distance of 15 m at a maximum speed of 19.2 Kbps.

SAR-PDU: The AAL service data unit (SDU) containing the user's data is accepted by the CS, which defines its own PDU by adding its CS-PDU header, CS-PDU trailer, and some padding octets. This CS-PDU is sent to the SAR sublayer, which defines its own PDU in the same way as CS. It adds its own SAR-PDU header and SAR-PDU trailer, and sends it to the ATM layer. The CS-PDU in general has a variable length. The SAR-PDU includes up to 44 octets of CS-PDU in its PDU and is defined by the SAR sublayer. The SAR-PDU also defines an indication about the number of CS-PDU octets included by the SAR payload indicator.

Satellite communication: A microwave link beyond the atmosphere, usually assigned a pair of frequencies at 4 and 6 GHz, 11 and 14 GHz, and 20 to 30 GHz. Usually, satellites at low altitude are known as *moving satellites*, which rotate rapidly with respect to a satellite earth station point on the earth's surface. On other sides, at higher altitudes (22,100 to 22,500 miles), the satellites rotate at the same speed as the earth rotates in 24 hours, and hence these satellites are fixed with respect to the satellite earth station on earth's surface. Such satellites are known as *geostationary satellites*.

Satellite frequency bands: There are different classes of frequency bands for satellite links, such as C band (6 and 4 GHz), Ku band (11/12/14 GHz), and K band (30/20 GHz). Each country is assigned a frequency band and the type of geostationary satellite they can use by United Nations International Telecommunications Union (ITU). For example, the U.S. Federal Communications Commission (FCC) uses C band for commercial communication and has divided this band into two distinct microwave frequencies: 3.7–4.2 GHz for downward microwaves and 5.925–6.425 GHz for upward waves. Currently, there are about 1500 satellites rotating in the earth's orbit and providing various services to subscribers.

Satellite link: A large number of voice-grade channels placed in microwave frequency bands can be accommodated on a satellite station (at a distance of 22,300 miles above the earth rotating at a speed of 7000 miles per hour). Due to its ability to cover a larger geographical area, it is used for broadcast transmission of analog signals, data, and also video signals. These satellites are known as satellites in synchronous orbit. At this orbit, the satellites cover large geographical areas and transmit signals at 6 and 4 GHz.

Secondary ring: The FDDI uses a pair of rings topology. The data usually flows in the primary ring in one direction. In the event this ring fails, the data is sent through the backup ring in the opposite direction. This backup ring is known as a *secondary ring*.

Segment: The user data part of the payload in DQDB.

Segmentation and reassembly sublayer (SAR): The layer higher to the ATM layer is known as the ATM adaptation layer (AAL). This layer provides an interface between ATM and higher layers. The AAL layer is divided into two sublayers — segmentation and reassembly (SAR) and convergence sublayer (CS). The SAR sublayer segments the PDUs of higher layers into fixed size of 48 octets (information fields of ATM cell) on the transmitting side.

Serial configuration: In this channel configuration, there is only one line between two nodes, and the transmission signal bits are transmitted bit by bit (as a function of time) or sequentially over the transmission media. The signaling rate is controlled by baud (rate of signaling per second on the circuit). The reciprocal of baud gives the time period of signaling for the transmission.

Serial line Internet protocol (SLIP): In situations where the user's computer is being used as an Internet host for obtaining a full Internet connection over telephone lines, we have to install an Internet application program (IAP) known as serial line internet protocol (SLIP) or point-to-point protocol (PPP). The user's computer can use the TCP/IP facilities after the telephone connection between the user's computer and the Internet host has been established. Now the computer can read the Internet host and can be identified by its electronic address. The users can also be connected to service providers such as UUNET (UNIX–UNIX Network) or PSI (Performance Systems International), which usually charge a nominal fee (a few hundred dollars) per minute for offering SLIP or PPP services.

Serial transmission: In this mode of transmission, message information is transmitted bit by bit over the link, and the transmission speed of the transmitting site depends on the signaling speed. Both types of digital signals (data and control) are transmitted serially. The control signals are used for providing synchronization between two sites. The signaling rate is the rate at which signaling per second is selected for the communication device and is usually expressed in baud. The time period of one signaling rate can be defined as a reciprocal of the baud.

Service access point (SAP): Each peer-to-peer protocol (defined for each layer separately in OSI-RM) carries information defined in terms of data units known as protocol data units (PDUs). The communication between layers takes place through a hardware address or port known as a service access point (SAP).

Service data unit (SDU): Peer protocol layers exchange messages which have commonly understood formats and such messages are known as protocol data units (PDUs). They also correspond to the unit of data in *m*-peer-to-peer protocol. The service data unit (SDU) associated with layer primitives is delivered to the higher layer.

SG COM: Study groups dealing with computers.

SG COM I: Definition and operations of telegraph, data transmission, etc.

SG COM II: Operation of telephone network and ISDN.
SG COM III: General tariff principles.
SG COM IV: Transmission of international lines and circuits, etc.
SG COM V: Protection against electromagnetic origin.
SG COM VI: None.
SG COM VIII: Terminal equipment for telemetric services.
SG COM IX: Telegraph networks and terminal equipment.
SG COM X: None.
SG COM XI: ISDN and telephone network switching and signaling.
SG COM XII: Transmission performance of telephone networks and terminals.
SG COM XV: Transmission systems.
SG COM XVII: Data transmission over the telephone networks.
SG COM XVIII: Digital networks including ISDN.

Shielded twisted pair: This type of twisted pair of wires includes extra insulation via jacket shielding to reduce the external interference. These wires are used for long-distance communication and are more expensive than unshielded twisted pair.

Signal distortion: Noises, distances, attenuation, etc., generally influence electrical signals transmitted over transmission media. Due to atmospheric noises, the signal-to-noise ratio may become smaller at the receiving side, while signals may become weak as they travel over a transmission line over a long distance, or if the communication media pose significant attenuation (due to capacitive and inductive effects, resistance, etc.) to the signals. Under all of these cases, signals received at the receiving side may not provide us with the exact message information transmitted. There are various types of distortions and unwanted signals (called noises) which contribute to signal distortion: attenuation, high frequency, and delay distortion.

Signaling: Control signaling provides a means for managing and controlling various activities in the networks. It is being exchanged between various switching elements and also between switches and networks. Every aspect of network behavior is controlled and managed by these signals, also known as *signaling functions*. Various signaling techniques being used in circuit-switching networks are in-channel in-band, in-channel out-of-band, and common channel. With circuit switching, the popular control signaling has been in-channel.

Signaling System No. 7 (SS7): A new signaling scheme based on digital signals recommended by CCITT and widely accepted and implemented by various telephone companies. This scheme was the first signaling scheme used in ISDNs and has already been accepted as a standard signaling system specification. Many telephone companies have already decided to switch to SS7. AT&T and Telecom Canada have implemented message transfer part (MTP), which is ISDN user part (ISDNUP), and signaling connection control part (SCCP) based on the SS7 signaling scheme.

Signal-to-noise (S/N): In order to receive an analog modulated signal, it needs to be amplified at a regular distance. The amplifier raises the amplitude to the level determined by the amplification factor and it amplifies the unidirectional signal and retransmits it over the link. If any noise or unwanted signal is present in the modulated signal, it will also be amplified by the same amplification factor. Thus, the quality received depends on the signal-to-noise ratio, which sometimes becomes a main design consideration in analog transmission.

Simple mail transfer protocol (SMTP): This protocol defined by DoD allows Internet users to send e-mail directly from one host to another host on TCP connection. The sender and receiver SMTP establish a transport connection on TCP and then transfer the data on IP.

Simple network management protocol (SNMP): WANs offer network services to users of a vast geographical area (thousands of miles) and make use of different vendor network products. The service providers have to manage the requests from different users and ensure that the network services are available to the users. Similarly, the system administrator of a WAN has to ensure the quality and availability of services to its service providers. With the introduction of simple network management protocols, it becomes very easy for the system administrators and services providers to manage from a remote location and also generate different statistical reports.

Single attachment station: In an FDDI ring, we use two rings — primary and secondary. Usually the data flows through the primary ring in one direction. In the event of ring failure, the secondary ring is used to carry the data in the opposite direction. Usually the nodes have access to both rings via attachment. Some stations that have access to only the primary ring via single attachment are known as *single attachment stations*.

Single mode fiber: Optical fiber is used in long-distance telecommunications, LANs, metropolitan trunks, and also in military applications. Two different types of light sources are being used in fiber optic systems — light emitting diode (LED) and injection laser diode (ILD). It supports two modes of propagation — single- and multimode. In the case of single mode, the radius of the core is reduced which will pass a single angle or a few angles. This will reduce the number of reflections, as fewer angles are there. It offers better performance than multi-mode, where multiple propagation paths with different lengths are created. These angle signals will travel through fiber, and they will also spread out in time.

Single sideband (SSB): In amplitude modulation, two sidebands are generated around a carrier frequency. One has a frequency lower than carrier frequency and is known as the lower sideband, while the one above it is known as the upper sideband. We can send either one sideband or both sidebands. Single sideband transmission will carry only one sideband.

Single sideband AM (SSB-AM): In single-sideband suppressed carrier (SSBSC-AM) or in short SSB-AM, only one sideband is transmitted and the other sideband and carrier are suppressed. By sending only one sideband, half the bandwidth is needed and also less power is required for the transmission. SSB modulation is used in telephony and also as an intermediate stage in constructing frequency division multiplexing of telephone channels before they can be sent using radio transmission (microwave link, satellite link), which uses another method for modulation.

Skin effect: With the increase in frequency, the flow of current in a metal conductor develops a tendency of going toward the shallow zone on the surface, thus causing an increase in effective resistance of wire. This is known as *skin effect*. In order to reduce the skin effect, the coaxial cable is defined as a skin with a center conductor and is suitable for high frequency.

Slotted ring: Another version of a ring-based system where time slots of fixed size or lengths are defined. These slots are represented as carriers for the data frames around the ring. Each slot has its own status indicator and, in the case of a free slot, the data frames are copied into the carriers of fixed sizes. If the length of data is greater than the length of the carrier, it will be partitioned into packets of the size of the carriers (slots) and transmitted. For each slot, a bit can be set/reset to indicate the acknowledgment and the source node must make it free for further transfer of data frames.

SMDS interface protocol: This protocol defines the functions of the lower three layers of OSI-RM that include the following functions of the SMDS subscriber-to-network interface: user frame, addressing, error control, and overall transport.

S/N ratio: This ratio represents the strength of signal and noise. Both signals and noise should have their strength expressed in the same unit. It is usually expressed in decibels (db).

Source routing: In this routing technique, the source address defined in the packet is used to route the packet through switching nodes of the network.

Specific application service element (SASE): Different classes of applications, such as terminal handling, file handling, and job transfer and manipulation, have been standardized and are usually termed as specific application service elements (SASEs). The terminal handling application is mainly concerned with basic graphics, images, etc.

Spectrum: A range of wavelength that is usually defined in radio frequency signals.

Spectrum analyzer: This device shows the frequency representation of the signal. Here the signal is shown on the horizontal axis of frequency representation.

Spread spectrum: This is defined as a modulation technique that allows the bandwidth of output signals to be spread over a larger frequency range.

Standards organization: Standards organizations are classified as government representatives, voluntary, non-treaty, and non-profit organizations. A standards organization is a hierarchical structure based on a bottom-up model for reviewing and approving the standards. There are two levels of standards organizations: international organizations and national (domestic) organizations.

Star: This LAN topology is defined by a central or common switching node (working as a primary or master) providing a direct connectivity to other nodes (known as secondary or slave). It defines one primary or central node which provides communication between different computers or secondary nodes. Each of these computers or secondary nodes are connected directly to the primary node on a separate transmission link. Each secondary node sends a data packet the primary node, which passes it on to the destination secondary node. The maximum number of hops (the number of paths or routes) between any pair of nodes is two. The failure of any node will not affect the communication networks other than that particular node. The star topology is useful for transmitting digital data switches and PBX switches. Thus, the star topology supports both digital or voice/digital services in a variety of applications.

Station management: The management operations of this layer are defined by the entity of this layer.

Statistical multiplexer: This type of multiplexer is in fact an improvement over the traditional TDM. In TDM, a time slot is preassigned to every node. The scanner goes through all of the nodes, even if they may not have any data to transmit. This wastes the capacity of the link. In a statistical multiplexer, the scanner will skip the node that does not have any data to send, thus reducing the capacity wastage.

Statistical time division multiplexing (STDM): In conventional TDM, each port is assigned a time slot, and if there is no information on a particular port, the TDM waits for that preassigned time at that port before going to the next port. But, if we can allocate the time slots to the ports in a dynamic mode, i.e., if any particular port is empty, the scanning process will slip and go to the next port and so on. Therefore, we can prevent the wastage of bandwidth and also reduce waiting time. This is the working principle of statistical time division multiplexing (STDM).

Stop-and wait protocol: This protocol is used for flow control in the network. The sender node, after sending the packet, has to wait for acknowledgment before it can send the next packet.

Strongly connected (fully connected): In this LAN topology, all the nodes are connected to each other by a separate link. Since each node is connected to each other directly, the number of hops (a simple path or route between two nodes) is always one between any pair of the nodes. This topology sometimes is also known as "crossbar" and has been very popular in various networking strategies for fast response requirements. It is very fast, and as such it offers very low (negligible) queue delays.

STS synchronous payload envelope (SPE): The size of an STS-1 SONET (standard) frame is 51.84 Mbps. This frame has 9 rows and 90 columns. There are 810 bytes in one frame, and the frames are generated at the rate of 125 μs. The transport overhead is

27 bytes, while payload capacity is 783 bytes. Thus the usable payload is given by 783 – 27 bytes. The envelope payload is defined as a usable bandwidth within the frame that carries the STS SPE. Combing several STSs can increase the bandwidth of the payload. The start of STS SPE can be defined anywhere within the STS frame.

Subnetwork: The Internet is a collection of networks that appears to be one network. All of the networks so connected are referred to as *subnetworks*.

Subscriber-to-network interface: This defines the service access point into the underlying network or MAN in DQDB networks.

Sub-split: The guard band is usually around 40 MHz in a broadband system, and sub-split offers four channels for a return path occupying the least amount of spectrum in the frequency range.

Switched multimegabit data service (SMDS): This network (defined by Bellcore) is based on the concept of fast packet switching and is used in LAN interconnectivity for transferring the data traffic. The network provides broadband services for LANs, intra-LATA, and WANs using connectionless mode of operation. The specifications related to interfaces, switching, etc., of SMDS are defined in the specification network interface (SNI). Three types of interfaces are interswitching system interface, intercarrier interface, and operation system interface.

Switched network: Interconnection between computers allows the computers to share data, and interact with others, and as such a computer network so defined is also known as a *switched network*. In these networks (e.g., telephone networks, packet-switched networks, etc.), switching for routing the data packets performs functions such as providing the routing information between source and destination across the network and the link utilization techniques which can transmit the traffic equally (depending on the capacity of the links) on all links.

Switched services: The communication channels are used to access the services of networks in switched and leased forms. Switched services are offered on public switched networks and are sometimes also known as *dial-up services*. They are controlled and managed by AT&T, Data-Phone system, and Western Union broadband channel.

Symmetric encryption: There are two ways for providing data and source integrity through encryption — asymmetric and symmetric. In asymmetric encryption, private and public keys are used by encryption and decryption, while in the case of symmetric encryption, the same key is used for both encryption and decryption.

Synchronization in transmission: One of the main problems with the transmission configurations (balanced or unbalanced) is the identification of characters being sent, and this problem becomes more serious in the case of serial configurations, as the bits are being sent continuously. On the receiving side, the node has to identify the bits for each of the transmitted characters. Further, the receiving node must be informed about the transmission of data from the transmitting node. These nodes have to use some synchronizing bits between the characters, not only to identify the boundary of each character, but also to provide synchronization between them.

Synchronous communication: This type of communication transmits character symbols over the communication link. We have to provide a synchronization between transmitting and receiving sites whereby the receiver is informed of the start and end of the information, and for this reason some control characters are always sent along with the character message information. The field corresponding to these control characters is defined as idle (IDLE) or synchronization (SYNC), which typically contains unique control characters or bit patterns and is attached before the message information.

Synchronous data link control (SDLC): This protocol for the data link layer of OSI-RM was defined by IBM and is based on bit-oriented protocol. It is interesting to note that most of the protocols for the data link layer were derived from the most popular protocol — high-level data link protocol (HDLC) — in one way or the other. The SDLC protocol was also derived from HDLC.

Synchronous digital hierarchy (SDH): Two popular interface data rates for B-ISDN based on ATM have been recommended — 155.520 Mbps and 622.080 Mbps — and these data bit rates are on the lower side of bandwidths defined by the CCITT G.700 recommendation, synchronous digital hierarchy (SDH). The SDH in fact represents a bandwidth range for transmission hierarchy. This hierarchy is based on the Synchronous Optical Network (SONET) concept adopted in North America. The main objectives of SDH are to provide transmission flexibility for adding or deleting capabilities for multiplexed signals (in contrast to plesiochronous digital hierarchy — PDH), support operation and maintenance applications of existing networks with little overhead, and support the standardization process in optical transmission systems. The SDH-based interfaces of 155.520 Mbps and 622.080 Mbps have different frame formats.

Synchronous line driver: Uses the twisted pair line or Telco line, which can be leased from the telephone company and supports RS-232/V.24 interfaces. The general configuration of these line drivers includes mainframe, front-end processor (FEP) for point-to-point or multi-point models. It offers higher distance than asynchronous line drivers.

Synchronous multiplexing (STM): SONET and SDH optical standards provide asynchronous transmission hierarchy signals. The DS series (SONET) is used in North America, while the E series (PDH/SDH) is used in Europe. The earlier digital hierarchy was defined as plesiochronous digital hierarchy (PDH), which support synchronous transmission. Another standard, STM-n, offers a much higher bit rate, and simple multiplexing and demultiplexing methods can obtain the multiplexing of PDH signals.

Synchronous optical network: This standard for fiber is very popular in North America, and another standard known as synchronous digital hierarchy (SDH) is popular in Europe. The high-speed networks based on B-ISDN, ATM, and frame relay all provide LAN interconnectivity for transmitting voice, video, and data over the network. The traffic coming from any LAN going to another LAN goes through these high-speed networks. All of these networks are based on optical fiber communications. Each of these networks provide different protocols for LAN interconnectivity. Some of the transmission media interfaces are T1 (1.544 Mbps), E1 (2.048 Mbps), T3 (45 Mbps), E3 (34 Mbps), E4 (140 Mbps), SONET STS-3c/SDH STM1 (155 Mbps), SONET STS 12c/SDH STM-4 (622 Mbps), and FDDI (100 Mbps).

Synchronous terminals: These terminals have a buffer capability and can store blocks of data (constituting more than one character). When there is no digital signal, a flag bit pattern of 01111110 is continuously transmitted from the terminals to the mainframe. When a character is transmitted from the keyboard (by hitting a particular key), the synchronizing bits are also transmitted, and the end of the block of data is recognized by the return key. Since the buffer stores a block of data, these terminals are also known as block mode terminals (BMTs), e.g., IBM 3270, etc.

Synchronous time-division multiplexing (STDM): In time-division multiplexing, various devices are assigned time slots, during which signals can be sent. In synchronous time division multiplexing, different I/O channels, devices, and others share the time slots on a fixed and predefined time basis.

Synchronous transfer mode (STM): CCITT study group XVIII focuses on defining standards for B-ISDN and has defined a transfer mode (TM) for switching and multiplexing operations put together. The transfer mode is important to both switching and transmission aspects of the network interface. There exist two types of transfer modes: synchronous transfer mode (STM) and asynchronous transfer mode (ATM). In STM, the network uses synchronous time division multiplexing (STDM) and circuit-switching technique for data communication on demand.

Synchronous transport module level: The SDH standard defines a base signal of STM-1 and byte interleaving around STM-1 will define N STM-1 signals.

Synchronous transport signal 1 (STS-1): A new multiplexing hierarchy for a base signal of 51.84 Mbps of the SONET standard. This standard defines an electrical signal equivalent to the optical signal defined by another standard as optical carrier (OC). The mapping between STS and OC is as follows: STS-1/OC-1 (51.84 Mbps), STS-3/OC-3 (155.52 Mbps), STS-12/OC-12 (622.08 Mbps), STS-24/OC-24 (1.244 Gbps), STS-48/OC-48 (2.488 Gbps), and STS-192/OC-192 (9.6 Gbps).

System network architecture (SNA): IBM has defined its own LAN (non-OSI) for interconnecting PCs and also SNA networks, known as system network architecture (SNA). Although SNA is a non-OSI architecture, its upper layers perform the functions of the application layer of OSI-RM. In this architecture, users can define their private networks on a host or even on subnetworks. The SNA architecture is a complete model of layers, and each layer has specific specifications and implementation guidelines.

System service control points (SSCP): The logical unit (LU) in an SNA LAN can be considered a port through which end users communicate with each other. Interconnecting LUs perform the network operations, including transportation. The SNA network contains different types of units/controls, which are collectively known as a network addressable unit (NAU). These units are physical units (PUs), logical units (LUs), and system service control points (SSCPs). The SSCP is mainly responsible for the management and control of the network. It resides in a host computer. There is only one SSCP in a hierarchy and it offers control over the network's operation.

Systeme Electronique Couleur Avec Memoire (SECAM): This standard was defined by France in 1967 and has 625 lines and 25 frames. Multisystems offering matching with other standards such as NTSC or PAL are available. Some of the countries using this standard include Albania, Egypt, France, Poland, Romania, and Saudi Arabia. Some of these countries also use PAL systems.

T1: Integrated services digital network (ISDN) became the first network to provide integrated services of data and video. One of the popular applications of ISDN has been videoconferencing that has been used for distance learning, face-to-face communications, on-line departmental discussions and meetings, etc. The ISDN offers a data rate of 1.544 Mbps via different channels: B (64 Kbps), D (16 or 64 Kbps), and H (384, 1536, and 1920 Kbps). With the advent of broadband services based on ISDN, ISDN is now known as N-ISDN. A network offering services at data rates above 1.544 Mbps is known as B-ISDN. The primary rate of ISDN is defined. The ISDN standard is defined by the CCITT I-recommendation series. The primary access must allow users to request higher capacities in applications such as LANs, digital PBX, and so on. The standards for primary access are T1 (1.544 Mbps) using DS1 format in North America, Canada, and Japan, and E1 (2.048 Mbps) in Europe.

T1 carrier: The standard multiplexing digital hierarchy T1 is used in North America, Canada, and Japan. This supports 24 channels of voice grade and offers a data rate of 1.544 Mbps over T1 carrier or trunk. Its equivalent carrier in Europe is E1 (2.054 Mbps) carrying 30 voice-grade channels.

T3: Usually represent DS3 signal operating at 44.736 Mbps.

Talk: A very useful resource service on Internet Talk (client application program) being offered to the users. Two users may be thousands of miles apart, but can send and receive messages back and forth simultaneously (as if they are sitting in front of each other and talking). These two users communicate with each other by typing the message at the same time, and this does not affect each other's typing of messages. The Internet Talk program partitions the screen into two sectors and the user can see the typed messages from the remote user on the top (or lower) sector while the local user's message will be displayed on the lower (or top) sector.

Tap: A device that provides an interface to the user to get data from the communication system link. This can be considered similar to a port and is also used for sending the data to the communication system.

Target token rotation time (TTRT): This represents a predefined word value that is stored at all the stations. Initially each station stores the same value of TTRT.

TAXI block: The physical layer of B-ISDN protocol supports a variety of physical media. For each medium, a separate convergence layer protocol (similar to a driver) is defined which provides mapping between different media and ATM speeds. The ATM cells are comprised of 53 bytes, of which 48 bytes represent the data while the remaining 5 bytes represent the header. A number of transmission convergence protocols have been defined to support different media with different specifications. Some of these include SMDS physical layer convergence procedure (PLCP) T1 and T3, ETSI PLCP E1 and E3, ATM TC, and TAXI block.

TDM channels cascading: In order to provide communications between remote ports on the host and various branch terminals located at a distance, we typically should require the number of lines for each terminal equal to the number of ports on the host. This number of lines can be reduced to one and still provide communication between them. TDMs can be cascaded. This configuration will provide communication between the host and remote terminals on one pair of lines. It offers various functions such as conversion of speed code and formats of the data information, polling, error control, routing, and many others, depending on a particular type of concentrator under consideration. The cascading of TDMs can reduce the number of telephone lines.

TDM standards: The TDM defines a time slot, and each time slot can carry a digital signal from different sources. The multiplexed TDM channel offers much higher data rates than an individual channel. The total data rate at the output of the multiplexer depends on the number of channels used. The number of channels defined for this type of multiplexer is different for different countries, e.g., North America and Japan define 24 voice channels, while European countries which comply with CCITT define 30 channels. The total data rate of TDM with 24 channels will be given by 1.544 Mbps ($8000 \times 24 \times 8 + 8000$), while 30 channels will offer data rate of 2.048 Mbps ($8000 \times 30 \times 8$). These TDM circuits are also known as E1 links.

Technical Committee (TC): ISO is a non-treaty, voluntary organization and has over 90 members who are the members of national standards organizations of their respective countries. The main objective of this organization is to define/issue various standards and other products to provide exchange of services and other items at the international level. It also offers a framework for interaction among various technical, economical, and scientific activities to achieve its purpose. Various Technical Committees (TCs) have been formed to define standards in each of the specific areas separately. ISO currently has nearly 2000 TCs, each dealing with a specific area/subject. Further, TCs are broken into subcommittees and subgroups that develop the standards for a very specific task. For example, TC 97 (Technical Committee 97) is concerned with the standardization of computers and information processing systems.

Technical Office Protocol (TOP): An attempt (based on factory automation by GM) was made for office automation by Boeing Company. The main objective of defining their own LAN was to support office automation and engineering aspects in real time domain to Boeing 747 fleets. Boeing defined its own LAN and set of protocols for office automation and called this Technical Office Protocol (TOP). The TOP is now under SME. Both GM and Boeing worked together to provide interoperability between their networks.

Telco line bridge: This unit provides bridging between leased lines and the terminals via one modem. This modem connects the bridge to each of the leased lines simultaneously.

Telephone lines or links: Telephone lines were introduced in the late 1890s and the concept of multiplexing was used over a pair of telephone wires to carry more than one voice channel in the 1920s. In the early 1940s, the coaxial cable was used for carrying 600 multiplexed voice channels and 1800 channels with multiple wire pair configuration. The total data rates offered by these 600 channels is about 100 Mbps. The number of channels which can be multiplexed over the same coaxial cable continuously increased with the technology changes in switching and transmission techniques, and hence the data rates were also higher for voice communication. Some popular number of channels over coaxial cable include 1860 voice channels (with data rates of about 600 Mbps), 10,800 voice channels (with data rates of 6 Gbps), and 13,200 voice channels (with data rates of 9 Gbps). Now with the introduction of optical fiber optics in 1990, we can get data rates of over 10 Gbps, with various advantages of fiber over cable such as high bandwidth, low error rate, most reliable, etc. At the same time, another communication link based on wireless transmission was introduced — microwave links — which offer a number of voice channels ranging from 2400 in 1960 to over 42,000 in the 1980s.

Telephone local area network: The telephone LAN is defined for private branch exchange (PBX) or private automatic branch exchange (PABX) and uses a star topology. Both data and voice signals can be transmitted over PBX, which is basically a switching device. PBX offers connectivity similar to that of baseband or broadband LANs, but offers slow data rates compared to them. If we try to compare PBX or PABX with standard LANs, we notice that both provide connectivity to both synchronous and asynchronous signals. In terms of wiring and installation charges, the digital PBX may be cheaper than standard LANs. But LANs will be preferred over digital PBX for the speed of transmission, as standard LANs can offer data rates of 1 to 100 Mbps, as opposed to 56 Kbps or maybe 64 Kbps in the case of PBX.

Telephone network system: Each country has its own telephone network that is managed either by the government or by a duly recognized company. In order to understand the infrastructure of telephone networks, we consider the U.S. The services of telephone network are provided to the customers by seven Regional Bell Holding Companies (RBHCs) which control various long-distance carriers such as AT&T, MCI, Sprint, etc. The main duties of RBHCs are to install and operate under about 165 local access and transport areas (LATAs), which are usually located in metropolitan areas. Inter-LATA services are provided by RBHCs, and they provide equal-access inter-LATA service to all long-distance carriers.

Teletex: This communication system is one-way transmission of signals/data, e.g., a television system.

Teletex services: Similar to videotex in that it offers services such as news summaries, sports news, financial newsletters, travel listings, stock market reports, etc. However, in videotex, users can select the use of services interactively by user interfaces (provided by information providers), while teletex service is one-way communication that we can view but cannot perform any operations on.

Telex communication: In order to transmit information in digital form (e.g., text, pictures, etc.), the services are offered through different forms of communication links. Each link is suited to a particular application and is obtained through appropriate devices; for example, telegraphy (transmission of alphanumeric messages using electrical signals), telex (transmission and routing of alphanumeric text using a teleprinter), teletex (comparable to a word processor), telecopying (facsimile transmission of pictures, text), telecontrol (concerned with telemonitoring, remote control, telemetry, etc.). These communication links provide services to the data communication from the point of view of integrated telecommunication computer-based communication. These are also known as telecomputing or telematics (videotex and teletex).

Telex network: These networks are similar to telephone networks, with only difference being in the devices which can be connected together. In telex, teleprinters are connected using point-to-point communication and can send the typed messages on the telex network (switched type networks). New services which have been introduced in these networks, are store-and-forward capability, automatic conference calls, set up facility, etc. Telex networks provide a speed of 40–60 bps (baud of code), while American TWX networks provide 100 bps (ASCII code). Many countries are connected to telex networks, and there are a few million telex sets around the world. Although they are used by many countries, telex networks are very slow and unreliable and produce very poor-quality outputs. Telex communication is being replaced by teletex communication that offers higher speed and better quality of outputs (using higher quality printers) on word processors. Teletex communication can provide transmission speeds of 2400 bps, 4800 bps, 9600 bps, and so on.

Telex protocol (TP): Due to incompatibility between various word processors (by different vendors), CCITT in 1982 defined an international standard for telex communication. This standard is based on the ISO open system interconnection seven-layered architecture model and includes protocols, definition of communication rules, character sets, error-control functions, etc., and is collectively known as *teletex protocol* (TP). This protocol provides interface between different types of word processors and also allows conversion of formats between telex and teletex networks. Different countries use different types of teletex networks, e.g., packet-switched networks, public switched telephone networks, or circuit-switched teletex networks.

Telnet: GTE Telenet Communication Corporation (U.S.), known as Telenet, introduced a public network based on ARPANET in 1974. It is a geographically distributed packet-switched network, which uses high-speed digital lines at data rates of 54 Kbps. One of its main objectives was to allow users to access their remote computers at a reasonable price and thus utilize resources effectively. This is a very popular protocol offered by TCP/IP communication interface. It is also one of the important services offered by the Internet which offers remote login on an Internet host. The Telnet is in fact a terminal emulation protocol that allows users to log onto other computers. It is important to know that this service protocol is one of over 100 service protocols defined by TCP/IP. It offers access to public services, libraries, and databases at remote computers (servers). This protocol defines a telnet connection similar to TCP for transferring user data packets, control packets, and telnet control information. It is also known as virtual terminal protocol (VTP) proving interfaces to terminal devices and terminal-oriented processes. It is based on the concept of network virtual terminal (NVT), which is nothing but a half-duplex, ASCII communication code-oriented device, a printer, and a keyboard. The main function of telnet is, in fact, to provide a mapping of real terminals into the virtual terminal and vice versa.

Terminal: Terminals are used for processing and communication of data and provide an interface between users and the computer systems. The terminals are a very useful component for processing, collecting, retrieving, and communicating data in the system and have become an integral requirement of any application system. The main function of terminals is to convert characters of the keyboard into electrical signal pulses expressed in the form of a code that is processed by computers. The interface (provided by terminals) for data communication takes input from input devices such as keyboard, card reader, magnetic tape, or any other input device and converts it into output for output devices such as display devices and paper tape.

Terminal eliminator: For controlling the data from asynchronous devices (CPU, data concentrators, data PABXs, data testers, PCs, mainframes, etc.) by one communication console, a terminal eliminator is used.

Terminal/line sharing interface: The line sharer allows the dial-up line or leased line to be shared, but the terminal/line sharing interface allows the sharing of both line and modem among the terminals. The modem and line can be shared by both synchronous and asynchronous devices and can also be cascaded.

Thermal noise: Due to a change in the temperature of transmission media, the number of electrons is generated randomly, affecting the normal current flow. Due to this sudden change in the number of electrons, the signal-to-noise ratio becomes low. This type of noise is known as *thermal noise*.

3270 port contender: This unit allows users to access an IBM 3174/3274 controller which offers many ports. This unit uses one port of this controller and provides connection to asynchronous devices, terminals, etc.

Throughput: This parameter indicates the number of packets that can be delivered over a period of time. It is very important, as it helps in determining utilization of the network. It depends on a number of factors, e.g., access control technique, number of stations connected, average waiting time, propagation time delay, connection, etc.

Time division multiplexing (TDM): This method allows users to share a high-speed line for sending the data from their low-speed keyboard computers. For each port, a time slot and a particular device is allocated, and each port when selected can provide bits, bytes, characters, or packets to the communication line from that device. At that time, the user can utilize the entire bandwidth and capacity of the channel. TDM scans all of these ports in a circular fashion, picks up the signals, separates them out and, after interleaving into a frame, transmits them on the communication link. It is important to note that only digital signals are accepted, transmitted, and received in TDM.

Time-sensitive: There are different types of traffic such as data, voice, graphics, video, multimedia, etc. There are some types of data for which delay may cause changes in the contents of the data, and as such no delay is acceptable in those types of data during transmission. Some of these include video, voice, real-time applications, multimedia, etc.

Token: This is a 3-byte frame that circulates around the ring of a token ring LAN. Any station connected to the ring needs to capture the token, append it, and send it back to the network. The appended token carries the data.

Token hold time (THT): When a node receives a token, it transmits the data from frames of the highest priority queues (i.e., class 6) until the transmission becomes equal to token hold time (THT) or there is no data to transmit. If the token rotation time (TRT) is less than the TRT of class i, then the data is transmitted in the order of priority. The token cycle time is defined as follows: If $n \times$ THT is greater than $n \times$ maximum TRT, then only class data will be transmitted. But if, $n \times$ THT is less than $n \times$ maximum TRT, the class 6 data is assured of the requested capacity, while other class data will get the remaining capacity.

Token passing topology: The token passing method for accessing data over the network determines which node gets an access right to the ring for data communication. In this method, a token of unique fixed bit pattern is defined, and this token always circulates around the ring topology. Any node wishing to send the frame must first capture the token. After it has captured the token, it attaches the data packet (frame) with the token and sends it back onto the ring. The destination node receives this token along with the frame, copies the frame onto its buffer, and changes the bit position in the token frame to indicate its acknowledgment to the sender node.

Token ring: This LAN was defined by IBM and uses a token passing access technique for LAN topology. The token ring topology (typically a bus) offers data rates of 4 and 16 Mbps.

Token rotation time (TRT): This is defined as the time taken by the token to go around the ring once. Both token hold and token notation times are controlled by respective timers. Initially each station is initialized with TRT which is set equal to TTRT.

Transceiver: The baseband-based LANs may use any cable (coaxial, twisted pair, etc.), and no modulation is required due to its support for only very short distances (typically a few kilometers). It uses 50-Ohm cable and supports multiple lines. A device known as a *transceiver* usually provides the access method to the baseband system (usually Ethernet). This device receives the data from the node and defines the packets for transmitting over the bus. The packet contains a source address, destination address, user's data, and some control information.

Transmission control protocol (TCP): The Internet is a collection of different types of computer networks and offers global connectivity for data communications between them. A common language through a standard protocol, transmission control protocol/Internet protocol (TCP/IP), defines the communication between these computer networks. This protocol was defined initially for ARPANET to provide communication between unreliable subnets. When this protocol is combined with IP, the user can transmit large files over the network reliably. It performs the same functions on the packets it receives from the session layer as those of the transport layer of OSI-RM.

Transmission convergence: The physical layer of the B-ISDN protocol is divided into two sublayers — physical medium and transmission convergence. The physical medium sublayer is the lowest B-ISDN protocol and is mainly concerned with medium-related functions, mapping between electrical and optical signals, timing, etc. The transmission convergence sublayer, on the other hand, is mainly concerned with functions related to cell transmission such as HEC header sequence generation, cell delineation, frame adaptation, and frame generation/recovery.

Transmission media: This is a physical link connecting different devices to the network. The sending and receiving nodes on the network can communicate with each other by establishing virtual circuits/logical channels over the transmission media. There are different types of media, which can be grouped into two categories: wired and wireless. The wired transmission media include twisted-pair wires, cables, fiber optics, etc., while wireless media include microwave, satellite, radio, etc.

Transport layer: This is the fourth layer (from the bottom) of OSI-RM. Its main functions are to provide end-to-end communication between two hosts and to provide error control.

Tree: This LAN topology is derived from bus topology and is characterized as having a root (headend) node (in contrast to the bus, where all the nodes are the same and can make a request to access the network at any time). We can also say that this topology is a generalization of bus topology without any loop or star-shaped topology. It is characterized as a tree structure, which defines certain intermediate nodes for controlling the information flow. There is one root node which is connected to other nodes over respective bus topologies based on the hierarchy of branches and subbranches coming out of the root node.

Twisted pair: This transmission media is comprised of two insulated wires twisted together. There are two types of twisted pairs: shielded and unshielded. These wires are very popular in local or end or subscriber loop connecting the subscriber's set with the local or end office. This local office is in fact an exchange (private branch exchange) located within the building or organization and is connected to the public telephone network via different media such as cables, fiber optics, and so on. The twisted pair is very popular for both analog and digital signals. The digital data switch or digital PBX within the building can offer data rates of 64 Kbps. These are also used in LANs offering data rates of 10 Mbps and 100 Mbps.

Tymnet: The first commercial public packet network was introduced in early 1970 and known as the McDonnell Douglas Tymnet network, after the introduction of X.25 as a basic interface to the public network by CCITT. The initial version of Tymnet offered support to only low-speed asynchronous terminal interfaces that were connected to its nodes (located at main locations) via dial-up connections. This version was then updated to also support X.25 and synchronous terminal interfaces. Tymnet uses a centralized routing technique that is based on the concept of a virtual circuit, which defines a fixed route or circuit during the establishment of connection.

Unbalanced transmission: Normally for transmission of a signal between two connected nodes, we need two physical paths. The current from the sender flows through one physical path to the destination and comes back on the second path. In unbalanced transmission, only one wire is used between the nodes. The current flows over the wire in each direction and returns back through earth in the other direction. Both sender and receiver nodes share the earth as another physical path between them.

Universal addressing: In any network, the user gets a unique address. For the large networks, the network interface card carries a built-in address and is used as the address of the user/PC. This address cannot be duplicated.

Universal asynchronous receiver and transmitter (UART): The data transmission between two DTEs usually takes place in serial mode, while the CPU within the DTE processes the data in parallel mode in power of 1 byte (8 bits). The CPU accepts the serial data from the DTE and, after processing, sends it back to the DTE. There must be a device between the DTE and CPU which will convert incoming serial data into parallel (for its processing by CPU), and then after processing, convert it back into serial mode for its transmission to DTE. The interface between DTE and CPU is a special integrated circuit (IC) chip known as a *universal asynchronous receiver and transmitter* (UART). This device is programmable and can be configured accordingly by the designers or users. The programming of UART requires certain controlling data signals which can be specified by the users, and the device will perform various functions such as conversion of incoming serial data into parallel, detecting parity and framing errors, and many other features.

Unshielded twisted pair: This medium has been used since the telephone system came into existence. It is in fact a telephone wire and can be found in almost any building. This is a very cheap medium, and external electromagnetic interference, external noises, etc., affect its performance. A metallic braid or sheathing around twisted pair can reduce the effects of external interference. It is 100-Ohm twisted pair and commonly used for telephone (voice-grade) applications. It is also used for data communications (LANs).

Upload: This method is used when a user wants to keep the documents in a shareable server. The user will transmit (upload) the data onto the server.

Upper side band (USB): The amplitude modulation (AM) method defines two side band frequency components around the carrier frequency as lower and upper side bands. The upper side band defines the frequency above the carrier frequency. The transmission may send one side band, both side bands, or suppress them, and as such define different types of modulation techniques.

User channel: The standard SONET channel allocates a part of its bandwidth to the user, who uses this for maintenance functions of the network.

User–network interface (UNI): This interface provides interconnectivity between LANs and high-speed networks (frame relay, ATM, B-IDDN, SMSD). In the case of ATM networks, the cell header defines two types of interfaces — user–network interface (UNI) and network–node interface (NNI). In UNI, the header format of the cell header defines an interface between user and network, while in NNI, the cell header format defines an interface between the switching nodes of the network. The frame relay network is more concerned with NNI for interconnecting different networks (for its operation) than UNI. The operations are basically aimed at providing switched virtual circuit services and fault tolerance.

User plane: The reference protocol architecture of B-ISDN consists of three planes: management (top level), user, and control (bottom level). The user plane deals with various functions for the transfer of messages between users through the network. Various functions include flow control, error recovery, and congestion control. Various services provided by this plane include connection-oriented and connectionless data, video, voice, and multimedia.

User-to-user protocols: This protocol defines an interface between the users and various applications that are transparent to the network, e.g., telnet, FTP, and others.

User–user interface: This interface protocol works between users and is usually transparent to the network.

V.35: This ITU standard defines a trunk interface between modem or any other network access device and the public switched network. It offers a data rate of 48 Kbps and uses a band of 60 to 108 KHz.

Value-added network: FCC defined this network as a mechanism which allows the carriers or any organization to lease lines for services. It defines a common carrier and uses an open policy that allows potential public network operators to apply for FCC approval. It does not provide any data processing or data communication, but offers the services to end users at a reasonable price. This network is based on packet switching and is typically owned by an organization.

Variable bit rate: The broadband networks support different types of traffic (data, voice, video, multimedia, etc.). A lot of memory and bandwidth are required when transmitting these signals. Usually these signals, requiring a very large bandwidth, are compressed using different compression techniques (MPEG, JPEG, etc.). The compression technique converts the traffic signal (which is typically a continuous bit rate) into variable bit rate format in a frame-by-frame mode. Due to the variable size of the frame, the transmission of traffic does not follow a uniform flow of data and hence there needs to be a mechanism which can handle this type of flow control.

Velocity of propagation: The speed at which the signals (electromagnetic waves) travel through transmission media (wired or wireless).

Vertical redundancy check (VRC): This code can only detect either odd or even number of bits error. It cannot detect both errors together. The geometric code includes one parity bit to each character and one parity bit to each level of bits for all the characters in the block. The parity bit added with character corresponds to vertical redundancy check (VRC), while the parity bit added with each bit level corresponds to longitudinal redundancy check (LRC).

Vestigial sideband: Amplitude modulation defines two side bands having frequencies around carrier frequency known as lower and upper side bands. In a form of modulation, we transmit one full side band (upper or lower) and a part of another side band.

Videoconferencing: This is a very popular application of ISDN that is finding more popularity with the advent of high-speed networks with broadband service capabilities such as multimedia, voice, etc. Videotex is an interactive text and graphic service and offers text and graphics over a videotex terminal, over a TV with a special decoder, over cable TV, or even over telephone lines attached to a PC via modems. The videotex vendor usually has contracts with information providers, which then make contents available to the users. Various services offered on videotex include request for any balance, transfer of funds, payment, purchase, database queries, purchase of tickets, travel documents, interactive games, etc.

Video on demand: Another application of broadband services where users can view the program from a remote VCR by making a request. Other similar applications include remote database access, banking transactions from home, and pay-per-view. All of these applications are available on a CATV network that uses high-speed switches such as ATM, frame relay, and so on.

Videotex: A very popular application of broadband services is videotex, a communication system that works in two-way mode such that both video and other related information can be transmitted simultaneously. Users can access the information (data, text, voice, video) stored in file servers. Other examples include distance learning and training, tele-software, tele-shopping, etc.

Video voice mail: The broadband services are available through different types of high-speed networks at varied data rates and large bandwidth. Video voice mail is one of the popular applications of broadband services where a remote video machine with voice can provide functions similar to those of an ordinary telephone answering machine but with video capability such as transmission of video moving pictures along with sound.

Virtual channel connection: ATM networks are based on fast packet switching, similar to the way frame relay uses it. X.25 is also a packet switched network. In X.25, a virtual circuit link is established before transmission, while a frame relay network establishes a data link connection. In the same way, in ATM networks a virtual circuit channel link is established before the transmission. This connection is defined between end users and carries the signal of both the interfaces — user–network and network–node — and of course ATM cells.

Virtual channel identifier: This is one of six fields in the header of an ATM cell. This field is required to establish connections across the switching nodes. It uses a mapping table to provide mapping between incoming VCI and outgoing VCI at each switch. This defines a virtual circuit. The sequence of virtual circuits between end points defines a virtual connection. It has 16 bits and does not use the bandwidth until the requested information is actually transmitted. Interestingly, this type of virtual connection establishment offers the same function as that of TDM in circuit-switched networks. A field within a segment header of DQDB defines the types of operations (read, write, copy) on a payload segment.

Virtual circuit: The packet-switching technique provides connection-oriented (reliable) and connectionless (unreliable) services. In reliable service, it establishes a dedicated virtual circuit and transmits the packets of fixed size. The packets are stored at intermediate switching nodes and forwarded to the next node until they reach the destination node. The overhead caused is due to call setup, overloading of packets, extra bits in each packet, sequencing of packets, etc. In unreliable service (datagram), no dedicated path is established before the transmission. The overhead is caused due to packet queuing, packet recovery, and an extra bit in each packet. The allocation of bandwidth is done to virtual circuits only when they are in use.

Virtual LAN: A concept of defining a logical network by connecting different LAN media and giving the appearance to the users that they are connected by the same physical LAN media. This will also give the appearance to the users that all the resources are local. It encourages connectivity, but we have to be careful as we are dealing with different LAN media and MAC protocols.

Virtual path connection: In ATM, a virtual path is defined which basically consists of different virtual channel connections having the same end points. Thus, a virtual path connection is a collection of all the VCCs that have the same end points. This way, the controlling cost of different VCCs can be reduced, since these VCCs are sharing the same common path. This virtual path concept offers various advantages, such as better performance of the network, simple architecture, smaller connection set-up time, and so on.

Virtual path identifier: One of the six fields in the header of an ATM cell. It is required to establish a virtual path similar to VCI. It is mainly concerned with routing and management issues. It has 8 or 12 bits and defines a virtual path between end points that is comprised of different virtual channels.

Virtual telecommunication access method (VTAM): IBM has defined an application package, virtual telecommunication access method (VTAM), that supports different types of data link protocols, e.g., bisynchronous, asynchronous, SDLC, etc. It also supports security and reliability. The front-end processor (FEP) defines the management layer via network control program (NCP)/virtual storage (VS) protocol.

Voltage controlled oscillator: This oscillator generates the frequency components that are proportional to input voltage.

Wavelength: The distance between any two points that have a phase difference of two π in a periodic representation of signal.

World Wide Web (WWW): A very popular application program on the Internet. It is based on hypertext, which provides a variety of other useful links. Since all the documents are defined in hypertext, these are easily searchable.

X.2: This recommendation of CCITT describes international user classes of services in public data networks (PDNs) and ISDNs.

X.21: This defines an interface between DTE and DCE of digital circuit-switching network. The X.21 bis (another standard) was introduced by CCITT (similar to RS-232 with added advantages) to be used in the existing telephone networks.

X.21 bis: This standard was introduced by CCITT as an alternative to RS-232 to be used in public telephone network before the deployment of public data networks.

X.25: This recommendation of CCITT defines an interface between data terminal equipment (DTE) and data circuit-terminating equipment (DCE) for terminals operating in packet mode. It further connects the interface to the public data network by dedicated circuit.

X.75: This recommendation of CCITT defines terminal and transit call procedures and data transfer system on international circuits between packet-switched data networks. It provides an interface between terminals and public packet-switched network (X.25). The terminals are usually asynchronous and hence the X.75 accepts asynchronous characters, converts them into synchronous characters, and transmits using X.25 standard protocols.

X.200: This recommendation of CCITT describes OSI-RM for CCITT applications.

X.210: This recommendation of CCITT defines various layer service definitions.

X.211: This standard defines the services between physical and data link layers of OSI-RM. It describes six primitives to be used by the physical layer to initialize the physical connection, data transfer, and termination of physical connection.

X.213: This recommendation describes the services offered within OSI-RM for CCITT applications.

X.214: This recommendation defines the services offered within OSI-RM for CCITT applications.

X.215: This recommendation describes services offered within OSI-RM for CCITT applications.

X.224: This recommendation discusses the transport protocol specification within OSI-RM for CCITT applications.

X.225: This recommendation describes the session protocol specification within OSI-RM for CCITT applications.

X.400 protocol: The recommendation of CCITT discusses the message handling system.

X.409: This recommendation of CCITT discusses the message handling system and its presentation transfer syntax and notation.

X.440 and X.441: These recommendations of CCITT describe the ISDN user–network data link layer general aspects and specification.

X.500: This recommendation of CCITT defines directory systems (ISO 9595).

XML (eXtensible Markup Language): This document description language, much like HTML but more versatile, is used to construct Web pages. It allows the user to define the tags but does not allow the description of presentation style of data. An associated language, Extensible Style Language (ESL), must be used for this.

Yagi antenna: This type of antenna uses half-wave dipoles as an active element for transmission and reception of electromagnetic signals.

Index

A

B

C

D

E

F

G

H

I

J

K

L

M

N

O

P

Q

R

S

T

U

V

W

X